Glioblastoma

Glioblastoma

State of the Art and Future Perspectives

Special Issue Editors

Ghazaleh Tabatabai
Hiroaki Wakimoto

MDPI • Basel • Beijing • Wuhan • Barcelona • Belgrade • Manchester • Tokyo • Cluj • Tianjin

Special Issue Editors
Ghazaleh Tabatabai
University Hospital Tübingen,
Eberhard Karls University Tübingen
Germany

Hiroaki Wakimoto
Massachusetts General Hospital
USA

Editorial Office
MDPI
St. Alban-Anlage 66
4052 Basel, Switzerland

This is a reprint of articles from the Special Issue published online in the open access journal *Cancers* (ISSN 2072-6694) (available at: https://www.mdpi.com/journal/cancers/special_issues/Glioblastoma_SAFP).

For citation purposes, cite each article independently as indicated on the article page online and as indicated below:

LastName, A.A.; LastName, B.B.; LastName, C.C. Article Title. *Journal Name* **Year**, *Article Number*, Page Range.

ISBN 978-3-03928-260-9 (Hbk)
ISBN 978-3-03928-261-6 (PDF)

Contents

About the Special Issue Editors

Ghazaleh Tabatabai, Neurologist and full Professor of Neuro-Oncology, University Hospital Tübingen and Eberahrd Karls University of Tübingen, Germany. Prof. Tabatabai's research interest is focussed on central nervous system tumors, molecular mechanisms of acquired resistance to therapy, modifications of the tumor-associated microenvironment by cell-based therapies, innovative early phase clinical trials.

Hiroaki Wakimoto, Associate professor of Neurosurgery at Harvard Medical School. Dr. Wakimoto's research interest is in developing novel biological and targeted treatment strategies for central nervous system malignancies using clinically relevant disease models.

cancers

MDPI

Editorial

Glioblastoma: State of the Art and Future Perspectives

Ghazaleh Tabatabai [1,*] and Hiroaki Wakimoto [2,*]

1 Interdisciplinary Division of Neuro-Oncology, Hertie Institute for Clinical Brain Research, Center for Neuro-Oncology, Comprehensive Cancer Center Tübingen Stuttgart, University Hospital Tübingen, Eberhard Karls University Tübingen, 72076 Tübingen, Germany

2 Department of Neurosurgery, Massachusetts General Hospital, Harvard Medical School Boston, Boston, MA 02114, USA

* Correspondence: ghazaleh.tabatabai@uni-tuebingen.de (G.T.); HWAKIMOTO@mgh.harvard.edu (H.W.)

Received: 29 July 2019; Accepted: 30 July 2019; Published: 31 July 2019

This special issue is dedicated to glioblastoma and elucidates this disease from different perspectives. Despite multimodal therapies, the prognosis is still dismal. Many features contribute to this therapeutic challenge including high intratumoral and intertumoral heterogeneity, resistance to therapy, migration and invasion, and immunosuppression. With the advent of novel high throughput technologies, significant progress has been made to understand molecular and immunological signatures underlying the pathology of glioblastoma. This special issue aimed at updating researchers on current topics and progress made in basic, preclinical, and clinical glioblastoma research.

The original articles in this special issue present novel findings on molecular mechanisms of radiosensitization, radioresistance and acquired resistance to therapy (Kessler et al., Müller-Längle et al., Shah et al., Llaguno-Munive et al.) [1–4], cell migration (Nowicki et al., Hübner et al.) [5,6], intracellular drug levels (Colin et al.) [7], strategies targeting renewal capacities of cancer stem-like cells (Gravina et al., Sansalone et al., Linder et al.) [8–10], mathematic modeling of synergy, machine learning, deep learning (Kim et al., Hana et al., Wong et al.) [11–13], distinct signaling pathways (Barbagallo et al., Akgül et al., Bensalma et al., Saito et al., Liu et al.) [14–18], prognostic and predictive effects of imaging patterns (Jungk et al., Puig et al.) [19,20], tumor-associated epilepsy (Berendsen et al.) [21], and novel models and experimental therapeutic approaches (Berthier et al., Offenhäuser et al., Privat-Maldonado et al., Lozada-Delgado et al.) [22–25].

Moreover, review articles summarize the current state of knowledge in the fields of platelets (Marx et al.) [26], subventricular zone (Altmann et al.) [27] and microenvironment (Schiffer et al.) [28], lentiviral vectors (Del Vecchio et al.) [29], allergic inflammation (Costanza and Finocchiaro) [30], elderly patients (Minniti et al.) [31], EMT and authophagy (Colella et al.) [32], notch signaling and CXCL12 signaling (Bazzoni et al., Giordano et al.) [33,34], protein phosphatases (Tomiyama et al.) [35], nitric oxide antagonism (Fahey and Girotti) [36], virus-based immunotherapy (Martikainen and Essand; Stepanenko et al.) [37,38], tumor-treating fields (Fabian et al.) [39], amino acid PET (Lohmann et al.) [40], cholesterol metabolism (Ahmad et al.) [41], preclinical modeling (Rominiyi et al.) [42], and EGFR as a therapeutic target (Gao et al.) [43].

It becomes clear that a holistic view from different angles is required to understand this complex disease and discover novel therapeutic targets and biomarkers. We are grateful for all the work the authors have included in this special issue.

Finally, we would like to emphasize the most important perspective, i.e., our patients' perspectives. From their point of view, the main readout for success in patient-centered research is prolonged survival with maintained quality of life. A culture of continued collaboration between disciplines and research teams will be necessary to meet this challenge.

Conflicts of Interest: Ghazaleh Tabatabai: Personal fees for lectures and advisory board participation from Bristol-Myers-Squibb, AbbVie, Novocure, Medac. Research and travel grants from Bristol-Myers-Squibb, Novocure, Roche Diagnostics, Medac. Member of steering committees of the non-interventional studies TIGER (Novocure) and ONTRk (Bayer). Hiroaki Wakimoto declares no conflict of interest.

References

1. Kessler, J.; Hohmann, T.; Güttler, A.; Petrenko, M.; Ostheimer, C.; Hohmann, U.; Bache, M.; Dehghani, F.; Vordermark, D. Radiosensitization and a Less Aggressive Phenotype of Human Malignant Glioma Cells Expressing Isocitrate Dehydrogenase 1 (IDH1) Mutant Protein: Dissecting the Mechanisms. *Cancers* **2019**, *11*, 889. [CrossRef] [PubMed]
2. Müller-Längle, A.; Lutz, H.; Hehlgans, S.; Rödel, F.; Rau, K.; Laube, B. NMDA Receptor-Mediated Signaling Pathways Enhance Radiation Resistance, Survival and Migration in Glioblastoma Cells—A Potential Target for Adjuvant Radiotherapy. *Cancers* **2019**, *11*, 503. [CrossRef] [PubMed]
3. Shah, S.; Rodriguez, G.; Musick, A.; Walters, W.; de Cordoba, N.; Barbarite, E.; Marlow, M.; Marples, B.; Prince, J.; Komotar, R.; et al. Targeting Glioblastoma Stem Cells with 2-Deoxy-D-Glucose (2-DG) Potentiates Radiation-Induced Unfolded Protein Response (UPR). *Cancers* **2019**, *11*, 159. [CrossRef] [PubMed]
4. Llaguno-Munive, M.; Romero-Piña, M.; Serrano-Bello, J.; Medina, L.; Uribe-Uribe, N.; Salazar, A.; Rodríguez-Dorantes, M.; Garcia-Lopez, P. Mifepristone Overcomes Tumor Resistance to Temozolomide Associated with DNA Damage Repair and Apoptosis in an Orthotopic Model of Glioblastoma. *Cancers* **2019**, *11*, 16. [CrossRef] [PubMed]
5. Nowicki, M.; Hayes, J.; Chiocca, E.; Lawler, S. Proteomic Analysis Implicates Vimentin in Glioblastoma Cell Migration. *Cancers* **2019**, *11*, 466. [CrossRef] [PubMed]
6. Hübner, M.; Hinske, C.; Effinger, D.; Wu, T.; Thon, N.; Kreth, F.; Kreth, S. Intronic miR-744 Inhibits Glioblastoma Migration by Functionally Antagonizing Its Host Gene MAP2K4. *Cancers* **2018**, *10*, 400. [CrossRef] [PubMed]
7. Colin, M.; Delporte, C.; Janky, R.; Lechon, A.; Renard, G.; Van Antwerpen, P.; Maltese, W.; Mathieu, V. Dysregulation of Macropinocytosis Processes in Glioblastomas May Be Exploited to Increase Intracellular Anti-Cancer Drug Levels: The Example of Temozolomide. *Cancers* **2019**, *11*, 411. [CrossRef]
8. Gravina, G.; Mancini, A.; Colapietro, A.; Delle Monache, S.; Sferra, R.; Vitale, F.; Cristiano, L.; Martellucci, S.; Marampon, F.; Mattei, V.; et al. The Small Molecule Ephrin Receptor Inhibitor, GLPG1790, Reduces Renewal Capabilities of Cancer Stem Cells, Showing Anti-Tumour Efficacy on Preclinical Glioblastoma Models. *Cancers* **2019**, *11*, 359. [CrossRef]
9. Sansalone, L.; Veliz, E.; Myrthil, N.; Stathias, V.; Walters, W.; Torrens, I.; Schürer, S.; Vanni, S.; Leblanc, R.; Graham, R. Novel Curcumin Inspired Bis-Chalcone Promotes Endoplasmic Reticulum Stress and Glioblastoma Neurosphere Cell Death. *Cancers* **2019**, *11*, 357. [CrossRef]
10. Linder, B.; Wehle, A.; Hehlgans, S.; Bonn, F.; Dikic, I.; Rödel, F.; Seifert, V.; Kögel, D. Arsenic Trioxide and (−)-Gossypol Synergistically Target Glioma Stem-Like Cells via Inhibition of Hedgehog and Notch Signaling. *Cancers* **2019**, *11*, 350. [CrossRef]
11. Kim, Y.; Lee, J.; Lee, D.; Othmer, H. Synergistic Effects of Bortezomib-OV Therapy and Anti-Invasive Strategies in Glioblastoma: A Mathematical Model. *Cancers* **2019**, *11*, 215. [CrossRef] [PubMed]
12. Hana, T.; Tanaka, S.; Nejo, T.; Takahashi, S.; Kitagawa, Y.; Koike, T.; Nomura, M.; Takayanagi, S.; Saito, N. Mining-Guided Machine Learning Analyses Revealed the Latest Trends in Neuro-Oncology. *Cancers* **2019**, *11*, 178. [CrossRef] [PubMed]
13. Wong, K.; Rostomily, R.; Wong, S. Prognostic Gene Discovery in Glioblastoma Patients using Deep Learning. *Cancers* **2019**, *11*, 53. [CrossRef] [PubMed]
14. Barbagallo, D.; Caponnetto, A.; Brex, D.; Mirabella, F.; Barbagallo, C.; Lauretta, G.; Morrone, A.; Certo, F.; Broggi, G.; Caltabiano, R.; et al. CircSMARCA5 Regulates VEGFA mRNA Splicing and Angiogenesis in Glioblastoma Multiforme Through the Binding of SRSF1. *Cancers* **2019**, *11*, 194. [CrossRef] [PubMed]
15. Akgül, S.; Patch, A.; D'Souza, R.; Mukhopadhyay, P.; Nones, K.; Kempe, S.; Kazakoff, S.; Jeffree, R.; Stringer, B.; Pearson, J.; et al. Intratumoural Heterogeneity Underlies Distinct Therapy Responses and Treatment Resistance in Glioblastoma. *Cancers* **2019**, *11*, 190. [CrossRef] [PubMed]

16. Bensalma, S.; Turpault, S.; Balandre, A.; De Boisvilliers, M.; Gaillard, A.; Chadéneau, C.; Muller, J. PKA at a Cross-Road of Signaling Pathways Involved in the Regulation of Glioblastoma Migration and Invasion by the Neuropeptides VIP and PACAP. *Cancers* **2019**, *11*, 123. [CrossRef] [PubMed]
17. Saito, N.; Hirai, N.; Aoki, K.; Suzuki, R.; Fujita, S.; Nakayama, H.; Hayashi, M.; Ito, K.; Sakurai, T.; Iwabuchi, S. The Oncogene Addiction Switch from NOTCH to PI3K Requires Simultaneous Targeting of NOTCH and PI3K Pathway Inhibition in Glioblastoma. *Cancers* **2019**, *11*, 121. [CrossRef] [PubMed]
18. Liu, H.; Su, Y.; Bamodu, O.; Hueng, D.; Lee, W.; Huang, C.; Deng, L.; Hsiao, M.; Chien, M.; Yeh, C.; et al. The Disruption of the β-Catenin/TCF-1/STAT3 Signaling Axis by 4-Acetylantroquinonol B Inhibits the Tumorigenesis and Cancer Stem-Cell-Like Properties of Glioblastoma Cells, In Vitro and In Vivo. *Cancers* **2018**, *10*, 491. [CrossRef] [PubMed]
19. Jungk, C.; Warta, R.; Mock, A.; Friauf, S.; Hug, B.; Capper, D.; Abdollahi, A.; Debus, J.; Bendszus, M.; von Deimling, A.; et al. Location-Dependent Patient Outcome and Recurrence Patterns in IDH1-Wildtype Glioblastoma. *Cancers* **2019**, *11*, 122. [CrossRef]
20. Puig, J.; Biarnés, C.; Daunis-i-Estadella, P.; Blasco, G.; Gimeno, A.; Essig, M.; Balaña, C.; Alberich-Bayarri, A.; Jimenez-Pastor, A.; Camacho, E.; et al. Macrovascular Networks on Contrast-Enhanced Magnetic Resonance Imaging Improves Survival Prediction in Newly Diagnosed Glioblastoma. *Cancers* **2019**, *11*, 84. [CrossRef]
21. Berendsen, S.; Spliet, W.; Geurts, M.; Van Hecke, W.; Seute, T.; Snijders, T.; Bours, V.; Bell, E.; Chakravarti, A.; Robe, P. Epilepsy Associates with Decreased HIF-1α/STAT5b Signaling in Glioblastoma. *Cancers* **2019**, *11*, 41. [CrossRef] [PubMed]
22. Berthier, S.; Larrouquère, L.; Champelovier, P.; Col, E.; Lefebvre, C.; Cottet-Rouselle, C.; Arnaud, J.; Garrel, C.; Laporte, F.; Boutonnat, J.; et al. A New Patient-Derived Metastatic Glioblastoma Cell Line: Characterisation and Response to Sodium Selenite Anticancer Agent. *Cancers* **2019**, *11*, 12. [CrossRef] [PubMed]
23. Offenhäuser, C.; Al-Ejeh, F.; Puttick, S.; Ensbey, K.; Bruce, Z.; Jamieson, P.; Smith, F.; Stringer, B.; Carrington, B.; Fuchs, A.; et al. EphA3 Pay-Loaded Antibody Therapeutics for the Treatment of Glioblastoma. *Cancers* **2018**, *10*, 519. [CrossRef] [PubMed]
24. Privat-Maldonado, A.; Gorbanev, Y.; Dewilde, S.; Smits, E.; Bogaerts, A. Reduction of Human Glioblastoma Spheroids Using Cold Atmospheric Plasma: The Combined Effect of Short- and Long-Lived Reactive Species. *Cancers* **2018**, *10*, 394. [CrossRef] [PubMed]
25. Lozada-Delgado, E.; Grafals-Ruiz, N.; Miranda-Román, M.; Santana-Rivera, Y.; Valiyeva, F.; Rivera-Díaz, M.; Marcos-Martínez, M.; Vivas-Mejía, P. Targeting MicroRNA-143 Leads to Inhibition of Glioblastoma Tumor Progression. *Cancers* **2018**, *10*, 382. [CrossRef]
26. Marx, S.; Xiao, Y.; Baschin, M.; Splittstöhser, M.; Altmann, R.; Moritz, E.; Jedlitschky, G.; Bien-Möller, S.; Schroeder, H.; Rauch, B. The Role of Platelets in Cancer Pathophysiology: Focus on Malignant Glioma. *Cancers* **2019**, *11*, 569. [CrossRef] [PubMed]
27. Altmann, C.; Keller, S.; Schmidt, M. The Role of SVZ Stem Cells in Glioblastoma. *Cancers* **2019**, *11*, 448. [CrossRef]
28. Schiffer, D.; Annovazzi, L.; Casalone, C.; Corona, C.; Mellai, M. Glioblastoma: Microenvironment and Niche Concept. *Cancers* **2019**, *11*, 5. [CrossRef]
29. Del Vecchio, C.; Calistri, A.; Parolin, C.; Mucignat-Caretta, C. Lentiviral Vectors as Tools for the Study and Treatment of Glioblastoma. *Cancers* **2019**, *11*, 417. [CrossRef]
30. Costanza, M.; Finocchiaro, G. Allergic Signs in Glioma Pathology: Current Knowledge and Future Perspectives. *Cancers* **2019**, *11*, 404. [CrossRef]
31. Minniti, G.; Lombardi, G.; Paolini, S. Glioblastoma in Elderly Patients: Current Management and Future Perspectives. *Cancers* **2019**, *11*, 336. [CrossRef]
32. Colella, B.; Faienza, F.; Di Bartolomeo, S. EMT Regulation by Autophagy: A New Perspective in Glioblastoma Biology. *Cancers* **2019**, *11*, 312. [CrossRef]
33. Bazzoni, R.; Bentivegna, A. Role of Notch Signaling Pathway in Glioblastoma Pathogenesis. *Cancers* **2019**, *11*, 292. [CrossRef]
34. Giordano, F.; Link, B.; Glas, M.; Herrlinger, U.; Wenz, F.; Umansky, V.; Brown, J.; Herskind, C. Targeting the Post-Irradiation Tumor Microenvironment in Glioblastoma via Inhibition of CXCL12. *Cancers* **2019**, *11*, 272. [CrossRef]
35. Tomiyama, A.; Kobayashi, T.; Mori, K.; Ichimura, K. Protein Phosphatases—A Touchy Enemy in the Battle Against Glioblastomas: A Review. *Cancers* **2019**, *11*, 241. [CrossRef]

36. Fahey, J.; Girotti, A. Nitric Oxide Antagonism to Anti-Glioblastoma Photodynamic Therapy: Mitigation by Inhibitors of Nitric Oxide Generation. *Cancers* **2019**, *11*, 231. [CrossRef]
37. Martikainen, M.; Essand, M. Virus-Based Immunotherapy of Glioblastoma. *Cancers* **2019**, *11*, 186. [CrossRef]
38. Stepanenko, A.; Chekhonin, V. Recent Advances in Oncolytic Virotherapy and Immunotherapy for Glioblastoma: A Glimmer of Hope in the Search for an Effective Therapy? *Cancers* **2018**, *10*, 492. [CrossRef]
39. Fabian, D.; Guillermo Prieto Eibl, M.; Alnahhas, I.; Sebastian, N.; Giglio, P.; Puduvalli, V.; Gonzalez, J.; Palmer, J. Treatment of Glioblastoma (GBM) with the Addition of Tumor-Treating Fields (TTF): A Review. *Cancers* **2019**, *11*, 174. [CrossRef]
40. Lohmann, P.; Werner, J.; Shah, N.; Fink, G.; Langen, K.; Galldiks, N. Combined Amino Acid Positron Emission Tomography and Advanced Magnetic Resonance Imaging in Glioma Patients. *Cancers* **2019**, *11*, 153. [CrossRef]
41. Ahmad, F.; Sun, Q.; Patel, D.; Stommel, J. Cholesterol Metabolism: A Potential Therapeutic Target in Glioblastoma. *Cancers* **2019**, *11*, 146. [CrossRef]
42. Rominiyi, O.; Al-Tamimi, Y.; Collis, S. The 'Ins and Outs' of Early Preclinical Models for Brain Tumor Research: Are They Valuable and Have We Been Doing It Wrong? *Cancers* **2019**, *11*, 426. [CrossRef]
43. Gao, Y.; Vallentgoed, W.; French, P. Finding the Right Way to Target EGFR in Glioblastomas; Lessons from Lung Adenocarcinomas. *Cancers* **2018**, *10*, 489. [CrossRef]

cancers

MDPI

Article

Targeting MicroRNA-143 Leads to Inhibition of Glioblastoma Tumor Progression

Eunice L. Lozada-Delgado [1,2,3], Nilmary Grafals-Ruiz [3,4], Miguel A. Miranda-Román [1,3], Yasmarie Santana-Rivera [1,3], Fatma Valiyeva [3], Mónica Rivera-Díaz [2,3], María J. Marcos-Martínez [5,6] and Pablo E. Vivas-Mejía [2,3,*]

1 Department of Biology, Rio Piedras Campus, University of Puerto Rico, San Juan, PR 00931, USA; eunice.lozada@upr.edu (E.L.L.-D.); mirandar.miguel@gmail.com (M.A.M.-R.); yasmarie.santana@upr.edu (Y.S.-R.)

2 Department of Biochemistry, Medical Sciences Campus, University of Puerto Rico, San Juan, PR 00936, USA; mrivera@bromediconllc.com

3 Comprehensive Cancer Center, University of Puerto Rico, San Juan, PR 00935, USA; nilmary.grafals1@upr.edu (N.G.-R.); fvaliyeva@cccupr.org (F.V.)

4 Department of Physiology, Medical Sciences Campus, University of Puerto Rico, San Juan, PR 00936, USA

5 Department of Pathology and Laboratory Medicine, Medical Sciences Campus, University of Puerto Rico, San Juan, PR 00936, USA; maria.marcos@upr.edu

6 Anatomic Pathology Laboratory, Puerto Rico Medical Services Administration, San Juan, PR 00936, USA

* Correspondence: pablo.vivas@upr.edu; Tel.: +1-787-772-8300

Received: 9 August 2018; Accepted: 8 October 2018; Published: 12 October 2018

Abstract: Glioblastoma (GBM) is the most common and aggressive of all brain tumors, with a median survival of only 14 months after initial diagnosis. Novel therapeutic approaches are an unmet need for GBM treatment. MicroRNAs (miRNAs) are a class of small non-coding RNAs that regulate gene expression at the post-transcriptional level. Several dysregulated miRNAs have been identified in all cancer types including GBM. In this study, we aimed to uncover the role of miR-143 in GBM cell lines, patient samples, and mouse models. Quantitative real-time RT-PCR of RNA extracted from formalin-fixed paraffin-embedded (FFPE) samples showed that the relative expression of miR-143 was higher in GBM patients compared to control individuals. Transient transfection of GBM cells with a miR-143 oligonucleotide inhibitor (miR-143-inh) resulted in reduced cell proliferation, increased apoptosis, and cell cycle arrest. SLC30A8, a glucose metabolism-related protein, was identified as a direct target of miR-143 in GBM cells. Moreover, multiple injections of GBM tumor-bearing mice with a miR-143-inh-liposomal formulation significantly reduced tumor growth compared to control mice. The reduced in vitro cell growth and in vivo tumor growth following miRNA-143 inhibition suggests that miR-143 is a potential therapeutic target for GBM therapy.

Keywords: glioblastoma; microRNAs; mouse model; cell proliferation

1. Introduction

Glioblastoma (GBM), also known as glioblastoma multiforme, is the most common and lethal form of brain tumor [1]. Currently, there are no optimal treatments for this disease, which accounts for about 14,000 annual deaths in the U.S., having an incidence ratio of 2 to 3 out of 100,000 adults per year [2]. GBM, or WHO Grade IV astrocytoma, can either develop de novo (primary, 90% of cases) or derive from WHO grade II or grade III astrocytomas (secondary) [1,3–5]. GBM tumors are known to be fast growing in the cerebral white matter and patients typically remain asymptomatic until advanced stages of the disease [6]. The current therapeutic strategy for GBM involves tumor resection surgery followed by radiotherapy (XRT) and/or radiosurgery in combination with temozolomide (TMZ)-based chemotherapy [1,7]. However, most GBM patients become resistant to a second round of TMZ treatment due to over-activation of the DNA repair enzyme O6-methylguanine-DNA methyltransferase (MGMT) [2,8]. Despite the aggressive treatment, the prognosis of GBM patients remains poor, with survival rates of only 12–14 months after initial diagnosis. Therefore, novel and more effective therapies for GBM are urgently needed.

MicroRNAs (miRNAs) are small non-coding RNAs of about 18–22 nucleotides in length that regulate gene expression post-transcriptionally by recognizing and binding preferably to the 3′-Untranslated Region (3′-UTR) of their target messenger RNAs (mRNAs) [9]. In addition, miRNAs binding to the 5′-UTR or coding regions have also been observed [9–11]. Evidence indicates that altered expression of miRNAs in GBM plays a central role in GBM initiation, progression, and tumor maintenance [12–14]. This has led to the proposal of several miRNAs as both diagnostic and prognostic markers, and as targets for GBM therapy [2].

MicroRNA-143 (miR-143) is part of a conserved miRNA cluster composed of miR-143/miR-145 on chromosome 5 (5q33) in humans [15]. MiR-143 has been shown to have a role in tumor progression, cancer cell growth, and invasiveness of cancer cells, including GBM cells [16,17]. As most miRNAs, miR-143 expression appears to be tissue-specific [17,18]. In normal tissues, miR-143 expression ranges from highest in the colon to lowest in the brain and liver [15]. Reports have shown that high levels of miR-143 sensitize cells to chemotherapeutic drugs including docetaxel in prostate cancer cells and TMZ in GBM cells [8,19]. However, others show an association between high levels of miR-143 with increased invasive potential of GBM cells compared with parental GBM cells, suggesting an oncogenic role [16]. Since the biological role of miR-143 in GBM is not well understood, in the present study we rigorously investigated the role of miR-143 in GBM cell lines, patient samples, and a subcutaneous GBM mouse model.

2. Results

2.1. Expression of MiR-143 in GBM Patients

First, we assessed the miR-143 expression levels in FFPE samples of GBM patients [14,20,21]. MiR-143 expression levels were found to be significantly increased in GBM patients compared to controls (* $p = 0.0208$) (Figure 1).

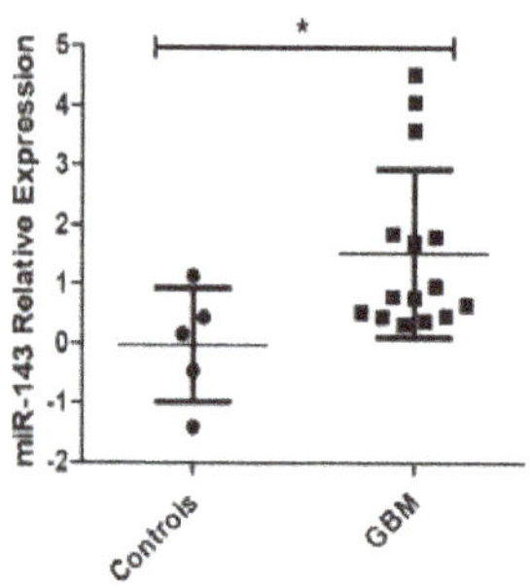

Figure 1. MiR-143 expression levels in GBM. Formalin-fixed paraffin-embedded (FFPE) tissue blocks from 19 newly diagnosed Glioblastoma (GBM) patients (13 females, 6 males) and 5 control patients (2 females, 3 males) were used in this study. GBM patients showed higher miR-143 expression compared to control patient samples (* $p < 0.05$); dots represent the means of triplicates $\pm$ SD.

2.2. Effect of MiR-143 Targeting on GBM Cell Proliferation

We measured miR-143 expression levels in a panel of three well-known GBM cell lines (U87-MG, T98G, A-172). MiR-143 was found to be expressed in higher levels in the U87-MG (U87) cell line, while the T98G cells expressed the lowest miR-143 levels (Figure 2A). Therefore, the U87 cell line was used for miR-143 inhibition experiments, while the T98G cell line was chosen for ectopic miR-143 overexpression. To examine the effect of targeting miR-143 on cell proliferation, U87 cells were transiently transfected with 100 nM of oligonucleotide-inhibitors (miR-143-inh or NC-inh). Following this treatment, miR-143 levels were significantly reduced in U87 cells by ~90% (* $p = 0.0211$; Figure 2B). These cells showed a reduced ability to proliferate as shown by the 78% reduction in the number of colonies (* $p = 0.0202$) compared to the NC-inh transfected cells (Figure 2C). Transient transfection of A-172 cells with 100 nM of miR-143-inh produced similar cell growth inhibitory effects (Figure 2D).

Taq-Man-based RT-PCR (qPCR) analysis with the total RNA extracted from stable transfected clones showed a 6.98-fold and 2.17-fold increase (compared with empty vector clones) in two of the miR-143 selected clones (Figure 2E). In a colony formation assay, the miR-143-overexpressed clone miR-143-1 grew significantly faster than cells expressing the empty vector clones (EV) (Figure 2F). Together, these results suggest that high miR-143 levels promote cell proliferation of GBM cells.

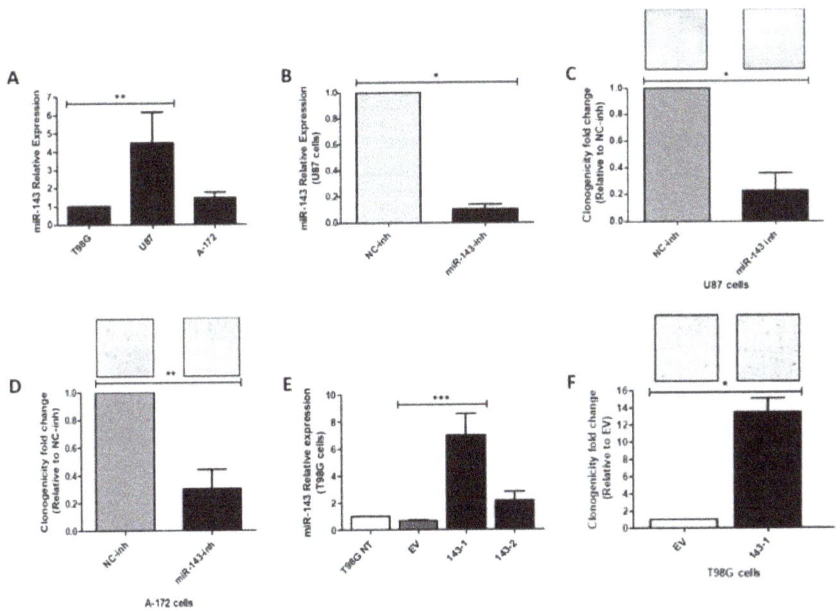

Figure 2. Effect of miR-143 inhibition or overexpression on cell proliferation. Total RNA was isolated, and qPCR was performed. (**A**) Relative miR-143 expression in a panel of Glioblastoma (GBM) cell lines, calculated relative to T98G cells; (**B**) Relative miR-143 expression after transient transfection of U87 with miR-inhibitors, calculated relative to the Negative control (NC) inhibitor. Colony formation assay after transfection of (**C**) U87 and (**D**) A-172 GBM cells with miR-143 inhibitor (miR-143-inh) or negative control inhibitor (NC-inh). (**E**) qPCR for relative miR-143 expression in empty vector (EV) and miR-143 T98G clones, calculated relative to the T98G non-treated (T98G NT) cells. (**F**) Colony formation assay of the T98G (143-1) miR-143 overexpressing clone and T98G Empty Vector (EV) clone. Columns represent the means of at least triplicates ± SEM (* $p < 0.05$, ** $p < 0.01$, *** $p < 0.001$).

2.3. *Effect of MiR-143 Targeting on Cell Cycle Progression and Apoptosis*

Next, we investigated whether the reduction of cell proliferation after miR-143 downregulation was due to activation of apoptosis and/or inhibition of cell cycle progression. Transient transfection of miR-143-inh in U87 GBM cells produced a cell cycle arrest in the G0/G1 to S phase 72 h post-transfection (Figure 3A). These results were further validated by western blot analysis of key proteins involved in G0/G1 to S phase transition. Particularly, cyclin A2, cyclin D3, and cyclin-dependent kinase 2 (Cdk2) protein levels decreased following miR-143 inhibition (Figure 3B,C).

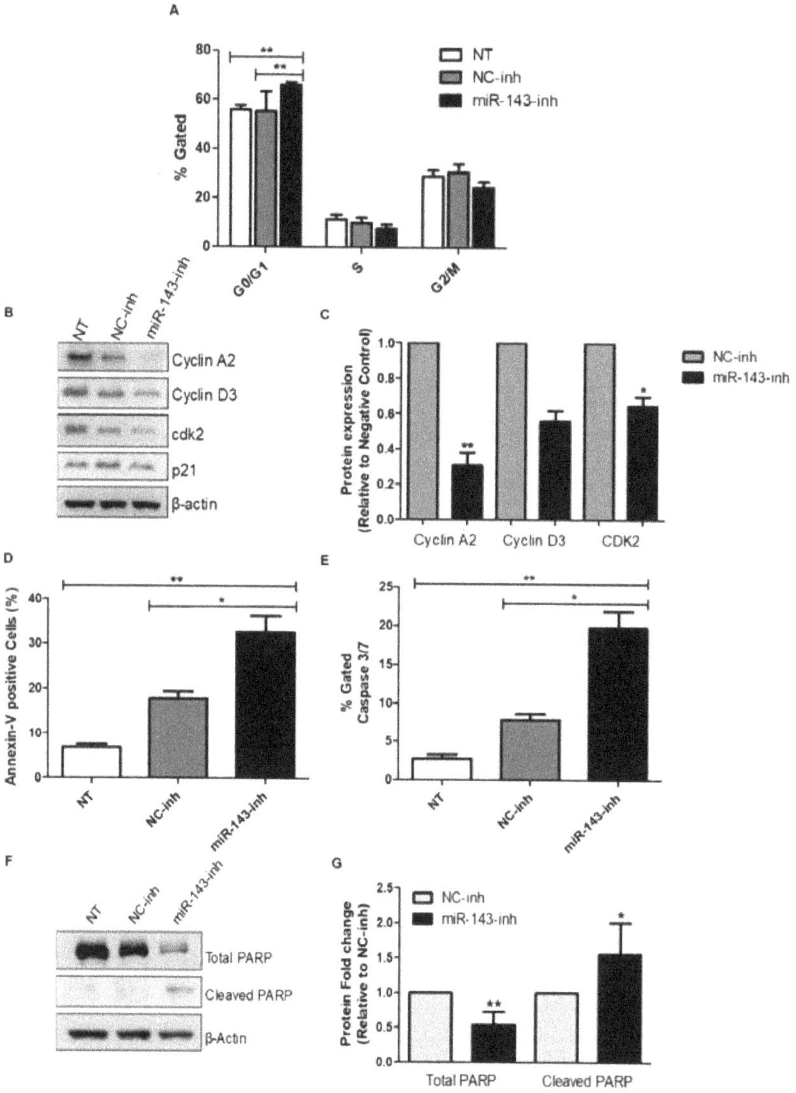

Figure 3. Inhibition of miR-143 induces apoptosis and cell cycle arrest. Apoptosis and cell cycle progression were measured by flow cytometry as described in the "Materials and Methods" section. U87 cells were transfected with 100 nM of negative control (NC-inh) or miR-143 inhibitor (miR-143-inh). (**A**) Seventy-two hours later, cells were fixed and cell cycle progression was assessed using the Muse Cell Analyzer. (**B**) Western blot analysis was performed 72 h after miR-inh transfection to detect changes in cell cycle-related proteins. (**C**) Densitometric analysis of the band intensities from (**B**) was performed and intensity values were expressed relative to NC-inh treated cells. U87 cells were treated as in (**A**), and 72 h later the Muse Cell Analyzer was used to measure apoptosis with (**D**) Annexin V and (**E**) Caspase 3/7 activity assays. (**F**) U87 cells were treated as in (**A**) and western blot analysis was performed to detect PARP-1 expression. (**G**) Densitometric analysis of the band intensities from (**F**) was performed and values were expressed relative to NC-inh treated cells. Columns represent the means of at least triplicates ± SEM (* $p < 0.05$, ** $p < 0.01$).

Furthermore, compared to NC-inh transfected cells, treatment of U87 cells with the miR-143-inh significantly increased apoptosis as evidenced by increased Annexin-V positive cells (Figure 3D) and caspase 3/7 levels (Figure 3E). Western blot analysis confirmed the induction of apoptosis showing a drastic reduction in the total poly (ADP-ribose) polymerase (PARP) protein levels accompanied by a significant increase in cleaved PARP (Figure 3F,G). Together, these results suggest that high levels of miR-143 protect GBM cells from apoptotic cell death.

2.4. MiR-143 Target Prediction and Validation

To identify further miR-143 target mRNAs in GBM cells, we performed in silico analysis using six different microRNA target identification softwares (Targetscan, Diana microT, miRPath, miRecords, miRDIP, miRGator). Over a hundred potential miR-143-regulated genes were identified by this approach (data not shown). Targets identified by at least three softwares generated a list of 21 potential miR-143-regulated genes (Figure 4A). Total RNA extracted from T98G cells transiently transfected with 100 nM of miR-143-inh or NC-inh was used for SYBR Green I-based qPCR analyses. Compared with the NC-inh transfected cells, transient transfection of miR-143-inh increased the mRNA levels of 6 out of the 21 miR-143 potential target genes (Figure 4A). These genes include SLC30A8, ITGA7, GXYLT1, ABL2, and ITM2B (Table 1). ELK-1 has been previously reported as a direct miR-143-regulated gene and was used as a positive control [22]. Western blot analysis was performed with protein lysates from miR-143 overexpressed clones 143-1 and 143-2 (Figure 2E). A reduction in the protein levels of ITM2B, SLC30A8, and ELK-1 was observed in the miR-143 overexpressed clones (143-1, 143-2) compared to EV or non-treated (NT) protein samples (Figure 4B). However, no observable changes in protein levels were detected for ABL2, GXYLT1, or ITGA7 in the samples tested (Figure 4B). Moreover, western blot analysis of U87 cells transiently transfected with 200 nM of miR-143-inh showed increases in expression of both ITM2B and SLC30A8 compared to the NC-inh U87 transfected cells (Figure 4C). These results suggest that both ITM2B and SLC30A8 are regulated by miR-143, either directly or indirectly.

Dual-luciferase reporter assays were performed to confirm that miR-143 binds directly to the 3′-UTR of the mRNA of ITM2B and SLC30A8. A plasmid containing the 3′-UTR region of each mRNA (ITM2B or SLC30A8) and a plasmid containing the pre-miR-143 were co-transfected, as described in the "Materials and Methods" section. The 3′-UTR of SLC30A8 was divided into two vectors due to its length (A and B). The sequence of the A and B 3′-UTR fragments are shown in Supplementary Figure S1. A reduction of 37% in the relative luciferase activity of the SLC30A8 (A) 3′-UTR (* $p < 0.05$) was observed, compared to the 3′-UTR control vector confirming direct binding of miR-143 to this target mRNA (Figure 4D). Supplementary Figure S2 shows the binding region of miR-143 to SLC30A8. However, changes in luciferase activity for ITM2B were not observed (Figure 4D). Ingenuity pathway analysis (IPA) shows that SLC30A8 is a zinc transporter protein associated with glucose metabolism (Figure 4E) [23]. Moreover, SLC30A8 has been shown to interact with CREB3L1 (cAMP responsive element binding protein 3-like 1), a transcription factor related to unfolded protein response of the Golgi apparatus and endoplasmic reticulum stress [24].

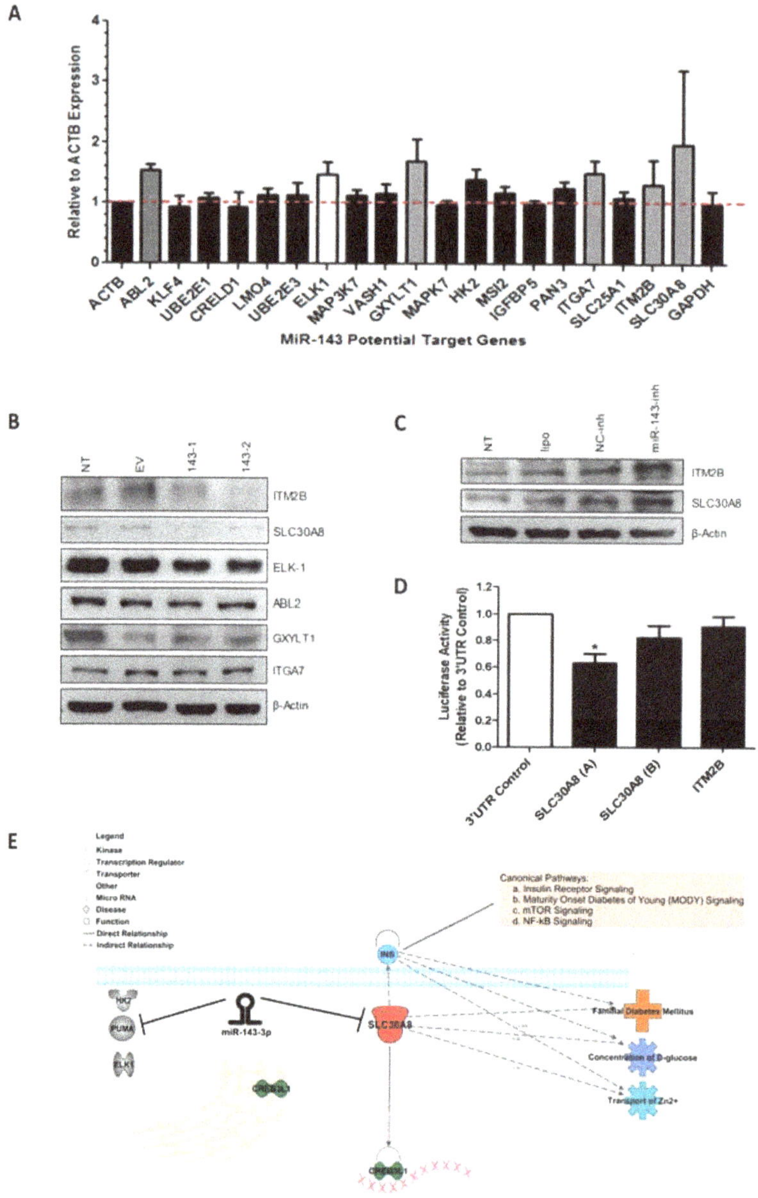

Figure 4. Identification of miR-143 target genes in GBM cells. (**A**) SYBR Green-based qPCR was performed with total RNA isolated from T98G cells transiently transfected with miR-143-inh or NC-inh. (**B**) Western blot analysis was performed with protein extracts from miR-143 overexpressed (143-1, 143-2) and EV clones. (**C**) Western blot analysis of protein extracts from U87 cells treated with 200 nM miR-143-inh or NC-inh, Non-treated cells (NT), and U87 cells with transfection reagent alone (lipo) as loading controls. (**D**) Dual-luciferase reporter assays were performed where luciferase activity was calculated relative to the NC-inh. (**E**) IPA analysis showing the interaction of miR-143 with its target genes and SLC30A8 association with glucose metabolism.

Table 1. List of the five miR-143 predicted target genes from qPCR results.

Full Name	Gene Symbol	Fold Change miR-143-Inh vs. NC-Inh in T98G Cells	Biological Role
Solute carrier family 30 (zinc transporter), member 8	SLC30A8	1.98	Role in glucose homeostasis; diabetes mellitus [25,26].
Integrin, alpha 7	ITGA7	1.51	Receptor for the basement membrane protein laminin-1 [27]; Tumor suppressor gene [28].
Glucoside xylosyltransferase 1	GXYLT1	1.69	Notch xylosyltransferase [29].
ABL proto-oncogene 2, non-receptor tyrosine kinase	ABL2	1.53	Oncogene related to cell migration and invasion in various cancers; role in cytoskeletal rearrangement, regulator of neuronal structural stability [30–32].
Integral membrane protein 2B	ITM2B	1.32	Role in triggering apoptosis P53 independent; tumor suppressor [33]. Transmembrane protein involved in negative regulation of amyloid processing. Related to dementia [34–36].

2.5. *In Vivo Targeting of MiR-143*

We then assessed the effect of targeting miR-143 on tumor progression in a GBM subcutaneous mouse model. MiRNA inhibitors were administered as a DOPC-PEG-cholesterol-based nanoliposome formulation [37]. The experimental design is illustrated in Figure 5A. Tumor-bearing mice were divided into two groups: DOPC-PEG-liposomal-miR-NC-Inh (N = 7) and DOPC-PEG-liposomal-miR-143-Inh (N = 9). Nine days after cell implantation, nude mice bearing U87 tumors were injected *i.p.* three times a week for three weeks. Each dose consisted of 2.5 µg of DOPC-PEG-liposomal-miR-NC-Inh or 2.5 µg of DOPC-PEG-liposomal-miR-143-Inh. Liposomal miR-143-inh treatment reduced tumor growth compared to the control group (Figure 5B). The differences among groups were statistically significant, particularly for the last two treatment sessions. To confirm that the reduction in tumor growth was caused by the inhibition of miR-143, we assessed the miR-143 expression levels in total RNA extracted from mouse tumor samples of both NC-inh and miR-143-inh treated groups. qPCR results showed a significant reduction in miR-143 expression levels in the miR-143-inh group compared to the NC-inh group (Figure 5C). Western blot analysis of protein extracted from mouse tumor samples showed increased SLC30A8 protein levels in three out of the four miR-143-inh samples (miR-143-inh-1, 3, 4) compared to the NC-inh group (Figure 5D). These results further confirm that SLC30A8 is a miR-143 target in GBM cells.

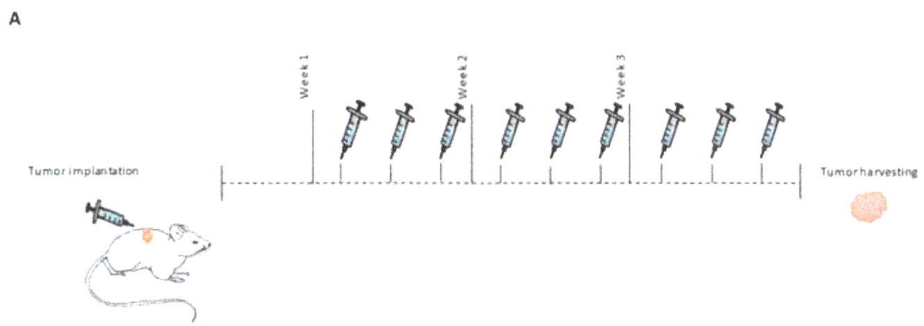

Figure 5. *Cont.*

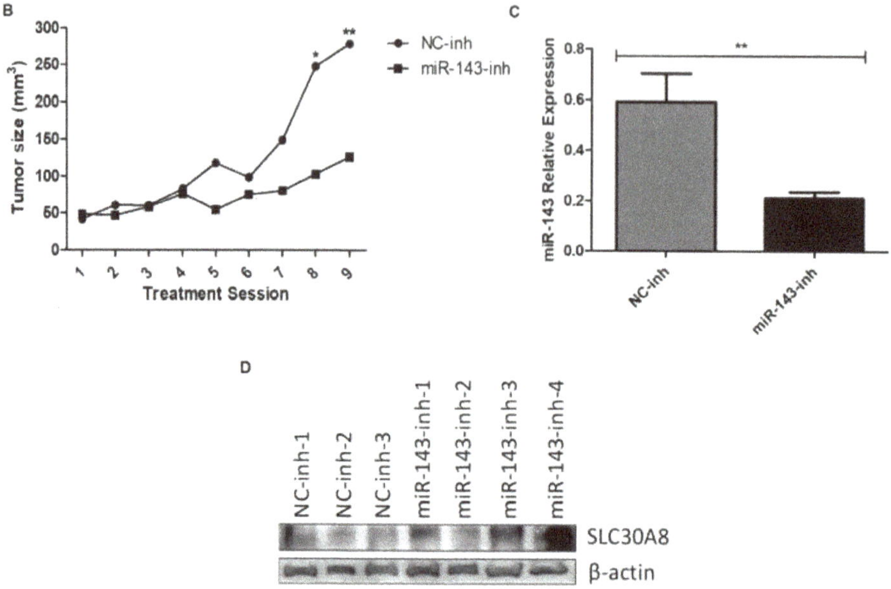

Figure 5. In vivo targeting of miR-143 reduces tumor growth. (**A**) Experimental design. (**B**) MiR-143 inhibition effect on tumor growth was calculated as described in the "Materials and Methods" section. (**C**) MiR-143 expression levels were measured by qPCR from total RNA extracted from mouse tumor tissues. Columns represent the means of N = 7 for NC-inh and N = 9 for miR-143-inh treatments ± SEM. (**D**) SLC30A8 expression levels were measured by western blot analysis in protein samples extracted from the mouse tumor tissues. * $p < 0.05$, ** $p < 0.01$.

3. Discussion

The major findings of this study are that miR-143 levels are increased in GBM patient samples compared to normal controls; and that targeting miR-143 reduced cell growth in vitro and tumor growth in a xenograft mouse model of GBM. Early studies showed divergences regarding the role of miR-143 in cancer [38]. In colorectal cancer, low levels of miR-143 have been associated with exacerbation of cancer proliferation in patient samples and cell lines [39]. Similar results were obtained in gastric cancer [40]. In contrast, a recent study showed that miR-143 expression in stromal cells of a mouse model of lung adenocarcinoma promoted tumor progression through neoangiogenesis [38]. Similarly, Koo et al. showed that serial selection for invasiveness in glioblastoma cell lines (U87, U251, U373, and rat glioma cell line C6) exhibited increased expression of miR-143 compared with parental GBM cells, suggesting an invasive oncogenic role of miR-143 in these cells [16].

While Wang and colleagues found that miR-143 is down-regulated in GBM tissues compared to normal brains [8], we obtained opposite results. Differences in the integrity of the tissue samples, the internal standard used for the real-time PCR experiments, and miRNA polymorphisms could contribute to these divergent results [41–43]. Furthermore, the presence of multiple cancer cell populations in a tumor (tumor heterogeneity) may result in diverse miRNA expression patterns, as evidenced by other miRNA expression profiles [2,44].

We observed a reduction in cell proliferation, activation of apoptosis, and cell cycle arrest following miR-143 inhibition. These results support the conclusion that high levels of miR-143 promote cell proliferation and protect GBM cells from apoptosis. In fact, PUMA (p53 upregulated modulator of apoptosis), a reported miR-143-target, is a proapoptotic protein that has been demonstrated to reduce the growth of subcutaneous U87 tumors in nude mice [45–47]. In our study, multiple injections of a liposomal-miR-143-Inh reduced tumor growth in a subcutaneous GBM mouse model. A similar cancer-driving role for miR-143 is supported by other studies, including Koo et al., where up-regulation of miR-143 increased the invasive ability of GBM cells [16,38]. These results corroborate the oncogenic role of miR-143 in GBM cells.

In this study, we identified SLC30A8 as a direct target of miR-143 in GBM cells. SLC30A8, or Solute carrier family 30 (zinc transporter) member 8, is a protein highly expressed in pancreatic beta cells. Evidence indicates that SLC30A8 is necessary for the transport of cytosolic zinc into insulin granules for further insulin maturation (Figure 5E) [25,48]. Interestingly, miR-143 has been reported to promote type 2 diabetes and adipocyte differentiation [49]. In different tumor types including breast, colon, and in GBM stem-like cells, miR-143 has been linked to glucose metabolism and regulation of cancer glycolysis via targeting hexokinase-2 [50–54]. As metabolic rewiring has been recently coined as a hallmark of cancer [55], further studies should be performed to uncover the role of miR-143 and its target gene SLC30A8, in metabolism pathways contributing to GBM progression and tumor maintenance. Again, as the expression of miRNA-143 (as most miRNAs) is tissue- and time-dependent, its role in the intracellular glucose accumulation and the progression of GBM should be carefully interpreted [2]. Of note, recent reports indicate that the circular RNA, DLGAP4 (circDLGAP4) [56]; and the Mir-143 host gene (MIR143HG) [57] act as miRNA sponges that sequester miRNA-143, thus, reducing the ability of this miRNA to bind to its cognate targets.

Moreover, miR-143 post-transcriptionally regulates other genes including HK2, PUMA, and ELK-1 [22,46,50–53]. The decreased expression of all these genes (including SLC30A8), as a result of the aberrant overexpression of miR-143, could contribute to the uncontrolled growth, proliferation, and tumor maintenance of GBM cells. Although direct binding of miR-143 to the mRNA of ITM2B was not observed, the changes in ITM2B protein levels upon miR-143 manipulation suggest that this protein is a miR-143 downstream effector in GBM cells. ITM2B is known as a tumor suppressor that triggers p53-independent apoptosis [33,36]. The molecular mechanism of how ITM2B is regulated by miR-143 in GBM cells should be further investigated.

4. Materials and Methods

4.1. Tumor Samples: RNA and Protein Isolation

Formalin-fixed paraffin-embedded (FFPE) tissue blocks (2008–2010) from 19 newly diagnosed (de novo) glioblastoma patients (12 females, seven males; median age 59) and five control patients (two females, three males; median age 38) were obtained from the Pathology Department of the University of Puerto Rico-Medical Sciences Campus. The control samples were from patients selected based on having non-neoplastic, non-infectious brain tissue with no metabolic conditions and without hemorrhagic or necrotic diathesis. The research protocol conducted was approved by the University of Puerto Rico-Medical Sciences Campus Institutional Research Board (IRB) on 26 January 2016 (protocol number: A9180112). A representative hematoxylin and eosin (H&E)-stained slide from each of the selected tissue blocks was evaluated by a pathologist and neurooncologist to corroborate diagnosis and delineate tumor tissue from normal brain tissue. RNA isolation followed the protocol previously described by Rivera et al. [14]. Briefly, for each FFPE tissue block, a 3-mm punch biopsy sample was obtained from the tumor area previously delineated by the pathologist. Samples were processed for total RNA isolation using the RecoverAll Total Nucleic Acid Isolation Kit (Ambion, Austin, TX, USA). RNA concentrations were measured using NanoDrop (Thermo Scientific, Wilmington, DE, USA). The threshold cycles (Ct) were used to calculate the relative miR-143 expression using U48 as the internal control [14,58]. RNA was isolated from mice tumor tissue with the RecoverAll Total Nucleic Acid Isolation kit as per the manufacturer instructions. Protein extraction from mice tumor tissue was performed using the same procedure of protein isolation described below. In both cases, the frozen tumor tissue was placed in liquid nitrogen, followed by tissue homogenization with a mortar and electric tissue homogenizer (D1000 Hand-held homogenizer, Benchmark Scientific, Sayreville, NJ, USA).

4.2. Cells and Culture Conditions

T98G, U87-MG (U87), and A-172 GBM human cancer cells (RRID: CVCL_0556, CVCL_0022, and CVCL_0131, respectively) were purchased from American Type Culture Collection (ATCC, (Manassas, VA, USA). These lines were authenticated by the ATCC on August 2016. The cells were maintained in adherent culture in Dulbecco's Modified Eagle Medium: Nutrient Mixture F-12 (DMEM/F-12; GIBCO) (Invitrogen Corporation) supplemented with 10% fetal bovine serum (FBS) (Thermo Scientific, Logan, UT, USA), 100 U/mL penicillin/streptomycin (Thermo Scientific) at 37 °C in a humidified chamber with 5% CO_2. In vitro assays were performed at 75−85% cell density in passages 1−8.

4.3. Transient and Stable Transfections

For transient transfections, U87 cells (3×10^4 cells/mL) were plated in 6-well plates or Petri dishes. After 24 h, a mixture of 100 nM (final concentration) of miRNA oligonucleotide-inhibitors (negative control and miR-143) (Life Technologies, Grand Island, NY, USA), lipofectamine RNAiMAX (Life Technologies) (1:1 ratio, v/v), and Opti-MEM I (Life Technologies) was added to the cells for 6−8 h. Then, the media was changed to regular DMEM with 10% FBS. For stable transfections, T98G cells were plated in a 6-well plate (3×10^4 cells/mL). After 24 h, a mixture of 3 µg of miR-143 (pCMV-MIR143) or Empty (pCMV-EV) OriGene vectors (OriGene Technologies, Inc. Rockville, MD), lipofectamine RNAiMAX (Life Technologies) (1:1 ratio, v/v), and Opti-MEM I (Life Technologies) was added to the cells for 6−8 h. These vectors contain a Neomycin resistance cassette, which was used for mammalian cell clone selection and maintenance. Independent clones were picked and grown individually. RNA was isolated using the mirVana miRNA Isolation Kit (Ambion, Austin, TX, USA), following manufacturer's instructions. To monitor miR-143 expression in empty vector and miR-143-overexpressing clones we used Taqman based qPCR (Applied Biosystems, Life Technologies, NY, USA). Threshold cycles (Ct) were used to calculate the relative miR-143 expression using U48 as an internal control [14,58].

4.4. Colony Formation Assay

Cell proliferation was assessed by a clonogenic assay. U87 cells were seeded in a 6-well plate (3.0×10^4 cells/mL). The next day cells were transiently transfected with miR-143 oligonucleotide-inhibitor (miR-143-inh) or negative-control inhibitor (NC-inh) as described above. The next day cells were collected, and 1000 cells were seeded in 10 cm-Petri dishes. Ten days later, colonies were stained with 0.5% crystal violet in methanol and counted using an Eclipse TS100 microscope (Nikon, Minato, Tokyo, Japan) [59]. Colony formation assay was also performed with T98G miR-143-overexpressed clones without the oligonucleotide-inhibitor transfection step.

4.5. Assessment of Cell Cycle and Apoptosis

U87 cells were transiently transfected with a 100 nM (final concentration) of miR-143-inh or NC-inh as described above. Apoptosis was evaluated by flow cytometry analysis 72 h post-transfection using both Muse Annexin V Dead Cell Kit and Caspase 3/7 Kits (EMD Millipore Headquarters, Burlington, MA, USA). Cell cycle progression was evaluated by flow cytometry 72 h post-transfection with the Cell Cycle Kit (EMD Millipore Headquarters). Data was analyzed and collected using the Muse Cell Analyzer (EMD Millipore Headquarters). For western blot analysis, U87 transiently transfected (miR-143-inh or NC-inh oligonucleotides) cells were collected for western blot analysis.

4.6. MiRNA Target Prediction and SYBR-Green I RT-PCR for Target Identification

Online target prediction softwares (microrna.org, Diana-microT, Target Scan Human, mirPath, miRecords, miRGator, and mRDB) were used to identify potential target genes of miR-143. Total RNA from miR-143-inh and NC-inh samples were subjected to reverse transcription using the iScript cDNA Synthesis Kit from Bio-Rad Laboratories (Hercules, CA, USA). Data was collected and analyzed using a StepOne Software v2.1 from Applied Biosystems following manufacturer's protocol (95 °C for 2 min followed by 40 cycles of 5 s at 95 °C and 30 s at 60 °C). β-actin was used as the internal standard for the gene expression values [59,60]. We used the multalin software (http://multalin.toulouse.inra.fr/multalin/) [61] for alignment analysis of miR-143-3p (mature strand) and the SLC30A8(A) fragment mRNA.

4.7. Western Blot Analysis

Cells were collected, washed with PBS 1X, and stored at −80 °C until processed. For protein extraction, cells were lysed on ice with lysis buffer (1% Triton X, 150 mM NaCl, 25 mM Tris HCl, 0.4 mM $NaVO_4$, 0.4 mM NaF and protease inhibitor cocktail from Sigma, St. Louis, MO, USA) for 30 min vortexing periodically. Lysates were centrifuged for 15 min at 4 °C, supernatants collected, and total protein concentration was determined using Bio-Rad DC Protein Assay reagents (Bio-Rad) following the manufacturer's protocol. Equal protein quantities for each sample (30 or 50 μg per lane) were separated by SDS-PAGE, blotted onto nitrocellulose membranes, blocked with 5% non-fat milk, and probed with the appropriate dilution of the corresponding primary antibody. Once incubated with the primary antibody, membranes were rinsed and incubated with the corresponding HRP-conjugated secondary antibody. Bound antibodies were detected using an enhanced chemiluminescence substrate followed by autoradiography using a FluorChemTM 8900 (Alpha Innotech Corporation, San Leandro, CA, USA). Primary antibodies used: anti-ITM2B (1C11) (3 0kDa), anti-ELK-1 (E277) (45 kDa), anti-ITGA7 (129 kDa), anti-ABL2 (EPR1222(2)) (128 kDa), anti-SLC30A8 (40 kDa), anti-GXYLT1 (51 kDa) (Abcam, Cambridge, UK); cdk2 (78B2] (33 kDa), Cyclin D3 (DCS22) (31 kDa), p21 (12D1) (21 kDa), Cyclin A2 (BF683) (55 kDa), PARP (89, 116 kDa) (Cell Signaling, Danvers, MA, USA); and anti-β-actin (42 kDa) (Sigma). Secondary antibodies used: anti-mouse and anti-rabbit IgG horseradish peroxidase (HRP) (Cell Signaling).

4.8. Dual-Luciferase Reporter Assays

T98G cells (3.5×10^4 cells/mL) were transiently transfected with 1.5 μg pre-miR vectors (miR-143 and empty vector control) and lipofectamine RNAiMax (1:1 ratio, v/v) in Opti-MEM I media. After 6 h of transfection cells were washed with PBS 1X and the second transfection of 1.5 μg of dual Firefly/Renilla luciferase reporter mammalian expression vectors (pEZX-MT06; GeneCopoeia, Rockville, MD, USA) was performed. Vectors included the 3′-UTR of SLC30A8 (in two portions-A and B), ITM2B, and an empty vector control. The transfection mix included the 3′-UTR vector, lipofectamine RNAiMax (Life Technologies) (1:1 ratio, v/v) and Opti-MEM I which was added for another 6 h. Then, the media was changed to fresh DME/F-12 (10% FBS and 0.1% penicillin/streptomycin). After 48 h firefly and renilla activity were measured in a Glomax 20/20 luminometer using the Dual-Luciferase Reporter Assay System kit (Promega, Madison, WI) following the manufacturer's protocol. The relative luciferase activity was calculated and graphed relative to the negative-control inhibitor samples with each 3′-UTR vector.

4.9. Tumor Implantation and Treatment

Male (10) and female (10) BALB/c nude mice of 4 to 6 weeks of age were purchased from Taconic Biosciences (Rensselaer, NY, USA). U87 cells were subcutaneously (*s.c.*) injected into the right dorsal flank (2.0×10^6 cells/200 μL in PBS/Matrigel mixture). Tumor size was measured each treatment session (every three days) with a caliper. Tumor volumes were calculated using the following formula: volume = (L × W × H) × 0.5, where L is the length (longest diameter), W is the weight (thickness), and H is the height (shorter diameter). Liposomal-miRNA-inh administration initiated once tumors were visible (after 7 days). Mice were injected intraperitoneally (*i.p.*) three times a week for three weeks with DOPC-PEG-liposomal-miR-NC-Inh or DOPC-PEG-liposomal-miR-143-Inh. Liposomes were composed of miRNA-inh; 1,2-dioleoyl-sn-glycero-3-phosphocholine (DOPC) (1:10 w/w ratio); (1,2-distearoyl-sn-glycero-3-phosphoethanolamine-*N*-(amino(polyethylene glycol)-2000) (ammonium salt) – (DSPE-PEG-2000) (5% mol/mol of DOPC); and cholesterol (25% w/w of DOPC). This liposome formulation has been previously described [37]. At the end of treatment, mice were euthanized, tumors were measured and weighted. Tumor samples were processed, and total RNA extraction was obtained with the mirVana miRNA Isolation Kit (Life Technologies, Thermo Fisher Scientific, Waltham, MA, USA) as per de protocol instructions. MiR-143 and U48 (internal control) were measured by real-time PCR with TaqMan specific probes (Applied Biosystems). Animal handling and research protocols were approved by the Institutional Animal Care and Use Committee (IACUC) of the University of Puerto Rico, Medical Sciences Campus on 10 April 2018 (protocol number: A870110).

4.10. Statistical Analysis

Statistical analysis was performed using GraphPad Prism 5 (GraphPad Software, Inc., La Jolla, CA, USA). Data was analyzed using Student's t-test for comparing two groups and ANOVA tests for multiple group comparisons, with $p < 0.05$ considered statistically significant (* $p < 0.05$, ** $p < 0.01$, *** $p < 0.001$). All experiments were performed at least in triplicates.

5. Conclusions

We provide evidence that miR-143 levels are increased in GBM patients and that miR-143 acts as an oncogene by promoting cell proliferation and survival of GBM cells. Further preclinical studies using orthotopic xenograft models and direct administration of the miR-143 inhibitor into an orthotopic brain tumor should be conducted to confirm that miR-143 is a viable molecular target for GBM treatment. We also identified SLC30A8 as a direct target mRNA of miR-143 in GBM cells in vitro and in vivo. Further studies should elucidate the role of this novel cancer-associated gene in GBM.

Supplementary Materials: Supplementary materials can be found at http://www.mdpi.com/2072-6694/10/10/382/s1, Figure S1: SCL30A8 3′UTR vectors A and B sequences, Figure S2: Sequence alignment of miR-143 and SLC30A8 3′UTR.

Author Contributions: Conceptualization: E.L.L.-D. and P.E.V.-M.; Methodology: E.L.L.-D. and P.E.V.-M.; Validation: E.L.L.-D. and P.E.V.-M.; Formal Analysis: E.L.L.-D.; Investigation: E.L.L.-D., N.G.-R., M.A.M.-R., Y.S.-R., F.V., M.R.-D. and M.J.M.-M.; Resources: M.J.M.-M. and P.E.V.-M.; Data Curation: E.L.L.-D.; Writing-Original Draft Preparation: E.L.L.-D.; Writing-Review & Editing: E.L.L.-D., M.A.M.-R. and P.E.V.-M.; Visualization: E.L.L.-D., N.G.-R. and P.E.V.-M.; Supervision: P.E.V.-M.; Project Administration: E.L.L.-D. and P.E.V.-M.; Funding Acquisition: E.L.L.-D. and P.E.V.-M. All authors reviewed the results and approved the final version of the manuscript.

Funding: This research was funded by RCMI grant U54 MD007600 (National Institute on Minority Health and Health Disparities) from the National Institutes of Health, institutional seed funds from the University of Puerto Rico Comprehensive Cancer Center (P.E.V.-M.), and the NIGMS-RISE Grant Number R25-GM061838 (E.L.L.-D., N.G.-R.).

Acknowledgments: This manuscript is a partial fulfillment of the doctoral dissertation of Eunice L. Lozada-Delgado. We would like to acknowledge Gwilz for the adapted mouse vector image used in Figure 4A (license: https://creativecommons.org/licenses/by-sa/4.0/legalcode). We would like to thank Surangani Dharmawardhane Flanagan, Ph.D., for critical reading of the manuscript.

Conflicts of Interest: The authors declare no conflict of interest.

References

1. Furnari, F.B.; Fenton, T.; Bachoo, R.M.; Mukasa, A.; Stommel, J.M.; Stegh, A.; Hahn, W.C.; Ligon, K.L.; Louis, D.N.; Brennan, C.; et al. Malignant astrocytic glioma: Genetics, biology, and paths to treatment. *Genes Dev.* **2007**, *21*, 2683–2710. [CrossRef] [PubMed]
2. Lozada-Delgado, E.L.; Grafals-Ruiz, N.; Vivas-Mejía, P.E. RNA interference for glioblastoma therapy: Innovation ladder from the bench to clinical trials. *Life Sci.* **2017**, *188*, 26–36. [CrossRef] [PubMed]
3. Shugg, D.; Allen, B.J.; Blizzard, L.; Dwyer, T.; Roder, D. Brain cancer incidence, mortality and case survival: Observations from two australian cancer registries. *Int. J. Cancer* **1994**, *59*, 765–770. [CrossRef] [PubMed]
4. Ohgaki, H.; Kleihues, P. Epidemiology and etiology of gliomas. *Acta Neuropathol.* **2005**, *109*, 93–108. [CrossRef] [PubMed]

5. Kleihues, P.; Burger, P.C.; Scheithauer, B.W. The new who classification of brain tumours. *Brain Pathol.* **1993**, *3*, 255–268. [CrossRef] [PubMed]
6. Adamson, C.; Kanu, O.O.; Mehta, A.I.; Di, C.; Lin, N.; Mattox, A.K.; Bigner, D.D. Glioblastoma multiforme: A review of where we have been and where we are going. *Expert Opin. Investig. Drugs* **2009**, *18*, 1061–1083. [CrossRef] [PubMed]
7. Sathornsumetee, S.; Reardon, D.A.; Desjardins, A.; Quinn, J.A.; Vredenburgh, J.J.; Rich, J.N. Molecularly targeted therapy for malignant glioma. *Cancer* **2007**, *110*, 13–24. [CrossRef] [PubMed]
8. Wang, L.; Shi, Z.M.; Jiang, C.F.; Liu, X.; Chen, Q.D.; Qian, X.; Li, D.M.; Ge, X.; Wang, X.F.; Liu, L.Z.; et al. Mir-143 acts as a tumor suppressor by targeting n-ras and enhances temozolomide-induced apoptosis in glioma. *Oncotarget* **2014**, *5*, 5416–5427. [CrossRef] [PubMed]
9. Bartel, D.P. Micrornas: Genomics, biogenesis, mechanism, and function. *Cell* **2004**, *116*, 281–297. [CrossRef]
10. Tay, Y.; Zhang, J.; Thomson, A.M.; Lim, B.; Rigoutsos, I. Micrornas to nanog, oct4 and sox2 coding regions modulate embryonic stem cell differentiation. *Nature* **2008**, *455*, 1124–1128. [CrossRef] [PubMed]
11. Gu, W.; Xu, Y.; Xie, X.; Wang, T.; Ko, J.H.; Zhou, T. The role of rna structure at 5′ untranslated region in microrna-mediated gene regulation. *RNA* **2014**, *20*, 1369–1375. [CrossRef] [PubMed]
12. Wang, H.; Xu, T.; Jiang, Y.; Yan, Y.; Qin, R.; Chen, J. Micrornas in human glioblastoma: From bench to beside. *Front. Biosci.* **2015**, *20*, 105–118.
13. Genovese, G.; Ergun, A.; Shukla, S.A.; Campos, B.; Hanna, J.; Ghosh, P.; Quayle, S.N.; Rai, K.; Colla, S.; Ying, H.; et al. Microrna regulatory network inference identifies mir-34a as a novel regulator of tgf-β signaling in glioblastoma. *Cancer Discov.* **2012**, *2*, 736–749. [CrossRef] [PubMed]
14. Rivera-Díaz, M.; Miranda-Román, M.A.; Soto, D.; Quintero-Aguilo, M.; Ortiz-Zuazaga, H.; Marcos-Martinez, M.J.; Vivas-Mejía, P.E. Microrna-27a distinguishes glioblastoma multiforme from diffuse and anaplastic astrocytomas and has prognostic value. *Am. J. Cancer Res.* **2015**, *5*, 201–218. [PubMed]
15. Iio, A.; Nakagawa, Y.; Hirata, I.; Naoe, T.; Akao, Y. Identification of non-coding rnas embracing microrna-143/145 cluster. *Mol. Cancer* **2010**, *9*, 136. [CrossRef] [PubMed]
16. Koo, S.; Martin, G.S.; Schulz, K.J.; Ronck, M.; Toussaint, L.G. Serial selection for invasiveness increases expression of mir-143/mir-145 in glioblastoma cell lines. *BMC Cancer* **2012**, *12*, 143. [CrossRef] [PubMed]
17. Kent, O.A.; McCall, M.N.; Cornish, T.C.; Halushka, M.K. Lessons from mir-143/145: The importance of cell-type localization of mirnas. *Nucleic Acids Res.* **2014**, *42*, 7528–7538. [CrossRef] [PubMed]
18. Mishra, S.; Yadav, T.; Rani, V. Exploring mirna based approaches in cancer diagnostics and therapeutics. *Crit. Rev. Oncol. Hematol.* **2016**, *98*, 12–23. [CrossRef] [PubMed]
19. Xu, B.; Niu, X.; Zhang, X.; Tao, J.; Wu, D.; Wang, Z.; Li, P.; Zhang, W.; Wu, H.; Feng, N.; et al. Mir-143 decreases prostate cancer cells proliferation and migration and enhances their sensitivity to docetaxel through suppression of kras. *Mol. Cell. Biochem.* **2011**, *350*, 207–213. [CrossRef] [PubMed]
20. Li, J.; Smyth, P.; Flavin, R.; Cahill, S.; Denning, K.; Aherne, S.; Guenther, S.M.; O'Leary, J.J.; Sheils, O. Comparison of mirna expression patterns using total rna extracted from matched samples of formalin-fixed paraffin-embedded (ffpe) cells and snap frozen cells. *BMC Biotechnol.* **2007**, *7*, 36. [CrossRef] [PubMed]
21. Xi, Y.; Nakajima, G.; Gavin, E.; Morris, C.G.; Kudo, K.; Hayashi, K.; Ju, J. Systematic analysis of microrna expression of rna extracted from fresh frozen and formalin-fixed paraffin-embedded samples. *RNA* **2007**, *13*, 1668–1674. [CrossRef] [PubMed]
22. Cordes, K.R.; Sheehy, N.T.; White, M.P.; Berry, E.C.; Morton, S.U.; Muth, A.N.; Lee, T.H.; Miano, J.M.; Ivey, K.N.; Srivastava, D. Mir-145 and mir-143 regulate smooth muscle cell fate and plasticity. *Nature* **2009**, *460*, 705–710. [CrossRef] [PubMed]
23. Krämer, A.; Green, J.; Pollard, J.; Tugendreich, S. Causal analysis approaches in ingenuity pathway analysis. *Bioinformatics* **2014**, *30*, 523–530. [CrossRef] [PubMed]
24. Murakami, T.; Kondo, S.; Ogata, M.; Kanemoto, S.; Saito, A.; Wanaka, A.; Imaizumi, K. Cleavage of the membrane-bound transcription factor oasis in response to endoplasmic reticulum stress. *J. Neurochem.* **2006**, *96*, 1090–1100. [CrossRef] [PubMed]
25. Chimienti, F.; Devergnas, S.; Favier, A.; Seve, M. Identification and cloning of a beta-cell-specific zinc transporter, znt-8, localized into insulin secretory granules. *Diabetes* **2004**, *53*, 2330–2337. [CrossRef] [PubMed]

26. Sladek, R.; Rocheleau, G.; Rung, J.; Dina, C.; Shen, L.; Serre, D.; Boutin, P.; Vincent, D.; Belisle, A.; Hadjadj, S.; et al. A genome-wide association study identifies novel risk loci for type 2 diabetes. *Nature* **2007**, *445*, 881–885. [CrossRef] [PubMed]
27. von der Mark, H.; Williams, I.; Wendler, O.; Sorokin, L.; von der Mark, K.; Pöschl, E. Alternative splice variants of alpha 7 beta 1 integrin selectively recognize different laminin isoforms. *J. Biol. Chem.* **2002**, *277*, 6012–6016. [CrossRef] [PubMed]
28. Ren, B.; Yu, Y.P.; Tseng, G.C.; Wu, C.; Chen, K.; Rao, U.N.; Nelson, J.; Michalopoulos, G.K.; Luo, J.H. Analysis of integrin alpha7 mutations in prostate cancer, liver cancer, glioblastoma multiforme, and leiomyosarcoma. *J. Natl. Cancer Inst.* **2007**, *99*, 868–880. [CrossRef] [PubMed]
29. Sethi, M.K.; Buettner, F.F.; Krylov, V.B.; Takeuchi, H.; Nifantiev, N.E.; Haltiwanger, R.S.; Gerardy-Schahn, R.; Bakker, H. Identification of glycosyltransferase 8 family members as xylosyltransferases acting on o-glucosylated notch epidermal growth factor repeats. *J. Biol. Chem.* **2010**, *285*, 1582–1586. [CrossRef] [PubMed]
30. Lin, Y.C.; Yeckel, M.F.; Koleske, A.J. Abl2/arg controls dendritic spine and dendrite arbor stability via distinct cytoskeletal control pathways. *J. Neurosci.* **2013**, *33*, 1846–1857. [CrossRef] [PubMed]
31. Procaccia, V.; Nakayama, H.; Shimizu, A.; Klagsbrun, M. Gleevec/imatinib, an abl2 kinase inhibitor, protects tumor and endothelial cells from semaphorin-induced cytoskeleton collapse and loss of cell motility. *Biochem. Biophys. Res. Commun.* **2014**, *448*, 134–138. [CrossRef] [PubMed]
32. Xing, Q.T.; Qu, C.M.; Wang, G. Overexpression of abl2 predicts poor prognosis in hepatocellular carcinomas and is associated with cancer cell migration and invasion. *Onco Targets Ther.* **2014**, *7*, 881–885. [CrossRef] [PubMed]
33. Fleischer, A.; Rebollo, A. Induction of p53-independent apoptosis by the bh3-only protein itm2bs. *FEBS Lett.* **2004**, *557*, 283–287. [CrossRef]
34. Vidal, R.; Frangione, B.; Rostagno, A.; Mead, S.; Révész, T.; Plant, G.; Ghiso, J. A stop-codon mutation in the bri gene associated with familial british dementia. *Nature* **1999**, *399*, 776–781. [CrossRef] [PubMed]
35. Tamayev, R.; Giliberto, L.; Li, W.; d'Abramo, C.; Arancio, O.; Vidal, R.; D'Adamio, L. Memory deficits due to familial british dementia bri2 mutation are caused by loss of bri2 function rather than amyloidosis. *J. Neurosci.* **2010**, *30*, 14915–14924. [CrossRef] [PubMed]
36. Fleischer, A.; Ayllon, V.; Rebollo, A. Itm2bs regulates apoptosis by inducing loss of mitochondrial membrane potential. *Eur. J. Immunol.* **2002**, *32*, 3498–3505. [CrossRef]
37. Reyes-González, J.M.; Armaiz-Peña, G.N.; Mangala, L.S.; Valiyeva, F.; Ivan, C.; Pradeep, S.; Echevarría-Vargas, I.M.; Rivera-Reyes, A.; Sood, A.K.; Vivas-Mejía, P.E. Targeting c-myc in platinum-resistant ovarian cancer. *Mol. Cancer Ther.* **2015**, *14*, 2260–2269. [CrossRef] [PubMed]
38. Dimitrova, N.; Gocheva, V.; Bhutkar, A.; Resnick, R.; Jong, R.M.; Miller, K.M.; Bendor, J.; Jacks, T. Stromal expression of mir-143/145 promotes neoangiogenesis in lung cancer development. *Cancer Discov.* **2016**, *6*, 188–201. [CrossRef] [PubMed]
39. Michael, M.Z.; O' Connor, S.M.; van Holst Pellekaan, N.G.; Young, G.P.; James, R.J. Reduced accumulation of specific micrornas in colorectal neoplasia. *Mol. Cancer Res.* **2003**, *1*, 882–891. [PubMed]
40. Takagi, T.; Iio, A.; Nakagawa, Y.; Naoe, T.; Tanigawa, N.; Akao, Y. Decreased expression of microrna-143 and -145 in human gastric cancers. *Oncology* **2009**, *77*, 12–21. [CrossRef] [PubMed]
41. Dang, W.; Sun, L. Determination of internal controls for quantitative real time rt-pcr analysis of the effect of edwardsiella tarda infection on gene expression in turbot (scophthalmus maximus). *Fish Shellfish Immunol.* **2011**, *30*, 720–728. [CrossRef] [PubMed]
42. Wang, X.; Ren, H.; Zhao, T.; Ma, W.; Dong, J.; Zhang, S.; Xin, W.; Yang, S.; Jia, L.; Hao, J. Single nucleotide polymorphism in the microrna-199a binding site of hif1a gene is associated with pancreatic ductal adenocarcinoma risk and worse clinical outcomes. *Oncotarget* **2016**, *7*, 13717–13729. [CrossRef] [PubMed]
43. Yin, Z.; Cui, Z.; Ren, Y.; Xia, L.; Wang, Q.; Zhang, Y.; He, Q.; Zhou, B. Association between polymorphisms in pre-mirna genes and risk of lung cancer in a chinese non-smoking female population. *Lung Cancer* **2016**, *94*, 15–21. [CrossRef] [PubMed]
44. Alves, T.R.; Lima, F.R.; Kahn, S.A.; Lobo, D.; Dubois, L.G.; Soletti, R.; Borges, H.; Neto, V.M. Glioblastoma cells: A heterogeneous and fatal tumor interacting with the parenchyma. *Life Sci.* **2011**, *89*, 532–539. [CrossRef] [PubMed]
45. Yu, J.; Zhang, L. Puma, a potent killer with or without p53. *Oncogene* **2008**, *27*, S71–S83. [CrossRef] [PubMed]

46. Bai, Y.; Zhang, Y.; Hua, J.; Yang, X.; Zhang, X.; Duan, M.; Zhu, X.; Huang, W.; Chao, J.; Zhou, R.; et al. Silencing microrna-143 protects the integrity of the blood-brain barrier: Implications for methamphetamine abuse. *Sci. Rep.* **2016**, *6*, 35642. [CrossRef] [PubMed]
47. Ito, H.; Kanzawa, T.; Miyoshi, T.; Hirohata, S.; Kyo, S.; Iwamaru, A.; Aoki, H.; Kondo, Y.; Kondo, S. Therapeutic efficacy of puma for malignant glioma cells regardless of p53 status. *Hum. Gene Ther.* **2005**, *16*, 685–698. [CrossRef] [PubMed]
48. Merriman, C.; Huang, Q.; Rutter, G.A.; Fu, D. Lipid-tuned zinc transport activity of human znt8 protein correlates with risk for type-2 diabetes. *J. Biol. Chem.* **2016**, *291*, 26950–26957. [CrossRef] [PubMed]
49. Li, B.; Fan, J.; Chen, N. A novel regulator of type ii diabetes: Microrna-143. *Trends Endocrinol. Metab.* **2018**, *29*, 380–388. [CrossRef] [PubMed]
50. Jiang, S.; Zhang, L.F.; Zhang, H.W.; Hu, S.; Lu, M.H.; Liang, S.; Li, B.; Li, Y.; Li, D.; Wang, E.D.; et al. A novel mir-155/mir-143 cascade controls glycolysis by regulating hexokinase 2 in breast cancer cells. *EMBO J.* **2012**, *31*, 1985–1998. [CrossRef] [PubMed]
51. Fang, R.; Xiao, T.; Fang, Z.; Sun, Y.; Li, F.; Gao, Y.; Feng, Y.; Li, L.; Wang, Y.; Liu, X.; et al. Microrna-143 (mir-143) regulates cancer glycolysis via targeting hexokinase 2 gene. *J. Biol. Chem.* **2012**, *287*, 23227–23235. [CrossRef] [PubMed]
52. Peschiaroli, A.; Giacobbe, A.; Formosa, A.; Markert, E.K.; Bongiorno-Borbone, L.; Levine, A.J.; Candi, E.; D'Alessandro, A.; Zolla, L.; Finazzi Agrò, A.; et al. Mir-143 regulates hexokinase 2 expression in cancer cells. *Oncogene* **2013**, *32*, 797–802. [CrossRef] [PubMed]
53. Gregersen, L.H.; Jacobsen, A.; Frankel, L.B.; Wen, J.; Krogh, A.; Lund, A.H. Microrna-143 down-regulates hexokinase 2 in colon cancer cells. *BMC Cancer* **2012**, *12*, 232. [CrossRef] [PubMed]
54. Zhao, S.; Liu, H.; Liu, Y.; Wu, J.; Wang, C.; Hou, X.; Chen, X.; Yang, G.; Zhao, L.; Che, H.; et al. Mir-143 inhibits glycolysis and depletes stemness of glioblastoma stem-like cells. *Cancer Lett.* **2013**, *333*, 253–260. [CrossRef] [PubMed]
55. Pavlova, N.N.; Thompson, C.B. The emerging hallmarks of cancer metabolism. *Cell Metab.* **2016**, *23*, 27–47. [CrossRef] [PubMed]
56. Bai, Y.; Zhang, Y.; Han, B.; Yang, L.; Chen, X.; Huang, R.; Wu, F.; Chao, J.; Liu, P.; Hu, G.; et al. Circular rna dlgap4 ameliorates ischemic stroke outcomes by targeting mir-143 to regulate endothelial-mesenchymal transition associated with blood-brain barrier integrity. *J. Neurosci.* **2018**, *38*, 32–50. [PubMed]
57. Du, C.; Shen, Z.; Zang, R.; Xie, H.; Li, H.; Chen, P.; Hang, B.; Xu, X.; Tang, W.; Xia, Y. Negative feedback circuitry between mir143hg and rbm24 in hirschsprung disease. *Biochim. Biophys. Acta* **2016**, *1862*, 2127–2136. [CrossRef] [PubMed]
58. Reyes-Gonzalez, J.M.; Pietri-Vazquez, F.M.; Vivas-Mejia, P.E. A general overview of the nano-sized carriers for cancer treatment. In *Nano Based Drug Delivery*; Naik, J., Ed.; IAPC Publishing: Zagreb, Croacia, 2015; pp. 317–333.
59. Echevarría-Vargas, I.M.; Valiyeva, F.; Vivas-Mejía, P.E. Upregulation of mir-21 in cisplatin resistant ovarian cancer via jnk-1/c-jun pathway. *PLoS ONE* **2014**, *9*, e97094. [CrossRef] [PubMed]
60. Vargas, I.M.; Vivas-Mejía, P.E. Assessment of mrna splice variants by qrt-pcr. *Methods Mol. Biol.* **2013**, *1049*, 171–186. [PubMed]
61. Corpet, F. Multiple sequence alignment with hierarchical clustering. *Nucleic Acids Res.* **1988**, *16*, 10881–10890. [CrossRef] [PubMed]

cancers

MDPI

Article

Intronic miR-744 Inhibits Glioblastoma Migration by Functionally Antagonizing Its Host Gene MAP2K4

Max Hübner [1,2,†], Christian Ludwig Hinske [1,†], David Effinger [1,2], Tingting Wu [1,2], Niklas Thon [3], Friedrich-Wilhelm Kreth [3] and Simone Kreth [1,2,*]

1 Department of Anesthesiology, University Hospital, LMU Munich, 81377 Munich, Germany; max.huebner@med.uni-muenchen.de (M.H.); christian.hinske@med.uni-muenchen.de (C.L.H.); david.effinger@med.uni-muenchen.de (D.E.); Tingting.wu@med.uni-muenchen.de (T.W.)

2 Walter-Brendel Center of Experimental Medicine, Faculty of Medicine, LMU Munich, 81377 Munich, Germany

3 Department of Neurosurgery, University Hospital, LMU Munich, 81377 Munich, Germany; Niklas.Thon@med.uni-muenchen.de (N.T.); friedrich-wilhelm.kreth@med.uni-muenchen.de (F.-W.K.)

* Correspondence: Simone.Kreth@med.uni-muenchen.de; Tel.: +49-89-4400-74567

† These authors contributed equally.

Received: 1 September 2018; Accepted: 24 October 2018; Published: 25 October 2018

Abstract: Background: The second intron of Mitogen-Activated Protein Kinase Kinase 4 (MAP2K4), an important hub in the pro-invasive MAPK pathway, harbors miR-744. There is accumulating evidence that intronic micro-RNAs (miRNAs) are capable of either supporting or restraining functional pathways of their host genes, thereby creating intricate regulative networks. We thus hypothesized that miR-744 regulates glioma migration by interacting with its host's pathways. Methods: Patients' tumor specimens were obtained stereotactically. MiR-744 was overexpressed in U87, T98G, and primary glioblastoma (GBM) cell lines. Cell mobility was studied using migration and Boyden chamber assays. Protein and mRNA expression was quantified by SDS-PAGE and qRT-PCR. Interactions of miR-744 and 3′UTRs were analyzed by luciferase reporter assays, and SMAD2/3, p38, and beta-Catenin activities by TOP/FOPflash reporter gene assays. Results: As compared to a normal brain, miR-744 levels were dramatically decreased in GBM samples and in primary GBM cell lines. Astrocytoma WHO grade II/III exhibited intermediate expression levels. Re-expression of miR-744 in U87, T98G, and primary GBM cell lines induced focal growth and impaired cell mobility. Luciferase activity of 3′UTR reporter constructs revealed the pro-invasive factors TGFB1 and DVL2 as direct targets of miR-744. Re-expression of miR-744 reduced levels of TGFB1, DVL2, and the host MAP2K4, and mitigated activity of TGFB1 and DVL2 downstream targets SMAD2/3 and beta-Catenin. TGFB1 knock-down repressed MAP2K4 expression. Conclusion: MiR-744 acts as an intrinsic brake on its host. It impedes MAP2K4 functional pathways through simultaneously targeting SMAD-, beta-Catenin, and MAPK signaling networks, thereby strongly mitigating pro-migratory effects of MAP2K4. MiR-744 is strongly repressed in glioma, and its re-expression might attenuate tumor invasiveness.

Keywords: glioblastoma; migration; microRNAs; MAP2K4

1. Introduction

Micro-RNAs (miRNAs) are short RNA molecules with an established role as important epigenetic regulators of the transcriptome via recognition of base-complementary signals in the 3′ untranslated regions of target mRNAs [1,2]. Interestingly, the majority of human miRNA genes are located within non-coding regions of protein-coding genes [3,4]. It also appears that this colocalization leads to coregulation in several instances, either through cotranscription and therefore coexpression, or via shared cis-regulatory elements [5]. Increasing evidence has begun to unravel the shades of a mechanism bearing an important role in the regulation of central cellular pathways, rather than just being a biological pendent of an information compression algorithm. However, we are only on the verge of understanding its impact on health and disease [6,7]. The Mitogen-Activated Protein Kinase (MAPK) pathway is one of such central signal transduction pathways, regulating a plethora of essential cellular functions, including proliferation, differentiation, and modulation of gene expression. One of its members, Mitogen-Activated Protein Kinase Kinase 4 (MAP2K4), is a well-known tumorigenic kinase with an established role in metastasis, invasion, and cancer progression [8–10]. *MAP2K4* in turn hosts the intronic microRNA *hsa-miR-744*, located in the second intron of its host. Even though MAP2K4 has long been recognized as a potent proto-oncogene, little is known about its intronic miRNA. Most importantly, the relationship between MAP2K4 and miR-744 is as yet completely uncharacterized. Due to the central and established role of MAP2K4 in tumor biology [11,12], we hypothesized miR-744 to be guilty by association, possibly augmenting or antagonizing MAP2K4s tumor-promoting effects.

In the following manuscript, we provide new and unprecedented evidence of a significant role of miR-744 in the regulation of its host gene MAP2K4 in human glioma. This is accomplished via controlling both expression and functional aspects of its host gene. We show how miR-744 inhibits cell migration and invasion, a key characteristic of glioblastoma (GBM), via targeting three essential cellular pathways. We finally validate our hypothesis in cell line experiments as well as patients´ tissue samples. We believe that our findings shed some more light on the as yet blurry contours of the mechanisms underlying tumor development through dysregulation of cellular signaling, and portray miR-744 as a central molecule in the formation of GBM.

2. Results

2.1. miR-744 Is Strongly Repressed in Human Glioma

To identify human tissues in which miR-744 may fulfill important regulatory functions, we used the intragenic microRNA database miRIAD (http://bmi.ana.med.uni-muenchen.de/miriad/) [4], and screened for tissues with high expression levels of this miRNA. As depicted in Figure 1A, among five different tissues deposited in miRIAD (heart, testis, kidney, cerebellum, and brain), human brain exhibited the highest expression levels of miR-744, which suggests its implication in the maintenance of homeostatic conditions in the central nervous system. To investigate our initial hypothesis, we next quantified miR-744 in stereotactically obtained GBM specimens and primary cell lines obtained from open GBM resections by qRT-PCR, and found a dramatic reduction of miR-744 as compared to normal brain tissue (Figure 1B; GBM samples: reduction by 90.3% ± 14.7%, primary GBM cell lines: reduction by 92.7% ± 7.3%; n = 9 for normal brain tissue, n = 21 for GBM samples, n = 8 for primary GBM cell lines; $p < 0.01$). Also, we could detect reduced expression of miR-744 in U87 cells, a human GBM cell line (Figure 1B; reduction by 97.7% ± 6%, n = 9, $p < 0.001$).

Collectively, this data shows that miR-744 is highly expressed in human brain tissue, whereas it is almost entirely repressed in GBM.

To assess the expression of miR-744 in human glioma of different grades, we quantified miR-744 in 15 stereotactically obtained WHO II/III tumors by qRT-PCR. As depicted in Figure 1C, we found miR-744 also to be repressed; however, expression levels were significantly higher as compared to GBM (48% ± 20%; WHO II/III: n = 15, GBM: n = 21, $p = 0.034$). It thus appears that miR-744 expression is inversely correlated with tumor grade and may contribute to increased tumor aggressiveness.

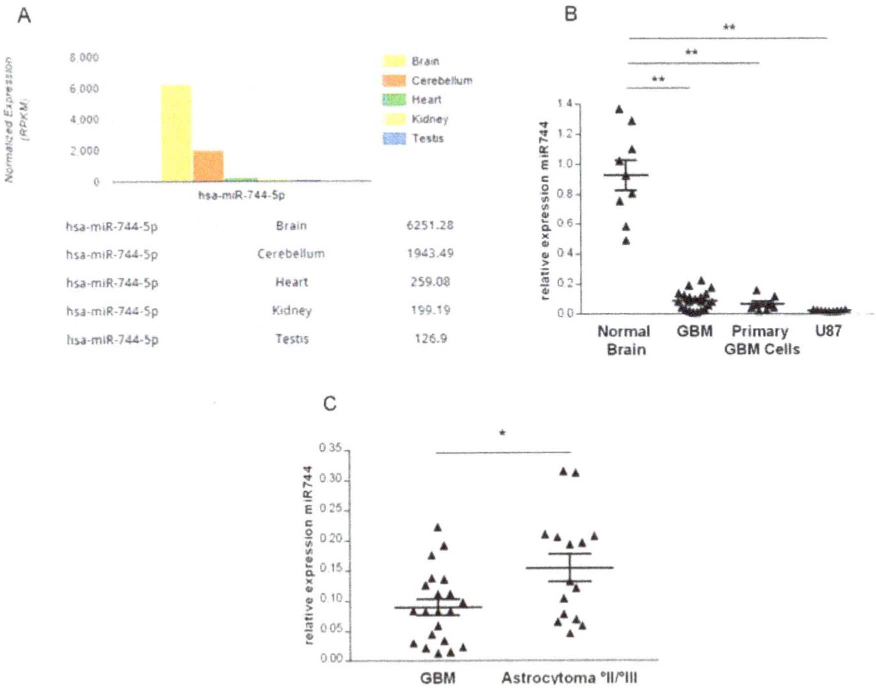

Figure 1. miR-744 is strongly repressed in glioma. MiR-744 expression was quantified by qRT-PCR. U47 served as the endogenous reference. (**A**) Expression levels of miR-744 in five different tissues. (**B**) Expression of miR-744 in normal brain tissue (NB) (n = 9), glioblastoma (GBM) (n = 21), primary GBM cell lines (n = 9), and U87 cells (n = 9), $p < 0.001$. (**C**) Expression of miR-744 in WHO grade °II/°III glioma (n = 15) compared to GBM (n = 21), $p = 0.034$. ** $p < 0.001$; * $p < 0.05$.

2.2. Overexpression of miR-744 Reduces Migration of GBM Cells

It is a frequently occurring phenomenon that tumors down-regulate, or even hamper, the expression of genes that are not useful for malignant transformation. Our next aim was to understand the reasons for miR-744 downregulation in human GBM, and thus we investigated the biological functions of miR-744 in glioma cells. To this end, we transiently re-expressed miR-744 in U87, T98G, and primary GBM cell lines by transfection of the respective premiR and assessed the resulting phenotype. Surprisingly, we could not detect any alterations of apoptosis or proliferation after transfection of miR-744 (data not shown). 2D wound closure assays however, revealed a strong impact of miR-744 on cellular migration, which was markedly attenuated in miR-744 transfected cells (Figure 2A). To study the long-term effects of miR-744 on cellular migration, we constructed a miR-744 delivering expression vector, and stably transfected U87 GBM cells (Figure 2D, left panel). As shown in Figure 2B, overexpression of miR-744 leads to a decrease in cellular spreading and induces focal growth, pointing towards an alteration of cellular mobility. Importantly, 2D migration and Boyden Chamber assays revealed a less invasive phenotype (Figure 2C,D, right panel; reduction of 46% ± 5.8%, n = 4, $p = 0.029$). Taken together, this data shows that miR-744 inhibits migration of GBM cells.

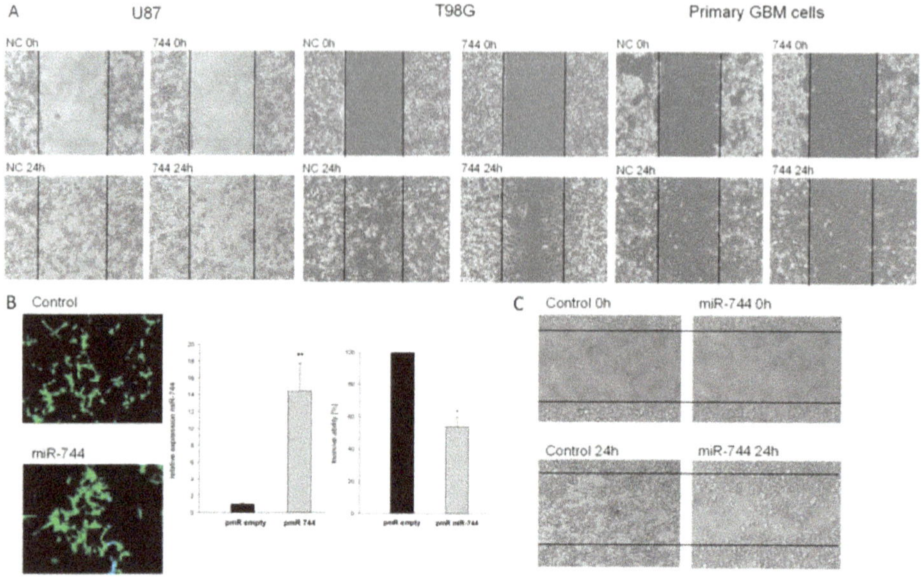

Figure 2. Overexpression of miR-744 induces focal cell growth and inhibits invasion and migration. (**A**) 2D migration assays of transiently miR-744 transfected cells (left panel: U87; middle panel: T98G; right panel: primary GBM cell lines) at start and after 24 h. Lines mark the initially cell-free area. A typical example of 3 experiments is shown. (**B**) Fluorescence microscopy of control and stably miR-744 overexpressing cells. (**C**) 2D migration assays of stably transfected cells; depicted are start state and after 24 h of incubation. Lines mark the cell-free area. A typical example of 3 similar experiments is shown. (**D**) Left panel: MiR-744 levels after stable transfection (induction 14.4-fold $\pm$ 6.0, n = 5, $p < 0.001$). Right panel: Invasion after stable transfection of U87 cells, measured with Collagen-coated Boyden Chamber invasion assays (n = 4, $p = 0.029$). ** $p < 0.001$; * $p < 0.05$.

2.3. TGFB1 and DVL2 are Direct Targets of miR-744, Regulate Migration in GBM Cells, and are Induced in Tissue of Human Malignant Glioma

To identify direct targets of miR-744 possibly accounting for the detected phenotypic alterations, we next combined in silico target prediction and pathway analysis to extract mRNAs (a) containing miR-744 binding sites in their 3′ untranslated region (3′UTR), and (b) being involved in the regulation of cellular mobility. This prompted us to investigate Transforming Growth Factor Beta 1 (TGFB1) and Dishevelled2 (DVL2) in detail, as these were the most promising predicted targets of miR-744 with respect to a supposed role in GBM migration [13,14]. To test this assumption, we quantified mRNA and protein expression levels of TGFB1 and DVL2 in U87 cells stably overexpressing miR-744, and indeed detected a marked decrease of TGFB1 and DVL2 mRNA (reduction of 38% $\pm$ 5.6% and 33.6% $\pm$ 4.9%, respectively, n = 5, $p < 0.05$), and protein expression (reduction of 35.6% $\pm$ 8.3% and 36.8% $\pm$ 5.3%; Figure 3A, left and middle panels). To provide experimental proof that both genes are bona fide targets of miR-744 in GBM, we performed luciferase reporter gene assays on vector constructs containing the full-length 3′-UTR of either TGFB1 or DVL2. As shown in Figure 3A (right panel), co-transfection of miR-744 mimic and reporter constructs diminished luciferase activity by 37% and 47%, respectively, compared to miR scrambled control (n = 5, $p < 0.05$), thereby proving that both genes are direct targets of miR-744. Transient knock-down of both TGFB1 and DVL2 by specific siRNAs (knock-down efficiency: Supplementary Figure S1) reduced the migratory capabilities of GBM cells in 2D migration assays (Figure 3B), thereby closing the anticipated functional loop; miR-744 impairs migration of human glioma cells by direct targeting of TGFB1 and DVL2.

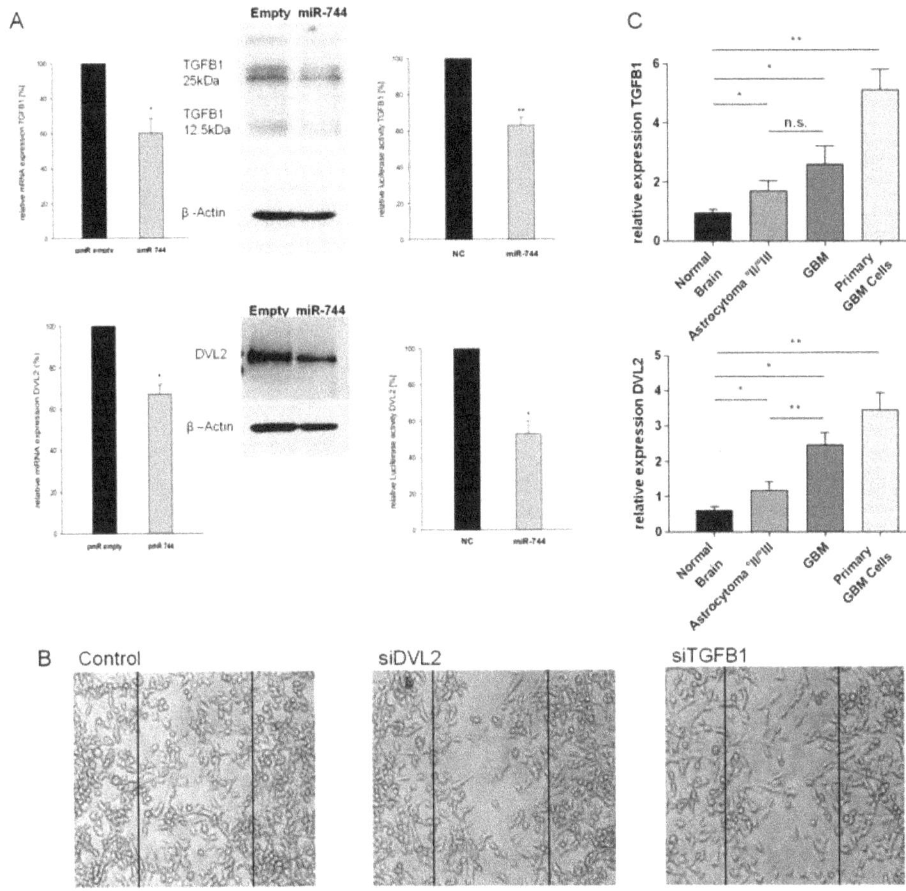

Figure 3. Transforming Growth Factor Beta 1 (TGFB1) and Dishevelled2 (DVL2) are direct targets of miR-744, regulate migration, and are induced in GBM and in Astrocytoma °II/°III. (**A**) Left and middle panels: mRNA and protein expression of TGFB1 and DVL2 in U87 cells stably transfected with miR-744 or empty vector control, respectively (n = 5, $p < 0.05$). Right panels: TGFB1 and DVL2 3′UTR Luciferase Reporter Gene activity after co-transfection of the respective reporter vectors with miR-744 or with scrambled control, respectively (DVL2: −47% ± 3,2%, TGFB1: −37% ± 2%, n = 5, $p < 0.05$). All Luciferase and mRNA experiments were performed in triplicates. (**B**) 2D migration assay after transient knock-down of DVL2 and TGFB1 in U87 cells. One representative example of 3 experiments is shown. (**C**) Expression of DVL2 and TGFB1 mRNA in Astrocytoma °II/°III biopsies (n = 10, $p < 0.05$), GBM biopsies (DVL: n = 17, $p = 0.001$; TGFB1: n = 39, $p = 0.015$), and primary GBM cell lines (n = 8, $p < 0.001$), as compared to normal brain tissue (n = 9). ** $p < 0.001$; * $p < 0.05$. n.s. = not significant.

In GBM and Astrocytoma °II/°III specimens, as well as in primary GBM cells, we found significantly increased expression levels of DVL2 and TGFB1. These results are in line with the detected down-regulation of miR-744 (Figure 3C; DVL2: GBM, induction 3.35-fold ± 0.14, $p = 0.002$; primary GBM cells, induction 5.4-fold ± 0.46, $p < 0.001$; normal brain tissue (NB), n = 9; GBM, n = 17; primary GBM cells, n = 8. TGFB1: GBM, induction 2.19-fold ± 0.09, $p = 0.015$; primary GBM cells, induction 4.33-fold ± 0.69, $p < 0.001$; NB, n = 9; GBM, n = 39; primary GBM cells, n = 8). Notably, GBM exhibited higher levels of TGFB1 and DVL2 as compared to Astrocytoma °II/°III (Figure 3; TGFB1:

induction 1.5-fold ± 0.36, p = n.s.; DVL2: induction 2.1-fold ± 0.3, $p < 0.001$; n = 10 for Astrocytoma °II/°III).

Hence, this data indicates that miR-744 puts the brake on the expression of DVL2 and TGFB1, which both play an important role as promoters of migration in GBM.

2.4. Via Repression of DVL2 and TGFB1, miR-744 Regulates Beta-Catenin and SMAD-Signaling Pathways

We next set out to gain insight into the molecular pathways underlying the observed phenotypic alterations. TGFB1 is assumed to enhance cellular mobility through SMAD-dependent induction of pro-invasive factors such as Matrix Metalloproteinases (MMPs) [15,16], while DVL2 represents a central inducer of beta-Catenin signaling [17]. Both pathways induce epithelial-to-mesenchymal transition (EMT), and enhance tumor cell migration [14,17]. Consequently, we next assessed the activation status of SMAD2/3 and beta-Catenin upon overexpression of miR-744, as these transcription factors represent important downstream effector molecules of TGFB1 and DVL2 [16,18], respectively. As depicted in Figure 4A, miR-744 overexpression resulted in a marked reduction of the transcriptionally active forms of SMAD2/3 and beta-Catenin (SMAD2/3: reduction of 52% ± 15.8%; active beta-Catenin: reduction of 38.6% ± 5.7%). In addition, immunohistochemistry showed a significantly weaker staining for transcriptional active beta-Catenin upon overexpression of miR-744 as compared to controls (Figure 4B). In line with these results, subsequent analysis of beta-Catenin-mediated transcriptional activity by Lef/Tcf luciferase reporter gene assay revealed a significantly reduced activity in cells stably overexpressing miR-744 (Figure 4C; −46.1% ± 10.7%, n = 5, $p < 0.01$).

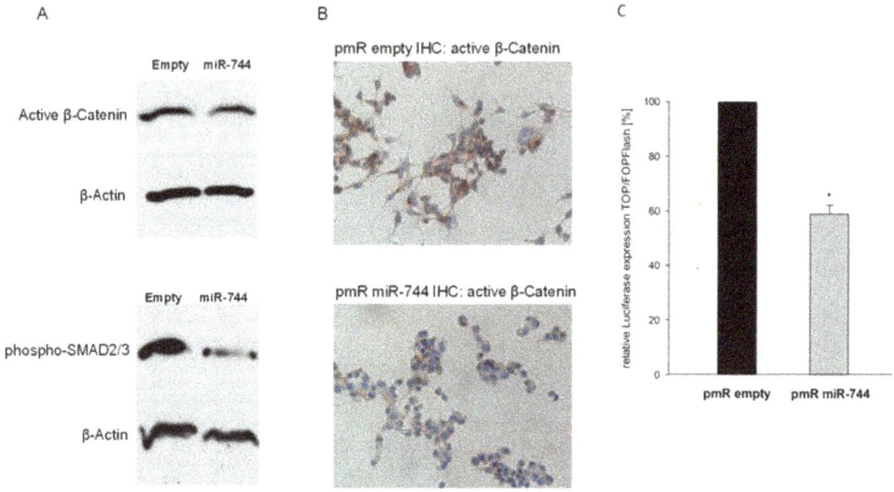

Figure 4. Stable overexpression of miR-744 represses activity of TGFB1 and DVL2 downstream effector molecules SMAD- and β-Catenin. Experiments were conducted after stable overexpression of miR-744 as compared to empty vector controls. (**A**) SDS-PAGE of active β-Catenin and phosphorylated SMAD 2/3. (**B**) Immunohistochemistry staining of active β-Catenin. (**C**) β-Catenin-dependent transcriptional activity of Lef/Tcf, measured with TOP/FOPFlash Luciferase Reporter Gene Assay (n = 4, $p = 0.007$). * $p < 0.05$.

Taken together, our results show that miR-744 via DVL2 and TGFB1 ameliorates invasive properties of GBM cells by down-regulation of beta-Catenin and SMAD-signaling.

2.5. *miR-744 Reduces It's Host Gene MAP2K4 Through TGFB1-Mediated Negative Feedback*

So far, our results have suggested that miR-744 acts as a counterpart of its host gene MAP2K4, an enhancer of cancer progression [8,10]. It thus appeared likely that a negative feedback relationship between miR and host might exist. To test this assumption, we assessed MAP2K4 expression in U87 cells stably overexpressing miR-744, and indeed found reduced mRNA and protein levels (Figure 5A; mRNA: −66.2% ± 7.9%; protein: −56.2% ± 9.6%; n = 5, $p < 0.05$). Concordantly, miR-744 overexpression attenuated activity of MAP2K4s downstream effector p38. As the 3′UTR of MAP2K4 does not contain any putative binding sites of miR-744, the host obviously is not directly targeted. We thus hypothesized that miR-744 indirectly represses its host by targeting an activator of the host gene. As TGFB1 is known to activate p38 MAPK, we speculated that TGFB1 might also fulfill a function as an enhancer of MAP2K4 signaling. Gene-specific knock-down of TGFB1 (Figure 5C, right panel) significantly decreased MAP2K4 mRNA and protein levels, thereby strongly supporting this assumption (Figure 5B; −47.7% ± 3.8%, n = 4, $p = 0.004$). Not unexpectedly, intronic miR-744 was also significantly affected by TGFB1 knock-down (Figure 5C, left panel; −38.1% ± 9.5%, n = 4, $p = 0.015$).

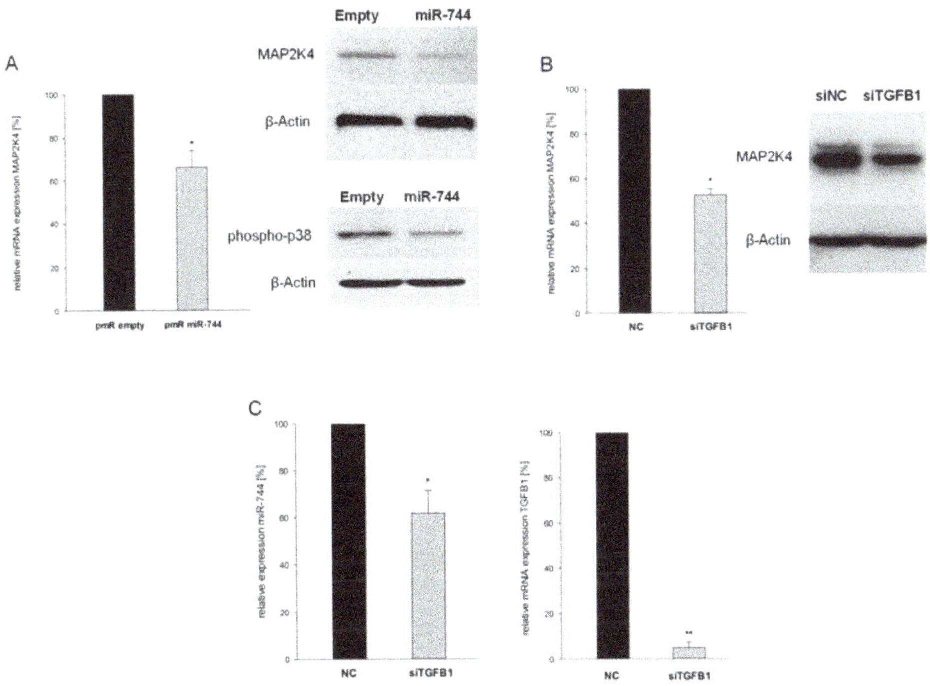

Figure 5. miR-744 reduces its host gene Mitogen-Activated Protein Kinase Kinase 4 (MAP2K4) through TGFB1-mediated negative feedback. (**A**) mRNA expression of MAP2K4 (left panel), and protein expression of MAP2K4 and its downstream target phospho-p38 MAPK (right panel), after stable overexpression of miR-744 as compared to empty vector controls (n = 5, $p = 0.003$). (**B**) mRNA (left panel) and protein expression (right panel) of MAP2K4 after knock-down of TGFB1 (n = 5, $p = 0.004$). (**C**) Expression of miR-744 (left panel) and TGFB1 (right panel) after knock-down of TGFB1 (miR-744: n = 5, $p = 0.015$; TGFB1: n = 5, $p < 0.001$). ** $p < 0.001$; * $p < 0.05$.

These results sketch a regulatory loop: TGFB1 simultaneously enhances transcription of MAP2K4 and miR-744. The latter directly targets TGFB1, which as a negative feedback results in concurrent repression of MAP2K4 and miR-744 (Figure 6). These results highlight the close regulatory relationship between intronically located miRs and their host genes.

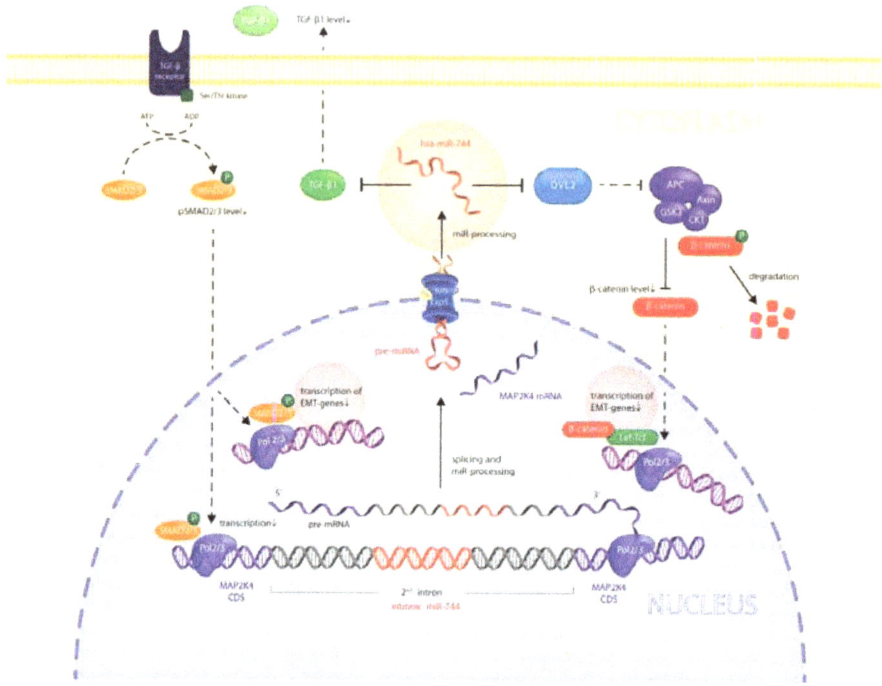

Figure 6. Model of miR-744 effects on it's host gene´s expression and on tumor cell migration. Direct repression of DVL2 and TGFB1 by miR-744 inhibits SMAD- and beta-Catenin-dependent transcriptional activity, respectively, thus reducing expression of EMT- and pro-invasive genes. Moreover, miR-744 constitutes a second-order negative feedback loop on its host gene MAP2K4 through direct repression of TGFB1 levels, concomitantly repressing host gene and miR-744 expression.

3. Discussion

In recent years, evidence has accumulated that colocalization of intronic miRNA and the host gene is not a random choice by nature, but rather fulfills important functional tasks within the host genes' pathways. Several studies have experimentally proven in different contexts that intronic miRNAs are capable of either supporting or restraining functional pathways of their host genes, thereby creating intricate regulative networks [19,20]. As about half of human miRNAs reside in introns of protein-coding genes, these examples may signify a more general biological principle. A very recent study however, has suggested low prevalence of functional association between host and intronic miRNAs in general, but has assumed a key role of this type of regulation in cellular signaling pathways requiring tight control [21]. To date, experimental evidence supporting this view is scarce. As an example of potential high interest, we investigated miR-744, located in the second intron of the tumorigenic kinase MAP2K4, and its impact on it's host gene's functional networks.

Expression patterns of miRNAs are highly heterogeneous among cell types, and pronounced expression levels of a certain miRNA within a specific tissue is likely to reflect functional relevance. After analysis of different tissues, we found brain tissue to be the most suitable for further analyses, due to its high miR-744 expression. With respect to the role of MAP2K4 as a driver of malignancy, we assessed miR-744 expression in human brain tissue, in Astrocytoma WHO grade II/III, and in the most malignant brain tumor (i.e., GBM). Interestingly, we detected a gradual tumor-grade-dependent loss of expression, with almost completely lost miR-744 expression levels in GBM. These findings made us assume that miR-744 is a gatekeeper of oncogenic signaling during the malignization process. Indeed, stable re-expression of miR-744 in GBM cells resulted in a significantly more benign phenotype, with tumor cells growing focally, and migrating less. To uncover the mechanisms underlying these phenotype changes, we stably overexpressed miR-744 in GBM cells, and evaluated potential miR-744 target genes fulfilling the criteria of (a) being involved in the regulation of cell motility and (b) harboring potential miR-744 binding sites in their 3′UTRs. We identified and experimentally validated DVL2 and TGFB1 to be directly regulated by miR-744 in GBM cells. We were further able to show that specific knock-down of these target genes impaired cellular migration similar to miR-744 overexpression. In line with these results, we found that down-regulation of miR-744 in human GBM is accompanied by a marked increase in DVL2 and TGFB1 expression levels, indicating clinical relevance of these functional networks.

DVL2 and TGFB1 have repeatedly been shown to act as inductors of epithelial-to-mesenchymal transition (EMT) [13,22,23], one of the hallmarks of cancer progression. During EMT, induction of Matrix-Metalloproteinases initiates breakdown of the extracellular matrix and reduces cellular adhesion, which results in enhanced migratory capacity and invasiveness [24,25]. Particularly in GBM, TGFB1 is one of the most powerful cytokines secreted by the tumor itself. It has repeatedly been shown that GBM-derived TGFB1 induces EMT via activation of transcription factors of the SMAD family, thereby increasing migration of GBM cells into the surrounding brain tissue. It thus represents a hallmark of GBM progression [26–28]. In addition, DVL2, as a key enhancer of beta-Catenin transcriptional activity, represents another orchestrator of EMT [29–31]. In GBM, the DVL2-beta-catenin signaling axis has been found to be markedly activated, which significantly contributes to the extremely invasive nature of these tumors [13,32,33]. To this end, we tested whether inhibition of DVL2 and TGFB1 by overexpression of miR-744 exerted the supposed impact on the respective downstream effector molecules, beta-Catenin and SMAD. Based on immunohistochemistry, protein analyses, and reporter gene assays, our results provided evidence that miR-744 hampers GBM cell migration via concurrent reduction of SMAD- and beta-Catenin signaling. Thus, downregulation of miR-744 during gliomagenesis may be an effective tumor intrinsic mechanism to support tumor progression by simultaneously affecting different oncogenic signaling pathways.

This negative impact of miR-744 on tumor cell migration strongly contrasts with the functions of its host gene, MAP2K4, which has been shown to profoundly enhance cancer cell migration and metastasis [8,10]. It can thus be concluded that that miR-744 acts as functional antagonist of its host, thereby keeping a molecular balance. As stable overexpression of miR-744 resulted in significantly reduced MAP2K4 levels, we assumed that miR-744 may not only control its host functionally, but also on the level of gene expression through negative feedback. The concept of intronic miRs regulating their host genes by direct or indirect feedback loops has been experimentally proven in several different contexts [6,34]. Effects are achieved either directly by targeting of the host's 3′UTR (first order negative feedback), or indirectly by targeting an interposed gene that subsequently affects the host gene's expression (second order negative feedback). As the 3′UTR of MAP2K4 does not contain any predicted miR-744 binding sites, we focused on the identification of a second order negative feedback. In this regard, our finding that stable overexpression of miR-744 attenuated the activity of MAP2K4's downstream effector p38 prompted us to come back to TGFB1 as an important activator of p38 MAPKs [35,36]. Indeed, we could show that gene-specific knock-down of TGFB1 significantly decreased MAP2K4 expression. Concomitantly, miR-744 levels were decreased, supporting the notion of a negative feedback loop acting as an "intrinsic transcriptional brake" to prevent inadequate transcription of MAP2K4. Inhibition of TGFB1 by miR-744 thus fulfills a dual role in mediating effects of this intronic miRNA on its host MAP2K4: (i) Functional antagonization via blocking SMAD-signaling, thereby reducing glioma migration and invasion; and (ii) control of the tumorigenic host's expression levels.

It is a limitation of the current study that the molecular mechanisms enabling this fundamental switch remain elusive. However, due to the assumable multi-layered nature of these processes, this question may be addressed in further research projects.

4. Materials and Methods

4.1. Bioinformatics

Analysis of potential miR-mRNA interactions was performed using the public databases TargetScan, PITA, miRIAD, and picTAR. Potential direct interactions were considered probable when 2 or more algorithms returned a positive target prediction. In-silico analysis of miR expression levels was conducted using the intragenic microRNA database miRIAD (http://bmi.ana.med.uni-muenchen.de/miriad/). Involvement of target genes in tumor-associated pathways was evaluated using the KEGG database (www.genome.jp/kegg/pathway.html).

4.2. Human Tissue Samples

Tissue samples (n = 21, GBM; n = 15, WHO II/III Astocytoma; n = 8, primary GBM cell lines; and n = 9, normal brain) were obtained and processed as described previously [7]. Written informed consent was given by all patients, and the study protocol was approved by the Institutional Review Board of the Ludwig-Maximilians University of Munich, Germany (approval number: 216/14). For this study, only specimens from patients diagnosed with primary GBM, LOH 1p/19q negative, and IDH wild type were used.

4.3. RNA Extraction and cDNA Synthesis

RNA was extracted using the RNAqueous or miRvana Isolation kit (Ambion, Waltham, MA, USA), followed by DNAse treatment (Turbo DNAse, Ambion) according to the manufacturer's instructions. RNA amount and quality was assessed using a NanoDrop 2000 spectrophotometer (Thermo Scientific, Waltham, MA, USA). Equal amounts of RNA were transcribed using Oligo-dT Primers, Random Hexamers (Qiagen, Venlo, Netherlands), dNTPs, RNAse OUT, and Superscript® III Reverse Transcriptase (Invitrogen, Waltham, MA, USA), following the manufacturer's instructions.

4.4. Quantitative RT-PCR

Quantitative analysis of mRNA levels was performed on a LightCycler 480 (Roche Diagnostics, Penzberg, Germany) using 10 ng of cDNA/well. Succinate Dehydrogenase Subunit A (SDHA) and TATA Box Binding Protein (TBP) were used as reference genes. Quantitative real-time PCR (qRT-PCR) was conducted using the primers (Metabion, Martinsried, Germany) and UPL Probes (Roche Diagnostics) provided in Supplementary Table S1. All assays were designed intron spanning. qRT-PCR conditions comprised initial denaturation for 10 Minutes (95 °C), and 50 cycles of 95 °C for 10 s, 60 °C for 30 s, and 72 °C for 1 s. Quantification cycle (C_q) values were calculated employing the "second derivative maximum" method computed by the LightCycler® software.

4.5. Quantification of miRNA Expression

Mir-744 expression was studied using TaqMan miRNA assays (Applied Biosystems, Waltham, MA, USA) according to manufacturer's instructions. MiR-744 and reference gene expression was measured in technical duplicates. U47 served as an endogenous reference. All patient samples were calibrated using a sample of normal brain tissue.

4.6. SDS-PAGE

Cells were lysed in cell lysis buffer containing protease and phosphatase inhibitors (Cell Signaling Technologies, Danvers, MA, USA). Protein concentrations were assessed through BCA assays (Thermo Fisher Scientific, Waltham, MA, USA) according to the manufacturer's instructions. Forty micrograms of the protein extracts were electrophoresed on 10% SDS-PAGE gels and electroblotted on PVDF-membranes. Nonspecific binding was blocked with 5% Bovine Serum Albumin (BSA) in TBS-Tween-20 (TBST) (Sigma, St.Louis, MO, USA). Antibodies for MAP2K4 (Cat. No. 9152),

TGFB1 (Clone 56E4, Cat. No. 3709), DVL2 (Clone 30D2, Cat. No. 3224), and β-Actin (Clone 13E5, Cat. No. 4970) (all Cell Signaling Technologies) were diluted in TBST with 1% BSA. β-Actin served as the loading control. Immunoreactivity was assessed using horseradish peroxidase-labeled goat anti-mouse or goat anti-rabbit antibodies (Cell Signaling Technologies).

4.7. Cell Culture

U87 and T98G GBM cells were purchased from the American Type Cell Culture Collection (ATCC). Cells were maintained in DMEM (Gibco) with 10% FCS (Biochrom AG), 2% L-Glutamine, 1% Penicillin/Streptomycin, 1% MEM NEAA (Invitrogen), and 1% Sodium Pyruvate (PAA). HEK-293 cells (ATCC) were maintained in DMEM with 10% FCS, 2% L-Glutamine, 1% Penicillin/Streptomycin, and 1% MEM NEAA. Primary GBM cell lines were obtained from patients undergoing open GBM resection, according to the study protocol mentioned above. Tumor tissue was dissociated using the Brain Tumor Dissociation Kit P (Miltenyi, Bergisch-Gladbach, Germany) according to the manufacturer's instructions. Primary GBM cells were cultivated in MACS Neuro Medium supplemented with Neuro Brew-21 without Vitamin A (Miltenyi).

4.8. Cloning of Reporter Constructs

The 3'UTRs of TGFB1 and DVL2 were amplified using genomic DNA and primer with XhoI and NotI, or PmEI and XhoI, restriction sites. Primer sequences are supplied in Supplementary Table S2. PCR products were cloned into the psiCHECK2 vector (Promega, Mannheim, Germany). Correct sequences were verified by Sanger sequencing (Eurofins Operon, Ebersberg, Germany). Plasmids were purified using the Qiaprep Spin Plasmid Miniprep Kit (Qiagen) and the Pure Yield Plasmid Midiprep System (Promega). DNA concentrations were determined using a NanoDrop 2000 spectrophotometer (Thermo Fisher Scientific).

4.9. Cloning of miRNA Expression Vector

The pre-miR sequence was amplified using genomic DNA and specific primers (sequences supplied in Supplementary Table S2). The resulting amplicon was cloned into the pmRZs-Green1 vector (Promega).

4.10. Transfections

Transfections were conducted using the NEON electroporation device (Life Technologies, Waltham, MA, USA). Transient transfection with microRNA precursors (premiR, Thermo Fisher Scientific) or siRNA (Dharmacon, Lafayette, CO, USA) was carried out at final concentrations of 50 nM (premiR) or 100 nM (siRNA), and 250,000 cells per well. Cells were incubated for 36 h at 37 °C and 5% CO_2 in an antibiotics-free medium. Stable transfection was performed using 1 million cells and 10 μg plasmid/well. After incubation for 12 h in antibiotics-free medium, cells were seeded in DMEM containing 750 μg/mL Geneticin (Life Technologies). Stable transfection was analyzed by flow cytometry (Attune, Life Technologies). Monoclonal cell lines were obtained by single-cell picking. Overexpression of miR-744 was assessed through TaqMan assays. Co-transfection of luciferase reporter plasmids and premiR™ was carried out using 100,000 HEK-293 cells and 1 μg of Psi-CHECK™2 plasmid. All transfection experiments were performed in triplicate.

4.11. Reporter Gene Assays

After 36 h of incubation, co-transfected cells were harvested, washed twice, and resuspended in 20μl medium. Luminescence was measured with the MicroLumat Plus (Berthold Technologies, Bad Wildbad, Germany) using the Dual-Glo Luciferase Assay system (Promega), according to manufacturer's instructions. All experiments were performed in triplicate.

4.12. Migration Assays

70,000 cells/well were seeded in 2-well culture inserts (Ibidi, Martinsried, Germany) and incubated for 24 h at 37 °C with 5% CO_2 in a humidified incubator. Inserts were removed, cells were washed with cell culture media, and pictures were obtained using an inverted microscope (Carl Zeiss, Jena, Germany). Cells were incubated for 24 h in a cell culture incubator. Pictures were obtained and cells were fixed with methanol.Invasive properties were studied with the Cytoselect Assay, Collagen I, Colorimetric Format (Cell Biolabs, San Diego, CA, USA). 100.000 Cells were resuspended in 250 μL FCS-free media and seeded into the upper compartment of the boyden chamber. DMEM containing 10% FCS was added to the lower compartment. After incubation for 24 h in a cell culture incubator, non-invasive cells were removed with cotton swabs, invasive cells were stained, lysed, and the optical density was determined. Experiments were performed in duplicate.

4.13. Immunohistochemistry

100,000 cells were seeded on glass slides (Falcon) and incubated for 2 days. Cells were washed with PBS, fixed with ice-cold acetone, and washed with TBST. Endogenous peroxidase was blocked (1% H_2O_2, 10 minutes), cells were washed twice with TBST, and incubated in 5% normal goat serum for 1h. Antibodies were diluted in Antibody Diluent (Cell Signaling Technologies) at a concentration of 1:800 and incubated at 4 °C overnight. Cells were washed twice with TBST and Signal Stain Boost IHC Detection Reagent was added (Cell Signaling Technologies). After washing with TBST, Signal Stain DAB substrate was added for 45 s. Thereafter, cells were washed with TBST. Counterstaining was performed with Haematoxylin (Sigma Aldrich).

4.14. Statistics

All data is presented as the mean ± SEM. p-values were calculated using student's t-tests. Statistical analyses were performed using SigmaPlot 12.0 (Systat Software). p-values below 0.05 were considered statistically significant (* $p < 0.05$; ** $p < 0.001$).

5. Conclusions

Taken together, we here uncovered a new regulatory circuit consisting of the tumorigenic host gene MAP2K4 and its intronically located miR-744; miR-744 acts as an intrinsic brake on its host by counterbalancing both its expression and function. In human glioma, this circuit is disrupted, leading to an invasion-promoting constellation where miR-744 is almost completely repressed while its host is induced.

Our data underscores the necessity to gain further profound insights into the networks of intronic miRNAs and their hosts, which is particularly important with respect to potential clinical approaches. Counteracting the shortfall of miR-744 in GBM pharmacologically, for example, might be an innovative clinical direction to pursue.

Supplementary Materials: The following are available online at http://www.mdpi.com/2072-6694/10/11/400/s1, Table S1: Primer sequences for qRT-PCR; Table S2: Primer sequences for molecular cloning; Figure S1: Knock-down efficiency after transient transfection of U87 GBM cells with DVL2 or TGFB1 siRNA, as analyzed by qRT-PCR.

Author Contributions: M.H. designed and performed experiments, collected, analyzed, and interpreted data, and wrote the manuscript; C.L.H. designed and performed research, analyzed and interpreted data, and wrote the manuscript; D.E. performed experiments and interpreted data; T.W. planned and performed research; N.T. contributed patient samples and helped to revise the manuscript. F.-W.K. contributed patient samples and interpreted data, and participated in writing the manuscript; S.K. designed the study, analyzed and interpreted data, wrote the manuscript, and supervised the study.

Funding: This research was funded by institutional grants of the LMU of Munich.

Acknowledgments: The authors are indebted to Jessica Rink, Gabriele Gröger, and Gudrun Prangenberg for their excellent technical assistance.

Conflicts of Interest: The authors declare no conflict of interest.

References

1. Kreth, S.; Hübner, M.; Hinske, L.C. MicroRNAs as Clinical Biomarkers and Therapeutic Tools in Perioperative Medicine. *Anesth. Analg.* **2018**, *126*, 670–681. [CrossRef] [PubMed]
2. Lewis, B.P.; Shih, I.-H.; Jones-Rhoades, M.W.; Bartel, D.P.; Burge, C.B. Prediction of mammalian microRNA targets. *Cell* **2003**, *115*, 787–798. [CrossRef]
3. Hinske, L.C.; Dos Santos, F.R.C.; Ohara, D.T.; Ohno-Machado, L.; Kreth, S.; Galante, P.A.F. MiRIAD update: using alternative polyadenylation, protein interaction network analysis and additional species to enhance exploration of the role of intragenic miRNAs and their host genes. *Database* **2017**. [CrossRef] [PubMed]
4. Hinske, L.C.; França, G.S.; Torres, H.A.M.; Ohara, D.T.; Lopes-Ramos, C.M.; Heyn, J.; Reis, L.F.L.; Ohno-Machado, L.; Kreth, S.; Galante, P.A.F. miRIAD-integrating microRNA inter- and intragenic data. *Database* **2014**. [CrossRef] [PubMed]
5. Kim, Y.-K.; Kim, V.N. Processing of intronic microRNAs. *EMBO J.* **2007**, *26*, 775–783. [CrossRef] [PubMed]
6. Schmitt, D.C.; Madeira da Silva, L.; Zhang, W.; Liu, Z.; Arora, R.; Lim, S.; Schuler, A.M.; McClellan, S.; Andrews, J.F.; Kahn, A.G.; et al. ErbB2-intronic microRNA-4728: a novel tumor suppressor and antagonist of oncogenic MAPK signaling. *Cell Death Dis.* **2015**, *6*, e1742. [CrossRef] [PubMed]
7. Hinske, L.C.; Heyn, J.; Hübner, M.; Rink, J.; Hirschberger, S.; Kreth, S. Intronic miRNA-641 controls its host Gene's pathway PI3K/AKT and this relationship is dysfunctional in glioblastoma multiforme. *Biochem. Biophys. Res. Commun.* **2017**, *489*, 477–483. [CrossRef] [PubMed]
8. Pavese, J.M.; Ogden, I.M.; Voll, E.A.; Huang, X.; Xu, L.; Jovanovic, B.; Bergan, R.C. Mitogen-activated protein kinase kinase 4 (MAP2K4) promotes human prostate cancer metastasis. *PLoS ONE* **2014**, *9*, e102289. [CrossRef] [PubMed]
9. Wang, L.; Pan, Y.; Dai, J.L. Evidence of MKK4 pro-oncogenic activity in breast and pancreatic tumors. *Oncogene* **2004**, *23*, 5978–5985. [CrossRef] [PubMed]
10. Wu, X.; Gong, Z.; Sun, L.; Ma, L.; Wang, Q. MicroRNA-802 plays a tumour suppressive role in tongue squamous cell carcinoma through directly targeting MAP2K4. *Cell Prolif.* **2017**, *50*. [CrossRef] [PubMed]
11. Khatlani, T.S.; Wislez, M.; Sun, M.; Srinivas, H.; Iwanaga, K.; Ma, L.; Hanna, A.E.; Liu, D.; Girard, L.; Kim, Y.H.; et al. c-Jun N-terminal kinase is activated in non-small-cell lung cancer and promotes neoplastic transformation in human bronchial epithelial cells. *Oncogene* **2007**, *26*, 2658–2666. [CrossRef] [PubMed]
12. Finegan, K.G.; Tournier, C. The mitogen-activated protein kinase kinase 4 has a pro-oncogenic role in skin cancer. *Cancer Res.* **2010**, *70*, 5797–5806. [CrossRef] [PubMed]
13. Pulvirenti, T.; Van Der Heijden, M.; Droms, L.A.; Huse, J.T.; Tabar, V.; Hall, A. Dishevelled 2 signaling promotes self-renewal and tumorigenicity in human gliomas. *Cancer Res.* **2011**, *71*, 7280–7290. [CrossRef] [PubMed]
14. Frei, K.; Gramatzki, D.; Tritschler, I.; Schroeder, J.J.; Espinoza, L.; Rushing, E.J.; Weller, M. Transforming growth factor-β pathway activity in glioblastoma. *Oncotarget* **2015**, *6*, 5963–5977. [CrossRef] [PubMed]
15. Golestaneh, N.; Mishra, B. TGF-beta, neuronal stem cells and glioblastoma. *Oncogene* **2005**, *24*, 5722–5730. [CrossRef] [PubMed]
16. Mehta, S.; Lo Cascio, C. Developmentally regulated signaling pathways in glioma invasion. *Cell. Mol. Life Sci.* **2018**, *75*. [CrossRef] [PubMed]
17. Lee, Y.; Lee, J.-K.; Ahn, S.H.; Lee, J.; Nam, D.-H. WNT signaling in glioblastoma and therapeutic opportunities. *Lab. Invest.* **2016**, *96*, 137–150. [CrossRef] [PubMed]
18. Sharma, M.; Castro-Piedras, I.; Simmons, G.E., Jr.; Pruitt, K. Dishevelled: A masterful conductor of complex Wnt signals. *Cell. Signal.* **2018**, *47*, 52–64. [CrossRef] [PubMed]
19. Dill, H.; Linder, B.; Fehr, A.; Fischer, U. Intronic miR-26b controls neuronal differentiation by repressing its host transcript, ctdsp2. *Genes Dev.* **2012**, *26*, 25–30. [CrossRef] [PubMed]
20. Hinske, L.C.; Galante, P.A.F.; Limbeck, E.; Möhnle, P.; Parmigiani, R.B.; Ohno-Machado, L.; Camargo, A.A.; Kreth, S. Alternative polyadenylation allows differential negative feedback of human miRNA miR-579 on its host gene ZFR. *PLoS ONE* **2015**, *10*, e0121507. [CrossRef] [PubMed]
21. Steiman-Shimony, A.; Shtrikman, O.; Margalit, H. Assessing the functional association of intronic miRNAs with their host genes. *RNA* **2018**. [CrossRef] [PubMed]

22. Ye, X.-Z.; Xu, S.-L.; Xin, Y.-H.; Yu, S.-C.; Ping, Y.-F.; Chen, L.; Xiao, H.-L.; Wang, B.; Yi, L.; Wang, Q.-L.; et al. Tumor-associated microglia/macrophages enhance the invasion of glioma stem-like cells via TGF-β1 signaling pathway. *J. Immunol.* **2012**, *189*, 444–453. [CrossRef] [PubMed]
23. Joseph, J.V.; Conroy, S.; Tomar, T.; Eggens-Meijer, E.; Bhat, K.; Copray, S.; Walenkamp, A.M.E.; Boddeke, E.; Balasubramanyian, V.; Wagemakers, M.; et al. TGF-β is an inducer of ZEB1-dependent mesenchymal transdifferentiation in glioblastoma that is associated with tumor invasion. *Cell Death Dis.* **2014**, *5*, e1443. [CrossRef] [PubMed]
24. Wick, W.; Platten, M.; Weller, M. Glioma cell invasion: regulation of metalloproteinase activity by TGF-beta. *J. Neurooncol.* **2001**, *53*, 177–185. [CrossRef] [PubMed]
25. Han, J.; Alvarez-Breckenridge, C.A.; Wang, Q.-E.; Yu, J. TGF-β signaling and its targeting for glioma treatment. *Am. J. Cancer Res.* **2015**, *5*, 945–955. [PubMed]
26. Shi, Y.; Massagué, J. Mechanisms of TGF-beta signaling from cell membrane to the nucleus. *Cell* **2003**, *113*, 685–700. [CrossRef]
27. Bruna, A.; Darken, R.S.; Rojo, F.; Ocaña, A.; Peñuelas, S.; Arias, A.; Paris, R.; Tortosa, A.; Mora, J.; Baselga, J.; Seoane, J. High TGFβ-Smad Activity Confers Poor Prognosis in Glioma Patients and Promotes Cell Proliferation Depending on the Methylation of the PDGF-B Gene. *Cancer Cell* **2007**, *11*, 147–160. [CrossRef] [PubMed]
28. Seystahl, K.; Tritschler, I.; Szabo, E.; Tabatabai, G.; Weller, M. Differential regulation of TGF-β-induced, ALK-5-mediated VEGF release by SMAD2/3 versus SMAD1/5/8 signaling in glioblastoma. *Neuro. Oncol.* **2015**, *17*, 254–265. [CrossRef] [PubMed]
29. Gan, X.-Q.; Wang, J.-Y.; Xi, Y.; Wu, Z.-L.; Li, Y.-P.; Li, L. Nuclear Dvl, c-Jun, β-catenin, and TCF form a complex leading to stabiLization of β-catenin–TCF interaction. *J. Cell Biol.* **2008**, *180*, 1087–1100. [CrossRef] [PubMed]
30. Zhu, Y.; Tian, Y.; Du, J.; Hu, Z.; Yang, L.; Liu, J.; Gu, L. Dvl2-dependent activation of Daam1 and RhoA regulates Wnt5a-induced breast cancer cell migration. *PLoS ONE* **2012**, *7*, e37823. [CrossRef] [PubMed]
31. Sharma, J.; Mulherkar, S.; Mukherjee, D.; Jana, N.R. Malin regulates Wnt signaling pathway through degradation of dishevelled2. *J. Biol. Chem.* **2012**, *287*, 6830–6839. [CrossRef] [PubMed]
32. Kahlert, U.D.; Maciaczyk, D.; Doostkam, S.; Orr, B.A.; Simons, B.; Bogiel, T.; Reithmeier, T.; Prinz, M.; Schubert, J.; Niedermann, G. Others Activation of canonical WNT/β-catenin signaling enhances in vitro motility of glioblastoma cells by activation of ZEB1 and other activators of epithelial-to-mesenchymal transition. *Cancer Lett.* **2012**, *325*, 42–53. [CrossRef] [PubMed]
33. Zhang, J.; Huang, K.; Shi, Z.; Zou, J.; Wang, Y.; Jia, Z.; Zhang, A.; Han, L.; Yue, X.; Liu, N.; et al. High β-catenin/Tcf-4 activity confers glioma progression via direct regulation of AKT2 gene expression. *Neuro. Oncol.* **2011**, *13*, 600–609. [CrossRef] [PubMed]
34. Truscott, M.; Islam, A.B.M.M.K.; Lightfoot, J.; López-Bigas, N.; Frolov, M.V. An intronic microRNA links Rb/E2F and EGFR signaling. *PLoS Genet.* **2014**, *10*, e1004493. [CrossRef] [PubMed]
35. Gomes, L.R.; Terra, L.F.; Wailemann, R.A.; Labriola, L.; Sogayar, M.C. TGF-β1 modulates the homeostasis between MMPs and MMP inhibitors through p38 MAPK and ERK1/2 in highly invasive breast cancer cells. *BMC Cancer* **2012**, *12*, 26. [CrossRef] [PubMed]
36. Dziembowska, M.; Danilkiewicz, M.; Wesolowska, A.; Zupanska, A.; Chouaib, S.; Kaminska, B. Cross-talk between Smad and p38 MAPK signalling in transforming growth factor β signal transduction in human glioblastoma cells. *Biochem. Biophys. Res. Commun.* **2007**, *354*, 1101–1106. [CrossRef] [PubMed]

cancers

MDPI

Article

PKA at a Cross-Road of Signaling Pathways Involved in the Regulation of Glioblastoma Migration and Invasion by the Neuropeptides VIP and PACAP

Souheyla Bensalma [1], Soumaya Turpault [1], Annie-Claire Balandre [2], Madryssa De Boisvilliers [1], Afsaneh Gaillard [3], Corinne Chadéneau [1] and Jean-Marc Muller [1,*]

1 Team Récepteurs, Régulations, Cellules Tumorales (2RCT), EA3842 CAPTuR, Pôle Biologie-Santé, Université de Poitiers, F-86022 Poitiers, France; souheyla.bensalma@gmail.com (S.B.); soumaya.turpault@gmail.com (S.T.); m.de.boisvilliers@outlook.fr (M.D.B.); corinne.chadeneau@univ-poitiers.fr (C.C.)

2 STIM Laboratory, CNRS ERL 7003-EA7349, Pôle Biologie-Santé, Université de Poitiers, F-86022 Poitiers, France; annie-claire.balandre@univ-poitiers.fr

3 Laboratoire de Neurosciences Expérimentales et Cliniques (LNEC)–INSERM UMR-S1084, Pôle Biologie-Santé, Université de Poitiers, F-86022 Poitiers, France; afsaneh.gaillard@univ-poitiers.fr

* Correspondence: jean.marc.muller@univ-poitiers.fr; Tel.: +33-054-945-3725

Received: 19 December 2018; Accepted: 10 January 2019; Published: 21 January 2019

Abstract: Glioblastoma (GBM) remains an incurable disease, mainly due to the high migration and invasion potency of GBM cells inside the brain. PI3K/Akt, Sonic Hedgehog (SHH), and PKA pathways play major regulatory roles in the progression of GBM. The vasoactive intestinal peptide (VIP) family of neuropeptides and their receptors, referred in this article as the "VIP-receptor system", has been reported to regulate proliferation, differentiation, and migration in a number of tumor cell types and more particularly in GBM cells. These neuropeptides are potent activators of the cAMP/PKA pathway. The present study aimed to investigate the cross-talks between the above cited signaling cascades. Regulation by VIP-related neuropeptides of GBM migration and invasion was evaluated ex vivo in rat brain slices explanted in culture. Effects of different combinations of VIP-related neuropeptides and of pharmacological and siRNA inhibitors of PKA, Akt, and of the SHH/GLI1 pathways were tested on GBM migration rat C6 and human U87 GBM cell lines using the wound-healing technique. Quantification of nuclear GLI1, phospho-Akt, and phospho-PTEN was assessed by western-immunoblotting. The VIP-receptor system agonists VIP and PACAP-38 significantly reduced C6 cells invasion in the rat brain parenchyma ex vivo, and C6 and U87 migration in vitro. A VIP-receptor system antagonist, VIP_{10-28} increased C6 cell invasion in the rat brain parenchyma ex vivo, and C6 and migration in vitro. These effects on cell migration were abolished by selective inhibitors of the PI3K/Akt and of the SHH pathways. Furthermore, VIP and PACAP-38 reduced the expression of nuclear GLI1 while VIP_{10-28} increased this expression. Selective inhibitors of Akt and PKA abolished VIP, PACAP-38, and VIP_{10-28} effects on nuclear GLI1 expression in C6 cells. PACAP-38 induced a time-dependent inhibition of phospho-Akt expression and an increased phosphorylation of PTEN in C6 cells. All together, these data indicate that triggering the VIP-receptor system reduces migration and invasion in GBM cells through a PKA-dependent blockade of the PI3K/Akt and of the SHH/GLI1 pathways. Therefore, the VIP-receptor system displays anti-oncogenic properties in GBM cells and PKA is a central core in this process.

Keywords: glioblastoma; protein kinase A; Sonic Hedgehog; Akt; PTEN; VIP-receptor system

---

1. Introduction

Glioblastoma multiforme (GBM) are the most frequent and malignant adult brain cancers. Their origin remains unclear, but it has been proposed that they could develop from the transformation of poorly differentiated glial progenitors or of neural stem cells [1–4]. Since 2005, the Stupp protocol has become the standard of care for the treatment of GBM. It consists of radiotherapy and concomitant chemotherapy with temozolomide, an alkylating agent [5]. Since then, in spite of constant progress of neurosurgical resection, radio- and chemotherapy, GBM remains an incurable disease with a median survival time less than two years after diagnosis. This is the consequence of the high migration and invasion potency of GBM cells, more particularly of the subset of so-called GBM stem cells or GSCs [6]. A strategy to cure GBM patients could thus be to inhibit tumor infiltration into the surrounding brain parenchyma. Clinical trials for GBM infiltration using the matrix metalloproteinase inhibitor *Marimastat* or the integrin antagonist *Cilengitide* have been attempted with no real success [7]. Numerous recent therapeutic trials targeting the pro-invasive role in GBM of Ephrin receptors, TGFβR1, Integrin β8 chain, Rho GTPases, and casein kinase 2 (CK2) are under development [8]. Recent immunotherapy early phase trials targeting the GBM stem cells led to a significant improvement of the median survival of patients [9].

The signaling pathways that play central roles in the invasive potential and in the radio- and chemo-resistance of GBM have been extensively studied. Among them are the PI3K/Akt/PTEN/mTOR and the SHH/GLI1 cascades [10]. In numerous GBM cases, PI3K/Akt is abnormally activated, due to amplification of EGFR, gene amplification, or activating mutations of the p110α catalytic or of the p85 regulatory subunits of PI3K. Almost half of GBM patients bear deletions, mutations, or epigenetic silencing of the PTEN gene leading to a loss of function of this anti-oncogenic factor associated with poor survival. Alterations of at least one of the EGFR, PTEN, or p110 PI3K genes is frequently detected in primary and or secondary GBM [11,12]. Effectors of this pathway have been targeted by a number of small molecules that demonstrated poor therapeutic benefit on GBM progression in clinical trials [13–17].

Another major cascade in GBM pathogenesis is triggered by the developmental protein Sonic Hedgehog (SHH) binding to the transmembrane glycoprotein Patched-1 (PTCH1), which releases its repressor activity on the smoothened (SMO) co-receptor, a member of the G-protein coupled receptors (GPCR) family. This causes the expression, activation, and nuclear import of glioma-associated oncogene homolog 1 (GLI1), a zinc finger transcription factor, regulating directly or indirectly the expression of numerous factors involved in GBM progression. Growth factors also activate GLI1 through the PI3K/Akt and Ras/MAP kinases cascades, while GPCR activation of PKA represses this process [18,19]. A number of small compounds that inhibit different effectors of this pathway have been developed. Despite their efficacy in vitro and in preclinical assays, SMO inhibitors like the plant alkaloid cyclopamine and its derivatives failed to improve the overall patient survival in clinical trials. This may be due to their limited bioavailability and to unintentional side effects, since the SHH pathway is involved in many physiological cell processes. Moreover, resistance to these inhibitors have been observed in animal models as a consequence of, for example, SMO activating or PTCH1 inactivating mutations, and PTCH1 suppression by the microRNA miR-9 [20–25].

The "VIP-receptor system" is composed of the 28-amino-acid neuropeptide VIP (vasoactive intestinal peptide) and VIP-related peptides, such as the 38-amino-acid PACAP-38 (pituitary adenylate-cyclase activating peptide) and their GPCR: VPAC1 and VPAC2, which display a high affinity for both VIP and PACAP-38, and PAC1 which is selective for PACAP-38. The pleiotropic functions of the VIP-receptor system in the body, particularly on glial and neuronal differentiation and on the progression of a number of cancer types, are at least partly mediated by a potent activation of the cAMP/PKA pathway [26–30]. We and others demonstrated that GBM generally express different combinations of components of the VIP-receptor system that are involved in the control of proliferation and migration of GBM cells [31–38]. VIP and PACAP act as anti-invasive factors in different GBM cell lines, a function mediated by VPAC1-dependent inhibition of AKT phosphorylation [36,38]. PACAP also acts as a strong tumor suppressor in medulloblastoma (MB), a highly aggressive tumor of the cerebellum, by repressing GLI1 expression and activity in a PKA-dependent manner [39,40]. It has been proposed that in MB, a compartmentalized pool of PKA in the vicinity of primary cilia could be involved in the regulation by PACAP of the SHH/GLI1 pathway [41]. PKA directly regulates GLI1 by phosphorylating a Threonine residue near the nuclear localization signal of this protein, resulting in its cytoplasmic retention [42]. Overexpression of Gαs in transgenic mice leads to increased cAMP-dependent signaling, inhibits ciliary trafficking of SHH components and SMO activation, leading to decreased MB cell proliferation [43]. These data indicate that, as an alternative to SMO inhibition, the SHH signaling could be inhibited by PACAP/PKA signaling in tumor cells.

Shortly after the discovery of the intracellular messenger cyclic adenosine monophosphate (cAMP) and of its synthetase adenylate cyclase (ADCY), a relationship between the levels of this nucleotide, cell differentiation and tumor biology has been evident. Decreased activation of the cAMP-dependent pathway and lower levels of cAMP are generally associated with increased tumor grade and decreased overall patient survival. An extensive and converging documentation supports the potent tumor suppressor function of the cAMP/PKA axis in GBM and MB cells [43–48].

All together, these observations encouraged us to investigate further the signaling pathways triggered by the VIP-receptor system involved in GBM migration and invasion, paying particular attention to the potential implication of PKA in the cross-talks between the PI3K/PTEN and the SHH/GLI1 cascades in these phenomena.

2. Results

2.1. Components of the VIP-Receptor System Are Expressed in C6 and U87 Cell Lines

Analysis by RT-qPCR of mRNA expression of the components of the VIP-receptor system reveals that C6 cells mainly express VIP and the VPAC2 receptor and very limited levels of PACAP, VPAC1 and PAC1 mRNAs. In U87 cells, the main expressed components are VIP and PAC1 with a very low level of VPAC1 and PACAP mRNAs, while VPAC2 mRNA was undetectable, as shown in Table S1. Thus, both cell lines expressed mainly the VIP mRNA and at least one mRNA encoding a receptor of the VIP-receptor system.

2.2. VIP, PACAP-38, and $VIP_{10\text{-}28}$ Effects on Invasion by C6-GFP Cells of the Rat Brain Parenchyma

Ex vivo invasion assays on rat brain slices were performed in order to assess the potential action of VIP and related peptides on this process in rat C6 GBM cells. After 48 h of treatment of the slices with 1 μM VIP, PACAP-38, or $VIP_{10\text{-}28}$, the width of invasion of GFP-C6 cells was determined using a fluorescent macroscope, as shown in Figure 1A,B. The depth of invasion of GFP-C6 cells was assessed by confocal microscopy, as shown in Figure 1C. The width of the invasive area was significantly reduced by about 50% by the VIP or PACAP-38 receptors agonists and the depth of invasion was decreased by about 35%. On the contrary, the VIP receptor antagonist $VIP_{10\text{-}28}$ increased by 2.3-fold the width of the invasive area and by about 1.4-fold the width of invasion by C6 cells. The VIP and PACAP-38 analog secretin, which does not interact with VIP/PACAP receptors, did not significantly modify invasion of brain slices by C6 cells, as shown in Figure 1B,C.

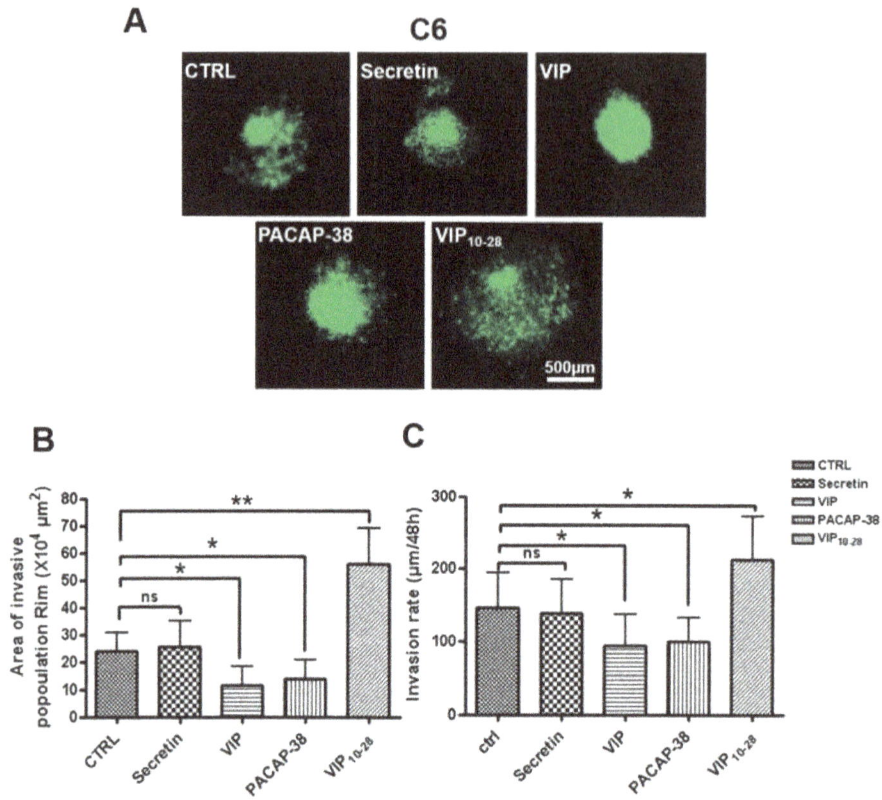

Figure 1. The rat brain slice model system was used to measure ex vivo C6 cell invasiveness. C6 cells stably expressing GFP were transplanted into the center of the putamen in brain slices. Then brain slices were treated or not (CTRL, control) with 10^{-6}M of secretin, vasoactive intestinal peptide (VIP), PACAP-38, or VIP_{10-28}, and the C6 cells were allowed to invade for 48 h. (**A**) Images were taken with a macroscope. (**B**) The area of GFP-stained cells in invasive population (RIM) in each section was measured. (**C**) The depth of invasion at 48 h was determined using the confocal microscopy. The values are the mean ± SD of quantification of glioblastoma (GBM) invasion in eight different brain slices, obtained from three independent experiments for each experimental condition. About 4–5 newborn rats were necessary to obtain 10 usable brain slices. Data were analyzed using the Kruskal–Wallis test and Dunn's post hoc test (* $p < 0.05$; ** $p < 0.01$).

2.3. *VIP, PACAP-38, and VIP_{10-28} Regulation of the Migration Process of C6 and U87 Cells is Affected by Different Signaling Pathways Inhibitors*

Neuropeptide effects on GBM cell migration were evaluated by the wound-healing technique, as described in the methods section and as previously reported [36]. In both C6 and U87 cells, VIP or PACAP-38 but not secretin, dose-dependently and robustly reduced the wound-healing closure process, as shown in Figure 2A,B. The VIP receptor antagonist VIP_{10-28} significantly increased the migration of C6 cells but had no significant effect in the U87 cell line. The PKA inhibitor H89 at 1 μM abolished the inhibition of migration induced by 10^{-7} M VIP or PACAP-38 in both cell types, as shown in Figure 2C,D.

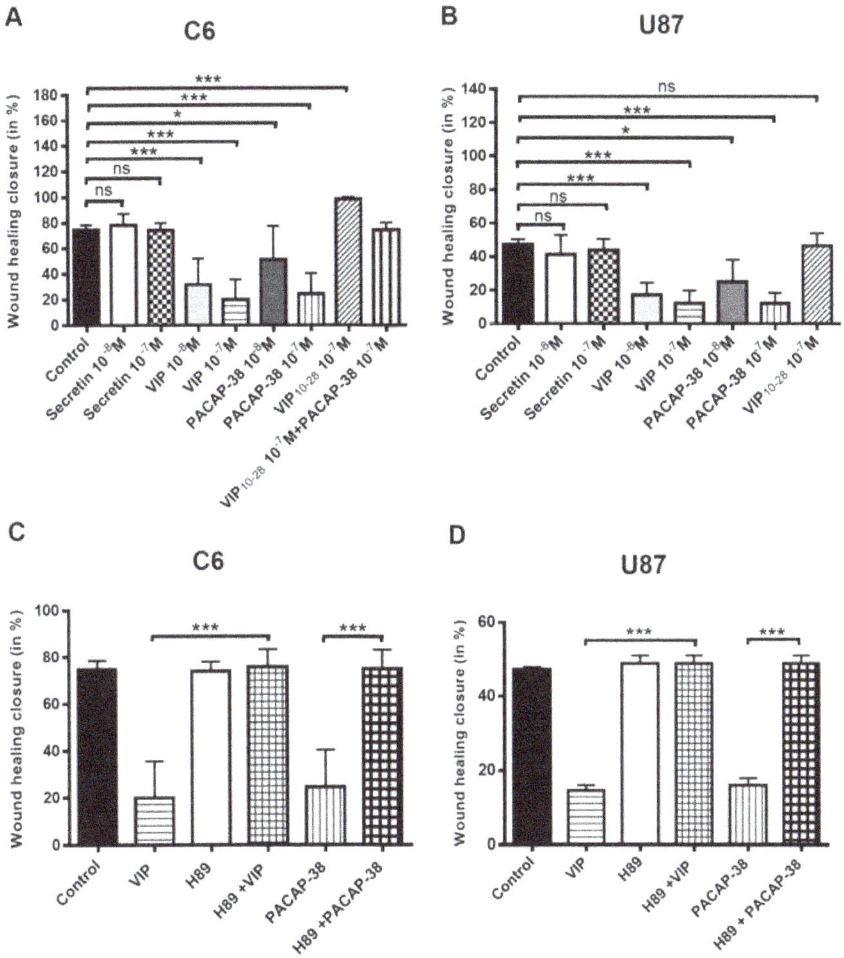

Figure 2. Effects of VIP, PACAP-38, $VIP_{10\text{-}28}$ and/or inhibitors of signaling pathways H89 (PKA inhibitor), IA (Akt inhibitor), on the migration process of C6 and U87 cells. Migration was analyzed by the technique of wound healing closure, as described in Materials and Methods. (**A**,**B**) Effect of neuropeptides on the migration of C6 or U87 cells. (**C**,**D**) Migration of C6 and U87 cells treated or not with VIP or PACAP-38 at 10^{-7} M in absence or presence of H89. Data represent the wound healing closure expressed in % and are the mean ± SD of at least three independent experiments, each performed in duplicate. Data were analyzed using the Kruskal–Wallis test and Dunn's post hoc test (* $p < 0.05$, *** $p < 0.001$, ns: not significant).

The increased migration of C6 cells by $VIP_{10\text{-}28}$ was blocked by 1 μM of the AKT protein kinase inhibitor or by the SMO co-receptor antagonist cyclopamine. None of these inhibitors affected the cell migration on their own at the concentration used, as shown in Figure 3A. When C6 cells were transfected with a siRNA targeting GLI1 (Gli-1 siRNA), an important reduction of wound closure by about 3-fold was obtained when compared with cells transfected with a control siRNA. The VIP-receptors antagonist $VIP_{10\text{-}28}$ significantly increased migration in the cells transfected with the control siRNA, but this effect could not be observed in the cells transfected with the GLI1-siRNA, as shown in Figure 3B. These results indicate that PKA is implicated in the regulation of C6 and U87 cell migration by the VIP-receptor system and that the component of the Hh pathway GLI1 participates also in this regulation in C6 cells.

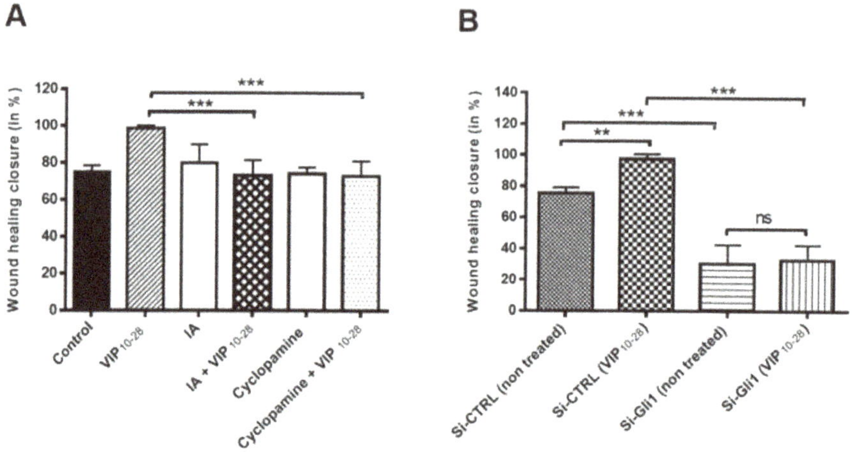

Figure 3. Effects of $VIP_{10\text{-}28}$ and/or inhibitors IA (Akt inhibitor), cyclopamine (Smo inhibitor), or Gli1 (Si-RNA) on the migration process of C6 cells. (**A**) Migration of C6 treated or not with $VIP_{10\text{-}28}$ at 10^{-7} M in absence or presence of 1 μM of IA or cyclopamine. Cells underwent incubation for 20 min with an inhibitor (1 μM of IA or cyclopamine) followed by treatment with $VIP_{10\text{-}28}$ at 10^{-7} M for 24 h. (**B**) Wound healing assay using confluent C6 cells transfected with 25 nM siRNA targeting GLI1 (GLI1-siRNA) or 25 nM negative control siRNA. Cells were wounded and treated or not with $VIP_{10\text{-}28}$ 10^{-7} M for 24 h. Data represent the wound healing closure expressed in % and are the mean ± SD of at least three independent experiments, each performed in duplicate. Data were analyzed using the Kruskal–Wallis test and Dunn's post hoc test (** $p < 0.01$, and *** $p < 0.001$, ns: not significant).

2.4. Inhibitors of PKA (H89) or of Akt (IA) Abolish the Effects of VIP-Related Peptides on the GLI1 Protein Nuclear Expression

Western immunoblotting detection of GLI1 in C6 or U87 lineages revealed the major expression of this protein in nuclear extracts of these GBM cells, as shown in Figure 4A. Expression of GLI1 was analyzed in cells treated for 24 h with neuropeptides at 10^{-7} M, following or not a 1 μM pretreatment with H89 or IA, as shown in Figure 4B–D. In both cell types, VIP or PACAP-38 induced an important reduction of nuclear expression of Gli1 protein. The effects of these neuropeptides were totally blocked by H89. In C6 cells, the $VIP_{10\text{-}28}$ antagonist triggered a robust rise of the expression of nuclear GLI1 protein, an effect which was completely abolished by IA.

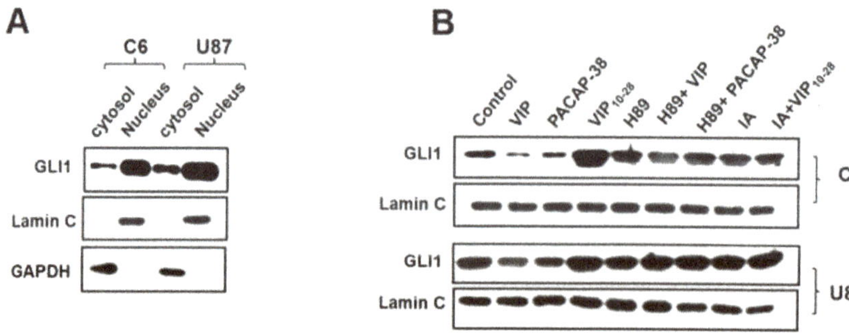

Figure 4. *Cont.*

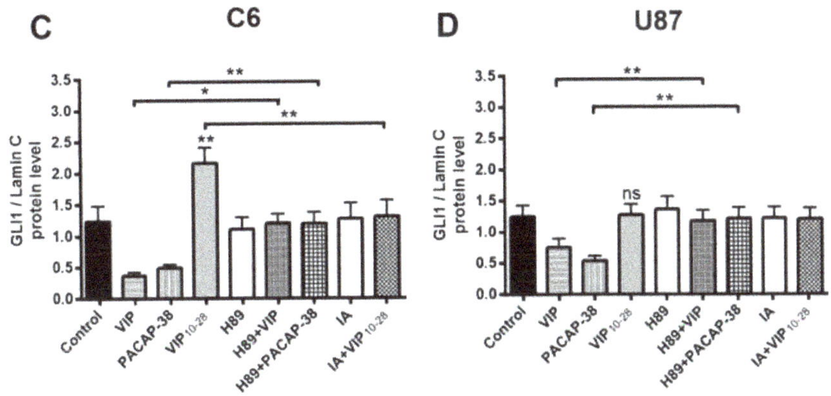

Figure 4. Effects of VIP, PACAP-38, $VIP_{10\text{-}28}$, and/or inhibitors H89 (PKA inhibitor), IA (Akt inhibitor) on the expression and intracellular localization of GLI1 protein in C6 and U87 cells. (**A**) Western immunoblotting of GLI1 proteins in C6 and U87 cells. Cytosolic and nuclear proteins of cells were resolved by SDS-PAGE, transferred to membranes, and further probed with antibody anti-GLI1, anti-GAPDH, or anti-Lamin-C. GAPDH and Lamin-C proteins are used as a control for cytosolic and nuclear extracts, respectively. (**B**) Western immunoblotting of GLI1 proteins in C6 and U87 cells treated or not with neuropeptides in absence or presence of pathways inhibitors. Nuclear proteins were extracted from cells treated or not with neuropeptides at 10^{-7} M for 24 h. Cells were incubated for 20 min with an inhibitor (1 μM of H89 or IA) followed by treatment with VIP, PACAP-38, or $VIP_{10\text{-}28}$ at 10^{-7} M for 24 h. Nuclear proteins of cells were resolved by SDS-PAGE, transferred to membranes and further probed with antibody anti-GLI1 or anti-Lamin-C. (**C**,**D**) Cumulative data from three independent experiments, in same conditions as in (**B**), after densitometry quantification of immunoblotting (WB) data. Quantification data are the mean ± SD. Data were analyzed using the Kruskal–Wallis test and Dunn's post hoc test (* $p < 0.05$ and ** $p < 0.01$, ns: not significant).

2.5. PACAP-38 Triggers a Time-Dependent Decrease of Phospho-Akt and Elevation of Phospho-PTEN Protein Expression in C6 Cells

C6 cells were treated with 10^{-7} M PACAP-38 for increasing periods of time ranging from 30 min to 24 h. Western immunoblotting detection of total and phosphorylated forms of Akt and phosphorylated PTEN was then carried out in cell lysates from these experiments, as shown in Figure 5A–C. PACAP-38 time-dependently decreased the phospho-Akt/total Akt ratio with a very strong but transient maximal effect after 6 h of treatment. At 24 h, this ratio returned to its initial level at the onset of the experiment. On the opposite, the phospho-PTEN/GAPDH ratio started to significantly increase after 2 h of treatment with a maximal increase of about 10-fold after 3 h. Then this ratio dropped back to its initial value after 24 h. Increased phosphorylation of PTEN by PACAP-38 was abolished in the presence of the PKA inhibitor H89, as shown in Figure 5D,E.

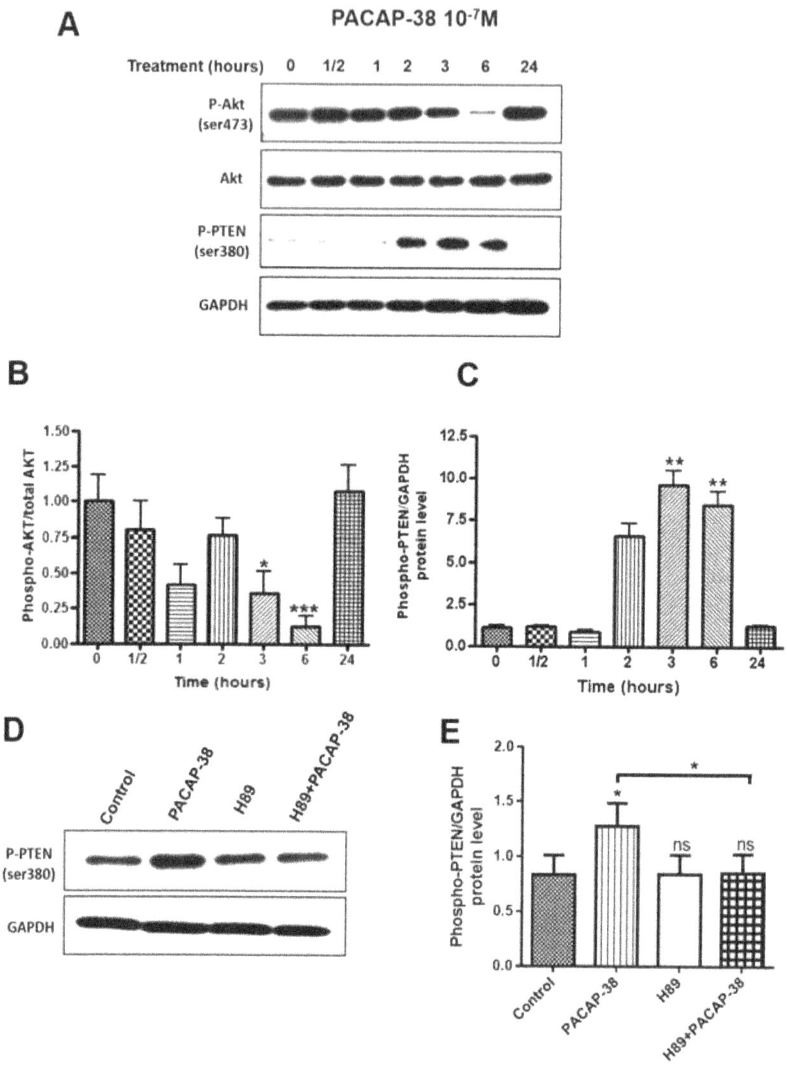

Figure 5. Effect of PACAP-38 on the phosphorylation of Akt and PTEN in C6 cells. Effect of the H89 PKA inhibitor on PACAP638 induced inhibition of PTEN phosphorylation. (**A–C**) Cells were treated with 10^{-7} M PACAP-38 for the indicated times. Total cell lysates were analyzed by immunoblotting (WB) with antibodies recognizing total Akt, phosphorylated Akt, phosphorylated PTEN or GAPDH. (**A**) corresponds to a representative WB experiment. (**B**,**C**) Cumulative data from three independent experiments after quantification by densitometry. Data were normalized by attributing a value of 1 for the control experiment. (**D**,**E**) Cells were treated for 3 h with 10^{-7} M PACAP-38, in the presence or not of 1 μM H89. (**D**) corresponds to a representative WB experiment. (**E**) Cumulative data from three independent experiments after quantification by densitometry. Quantification data are the mean ± SD. Data were analyzed using the Kruskal–Wallis test and Dunn's post hoc test (* $p < 0.05$, ** $p < 0.01$, and *** $p < 0.001$, ns: not significant).

3. Discussion

Previous work from our group demonstrated that the VIP-receptor system displayed anti-oncogenic properties in different human GBM cell lines. VIP receptor antagonists and a PACAP antibody enhanced migration, suggesting an autocrine or paracrine action of VIP-related peptides on this process [36,38]. The aim of the present study was to extend the analysis of GBM migration and invasion with the objective to investigate the signaling pathways involved in these processes. The C6 rat GBM cells were chosen because they have been long recognized as a valuable experimental model system for studying ex vivo or in vivo GBM progression after injection in Wistar rat brain [49]. This animal model was utilized in our group to test the in vivo antitumor activity of a cyclopamine glucuronide prodrug on GBM progression [50]. The U87 cell line has been demonstrated to express the PAC1 receptor and this cell line has been utilized to evaluate the effect of a VIP receptors antagonist on xenograft proliferation in nude mice [31]. We have generated C6 cells expressing the GFP protein to better follow the extent of tumor cell invasion after implantation of these cells inside the brain parenchyma. This allowed us to precisely determine the width and depth of invasion in rat brain slices by C6 cells and to demonstrate that both VIP and PACAP-38 were able to limit GBM cell invasion in this tissue. Of course, these ex vivo conditions do not reflect the real conditions of GBM development inside the brain. In the future, it will be of chief importance to study the effects of these neuropeptides on GBM progression on in vivo developing tumors in animal models. The wound healing technique was carried out to assess the efficiency of inhibitors of PKA (H89) and Akt (IA) protein kinases on the VIP and PACAP-38 inhibition of C6 and U87 migration process. The data indicated that the effects of these neuropeptides were abolished by these compounds, suggesting an involvement of the corresponding pathways in the migration process of GBM cells. In the C6 cells that are expressing important levels of VIP mRNA, the VIP receptors antagonist $VIP_{10\text{-}28}$ increased cell migration, arguing for an autocrine or paracrine reduction by endogenous VIP of the C6 cells migration. The $VIP_{10\text{-}28}$ antagonist has been demonstrated to be more selective for the VPAC1 and VPAC2 receptor subtypes [51,52], indicating their possible implication in the VIP and PACAP-38 inhibition of the C6 and U87 migration process. Interestingly, the fact that both VIP and PACAP-38 reduced U87 GBM migration also indicates that these peptides could act on the low amount of VPAC1 receptors expressed in this cell line or perhaps on certain splice variants of the PAC1 receptor that have been demonstrated to display high-affinity for both VIP and PACAP [53]. Increased migration by $VIP_{10\text{-}28}$ was not observed in U87 cells, which may be due to a lesser expression of VIP, as observed by real-time RT-PCR analysis. In the C6 cells, H89, IA, GLI1 siRNA, and cyclopamine, were all capable to totally abolish the increased cell migration caused by $VIP_{10\text{-}28}$. Migration was potently reduced in C6 transfected with a GLI1 siRNA and in these cells, $VIP_{10\text{-}28}$ failed to increase cell migration. Taken together, these data argue for an implication of the PKA, PI3K/Akt, and SHH/GLI1 pathways in the migration process of GBM cells.

To confirm these hypotheses, effects of the neuropeptides were evaluated on nuclear GLI1 protein expression in C6 and U87 cells. The data indicated that VIP or PACAP-38 strongly reduced the nuclear level of GLI1 protein, a process which was abolished by the PKA inhibitor H89. In C6 cells, the $VIP_{10\text{-}28}$ increase of nuclear GLI1 expression was completely blocked by IA, but the Akt inhibitor had no effect on VIP or PACAP-38 reduction of nuclear GLI1 level. These observations clearly demonstrate that regulation of GLI1 localization and activity by the VIP-receptor system are dependent on the PKA and the PI3K/Akt pathways. Regulation of the nuclear localization and transcriptional activity of GLI1 by Akt is well documented in a number of malignancies in general and in GBM in particular [54–57]. Furthermore, it has been described that phosphorylation of GLI1 near its nuclear localization signal (NLS) by PKA retains GLI1 in the cytoplasm [42].

A time-dependent treatment of C6 cells by PACAP-38 revealed that this neuropeptide could reduce the expression of the phosphorylated form of Akt with a very strong effect at 6 h. Conversely, PACAP-38 triggered a potent elevation of phospho-PTEN protein with the highest amplitude after 3 h of treatment, i.e., 3 h before the maximal inhibition of phospho-Akt, suggesting a close link between these two mechanisms. These experiments have been conducted in C6 cells because they express a wild-type PTEN protein, contrary to U87 presenting an in-frame deletion in the PTEN gene [58]. PKA could play a major role in the phosphorylation events observed in our studies because it has been shown that the p85 regulatory subunit of PI3K is phosphorylated in vivo and in vitro by PKA on its serine residue 83, a process which was demonstrated to play a major role on G1 cell cycle arrest [59]. Hence, the strong inhibition of Akt phosphorylation by PACAP-38 could be due to the combined phosphorylation by PKA of the P85 regulatory PI3K subunit and potentially of the phosphatase PTEN, leading to PIP3 dephosphorylation and to inhibition of the p110 catalytic PI3K subunit. The latter process remains, however, to be demonstrated in GBM cells. Subsequently, the decreased nuclear localization and activity of GLI1 observed in GBM cells could result both from the above discussed PKA-dependent inhibition of PI3K/Akt activity and from GLI1 phosphorylation by PKA (Figure 6). Hence PKA appears like a possible central core in the complex network linking these different pathways.

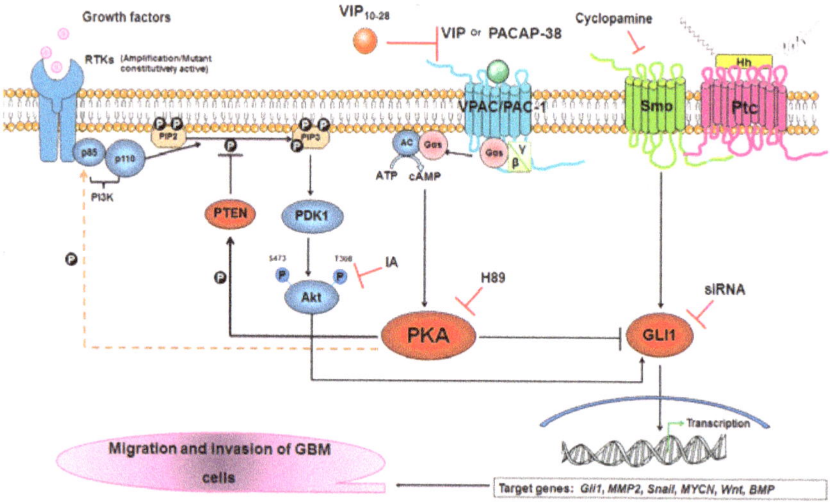

Figure 6. Potential molecular pathways and mechanisms regulated by the VIP-receptor system in glioblastoma (GBM) cells. Binding of VIP and related peptide agonists to their G-protein coupled receptors (GPCR) VPAC-1/2 and PAC1, leads to adenylate cyclase (AC) and cAMP-dependent Protein Kinase (PKA) activation. PKA is a central enzyme in these mechanisms because it could potentially: (i) inhibit glioma-associated oncogene homolog 1 (GLI1) nuclear translocation and activity; (ii) phosphorylate the phosphatase tensin-homolog deleted on chromosome ten (PTEN) and activate its Phosphatidylinositol (3,4,5)-trisphosphate (PIP3) phosphatase activity, leading to an inhibition of Akt phosphorylation and activation and consequently to a reduced nuclear localization and transcriptional activity of GLI1; (iii) phosphorylate the p85 regulatory subunit of PI3K on its serine residue 83. This hypothetic process (dashed line) that was described in vitro and in vivo to play a major role on G1 cell cycle arrest (see the discussion section) remains however to be demonstrated in GBM. Hh: Sonic Hedgehog; Ptc: Patched, the Hh receptor; RTK: receptor tyrosine kinase; Smo: smoothened, the Ptc co-receptor.

These data all together indicate that triggering the VIP-receptor system, a potent activator of the cAMP/PKA axis, reduces migration and invasion in GBM cells through a blockade of the PI3K/Akt and of the Shh/GLI1 pathways. These effects are efficiently abolished by H89, a PKA inhibitor. One has to keep in mind that pharmacological inhibitors can display off-target effects, as reported for H89, which besides PKA, can act on other protein kinases and signaling mechanisms [60,61]. In spite of these limitations, the observations reported in the present study support the hypothesis of a central role of PKA at a cross-road of signaling pathways involved in the regulation of GBM migration and invasion. Ten adenylate cyclases (ADCYs) have been identified in mammals. They can be activated by the plant diterpene Forskolin. Some of these isoforms are often overexpressed in brain cancers. ADCY3 is localized in the primary cilia of neural cells. Development of specific activators of certain ADCY isoforms expressed in GBM cells could thus be an interesting paradigm. Another strategy could be to block the phosphodiesterase PDE4 by Rolipram or other inhibitors of this enzyme. PDE4, which specifically hydrolyzes cAMP, is often overexpressed in cancer cells [43–48].

As a powerful activator of the cAMP/PKA pathway, the VIP-receptor system should also be considered as an alternative strategy to inhibit the interrelated signaling pathways involved in GBM cells in particular and in cancer progression in general. In this respect, a number of VIP or PACAP more stable derivatives or with improved receptor subtype specificity have been developed. Some of them have been utilized for in vivo studies on their neuroprotective properties. They really deserve to be considered for future in vitro and in vivo preclinical studies targeting the VIP-receptor system in cancer progression [62–65].

4. Materials and Methods

4.1. Materials

Synthetic VIP, PACAP-38, and antagonist VIP_{10-28} were purchased from Polypeptide Laboratories. H89 (a PKA antagonist) and IA (Akt Inhibitor VIII, Isozyme-Selective) were obtained from Calbiochem Merck, Guyancourt France. The Smo antagonist cyclopamine (an inhibitor of SHH pathway) was obtained from Sigma Aldrich, Saint-Quentin Fallavier, France).

4.2. Cell Culture

The rat C6 glioma cell line [66] was purchased from the European Collection of Animal Cell Culture (ECACC) (Salisbury, UK). The human U87 MG glioblastoma cell line was obtained from the American Type Culture Collection (Manassas, VA, USA). C6 cells were utilized from passage 9 to 20, to avoid the possible loss in these cells of efficient coupling of the VIP and PACAP receptors to adenylate cyclase at late passages reported by other investigators [33]. In standard monolayer conditions, the cells were cultured in Dulbecco's modified Eagle's medium (DMEM) with GlutamaxTM I, and supplemented with 10% fetal bovine serum, 100 U/mL penicillin, and 100 µg/mL streptomycin. Cells were incubated in humidified 95% air, 5% CO_2 at 37 °C. Medium was changed twice a week, and cells were subcultured once a week using trypsin-EDTA solution. Cell culture media, supplements and reagents were from Lonza, Levallois-Perret, France.

4.3. Transfection with a GFP-Encoding Vector

C6 cells were stably transfected with copGFP plasmid (Santa Cruz Biotechnology, Clinisciences, Nanterre, France) using FuGENE (Promega, Charbonnières-les-Bains, France) according to the manufacturer's instructions. Twenty-four hours before transfection experiments, 150,000 C6 cells were plated per well in 6-well dishes and grown in serum-containing medium without antibiotics. FuGENE (3 µL) and 1 µg of a copGFP plasmid was mixed with Opti-MEM and the mixture was incubated for 15 min at room temperature. The combined mixtures were added to the cells. After a 24-h culture period, the medium was changed and puromycin selection (12 µg/mL) of transfected cells was started.

4.4. siRNA Transfection

C6 cells were transfected with rat GLI-1 siRNA (Santa Cruz Biotechnology) using Lipofectamine RNAiMax (Invitrogen, Paris, France) according to the manufacturer's instructions. Four microliters of Lipofectamine RNAiMax and 25 nM of GLI-1 siRNA or a control siRNA FITC Conjugate-A (Santa Cruz Biotechnology, Clinisciences, Nanterre, France) were each mixed with 100 µL of Opti-MEM (Invitrogen). The RNAiMax/Opti-MEM mixture was incubated for 5 min at room temperature and next combined with the vector/Opti-MEM mixture. The combined mixtures were incubated for 20 min at room temperature and then added to the cells. After a 24-h culture period, the medium was changed.

4.5. Ex Vivo Invasion Assay on Rat Brain Slices

The ex vivo invasion assay on rat brain slices was performed as described in Nakada and collaborators [67]. The experimental protocol was approved by the local Animal Care Committee (Comité d'éthique en expérimentation animale COMETHEA). Briefly, 400-µm-thick sections were prepared from newborn Wistar rat brain. Brain slices then were laid down on a Millicell-CM membrane insert (Millipore, Molsheim, France) and the insert was placed in individual wells of six-well plates. Medium (1 mL) was added to the bottom of culture plates. Slices were cultured in the same culture medium as C6 cells at 37 °C in a 5% CO_2 incubator. After 24 h, a volume of 0.1 µL containing 5×10^3 C6 cells stably expressing GFP protein was injected with a Hamilton syringe in the putamen of the brain slices. Two hours after injection, the culture medium was changed, and the slices were treated with 10^{-6}M of VIP, PACAP-38, or $VIP_{10\text{-}28}$. After 48 h of treatment, the slices were incubated for 4 h with 4% of paraformaldehyde. Imaging of specimens was performed using a macro-fluorescent imaging system (Olympus, Rungis, France) equipped with a GFP barrier filter (DP50, Olympus). The area of GFP-stained cells in invasive population (RIM) in each section was measured and the depth of invasion at 48 h was determined using a confocal FV-1000 station installed on an inverted microscope IX-81 (Olympus, Rungis, France).

4.6. Wound Healing

Cells were cultured in 24-well dishes and a wound was made in the confluent monolayer by mechanical scraping using a p200 pipette tip. After a wash with serum-free medium, the cells were incubated in the absence or presence of increasing concentrations of VIP, PACAP-38, or $VIP_{10\text{-}28}$ in serum-free medium supplemented with 0.1% BSA. Cells underwent an incubation for 20 min with an inhibitor (1 µM of H89, IA, or cyclopamine) followed by a treatment with VIP, PACAP-38, or $VIP_{10\text{-}28}$. Phase-contrast micrographs were collected after 24 h and the width of the wound was measured. The data in percent represented the wound healing closure according to the calculation: ((Initial width − final width)/initial width) × 100.

4.7. Western Immunoblotting and Antibodies

C6 and U87 cells were cultured in 25-cm^2 flasks to about 75% confluence and then incubated for 0, 1/2, 1, 2, 3, 6, or 24 h, in the presence or absence of VIP or $VIP_{10\text{-}28}$ 10^{-7} M in serum-free medium supplemented with 0.1% BSA. After treatments, cells were harvested with 1× trypsin/EDTA, washed twice with cold Phosphate Buffered Saline (PBS) and resuspended in 10 µL per 10^6 cells of ice-cold lysis buffer (10 mM Tris, pH 7.5, 0.5 mM EDTA, pH 8.0, 0.5% CHAPS, 10% glycerol) supplemented with a cocktail of protease inhibitors (Calbiochem Merck, Guyancourt, France). After 30 min on ice, the samples were centrifuged at 4 °C for 20 min at 10,000× *g* and stored at −80 °C.

For extraction of nuclear and cytosolic proteins, cells were resuspended in 1 mL per 5×10^6 cells of lysis buffer (10 mM Tris, pH 7.4, 3 mM $MgCl_2$, 10 mM NaCl, 0.5% NP40). After 5 min on ice and centrifugation for 10 min at 2000× *g* at 4 °C, the supernatant containing the cytosolic fraction was collected. The pellet containing nuclei was resuspended in 30 µL per 5×10^6 cells of lysis buffer (20 mM HEPES, pH 7.9, 0.3 M NaCl, 1.5 mM $MgCl_2$, and the cocktail of protease inhibitors mentioned above) and incubated 30 min on ice under vigorous shaking. Samples were then centrifuged 20 min at 10,000× *g* at 4 °C. The supernatants were stored at −80 °C.

For all protein extracts, protein concentration was determined using the Bio-Rad DC Protein Assay (Bio-Rad, Marnes-La-Coquette, France). Total, cytosolic, and nuclear proteins were resolved in 8% SDS-PAGE and electroblotted for 1 h at 200 mA onto Immobilon-P membranes (Millipore). Membranes were blocked overnight at 4 °C using 5% nonfat milk in Tris-buffered saline (TBS) containing 0.1% Tween-20 (TBST) and incubated for 1 h at room temperature with a polyclonal rabbit anti-Akt antibody (1:1000; Cell Signaling Ozyme, San Quentin Yvelines, France), a polyclonal rabbit anti-phospho-Akt antibody (1:1000; Cell Signaling Ozyme, San Quentin Yvelines, France), a polyclonal rabbit anti-phospho-PTEN antibody (1:1000; Cell Signaling, Saint Quentin Yvelines, France), a polyclonal rabbit anti-Gli1 antibody (1:200; Santa-Cruz Clinisciences, Nanterre, France) or 1 h at room temperature with a monoclonal mouse anti-GAPDH antibody (1:80,000; Abcam, Paris, France) or monoclonal mouse anti-Lamin-C (1:1000; Abcam, Paris, France) used as a loading control, in blocking solution. After three washes, the membrane was next incubated 1 h at room temperature with goat anti-rabbit or goat anti-mouse secondary antibodies (1:20,000) (Calbiochem, Beeston Nottingham, UK) conjugated to horseradish peroxydase. The enhanced chemiluminescence (ECL) stainings obtained using ECL prime western blotting detection reagent (GE Healthcare, Buc, France) were quantified by densitometry with Image J software (National Institutes of Health, USA).

4.8. cDNA Synthesis

Total RNA was isolated using the GenElute$_{TM}$ Mammalian Total RNA kit (Sigma-Aldrich, Saint-Quentin Fallavier, France) following the manufacturer's instructions. Total RNA (1 µg) was treated with 1 U/µg RNA of DNase I Amplification Grade (Invitrogen) according to the manufacturer's instructions, and in the presence of 10 U/µg RNA of RNaseOUT (Invitrogen). After DNase inactivation, RNA was reverse transcribed using random hexamers (Promega) and M-MLV Reverse Transcriptase H Minus (Promega) according to the manufacturer's instructions.

4.9. Quantitative Polymerase Chain Reaction of Reverse Transcribed mRNA

Real-time polymerase chain reaction quantification was carried out with the LightCycler System (Roche Diagnostics, Basel, Switzerland) using the SYBR qPCR Premix Ex Taq (Takara, Saint-Germain-en-Laye, France). Sequences of primers used are listed in Table S2 (Supplementary data). The cDNA amplification program was 10 s at 95 °C to activate ExTaq polymerase followed by at least 45 cycles of 5 s at 95 °C, 5 s at 60 °C, and 10 s at 72 °C.

5. Conclusions

Triggering the VIP-receptor system reduces migration and invasion in GBM cells through a PKA-dependent blockade of the PI3K/Akt and of the SHH/GLI1 pathways. Therefore, the VIP-receptor system displays anti-oncogenic properties in GBM cells and PKA appears like a central core in this process. As a powerful activator of the cAMP/PKA pathway, targeting the VIP-receptor system should be considered as an alternative strategy to inhibit the interrelated signaling pathways involved in GBM and possibly other cancers progression.

Supplementary Materials: The following are available online at http://www.mdpi.com/2072-6694/11/1/123/s1, Table S1: Expression of mRNAs encoding the components of the VIP-receptor system in C6 and U87 glioblastoma cell lines. Table S2: Primer sequences used for RT-qPCR analysis of expression of mRNAs encoding the components of the VIP-receptor system in C6 and U87 glioblastoma cell lines.

Author Contributions: Investigation, S.B., S.T., A.-C.B. and M.D.B.; Writing–original draft, S.B.; Conceptualization of animal experimentation, A.G.; Conceptualization and supervision, C.C. and J.-M.M.

Funding: S.B. was the recipient of a Ph.D. fellowship from the "Region Poitou–Charentes," France. This work was supported by grants from the "Ligue contre le Cancer du Grand-Ouest, comité de la Vienne et comité des deux-Sèvres" and from the "Lions Club de Melle".

Acknowledgments: The authors would like to thank Anne Cantereau, and Emile Béré for their technical support and Marianne Benoit-Marand for her skillful assistance to establish the organotypic brain slice cultures.

Conflicts of Interest: The authors declare no conflicts of interest.

References

1. Zong, H.; Verhaak, R.G.W.; Canoll, P. The cellular origin for malignant glioma and prospects for clinical advancements. *Expert Rev. Mol. Diagn.* **2012**, *12*, 383–394. [CrossRef] [PubMed]
2. Muñoz, D.M.; Tung, T.; Agnihotri, S.; Singh, S.; Guha, A.; Zadeh, G.; Hawkins, C. Loss of p53 cooperates with K-ras activation to induce glioma formation in a region-independent manner. *Glia* **2013**, *61*, 1862–1872. [CrossRef]
3. Li, Q.-J.; Cai, J.-Q.; Liu, C.-Y. Evolving Molecular Genetics of Glioblastoma. *Chin. Med. J.* **2016**, *129*, 464–471. [CrossRef] [PubMed]
4. Xu, H.-S.; Qin, X.-L.; Zong, H.-L.; He, X.-G.; Cao, L. Cancer stem cell markers in glioblastoma—An update. *Eur. Rev. Med. Pharmacol. Sci.* **2017**, *21*, 3207–3211. [PubMed]
5. Stupp, R.; Mason, W.P.; van den Bent, M.J.; Weller, M.; Fisher, B.; Taphoorn, M.J.B.; Belanger, K.; Brandes, A.A.; Marosi, C.; Bogdahn, U.; et al. Radiotherapy plus Concomitant and Adjuvant Temozolomide for Glioblastoma. *N. Engl. J. Med.* **2005**, *352*, 987–996. [CrossRef] [PubMed]
6. Kwiatkowska, A.; Symons, M. Signaling determinants of glioma cell invasion. *Adv. Exp. Med. Biol.* **2013**, *986*, 121–141. [PubMed]
7. Vehlow, A.; Cordes, N. Invasion as target for therapy of glioblastoma multiforme. *Biochim. Biophys. Acta* **2013**, *1836*, 236–244. [CrossRef] [PubMed]
8. De Gooijer, M.C.; Guillén Navarro, M.; Bernards, R.; Wurdinger, T.; van Tellingen, O. An Experimenter's Guide to Glioblastoma Invasion Pathways. *Trends Mol. Med.* **2018**, *24*, 763–780. [CrossRef]
9. Swartz, A.M.; Shen, S.H.; Salgado, M.A.; Congdon, K.L.; Sanchez-Perez, L. Promising vaccines for treating glioblastoma. *Expert Opin. Biol. Ther.* **2018**, *18*, 1159–1170. [CrossRef]
10. Paw, I.; Carpenter, R.C.; Watabe, K.; Debinski, W.; Lo, H.-W. Mechanisms regulating glioma invasion. *Cancer Lett.* **2015**, *362*, 1–7. [CrossRef]
11. Höland, K.; Salm, F.; Arcaro, A. The phosphoinositide 3-kinase signaling pathway as a therapeutic target in grade IV brain tumors. *Curr. Cancer Drug Targets* **2011**, *11*, 894–918. [CrossRef] [PubMed]
12. Wen, P.Y.; Lee, E.Q.; Reardon, D.A.; Ligon, K.L.; Alfred Yung, W.K. Current clinical development of PI3K pathway inhibitors in glioblastoma. *Neuro-Oncology* **2012**, *14*, 819–829. [CrossRef] [PubMed]
13. Fan, Q.-W.; Weiss, W.A. Targeting the RTK-PI3K-mTOR axis in malignant glioma: Overcoming resistance. *Curr. Top. Microbiol. Immunol.* **2010**, *347*, 279–296. [PubMed]
14. McDowell, K.A.; Riggins, G.J.; Gallia, G.L. Targeting the AKT pathway in glioblastoma. *Curr. Pharm. Des.* **2011**, *17*, 2411–2420. [CrossRef] [PubMed]
15. Narayan, R.S.; Fedrigo, C.A.; Stalpers, L.J.A.; Baumert, B.G.; Sminia, P. Targeting the Akt-pathway to improve radiosensitivity in glioblastoma. *Curr. Pharm. Des.* **2013**, *19*, 951–957. [CrossRef] [PubMed]
16. Sami, A.; Karsy, M. Targeting the PI3K/AKT/mTOR signaling pathway in glioblastoma: Novel therapeutic agents and advances in understanding. *Tumour Biol. J. Int. Soc. Oncodev. Biol. Med.* **2013**, *34*, 1991–2002. [CrossRef]
17. Majewska, E.; Szeliga, M. AKT/GSK3β Signaling in Glioblastoma. *Neurochem. Res.* **2017**, *42*, 918–924. [CrossRef]
18. Katoh, Y.; Katoh, M. Hedgehog target genes: Mechanisms of carcinogenesis induced by aberrant hedgehog signaling activation. *Curr. Mol. Med.* **2009**, *9*, 873–886. [CrossRef]
19. Santoni, M.; Burattini, L.; Nabissi, M.; Morelli, M.B.; Berardi, R.; Santoni, G.; Cascinu, S. Essential role of Gli proteins in glioblastoma multiforme. *Curr. Protein Pept. Sci.* **2013**, *14*, 133–140. [CrossRef]
20. Ng, J.M.Y.; Curran, T. The Hedgehog's tale: Developing strategies for targeting cancer. *Nat. Rev. Cancer* **2011**, *11*, 493–501. [CrossRef]
21. Sahebjam, S.; Siu, L.L.; Razak, A.A. The utility of hedgehog signaling pathway inhibition for cancer. *Oncologist* **2012**, *17*, 1090–1099. [CrossRef] [PubMed]
22. Sandhiya, S.; Melvin, G.; Kumar, S.S.; Dkhar, S.A. The dawn of hedgehog inhibitors: Vismodegib. *J. Pharmacol. Pharmacother.* **2013**, *4*, 4–7. [CrossRef] [PubMed]
23. Tanaka, S.; Louis, D.N.; Curry, W.T.; Batchelor, T.T.; Dietrich, J. Diagnostic and therapeutic avenues for glioblastoma: No longer a dead end? *Nat. Rev. Clin. Oncol.* **2013**, *10*, 14–26. [CrossRef] [PubMed]
24. Munoz, J.L.; Rodriguez-Cruz, V.; Walker, N.D.; Greco, S.J.; Rameshwar, P. Temozolomide resistance and tumor recurrence: Halting the Hedgehog. *Cancer Cell Microenviron.* **2015**, *2*, e747. [PubMed]

25. Skoda, A.M.; Simovic, D.; Karin, V.; Kardum, V.; Vranic, S.; Serman, L. The role of the Hedgehog signaling pathway in cancer: A comprehensive review. *Bosn. J. Basic Med. Sci.* **2018**, *18*, 8–20. [CrossRef] [PubMed]
26. Muller, J.M.; Lelievre, V.; Becq-Giraudon, L.; Meunier, A.C. VIP as a cell-growth and differentiation neuromodulator role in neurodevelopment. *Mol. Neurobiol.* **1995**, *10*, 115–134. [CrossRef] [PubMed]
27. Muller, J.-M.; Debaigt, C.; Goursaud, S.; Montoni, A.; Pineau, N.; Meunier, A.-C.; Janet, T. Unconventional binding sites and receptors for VIP and related peptides PACAP and PHI/PHM: An update. *Peptides* **2007**, *28*, 1655–1666. [CrossRef]
28. Gozes, I. VIP, from gene to behavior and back: Summarizing my 25 years of research. *J. Mol. Neurosci.* **2008**, *36*, 115–124. [CrossRef]
29. Moody, T.W.; Chan, D.; Fahrenkrug, J.; Jensen, R.T. Neuropeptides as autocrine growth factors in cancer cells. *Curr. Pharm. Des.* **2003**, *9*, 495–509. [CrossRef]
30. Vaudry, D.; Falluel-Morel, A.; Bourgault, S.; Basille, M.; Burel, D.; Wurtz, O.; Fournier, A.; Chow, B.K.C.; Hashimoto, H.; Galas, L.; et al. Pituitary adenylate cyclase-activating polypeptide and its receptors: 20 years after the discovery. *Pharmacol. Rev.* **2009**, *61*, 283–357. [CrossRef]
31. Sharma, A.; Walters, J.; Gozes, Y.; Fridkin, M.; Brenneman, D.; Gozes, I.; Moody, T.W. A vasoactive intestinal peptide antagonist inhibits the growth of glioblastoma cells. *J. Mol. Neurosci.* **2001**, *17*, 331–339. [CrossRef]
32. D'Amico, A.G.; Scuderi, S.; Saccone, S.; Castorina, A.; Drago, F.; D'Agata, V. Antiproliferative Effects of PACAP and VIP in Serum-Starved Glioma Cells. *J. Mol. Neurosci.* **2013**, *51*, 503–513. [CrossRef] [PubMed]
33. Sokolowska, P.; Nowak, J.Z. Cyclic AMP formation in C6 glioma cells: Effect of PACAP and VIP in early and late passages. *Ann. N. Y. Acad. Sci.* **2006**, *1070*, 566–569. [CrossRef] [PubMed]
34. Sokolowska, P.; Nowak, J.Z. Effects of PACAP and VIP on cAMP-generating system and proliferation of C6 glioma cells. *J. Mol. Neurosci.* **2008**, *36*, 286–291. [CrossRef] [PubMed]
35. Robberecht, P.; Woussen-Colle, M.C.; Vertongen, P.; De Neef, P.; Hou, X.; Salmon, I.; Brotchi, J. Expression of pituitary adenylate cyclase activating polypeptide (PACAP) receptors in human glial cell tumors. *Peptides* **1994**, *15*, 661–665. [CrossRef]
36. Cochaud, S.; Chevrier, L.; Meunier, A.-C.; Brillet, T.; Chadéneau, C.; Muller, J.-M. The vasoactive intestinal peptide-receptor system is involved in human glioblastoma cell migration. *Neuropeptides* **2010**, *44*, 373–383. [CrossRef] [PubMed]
37. Barbarin, A.; Séité, P.; Godet, J.; Bensalma, S.; Muller, J.-M.; Chadéneau, C. Atypical nuclear localization of VIP receptors in glioma cell lines and patients. *Biochem. Biophys. Res. Commun.* **2014**, *454*, 524–530. [CrossRef]
38. Cochaud, S.; Meunier, A.-C.; Monvoisin, A.; Bensalma, S.; Muller, J.-M.; Chadéneau, C. Neuropeptides of the VIP family inhibit glioblastoma cell invasion. *J. Neurooncol.* **2015**, *122*, 63–73. [CrossRef]
39. Lelievre, V.; Seksenyan, A.; Nobuta, H.; Yong, W.H.; Chhith, S.; Niewiadomski, P.; Cohen, J.R.; Dong, H.; Flores, A.; Liau, L.M.; et al. Disruption of the PACAP gene promotes medulloblastoma in ptc1 mutant mice. *Dev. Biol.* **2008**, *313*, 359–370. [CrossRef]
40. Cohen, J.R.; Resnick, D.Z.; Niewiadomski, P.; Dong, H.; Liau, L.M.; Waschek, J.A. Pituitary adenylyl cyclase activating polypeptide inhibits gli1 gene expression and proliferation in primary medulloblastoma derived tumorsphere cultures. *BMC Cancer* **2010**, *10*, 676. [CrossRef]
41. Niewiadomski, P.; Zhujiang, A.; Youssef, M.; Waschek, J.A. Interaction of PACAP with Sonic hedgehog reveals complex regulation of the hedgehog pathway by PKA. *Cell. Signal.* **2013**, *25*, 2222–2230. [CrossRef] [PubMed]
42. Sheng, T.; Chi, S.; Zhang, X.; Xie, J. Regulation of Gli1 localization by the cAMP/protein kinase A signaling axis through a site near the nuclear localization signal. *J. Biol. Chem.* **2006**, *281*, 9–12. [CrossRef] [PubMed]
43. He, X.; Lu, Q.R. G-Protein Gαs controls medulloblastoma initiation by suppressing sonic hedgehog signaling. *Mol. Cell. Oncol.* **2015**, *2*, e975070. [CrossRef] [PubMed]
44. Warrington, N.M.; Sun, T.; Rubin, J.B. Targeting brain tumor cAMP: The case for sex-specific therapeutics. *Front. Pharmacol.* **2015**, *6*, 153. [CrossRef] [PubMed]
45. Rao, R.; Salloum, R.; Xin, M.; Lu, Q.R. The G protein Gαs acts as a tumor suppressor in sonic hedgehog signaling-driven tumorigenesis. *Cell Cycle* **2016**, *15*, 1325–1330. [CrossRef] [PubMed]

46. Mucignat-Caretta, C.; Denaro, L.; D'Avella, D.; Caretta, A. Protein Kinase A Distribution Differentiates Human Glioblastoma from Brain Tissue. *Cancers* **2017**, *10*, 2. [CrossRef] [PubMed]
47. Sapio, L.; Di Maiolo, F.; Illiano, M.; Esposito, A.; Chiosi, E.; Spina, A.; Naviglio, S. Targeting protein kinase A in cancer therapy: An update. *EXCLI J.* **2014**, *13*, 843–855.
48. Sapio, L.; Gallo, M.; Illiano, M.; Chiosi, E.; Naviglio, D.; Spina, A.; Naviglio, S. The Natural cAMP Elevating Compound Forskolin in Cancer Therapy: Is It Time? *J. Cell. Physiol.* **2017**, *232*, 922–927. [CrossRef]
49. Grobben, B.; De Deyn, P.P.; Slegers, H. Rat C6 glioma as experimental model system for the study of glioblastoma growth and invasion. *Cell Tissue Res.* **2002**, *310*, 257–270. [CrossRef]
50. Bensalma, S.; Chadeneau, C.; Legigan, T.; Renoux, B.; Gaillard, A.; de Boisvilliers, M.; Pinet-Charvet, C.; Papot, S.; Muller, J.M. Evaluation of cytotoxic properties of a cyclopamine glucuronide prodrug in rat glioblastoma cells and tumors. *J. Mol. Neurosci.* **2015**, *55*, 51–61. [CrossRef]
51. Lamouche, S.; Yamaguchi, N. Role of PAC(1) receptor in adrenal catecholamine secretion induced by PACAP and VIP in vivo. *Am. J. Physiol. Regul. Integr. Comp. Physiol.* **2001**, *280*, R510–R518. [CrossRef]
52. Summers, M.A.; O'Dorisio, M.S.; Cox, M.O.; Lara-Marquez, M.; Goetzl, E.J. A lymphocyte-generated fragment of vasoactive intestinal peptide with VPAC1 agonist activity and VPAC2 antagonist effects. *J. Pharmacol. Exp. Ther.* **2003**, *306*, 638–645. [CrossRef] [PubMed]
53. Lutz, E.M.; Ronaldson, E.; Shaw, P.; Johnson, M.S.; Holland, P.J.; Mitchell, R. Characterization of novel splice variants of the PAC1 receptor in human neuroblastoma cells: Consequences for signaling by VIP and PACAP. *Mol. Cell. Neurosci.* **2006**, *31*, 193–209. [CrossRef] [PubMed]
54. Wang, Y.; Ding, Q.; Yen, C.-J.; Xia, W.; Izzo, J.G.; Lang, J.-Y.; Li, C.-W.; Hsu, J.L.; Miller, S.A.; Wang, X.; et al. The crosstalk of mTOR/S6K1 and Hedgehog pathways. *Cancer Cell* **2012**, *21*, 374–387. [CrossRef] [PubMed]
55. Ruiz i Altaba, A. Hedgehog signaling and the Gli code in stem cells, cancer, and metastases. *Sci. Signal.* **2011**, *4*, pt9. [CrossRef] [PubMed]
56. Wei, L.; Xu, Z. Cross-signaling among phosphinositide-3 kinase, mitogen-activated protein kinase and sonic hedgehog pathways exists in esophageal cancer. *Int. J. Cancer* **2011**, *129*, 275–284. [CrossRef] [PubMed]
57. Stecca, B.; Mas, C.; Clement, V.; Zbinden, M.; Correa, R.; Piguet, V.; Beermann, F.; Ruiz i Altaba, A. Melanomas require HEDGEHOG-GLI signaling regulated by interactions between GLI1 and the RAS-MEK/AKT pathways. *Proc. Natl. Acad. Sci. USA* **2007**, *104*, 5895–5900. [CrossRef] [PubMed]
58. Furnari, F.B.; Lin, H.; Huang, H.-J.S.; Cavenee, W.K. Growth suppression of glioma cells by PTEN requires a functional phosphatase catalytic domain. *Proc. Natl. Acad. Sci. USA* **1997**, *94*, 12479–12484. [CrossRef]
59. Cosentino, C.; Di Domenico, M.; Porcellini, A.; Cuozzo, C.; De Gregorio, G.; Santillo, M.R.; Agnese, S.; Di Stasio, R.; Feliciello, A.; Migliaccio, A.; et al. p85 regulatory subunit of PI3K mediates cAMP-PKA and estrogens biological effects on growth and survival. *Oncogene* **2007**, *26*, 2095–2103. [CrossRef]
60. Lochner, A.; Moolman, J.A. The many faces of H89: A review. *Cardiovasc. Drug Rev.* **2006**, *24*, 261–274. [CrossRef]
61. Murray, A.J. Pharmacological PKA Inhibition: All May Not Be What It Seems. *Sci. Signal.* **2008**, *1*, re4. [CrossRef] [PubMed]
62. Bourgault, S.; Vaudry, D.; Botia, B.; Couvineau, A.; Laburthe, M.; Vaudry, H.; Fournier, A. Novel stable PACAP analogs with potent activity towards the PAC1 receptor. *Peptides* **2008**, *29*, 919–932. [CrossRef] [PubMed]
63. Bourgault, S.; Vaudry, D.; Dejda, A.; Doan, N.D.; Vaudry, H.; Fournier, A. Pituitary adenylate cyclase-activating polypeptide: Focus on structure-activity relationships of a neuroprotective Peptide. *Curr. Med. Chem.* **2009**, *16*, 4462–4480. [CrossRef] [PubMed]
64. Doan, N.-D.; Bourgault, S.; Dejda, A.; Létourneau, M.; Detheux, M.; Vaudry, D.; Vaudry, H.; Chatenet, D.; Fournier, A. Design and in vitro characterization of PAC1/VPAC1-selective agonists with potent neuroprotective effects. *Biochem. Pharmacol.* **2011**, *81*, 552–561. [CrossRef] [PubMed]
65. De Boisvilliers, M.; Perrin, F.; Hebache, S.; Balandre, A.-C.; Bensalma, S.; Garnier, A.; Vaudry, D.; Fournier, A.; Festy, F.; Muller, J.-M.; et al. VIP and PACAP analogs regulate therapeutic targets in high-risk neuroblastoma cells. *Peptides* **2016**, *78*, 30–41. [CrossRef] [PubMed]

66. Benda, P.; Lightbody, J.; Sato, G.; Levine, L.; Sweet, W. Differentiated rat glial cell strain in tissue culture. *Science* **1968**, *161*, 370–371. [CrossRef] [PubMed]
67. Nakada, M.; Anderson, E.M.; Demuth, T.; Nakada, S.; Reavie, L.B.; Drake, K.L.; Hoelzinger, D.B.; Berens, M.E. The phosphorylation of ephrin-B2 ligand promotes glioma cell migration and invasion. *Int. J. Cancer* **2010**, *126*, 1155–1165. [CrossRef]

cancers

MDPI

Article

Proteomic Analysis Implicates Vimentin in Glioblastoma Cell Migration

Michal O. Nowicki, Josie L. Hayes, E. Antonio Chiocca and Sean E. Lawler *

Harvey W. Cushing Neurooncology Laboratories, Department of Neurosurgery, Brigham and Women's Hospital, Harvard Medical School, Boston, MA 02115, USA; mnowicki@bwh.harvard.edu (M.O.N.); jlhayes1982@gmail.com (J.L.H.); eachiocca@bwh.harvard.edu (E.A.C.)

* Correspondence: slawler@bwh.harvard.edu; Tel.: +1-617-525-5650

Received: 25 February 2019; Accepted: 29 March 2019; Published: 3 April 2019

Abstract: We previously showed lithium chloride (LiCl) and other inhibitors of glycogen synthase kinase-3 (GSK-3) including 6-bromo-indirubin-3-oxime (BIO), can block glioblastoma (GBM) cell migration. To investigate the mechanisms involved we used two-dimensional difference in-gel electrophoresis (2D-DIGE) and mass spectrometry to identify proteins altered after treatment of U251 GBM cells with 20 mM LiCl. Downregulation of the intermediate filament protein vimentin was the most significant change identified. Analysis of patient tumor samples revealed that vimentin is expressed abundantly in GBM, and is prognostic especially in lower grade tumors. Additionally, siRNA-mediated vimentin knockdown impaired GBM migration. Western blotting showed that treatment with LiCl or small molecule GSK-3 inhibitors led to the rapid downregulation of detergent soluble vimentin levels across a panel of GBM-derived cells. Fluorescence reactivation after photobleaching (FRAP) microscopy studies showed a significant reduction in the ability of the vimentin cytoskeleton to recover from photo-bleaching in the presence of LiCl or BIO. Biochemical studies revealed that GSK-3 and vimentin directly interact, and analysis of vimentin revealed a GSK-3 consensus phosphorylation site. We conclude that anti-migratory compounds with the ability to inhibit GSK-3 have effects on vimentin cytoskeletal dynamics, which may play a role in their anti-invasive activity.

Keywords: vimentin; GSK-3; glioblastoma; motility; cytoskeleton

1. Introduction

Invasiveness is one of the main hallmarks of glioblastoma (GBM) and other glial tumors, in which malignant cells diffusely infiltrate normal brain tissue and migrate along blood vessels and defined structures of the brain [1,2]. This has important clinical consequences as it prevents complete surgical tumor resection, contributes to therapeutic resistance and helps to drive rapid tumor growth. Invasion is facilitated by multiple cellular processes including extracellular matrix remodeling [3], interactions of malignant cells with the normal cells of the brain and the host immune system [4], changes in cell adhesion properties [5] and cytoskeletal dynamics [6,7].

GBM is the most common malignant brain tumor and has a dismal prognosis [8,9]. At present there are no clinical approaches to block GBM invasion. We previously showed that lithium chloride (LiCl) potently and specifically blocks GBM cell migration in vitro [8]. LiCl has multiple targets within the cell [9–11], including the protein kinase glycogen synthase kinase-3 (GSK-3). GSK-3 is ubiquitously expressed in mammalian cells, with two closely related isoforms, GSK-3α and GSK-3β which have distinct and overlapping functions [12–14]. GSK-3 is a multi-tasking enzyme, known for its role in the regulation of glycogen synthesis, via inactivation by Akt, and also for its role as a critical mediator of Wnt signaling [15]. GSK-3 has been associated with a number of processes involved in cell motility via interactions with cytoskeletal components [11,16].

In pre-clinical murine GBM xenograft models we showed that protein kinase inhibitors of the indirubin family, which are known to potently inhibit GSK-3, block invasion and angiogenesis improving survival [17,18]. However, we have not yet defined the mechanisms underlying our observations in GBM cells. Here, we performed a proteomic study, which revealed consistent and highly significant downregulation of the intermediate protein vimentin after treatment of GBM cells with LiCl, and the indirubin derivative 6-bromo-indirubin-3-oxime (BIO).

Vimentin is a major component of the intermediate filament cytoskeleton and is best known in cancer as a marker of cellular epithelial to mesenchymal transition (EMT) [9], a phenomenon associated with cancer cell invasion and metastatic tumor spread. Vimentin is known to play an important role in maintenance of cellular integrity and resistance to stress, and vimentin knockout mice show impaired wound healing ability [19]. Vimentin has been shown to regulate cell migration via interaction with paxillin, integrin α6β4 and myosin II [20]. The role of vimentin in GBM has not been explored in detail regardless of the fact that it has been used as biomarker [21].

Here we show that LiCl and BIO, which block GBM cell migration, both affect the dynamic regulation of vimentin in GBM cells, and that vimentin knockdown partially blocks GBM migration. We also show that vimentin and GSK-3 physically associate. Thus, vimentin may affect GBM invasion downstream of anti-invasive compounds LiCl and indirubin derivatives.

2. Results

2.1. Two-Dimensional Fluorescence DIGE Proteomics Analysis Identifies Vimentin Downregulation by LiCl in GBM Cells

Our previous studies showed that LiCl specifically and reversibly blocks GBM cell migration with minimal cytotoxicity in a dose-dependent manner with maximal effects at a concentration of 20 mM [8]. To better understand these observations, proteomic analysis was performed on U251 GBM cells treated with 20 mM LiCl. Image analysis after 2-D fluorescence difference gel electrophoresis (DIGE) revealed 22 significantly altered spots on the gel comparing LiCl treated U251 cells with untreated controls ($\geq$1.5 fold average increase/decrease and $p \leq 0.05$, Students t-test) (Figure 1a). Individual spots were cut from the gels, and mass spectrometry was used to determine the identity of the most significant hits identified by 2D-DIGE (Figure 1b). Interestingly, the six top primary targets identified (LMNA, ARF1, AIF1L, VIM, TUBB and CRMP4) are all components of the cytoskeleton, with the intermediate filament VIM (vimentin) being the top scoring hit. However, the role of intermediate filaments in GBM is poorly characterized. In this study, we therefore further explored the potential link between LiCl and vimentin in GBM as suggested by the proteomic data.

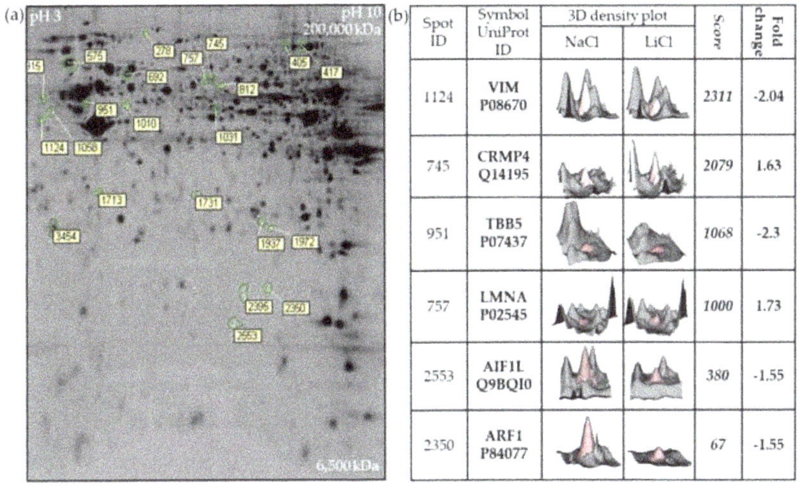

Figure 1. Two-dimensional fluorescence difference gel electrophoresis (DIGE) proteomics analysis identified 22 protein spots significantly altered by 20 mM lithium chloride (LiCl) treatment of glioblastoma (GBM) cells for 24 h. The experiment was performed in triplicate. (**a**) Representative Cy3/Cy5 stained (presented in gray) gel showing differentially regulated spots after 24 h treatment of U251 cells with 20 mM LiCl. (**b**) The top six targets were excised from the gel based on significance after analysis of alterations in fluorescence intensity post-LiCl treatment and identified by mass spectrometry peptide analysis.

2.2. Vimentin Is Highly Expressed in GBM Patient Tumor Specimens and Is Associated with Cell Migration

Initially, to confirm the relevance of vimentin in GBM, we examined its expression by Western blotting of patient tumor specimens. This revealed high levels of vimentin in GBM, anaplastic astrocytoma (AA) and oligodendroglioma (OD) specimens compared with normal brain controls (Figure 2a). This was confirmed by examination of gene expression data which showed significantly elevated median vimentin expression in GBM ($p = 3.5 \times 10^{-23}$) compared with normal brain (Figure 2b) and also that vimentin expression is a prognostic factor in glioma analyzed in the Rembrandt glioma database (Figure 2c). Analysis of vimentin mRNA transcript levels in the TCGA database shows that it is present in all types of malignancies, where most invasive cancer types like melanoma and glioma show the highest levels (Figure S1a). The TCGA database allows for survival analysis of combined GBM and Low-Grade Glioma (LGG) datasets, as shown in Figure S1b; this data also suggest that vimentin is a prognostic factor in both GBM and LGG.

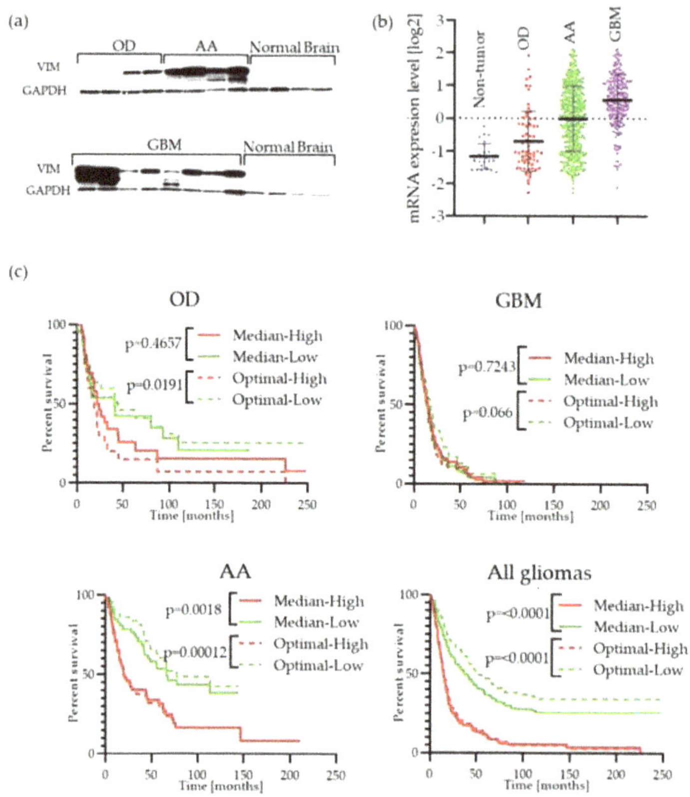

Figure 2. Expression of vimentin in malignant gliomas: (**a**) Western blot analysis of vimentin expression in human patient malignant glioma surgical specimens: OD—oligodendroglioma, AA—anaplastic astrocytoma, GBM—glioblastoma multiforme; compared with normal brain; (**b**) the mRNA transcript levels of vimentin according REMBRANDT database (http://www.betastasis.com/glioma/rembrandt), sample size: non-tumor n = 29, OD n = 67, AA n = 460, GBM n = 220; (**c**) survival curves based on high versus low vimentin mRNA expression obtained from the REMBRANDT database. The solid lines present data for samples with levels higher (red) or lower (green) than median of dataset; the dashed lines presents data for samples divided into two groups (higher—red or lower—green) based on the "optimal cut-off" algorithm [22].

2.3. Vimentin Immunofluorescence Staining in GBM cells and in Patient Specimens

To further understand vimentin in GBM we performed immunofluorescence staining in U251 GBM cells grown as monolayers. The vimentin cytoskeleton could clearly be seen ramifying throughout the cytosol, with a concentrated perinuclear area associated with the aggresome as previously observed [23] (Figure 3a). In patient specimens variable staining was observed, but in general a very strong signal could be seen throughout the tumor (Figure 3b). The small subset of glioma tissues in the Human Protein Atlas (www.proteinatlas.org) [24] are shown in Figure S2.

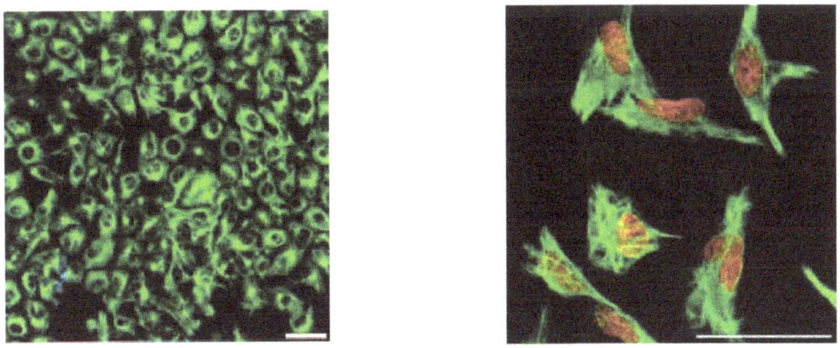

(a)

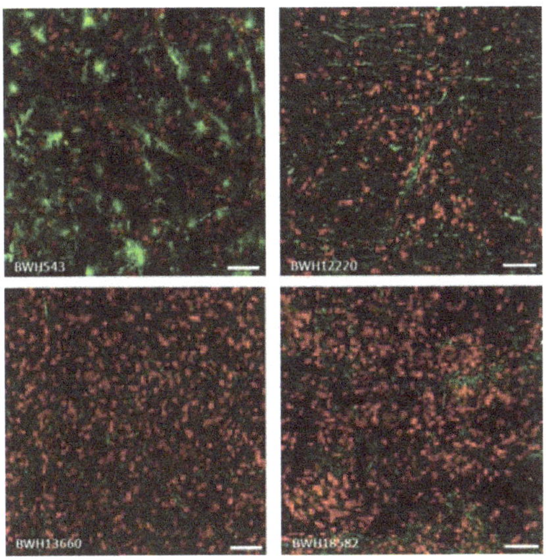

(b)

Figure 3. Vimentin staining in human GBM: (**a**) U251 cells grown as monolayers (left panel, field of view 400 by 400 μm, scale 50 μm) and high magnification (right, scale bar = 50 μm) of single cells, vimentin staining is presented here as a high dynamic range (HDR) image showing the perinuclear vimentin-rich region and fine network of filaments. (**b**) Surgically obtained human tumor specimens were stained for vimentin, field of view 400 by 400 μm, scale 50 μm.

2.4. Vimentin Knockdown Reduces GBM Cell Migration

To establish further the relevance of vimentin in the process of GBM migration we performed siRNA knockdown. Even though we were not able to fully knockdown endogenous vimentin, similar to other reports [18,20], a significant reduction in transwell migration was observed in three distinct GBM cell lines (Figure 4).

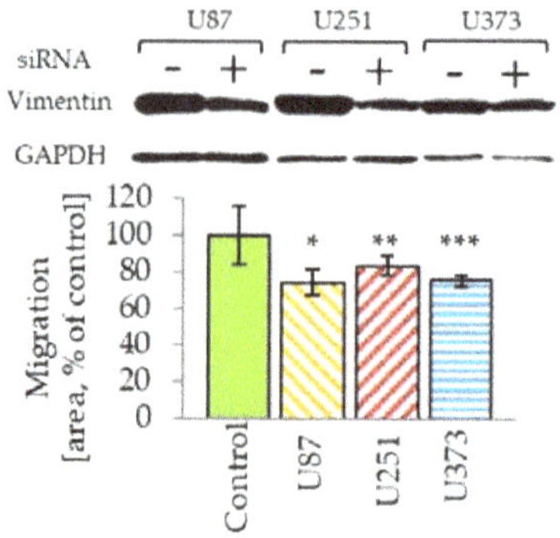

Figure 4. siRNA vimentin downregulation reduces the ability of cells to migrate across a transwell membrane in the modified Boyden chamber assay at 24 h post-transfection. The number of cells measured on the bottom side of membrane was normalized to the performance of the respective scrambled siRNA control (=100%) for each cell line; * $p = 0.010$, ** $p = 0.015$, *** $p = 0.037$. Western blot (upper) shows the actual degree of vimentin downregulation obtained for each cell line compared with scrambled siRNA control.

2.5. Downregulation of the Triton-x-100 Soluble Vimentin Fraction in GBM Cell Lines by GSK-3 Inhibitors

Having established the potential role of vimentin in GBM cell migration, we further investigated the effects of LiCl on vimentin levels in a range of GBM cell lines. Initially we used Western blotting to validate alterations in vimentin levels observed in the proteomics study. This confirmed the effects seen in our proteomics screen, with downregulation of vimentin expression observed in both U251 and U373 GBM cell lines, and was detected very rapidly, minutes after LiCl administration (Figure 5a). The effects were much more pronounced at earlier time points, with approximately 3-fold downregulation typically observed three hours post LiCl treatment. This effect was also observed in a panel of patient-derived GBM stem-like cells (GSCs) grown as neurospheres and treated with 20 mM LiCl for three hours (Figure 5b). The reduction of vimentin levels in the detergent soluble fraction was also observed in correlation with concentration dependent inhibition of GSK-3 by the indirubin derivative, 6-Bromoindirubin-3′-oxime (BIO) and other GSK-3 inhibitors (Figure 5c). The effect of BIO on vimentin levels was dose dependent (Figure 5d). These data suggest that vimentin downregulation is a common feature of these anti-invasive drugs. However, in all of these experiments lysates created in parallel using 8 M urea to solubilize all intracellular proteins did not show a difference in vimentin levels (shown in Figure 5a). Thus, there was in fact not a change in absolute vimentin levels, rather a change in its physical state, likely polymerized versus depolymerized within the cell. These data therefore suggest a change in the distribution of vimentin between intracellular triton-x-100 soluble and insoluble pools after treatment with GSK-3 inhibitors including LiCl and BIO.

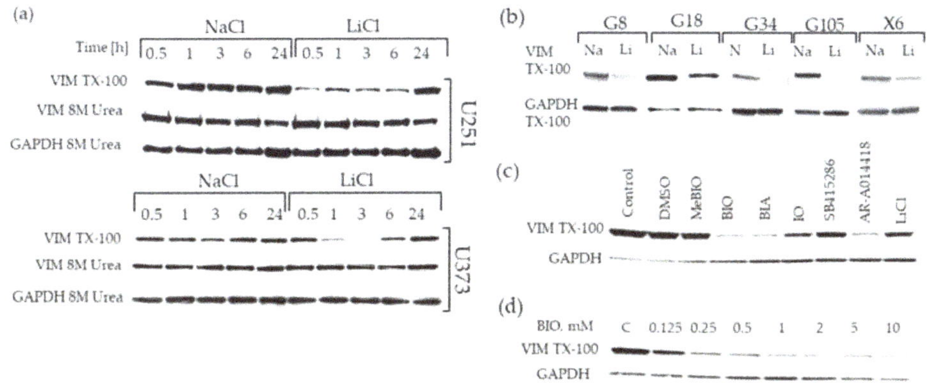

Figure 5. Downregulation of vimentin in detergent soluble extracts from GBM cells treated with LiCl, and a panel of small molecular glycogen synthase kinase-3 (GSK-3) inhibitors: (**a**) Validation in GBM cell lines. U251 and U373 glioma cells were treated with 20 mM LiCl or NaCl, for the time (hours) as indicated. Vimentin levels were determined in triton-x-100 extracts and in "total" extracts in 8 M urea. (**b**) Effects of 20 mM LiCl across a panel of GBM cell lines 3 h post-treatment. (**c**) Effects of a panel of GSK-3 inhibitors on vimentin levels in U251 cells 3 h post-treatment. (**d**) Dose-dependent effects of 6-bromo-indirubin-3-oxime (BIO) on triton-x-100 soluble vimentin levels 3 h post-treatment.

2.6. Changes in Vimentin Dynamics in LiCl and BIO Treated GBM cells

We used microscopic approaches to investigate these observations further using U251 cells stably expressing a green fluorescent protein (GFP)-vimentin fusion protein to visualize the effects of LiCl on the vimentin intermediate filament network. Exposure to 20 mM LiCl or 5 μM BIO for 3 h induced a reduction in the amount and GFP-intensity of small size vimentin fibers located at the cell periphery, and increased the GFP signal in the perinuclear area (Figure 6a). Time-lapse observations of GFP-vimentin fiber movement showed that in our cellular model, fibers are highly mobile laterally and possibly also have a dynamic rebuilding rate. We therefore utilized fluorescence recovery after photobleaching (FRAP) in which real-time fluorescence microscopy is used to determine the time taken for a cell to recover fluorescence levels after bleaching a small area in the presence and absence of drug treatment. A photobleached area was made in GFP-vimentin expressing U251 cells (Figure 6b), and recovery-time was measured. This showed a significant reduction in the time necessary for fibers to recover their fluorescent signal after LiCl or BIO treatment (Figure 6b, plot). We also utilized a time-lapse based approach to further assess effects of BIO treatment on fine vimentin network movement, which leads to up to 40% reduction in measured values compared with controls (Figure 6c). We also verified our previously reported observations [8,17] that LiCl and BIO treatments lead to overall reduction in cell motility in all tested cell lines and assays (Figure S3).

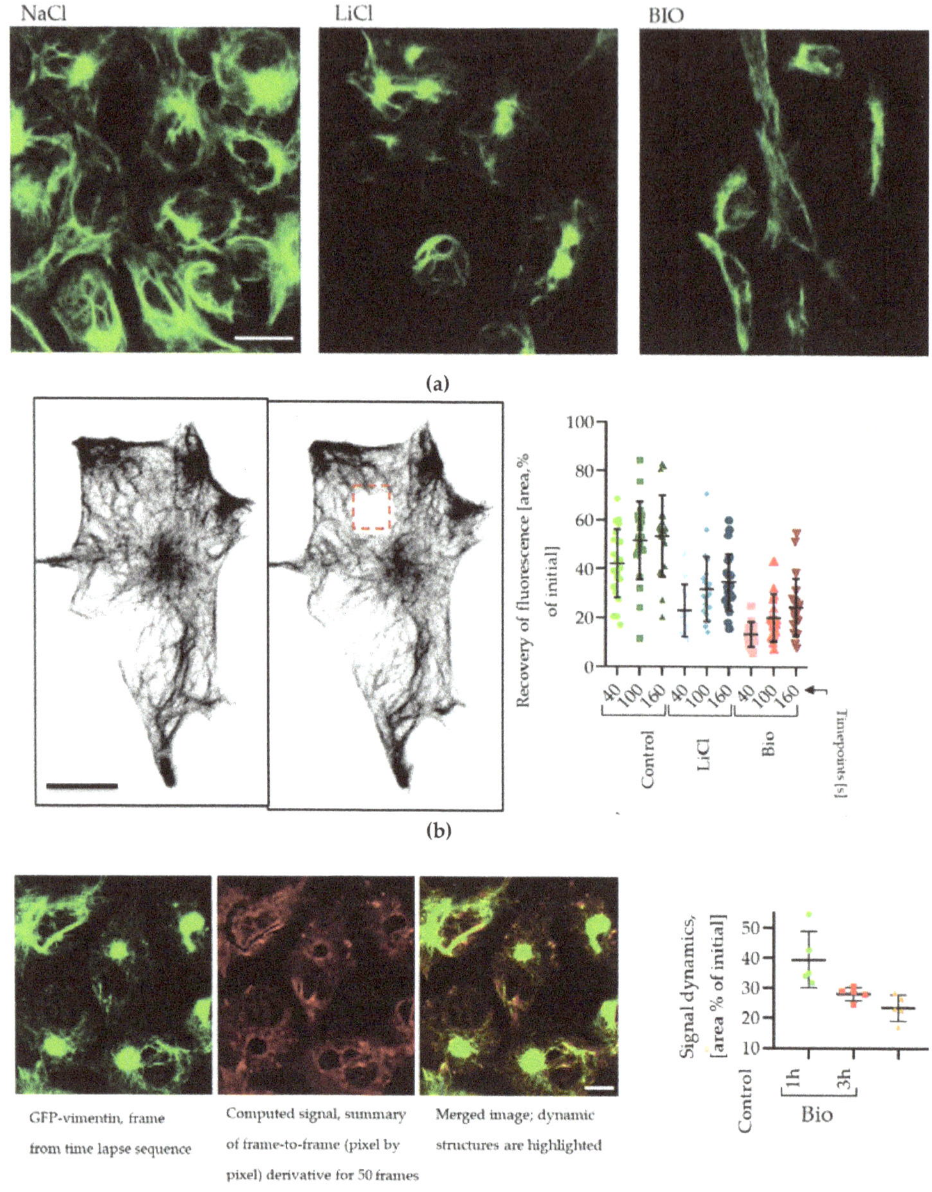

Figure 6. (**a**) Appearance of green fluorescent protein (GFP)-vimentin filaments after 20 mM LiCl or 5 μM BIO treatment (3 h, scale bar = 10 μm). (**b**) Recovery of photo-bleached GFP-vimentin in time points, scale bar = 10 μm. (**c**) Quantification of dynamic of vimentin skeleton movement by frame-to-frame derivative (μm^2) in 50 frames (200 s) time lapse series; scale = 10 μm.

2.7. GSK-3 and Vimentin Colocalize in GBM Cells

Both LiCl and BIO are known to inhibit the protein kinase GSK-3. Therefore, to investigate the relationship of vimentin and GSK-3, co-precipitation studies were performed. We used U251 cells transiently overexpressing glutathione-s-transferase (GST)-GSK-3α and GSK-3β fusion proteins. The cell lysate was incubated with agarose-glutathione beads, washed and subsequently assayed by Western blotting. We found that both isoforms of GSK-3 are able to interact with vimentin isolated from triton-x-100 cell lysates (Figure 7a). We then used a co-sedimentation assay to further investigate GSK-3 interaction with vimentin. In this assay, vimentin is solubilized and precipitated by a series of buffer exchanges in two cycles of dialysis, followed by Western blotting (Figure 7b). Additionally, Fluorescence Resonance Energy Transfer (FRET) measurements showed co-localization of cyan-fluorescent protein (CFP)-GSK3 and yellow-fluorescent protein (YFP)-vimentin fusion proteins in live cells (Figure 7c). Thus GSK-3 and vimentin are physically associated in GBM cells. These data identify a novel relationship between GSK-3 and vimentin for the first time and suggest that alterations in vimentin dynamics may contribute to the inhibition of GBM cell migration in response to LiCl and BIO.

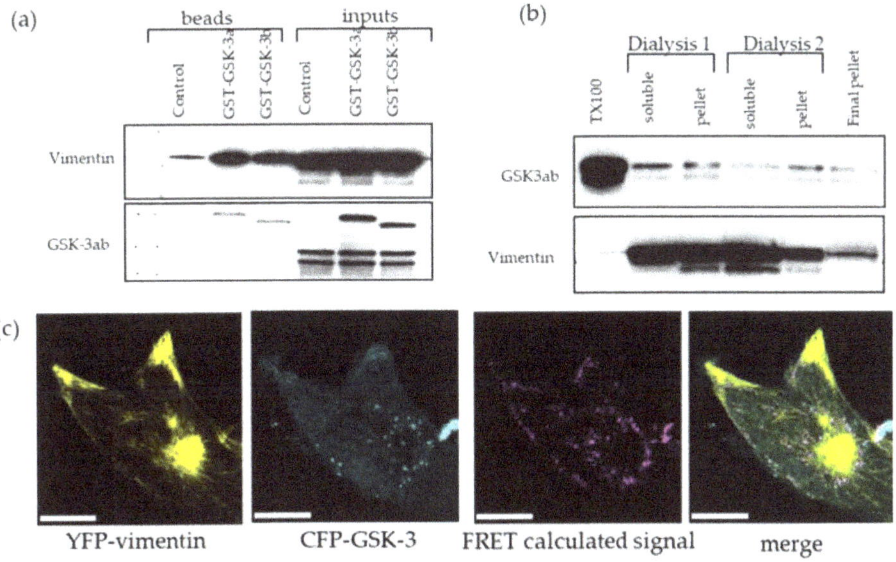

Figure 7. GSK-3 interaction with vimentin shown by: (**a**) glutathione-s-transferase (GST)-GSK-3 pull down assay; and (**b**) co-precipitation assay. GSK3 migrates as a double band with GSK-3α showing a slightly higher molecular weight than GSK-3β. (**c**) Yellow-fluorescent protein (YFP)-vimentin and cyan-fluorescent protein (CFP)-GSK-3β fusion proteins interaction as shown by Fluorescence REsonance Energy Transfer (FRET) colocalization, scale bar = 10 μm.

3. Discussion

Our previous studies implicated the conserved, ubiquitous and multi-tasking protein kinase GSK-3 in GBM cell migration [8,17]. Here we performed a proteomics screen to investigate the mechanisms underlying these observations, and show for the first time that GSK-3 plays a role in regulation of the intermediate cytoskeleton component vimentin. Specifically, we establish that (1) GSK-3 regulates vimentin dynamics, (2) GSK-3 interacts with and phosphorylates vimentin, (3) vimentin facilitates GBM cell migration, and (4) vimentin is highly expressed in GBM patient specimens, and is prognostic, particularly in lower grade gliomas.

This study was performed as a development from our previously published observations on small molecules with potent inhibitory effects on GBM cell migration [8,17]. First, we showed that LiCl was a potent and reversible inhibitor of GBM cell migration in vitro [8]. These effects were observed at 10–20 mM LiCl, which exceeds the maximum tolerated dose of around 1 mM. LiCl is known to inhibit GSK-3 [25], therefore we went on to show that GSK-3α/β siRNA could block GBM migration and that other small molecule GSK-3 inhibitors could block GBM cell migration. During these studies, we examined derivatives of the Chinese medicine indirubin, which are known to inhibit GSK-3 [26]. These molecules blocked migration and angiogenesis in vivo in murine GBM models, improving animal survival [17].

To further study the underlying mechanisms, here we examined the impact of LiCl on GBM cells using proteomics-based approach. 2D-gel electrophoresis/mass spectrometry identified several proteins significantly altered by LiCl treatment. Interestingly these were all involved with cytoskeleton regulation and merit further study in the context of GBM cell migration. Indeed, one of these proteins, CRMP-4, regulates microtubule dynamics in axon outgrowth, and is also a known GSK-3 substrate [14]. Vimentin was the most significantly altered protein and the focus of this study. Additionally, by analysis of patient specimens by Western blotting and immunofluorescence, as well gene expression databases and the human protein atlas, we showed that vimentin is highly abundant in most GBM cases. At the mRNA level it is significantly prognostic in GBM and even more so in astrocytoma where the distribution of vimentin expression appears much broader than in GBM. Further studies are needed to determine whether the reduced expression in astrocytoma compared with GBM is at the tumor cell level, or rather reflects lower cellularity in lower grade tumors. To determine whether vimentin is truly a useful prognostic marker, much more detailed studies are required with larger well-characterized patient populations.

Our data suggest that a primary target of the anti-invasive drugs is vimentin dynamics. This is supported by the observations that (1) detergent soluble vimentin levels are rapidly reduced upon exposure to either LiCl or BIO, and (2) FRAP studies which showed a reduction in the ability of the vimentin cytoskeleton to recover after photobleaching of a GFP-vimentin fusion protein. Presumably this prevents the cell undergoing the restructuring and dynamic alterations required for effective migration and invasion.

Structurally, vimentin is a 57 kDa protein with a highly conserved α−helical rod domain flanked by N- and C-terminal domains which associate to ultimately form intermediate filament networks. Phosphorylation of vimentin, largely at the N-terminus, regulates its ability to form filaments and several kinases have been implicated in this process including protein kinase A (PKA), aurora kinase B and Akt which phosphorylate vimentin at various N-terminal sites [11,12]. Our data strongly implicates GSK-3 as a candidate vimentin kinase, although we have not yet definitively shown this. Analysis of the vimentin amino acid sequence reveals the presence of GSK-3 phosphorylation sites (Ser 50) [14,27] at the N-terminus, which are under investigation.

Further studies on the role of vimentin in GBM are warranted. This protein is highly abundant in GBM and is involved in multiple cellular processes including cell migration. However, one challenge we faced throughout this study was that vimentin manipulation in GBM cells is difficult; sustained knockdown was not straightforward due to the high levels of vimentin in GBM cells, and we have so far been unable to generate CRISPR gene edited GBM cell lines null for vimentin. Additionally, further detailed molecular studies are needed to understand better the effects of vimentin on GBM cell migration. Our data also shows that proteomics can be a very effective approach to identifying molecular changes relevant in GBM cell migration, revealing new targets and mechanisms.

4. Materials and Methods

4.1. Cell Lines, Tissue Samples and Chemicals

The U87 cell line was provided by Dr Webster Cavenee (Ludwig Institute for Cancer Research, La Jolla, CA, USA). U251 and U373 cells were purchased from ATCC (Manassas, VA, USA). All cells were grown in Dulbecco's Modified Eagles Medium (DMEM) with 10% FBS and 1% penicillin-streptomycin. The GBM stem-like cells (GSCs) were derived from fresh samples of surgically resected brain tumors obtained from The Ohio State University Tissue Procurement Shared Resources under a protocol approved by OSU IRB. GSCs were maintained as neurosphere suspension cultures in Neurobasal medium with 20 ng/mL fibroblast growth factor (FGF) and epidermal growth factor (EGF). Unless otherwise stated, reagents, chemicals and supplies were purchased from Thermo-Fisher Scientific (Waltham, MA, USA).

4.2. Western Blotting

Proteins from 5×10^6 cells were isolated by 15 min extraction in TX100 lysis buffer: 1% Triton X100, 50 mM Tris pH 7.5, 100 mM NaCl, 1 mM EDTA, 1 mM EGTA, 50 mM β-glycerophosphate supplemented with 50 mM NaF, 1 mM PMSF, 1 mM DTT, 2 mM Na_3VO_4, 10 μg/mL Leupeptin, 10 μg/mL Aprotinin, 10 μg/mL Pepstatin, 0.25 ng/mL Microcystin LR (EMD Millipore, Billerica, MA, USA). To achieve total solubilization of vimentin fibers, TX100 lysis buffer was supplemented with 8 M urea. Tissue samples were solubilized by sonication in urea containing TX100 lysis buffer. For Western blotting, proteins (100 μg) were separated by SDS-PAGE (Criterion system, Bio-Rad, Hercules, CA, USA) and transferred to nitrocellulose (Bio-Rad). Antibodies used for Western blotting were: mouse anti-vimentin V9 (MS-129-P, Thermo Scientific, (Waltham, MA, USA) mouse anti-GSK-3αβ (368662, EMD Millipore), mouse anti-α-tubulin (T6074, Sigma-Aldrich, St. Louis, MO, USA), mouse anti-GAPDH (ab9484-200, Abcam, Cambridge, MA, USA) and peroxidase-conjugated secondary antibodies (Jackson Laboratories, Bar Harbor, ME, USA).

4.3. Cell Migration Assays

For vimentin knockdown, 200 pmols of Smartpool siRNA, or scrambled siRNA control (Dharmacon, Chicago, IL, USA) was introduced into cells with Lipofectamine 2000 (Thermo Scientific) in a six-well plate 24 h prior to the transwell assay (modified Boyden chamber assay [28]). Assays were performed using 8 μm pore size inserts (ISC Bioexpress, Kaysville, UT, USA), and 50,000 cells per well. Cells were allowed to migrate for 6 h prior to fixation in 1% glutaraldehyde and cell visualization was performed by staining with 10 μg/mL 4′,6-diamidino-2-phenylindole (DAPI). Images of the whole membrane were captured with a motorized Nikon Eclipse Ti microscope system. The time-lapse version of the transwell assay was performed with FluoroBlock inserts (Corning cat# 351152 & 353504) and U251pCDH (copGFP labeled cells, System Biosciences plasmid pCDH-EF1-MCS-T2A-copGFP). Images of the whole insert were captured with a motorized Nikon Eclipse TE2000 microscope system (Nikon, Tokyo, Japan) equipped with an on-stage incubator with CO_2 and temperature control. The invasion/migration assay performed in collagen gel was performed as described previously [8]. Analysis and quantification of signal was performed in ImageJ (https://imagej.nih.gov/ij) [29], Microsoft Excel was used for data handling and Prism Graphpad version 8 (San Diego, CA, USA) was used for data presentation.

4.4. Two-Dimensional Fluorescence Difference Gel Electrophoresis

Whole cell proteomic analysis was performed by using 2-D fluorescence difference gel electrophoresis (DIGE) recommended by The Ohio State University Proteomics Shared Resource https://www.ccic.osu.edu/msp-proteomics [30]. U251 cells (1×10^7) were incubated with 20 mM NaCl or 20 mM LiCl for 24 h. Cell pellets were snap frozen and processed for DIGE analysis by The Ohio State University Mass Spectrometry and Proteomics Facility using the Ettan DIGE system. After 2D-SDS-PAGE, gels were imaged with the Typhoon Variable Mode Imager. Image analysis and quantitation were performed using DeCyder Differential Analysis Software (GE Healthcare, Atlanta, GA, USA). Selected spots were excised and processed for protein/peptide identification.

4.5. Fluorescence Recovery After Photobleaching (FRAP) Assay

U251 cells were stably transfected with a pEGFP-vimentin expression vector, kindly provided by Professor Robert Goldman, Northwestern University [31]. High magnification, time-lapse image series of GFP-vimentin-labeled cells were collected using a Zeiss LSM510 confocal microscope system

(Carl Zeiss Inc., Thornwood, NY, USA), equipped with an in vitro cell culture on-stage incubator (PeCon GmbH, Erbach, Germany). Images were collected every 10 s, for 20 min with minimal laser power (1%). A 5 μm^2 area was burned with 100% laser power for 2 s (until the fluorescence from the selected area was equal to background levels). The half-time of fluorescence recovery T1/2 was analyzed using Zeiss LSM510 control software v3.2 (Jena, Germany). For image dynamic measurement 200-seconds-long, 50 frames sequences were collected. A frame-to-frame difference was calculated with ImageJ/Image Calculator function; data is presented as percent value of first frame of time sequence, accounting for background/noise. Co-localization of GSK-3 with vimentin was assayed on the same microscope system; U251 cells were transiently transfected with pcDNA4TO/CFP-hVimentin (NM_003380) and pcDNA4TO/YFP-GSK-3α/β plasmids. The pcDNA4TO plasmid was purchased from Thermo-Fisher/Life Technologies (cat# V1020), human vimentin was obtained from OriGene (SC111054), GSK-3α/β coding sequences were provided by Dr. Chris Phiel, Nationwide Children's Hospital, Columbus, OH, USA.

4.6. GST Pull-Down Assay

U251 cells were transiently transfected with pDEST27/GST (control) and pDEST27/GST-GSK-3 (either α or β isoform) expression vector. A total of 500 μg of TX100 protein extract was incubated with 50 μL of glutathione agarose beads (Invitrogen, Carlsbad, CA, USA) at 4 °C for 1 h. Beads were extensively washed with TX100 lysis buffer and pulled-down proteins were extracted by boiling in 2× Laemmli Western blot sample buffer (Bio-Rad) and analyzed by Western blotting.

4.7. Digital Data Resources

Digital data presenting vimentin expression levels and Kaplan-Meier survival plots were obtained from The Cancer Genome Atlas (TCGA) (http://cancergenome.nih.gov/), and Rembrandt (https://caintegrator.nci.nih.gov/rembrandt/) databases.

5. Conclusions

Proteomic analysis of GBM cells treated with LiCl and other GSK-3 inhibitors showed alterations in multiple cytoskeletal proteins which may play a role in GBM migration. The intermediate filament protein vimentin was the most significantly altered of these, is highly expressed in GBM, and is prognostic, particularly in LGG. Knockdown of vimentin impairs GBM migration and its dynamics appear to be altered by LiCl treatment. Vimentin physically associates with GSK-3 and this may represent a novel mechanism that mediates intermediate filament dynamics and cell migration.

Supplementary Materials: The following are available online at http://www.mdpi.com/2072-6694/11/4/466/s1, Figure S1: (a) The mRNA levels of vimentin collected from the TCGA database, Levels of transcript are presented as Fragments Per Kilobase of exon per Million reads (FPKM), Figure S2: The small subset of tissues for vimentin and labeled with DAB, Figure S3: LiCl and BIO Treated GBM cells reduces cell motility.

Author Contributions: Methodology, M.O.N.; investigation, M.O.N., J.L.H.; visualization, M.O.N.; resources M.O.N.; data curation, M.O.N., S.E.L., J.L.H.; writing—original draft preparation, S.E.L., M.O.N.; writing—review and editing, S.E.L., M.O.N.; supervision, S.E.L., E.A.C.; project administration, S.E.L., E.A.C.; funding acquisition, E.A.C.

Funding: This research was partially funded by American Brain Tumor Association basic research fellowship for MON and NCI R01 CA166172 (EAC)

Conflicts of Interest: The authors declare no conflict of interest.

References

1. Sattiraju, A.; Sai, K.K.S.; Mintz, A. Glioblastoma Stem Cells and Their Microenvironment. *Adv. Exp. Med. Biol.* **2017**, *1041*, 119–140.
2. Bellail, A.C.; Hunter, S.B.; Brat, D.J.; Tan, C.; Van Meir, E.G. Microregional extracellular matrix heterogeneity in brain modulates glioma cell invasion. *Int. J. Biochem. Cell Biol.* **2004**, *36*, 1046–1069. [CrossRef] [PubMed]

3. Roomi, M.W.; Kalinovsky, T.; Rath, M.; Niedzwiecki, A. Modulation of MMP-2 and MMP-9 secretion by cytokines, inducers and inhibitors in human glioblastoma T-98G cells. *Oncol. Rep.* **2017**, *37*, 1907–1913. [CrossRef] [PubMed]
4. Matias, D.; Predes, D.; Niemeyer Filho, P.; Lopes, M.C.; Abreu, J.G.; Lima, F.R.S.; Moura Neto, V. Microglia-glioblastoma interactions: New role for Wnt signaling. *Biochim. Biophys. Acta Rev. Cancer* **2017**, *1868*, 333–340. [CrossRef] [PubMed]
5. Turaga, S.M.; Lathia, J.D. Adhering towards tumorigenicity: Altered adhesion mechanisms in glioblastoma cancer stem cells. *CNS Oncol.* **2016**, *5*, 251–259. [CrossRef] [PubMed]
6. Kłopocka, W.; Korczyński, J.; Pomorski, P. Cytoskeleton and nucleotide signaling in glioma C6 cells. *Adv. Exp. Med. Biol.* **2013**, *986*, 103–119.
7. Schoumacher, M.; Goldman, R.D.; Louvard, D.; Vignjevic, D.M. Actin, microtubules, and vimentin intermediate filaments cooperate for elongation of invadopodia. *J. Cell Biol.* **2010**, *189*, 541–556. [CrossRef]
8. Nowicki, M.O.; Dmitrieva, N.; Stein, A.M.; Cutter, J.L.; Godlewski, J.; Saeki, Y.; Nita, M.; Berens, M.E.; Sander, L.M.; Newton, H.B.; et al. Lithium inhibits invasion of glioma cells; possible involvement of glycogen synthase kinase-3. *Neuro-Oncology* **2008**, *10*, 690–699. [CrossRef]
9. Mendez, M.G.; Kojima, S.-I.; Goldman, R.D. Vimentin induces changes in cell shape, motility, and adhesion during the epithelial to mesenchymal transition. *FASEB J.* **2010**, *24*, 1838–1851. [CrossRef]
10. Lowery, J.; Kuczmarski, E.R.; Herrmann, H.; Goldman, R.D. Intermediate Filaments Play a Pivotal Role in Regulating Cell Architecture and Function. *J. Biol. Chem.* **2015**, *290*, 17145–17153. [CrossRef]
11. Köster, S.; Weitz, D.A.; Goldman, R.D.; Aebi, U.; Herrmann, H. Intermediate filament mechanics in vitro and in the cell: From coiled coils to filaments, fibers and networks. *Curr. Opin. Cell Biol.* **2015**, *32*, 82–91. [CrossRef] [PubMed]
12. Cole, A.R.; Knebel, A.; Morrice, N.A.; Robertson, L.A.; Irving, A.J.; Connolly, C.N.; Sutherland, C. GSK-3 phosphorylation of the Alzheimer epitope within collapsin response mediator proteins regulates axon elongation in primary neurons. *J. Biol. Chem.* **2004**, *279*, 50176–50180. [CrossRef] [PubMed]
13. Nagini, S.; Sophia, J.; Mishra, R. Glycogen synthase kinases: Moonlighting proteins with theranostic potential in cancer. In *Seminars in Cancer Biology*; Academic Press: Cambridge, MA, USA, 2018.
14. Patel, P.; Woodgett, J.R. Glycogen Synthase Kinase 3: A Kinase for All Pathways? *Curr. Top. Dev. Biol.* **2017**, *123*, 277–302. [PubMed]
15. Tejeda-Muñoz, N.; Robles-Flores, M. Glycogen synthase kinase 3 in Wnt signaling pathway and cancer. *IUBMB Life* **2015**, *67*, 914–922. [CrossRef]
16. Zhu, Q.-S.; Rosenblatt, K.; Huang, K.-L.; Lahat, G.; Brobey, R.; Bolshakov, S.; Nguyen, T.; Ding, Z.; Belousov, R.; Bill, K.; et al. Vimentin is a novel AKT1 target mediating motility and invasion. *Oncogene* **2011**, *30*, 457–470. [CrossRef]
17. Williams, S.P.; Nowicki, M.O.; Liu, F.; Press, R.; Godlewski, J.; Abdel-Rasoul, M.; Kaur, B.; Fernandez, S.A.; Chiocca, E.A.; Lawler, S.E. Indirubins decrease glioma invasion by blocking migratory phenotypes in both the tumor and stromal endothelial cell compartments. *Cancer Res.* **2011**, *71*, 5374–5380. [CrossRef] [PubMed]
18. Majewska, E.; Szeliga, M. AKT/GSK3β Signaling in Glioblastoma. *Neurochem. Res.* **2017**, *42*, 918–924. [CrossRef]
19. Cheng, F.; Shen, Y.; Mohanasundaram, P.; Lindström, M.; Ivaska, J.; Ny, T.; Eriksson, J.E. Vimentin coordinates fibroblast proliferation and keratinocyte differentiation in wound healing via TGF-β-Slug signaling. *Proc. Natl. Acad. Sci. USA* **2016**, *113*, E4320–E4327. [CrossRef]
20. Menko, A.S.; Bleaken, B.M.; Libowitz, A.A.; Zhang, L.; Stepp, M.A.; Walker, J.L. A central role for vimentin in regulating repair function during healing of the lens epithelium. *Mol. Biol. Cell* **2014**, *25*, 776–790. [CrossRef] [PubMed]
21. Lin, L.; Wang, G.; Ming, J.; Meng, X.; Han, B.; Sun, B.; Cai, J.; Jiang, C. Analysis of expression and prognostic significance of vimentin and the response to temozolomide in glioma patients. *Tumour Biol. J. Int. Soc. Oncodev. Biol. Med.* **2016**, *37*, 15333–15339. [CrossRef]
22. Hothorn, T.; Lausen, B. On the exact distribution of maximally selected rank statistics. *Comput. Stat. Data Anal.* **2003**, *43*, 121–137. [CrossRef]
23. Ádám, C.; Fekete, A.; Bőgel, G.; Németh, Z.; Tőkési, N.; Ovádi, J.; Liliom, K.; Pesti, S.; Geiszt, M.; Buday, L. Accumulation of the PX domain mutant Frank-ter Haar syndrome protein Tks4 in aggresomes. *Cell Commun. Signal.* **2015**, *13*, 33. [CrossRef] [PubMed]

24. Uhlén, M.; Fagerberg, L.; Hallström, B.M.; Lindskog, C.; Oksvold, P.; Mardinoglu, A.; Sivertsson, Å.; Kampf, C.; Sjöstedt, E.; Asplund, A.; et al. Proteomics. Tissue-based map of the human proteome. *Science* **2015**, *347*, 1260419. [CrossRef] [PubMed]
25. Jope, R.S. Lithium and GSK-3: One inhibitor, two inhibitory actions, multiple outcomes. *Trends Pharmacol. Sci.* **2003**, *24*, 441–443. [CrossRef]
26. Meijer, L.; Skaltsounis, A.-L.; Magiatis, P.; Polychronopoulos, P.; Knockaert, M.; Leost, M.; Ryan, X.P.; Vonica, C.A.; Brivanlou, A.; Dajani, R.; et al. GSK-3-selective inhibitors derived from Tyrian purple indirubins. *Chem. Biol.* **2003**, *10*, 1255–1266. [CrossRef] [PubMed]
27. Gnad, F.; Ren, S.; Cox, J.; Olsen, J.V.; Macek, B.; Oroshi, M.; Mann, M. PHOSIDA (phosphorylation site database): Management, structural and evolutionary investigation, and prediction of phosphosites. *Genome Biol.* **2007**, *8*, R250. [CrossRef] [PubMed]
28. Boyden, S. The chemotactic effect of mixtures of antibody and antigen on polymorphonuclear leucocytes. *J. Exp. Med.* **1962**, *115*, 453–466. [CrossRef]
29. Chen, W.; Wong, C.; Vosburgh, E.; Levine, A.J.; Foran, D.J.; Xu, E.Y. High-throughput image analysis of tumor spheroids: A user-friendly software application to measure the size of spheroids automatically and accurately. *J. Vis. Exp.* **2014**, *89*, 51639. [CrossRef] [PubMed]
30. Blundon, M.; Ganesan, V.; Redler, B.; Van, P.T.; Minden, J.S. Two-Dimensional Difference Gel Electrophoresis. *Methods Mol. Biol.* **2019**, *1855*, 229–247.
31. Yoon, M.; Moir, R.D.; Prahlad, V.; Goldman, R.D. Motile properties of vimentin intermediate filament networks in living cells. *J. Cell Biol.* **1998**, *143*, 147–157. [CrossRef] [PubMed]

cancers

MDPI

Article

CircSMARCA5 Regulates VEGFA mRNA Splicing and Angiogenesis in Glioblastoma Multiforme Through the Binding of SRSF1

Davide Barbagallo [1,2], Angela Caponnetto [1], Duilia Brex [1], Federica Mirabella [1], Cristina Barbagallo [1], Giovanni Lauretta [1], Antonio Morrone [3], Francesco Certo [2,3], Giuseppe Broggi [3], Rosario Caltabiano [3], Giuseppe M. Barbagallo [2,3], Vittoria Spina-Purrello [4], Marco Ragusa [1,2,5,†], Cinzia Di Pietro [1,†], Thomas B. Hansen [6,7,†,*] and Michele Purrello [1,2,†,*]

1 Department of Biomedical and Biotechnological Sciences—Section of Biology and Genetics "Giovanni Sichel", University of Catania, 95123 Catania, Italy; dbarbaga@unict.it (D.B.); caponnettoangela@gmail.com (A.C.); duiliabrex@gmail.com (D.B.); mirabella.federica.91@gmail.com (F.M.); barbagallocristina@gmail.com (C.B.); giovannilau91@hotmail.it (G.L.); mragusa@unict.it (M.R.); dipietro@unict.it (C.D.P.)

2 Multidisciplinary Research Center on Brain Tumors Diagnosis and Therapy, University of Catania, 95123 Catania, Italy; cicciocerto@yahoo.it (F.C.); gbarbagallo@unict.it (G.M.B.)

3 Department of Medical, Surgical and Advanced Technological Sciences "G.F. Ingrassia", University of Catania, 95123 Catania, Italy; morant592@libero.it (A.M.); giuseppe.broggi@gmail.com (G.B.); rosario.caltabiano@unict.it (R.C.)

4 Department of Biomedical and Biotechnological Sciences—Section of Medical Biochemistry, University of Catania, 95123 Catania, Italy; spinavit@unict.it

5 Oasi Research Institute-IRCCS, 94018 Troina, Italy

6 Department of Molecular Biology and Genetics (MBG), Aarhus University, 8000 Aarhus C, Denmark

7 Interdisciplinary Nanoscience Center (iNANO), Aarhus University, 8000 Aarhus C, Denmark

* Correspondence: tbh@mbg.au.dk (T.B.H.); purrello@unict.it (M.P.); Tel.: +39-095-378-2078 (M.P.)

† Senior authors.

Received: 30 December 2018; Accepted: 6 February 2019; Published: 7 February 2019

Abstract: Circular RNAs are a large group of RNAs whose cellular functions are still being investigated. We recently proposed that circSMARCA5 acts as sponge for the splicing factor Serine and Arginine Rich Splicing Factor 1 (SRSF1) in glioblastoma multiforme (GBM). After demonstrating by RNA immunoprecipitation a physical interaction between SRFS1 and circSMARCA5, we assayed by real-time PCR in a cohort of 31 GBM biopsies and 20 unaffected brain parenchyma controls (UC) the expression of total, pro-angiogenic (Iso8a) and anti-angiogenic (Iso8b) mRNA isoforms of Vascular Endothelial Growth Factor A (VEGFA), a known splicing target of SRSF1. The Iso8a to Iso8b ratio: (i) increased in GBM biopsies with respect to UC (p-value < 0.00001); (ii) negatively correlated with the expression of circSMARCA5 (r-value = −0.46, p-value = 0.006); (iii) decreased in U87-MG overexpressing circSMARCA5 with respect to negative control (p-value = 0.0055). Blood vascular microvessel density, estimated within the same biopsies, negatively correlated with the expression of circSMARCA5 (r-value = −0.59, p-value = 0.00001), while positively correlated with that of SRSF1 (r-value = 0.38, p-value = 0.00663) and the Iso8a to Iso8b ratio (r-value = 0.41, p-value = 0.0259). Kaplan-Meier survival analysis showed that GBM patients with low circSMARCA5 expression had lower overall and progression free survival rates than those with higher circSMARCA5 expression (p-values = 0.033, 0.012, respectively). Our data convincingly suggest that circSMARCA5 is an upstream regulator of pro- to anti-angiogenic VEGFA isoforms ratio within GBM cells and a highly promising GBM prognostic and prospective anti-angiogenic molecule.

Keywords: circular RNA; hsa_circ_0001445; RNA binding proteins; alternative splicing; glioblastoma multiforme; angiogenesis; VEGFA

1. Introduction

Circular RNAs (circRNAs) are a recently discovered large group of RNAs, of which biogenesis and functions within cells remain mostly to be investigated [1–4]. Notwithstanding the limited number of studies describing the mechanisms through which they act, the expression of several circRNAs has been found deregulated in pathological phenotypes from cancer to degenerative conditions: this suggests an active role of these molecules in their pathogenesis [5–10]. Splicing is thought to represent the main mechanism by which circRNAs originate [11–13]. Glioblastoma multiforme (GBM) is the most deadly human brain cancer to date: its median overall survival is 14.6 months after diagnosis and the prognosis is dismal in the majority of cases [14]. Notwithstanding the published data on the involvement of circRNAs and splicing in GBM, many questions remain without an answer [15–18]. Recently, we hypothesized that circSMARCA5 may carry out its function by tethering the splicing factor Serine and Arginine Rich Splicing Factor 1 (SRSF1) and that through this mechanism it may regulate GBM cells migration [19]. In addition to splicing, SRSF1 is involved in a plethora of other cellular functions as: (i) regulation of RNA metabolism; (ii) mRNA translation; (iii) miRNA processing; (iv) protein sumoylation; (v) stress response [20]. SRSF1 is also upregulated and acts as oncoprotein in several cancers [21]. Here, we demonstrate the physical interaction between circSMARCA5 and the splicing factor SRSF1 in GBM cells; based on our data, we propose that this interaction regulates the switch between pro- and anti-angiogenic isoforms generated by the splicing of the clinically relevant Vascular Endothelial Growth Factor A (VEGFA) pre-mRNA in GBM cells.

2. Results

2.1. SRSF1 and CircSMARCA5 Physically Interact within GBM Cells

CircSMARCA5 was predicted to be bound by SRSF1 in at least seven different evolutionarily conserved sites (see [19] and Figure 1A). SRSF1 RNA immunoprecipitation (RIP) allowed us to validate this interaction, showing a significant enrichment of circSMARCA5 in addition to SRSF3 mRNA, a known splicing target of SRSF1 (positive control), in immunoprecipitated U87-MG cells lysate with respect to a negative control (Figure 1B,C).

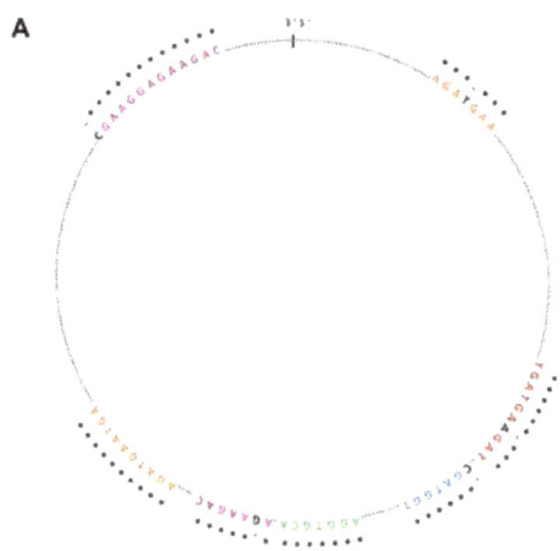

Figure 1. *Cont.*

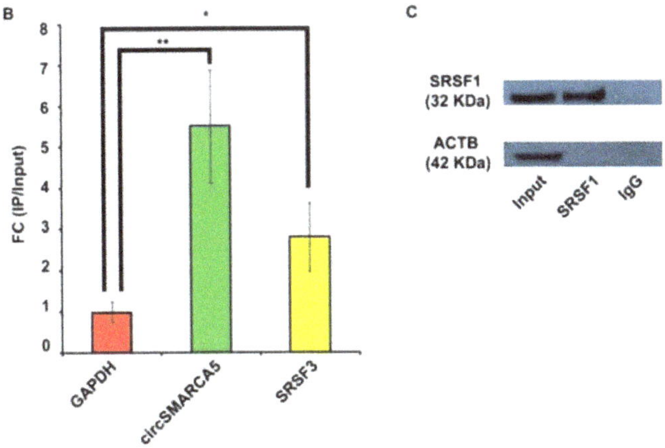

Figure 1. CircSMARCA5 and SRSF1 physically interact within U87-MG cells. (**A**) SRSF1 binding sites predicted by RBPMap database (http://rbpmap.technion.ac.il/) are highlighted with different colors within circSMARCA5 sequence. * indicates sequence conservation among twelve primates, including *Homo sapiens* (see [19]). (**B**) Fold enrichment (FC) of circSMARCA5, SRSF3 and GAPDH are shown as IPed samples/input. See [22] and Section 4 Materials and Methods for further details (* p-value < 0.05; ** p-value < 0.01, $N = 4$, two-samples t-test). (**C**) Representative western blot of U87-MG Input, SRSF1 and normal IgG RIPed samples. Antibodies against SRSF1 and Actin Beta (ACTB) were used as described in Section 4 Materials and Methods.

2.2. SRSF1 and VEGFA Are Upregulated While circSMARCA5 Is Downregulated in GBM Biopsies

SRSF1 was upregulated in GBM biopsies with respect to unaffected brain parenchyma (UC), both as mRNA (p-value = 0.0009, Mann–Whitney test) and protein (p-value = 0.022, two-sample t-test) (Figure 2A–C). After a literature review of SRSF1's splicing targets (Table S1, Supplementary Materials), we focused on VEGFA, a molecule of high clinical interest in GBM (see Discussion). We demonstrated that total VEGFA mRNA was upregulated in the same cohort (p-value < 0.00001, Mann–Whitney test), while circSMARCA5 was downregulated (p-value < 0.00001, Mann–Whitney test), confirming our data previously obtained in an independent GBM cohort (Figure 2A). Upregulation of SRSF1 and VEGFA mRNA, as well as their positive correlation, was also observed in the extended cohort of GBM biopsies analyzed in REpository for Molecular BRAin Neoplasia DaTa (REMBRANDT) (Figure S1, Supplementary Materials) and The Cancer Genome Atlas (TCGA) (Figure S2, Supplementary Materials) databases. Most specifically, based on REMBRANDT data, SRSF1 mRNA was significantly upregulated in all glioma grades with respect to normal brain (p-value < 0.0001, ANOVA test) (Figure S1A, Supplementary Materials), while VEGFA was significantly upregulated specifically in GBM (p-value < 0.0001, ANOVA test) (Figure S1B, Supplementary Materials). In more detail, SRSF1 and VEGFA mRNAs were both significantly upregulated in all GBM subtypes, except for the neural subtype, with respect to unaffected brain samples (see significant p-values within Figure S2, Supplementary Materials) [23,24].

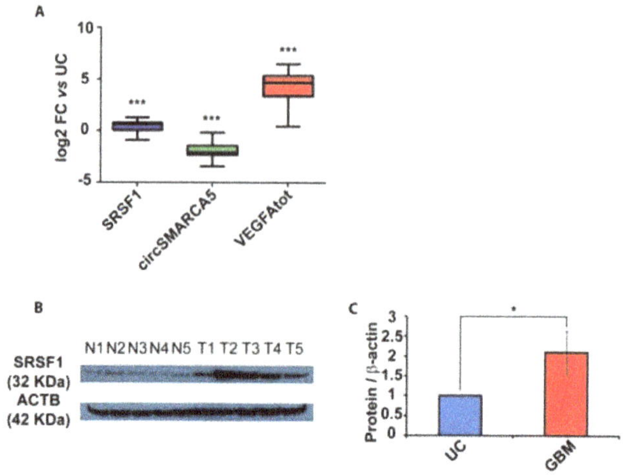

Figure 2. SRSF1, circSMARCA5 and VEGFA expression in GBM biopsies and unaffected controls (UC). (**A**) Box-and-whisker plots, representing the expression of SRSF1, circSMARCA5 and total VEGFA RNA in the studied cohort. Data are represented as $\log_2$ fold change (FC) values versus UC. (*** *p*-value < 0.001, $N_{(GBM)} = 31$, $N_{(UC)} = 20$, Mann–Whitney test). (**B**) Western blot of SRSF1 in a selection of UC (N) and GBM (T) samples. ACTB was used as a loading control. (**C**) Bar graph representing densitometric quantification of SRSF1. Data, shown as mean ± standard deviation, represent fold change versus UC (* *p*-value < 0.05, $N_{(GBM)} = 14$, $N_{(UC)} = 8$, two-sample *t*-test).

2.3. *The Ratio of Pro- to Anti-Angiogenic VEGFA mRNA Isoforms Increases in GBM Cells as Compared to Unaffected Brain Parenchyma and Decreases in U87-MG Overexpressing circSMARCA5*

In order to define the role of circSMARCA5 in the control of VEGFA mRNA splicing, the expression of total VEGFA, pro- and anti-angiogenic VEGFA mRNA isoforms (Iso8a and Iso8b, respectively) were assayed in the same cohort previously analyzed. Iso8a isoforms were most expressed in all biopsies, as shown by real-time PCR data. Most specifically, the Iso8a-to-Iso8b ratio was significantly higher in GBM biopsies with respect to UC (*p*-value < 0.00001, Mann–Whitney test) (Figure 3A and Figure S3) (see Section 4 Materials and Methods for details on data analysis). In the same cohort, the expression of circSMARCA5 negatively correlated with that of SRSF1 mRNA (*r*-value = −0.36, *p*-value = 0.011, Spearman correlation test) and with the Iso8a-to-Iso8b ratio (*r*-value = −0.47, *p*-value = 0.006, Spearman correlation test) (Figure 3B). The Iso8a-to-Iso8b ratio was also higher in three GBM cell lines (A172, CAS-1 and U87MG) with respect to unaffected brain (Figure S4, Supplementary Materials). Interestingly, U87-MG overexpressing circSMARCA5 showed a significant decrease in the Iso8a-to-Iso8b ratio with respect to the same cells transfected with the empty vector (negative control, NC) (*p*-value = 0.0055, two-sample *t*-test) (Figure 3C; Figures S3 and S5, Supplementary Materials).

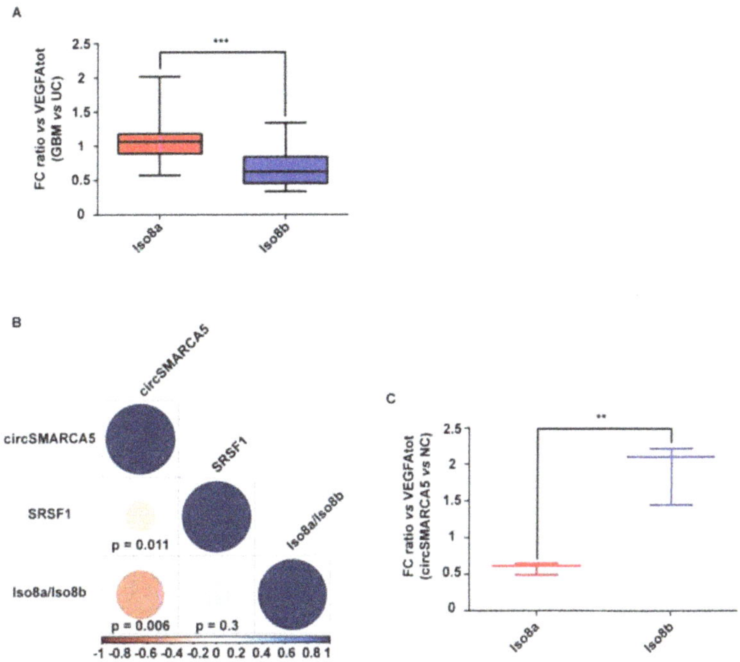

Figure 3. Iso8a-to-Iso8b ratio in GBM biopsies vs. UC. (**A**) Box-and-whisker plots, representing the ratios between fold changes of Iso8a and total VEGFA ($VEGFA_{tot}$) and Iso8b and $VEGFA_{tot}$ in GBM compared to UC (*** p-value < 0.00001, N = 27, Mann–Whitney test) (see Section 4 Materials and Methods for details on data analysis). (**B**) Correlation matrix among the expression of circSMARCA5 and SRSF1 and Iso8a-to-Iso8b ratio. Positive and negative correlations are displayed in blue and red colors, respectively. The color scale bar indicates r values. Color intensity and the size of the circle are proportional to the correlation coefficients into the correlogram. (**C**) Box-and-whisker plots, representing the ratios between fold changes of Iso8a and $VEGFA_{tot}$ and Iso8b and $VEGFA_{tot}$ in U87-MG overexpressing circSMARCA5 with respect to U87-MG transfected with the empty vector (NC) (** p-value < 0.001, N = 3, two-sample t-test) (see Section 4 Materials and Methods for details on data analysis).

2.4. Blood Vascular Microvessel Density (MVD) Negatively Correlates with the Expression of circSMARCA5, While Positively Correlating with SRSF1 mRNA Expression and Pro/Anti-Angiogenic VEGFA Isoforms Ratio

As expected, MVD evaluated in the same cohort previously characterized was significantly higher in GBM samples than in UC (p-value < 0.00001, two-sample t-test) (Figure 4A,B). MVD positively correlated with the SRSF1 mRNA expression and the Iso8a-to-Iso8b ratio (r-values = 0.38, and 0.41, respectively; p-values = 0.00663, and 0.026, respectively, Spearman correlation test), while negatively correlating with circSMARCA5 expression (r-value = −0.59, p-value = 0.00001, Spearman correlation test) (Figure 4C).

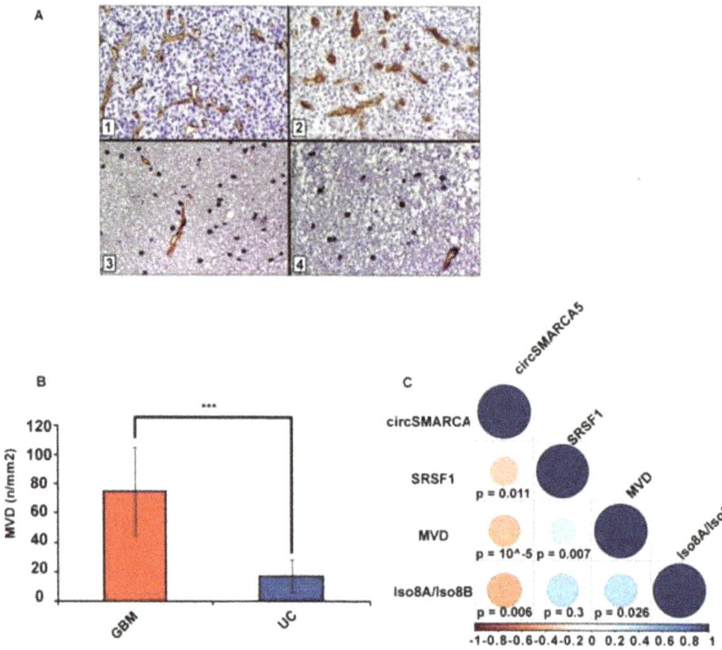

Figure 4. MVD in GBM and UC samples. (**A**) Representative immunohistochemical staining for CD31 showing areas of high MVD with multiple branching vessels in glioblastoma tissue (1 and 2) and a lower MVD in unaffected brain tissue (3 and 4). Immunoperoxidase staining; 400× magnification (**B**) Bar graph representing the mean MVD in GBM and UC samples. Data are represented as mean ± standard deviation (*** p-value < 0.00001, $N_{(GBM)} = 31$, $N_{(UC)} = 18$, two sample t-test). (**C**) Correlation matrix among MVD, circSMARCA5 and SRSF1 expression and Iso8a to Iso8b ratio. Positive and negative correlations are displayed in blue and red color, respectively. Color scale bar indicates r values. Color intensity and the size of the circle are proportional to the correlation coefficients into the correlogram.

2.5. CircSMARCA5 Expression Negatively Correlates with GBM Patients' Overall Survival (OS) and Progression-Free Survival (PFS)

We found that lower circSMARCA5 expression was associated with poorer OS and PFS in GBM (p-values = 0.033, and 0.012, respectively, logrank test) (Figure 5A,B). Kaplan–Meier analysis performed on TCGA data showed also a negative correlation between mesenchymal GBM patients' OS and SRSF1 expression (p-value = 0.0172, logrank test) (Figure S6, Supplementary Materials).

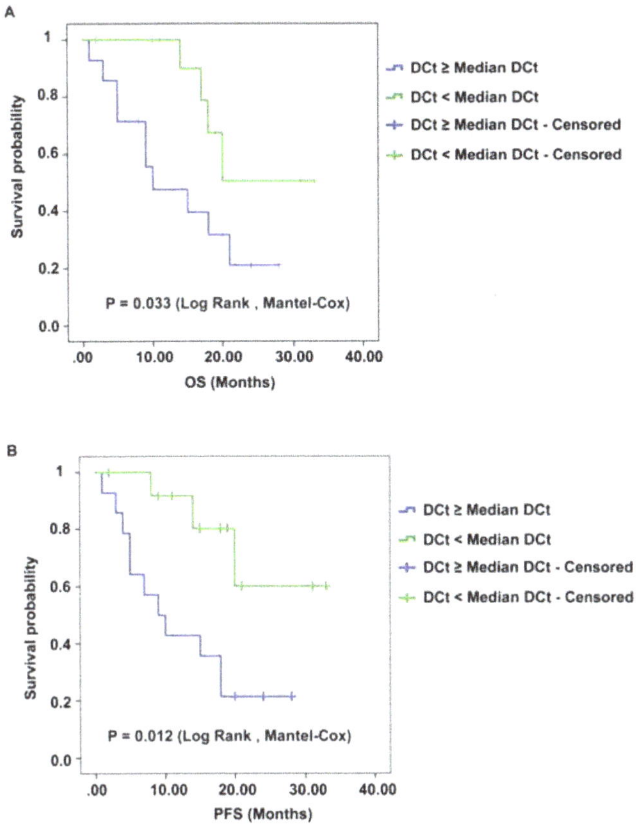

Figure 5. Kaplan–Meier overall survival (OS) and progression-free survival (PFS) curves of GBM patients, based on the expression of circSMARCA5. Patients having a lower expression of circSMARCA5 (DCt ≥ Median DCt) survive less (**A**) show a shorter PFS (**B**) than patients with a higher expression of circSMARCA5 (DCt < Median DCt).

3. Discussion

As previously described by our group, circSMARCA5 sequence is predicted to be enriched in SRSF1 protein-binding sites: accordingly, a physical interaction between these two molecules has been identified by high-throughput enhanced UV crosslinking and immunoprecipitation (eCLIP) analysis in K562 cells [19,25]. Deeper knowledge on the interactions between RNAs and RNA-binding proteins (RBPs) in specific cell contexts is a critical piece of the cell function puzzle [26]. For instance, the ability of specific viral RNAs to function as a sponge for several host cell's RBPs has been extensively analyzed [27]; similarly, many coding and non-coding RNAs have been described to behave as miRNAs- and RBPs-sponges in eukaryotic cells [28,29]. Here, we validate, through RIP data, the physical interaction between circSMARCA5 and SRSF1 in human GBM cells. SRSF1 is a protein involved in many biomolecular functions [20] and its expression is known to be upregulated in several cancers [30,31]. In order to understand if the interaction between circSMARCA5 and SRSF1 may alter the splicing pattern of SRSF1's targets, we focused on VEGFA. It is known that VEGFA pre-mRNA can be alternatively spliced generating both pro-angiogenic isoforms (VEGF-A_{xxx}a) and anti-angiogenic isoforms (VEGF-A_{xxx}b), depending on the recognition of a proximal splicing site (PSS) within the eighth exon of VEGFA pre-mRNA by SRSF1: specifically, the higher the amount of SRSF1 binding this PSS, the higher the retention of full-length eighth exon is, improving the synthesis of VEGF-A_{xxx}a isoforms [32–34]. Aberrant splicing of VEGFA, leading to an alteration in the pro/anti-angiogenic ratio was described in human colon cancer [35]. Data from biopsies and our in vitro model following overexpression of circSMARCA5, together with our previous data [19], support the hypothesis that this circRNA performs a *trans*-acting splicing function within GBM cells, which it carries out by tethering SRSF1. Moreover, this pathway seems to be specifically deregulated in GBM cells, among the other glioma grades, based on our previous observations and on REMBRANDT and TCGA data. Additionally, the hypothesis of a functional involvement of circSMARCA5 in the regulation of angiogenesis is supported by correlations observed among the circSMARCA5/SRSF1/VEGFA expression and the blood microvascular vessel density. This is an important aspect of GBM biology: ultimately, the control of the pro- to anti-angiogenic switch may represent a valid alternative to the therapy based on monoclonal anti-VEGFA antibodies that did not reach the expected results to date [36]. Finally, even if on a limited cohort of patients, the perspective use of circSMARCA5 as a prognostic biomarker and a therapeutic target [33,37,38] in GBM cells appears to be promising.

4. Materials and Methods

4.1. GBM Biopsies

Each specimen obtained during intraoperative navigation-assisted neurosurgery was divided in two halves: one half was supplied to pathologists for fixation and tissue embedded into paraffin, and the other was immediately frozen and stored at −80 °C until use. All patients enrolled in this study supplied their written informed consent before surgery. The study was conducted in accordance with the declaration of Helsinki and the protocol was approved by the ethical committee of Azienda Ospedaliero-Universitaria "Policlinico-Vittorio Emanuele", Catania, Italy. Only biopsies from patients with a confirmed pathological diagnosis of GBM were included into the study (N = 31). Unaffected brain parenchyma was obtained, when possible, from a non-eloquent region of the brain, adjacent to the tumor and negative to 5-aminolevulinic acid (5-ALA) fluorescence: this type of sample has been defined as unaffected control and used as calibrator tissue in this study only after pathologists observed no infiltration of cancer cells (N = 20). Clinical data from patients enrolled in the study are summarized in Table 1.

Table 1. Clinical data of Glioblastoma (GBM) and control samples.

Sample	N	Mean Age (Years ± Std. Dev.)	Sex		Mean OS (Months ± Std. Dev.)	Mean PFS (Months ± Std. Dev.)
			M	F		
Fresh frozen GBM biopsies	31	63.6 ± 10.9	15	16	15 ± 8.2	13.8 ± 8.7
Fresh frozen unaffected brain parenchyma	20	64 ± 10.3	8	12		
FirstChoice® Human Brain Reference RNA	1 (commercially available)	68.3 ± 15	13	10		

4.2. Cell Cultures and Transfection

GBM cell lines A172, CAS-1 and U87-MG culture and transfection with pcDNA3-circSMARCA5 or empty pcDNA3 vectors were performed as previously described [10,19].

4.3. RNA Immunoprecipitation (RIP)

Briefly, cells were seeded in 10 cm dishes at a density of 3.6 × 10^6 and cultured for 72 hours. RIP was performed as previously described by Peritz et al. [39], with some modifications. More specifically, RIP was performed without cross-linking. Immunoprecipitation was performed using 5 µg of mouse monoclonal IgG2b antibody against SRSF1 (Santa Cruz Biotechnology, Inc., Heidelberg, Germany, Cat.

n. sc-73026) or isotype control IgG from mouse (negative control) (Santa Cruz Biotechnology, Inc., Cat. n. sc-2025). Data were analyzed as described by Ratnadiwakara et al. [22]. RIP methodology and data analysis are fully described in Supplementary Materials.

4.4. RNA Extraction and Real-Time PCR

RNA was extracted by using Trizol (ThermoFisher Scientific, Waltham, MA USA), according to manufacturer's instruction and quantified both by spectrophotometer and Qubit™ fluorometer (ThermoFisher Scientific). A commercially available RNA from human brain (Ambion, Austin, TX, USA) has been used as further unaffected control. Real-time PCR was performed as previously described [40] and relative RNA amounts were estimated by using 2^{-DDCt} method [41]. For a fully description of real-time PCR data analysis within this manuscript, see Supplementary Materials. Linear and circular RNAs were amplified by using convergent and divergent primers, respectively, as described in Table S2 and in Figure S7 (Supplementary Materials).

4.5. Protein Extraction and Immunoblotting

Proteins from biopsies were extracted by using RIPA buffer (Abcam, Cambridge, UK) and quantified by Qubit™ fluorometer (ThermoFisher Scientific). Human Brain Cerebral Cortex Protein Medley (Takara Clontech®, Mountain View, CA, USA) was used as further unaffected control. Western blot analysis was performed as previously described [42]. Primary antibodies against the following proteins were used: SRSF1 (mouse monoclonal antibody from Santa Cruz Biotechnology, Inc., Cat. n. sc-73026) and ACTB (rabbit polyclonal antibody from Abcam, Cat. n. ab16039). Secondary antibodies were HRP-conjugated anti-mouse (for SRSF1) or anti-rabbit (for ACTB) (Santa Cruz Biotechnology, Inc, Cat. n. sc-516102 and sc-2004, respectively) for chemiluminescent detection. Gel bands were quantified by ImageJ software (https://imagej.nih.gov/ij/index.html).

4.6. Immunohistochemistry

For each case, all Hematoxylin and Eosin (H&E) stained sections were assessed by two pathologists and one representative sample was identified. Sections were processed as previously described [43]. Briefly, slides were cut at 4–5 μm, dried, deparaffinized and rehydrated. Then, sections were incubated for 30 min at 4 °C with mouse monoclonal anti-Human CD31, Endothelial Cell antibody (JC70A, Dako Corporation, Glostrup, Denmark), diluted 1/40 in PBS (Sigma, Milan, Italy). The biotinylated anti-rabbit secondary antibody was applied for 30 min at 20 °C, followed by the avidin–biotin–peroxidase complex (Vector Laboratories, Burlingame, CA, USA) for a further 30 min at 20 °C. The immunoreaction was visualized by incubating the sections for 4 min in a diaminobenzidine (DAB) and 0.02% hydrogen peroxide solution (DAB substrate kit, Vector Laboratories, CA, USA).

4.7. Assessment of Blood Vascular Microvessel Density (MVD)

MVD of CD31 was evaluated by two pathologists. Immunohistochemical sections were observed with a Zeiss Axioplan light microscope (CarlZeiss, Oberkochen, Germany). MVD assessment was performed as described by Weidner et al. and Mikkelsen et al., with some modifications [44,45]. Briefly, vascular hotspots were identified on CD31 sections by a light microscope at 4× and 10× magnifications. MVD was evaluated as the total number of vessels per mm^2 with a conversion factor of 1 mm^2 equaling 4 high power fields (HPFs) at 40× magnification of highest vascular densities [46]. Areas with ≥50 of viable tumor cells were counted; tissues with extensive necrosis, hemorrhage and desmoplasia were excluded. Every single stained endothelial cell and each lumen for long branched vessels and glomeruloid tufts were counted. Moreover, small clusters of ≥2 staining endothelial cells within the same vessel was counted as one vascular structure.

4.8. Statistics

All the statistical tests used in this study are described throughout the text and in figure legends. p-values lower than 0.05 were considered significant.

5. Conclusions

We proposed that our data convincingly suggest that circSMARCA5 is an upstream regulator of the ratio of the pro- and anti-angiogenic VEGFA isoforms within GBM cells and is a promising prognostic biomarker and a prospective therapeutic target in GBM cells.

Supplementary Materials: The following are available online at http://www.mdpi.com/2072-6694/11/2/194/s1, Table S1, List of SRSF1's splicing targets retrieved by literature. Table S2, Sequences of primers used in the study. Figure S1, SRSF1 and VEGFA mRNA expression in different types of glioma (REMBRANDT database). Figure S2, SRSF1 and VEGFA mRNA expression in GBM subtypes (TCGA database). Figure S3, Box-and-whisker plots, representing the Iso8a/Iso8b ratio in GBM and UC. Figure S4, Bar graph showing Iso8A vs Iso8B ratio in three different GBM cell lines. Figure S5, CircSMARCA5 but not linear SMARCA5 is overexpressed in U87-MG transfected with pcDNA3-circSMARCA5 vector with respect to NC (U87-MG transfected with the empty vector). Figure S6, Kaplan-Meier overall survival curves of mesenchymal GBM patients, based on the expression of SRSF1. Figure S7, Diagram of primers used to amplify Iso8a and Iso8b VEGFA specific isoforms by qRT PCR.

Author Contributions: M.P., T.B.H., C.D.P., M.R. and D.B. conceived and designed the experiments. D.B. and A.C. performed the experiments. G.M.B., R.C., F.C., G.B. and A.M. obtained, characterized and curated clinical data from patients' biopsies. D.B., F.M., C.B., G.L. and V.S.-P. contributed to the acquisition, analysis and interpretation of data. M.P. and D.B. wrote the paper.

Funding: This project was financially supported by Lega Italiana per la Lotta contro i Tumori (LILT), section of Catania, by Ministero dell'Università e della Ricerca Scientifica e Tecnologica (MIUR) and by Università degli Studi di Catania.

Acknowledgments: We wish to thank Massimo Libra, Massimo Romani, Maria Angela Sortino and their collaborators for scientific and logistic support to the project.

Conflicts of Interest: The authors declare no conflicts of interest.

References

1. Hansen, T.B.; Jensen, T.I.; Clausen, B.H.; Bramsen, J.B.; Finsen, B.; Damgaard, C.K.; Kjems, J. Natural RNA circles function as efficient microRNA sponges. *Nature* **2013**, *495*, 384–388. [CrossRef] [PubMed]
2. Memczak, S.; Jens, M.; Elefsinioti, A.; Torti, F.; Krueger, J.; Rybak, A.; Maier, L.; Mackowiak, S.D.; Gregersen, L.H.; Munschauer, M.; et al. Circular RNAs are a large class of animal RNAs with regulatory potency. *Nature* **2013**, *495*, 333–338. [CrossRef]
3. Ebbesen, K.K.; Hansen, T.B.; Kjems, J. Insights into circular RNA biology. *RNA Biol.* **2017**, *14*, 1035–1045. [CrossRef] [PubMed]
4. Zhang, Z.; Yang, T.; Xiao, J. Circular RNAs: Promising Biomarkers for Human Diseases. *EBioMedicine* **2018**, *34*, 267–274. [CrossRef] [PubMed]
5. Dahl, M.; Daugaard, I.; Andersen, M.S.; Hansen, T.B.; Gronbaek, K.; Kjems, J.; Kristensen, L.S. Enzyme-free digital counting of endogenous circular RNA molecules in B-cell malignancies. *Lab. Invest.* **2018**, *98*, 1657–1669. [CrossRef]
6. Kristensen, L.S.; Hansen, T.B.; Veno, M.T.; Kjems, J. Circular RNAs in cancer: Opportunities and challenges in the field. *Oncogene* **2018**, *37*, 555–565. [CrossRef]
7. Dragomir, M.; Calin, G.A. Circular RNAs in Cancer—Lessons Learned From microRNAs. *Front. Oncol.* **2018**, *8*, 179. [CrossRef]
8. Floris, G.; Zhang, L.; Follesa, P.; Sun, T. Regulatory Role of Circular RNAs and Neurological Disorders. *Mol. Neurobiol.* **2017**, *54*, 5156–5165. [CrossRef]
9. Altesha, M.A.; Ni, T.; Khan, A.; Liu, K.; Zheng, X. Circular RNA in cardiovascular disease. *J. Cell Physiol.* **2018**. [CrossRef]
10. Barbagallo, D.; Condorelli, A.; Ragusa, M.; Salito, L.; Sammito, M.; Banelli, B.; Caltabiano, R.; Barbagallo, G.; Zappala, A.; Battaglia, R.; et al. Dysregulated miR-671-5p/CDR1-AS/CDR1/VSNL1 axis is involved in glioblastoma multiforme. *Oncotarget* **2016**, *7*, 4746–4759. [CrossRef]

11. Pajares, M.J.; Ezponda, T.; Catena, R.; Calvo, A.; Pio, R.; Montuenga, L.M. Alternative splicing: An emerging topic in molecular and clinical oncology. *Lancet Oncol.* **2007**, *8*, 349–357. [CrossRef]
12. Wang, B.D.; Lee, N.H. Aberrant RNA Splicing in Cancer and Drug Resistance. *Cancers (Basel)* **2018**, *10*, 458. [CrossRef] [PubMed]
13. Montes, M.; Sanford, B.L.; Comiskey, D.F.; Chandler, D.S. RNA Splicing and Disease: Animal Models to Therapies. *Trends Genet.* **2019**, *35*, 68–87. [CrossRef]
14. Ostrom, Q.T.; Gittleman, H.; Liao, P.; Vecchione-Koval, T.; Wolinsky, Y.; Kruchko, C.; Barnholtz-Sloan, J.S. CBTRUS Statistical Report: Primary brain and other central nervous system tumors diagnosed in the United States in 2010-2014. *Neuro-Oncology* **2017**, *19*, v1–v88. [CrossRef]
15. Eger, N.; Schoppe, L.; Schuster, S.; Laufs, U.; Boeckel, J.N. Circular RNA Splicing. *Adv. Exp. Med. Biol.* **2018**, *1087*, 41–52. [CrossRef]
16. Qin, M.; Wei, G.; Sun, X. Circ-UBR5: An exonic circular RNA and novel small nuclear RNA involved in RNA splicing. *Biochem. Biophys. Res. Commun.* **2018**, *503*, 1027–1034. [CrossRef]
17. Ashwal-Fluss, R.; Meyer, M.; Pamudurti, N.R.; Ivanov, A.; Bartok, O.; Hanan, M.; Evantal, N.; Memczak, S.; Rajewsky, N.; Kadener, S. circRNA biogenesis competes with pre-mRNA splicing. *Mol. Cell* **2014**, *56*, 55–66. [CrossRef] [PubMed]
18. Conn, S.J.; Pillman, K.A.; Toubia, J.; Conn, V.M.; Salmanidis, M.; Phillips, C.A.; Roslan, S.; Schreiber, A.W.; Gregory, P.A.; Goodall, G.J. The RNA binding protein quaking regulates formation of circRNAs. *Cell* **2015**, *160*, 1125–1134. [CrossRef] [PubMed]
19. Barbagallo, D.; Caponnetto, A.; Cirnigliaro, M.; Brex, D.; Barbagallo, C.; D'Angeli, F.; Morrone, A.; Caltabiano, R.; Barbagallo, G.M.; Ragusa, M.; et al. CircSMARCA5 Inhibits Migration of Glioblastoma Multiforme Cells by Regulating a Molecular Axis Involving Splicing Factors SRSF1/SRSF3/PTB. *Int. J. Mol. Sci.* **2018**, *19*, 480. [CrossRef] [PubMed]
20. Das, S.; Krainer, A.R. Emerging functions of SRSF1, splicing factor and oncoprotein, in RNA metabolism and cancer. *Mol. Cancer Res.* **2014**, *12*, 1195–1204. [CrossRef]
21. Karni, R.; de Stanchina, E.; Lowe, S.W.; Sinha, R.; Mu, D.; Krainer, A.R. The gene encoding the splicing factor SF2/ASF is a proto-oncogene. *Nat. Struct. Mol. Biol.* **2007**, *14*, 185–193. [CrossRef] [PubMed]
22. Ratnadiwakara, M.; Archer, S.K.; Dent, C.I.; Ruiz De Los Mozos, I.; Beilharz, T.H.; Knaupp, A.S.; Nefzger, C.M.; Polo, J.M.; Anko, M.L. SRSF3 promotes pluripotency through Nanog mRNA export and coordination of the pluripotency gene expression program. *Elife* **2018**, *7*. [CrossRef]
23. Madhavan, S.; Zenklusen, J.C.; Kotliarov, Y.; Sahni, H.; Fine, H.A.; Buetow, K. Rembrandt: helping personalized medicine become a reality through integrative translational research. *Mol. Cancer Res.* **2009**, *7*, 157–167. [CrossRef] [PubMed]
24. Tomczak, K.; Czerwinska, P.; Wiznerowicz, M. The Cancer Genome Atlas (TCGA): an immeasurable source of knowledge. *Contemp. Oncol. (Pozn)* **2015**, *19*, A68–A77. [CrossRef] [PubMed]
25. Van Nostrand, E.L.; Pratt, G.A.; Shishkin, A.A.; Gelboin-Burkhart, C.; Fang, M.Y.; Sundararaman, B.; Blue, S.M.; Nguyen, T.B.; Surka, C.; Elkins, K.; et al. Robust transcriptome-wide discovery of RNA-binding protein binding sites with enhanced CLIP (eCLIP). *Nat. Methods* **2016**, *13*, 508–514. [CrossRef]
26. Zang, J.; Lu, D.; Xu, A. The interaction of circRNAs and RNA binding proteins: An important part of circRNA maintenance and function. *J. Neurosci. Res.* **2018**. [CrossRef]
27. Charley, P.A.; Wilusz, J. Sponging of cellular proteins by viral RNAs. *Curr. Opin. Virol.* **2014**, *9*, 14–18. [CrossRef]
28. Barbagallo, C.; Brex, D.; Caponnetto, A.; Cirnigliaro, M.; Scalia, M.; Magnano, A.; Caltabiano, R.; Barbagallo, D.; Biondi, A.; Cappellani, A.; et al. LncRNA UCA1, Upregulated in CRC Biopsies and Downregulated in Serum Exosomes, Controls mRNA Expression by RNA-RNA Interactions. *Mol. Ther. Nucleic Acids* **2018**, *12*, 229–241. [CrossRef]
29. Kim, J.; Abdelmohsen, K.; Yang, X.; De, S.; Grammatikakis, I.; Noh, J.H.; Gorospe, M. LncRNA OIP5-AS1/cyrano sponges RNA-binding protein HuR. *Nucleic Acids Res.* **2016**, *44*, 2378–2392. [CrossRef]
30. Anczukow, O.; Akerman, M.; Clery, A.; Wu, J.; Shen, C.; Shirole, N.H.; Raimer, A.; Sun, S.; Jensen, M.A.; Hua, Y.; et al. SRSF1-Regulated Alternative Splicing in Breast Cancer. *Mol. Cell.* **2015**, *60*, 105–117. [CrossRef]
31. Martinez-Terroba, E.; Ezponda, T.; Bertolo, C.; Sainz, C.; Remirez, A.; Agorreta, J.; Garmendia, I.; Behrens, C.; Pio, R.; Wistuba, I.I.; et al. The oncogenic RNA-binding protein SRSF1 regulates LIG1 in non-small cell lung cancer. *Lab. Invest.* **2018**, *98*, 1562–1574. [CrossRef] [PubMed]

32. Amin, E.M.; Oltean, S.; Hua, J.; Gammons, M.V.; Hamdollah-Zadeh, M.; Welsh, G.I.; Cheung, M.K.; Ni, L.; Kase, S.; Rennel, E.S.; et al. WT1 mutants reveal SRPK1 to be a downstream angiogenesis target by altering VEGF splicing. *Cancer Cell* **2011**, *20*, 768–780. [CrossRef]
33. Mavrou, A.; Brakspear, K.; Hamdollah-Zadeh, M.; Damodaran, G.; Babaei-Jadidi, R.; Oxley, J.; Gillatt, D.A.; Ladomery, M.R.; Harper, S.J.; Bates, D.O.; et al. Serine-arginine protein kinase 1 (SRPK1) inhibition as a potential novel targeted therapeutic strategy in prostate cancer. *Oncogene* **2015**, *34*, 4311–4319. [CrossRef] [PubMed]
34. Ye, X.; Abou-Rayyah, Y.; Bischoff, J.; Ritchie, A.; Sebire, N.J.; Watts, P.; Churchill, A.J.; Bates, D.O. Altered ratios of pro- and anti-angiogenic VEGF-A variants and pericyte expression of DLL4 disrupt vascular maturation in infantile haemangioma. *J. Pathol.* **2016**, *239*, 139–151. [CrossRef] [PubMed]
35. Hamdollah Zadeh, M.A.; Amin, E.M.; Hoareau-Aveilla, C.; Domingo, E.; Symonds, K.E.; Ye, X.; Heesom, K.J.; Salmon, A.; D'Silva, O.; Betteridge, K.B.; et al. Alternative splicing of TIA-1 in human colon cancer regulates VEGF isoform expression, angiogenesis, tumour growth and bevacizumab resistance. *Mol. Oncol.* **2015**, *9*, 167–178. [CrossRef]
36. Gilbert, M.R.; Dignam, J.J.; Armstrong, T.S.; Wefel, J.S.; Blumenthal, D.T.; Vogelbaum, M.A.; Colman, H.; Chakravarti, A.; Pugh, S.; Won, M.; et al. A randomized trial of bevacizumab for newly diagnosed glioblastoma. *N. Engl. J. Med.* **2014**, *370*, 699–708. [CrossRef] [PubMed]
37. Holdt, L.M.; Kohlmaier, A.; Teupser, D. Circular RNAs as Therapeutic Agents and Targets. *Front. Physiol.* **2018**, *9*, 1262. [CrossRef]
38. Fukuhara, T.; Hosoya, T.; Shimizu, S.; Sumi, K.; Oshiro, T.; Yoshinaka, Y.; Suzuki, M.; Yamamoto, N.; Herzenberg, L.A.; Herzenberg, L.A.; et al. Utilization of host SR protein kinases and RNA-splicing machinery during viral replication. *Proc. Natl. Acad. Sci. USA* **2006**, *103*, 11329–11333. [CrossRef]
39. Peritz, T.; Zeng, F.; Kannanayakal, T.J.; Kilk, K.; Eiriksdottir, E.; Langel, U.; Eberwine, J. Immunoprecipitation of mRNA-protein complexes. *Nat. Protoc.* **2006**, *1*, 577–580. [CrossRef]
40. Di Pietro, C.; Vento, M.; Guglielmino, M.R.; Borzi, P.; Santonocito, M.; Ragusa, M.; Barbagallo, D.; Duro, L.R.; Majorana, A.; De Palma, A.; et al. Molecular profiling of human oocytes after vitrification strongly suggests that they are biologically comparable with freshly isolated gametes. *Fertil. Steril.* **2010**, *94*, 2804–2807. [CrossRef]
41. Livak, K.J.; Schmittgen, T.D. Analysis of relative gene expression data using real-time quantitative PCR and the 2(-Delta Delta C(T)) Method. *Methods* **2001**, *25*, 402–408. [CrossRef]
42. Urbano, F.; Filippello, A.; Di Pino, A.; Barbagallo, D.; Di Mauro, S.; Pappalardo, A.; Rabuazzo, A.M.; Purrello, M.; Purrello, F.; Piro, S. Altered expression of uncoupling protein 2 in GLP-1-producing cells after chronic high glucose exposure: Implications for the pathogenesis of diabetes mellitus. *Am. J. Physiol. Cell Physiol.* **2016**, *310*, C558–C567. [CrossRef]
43. Caltabiano, R.; Puzzo, L.; Barresi, V.; Ieni, A.; Loreto, C.; Musumeci, G.; Castrogiovanni, P.; Ragusa, M.; Foti, P.; Russo, A.; et al. ADAM 10 expression in primary uveal melanoma as prognostic factor for risk of metastasis. *Pathol. Res. Pract.* **2016**, *212*, 980–987. [CrossRef] [PubMed]
44. Weidner, N.; Semple, J.P.; Welch, W.R.; Folkman, J. Tumor angiogenesis and metastasis—Correlation in invasive breast carcinoma. *N. Engl. J. Med.* **1991**, *324*, 1–8. [CrossRef] [PubMed]
45. Mikkelsen, V.E.; Stensjoen, A.L.; Granli, U.S.; Berntsen, E.M.; Salvesen, O.; Solheim, O.; Torp, S.H. Angiogenesis and radiological tumor growth in patients with glioblastoma. *BMC Cancer* **2018**, *18*, 862. [CrossRef] [PubMed]
46. Burton, A.L.; Egger, M.E.; Gilbert, J.E.; Stromberg, A.J.; Hagendoorn, L.; Martin, R.C.; Scoggins, C.R.; McMasters, K.M.; Callender, G.G. Assessment of mitotic rate reporting in melanoma. *Am. J. Surg.* **2012**, *204*, 969–974. [CrossRef] [PubMed]

cancers

MDPI

Article

The Disruption of the β-Catenin/TCF-1/STAT3 Signaling Axis by 4-Acetylantroquinonol B Inhibits the Tumorigenesis and Cancer Stem-Cell-Like Properties of Glioblastoma Cells, In Vitro and In Vivo

Heng-Wei Liu [1,2,3,4], Yu-Kai Su [1,2,3,4], Oluwaseun Adebayo Bamodu [4,5], Dueng-Yuan Hueng [6], Wei-Hwa Lee [7], Chun-Chih Huang [8], Li Deng [9,10], Michael Hsiao [11], Ming-Hsien Chien [1], Chi-Tai Yeh [1,4,5,*] and Chien-Min Lin [1,2,3,4,*]

1 Graduate Institute of Clinical Medicine, College of Medicine, Taipei Medical University, Taipei City 11031, Taiwan; henryway0404@hotmail.com (H.-W.L.); yukai.su@gmail.com (Y.-K.S.); mhchien1976@gmail.com (M.-H.C.)

2 Department of Neurology, School of Medicine, College of Medicine, Taipei Medical University, Taipei City 11031, Taiwan

3 Division of Neurosurgery, Department of Surgery, Taipei Medical University-Shuang Ho Hospital, New Taipei City 23561, Taiwan

4 Taipei Neuroscience Institute, Taipei Medical University, Taipei 11031, Taiwan; dr_bamodu@yahoo.com

5 Department of Medical Research & Education, Taipei Medical University-Shuang Ho Hospital, New Taipei City 23561, Taiwan

6 Department of Neurological Surgery, Tri-Service General Hospital, National Defense Medical Center, Taipei 11490, Taiwan, ROC; hondy2195@yahoo.com.tw

7 Department of Pathology, Taipei Medical University-Shuang Ho Hospital, Taipei 23561, Taiwan; whlpath97616@s.tmu.edu.tw

8 Department of Applied Chemistry, Chaoyang University of Technology, Taichung 41147, Taiwan; john@newbellus.com.tw

9 Beijing Bioprocess Key Laboratory, College of Life Science and Technology, Beijing University of Chemical Technology, Beijing 100029, China; dengli@mail.buct.edu.cn

10 Amoy-BUCT Industrial Bio-technovation Institute, Amoy 361022, China

11 Genomics Research Center, Academia Sinica, Taipei 11529, Taiwan; mhsiao@gate.sinica.edu.tw

* Correspondence: ctyeh@s.tmu.edu.tw (C.-T.Y.); m513092004@s.tmu.edu.tw (C.-M.L.); Tel.: +886-2-2490088 (ext. 8881) (C.-T.Y.); +886-2-2490088 (ext. 8885) (C.-M.L.); Fax: 886-2-2248-0900 (C.-T.Y.); +886-2-2248-0900 (C.-M.L.)

Received: 3 November 2018; Accepted: 4 December 2018; Published: 5 December 2018

Abstract: Background: Glioblastoma (GBM), a malignant form of glioma, is characterized by resistance to therapy and poor prognosis. Accumulating evidence shows that the initiation, propagation, and recurrence of GBM is attributable to the presence of GBM stem cells (GBM-CSCs). Experimental approach: Herein, we investigated the effect of 4-Acetylantroquinonol B (4-AAQB), a bioactive isolate of *Antrodia cinnamomea*, on GBM cell viability, oncogenic, and CSCs-like activities. Results: We observed that aberrant expression of catenin is characteristic of GBM, compared to other glioma types ($p = 0.0001$, log-rank test = 475.2), and correlates with poor prognosis of GBM patients. Lower grade glioma and glioblastoma patients ($n = 1152$) with low catenin expression had 25% and 21.5% better overall survival than those with high catenin expression at the 5 and 10-year time-points, respectively ($p = 3.57e\text{-}11$, log-rank test = 43.8). Immunohistochemistry demonstrated that compared with adjacent non-tumor brain tissue, primary and recurrent GBM exhibited enhanced catenin expression (~10-fold, $p < 0.001$). Western blot analysis showed that 4-AAQB significantly downregulated β-catenin and dysregulated the catenin/LEF1/Stat3 signaling axis in U87MG and DBTRG-05MG cells, dose-dependently. 4-AAQB–induced downregulation of catenin positively correlated with reduced Sox2 and Oct4 nuclear expression in the cells. Furthermore, 4-AAQB markedly reduced the viability of U87MG and DBTRG-05MG cells with 48 h IC_{50} of 9.2 M and 12.5 M, respectively, effectively inhibited the nuclear catenin, limited the migration and invasion of GBM cells, with concurrent downregulation of catenin, vimentin, and slug; similarly, colony and tumorsphere formation was significantly attenuated with reduced expression of c-Myc and KLF4 proteins. Conclusions: Summarily, we show for the first time that 4-AAQB suppresses the tumor-promoting catenin/LEF1/Stat3 signaling, and inhibited CSCs-induced oncogenic activities in GBM in vitro, with in vivo validation; thus projecting 4-AAQB as a potent therapeutic agent for anti-GBM target therapy.

Keywords: glioblastoma; β-catenin; cancer stem cell; 4-AAQB; chemoresistance; prognosis; survival

1. Introduction

Glioblastoma (GBM), one of the most lethal malignancies in adults, with an average incidence of 3.2/100 000, median survival of about 15 months, and median age at diagnosis of 64 years, is the most common and highly aggressive brain tumor [1]. Histopathologically-defined by necrosis and endothelial proliferation, GBM are high-grade (World Health Organization (WHO) grade IV) gliomas which are characteristically resistant to anticancer therapy, resulting in death within 12 months of diagnosis or initiation of therapy [2]. Glioblastoma is characterized by sustained self-renew potential, enhanced tumorigenicity and invasiveness, high likelihood to relapse, and increased resistance to chemotherapy; all features associated with the presence and activities of a small tumor subpopulation called cancer stem cells (CSCs) or tumor-initiating cells (TICs) and referred to herein as GBM stem cells (GBM-SCs). Current anti-GBM therapeutic strategy consisting of concurrent chemoradiotherapy (CCRT) with the DNA alkylating agent, temozolomide (TMZ), is plagued by eventual disease relapse and only prolongs the median survival to 14.6 months [3], necessitating the discovery and/or development of new therapeutic strategy with high anti-GBM efficacy.

In their published review, Vogelstein et al. [4] analyzed several alterations in the human cancer genome and identified tumor driver or promoter genes, as well as outlined twelve associated signaling pathways including Notch, Hedgehog (Hh), and Wnt/β-catenin, known to regulate the determination of cell fate, survival, proliferation, and genome stability. Though infrequently mutated in GBM, these pathways are often aberrantly activated and enhance the CSC-like phenotypes of GBM cells [5]. While our understanding of the molecular mechanism underlying the pathogenesis of GBM is incomplete and continues to evolve, as suggested above, there is increasing evidence that GBM-SCs play an important role in gliomagenesis, tumor maintenance and subsequent disease progression [5], and that Wnt/β-catenin signaling is implicated in the modulation of GBM-SCs [6]. Consistent with this, in a recent study, FoxM1 was shown to enhance the nuclear localization of β-catenin, regulate the expression of target genes of Wnt and gliomagenesis; similarly, close correlation was demonstrated between Wnt/β-catenin signaling, GBM progression, and patient prognosis [7], as well as cancer stemness, self-renewal, and resistance to therapy [8,9], thus, reinforcing the notion that the targeting of Wnt/β-catenin-mediated oncogenicity, self-renewal, and therapy-resistance may represent a critically efficacious anti-GBM treatment strategy.

The principal factor in the canonical Wnt signaling pathway is β-catenin. Nuclear β-catenin facilitates activation of T-cell factor/lymphoid enhancing factor (TCF/LEF) transcription factors, consequently modulating tumor formation, cell-cycle progression, cell survival, and stem-cell like activities [5], thus, making β-catenin a potential therapeutic target in anti-GBM therapy.

In the present study, understanding that the aberrant activation of the Wnt signaling is implicated in gliomagenesis, mediates the maintenance of GBM-SCs and enhances GBM invasive potential [6], as well as building on the demonstrated anticancer therapeutic efficacy of the relatively novel phytochemical 4-acetylantroquinonol B (4-AAQB), a derivative of mono-acetylated antroquinonol, which is a bioactive extract of *Antrodia camphorata* [10,11], we hypothesized and investigated if and how 4-AAQB through the downregulation of β-catenin expression and/or activity, can inhibit the activation of β-catenin-modulated genes in GBM cell lines, U87MG and DBTRG-05MG. The plausibility of this hypothesis is rooted in our previous demonstration of the ability of 4-AAQB to suppress autophagic flux and improves cisplatin sensitivity in highly aggressive epithelial cancer through the PI3K/Akt/mTOR/p70S6K signaling pathway [11], coupled with our understanding that the inhibition of autophagy by mTORC1, a complex of mTOR, rescues disheveled (Dvl), which is key component of Wnt signaling, and thus lead to the activation of Wnt/β-catenin pathway [12]. In addition, we examined the role of 4-AAQB in modulating the responsiveness of GBM cells to anticancer therapy. Thus, we present a novel anti-GBM therapeutic approach by inhibiting the oncogenic Wnt signaling through direct β-catenin targeting by the 4-AAQB.

2. Materials and Methods

2.1. Drugs and Chemicals

4-acetylantroquinonol B (4-AAQB; >99% HPLC purity) was purchased from New Bellus Enterprises Co., Ltd (Tainan, Taiwan). Stock solution of 1 mM dissolved in dimethyl sulfoxide (DMSO; Sigma–Aldrich Co., St. Louis, MO, USA) was stored at −20 °C then further diluted in sterile culture medium immediately prior to use. Gibco® RPMI-1640, fetal bovine serum (FBS), Trypsin/EDTA, DMSO, phosphate buffered saline (PBS), sulforhodamine B (SRB) medium, Acetic acid, and TRIS base were also purchased from Sigma-Aldrich Co. (St. Louis, MO, USA).

2.2. Cell Lines and Culture

The human GBM cell lines U87MG and DBTRG-05MG were obtained from American Type Culture Collection (ATCC, Manassas, VA, USA). Cells were cultured in RPMI-1640, supplemented with 10% FBS and 1% penicillin/streptomycin (Invitrogen, Life Technologies, Carlsbad, CA, USA) and incubated in 5% humidified CO_2 incubator at 37 °C. The cells were passaged at 95% confluence or culture medium changed every 72 h. For drug cytotoxicity assays, the cells were treated with different concentrations of 4-AAQB for different duration.

2.3. Sulforhodamine B (SRB) Cell Viability Assay

U87MG or DBTRG-05MG cells were seeded in supplemented media at a density of 3.5×10^3 cells/well in triplicates in 96-well plates. After 24 h incubation, cells were treated with different concentrations of 4-AAQB. After 24 h or 48 h of treatment, the treated cells were washed in PBS, fixed with 10% trichloroacetic acid (TCA) for 1 h, washed with distilled water, and the viable cells incubated in 0.4% SRB (w/v) in 1% acetic acid at room temperature for 1 h. The unbound dye was removed by 1% acetic acid washing thrice and the plates air-dried. Attached dye was dissolved in 10mM trizma base, and absorbance was read in a microplate reader at a wavelength of 570 nm.

2.4. Western Blot Analysis

Ten-μg protein samples were run in 10% SDS-PAGE gel and transferred onto a polyvinylidene fluoride (PVDF) membrane using the Bio-Rad Mini-Protean system (Bio-Rad Laboratories, Inc, Hercules, CA, USA). Non-specific binding was blocked by incubating the membranes in 5% skimmed milk in tris-buffered saline with Tween 20 (TBST) for 1 h and then overnight at 4 °C with the antibodies against total β-catenin (1:1000, Cell Signaling Technology, Danvers, MA, USA), free β-catenin (1:1000, Cell Signaling Technology), p-GSK-3β (1:1000, Cell Signaling Technology), GSK-3β (1: 1000, Cell Signaling Technology), TCF1/TCF7 (1:1000, Cell Signaling Technology), LEF1 (1:1000, Cell Signaling Technology), p-Stat3 (1:1000, Santa Cruz, California, USA), Stat3 (1:1000, Santa Cruz), KLF4 (1:1000, Santa Cruz), c-Myc (1:1000, Santa Cruz), vimentin (1:1000, Cell Signaling Technology), Slug (1:1000, Cell Signaling Technology), and β-actin (1:500, Santa Cruz). After overnight probing with primary antibody, the membranes were incubated with horseradish peroxidise (HRP)-linked secondary antibodies for 1 h, then washed with PBS thrice. The protein band signals were detected and developed using an enhanced chemiluminescence (ECL) detection system (Thermo Fisher Scientific Inc, Waltham, MA, USA). Protein bands were quantified using ImageJ software.

2.5. *β-catenin siRNA Infection*

For β-catenin loss of function assays, U87MG or DBTRG-05MG cells were infected with stealth siRNAs targeting CTNNB1 (HSS102461; Thermo Fisher Scientific Inc., Lai Fu Life Science and Technology Co., Ltd. Taipei, Taiwan) according to the manufacturers' protocol. Briefly, the cells at 70–75% confluence per well in 6-well plates were infected with 30 nmol/L siRNA or control/mock siRNA and 3 μL RNAiMAX reagent (Thermo Fisher Scientific Inc., Lai Fu Life Science and Technology Co., Ltd. Taipei, Taiwan). The infection efficiency was confirmed using microscopy and Western blot assays before being used in this study.

2.6. *Immunohistochemical Staining and Quantification*

Paraffin-embedded primary or recurrent GBM tissue sections were used for examination of β-catenin expression, as well as hematoxylin and eosin (H&E) staining. For Immunohistochemistry (IHC) assay, sections were incubated with primary anti- β-catenin (Cell Signaling Technology 1:200 dilutions) at 4 °C overnight, followed by a biotin-labeled secondary antibody (1:100 dilutions) at room temperature for 1 h. Sections were incubated in ABC-peroxidase and diaminobenzidine (DAB), counterstained with hematoxylin and visualized using light microscopy. For Immunofluorescence staining and analysis, the cells were plated in 6-well chamber slides (Nunc, Thermo Fisher Scientific, Taipei, Taiwan) at 4 °C overnight, fixed in 2% paraformaldehyde at room temperature for 10 min, then permeabilized with 0.1% Triton X-100 in 0.01 M phosphate-buffered saline (PBS), pH 7.4 containing 0.2% bovine serum albumin (BSA). Thereafter, they were air-dried and rehydrated in PBS. Followed by incubating the cells with antibody against β-catenin (D10A8; XP Rabbit mAb #8480; Cell Signaling Technology), Sox2 (D6D9; XP Rabbit mAb #3579; Cell Signaling Technology), Oct4 (C30A3; Rabbit mAb #2840; Cell Signaling Technology), or F-actin (13E5; Rabbit mAb #4970; Cell Signaling Technology) diluted 1:500, in PBS containing 3% normal goat serum at room temperature for 2 h. For negative controls we omitted the primary antibody. After PBS washing twice for 10 min each, anti-rabbit IgG fluorescein isothiocyanate-conjugated secondary antibody (Jackson Immunoresearch Lab. Inc., West Grove, PA, USA) diluted 1:500 in PBS was added, and the cells incubated at room temperature for 1 h. This was followed by PBS washing and cell mounting using the Vectashield mounting medium and 4′, 6′-diamidino-2-phenylindole (DAPI) to counter stain DNA for nucleus visualization. Cells were observed under a Zeiss Axiophot (Carl Zeiss, Strasse, Oberkochen Germany) fluorescence microscope. Immunohistochemistry in the evaluation was performed by 2 independent pathologists who were blinded to the study. Immunohistochemistry in the evaluation was performed by 2 independent pathologists who were blinded to the study. The staining results were classified into 3 categories based on β-catenin subcellular localization, namely nuclear, cytoplasmic or membranous. All slides were assessed under microscope and initial score representing the estimated proportion (0: $\leq$ 5%, 1: 5%–25%, 2: 25%–75%, and 3: $\geq$ 75%) of positively stained (P) cancer cells, while the intensity (I) scores 1, 2, or 3 were assigned to weak, moderate, and strong stainings, respectively. Slides with indeterminate evaluation were re-assessed until consensus was reached. The Quick score (Q) was calculated by multiplying the percentage of positive cells (P) by the intensity (I) based on the formula below:

$$Q = P \times I;\ maximum = 300$$

2.7. *Tumorsphere Formation Assay*

For tumorsphere generation, U87MG and DBTRG-05MG cells were cultured in HEScGRO^{TM} serum-free medium for human embryonic stem cell culture (Chemicon, SCM020; Merck KGaA,

Darmstadt, Germany), supplemented with 10 ng/mL hbFGF (Invitrogen, Carlsbad, CA), 20 ng/mL hEGF (Millipore, Bedford, MA), B27 supplement (Invitrogen, Carlsbad, CA), heparin (#07980; STEMCELL Technologies Inc., Interlab Co., Ltd, Taipei, Taiwan), and NeuroCult^{TM} NS-A proliferation supplement (Human; #05753; STEMCELL Technologies Inc. Interlab Co., Ltd, Taipei, Taiwan). Cells were seeded at 1000 cells/mL/well in 6-well ultra-low adhesion plates (Corning Inc., Corning, NY, USA) and cultured for 7–10 days. The non-adherent tumorspheres ($\geq$ 90 m in diameter) were viewed, counted, and photographed under inverted phase contrast microscope.

2.8. Colony Formation Assay

Two $\times$ 10^4 U87MG or DBTRG-05MG cells were seeded per well in triplicate in 6-well culture plates with or without indicated concentration of 4-AAQB and incubated at 37 °C in a 5% CO_2 atmosphere incubator for 12–14 days. Generated colonies with diameter $\geq$ 100 μm and consisting of $\geq$50 cells, were washed two times with PBS, fixed with methanol for 15 min, and stained with 0.005% crystal violet for 15 min at room temperature for visualization and counting of colonies. The number and size of colonies formed were estimated with the ChemiDoc-XRS imager (QuantityOne software package; Bio-Rad, Hercules, CA, USA).

2.9. Invasion Assay

For the invasion assay, 24-well plate Transwell systems were used; 3 $\times$ 10^4 U87MG or DBTRG-05MG cells were seeded into the upper chamber of the insert (BD Bioscience, 8 μm pore size) containing media without serum, while media containing 10% FBS in the lower chamber was used as chemo-attractant. Media was discarded after incubating for 24 h, and the GBM cells on the filter membrane were fixed for 1 h with 3.7% formaldehyde, before staining with crystal violet dye, and cells lying on the upper surface of the insert were cleared off using a cotton swab. Visualization of the migrated cells and evaluation of their migratory capacity based on the total number of cells on the lower surface of the filter membrane was performed under microscope.

2.10. Wound-healing Migration Assay

The GBM cell lines, U87MG or DBTRG-05MG, were seeded in 6-well plates and incubated in 5% CO_2 atmosphere incubator for 48 h until they were 100% confluent. Scratch wounds of similar sizes were then made along the median axis of the adherent cells using sterile 200 μL micropipette tips. After PBS washing to remove detached cells, adherent cells were incubated in new culture medium in 5% CO_2 humidified incubator at 37 °C to allow wound to close. Healing/closure of wound-gap was monitored and photographed at the indicated time points.

2.11. Animal Studies

Six–8-weeks old female Non-Obese Diabetic (NOD)/Severe-Combined Immunodeficient Mice (SCID) mice (n = 15) obtained from BioLASCO Taiwan Co., Ltd (Taipei, Taiwan) were bred under standard experimental animals specific-pathogen-free conditions. The mice (5/treatment group) were subcutaneously inoculated on the right flank with 0.5 $\times$ 10^6 U87MG cells in 0.5 ml PBS. Treatment was started on day 7–10 when tumors reached an average size of $\geq$150 mm^3. Treatment group 1 consisted of thrice weekly intraperitoneal (i.p.) injections of 5 mg/kg 4-AAQB in 0.5 ml PBS for 4 weeks; Treatment group 2 consisted of thrice weekly oral (p.o.) gavage of 5 mg/kg 4-AAQB in 0.5 ml PBS for 4 weeks; while the control group were injected with PBS. Tumor growth was measured two times per week using calipers, and tumor volume (v) calculated using the formula: $v = (\text{width})^2 \times \text{length}/2$. Animals were followed for another 3 weeks after the final 4-AAQB treatment (i.e., 7 weeks after tumor inoculation), then the treated and control with extremely large tumors were humanely sacrificed. Mice with small tumors were allowed for an additional 8–12 weeks. Assessment of tumor growth for each of the studies was carried out and statistical analyses was determined by Student's t-test using Sigma plot v13 (Stystat Software, CA, U.S.A.). $p < 0.05$ was considered statistically significant. All the experimental animal procedures were approved by and performed in accordance to the Institutional Animal Care and Use Committee/Panel (IACUC/P) protocol approval LAC-2015-0386.

2.12. Statistical Analysis

Each experiment was performed at least 4 times in triplicates. Shown data represent means ± SD. Comparison between two groups was estimated using the 2-sided Student's *t*-test, while the one-way analysis of variance (ANOVA) was used for comparison between 3 or more groups. p-value < 0.05 was considered statistically significant.

3. Results

3.1. Aberrant Expression of β-Catenin Is Characteristic of GBM and Correlates with Poor Prognosis

To determine the clinical relevance of the Wnt/β-catenin signaling in GBM, we accessed and analyzed the cancer genome atlas (TCGA) lower grade glioma and glioblastoma cohort (GBMLGG, n = 1152) dataset. We observed that of all four histological sub-types of glioma brain tumor in the GBMLGG cohort, namely astrocytoma (n = 196), oligoastrocytoma (n = 134), oligodendroglioma (n = 195), and GBM (n = 604), GBM in its various forms exhibit the poorest survival rates (p = 0.001, log-rank test = 475.2), with a 100% mortality rate by 4.1, 8.2, and 10 year post-diagnosis for GBM (n = 30), treated primary GBM (n =20), and untreated primary (de novo) GBM (n = 554), respectively (Figure 1A). Furthermore, our analyses revealed the histological subtypes of glioma with the poorest overall survival were characterized by the most aberrantly expressed β-catenin gene expression as demonstrated by a median β-catenin expression of 13.125 in patients with GBM, treated primary GBM and untreated primary (de novo) GBM, compared to 12.80, 12.69, and 12.70 in those with astrocytoma, oligoastrocytoma and oligodendroglioma, respectively (Figure 1B). This is consistent with the high grade, highly malignant, and aggressive character of GBM, compared to the lower grade gliomas. In parallel analyses, we observed that 23.75% and 18.75% of GBM patients with low expression of β-catenin (n = 301) were more likely to be alive at the 5- and 10-yr time-point post-diagnosis, respectively, compared to patients with high β-catenin expression (n = 303) (p = 3.57e-11; log-rank test = 43.8; Figure 1C). These data not only indicate a functional relationship between patients' survival and the expression levels of β-catenin, but also points to a probable causative or at least participatory role for β-catenin overexpression in the poor prognosis of GBM patients (also see Figure S1).

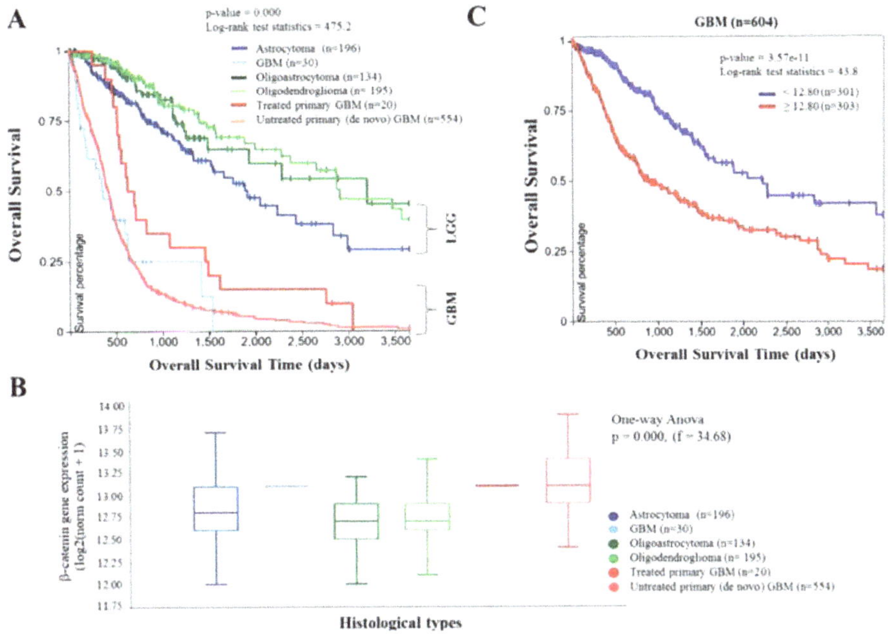

Figure 1. Aberrant expression of β-catenin is characteristic of glioblastoma (GBM) and correlates with poor prognosis. (**A**) Kaplan–Meier plot showing the overall survival of The Cancer Genome Atlas (TCGA) lower grade glioma and glioblastoma cohort (Glioma (GBMLGG), n = 1152) according to their histological type. (**B**) β-catenin expression across the different histological type of glioma and glioblastoma in the TCGA GBMLGG cohort. (**C**) Kaplan–Meier plot based on the β-catenin expression level in only the glioblastoma component of the TCGA GBMLGG dataset using the gene expression RNAseq-illuminaHiSeq. Expression cutoff is based on median expression. LGG, low grade glioma; GBM, glioblastoma.

3.2. β-Catenin Facilitates GBM Oncogenicity and Disease Recurrence, as well as Their Cancer Stem Cell-Like Traits

To better appreciate the oncogenic role of β-catenin in GBM and investigate its likely causative or participatory role in gliomagenesis and disease course, we accessed and analyzed the TCGA GBM dataset (n = 538), for differential β-catenin expression profile. Using the Human Genome U133A Array, compared to the normal brain tissues (n = 10), β-catenin (reporter ID 201533_at), was overexpressed in the GBM tissues (n = 528; p = 9.4e-6), with a Tukey's honest significant difference (HSD) of 0.47 and 95% confidence interval of 0.26–0.68 (Figure 2A). This data does suggest an initiating or enhancing role for β-catenin in GBM. In confirmatory assays using clinical samples from our GBM cohort, we evaluated and compared the expression of β-catenin in primary GBM, recurrent GBM and non-tumor ("normal") brain tissue. Our immunohistochemical staining results showed enhanced β-catenin expression in the primary (8-fold, $p < 0.001$) and recurrent GBM (8.1-fold, $p < 0.001$) tissues, compared to the non-tumor tissues (Figure 2B). To further provide insight into the mechanistic underlining of GBM oncogenicity and recurrence, using the SE20736/GPL570/GDS4535 (HG-U133_Plus_2) Affymetrix Human Genome U133 Plus 2.0 Array dataset, we examined the expression profile of β-catenin (1554411_at) in cultures of progenitor (GBM-SCs) and four-day differentiated GBM cells derived from surgical specimens; our data show that compared to the differentiated GBM cells in samples GSM520521, GSM520523, and GSM520525, the cumulative β-catenin expression in the GBM-SCs samples GSM520522, GSM520524, and GSM520526 was significantly higher (Figure 2C). Specifically, as shown in Figure 2C, while the differential β-catenin expression was equivocal between GSM520521 and GSM520522, a 30% and 32.25% enhancement in β-catenin expression was observed in GSM520524 and GSM520526 compared to GSM520523 and GSM520525, respectively. These data indicate the probable involvement of β-catenin expression in facilitating GBM oncogenicity and disease recurrence, as well as their cancer stem cell-like traits.

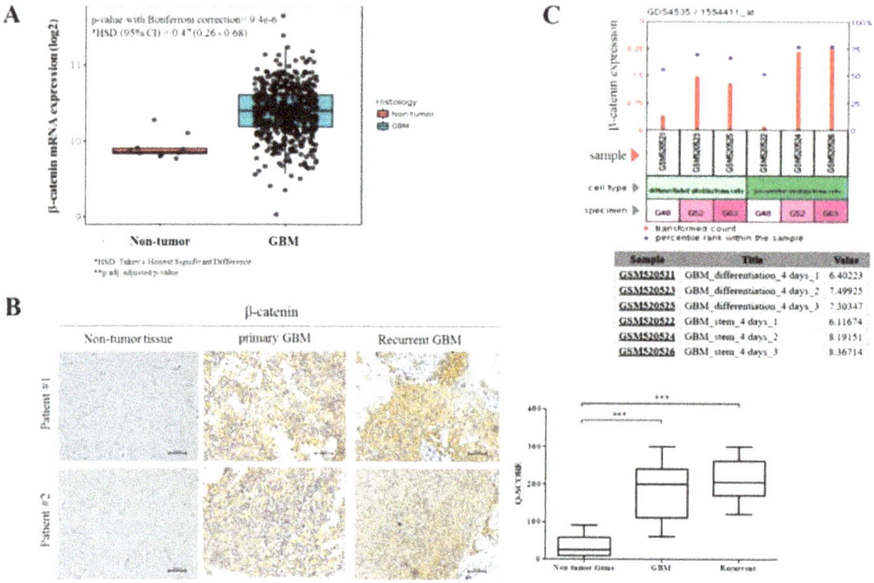

Figure 2. β-catenin facilitates GBM oncogenicity and disease recurrence, as well as their cancer stem cell-like traits. (**A**) TCGA brain dataset (n = 557) using the Human Genome U133A Array showing the differential β-catenin mRNA expression in GBM (n = 547) and normal brain (n = 10) samples. (**B**) Immunohistochemistry (IHC) staining comparing expression of β-catenin in GBM, recurrent and adjacent non-tumor brain tissues in 2 Taipei Medical University Shuang Ho Hospital (TMU-SHH) brain tumor patients. * $p < 0.05$, ** $p < 0.01$, *** $p < 0.001$; GBM, glioblastoma. Scale bar: 50 μm (**C**) GEO dataset GSE20736/GPL570/GDS4535 shows time-dependent enhancement of β-catenin expression levels in human progenitor GBM cells compared to their differentiated counterparts.

3.3. 4-AAQB Disrupts the CSC-Associated Oncogenic β-Catenin/TCF-1/Stat3 Signaling Axis in GBM Cells

Having demonstrated the efficacy of 4-AAQB alone or in combination with cisplatin in highly aggressive epithelial cancers, in vitro and in vivo, in earlier work [11], we thus investigated the therapeutic effect and probable β-catenin—mediated efficacy of 4-AAQB (Figure 3A) in the human GBM cell lines—U87MG and DBTRG05MG. Using the Western blot assay, we demonstrated that similar to the inhibitory effect of the transient silencing of β-catenin using short interfering RNA (siRNA) on the protein expression of β-catenin, its upstream modulator p-GSK-3, and downstream effectors TCF1/TCF7 and LEF1 in U87MG cells (Figure 3B), treatment of U87MG and DBTRG05MG cells with 5 and 10 μM 4-AAQB significantly and dose-dependently downregulated the expression of β-catenin, p-GSK-3β, TCF1/TCF7 and LEF1, as well as p-Stat3 (Figure 3C). These data not only corroborate demonstrated convergence of the canonical Wnt/β-catenin and Stat3 signaling pathways [13–15], but also is indicative of the therapeutic efficacy of 4-AAQB in GBM cells through disruption of the CSC-associated oncogenic β-catenin/TCF-1/Stat3 signaling axis.

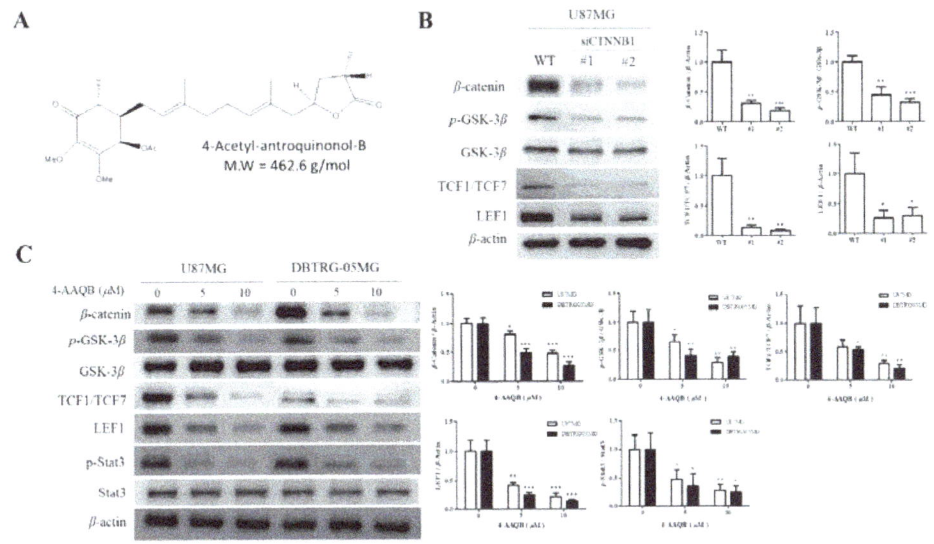

Figure 3. 4-AAQB disrupts β-catenin/TCF-1/Stat3 signaling axis in GBM cells. (**A**) Chemical structure of 4-Acetylantroquinonol B; C26H38O7, 462.58 g/mol. Representative Western blot data and graphical quantification showing (**B**) the knock-down efficiency of β-catenin in human GBM cell line U87MG, with corresponding effect on the expression levels of pGSK3β, GSK-3β, TCF1/TCF7, and LEF1 proteins; and (**C**) the dose-dependent downregulation of β-catenin, pGSK3β, TCF1/TCF7, LEF1, and p-Stat3 proteins in U87MG and DBTRG05MG cells treated with increasing concentrations of 4-AAQB for 24 h. β-actin served as loading control. siCTNNB1, short interfering RNA directed specifically against β-catenin; WT, wild type; * $p < 0.05$, ** $p < 0.01$, *** $p < 0.001$. All data is representative of experiments performed 4 times in triplicates.

3.4. 4-AAQB Inhibits the Nuclear Localization of β-Catenin, Sox2, and Oct4 in GBM Cells

Since there is evidence that the key stemness genes *Nanog*, *Oct4*, and *Sox2* are directly or indirectly regulated in a context-specific and TCF1/TCF3-involved manner by *β-catenin* [16], and that the blocking of β-catenin nuclear localization with small-molecule inhibitors significantly enhances reprogramming efficiency of stem cells [17], we then evaluated the effect of 4-AAQB on the nuclear localization of β-catenin, Sox2, and Oct4 using the dual-color immunofluorescent staining. Our results demonstrated that 48 h exposure to 5–10 μM 4-AAQB significantly reduced the nuclear expression and co-localization of Sox2 and Oct4 in U87MG (Figure 4A) and DBTRG05MG (Figure 4B) cells, and this correlates with the concurrent decrease in β-catenin and F-actin expression after treatment with 10 μM 4-AAQB (Figure 4C). These data are corroboratory of previous results and demonstrate the therapeutic efficacy of 4-AAQB in GBM cells through disruption of F-actin-mediated essential cellular bioactivities, pluripotency, and CSCs-associated oncogenic β-catenin signaling.

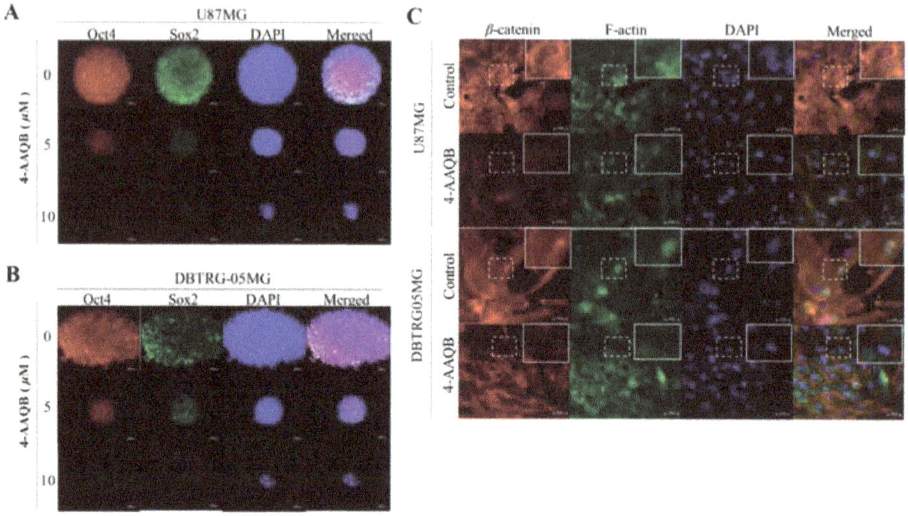

Figure 4. 4-AAQB inhibits the nuclear localization of β-catenin, Sox2, and Oct4 in GBM cells. U87MG and DBTRG-05MG cells treated with or without 5–10 μM 4-AAQB were immunostained with Sox2 (green), Oct4 (red), β-catenin (red), and F-actin (green) antibodies, then image visualization and analysis carried out by fluorescence microscopy. Treatment with 5 or 10 μM 4-AAQB markedly decreased nuclear Sox2 and Oct4 protein expression in (**A**) U87MG or (**B**) DBTRG-05MG cells. (**C**) Immunohistochemistry show reduced nuclear expression of β-catenin and F-actin after treatment with 10 μM 4-AAQB. Original magnification ×200. DAPI (blue) served as nuclear marker.

3.5. 4-AAQB Significantly Suppresses the Viability and Oncogenicity of GBM Cells

To further understand the therapeutic relevance of 4-AAQB in GBM, we examined the effect and therapeutic efficacy of 2.5–15 M 4-AAQB in U87MG and DBTRG05MG cells. Our drug cytotoxicity assay results demonstrate that the viability of U87MG and DBTRG05MG cells treated with 4-AAQB was markedly reduced in a dose- and time-dependent manner, compared to the untreated control cells (Figure 5A). Since oncogenic signaling and/or oncogenic transformation of cancer cells is associated with and underlies the acquisition of migratory, invasive, and metastatic phenotypes; we evaluated the effect of 4-AAQB on the migration of GBM cells using the scratch wound healing assay. Confluent U87MG or DBTRG05MG monolayer adherent cells were scratched along the median axis and 4-AAQB- or vehicle-containing culture medium was added, and wound closure monitored over 24 h. The untreated control U87MG or DBTRG05MG cells migrated significantly faster into the scratched area, compared to the cells treated with 5 M or 10 M 4-AAQB (Figure 5B). Using the transwell invasion assay, we demonstrated that when treated with 5 M or 10 M 4-AAQB, the GBM cells were less invasive than their untreated counterparts, with a 52% ($p < 0.01$) or 70% ($p < 0.001$) reduction in number of 5 M or 10 M 4-AAQB-treated invaded U87MG cells, respectively, compared to the untreated control cells. The number of 5 M or 10 M 4-AAQB-treated invaded DBTRG05MG cells was reduced by 35% ($p < 0.01$) or 48% ($p < 0.001$), respectively, in comparison to their untreated counterpart (Figure 5C). To characterize the underlying mechanism for the observed reduction in cell viability, migration and invasion, we evaluated the expression of a selected panel of oncogenic proteins and observed that 5 μM or 10 μM 4-AAQB significantly downregulated the expression levels of β-catenin, vimentin, and slug proteins, in a dose-dependent manner (Figure 5D). These results are indicative of 4-AAQB ability to inhibit GBM cell viability and/or proliferation, migration, invasion, and consequently, deter metastasis in GBM cells.

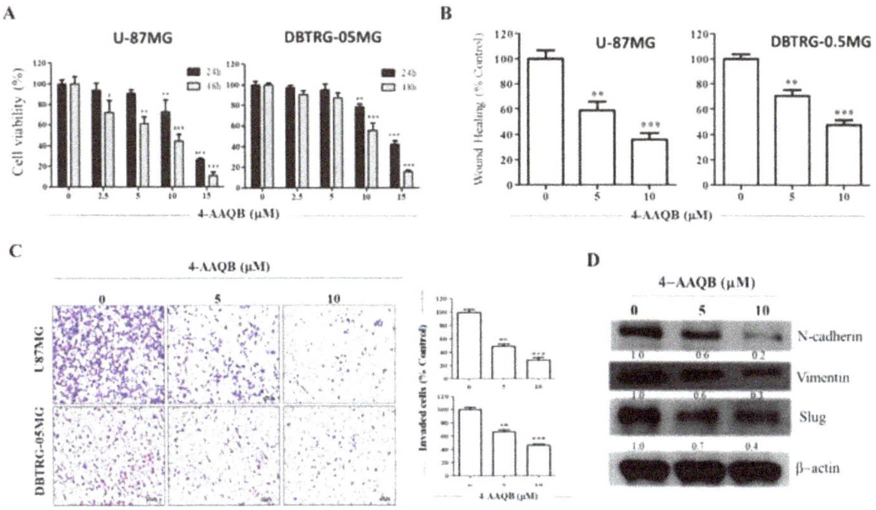

Figure 5. 4-AAQB significantly suppresses the viability and oncogenicity of GBM cells. (**A**) The dose-dependent cytotoxic effect of 4-AAQB on the cell viability of U-87MG and DBTRG-05MG. (**B**) Representative photo-images and graphical quantification showing the effect of 5 μM and 10 μM of 4-AAQB on the migration of U-87MG and DBTRG-05MG over 24 h duration. (**C**) 4-AAQB dose-dependently inhibit the number of invaded U-87MG and DBTRG-05MG cells. Scale bar: 50 μm. (**D**) The inhibitory effect of 5 M and 10 M 4-AAQB on the expression levels of β-catenin, vimentin, and slug proteins in DBTRG-05MG cells as shown by Western blot assay. β-actin served as loading control. * $p < 0.05$, ** $p < 0.01$, *** $p < 0.001$; All data is representative of experiments performed 4 times in triplicates.

3.6. 4-AAQB Markedly Inhibits the Stem Cell-Like Phenotype of U87MG and DBTRG-05MG Cells by Modulation of β-Catenin Expression

Since undifferentiated GBM-SCs are highly proliferative, invasive, and drug-resistant cells with enhanced colony- and tumorsphere-formation efficiency [18–20], compared to the remaining cancer cell population, and are implicated in tumor formation, reduced treatment response, and recurrence [20], we evaluated the effect of 4-AAQB on these cancer stem-cell-like phenotypes in human GBM cell lines, U87MG and DBTRG05MG. Results of our anchorage-independent tumorsphere formation assay demonstrated that treatment with 5 M or 10 M 4-AAQB significantly reduced the tumorsphere-forming capacity of U87MG and DBTRG05MG cells (79–96% reduction, $p < 0.001$) (Figure 6A,B). More interestingly, consistent with our portended self-renewal-limiting potential of 4-AAQB, U87MG- or DBTRG-05MG—derived tumorspheres pre-treated with 5 M 4-AAQB significantly lost their ability to form secondary or tertiary generation of tumorspheres, qualitatively and quantitatively ($p < 0.001$) (Figure 6A,B). Similarly, using the colony formation assay, we demonstrated that the treatment of GBM cells with 4-AAQB inhibits the formation of GBM colonies in a dose-dependent manner, as we observed a 39% ($p < 0.01$) and 80% ($p < 0.001$) reduction in number of colonies formed by U87MG cells treated with 5 or 10 M 4-AAQB, respectively, compared to the untreated cells. For the treated DBTRG05MG cells, a 58% ($p < 0.01$) and 64% ($p < 0.001$) inhibition in capacity to form colonies was noted after treatment with 5 or 10 μM 4-AAQB, respectively, compared to their untreated counterparts (Figure 6C,D). These findings were corroborated by Western blot data showing marked downregulation in the expression levels of β-catenin, and the stem cell markers, Oct4, Sox2, c-Myc and KLF4 proteins in a dose-dependent manner when treated with 5 and 10 M 4-AAQB (Figure 6E). These results do demonstrate that 4-AAQB effectively inhibits the stem cell-like phenotype and attenuates the self-renewal capacity of GBM cells and is consistent with the significant downregulation in expression or marked decrease in nuclear localization of β-catenin, and the stemness marker, Sox2 and Oct4 demonstrated earlier in Figures 3 and 4.

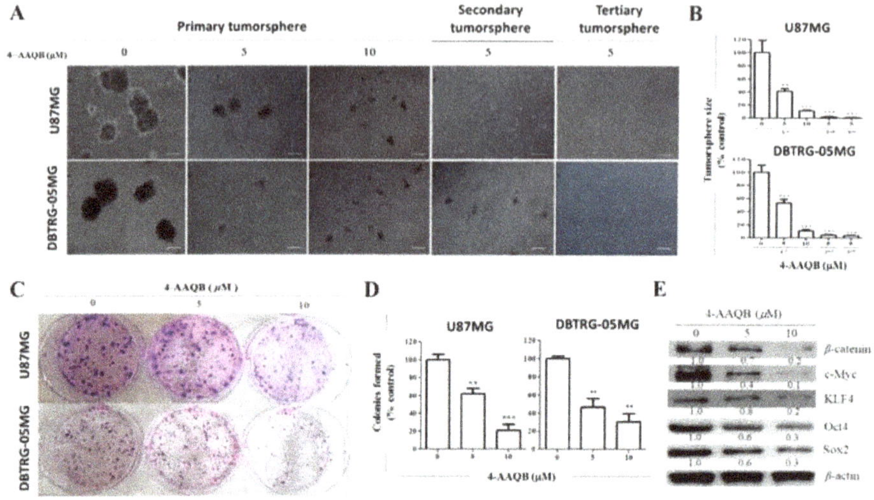

Figure 6. 4-AAQB markedly inhibits the stem cell-like phenotype of U87MG and DBTRG-05MG cells. (**A**) Representative photo-images and (**B**) graphical quantification showing 4-AAQB significantly reduce the size and number of U87MG and DBTRG-05MG primary, secondary, and tertiary generation tumorspheres formed. Original magnification = 200×. (**C**) Representative photo-images and (**D**) graphical representations of 4-AAQB effect on the ability of U87MG and DBTRG-05MG cells to form colonies; (**E**) The dose-dependent inhibitory effect of 5–10 M 4-AAQB treatment on the expression levels of β-catenin, c-Myc, KLF4, Oct4, and Sox2 proteins in DBTRG05MG cells shown in a representative Western blot data. β-actin served as loading control. * $p < 0.05$, ** $p < 0.01$, *** $p < 0.001$; All data is representative of experiments performed 4 times in triplicates.

3.7. Compared to Oral Gavage, Intraperitoneal 4-AAQB Significantly and More Efficiently Suppresses GBM Stem Cell-Induced Tumor Growth, in Vivo

Considering the enhanced tumor-initiating capacity of undifferentiated GBM-SCs and their implication in poor treatment response and tumor recurrence [18–20], we evaluated the effect of 4-AAQB administered through different routes, namely intraperitoneally (i.p) or orally (p.o) on these CSC-like phenotypes of GBM cells using U87MG single cell solution dissociated from formed tumorspheres, with the aim of validating our in vitro results in vivo. Subcutaneous injection of U87MG cells resulted in tumor formation in all 15 nude mice, which were thereafter treated with PBS, 4-AAQB (i.p), or 4-AAQB (p.o) from day 7 after tumor cell inoculation. No animals died during the treatment. We observed no dyscrasia nor significant difference in body weight between the 3 treatment groups before and after the experiments, neither was any metastases found in organs of the thoracic or abdominal cavity. Tumor growth curves showed that the average size of tumors in the 4-AAQB (i.p) group was significantly smaller compared to the 4-AAQB (p.o) (163.6 ± 78.4 mm^3 vs. 196.4 ± 108.9 mm^3, $p = 0.0095$) or PBS-treated control groups (163.6 ± 78.4 mm^3 vs. 307.0 ± 231.4 mm^3, $p = 0.0001$) (Figure 7A,B). Similarly, our immunoflourescence staining showed that compared with data from mice subjected to oral gavage of 4-AAQB (p.o), intraperitoneal injection of 4-AAQB (i.p) significantly decreased the nuclear and cytomenbranous expression of both TCF-1 and β-catenin, as well as inhibited their nuclear co-localization in the xenograft-derived GBM primary culture (Figure 7C). These findings indicate that 4-AAQB treatment, especially the intraperitoneal administration, inhibits tumor growth effectively.

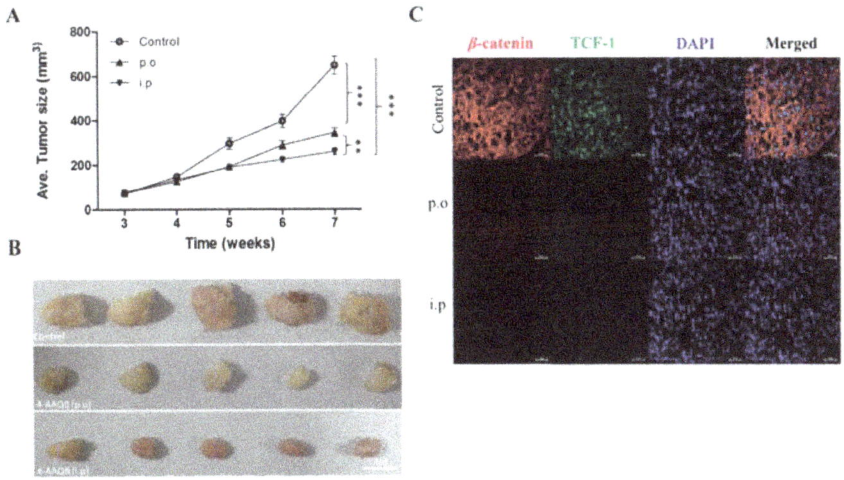

Figure 7. Compared to oral gavage, intraperitoneal 4-AAQB significantly and more efficiently suppresses GBM stem cell-induced tumor growth, in vivo. (**A**) Tumor size vs time curve show the inhibitory effect of 4-AAQB on U87MG tumor growth either via p.o or i.p route, as compared to the control group. (**B**) Photographs of tumor samples harvested from the in vivo studies. (**C**) The differential effects of oral and intraperitoneal administration of 4-AAQB on the expression and localization of TCF-1 and β-catenin in xenograft-derived GBM primary culture. * $p < 0.05$, ** $p < 0.01$, *** $p < 0.001$; 4′, 6′-diamidino-2-phenylindole (DAPI) served as nuclear marker.

4. Discussion

β-catenin is an essential transcription co-activator of the Wnt/β-catenin signaling pathway. Inappropriate activation and/or expression of the β-catenin induces aberrant activation of the canonical Wnt pathway in various cancer types including brain tumors and are in fact implicated in the enhanced proliferation and evasion of cell death of GBM cells [21], and poor prognosis in GBM patients [22]. While it has been suggested that β-catenin expression is a prognostic marker in GBM, for the first time, in a comparative analysis of different histological types of glioma, we showed that the aberrant expression of β-catenin is most characteristic of GBM cells and correlates with shorter overall and relapse-free survival (Figure 1). These findings are interesting in the context of our data showing overexpression of β-catenin in primary and recurrent GBM relative to normal brain tissues (Figure 2), which is corroboratory to the findings of Liu X, et al. [21] in which higher expression level of β-catenin was found in astrocytic glioma patients with high grade in comparison with the normal controls, while siRNA-mediated silencing of β-catenin inhibited the proliferation and resulted in apoptosis of human U251 cells, with arrested cell cycle in G_0/G_1. Against the background of β-catenin overexpression in recurrent GBM cells, of particular clinical relevance is our data showing cumulative enhanced expression of β-catenin in undifferentiated (progenitor) GBM-SCs compared to the differentiated GBM (Figure 2), as it not only suggests a critical role for β-catenin in the resistance of GBM cells to therapy and disease recurrence, which are characteristic of CSCs, thus projecting β-catenin as a putative biomarker for GBM initiating cells or GBM-SCs. GBM-SCs exhibit distinct phenotypic, histopathologic and molecular features; they are highly tumorigenic, differentiate into heterogeneous glioma cell types or self-renew to sustain the GBM-SC pool, thus, facilitating and enhancing gliomagenesis [5]. These features make the eradication of these β-catenin-rich GBM-SCs a crucial therapeutic strategy for the effective treatment of GBM [23].

4-Acetylantroquinonol B (4-AAQB), an acetylated form of antroquinonol, a bioactive mycelial isolate of *Antrodia cinnamomea*, which is a Taiwanese mushroom with documented anti-inflammatory, blood sugar-lowering, vascular tone-relaxing, anti-proliferative, anti-metastasis, and autophagy-modulating activity [10,11,24]. Following the demonstrated CSCs-limiting anticancer efficacy of 4-AAQB in different cancer types, including colorectal, hepatocellular, breast and ovarian carcinomas, by our team and others [10,11,24], we herein investigated and demonstrated for the first time to the best of our knowledge, the therapeutic effect of 4-AAQB against GBM cells, by targeting GSM-SCs through the dysregulation of the canonical Wnt signaling pathway in a β-catenin-mediated manner. The canonical Wnt signaling pathway, especially mediated by aberrant or ectopic expression of a constitutively-active β-catenin, is broadly known to be involved in epithelial-to-mesenchymal transition (EMT), increased cell motility and tumor invasion, which are elements of tumor metastasis and contribute largely to cancer-related deaths [6], thus, our findings showing that treatment with 4-AAQB significantly inhibit the viability, migration, and invasive potential of GBM cells, while concurrently suppressing the expression levels of β-catenin, p-Stat3, vimentin, and slug proteins (Figures 3 and 5) has clinical implications and is consistent with research reports that the inhibition of β-catenin signaling disrupts the Wnt pathway, inhibits the downstream TCF/LEF transcriptional activity and suppress oncogenic activity [25,26]; more so we demonstrated that this suppression of GBM cell oncogenicity is associated with the concurrent marked reduction in the expression of nuclear β-catenin, Sox2, and Oct4, as well as inhibited nuclear co-localization of β-catenin and F-actin in the GBM cells (Figures 3 and 4), which is consistent with contemporary knowledge that the Wnt/β-catenin signaling plays with vital role in embryogenesis, activates embryonic stem cells (ESCs), regulates adult stem cells (ASCs), and modulates CSC biology by modulating the expression and activity of pluripotency and self-renewal transcription factor, namely Nanog, Oct4, Sox2, c-Myc [27].

Interestingly, in corroborating assays, we showed that 4-AAQB significantly diminished the colony-forming and tumorsphere-forming population, as well as the self-renewal capacity of GBM cells in a dose-dependent manner (Figure 6). Tumorspheres are characterized by enhanced anchorage-independent clonogenicity and clonal expansivity [28], thus their use as in vitro representation of the human CSC model. The GBM-SCs are implicated in tumorigenesis, therapy resistance, metastasis, and GBM recurrence, making them crucial molecular targets for effective anti-GBM therapeutic strategy, thus, the ability of 4-AAQB to effectively target this cellular subset is therapeutically-relevant in the light of the short-lived response of GBM patients to standard anti-GBM therapy consisting of temozolomide (TMZ) and radiation, which is almost always followed by recurrence, and not unconnected with the presence of unscathed treatment-insensitive GBM-SCs [5,18–20,23,27]. Results of our tumor xenograft in vivo studies are corroboratory of our in vitro findings, as they showed that treatment with 4-AAQB significantly suppress GBM-SC-induced tumor growth, in vivo. More so, we observed that compared to oral gavage, the intraperitoneal administration of 4-AAQB not only significantly, but also more efficiently suppressed the formation and growth of tumors in the GBM mice models (Figure 7). This is clinically relevant as it informs clinical decision making on the optimal route of drug administration in the treatment of GBM. As with the experimental design of in vivo animal studies, the administration of therapeutic compound in human is a critical component of every therapeutic strategy, as the optimization of drug delivery while minimizing likely adverse effects may constitute a deciding factor in treatment failure or response, and the degree or extent of such response [29]. Consistent with our observation, parenteral administration of therapeutics, specifically intraperitoneally, typically exhibits enhanced bioavailability of the drug because this route avoids the first-pass effect of liver metabolism, as commonly witnessed with oral administration of therapeutics; Similarly, the i.p route circumvents the unpredictability often associated with enteral absorptive processes [30]. Our results highlight the therapeutic potential of 4-AAQB for anti-GBM treatment, especially for reversal of GBM stem cell-associated TMZ resistance and lay the groundwork for further pre-clinical exploration and clinical utility of 4-AAQB for enhancement of TMZ efficacy in hitherto therapy-resistant GBM cells by either sensitizing GBM-SCs to TMZ or potentiating the anticancer effect of TMZ.

5. Conclusions

In conclusion, for the first time to the best of our knowledge, as shown in our Graphical Abstract, we demonstrate that 4-AAQB effectively targets the often unscathed chemoradiotherapy, insensitive GBM-SCs, and inhibits the not-too-innocent differentiated bystander GBM cells by suppressing β-catenin, inhibiting Stat3 activation and disrupting the canonical Wnt signaling pathway, thus projecting 4-AAQB as a novel small-molecule inhibitor of the Wnt/β-catenin signaling pathway with GBM-SC -targeting potentials, Thus projecting 4-AAQB as a putative therapeutic agent for effective anti-GBM target therapy with the promise of improving survival.

Supplementary Materials: The following are available online at http://www.mdpi.com/2072-6694/10/12/491/s1, Figure S1: Aberrant expression of β-catenin is characteristic of GBM and correlates with poor prognosis.

Author Contributions: H.-W.L., Y.-K.S.: Project conception, Collation and/or assembly of data, Data analysis and interpretation, Manuscript writing. H.-W.L., O.A.B.: Project conception, Experimental design, Collation and/or assembly of data, Data analysis and interpretation, Bioinformatics, Manuscript writing and revision. D.-Y.H., W.-H.L., M.-H.C. and C.-C.H.: Data assembly and analysis. C.-M.L., C.-T.Y., L.D. and M.H.: Experimental design, Data analysis and interpretation, Provision of useful research insight and essential reagents, Administrative oversight. All authors read and approved the final submitted version.

Funding: This work was supported by the National Science Council of Taiwan grant to Chien-Min Lin (MOST 107-2314-B-038 -056 -MY3) and grants to Yu-Kai Su (MOST 107-2314-B-038-022 -). This study was also supported by grants from Taipei Medical University, Taiwan (106-FRP-03) to Chien-Min Lin. The authors thank the laboratory assistant Mr. Iat-Hang Fong (Department of Medical Research and Education, Taipei Medical University-Shuang Ho Hospital) for his technical assistance.

Conflicts of Interest: The authors declare that there are no potential conflict of interest.

References

1. Thakkar, J.P.; Dolecek, T.A.; Horbinski, C.; Ostrom, Q.T.; Lightner, D.D.; Barnholtz-Sloan, J.S.; Villano, J.L. Epidemiologic and Molecular Prognostic Review of Glioblastoma. *Cancer Epidemiol. Biomarkers Prev.* **2014**, *23*, 1985–1996. [CrossRef] [PubMed]
2. Wirsching, H.G.; Galanis, E.; Weller, M. Glioblastoma. *Handb. Clin. Neurol.* **2016**, *134*, 381–397. [PubMed]
3. Stupp, R.; Mason, W.P.; Bent, M.J.V.D. Radiotherapy plus Concomitant and Adjuvant Temozolomide for Glioblastoma. *Oncol. Times* **2005**, *27*, 15–16. [CrossRef]
4. Vogelstein, B.; Papadopoulos, N.; Velculescu, V.E.; Zhou, S.; Diaz, L.A.; Kinzler, K.W. Cancer Genome Landscapes. *Science* **2013**, *339*, 1546–1558. [CrossRef] [PubMed]
5. Wang, J.; Ma, Y.; Cooper, M.K. Cancer stem cells in glioma: Challenges and opportunities. *Transl. Cancer Res.* **2013**, *2*, 429–441. [PubMed]
6. Lee, Y.; Lee, J.K.; Ahn, S.H.; Lee, J.; Nam, D.H. WNT signaling in glioblastoma and therapeutic opportunities. *Lab Investig.* **2016**, *96*, 137–150. [CrossRef] [PubMed]
7. Liu, C.; Tu, Y.; Sun, X.; Jiang, J.; Jin, X.; Bo, X.; Li, Z.; Bian, A.; Wang, X.; Liu, D.; et al. Wnt/β-Catenin pathway in human glioma: Expression pattern and clinical/prognostic correlations. *Clin. Exp. Med.* **2010**, *11*, 105–112. [CrossRef] [PubMed]
8. Lee, Y.; Kim, K.H.; Kim, D.G.; Cho, H.J.; Kim, Y.; Rheey, J.; Shin, K.; Seo, Y.J.; Choi, Y.S.; Lee, J.I.; et al. FoxM1 Promotes Stemness and Radio-Resistance of Glioblastoma by Regulating the Master Stem Cell Regulator Sox2. *PloS ONE* **2015**, *10*, e0137703. [CrossRef] [PubMed]
9. Rheinbay, E.; Suvà, M.L.; Gillespie, S.M.; Wakimoto, H.; Patel, A.P.; Shahid, M.; Oksuz, O.; Rabkin, S.D.; Martuza, R.L.; Rivera, M.N.; et al. An Aberrant Transcription Factor Network Essential for Wnt Signaling and Stem Cell Maintenance in Glioblastoma. *Cell Rep.* **2013**, *3*, 1567–1579. [CrossRef] [PubMed]
10. Chang, T.C.; Yeh, C.T.; Adebayo, B.O.; Lin, Y.C.; Deng, L.; Rao, Y.K.; Huang, C.C.; Lee, W.H.; Wu, A.T.; Hsiao, M.; et al. 4-Acetylantroquinonol B inhibits colorectal cancer tumorigenesis and suppresses cancer stem-like phenotype. *Toxicol. Appl. Pharmacol.* **2015**, *288*, 258–268. [CrossRef]
11. Liu, M.; Bamodu, O.A.; Huang, W.C.; Zucha, M.A.; Lin, Y.K.; Wu, A.T.H.; Huang, C.C.; Lee, W.H.; Yuan, C.C.; Hsiao, M.; et al. 4-Acetylantroquinonol B suppresses autophagic flux and improves cisplatin sensitivity in highly aggressive epithelial cancer through the PI3K/Akt/mTOR/p70S6K signaling pathway. *Toxicol. Appl. Pharmacol.* **2017**, *325*, 48–60. [CrossRef] [PubMed]

12. Vadlakonda, L.; Pasupuleti, M.; Pallu, R. Role of PI3K-AKT-mTOR and Wnt signaling pathways in transition of G1-S phase of cell cycle in cancer cells. *Front. Oncol.* **2013**, *3*, e85. [CrossRef] [PubMed]
13. Ye, S.; Zhang, D.; Cheng, F.; Wilson, D.; Mackay, J.; He, K.; Ban, Q.; Lv, F.; Huang, S.; Liu, D.; et al. Wnt/β-catenin and LIF–Stat3 signaling pathways converge on Sp5 to promote mouse embryonic stem cell self-renewal. *Development* **2016**, *129*, 269–276.
14. Hao, J.; Li, T.-G.; Qi, X.; Zhao, D.-F.; Zhao, G.-Q. WNT/β-catenin pathway up-regulates Stat3 and converges on LIF to prevent differentiation of mouse embryonic stem cells. *Development. Biol.* **2006**, *290*, 81–91. [CrossRef] [PubMed]
15. Fragoso, M.A.; Patel, A.K.; Nakamura, R.E.I.; Yi, H.; Surapaneni, K.; Hackam, A.S. The Wnt/β-Catenin Pathway Cross-Talks with STAT3 Signaling to Regulate Survival of Retinal Pigment Epithelium Cells. *PLoS ONE* **2012**, *7*, e46892. [CrossRef] [PubMed]
16. Raggioli, A.; Junghans, D.; Rudloff, S.; Kemler, R. β-Catenin Is Vital for the Integrity of Mouse Embryonic Stem Cells. *PLoS ONE* **2014**, *9*, e86691. [CrossRef] [PubMed]
17. Murayama, H.; Masaki, H.; Sato, H.; Hayama, T.; Yamaguchi, T.; Nakauchi, H. Successful Reprogramming of Epiblast Stem Cells by Blocking Nuclear Localization of β-Catenin. *Stem Cell Rep.* **2015**, *4*, 103–113. [CrossRef]
18. Bradshaw, A.; Wickremsekera, A.; Tan, S.T.; Peng, L.; Davis, P.F.; Itinteang, T. Cancer Stem Cell Hierarchy in Glioblastoma Multiforme. *Front. Surg.* **2016**, *3*, e21. [CrossRef]
19. Iacopino, F.; Angelucci, C.; Piacentini, R.; Biamonte, F.; Mangiola, A.; Maira, G.; Grassi, C.; Sica, G. Isolation of Cancer Stem Cells from Three Human Glioblastoma Cell Lines: Characterization of Two Selected Clones. *PLoS ONE* **2014**, *9*, e105166. [CrossRef]
20. Beier, D.; Schulz, J.B.; Beier, C.P. Chemoresistance of glioblastoma cancer stem cells—much more complex than expected. *Mol. Cancer* **2011**, *10*, e128. [CrossRef]
21. Liu, X.; Wang, L.; Zhao, S.; Ji, X.; Luo, Y.; Ling, F. β-Catenin overexpression in malignant glioma and its role in proliferation and apoptosis in glioblastma cells. *Med. Oncol.* **2010**, *28*, 608–614. [CrossRef] [PubMed]
22. Rossi, M.; Magnoni, L.; Miracco, C.; Mori, E.; Tosi, P.; Pirtoli, L.; Tini, P.; Oliveri, G.; Cosci, E.; Bakker, A. β-catenin and Gli1 are prognostic markers in glioblastoma. *Cancer Biol. Ther.* **2011**, *11*, 753–761. [CrossRef]
23. Dietrich, J.; Diamond, E.L.; Kesari, S. Glioma stem cell signaling: Therapeutic opportunities and challenges. *Exp. Rev. Anticancer Ther.* **2010**, *10*, 709–722. [CrossRef] [PubMed]
24. Chang, C.H.; Huang, T.F.; Lin, K.T.; Hsu, C.C.; Chang, W.L.; Wang, S.W.; Ko, F.N.; Peng, H.C.; Chung, C.H. 4-Acetylantroquinonol B Suppresses Tumor Growth and Metastasis of Hepatoma Cells via Blockade of Translation-Dependent Signaling Pathway and VEGF Production. *J. Agric. Food Chem.* **2014**, *63*, 208–215. [CrossRef] [PubMed]
25. Grossmann, T.N.; Yeh, J.T.-H.; Bowman, B.R.; Chu, Q.; Moellering, R.E.; Verdine, G.L. Inhibition of oncogenic Wnt signaling through direct targeting of β-catenin. *Proc. Natl. Acad. Sci. USA* **2012**, *109*, 17942–17947. [CrossRef] [PubMed]
26. Hwang, S.Y.; Deng, X.; Byun, S.; Lee, C.; Lee, S.J.; Suh, H.; Zhang, J.; Kang, Q.; Zhang, T.; Westover, K.D.; et al. Direct Targeting of β-Catenin by a Small Molecule Stimulates Proteasomal Degradation and Suppresses Oncogenic Wnt/β-Catenin Signaling. *Cell Rep.* **2016**, *16*, 28–36. [CrossRef]
27. Nager, M.; Bhardwaj, D.; Cantí, C.; Medina, L.; Nogués, P.; Herreros, J. β-Catenin Signalling in Glioblastoma Multiforme and Glioma-Initiating Cells. *Chemother. Res. Practice* **2012**, *2012*, 1–7. [CrossRef]
28. Rajendran, V.; Jain, M.V. In Vitro Tumorigenic Assay: Colony Forming Assay for Cancer Stem Cells. *Cancer Stem Cells* **2017**, *1692*, 89–95.
29. Agoram, B.M. Use of pharmacokinetic/pharmacodynamic modelling for starting dose selection in first-in-human trials of high-risk biologics. *Br. J. Clin. Pharmacol.* **2009**, *67*, 153–160. [CrossRef]
30. Turner, P.V.; Brabb, T.; Pekow, C.; Vasbinder, M.A. Administration of Substances to Laboratory Animals: Routes of Administration and Factors to Consider. *J. Am. Assoc. Lab. Anim. Sci.* **2011**, *50*, 600–613.

cancers

MDPI

Article

The Oncogene Addiction Switch from NOTCH to PI3K Requires Simultaneous Targeting of NOTCH and PI3K Pathway Inhibition in Glioblastoma

Norihiko Saito *, Nozomi Hirai, Kazuya Aoki, Ryo Suzuki, Satoshi Fujita, Haruo Nakayama, Morito Hayashi, Keisuke Ito, Takatoshi Sakurai and Satoshi Iwabuchi

Department of Neurosurgery, Toho University Ohashi Medical Center, Tokyo 153-8515, Japan; nozomi.hirai@med.toho-u.ac.jp (N.H.); kaoki@med.toho-u.ac.jp (K.A.); ryo.suzuki@med.toho-u.ac.jp (R.S.); satoshi.fujita@med.toho-u.ac.jp (S.F.); haruonakayama@med.toho-u.ac.jp (H.N.); morito@med.toho-u.ac.jp (M.H.); keisuke@med.toho-u.ac.jp (K.I.); cherry@med.toho-u.ac.jp (T.S.); iwabuchi@med.toho-u.ac.jp (S.I.)

* Correspondence: bsaitob@med.toho-u.ac.jp; Tel.: +81-3-3468-1251

Received: 16 December 2018; Accepted: 19 January 2019; Published: 20 January 2019

Abstract: The NOTCH pathway regulates neural stem cells and glioma initiating cells (GICs). However, blocking NOTCH activity with γ-secretase inhibitors (GSIs) fails to alter the growth of GICs, as GSIs seem to be active in only a fraction of GICs lines with constitutive NOTCH activity. Here we report loss of *PTEN* function as a critical event leading to resistance to NOTCH inhibition, which causes the transfer of oncogene addiction from the NOTCH pathway to the PI3K pathway. Drug cytotoxicity testing of eight GICs showed a differential growth response to GSI, and the GICs were thus stratified into two groups: sensitive and resistant. In the sensitive group, GICs with loss of PTEN function appeared less sensitive to GSI treatment. Here we show that NOTCH regulates PTEN expression and the activity of the PI3K pathway in GICs, as treatment with GSI attenuated the NOTCH pathway and increased PTEN expression. NOTCH regulates PTEN expression via Hes-1, as knockdown of Notch or Hes1 increased expression of PTEN. This novel observation suggests that both pathways must be simultaneously inhibited in order to improve therapeutic efficacy in human glioblastomas (GBMs).

Keywords: NOTCH; glioma initiating cell; PTEN; PI3K pathway; glioblastoma

1. Introduction

Glioblastoma (GBM), the most common adult glioma, has a poor prognosis. Genetic heterogeneity between patients and even within tumors is high, and GBM is characterized by evolving genetic aberration resulting from dynamic genetic instability. The genes most commonly affected are *CDKN2A*, *TP53*, *EGFR*, *PTEN* and *RB* [1]. By combining sequencing data with other types of genomic information, the Cancer Genome Atlas team produced a tentative overview of the main biological pathways involved in GBM. Each of the three pathways (namely, the CDK/RB, p53 and RTK/RAS/PI3K pathways) was disrupted in more than three-quarters of GBM tumors. Signal transduction pathways are complex and exhibit overlap and crosstalk [2]. The complexity of these pathways may allow for compensatory effects in alternative pathways, which could lead to resistance to single agents that regulate only one target. Successful novel therapeutic strategies for GBMs may thus require simultaneous targeting of multiple dysregulated molecules.

The NOTCH signaling pathway is an evolutionarily conserved system that is important in most multicellular processes such as neural differentiation, proliferation, survival, angiogenesis and stemness [3–5]. About 45% of proneural GBMs exhibit a high expression of representative NOTCH pathway genes, which has been implicated in the pathogenesis of solid tumors [6]. When the NOTCH receptor is triggered by a ligand, it promotes two proteolytic cleavage events at the NOTCH receptor: by means of an ADAM metalloprotease and γ-secretase complex. The cleavage can release the NOTCH intracellular domain (NICD), which translocates to the nucleus and interacts with the CSL-binding protein to activate expressions of NOTCH targeting genes [3,4].

Recent studies suggest that PTEN is regulated through the NOTCH pathway in a variety of settings, such as fibroblasts [7,8], T-cell acute lymphoblastic leukemia cells [9] and prostate tumor cells [10]. NOTCH interaction with PTEN has been well characterized in T-cell leukemia, in which NOTCH and PTEN induce resistance to γ-secretase inhibition. Here we report that PTEN regulates GBM sensitivity to γ-secretase inhibitors (GSIs), thereby highlighting the need for simultaneous inhibition of the PI3K/AKT and NOTCH pathways in PTEN-mutant GBMs. Thus, PTEN may be an important factor of GSI-induced attenuation of cell growth through a regulatory circuit linking NOTCH signaling with PTEN expression. This finding supports a need for combination therapeutic strategies in the treatment of GBM.

2. Results

2.1. GICs Show Differential Growth in Response to GSIs

We quantified sensitivity to three GSIs, as seen in Figure S1, in a panel of eight glioma initiating cell lines (GICs) and four glioma cell lines by measuring the IC50 or half-maximal inhibitory concentration after 72 h of continuous exposure. GSIs showed a dose-dependent growth inhibition of GICs and glioma cell lines (Figure 1a,b). Expression of the Notch signaling, PTEN and AKT are shown in Figure 1c [11]. NICD and Hes1—a NOTCH-1 pathway component—were expressed in U87, A172 and LN18. PTEN expression was absent in U87 and U251, suggesting that loss of PTEN function (Figure 1c). Figure 1d shows representative waterfall plots of the differential responses to GSIs, which were used to classify GICs as sensitive and resistant. Sensitive cell lines were those with IC50 values of 3–18 μmol/L for N-[N-(3,5-difluorophenacetyl)-L-alanyl]-S-phenylglycine t-butyl ester (DAPT) and 0.5–2 μmol/L for BMS-708163 and RO4929097. Resistant cell lines were those with IC50 values greater than 20 μmol/L for DAPT and greater than 3 μmol/L for BMS-708163 and RO4929097 (Figure 1d).

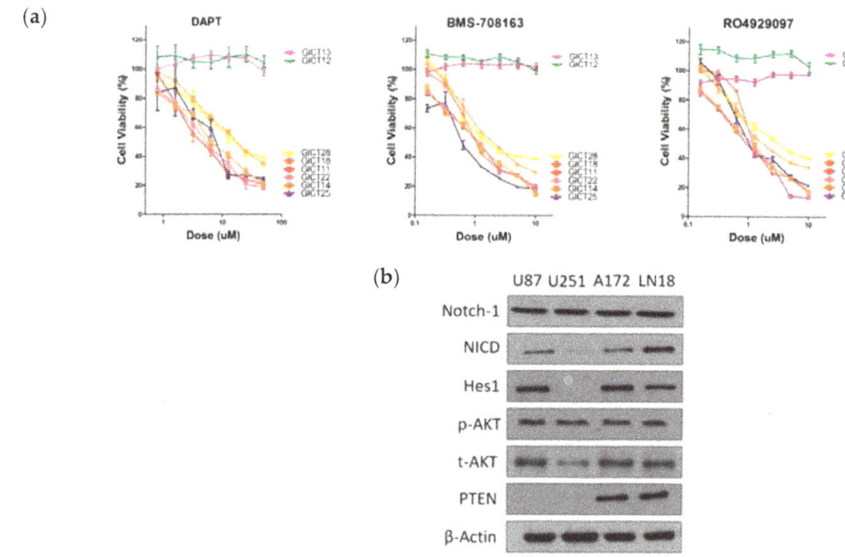

Figure 1. *Cont.*

(c)

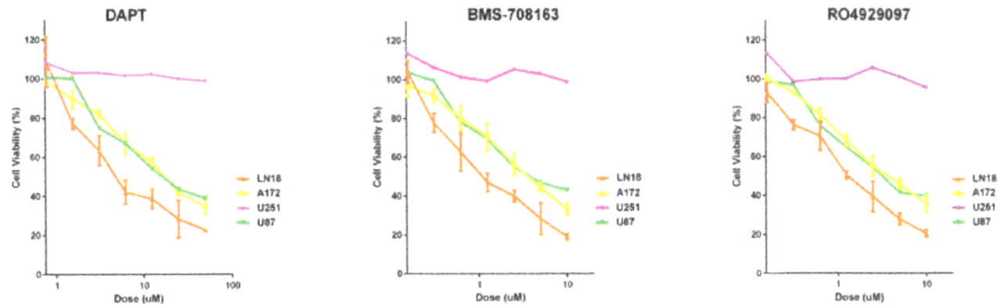

(d)

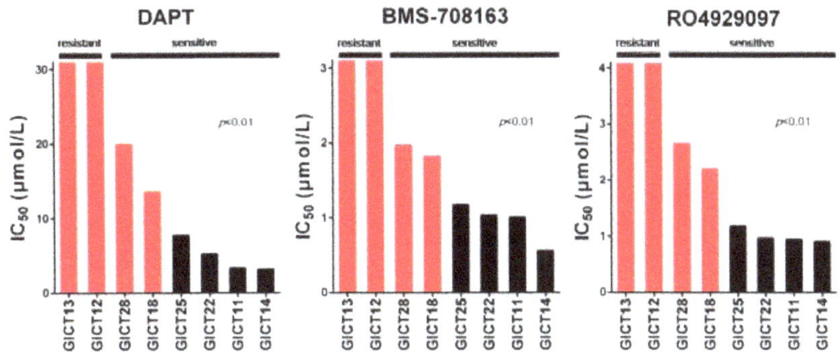

Figure 1. γ-Secretase inhibitors (GSIs) showed dose-dependent growth inhibition of glioma tumor-initiating cells (GICs). (**a**) A panel of GIC lines was treated with various concentrations of the GSIs. Cells were treated with increasing concentrations of GSIs in triplicate wells for 72 h, and cell viability was assessed with the CellTiter-Blue assay. Cell viability in the vehicle control was considered to be 100%; (**b**) GSIs showed dose-dependent growth inhibition of glioma cells. A panel of glioma cell lines was treated with various concentrations of the GSIs. Cells were treated with increasing concentrations of GSIs in triplicate wells for 72 h, and cell viability was assessed with the CellTiter-Blue assay. Cell viability in the vehicle control was considered to be 100%; (**c**) Western blotting of the Notch signaling, AKT and PTEN in glioma cell lines. β-Actin was used as loading control; (**d**) Waterfall plot of IC50 values for eight GICs. These figures show that GSIs have a particular growth inhibition signature: some cell lines are very sensitive and others are relatively resistant.

2.2. Expression of the NOTCH Signaling Pathway and PTEN Status in GICs

To identify associations between the activation of the NOTCH signaling pathway, PTEN status and the sensitivity of GSIs, Western blot analysis was used to evaluate gene expression in the NOTCH signaling pathway in eight GIC lines, as seen in Figure 2. NICD and Hes1—a NOTCH-1 pathway component—were all expressed in sensitive and resistant GICs. PTEN expression was absent, or very low, in resistant GICs, which suggests that the response may be related to PTEN status in GICs.

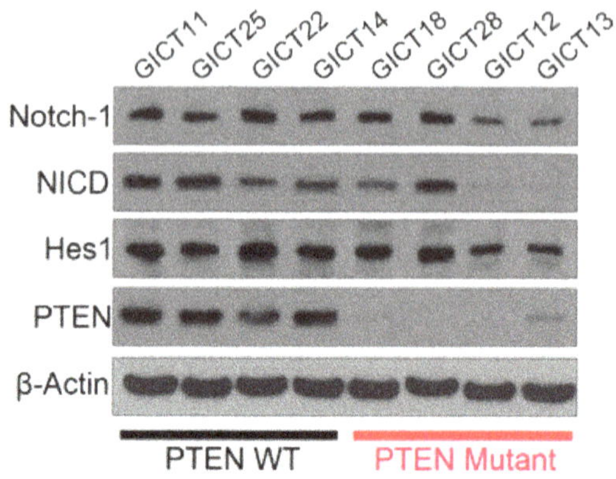

Figure 2. Western blotting of the NOTCH signaling pathway in eight GICs. Activated Notch-1 (NICD) and Hes1 were expressed in sensitive GICs. β-Actin was used as the loading control.

2.3. NOTCH Regulates Activity of the PI3K/AKT Pathway via Hes1 in PTEN–Wild-Type GICs

The close association between the presence of PTEN mutations and GSI resistance in GICs prompted us to ask whether PTEN might be functionally linked to NOTCH-1 signaling. Real-time PCR analysis of PTEN transcript levels upon NOTCH-1 inhibition by GSI showed PTEN relative expression levels were increased by GSI treatment in GICT25, i.e., GSI-sensitive/PTEN–wild-type cells (Figure 3a). In GICT25, NICD and HES1 expressions decreased and PTEN expression increased after treatment with GSIs, as seen in Figure 3b. In NOTCH knockdown experiments, NOTCH-1, NICD and HES1 expressions were decreased and PTEN expression was increased. siRNA knockdown of Hes1 decreased HES1 expression and increased PTEN expression, thus suggesting that Hes1 caused the down-regulation of PTEN expression by transcriptional repression of the PTEN promoter (Figure 3c). In GICT18—a PTEN-mutant GIC—Western blotting was used to investigate the expression of NICD, HES1, PTEN, p-AKT and total AKT proteins. Transfection with PTEN resulted in the up-regulation of PTEN expression and down-regulation of p-AKT expression in GICT18. After transfection with PTEN, PTEN expression increased after GSI treatment, as seen in Figure 3d. These results indicate that GSI might affect the PTEN and AKT pathway and that PTEN is a downstream target of the NOTCH pathway via HES1 in PTEN–wild-type GICs.

(a)

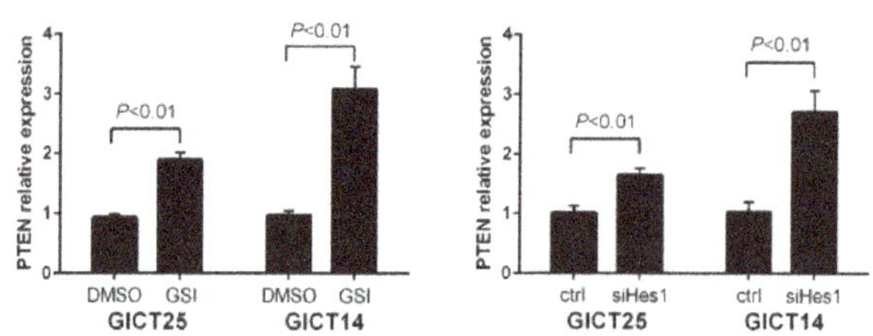

Figure 3. *Cont.*

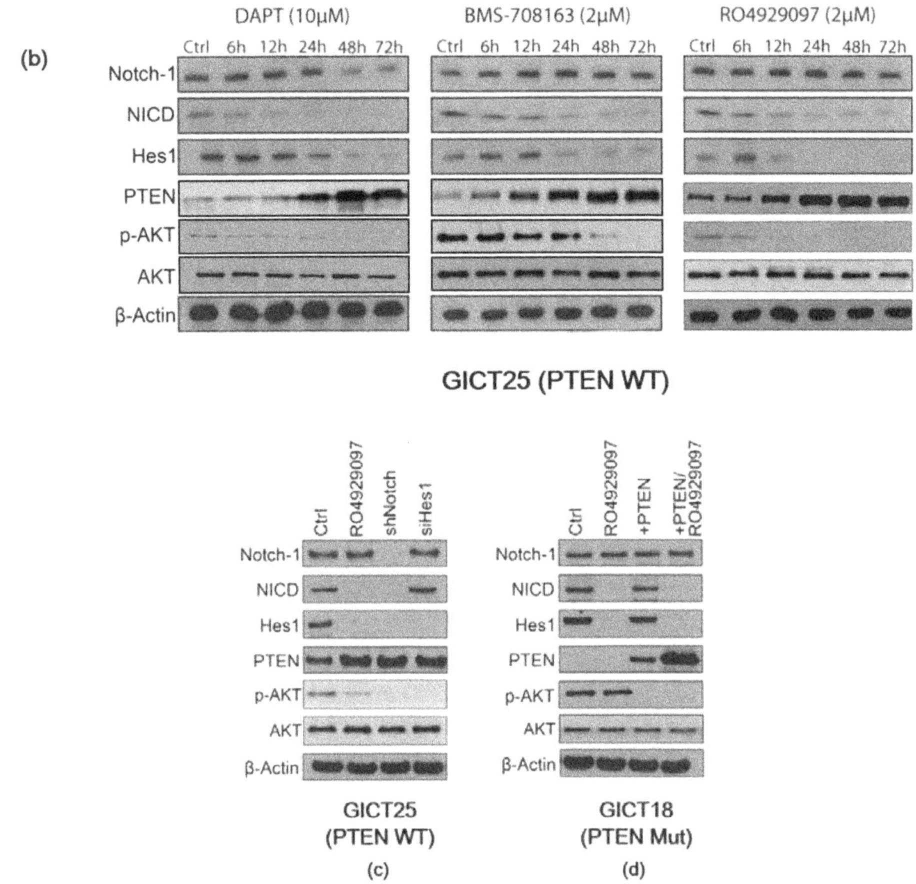

Figure 3. (**a**) Real-time PCR analysis of PTEN transcript levels upon NOTCH-1 inhibition by GSI in GICT25 and GICT14 relative to (DMSO) controls. GAPDH levels were used as the reference control; (**b**) A GICT25 panel was treated with the indicated doses of DAPT, BMS-708163 and RO4929097 for the indicated time intervals. All GSIs inhibited the expression of NICD, Hes1 and p-AKT in a time-dependent manner. The decrease in NICD was followed by a decrease in Hes1 expression, whereas Notch-1 expression did not change; (**c**) Western blotting confirmed the knockdown effect of Notch shRNA lentivirus on GICT25. siRNA knockdown of Hes1 decreased HES1 expression and increased PTEN expression; (**d**) After transfection with the PTEN gene, PTEN expression increased with GSI treatment.

2.4. The Combination of GSI and PI3K Inhibitor had a Synergistic Effect in PTEN Mutant GICs

We performed drug sensitivity testing and compared the control with PTEN knockdown in GICT25. The IC50 of GICT25 increased from 7.69 to 19.8 μM in DAPT, from 0.55 to 2.00 μM in BMS-708163 and from 0.96 to 4.67 μM in RO4929097, indicating a decrease in sensitivity to GSI treatment, as seen in Table 1a and Figure 4a. We also performed drug sensitivity testing and compared the control with PTEN-transfected GICT18. The IC50 of GICT18 decreased from 13.50 to 5.78 μM in DAPT, from 1.82 to 0.84 μM in BMS-708163 and from 2.21 to 1.11 μM in RO4929097, indicating an increase of sensitivity to GSI treatment, as seen in Table 1b and Figure 4b. We tested the efficacy of combining RO4929097 with two inhibitors of the PI3K/AKT/mTOR pathway, namely BEZ235 and BKM120. For BKM120 and RO4929097, the maximum concentration was 2 μM for BKM120 and

1 μM for RO4929097, which were diluted to a 2:1 ratio. For BEZ235 and RO4929097, the maximum concentration was 200 nM for BEZ235 and 1 μM for RO4929097, which were diluted to a 1:5 ratio. We calculated the Combination Index by using the Calcusyn software package and observed the synergistic inhibition of cell proliferation in GICT18, as seen in Figure 5 and Table 2.

Table 1. PTEN expression is required for the GSI response in cell growth inhibition. (**a**) PTEN knockdown decreased the response to GSI in PTEN–wild-type GIC. (**b**) PTEN expression increased the response to GSI in PTEN-mutant GIC.

IC_{50}	GICT25	shPTEN
	(a)	
DAPT (μM)	7.69	19.8
BMS-708163 (μM)	0.55	2.00
RO4929097 (μM)	0.96	4.67
	(b)	
DAPT (μM)	13.50	5.78
BMS-708163 (μM)	1.82	0.84
RO4929097 (μM)	2.21	1.11

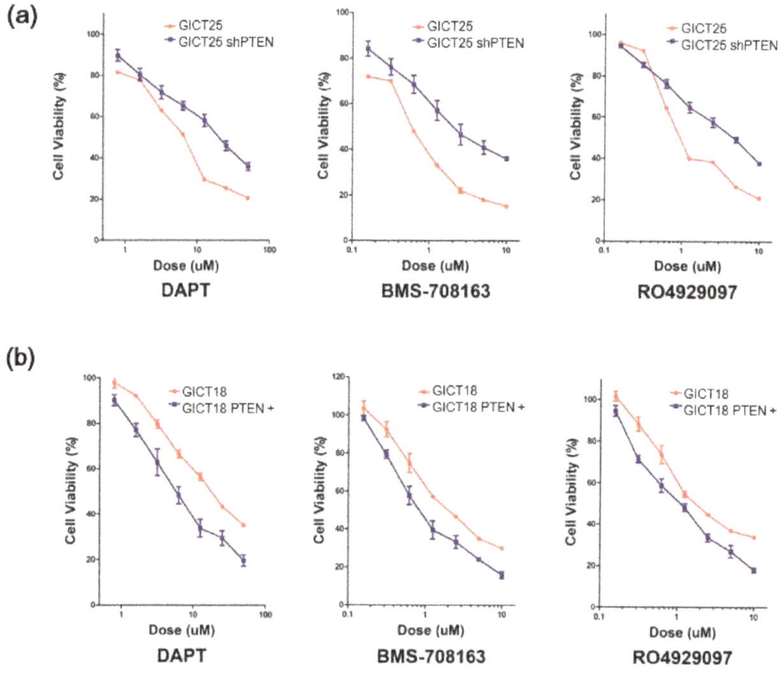

Figure 4. PTEN expression is required for the GSI response in cell growth inhibition. (**a**) PTEN knockdown decreased the response to GSI in PTEN–wild-type GIC; (**b**) PTEN expression increased the response to GSI in PTEN-mutant GIC.

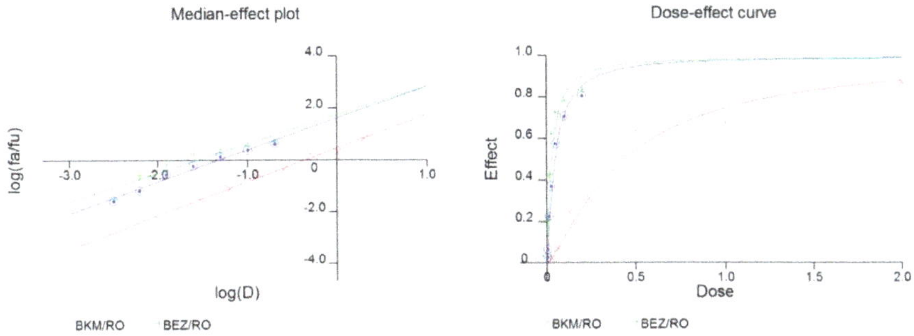

Figure 5. The combination of a GSI and PI3K inhibitor had a synergistic effect on PTEN-mutant GIC.

Table 2. Combination Index of GSI and PI3K inhibitor in GICs cells in vitro.

Reagents	CI Values at ED50
BKM120/GSI (2:1)	0.47
BEZ235/GSI (1:5)	0.58

2.5. The Combination of GSI and PI3K Inhibitor Regulated Survival in an Orthotropic Mouse Model

To assess the anti-glioma efficacy of the combination of GSI and PI3K inhibitor in vivo, we used a GICT25 orthotopic model of human glioma intracranial xenografts in nude mice. On day 4 after tumor cell implantation, the animals were treated with vehicle or with 10 mg/kg RO4929097 in methyl cellulose alone, 20 mg/kg BKM120 in methyl cellulose alone, a combination of RO4929097 and BKM120 in methyl cellulose or methyl cellulose alone (control) once a week for a total of 5 weeks, as described in the Materials and Methods section. The median survival duration for animals treated with methylcellulose (control) was 64 days, as seen in Figure 6. Treatment with 10 mg/kg RO4929097 and 20 mg/kg BKM120 alone extended survival to a median of 109.5 and 117.5 days, respectively ($p < 0.05$, log-rank test for both experiments). Combination treatment with RO4929097 and BKM120 significantly increased the survival of mice, as compared with either agent alone (median 168.5 days, $p < 0.05$). These results show that the combination of RO4929097 and BKM120 can inhibit tumor growth and prolong the survival of mice with xenograft tumors, which suggests that the combination of RO4929097 with BKM120 has potential therapeutic efficacy in vivo.

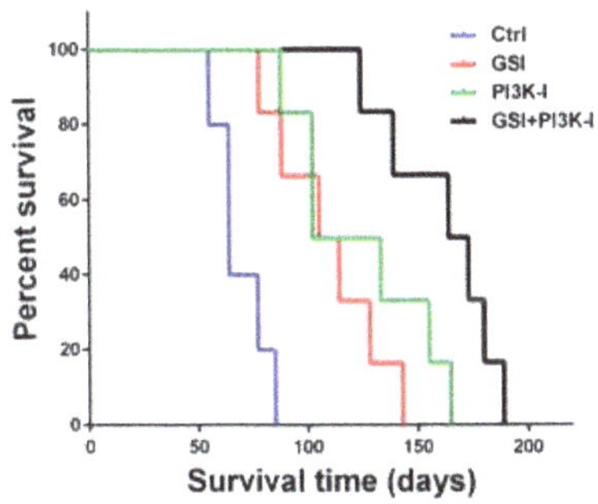

Figure 6. The combination of a GSI and PI3K inhibitor had a synergistic effect in PTEN-mutant GIC. Kaplan–Meier survival probability plots of tumor-bearing mice in vehicle or RO4929097, BKM120 and combined RO4929097 and BKM120 treatment groups were graphed, and the log-rank test was used to compare groups. All treatments (different colored lines) showed a statistically significant improvement versus control ($p < 0.05$).

3. Discussion

The NOTCH signaling pathway is involved in cell-fate decisions during normal development and in multicellular processes and has been implicated in the maintenance of neural progenitors during brain development [12,13]. In gliomas, NOTCH signaling seems to confer radioresistance to glioma initiating cells [14] and the inhibition of NOTCH through GSIs [15] or Delta-4 monoclonal antibodies [16] decreased the numbers of glioma initiating cells and/or their tumorigenicity in some preclinical models. This suggests that a therapeutic strategy that includes NOTCH inhibitors might be used clinically to target glioma initiating cells and overcome chemoresistance and radioresistance. We previously reported that proneural GICs with high NOTCH pathway activation responded to GSIs and that PTEN status was an important factor in the sensitivity to GSI treatment [6]. Amplifying or activating mutations of PIK3CA were found in about 15% of patients with GBM [17–19]. Similarly, loss-of-function mutations, chromosomal deletions or epigenetic gene silencing of PTEN, which are associated with poor survival [19], were found in approximately 40% of GBM cases [1].

The close association between PTEN mutation and GSI resistance in GICs suggests that PTEN might be linked to NOTCH signaling functionally. Analysis of the transcriptional responses of GSI-sensitive GICs with PTEN wildtype to NOTCH inhibition showed significant upregulation of the PTEN expression level. However, the mechanism responsible for PI3K-AKT upregulation via NOTCH activation in GICs remains unknown. The inhibitory effect of NOTCH signaling on PTEN expression is inconsistent with the established role of NOTCH as a transcriptional activator [20–23]. Thus, we hypothesized that the inhibition of PTEN by NOTCH could be mediated by HES1 regulated by NOTCH1.

"Oncogene addiction" refers to the need that tumor cells have for sustained abundant oncogene signaling, which maintains their viability [24–26], and explains the requirement for continuous NOTCH signaling in GICs with mutations in NOTCH [27,28]. Our data indicate that loss of function with PTEN in GICs may result in resistance to NOTCH inhibition by GSI treatment. Thus, we hypothesized that GSI resistance results from a change in oncogene addiction, from NOTCH to constitutive AKT signaling. Our preliminary data reveal a synergistic attenuation of cell growth by the combination of GSIs and PI3K inhibitors in PTEN mutant GICs. Thus, a future study should evaluate the efficacy of combined NOTCH and PI3K/AKT inhibition therapies in glioma.

4. Materials and Methods

4.1. Cell Lines and Reagents

The four glioma cell lines used in this study were obtained from the JCRB cell bank. The eight glioma initiating cell lines were maintained in neurosphere medium by using a previously described method [29] to isolate neurosphere-forming cells from surgical specimens of human GBM. The study was approved by the Institutional Review Board of Toho University (H22-62). These GIC lines were cultured as GBM neurospheres in DMEM/F12 medium supplemented with B27 (Invitrogen, Grand Island, NY, USA), L-glutamine (GIBCO), penicillin/streptomycin and growth factors (20 ng/mL EGF and 20 ng/mL FGF-2; Invitrogen). DAPT, BMS-708163, RO4929097 and NVP-BKM120 were purchased from Selleck Chemicals. For in vitro use, all inhibitors were dissolved in dimethyl sulfoxide (DMSO; Sigma-Aldrich, St. Louis, MO, USA) to a concentration of 10 mmol/L, stored at −20 °C and further diluted to an appropriate final concentration in DMEM/F12 medium at the time of use. The DMSO in the final solution did not exceed 0.1% (v/v).

4.2. Cell Proliferation Assay

Cells were seeded in 96-well plates (2000 cells/well) and incubated at 37 °C for 24 h before addition of serial dilutions of GSIs DAPT, BMS-708163 and RO4929097. Growth inhibition was estimated using the CellTiter-Blue (Promega, Madison, WI, USA) viability assay. The IC50 value was calculated as the mean drug concentration required in order to inhibit cell proliferation by 50% compared with vehicle controls. The data are expressed as percentages of the vehicle-treated controls, and IC50 values were calculated with CalcuSyn 2.0 software (BIOSOFT, Great Shelford, Cambridge, UK).

4.3. Western Blot Analysis

Cells were harvested in lysis solution, as previously described [30], and subjected to Western blotting. Membranes were probed with the following primary antibodies: NOTCH-1, PTEN, phospho-specific AKT, total AKT (Cell Signaling, Boston, MA, USA) and Hes-1 (Millipore, Burlington, MA, USA). Anti–β-actin antibody was purchased from Sigma and used as a loading control.

4.4. RNA Extraction and cDNA Synthesis

Total RNAs were extracted from each sample and placed in 350 μL of RLT buffer (Qiagen, Hilden, Germany) supplemented with 1% β-mercaptoethanol. Next, the total RNAs were purified with an RNeasy Micro Kit (Qiagen, Valencia, CA, USA) according to the manufacturer's protocol. After RNA extraction, we synthesized first-strand cDNA by using random primers and TaqMan reverse-transcription reagents (Applied Biosystems, Foster City, CA, USA).

4.5. Quantitative Real-Time PCR Analysis

Quantitative real-time PCR was performed with an ABI PRISM 7000 sequence detection system (Applied Biosystems) according to the manufacturer's instructions. The relative expressions of PTEN mRNAs were normalized to the amount of GAPDH in the same cDNA by using the delta Ct method, as described by the manufacturer. Each sample was assayed in triplicate and analyzed with SDS software (Applied Biosystems).

4.6. Knockdown of Notch-1 by Lentiviral shRNA

Lentiviral vector encoding shRNA for PTEN was purchased from GeneCopoeia Inc. Lentiviral vector with pLKO.1-mediated expression of shRNA for targeting human PTEN was performed according to the manufacturer's instructions. Lentiviral particles, which expressed targeting and control scramble, were produced in HEK293 cells with the mixed set of packing plasmids, and the viruses were concentrated and titered as previously described [31]. GICT25 cells were infected with the PTEN shRNA. The produced lentiviruses were concentrated with a Centricon Plus-20 centrifugal filter device (Millipore). To ensure that the same number of lentiviral particles was used in each experiment, the produced lentiviral stock was stored at −80 °C. For the in vitro infection of GICs with the lentivirus, we disaggregated cultured tumorspheres before infection, to increase infection efficiency and uniformity. To infect target cells with lentiviruses, we exposed GICT25 for 24 h. Cells were washed and then cultured with regular complete medium for two additional days in the 2.5 μg/μL puromycin. Lastly, the cells were washed and analyzed for protein expression with the Western blotting protein assay.

4.7. Transient RNA Interference

Small interfering RNA (siRNA) duplexes targeting human Hes1 sequences and a scrambled siRNA were purchased from Sigma-Aldrich. Transfection of the siRNA duplexes was performed by TransIT-TKO Transfection Reagent (Takara, Kusatsu, Shiga, Japan) according to the manufacturer's instructions.

4.8. Combination Studies

In the in vitro combination studies, cells were seeded in 96-well plates (2000 cells/well) and incubated at 37 °C for 24 h before addition. Then cells were treated with RO4929097, BKM120 and BEZ235. Cell viability was quantified with the CellTiter-Blue assay. Drug synergy was analyzed by calculating the combination index (CI) as a measure of the interaction between two drugs.

The Combination Index (CI) was calculated according to the median-effect principle of the Chou and Talalay method, using the CalcuSyn software, version 2.1 (BioSoft, Great Shelford, Cambridge, UK). CI values are generated over a range of Fa levels, from 0.05–0.90 (5–90% growth inhibition). A CI of 1 indicates an additive effect between two agents, whereas a CI of <1 or >1 indicates synergism or antagonism, respectively [32].

4.9. Animal Studies

The mice were housed and cared for at the animal care facility of Toho University School of Medicine in accordance with the institution's guidelines for the care and use of laboratory animals. The experimental protocol was approved by the Animal Research Committee, Toho University School of Medicine (ARC/TUSM-R16-14). We examined the antitumor efficacy of RO4929097 and BKM120 in intracranial xenografts, using GICs. Nude (nu/nu) 6–8 week-old mice (n = 24) were purchased from Charles River Co. (Japan). In this study, we used a guide-screw system to implant 5 $\times$ 10^5 GICT25 cells in DMEM/F-12 serum-free media (5 mL), as described previously [6], and then randomly divided the mice into four groups of six mice each. Starting on day 4 after the tumor cells were implanted, mice were treated by oral gavage with 10 mg/kg RO4929097 in methyl cellulose alone, 20 mg/kg BKM120 in methyl cellulose alone, a combination of RO4929097 and BKM120 in methyl cellulose or methyl cellulose alone (control) once a week for a total of 5 weeks. Mice were monitored daily and euthanized when they became moribund. At necropsy, all organs were analyzed grossly and microscopically to assess toxicity.

4.10. Statistical Analysis

The data were analyzed with the Student unpaired t–test. The results are presented as the mean of at least three independent experiments. Differences were considered significant at a p value of <0.05, in all comparisons.

5. Conclusions

Our data show that PTEN is an important mediator of GSI-induced attenuation of cell growth and suggest the presence of a regulatory circuit linking NOTCH signaling with the PI3K/PTEN/AKT pathway. This finding may yield new therapeutic strategies and indicates that both pathways should be simultaneously inhibited in order to improve therapeutic efficacy in human GBMs.

Supplementary Materials: The following are available online at http://www.mdpi.com/2072-6694/11/1/121/s1, Figure S1: γ-Secretase inhibitors. DAPT, BMS-708163 and RO4929097 were purchased from Selleck Chemicals.

Author Contributions: N.S. designed and performed experiments, collected, analyzed, and interpreted data, and wrote the manuscript; N.H., K.A., R.S., S.F., H.N., M.H., K.I. and T.S. performed experiments and interpreted data; S.I. supervised the study.

Funding: This work was supported by JSPS KAKENHI Grant Number 15K19980.

Conflicts of Interest: The authors declare no conflict of interest.

References

1. The Cancer Genome Atlas Research Network. Comprehensive genomic characterization defines human glioblastoma genes and core pathways. *Nature* **2008**, *23*, 1061–1068.
2. Brennan, C.; Momota, H.; Hambardzumyan, D.; Ozawa, T.; Tandon, A.; Pedraza, A.; Holland, E. Glioblastoma subclasses can be defined by activity among signal transduction pathways and associated genomic alterations. *PLoS ONE* **2009**, *4*, e7752. [CrossRef] [PubMed]
3. Kopan, R.; Ilagan, M.X. The canonical Notch signaling pathway: Unfolding the activation mechanism. *Cell* **2009**, *137*, 216–233. [CrossRef] [PubMed]
4. Fortini, M.E. Notch signaling: The core pathway and its posttranslational regulation. *Dev. Cell* **2009**, *16*, 633–647. [CrossRef] [PubMed]

5. Kageyama, R.; Ohtsuka, T. The Notch-Hes pathway in mammalian neural development. *Cell Res.* **1999**, *9*, 179–188. [CrossRef] [PubMed]
6. Saito, N.; Fu, J.; Zheng, S.; Yao, J.; Wang, S.; Liu, D.D.; Yuan, Y.; Sulman, E.P.; Lang, F.F.; Colman, H.; et al. A high Notch pathway activation predicts response to γ secretase inhibitors in proneural subtype of glioma tumor-initiating cells. *Stem Cells* **2014**, *32*, 301–312. [CrossRef]
7. Chappell, W.H.; Green, T.D.; Spengeman, J.D.; McCubrey, J.A.; Akula, S.M.; Bertrand, F.E. Increased protein expression of the PTEN tumor suppressor in the presence of constitutively active Notch-1. *Cell Cycle* **2005**, *4*, 1389–1395. [CrossRef]
8. Wu, X.; Xu, K.; Zhang, L.; Deng, Y.; Lee, P.; Shapiro, E.; Monaco, M.; Makarenkova, H.P.; Li, J.; Lepor, H.; et al. Differentiation of the ductal epithelium and smooth muscle in the prostate gland are regulated by the Notch/PTEN-dependent mechanism. *Dev. Biol.* **2011**, *356*, 337–349. [CrossRef]
9. Palomero, T.; Sulis, M.L.; Cortina, M.; Real, P.J.; Barnes, K.; Ciofani, M.; Caparros, E.; Buteau, J.; Brown, K.; Perkins, S.L.; et al. Mutational loss of PTEN induces resistance to NOTCH1 inhibition in T-cell leukemia. *Nat. Med.* **2007**, *13*, 1203–1210. [CrossRef]
10. Bertrand, F.E.; McCubrey, J.A.; Angus, C.W.; Nutter, J.M.; Sigounas, G. NOTCH and PTEN in prostate cancer. *Adv. Biol. Regul.* **2014**, *56*, 51–65. [CrossRef]
11. Barretina, J.; Caponigro, G.; Stransky, N.; Venkatesan, K.; Margolin, A.A.; Kim, S.; Wilson, C.J.; Lehár, J.; Kryukov, G.V.; Sonkin, D.; et al. The Cancer Cell Line Encyclopedia enables predictive modelling of anticancer drug sensitivity. *Nature* **2012**, *483*, 603–607. [CrossRef] [PubMed]
12. Kageyama, R.; Ohtsuka, T.; Hatakeyama, J.; Ohsawa, R. Roles of bHLH genes in neural stem cell differentiation. *Exp. Cell Res.* **2005**, *306*, 343–348. [CrossRef] [PubMed]
13. Hatakeyama, J.; Sakamoto, S.; Kageyama, R. Hes1 and Hes5 regulate the development of the cranial and spinal nerve systems. *Dev. Neurosci.* **2006**, *28*, 92–101. [CrossRef] [PubMed]
14. Wang, R.F. Regulatory T cells and innate immune regulation in tumor immunity. *Springer Semin. Immunopathol.* **2006**, *28*, 17–23. [CrossRef] [PubMed]
15. Fan, X.; Khaki, L.; Zhu, T.S.; Soules, M.E.; Talsma, C.E.; Gul, N.; Koh, C.; Zhang, J.; Li, Y.M.; Maciaczyk, J.; et al. NOTCH pathway blockade depletes CD133-positive glioblastoma cells and inhibits growth of tumor neurospheres and xenografts. *Stem Cells* **2010**, *28*, 5–16. [CrossRef] [PubMed]
16. Hoey, T.; Yen, W.C.; Axelrod, F.; Basi, J.; Donigian, L.; Dylla, S.; Fitch-Bruhns, M.; Lazetic, S.; Park, I.K.; Sato, A.; et al. DLL4 blockade inhibits tumor growth and reduces tumor-initiating cell frequency. *Cell Stem Cell* **2009**, *5*, 168–177. [CrossRef] [PubMed]
17. Holland, E.C.; Celestino, J.; Dai, C.; Schaefer, L.; Sawaya, R.E.; Fuller, G.N. Combined activation of Ras and Akt in neural progenitors induces glioblastoma formation in mice. *Nat. Genet.* **2000**, *25*, 55–57. [CrossRef]
18. Koul, D.; Fu, J.; Shen, R.; LaFortune, T.A.; Wang, S.; Tiao, N.; Kim, Y.W.; Liu, J.L.; Ramnarian, D.; Yuan, Y.; et al. Antitumor activity of NVP-BKM120–a selective pan class I PI3 kinase inhibitor showed differential forms of cell death based on p53 status of glioma cells. *Clin. Cancer Res.* **2012**, *18*, 184–195. [CrossRef]
19. Masica, D.L.; Karchin, R. Correlation of somatic mutation and expression identifies genes important in human glioblastoma progression and survival. *Cancer Res.* **2011**, *71*, 4550–4561. [CrossRef]
20. Palomero, T.; Lim, W.K.; Odom, D.T.; Sulis, M.L.; Real, P.J.; Margolin, A.; Barnes, K.C.; O'Neil, J.; Neuberg, D.; Weng, A.P.; et al. NOTCH1 directly regulates c-MYC and activates a feed-forward-loop transcriptional network promoting leukemic cell growth. *Proc. Natl. Acad. Sci. USA* **2006**, *103*, 18261–18266. [CrossRef]
21. Jarriault, S.; Brou, C.; Logeat, F.; Schroeter, E.H.; Kopan, R.; Israel, A. Signalling downstream of activated mammalian Notch. *Nature* **1995**, *377*, 355–358. [CrossRef] [PubMed]
22. Satoh, Y.; Matsumura, I.; Tanaka, H.; Ezoe, S.; Sugahara, H.; Mizuki, M.; Shibayama, H.; Ishiko, E.; Ishiko, J.; Nakajima, K.; et al. Roles for c-Myc in self-renewal of hematopoietic stem cells. *J. Biol. Chem.* **2004**, *279*, 24986–24993. [CrossRef] [PubMed]
23. Weng, A.P.; Millholland, J.M.; Yashiro-Ohtani, Y.; Arcangeli, M.L.; Lau, A.; Wai, C.; DelBianco, C.; Rodriguez, C.G.; Sai, H.; Tobias, J.; et al. c-Myc is an important direct target of Notch1 in T-cell acute lymphoblastic leukemia/lymphoma. *Genes Dev.* **2006**, *20*, 2096–2109. [CrossRef] [PubMed]
24. Pagliarini, R.; Shao, W.; Sellers, W.R. Oncogene addiction: Pathways of therapeutic response.; resistance.; and road maps toward a cure. *EMBO Rep.* **2015**, *16*, 280–296. [CrossRef] [PubMed]
25. Yan, W.; Zhang, W.; Jiang, T. Oncogene addiction in gliomas: Implications for molecular targeted therapy. *J. Exp. Clin. Cancer Res.* **2011**, *17*, 30–58. [CrossRef] [PubMed]

26. Aravindan, N.; Aravindan, S.; Manickam, K.; Natarajan, M. High Energy Particle Radiation-associated Oncogenic Transformation in Normal Mice: Insight into the Connection between Activation of Oncotargets and Oncogene Addiction. *Sci. Rep.* **2016**, *6*, 37623. [CrossRef] [PubMed]
27. DeGraffenried, L.A.; Fulcher, L.; Friedrichs, W.E.; Grünwald, V.; Ray, R.B.; Hidalgo, M. Reduced PTEN expression in breast cancer cells confers susceptibility to inhibitors of the PI3 kinase/Akt pathway. *Ann. Oncol.* **2004**, *15*, 1510–1516. [CrossRef]
28. She, Q.B.; Solit, D.; Basso, A.; Moasser, M.M. Resistance to gefitinib in PTEN-null HER-overexpressing tumor cells can be overcome through restoration of PTEN function or pharmacologic modulation of constitutive phosphatidylinositol 3′-kinase/Akt pathway signaling. *Clin. Cancer Res.* **2003**, *9*, 4340–4346.
29. Bhat, K.P.; Salazar, K.L.; Balasubramaniyan, V.; Wani, K.; Heathcock, L.; Hollingsworth, F.; James, J.D.; Gumin, J.; Diefes, K.L.; Kim, S.H.; et al. The transcriptional coactivator TAZ regulates mesenchymal differentiation in malignant glioma. *Genes Dev.* **2011**, *25*, 2594–2609. [CrossRef]
30. Koul, D.; Jasser, S.A.; Lu, Y.; Davies, M.A.; Shen, R.; Shi, Y.; Mills, G.B.; Yung, W.K. Motif analysis of the tumor suppressor gene MMAC/PTEN identifies tyrosines critical for tumor suppression and lipid phosphatase activity. *Oncogene* **2002**, *21*, 57–64. [CrossRef]
31. Bao, S.; Wu, Q.; Li, Z.; Sathornsumetee, S.; Wang, H.; McLendon, R.E.; Hjelmeland, A.B.; Rich, J.N. Targeting cancer stem cells through L1CAM suppresses glioma growth. *Cancer Res.* **2008**, *68*, 6043–6048. [CrossRef] [PubMed]
32. Damaraju, V.L.; Bouffard, D.Y.; Wong, C.K.; Clarke, M.L.; Mackey, J.R.; Leblond, L.; Cass, C.E.; Grey, M.; Gourdeau, H. Synergistic activity of troxacitabine (Troxatyl) and gemcitabine in pancreatic cancer. *BMC Cancer.* **2007**, *7*, 121. [CrossRef] [PubMed]

cancers

MDPI

Article

EphA3 Pay-Loaded Antibody Therapeutics for the Treatment of Glioblastoma

Carolin Offenhäuser [1,†], Fares Al-Ejeh [1,†], Simon Puttick [2,3], Kathleen S. Ensbey [1], Zara C. Bruce [1], Paul R. Jamieson [1], Fiona M. Smith [1], Brett W. Stringer [1], Benjamin Carrington [1], Adrian V. Fuchs [2,4], Craig A. Bell [2,4], Rosalind Jeffree [5], Stephen Rose [3], Kristofer J. Thurecht [2,4], Andrew W. Boyd [1,‡] and Bryan W. Day [1,6,‡,*]

1 Department of Cell and Molecular Biology, QIMR Berghofer Medical Research Institute, Brisbane 4006, Australia; Carolin.Offenhauser@qimrberghofer.edu.au (C.O.); Fares.Al-Ejeh@qimrberghofer.edu.au (F.A.-E.); Kathleen.Ensbey@qimrberghofer.edu.au (K.S.E.); Zara.Bruce@qimrberghofer.edu.au (Z.C.B.); Jamieson.P@wehi.edu.au (P.R.J.); Fiona.Smith@qimrberghofer.edu.au (F.M.S.); Brett.W.Stringer@gmail.com (B.W.S.); Ben.Carrington@hansonwade.com (B.C.); Andrew.Boyd@qimrberghofer.edu.au (A.W.B.)

2 Australian Institute for Bioengineering and Nanotechnology, The University of Queensland, Brisbane 4072, Australia; S.Puttick@uq.edu.au (S.P.); adrianfuchs1@hotmail.com (A.V.F); C.Bell1@uq.edu.au (C.A.B.); K.Thurecht@uq.edu.au (K.J.T.)

3 Commonwealth Scientific and Industrial Research Organisation (CSIRO), Brisbane 4006, Australia; Stephen.Rose@csiro.au

4 Centre for Advanced Imaging, The University of Queensland, Brisbane 4072, Australia

5 Kenneth G. Jamieson Department of Neurosurgery, Royal Brisbane and Women's Hospital, Brisbane 4029, Australia; Lindy.Jeffree@health.qld.gov.au

6 School of Biomedical Sciences, Faculty of Health, Queensland University of Technology, Brisbane 4059, Australia

* Correspondence: Bryan.Day@qimrberghofer.edu.au; Tel.: +61-7-3845-3885

† These authors contributed equally to the manuscript.

‡ Co-Senior authors.

Received: 13 November 2018; Accepted: 11 December 2018; Published: 17 December 2018

Abstract: The EphA3 receptor has recently emerged as a functional tumour-specific therapeutic target in glioblastoma (GBM). EphA3 is significantly elevated in recurrent disease, is most highly expressed on glioma stem cells (GSCs), and has a functional role in maintaining self-renewal and tumourigenesis. An unlabelled EphA3-targeting therapeutic antibody is currently under clinical assessment in recurrent GBM patients. In this study, we assessed the efficacy of EphA3 antibody drug conjugate (ADC) and radioimmunotherapy (RIT) approaches using orthotopic animal xenograft models. Brain uptake studies, using positron emission tomography/computed tomography (PET/CT) imaging, show EphA3 antibodies are effectively delivered across the blood-tumour barrier and accumulate at the tumour site with no observed normal brain reactivity. A robust anti-tumour response, with no toxicity, was observed using EphA3, ADC, and RIT approaches, leading to a significant increase in overall survival. Our current research provides evidence that GBM patients may benefit from pay-loaded EphA3 antibody therapies.

Keywords: EphA3; antibody drug conjugate; radioimmunotherapy; glioblastoma; stem cells

1. Introduction

Glioblastoma (GBM) is the most common and aggressive adult malignant brain cancer. Despite surgical resection, radiation, and temozolomide chemotherapy, median survival is approximately 15 months [1,2]. Brain cancer sufferers have not seen meaningful increases in overall survival for decades. This is, in large part, due to the highly infiltrative and heterogeneous nature of these aggressive tumours [3]. GBM also displays complex cellular hierarchies containing self-renewing glioma stem cells (GSCs) with the ability to re-populate the tumour post-therapy.

Eph receptors are the largest family of receptor tyrosine kinases (RTKs). Whilst they have critical functions during embryonic development, they are typically expressed at low levels in normal adult tissues [4]. It is now established that numerous Eph receptors are re-expressed and functional in human cancers, making them attractive, relatively tumour-specific targets [5,6]. The ephrin type-A receptor 3 (EphA3) (formerly known as HEK) was first identified in 1992 as a surface antigen on pre-B lymphoblastic leukaemia cells [7]. Since then, EphA3 has been widely implicated in a number of developmental processes involving cell adhesion, cell migration, and tissue boundary formation. EphA3 has been shown to be elevated in a number of haematological cancers and solid tumours, with both oncogenic and tumour suppressive functions being described [8]. We recently reported that EphA3 has oncogenic functions in GBM [9,10]. Critically, we showed that EphA3 is expressed at low levels in a normal brain, but is most highly expressed on GSCs, where it has a functional role in survival and self-renewal. EphA3 knockdown (KD) induced tumour cell differentiation, apoptosis, and reduced tumour formation in vivo. More recently in GBM, EphA3 was shown to be significantly elevated in recurrent post-treatment versus primary treatment naïve disease. These findings further establish this receptor as an attractive tumour-specific target for brain cancer therapy [9,11,12]. Moreover, EphA3 has been shown to be over-expressed and functional on mesenchymal stromal cells in a number of human cancers, and EphA3 antibody targeting inhibited tumour growth by disrupting newly formed tumour microvasculature [13]. We previously generated the EphA3 monoclonal antibody (mAb) IIIA4 [7]. IIIA4 binds the EphA3 globular ephrin-binding domain with high affinity ($K_D \sim 5 \times 10^{-10}$ mol/L) with a low dissociation constant ($K_D = 3 \times 10^{-4}$/s) [14–16]. IIIA4, similar to how the high affinity ligand ephrin-A5, triggers rapid EphA3 activation and internalisation by inducing a conformation change allowing for the assembly of EphA3/ephrin-A5 signal clusters [17]. An investigator-sponsored Phase 0/1 clinical trial is currently underway using the EphA3-targeting monoclonal antibody (mAb) Ifabotuzumab [18] (Humanigen Inc., Brisbane, CA, USA), a humanised version of IIIA4, in patients with recurrent GBM. This study will confirm the safety and recommended dose to achieve optimal tumour penetration using positron emission tomography/computed tomography (PET/CT) imaging. This study is in part motivated by reports of the positive results of other antibody drug conjugate (ADC) therapy in the treatment of GBM. Depatuxizumab mafodotin (Depatux-m, formerly ABT-414) is an ADC against activated epidermal growth factor receptor (EGFR), conjugated, via a non-cleavable linker, to the cytotoxic microtubule-targeting agent monomethyl auristatin F (MMAF). Efficacy has been demonstrated as both a single agent and in combination with temozolomide (TMZ) [19,20].

EphA3 is predominantly expressed on the stem cell fraction in GBM and hence represents a logical target for further exploitation of immunotherapeutic approaches to this disease. To this end, we assessed the efficacy of EphA3 pay-loaded antibody strategies in GBM, using both EphA3 ADC and radioimmunotherapy (RIT) approaches in orthotopic GBM xenograft models. Our results shed light on further clinical evaluation of EphA3 targeting in GBM, and highlight this receptor as a tumour-specific candidate suitable for ADC or RIT strategies.

2. Results

2.1. The EphA3-Maytansine ADC Induces Potent Cell Killing In Vitro

The IIIA4 antibody was deemed to be an excellent candidate for ADC therapy, as previous studies have shown that the IIIA4 antibody-EphA3 interaction induces rapid internalisation [17]. As a first step, we prepared an EphA3-ADC by conjugating the IIIA4 antibody to the cytotoxic microtubule-targeting agent maytansine (USAN), using a non-cleavable succinimidyl 4-(*N*-maleimidomethyl) cyclohexane-1-carboxylate (SMCC) linker, with a drug-to-antibody ratio (DAR) of 2.81 (Figure 1A and Figure S1A). A DAR of ~2–4 has been shown to be optimal for in vivo efficacy, as a low DAR reduces potency while a high DAR (~8) negatively affects antibody clearance and pharmacokinetics (PK) [21]. Bio-Layer Interferometry (BLI) Octet®, analysis showed that IIIA4 binding kinetics were unaffected following USAN conjugation (Figure S1B). Immunofluorescence (IF) analysis of EphA3-positive primary GBM cell lines, labelled with IIIA4 at 4 °C to prevent internalisation of receptor-antibody complexes, confirmed expression of the receptor on the cell membrane. To confirm receptor complexes were internalised following IIIA4 treatment, we performed IF labelling of EphA3 post-antibody treatment (IIIA4, 1 µg/mL at 37 °C for 20 min) (Figure 1B and Figure S1C). To assess IIIA4-USAN efficacy, we selected the EphA3-positive lines, U251 and SJH1, and an EphA3-negative line PB1. SJH1 and PB1 are primary GBM cultures, generated in-house (Q-Cell), and grown as glioma neural stem (GNS) cell cultures (Figure 1C) [22,23]. IIIA4-USAN showed a significant and specific reduction of cell viability in the EphA3-positive U251 and SJH1 lines 7 days post-treatment, which was accompanied with apoptotic cell death as judged by induction of caspase-3 cleavage (Figure 1D,E).

We expanded our cytotoxicity studies to a panel of four Q-Cell EphA3-positive primary GBM GNS cultures, representing all GBM subtypes, classical (CL), proneural (PN) and mesenchymal (MES), (WK1-CL, JK2-PN, BAH1-PN, SB2b-MES) (Figure 2A) [24–26]. IIIA4-USAN was potent across these GBM cultures with a half maximal inhibitory concentration (IC_{50}) of less than 0.7 µg/mL (Figure 2B).

2.2. IIIA4 Crosses the Blood-Tumour Barrier (BTB) and Accumulates Specifically in GBM

IIIA4 binds both mouse and human EphA3 with high affinity [17]. To assess IIIA4 brain penetration and tumour uptake, we engrafted the EphA3-positive U251 cell line into the right striatum of a NOD/SCID mouse and allowed the tumour to form orthotopically for 30 days. Biotin-labelled IIIA4 was then injected intravenously (IV) via the tail vein and allowed to circulate for two hours. The animal was then euthanised and coronal brain sections prepared, followed by IF analysis using streptavidin-Alexa Fluor 488. Results showed that the IIIA4-biotin accumulated specifically at the tumour site, while no staining was detected in surrounding normal brain (Figure 3A). To further assess tumour penetration, we engrafted the EphA3-positive primary model WK1 into the striatum of a NOD/SCID mouse. WK1 tumours were allowed to form for 60 days before PET/CT imaging was conducted 12 h post ^{64}Cu-IIIA4 IV injection, followed by magnetic resonance imaging (MRI). PET/CT revealed specific tumour uptake with a minimal signal detected in contralateral normal brain (Figure 3B).

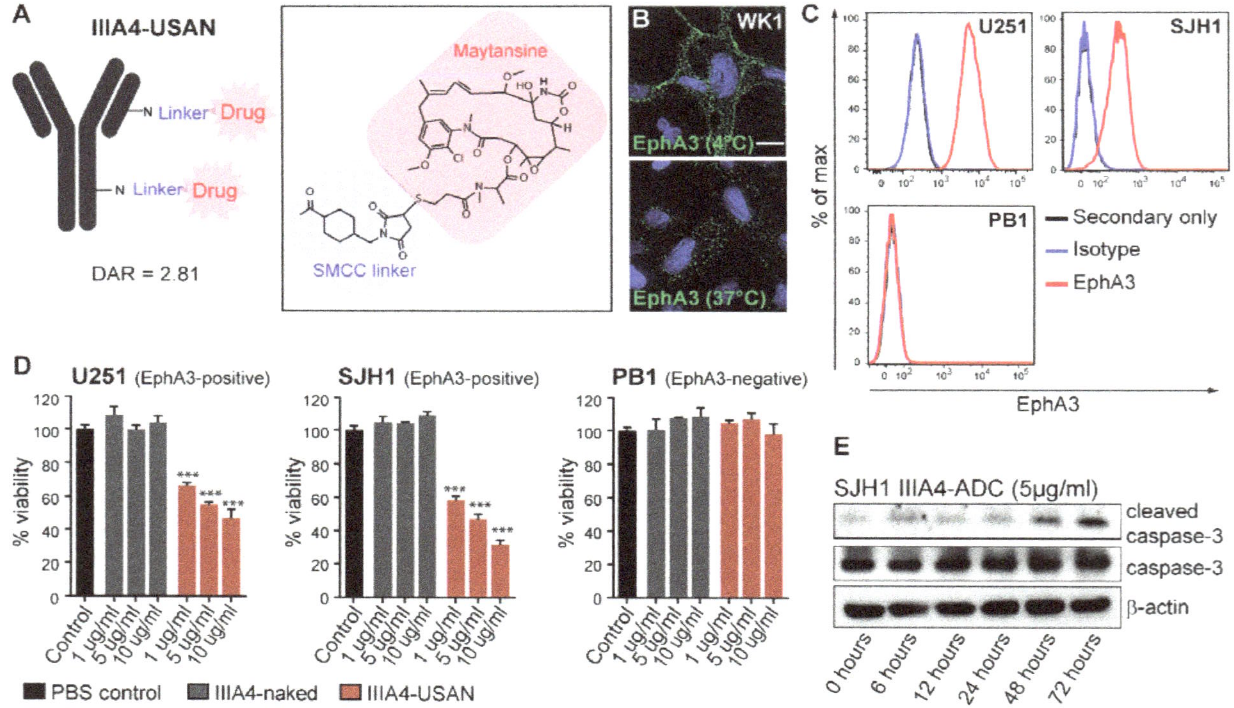

Figure 1. The EphA3-maytansine antibody drug conjugate (ADC) induces potent killing in vitro. (**A**) Schematic of IIIA4-USAN, an EphA3-specific monoclonal antibody (mAb) (IIIA4) conjugated to the tubulin-inhibitor maytansine (USAN) via a non-cleavable succinimidyl 4-(*N*-maleimidomethyl) cyclohexane-1-carboxylate (SMCC) linker. (**B**) Representative immunofluorescence images of EphA3-labelling with IIIA4 mAbs (shown in green) on the cell surface (top panel) and after antibody internalization (bottom panel) in the patient-derived glioblastoma (GBM) cell line WK1. (**C**) Flow cytometry analysis of EphA3 cell surface expression compared to isotype control in U251 and two primary GBM cell lines. (**D**) Cell viability of the EphA3$^+$ U251 and SJH1 and EphA3$^-$ PB1 GBM cell lines was measured by MTS colorimetric assay 7 days post IIIA4-USAN treatment. (**E**) SJH1 cells were treated with IIIA4-ADC (5 µg/mL) for the indicated times, followed by Western blot analysis. Scale bar represents 10 µm.

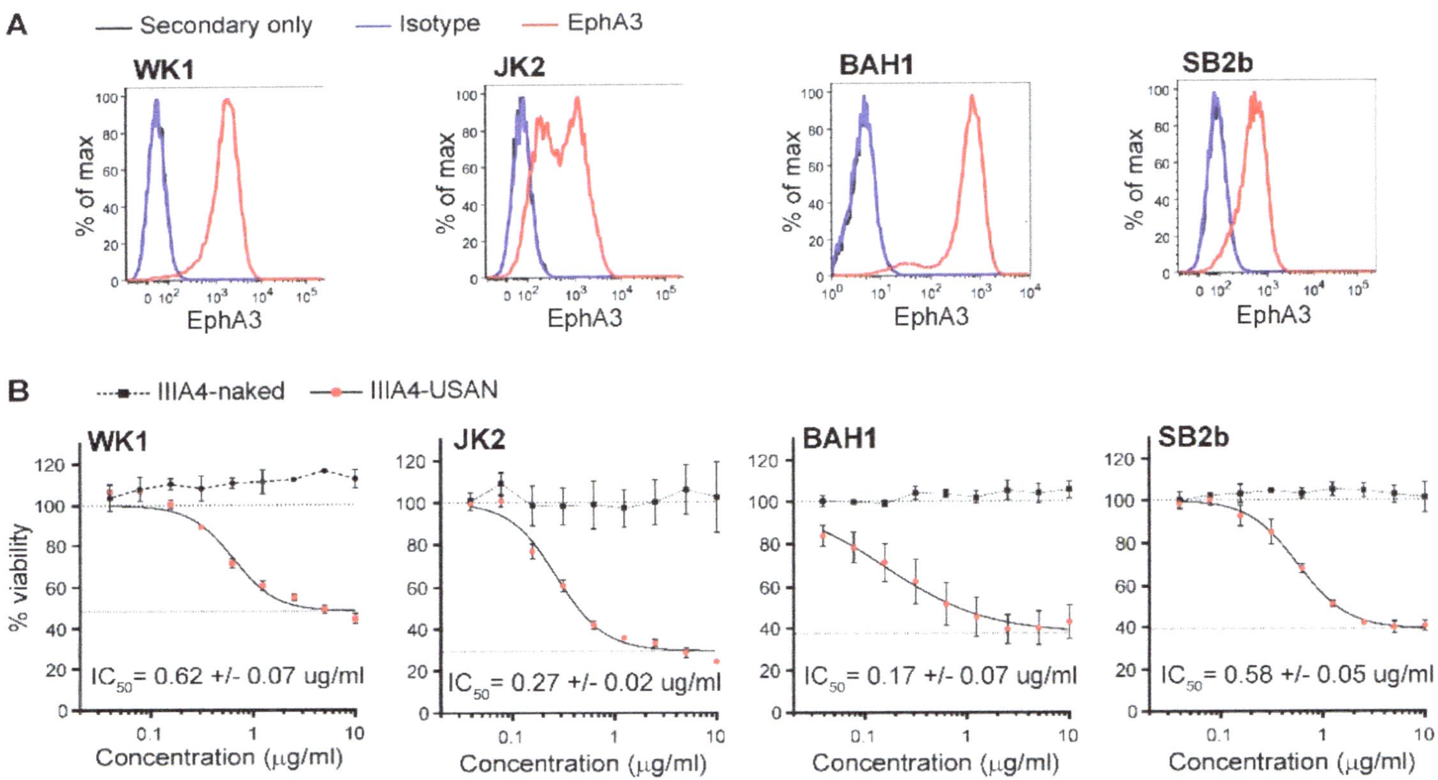

Figure 2. IC_{50} concentrations of IIIA4-USAN in primary models of GBM. (**A**) EphA3 cell surface flow cytometry profiles of four primary GBM cell lines. (**B**) Cell viability relative to the control was measured by MTS colorimetric assay and IC_{50} concentrations determined after 4 days of treatment with increasing doses of IIIA4-USAN versus IIIA4-naked.

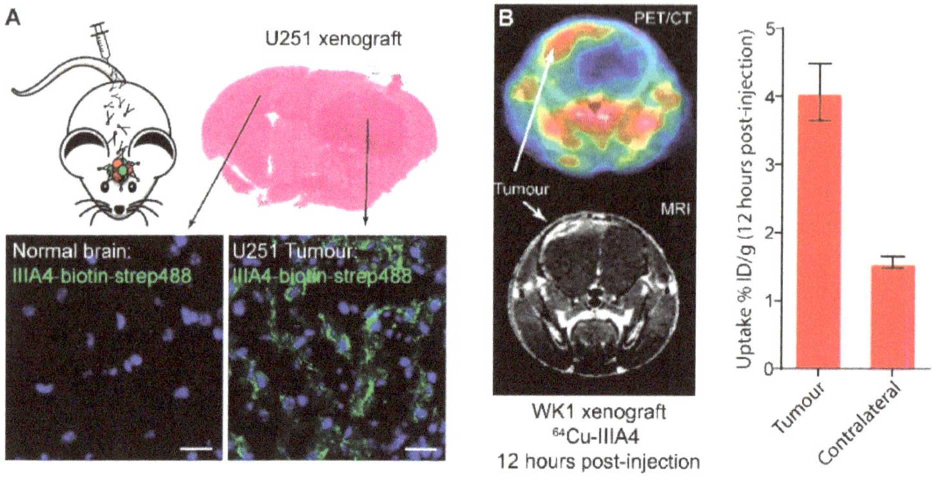

Figure 3. IIIA4 crosses the blood-tumour barrier (BTB) and accumulates specifically in GBM. (**A**) Tumour uptake of biotin-labelled IIIA4 (shown in green) 2 h post IV injection in a U251 orthotopic tumour xenograft, compared to adjacent normal brain was visualized with streptavidin-Alexa Fluor 488. (**B**) positron emission tomography/computed tomography (PET/CT) imaging and image quantification of ^{64}Cu-labelled IIIA4 uptake into MRI-confirmed tumour tissue in an EphA3^{+} intracranial primary GBM tumour model (WK1). Images were acquired 12 h after IV injection of the labelled antibody. Scale bars represent 50 µm.

2.3. The EphA3-Maytansine ADC Induces A Potent GBM Anti-Tumour Response In Vivo

To assess the efficacy of the IIIA4-USAN ADC, we first selected the EphA3-positive U251 model. Three weeks post-orthotopic injection, large tumours were detected via bioluminescent imaging. IIIA4-USAN therapy (10 mg/kg IV twice weekly) was commenced and compared to treatment with unlabelled IIIA4 (IIIA4-naked) or vehicle (PBS)-only arms. No signs of toxicity were observed at any time during the course of the up to 22 IIIA4-USAN injections. Weekly imaging revealed an initial anti-tumour response, which appeared to reduce tumour burden to near completion in some animals after eight treatments (Figure 4A and Figure S2A). Tumours rapidly progressed in both control arms and all control animals had been euthanised five weeks after treatment initiation from advanced tumour burden. Despite continuation of IIIA4-USAN injections, tumours recurred and progressed in all animals. IIIA4-USAN led to a significant ($p = 0.0007$, $n = 6$) increase in overall survival compared to controls. Notably, IIIA4-USAN almost doubled overall survival. Median survival in days was 46.5 for PBS and 47.5 for IIIA4-naked, versus 90.5 days for IIIA4-USAN (Figure 4B). Post mortem, we examined EphA3 mRNA levels in controls versus treated animals by qPCR. Tumours were collected from four animals per group and expression analysed. IIIA4-USAN treatment led to a significant ($p < 0.01$) reduction in EphA3 mRNA levels in these tumours (Figure 4C). We repeated our initial U251 IIIA4-USAN animal study, on this occasion we allowed tumours to form for six weeks before treatment was commenced. Weekly imaging revealed a similar initial robust anti-tumour response, while tumours rapidly progressed in both control arms. ADC treatment significantly increased overall survival ($p = 0.0024$, $n = 5$). Median survival in days was 56 for PBS and 60 for IIIA4-naked, versus 92 days for IIIA4-USAN (Figure S2B). We have previously shown that shRNA-mediated knockdown (KD) of EphA3 in U251 cells prolongs survival following orthotopic engraftment [9]. To confirm that EphA3 has a similar oncogenic role in the WK1 primary model, we conducted a comparable orthotopic animal study. WK1 control shRNA cells were compared to constitutive EphA3 shRNA expressing cells. Effective EphA3 KD in WK1 was confirmed as part of a previous study [9]. EphA3 KD led to a significant increase in overall survival ($p = 0.0225$, $n = 6$). Median survival in days was control 109.5

for shRNA versus 148.5 for EphA3 KD (Figure 4D). We next assessed IIIA4-USAN, using the WK1 model as a single agent (10 mg/kg IV twice weekly). A significant increase ($p = 0.0002$, $n = 6$) in overall survival was observed from the IIIA4-USAN compared to PBS and IIIA4-naked (Figure 4E).

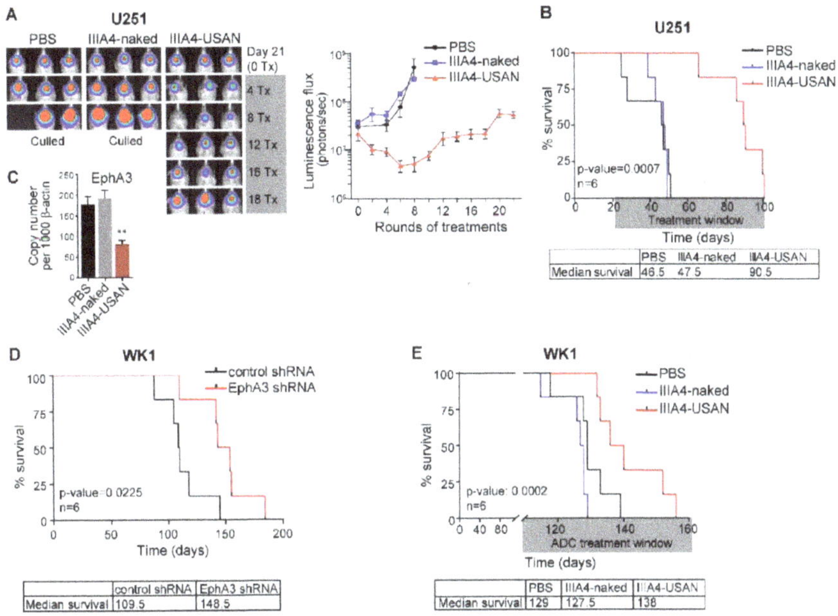

Figure 4. The EphA3-maytansine ADC induces a potent GBM anti-tumour response in vivo. (**A**) Luciferase-expressing U251 cells were engrafted into NRG mice. From day 21 after engraftment until endpoint, animals received twice weekly IV injections of IIIA4-USAN (10 mg/kg), IIIA4-naked (10 mg/kg), or PBS vehicle control. Representative images of tumour signals are shown. For complete analysis, see Figure S2A. (**B**) Kaplan-Meier survival analysis showing a significant increase in overall survival following IIIA4-USAN treatment. (**C**) qPCR analysis showing a significant (** $p < 0.01$) reduction in EphA3 mRNA levels in IIIA4-USAN-treated animals compared to controls. Tumors tissue from 4 animals per arm were analysed. (**D**) Kaplan Meier survival analysis showing a significant increase in overall survival of animals engrafted orthotopically with WK1 EphA3 shRNA-expressing cells compared to control shRNA-expressing cells. (**E**) Kaplan Meier survival analysis showing a significant increase in overall survival in NRG mice engrafted with the WK1 cells following IIIA4-USAN (10 mg/kg) treatment compared to controls.

2.4. EphA3-Targeting RIT Induces A GBM Anti-Tumour Response In Vivo

To examine if EphA3-targeting RIT could have benefit as a GBM therapy we radiolabelled IIIA4-DOTA with the β particle-emitting radionuclide lutetium-177 (^{177}Lu-IIIA4). Established U251 xenografts in BALB/c nude mice were treated with a single dose of ^{177}Lu-IIIA4 (450 MBq/kg, equivalent to ~9 MBq per animal), administered IV. No signs of toxicity were observed at any time during the course of the experiment. Compared to the unlabelled IIIA4-DOTA control, a significant anti-tumour response and a delayed onset of tumour progression, up to 32 days, was detected by imaging (Figure 5A). This delay is significant considering the 6.7 days half-life of ^{177}Lu, and that by day 32 approximately five half-lives have passed. Tumours rapidly progressed in the control arm and all control animals had been euthanised 32 days after commencing treatment, due to advanced tumour burden. ^{177}Lu-IIIA4 led to a significant ($p = 0.0329$) increase in overall survival compared to the control arm. Median survival in days was 47 for IIIA4-DOTA-control versus 64.5 for ^{177}Lu-IIIA4 (Figure 5B).

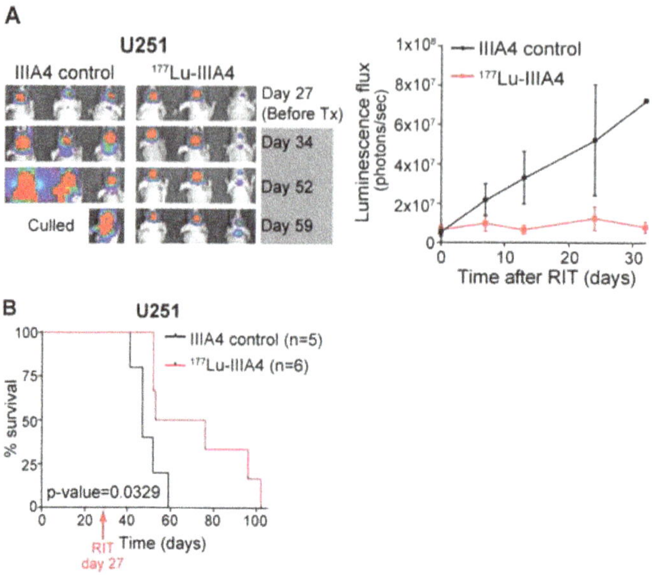

	IIIA4 control	^{177}Lu-IIIA4
Median survival	47	64.5

Figure 5. EphA3-targeting RIT induces a GBM anti-tumour response in vivo. (**A**) Luciferase-expressing U251 cells were engrafted orthotopically and, on day 27, animals received a single dose of 450 MBq/kg dose of IIIA4-DOTA-^{177}Lu (^{177}Lu-IIIA4) or IIIA4-DOTA control (IIIA4 control). Representative images of tumour signals are shown. (**B**) Kaplan-Meier survival analysis showing a significant increase in overall survival following a single dose of ^{177}Lu-IIIA4 (450 MBq/kg) treatment.

3. Discussion

Since 2005, when TMZ was adopted into the clinical management of GBM, no therapy has significantly improved overall patient survival, prompting a search for alternative therapeutic modalities [1,2]. ADC therapy has emerged as an effective part of anti-cancer treatment. A recent example is the HER2 mAb Trastuzumab coupled to the maytansinoid anti-microtubule agent (DM1) for the treatment of HER2-positive breast cancer [27–29]. In GBM, Depatux-m (ABT-414), an ADC against activated EGFR is showing promise as an effective therapy [19,20]. A large body of data on the role of EphA3 biology in cancer has identified this receptor as one of the most promising targets for therapy, based on low expression levels in normal adult tissues and the tumour-promoting properties, including stem cell maintenance, angiogenesis, metastatic functions, and enriched expression in recurrent disease [30]. Here, we build upon our existing body of work and the work of others to characterize EphA3 as a tumour-specific therapeutic target for the treatment of GBM [9–12]. Our findings show that EphA3 is an effective target for antibody-based pay-loaded therapy in GBM. PET/CT imaging showed that the IIIA4 antibody readily enters the tumour in our orthotopic models, accumulating specifically at the tumour site with minimal uptake in normal brain. Importantly, we observed no toxicity from either EphA3 ADC or RIT treatment. IIIA4 has equal affinity for both mouse and human EphA3, allowing accurate assessment of off-target toxicity in our animal studies. EphA3 ADC or RIT were both very well tolerated, in the case of one animal study, up to 22 doses of IIIA4-USAN were administered over an eleven-week period with no toxic response. In all our animal studies, therapy was commenced only after large tumours had already formed in the brain. Despite this, both ADC and RIT therapies were able to significantly reduce tumour burden. In one experiment, although ADC therapy was not commenced until six weeks post engraftment, efficacy was still achieved, leading to a significant increase in overall survival in the treated group. Thus, both pay-loading strategies were shown to be effective, whereas the IIIA4 antibody alone showed no effect on tumour progression. The EphA3-targeting mAb Ifabotuzumab, a humanised version of IIIA4, has been trialed in its naked form in patients with advanced haematologic malignancies [18]. Results from this study found that the antibody was well-tolerated. While some clinical responses were observed, overall responses were modest. An investigator-sponsored Phase 0/1 clinical trial has been instigated to assess Ifabotuzumab in patients with recurrent GBM (NCT03374943). This study will confirm safety in a solid tumour setting and determine the biodistribution and pharmacokinetics of ^{89}Zr-Ifabotuzumab using PET/CT imaging. In addition, PET/CT imaging could be used to screen EphA3 levels in patient tumours prior to commencing Ifabotuzumab therapy. Though efficacy of naked Ifabotuzumab is yet to be determined in patients with GBM, it might be expected that greater responses could be achieved using pay-loaded approaches in these aggressive diseases. With this in mind, we chose to assess EphA3 targeting antibody pay-loaded strategies in pre-clinical GBM models to inform potential future trial design.

Ultimately, all tumours recurred using both pay-loaded strategies. This was despite ADC therapy being administered throughout the course of the experiments. In seeking to understand the escape from the initial response, it was notable that ADC therapy also significantly reduced EphA3 receptor levels in recurrent tumours, indicating that escape might be partly mediated by down-regulation of EphA3. Alternatively, this finding could indicate that therapy was specifically targeting the EphA3 positive tumour compartment, reducing overall receptor levels. In recent years, acquired resistance to ADCs has emerged as a challenge for this class of cancer therapeutics. Most patients treated with Trastuzumab-maytansine (T-DM1), while initially responsive, develop acquired drug resistance to this therapy [31]. Preclinical studies have recently revealed several possible resistance mechanisms, including decreased cell surface antigen expression, altered intracellular trafficking, impaired lysosomal degradation, and payload removal via multiple drug resistance transporters [32]. This highlights the need for drugs which target multiple aspects of tumour biology in order to develop effective combinations of therapies, rather than single agents alone. However, despite these limitations, our pre-clinical study indicates that GBM sufferers may benefit from pay-loaded EphA3 antibody targeting strategies and supports further clinical evaluation of these therapies.

4. Materials and Methods

4.1. Primary Cell Cultures

We have developed a characterised GBM patient-derived cell line resource (Q-Cell) [23], in which GBM lines are maintained as glioma neural stem cell (GNS) cultures [22] or as neurosphere cultures using StemPro NSC SFM (Invitrogen, Carlsbad, CA, USA) as per manufacturer's guidelines. Characterisation data is freely available from: https://www.qimrberghofer.edu.au/q-cell/. U251-MG GBM cells were obtained from the ATCC (Manassas, VA, USA) and cultured as GNS cultures as described above. Luciferase-expressing lines were generated by transducing U-251-MG and WK1 cells with Firefly Luciferase Lentifect™ Purified Lentiviral Particles (GeneCopoeia, Rockville, MD, USA). 70–80% confluent cells were transduced with 2 µL of Lentifect particles in 2 mL of culture medium, supplemented with 8 µg/mL polybrene per well of a 6-well plate. After adding the virus, cells were centrifuged at 500× g in a swing-out rotor for 45 min at room temperature and subsequently cultured for 24 h before the medium was replaced. Puromycin was added 72 h after transduction to select successfully transduced cells.

4.2. ADC Preparation

IIIA4 was purified in-house as previously described [17]. To conjugate IIIA to the non-cleavable linker SMCC, sulfo-SMCC (2.2 mg in 220 µL DMSO) was added dropwise to IIIA4 in PBS (8 mL at 9.17 mg/mL) at 4 °C and stirred overnight. IIIA4-SMCC was purified by centrifugal filtration (Amicon Ultra-15 centrifugal filters, 3 kDa MWCO). The SMCC linker to antibody ratio of 7.2

SMCC units per IIIA4 was determined by Matrix Assisted Laser Desorption Ionisation-Time of Flight Mass Spectroscopy (MALDI-ToF MS). Maytansine DM1 (1.62 mg in 162 μL DMSO, Abcam, Cambridge, UK) was added dropwise to IIIA4-SMCC (8.2 mL at 9.26 mg/mL) in PBS at 4 °C, stirred overnight and purified by centrifugal filtration (Amicon Ultra-15 centrifugal filters, 3 kDa MWCO). A drug-to-antibody (DAR) ratio of 2.81 USAN units per IIIA4-SMCC was determined by MALDI-ToF. MALDI-ToF MS was conducted on a Bruker Autoflex III equipped with a nitrogen laser delivering 3 ns laser pulses at 337 nm. A matrix solution was prepared consisting of sinapinic acid (saturated) in MeCN containing 0.1% *v*/*v* trifluoroacetic acid. The antibody solutions (in PBS) were diluted 10× with Milli-Q water, then mixed in an Eppendorf tube with the matrix solution in a 1:1 ratio. The resulting solutions were then applied to a polished steel target (1 μL) followed be solvent evaporation to prepare a thin matrix/analyte film. Samples were analysed in positive linear mode over a mass range of 20–200 kDa with an average of 10,000 shots. The average number of maleimide reactive groups per IIIA4-SMCC for USAN attachment was calculated from the mass difference between IIIA4-SMCC and IIIA4 divided by MW_{SMCC} after conjugation (220.1 Da). Subsequently, the DAR was calculated from the mass difference between IIIA4-USAN and IIIA4-SMCC divided by MW_{USAN} after conjugation (738.3 Da).

4.3. Bio-Layer Interferometry (BLI)

Antibody binding kinetics of IIIA4-naked and IIIA4-USAN to EphA3-Fc-loaded biosensors was compared by Bio-Layer Interferometry using the Octet® RED system (ForteBio, Fremont, CA, USA). EphA3-Fc protein (in-house) was conjugated to biotin using a 1:1 molar ratio of protein to EZ-Link Sulfo NHS-LC-LC-Biotin (ThermoFisher Scientific, Waltham, MA, USA) following the manufacturer's instructions. Excess biotin was removed using Zeba Spin desalting columns 7K MWCO (ThermoFisher Scientific). Biotinylated EphA3-Fc protein (50 μg/mL) was loaded onto streptavidin biosensors (ForteBio) as per the manufacturer's instructions, and binding of IIIA4-naked and IIIA4-USAN (30 μg/mL) to individual sensors recorded. On reaching saturation, the probes were exchanged into 1× kinetics buffer (1 mM Phosphate, 15 mM NaCl, 0.005% Tween 20 and 0.1 mg/mL BSA) to allow dissociation.

4.4. Cell Viability Assay

To assess cell viability, 5×10^3 cells per well were seeded in 96-well plates in GNS conditions and incubated overnight. Cells were incubated with IIIA4-naked and IIIA4-USAN at the indicated concentrations or medium alone (control) for seven days before cell viability was assessed by MTS assay as per manufacturer's instructions (CellTiter 96 Aqueous One Cell Proliferation Assay; Promega, Madison, WI, USA). To measure dose response curves, cells were incubated with nine 1:2 serial dilutions, starting from 10 μg/mL, for 96 h before cell viability was assessed. Responses were normalised to percent viability of control and maximal effect and half-maximal inhibitory concentration (IC_{50}) determined by four-parameter nonlinear regressing analysis (top plateau constrained to 100%) using GraphPad Prism 7 software (GraphPad, San Diego, CA, USA).

4.5. Caspase-3 Activation

To assess caspase activation, cells were cultured in GNS conditions and treated with 5 ug/mL IIIA4-ADC for the indicated times, or incubated in medium alone. Cells were lysed and protein expression of cleaved caspase-3 (Cell Signaling Technology, Danvers, MA, USA, #9661, 1:1000), total caspase-3 (Cell Signaling Technology, #9662, 1:1000) and β-actin (Cell Signaling Technology, #9700, 1:5000) determined by Western blot analysis as previously described [33].

4.6. Immunofluorescence Analysis

Cells were seeded on coverslips and cultured overnight in GNS conditions. Cells were incubated with IIIA4 antibody at 4 °C (to prevent internalisation of antibody receptor complexes) or 37 °C for

20 min. Cells were washed twice in PBS and fixed in 4% paraformaldehyde in PBS for 30 min at room temperature. Cells were permeabilised with 0.25% Triton X-100 in PBS for 10 min at room temperature and EphA3 receptor-antibody complexes visualised with Alexa Fluor 488-conjugated secondary antibodies (ThermoFisher Scientific). Nuclei were labelled with DAPI. For in vivo uptake of IIIA4 antibody into intracranially engrafted tumours, a U251-MG tumour-bearing mouse was injected with a single dose of biotin-conjugated IIIA4 antibody (10 mg/kg). The antibody was allowed to circulate for 2 h before the mouse was euthanised and the brain removed and fixed in 10% neutral buffered formalin. IIIA4 antibody uptake was visualised in coronal cross-sections using Alexa Fluor 488 streptavidin (ThermoFisher Scientific), and nuclei were labelled with DAPI.

4.7. Flow Cytometry

Cell surface expression of EphA3 was analysed using the IIIA4 antibody (in-house, 1:100). IgG1 isotype control antibody (Abcam, 1:100) was used to indicate background staining. Cells were washed twice in ice-cold PBS and blocked for 20 min in 1% bovine serum albumin (BSA) in PBS on ice. After blocking, cells were incubated with primary antibodies for 20 min on ice, washed and incubated with secondary Alexa Fluor 488-conjugated antibodies for 20 min on ice in the dark (ThermoFisher Scientific). EphA3 expression levels were examined using a LSR Fortessa flow cytometer (BD, Franklin Lakes, NJ, USA) and acquired data analysed using FlowJo (FlowJo, LLC, CA, USA) software.

4.8. Animal Studies

All mice experiments were performed according to the National Health and Medical Research Council (2013) Australian code for the care and use of animals for scientific purposes, under experimental protocols approved by the QIMR Berghofer Animal Ethics Committee, P1572-A1412-613M approved 17 December 2014; P1173-A0812-608M approved 12 December 2008. For intracranial (orthotopic) xenograft studies, five week-old female NOD/SCID (NOD.CB17-Prkdc scid/Arc), BALB/c Nude (BALB/c-Foxn1^{nu}/Arc) or NRG mice (NOD.Cg-Rag1^{tm1MOM} IL2rg$^{tm1Wjl/SzJ}$) animals were sourced from the Animal Resources Centre (ARC) in Canning Vale, Western Australia). Mice were anaesthetised using 2% isoflurane in oxygen and engrafted intracranially using a small animal stereotactic device. Mice were given analgesia (Meloxicam (Ilium) 5 mg/kg, delivered subcutaneously) 30 min prior to surgery and again the following day. Tumour growth was monitored weekly by bioluminescence imaging. To generate bioluminescence signals, mice were anaesthetised with 2% isoflurane in oxygen, injected IP with 100 µL of 5 mg/mL D-Luciferin (Pure Science Ltd, Porirua, New Zealand) in PBS and imaged using the IVIS Spectrum in vivo imaging system (Perkin Elmer, Waltham, MA, USA). Following tumour formation, mice were randomised and tumour-matched based on bioluminescence imaging results. A non-specific antibody (1A7, in-house, 500 µg/mouse) was injected into the peritoneum 2 h prior to ADC therapy to block unspecific tissue uptake by unoccupied Fc receptors in immune-compromised mice. IIIA4-naked and IIIA4-USAN in PBS were administered intravenously at 10mg/kg twice weekly via the lateral tail vain. Mice were monitored daily for signs of illness or tumour burden as per our ethical guidelines, animal monitoring criteria, and scoring. At the endpoint, animals were euthanised by cervical dislocation.

4.9. Imaging

4.9.1. Generation of IIIA4-NOTA

IIIA4 (1 mg) was buffer exchanged into 0.1 M Na_2CO_3 (pH 9.2) and pSCN-Bn-NOTA (35 µg in 3.5 µL DMSO, Macrocyclics) added. The reaction mixture was incubated for 1 h at 37 °C with gentle agitation followed by purification by size exclusion chromatography (ÄKTA PrimePlus fitted with a 5 mL, HiTrap Desalting G25 column running a mobile phase of 0.1 M PBS at 0.5 mL/min, GE Life Sciences, NSW, Australia).

4.9.2. Radiolabelling

IIIA4-NOTA (200 µg) was buffer exchanged into 0.1 M NH_3OAc (pH 5.5) and incubated with 100 MBq ^{64}CuOAc for 1 h at 37 °C with gentle agitation followed by purification by centrifugal filtration (Zeba Spin Desalting Column, 7K MWCO, 0.5 mL equilibrated with 1× PBS, ThermoFisher). Radiochemical purity was assessed by iTLC (50 mM EDTA mobile phase using silica-gel impregnated glass microfiber paper strips (iTLC-SG, Agilent, Santa Clara, CA, USA). Radiochemical purity was >98%, radiochemical yield was 70% and specific activity was 350 MBq/mg.

4.9.3. PET/CT Imaging

NOD/SCID (n = 3) mice bearing intracranial WK1 tumours were injected with IIIA4-NOTA-^{64}Cu (5–7 MBq) via the lateral tail vein. At 12 h post injection, mice were anaesthetised and maintained using 2% isoflurane in oxygen at a flow rate of 2 L/min and positioned on the scanning bed of an Inveon Preclinical PET/CT system (Siemens, Munich, Germany). A 30 min PET image was acquired followed by a CT for anatomical registration and attenuation correction. PET images were reconstructed using the OSEM-2D reconstruction algorithm in Inveon Acquisition Workspace (IAW, Siemens) correcting for isotope half-life, attenuation, and ^{64}Cu detector efficiency.

4.9.4. Magnetic Resonance Imaging (MRI)

MRI images were acquired 24 h following PET/CT imaging. Mice were anaesthetised and maintained using 2% isoflurane in oxygen at a flow rate of 2 L/min. Immediately prior to acquisition of the MRI image, mice were injected with 50 µL of a 0.2 M solution of Magnevist (Bayer, Leverkusen, Germany) in 1× PBS. Mice were placed on the scanning bed of a Brüker 7T ClinScan (Brüker, Billerica, MA, USA) fitted with a 23 mm volume coil and 3D T_1 weighted contrast enhanced images were acquired using a VIBE sequence (TR = 12 ms, TE = 1.78 ms, FA = 21°, 0.12 mm isotropic resolution).

4.9.5. Image Analysis

Decay-corrected PET images and contrast-enhanced MRI images were co-registered to the CT image space using Inveon Research Workspace (Siemens). The brain was segmented from the CT image and applied as a mask to the PET and MRI images. Contrast-enhancing tumour was segmented from the MRI image by thresholding the brain volume voxel histogram and the segmented volume duplicated and horizontally reflected to form tumour and contralateral brain volumes of interest (VOIs). These VOIs were transformed onto the co-registered PET image and mean %ID/g within each VOI was measured.

4.10. Radioimmunotherapy (RIT)

The EphA3 mAb IIIA4 was conjugated to DOTA-NHS (Macrocyclics, Plano, TX, USA), radiolabelled with $^{177}LuCl_3$, and purified using 100 kDA-cutoff microconcentrators (Millipore, Burlington, MA, USA) as previously described [9,34,35]. The specific activity of ^{177}Lu-IIIA4 ranged from 225 and 260 MBq/mg, and radiolabelling efficiency was >98% as judged by standard, instant iTLC-SG [34]. Mice were injected intravenously with ^{177}Lu-IIIA4 once only at 450 MBq/kg dose (equivalent to 1.7–2.0 mg/kg of IIIA4 mAb dose).

4.11. Statistical Analysis

In vivo experiments—sample size or replicate number (designated as "n")—for each experiment is indicated in the figure. Charts and survival plots were generated using GraphPad Prism 7 software. A log-rank (Mantel-Cox) test was used to determine the significance between experimental groups. p-values are as indicated and * $p \leq 0.05$ was considered to be statistically significant. In vitro experiments—a Student's t-test determined the probability of difference—* $p < 0.05$ was considered significant; all statistical tests were two-sided.

5. Conclusions

Our data demonstrate that EphA3 can be effectively targeted using both ADC or RIT approaches, and that EphA3 mAbs cross the BTB to allow tumour-specific targeting. Our pre-clinical study supports further clinical evaluation of pay-loaded EphA3 antibody therapies in patients with GBM.

Supplementary Materials: The following are available online at http://www.mdpi.com/2072-6694/10/12/519/s1, Figure S1: Validation of the EphA3-maytansine ADC IIIA4-USAN, Figure S2: In vivo efficacy of the EphA3-maytansine ADC IIIA4-USAN.

Author Contributions: Conceptualization, C.O.; F.A.-E.; A.W.B. and B.W.D.; Methodology, C.O.; F.A.-E.; S.P.; K.J.T.; S.R.; A.W.B. and B.W.D. Formal Analysis and Investigation, C.O.; F.A.-E.; S.P.; B.C.; K.S.E.; Z.C.B.; P.R.J.; F.M.S.; A.V.F.; and C.A.B.; Resources, B.W.S and R.J.; Data Curation, C.O.; F.A.-E.; S.P.; C.A.B.; K.J.T.; A.W.B. and B.W.D.; Writing-Original Draft Preparation, C.O.; A.W.B. and B.W.D.; Supervision, A.W.B. and B.W.D. Project Administration, B.W.D. Funding Acquisition, A.W.B. and B.W.D.

Funding: This research was funded by the National Health & Medical Research Council (NHMRC) of Australia grant application I.D.1078893 and generous donation support from the Sid Faithfull Group and Cure Brain Cancer Foundation.

Acknowledgments: We acknowledge the QIMR Berghofer histology, microscopy, flow cytometry and animal facilities for their assistance and expertise in enabling us to complete this study.

Conflicts of Interest: A.W.B. is a member of a consortium of academic institutions who have licensed the IIIA4 antibody to Humanigen Inc., Brisbane, CA USA. The other authors declare no conflict of interest.

References

1. Stupp, R.; Mason, W.P.; van den Bent, M.J.; Weller, M.; Fisher, B.; Taphoorn, M.J.; Belanger, K.; Brandes, A.A.; Marosi, C.; Bogdahn, U.; et al. Radiotherapy plus concomitant and adjuvant temozolomide for glioblastoma. *N. Engl. J. Med.* **2005**, *352*, 987–996. [CrossRef]
2. Stupp, R.; Hegi, M.E.; Mason, W.P.; van den Bent, M.J.; Taphoorn, M.J.; Janzer, R.C.; Ludwin, S.K.; Allgeier, A.; Fisher, B.; Belanger, K.; et al. Effects of radiotherapy with concomitant and adjuvant temozolomide versus radiotherapy alone on survival in glioblastoma in a randomised phase III study: 5-year analysis of the EORTC-NCIC trial. *Lancet Oncol.* **2009**, *10*, 459–466. [CrossRef]
3. Hanahan, D.; Weinberg, R.A. Hallmarks of cancer: The next generation. *Cell* **2011**, *144*, 646–674. [CrossRef] [PubMed]
4. Pasquale, E.B. Eph receptor signalling casts a wide net on cell behaviour. *Nat. Rev. Mol. Cell Biol.* **2005**, *6*, 462–475. [CrossRef] [PubMed]
5. Pasquale, E.B. Eph receptors and ephrins in cancer: Bidirectional signalling and beyond. *Nat. Rev. Cancer* **2010**, *10*, 165–180. [CrossRef] [PubMed]
6. Boyd, A.W.; Bartlett, P.F.; Lackmann, M. Therapeutic targeting of EPH receptors and their ligands. *Nat. Rev. Drug Discov.* **2013**, *13*, 39–62. [CrossRef]
7. Boyd, A.W.; Ward, L.D.; Wicks, I.P.; Simpson, R.J.; Salvaris, E.; Wilks, A.; Welch, K.; Loudovaris, M.; Rockman, S.; Busmanis, I. Isolation and characterization of a novel receptor-type protein tyrosine kinase (hek) from a human pre-B cell line. *J. Biol. Chem.* **1992**, *267*, 3262–3267. [PubMed]
8. Pasquale, E.B. Eph-ephrin bidirectional signaling in physiology and disease. *Cell* **2008**, *133*, 38–52. [CrossRef] [PubMed]
9. Day, B.W.; Stringer, B.W.; Al-Ejeh, F.; Ting, M.J.; Wilson, J.; Ensbey, K.S.; Jamieson, P.R.; Bruce, Z.C.; Lim, Y.C.; Offenhauser, C.; et al. EphA3 maintains tumorigenicity and is a therapeutic target in glioblastoma multiforme. *Cancer Cell* **2013**, *23*, 238–248. [CrossRef]
10. Day, B.W.; Stringer, B.W.; Boyd, A.W. Eph receptors as therapeutic targets in glioblastoma. *Br. J. Cancer* **2014**, *111*, 1255–1261. [CrossRef]
11. Qazi, M.A.; Vora, P.; Venugopal, C.; Adams, J.; Singh, M.; Hu, A.; Gorelik, M.; Subapanditha, M.K.; Savage, N.; Yang, J.; et al. Cotargeting Ephrin Receptor Tyrosine Kinases A2 and A3 in Cancer Stem Cells Reduces Growth of Recurrent Glioblastoma. *Cancer Res.* **2018**, *78*, 5023–5037. [CrossRef]
12. Ferluga, S.; Tome, C.M.; Herpai, D.M.; D'Agostino, R.; Debinski, W. Simultaneous targeting of Eph receptors in glioblastoma. *Oncotarget* **2016**, *7*, 59860–59876. [CrossRef] [PubMed]

13. Vail, M.E.; Murone, C.; Tan, A.; Hii, L.; Abebe, D.; Janes, P.W.; Lee, F.T.; Baer, M.; Palath, V.; Bebbington, C.; et al. Targeting EphA3 Inhibits Cancer Growth by Disrupting the Tumor Stromal Microenvironment. *Cancer Res.* **2014**, *74*, 4470–4481. [CrossRef] [PubMed]
14. Lackmann, M.; Mann, R.J.; Kravets, L.; Smith, F.M.; Bucci, T.A.; Maxwell, K.F.; Howlett, G.J.; Olsson, J.E.; Vanden Bos, T.; Cerretti, D.P.; et al. Ligand for EPH-related kinase (LERK) 7 is the preferred high affinity ligand for the HEK receptor. *J. Biol. Chem.* **1997**, *272*, 16521–16530. [CrossRef] [PubMed]
15. Lackmann, M. Isolation and characterization of "orphan-RTK" ligands using an integrated biosensor approach. *Methods Mol. Biol.* **2001**, *124*, 335–359.
16. Smith, F.M.; Vearing, C.; Lackmann, M.; Treutlein, H.; Himanen, J.; Chen, K.; Saul, A.; Nikolov, D.; Boyd, A.W. Dissecting the EphA3/Ephrin-A5 interactions using a novel functional mutagenesis screen. *J. Biol. Chem.* **2004**, *279*, 9522–9531. [CrossRef] [PubMed]
17. Vearing, C.; Lee, F.T.; Wimmer-Kleikamp, S.; Spirkoska, V.; To, C.; Stylianou, C.; Spanevello, M.; Brechbiel, M.; Boyd, A.W.; Scott, A.M.; et al. Concurrent binding of anti-EphA3 antibody and ephrin-A5 amplifies EphA3 signaling and downstream responses: Potential as EphA3-specific tumor-targeting reagents. *Cancer Res.* **2005**, *65*, 6745–6754. [CrossRef]
18. Swords, R.T.; Greenberg, P.L.; Wei, A.H.; Durrant, S.; Advani, A.S.; Hertzberg, M.S.; Jonas, B.A.; Lewis, I.D.; Rivera, G.; Gratzinger, D.; et al. KB004, a first in class monoclonal antibody targeting the receptor tyrosine kinase EphA3, in patients with advanced hematologic malignancies: Results from a phase 1 study. *Leuk. Res.* **2016**, *50*, 123–131. [CrossRef]
19. Gan, H.K.; Reardon, D.A.; Lassman, A.B.; Merrell, R.; van den Bent, M.; Butowski, N.; Lwin, Z.; Wheeler, H.; Fichtel, L.; Scott, A.M.; et al. Safety, pharmacokinetics, and antitumor response of depatuxizumab mafodotin as monotherapy or in combination with temozolomide in patients with glioblastoma. *Neuro Oncol.* **2018**, *20*, 838–847. [CrossRef]
20. Lassman, A.B.; van den Bent, M.J.; Gan, H.K.; Reardon, D.A.; Kumthekar, P.; Butowski, N.; Lwin, Z.; Mikkelsen, T.; Nabors, L.B.; Papadopoulos, K.P.; et al. Safety and efficacy of depatuxizumab mafodotin + temozolomide in patients with EGFR-amplified, recurrent glioblastoma: Results from an international phase I multicenter trial. *Neuro Oncol.* **2018**. [CrossRef]
21. Hamblett, K.J.; Senter, P.D.; Chace, D.F.; Sun, M.M.; Lenox, J.; Cerveny, C.G.; Kissler, K.M.; Bernhardt, S.X.; Kopcha, A.K.; Zabinski, R.F.; et al. Effects of drug loading on the antitumor activity of a monoclonal antibody drug conjugate. *Clin. Cancer Res.* **2004**, *10*, 7063–7070. [CrossRef] [PubMed]
22. Pollard, S.M.; Yoshikawa, K.; Clarke, I.D.; Danovi, D.; Stricker, S.; Russell, R.; Bayani, J.; Head, R.; Lee, M.; Bernstein, M.; et al. Glioma stem cell lines expanded in adherent culture have tumor-specific phenotypes and are suitable for chemical and genetic screens. *Cell Stem Cell* **2009**, *4*, 568–580. [CrossRef] [PubMed]
23. Day, B.W.; Stringer, B.W.; Wilson, J.; Jeffree, R.L.; Jamieson, P.J.; Ensbey, K.S.; Bruce, Z.C.; Inglis, P.; Allan, S.; Winter, C.; et al. Glioma Surgical Aspirate: A Viable Source of Tumor Tissue for Experimental Research. *Cancers* **2013**, *5*, 357–371. [CrossRef] [PubMed]
24. Verhaak, R.G.; Hoadley, K.A.; Purdom, E.; Wang, V.; Qi, Y.; Wilkerson, M.D.; Miller, C.R.; Ding, L.; Golub, T.; Mesirov, J.P.; et al. Integrated genomic analysis identifies clinically relevant subtypes of glioblastoma characterized by abnormalities in PDGFRA, IDH1, EGFR, and NF1. *Cancer Cell* **2010**, *17*, 98–110. [CrossRef] [PubMed]
25. Brennan, C.; Momota, H.; Hambardzumyan, D.; Ozawa, T.; Tandon, A.; Pedraza, A.; Holland, E. Glioblastoma subclasses can be defined by activity among signal transduction pathways and associated genomic alterations. *PLoS ONE* **2009**, *4*, e7752. [CrossRef] [PubMed]
26. Wang, Q.; Hu, B.; Hu, X.; Kim, H.; Squatrito, M.; Scarpace, L.; de Carvalho, A.C.; Lyu, S.; Li, P.; Li, Y.; et al. Tumor Evolution of Glioma-Intrinsic Gene Expression Subtypes Associates with Immunological Changes in the Microenvironment. *Cancer Cell* **2017**, *32*, 42–56. [CrossRef] [PubMed]
27. Hudis, C.A. Trastuzumab—Mechanism of action and use in clinical practice. *N. Engl. J. Med.* **2007**, *357*, 39–51. [CrossRef] [PubMed]
28. Burris, H.A., 3rd; Rugo, H.S.; Vukelja, S.J.; Vogel, C.L.; Borson, R.A.; Limentani, S.; Tan-Chiu, E.; Krop, I.E.; Michaelson, R.A.; Girish, S.; et al. Phase II study of the antibody drug conjugate trastuzumab-DM1 for the treatment of human epidermal growth factor receptor 2 (HER2)-positive breast cancer after prior HER2-directed therapy. *J. Clin. Oncol.* **2011**, *29*, 398–405. [CrossRef] [PubMed]

29. Krop, I.E.; Beeram, M.; Modi, S.; Jones, S.F.; Holden, S.N.; Yu, W.; Girish, S.; Tibbitts, J.; Yi, J.H.; Sliwkowski, M.X.; et al. Phase I study of trastuzumab-DM1, an HER2 antibody-drug conjugate, given every 3 weeks to patients with HER2-positive metastatic breast cancer. *J. Clin. Oncol.* **2010**, *28*, 2698–2704. [CrossRef]
30. Janes, P.W.; Slape, C.I.; Farnsworth, R.H.; Atapattu, L.; Scott, A.M.; Vail, M.E. EphA3 biology and cancer. *Growth Factors* **2014**, *32*, 176–189. [CrossRef]
31. Barok, M.; Joensuu, H.; Isola, J. Trastuzumab emtansine: Mechanisms of action and drug resistance. *Breast Cancer Res.* **2014**, *16*, 209. [CrossRef] [PubMed]
32. Loganzo, F.; Sung, M.; Gerber, H.P. Mechanisms of Resistance to Antibody-Drug Conjugates. *Mol. Cancer Ther.* **2016**, *15*, 2825–2834. [CrossRef] [PubMed]
33. Day, B.W.; Stringer, B.W.; Spanevello, M.D.; Charmsaz, S.; Jamieson, P.R.; Ensbey, K.S.; Carter, J.C.; Cox, J.M.; Ellis, V.J.; Brown, C.L.; et al. ELK4 neutralization sensitizes glioblastoma to apoptosis through downregulation of the anti-apoptotic protein Mcl-1. *Neuro Oncol.* **2011**, *13*, 1202–1212. [CrossRef] [PubMed]
34. Al-Ejeh, F.; Shi, W.; Miranda, M.; Simpson, P.T.; Vargas, A.C.; Song, S.; Wiegmans, A.P.; Swarbrick, A.; Welm, A.L.; Brown, M.P.; et al. Treatment of triple-negative breast cancer using anti-EGFR-directed radioimmunotherapy combined with radiosensitizing chemotherapy and PARP inhibitor. *J. Nucl. Med. Off. Publ. Soc. Nucl. Med.* **2013**, *54*, 913–921. [CrossRef] [PubMed]
35. Al-Ejeh, F.; Pajic, M.; Shi, W.; Kalimutho, M.; Miranda, M.; Nagrial, A.M.; Chou, A.; Biankin, A.V.; Grimmond, S.M.; Australian Pancreatic Cancer Genome Initiative; et al. Gemcitabine and CHK1 inhibition potentiate EGFR-directed radioimmunotherapy against pancreatic ductal adenocarcinoma. *Clin. Cancer Res.* **2014**, *20*, 3187–3197. [CrossRef] [PubMed]

cancers MDPI

Article

The Small Molecule Ephrin Receptor Inhibitor, GLPG1790, Reduces Renewal Capabilities of Cancer Stem Cells, Showing Anti-Tumour Efficacy on Preclinical Glioblastoma Models

Giovanni Luca Gravina [1], Andrea Mancini [2], Alessandro Colapietro [2], Simona Delle Monache [3], Roberta Sferra [4], Flora Vitale [5], Loredana Cristiano [6], Stefano Martellucci [3,7], Francesco Marampon [8], Vincenzo Mattei [7], Filip Beirinckx [9], Philippe Pujuguet [10], Laurent Saniere [10], Giocondo Lorenzon [9], Ellen van der Aar [9] and Claudio Festuccia [2,*]

1. Division of Radiation Oncology, Department of Biotechnological and Applied Clinical Sciences, University of L'Aquila, Via Vetoio (Coppito II), 67100 L'Aquila, Italy; giovanniluca.gravina@univaq.it
2. Laboratory of Radiobiology, Department of Biotechnological and Applied Clinical Sciences, University of L'Aquila, Via Vetoio (Coppito II), 67100 L'Aquila, Italy; mancio_1982@hotmail.com (A.M.); alecolapietro@gmail.com (A.C.)
3. Laboratory of cellular pathology, Department of Biotechnological and Applied Clinical Sciences, University of L'Aquila, Via Vetoio 1 (Coppito II), 67100 L'Aquila, Italy; simona.dellemonache@univaq.it (S.D.M.); s.martellucci@sabinauniversitas.it (S.M.)
4. Laboratory of Human Anatomy; Department of Biotechnological and Applied Clinical Sciences, University of L'Aquila, Via Vetoio (Coppito II), 67100 L'Aquila, Italy; Roberta.sferra@univaq.it
5. Laboratory of Neurophysiology, Department of Biotechnological and Applied Clinical Sciences, University of L'Aquila, Via Vetoio (Coppito II), 67100 L'Aquila, Italy; floravitale86@Hotmail.it
6. Department of Life, Health and Environmental Sciences, University of L'Aquila, Via Vetoio (Coppito II), 67100 L'Aquila, Italy; Loredana.cristiano@univaq.it
7. Laboratory of Experimental Medicine and Environmental Pathology, University Hub "Sabina Universitas", 02100 Rieti, Italy; vincenzo.mattei@uniroma1.it
8. Department of Radiological, Oncological and Pathological Sciences, Sapienza University of Rome, 00161 Rome, Italy; f.marampon@gmail.com
9. Galapagos NV, Industriepark Mechelen Noord, General De Wittelaan L11 A3, 2880 Mechelen, Belgium; filip.beirinckx@glpg.com (F.B.); Giocondo.lorenzon@glpg.com (G.L.); Ellen.vanderAar@glpg.com (E.v.d.A.)
10. Galapagos SASU, 102 avenue Gaston Roussel, 93230 Romainville, France; philippe.pujuguet@glpg.com (P.P.); laurent.saniere@glpg.com (L.S.)

* Correspondence: claudio.festuccia@univaq.it; Tel.: +39-0862-433585; Fax: +39-0862-433523

Received: 8 January 2019; Accepted: 5 March 2019; Published: 13 March 2019

Abstract: Therapies against glioblastoma (GBM) show a high percentage of failure associated with the survival of glioma stem cells (GSCs) that repopulate treated tumours. Forced differentiation of GSCs is a promising new approach in cancer treatment. Erythropoietin-producing hepatocellular (Eph) receptors drive tumourigenicity and stemness in GBM. We tested GLPG1790, a first small molecule with inhibition activity versus inhibitor of various Eph receptor kinases, in preclinical GBM models using in vitro and in vivo assays. GLPG1790 rapidly and persistently inhibited Ephrin-A1-mediated phosphorylation of Tyr^{588} and Ser^{897}, completely blocking EphA2 receptor signalling. Similarly, this compound blocks the ephrin B2-mediated EphA3 and EphB4 tyrosine phosphorylation. This resulted in anti-glioma effects. GLPG1790 down-modulated the expression of mesenchymal markers CD44, Sox2, nestin, octamer-binding transcription factor 3/4 (Oct3/4), Nanog, CD90, and CD105, and up-regulated that of glial fibrillary acidic protein (GFAP) and pro-neural/neuronal markers, βIII tubulin, and neurofilaments. GLPG1790 reduced tumour growth in vivo. These effects were larger compared to radiation therapy (RT; U251 and T98G xenografts) and smaller than those of temozolomide (TMZ; U251 and U87MG cell models). By contrast, GLPG1790 showed effects that were higher than Radiotherapy (RT) and similar to Temozolomide (TMZ) in orthotopic U87MG and CSCs-5 models in terms of disease-free survival (DFS) and overall survival (OS). Further experiments were necessary to study possible interactions with radio- and chemotherapy. GLPG1790 demonstrated anti-tumor effects regulating both the differentiative status of Glioma Initiating Cells (GICs) and the quality of tumor microenvironment, translating into efficacy in aggressive GBM mouse models. Significant common molecular targets to radio and chemo therapy supported the combination use of GLPG1790 in ameliorative antiglioma therapy.

Keywords: glioblastoma; ephrin system; glioma stem like cells

---

1. Introduction

Glioblastoma multiforme (GBM) is the most malignant human brain tumour [1]. This neoplasia is characterised by high cell heterogeneity and the presence of a complex tumour microenvironment [2]. Its aggressiveness is due to the high capacity of tumour cells to move and infiltrate healthy brain tissue [3] as well as to produce soluble factors capable of recruiting blood vessels and inflammatory cells [4,5]. This means that aggressive/infiltrating tumour cells are not eradicated with surgery or standard chemo and radio therapy [6]. Abnormal angiogenesis is also supported by vasculogenic mimicry [4–7]. Despite the abundant tumour-associated vasculogenesis, tumour necrosis is increased and this renders some tumour areas strongly hypoxic [8]. This is associated with the recruitment of precursors of endothelial cells/pericytes, bone marrow derived myeloid circulating cells and glioma stem cells (GSCs) [9–11]. Recruited cancer stem cells grow rapidly, they determine tumour recurrence [12] and are frequently resistant to chemotherapeutic agents [13]. The inflammatory state of gliomas is constantly maintained to support vasculogenesis and hypoxia [9,14]. Median overall survival (OS) for newly diagnosed GBM is 14.6 months [15]. Failure of standard chemo and radio therapy is attributed to multiple factors, such as microenvironment protection, -de novo and/or acquired tumour resistance, limitations in drug deliver, increased angiogenesis and/or vasculogenesis and GSCs.

Erythropoietin-producing hepatocellular (EPH) proteins are a large family of receptors with tyrosine kinases (TKs) moieties widely considered targets for glioblastoma [16]. These are grouped as Eph-A and Eph-B based on binding cell associated ligands, Ephrin-A and Ephrin-B, respectively. Binding of the ligand to the receptor starts a complex network of signals through "trans" and "cis" mechanisms [17] as a function of Eph and Ephrin expression in different cells or in the same cell, respectively. For the trans-modality, a "forward" activation, which is dependent on TK activity, propagates signals in the receptor-expressing cell, while the "reverse" signals are propagated in the Ephrin-expressing cell. "Forward" signaling includes the activation of Rho and Ras family GTPases and Akt/mTORC1, whereas "reverse" signaling involves Src family kinase activation [16,17]. The Eph/Ephrin network plays a key role in tissue remodeling during embryogenesis, cardiovascular and skeletal development, axon guidance, tissue patterning (for reviews, see Saha et al 2018 [18]). The major members of this receptor family involved in GBM progression and glial/neuronal differentiation are EphA2, EphA3 and EphB4 [19–25]. In particular, EphA3 is a protein frequently overexpressed in the most aggressive mesenchymal subtype [22]. Importantly, EphA3 is highly expressed on the tumor-initiating cell population in glioma, and appears critically involved in maintaining tumor cells in a less differentiated state by modulating mitogen-activated protein kinase signalling. EphA2 is co-expressed with EphA3 and promotes infiltrative invasion of Glioma Initiating Cells (GICs) in vivo through cross-talk with Akt and regulates stem cell properties [20]. Several processes correlated to tumour progression, such as neovascularization/angiogenesis, local invasion and metastatization are regulated by these receptors [23–25]. Many recent reports also indicate that the signals triggered by the EPH/Ephrin system modulate aspects of cancer stem cell self-renewal [20]. Different compounds targeting the Eph receptor are developed in cancer research [26–28]. GLPG1790 is one of these compounds. This compound is a small molecule with a pan-specific pharmacological compound with inhibition activity versus various Eph receptors kinases and with oral bioavailability [27,28]. This compound shows remarkable inhibition activity with nanomolar potency against EphB2, EphA2, EphB4, EphA4 and EphA1, but was not able to block EphA7 (if not in high doses). GLPG1790 was not previously tested for EpHA3 inhibition. Therefore, we are aware that different members of Eph receptors may be involved in tumor progression, stemness, differentiation of glioma stem like cells and angiogenesis [29].

It has also been demonstrated that GLPG1790 inhibits EphA2 receptor phosphorylation after a single oral administration of 30 and 100 mg/kg doses. Previously, we observed that pharmacological doses of GLPG1790 reduced tumour cell proliferation in rabdo-myosarcoma, modulating muscle cell differentiation [28]. Molecular action mechanisms are under investigation in tumour histotypes over-expressing Eph receptors. In this current study, the effectiveness of GLPG1790 was evaluated in GBM. For this purpose, we selected three cell lines (U87MG, U251 and T98G), which were subcutaneously xenografts in nude mice, and luciferase tagged U87MG and GSCs, which were orthotopically injected into the brains of female cd1 nude mice. We assessed changes in tumour growth, disease-free survival time (DSF) and OS after the GLPG1790 treatment, and compared these parameters with those collected by standard radio-, chemo- and chemo and radio therapies.

2. Material and Methods

2.1. Cell Lines and Cell Cultures

All the materials for the tissue culture were purchased from the Italian distributor of Euroclone (Euroclone S.p.A, Milan, Italy). Six human glioma cell lines (U251MG, A172, U87MG, SW1783, LN229 and T98G) were cultured at 37 °C in 5% CO_2 in Dulbecco's modified Eagle medium (DMEM, Euroclone) containing 10% (v/v) fetal bovine serum, 4 mM glutamine, 100 IU/mL penicillin, 100 μg/mL streptomycin and 1% nonessential amino acids (Thermo Fisher Scientific Inc., Carlsbad, CA, USA). The risk of working with misidentified and/or contaminated cell lines was minimised by using GBM cells at very low passages and periodic short tandem repeat (STR) DNA profiling. Luciferase-tagged U87MG cells were generated and provided by Jari E. Heikkila (Abo Akademi University, Turku, Finland). Five GBM patient-derived stem cell lines (BT12M, BT25M, BT48EF, BT50EF and BT53M), were provided by J. Gregory Cairncross, and Samuel Weiss (University of Calgary, Calgary, Canada) [30], and GSCs-5 and CSCs-7 were obtained from Marta Izquierdo (Universidad Autónoma de Madrid, Spain) [31]. These cells were maintained in neurosphere medium DMEM:F12 (1:1) supplemented with B-27, epidermal growth factor (EGF; 20 ng/mL) and basic fibroblast growth factor (bFGF; 10 ng/mL). Luciferase was inserted in the genome of GSCs-5 cells using the pGL4.13 vector (Promega, Milan, Italy) and the jetPEI DNA transfection method (Polyplus, Illkirch, France). Isolated neurospheres were assayed for stemness properties in terms of clonogenic capacity and positivity for stem cell markers.

2.2. Reagents and Enzymatic Activities

Antibodies against total Akt (Sc-377457), p-AktSer473 (sc-135651), total ERK 1/2 (H-72, sc-292838), p-ERK 1/2 (Thr 202/Tyr 204, sc-16982), p38α (clone C-20, sc-535), p-p38 (Tyr 182, sc-101759), EphA2 (clone C-20, sc-924), EphB4 (clone H-200, sc-5536), EphB2 (clone H-80, sc-28980), OCT3/4 (C-10, sc-5279), CA-IX (clone H11, sc-365900), anti-human CD31 (PECAM-1, clone M-20, sc-1506), nestin (clone 10c2, sc-23927), βIII tubulin (clone 3H3091, sc-69966), NFH (clone 2D2, sc-20014), GFAP (sc-6170) and SOX2 (clone A-5, sc-365964) were purchased from Santa Cruz Biotechnology (Santa Cruz, CA, USA). Antibodies against Phospho-EphA2 (Ser897, 6347) and Phospho-EphA2 (Tyr594, 3970) were purchased from Cell Signaling Technology Europe, B.V. (Leiden, The Netherland). Antibodies against Ki67 (Clone MIB-1, M7240) were purchased from Dako (Agilent Technologies Italia S.p.A., Cernusco sul Naviglio, Milan, Italy). Antibodies against CD44 (ab157107), Stro-1 (ab214086), murine CD31 (clone MEC 7.46, ab7388) and Ephrin A1 (ab199697) were purchased from Abcam (Cambridge, UK). Antibodies against CD105, conjugated with phycoerythrin (PE; MEM226) or un-conjugated (MEM-229), and CD90 (5E10) were purchased from Immunotools (Friesoythe, Germany). Antibodies against CXCR4 (A00995) and VEGF-A (251575) were purchased from GenScript (Piscataway, NJ, USA) and Abbiotec (San Diego, CA, USA), respectively. Cell-based enzyme-linked immunosorbent assays (ELISAs) for total and phosphorylated isoforms in Ser 897 and Tyr 588 EphA2 as well as total and p-Tyrosine779 EphA3 and total and p-Tyrosine987 EphB4 were used for detecting and quantifying target proteins in cultured cells following the "In-Cell ELISA protocol" on the Abcam website (http://www.abcam.com/protocols/in-cell-elisa-ice). Recombinant Human Ephrin-A1 (EFNA1-150H) and Recombinant Human Ephrin-B2 (EFNB2-480H) were purchased from Creative Biomart (Shirley, NY, USA).

2.3. Immunofluorescence Studies

BT48EF and BT50EF GSC cells were used for immuno-fluorescence analyses. Spheres were seeded at a density of 10,000 cells/cm^2 on glass coverslips pretreated with 30 μg/mL poly-L lysine to promote adherence. The slides were then washed twice with phosphate-buffered saline (PBS) and fixed with 4% paraformaldehyde for 20 min at room temperature. To stain cytoplasmic markers, slides were permeabilized with 0.3% Triton-X-100 for 5 min at room temperature. Spheres were then incubated overnight at 4 °C with the following primary antibodies accordingly to their data sheets: anti-OCT3/4, Ki67, EphA2, nestin, βIII tubulin, NFH, GFAP, SOX2, Stro-1 and CD44. After washing with PBS, cells were incubated for 30 min at RT with AlexaFluor 488 anti-rabbit IgG, AlexaFluor 595 anti-goat IgG or AlexaFluor 633 anti-mouse IgG secondary antibody (1:2000; Molecular Probes, Invitrogen, Carlsbad, CA, USA). Controls were performed by omitting the primary antibody. Cell nuclei were stained with DAPI (0.5 μg/mL). Coverslips were mounted with Vectashield Mounting Medium and examined with a Leica TCS SP5 confocal microscope (Leica Microsystems Inc., Mannheim, Germany).

2.4. Fluorescence-Activated Cell Sorter (FACS) Analyses

Expression of surface antigens in BT48EF and BT12M cell lines, treated or untreated with 0.5 μM GLPG 1790, was quantified by flow cytometry. BT48EF and BT12M cells were fixed with 4% paraformaldehyde for 10 min at 4 °C and, after washing, cells were incubated for 1 h at room temperature with anti-CD44, anti-CD105 and anti-EphA2, followed by additional 30 min with CY5-conjugated anti-rabbit IgG H&L or PE-conjugated anti-mouse IgG purchased from Abcam (Cambridge, UK). Moreover, we quantified the expression of cytoplasmatic antigens by flow cytometry in the same cell lines treated as above. The cells were fixed with 4% paraformaldehyde for 10 min at 4 °C, and after washing, were permeabilized with 0.1% (v/v) Triton X-100 for 10 min at room temperature. The cells were washed in PBS and incubated for 1 h at RT with anti-NFH, anti-Nestin and anti-GFAP, followed by an additional 30 min with Cyanine dye 5 (CY5)-conjugated anti-rabbit IgG H&L or Phycoerythrin PE-conjugated anti-mouse IgG. All samples were analysed using a BD AccuriTM C6 Plus Flow cytometer (Becton Dickinson Italia SpA, Milan, Italy) equipped with a blue laser (488 nm) and a red laser (640 nm). At least 10,000 events were acquired. Negative controls were obtained by analysing samples treated without the primary antibody. Although Fluorescent-Activated Cell Sorter (FACS) should be made on live cells, especially for membrane associated antigens, we choose to use a fixation in 4% paraformaldehyde to avoid the internalisation and degradation of membrane markers as demonstrated for different membrane targets such as EphA2 [32] or CD44 [33].

2.5. Growth Assays

Twenty-four-well plates were seeded with 2 × 10^4 cells/mL GBM cells. After cell adhesion and growth in 5% fetal calf serum (FCS) DMEM for 24 h, different concentrations of GLPG1790 were added in appropriate culture conditions. A Nikon Diaphot inverted phase-contrast photomicroscope (Nikon Corp., Tokyo, Japan) was used before cell trypsinization and counting. Cell counts were made with a NucleoCounter NC-100 (Chemotec, Gydevang, Denmark) as previously described [34]. IC50 values, the concentration of drug required for a 50% reduction in growth/viability, were calculated using the Dojindo Cell Counting Kit-8 (Dojindo EU GmbH, Munich, Germany). For neurosphere proliferation, we used two different modalities of study: (i) a direct count and sizing of neurospheres at 1 week of culture from pre-formed spheres, and (ii) the evaluation of the clonal capacity of cancer stem cells (CSCs) cultured as single cells after 14–30 days. For the analysis of sphere growth, pre-formed neurospheres were treated with different doses of GLPG1790 for 72 h. After treatment, spheres were photographed and counted at contrast phase microscopy. Spheres were recorded as either large colonies (>50 cells) or small colonies (<50 cells). Single cells were also manually counted per microscopic field at 100× magnification. For the clonogenic assay, glioma tumour-initiating cells (GICs) were seeded in 96-well plates as a single cell suspension at a density of 2 cells/mL (equivalent of 1 cell every 3 wells). Cells were maintained for 14–30 days in their culturing media and then the wells were visually scanned using light microscopy to identify and count the clones (spheres) produced.

2.6. Cell Viability and Apoptosis Assay

Apoptosis was evaluated using the Tali® Apoptosis Kit—Annexin V Alexa Fluor® 488 & Propidium Iodide-based (Life Technologies Italia, Monza, Italy), as well as caspase-specific chromogenic substrates at 450 nm in an ELISA plate reader. In addition, we used Ac-DEVD-pNA (caspase-3), Ac-IETD-pNA (caspase 8) and Ac-LEHD-pNA purchased from Kaneka Eurogentec SA (Seraing, Belgium). N-Succinyl-Leu-Tyr-7-amido-4-methylcoumarin (Sigma Aldrich, St. Louis, MO, USA) was used to test calpain I activity in cell extracts in order to evaluate necrosis.

2.7. Western Blot

Cell extracts were obtained from treated or untreated cultures washed with cold PBS and lysed with lysis buffer containing proteinase and phosphatase inhibitor cocktails. Proteins were subjected to 7% or 15% sodium dodecyl sulphate-polyacrylamide gel electrophoresis (SDS-PAGE), transferred to nitrocellulose and probed with appropriate antibodies based on the recommendations of the suppliers. Reactive bands were visualised with a chemiluminescent detection kit (Perbio Science, Tattenhall, UK) in a Bio-Rad Gel Doc system (Bio-Rad Laboratories S.r.l., Milan, Italy). Normalization of specific bands was performed using an anti β-actin antibody.

2.8. In Vivo Experiments: Xenograft Model

Six-week-old female CD1 nu/nu mice were purchased from Charles River Laboratories Italia, SRL (Calco, Italy) and maintained under European Community (EC) guidelines (2010/63/UE and DL 26/2014 for the use of laboratory animals). All mice received subcutaneous flank injections of 1×10^6 U87MG, U251 or T98G cells. Tumour growth was assessed by measuring bi-weekly tumour diameters with Vernier calipers. Tumour volumes were calculated according to the formula: tumour volume (mm^3) = $4/3\pi \times R1 \times R2 \times R3$, where R1, R2 and R3 were the three rays of the ellipsoid, as previously described [31] At about 20 days after the tumour injection, 60 mice with tumour volumes of 85–100 mm^3 were randomised to receive the vehicle (methylcellulose 0.5% used to dissolve GLPG1790) or 30 mg/kg po qd GLPG1790, and for comparison with standard therapeutic procedures, RT, 4 Gy in a single administration), TMZ (16 mg/kg/5 consecutive days), or RT plus TMZ. The duration of treatments was 35 days or when control/untreated tumours reached critical volumes with regard to animal welfare laws. However, when an experiment was stopped, treated and control animals were sacrificed via carbon dioxide inhalation, and tumours were removed for biochemical (frozen tissue) and histological (paraffin fixed tissue) analyses. To compare different treatments, we considered as previously described [34]: (1) tumour volume, measured at different times; (2) tumour weight, measured at the end of experiment; (3) tumour progression (TP), defined as an increase >50% of tumour volume with respect to baseline; and (4) time to progression (TTP) which was the time necessary for TP.

2.9. Immunohistochemical Analyses

Indirect immunoperoxidase staining was performed on 4 μm paraffin-embedded tissue sections. A consensus judgment as indicated in our previous report [35] was adopted for immunohistochemical scoring of tumours based on the strength of positivity: negative (score 0), weak (score 1), moderated (score 2), or strong staining (score 3). In each category, the percentage of positive cells was assessed by scoring at least 1000 cells in the area with the highest density of antigen positive cells. Cytoplasmic/membrane staining intensity was graded as follow: 0 = negative; 1 $\leq$ 10% positive cells; 2 = positive cells in a range of 10–50%; and 3 $\geq$ 50% positive cells. Overall expression was defined by the staining index (SI) and ranged between 0 and 9, with an SI $\leq$ 4 indicating a low expression. Proliferation index (labeling index) was determined though the evaluation of the percentage of Ki67 positive cells by analysing 500 cells at 100× magnification. A TACS Blue Label kit (R&D Systems, Inc., Minneapolis, MN, USA) was used for in situ apoptosis determination and the percentage of terminal deoxynucleotidyl transferase dUTP nick end labeling (TUNEL) positive cells was determined in five random fields evaluated at 400× magnification. In order to count the number of CD31+ microvessels, five arbitrarily selected fields were analysed for each group at 100× magnification (tumour microvessels). Martius yellow-brilliant crystal scarlet blue stain was used to stain erythrocytes, and consequently, the presence of micro-thrombi and bleeding zones.

2.10. Orthotopic Intra-Brain Model

Nude mice were inoculated intra-cerebrally as previous described [36,37]. We injected 3 μL of cell suspension containing 3 × 10^3 U87MG or CSCs-5 luciferase tagged cells directly into the brain; the injection rate was set to 1 μL/min. Therapeutic administrations were started 5 days after cell injection when no luciferase activity was detectable intracranially. At this time, animals were randomised to receive vehicle (methylcellulose 0.5%) or 30 mg/kg GLPG1790 (po qd), and for comparison with standard therapeutic procedures, radiotherapy (RT, 4 Gy in a single administration), temozolomide (TMZ, 32 mg/kg/5 consecutive days), or RT plus TMZ. Each experimental group was formed using 10 mice. Animals were tested weekly via bioluminescence assay. The time at which a visible bioluminescence lesion was observed defined the parameter called disease free survival (DFS). Repeated bioluminescence assays were performed in order to also have data on tumour progression. Bioluminescence imaging was performed using an Alliance Mini HD6 system (Uvitec Ltd., Cambridge, UK) after intra-peritoneal (i.p.) injection with 150 μg/g D-luciferin (Synchem UG & Co. Altenburg, Germany) in pre-anesthetized animals. Treatments were completed after 35 days when a period without drug administration was started. A similar therapeutic approach could mimic the progression of surgically treated tumours in which a small portion of cells remain in the surgical bed from which the tumour occurred. DSF was the first oncological outcome analysed and defined the time during which no bioluminescence evidence of the tumour was recorded. If progression was observed, this might be recorded during or after the discontinuation of treatment. Mice were euthanised when they displayed neurological signs, such as altered gait, tremors/seizures and lethargy, or weight loss of $\geq$20% of pre-surgical weight. A follow-up provided us with data on OS, defined as the time (days) prior to which an animal did not show the distress signs given above, and was equal to the time of euthanasia. Brains were collected, fixed with 4% paraformaldehyde and paraffin was embedded.

2.11. Statistical Details

Continuous variables were summarised as mean and standard deviation (SD) or as median and 95% confidence intervals (CI). For continuous variables that were not normally distributed, statistical comparisons between control and treated groups were established by carrying out the Kruskal–Wallis test and Dwass–Steel–Chritchlow–Fligner method. For continuous variables that were normally distributed, statistical comparisons between control and treated groups were established by carrying out an analysis of variance (ANOVA) test or by Student t-test for unpaired data (for two comparisons).

When the ANOVA test revealed a statistical difference, pair-wise comparisons were made using Tukey's honestly significant difference (HSD) test. TTP was analysed using Kaplan–Meier curves and Gehan's generalized Wilcoxon test. When more than two survival curves were compared, the logrank test for trend was used. This tested the probability that there was a trend in survival scores across the groups. All tests were two-sided and were determined using Monte Carlo significance. Statistical significance occurred when p-values were at least < 0.05. MedCalc (MedCalc Software, Ostend, Belgium) was used as a complete statistical program. We analysed the Kaplan–Meyer curves [36] in terms of hazard ratios (HR), an expression of the ratio of the chance of an event occurring in the treatment arm compared to it occurring in the control arm. For comparisons, HRs are displayed with forest plot graphs.

3. Results

3.1. GLPG1790 Inhibits Phosphorylation of Both Tyrosine and Serine in U87MG Cells

Semi-quantitative determinations for basal expression of EphA2, EphA3 and EphB4 were obtained using cell-based ELISAs, whereas the affinities of GLPG1790 for EphA2, EphA3 and EphB4 were calculated by treating 80% sub-confluent GBM cultures with different doses of GLPG1790 that were administered 5 min before the addition of 0.25 µg/mL Ephrin-A1 (ligand of EphA2), and Ephrin-b2 (ligand for both EphA3 and EphB4). IC50 determinations were calculated 1 h after GLPG1790 administration. We verified the IC50 values for p-Tyrosine588 and p-Serine897 of EphA2, p-Tyrosine779 of EphA3 and p-Tyrosine987 of EphB4 (Figure 1A). We observed that the values of $OD_{450\ nm}$ (and then the amount of proteins) for EphA2, EphA3 and EphB4 were higher in GICs versus GBM cells. Similarly, the expression levels of EphA2 were higher when compared to those observed for EphA3 and EphB4 in the same cells. The growth of GBM cells lines was markedly reduced by GLPG1790 administration with an IC50 that ranged between 491 nM (A172) and 653 nM (U87MG; Figure 1B). IC50 values calculated for GBM cell lines considered in our study are summarised in this panel. We also showed that GLPG1790 increased the neuron-like cell differentiation in the mixed neuron-like/glial (epithelioid cells) U251 cell line (Figure 1C). Next, we tested the effects of GLPG1790 on seven patient-derived and proliferating GSC lines. Figure 2A shows the growth of BT12M cells. These cells were highly sensitive to GLPG1790 administration. This compound induced cell detachment from the spheres with a reduction of both the number and size of BT12M spheres. IC50 values were calculated for CSCs ranged between 109 nM (BT12M) and 420 nM (BT53M), as summarized in Figure 2B for single cultures. GLPG1790 was also able to reduce dose-dependent sphere generation from single cells in a clonogenic assay (Figure 2C). This indicated that GLPG1790, inhibiting the spheres formation, might reduce the self-renewal potential of CSCs, which in turn, could cause a reduction in tumour recurrence capacity, when these data are extrapolated to an in vivo situation. In order to have further information regarding the cell proliferation and differentiation status of GSC cultures, we performed an immuno-fluorescent analysis using a double immunostaining with Sox2 and Ki67. In BT12M cells, GLPG1790 reduced the percentage of Ki67 (proliferating antigen) positive cells (Figure 2D). These results were also quantified in other GSCs (Figure 2D). Confocal analysis also showed that double positivity for Sox2 and Ki67 cells was basally distributed to the peripheral layers of spheres, whereas lower expression of Ki67 was demonstrated in the innermost cells of the sphere that are positive for Sox-2. The nuclear Ki67 localisation was demonstrated to be cell cycle-dependent [38] and linked to rRNA transcription [39]. Quiescent cells blocked in the G0/G1 phase of the cell cycle were negative or exhibited very faint Ki67 nucleoplasm staining. Late G1 and early S-phase patterns were very similar with intense dot-like, Ki67-stained cells (Figure 2D). In the G2 phase, the nuclear staining intensity was reduced and accompanied by well-defined and bright nucleoli; four cells with this appearance were also observed in the control group (Figure 2D). GLPG1790 treated spheres were smaller when compared to untreated cultures (Figure 2D,E). The percentage of cells with an intense nuclear background (mid and late S-phase) was significantly reduced in a dose-dependent manner, whereas the percentage of cells with no Ki67 expression (negative) or exhibiting very faint Ki-67 nucleoplasm staining was increased. Spheres were compacted in untreated cultures, whereas spheres derived from GLPG1790-treated cultures showed less adhesive cells. As indicated in Figure 2A, GLPG1790 could induce cell detachment from the outer/peripheral cell layers of spheres; therefore, the proliferating cells from spheres could be lost after the GLPG1790 treatment. Next, we analysed the expression of Sox2/Ki67 in single cells or in very small cell aggregates from treated and untreated cultures. As shown in Figure 2F, some cells from this population were proliferating (Ki67 positive). GLPG1790 was also able to reduce the percentage of Ki67 expressing cells suggesting that this compound might influence cell proliferation, also through detached cells. Taken together, these data suggested that GLPG1790 reduced both the recruitment and induction of cell proliferation (commitment) of CSCs.

A (colorimetric cell based ELISA)

Cell line	EphA2 OD	p-Ser-897 EphA2 (IC50, nM)	p-Tyr-588 EphA2 (IC50, nM)	EphA3 OD	p-Tyr EphA3 (IC50, nM)	EphB4 OD	p-Tyr EphB4 (IC50, nM)
U87MG	0.855 ± 0.092	289 ± 40	165 ± 17	0.325 ± 0.044	348 ± 37	0.558 ± 0.059	245 ± 42
U251	1.044 ± 0.169	215 ± 57	188 ± 21	0.412 ± 0.078	285 ± 24	0.677 ± 0.092	204 ± 28
BT48EF	1.446 ± 0.188	157 ± 17	189 ± 22	1.454 ± 0.038	224 ± 12	0.853 ± 0.122	210 ± 16
BT12M	1.610 ± 0.132	135 ± 10	115 ± 17	0.868 ± 0.084	200 ± 24	0.974 ± 0.105	233 ± 27

B (GLPG179 antiproliferation effects IC50 values calculated in four different glioblastoma cell lines)

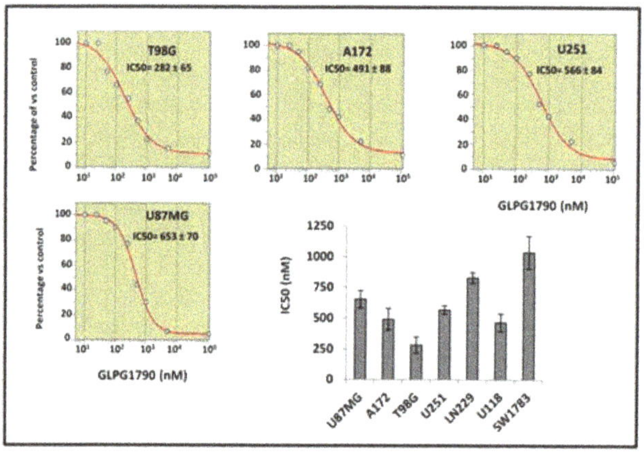

C (Time-dependent morphologic changes, U251 cells)

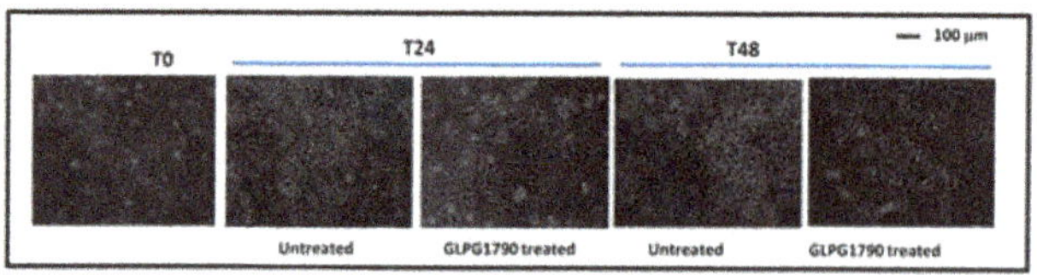

Figure 1. Biochemical and cellular effects of different doses of GLPG1790. (**A**) Colorimetric cell-based ELISA determinations for total and phosphorylated Tyr588 and Ser897 EphA2, total and phosphorylated Tyr779 EphA3 and total and phosphorylated Tyr987 EphB4 in GLPG1790-treated U87MG cells. Data show also the IC50 values calculated for the inhibition of phosphorylation for EphA2, EphA3 and EphB4 isoforms. (**B**) Growth curves and IC50 determination for GLPG1790-treated A172, U251, T98G and U87MG and GBM cell lines as well as in BT48EF and BT12M GICs. (**B**) antiproliferative effects and IC50 values calculated for GBM cell lines. (**C**) Morphology of GLPG1790-treated U251 cells.

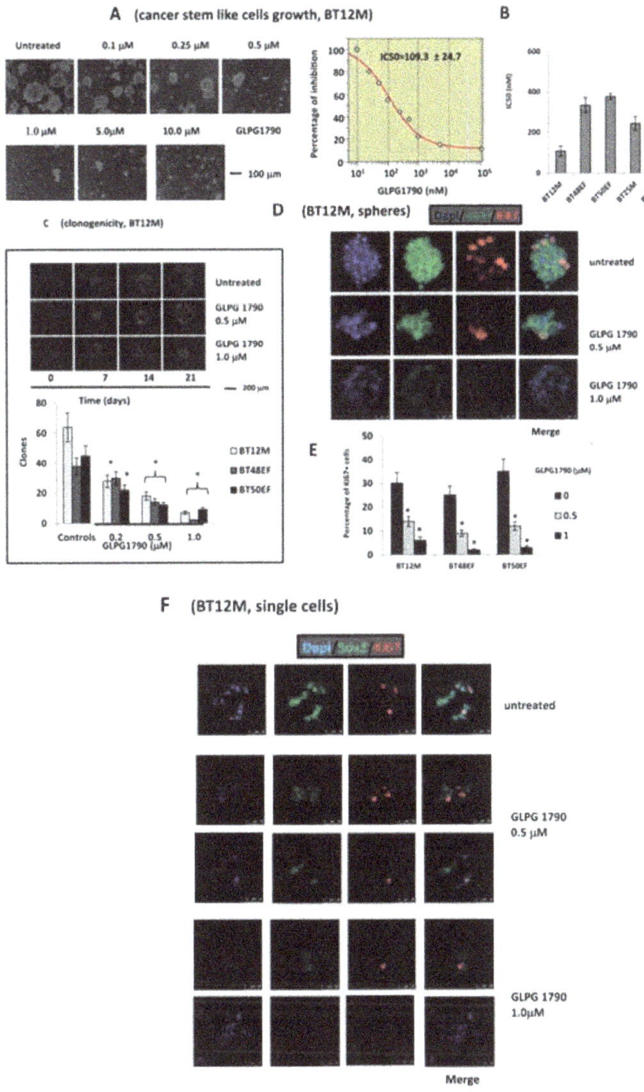

Figure 2. Proliferation of glioma stem cells cultured with different doses of GLPG1790. (**A**) Growth changes of BT12M CSCs; GLPG1790 reduced the number and size of BT12M cells. (**B**) IC50 values calculated for singles GIC cultures. (**C**) GLPG1790 reduced the sphere generation dose-dependently in clonogenic assays in BT12M, BT48EF and BT50EF cells. (**D**) Reduced Ki67 expression after the GLPG1790 treatment, both as a protein amount and percentage of positive cells; the reduction was dose-dependent in the BT48EF culture. Scale bar: 25 μm. (**E**) Dual Sox2 and Ki67 determination in single cells or in very small cell aggregates from treated and untreated cultures. (**F**) Dual Sox2/ki67 determination in single cells. Scale bar: 25 μm. Data are representative of three similar results performed in triplicate. Confocal images were collected and shown as maximal projection of about 20 analysed spheres observed with 0.29-μm size serial sections. * $p < 0.05$.

3.2. GLPG1790 Reduces Mesenchymal/Stem Cell Marker Expression in GICs

Of all the cancer stem cell markers identified to date, our attention was focused on CD44, CD90, CD105, Nestin, Sox2, Oct3/4, GFAP, βIII tubulin and neuro-filaments (NFH/Tuj1). In Figure 3A,B the representative cyto-fluorimetric analyses (BT48EF and BT12M cells) and western blots (BT48EF alone) are shown. Confocal immuno-fluorescence analyses (Figure 3C–I) were also performed to verify possible changes in expression and localisation of CD44 (Figure 3C,D), Sox2 (Figure 3E,F), NFH (Figure 3E), Oct3/4 (Figure 3H), GFAP (Figure 3I), Nestin (Figure 3F) and EphA2 (Figure 3C,D). Figure 3H shows the co-expression of actins and integrin-linked kinase (ILK) in the semi-adherent cultures. Notably, the CD44-positive cell percentage was reduced by approximately 40% (79.4 ± 2.5 vs. 48.0 ± 3.7 in untreated and GLPG1790 treated cultures, respectively) in BT12M cells and by 20% (68.5 ± 3.9 vs. 54.8 ± 4.2 in untreated and GLPG1790 treated cultures, respectively) in BT48EF. GLPG1790 administration reduced the expression of the CD44 standard isoform (CD44s) as indicated via western blot; however, as the difference observed between 0.5 and 1.0 μM treatments were minimal, it suggested this effect was not dose-dependent. CD44 positive cells were also EphA2-positive as suggested by the confocal data. The percentage of EphA2 positive cells was very high in both control GSC cultures. EphA2 was immuno-detected in 83.0 ± 7.0% of BT48EF cells and 92.5 ± 2.4% of BT12M cells.

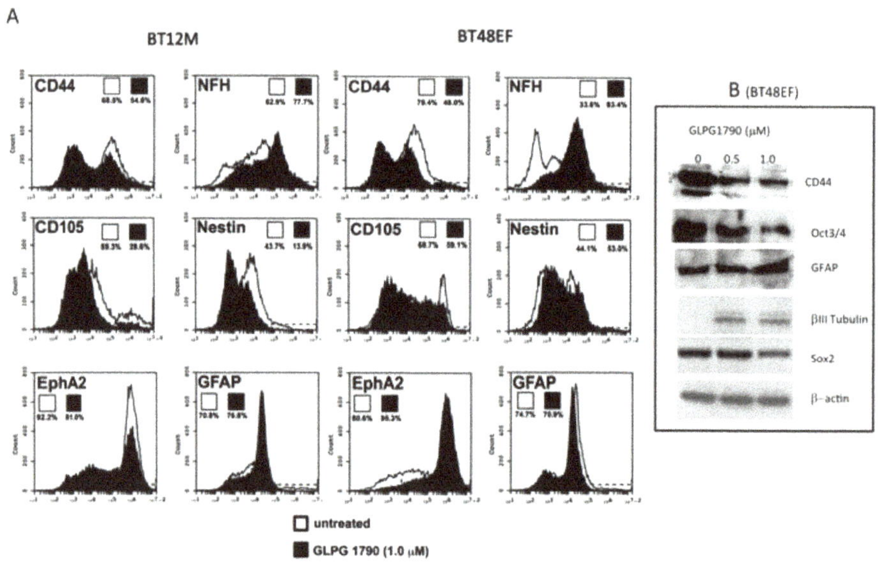

Figure 3. *Cont.*

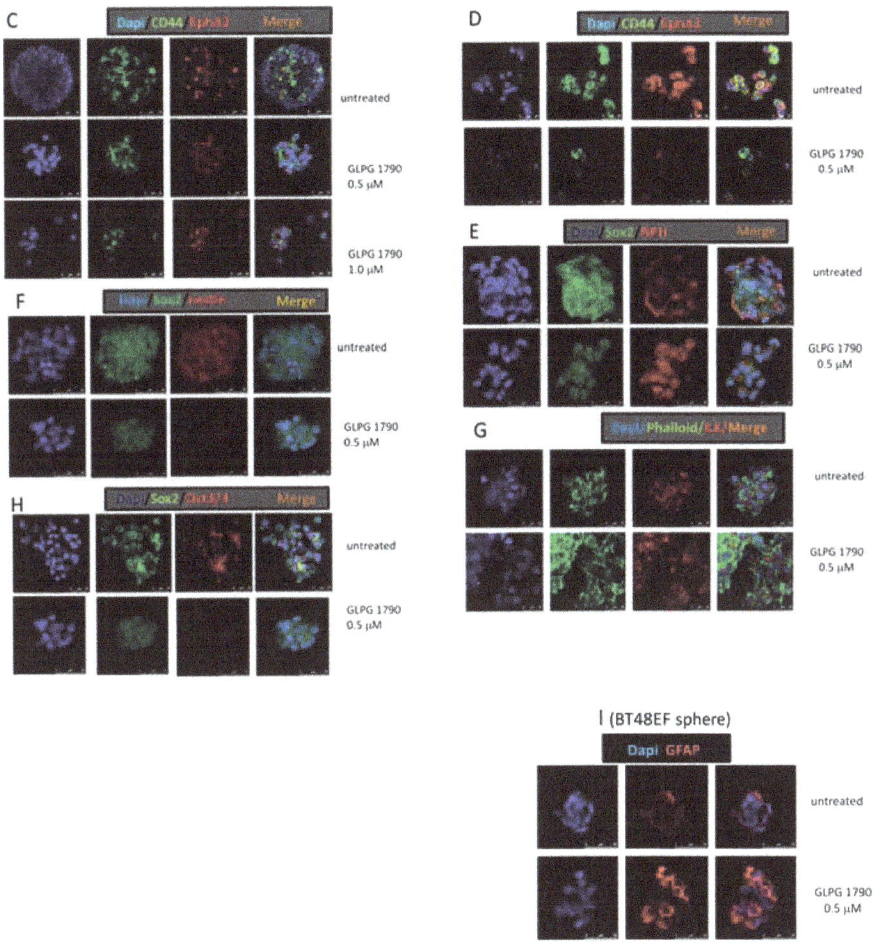

Figure 3. Phenotypic modifications in GLPG1790-treated GICs: changes in mesenchymal/stem cell marker expression. (**A**) FACS analysis performed in controls and GLPG1790-treated BT12 and BT48EF cultures. Data are representative of three separated experiments performed in triplicate and values are expressed as a percentage of positive cells present in the analyzed cell suspension. (**B**) Western blot determinations performed in control or treated BT48EF cultures. Data are representative of three different gels/experiments and lanes were charged with 40 μg of proteins. (**C–I**) Confocal immuno-fluorescence analyses performed in BT48EF: dual CD44/EphA2 expression in cell spheres (**C**) and in single or small cell aggregates (**D**), dual Sox2/NFH expression in cell sphere cultures (**E**), dual Sox2/nestin expression in cell sphere cultures (**F**), dual phalloidin/FAK expression in adherent cells (**G**), dual Sox2/Oct 3/4 expression in cell sphere cultures (**H**), and GFAP expression in BT48EF spheres (**I**). Confocal images were collected and shown as a maximal projection of about 20 analysed spheres observed with 0.29-μm size serial sections. Scale bar: 25 μm.

GLPG1790 administration induced a significant decrease in EphA2 expression in BT12M cells (81.3 ± 3.4%, p = 0.0016, with a reduction of 12%), whereas no significant variation was observed in BT48EF lines (92.7 ± 5.2%, p = 0 0670). However, confocal immuno-fluorescence analysis showed a reduction of the EphA2 signal in BT48EF treated cells suggesting that GLPG1790 might reduce EphA2 expression in single cells. As GLPG1790 might induce cell detachment from outer/peripheral layers of cells from spheres, we also analysed EphA2 expression in this GIC population. Co-expression of CD44 and EphA2 was reduced after the GLPG1790 administration (Figure 3D), and significant changes were observed for CD105 expression. This antigen was basally detected in 68.7 ± 2.8% and 59.3 ± 2.7% of cells in BT48EF and BT12M cultures, respectively. The percentage of CD105 positive cells was significantly reduced after GLPG1790 administration, being detected in 52.2 ± 3.2% (reduction of 24%, $p < 0.005$) in BT48EF cells and 28.6 ± 2.5 (reduction of 51.8%, $p < 0.0001$) in BT12M cells. The percentage of CD90 expressing cells was not modified by GLPG1790 administration in BT48EF cells (65.9 ± 3.5% vs. 64.8 ± 2.7%), whereas it was statistically reduced in BT12M cells (77.1 ± 4.2% vs. 61.7 ± 3.6%, $p < 0.005$ with a reduction of 20%). Similarly, nestin was significantly reduced in BT12M cells (43.7 ± 2.8% vs. 13.9 ± 2.7%, $p < 0.001$ with a reduction of 68.2%), whereas the expression of this antigen was not statistically modified in BT48EF cells via the GLPG1790 treatment (45.5 ± 3.8% vs. 52.2 ± 5.3%, $p = 0.0947$). Nevertheless, confocal immuno-fluorescence analysis showed a reduction of the nestin signal in BT48EF-treated cells suggesting that GLPG1790 induced a reduced expression rather than a modification in the percentage of positive cells. This reduced signal was also confirmed by confocal immuno-fluorescence analyses performed in BT50EF cells. Next, we analysed the variation of differentiating marker after GLPG1790 administration and found that the expression of the neuronal marker NFH/Tuj1 was strongly increased in BT48EF cells (33.6 ± 2.4% vs. 83.4 ± 6.5%, $p < 0.0001$ with an increase of 2.48-fold), and was moderately increased in BT12M cells (62.9 ± 4.2% vs. 91.3 ± 6.4%, $p < 0.0001$ with an increment of 1.45-fold). These effects were particularly evident in the confocal immunofluorescence analysis in which NFH positive cells were strongly increased, as well as the signal that was increased after GLPG1790 administration (Figure 3C). The astrocyte marker GFAP was increased by 1.22-fold (70.0 ± 4.3% vs. 77.5 ± 3.4%, $p = 0.0345$) in BT12M cells but was not modified in BT48EF cells (74.7 ± 3.7% vs. 71.5 ± 4.3%). However, from western blots, the levels of GFAP seemed to be increased in extracts from BT48EF cells treated with higher doses of GLPG1790. For other stem cell markers, including Oct3/4, the signal on confocal immuno-fluorescence was dose-dependent and significantly reduced after GLPG1790 administration (Figure 3C). Western blot analysis confirmed the decrease in CD44, Oct3/4 and Sox2, as well as the increase of GFAP and βIII tubulin.

3.3. GLPG1790 Modifies Tumour Growth of GBM Cells Subcutaneously Injected in Female nu/nu Mice (Xenograft Model)

The efficacy of GLPG1790 as a single therapy was verified in comparison with standard therapeutic approaches (RT, TMZ or RT+TMZ) in Cd1 nu/nu mice with experimental brain tumours by using subcutaneous xenograft models of U87MG, U251 and T98G cells. Luciferase transfected U87MG and patient derived glioma stem-like CSCs-5 cells were injected in the striatum for the orthotopic models. For the first evaluation, we verified the in vivo efficacy of single doses of 30 mg/Kg/day GLPG1790 in mice by using three human models of high-grade glioma cells (U87MG, U251 and T98G) grown as subcutaneous xenografts and showing different levels of EphA2. GLPG1790 efficacy was compared with standard radio (RT, 1 doses of 4 Gy), chemo (temozolomide, TMZ, 16 mg/kg for 5 consecutive days) and chemo and radio (4 Gy + 16 mg/Kg TMZ) therapy. Figure 4A–D shows the percentage of mice in progression plotted with time of progression using Kaplan–Meyer analysis. In U87MG and U251 xenografts (Figure 4A,B), the anti-tumour effect of GLPG1790 was intermediate between that of RT and temozolomide single therapies, whereas the effects of GLPG1790 were similar to TMZ in T98G cells (Figure 4C).

GLPG1790 effects were not significantly different to RT in U87MG (Figure 4A), whereas this agent was more active than RT in U251 (HR = 2.87, $p < 0.005$; Figure 4B) and T98G (HR = 2.23, $p < 0.05$; Figure 4C). GLPG1790 effects were lower in U87MG (HR = 2.10, $p < 0.05$; Figure 4A) and U251 (HR = 2.88, $p < 0.005$; Figure 4B), and similar in T98G (HR = 1.49, $p > 0.05$; Figure 4C). GLPG1790 effects were significantly lower in comparison with standard chemo and radio therapy (RT+TMZ) in all cell systems (Figure 4A–C). In Table S1, we summarised the statistical analysis of xenograft experiments.

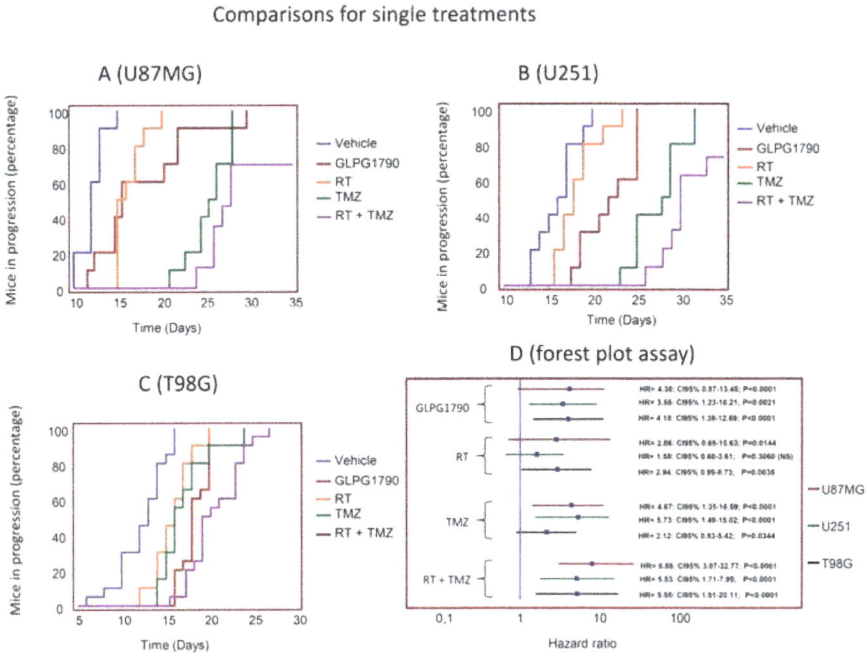

Figure 4. In vivo experiments: GLPG1790-modified tumour growth of GBM cells subcutaneously injected in female nu/nu mice (xenograft model). In vivo efficacy of single dose (30 mg/Kg/day) of GLPG1790 compared with standard RT (1 dose of 4 Gy), chemo (TMZ, 16 mg/kg for 5 consecutive days) and chemo and radio (4 Gy + 16 mg/kg TMZ) therapy. (**A**) Percentage of mice in progression plotted for the time by using Kaplan–Meyer analysis for U87MG cell model. (**B**) Percentage of mice in progression plotted for the time by using Kaplan–Meyer analysis for U251 cell model. (**C**) Percentage of mice in progression plotted for the time by using Kaplan–Meyer analysis for T98G cell model. (**D**) Forrest plot analysis. In Table S1, we summarised the statistical analysis of xenograft experiments.

3.4. Histological Evaluation of GLPG1790 Treated Xenografts: Involvement of Tumour Microenvironment in the GLPG1790-Mediated Anti-Tumour Effects

Previously, we observed that different experimental xenograft models including GBMs were characterised by the presence of heterogeneous populations of tumour and inflammatory cells [40–43]. In addition, GBM xenografts contained giant cells, multinucleated cells and polygonal or spindle-shaped cells with abundant and intensely eosinophilic cytoplasm and hypochromatic nuclei. Tumour cells in rapid growth were dispersed on a fibrillar collagen background that enveloped abundant vasculature. From a morphological point of view, glioma xenografts did not undergo noticeable cellular changes after the administration of GLPG1790. Instead, we observed deep changes in the quality of the tumour microenvironment as modifications of: (i) angiogenesis and vasculomimicry (Figure 5A–D); (ii) necrosis (Figure 5E) and hypoxia-inducible factor 2α (HIF2α) immunostaining (Figure 5F); as well as (iii) tumour infiltration by leukocytes (Figure 5G). In particular, GLPG1790 administration reduced angiogenesis and vasculomimicry as demonstrated by periodic acid-Schiff (PAS) staining (Figure 5A), Masson's trichrome staining (Figure 5B) and CD31 immuno-localization (Figure 5C), when compared to controls. Next, we used PAS staining to analyse angiogenesis and vasculogenic mimicry. We showed that vessels were equally distributed in control xenografts. By contrast, RT caused a switch from angiogenesis to vasculogenic mimicry. This event started with single medium-sized vessels and resulted in a significant reduction of overall vessel number (Figure 5C,D), as recently also observed in bevacizumab treated tumours [35]. This could be associated with endothelial cell damage in different portions of the vessel wall producing endothelial cell discontinuity. In this case, tumour cells oriented themselves towards the lumen of the vessel (intra-vascular mimicry, Figure 5A,B). Therefore, tumour cells were trying to give hardness to the vascular structure, which was straightened by a dense PAS-positive connective matrix. Small endothelial cell-free vascular lacunae were also present. GLPG1790 administration significantly reduced the number of medium/large-sized vessels as well as vascular lacunae distributed in PAS-positive regions. The vascular pattern was significantly reduced after GLPG1790 administration when compared to those observed for RT or TMZ; whereas, the number of vessels (Figure 5D) was lower with respect to tissues treated with RT+TMZ. These morphological results were related to hemoglobin content (Figure 5D). Table S2 summarises these angiogenetic and vasculomimicry differences. Tumour tissues from experimental GBM xenografts also showed areas of necrosis dispersed in the tumour parenchyma (Figure 5E), both in the central and peripheral portions. This appearance was not completely comparable to human gliomas since evident palisating necrosis was absent in experimental gliomas. This necrosis showed elevated leukocyte infiltration (Figure 5E). We also observed that GLPG1790 administration reduced the necrosis percentage, as indicated in Table S3. This seemed to be due to reduced tumour masses. RT, TMZ and RT+TMZ administration, being much more cytotoxic therapeutic procedures, showed higher necrosis due to increased tissue damage. The anti-vascular properties of GLPG1790 were tested in vitro by evaluating the vasculomimicry (VM) of U87MG cells and tubule formation of human brain derived endothelial cells (HBMVECs), as shown in Figure 5F. GLPG1790 reduced both phenomena in a dose-dependent manner. It had been widely demonstrated that necrosis was associated with increased hypoxia and here we showed high HIF2α immunostaining in peri-necrotic areas (Figure 5G), suggesting an involvement of hypoxia. In Table S4, we summarised immunostaining scores of both necrosis and HIF2α. Further molecular evaluations were in progress. One area of this study was to elucidate the association of GLPG1790 administration and the control of hypoxia in stem cell recruitment and recurrence. Temozolomide-treated tumours showed a lower level of leukocyte infiltration but higher necrosis and fibrosis compared to those treated with RT and GLPG1790. Infiltration of granulocytes (both neutrophils and eosinophils) and mono-nucleated blood cells (mainly monocytes) participated in the establishment and maintenance of local inflammation.

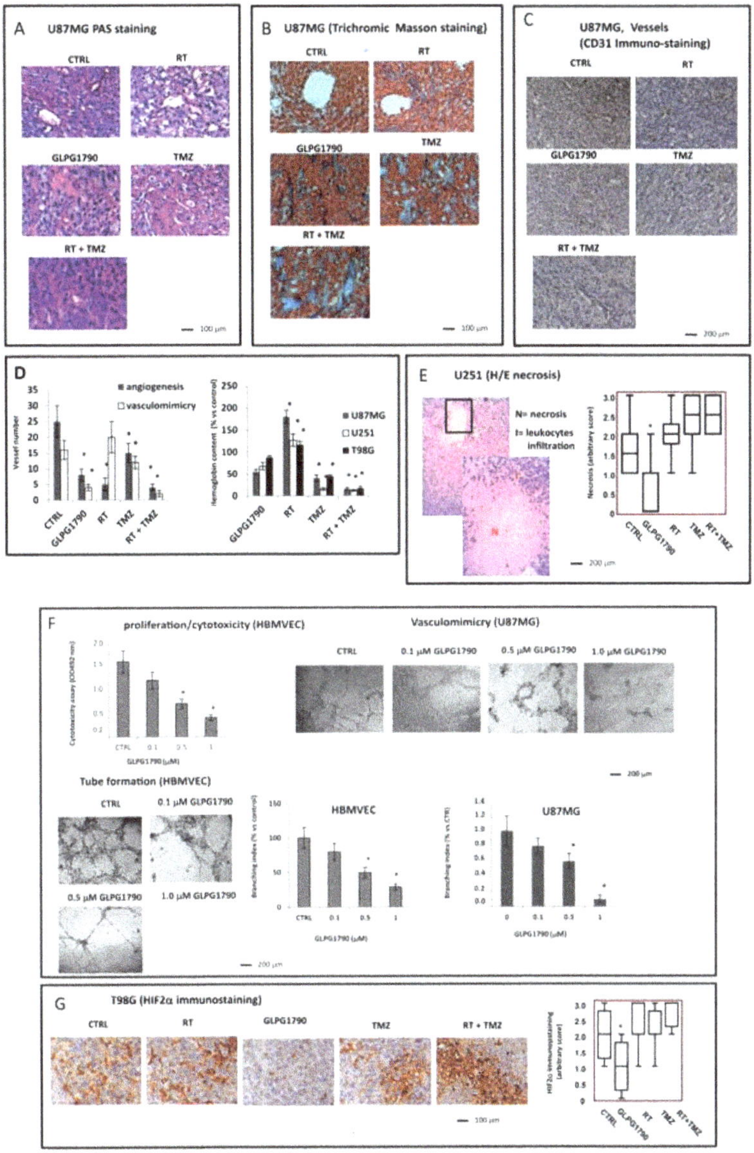

Figure 5. Histological evaluation of GLPG1790-treated xenografts: involvement of tumour microenvironment in the GLPG1790-mediated anti-tumour events. GBM xenografts contained giant cells, multinucleated cells and polygonal or spindle-shaped cells with abundant and intensely eosinophilic cytoplasm and hypochromatic nuclei. Areas with elevated tumour cell proliferation show a fibrillar collagen background that envelopes abundant vasculature and tumour cells. Deep changes in the quality of tumour microenvironment as modifications on: angiogenesis and vasculomimicry: (**A**) PAS staining; (**B**) trichromic Masson staining; (**C**) CD31 immunostaining; (**D**) vessel counts and hemoglobin content. (**E**) Necrosis, which is functionally associated to (**F**) increased in vitro cytotoxicity of HBMVEC, reduced in vitro vasculomimicry of U87MG cells and tubule formation of HBMVEC cells, as well as HIF2α-immunostaining (**G**). * $p < 0.05$.

Leukocyte infiltration was concentrated both along the front of tumour cell growth, as well as dispersed among the tumour cell sheets (Figure 6A,B). The grade of this infiltration is shown in Table S4. Further molecular evaluations are in progress looking specifically at elucidating the association of GLPG1790 administration and the control of inflammation in stem cell recruitment and recurrence.

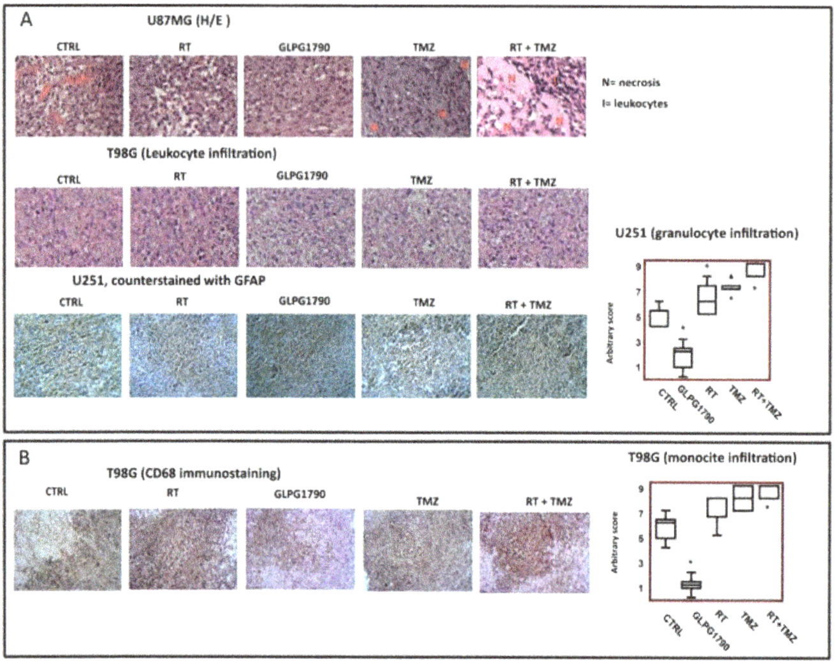

Figure 6. Tumour infiltration of inflammatory cells. (**A**) Infiltration of granulocyte (neutrophils and eosinophils) and (**B**) mono-nucleated blood cells (monocytes). Leukocyte infiltration was concentrated both along the front of tumour cell growth, as well as dispersed among the tumour cell sheets. * $p < 0.05$ vs. CTRL. Scale bar: 1 cm = 100 µm.

3.5. GLPG1790 Increased Disease-Free and Overall Survival in Orthotopic Intra-Brain Tumours as Determined by Using Differentiated U87MG Cells

The efficacy of GLPG1790 was investigated in mice with experimental brain tumours. We wanted to inoculate a small number of cells (3×10^3) into the brain to simulate chemo and radio therapy treatment made after surgery. In such a case, a few tumour cells might remain in the operating bed. These cells would be able to re-grow and recur. We administered treatments after 5 days when no bioluminescence could be detected intracranially. Next, we treat animals for 35 days, with a maximum of 200 days of follow-up without drug administration (total 240 days).

In Figure 7A, we show a schedule of treatments with the animals having different amounts of bioluminescent signal. Bioluminescence was associated with low, intermediate and large intra-brain tumors at necroscopy. In Figure 7B, we show the signals obtained by using Alliance Mini HD6 system (Uvitec Ltd., Cambridge, UK), issued at 42 days from intra-brain cell inoculation in 7/9 mice (one animal had just died) of control and 3/10 of mice treated with GLPG1790. Parameters pertinent to our analyses were recorded, including: (i) tumour growth delay (the time at which a luciferase activity was intracranially detectable), indicating the recurrence time and equivalent to human parameter DFS; (ii) tumour progression through the analysis of bioluminescence imaging (BLI) photon counts and tumour volumes (calculated by magnetic resonance imaging, MRI); and (iii) the survival time (equivalent to human parameter OS) as indicated above. In Figure 7C we analyzed the time of appearance of luciferase signal equivalent of the Disease Free Survival (DFS) in mice injected with luciferase transfected U87MG. In particular control mice developed a bioluminescent lesion after 20–50 days with a mean of 29.0 ± 2.53 days and a median of 30.0 days (20.0–35.0 days, 95% CI); whereas, in GLPG1790-treated animals, the bioluminescence appeared after 20–65 days with a mean of 46.5 ± 4.53 (p = 0.0050 vs. the control) and a median of 45.0 days (35–65 days, 95% CI). RT showed a bioluminescent lesion from 20 to 50 days with a mean of 36.0 ± 3.15 (p = 0.1174 vs. the control and p = 0.0877 vs. GLPG1790) and a median of 32.5 days (30–50 days, 95% CI). The animals treated with TMZ showed recurrent tumours from 30 to 80 days with a mean of 52 ± 5.06 days (p = 0.0012 vs. control, p = 0.0202 vs. RT and p = 0.4523 vs. GLPG1790) and a median of 47.5 (45.0–70.0 days, 95% CI). Standard RT plus TMZ administration significantly delayed the insurgence of recurrent tumours from 50 to 240 days with a mean of 128 ± 25.1 (p = 0.0016 vs. control, p = 0.0028 vs. RT, p = 0.0114 vs. TMZ and p = 0.0072 vs. GLPG1790) and a median of 90 days (50.0–160.0 days, 95% CI). Next, we generated Kaplan–Meier curves (Figure 7D) and calculated the HRs for each experimental group. RT treatment showed a non-significant increase in terms of DFS with an HR = 1.67 (p = 0.1379) compared to untreated animals, whereas GLPG1790 (HR = 2.85, p = 0.0022), TMZ (HR = 3.38, p = 0.0003) and RT+TMZ (HR = 5.21, $p < 0.0001$) treatments significantly reduced the percentage of animals in progression. Although 30% of GLPG1790-treated animals showed bioluminescence during the course of treatment compared to 70% of RT-treated animals, Kaplan–Meyer curves indicated that the effects of GLPG1790 and RT were not significant different (Table S5). Similarly, no statistically significant differences were observed in the comparison between GLPG1790 and TMZ, where 10% of animals showed progression during the cycle of pharmacological treatment. By contrast, the combination RT+TMZ produced no animals in progression during the cycle of treatments, being statistically more active when compared with other single treatments, including GLPG1790. These results could indicate that: (i) the tumour growth rate in each tumour was not different between RT, TMZ or GLPG1790; and (ii) if the administration time had been repeated with further therapy cycles, the differences between these individual administrations could have become significant. Next, we assessed Overall Survival (OS) values (Figure 7E). Control mice were euthanised from 40 to 80 days with a mean of 61.5 ± 3.4 days and a median of 60.0 (60–65 days, 95% CI). GLPG1790-treated animals survived from 66 to 130 days with a mean of 88.0 ± 6.1 (p = 0.0021 vs. control) and a median of 82.5 days (75–100 days, 95% CI). RT-treated mice were euthanised from 55 to 90 days with a mean of 75.5 ± 3.1 (p = 0.0099 vs. control and p = 0.1019 vs. GLPG1790) and a median of 75 days (70–80 days, 95% CI). TMZ showed a significantly increased survival of animals from 60 to 140 days with a mean of 102.5 ± 8.64 days (p = 0.0005 vs. control, p = 0.0121 vs. RT and p = 0.2105 vs. GLPG1790) and a median of 100 days (90–140 days, 95% CI). RT plus TMZ administration showed increased survival of animals from 90 to 200 days with a mean of 170.0 ± 18.0 ($p < 0.0001$ vs. control and RT, p = 0.0049 vs. TMZ and p = 0.0007 vs. GLPG1790) and a median of 180 days (100–200 days, 95% CI). The analysis performed on Kaplan–Meier curves (Figure 7F) showed that a HR = 2.44 (p = 0.0147) for RT administration in the comparison with controls. Similarly, the efficacy of GLPG1790 (HR = 3.38; p = 0.0010), TMZ (HR = 3.65; $p < 0.0001$) and RT+TMZ (5.35; $p < 0.0001$) were significantly increased in term of OS versus controls. When GLPG1790 was compared with standard therapies, we found no significant differences. The combination therapy RT+TMZ was statistically more active when compared with the other single treatments, including GLPG1790. Once again, this suggested an advantage in using GLPG1790 in combination with RT or TMZ. Table S5 summarises the statistical analyses of orthotopic results with U87MG cells.

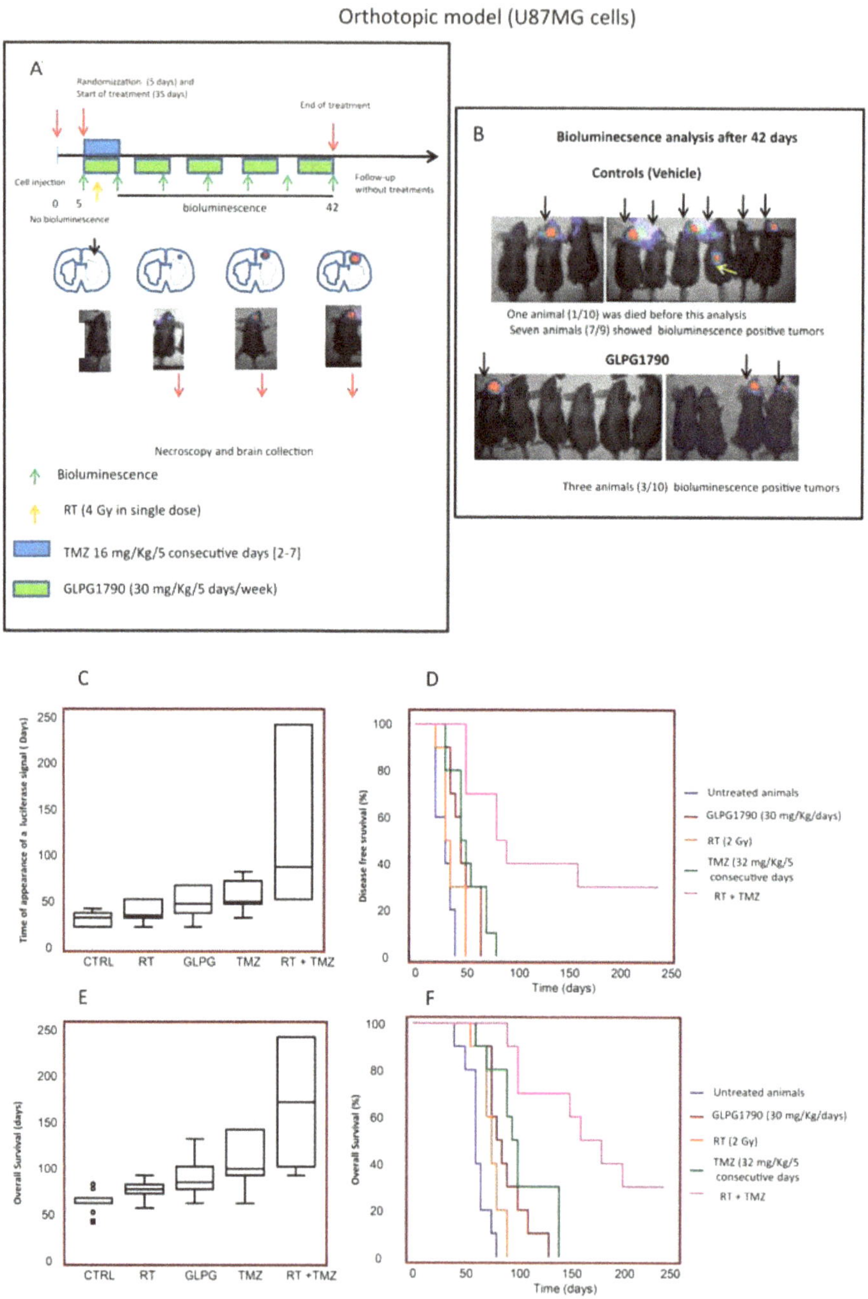

Figure 7. GLPG1790 increased Disease Free Survival (DFS) and Overall Survival (OS) in orthotopic intra-brain tumours using differentiated U87MG cells: a comparison with standard therapies. A small number of cells (3×10^3) were inoculated into the brain to simulate a chemo and radio therapy treatment made with tumours in recurrent phase after surgery. Such a low number of cells remaining in the operatory bed could re-grow and give rise to a recurrence. Five days after cell injection, when

no luciferase activity was detectable intracranially, treatments were started for a maximum of 35 days with 200 days of follow-up without drug administration (total 240 days). Ten animals per group were considered. DFS and OS were recorded as indicated in the Materials and Methods section. (**A**) Graphical methodological approach regarding protocols and type of analyses. (**B**) Representative images for bioluminescence analysis performed after 42 days from tumor injection and showing the number and different sizes of positive (bioluminescent) animals for controls (vehicle treated animals) and GLPG1790. (**C**) DFS: comparison of GLPG1790 administration with standard therapies. (**D**) Percentage of animals in progression analysed using Kaplan–Meyer curves. (**E**) OS (the time of animal's death). (**F**) Percentage of dead animals analysed using Kaplan–Meyer curves.

3.6. GLPG1790 Increased DFS and OS in Orthotopic Intra-Brain Tumours Generated from Patient Derived Stem-Like GSCs-5 Cells

We first evaluated the DFS in GCS-5 brain tumours. Controls developed a bioluminescent lesion from 30 to 50 days with a mean of 33.5 ± 1.9 days and a median of 30 days (30–35 days, 95% CI; Figure 8A,B). GLPG1790-treated animals developed a visible lesion between 30 and 70 days with a mean of 48.5 ± 4.5 days ($p = 0.0324$ vs. control) and a median of 45 days (35–65 days, 95% CI). RT showed a bioluminescent lesion from 30 to 50 days with a mean of 35.5 ± 2.2 days ($p = 0.5172$ vs. control and $p = 0.0818$ vs. GLPG1790) and a median of 32.5 days (30–40 days, 95% CI). TMZ showed a significant increase in the insurgency of recurrent tumours from 30 and 80 days with a mean of 58.5 ± 5.1 days ($p = 0.0123$ vs. control, $p = 0.0290$ vs. RT and $p = 0.5140$ vs. GLPG1790) and a median of 60 days (45–75 days, 95% CI).

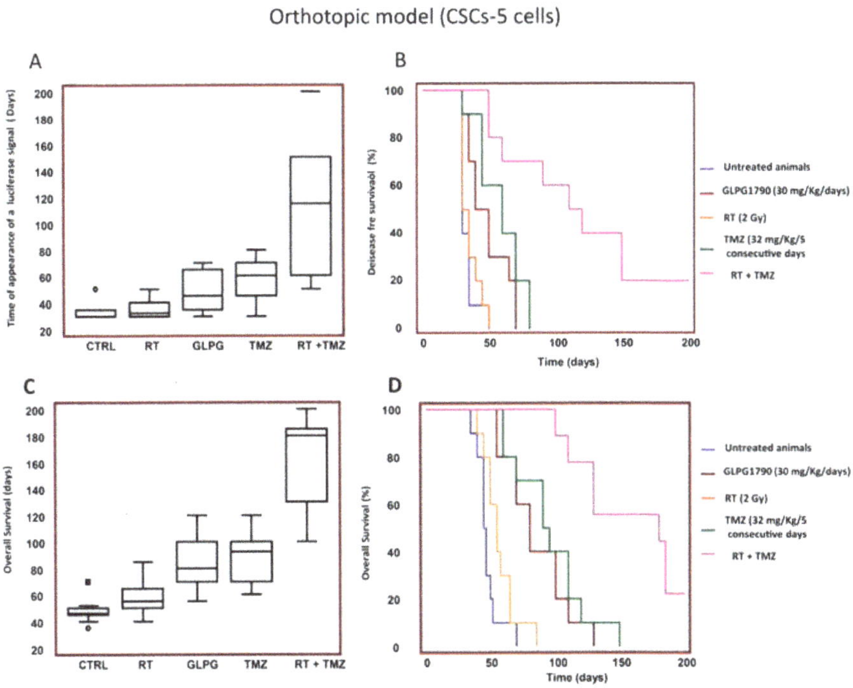

Figure 8. GLPG1790 increases DFS and OS in orthotopic intra-brain tumours using differentiated patient derived GICs (GSCs-5). GICs were inoculated into the brain and treated as indicated in Figure 7. (**A**) DFS: comparison of GLPG1790 administration with standard therapies. (**B**) Percentage of animals in progression analysed using Kaplan–Meyer curves. (**C**) OS (the time of animal's death). (**D**) Percentage of dead animals analysed using Kaplan–Meyer curves.

RT plus TMZ administration increased the insurgency of recurrent tumours from 50 and 200 days with a mean of 118 ± 17.0 (p = 0.0002 vs. control and RT, p = 0.0016 vs. TMZ and p = 0.0009 vs. GLPG1790). Kaplan–Meier curves (Figure 8B) were generated and HR values were calculated. We observed that the RT treatment showed a non-significant increase in terms of DFS with an HR = 1.45 (p = 0.6491) compared to untreated animals, whereas GLPG1790 (HR = 2.45, p = 0.0060), TMZ (HR = 3.11, p = 0.0006) and RT+TMZ (HR = 4.20, p < 0.0001) significantly reduced the percentage of animals in progression. When GLPG1790 was compared with RT, we found that 30% of GLPG1790-treated animals showed a bioluminescence signal during the course of the treatment compared to 80% of RT-treated animals, and Kaplan–Meyer curves indicated the GLPG1790 effects (HR = 2.27, p = 0.0192) were significantly different to RT (Table S2). TMZ showed significantly higher effects in the comparison with RT treated animals (HR = 3.14, p = 0.0006), but not in comparison with GLPG1790 treated animals (HR = 1.75, p = 0.1421). The combination RT+TMZ showed no animals in progression during the cycle of treatments and this resulted in it being statistically much more active than any other single treatment, including GLPG1790. Figure 8C,D shows the OS of animals treated with single treatments. Controls were euthanised from 35 and 70 days with a mean of 47.6 ± 9.2 days and a median of 46 days (45–50 days, 95% CI). GLPG1790-treated animals survived from 55 to 130 days with a mean of 85.0 ± 7.3 (p = 0.0002 vs. control) and a median of 80 days (70–100 days, 95% CI). RT-treated mice were euthanised from 40 to 85 days with a mean of 56.8 ± 12.7 (p = 0.0802 vs. control and p = 0.0038 vs. GLPG1790) and a median of 55 days (50–65 days, 95% CI). TMZ showed a significant increase in survival of animals from 60 and 130 days with a mean of 95.5 ± 8.5 days (p < 0.0001 vs. control, p = 0.0004 vs. RT and p = 0.3742 vs. GLPG1790) and a median of 92.5 days, 95% CI). RT plus TMZ administration increased survival of animals to 90–240 days with a mean of 160.0 ± 12.1 (p < 0.0001 vs. control, RT, TMZ and GLPG1790) and a median of 180 days (130–185 days, 95% CI) Next, Kaplan–Meier curves were generated and HR values were calculated (Figure 8D). We observed that the RT treatment increased the OS with a HR = 2.00 (p = 0.0733) compared to untreated animals. Similarly, the efficacy of GLPG1790 (HR = 4.15; p < 0.0001), TMZ (HR = 4.30; p < 0.0001) and RT+TMZ (5.25; p < 0.0001) were significantly increased in term of OS. OS observed for GLPG1790 was significantly higher when compared to RT (HR = 3.04, p = 0.0027); however, it was similar when compared to TMZ (HR = 1.40, p = 0.4006) and lower when compared with RT plus TMZ administration (HR = 4.11, p = 0.0002).

A comparison of HRs showed that: (i) HR values for DFS were low and not significantly different (1.67 and 1.45, respectively), both in RT-treated patient-derived glioma, stem-like cells and differentiated U87MG cells; (ii) HR values for OS were slightly higher but not significant higher (2.44 and 2.00, respectively) in the same groups; (iii) GLPG1790 effects were higher, in terms of DFS (HR of 2.85 and 2.45, respectively) or OS (3.38 and 4.15, respectively), when compared to RT, in patient-derived glioma stem-like cells and differentiated U87MG cells; and (iv) TMZ showed higher HR values, though not significantly higher than GLPG1790, mainly in patient-derived, stem-like cells. Table S6 data summarises statistical analyses of orthotopic results with CSCs-5 cells. In addition, results on the effects of GLPG1790 obtained in combination with RT or temozolomide are collected for a further report that is in preparation.

4. Discussion

Recent studies indicate that GBMs initiate from a small population of GSCs that are committed to proliferate, namely GICs. This population maintains the stem cell's ability of self-renewal, as well as the ability to modulate, through stem cell signal transduction pathways, different aspects of the malignancy, such as tumourigenicity, local invasion/metastasis and resistance to anticancer drugs [26,27]. Conventional therapies mainly induce death of the differentiated cancer cells. Therefore, a forced differentiation of GICs could increase the percentage of cells that can be targeted by these therapies. This is an interesting and new approach for cancer treatment today. As such, the reactivation of endogenous differentiation programs tends to eliminate stem-like cell phenotypes in a tumour mass, reducing resistance to therapy and tumour recurrence. In addition, differentiation agents show less toxicity than conventional cancer treatments, though their pharmacological efficacy seems to be limited. However, novel compounds targeting cancer stem cells are in preclinical and clinical trials. It has been demonstrated that EphA2 [21], EphA3 [22] and EphB4 [23] are associated with GBM progression, and modulate glial and neuronal differentiation. In addition, overexpression of EphA2 promotes glioma cell migration and local invasion [44], and coincides with cancer stem cell renewal [28,44,45]. Small molecule inhibitors [46,47], monoclonal antibodies [48] or antagonists [49,50] of these receptors have been developed and they have shown encouraging results.

In a previous report, we described the anti-tumour effects of a pan Eph receptor antagonist, UniPr1331 [35], which is able to block both the reverse and forward signaling. In this current study, we studied GLPG1790, a novel small molecule targeting the intracellular kinase domain, as a forced differentiation agent in GBM preclinical models. GLPG1790 shows a pan Eph receptor activity with a range of 1 (EphB2) to 77 (EphA1) nM for affinities on isolated proteins [29] and a good oral bio-availability, being able to cross the blood brain barrier. GLPG1790 was not previously tested for EpHA3 inhibition, whereas it shows an affinity for EphB4 of 22 nM and for EphA2 of 11 nM [39]. Nevertheless, here we analysed the basal expression of EphA3, EphA2 and EphB4 in two different GIC cultures corresponding to mesenchymal phenotype (BT48EF and BT12M), as well as in two glioma cell models (U87MG and U251). Next, we also analysed the inhibition activity versus phosphotyrosines on whole cells treated with different doses of GLPG1790 and 0.25 μg/mL of ephrin-A1 (ligand of EphA2) and ephrin-B2 (ligand of both EphA3 and EphB4) with IC50 values ranged between 200 nM (BT12M) and 348 nM (U87MG) while EphA2 (p-Tyr) activity was inhibited with IC50 values that ranged between 115 nM (BT12M) and 189 nM (BT48EF) in the same assays.

Previously, this compound was shown to reduce the proliferation of human rabdo-myosarcoma [28]. where it induces muscle cell differentiation. Here, we hypothesized that GLPG1790 would be able to revert GICs towards a less aggressive and easier to treat tumour phenotype. Therefore, we analysed the expression of stem and differentiation markers in patient-derived glioma stem-like cells after the administration of GLPG1790. We observed that GLPG1790 was able to reduce stem cell renewal and the growth of glioma spheres. GLPG1790 also induced cell detachment from sphere cultures. The number of single cells was increased in a dose-dependent manner, and maximally between 1.0 and 5.0 μM. A cohort of stem cell markers (CD44, Sox2, Oct3/4, Nestin, Stro-1, CD90 and CD105) was down-modulated using GLPG1790 at low doses (0.5 and 1.0 μM). By contrast, differentiation markers (βIII Tubulin/Tuj1, neurofilaments and GFAP) were increased. In particular, CD44 expression was reduced as assessed using western blotting and the percentage of cells was reduced as determined by FACS. Several data present in the literature indicate that CD44 is essential for both cell stemness and the proliferation of GICs [51,52]. CD44 signaling, which is obtained via ligation of different substrates including osteopontin (OPN) and $\alpha v\beta 3$ and $\alpha v\beta 5$ integrins, has recently been implicated in the stem cell phenotype associated with a lower expression of differentiation marker (GFAP and/or βIII tubulin/Tuj1) and a higher level of sphere formation and expression of stem cell markers (CD133, nestin and Oct4). Therefore, CD44 might be considered a true stem cell marker since reduction of its expression reduced self-renewal and induced cell proliferation and differentiation of committed cells. In addition, CD44 ligation increases DNA repair and impacts GBM radio-sensitivity [53].

A reduction of stemness is also associated with a reduction of CD90 (Thy1). This marker is responsible for cell pluripotency and exhibits key properties of cancer initiating cells (CICs), such as proliferation, differentiation, spheroid formation and tumourigenicity in immuno-deficient mice. CD90 expression is high in high-grade gliomas, whereas it is extremely low in low-grade gliomas or normal brains [54]. CD90 expression is reduced during cell differentiation. In GBM, CD90 positive cells cluster proximally to the tumour vasculature. In addition, this marker is associated with therapy-resistant, quiescent and pluripotent cell populations. CD90 may be considered as a potential prognostic marker mainly because it resides within an endothelial niche where it may play critical roles in vasculature maintenance via VM or differentiation into endothelial cells. Here, we observed that the expression of different stem cell markers was reduced after GLPG1790 administration, including SOX-2, a marker necessary for the maintenance of stem cell features [55], and Oct3/4, another self-renewal regulator in stem cells in GBM [56,57], which up regulates focal adhesion kinase (FAK) and c-Src expression. Nestin characterises immature neural stem cells and promotes tumour cell proliferation, migration, and invasion of GBM, correlating with histological grade and poor survival [58]. We observed that after GLPG1790 treatment, nestin expression was significantly reduced in BT12M cells, whereas the changes were minimal in BT48EF cells. The expression of CD105 [59] (endoglin) is often associated to increased VM. This marker is overexpressed in the endothelial niche. We observed that GLPG1790-induced variations in CD105 expression were significantly related to reduced VM. Next, we found that expression of β-tubulin III/Tuj1 and neuro-filaments, as well as GFAP, was significantly increased after GLPG1790 administration. Although these markers are considered to be associated to neuronal and astrocyte differentiation, respectively, some reports indicated that Tuj1 correlates with malignancy, high proliferative rates and poor prognosis of GBM [60]. In addition, consistent with other reports, GBM neurospheres express high levels of GFAP [61].

In vivo experiments indicated that GLPG1790 reduced tumour growth of three different GBM xenografts. The comparison with standard radio-, chemo- or radio and chemo therapy shows that GLPG1790 effectiveness was higher than that observed for RT, lower with respect to temozolomide in U87M and U251, and higher compared to TMZ in O^6-methylguanine-DNA methyltransferase (MGMT)-positive T98G cells. In addition, GLPG1790 increased DFS and OS in orthotopic intra-brain tumours generated from patient-derived luciferase tagged stem-like GSCs-5 cells. In comparison with RT, GLPG1790 effectiveness in the intra-brain model was much more elevated compared to that observed in xenografts. Compared with TMZ, GLPG1790 administration showed similar efficacy. Nevertheless, the effectiveness of GLPG1790 was lower to that observed for the RT plus TMZ administration. This difference between the treatments was more reduced in the cerebral model when compared to the subcutaneous model. In addition, immunohistochemical analyses indicated that GLPG1790 affected the quality of the tumour microenvironment, which showed to also have a tumour-supportive character. Tumours with necrotic regions had an inadequate blood supply, promoting a local increase in vasculogenesis. This was also associated with impaired blood vessels that had elevated vascular blending, hemorrhage and thrombosis. Although EphrinB2/EphB4 pathway is important in the regulation of postnatal angiogenesis showing a potential therapeutic target for ischemic cardiovascular disease [62], vasculogenic processes including the vascular mimicry are process in which the major role is played by EphA2. In aggressive solid tumors including GBM, it has been demonstrated that vasculogenic mimicry is dependent on the VE-cadherin/EphA2/MMP9/MMP2 axis [63], as well as being mediated by the activity of the EphA2/focal adhesion kinase/paxillin signaling pathway [64].

As a result, these effects amplify GBM tissue damage contributing to inflammation and ischemia. It is widely considered that human GBM trigger strong pro-immunogenic effects with elevated recruitment of granulocytes. Neutrophils are the first cells recruited to inflammatory sites, constituting an important component of the tumour stroma and where they may have both anti-tumoral and pro-tumoral activities. The imbalance of these activities modulates the appearance of drug resistance to anti-angiogenic therapies [40–43]. Similarly, eosinophils are able to produce neuro-mediators, pro-inflammatory cytokines and pro-fibrotic and angiogenic factors involved in tissue remodeling and repair. Resident macrophages and circulating monocytes may also infiltrate the surrounding necrotic areas, continuing to release inflammatory cytokines and sustaining inflammation.

Previous reports indicate that the Eph/ephrin system is involved in the regulation of inflammation [65,66]. Indeed, EphA2 expression is associated in atherosclerotic plaque formation [67]. Ephrin-A1 is upregulated in inflammatory vasculature and EphA2/ephrin-A1, regulates transendothelial migration/tissue infiltration of monocytes/macrophages. Therefore, EphA2 inhibition might reduce chronic inflammation [66]. In our experience, GLPG1790 reduced angiogenesis, necrosis and inflammation. By contrast, RT and TMZ-based chemotherapy, being much more cytotoxic procedures, increased tissue damage of vascular structures, tumour mass and healthy tissue. As such, this resulted in the establishment and maintenance of local inflammation, which in turn, sustained further necrosis and hypoxia, recruiting GSCs glioma stem cells and may have participated in tumour recurrence. Further molecular evaluations aim to elucidate the association of GLPG1790 administration with the control of hypoxia in stem cell recruitment as well as with the control of inflammation in tumour recurrence. No specific experiments to record toxicology data were performed. However, in the experiments presented here, we found no differences in weight wasting. The animals with intracranial tumors treated with GLPG showed a better quality of life linked to the reduced cachecsia associated with reduced tumor growth and brain edema. However, when tumors reached critical values, the quality of life parameters dropped and animals were euthanatised.

5. Conclusions

Our data suggested that GLPG1790 possessed satisfactory anti-tumour effects, regulating both the differentiation status of GICs and the quality of the tumour microenvironment. Of particular importance was its ability to reduce tissue damage that was the main cause of the maintenance of local inflammation, recruitment of cancer stem cells and tumour recurrence. Lower local inflammation may have also been important for the reduction of edema, which is the cause of the main symptoms of GBM. In addition, GLPG1790 administration showed low toxic side effects. Altogether, these data support the use of GLPG1790 for clinical trials in GBM patients. Further experimentation is needed to study possible interactions with radio- and chemotherapy, and preliminary results from our studies indicate that GLPG1790 showed sensitizing effects with RT and temozolomide.

Supplementary Materials: The following are available online at http://www.mdpi.com/2072-6694/11/3/359/s1, Table S1: Hazard ratios for the single therapies (subcutanous xenografts), Table S2: GLPG1790 reduced vasculogenesis, Table S3: GLPG1790 reduced hypoxic areas, Table S4: GLPG1790 reduced leuocytes infiltration, Table S5: Hazard ratios for the single therapies (subcutanous xenografts), Table S6: Hazard ratios for the single therapies (CSCs5 cell orthotopic model).

Author Contributions: C.F.: Conceptualisation, formal analysis, investigation, performed the experiments, data evidence collection, provided resources, data care, writing, review and editing, visualisation, supervision, project administration and funding acquisition. G.L.G.: Conceptualisation, formal analysis, data evidence collection, provided resources, data care, writing, review and editing, visualisation, supervision and funding acquisition. A.M.: formal analysis, investigation (animal manipulation and drug administration, data evidence collection and data care). A.C.: cell cultures, biochemical analyses and immunohistochemical evaluations. S.D.M.: angiogenetic assays and analyses of data. F.V.: Investigation (orthotopic cell injection). S.M.: FACS analyses and Data care. L.C.: confocal analyses and Data care. F.M.: formal analysis and data care. V.M.: Resources, writing, review and editing and funding acquisition. F.B.: Investigation and data evidence collection. P.P.: Investigation and data evidence collection. L.S.: Conceptualisation, data evidence collection, data care, writing, review and editing. G.L.: Resources, writing, review and editing. E.v.d.A.: Conceptualisation, writing, review and editing.

Funding: This work was partially supported by a research contract with Galapagos Sasu and by the ALCLI "Giorgio e Silvia" Non-profit Association.

Acknowledgments: We are grateful to the Umberto Veronesi Foundation for awarding a post-doctoral fellowship to Francesco Marampon.

Conflicts of Interest: Filip Berincx, Philippe Pujuguet, Laurent Saniere, Ellen Vab der AAR and Giocondo Lorenzon are employers of Galapagos Sasu.

References

1. Gittleman, H.; Boscia, A.; Ostrom, Q.T.; Truitt, G.; Fritz, Y.; Kruchko, C.; Barnholtz-Sloan, J.S. Survivorship in Adults with Malignant Brain and other Central Nervous System Tumor from 2000–2014. *Neuro Oncol.* **2018**. [CrossRef] [PubMed]
2. Chen, Z.; Hambardzumyan, D. Immune Microenvironment in Glioblastoma Subtypes. *Front. Immunol.* **2018**, *9*, 1004. [CrossRef] [PubMed]
3. Bougnaud, S.; Golebiewska, A.; Oudin, A.; Keunen, O.; Harter, P.N.; Mäder, L.; Azuaje, F.; Fritah, S.; Stieber, D.; Kaoma, T.; et al. Molecular crosstalk between tumour and brain parenchyma instructs histopathological features in glioblastoma. *Oncotarget* **2016**, *7*, 31955–31971. [CrossRef] [PubMed]

4. Huang, Y.; Rajappa, P.; Hu, W.; Hoffman, C.; Cisse, B.; Kim, J.H.; Gorge, E.; Yanowitch, R.; Cope, W.; Vartanian, E.; et al. A proangiogenic signaling axis in myeloid cells promotes malignant progression of glioma. *J. Clin. Investig.* **2017**, *127*, 1826–1838. [CrossRef]
5. Zanotto-Filho, A.; Gonçalves, R.M.; Klafke, K.; de Souza, P.O.; Dillenburg, F.C.; Carro, L.; Gelain, D.P.; Moreira, J.C. Inflammatory landscape of human brain tumors reveals an NFκB dependent cytokine pathway associated with mesenchymal glioblastoma. *Cancer Lett.* **2017**, *390*, 176–187. [CrossRef] [PubMed]
6. Li, M.; Song, X.; Zhu, J.; Fu, A.; Li, J.; Chen, T. The interventional effect of new drugs combined with the Stupp protocol on glioblastoma: A network meta-analysis. *Clin. Neurol. Neurosurg.* **2017**, *159*, 6–12. [CrossRef] [PubMed]
7. Guo, X.; Xu, S.; Gao, X.; Wang, J.; Xue, H.; Chen, Z.; Zhang, J.; Guo, X.; Qian, M.; Qiu, W.; et al. Macrophage migration inhibitory factor promotes vasculogenic mimicry formation induced by hypoxia via CXCR4/AKT/EMT pathway in human glioblastoma cells. *Oncotarget* **2017**, *8*, 80358–80372. [CrossRef]
8. Huang, W.J.; Chen, W.W.; Zhang, X. Glioblastoma multiforme: Effect of hypoxia and hypoxia inducible factors on therapeutic approaches. *Oncol. Lett.* **2016**, *12*, 2283–2288. [CrossRef] [PubMed]
9. Du, R.; Lu, K.V.; Petritsch, C.; Liu, P.; Ganss, R.; Passegué, E.; Song, H.; Vandenberg, S.; Johnson, R.S.; Werb, Z.; et al. HIF1alpha induces the recruitment of bone marrow-derived vascular modulatory cells to regulate tumor angiogenesis and invasion. *Cancer Cell* **2018**, *13*, 206–220. [CrossRef]
10. Holohan, C.; Van Schaeybroeck, S.; Longley, D.B.; Johnston, P.G. Cancer drug resistance: An evolving paradigm. *Nat. Rev. Cancer.* **2013**, *13*, 714–726. [CrossRef]
11. Lim, E.J.; Suh, Y.; Yoo, K.C.; Lee, J.H.; Kim, I.G.; Kim, M.J.; Chang, J.H.; Kang, S.G.; Lee, S.J. Tumor-associated mesenchymal stem-like cells provide extracellular signaling cue for invasiveness of glioblastoma cells. *Oncotarget* **2017**, *8*, 1438–1448. [CrossRef] [PubMed]
12. Kioi, M.; Vogel, H.; Schultz, G.; Hoffman, R.M.; Harsh, G.R.; Brown, J.M. Inhibition of vasculogenesis, but not angiogenesis, prevents the recurrence of glioblastoma after irradiation in mice. *J. Clin. Investig.* **2010**, *120*, 694–705. [CrossRef] [PubMed]
13. Sattiraju, A.; Sai, K.K.S.; Mintz, A. Glioblastoma Stem Cells and Their Microenvironment. *Adv. Exp. Med. Biol.* **2017**, *1041*, 119–140. [PubMed]
14. Wang, P.; Lan, C.; Xiong, S.; Zhao, X.; Shan, Y.; Hu, R.; Wan, W.; Yu, S.; Liao, B.; Li, G.; et al. HIF1α regulates single differentiated glioma cell dedifferentiation to stem-like cell phenotypes with high tumorigenic potential under hypoxia. *Oncotarget* **2017**, *8*, 28074–28092. [PubMed]
15. Gauden, A.J.; Hunn, A.; Erasmus, A.; Waites, P.; Dubey, A.; Gauden, S.J. Combined modality treatment of newly diagnosed glioblastoma multiforme in a regional neurosurgical centre. *J. Clin. Neurosci.* **2009**, *16*, 1174–1179. [CrossRef] [PubMed]
16. Day, B.W.; Stringer, B.W.; Boyd, A.W. Eph receptors as therapeutic targets in glioblastoma. *Br. J. Cancer* **2014**, *111*, 1255–1261. [CrossRef] [PubMed]
17. Yin, Y.; Yamashita, Y.; Noda, H.; Okafuji, T.; Go, M.J.; Tanaka, H. EphA receptor tyrosine kinases interact with co-expressed ephrin—A ligands in cis. *Neurosci. Res.* **2004**, *48*, 285–296. [CrossRef] [PubMed]
18. Saha, N.; Robev, D.; Mason, E.O.; Himanen, J.P.; Nikolov, D.B. Therapeutic potential of targeting the Eph/ephrin signaling complex. *Int. J. Biochem. Cell Biol.* **2018**, *105*, 123–133. [CrossRef]
19. Tognolini, M.; Incerti, M.; Pala, D.; Russo, S.; Castelli, R.; Hassan-Mohamed, I.; Giorgio, C.; Lodola, A. Target hopping as a useful tool for the identification of novel EphA2 protein-protein antagonists. *ChemMedChem* **2014**, *9*, 67–72. [CrossRef]
20. Day, B.W.; Stringer, B.W.; Al-Ejeh, F.; Ting, M.J.; Wilson, J.; Ensbey, K.S.; Jamieson, P.R.; Bruce, Z.C.; Lim, Y.C.; Offenhäuser, C.; et al. EphA3 maintains tumorigenicity and is a therapeutic target in glioblastoma multiforme. *Cancer Cell* **2013**, *23*, 238–248. [CrossRef]
21. Tu, Y.; He, S.; Fu, J.; Li, G.; Xu, R.; Lu, H.; Deng, J. Expression of EphrinB2 and EphB4 in glioma tissues correlated to the progression of glioma and the prognosis of glioblastoma patients. *Clin. Transl. Oncol.* **2012**, *14*, 214–220. [CrossRef] [PubMed]
22. Chen, J.; Song, W.; Amato, K. Eph receptor tyrosine kinases in cancer stem cells. *Cytokine Growth Factor Rev.* **2015**, *26*, 1–6. [CrossRef] [PubMed]
23. Lodola, A.; Giorgio, C.; Incerti, M.; Zanotti, I.; Tognolini, M. Targeting Eph/ephrin system in cancer therapy. *Eur. J. Med. Chem.* **2017**, *142*, 152–162. [CrossRef] [PubMed]

24. Miao, H.; Gale, N.W.; Guo, H.; Qian, J.; Petty, A.; Kaspar, J.; Murphy, A.J.; Valenzuela, D.M.; Yancopoulos, G.; Hambardzumyan, D.; et al. EphA2 promotes infiltrative invasion of glioma stem cells in vivo through cross-talk with Akt and regulates stem cell properties. *Oncogene* **2015**, *34*, 558–567. [CrossRef] [PubMed]
25. Ferluga, S.; Tomé, C.M.; Herpai, D.M.; D'Agostino, R.; Debinski, W. Simultaneous targeting of Eph receptors in glioblastoma. *Oncotarget* **2016**, *7*, 59860–59876. [CrossRef] [PubMed]
26. Hassan-Mohamed, I.; Giorgio, C.; Incerti, M.; Russo, S.; Pala, D.; Pasquale, E.B.; Zanotti, I.; Vicini, P.; Barocelli, E.; Rivara, S.; et al. UniPR129 is a competitive small molecule Eph-ephrin antagonist blocking in vitro angiogenesis at low micromolar concentrations. *Br. J. Pharmacol.* **2014**, *171*, 5195–5208. [CrossRef] [PubMed]
27. Pujuguet, P.; Beirinckx, F.; Delachaume, C.; Shenton, D.D.; Huck, J.; van der Aar, E.; Brys, R.; van Rompaey, L.; Wigerinck, P.; Saniere, L. Abstract 1753: GLPG1790: The first ephrin (EPH) receptor tyrosine kinase inhibitor for the treatment of triple negative breast cancer. In Proceedings of the Annual Meeting Cancer Research, San Diego, CA, USA, 5–9 April 2014.
28. Megiorni, F.; Gravina, G.L.; Camero, S.; Ceccarelli, S.; Del Fattore, A.; Desiderio, V.; Papaccio, F.; McDowell, H.P.; Shukla, R.; Pizzuti, A.; et al. Pharmacological targeting of the ephrin receptor kinase signalling by GLPG1790 in vitro and in vivo reverts oncophenotype, induces myogenic differentiation and radiosensitizes embryonal rhabdomyosarcoma cells. *J. Hematol. Oncol.* **2017**, *10*, 161. [CrossRef] [PubMed]
29. Krusche, B.; Ottone, C.; Clements, M.P.; Johnstone, E.R.; Goetsch, K.; Lieven, H.; Mota, S.G.; Singh, P.; Khadayate, S.; Ashraf, A.; et al. EphrinB2 drives perivascular invasion and proliferation of glioblastoma stem-like cells. *eLife* **2016**, *5*, e14845. [CrossRef]
30. Luchman, H.A.; Stechishin, O.D.; Dang, N.H.; Blough, M.D.; Chesnelong, C.; Kelly, J.J.; Nguyen, S.A.; Chan, J.A.; Weljie, A.M.; Cairncross, J.G.; et al. An in vivo patient-derived model of endogenous IDH1-mutant glioma. *Neuro Oncol.* **2012**, *14*, 184–191. [CrossRef]
31. Mendiburu-Eliçabe, M.; Gil-Ranedo, J.; Izquierdo, M. Efficacy of rapamycin against glioblastoma cancer stem cells. *Clin. Transl. Oncol.* **2014**, *16*, 495–502. [CrossRef] [PubMed]
32. Bruckheimer, E.M.; Fazenbaker, C.A.; Gallagher, S.; Mulgrew, K.; Fuhrmann, S.; Coffman, K.T.; Walsh, W.; Ready, S.; Cook, K.; Damschroder, M.; et al. Antibody-dependent cell-mediated cytotoxicity effector-enhanced EphA2 agonist monoclonal antibody demonstrates potent activity against human tumors. *Neoplasia* **2009**, *11*, 509–517. [CrossRef] [PubMed]
33. Masuko, K.; Okazaki, S.; Satoh, M.; Tanaka, G.; Ikeda, T.; Torii, R.; Ueda, E.; Nakano, T.; Danbayashi, M.; Tsuruoka, T.; et al. Anti-tumor effect against human cancer xenografts by a fully human monoclonal antibody to a variant 8-epitope of CD44R1 expressed on cancer stem cells. *PLoS ONE* **2012**, *7*, e29728. [CrossRef]
34. Gravina, G.L.; Mancini, A.; Marampon, F.; Colapietro, A.; Delle Monache, S.; Sferra, R.; Vitale, F.; Richardson, P.J.; Patient, L.; Burbidge, S.; et al. The brain-penetrating CXCR4 antagonist, PRX177561, increases the antitumor effects of bevacizumab and sunitinib in preclinical models of human glioblastoma. *J. Hematol. Oncol.* **2017**, *10*, 5. [CrossRef] [PubMed]
35. Festuccia, C.; Gravina, G.L.; Giorgio, C.; Mancini, A.; Pellegrini, C.; Colapietro, A.; Delle Monache, S.; Maturo, M.G.; Sferra, R.; Chiodelli, P.; et al. UniPR1331, a small molecule targeting Eph/ephrin interaction, prolongs survival in glioblastoma and potentiates the effect of antiangiogenic therapy in mice. *Oncotarget* **2018**, *9*, 24347–24363. [CrossRef] [PubMed]
36. Gravina, G.L.; Mancini, A.; Mattei, C.; Vitale, F.; Marampon, F.; Colapietro, A.; Rossi, G.; Ventura, L.; Vetuschi, A.; Di Cesare, E.; et al. Enhancement of radiosensitivity by the novel anticancer quinolone derivative vosaroxin in preclinical glioblastoma models. *Oncotarget* **2017**, *8*, 29865–29886. [CrossRef] [PubMed]
37. Gravina, G.L.; Tortoreto, M.; Mancini, A.; Addis, A.; Di Cesare, E.; Lenzi, A.; Landesman, Y.; McCauley, D.; Kauffman, M.; Shacham, S.; et al. XPO1/CRM1-selective inhibitors of nuclear export (SINE) reduce tumor spreading and improve overall survival in preclinical models of prostate cancer (PCa). *J. Hematol. Oncol.* **2014**, *7*, 46. [CrossRef]
38. Yuan, J.P.; Wang, L.W.; Qu, A.P.; Chen, J.M.; Xiang, Q.M.; Chen, C.; Sun, S.; Pang, D.; Liu, J.; Li, J. Quantum Dots-Based Quantitative and In Situ Multiple Imaging on Ki67 and Cytokeratin to Improve Ki67 Assessment in Breast Cancer. *PLoS ONE* **2015**, *10*, e0122734. [CrossRef] [PubMed]

39. Sysel, A.M.; Valli, V.E.; Bauer, J.A. Immunohistochemical quantification of the cobalamin transport protein, cell surface receptor and Ki-67 in naturally occurring canine and feline malignant tumors and in adjacent normal tissues. *Oncotarget* **2015**, *6*, 2331–2348. [CrossRef] [PubMed]
40. Mason, M.; Maurice, C.; McNamara, M.G.; Tieu, M.T.; Lwin, Z.; Millar, B.A.; Menard, C.; Laperriere, N.; Milosevic, M.; Atenafu, E.G.; et al. Neutrophil-lymphocyte ratio dynamics during concurrent chemo-radiotherapy for glioblastoma is an independent predictor for overall survival. *J. Neurooncol.* **2017**, *132*, 463–471. [CrossRef] [PubMed]
41. Lopes, M.; Carvalho, B.; Vaz, R.; Linhares, P. Influence of neutrophil-lymphocyte ratio in prognosis of glioblastoma multiforme. *J. Neurooncol.* **2018**, *136*, 173–180. [CrossRef]
42. Saito, T.; Sugiyama, K.; Hama, S.; Yamasaki, F.; Takayasu, T.; Nosaka, R.; Muragaki, Y.; Kawamata, T.; Kurisu, K. Prognostic importance of temozolomide-induced neutropenia in glioblastoma, IDH-wildtype patients. *Neurosurg. Rev.* **2018**, *41*, 621–628. [CrossRef] [PubMed]
43. Curran, C.S.; Bertics, P.J. Eosinophils in glioblastoma biology. *J. Neuroinflamm.* **2012**, *9*, 11. [CrossRef] [PubMed]
44. Hamaoka, Y.; Negishi, M.; Katoh, H. Tyrosine kinase activity of EphA2 promotes its S897 phosphorylation and glioblastoma cell proliferation. *Biochem. Biophys. Res. Commun.* **2018**, *499*, 920–926. [CrossRef] [PubMed]
45. Binda, E.; Visioli, A.; Giani, F.; Lamorte, G.; Copetti, M.; Pitter, K.L.; Huse, J.T.; Cajola, L.; Zanetti, N.; DiMeco, F.; et al. The EphA2 receptor drives self-renewal and tumorigenicity in stem-like tumor-propagating cells from human glioblastomas. *Cancer Cell* **2012**, *22*, 765–780. [CrossRef]
46. El-Khoueiry, A.; Gitlitz, B.; Cole, S.; Tsao-Wei, D.; Goldkorn, A.; Quinn, D.; Lenz, H.J.; Nieva, J.; Dorff, T.; Oswald, M.; et al. A first-in-human phase I study of sEphB4-HSA in patients with advanced solid tumors with expansion at the maximum tolerated dose (MTD) or recommended phase II dose (RP2D). *Eur. J. Cancer* **2016**, *69*, S11. [CrossRef]
47. Swords, R.T.; Greenberg, P.L.; Wei, A.H.; Durrant, S.; Advani, A.S.; Hertzberg, M.S.; Lewis, I.D.; Rivera, G.; Gratzinger, D.; Fan, A.C.; et al. KB004, a first in class monoclonal antibody targeting the receptor tyrosine kinase EphA3, in patients with advanced hematologic malignancies: Results from a phase 1 study. *Leuk Res.* **2016**, *50*, 123–131. [CrossRef]
48. Incerti, M.; Tognolini, M.; Russo, S.; Pala, D.; Giorgio, C.; Hassan-Mohamed, I.; Noberini, R.; Pasquale, E.B.; Vicini, P.; Piersanti, S.; et al. Amino Acid Conjugates of Lithocholic Acid As Antagonists of the EphA2 Receptor. *J. Med. Chem.* **2013**, *56*, 2936–2947. [CrossRef]
49. Castelli, R.; Tognolini, M.; Vacondio, F.; Incerti, M.; Pala, D.; Callegari, D.; Bertoni, S.; Giorgio, C.; Hassan-Mohamed, I.; Zanotti, I.; et al. Δ(5)-Cholenoyl-amino acids as selective and orally available antagonists of the Eph-ephrin system. *Eur. J. Med. Chem.* **2015**, *103*, 312–324. [CrossRef]
50. Lamour, V.; Henry, A.; Kroonen, J.; Nokin, M.J.; von Marschall, Z.; Fisher, L.W.; Chau, T.L.; Chariot, A.; Sanson, M.; Delattre, J.Y.; et al. Targeting osteopontin suppresses glioblastoma stem-like cell character and tumorigenicity in vivo. *Int. J. Cancer* **2015**, *137*, 1047–1057. [CrossRef]
51. Tanaka, S.; Nakada, M.; Yamada, D.; Nakano, I.; Todo, T.; Ino, Y.; Hoshii, T.; Tadokoro, Y.; Ohta, K.; Ali, M.A.; et al. Strong therapeutic potential of γ-secretase inhibitor MRK003 for CD44-high and CD133-low glioblastoma initiating cells. *J. Neurooncol.* **2015**, *121*, 239–250. [CrossRef]
52. Shankar, A.; Kumar, S.; Iskander, A.S.; Varma, N.R.; Janic, B.; deCarvalho, A.; Mikkelsen, T.; Frank, J.A.; Ali, M.M.; Knight, R.A.; et al. Subcurative radiation significantly increases cell proliferation, invasion, and migration of primary glioblastoma multiforme in vivo. *Chin. J. Cancer* **2014**, *33*, 148–158. [CrossRef]
53. He, J.; Liu, Y.; Zhu, T.; Zhu, J.; Dimeco, F.; Vescovi, A.L.; Heth, J.A.; Muraszko, K.M.; Fan, X.; Lubman, D.M. CD90 is identified as a candidate marker for cancer stem cells in primary high-grade gliomas using tissue microarrays. *Mol. Cell. Proteom.* **2012**, *11*, M111.010744. [CrossRef] [PubMed]
54. Song, W.S.; Yang, Y.P.; Huang, C.S.; Lu, K.H.; Liu, W.H.; Wu, W.W.; Lee, Y.Y.; Lo, W.L.; Lee, S.D.; Chen, Y.W.; et al. Sox2, a stemness gene, regulates tumor-initiating and drug-resistant properties in CD133-positive glioblastoma stem cells. *J. Chin. Med. Assoc.* **2016**, *79*, 538–545. [CrossRef] [PubMed]
55. Bhagat, M.; Palanichamy, J.K.; Ramalingam, P.; Mudassir, M.; Irshad, K.; Chosdol, K.; Sarkar, C.; Seth, P.; Goswami, S.; Sinha, S.; et al. HIF-2α mediates a marked increase in migration and stemness characteristics in a subset of glioma cells under hypoxia by activating an Oct-4/Sox-2-Mena (INV) axis. *Int. J. Biochem. Cell Biol.* **2016**, *74*, 60–71. [CrossRef]

56. Karmakar, S.; Seshacharyulu, P.; Lakshmanan, I.; Vaz, A.P.; Chugh, S.; Sheinin, Y.M.; Mahapatra, S.; Batra, S.K.; Ponnusamy, M.P. hPaf1/PD2 interacts with OCT3/4 to promote self-renewal of ovarian cancer stem cells. *Oncotarget* **2017**, *8*, 14806–14820. [CrossRef] [PubMed]
57. Matsuda, Y.; Ishiwata, T.; Yoshimura, H.; Hagio, M.; Arai, T. Inhibition of nestin suppresses stem cell phenotype of glioblastomas through the alteration of post-translational modification of heat shock protein HSPA8/HSC71. *Cancer Lett.* **2015**, *357*, 602–611. [CrossRef] [PubMed]
58. Smith, S.J.; Tilly, H.; Ward, J.H.; Macarthur, D.C.; Lowe, J.; Coyle, B.; Grundy, R.G. CD105 (Endoglin) exerts prognostic effects via its role in the microvascular niche of paediatric high grade glioma. *Acta Neuropathol.* **2012**, *124*, 99–110. [CrossRef] [PubMed]
59. Kawamura, Y.; Takouda, J.; Yoshimoto, K.; Nakashima, K. New aspects of glioblastoma multiforme revealed by similarities between neural and glioblastoma stem cells. *Cell Biol. Toxicol.* **2018**. [CrossRef] [PubMed]
60. Gállego Pérez-Larraya, J.; Paris, S.; Idbaih, A.; Dehais, C.; Laigle-Donadey, F.; Navarro, S.; Capelle, L.; Mokhtari, K.; Marie, Y.; Sanson, M.; et al. Diagnostic and prognostic value of preoperative combined GFAP, IGFBP-2, and YKL-40 plasma levels in patients with glioblastoma. *Cancer* **2014**, *120*, 3972–3980. [CrossRef] [PubMed]
61. Yang, D.; Jin, C.; Ma, H.; Huang, M.; Shi, G.P.; Wang, J.; Xiang, M. EphrinB2/EphB4 pathway in postnatal angiogenesis: A potential therapeutic target for ischemic cardiovascular disease. *Angiogenesis* **2016**, *19*, 297–309. [CrossRef]
62. Guo, J.Q.; Zheng, Q.H.; Chen, H.; Chen, L.; Xu, J.B.; Chen, M.Y.; Lu, D.; Wang, Z.H.; Tong, H.F.; Lin, S. Ginsenoside Rg3 inhibition of vasculogenic mimicry in pancreatic cancer through downregulation of VE-cadherin/EphA2/MMP9/MMP2 expression. *Int. J. Oncol.* **2014**, *45*, 1065–1072. [CrossRef] [PubMed]
63. Wang, H.; Sun, W.; Zhang, W.Z.; Ge, C.Y.; Zhang, J.T.; Liu, Z.Y.; Fan, Y.Z. Inhibition of tumor vasculogenic mimicry and prolongation of host survival in highly aggressive gallbladder cancers by norcantharidin via blocking the ephrin type a receptor 2/focal adhesion kinase/paxillin signaling pathway. *PLoS ONE* **2014**, *9*, e96982. [CrossRef] [PubMed]
64. Shiuan, E.; Chen, J. Eph Receptor Tyrosine Kinases in Tumor Immunity. *Cancer Res.* **2016**, *76*, 6452–6457. [CrossRef] [PubMed]
65. Nguyen, T.M.; Arthur, A.; Hayball, J.D.; Gronthos, S. EphB and Ephrin-B interactions mediate human mesenchymal stem cell suppression of activated T-cells. *Stem Cells Dev.* **2013**, *22*, 2751–2764. [CrossRef] [PubMed]
66. Ende, G.; Poitz, D.M.; Strasser, R.H.; Jellinghaus, S. The role of the Eph/ephrin-system in atherosclerotic plaque development: A complex puzzle. *Cardiovasc. Pathol.* **2014**, *23*, 251. [CrossRef] [PubMed]
67. Konda, N.; Saeki, N.; Nishino, S.; Ogawa, K. Truncated EphA2 likely potentiates cell adhesion via integrins as well as infiltration and/or lodgment of a monocyte/macrophage cell line in the red pulp and marginal zone of the mouse spleen, where ephrin-A1 is prominently expressed in the vasculature. *Histochem. Cell Biol.* **2017**, *147*, 317–339. [CrossRef]

cancers

MDPI

Article

A New Patient-Derived Metastatic Glioblastoma Cell Line: Characterisation and Response to Sodium Selenite Anticancer Agent

Sylvie Berthier [1,†], Louis Larrouquère [2,3,†], Pierre Champelovier [1], Edwige Col [4], Christine Lefebvre [5], Cécile Cottet-Rouselle [6], Josiane Arnaud [6,7], Catherine Garrel [7], François Laporte [7], Jean Boutonnat [4], Patrice Faure [7,8] and Florence Hazane-Puch [7,*]

1 Cytometry Platform, Institute of Biology and Pathology, Grenoble Alpes Hospital, CS10217, Grenoble CEDEX 9, France; SBerthier@chu-grenoble.fr (S.B.); pierre.champelovier@wanadoo.fr (P.C.)
2 BrainTech Lab, INSERM U1205, 38000 Grenoble, France; LLarrouquere@chu-grenoble.fr
3 Medical Oncology Department, Grenoble Alpes Hospital, CS10217, Grenoble CEDEX 9, France
4 Unit of Anatomopathology, Institute of Biology and Pathology, Grenoble Alpes Hospital, CS10217, Grenoble CEDEX 9, France; ECol@chu-grenoble.fr (E.C.); JBoutonnat@chu-grenoble.fr (J.B.)
5 Laboratory of Hematology, Onco-Genetic and Immunology, Institute of Biology and Pathology, Grenoble Alpes Hospital, CS10217, Grenoble CEDEX 9, France; CLefebvre@chu-grenoble.fr
6 Laboratory of Fundamental and Applied Bioenergetics (LBFA) and SFR BEeSy, University Grenoble Alpes, Inserm U1055, 38000 Grenoble, France; cecile.cottet@univ-grenoble-alpes.fr (C.C.-R.); JArnaud@chu-grenoble.fr (J.A.)
7 Unit Nutritional and Hormonal Biochemistry, Institute of Biology and Pathology, Grenoble Alpes Hospital, CS10217, 38043 Grenoble CEDEX 9, France; J.A, CGarrel@chu-grenoble.fr (C.G.); francois.laporte@cegetel.net (F.L.); PFaure@chu-grenoble.fr (P.F.)
8 Hypoxia-Physiopathology Laboratory (HP2), Inserm U1042, University Grenoble Alpes, 38000 Grenoble, France
* Correspondence: FPuch@chu-grenoble.fr; Tel.: +33-476769316; Fax: +33-476765664
† Co-first authors.

Received: 9 November 2018; Accepted: 14 December 2018; Published: 21 December 2018

Abstract: Glioblastoma multiform (GBM) tumors are very heterogeneous, organized in a hierarchical pattern, including cancer stem cells (CSC), and are responsible for development, maintenance, and cancer relapse. Therefore, it is relevant to establish new GBM cell lines with CSC characteristics to develop new treatments. A new human GBM cell line, named R2J, was established from the cerebro-spinal fluid (CSF) of a patient affected by GBM with leptomeningeal metastasis. R2J cells exhibits an abnormal karyotype and form self-renewable spheres in a serum-free medium. Original tumor, R2J, cultured in monolayer (2D) and in spheres showed a persistence expression of CD44, CD56 (except in monolayer), EGFR, Ki67, Nestin, and vimentin. The R2J cell line is tumorigenic and possesses CSC properties. We tested in vitro the anticancer effects of sodium selenite (SS) compared to temozolomide TMZ. SS was absorbed by R2J cells, was cytotoxic, induced an oxidative stress, and arrested cell growth in G2M before inducing both necrosis and apoptosis via caspase-3. SS also modified dimethyl-histone-3-lysine-9 (H3K9m2) levels and decreased histone deacetylase (HDAC) activity, suggesting anti-invasiveness potential. This study highlights the value of this new GBM cell line for preclinical modeling of clinically relevant, patient specific GBM and opens a therapeutic window to test SS to target resistant and recurrent GBM.

Keywords: Glioblastoma; cancer stem cells; new cell line; sodium selenite; xenograft; cell death; epigenetics

1. Introduction

Glioblastoma represents the most common type of primary tumors of the central nervous system (CNS) and has a poor prognosis [1]. Recent studies have clarified the common somatic genetic alterations that occur in human GBM. The cellular signaling pathways, including phosphatidylinositol-3 kinase (PI3K) and cell cycle deregulation, are often related to the malignant transformation [2]. Indeed, cyclin-dependent kinase inhibitors, including p15/INK4b, p16/INK4A, p21/Waf1, p27/Kip1, and Mouse double minute 2 homolog (MDM2), play a primordial role in this process [3]. Specific alterations affecting these pathways that include aberrant expression of oncogenes and tumor suppressor genes are shown to be correlated to tumor prognosis, disease progression, and cancer metastasis [4]. Among them, amplification of several growth factor receptors, including EGFR, have been associated to a poor prognosis in GBM [5]. Moreover, Ceccarelli et al. classified GBM in different subtypes, especially dependent on their transcriptional signature and with a correlation with therapeutic response [6].

Recent findings have identified a subpopulation of stem-like cells within tumors, known as cancer stem cells (CSC), exhibiting characteristics of both stem cells and cancer cells. In addition to self-renewal and differentiation capacities, CSCs have the ability to seed tumors when transplanted into the animal host [7,8]. Concerning the origin of CSCs, one hypothesis is that these cells arise from normal somatic cells acquiring stem-like characteristics and malignant behavior, for example, via a glial to mesenchymal transition (GMT) [8]. CSCs are highly resistant to current therapies and could be responsible for repopulating the initial GBM tumor, which explain the high recurrence of tumors [7]. To date, there is no single marker for CSCs, rather a combination of different markers is used to identify CSCs in an isolated cell line. Indeed, the expression of Nestin and Sox-2 has been associated with both the GMT process and CSCs [8–10]. CD34 and CD133 [11] have been used to isolate CSCs from different tumors [12]. The GMT, a crucial process in glial tumor progression, is characterized by the reduction of glial markers (GFAP, which is also used as an astrocytic lineage-specific marker [13]) in exchange for increased expression of mesenchymal markers, such as vimentin [14]. Moreover, the expression of several markers, including Snail, Sox-2, EGFR, CD44, and Six-1, has been associated with the GMT process [10,15]. It should be noted that the induction of epithelial (E)-MT also significantly led to the expression of markers associated with breast CSCs (for a review see [16]). In GBM, the malignant cells are thought to acquire motility by degrading or interacting with extracellular matrix (ECM) components through matrix metalloproteases (MMPs) and specific cell surface molecules, respectively. In this process, CD44, highly expressed during GMT [17], plays an important role in the implantation of tumor cells, allowing the initiation of a metastatic cascade related to the modulation of several cellular characteristics, including adhesion, motility, and matrix degradation [18]. Epithelial markers include adherens and tight junction proteins, such as E-cadherin and ZO-1, whereas mesenchymal markers include, for example, the extracellular matrix component, fibronectin, and the intermediate filament protein, vimentin (for a review see [17]).

Epigenetic modifications are implied in GBM progression [19]. Histone (H) methylation is a complex process implying several histone lysine demethylases (KDM) associated to a variety of physiological and pathological conditions, including cancer. Histone methylation involves specific lysine and arginine of H3 and H4 implied in the regulation of gene transcription and in the support of gene maintenance either on activation or on a repressive dynamic process [19]. Lower global levels of H3K9m2 predict poor prognosis in kidney and prostate cancers [20], but not in lung cancer [21]. Concerning H3K9m2 level regulation, some KDM has been studied, such as LSD1 (KDM1), associated with demethylation of H3K9m1/m2, lysine-specific demethylase 4C (KDM4C/JMJD2C), and G9a, which demethylase H3K9m2/m3 [22].

Temozolomide (TMZ) is the standard of care for GBM, but it still has limitations and unsatisfactory outcomes, particularly in patients with tumor cells expressing the O(6)-methyguanine-DNA-methyltransferase (MGMT) [1,23]. Moreover, resistance to TMZ has become a major concern in GBM treatment, thus new therapies are urgently needed.

In this way, we study sodium selenite's (SS) anticancer properties. We have previously showed that the chemical form and doses influenced its ways of actions [24,25] and we investigated SS anticancer properties in human GBM cell lines [26,27]. We, and others, indicated that SS induces cell death via oxidative stress. Indeed, in the T98G GBM cell line, 24h SS (5, 10 μM) treatment induced a significant decrease of thiol groups and glutathione [27].

The susceptibility of the cultures to the effects of SS depends on the cell type. Indeed, several authors showed that SS was preferentially toxic to malignant glioma cells over normal astrocytes. For example, Rooprai et al. [28] found an IC50 of 28.9 μM for SS in anaplastic astrocytoma within 8 h of treatment whereas it was not reached for 69.36 μM in normal human astrocytes. Kim et al. [29] reported an IC50 comprised between 3 and 5 μM for the established human GBM cell lines (U87MG, U343, A172, and U251) treated during 24 h, whereas, interestingly, no toxicity was reported with 7 μM of SS in normal astrocytes [29]. In vivo, SS accumulates preferentially in the tumor rather than the normal brain tissue in Wistar rats bearing C6 glioma and treated for 4 weeks with 2 ppm to 5 ppm of SS in drinking water [30].

This last point is of great relevance for GBM therapy as crossing the blood brain barrier (BBB) is one of the challenge of new drugs: SS crosses the BBB [31].

Anti-cancer properties of sodium selenite in vitro are so encouraging that SS deserves to be deeply studied for chemotherapeutic responses, before being given up. The clinical trial, SECAR (Sodium Selenite as a Cytotoxic Agent in Advanced Carcinoma, https://clinicaltrials.gov/ct2/show/study/NCT01959438), is ongoing. The phase I has allowed determination of the maximal tolerated dose (MTD) as 10.2 mg/m^2 (phase I). The aim of the phase II, which is in progress, is to use MTD and to study responses, if any, in malignant tumor and in treatment resistant tumors.

The aim of our study was first to characterize this new cell line, named R2J, which expresses the MGMT transcript and exhibits CSC properties associated with in vivo tumor growth, and, second, to evaluate in vitro SS effects.

2. Results

2.1. Immunohistochemistry of the Original Tumor

The resected tumor was positive for GFAP, Ki67, vimentin, CD56, and nestin. Only few cells were positive for EGFR and CD44 (Figure 1, Table 1). It is worth noting that Ki67 staining can reach 30% on some focus. Neuronal markers, NF70, synaptophysin, and NeuN, were also tested and were not detected.

Table 1. Immunohistochemical characterisation of the original tumor compared to R2J monolayer cells and to R2J forming spheres. The expression level is expressed as followed: -: no expression, +: positive expression, nd: Not determined.

Related Family	Proteins	Tumor	R2J Monolayer (2D)	R2J-Gliospheres
Conventional GBM markers and subtypes	ATRX	nd	+	+
	GFAP	+	−	−
	IDH1	nd	−	−
	Ki67	+	+	+
	Olig2	−	−	+ rare cells
	TP53 mutated	nd	−	−
	Vimentin	+	+	+
Neuronal markers	CD56	+	−	+
	neuN	−	−	−
	NF70	−	−	−
Growth factor receptors	EGFR	+ low expression	+ few cells	+ few cells
Cell cycle markers	P16/INK4	nd	+	+
	MDM2	nd	−	−
Cancer stem cell markers	CD34	−	−	−
	Nestin	+	+	+
Mesenchymal shift	CD44	+ rare cells	+ (>90%)	+ few cells
	E-cadherin	nd	−	−

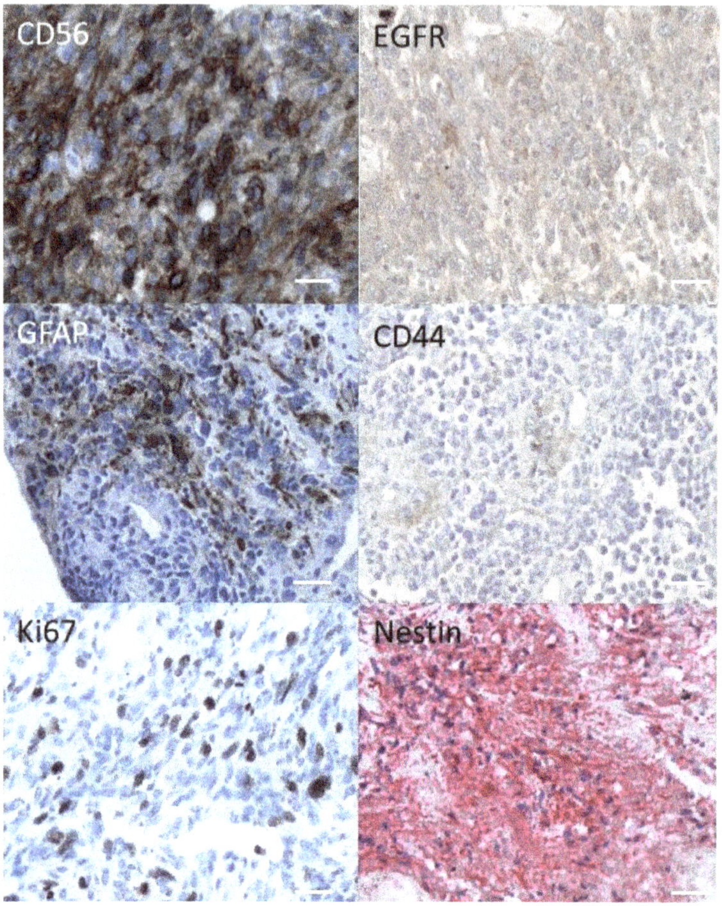

Figure 1. Original tumor was labelled with different markers (see Table 1) as described in the Materials and Methods, using a revelation DAB (peroxidase) for CD56, CD44, GFAP, EGFR, and Ki67 and a revelation Fast Red (alcaline phosphatase) for Nestin. Only positive labellings are shown. Pictures were captured using Leica ICC50 camera connected to a Leica DM2500 microscope (objective ×20). Scale bar: 100 μm.

2.2. *Phenotypic Characterisation of Tumor and R2J Cells*

Original tumor cells were heterogeneous with a big nucleolus (Figure 2a). Numerous mitosis was shown with a high mitotic index (MI) > 5%. More than 90% of the cells from CSF were round and uniform (10–15 μm) (Figure 2b); the nucleus was round with numerous nucleoli. Most of mitotic figures appeared morphologically normal (MI = 2%). Less than 5% of the cells were multinucleated.

Currently, R2J cells are maintained under passage 50 and retained cellular phenotypes' diversity. Indeed, more than 70% of the cells were round and small (15–20 μm), 10% of the cells were round, giant (30–50 μm), and multinucleated (2–5 nuclei), and less than 20% of the cells exhibited fibroblastic- or glial-like morphology (Figure 2c). Cytospin slides stained with MGG essentially showed small and giant cells.

R2J cells were able to form colonies with a PE = 12.4% ± 3.1 (PE = (counted colonies/cells plated) × 100).

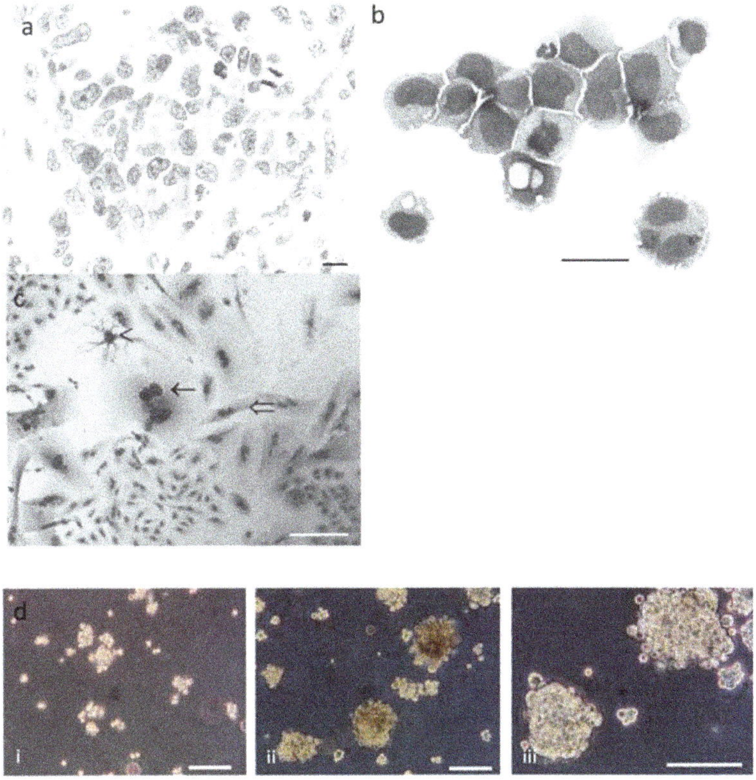

Figure 2. Morphological and phenotypic analysis of (**a**) the biopsy of the original tumor: HES staining shows the heterogeneity of the tumor cells and mitosis. Scale bar = 100 µm (**b**) cerebro-spinal fluid (CSF): May Grunwald Giemsa (MGG) staining shows mitosis and a giant cell. Scale bar = 25 µm (**c**) R2J cells in 2D culture: MGG staining shows giant cells (←), fibroblastic-like cells (⇐) and glial-like cells (<). Scale bar = 100 µm (d) R2J forming spheres in a medium without serum, 7 days (**i**) and 25 days (**ii**: ×100 and iii: ×200) after the seeding. Pictures are representative of more than four independent experiments. Scale bar = 100 µm.

2.3. R2J Cells are Able to Form Gliospheres

When R2J are plated in a cell repellent 10 cm-dish, small spheres containing 10–20 cells were formed 5–6 days after the plating. The sphere diameter increased each day and tumor spheres can be passaged every week for many generations in fresh medium without serum (Figure 2d). The percentage of TS-ICs to examine the efficacy of clonogenesis was 1.9% ± 0.1 (n = 3 independent experiments).

2.4. Immunohistochemistry of R2J Cells Cultured in 2D and in Gliospheres

Compared to the original tumor, R2J cells in culture (2D and spheres) lost the GFAP and CD56 expressions (only 2D) whereas Ki67, vimentin and nestin expressions were conserved as well as mesenchymal shift markers, such as CD44 (Figure 3, Table 1).

Comparing 2D vs. spheres, it appears that only olig2 and CD56 were expressed in spheres. E-Cad transcript was tardily detected in RT-q-PCR (Ct = 37.1 ± 0.9) and the protein was not detected (Figure 3). Concerning Sox2 transcript, it was detected early by RT-q-PCR both in 2D and spheres cells (Ct = 21.4 ± 0.9 and 24.5 ± 2.3, respectively). Moreover, N-Cad transcript was neither detected in adherent R2J cells nor in spheres.

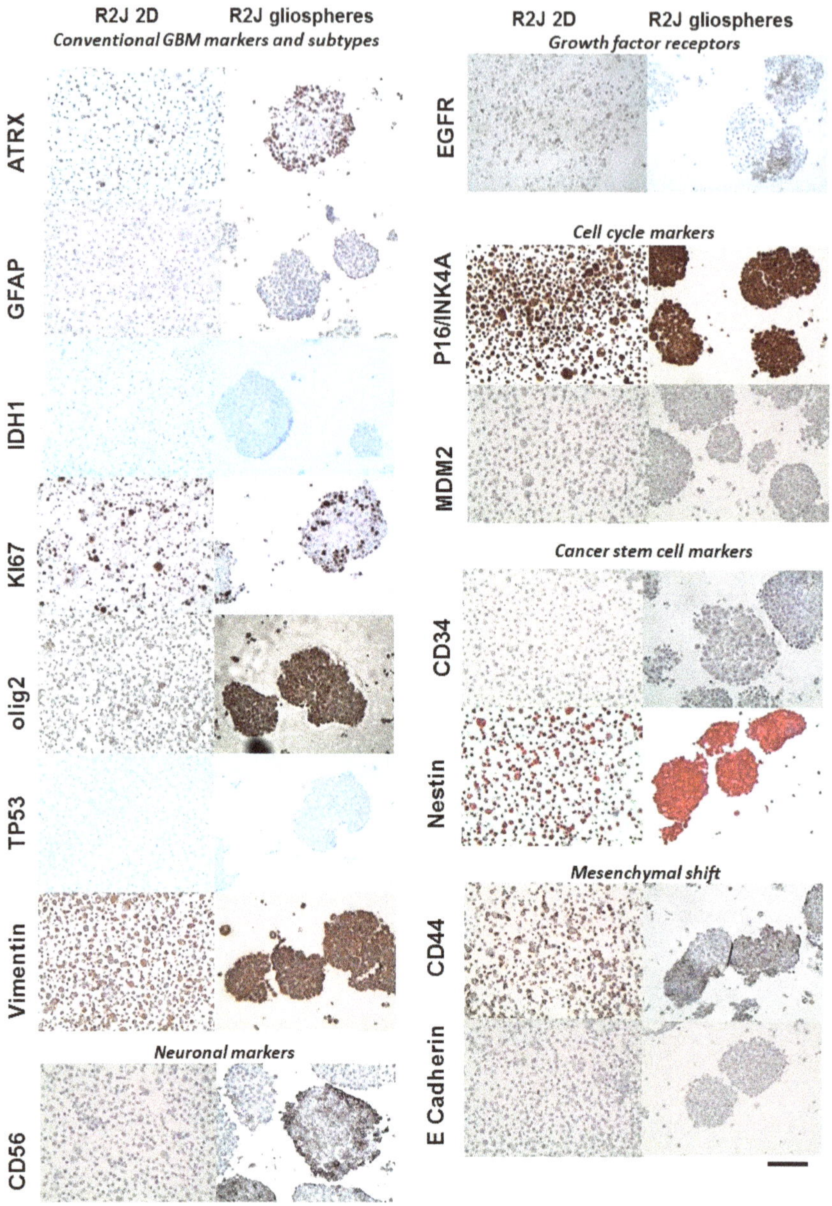

Figure 3. R2J cells cultured in monolayer or in spheres were labelled with different markers as described in the Materials and Methods. Scale bar = 100 μM.

2.5. MGMT Status of R2J Cells

R2J cells expressed MGMT transcript (evaluated by RT-q-PCR) with a cycle threshold (Ct) value=34.8 ± 4.1 (n = three independent experiments). U251 cell line was used as a negative control for MGMT status (no Ct) and T98G was used as a positive control with Ct = 26.11 ± 0.04 (n = three independent experiments).

2.6. Chromosome Analysis

Karyotype analysis, at passages 5 and 35, showed that proliferative R2J cells possess an abnormal karyotype (Supplementary Material, Figure S1). R2J cells are hypotriploid (modal number 64) and showed a large number of numerical abnormalities: A recurrent loss of chromosomes (chr-) 6, 8, 9, 10, 11, 13, 21, 22, and X, a gain of chr-7 (five copies), chr-14 (four copies), and chr-19 (four copies). The chr-Y was not observed whereas R2J was from a male patient. One recurrent structural change (add 7q11) was always present. This was consistent with the degree of malignancy of the original tumor (diagnosed GBM). Moreover, analysis of DNA content by flow cytometry confirmed the polyploidy of the R2J cells.

2.7. R2J Cells are Tumorigenic and Cancer Stem Cells

All the nude mice intracranially implanted with R2J cells cultivated in monolayer (2×10^5 cells, $n = 4$) and in spheres (2×10^5 cells, $n = 4$ and 1000 cells, $n = 4$) were tumor bearing (Figure 4). Two weeks after the implantations, MRI revealed the presence of tumors in mice, which was confirmed 56 days post implantation (PI) for monolayer cells (Figure 4a) and 32 days PI for spheres (Figure 4b,c).

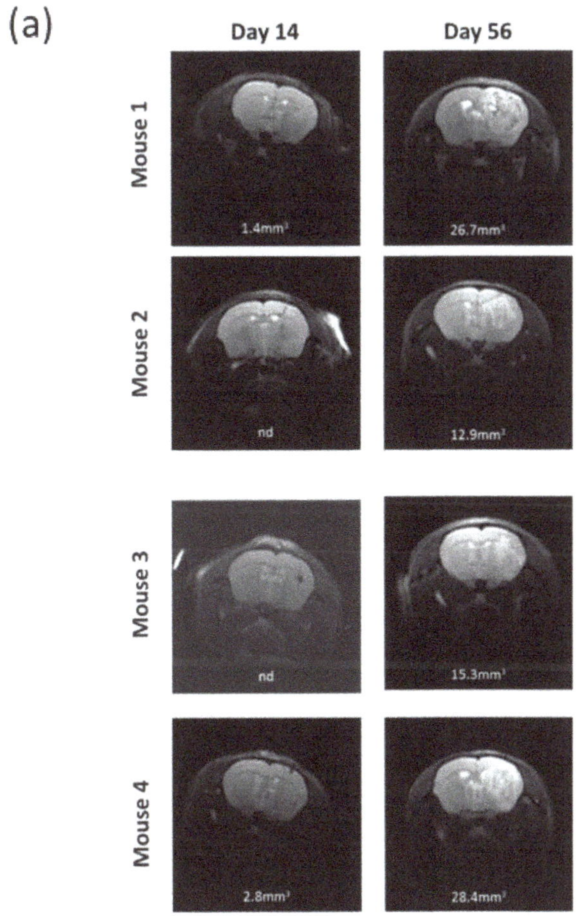

Figure 4. *Cont.*

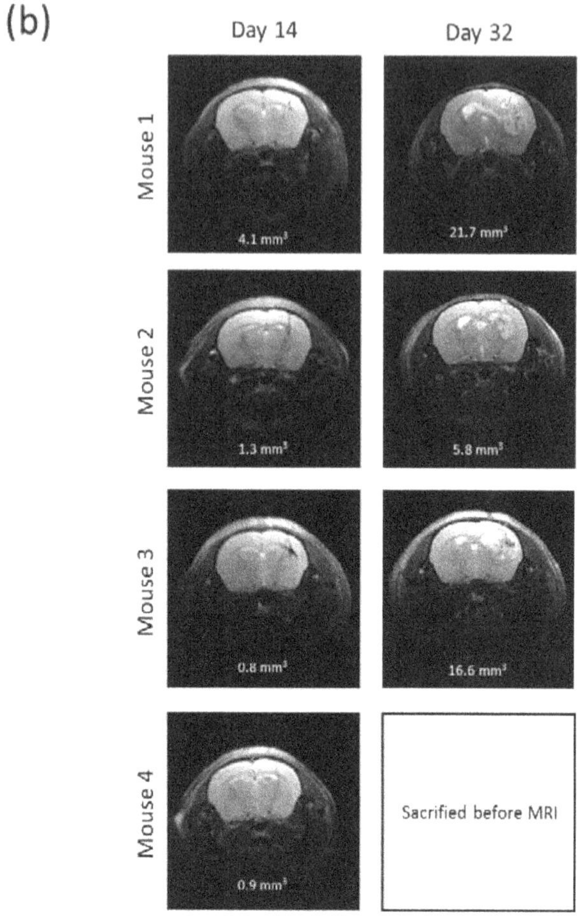

Figure 4. *Cont.*

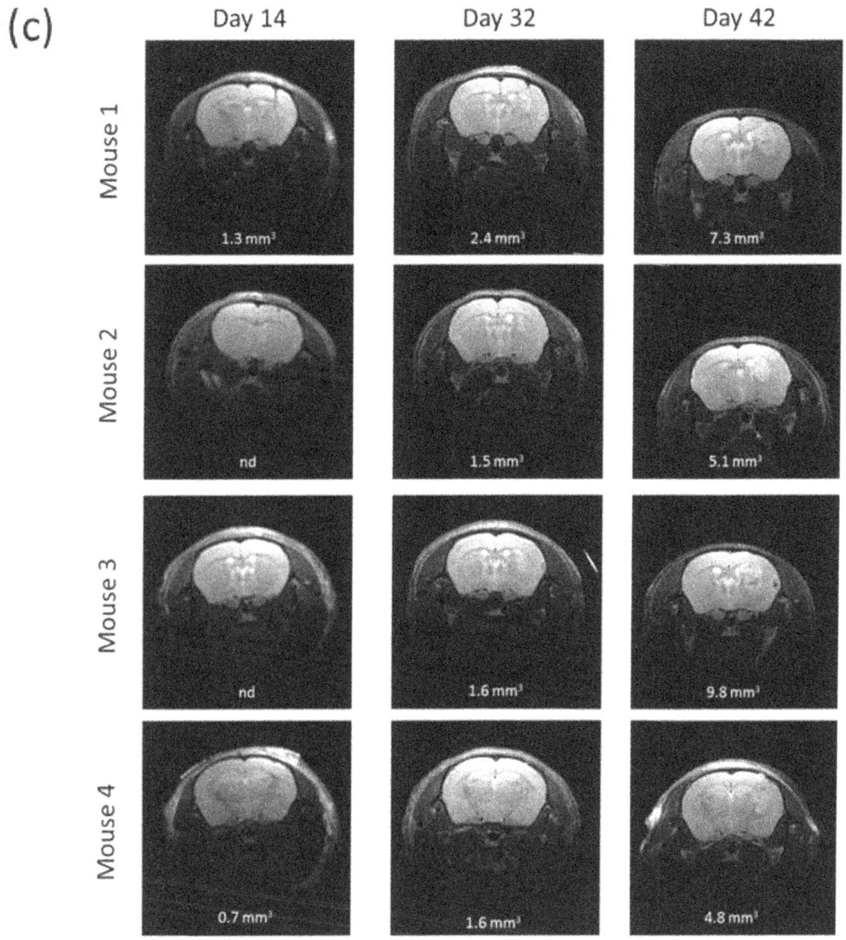

Figure 4. In vivo tumorigenicity of R2J cells after intracranial implantation in nude mice of (**a**) 2×10^5 cells cultivated in the monolayer (**b**) 2×10^5 or (**c**) 1000 cells cultivated in spheres. MRI acquisitions were performed post implantation at the times indicated. Mice were sacrificed after the last MRI. Tumor volumes were calculated by adding each tumor x slice thickness (0.5 mm^2). (**a**) Implantation of 2×10^5 R2J monolayer cultivated cells. (**b**) Implantation of 2×10^5 R2J sphere cells. (**c**) Implantation of 1000 R2J sphere cells.

2.8. SS Absorption

Se was measured by Inductively Coupled Plasma Mass Spectrometry ICP-MS both in lysates and medium in R2J-2D cells treated with SS. The quantity of Se absorbed significantly increased with the SS concentration added. Indeed, at 2.5 µM, the percentage of Se measured vs. Se added was 0.6% ± 0.2 vs. 2.8% ± 0.7 at 5 µM ($p < 0.05$ vs. 2.5 µM) vs. 3.7% ± 1.3 at 10 µM ($p < 0.005$ vs. 2.5 µM) for n = three independent experiments.

Se recovery was 102.7% ± 1.1 at 2.5 µM vs. 83.2%±4.1 at 5 µM ($p < 0.0001$ vs. 2.5 µM) and vs. 79.9% ± 7.0 at 10 µM ($p < 0.0001$ vs. 2.5 µM) for n = three independent experiments. It means that a loss of Se depended on the concentration added.

2.9. SS Triggered both Apoptosis and the Necrosis Cell Death Process

For SS treatment in R2J-2D cells, IC50 values were 3.4 μM ± 0.2 and 2.6 μM ± 0.2 at 24 h and 72 h, respectively (Figure 5a), whereas IC50 was not reached after TMZ acute treatment, until 1mM for 72 h [32] or after the long term treatment, although TMZ was significantly cytotoxic at 40 μM (Figure 5b). Our results suggest that R2J cells are resistant to TMZ. It should be noted that upper TMZ doses were not testable due to DMSO cytotoxicity. U251, which are MGMT negative and p53 mutant, exhibited comparable IC50, i.e., 5.44 μM ± 0.02 and 3.50 μM ± 0.03 at 24 h and 72 h, respectively (Figure 5c), but were more sensitive to the TMZ long term treatment with a significant cell mortality at 10 μM (Figure 5d).

Flow cytometry analysis showed that SS significantly triggered both necrosis and apoptosis at 5 μM for 24 h which was amplified at 48 h (Figure 5e). These results were confirmed by confocal analysis with increased PI staining, a nuclear bean shape specific of apoptosis, and loss of fluorescein diacetate (FDA) at 5 μM-24 h (Figure 5f).

The induction of apoptosis by SS was finally confirmed by a significant increase of caspase-3 activity at 5 μM-24 h (Figure 5g).

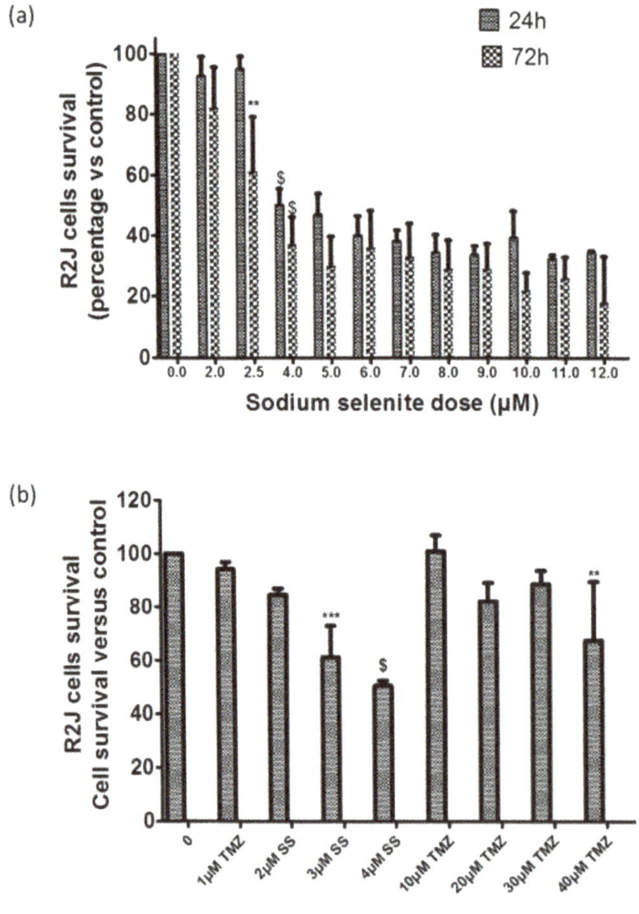

Figure 5. *Cont.*

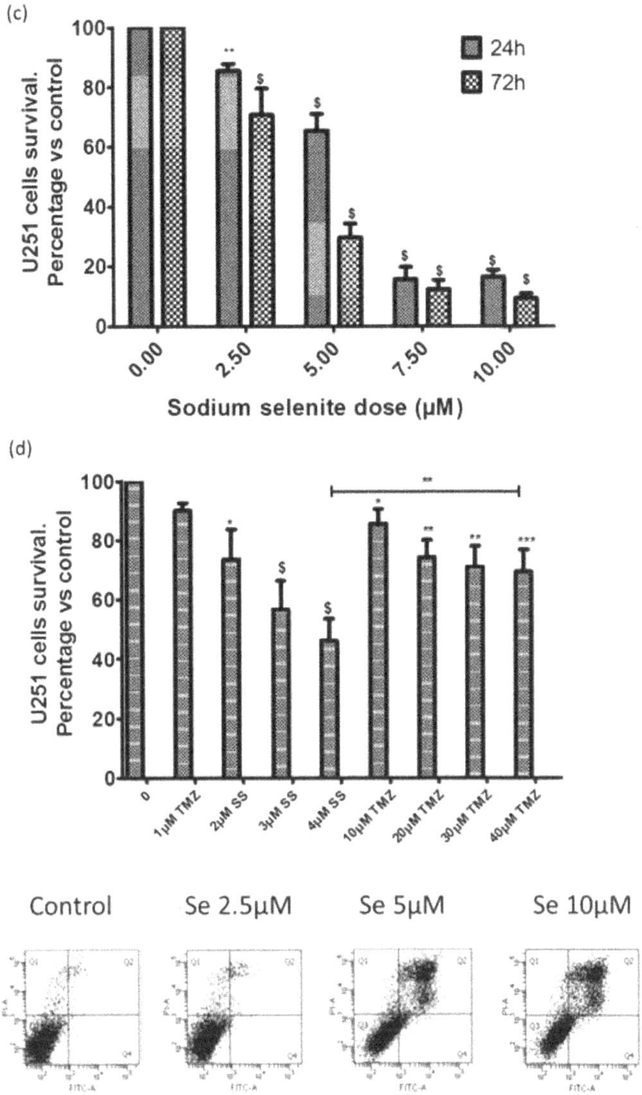

	Control	Se 2.5µM	Se 5µM	Se 10µM
Q1 IP+FITC-	3.3 ± 0.4	4.1 ± 2.0	4.2 ± 1.3	5.4 ± 2.5
Q2 IP+FITC+	4.2 ± 1.8	8.7 ± 4.9	29..2 ± 11.3* $p = 0.0002$	49.8 ± 13.6* $p < 0.0001$
Q3 IP-FITC-	91.8 ± 2.0	84.9 ± 7.1	52.8 ± 8.8* $p < 0.0001$	40.2 ± 15.1* $p < 0.0001$
Q4 IP-FITC+	0.8 ± 0.3	2.4 ± 1.5	7.2 ± 2.5* $p = 0.0065$	4.6 ± 1.5

Figure 5. *Cont.*

48H

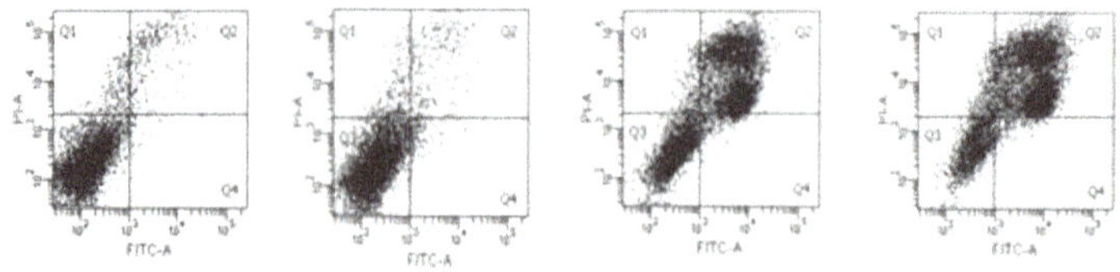

Q1 IP+FITC-	2.8 ± 0.7	2.7 ± 0.8	7.7 ± 6.5	5.9 ± 2.6
Q2 IP+FITC+	4.1 ± 1.3	8.2 ± 1.8	55.3 ± 13.0* *p<0.0001*	65.0 ± 10.1* *p<0.0001*
Q3 IP-FITC-	91.5 ± 2.2	85.7 ± 1.8	32.8 ± 15.0* *p<0.0001*	25.4 ± 12.1* *p<0.0001*
Q4 IP-FITC+	0.9 ± 0.4	3.4 ± 0.6* *p=0.0047*	4.2 ± 1.4* *p=0.0005*	3.7 ± 1.0* *p=0.0045*

(f)

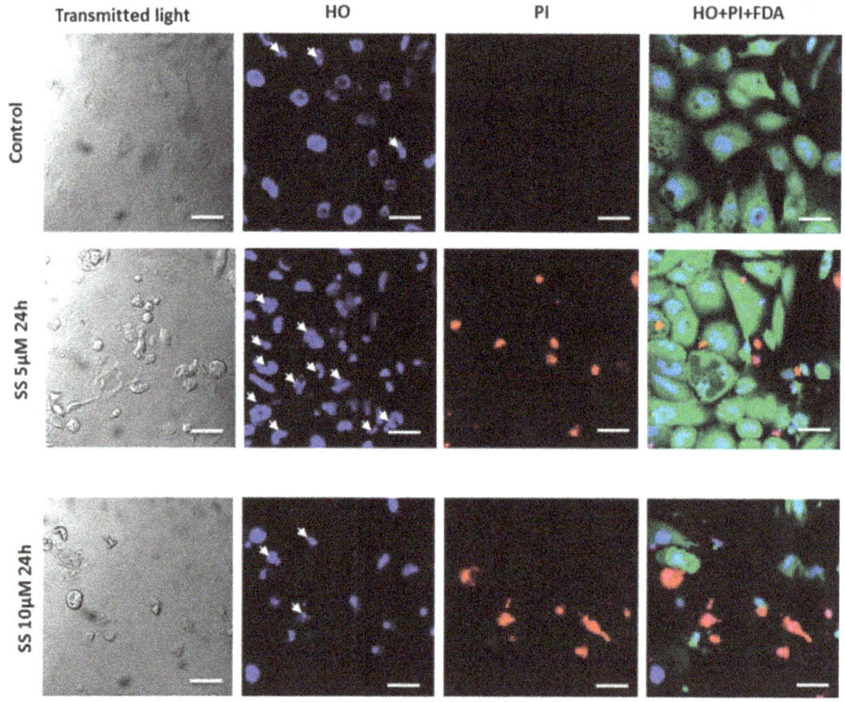

Figure 5. *Cont.*

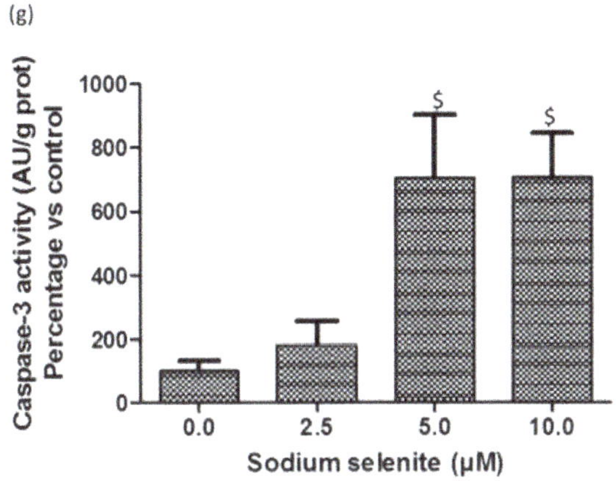

Figure 5. Sodium selenite (SS) cytotoxicity and cell death triggered. SS is more cytotoxic than TMZ in R2J ((**a**) and (**b**)) and in U251 ((**c**) and (**d**)) cells: Cell survival was evaluated by an MTT assay performed on cell growing without (control = 100%) or with variable doses of sodium selenite for 24 h and 72 h (**a**,**c**) or sodium selenite or TMZ for 72 h, followed by 72 h of wash-out (**b**,**d**). Results, expressed in percentage of cell survival vs. the control, are the mean ± SD of three independent experiments with * $p < 0.05$, ** $p < 0.005$, *** $p < 0.0005$, and $ $p < 0.0001$ vs. control. In (**a**), only the first statistical significance was specified to avoid an overloading of the graph. (**e**) Apoptosis (Annexin-V labelling) and necrosis (PI labelling) were evaluated by flow cytometry analysis, 24 h and 48 h after SS treatment. Results are the mean ± SD of four independent experiments. The percentage of cells in each quadrant was compared as a function of the SS concentration and when significant, the p value was informed. Q3 represents viable cells, Q1: Necrotic cells, Q2: Both necrotic and apoptotic cells, and Q4: Apoptotic cells. (**f**) Confocal microscopy: R2J were exposed to different concentrations of sodium selenite for 24 h and labelling was compared to the controls. Intact cells were revealed by intracellular green fluorescence of FDA (no loss of plasma membrane integrity) whereas the nuclei of necrotic cells were labelled with propidium iodide (PI) (red fluorescence). The Hoechst blue fluorescence allowed the discrimination of apoptotic nuclei with an irregular shape (informed white arrows) that exhibited bean-like morphology or were fragmented whereas the intact nuclei displayed a regular shape. Pictures are representative of three independent experiments. Scale Bar: 40 µm. (**g**) SS triggered apoptosis via caspase-3. Caspase 3 assay was performed in R2J cells seeded at 6000 cells/cm^2 and treated with SS for 24 h. Cells and medium were then recovered, centrifuged for 3 min, 360 g, room temperature and rinsed twice with PBS. Cells were lysed in 50 µL of the Caspase-3 lysis buffer. Results are mean ± SD of three independent experiments with $ $p < 0.0001$ versus the control.

2.10. SS Induced DNA Fragmentation and Cell Cycle Arrest in G2

R2J-2D cell cycle was evaluated by flow cytometry after SS treatment. At 24 h-2.5 µM, SS significantly moved the cells from the G0/G1 to the G2M phase. This arrest in G2 was dose-dependent and was amplified at 48 h (Figure 6).

The subG0G1 analysis allowed evaluation of the DNA fragmentation, which was significantly induced by SS-24 h for the 5 and 10 µM treatment. For the higher doses used at 48 h, i.e., 5 and 10 µM, only a third and a quarter of the R2J cells remained, respectively, alive (Figure 5e), making the cell cycle analysis difficult.

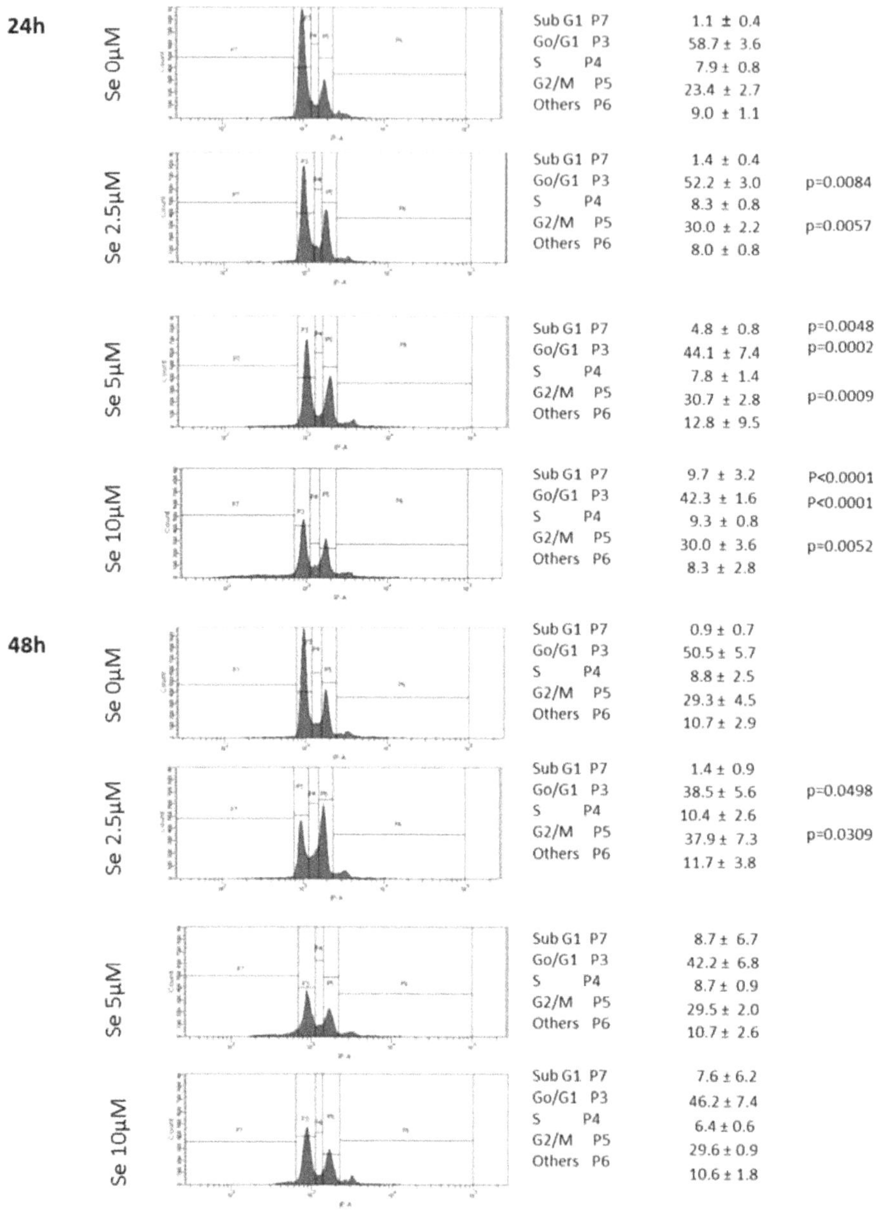

Figure 6. SS blocked the cell cycle in G2M. SS effects on R2J cell cycle (PI staining) was evaluated by flow cytometry analysis, 24 h and 48 h after SS treatment. Results are the mean ± SD of four independent experiments. Percentage of cells in each cell cycle phase was compared as a function of the SS concentration and when significant, the p value was informed. The subG1 was used to evaluate DNA fragmentation.

2.11. SS Acted at the Epigenetic Level

HDAC activity was determined using a fluorometric analysis, after 24 h and 48 h of SS-treatment in R2J-2D cells. SS significantly altered the HDAC activity at 48 h-2.5 μM (Figure 7a).

H3K9m2 level was evaluated by flow cytometry after 24 h SS-treatment. SS treatment for 24 h increased H3K9m2 level in R2J cells as a tendency at 2.5 μM (p = 0.055) and significantly at 5 μM (p = 0.0318) (Figure 7b).

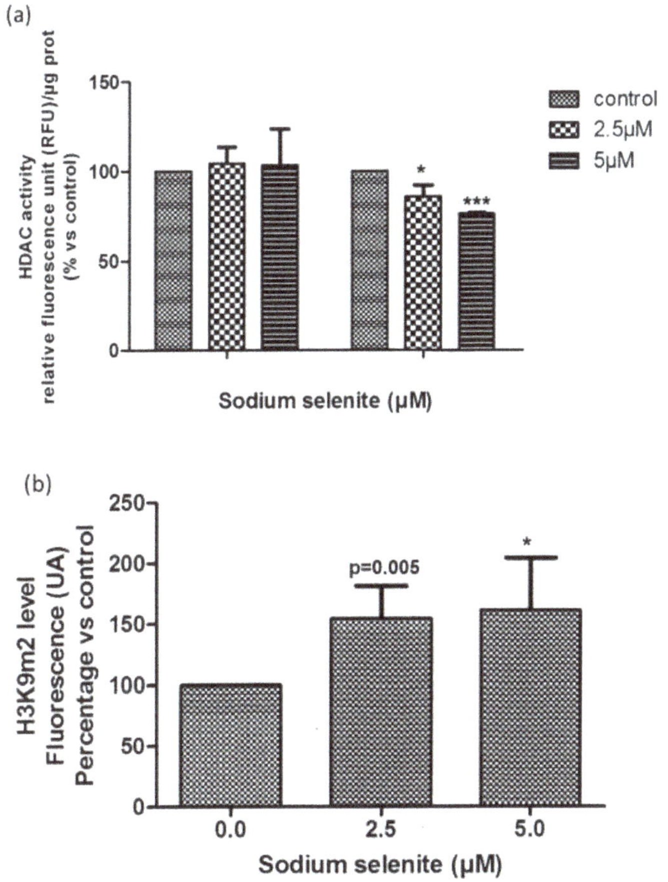

Figure 7. Epigenetic regulations under SS treatment. (**a**) SS altered HDAC activity. R2J-2D cells were not treated (control = 100%) or treated with 2.5 and 5 μM SS for 24 h, 48 h, and 72 h, and harvested. Cell lysates were obtained after five thaw-freeze cycles. HDAC activity (expressed in RFU) was determined using a fluorometric kit and results were normalized by the protein concentration in each sample. Results, expressed in percentage vs. control, are the mean ± SD of three independent experiments with * $p < 0.05$ and *** $p < 0.0005$ vs. control. (**b**) SS modulated the H3K9m2 level. 1×10^6 R2J-2D cells fixed and permabilized were incubated with a monoclonal anti-dimethyl-H3K9 antibody to determine the methylation level of H3K9 after 24 h culture in the presence of SS at 2.5 and 5 μM versus not treated cells. A goat anti rabbit-FITC labelled antibody allowed the flow cytometric analyses with FACSCanto II. Results expressed in percentage vs. the control (not treated) are mean ± SD of five independent experiments, with * $p < 0.05$ vs. control.

2.12. SS Induced Oxidative Stress, Autophagy, and Reticulum Endoplasmic Stress

Oxidative stress was evaluated by quantifying thiol groups concentration in R2J and U251 cells, grown in the monolayer, after 24 h of SS treatment. SS significantly decreased thiol groups in both cell lines whereas TMZ did not (Figure 8a).

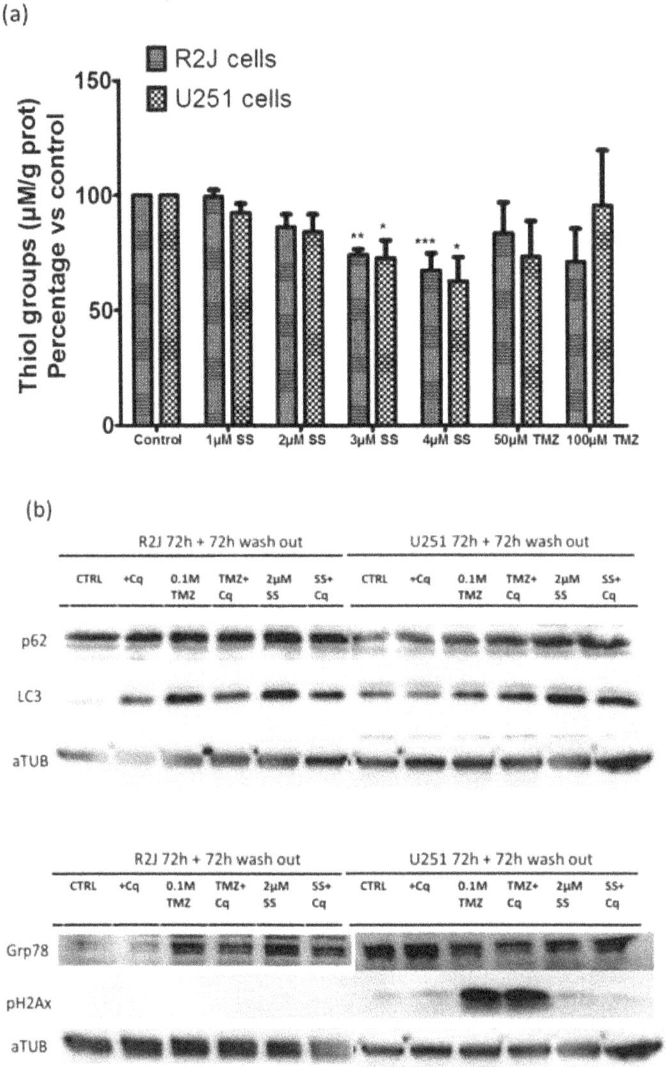

Figure 8. SS's potential way of action was studied by evaluating (**a**) oxidative stress by thiol group levels in R2J and U251 cell lines after being treated for 24h with SS or TMZ. Cell lysates were obtained after five thaw-freeze cycles. Thiol groups (µM) were normalized by the protein concentration in each sample. Results, expressed in percentage vs. control (not treated cells), are the mean ± SD of three independent experiments with * $p < 0.05$, ** $p < 0.005$, and *** $p < 0.0005$ vs. control. (**b**) Autophagy (p62 and LC3-II), reticulum stress (Grp78), and DNA damage (pH2AX) by Western-blots analysis, 72 h after SS or TMZ treatments followed by 72 h of wash-out. Chloroquine (Cq) at 20 µM was used as a positive control to block autophagy 1 h before SS or TMZ treatments.

Western-blots showed that SS at 2 μM and TMZ at 100 μM repressed autophagy with an increase of p62 and LC3-II both in R2J and in U251 cells and induced a reticulum stress with an increase of Grp78 only in R2J cells (Figure 8b). It should be noted that chloroquine (Cq) at 20 μM was used as a blocker of autophagy. Interestingly, SS at 2 μM did not cause pH2AX expression (DNA damage) whereas TMZ did (Figure 8b). All these results suggest that SS and TMZ do not have the same mechanism of action.

3. Discussion

Despite the great efforts made to fight against GBM and the advent of TMZ as the first line CT agent available for long term application, the prognosis of patient suffering for GBM remains poor [1]. One problem is chemo-resistance, supposedly related to CSC, in which cells escape CT-induced cell death [33]. One hypothesis is that CSC can later differentiate into proliferative progenitor-like and more differentiated tumor cells that characterize the histological traits of the tumor entity [8]. New therapeutic approaches are likely to generate new drugs able to destroy the CSC of the tumor to stop the tumor progression.

In vitro tests are essential to raise these questions with the development of new cancer cell lines with CSC characteristics. Their individual phenotypic and/or genotypic characteristics provides a valuable means for this search [34–36]. The R2J cell line described herein was established from a CSF from a patient with a GBM. The karyotype and cell cycle analysis revealed a mixture of diploid (2n) and tetraploid (4n), but also 8n and 16n cells, a gain of chr-7 and -19, which may be due to an endoreplicative process frequently occurring in malignant cells and commonly associated to GBM [37].

There is a diversity of cellular phenotypes within the R2J-2D cell population that may reflect mixtures of progenitors, differentiated cells, and presumed cancer stem cells. Indeed, our data supports that R2J cells are derived from a tumor containing multipotent neural stem cell-like cells where a population of CSC persists. Indeed, R2J cells expressed Nestin and CD44, which is different from other usual reported GBM MGMTt+ cell lines, like T98G [13]. R2J cells are able to form spheres (1.9%) in a serum-free medium, with self-renewal capability, as reported for U87 (2.5%) [38] and U251 (1.15%) [39].

The most prevalent form of glioma is referred to as astrocytoma, based on the predominance of GFAP+ astrocytes-like cells within the tumor mass. GFAP is then used as an astrocytic lineage-specific marker [13]. The original tumor expressed GFAP, which was lost in adherent and R2J spheres. Furthermore, adherent R2J did not express CD56, a neuronal lineage marker, whereas the spheres expressed it, suggesting that R2J cells are more likely to differentiate into neurons than in astrocytes. Moreover, in R2J spheres, the strong expression of Nestin, Sox2, and CD44 also suggests that R2J cells are engaged in an incomplete GMT process that may induce a partial acquisition of some CSC markers. Indeed, recent studies have established a possible link between a passage through GMT and the acquisition of GSC properties. Thus, the expression of Nestin and Sox-2 has been associated with both the GMT process and CSC [8–10]. In mammary cells, the induction of EMT has also significantly led to the expression of markers associated with breast CSCs (for a review see [16]).

Proliferation may be altered by cellular signaling pathways modifications related to growth factor receptor expressions [2,3]. EGFR, associated to a poor prognosis [40], is expressed in some R2J cells, which are denoted with the polysomy of chr-7, but the abnormal karyotype of R2J cells is a common feature of GBM [41]. From all our results and on the basis of the molecular profiling established by Verhaak et al. [6,42] and on the new WHO classification [43], the R2J cell line could be classified into the mesenchymal-like subtype into the isocitrate deshydrogenase IDH-wildtype glioma subgroup. Additionally, the combination of a high expression of mesenchymal (CD44), progenitor cell markers (nestin, SOX2) is evocative of an epithelial (glial)-to-mesenchymal transition that has been linked to dedifferentiated and transdifferentiated tumors [17]. Moreover, the R2J cell line presents some similarities with the mesenchymal described USC02 cell line [44], with SOX2, CD44, and nestin expressions, and the capability to form spheres when cultured with growth factors, and to form tumors after intracranial implantation into nude mice.

The xenograft of R2J cells cultivated in 2D or in spheres resulted in the formation of a detectable tumor within 14 days post implantation into the striatum of nude mice. These results allowed conclusions on the tumorigenicity of this cell line and on its CSC properties [7]. Indeed, only 1000 spheres isolated from the total R2J-2D population were able to trigger a highly proliferative tumor growth.

As a first conclusion, although numerous GBM cell lines have been published, few of them have been established from tumor cells expressing MGMT transcript and having resistance to TMZ. Indeed, T98G is p53 mutation, has a positive MGMT status, and is resistant to TMZ [45]. Only U251 [39] and U87 [38] have been reported to contain a CSC population, but these are both MGMT negative. Moreover, LN229 and U251 are different from T98G and U87 by expressing nestin, SOX2, and CD44 [13]. Then, the R2J cell line really has added value because in addition to being MGMTt+, they exhibit CSC characteristics with in vivo tumorigenesis that are relevant to the testing of new drugs.

In this line, as we reported SS-anticancer effects in GBM cell lines cultivated in 2D [26] and in cell-cluster spheroids [27], we in vitro tested SS on the R2J-2D cells.

SS was cytotoxic in the micromolar range and much more toxic than TMZ, for which the IC50 was not reached at a higher dose and after long term treatment. This is consistent with the fact that the patient did not respond to TMZ that may be correlated with a MGMT transcript positive expression. Similar results were obtained in the T98G cell line, also positive for MGMT transcript expression [46], which exhibited a high TMZ resistance (IC50 was not reached at 1mM-72 h [27]).

R2J cells absorbed SS, and an uncomplete recovery suggest the generation of volatiles Se metabolites, as reported in prostate, colon, and Jurkat cell lines [47]. The SS absorption triggered DNA fragmentation and cell cycle arrest in G2, leading to both necrosis and apoptosis via caspase-3 activation, as reported in three GBM cell lines that were SS-treated [26]. These results are essential as R2J cells expressed a high level of HDAC that plays an important role in gene expression associated to cell proliferation (p21/Waf1, p27/kip1, p16/ink4a) and apoptosis via Bad, Trail, and FasL in glioma cells [48].

SS metabolism generated Se intermediates, which are also responsible for SS toxicity, such as selenide, selenocystine, and elemental Se [49]. Oxidative stress via the generation of anion superoxide is considered an essential process by which selenite applies its cytotoxicity [47,49].

To sustain an oxidative stress way of action and subsequent pathways, we showed that SS depleted thiol group concentration, in both R2J (p53 wild–type) and U251 (p53 mutant) cells, contrary to TMZ. Indeed, reactive oxygen species ROS production is dependent on the reduction of selenite to selenide through the consumption of thiols [50]. It was also shown that the changes in the redox status induced cell death via apoptosis and via mitochondria-selective autophagy (mitophagy) in glioma cells [29]. Western-blots experiments showed that SS repressed autophagy in both cell lines and induced reticulum stress only in R2J cells. Interestingly, SS at 2 μM did not cause double strand breaks whereas TMZ did. In parallel, TMZ did not deplete thiol groups, suggesting that SS and TMZ do not have the same mechanism of action and thus could have complementary effects to target GBM cells, if used in combination.

Epigenetic regulation of chromatin organization is a key regulator of gene expression that can significantly impact tumor biology and tumor response to therapies. In many tumor types, mutated or modified epigenetic regulators have been clearly associated to tumorigenesis because of their effects on gene expression (reviewed in [19]), and this has led to the development of pharmacologic inhibitors specific of these actors. At the epigenetic level, we focused on histone acetylation and methylation, both modulated by SS in R2J cells. Indeed, HDAC activity was significantly reduced by SS, as already reported in GBM cell lines [26]. In R2J cells, the inhibition of HDAC concomitant with cell proliferation braking and apoptosis induction is consistent with the role of HDAC in these two processes. Therefore, our results led to the intriguing question: How do both these epigenetic modifications interact and would they be able to re-express suppressor tumor genes? Some recent data have revealed an interplay between histone methylation, DNA methylation, the autophagy process, and apoptosis. An interesting study [51] showed that the histone methyltransferase, G9a, inhibits the expression of genes involved in the autophagy process. Indeed, the pharmacologic inhibition or genetic depletion of G9a triggers the formation of LC3B-II, the aggregation of p62, and the formation of autophagosomes. This loss of G9a leads to a reduction of H3K9m2 levels. In addition, a recent study [52] showed that BIX01294, an inhibitor of G9a, induced autophagy in glioma cells and stimulated differentiation of glioma stem cells (spheres cultivated without serum), associated with lower levels of H3K9m2 as the promoter of autophagy related genes. Taken together, our data suggest that SS could cause a regional remodeling of chromatin that may trigger apoptosis and cell cycle arrest.

4. Materials and Methods

4.1. Clinical Data of the Patient

In October 2009, the patient (a 49-year old white male without any medical history) felt faint and lost consciousness. Magnetic resonance imaging (MRI) showed a right tempo extensive process compatible with a GBM. In November 2009, the tumor was resected and histological analysis confirmed the diagnosis of GBM with a leptomeningeal extension (Figure 1). After the surgery, hydrocephalia and an unmanageable headache appeared. The patient was unfit to receive any conventional treatment, such as chemotherapy or/and radiotherapy.

Instead, the patient received weekly intrathecal injections of Trastuzumab because CSF glioma cells overexpressed HER2. In May 2010, symptoms and the leptomeningeal extension decreased, but the patient presented a local tumor progression on the MRI. So, the patient was treated with radiotherapy plus concomitant TMZ. In August 2010, because of tumor progression and leptomeningeal carcinomatosis progression, no more treatment was compatible with the patient's health condition and he received standard palliative care. The patient died in October 2010.

4.2. Origin of R2J Cells

The tumor cells were isolated from the CSF of the patient (November 2009). The analysis of CSF showed numerous tumor cells with a high mitotic index, indicating a leptomeningeal carcinomatosis. These cells were cultured and the derived R2J cell line was successfully cryopreserved at different passages.

4.3. Morphological Analysis

The primary biopsy (original tumor) was processed for histopathologic study using standard methods and sections were stained with hematoxylin and eosin safran (HES).

R2J cells from the CSF were isolated by cytocentrifugation (Cytospin, Shandon, Pittsburgh, PA, USA) and stained using May Grunwald Giemsa (MGG) using a Zeiss microscope (Oberkochen, Germany). Pictures were taken using a Canon Power Shot S50 digital camera (Courbevoie, France).

4.4. Cell Culture

U251 cells were purchased from the American Type Culture Collection ATCC. They were cultured in DMEM medium containing 10% fetal calf serum (FCS) supplemented with penicillin (100 IU/mL), streptomycin (100 μg/mL, PS) and L-Glutamine (2mM) (Life Technology, Walthma, MA), in humidified normoxia incubator. U251 were harvested using 0.5%-Trypsin-EDTA (10×) (#15400-054, LifeTechnologies). U251 was chosen due to its MGMT negative and p53 mutant status.

R2J were cultured in RPMI 1640 medium enriched as described for U251, in humidified hypoxia incubator (3% O_2, 5% CO_2, 37 °C). R2J were harvested as described for U251.

For sphere formation, the R2J cells were grown in non-adherent 10-cm diameter Petri dishes (Greiner, les Ulis, France) coated with PolyHema (#P3932, Sigma, St Louis, MO, USA,) in DMEM-F12 medium (Life Technologies,) without serum containing 20 ng/mL human EGF, FGF-2, and 1× NeuroBrew-21 (Miltenyi Biotec, Bergirsch Gladbach, Germany), and PS (as above) in hypoxia.

To conform with already published methods, the cells were plated at low density (10 per µL) to allow clonal sphere formation [53,54].

PolyHema was prepared at 10 mg/mL in ethanol 95% and sterilized by filtration on a Stericup (Millipore Express PLUS 1070, Temecula, CA, USA). To coat the 10 cm Petri-dish, 4mL of this solution was applied. The dishes were put in a dry incubator overnight, at 60 °C to dry the PolyHema.

4.5. Immunohistochemistry

Embedded paraffin tumor was analyzed in 2009 for CD56, Ki67, GFAP, NeuN, NF70, and synaptophysin, according to the current procedure of the laboratory. The Bouin's fluid fixative used at this time is not compatible with the current methods of the laboratory and the immuno-histo staining obtained for some markers (IDH1 and olig2, for example) was unsatisfactory except for EGFR, CD44, CD34, and Nestin, performed after 2009.

R2J cells cultured in 2D or in spheres were fixed in formol 4%, 30 min before being included in HistogelTM (Microm Microtech, Francheville, France) and paraffin embedded. HES and IHC analyses were performed from 3 µm paraffin sections using a Histostain® plus kit (Thermo Fisher Scientific, Waltham, MA) and Vector novaRED™ substrate kit for peroxidase or Ventana Benchmark® XT platform (Roche, Rotkreug, Switzerland). Only nestin was revealed using Fast Red Bond Polymer Red Refine Detection (#DS93390, Leica Biosystems, Newcastle, UK).

CD-56 (#MAB24081) and Olig2 (#AF2418) antibodies are from R&D Systems (Temecula, CA, USA), MIB1/KI67 (#M7240), p53 (#M7001) and Vimentin (#M0725) were from Dako (Glostrup, Danemark), GFAP (#2202MGF, Eurodiagnostica, Malmö, Sweden), EGFR (#28-0005), E-Cadherin (#18-0223), and MDM2 (#33-7100) were from Zymed Lab, (South San Francisco, CA, USA), IDH1 (#DIA-H09, Histonova, Clinisciences, Nanterre, France), Nestin (#MAB5326), and NeuN (#MAB377) were from Zymed Lab (South San Francisco, CA, USA), ATRX (#HPA001906) and CD44 (#144M-95) were from Sigma Aldrich, P16 (#9511, MTM Roche, France), Neurofilament 70-200 (#Mob080, Diagnostic BioSystems, Pleasanton, CA, USA), and CD34 (#PN IM0786, Beckman Coulter, Brea, CA, USA).

4.6. Flow Cytometry Analysis

R2J-2D cells were seeded at 6000 cell/cm^2 in 10 cm Petri-dishes 48 h before treatment with SS (Sigma) for 24 h or 48 h, after which cells and medium were recovered, centrifuged (3 min, 360 g, RT), and cells were rinsed twice with PBS1X.

Apoptosis and the cell-cycle were evaluated with the FITC-Annexin V Apoptosis Detection kit and the Cycle Test Plus DNA reagent kit (BD Biosciences, San Jose, CA) following the manufacturer's instructions. The analysis of cells in the subG1 phase was used to determine DNA fragmentation.

For dimethyl-Histone H3-Lysine-9 (H3K9m2) analysis, 10^6 cells were fixed 30 min in PFA 1% on ice and then permeabilized with 0.1% Triton × 100 in PBS, 30 min on ice. R2J cells were washed in PBS/BSA 1% twice before adding 1 µL of the rabbit monoclonal antibody, anti-H3K9m2 (#MC554, Millipore), 45 min. The H3K9m2 level was revealed using a (FITC)-conjugated F(ab)′2 fragment goat anti-rabbit IgG(H+L) (Beckman Coulter), 30 min, 4 °C.

Cell fluorescence was detected with FACSCantoII (Becton Dickinson) and analysed with FACS Diva Software.

4.7. Chromosome Analysis

The chromosomes were analyzed according to the International System for Human Cytogenetic Nomenclature. Culture was exposed to 10 µg/mL colchicine for 60 min. Slides were processed for R-binding using heat and the Giemsa-RHG method [55].

4.8. Limiting Dilution/Tumor Sphere Initiation Assay (LDA)

An LDA was performed to quantify in R2J cell lines, the frequency of tumor spheres initiation cells (TS-ICs). R2J parental cells were calibrated to 1000 cells/mL (corresponding to 200 cells/200 µL/well) and then diluted into gradients of cell titers at 500, 250, 125, 62, 100, 50, 25, and 12.5 /mL and transferred into the wells of the 96-well microplate. Clonal spheres (non-adherent, tight, and spherical masses >5 cells) were counted under an inverted microscope (Leica, Wetzlar, Germany), at the end of two weeks. The results are expressed as the percentage of TS-ICs to examine the efficacy of clonogenesis [38].

4.9. Intracranial Cell Transplantation into Immunocompromised Mice

All animal protocols were approved by the GIN Animal Care and Use Committee (number 2017121517085709). R2J-2D cells (2×10^5) or spheres (2×10^5 or 1000) were implanted into the striatum of 4–6 week male athymicNudeFoxn1nu mice (n = 4, Envigo, Gannat, France) to analyze the tumor initiating capacity. Experimental details are available in Appendix A.

Mice were perfused; brains were removed and immediately fixed in formol 4% for subsequent histopathological analyses.

4.10. Magnetic Resonance Imaging

MRI sessions were performed at 4.7 T (Avance III console; Bruker, Ettlingen, Germany; IRMaGe MRI facility, Grenoble, France) using an actively decoupled cross-coil setup (volume coil for radiofrequency transmission and surface coil for signal reception). All MR experiments were performed under anaesthesia: 5% isoflurane for induction and 2% for maintenance in air. Mice were maintained at 37.0 °C and breath rate was cared for 60 breath/min throughout the acquisition. After shimming, MRI sequences were designed for anatomical T2-weighted (T2W) images acquisition using a spin-echo MRI sequence (TR /TE = 2500/33 ms, NA = 14, 21 slices with field of view (FOV) 20×20 mm^2, matrix = 256×256 and voxel size = $156 \times 156 \times 500$ µm^3). Acquisition duration was 9 min 20 s. Tumor volume was calculated by adding each tumor surface × slice thickness.

4.11. Selenium Absorption

R2J-2D cells were seeded at 6000 cell/cm^2 in 10 cm Petri-dishes, 24 h before SS supplementation (2.5, 5, and 10 µM).

Se concentration was determined both in cell lysates and medium using an ICP-MS (X series II, Thermo Fisher Scientific, Waltham, Massachusetts, USA) equipped with Collision Cell interference elimination Technology (CCT) after dilution in 1% nitric acid containing Gallium (Ga) as an internal standard. The gas introduced in the CTT was a mix of He and H2. The isotope ratio ^{78}Se/^{71}Ga and an external calibration curve (16, 80, 160 nmol Se/L) allowed the determination of the Se concentration. The results (mean ± SD) were calculated in nmoles of Se per gram of protein and expressed as a percentage of the Se measured vs. Se added. The Se recovery was calculated by adding the quantity of Se both measured in the medium and in the lysates and compared to the Se quantity added (referred as 100%).

4.12. MTT Assay

A MTT (3-(4,5-Dimethylthiazol-2-yl)-2,5-Diphenyltetrazolium Bromide) assay (Sigma) was performed to determine R2J-2D cell survival after SS treatment at 2.5 to 12 µM for 24 h, 48 h, or 72 h. 1×10^5 cells/well were seeded in 96-wells plate 24h before the addition of SS. The results (mean ± SD) are reported as the cell survival percentage versus the control (not treated). IC50 (concentrations inducing 50% cell death) values were determined by curve-fitting (using Prism 5, GraphPad).

In parallel, TMZ (Temodal, MSD, France) was dissolved in dimethyl sulfoxide (DMSO) and R2J cells were treated for 24 h, 48 h, and 72 h with 1, 10, 100, 500, and 1000 µM of TMZ. Cells were also treated with DMSO alone at the corresponding volume. A long term treatment was also performed by exposing R2J-2D cells to 400 µM TMZ for 72 h followed by a wash-out of 72 h. Results are expressed as the percentage of cell survival versus control (DMSO concentrations used were not cytotoxic).

4.13. Confocal Microscopy Analysis of Cell Viability

R2J-2D cells were set on a Lab-Tek-Chamber Slide system (Nalge Nunc International, Rochester, NY, USA) and were incubated with 5 μM of Fluorescein Diacetate (FDA, Thermofisher Scientific, Courtaboeuf, France), 1μM Hoechst 33342 (Interchim, Montluçon, France), and 10mg/mL Propidium Iodide (PI) (Interchim) in a 5% CO_2 incubator, 37 °C, 20 min. After staining, cells were installed on the microscope incubation chamber with a controlled atmosphere at 37 °C and 5% CO_2 (POC Chamber, Pecom, Erbach, Germany). Images were collected with a Leica TCS SP2 AOBS (Acoustico Optical Beam Splitter) inverted laser scanning confocal microscope equipped with an ×63 oil immersion objective (HCX PL APO 63.0× 1.40). Laser excitation and emission (adjusted with AOBS) were, respectively, 351–364/425–485 nm for Hoechst, 488/500–540 nm for FDA, and 543/600–650 nm for PI. Confocal pinhole (Airy units) was 1 for all channels. Each experiment was performed on a randomly chosen field. Raw image merging was obtained by LCS Lite software (Leica LCS, version 2.61, Overlay).

4.14. Caspase-3 Activity Evaluation

The caspase-3 Fluorometric Assay (BioVision, Mountain View, CA, USA) was used. R2J-2D cells were harvested and lysed into 50 μL of chilled Cell Lysis Buffer at 4 °C, 10 min, and centrifuged for 10 min, 10000 g at 4 °C. Protein concentration was determined (see below) and 50μL of supernatant containing 50 μg to 200 μg of protein was mixed with 50 μL of 2× reaction buffer and 5 μL of DEVD-AFC substrate (1 mM), in a 96-wells plate. Reaction was achieved for 1h at 37 °C. The plate was read on a fluorometer (Ex/Em = 400/505 nm). Caspase-3 activity is expressed in AU/g prot.

4.15. Determination of HDAC Activity

R2J-2D cells were treated for 24 h, 48 h, and 72 h with 2.5 and 5 μM SS and harvested. Pellets were rinsed twice in PBS and lysed in 400 μL sterile water, 10 min. HDAC activity was investigated in cell lysates from 50 μg total protein using a Fluorometric Assay (BioVision, Mountain View, CA, USA) as described by the manufacturer. HDAC activity is expressed in relative fluorescence units (RFU) per μg protein.

4.16. Real Time PCR

Total RNA extraction was performed with an RNeasy Mini Kit, following the manufacturer's recommendations, with RNase-free DNase I treatment (Qiagen, Courtaboeuf, France). cDNA was reverse transcribed from 1 μg of total RNA with the SuperScriptIII First-Strand Synthesis (Life Technologies) followed by RNaseH (1 μL) treatment. Real time PCR was conducted using the QuantiTect SYBR Green RT-PCR kit (Qiagen) and the Stratagene 3005MxPro (Santa Clara, CA, USA). The primers (sequences in Supplementary Material Table S1) (Life Technologies) were all used at 400 nM, Tm at 60 °C, with the following qPCR program: 1 cycle: 15 min-95 °C, 40 cycles: 15 s-94 °C, 30 s-60 °C, 30 s-72 °C, 1 cycle: 1 min-95 °C, 30 s-60 °C, 30 s-95 °C.

Gene expression was quantified using the comparative threshold cycle (Ct) method [56] with a normalization against HPRT1, RPL27, and RPL32.

4.17. Thiol Group Determination

Thiol groups were determined by a colorimetric method using the reducing properties of the SH groups as previously described [25]. The thiol group levels were determined as micromoles per gram of total cell proteins and expressed in percentage versus the control.

4.18. Western Blot Analysis

U251 and R2J, after 72 h treatment followed by 72 h of wash-out with TMZ or SS or chloroquine (Cq, used to block autophagy 1 h before SS or TMZ treatment at 20 μM, Sanofi-Aventis), were harvested in protein extraction buffer (Tris-HCl 20 mM, pH 7.4, NaCl 137 mM, EDTA 2 mM, Triton X-100 1%, supplemented with proteases' and phosphatases' inhibitors (1/100, Sigma-Aldrich). After pipetting up and down with a P1000, supernatants were collected after centrifugation at 10,000 g for 10 min, at 4 °C. Protein concentration was measured by the BCA method. Proteins (20 µg total) were treated with SDS-PAGE sample buffer [6× concentrated: 350 mM Tris, 10% (w/v) SDS, 30% (v/v) glycerol, 0.6 M DTT, 0.06% (w/v) bromophenol blue], boiled for 5 min at 95 °C, and loaded on 15% (for LC3, p62, and phospho-H2Ax) and 8% (for Grp78) acrylamide/bisacrylamide gels. Proteins were transferred onto nitrocellulose membranes, blocked with 5% bovine serum albumin containing 0.5% Tween (Euromedex) (PBST) for 1 h at room temperature, and probed overnight at 4 °C with the following primary antibodies diluted in 5% bovine serum albumin-PBST: 1:1000 rabbit anti-p62 antibody (Sigma-Aldrich), 1:200 goat anti-Grp78 antibody (Santa Cruz Biotechnology, Dallas, TX), 1:1000 anti-LC3-II antibody, 1:1000 rabbit anti-phospho-H2Ax antibody (both from Cell Signalling Technology), and 1:5000 mouse anti-alpha tubulin (Sigma-Aldrich). Membranes were washed with PBST, incubated for 1 h at room temperature with horseradish peroxidase-conjugated secondary antibodies (from Jackson Immunoresearch), and visualized using electrochemiluminescence (ECL) western blotting substrate (ThermoFisher Scientific) on a Chemidoc imaging system (BioRad, Hercules, CA).

4.19. Protein Concentration Determination

The concentration of protein in cell lysates (obtained as mentioned above) was determined using the BCA assay (Interchim, Montluçon, France) as previously described [24]. The absorption at 562 nm was measured using the Varioskan Flash (Thermo Fisher Scientific) associated to the SkanIt Software for the quantification.

4.20. Statistical Analysis

Results are expressed as means ± SD for the number of experiments indicated. All statistical analysis of data was computed using StatView® (SAS Institute, CA, USA). The sources of variation for multiple comparisons were assessed by analysis of variance ANOVA. The differences were considered statistically significant at $p < 0.05$ versus the control.

5. Conclusions

In conclusion, we described a newly established glioblastoma cell line (R2J), from a CSF from a patient affected by a metastatic GBM. R2J were MGMT+ and resistant to TMZ. In the presence of FGFb and EGF, the R2J forming spheres were phenocopied on 2D cultured R2J cells except for CD56, a neuronal marker, suggesting that R2J may differentiate into neuronal-like cells. The cell line could form a tumor when intracranially implanted into nude mice even at 1000 spheres, providing conclusions on their CSC properties. In vitro, SS could exert its anti-cancer properties in R2J cells, and is thus relevant for testing in this new GBM bearing mouse model.

Supplementary Materials: The following are available online at http://www.mdpi.com/2072-6694/11/1/12/s1, Figure S1: A representative G-banded karyotype of R2J cells. The R2J cell line was identified as near-triploid (hypotriploid) with 64 chromosomes, Table S1: primer sequences used for RT-q-PCR.

Author Contributions: Conceptualization: P.C., L.L. and F.H.-P.; Methodology, S.B., P.C., E.C., C.L., C.C., C.C.-R., J.A., C.G., L.L. and F.H.-P.; Validation, S.B., P.C., C.C.-R., J.A., F.L., J.B., L.L. and F.H.-P.; Formal Analysis, P.C., C.C.-R., J.A. and F.H.-P.; Resources, F.L., P.F. and F.H.-P.; Data Curation, S.B. and F.H.-P.; Writing-Original Draft Preparation, F.H.-P.; Writing-Review & Editing, S.B., P.C., C.C.-R., J.A., F.L., J.B., P.F. and F.H.-P.; Visualization, F.H.-P.; Supervision, L.L. and F.H.-P.; Project Administration, S.B., P.C., J.B., L.L. and F.H.-P.; Funding Acquisition, L.L. and F.H.-P.

Funding: This research was funded by grants from the Direction de la Recherche Clinique (DRC) of the Centre Hospitalier Universitaire Grenoble Alpes (CHUGA, France) and by the Groupement des Entreprises Françaises de Lutte contre le Cancer (GEFLUC).

Acknowledgments: We express our thanks to Laurence David-Boudet, Yveline Maffren, Aurélie Soldini, Martine Micheloni and Véronique Curri for technical assistance. We are grateful to Fabienne Agasse for the careful reading of the manuscript, Emmanuelle Grillon and Emmanuel Barbier for MRI assistance.

Conflicts of Interest: The authors declare no conflict of interest.

Abbreviations

2D	Monolayer
BSA	Bovine Serum Albumin
CSC	Cancer Stem Cell
CSF	Cerebro-Spinal Fluid
Ct	Cycle threshold
CT	ChemoTherapy
ECM	ExtraCellular Matrix
EGFR	Epithelial Growth Factor-Receptor
GMT	Glial to Mesenchymal Transition
FACS	Fluorescence Activated Cell Sorter
FCS	Foetal Calf Serum
FDA	Fluorescein DiAcetate
FISH	Fluorescence In Situ Hybridization
FITC	Fluorescein IsoThioCyanate
G-CSF-R	Granulocyte-Colony Stimulating Factor-Receptor
GFAP	Glial Fibrillary Acidic Protein
H3K9m2	dimethyl-Histone-3-lysine-9
HDAC	Histone Deacetylase
HES	Hematoxylin Eosin Safran
ICP-MS	Inductively Coupled Plasma Mass Spectrometry
IDH	Isocitrate DesHydrogenase
KDM	Histone Lysine Demethylase
MDM2	Mouse Double Minute 2 homolog
MGG	May Grunwald Giemsa
MGMTt+	positive transcript expression of O(6)-MethyGuanine-DNA-MethylTransferase
MMPs	Matrix Metallo-Proteases
MRI	Magnetic Resonance Imaging
N-CAM	Neural Cell Adhesion Molecule
neuN	NEUronal Nuclear antigen
p15/INK4B	Inhibitor of CDK4B
p16/INK4A	Inhibitor of CDK4A
p21/WAF1	Cyclin-dependent kinase inhibitor 1
p27/KIP1	Cyclin-dependent kinase inhibitor 1B
PBS	Phosphate Buffer Saline
PI	Propidium iodide
PI3K	Phosphatidylinositol-3 kinase
PBS	Phosphate buffer salt
PFA	ParaFormAldehyde
ROS	Reactive Oxygen Species
RT	Radiation Therapy
TMZ	Temozolomide
SS	Sodium selenite

Appendix A

Cells were harvested using Accutase (Sigma), counted, and suspended in 10 mL PBS. Cell viability and counting were assessed before and after the procedure using Scepter® (Millipore). All surgery was performed using sterile techniques within a tissue culture hood. Mice were anesthetized with isoflurane 5% and maintained under 2.5% during the surgery. NaCl 0.9% was put on eyes throughout the procedure and mice were kept on a warm pad throughout. Cells were injected into the right striatum using a stereotaxic apparatus. Injection coordinates were 2.6 mm lateral to Bregma and a

3.6mm depth. A small bore hole was made using a 25 g needle. The cell suspension was mixed and loaded into the syringe just prior to injection. 3 µL of cells was delivered using a 26 g Hamilton microsyringe (701RN, 10 µL). The syringe was stabilized for 1 min before injection and 0.5 µL of cell suspension was injected each 30 s. The needle was kept in place for a further 4 min to prevent backflow and was then removed slowly (1 µL/min). After cell injection, the skin incisions were closed. To avoid suffering, Buprecare® was intracutaneously injected (0.1 mg/kg) to each mouse. The animals were then revived under an infrared lamp and then returned with their congeners.

References

1. Stupp, R.; Hegi, M.E.; Mason, W.P.; van den Bent, M.J.; Taphoorn, M.J.; Janzer, R.C.; Ludwin, S.K.; Allgeier, A.; Fisher, B.; Belanger, K.; et al. Effects of radiotherapy with concomitant and adjuvant temozolomide versus radiotherapy alone on survival in glioblastoma in a randomised phase III study: 5-year analysis of the EORTC-NCIC trial. *Lancet Oncol.* **2009**, *10*, 459–466. [CrossRef]
2. Vivanco, I.; Sawyers, C.L. The phosphatidylinositol 3-Kinase AKT pathway in human cancer. *Nat. Rev. Cancer* **2002**, *2*, 489–501. [CrossRef] [PubMed]
3. Vermeulen, K.; Van Bockstaele, D.R.; Berneman, Z.N. The cell cycle: A review of regulation, deregulation and therapeutic targets in cancer. *Cell Prolif.* **2003**, *36*, 131–149. [CrossRef] [PubMed]
4. Bonomi, S.; Gallo, S.; Catillo, M.; Pignataro, D.; Biamonti, G.; Ghigna, C. Oncogenic alternative splicing switches: Role in cancer progression and prospects for therapy. *Int. J. Cell Biol.* **2013**, *2013*, 962038. [CrossRef] [PubMed]
5. Mao, X.; Hamoudi, R.A. Molecular and cytogenetic analysis of glioblastoma multiforme. *Cancer Genet. Cytogenet.* **2000**, *122*, 87–92. [CrossRef]
6. Ceccarelli, M.; Barthel, F.P.; Malta, T.M.; Sabedot, T.S.; Salama, S.R.; Murray, B.A.; Morozova, O.; Newton, Y.; Radenbaugh, A.; Pagnotta, S.M.; et al. Molecular Profiling Reveals Biologically Discrete Subsets and Pathways of Progression in Diffuse Glioma. *Cell* **2016**, *164*, 550–563. [CrossRef] [PubMed]
7. Chen, J.; Li, Y.; Yu, T.S.; McKay, R.M.; Burns, D.K.; Kernie, S.G.; Parada, L.F. A restricted cell population propagates glioblastoma growth after chemotherapy. *Nature* **2012**, *488*, 522–526. [CrossRef]
8. Yu, Z.; Pestell, T.G.; Lisanti, M.P.; Pestell, R.G. Cancer stem cells. *Int. J. Biochem. Cell Biol.* **2012**, *44*, 2144–2151. [CrossRef]
9. Jin, X.; Jin, X.; Jung, J.E.; Beck, S.; Kim, H. Cell surface Nestin is a biomarker for glioma stem cells. *Biochem. Biophys. Res. Commun.* **2013**, *433*, 496–501. [CrossRef]
10. Mimeault, M.; Batra, S.K. Molecular biomarkers of cancer stem/progenitor cells associated with progression, metastases, and treatment resistance of aggressive cancers. *Cancer Epidemiol. Prev. Biomark.* **2013**, *23*, 234–254. [CrossRef]
11. Choi, S.A.; Wang, K.C.; Phi, J.H.; Lee, J.Y.; Park, C.K.; Park, S.H.; Kim, S.K. A distinct subpopulation within CD133 positive brain tumor cells shares characteristics with endothelial progenitor cells. *Cancer Lett.* **2012**, *324*, 221–230. [CrossRef] [PubMed]
12. Singh, S.K.; Clarke, I.D.; Terasaki, M.; Bonn, V.E.; Hawkins, C.; Squire, J.; Dirks, P.B. Identification of a cancer stem cell in human brain tumors. *Cancer Res.* **2003**, *63*, 5821–5828.
13. Hong, X.; Chedid, K.; Kalkanis, S.N. Glioblastoma cell line-derived spheres in serumcontaining medium versus serum-free medium: A comparison of cancer stem cell properties. *Int. J. Oncol.* **2012**, *41*, 1693–1700. [CrossRef] [PubMed]
14. Thiery, J.P.; Sleeman, J.P. Complex networks orchestrate epithelial-mesenchymal transitions. *Nat. Rev. Mol. Cell Biol.* **2006**, *7*, 131–142. [CrossRef] [PubMed]
15. Mahabir, R.; Tanino, M.; Elmansuri, A.; Wang, L.; Kimura, T.; Itoh, T.; Ohba, Y.; Nishihara, H.; Shirato, H.; Tsuda, M.; et al. Sustained elevation of Snail promotes glial-mesenchymal transition after irradiation in malignant glioma. *Neuro Oncol.* **2013**, *16*, 671–685. [CrossRef]
16. May, C.D.; Sphyris, N.; Evans, K.W.; Werden, S.J.; Guo, W.; Mani, S.A. Epithelial-mesenchymal transition and cancer stem cells: A dangerously dynamic duo in breast cancer progression. *Breast Cancer Res.* **2011**, *13*, 202. [CrossRef]

17. Singh, A.; Settleman, J. EMT, cancer stem cells and drug resistance: An emerging axis of evil in the war on cancer. *Oncogene* **2010**, *29*, 4741–4751. [CrossRef]
18. Harabin-Slowinska, M.; Slowinski, J.; Konecki, J.; Mrowka, R. Expression of adhesion molecule CD44 in metastatic brain tumors. *Folia Neuropathol.* **1998**, *36*, 179–184.
19. Wilting, R.H.; Dannenberg, J.H. Epigenetic mechanisms in tumorigenesis, tumor cell heterogeneity and drug resistance. *Drug Resist. Updat.* **2012**, *15*, 21–38. [CrossRef]
20. Seligson, D.B.; Horvath, S.; McBrian, M.A.; Mah, V.; Yu, H.; Tze, S.; Wang, Q.; Chia, D.; Goodglick, L.; Kurdistani, S.K. Global levels of histone modifications predict prognosis in different cancers. *Am. J. Pathol.* **2009**, *174*, 1619–1628. [CrossRef]
21. Chen, M.W.; Hua, K.T.; Kao, H.J.; Chi, C.C.; Wei, L.H.; Johansson, G.; Shiah, S.G.; Chen, P.S.; Jeng, Y.M.; Cheng, T.Y.; et al. H3K9 histone methyltransferase G9a promotes lung cancer invasion and metastasis by silencing the cell adhesion molecule Ep-CAM. *Cancer Res.* **2010**, *70*, 7830–7840. [CrossRef] [PubMed]
22. Zang, L.; Kondengaden, S.M.; Che, F.; Wang, L.; Heng, X. Potential Epigenetic-Based Therapeutic Targets for Glioma. *Front. Mol. Neurosci.* **2018**, *11*, 408. [CrossRef] [PubMed]
23. Park, C.K.; Lee, S.H.; Kim, T.M.; Choi, S.H.; Park, S.H.; Heo, D.S.; Kim, I.H.; Jung, H.W. The value of temozolomide in combination with radiotherapy during standard treatment for newly diagnosed glioblastoma. *J. Neurooncol.* **2013**, *112*, 277–283. [CrossRef] [PubMed]
24. Hazane-Puch, F.; Champelovier, P.; Arnaud, J.; Garrel, C.; Ballester, B.; Faure, P.; Laporte, F. Long-term selenium supplementation in HaCaT cells: Importance of chemical form for antagonist (protective versus toxic) activities. *Biol. Trace Elem. Res.* **2013**, *154*, 288–298. [CrossRef] [PubMed]
25. Hazane-Puch, F.; Champelovier, P.; Arnaud, J.; Trocme, C.; Garrel, C.; Faure, P.; Laporte, F. Six-day selenium supplementation led to either UVA-photoprotection or toxic effects in human fibroblasts depending on the chemical form and dose of Se. *Met. Integr. Biomet. Sci.* **2014**, *6*, 1683–1692. [CrossRef] [PubMed]
26. Hazane-Puch, F.; Arnaud, J.; Trocme, C.; Faure, P.; Laporte, F.; Champelovier, P. Sodium Selenite Decreased HDAC Activity, Cell Proliferation and Induced Apoptosis in Three Human Glioblastoma Cells. *Anti-Cancer Agents Med. Chem.* **2016**, *16*, 490–500. [CrossRef]
27. Berthier, S.; Arnaud, J.; Champelovier, P.; Col, E.; Garrel, C.; Cottet, C.; Boutonnat, J.; Laporte, F.; Faure, P.; Hazane-Puch, F. Anticancer properties of sodium selenite in human glioblastoma cell cluster spheroids. *J. Trace Elem. Med. Biol.* **2017**, *44*, 161–176. [CrossRef] [PubMed]
28. Rooprai, H.K.; Kyriazis, I.; Nuttall, R.K.; Edwards, D.R.; Zicha, D.; Aubyn, D.; Davies, D.; Gullan, R.; Pilkington, G.J. Inhibition of invasion and induction of apoptosis by selenium in human malignant brain tumour cells in vitro. *Int. J. Oncol.* **2007**, *30*, 1263–1271. [CrossRef]
29. Kim, E.H.; Sohn, S.; Kwon, H.J.; Kim, S.U.; Kim, M.J.; Lee, S.J.; Choi, K.S. Sodium selenite induces superoxide-mediated mitochondrial damage and subsequent autophagic cell death in malignant glioma cells. *Cancer Res.* **2007**, *67*, 6314–6324. [CrossRef] [PubMed]
30. Zhang, Z.; Chinen, Y.; Zhu, Z.; Kimura, M.; Itokawa, Y. Uptake and distribution of sodium selenite in rat brain tumor. *Biol. Trace Elem. Res.* **1995**, *48*, 45–50. [CrossRef] [PubMed]
31. Cavalieri, R.R.; Scott, K.G.; Sairenji, E. Selenite (75Se) as a tumor-localizing agent in man. *J. Nucl. Med.* **1966**, *7*, 197–208. [PubMed]
32. Hazane-Puch, F.; Soldini, A. Unit Nutritional and Hormonal Biochemistry, Institute of Biology and Pathology, Grenoble Alpes Hospital, CS10217, 38043 Grenoble Cedex 9, France. Unpublished Work. 2017.
33. Beier, D.; Schulz, J.B.; Beier, C.P. Chemoresistance of glioblastoma cancer stem cells—Much more complex than expected. *Mol. Cancer* **2011**, *10*, 128. [CrossRef] [PubMed]
34. Loja, T.; Chlapek, P.; Kuglik, P.; Pesakova, M.; Oltova, A.; Cejpek, P.; Veselska, R. Characterization of a GM7 glioblastoma cell line showing CD133 positivity and both cytoplasmic and nuclear localization of nestin. *Oncol. Rep.* **2009**, *21*, 119–127. [PubMed]
35. Onda, K.; Nagai, S.; Tanaka, R.; Morii, K.; Yoshimura, J.I.; Tsumanuma, I.; Kumanishi, T. Establishment of two glioma cell lines from two surgical specimens obtained at different times from the same individual. *J. Neuro-Oncol.* **1999**, *41*, 247–254. [CrossRef]

36. Pollard, S.M.; Yoshikawa, K.; Clarke, I.D.; Danovi, D.; Stricker, S.; Russell, R.; Bayani, J.; Head, R.; Lee, M.; Bernstein, M.; et al. Glioma stem cell lines expanded in adherent culture have tumor-specific phenotypes and are suitable for chemical and genetic screens. *Cell Stem Cell* **2009**, *4*, 568–580. [CrossRef]
37. Therman, E.; Buchler, D.A.; Nieminen, U.; Timonen, S. Mitotic modifications and aberrations in human cervical cancer. *Cancer Genet. Cytogenet.* **1984**, *11*, 185–197. [CrossRef]
38. Yu, S.C.; Ping, Y.F.; Yi, L.; Zhou, Z.H.; Chen, J.H.; Yao, X.H.; Gao, L.; Wang, J.M.; Bian, X.W. Isolation and characterization of cancer stem cells from a human glioblastoma cell line U87. *Cancer Lett.* **2008**, *265*, 124–134. [CrossRef] [PubMed]
39. Iacopino, F.; Angelucci, C.; Piacentini, R.; Biamonte, F.; Mangiola, A.; Maira, G.; Grassi, C.; Sica, G. Isolation of cancer stem cells from three human glioblastoma cell lines: Characterization of two selected clones. *PLoS ONE* **2014**, *9*, e105166. [CrossRef] [PubMed]
40. Kong, D.S.; Song, S.Y.; Kim, D.H.; Joo, K.M.; Yoo, J.S.; Koh, J.S.; Dong, S.M.; Suh, Y.L.; Lee, J.I.; Park, K.; et al. Prognostic significance of c-Met expression in glioblastomas. *Cancer* **2009**, *115*, 140–148. [CrossRef] [PubMed]
41. Westphal, M.; Hansel, M.; Hamel, W.; Kunzmann, R.; Holzel, F. Karyotype analyses of 20 human glioma cell lines. *Acta Neurochir.* **1994**, *126*, 17–26. [CrossRef]
42. Verhaak, R.G.; Hoadley, K.A.; Purdom, E.; Wang, V.; Qi, Y.; Wilkerson, M.D.; Miller, C.R.; Ding, L.; Golub, T.; Mesirov, J.P.; et al. Integrated genomic analysis identifies clinically relevant subtypes of glioblastoma characterized by abnormalities in PDGFRA, IDH1, EGFR, and NF1. *Cancer Cell* **2010**, *17*, 98–110. [CrossRef] [PubMed]
43. Louis, D.N.; Perry, A.; Reifenberger, G.; von Deimling, A.; Figarella-Branger, D.; Cavenee, W.K.; Ohgaki, H.; Wiestler, O.D.; Kleihues, P.; Ellison, D.W. The 2016 World Health Organization Classification of Tumors of the Central Nervous System: A summary. *Acta Neuropathol.* **2016**, *131*, 803–820. [CrossRef] [PubMed]
44. Jhaveri, N.; Agasse, F.; Armstrong, D.; Peng, L.; Commins, D.; Wang, W.; Rosenstein-Sisson, R.; Vaikari, V.P.; Santiago, S.V.; Santos, T.; et al. A novel drug conjugate, NEO212, targeting proneural and mesenchymal subtypes of patient-derived glioma cancer stem cells. *Cancer Lett.* **2016**, *371*, 240–250. [CrossRef] [PubMed]
45. Lee, S.Y. Temozolomide resistance in glioblastoma multiforme. *Genes Dis.* **2016**, *3*, 198–210. [CrossRef] [PubMed]
46. Okamoto, R.; Takano, H.; Okamura, T.; Park, J.S.; Tanimoto, K.; Sekikawa, T.; Yamamoto, W.; Sparreboom, A.; Verweij, J.; Nishiyama, M. O(6)-methylguanine-DNA methyltransferase (MGMT) as a determinant of resistance to camptothecin derivatives. *Jpn. J. Cancer Res.* **2002**, *93*, 93–102. [CrossRef]
47. Lunoe, K.; Gabel-Jensen, C.; Sturup, S.; Andresen, L.; Skov, S.; Gammelgaard, B. Investigation of the selenium metabolism in cancer cell lines. *Met. Integr. Biomet. Sci.* **2011**, *3*, 162–168. [CrossRef]
48. Yin, D.; Ong, J.M.; Hu, J.; Desmond, J.C.; Kawamata, N.; Konda, B.M.; Black, K.L.; Koeffler, H.P. Suberoylanilide hydroxamic acid, a histone deacetylase inhibitor: Effects on gene expression and growth of glioma cells in vitro and in vivo. *Clin. Cancer Res.* **2007**, *13*, 1045–1052. [CrossRef]
49. Weekley, C.M.; Aitken, J.B.; Vogt, S.; Finney, L.A.; Paterson, D.J.; de Jonge, M.D.; Howard, D.L.; Witting, P.K.; Musgrave, I.F.; Harris, H.H. Metabolism of selenite in human lung cancer cells: X-ray absorption and fluorescence studies. *J. Am. Chem. Soc.* **2011**, *133*, 18272–18279. [CrossRef]
50. Olm, E.; Fernandes, A.P.; Hebert, C.; Rundlof, A.K.; Larsen, E.H.; Danielsson, O.; Bjornstedt, M. Extracellular thiol-assisted selenium uptake dependent on the x(c)-cystine transporter explains the cancer-specific cytotoxicity of selenite. *Proc. Natl. Acad. Sci. USA* **2009**, *106*, 11400–11405. [CrossRef]
51. Artal-Martinez de Narvajas, A.; Gomez, T.S.; Zhang, J.S.; Mann, A.O.; Taoda, Y.; Gorman, J.A.; Herreros-Villanueva, M.; Gress, T.M.; Ellenrieder, V.; Bujanda, L.; et al. Epigenetic regulation of autophagy by the methyltransferase G9a. *Mol. Cell Biol.* **2013**, *33*, 3983–3993. [CrossRef]
52. Ciechomska, I.A.; Przanowski, P.; Jackl, J.; Wojtas, B.; Kaminska, B. BIX01294, an inhibitor of histone methyltransferase, induces autophagy-dependent differentiation of glioma stem-like cells. *Sci. Rep.* **2016**, *6*, 38723. [CrossRef] [PubMed]
53. Coles-Takabe, B.L.; Brain, I.; Purpura, K.A.; Karpowicz, P.; Zandstra, P.W.; Morshead, C.M.; van der Kooy, D. Don't look: Growing clonal versus nonclonal neural stem cell colonies. *Stem Cells* **2008**, *26*, 2938–2944. [CrossRef] [PubMed]
54. Singh, S.K.; Hawkins, C.; Clarke, I.D.; Squire, J.A.; Bayani, J.; Hide, T.; Henkelman, R.M.; Cusimano, M.D.; Dirks, P.B. Identification of human brain tumour initiating cells. *Nature* **2004**, *432*, 396–401. [CrossRef] [PubMed]

55. Sehested, J. A simple method for R banding of human chromosomes, showing a pH-dependent connection between R and G bands. *Humangenetik* **1974**, *21*, 55–58. [CrossRef] [PubMed]
56. Giulietti, A.; Overbergh, L.; Valckx, D.; Decallonne, B.; Bouillon, R.; Mathieu, C. An overview of real-time quantitative PCR: Applications to quantify cytokine gene expression. *Methods* **2001**, *25*, 386–401. [CrossRef] [PubMed]

cancers

MDPI

Article

Dysregulation of Macropinocytosis Processes in Glioblastomas May Be Exploited to Increase Intracellular Anti-Cancer Drug Levels: The Example of Temozolomide

Margaux Colin [1], Cédric Delporte [2], Rekin's Janky [3], Anne-Sophie Lechon [1], Gwendoline Renard [1], Pierre Van Antwerpen [2], William A. Maltese [4] and Véronique Mathieu [1,5,*]

1 Department of Pharmacotherapy and Pharmaceuticals, Faculty of Pharmacy, Université Libre de Bruxelles (ULB), 1050 Brussels, Belgium; Margaux.Colin@ulb.ac.be or macolin@ulb.ac.be (M.C.); alechon@ulb.ac.be (A.-S.L.); renardgwendo@hotmail.com (G.R.)
2 RD3-Pharmacognosy, Bioanalysis and Drug Discovery Unit and Analytical Platform, Faculty of Pharmacy, Université libre de Bruxelles (ULB), 1050 Brussels, Belgium; Cedric.Delporte@ulb.ac.be (C.D.); pvantwer@ulb.ac.be (P.V.A.)
3 VIB Nucleomics Core, VIB, 3000 Leuven, Belgium; Rekins.Janky@vib.be
4 Department of Cancer Biology, University of Toledo College of Medicine, Toledo, OH 43614, USA; william.maltese@utoledo.edu
5 ULB Cancer Research Center, Université libre de Bruxelles (ULB), 1050 Bruxelles, Belgium
* Correspondence: veronique.mathieu@ulb.ac.be or vemathie@ulb.ac.be; Tel.: +32-265-051-79

Received: 27 December 2018; Accepted: 20 March 2019; Published: 22 March 2019

Abstract: Macropinocytosis is a clathrin-independent endocytosis of extracellular fluid that may contribute to cancer aggressiveness through nutrient supply, recycling of plasma membrane and receptors, and exosome internalization. Macropinocytosis may be notably triggered by epidermal growth factor receptor (EGFR) and platelet-derived growth factor receptor (PDGFR), two well-known markers for glioblastoma aggressiveness. Therefore, we studied whether the expression of key actors of macropinocytosis is modified in human glioma datasets. Strong deregulation has been evidenced at the mRNA level according to the grade of the tumor, and 38 macropinocytosis-related gene signatures allowed discrimination of the glioblastoma (GBM) samples. Honokiol-induced vacuolization was then compared to vacquinol-1 and MOMIPP, two known macropinocytosis inducers. Despite high phase-contrast morphological similarities, honokiol-induced vacuoles appeared to originate from both endocytosis and ER. Also, acridine orange staining suggested differences in the macropinosomes' fate: their fusion with lysosomes appeared very limited in 3-(5-methoxy -2-methyl-1H-indol-3-yl)-1-(4-pyridinyl)-2-propen-1-one (MOMIPP)-treated cells. Nevertheless, each of the compounds markedly increased temozolomide uptake by glioma cells, as evidenced by LC-MS. In conclusion, the observed deregulation of macropinocytosis in GBM makes them prone to respond to various compounds affecting their formation and/or intracellular fate. Considering that sustained macropinocytosis may also trigger cell death of both sensitive and resistant GBM cells, we propose to envisage macropinocytosis inducers in combination approaches to obtain dual benefits: increased drug uptake and additive/synergistic effects.

Keywords: glioma; macropinocytosis; methuosis; honokiol; vacquinol-1; MOMIPP; temozolomide

1. Introduction

Glioblastoma (GBM) remains associated with a dismal prognosis, partly due to a high level of resistance to proapoptotic stimuli and a propensity for cell migration and invasion of normal brain tissue [1,2]. Although the majority of chemotherapeutic drugs operate by triggering caspase-dependent apoptotic cell death, 17 cell death types have been described to date [3–5], including autophagy. The latter is a programmed pathway that can either promote or hinder tumor development. It is triggered by several anti-tumor agents, including temozolomide (TMZ), particularly in cancer cells lacking essential apoptotic modulators, such as GBM [6]. The autophagosomes arise under the control of the PI3K-Akt pathway—probably from diverse sources, including endoplasmic reticulum (ER), Golgi and plasma membrane—and further fuse with lysosomes to digest their contents [5]. To date, autophagy remains the best way to combat GBM, since it is induced by the standard-of-care drug TMZ [7]. Other forms of non-apoptotic cell death include methuosis, paraptosis, necroptosis, oncosis and lysosomal membrane permeabilization plus senescence. All these forms of cell death are associated with accumulation of cytoplasmic vacuoles (vacuolization) [3,5]. Irreversible vacuolization marks cytopathological conditions leading to cell death; however, the size, content, function and origin of the vacuoles may differ markedly from one type of cell death to another.

One mechanism that can lead to cellular vacuolization is dysregulation of macropinocytosis. Macropinocytosis is a clathrin-independent endocytic mechanism that was first described by Warren Lewis in 1931 [8]. This process is characterized by the nonselective uptake of extracellular fluid [9]. Macropinosomes consist of large vesicles (0.2 μm < diameter < 5 μm), formed by actin-mediated ruffling of the plasma membrane. The nascent macropinosomes migrate in a centripetal manner and rapidly acquire markers of late endosomes, such as Rab7, before their fusion with the lysosomal compartment [9]. Cancer cells generally employ macropinocytosis to internalize cell surface receptors and certain nutrients [10]. Starving tumors by inhibiting macropinocytosis was recently suggested as a potential therapeutic strategy to combat Ras-driven cancers [11]. Deregulation of macropinocytosis in cancer cells seems to be correlated with oncogenic *RAS* [12], and more specifically *K-RAS* in clinical samples [13,14], even if induction of Src may also trigger macropinocytosis [15,16]. With respect to glioma, high levels of active wild-type Ras have been reported in high grade astrocytomas, but oncogenic mutations in *H-RAS*, *K-RAS* or *N-RAS* are rare (occurring in less than 2% of the cases) [17]. Another major pathway known to stimulate macropinosome formation is under the control of the epidermal growth factor receptor (EGFR) and platelet-derived growth factor receptor (PDGFR) [18]. Amplification of wild-type or mutant *EGFR* is relatively common (43% of GBM cases), with the most frequent mutation being *EGFRvIII*. Mutation and/or overexpression of *EGFR* or *PDGFR* are very well known markers of glioma malignancy [19,20]. Importantly, EGF-mediated macropinocytosis appeared dependent on c-Src activity [16].

Despite the potential importance of macropinocytosis for survival of glioma cells, little information is currently available about the possible deregulation of macropinocytosis in these tumors. The possible relevance of macropinocytosis in gliomas is underscored when considering that it contributes, at least partly, to the uptake of extracellular vesicles (particularly exosomes) produced by glioma cells to enhance cancer progression and promote angiogenesis, metastasis, immunosuppression and chemoresistance [21,22]. While active macropinocytosis may represent a survival strategy for gliomas, it may also be exploited to increase the intracellular delivery of therapeutics (soluble as well as vesicular-formulated ones) [22–25]. Moreover, there is strong evidence that hyperstimulation of macropinocytosis by Ras overexpression or by treatment with small molecules can be detrimental to glioma cells, leading to a form of cell death termed as methuosis [26]. Normally, macropinosomes are formed from plasma membrane ruffling via Rac1-dependent rapid polymerization/depolymerisation of branching actin filaments. The macropinosomes then undergo a maturation process wherein they are either degraded via a late endosome/lysosome process or recycled back to the plasma membrane [23]. During abnormal macropinocytosis, as observed in methuosis, macropinosomes do not fuse with lysosomes or recycle to the plasma membrane. Instead, they fuse with each other to form multiple larger vacuoles that exhibit properties of non-functional late endosomes [27]. Through mechanisms that are yet to be elucidated, the extreme vacuolization ultimately leads to death and rupture of the cell. While Ras-induced methuosis is accompanied by caspase activation, cell death cannot be prevented by caspase inhibitors. Methuosis is therefore considered to be caspase-independent [28]. Although early studies demonstrated that methuosis can be induced in glioma by overexpressing active forms of Ras or Rac1, more recent studies have identified small molecules that can trigger this form of death in a manner that is independent from Ras or Rac1. Specifically, the indolyl chalcone 3-(5-methoxy -2-methyl-1H-indol-3-yl)-1-(4-pyridinyl)-2-propen-1-one (MOMIPP), which was used in the present study, can induce intense macropinocytosis, leading to methuosis in cultured GBM cells at low micromolar concentrations [28]. Similar results were obtained in TMZ-resistant glioma cells [29].

Thus, this study aims to (i) evaluate in a systematic manner whether genes implicated in macropinocytosis are deregulated in gliomas, employing analyses of mRNA expression levels in clinical sample datasets and (ii) evaluate whether the induction of severe vacuolization and macropinocytosis with small molecules may represent an interesting strategy to enhance the uptake and/or efficiency of chemotherapeutic agents by gliomas as detailed in the Discussion section.

2. Results

2.1. Evaluation of The Deregulation of Macropinocytosis in Human Gliomas

To evaluate whether macropinocytosis is deregulated in gliomas, we made use of mRNA expression databases of human samples. Table A1 (Appendix A) summarizes the 38 selected main proteins participating in either the initiation/regulation or the formation and recycling of macropinosomes, with their detailed function in that process. The GDS4290 Henry Ford database regroups the mRNA expression array data of 180 patients: 23 non-tumoral samples (from epilepsy patients) and 157 glioma samples. We first performed a systematic, statistical comparison of the mRNA expression of the 38 targets between normal brain samples ($n = 23$), grade II ($n = 45$), grade III ($n = 31$) and grade IV ($n = 81$) glioma samples through non-parametric Kruskall–Wallis comparison and two-tailed tests. Detailed results per probe, provided in Table A1, highlighted that 14/38 targets were overexpressed in glioma while 9/38 were downregulated. This means that more than 60% (23/38) of genes playing key roles in macropinocytosis are deregulated in glial tumors.

As expected, *EGFR* and *PDGFR* were upregulated according to the malignancy grade of the tumor (Figure 1a,b), but other key players in macropinocytosis and methuosis were also upregulated according to the grade of the tumor, including actin dynamic-related gene products (*SWAP70*, several *ARPC* involved in the Arp2/3 complex), their upstream activator adenosine diphosphate ribosylation factors 6 (*ARF6*), and actin-rich membrane ruffles associated protein Rab34 (Figure 1d,f,h). In contrast, the expression level of GIT1 that mediates the deactivation of Arf6 is decreased (Figure 1c), along with the other genes involved in the closure of the macropinosomes, such as *PAK1*, *CTBP1* and *PDL1* (Table A1, Figure 2). *RAB20* and *SNX5*, two markers of initial macropinosome maturation, are both upregulated (Table A1; Figure 1e,g; Figure 2). Thus, the expression of genes encoding numerous proteins associated with macropinocytosis is deregulated in glioma in comparison to normal brain tissues; these changes are additionally correlated with the grade of the tumor (Figure 1). Figure 2 schematically illustrates the macropinocytosis process, noting the proteins analyzed and the changes observed in their expression in glioma tumors. According to the high number of targets whose mRNA levels were found deregulated, we proceeded with an unsupervised hierarchical clustering of the 180 patient samples on the basis of those 38 gene expression signatures, and we observed a clear tendency of grouping according to the grade of the sample (Figure 3a). While normal brain tissue samples grouped on the left, the GBM samples were distinctly clustered on the right of the Euclidian-based tree (Figure 3a). Note that while grade II samples also appeared grouped between non-tumoral brain samples and grade IV samples, grade III samples were spread among grade II or grade IV samples (Figure 3a).

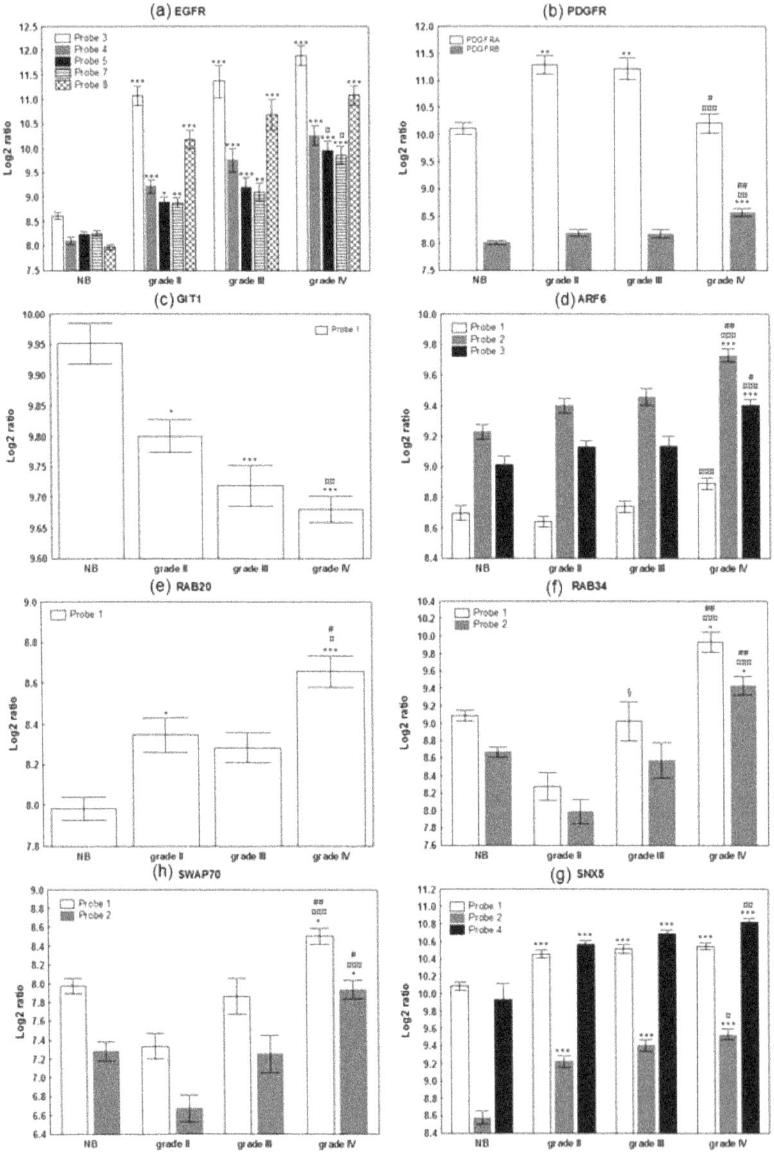

Figure 1. mRNA expression level of selected targets involved in macropinocytosis in patient brain samples. Data are expressed as mean ± SEM of mRNA intensities of each probe expressed in log 2 ratio. NB: normal brain; grade II, III and IV: glioma samples according to the histopathological grade of the tumor. (**a**) epidermal growth factor receptor (*EGFR*), (**b**) platelet-derived growth factor receptor A (*PDGFRA*) and *PDGFRB*, (**c**) *GIT1*, (**d**) *ARF6*, (**e**) *RAB20*, (**f**) *RAB34*, (**g**) *SWAP70* and (**h**) *SNX5*. Note that only probes with the highest intensities are presented for clarity of the figure with respect to *EGFR* and *SNX5*. Statistical comparisons with the non-tumoral brain samples are represented by (*), statistical comparisons between grade II and IV by (¤), between grade III and IV by (#) and between grade II and III by (§). All statistics are based on two-tailed tests according to conventional thresholds: $p < 0.05$ (*, ¤, # or §), $p < 0.01$ (**, ¤¤, ## or §§) and $p < 0.001$ (***, ¤¤¤, ### or §§§).

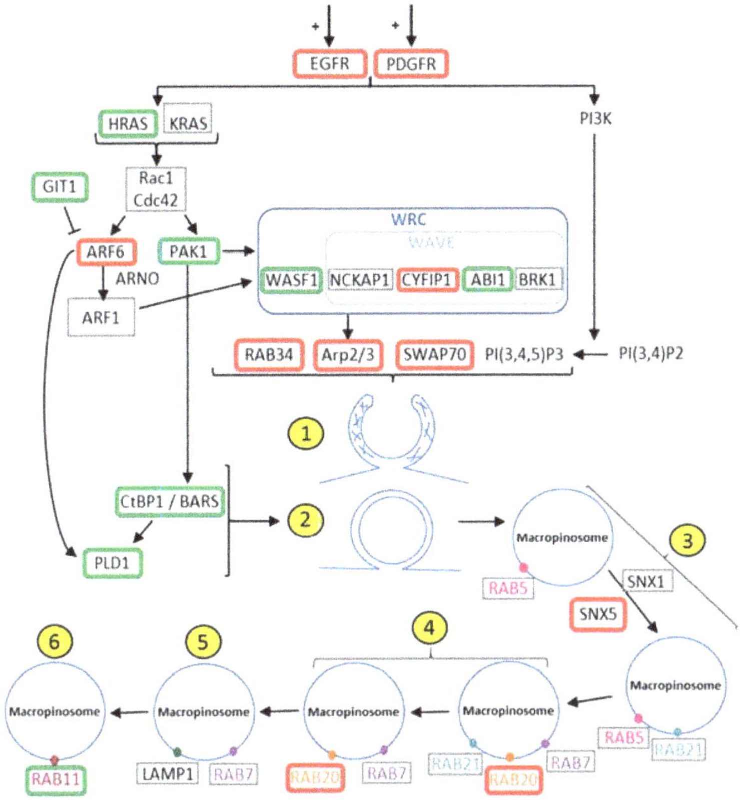

Figure 2. Macropinocytosis process and its deregulation in glioma. Upregulated genes are framed in red, downregulated ones in green, and those genes for which no change was observed are framed in grey. The stimulation of EGFR and PDGFR induces the activation of K-Ras and H-Ras. These Ras GTPases activate both Arf6 and Pak1 to initiate the actin polymerization. Arf6 recruits ARNO (Arf guanine nucleotide exchange factor) for the activation of Arf1 to enable the WAVE regulatory complex (WRC), composed by Wasf1 and the WAVE complex, which regroups Nckap1, Brk1, Cyfip and Abi1. WRC activates Arp2/3, which induces actin branch formation and actin polymerization (1) for the formation of macropinosomes with the help of Rab34, Swap70 and PI(3,4,5)P3. Both Ctbp1 (activated by Pak1) and PLD1 (activated by Arf6 and Ctbp1) are essential for the closure of the macropinocytic cup and the final fission from the plasma membrane (2). Rab5 is a marker of early endosome as well as Rab21, which remains on the intermediate endosome (3). SNX1 and SNX5 play a role in macropinosome maturation (3). Markers of intermediate endosomes are Rab21, Rab20 and Rab7 (4). Rab7 is also a marker of late endosome as well as Lamp1 (5). Rab11 is a marker of endosome recirculating to the plasma membrane (6).

In order to evaluate whether this deregulation has any impact on the survival of the patient, we carried out the same analysis on the data from the GDS53733 database of 70 samples of grade IV GBM patients (16 short term survival patients (<12 months); 23 long term survival patients (>36 months); and 31 intermediate survival patients). We found only very limited statistically significant changes, i.e., with respect to probes for *SWAP70* and *CYFIP1* and, to a lesser extent, *PDGFRA*, *EGFR* and *ABI1* (Table S1). Nevertheless, the unsupervised heatmap clustering revealed a tendency of grouping of the samples from long surviving GBM patients into a subcluster (Figure 3b). In addition, we noticed that many targets overexpressed in GBM samples of the 180 series were downregulated in long surviving patients, including *RAB20, RAB34, SWAP70, CYFIP1, ARPC1B, ARPC2* and *ARPC4*. The dataset is certainly limited in size and further investigation integrating all clinical aspects should be conducted to determine whether deregulated macropinocytosis could be a predictive marker for clinical outcome of the patient.

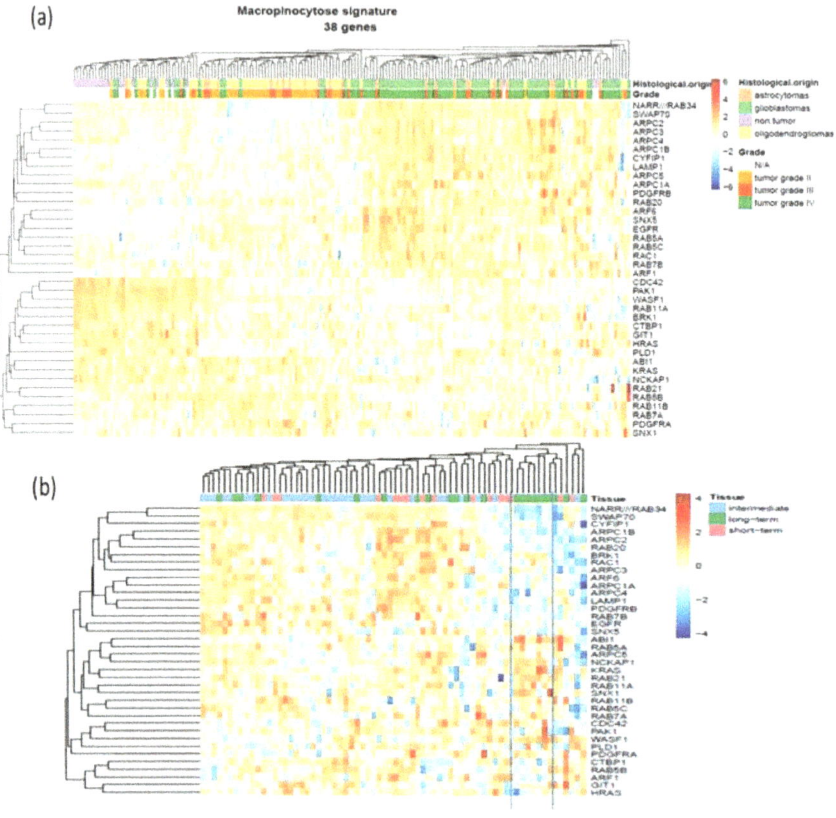

Figure 3. Heatmaps of 38 macropinocytosis gene signatures. (**a**) Dataset GSE4290 regrouping 180 samples (columns) of non-tumoral tissues (N/A) or glial tumors of various histological origin and grade according to the legend. For each gene, the data of the different probes have been averaged to be presented as a single line. Clustering has been made on the basis of the Euclidian distance. (**b**) Similar heatmap made on the dataset from GSE53733 regrouping 70 samples of GBM patients with different overall survival according to the legend. Overexpressed genes are in red, and downregulated genes are in blue (see scale next to the figures).

2.2. Morphological Comparison of Glioma Cells Treated by Honokiol, 3-(5-methoxy -2-methyl-1H-indol-3-yl)-1-(4-pyridinyl)-2-propen-1-one (MOMIPP) and Vacquinol-1

Next, we examined the effects of three compounds that alter vesicle trafficking or autophagy in glioma cells. Two are known for their propensity to induce macropinocytosis—i.e., vacquinol-1 [27] and MOMIPP [29]—while the last one, honokiol, is a natural bioactive polyphenol extracted from several parts of Magnolia genus tree [30] (Figure 4). The latter displays various well-known pharmacological properties, such as anti-oxidant [31], neuro-protective [32] and anti-inflammatory effects in both microglia and astrocytes [33]. These properties explain the traditional use of honokiol for the treatment of thrombotic stroke, gastrointestinal complaints, anxiety and nervous disturbance [34]. Honokiol has also been shown to display interesting anti-tumoral properties against glioma cells, where it may trigger p53-mediated cell cycle arrest and apoptosis [35] or alternative autophagic cell death [36].

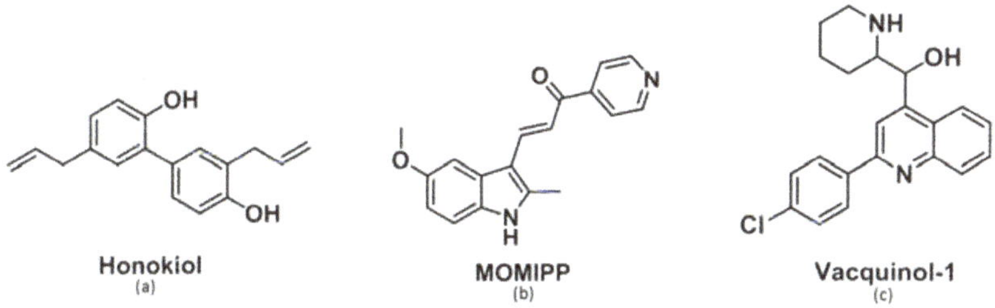

Figure 4. Chemical structure of the three compounds used in this study.

Surprisingly, we observed striking cytoplasmic vacuolization in cells exposed to honokiol at concentrations slightly lower than its IC_{50} in Hs683 (40 µM) and U373 (35 µM) glioma cells (Figure 5 with respect to the U373 cell line; Figure S1 for Hs683 cells). Although the vacuolization appeared delayed with honokiol in comparison to MOMIPP and vacquinol-1, the progressive increase in the number and size of highly refringent vacuoles and their apparent homotypic fusion resembled the morphological features observed with the macropinocytosis inducers at equipotent concentrations (close to their own IC_{50} as determined by means of MTT colorimetric assay, i.e., 3 µM for MOMIPP and 5 µM for vacquinol-1; see Figure S2 for the MTT curves).

2.3. Characterization of the Vacuoles Induced by Honokiol, Vacquinol-1 and MOMIPP

To obtain more insight regarding these vacuolization processes, we utilized different inhibitors of vacuolization, i.e., bafilomycin A1, ethyl-isopropyl-amiloride (EIPA), EHT1864, PP2 and filipin.

Bafilomycin A1 is an inhibitor of the vacuolar H^+-ATPase that plays crucial roles in maintaining low pHs of late endosomes and lysosomes. Bafilomycin A1 was accordingly shown to block the endosomal and endosome-lysosomal fusion during macropinocytosis [37]. Additionally, bafilomycin A1 has been suggested to also inhibit nascent macropinosome formation, similarly to the Na^+/H^+ exchanger inhibitor, by disrupting the fine tuning of submembranous pH needed for the recruitment and activation of Rac1 and Cdc42 to membrane ruffles [38]. As illustrated in Figure 6 with respect to the U373 cell line (Figure S3 for HS683 cells), bafilomycin A1 almost completely inhibited the vacuolization induced by each of the three compounds. Consistently, similar results were obtained with respect to the Na^+/H^+ exchanger inhibitor EIPA [9] when its concentration adapted to the cell line and the compound under investigation in accordance with the timing of the vacuolization that they induced (Figure 6 and Figure S3).

In contrast, the Rac1 inhibitor, EHT1864 [26], has only slight effects on honokiol and vacquinol-1-induced vacuolizations and no effect on MOMIPP–induced vacuolization (Figure 6 and Figure S3). Alternatively to oncogenic RAS stimulation, Src was also demonstrated to participate in macropinosome membrane ruffling via PI3K [15,16]. We therefore made use of the Src inhibitor PP2 [39]. This later inhibited vacuolization induced by honokiol at 25 µM (Figure S4). We tried higher concentrations to possibly inhibit MOMIPP and vacquinol-1 vacuoles but encountered solubility issues making us impossible to conclude at this stage whether Src may or not be involved in their vacuolization processes.

The last inhibitor, filipin, is a cholesterol binding agent [26] known to inhibit clathrin-independent endocytosis. Consistently, macropinocytosis occurs in cholesterol-rich membrane domains [40]. When used at nontoxic concentrations, i.e., 1 µg/mL, filipin had no effect on the vacuole formation induced by each of the three compounds (Figure S5). Previous studies demonstrated that filipin effectively impaired MOMIPP-induced vacuolization in a different cell line, but it was at higher concentrations, i.e., 12 µg/mL. However, in our cellular models, such high concentrations proved to be too toxic.

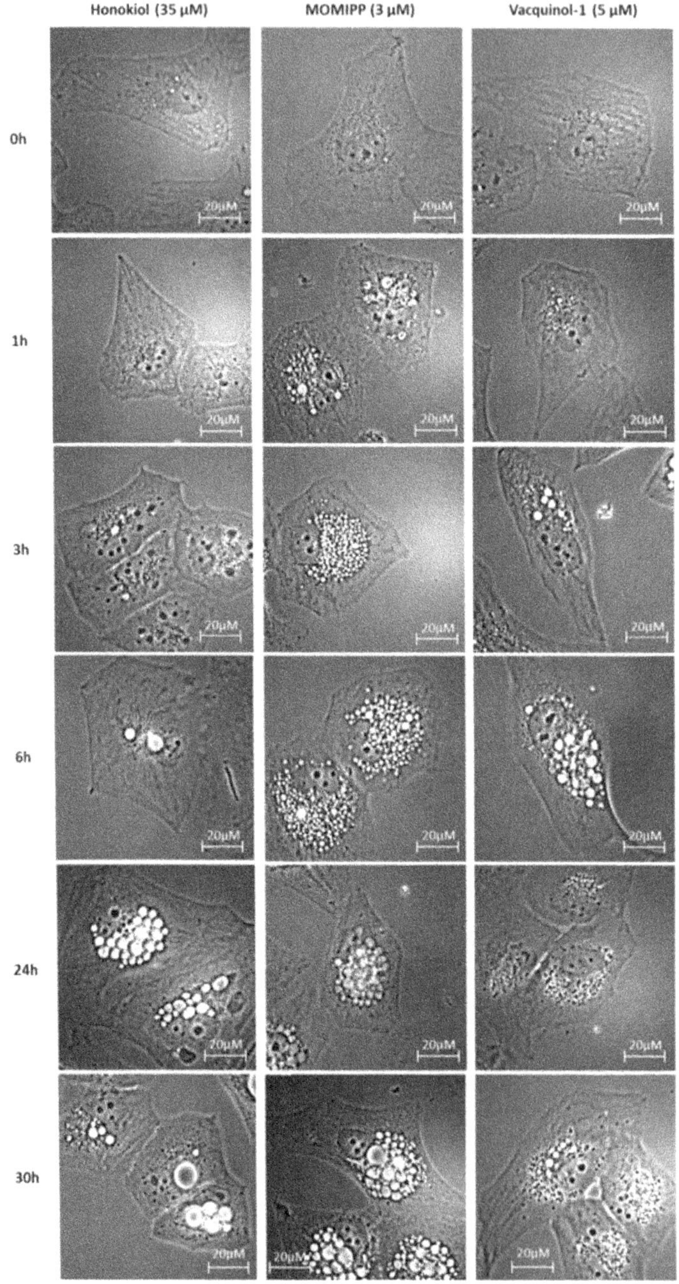

Figure 5. Morphological illustration of the vacuoles formed after treatment of the U373 cell line with honokiol (35 µM), 3-(5-methoxy-2-methyl-1H-indol-3-yl)-1-(4-pyridinyl)-2-propen-1-one (MOMIPP) (3 µM), or vacquinol-1 (5 µM) over time. The accumulation of vacuoles is only visible after 24 h of treatment with honokiol, but those induced by treatment with vacquinol-1 occurs after 3 h and after only 1 h of treatment with MOMIPP. The experiment has been conducted two times in triplicate.

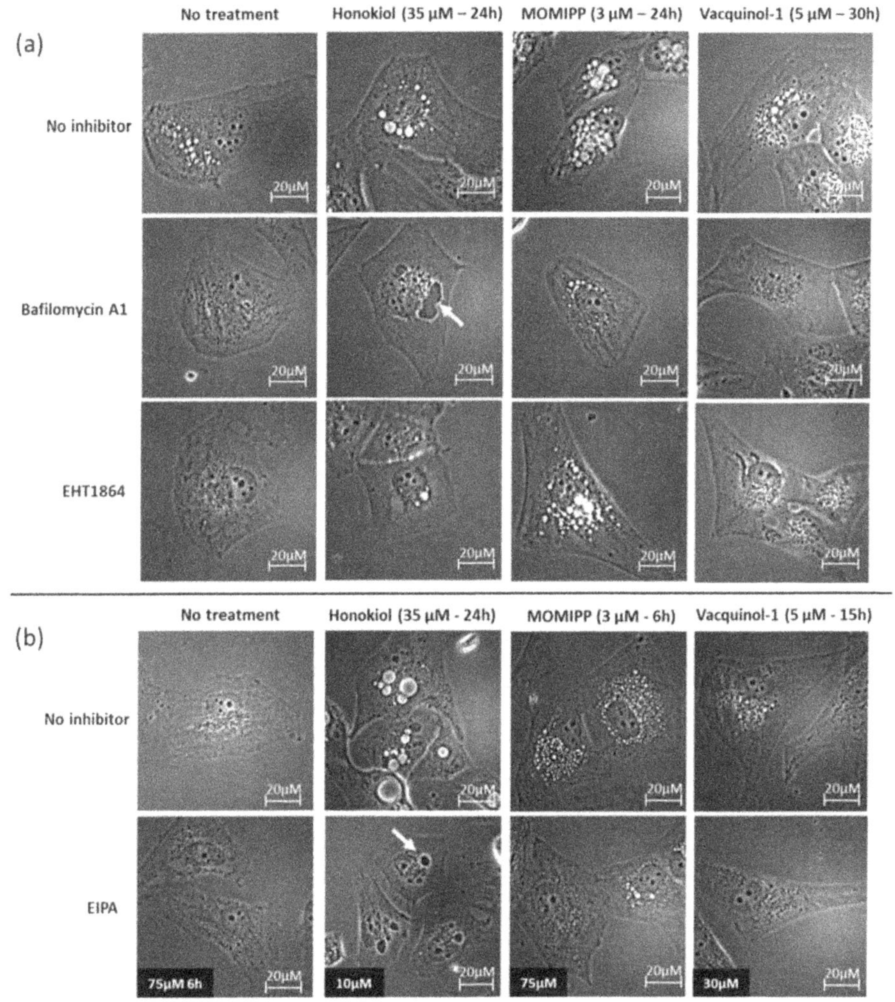

Figure 6. Effects of various inhibitors on vacuole formation following honokiol, vacquinol-1, and MOMIPP treatment in U373 cells. Representative brightfield illustrations of U373 cells treated with or without the inhibitors are as follows: (**a**) pre-treated 1 h with bafilomycin A1 (100 nM) before honokiol (35 µM), MOMIPP (3 µM) or vacquinol-1 (5 µM) treatment, co-treatment with EHT1864 (25 µM) and the compound. Bafilomycin A1 inhibits the vacuoles induced by each compound but not ER-derived dark dense vacuoles induced by honokiol (white arrow). EHT1864 slightly inhibits vacuolization induced by honokiol and vacquinol-1 but not for vacuolization induced by MOMIPP. (**b**) Cells were co-treated with EIPA (10, 30 and 75 µM) and each compound. EIPA inhibits the vacuoles induced by each compound except the ER-derived dark dense vacuoles induced by honokiol (white arrow; similar than bafilomycin A1 effects). Each experiment has been conducted at least twice in triplicate.

Finally, we utilized fluorescent probes to further decipher the origin of vacuoles. Lucifer yellow is a fluid-phase tracer that is internalized intracellularly by endocytic processes including macropinocytosis [12]. We observed, as expected, that both MOMIPP and vacquinol-1 increased the number of positive lucifer-yellow vacuoles—a feature that was also observed with honokiol, but to a lesser extent, with several vacuoles remaining negative (Figure 7a and Figure S6a).

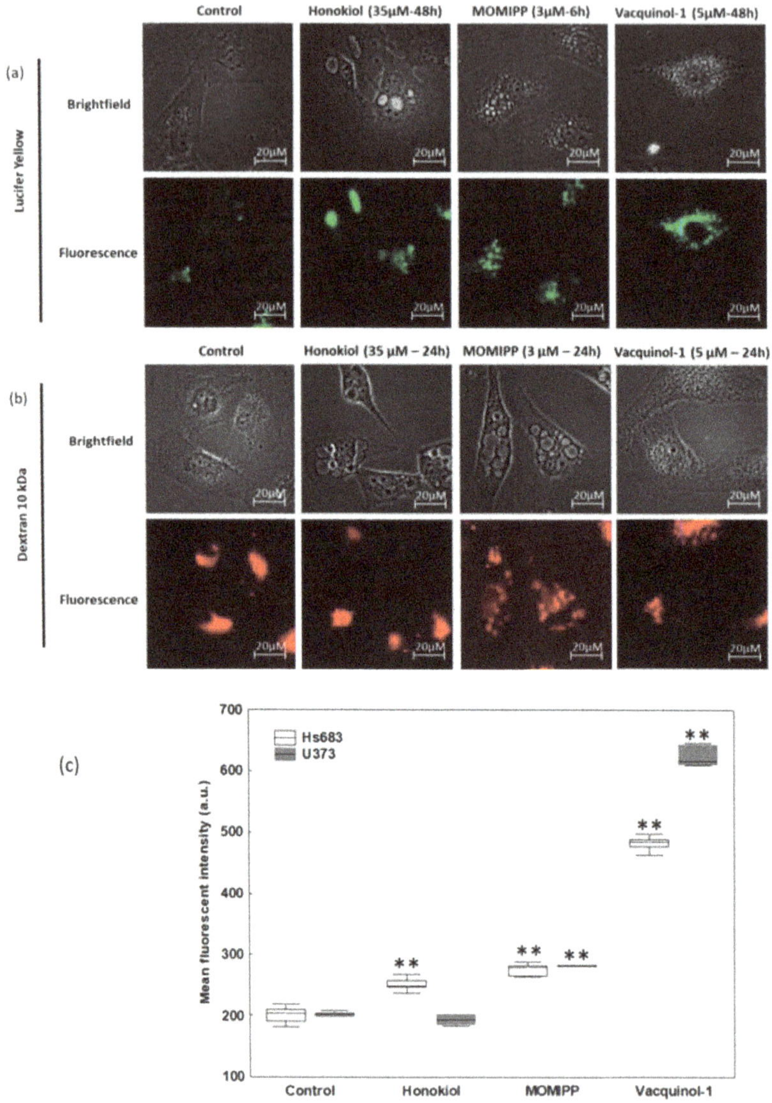

Figure 7. Lucifer yellow and fluorescent 10 kDa dextran uptake by U373 cells treated with honokiol, MOMIPP, or vacquinol-1. (**a**) Illustration of Lucifer yellow staining (100 µg/mL during the whole treatment period). Exposure time is the same for all conditions. The experiment has been conducted three times in duplicate. (**b**) Illustration of fluorescent 10 kDa dextran uptake. Again, cells were treated with each compound in presence of 10 kDa Texas-Red labeled dextran (125 µg/mL) for 24 h. Some vacuoles are positive with MOMIPP and vacquinol-1 but none with honokiol. Exposure time has been adjusted for each condition (550 ms for the control, 490 ms for honokiol, 650 ms for MOMIPP and 160 ms for vacquinol-1). The experiment has been conducted twice in duplicate. (**c**) Quantitative dextran 10 kDa uptake by U373 and HS683 determined by flow cytometry after 24 h and 30 h, respectively. Data are expressed as box plots: line, median; boxes, percentiles 25–75; and whiskers, non-outlier ranges of six experiments. Statistical comparisons with untreated cells are based on Mann–Whitney tests according to conventional thresholds: $p < 0.05$ (*), $p < 0.01$ (**) and $p < 0.001$ (***).

As macropinocytosis differs from other endocytic processes by its capacity to internalize large extracellular volumes and high molecular weight molecules, we evaluated this feature by means of 10 kDa and 70 kDa fluorescent dextran staining. Figure 7b and Figure S6b show that each of the three compounds seemed to trigger the uptake of 10 kDa dextrans that was observed in small vacuoles, but rarely in the most enlarged ones. MOMIPP and vacquinol-1 seemed to uptake 70 kDa dextrans, contrary to honokiol, which had no positive vacuole (Figure S7). A quantitative assay was conducted with 10 kDa Texas Red-labeled dextran. Significant increased uptake under treatment with each of the three compounds was confirmed, except in the case of U373 cells treated 24 h with honokiol (Figure 7c), and the most marked increase was obtained with vacquinol-1. By contrast, acridine orange labelled most of the enlarged vacuoles induced by honokiol in red, suggesting acidic content (Figure 8 and Figure S8). Interestingly, while only few of MOMIPP-induced vacuoles were also stained red (Figure 8 and Figure S8), vacquinol-1-induced vacuoles were positive for both red and green fluorescence after acridine orange staining [41]. When looking for lysosomal staining, these organelles were readily observed in cells treated either with honokiol or MOMIPP (Figure S9). Thus, the principal difference between these was that most of the large vacuoles remained unstained by acridine orange with MOMIPP, whereas most of them appeared acidic in honokiol-treated cells while being lyso-Tracker negative (acidic; Figure S9). This could suggest that the large vacuoles induced by honokiol might result from fusion of endosomes with lysosomes or autophagolysosomes at some point during their biogenesis, while MOMIPP completely blocks vacuole fusion with lysosomes (Figure 8 and Figures S8 and S9). Finally, although vacquinol-1 appears to trigger the strongest increase in macropinocytosis tracers (Lucifer Yellow and high molecular weight dextrans), few enlarged vacuoles have been observed, suggesting different effects on macropinosome maturation and recycling.

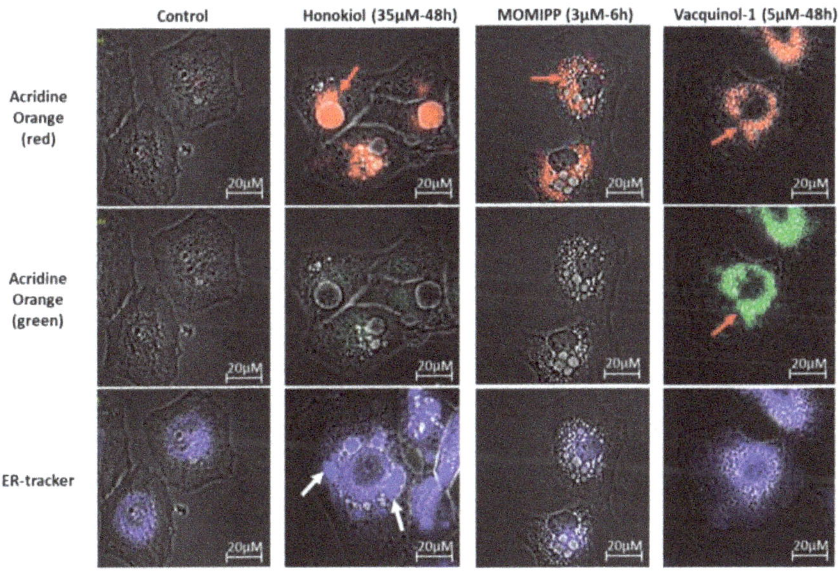

Figure 8. Illustration of the fluorescent characterization of the vacuoles using different fluorescent cell compartment trackers in U373 cells. Two kinds of vacuoles can be observed after treatment with honokiol: red arrows show acridine orange positive vacuoles possibly derived from macropinocytosis, and white arrows show ER-positive vacuoles. Some vacuoles induced by MOMIPP are red-positive with acridine orange (red arrow) but none to the ER-tracker. Vacquinol-1-induced vacuoles are red-positive (red arrow), but also green-positive stained by acridine orange. This experiment has been conducted three times in duplicate.

We also used probes for mitochondria and ER to assess the contributions made by these compartments to the vacuoles induced by honokiol. We observed some ER-positive vacuoles only in the honokiol-treated cells (Figure 8 and Figure S8). Both mitochondria and ER are thought to be a source of vacuolization during paraptosis [42]. Thus, honokiol differed from both MOMIPP and vacquinol-1 and could induce paraptosis features in these glioma cell lines, similar to what was previously described in leukemia cells [43]. ER swelling does not play any role in macropinocytosis. Accordingly, ER-positive vacuoles were still present after treatment with the inhibitors of macropinocytosis (Bafilomycin A1, EIPA and PP2; Figure S4). Note that none of the treatments triggered mitochondrial swelling (Figure S9). This supports the hypothesis that honokiol may induce paraptosis in addition to macropinocytosis according to the other organelle markers detailed above (see Discussion section).

2.4. *Evaluation of the Effects of Honokiol, Vacquinol-1 and MOMIPP on Intracellular Temozolomide (TMZ) Concentration*

To highlight the possibility of using these inducers of vacuolization to enhance the penetration of chemotherapeutic agents in cancer cells, we pre-treated glioma cells with the three compounds of interest before the addition of the treatment with TMZ. TMZ was chosen as the current first-line chemotherapeutic agent against newly diagnosed GBM [44,45] and for its pro-autophagic effects in glioma cells [46,47]. Honokiol was used 22 h, vacquinol-1 15 h and MOMIPP 3 h before the addition of TMZ. These timepoints were selected according to the time required to visualize the beginning of the vacuolization with each compound. The quantification of intracellular TMZ in Hs683 and U373 cells was achieved through HPLC-MS after two hours of subsequent exposure to TMZ at 200 μM with or without the macropinocytosis inducers. Figure 9 shows that all three compounds significantly increased the intracellular TMZ concentration in both Hs683 and U373 cells, except in the case of MOMIPP-pretreated Hs683 cells. Note that we pre-treated cells only for 3 h with MOMIPP, and longer incubation periods may be required in these cells.

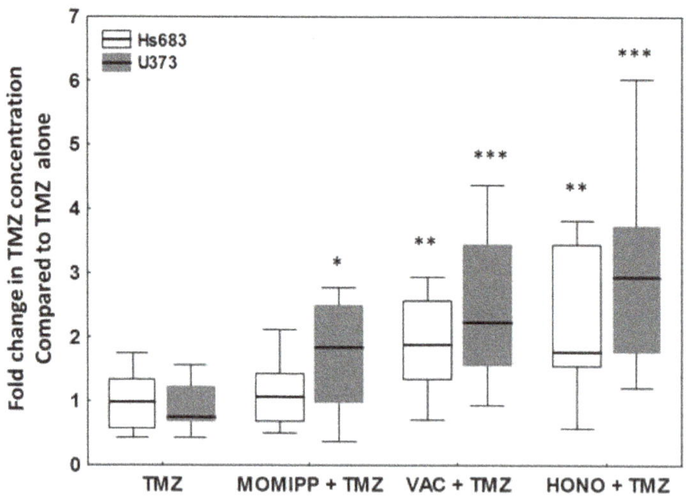

Figure 9. Temozolomide (TMZ) concentrations in U373 and Hs683 cell lysates after two hours of treatment with TMZ (200 μM). Cells were pre-treated with honokiol for 22 h (HONO + TMZ), vacquinol-1 for 15 h (VAC + TMZ) or MOMIPP for 3 h (MOMIPP + TMZ). Data are expressed as fold-change TMZ concentrations compared to its own control (TMZ alone) by box plots: line, median fold change; boxes, percentiles 25–75; whiskers, non-outlier ranges of 12 replicates. TMZ concentration of the cells treated with TMZ alone was normalized to 1. Statistical comparison has been made by the Mann–Whitney test. $p < 0.05$ (*), $p < 0.01$ (**) and $p < 0.001$ (***).

Despite apparent differences in their effect, certain considerations, including the concentration of the compound itself, the duration of the pretreatment required, or their ability to cross the blood brain barrier (BBB), are all crucial to be taken into account for future in vitro and in vivo combinations. Furthermore, this experiment only highlighted that the use of those compounds could help to increase the intracellular concentration of TMZ. Whether these effects actually related to macropinocytosis induction and/or other mechanisms should be evaluated in more depth. Such investigations would help decipher whether this strategy could be used with other kinds of drugs and pharmaceuticals. Other aspects and advantages of combinations (pro-methuotic or pro-paraptotic effects, additive or synergisctic effects) should also be considered, as discussed below.

3. Discussion

In this study, we showed for the first time that genes associated with macropinocytosis were deregulated in human glioma brain tumors. This conclusion is based on a systematic statistical analysis of the mRNA dataset from Henry Ford Hospital by regrouping non-tumoral human samples in comparison to gliomas samples of grades II, III and IV. Figure 2 summarizes the genes of the macropinocytosis process whose expression levels were evaluated in this study and whether they were up- or downregulated. Hereafter, we discuss these results briefly in relation to the scientific knowledge currently available regarding the main deregulated targets in glioma biology.

Consistent with existing literature, we found *EGFR* and *PDGFR* to be upregulated in gliomas [19,20,48]. These are potent inducers of macropinocytosis [18]. *H-RAS*, by contrast, was found to be downregulated, but only the mRNA level was investigated in the present study. Indeed, increased active wild-type Ras activity has been reported in high-grade glioma and may contribute to increased macropinocytosis as well [17]. Interestingly, Src plays key roles in EGF-triggered macropinocytosis associated with enhanced migration and further fusion of macropinosomes with lysosomes [16]. In this study we observed that the Src inhibitor PP2 impaired acid vacuolization induced by honokiol (see below).

We also found that several other key actors of macropinocytosis, such as Rab34, Arp2/3 complex and SWAP70, were significantly overexpressed according to the grade of the tumor (Figure 1, Table A1). These are all implicated in the organization of the actin cytoskeleton and ruffling of the plasma membrane during macropinocytosis [49–51]. Our results are consistent with previous data suggesting that overexpression of these proteins contributes to glioma cell invasion [52]. In this context, expression levels of *ARP2/3* and *RAB34* correlate with the grade of the tumor and/or the survival of the patient, with both being upregulated from low-grade to high-grade gliomas [53,54]. *RAB34* was additionally associated with poor patient survival [54]. Regarding other Rab proteins, only few studies have been published regarding their roles and/or deregulation in glial tumors. The increased expression of *RAB20* found in the present study could suggest a role in macropinocytosis in glioblastoma. Downregulation of *RAB21* by siRNA significantly inhibited cell proliferation and remarkably induced cell apoptosis [55], but its overexpression in glioblastoma was not reported, and no link to enhanced macropinocytosis was suggested in those tumors. In addition, mRNA coding for Arf6 was also found to be overexpressed in GBM samples (Figure 1d). During macropinocytosis, active Arf6 recruits the Arf nucleotide binding site opener (ARNO) that activates Arf1 via its guanine exchanger activity and allows the recruitment of the WAVE protein regulatory complex (WRC). This heteropentameric complex of WASP family proteins composed of WAVE, Cyfip1, Nap1 (*NCKAP1*), Abi1 and Brk1, in turn activates Arp2/3 to initiate actin polymerization [51,56]. Recently, a potential role for Arf1 in glioblastoma progression was suggested [48], and over-expression of Arf6 was shown to enhance glioma cell migration both in vitro and in vivo [48,57,58].

While we did not observe significant changes in *LAMP1* expression, Jensen et al. [59] found that *LAMP1* is more highly expressed in glioblastoma than in diffuse and anaplastic astrocytomas, even though its expression does not correlate with the overall survival of the patient [59,60]. *CTBP1/BARS* and its downstream target *PLD1*, whose activation triggers the closure and the final fission of macropinosomes from the plasma membrane [9,61,62], were found downregulated in high-grade tumors at the mRNA level herein. Those results are not consistent with previous data indicating positive correlation of Ctbp protein antigen expression with the histopathologic grade of the glioma [63] and worse survival of the patient [64]. The discrepancies could be related to evaluation at the mRNA versus protein expression level. Further, the downregulation of *PAK1* mRNA is not easy to integrate when considering that both protein phosphorylation and localization (cytoplasmic versus nuclear) appeared essential for its function and prognosis value in glioblastoma [65].

Nevertheless, as highlighted by Figure 2, the macropinocytosis process appeared obviously deregulated in glioma, and particularly in GBM (Figure 1). Accordingly, we showed that the mRNA expression signature of these 38 genes taken together, and covering most of the macropinosome formation, maturation and turn-over processes, enabled discriminating GBM from non-tumoral samples and lower grade glioma on the basis of unsupervised analysis (Figure 3). Macropinocytosis could, thus, participate in GBM aggressiveness, notably when considering its contribution for nutrient uptake and exosome GBM crosstalk [21,22]. Even if a systematic comparison of each target alone, according to the survival of GBM patient, did not reveal any significant changes (Table S1), long surviving GBM patients displayed a trend of grouping together on the basis of these 38 genes' signatures. A study at the protein level, and with full patient characterizations and follow-ups, should be conducted to further confirm an upregulation of macropinocytosis in GBM, and whether it might be of prognosis value for the patient.

Recent findings have suggested that stimulation of macropinocytosis in cancer cells can lead to increased chemotherapeutic efficiency [22–25,66]. Thus, we herein propose taking advantage of this possible Achilles' heel by the use of macropinocytosis-stimulating agents to combat GBM. Hyperstimulation of macropinocytosis with small molecules may ultimately induce methuotic cell death in glioma cells, as previously shown [26]. However, when used for short periods of time or at sub-lethal concentrations, methuosis-inducing compounds may be useful for increasing the intracellular concentrations of chemotherapeutic agents.

Therefore, we compared the effects of three potential compounds of interest—honokiol, vacquinol-1 and MOMIPP. We observed high morphological similarities between the vacuolization processes induced by those three compounds in terms of refringency, increase in number followed by apparent fusion, and cytoplasmic accumulation (Figure 5 and Figure S1). Although they are all inhibited by both bafilomycin A1 and EIPA, one unique inhibitor of macropinocytosis, our results indicate that the vacuoles they induce may differ in terms of origin and capacity to fuse with lysosomes. Vaquinol-1 actually appeared to stimulate macropinosome formation according to the marked increase in the uptake of Lucifer Yellow and dextrans (Figure 7 and Figure S6). The fact that it induces vacuoles of intermediate pH, as revealed by acridine orange staining (positive in both red and green fluorescences; Figure 8 and Figure S8) and the absence of increase in lysosomal content, are two features that differentiate vacquinol-1 from MOMIPP- and honokiol-induced effects. Indeed, these two latter display slight increases in lysosomal content assessed by Lyso-tracker. However, the enlarged vacuoles are not stained with that tracker. (Figure S9). Such enlarged vacuoles were not obtained with vacquinol-1 (Figure 5 and Figure S1). Important differences were also obtained with respect to uptake and acidity evaluated by acridine orange staining. In the case of honokiol, most of those large vacuoles appeared acidic as revealed by their red staining with acridine orange, while MOMIPP-induced enlarged vacuoles remained negative. We conclude that the three compounds affect macropinocytosis process and/or endosomal traffic in high-grade glioma cells, but that the mechanisms underlying those effects markedly differ from one compound to the other. In particular, the fusion of the vacuoles with lysosomal compartments seems to still occur with respect to honokiol, and partly to vacquinol-1 treatments, but not with MOMIPP. In addition, honokiol is the only compound in the current study that also induces ER-derived vacuolization (Figure 6 and Figure S8). Note that ER stress has been shown previously to inhibit endocytic pathways, including macropinocytosis [67,68]. This effect may also be linked to the ability of honokiol to induce paraptosis-like cell death, as previously reported in leukemia cells [43], and autophagy in glioblastoma [36,69]. Interestingly, vacquinol-1 was also recently shown to trigger mitophagy in GBM cells [70]. Although both macropinocytosis and autophagy represent two "opposite" major mechanisms to provide nutrient supply and recycling from extracellular versus intracellular sources respectively, they seem to share upstream and downstream regulation whose links remain to be fully deciphered [71]. In contrast, methuosis results from the accumulation of unprocessed macropinosomes that fuse together, rather than fusing with lysosomes or recycling to the plasma membrane [27], as described previously with respect to the MOMIPP chalcone.

Obviously, further studies are required to clarify if and how each of these compounds affect membrane ruffling, macropinosome formation, trafficking, fusion, and recycling independently to other effects such as ER swelling or autophagy.

Nevertheless, we observed that all three compounds were able to significantly increase intracellular TMZ concentration, even if TMZ uptake was excellent [47]. This study provides a proof of concept, and it encourages further investigations with drugs whose intracellular and/or BBB penetrations are more problematic. This could be the case with respect to antibody-based therapies, including depatuxizumab mafodotin, a new antibody-drug conjugate with promising clinical results notably in EGFR-amplified GBM cases [72,73]. Typically, patients with EGFR amplification may benefit from macropinocytosis-interfering drugs, such as those studied herein.

Honokiol and MOMIPP have been proven to cross the BBB [35,74,75]. Although vacquinol-1 also displays an adequate BBB penetration, its systemic toxicity requires a lowering dose or local delivery [76]. At a tolerable dose, vacquinol-1 alone allowed a reduction in tumor size, but it did not increase the overall survival of the GBM preclinical model [76]. Thus, toxicological issues have to be taken into account for future development. Honokiol is widely used and freely available as a phytotherapeutic complement in several countries for medicinal properties against anxiety, for facilitation of sleep, support of cognitive functions, and antioxidant effects. Further, possible additional and/or synergistic effects between macropinocytosis inducers and chemotherapeutic agents may be expected. Accordingly, honokiol has already been shown to increase TMZ cytotoxic effects in vitro [77]. Both honokiol and MOMIPP have been shown to kill drug-sensitive and -resistant glioma cells [29,77]. Recent studies pointed JNK Kinase as an important signal for both honokiol-induced effects on stem cells [78] and cytotoxic effects of indoylchalcones such as MOMIPP [75].

Numerous perspectives remain to be addressed, including the following: (1) evaluation of whether macropinocytosis deregulation may be linked to patient survival and response to chemotherapy, (2) deciphering how each compound affects macropinocytosis and other endocytosis processes, and (3) testing the efficiency of combined treatments for the proposed molecules investigated herein with chemotherapeutic agents, with a particular attention to new therapeutic drugs characterized by limited intracerebral brain pharmacokinetics.

4. Materials and Methods

4.1. Cell Lines and Compounds

The Hs683 human oligodendroglioma cell line was obtained from the American Type Culture Collection (ATCC, code HTB-138) and the human glioblastoma U373 cell line from the European Collection of Cell Culture (ECACC, code 08061901). Cells were cultivated in RPMI 1640 (Gibco, Thermofisher, Dilbeek, Belgium) culture medium supplemented with 10% heat-inactivated fetal bovine serum (Gibco), 0.6 mg/mL L-glutamine (Gibco), 200 IU/mL penicillin–streptomycin (Gibco), and 0.1 mg/mL gentamicin (Gibco) at 37 °C with 5% CO_2. Cultures were checked twice a month for mycoplasma.

Honokiol was purchased from Sigma-Aldrich (St. Louis, MO, USA) as well as EIPA and PP2, Vaquinol-1 and EHT1864 from Selleckchem (Houston, TX, USA), Bafilomycin A1 from Cayman chemicals (Selleckchem), and filipin complex from AG Scientific (San Diego, CA, USA). MOMIPP was synthesized and characterized as previously described [29].

4.2. MTT Colorimetric Assay

Cell viability was determined using a colorimetric assay as described previously [79]. Cells were seeded in 96 well plates (Sarstedt AG & CO, Nümbrecht, Germany) and were grown for 24 h. They were then treated with honokiol, vacquinol-1, or MOMIPP at concentrations ranging from 0.01 to 100 µM or left untreated for 72 h. The viability was estimated by using 3-(4,5-dimethylthiazol-2-yl)-2,5-diphenyl tetrazolium bromide (MTT, Sigma-Aldrich) mitochondrial reduction into formazan measured at 570 nm with a spectrophotometer (680XR, Bio-Rad Laboratories, Berkeley, CA, USA; reference wavelength 610 nm). The experiment was realized two times in sextuplicate for each cell line and each compound.

4.3. Characterization of Vacuoles

4.3.1. Phase Contrast Microscopy for Morphological Observations

In order to observe the morphological changes induced by treatments, pictures of living cells were taken with the Imager M2 fluorescence microscope (Carl Zeiss, Zaventem, Belgium) coupled with the AxioCam ICm1 and AxioImager software (Carl Zeiss). Cells were seeded on glass coverslips in six-well plates (Sarstedt AG & CO) and allowed to attach and start growing for 24 h. Afterwards, cells were either left untreated or treated with each compound as follows: 35 µM and 40 µM of honokiol for U373 and Hs683 cell lines, respectively; and 5 µM of vacquinol-1 and 3 µM of MOMIPP for both cell lines.

For the characterization of the vacuolization by using inhibitors, cells were either pre-treated 1 h with bafilomycin A1 (100 nM), filipin (1 µg/mL), or co-treated with EHT1864 (25 µM), EIPA (10, 30 or 75 µM) or PP2 (25 µM). At different time points, coverslips were washed twice in PBS, transferred onto microscope slides, and four pictures were taken per slide. Each experimental condition was tested twice in triplicate.

4.3.2. Fluorescent Microscopy Assays

Fluorescent probes that stained different cellular compartments were used to characterize the origin of the vacuoles. The dapoxyl ER-tracker blue-white dye, the Lyso-tracker red dye, the Mito-tracker green dye, and 10 kDa and 70 kDa Texas-Red labeled dextran were all obtained from Molecular Probes (Life Technologies, Merelbeke, Belgium). We also used Lucifer yellow (Lucifer Yellow CH, lithium salt) from Biotium (Fremont, CA, USA) and acridine orange from Sigma-Aldrich. Briefly, the cell seeding and treatment procedures were similar to the ones used for the phase contrast microscopy (Section 4.3.1). The dye solutions were simply added to the culture medium 1 h before the end of the treatment periods, excepted for the Lucifer yellow and both dextrans 10 kDa and 70 kDa, which were added for the whole duration of the treatment. The concentrations of the dyes were as follows: ER tracker, 0.5 µM; Lyso tracker, 75 nM; Mito tracker, 200 µM; Lucifer yellow, 100 µg/mL; acridine orange, 1 µg/mL; and dextrans 10 kDa and 70 kDa, 125 µg/mL; At the end of the treatment period, the procedure was similar to that of phase-contrast microscopy to take pictures of living cells with the Imager M2 fluorescence microscope (Carl Zeiss) coupled with the AxioCam ICm1 and AxioImager software (Carl Zeiss). The experiment was realized at least two times in duplicate.

4.3.3. Flow Cytometry

In order to quantify the internalization of Texas-red labeled dextran 10 kDa (Molecular Probes), cells were treated with honokiol (35 µM for U373 and 40 µM for Hs683 cells), MOMIPP (3 µM), or vacquinol-1 (5 µM), or left untreated in the presence of Texas-red labeled dextran 10 kDa (125 µg/mL). After 24 h for U373 and 30 h for Hs683, cells were washed twice with PBS, detached with trypsin-EDTA (Gibco) and centrifuged. The supernatant was removed and cells were resuspended in 250 µL of PBS for flow cytometry analysis with the Beckmann Gallios apparatus (Beckmann Coulter, Analis, Suarlee, Belgium). Each sample recorded 10,000 events, and the experiment was conducted once in sextuplicate.

4.4. *Quantification of Intracellular TMZ*

4.4.1. Sample Preparation

For the experiment, cells were seeded in T75 flasks (Sarstedt AG & CO). When the confluence of the flasks was around 75%, cells were pre-treated with honokiol (35 μM for U373 and 40 μM for Hs683), MOMIPP (3 μM), vacquinol-1 (5 μM), or left untreated for different periods of time (3, 15 and 22 h) before the addition of TMZ (200 μM) for 2 h. After the treatment, the culture medium was removed, the cells were washed twice with cold PBS (Gibco)), scrapped in 200 μL of ice-cold methanol (VWR International, Oud-Heverlee, Belgium), and sonicated for 30 s. As TMZ is stable at pH < 4 [80], 50 μL of cell lysate was diluted in 50 μL of an acid internal standard solution (2 μM theophyline and 0.1% formic acid in methanol).

4.4.2. LC-MS Process, Data Acquisition and Analysis

Ten microliters from the prepared sample were injected into the liquid chromatography (LC) system. Analyses were performed with a rapid resolution LC (RRLC) 1200 series from Agilent Technologies (Santa Clara, CA, USA). Separation was performed on a Zorbax Eclipse XDB-C18 Rapid Resolution HT column (4.6 × 50 mm, 1.8 μm particle size) from Agilent Technologies, preceded by a Zorbax Eclipse XDB-C18 pre-column (4.6 × 5 mm, 1.8 μm particle size) using water supplemented with a 0.1% formic acid/acetonitrile gradient. A 6520 series electrospray ion source (ESI)–quadrupole time-of-flight (QTOF) from Agilent Technologies was used for the MS analyses. Initial ESI–Q-TOF parameters were as follows: positive mode; capillary voltage of 4500 V; dynamic high resolution acquisition mode (2 Hz); gas temperature of 350 °C; drying gas flow of 9 L/min; nebulizer pressure of 50 psig; fragmentor voltage of 130; and skimmer voltage of 65 V. Data was acquired using the Mass Hunter Acquisition software (Agilent Technologies, version B.04 SP3) and analyzed by Mass Hunter Quantitative Analysis software (Agilent Technologies, version B.07 SP1). TMZ was monitored at an m/z value of 195.0625 and theophyline at an m/z value of 181.0720. A quantitative curve was performed over the range 0.1 to 10 μM of TMZ.

4.5. *Statistical Analyses*

The mRNA expression analyses were conducted on two datasets of human samples publicly available on the NCBI GEO repository. The GSE4290 dataset, from the Henry Ford Hospital, was published in 2006 [81]. The GSE53733 dataset regroups the mRNA expression data sets of 70 human glioblastoma (grade IV) primary tumors from the German Glioma Network and was published in 2014 [73]. Both expression data were generated using the same microarray platform (Affymetrix Human Genome U133 Plus 2.0 Array).

The microarray analysis was based on the robust multi-array average (RMA) expression values, which were obtained with affy package v1.56 of Bioconductor/R packages (http://www.bioconductor.org). The heatmaps were made using the pheatmap R package v1.0.10. The samples were grouped by annotation or clustered using hierarchical clustering using the average linkage on the Euclidian distance.

In both databases, we collected the data corresponding to the probes of 38 target genes involved in macropinocytosis (see list in Table A1). Comparison of the expression level for each probe was conducted using the Kruskal–Wallis test (a nonparametric, one-way analysis of variance), followed by two-tailed tests using Statistica (Statsoft, Tulsa, OK, USA).

For in vitro biological assay comparisons (flow cytometry and TMZ quantification), we conducted non-parametric Mann–Whitney tests using Statistica.

5. Conclusions

In conclusion, we found that more than 60% of the 38 macropinocytosis-related genes studied herein are overexpressed or down-regulated in GBM patient samples. These 38 genes may constitute a signature discriminating GBM from non-tumoral samples and lower grade gliomas. Those results suggest that macropinocytosis may play important roles in GBM aggressiveness.

We then proposed to make use of compounds that interfere with this endocytotic process to increase anti-cancer drug uptake. The three compounds selected for this purpose, i.e., honokiol, vacquinol-1 and MOMIPP indeed allowed significant increase in intracellular TMZ concentration in vitro. This study provides with a first proof of concept and paves the way to use macropinocytosis deregulators in combination with chemotherapeutic agents. Considering that EGFR is an activator of macropinocytosis, patients with amplified or mutated EGFR may represent better candidate for this approach.

Supplementary Materials: The following are available online at http://www.mdpi.com/2072-6694/11/3/411/s1, Table S1: Statistical comparison of mRNA expression data of macropinocytosis gene probes in the GSE53733 dataset, Figure S1: Morphological illustration of the vacuoles formed after treatment of Hs683 cell line with honokiol (40 μM), MOMIPP (3 μM) or vacquinol-1 (5 μM) over time, Figure S2: MTT curves of the three products used in this study, Figure S3: Effects of various inhibitors on vacuoles formation following honokiol, vacquinol-1 and MOMIPP treatment in Hs683 cells, Figure S4: Illustration of the ER-Tracker staining of U373 (a) and Hs683 (b) in presence of honokiol with or without the various vacuolization inhibitors, Figure S5: Effects of filipin on vacuoles formation following honokiol, vacquinol-1 and MOMIPP treatment in U373 and Hs683 cells, Figure S6: Illustration of lucifer yellow (a) and fluorescent 10 kDa dextran (b) uptake in Hs683 cells treated with honokiol, MOMIPP or vacquinol-1, Figure S7: Illustration of fluorescent 70 kDa dextran internalization in U373 (a) and Hs683 (b) cells treated with honokiol, MOMIPP, vacquinol-1, Figure S8: Illustrations of the fluorescent characterization of the vacuoles using different fluorescent cell compartment trackers in Hs683 cells, Figure S9: Illustration of the fluorescent marking of the vacuoles using mitochondrial and lysosomal cell compartment trackers in U373 (a) and Hs683 (b) cells.

Author Contributions: Conceptualization, V.M.; Cellular experiments, M.C., G.R. and A.-S.L.; mRNA dataset management and analysis: R.J.; Analytical HPLC-MS methodology and results analyses, C.D. and P.V.A.; Materials supply, W.A.M. (MOMIPP); Data interpretation, M.C., V.M. and W.A.M.; Writing—Original Draft Preparation, M.C. and V.M..; Critical Review & Editing, W.A.M.

Funding: This research was supported by the Belgian Brain Tumor Support (BBTS) and the United States National Institutes of Health (R01 CA 115495).

Conflicts of Interest: MOMIPP is included in a patent licensed to Systems Oncology, Inc. As a co-inventor, W.A.M. receives a portion of licensing fees and royalties. The other authors declare no conflict of interest.

Appendix A

Table A1. Statistical comparison of mRNA expression data of macropinocytosis gene probes in the GSE4290 dataset. Kruskall–Wallis results are provided per probe (probe 1/probe 2/ …). NS: $p > 0.05$; *: $p < 0.05$; **: $p < 0.01$; ***: $p < 0.001$. The number of two-tailed significant results per probe is then provided in a similar way: e.g., 0/1/0 means for probe 1 that no two-tailed comparison reached significant threshold ($p < 0.05$), for probe 2, one two-tailed comparison turned out significant, and again no significance was found between groups for probe 3. Note that the maximum comparison number is 5. ?: inconsistent results depending on the probe considered.

Gene	Name	Number of Probe	Kruskall - Wallis	2 Tailed	Up or Down Regulation	Role in Macropinocytosis
EGFR	Epidermal growth factor receptor	8	*/**/***/***/***/***/***/***	1/2/3/3/4/1/4/3	Up	• Stimulation of macropinocytosis [38].
PDGFRA	Platelet derived growth factor receptor alpha	1	***	4	Up	
PDGFRB	Platelet derived growth factor receptor beta	1	***	3	Up	
H-RAS	H-Ras proto-oncogene, GTPase	1	***	3	Down	• K-RAS induces membrane ruffles and macropinosomes more than H-RAS probably because it activates Rac1 more than H-RAS [82].
K-RAS	K-RAS proto-oncogene, GTPase	5	NS/NS/***/NS/NS	0/0/3/0/1	-	
CDC42	Cell division cycle 42	5	***/***/***/***/***	2/3/4/4/3	?	• These Rho GTPases work with PI-4,5-biP to initiate actin polymerization via PAK1 [9,38,51]. • Rac1 also activates WAVE regulatory complex (WRC). • Rac1 induces a decline in the pool of active ARF6 (ARF6-GTP), which is important for the vacuolization [83].
RAC1	Ras-related C3 botulinum toxin substrate 1 (rho family, small GTP binding protein Rac1)	3	***/**/***/NS	1/1/2/0	-	
GIT1	GIT ArfGAP 1	1	***	4	Down	• The presence of GIT1 is required for the decline of active ARF6. • This ARF6 GAP can interact with RAC1 [83].
ARF1	ADP ribosylation factor 1	3	***/***/*	4/1/1	?	• ARF6 recruits ARNO, an Arf guanine nucleotide exchange factor, which activates ARF that activates in turn the WAVE regulatory complex (WRC) for membrane ruffling [51,56]. • ARF6 also activates PLD1, which is required for the formation and the closure of macropinocytic cups [51,61].
ARF6	ADP ribosylation factor 6	3	***/***/***	1/3/3	Up	
PAK1	p21 (RAC1) activated kinase 1	4	**/***/***/***	3/3/3/3	Down	• PAK1 is activated by Rac1 and CDC42 [38]. • PAK1 regulates actin cytoskeleton organization through WASP/WAVE -Arp2/3 activation • PAK1 phosphorylates CtBP1/BARS, which is essential for the fission of macropinosomes from the plasma membrane [51,62,84].

Table A1. *Cont.*

Gene	Name	Number of Probe	Kruskall - Wallis	2 Tailed	Up or Down Regulation	Role in Macropinocytosis
WASF1	WAS protein family member 1	1	***	4	Down	• Activates Arp2/3 complex to stimulate actin polymerisation [50].
CYFIP1	Cytoplasmic FMR1 interacting protein 1	1	***	3	Up	• CYFIP1, NCKAP1, ABI1 et BRK1 compose the WRC with Wasf1. • The recruitment of the WRC to the membrane and its activation depends on Rac1 and an Arf GTPase (mainly Arf1 but also Arf5) [56,85].
NCKAP1	NCK associated protein 1	2	***/NS	2/0	-	
ABI1	Abl-interactor 1	2	***/***	2/4	Down	
BRK1	BRICK1, SCAR/WAVE actin nucleating complex subunit	1	NS	0	-	
ARPC1A	Actin related protein 2/3 complex subunit 1A	1	***	2	-	• The Arp2/3 complex is activated by WAVE [84]. • Arp2/3 complex mediates actin branch formation [50].
ARPC1B	Actin related protein 2/3 complex subunit 1B	1	***	4	Up	
ARPC2	Actin related protein 2/3 complex subunit 2	3	***/***/***	1/1/1	Up	
ARPC3	Actin related protein 2/3 complex subunit 3	1	***	1	Up	
ARPC4	Actin related protein 2/3 complex subunit 4	2	*/***	1/1	-	
ARPC5	Actin related protein 2/3 complex subunit 5	4	***/**/***/**	3/1/3/2	Up	
SWAP70	SWAP switching B-cell complex 70kDa subunit	2	***/***	3/3	Up	• SWAP70 regulates cellular actin dynamics and organization [49].
RAB34	RAB34, member RAS oncogene family	2	***/***	4/3	Up	• Rab34 is associated with actin-rich membrane ruffles [51].
CTBP1	C-terminal binding protein 1	4	***/***/NS	3/2/1	Down	• CtBP1 is phosphorylated by PAK1. This phosphorylation is essential for the final fission of macropinosomes from the plasma membrane [9]. • CtBP1/BARS is involved in the activation of PLD1, which is required for the formation and the closure of macropinocytic cups [61,62].
PLD1	Phospholipase D1	6	***/***/**/***/***/***	3/2/2/4/3/3	Down	• PLD1 regulates macropinocytosis at the earlier steps presumably at the formation and the closure of macropinocytic cups with the aid of CTBP1/BARS [61,62].

Table A1. *Cont.*

Gene	Name	Number of Probe	Kruskall - Wallis	2 Tailed	Up or Down Regulation	Role in Macropinocytosis
RAB5A	RAB5A, member RAS oncogene family	3	NS/NS/***	0/0/4	-	• Rab5 is located in the membrane of early endosomes. • It stabilizes macropinosomes [86].
RAB5B	RAB5B, member RAS oncogene family	1	***	3	Down	
RAB5C	RAB5C, member RAS oncogene family	2	NS/**	0/1	Up	
RAB20	RAB20, member RAS oncogene family	1	***	4	Up	• Closed homologs of Rab5 [87]. • Rab21 is colocalized with Rab5 in macropinosome membranes • Rab21 is recruited after Rab5 but before Rab7 [51]. • Rab20 is an intermediate macropinosome marker colocalized with Rab7 and LAMP1 [51,88].
RAB21	RAB21, member RAS oncogene family	4	*/**/*/**	1/2/2/2	?	
RAB7A	RAB7A, member RAS oncogene family	6	**/NS/***/NS/NS/*	2/0/2/0/0/1	-	• Rab7 is a marker of late endosomes [82].
RAB7B	RAB7B, member RAS oncogene family	2	***	1	-	
LAMP1	Lysosomal associated membrane protein 1	3	*/NS/NS	1/0/0	-	• Late endosomal/lysosomal protein [51].
SNX1	Sorting nexin 1	3	NS/***/NS	0/3/0	?	• The sorting nexins are a group of hydrophilic proteins characterized by a PX domain recruited to the endosomal system through binding with phosphoinositides [89]. • SNX are recruited to actin-rich regions of the plasma membrane in response to EGF. • Recruited to discrete subdomains on the cytoplasmic face of the macropinosomes. • SNX5 is transiently involved in microtubule formation and extension for macropinosome traffic to early endosome. • SNX5 activity is restricted to the initial stages of maturation [90,91].
SNX5	Sorting nexin 5	5	***/***/***/***/***	3/4/4/4/3	Up	
RAB11A	RAB11A, member RAS oncogene family	3	***/NS/***	2/0/4	Down	• RAB11 is a marker of endosomes recirculating to the plasma membrane [27].
RAB11B	RAB11B, member RAS oncogene family	2	*/NS	1/0	-	

References

1. Eisele, G.; Weller, M. Targeting apoptosis pathways in glioblastoma. *Cancer Lett.* **2013**, *332*, 335–345. [CrossRef]
2. Lefranc, F.; Le Rhun, E.; Kiss, R.; Weller, M. Glioblastoma quo vadis: Will migration and invasiveness reemerge as therapeutic targets? *Cancer Treat. Rev.* **2018**, *68*, 145–154. [CrossRef] [PubMed]
3. Galluzzi, L.; Bravo-San Pedro, J.M.; Vitale, I.; Aaronson, S.A.; Abrams, J.M.; Adam, D.; Alnemri, E.S.; Altucci, L.; Andrews, D.; Annicchiarico-Petruzzelli, M.; et al. Essential versus accessory aspects of cell death: Recommendations of the NCCD 2015. *Cell Death Differ.* **2015**, *28*, 58–73. [CrossRef] [PubMed]
4. Kornienko, A.; Mathieu, V.; Rastogi, S.K.; Lefranc, F.; Kiss, R. Therapeutic agents triggering nonapoptotic cancer cell death. *J. Med. Chem.* **2013**, *56*, 4823–4839. [CrossRef] [PubMed]
5. Galluzzi, L.; Vitale, I.; Abrams, J.M.; Alnemri, E.S.; Baehrecke, E.H.; Blagosklonny, M.V.; Dawson, T.M.; Dawson, V.L.; El-Deiry, W.S.; Fulda, S.; et al. Molecular definitions of cell death subroutines: Recommendations of the Nomenclature Committee on Cell Death 2012. *Cell Death Differ.* **2012**, *19*, 107–120. [CrossRef] [PubMed]
6. Yan, Y.; Xu, Z.; Dai, S.; Qian, L.; Sun, L.; Gong, Z. Targeting autophagy to sensitive glioma to temozolomide treatment. *J. Exp. Clin. Cancer Res.* **2016**, *35*, 23. [CrossRef] [PubMed]
7. Hart, M.G.; Garside, R.; Rogers, G.; Stein, K.; Grant, R. Temozolomide for high grade glioma. *Cochrane Database Syst. Rev.* **2013**, *4*, CD007415. [CrossRef] [PubMed]
8. Lewis, W.H. Pinocytosis. *Bull. Johns Hopkins Hops.* **1931**, *49*, 17–26.
9. Kerr, M.C.; Teasdale, R.D. Defining macropinocytosis. *Traffic* **2009**, *10*, 364–371. [CrossRef] [PubMed]
10. Maltese, W.A.; Overmeyer, J.H. Non-apoptotic cell death associated with perturbations of macropinocytosis. *Front. Physiol.* **2015**, *6*, 38. [CrossRef] [PubMed]
11. Redelman-Sidi, G.; Binyamin, A.; Gaeta, I.; Palm, W.; Thompson, C.B.; Romesser, P.B.; Lowe, S.W.; Bagul, M.; Doench, J.G.; Root, D.E.; et al. The Canonical Wnt Pathway Drives Macropinocytosis in Cancer. *Cancer Res.* **2018**, *78*, 4658–4670. [CrossRef] [PubMed]
12. Overmeyer, J.H.; Kaul, A.; Johnson, E.E.; Maltese, W.A. Active Ras Triggers Death in Glioblastoma Cells through Hyperstimulation of Macropinocytosis. *Mol. Cancer Res.* **2008**, *6*, 965–977. [CrossRef]
13. Commisso, C.; Davidson, S.M.; Soydaner-Azeloglu, R.G.; Parker, S.J.; Kamphorst, J.J.; Hackett, S.; Grabocka, E.; Nofal, M.; Drebin, J.A.; Thompson, C.B.; et al. Macropinocytosis of protein is an amino acid supply route in Ras-transformed cells. *Nature* **2013**, *497*, 633–637. [CrossRef] [PubMed]
14. Qian, Y.; Wang, X.; Liu, Y.; Li, Y.; Colvin, R.A.; Tong, L.; Wu, S.; Chen, X. Extracellular ATP is internalized by macropinocytosis and induces intracellular ATP increase and drug resistance in cancer cells. *Cancer Lett.* **2014**, *351*, 242–251. [CrossRef]
15. Veithen, A.; Cupers, P.; Baudhuin, P.; Courtoy, P.J. v-Src induces constitutive macropinocytosis in rat fibroblasts. *J. Cell Sci.* **1996**, *109*, 2005–2012.
16. Kasahara, K.; Nakayama, Y.; Sato, I.; Ikeda, K.; Hoshino, M.; Endo, T.; Yamaguchi, N. Role of Src-Family Kinases in Formation and Trafficking of Macropinosomes. *J. Cell. Physiol.* **2007**, *211*, 220–232. [CrossRef] [PubMed]
17. Kodaz, H.; Kostek, O.; Hacioglu, M.B.; Erdogan, B.; Kodaz, C.E.; Hacibekiroglu, I.; Turkmen, E.; Uzunoglu, S.; Cicin, I. Frequency of RAS Mutations (KRAS, NRAS, HRAS) in Human Solid Cancer. *Eurasian J. Med. Oncol.* **2017**, *1*, 1–7. [CrossRef]
18. Lim, J.P.; Gleeson, P.A. Macropinocytosis: An endocytic pathway for internalising large gulps. *Immunol. Cell Biol.* **2011**, *89*, 836–843. [CrossRef]
19. Nazarenko, I.; Hede, S.M.; He, X.; Hedrén, A.; Thompson, J.; Lindström, M.S.; Nistér, M. PDGF and PDGF receptors in glioma. *Ups. J. Med. Sci.* **2012**, *117*, 99–112. [CrossRef] [PubMed]
20. Westphal, M.; Maire, C.L.; Lamszus, K. EGFR as a Target for Glioblastoma Treatment: An Unfulfilled Promise. *CNS Drugs* **2017**, *31*, 723–735. [CrossRef] [PubMed]
21. Quezada, C.; Torres, Á.; Niechi, I.; Uribe, D.; Contreras-Duarte, S.; Toledo, F.; San Martín, R.; Gutiérrez, J.; Sobrevia, L. Role of extracellular vesicles in glioma progression. *Mol. Asp. Med.* **2018**, *60*, 38–51. [CrossRef]
22. Gourlay, J.; Morokoff, A.P.; Luwor, R.B.; Zhu, H.J.; Kaye, A.H.; Stylli, S.S. The emergent role of exosomes in glioma. *J. Clin. Neurosci.* **2017**, *35*, 13–23. [CrossRef]
23. Ha, K.D.; Bidlingmaier, S.M.; Liu, B. Macropinocytosis exploitation by cancers and cancer therapeutics. *Front. Physiol.* **2016**, *7*, 1–10. [CrossRef]

24. Giusti, I.; Francesco, M.; Dolo, V. Extracellular Vesicles in Glioblastoma: Role in Biological Processes and in Therapeutic Applications. *Curr. Cancer Drug Targets* **2017**, *17*, 221–235. [CrossRef]
25. Costa Verdera, H.; Gitz-Francois, J.J.; Schiffelers, R.M.; Vader, P. Cellular uptake of extracellular vesicles is mediated by clathrin-independent endocytosis and macropinocytosis. *J. Control. Release* **2017**, *266*, 100–108. [CrossRef] [PubMed]
26. Maltese, W.A.; Overmeyer, J.H. Methuosis: Nonapoptotic cell death associated with vacuolization of macropinosome and endosome compartments. *Am. J. Pathol.* **2014**, *184*, 1630–1642. [CrossRef] [PubMed]
27. Shubin, A.V.; Demidyuk, I.V.; Komissarov, A.A.; Rafieva, L.M.; Kostrov, S.V. Cytoplasmic vacuolization in cell death and survival. *Oncotarget* **2016**, *7*, 55863–55889. [CrossRef]
28. Trabbic, C.J.; Dietsch, H.M.; Alexander, E.M.; Nagy, P.I.; Robinson, M.W.; Overmeyer, J.H.; Maltese, W.A.; Erhardt, P.W. Differential induction of cytoplasmic vacuolization and methuosis by novel 2-indolyl-substituted pyridinylpropenones. *ACS Med. Chem. Lett.* **2014**, *5*, 73–77. [CrossRef] [PubMed]
29. Robinson, M.W.; Overmeyer, J.H.; Young, A.M.; Erhardt, P.W.; Maltese, W.A. Synthesis and evaluation of indole-based chalcones as inducers of methuosis, a novel type of nonapoptotic cell death. *J. Med. Chem.* **2012**, *55*, 1940–1956. [CrossRef]
30. Rauf, A.; Patel, S.; Imran, M.; Maalik, A.; Arshad, M.U.; Saeed, F.; Mabkhot, Y.N.; Al-Showiman, S.S.; Ahmad, N.; Elsharkawy, E. Honokiol: An anticancer lignan. *Biomed. Pharmacother.* **2018**, *107*, 555–562. [CrossRef]
31. Liou, K.T.; Shen, Y.C.; Chen, C.F.; Tsao, C.M.; Tsai, S.K. The anti-inflammatory effect of honokiol on neutrophils: Mechanisms in the inhibition of reactive oxygen species production. *Eur. J. Pharmacol.* **2003**, *475*, 19–27. [CrossRef]
32. Talarek, S.; Listos, J.; Barreca, D.; Tellone, E.; Sureda, A.; Nabavi, S.F.; Braidy, N.; Nabavi, S.M. Neuroprotective effects of honokiol: From chemistry to medicine. *Biofactors* **2017**, *43*, 760–769. [CrossRef] [PubMed]
33. Uta, R.; François, C.; Marvin, H.; Philipp, A.; Andrea, P.P.; Henrik, W.; Ralph, L. Anti-inflammatory properties of Honokiol in activated primary microglia and astrocytes. *J. Neuroimmunol.* **2018**, *323*, 78–86. [CrossRef]
34. Lee, Y.J.; Lee, Y.M.; Lee, C.K.; Jung, J.K.; Han, S.B.; Hong, J.T. Therapeutic applications of compounds in the Magnolia family. *Pharmacol Ther.* **2011**, *130*, 157–176. [CrossRef] [PubMed]
35. Lin, J.W.; Chen, J.T.; Hong, C.Y.; Lin, Y.L.; Wang, K.T.; Yao, C.J.; Lai, G.M.; Chen, R.M. Honokiol traverses the blood-brain barrier and induces apoptosis of neuroblastoma cells via an intrinsic bax-mitochondrion-cytochrome c-caspase protease pathway. *Neuro Oncol.* **2012**, *14*, 302–314. [CrossRef] [PubMed]
36. Lin, C.J.; Chen, T.L.; Tseng, Y.Y.; Wu, G.J.; Hsieh, M.H.; Lin, Y.W.; Chen, R.M. Honokiol induces autophagic cell death in malignant glioma through reactive oxygen species-mediated regulation of the p53/PI3K/Akt/mTOR signaling pathway. *Toxicol. Appl. Pharmacol.* **2016**, *304*, 59–69. [CrossRef] [PubMed]
37. Shubin, A.V.; Demidyuk, I.V.; Lunina, N.A.; Komissarov, A.A.; Roschina, M.P.; Leonova, O.G.; Kostrov, S.V. Protease 3C of hepatitis A virus induces vacuolization of lysosomal/endosomal organelles and caspase-independent cell death. *BMC Cell Biol.* **2015**, *16*, 1–18. [CrossRef] [PubMed]
38. Recouvreux, M.V.; Commisso, C. Macropinocytosis: A Metabolic Adaptation to nutrient Stress in Cancer. *Front. Endocrinol.* **2017**, *8*, 1–7. [CrossRef]
39. Hanke, J.H.; Gardner, J.P.; Dow, R.L.; Changelian, P.S.; Brissette, W.H.; Weringer, E.J.; Pollok, B.A.; Connelly, P.A. Discovery of a Novel, Potent, and Src Family-selective. *J. Biol. Chem.* **1996**, *271*, 695–701. [CrossRef] [PubMed]
40. Castro-Obregón, S.; Rao, R.V.; Del Rio, G.; Chen, S.F.; Poksay, K.S.; Rabizadeh, S.; Vesce, S.; Zhang, X.K.; Swanson, R.A.; Bredesen, D.E. Alternative, nonapoptotic programmed cell death: Mediation by arrestin 2, ERK2, and Nur77. *J. Biol. Chem.* **2004**, *279*, 17543–17553. [CrossRef]
41. Thomé, M.P.; Filippi-Chiela, E.C.; Villodre, E.S.; Migliavaca, C.B.; Onzi, G.R.; Felipe, K.B.; Lenz, G. Ratiometric analysis of acridine orange staining in the study of acidic organelles and autophagy. *J. Cell Sci.* **2016**, *129*, 4622–4632. [CrossRef] [PubMed]
42. Sperandio, S.; de Belle, I.; Bredesen, D.E. An alternative, nonapoptotic form of programmed cell death. *Proc. Natl. Acad. Sci. USA* **2000**, *97*, 14376–14381. [CrossRef] [PubMed]
43. Wang, Y.; Zhu, X.; Yang, Z.; Zhao, X. Honokiol induces caspase-independent paraptosis via reactive oxygen species production that is accompanied by apoptosis in leukemia cells. *Biochem. Biophys. Res. Commun.* **2013**, *430*, 876–882. [CrossRef] [PubMed]

44. Weller, M.; van den Bent, M.; Hopkins, K.; Tonn, J.C.; Stupp, R.; Falini, A.; Cohen-Jonathan-Moyal, E.; Frappaz, D.; Henriksson, R.; Balana, C.; et al. EANO guideline for the diagnosis and treatment of anaplastic gliomas and glioblastoma. *Lancet Oncol.* **2014**, *15*, 395–403. [CrossRef]
45. Serventi, J.; Behr, J. Surgery and Evidence-based Treatments in Patients with Newly Diagnosed High-grade Glioma. *Semin. Oncol. Nurs.* **2018**, *34*, 443–453. [CrossRef]
46. Kanzawa, T.; Germano, I.M.; Komata, T.; Ito, H.; Kondo, Y.; Kondo, S. Role of autophagy in temozolomide-induced cytotoxicity for malignant glioma cells. *Cell Death Differ.* **2004**, *11*, 448–457. [CrossRef] [PubMed]
47. Ramirez, Y.P.; Weatherbee, J.L.; Wheelhouse, R.T.; Ross, A.H. Glioblastoma multiforme therapy and mechanisms of resistance. *Pharmaceuticals (Basel)* **2013**, *6*, 1475–1506. [CrossRef] [PubMed]
48. López-Ginés, C.; Navarro, L.; Muñoz-Hidalgo, L.; Buso, E.; Morales, J.M.; Gil-Benso, R.; Gregori-Romero, M.; Megías, J.; Roldán, P.; Segura-Sabater, R.; et al. Association between epidermal growth factor receptor amplification and ADP-ribosylation factor 1 methylation in human glioblastoma. *Cell. Oncol.* **2017**, *40*, 389–399. [CrossRef]
49. Oberbanscheidt, P.; Balkow, S.; Kühnl, J.; Grabbe, S.; Bähler, M. SWAP-70 associates transiently with macropinosomes. *Eur. J. Cell Biol.* **2007**, *86*, 13–24. [CrossRef] [PubMed]
50. Smith, B.A.; Padrick, S.B.; Doolittle, L.K.; Daugherty-Clarke, K.; Corrêa, I.R.; Xu, M.Q.; Goode, B.L.; Rosen, M.K.; Gelles, J. Three-color single molecule imaging shows WASP detachment from Arp2/3 complex triggers actin filament branch formation. *Elife* **2013**, *2*, 1–25. [CrossRef] [PubMed]
51. Egami, Y.; Taguchi, T.; Maekawa, M.; Arai, H.; Araki, N. Small GTPases and phosphoinositides in the regulatory mechanisms of macropinosome formation and maturation. *Front. Physiol.* **2014**, *5*, 1–11. [CrossRef] [PubMed]
52. Liu, Z.; Yang, X.; Chen, C.; Liu, B.; Ren, B.; Wang, L.; Zhao, K.; Yu, S.; Ming, H. Expression of the Arp2/3 complex in human gliomas and its role in the migration and invasion of glioma cells. *Oncol. Rep.* **2013**, *30*, 2127–2136. [CrossRef] [PubMed]
53. Mariani, L.; Beaudry, C.; McDonough, W.S.; Hoelzinger, D.B.; Demuth, T.; Ross, K.R.; Berens, T.; Coons, S.W.; Watts, G.; Trent, J.M.; et al. Glioma cell motility is associated with reduced transcription of proapoptotic and proliferation genes: A cDNA microarray analysis. *J. Neurooncol.* **2001**, *53*, 161–176. [CrossRef] [PubMed]
54. Wang, H.J.; Gao, Y.; Chen, L.; Li, Y.L.; Jiang, C.L. RAB34 was a progression- and prognosis-associated biomarker in gliomas. *Tumor Biol.* **2014**, *36*, 1573–1578. [CrossRef] [PubMed]
55. Ge, J.; Chen, Q.; Liu, B.; Wang, L.; Zhang, S.; Ji, B. Knockdown of Rab21 inhibits proliferation and induces apoptosis in human glioma cells. *Cell. Mol. Biol. Lett.* **2017**, *22*, 1–11. [CrossRef] [PubMed]
56. Humphreys, D.; Davidson, A.C.; Hume, P.J.; Makin, L.E.; Koronakis, V. Arf6 coordinates actin assembly through the WAVE complex, a mechanism usurped by Salmonella to invade host cells. *Proc. Natl. Acad. Sci. USA* **2013**, *110*, 16880–16885. [CrossRef]
57. Hu, B.; Shi, B.; Jarzynka, M.J.; Yiin, J.J.; D'Souza-Schorey, C.; Cheng, S.Y. ADP-ribosylation factor 6 regulates glioma cell invasion through the IQ-domain GTPase-activating protein 1-Rac1-mediated pathway. *Cancer Res.* **2009**, *69*, 794–801. [CrossRef]
58. Li, M.; Ng, S.S.M.; Wang, J.; Lai, L.; Leung, S.Y.; Franco, M.; Peng, Y.; He, M.L.; Kung, H.F.; Lin, M.C.M. EFA6A enhances glioma cell invasion through ADP ribosylation factor 6/extracellular signal-regulated kinase signaling. *Cancer Res.* **2006**, *66*, 1583–1590. [CrossRef] [PubMed]
59. Jensen, S.S.; Aaberg-Jessen, C.; Christensen, K.G.; Kristensen, B. Expression of the lysosomal-associated membrane protein-1 (LAMP-1) in astrocytomas. *Int. J. Clin. Exp. Pathol.* **2013**, *6*, 1294–1305. [CrossRef]
60. Sarafian, V.S.; Koev, I.; Mehterov, N.; Kazakova, M.; Dangalov, K. LAMP-1 gene is overexpressed in high grade glioma. *Apmis* **2018**, *126*, 657–662. [CrossRef]
61. Haga, Y.; Miwa, N.; Jahangeer, S.; Okada, T.; Nakamura, S.I. CtBP1/BARS is an activator of phospholipase D1 necessary for agonist-induced macropinocytosis. *EMBO J.* **2009**, *28*, 1197–1207. [CrossRef]
62. Liberali, P.; Kakkonen, E.; Turacchio, G.; Valente, C.; Spaar, A.; Perinetti, G.; Böckmann, R.A.; Corda, D.; Colanzi, A.; Marjomaki, V.; et al. The closure of Pak1-dependent macropinosomes requires the phosphorylation of CtBP1/BARS. *EMBO J.* **2008**, *27*, 970–981. [CrossRef] [PubMed]
63. Liu, B.; Di, G. C-Terminal Binding Protein is Involved in Promoting to the Carcinogenesis of Human Glioma. *Mol. Neurobiol.* **2017**, *54*, 6121–6132. [CrossRef]

64. Zhao, C.; Shen, Y.; Tao, X.; Xu, J.; Lu, J.; Liu, C.; Xu, Z.; Tang, Q.; Tao, T.; Zhang, X. Silencing of CtBP1 suppresses the migration in human glioma cells. *J. Mol. Histol.* **2016**, *47*, 297–304. [CrossRef] [PubMed]
65. Aoki, H.; Yokoyama, T.; Fujiwara, K.; Tari, A.M.; Sawaya, R.; Suki, D.; Hess, K.R.; Aldape, K.D.; Kondo, S.; Kumar, R.; et al. Phosphorylated Pak1 level in the cytoplasm correlates with shorter survival time in patients with glioblastoma. *Clin. Cancer Res.* **2007**, *13*, 6603–6609. [CrossRef] [PubMed]
66. Gilbertson, R.J. Driving glioblastoma to drink. *Cell* **2014**, *157*, 289–290. [CrossRef] [PubMed]
67. Noh, J.; Lee, H.; Song, S.; Kim, N.S.; Im, W.; Kim, M.; Seo, H.; Chung, C.; Chang, J.; Ferrante, R.J.; et al. SCAMP5 Links Endoplasmic Reticulum Stress to the Accumulation of Expanded Polyglutamine Protein Aggregates via Endocytosis Inhibition. *J. Biol. Chem.* **2009**, *284*, 11318–11325. [CrossRef]
68. Gunduz, N.; Ceylan, H.; Guler, M.O.; Tekinay, A.B. Intracellular Accumulation of Gold Nanoparticles Leads to Inhibition of Macropinocytosis to Reduce the Endoplasmic Reticulum Stress. *Sci. Rep.* **2017**, 1–10. [CrossRef] [PubMed]
69. Wu, G.; Lin, C.; Lin, Y. Data in Brief Data analyses of honokiol-induced autophagy of human glioma cells in vitro and in vivo. *Data Brief* **2016**, *9*, 667–672. [CrossRef]
70. Sander, P.; Walther, P.; Moepps, B.; Hinz, M.; Mostafa, H.; Schaefer, P.; Pala, A.; Wirtz, R.; Georgieff, M.; Schneider, M. Mitophagy-related cell death mediated by vacquinol-1 and TRPM7 blockade in glioblastoma IV. *IntechOpen* **2019**, *5*, 81–93. [CrossRef]
71. Florey, O.; Overholtzer, M.; Overholtzer, M. Macropinocytosis and autophagy crosstalk in nutrient scavenging. *Philos. Trans. R. Soc. Lond. B Biol. Sci.* **2019**, *374*. [CrossRef]
72. Gan, H.K.; Reardon, D.A.; Lassman, A.B.; Merrell, R.; Van Den Bent, M.; Butowski, N.; Lwin, Z.; Wheeler, H.; Fichtel, L.; Scott, A.M.; et al. Safety, Pharmacokinetics and Antitumor Response of Depatuxizumab Mafodotin as Monotherapy or in Combination with Temozolomide in Patients with Glioblastoma. *Neuro Oncol.* **2018**, *20*, 838–847. [CrossRef] [PubMed]
73. Lassman, A.B.; Van Den Bent, M.J.; Gan, H.K.; Reardon, D.A.; Kumthekar, P.; Butowski, N.; Lwin, Z.; Mikkelsen, T.; Nabors, L.B.; Kyriakos, P.; et al. Safety and efficacy of depatuxizumab mafodotin + temozolomide in patients with EGFR-amplified, recurrent glioblastoma: Results from an international phase I multicenter trial. *Neuro Oncol.* **2019**, *21*, 106–114. [CrossRef]
74. Wang, X.; Duan, X.; Yang, G.; Zhang, X.; Deng, L.; Zheng, H.; Deng, C.; Wen, J.; Wang, N.; Peng, C.; et al. Honokiol Crosses BBB and BCSFB, and Inhibits Brain Tumor Growth in Rat 9L Intracerebral Gliosarcoma Model and Human U251 Xenograft Glioma Model. *PLoS ONE.* **2011**, *6*, e18490. [CrossRef]
75. Li, Z.; Mbah, N.E.; Overmeyer, J.H.; Sarver, J.G.; George, S.; Trabbic, C.J.; Erhardt, P.W.; Maltese, W.A. The JNK signaling pathway plays a key role in methuosis (non-apoptotic cell death) induced by MOMIPP in glioblastoma. *BMC Cancer* **2019**, *19*, 77. [CrossRef] [PubMed]
76. Ahlstedt, J.; Förnvik, K.; Zolfaghari, S.; Kwak, D.; Lars, G.J.; Ernfors, P.; Salford, L.G.; Redebrandt, H.N. Evaluating vacquinol-1 in rats carrying glioblastoma models RG2 and NS1. *Oncotarget* **2018**, *9*, 8391–8399. [CrossRef] [PubMed]
77. Chio, C.C.; Tai, Y.T.; Mohanraj, M.; Liu, S.H.; Yang, S.T.; Chen, R.M. Honokiol enhances temozolomide-induced apoptotic insults to malignant glioma cells via an intrinsic mitochondrion-dependent pathway. *Phytomedicine* **2018**, *49*, 41–51. [CrossRef] [PubMed]
78. Fan, Y.; Xue, W.; Schachner, M.; Zhao, W. Honokiol Eliminates Glioma/Glioblastoma Stem Cell-Like Cells via JAK-STAT3 Signaling and Inhibits Factor Receptor. *Cancer* **2018**, *11*, 22. [CrossRef] [PubMed]
79. Mosmann, T. Rapid colorimetric assay for cellular growth and survival: Application to proliferation and cytotoxicity assays. *J. Immunol. Methods* **1983**, *65*, 55–63. [CrossRef]
80. Ballesta, A.; Zhou, Q.; Zhang, X.; Lv, H.; Gallo, J. Multiscale Design of Cell-Type–Specific Pharmacokinetic/Pharmacodynamic Models for Personalized Medicine: Application to Temozolomide in Brain Tumors. *CPT Pharmacomet. Syst. Pharmacol.* **2014**, *4*, 1–11. [CrossRef]
81. Sun, L.; Hui, A.; Su, Q.; Vortmeyer, A.; Kotliarov, Y.; Pastorino, S.; Passaniti, A.; Menon, J.; Walling, J.; Bailey, R.; et al. Neuronal and glioma-derived stem cell factor induces angiogenesis within the brain. *Cancer Cell* **2006**, *9*, 287–300. [CrossRef] [PubMed]
82. Sun, P.; Yamamoto, H.; Suetsugu, S.; Miki, H.; Takenawa, T.; Endo, T. Small GTPase Rah/Rab34 is associated with membrane ruffles and macropinosomes and promotes macropinosome formation. *J. Biol. Chem.* **2003**, *278*, 4063–4071. [CrossRef] [PubMed]

83. Bhanot, H.; Young, A.M.; Overmeyer, J.H.; Maltese, W.A. Induction of Nonapoptotic Cell Death by Activated Ras Requires Inverse Regulation of Rac1 and Arf6. *Mol. Cancer Res.* **2010**, *8*, 1358–1374. [CrossRef]
84. Fujii, M.; Kawai, K.; Egami, Y.; Araki, N. Dissecting the roles of Rac1 activation and deactivation in macropinocytosis using microscopic photo-manipulation. *Sci. Rep.* **2013**, *3*, 1–10. [CrossRef] [PubMed]
85. Koronakis, V.; Hume, P.J.; Humphreys, D.; Liu, T.; Horning, O.; Jensen, O.N.; McGhie, E.J. WAVE regulatory complex activation by cooperating GTPases Arf and Rac1. *Proc. Natl. Acad. Sci. USA* **2011**, *108*, 14449–14454. [CrossRef] [PubMed]
86. Feliciano, W.D.; Yoshida, S.; Straight, S.W.; Swanson, J.A. Coordination of the Rab5 Cycle on Macropinosomes. *Traffic* **2011**, *12*, 1911–1922. [CrossRef] [PubMed]
87. Schwartz, S.L.; Cao, C.; Pylypenko, O.; Rak, A.; Wandinger-Ness, A. Rab GTPases at a glance. *J. Cell Sci.* **2008**, *121*, 246. [CrossRef]
88. Egami, Y.; Araki, N. Rab20 regulates phagosome maturation in RAW264 macrophages during Fc gamma receptor-mediated phagocytosis. *PLoS ONE* **2012**, *7*, e35663. [CrossRef]
89. Teasdale, R.D.; Loci, D.; Houghton, F.; Karlsson, L.; Gleeson, P.A. A large family of endosome-localized proteins related to sorting nexin 1. *Biochem. J.* **2001**, *358*, 7–16. [CrossRef]
90. Merino-Trigo, A.; Kerr, M.; Houghton, F.; Lindberg, A.; Mitchell, C.; Teasdale, R.; Gleeson, P. Sorting nexin 5 is localized to a subdomain of the early endosomes and is recruited to the plasma membrane following EGF stimulation. *J. Cell Sci.* **2004**, *117*, 6413–6424. [CrossRef]
91. Kerr, M.C.; Lindsay, M.; Luetterforst, R.; Hamilton, N.; Simpson, F.; Parton, R.; Gleeson, P.; Teasdale, R. Visualisation of macropinosome maturation by the recruitment of sorting nexins. *J. Cell Sci.* **2006**, *119*, 3967–3980. [CrossRef] [PubMed]

cancers

MDPI

Article

Mifepristone Overcomes Tumor Resistance to Temozolomide Associated with DNA Damage Repair and Apoptosis in an Orthotopic Model of Glioblastoma

Monserrat Llaguno-Munive [1,2], Mario Romero-Piña [1], Janeth Serrano-Bello [3], Luis A. Medina [4], Norma Uribe-Uribe [5], Ana Maria Salazar [6], Mauricio Rodríguez-Dorantes [7] and Patricia Garcia-Lopez [1,*]

1 Laboratorio de Farmacología, Subdirección de Investigación Básica, Instituto Nacional de Cancerología, Ciudad de México 14080, Mexico; muniv1250@hotmail.com (M.L.-M.); esau1708@gmail.com (M.R.-P.)
2 Posgrado en Ciencias Biomédicas, Universidad Nacional Autónoma de México (UNAM), Ciudad de México 04510, Mexico
3 Facultad de Odontología, Universidad Nacional Autónoma de México (UNAM), Ciudad de México 04510, Mexico; janserbe@hotmail.com
4 Instituto de Física, Universidad Nacional Autónoma de México (UNAM), Unidad de Investigación Biomédica en Cáncer INCan-UNAM, Ciudad de México, 14080, Mexico; medina@fisica.unam.mx
5 Instituto Nacional de Ciencias Médicas y de la Nutrición Salvador Zubirán, Ciudad de México 14080, Mexico; nofeliauribe@yahoo.com.mx
6 Instituto de Investigaciones Biomédicas, Universidad Nacional Autónoma de México (UNAM), Ciudad de México 04510, Mexico; anamsm@biomedicas.unam.mx
7 Instituto de Medicina Genómica, Ciudad de México 14080, Mexico; mrodriguez@inmegen.gob.mx
* Correspondence: pgarcia_lopez@yahoo.com.mx; Tel.: +52-55-56280400 (ext. 32085)

Received: 10 November 2018; Accepted: 20 December 2018; Published: 22 December 2018

Abstract: The standard treatment for glioblastoma multiforme (GBM) is surgery followed by chemo/radiotherapy. A major limitation on patient improvement is the high resistance of tumors to drug treatment, likely responsible for their subsequent recurrence and rapid progression. Therefore, alternatives to the standard therapy are necessary. The aim of the present study was to evaluate whether mifepristone, an antihormonal agent, has a synergistic effect with temozolomide (used in standard therapy for gliomas). Whereas the mechanism of temozolomide involves damage to tumor DNA leading to apoptosis, tumor resistance is associated with DNA damage repair through the O^6-methylguanine-DNA-methyltransferase (MGMT) enzyme. Temozolomide/mifepristone treatment, herein examined in Wistar rats after orthotopically implanting C6 glioma cells, markedly reduced proliferation. This was evidenced by a decreased level of the following parameters: a proliferation marker (Ki-67), a tumor growth marker (^{18}F-fluorothymidine uptake, determined by PET/CT images), and the MGMT enzyme. Increased apoptosis was detected by the relative expression of related proteins, (e.g. Bcl-2 (B-cell lymphoma 2), Bax (bcl-2-like protein 4) and caspase-3). Thus, greater apoptosis of tumor cells caused by their diminished capacity to repair DNA probably contributed significantly to the enhanced activity of temozolomide. The results suggest that mifepristone could possibly act as a chemo-sensitizing agent for temozolomide during chemotherapy for GBM.

Keywords: glioblastoma; temozolomide; mifepristone; MGMT; drug resistance; apoptosis

1. Introduction

Glioblastoma multiforme (GBM), the most common tumor of the central nervous system, is a highly aggressive cancer with a prognosis of 14.6 months median survival upon diagnosis [1]. Agents for GBM therapy encounter obstacles nonexistent in the treatment of non-neurological cancers. For instance, the blood-brain barrier restricts the passage of many chemotherapeutic agents due to certain well-known characteristics, such as the presence of tight junctions between cells and the lack of fenestrae [2,3].

The current treatment for GBM consists of surgery followed by radiotherapy and chemotherapy with temozolomide. Unfortunately, the great tumor resistance to this drug limits its effectiveness and often leads to recurrence and rapid tumor progression [4]. Subsequent to the oral administration of temozolomide, its spontaneous breakdown at physiological pH produces monomethyltriazene 5-(3-methyltriazen-1-yl)-imidazole-4-carboxamide (MTIC), which in turn reacts with water to liberate 5-aminoimidazole-4-carboxamide (AIC) and the highly reactive methyldiazonium cation. The latter methylates DNA purine residues, preferentially O-6 guanine (O^6-MeG, 6%) in guanine-rich regions [5], but also N-7 guanine (N^7-MeG, 70%) and N-3 adenine (N^3-MeA, 9%). Alkylation of the O^6 site on guanine promotes the insertion of a thymine instead of a cytosine during DNA replication, which can result in cell death [5].

Tumor cells often show chemoresistance to temozolomide, continuing to proliferate after treatment. Drug resistance is due in part to the overexpression of MGMT (O^6-methylguanine-DNA-methyltransferase), an enzyme involved in the repair of temozolomide-induced DNA damage. There is an inverse relationship between the levels of MGMT expression and the cellular response to temozolomide [6]. Hence, MGMT expression is a relevant prognostic factor for the response to temozolomide therapy, with an elevated expression linked to drug resistance [7].

Clinical and experimental data have established the resistance of GBM to apoptosis. Defects in apoptotic mechanisms foster tumorigenesis and contribute to the resistance to temozolomide, since its cytotoxic activity is exerted in part through the triggering of apoptosis [8]. In recent years, new molecules have been developed to attempt to overcome the problem of tumor resistance to temozolomide, but with little impact on the prognosis and survival of GBM patients [9–11]. One strategy for improving GBM treatment is to seek new chemo-sensitizing agents that enhance the effectiveness of temozolomide by increasing the apoptosis of tumor cells through a decrease in the activity of the DNA repair enzyme.

One possibility is mifepristone, which blocks the capacity of progesterone to stimulate the growth, migration and invasion of human astrocytoma cells lines (such as U373, U87 and D54) [12,13]. This synthetic steroid is used as an abortifacient drug because of its anti-progestational and anti-glucocorticoid action. Our group previously demonstrated an important role of mifepristone in chemo-radio-sensitization, describing its synergistic effect with cisplatin and radiotherapy to act against proliferation of cervical cancer cell lines and cervix xenografts [14]. We also studied the addition of mifepristone to temozolomide-based chemotherapy (accompanied by radiation), observing improved antiproliferative activity on a glioma xenograft in the flank of nude mice. The tumor growth rate was slower than that found with radiation alone or temozolomide alone, indicating a chemo- and radio-sensitizing effect [15].

However, the flank tumor model has drawbacks, especially the absence of the environment provided by the normal brain parenchyma and blood brain barrier. Since these factors are known to influence drug delivery to the tumor [16], orthotopic models are preferred. The aim of the present study was to evaluate, with an immunocompetent orthotopic model based on the intracranial implantation of a glioma cell line in rats, whether mifepristone is able to modulate the growth of temozolomide-treated tumors. The possible mechanism of action of mifepristone as a chemo-sensitizing agent was explored by determining the levels of MGMT (a DNA repair enzyme) as well as the expression of proteins as Bcl-2 (B-cell lymphoma 2), Bax (bcl-2-like protein 4) and caspase that participate in the apoptotic pathway.

2. Results

2.1. Tumor Growth Assessed by Molecular Imaging

Rats were subjected to four treatments: the vehicle only, temozolomide only, mifepristone only, and mifepristone/temozolomide. PET/CT scans were performed at baseline (day 0 of drug treatment, 2 weeks post-implantation) and at the end of the experiment (day 21). In the resulting images (Figure 1a), the presence of red reflects the uptake of ^{18}F-fluorothymidine (18F-FLT), a tumor cell growth marker. It is measured as total lesion proliferation (TLP).

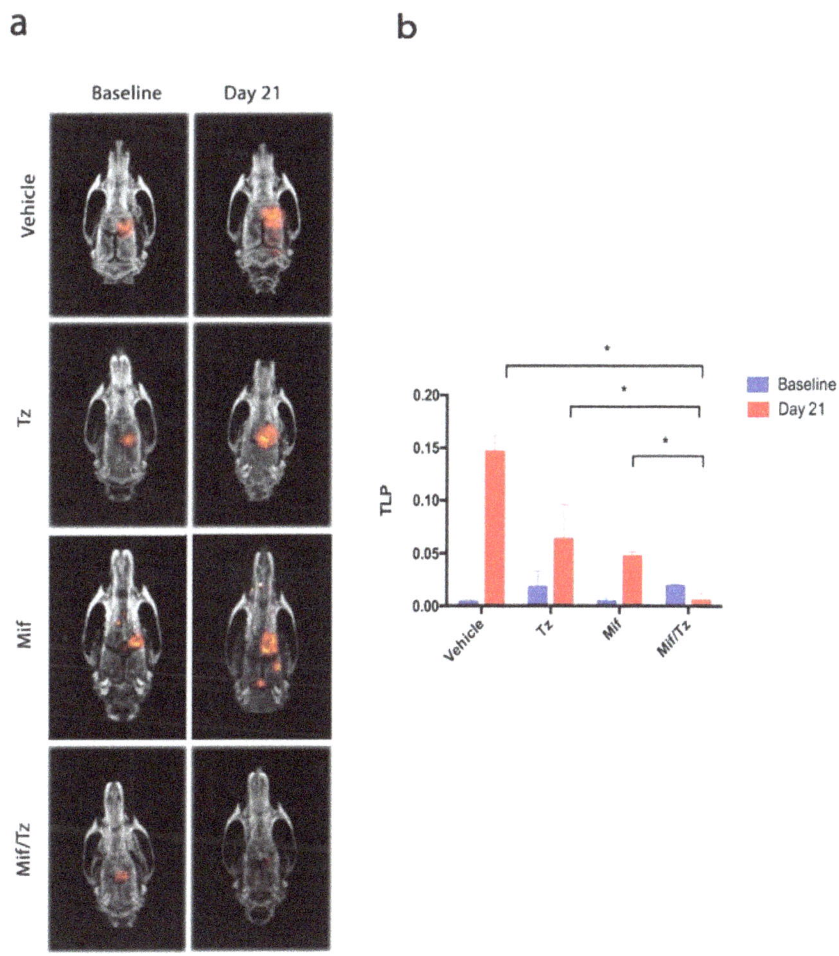

Figure 1. Antitumor activity in the orthotopic rat model of a glioma. (**a**) PET/CT images showing ^{18}F-FLT (^{18}F-fluorothymidine) tumor uptake for the distinct groups. The images on the left represent the beginning of drug treatment (baseline, two weeks after tumor cell implantation, considered as day 0) and those on the right depict the end of the third week of treatment (day 21). Red reflects the uptake of (^{18}F-FLT). (**b**) The proliferative activity of tumors, measured as total lesion proliferation (TLP). Data are expressed as the mean ± SEM of three animals. * A significant difference ($p < 0.05$) between the Mifeprostone/Temozolomide (Mif/Tz) group and the other groups with implantation of glioma cells (the vehicle only, Tz only and Mif only).

After 21 days of treatment, there was an increase in TLP and consequently in the tumor size for the control, temozolomide and mifepristone groups, and a decrease in this parameter for the mifepristone/temozolomide group (Figure 1b). At day 21, both the increase for the vehicle only (control) animals and the decrease for the mifepristone/temozolomide group were significant compared to baseline. Moreover, a significantly lower rate of proliferation was found for the combination treatment than for each of the other three groups (the control, temozolomide only and mifepristone only).

2.2. Determination of Body Weight and Overall Survival

Tumor growth was also assessed by means of weight loss (Figure 2a). All animals continued to gain weight during the first two weeks post-implantation of tumor cells. Subsequently, the rats treated with the vehicle or temozolomide only rapidly lost weight, evidencing accelerated tumor proliferation.

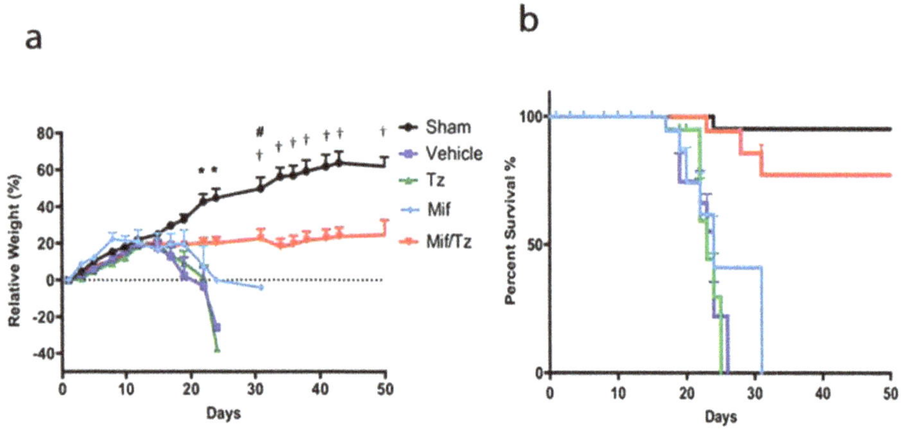

Figure 2. The effect of tumor growth in the orthotopic rat model of a glioma, evaluated by weight loss and overall survival. (**a**) Relative weight of rats in the five groups: sham surgery (without implanting glioma cells) (●), and the implantation of glioma cells followed by each of the treatments: vehicle only (•), temozolomide only (•), mifepristone only (•), and the mifepristone/temozolomide combination (•). Each point represents the mean ± SEM of six animals. * A significant difference ($p < 0.05$) between Mif/Tz and the other groups (Sham, Control, Tz only and Mif only) on day 24. # A significant difference ($p < 0.05$) between Mif/Tz and Mif only on day 31. † A significant difference ($p < 0.05$) between the Mif/Tz and the Sham group on day 31. (**b**) Survival analysis of the same groups for up to 50 days after implantation (day 0 = day of surgery).

The rats in the group given mifepristone only displayed a less drastic weight loss. All animals in these three groups survived between 17 and 32 days (Figure 2b). In contrast, the rats receiving mifepristone/temozolomide maintained their weight throughout the study (Figure 2a), and those having no tumor cell implantation (sham operation) gained weight. The latter two groups were still alive when all the rats of the other three groups had died, and most survived to the end of the study (50 days post-implantation) (Figure 2b).

2.3. Histological and Immunohistochemical Analysis

At the end of the treatments, the brain tissue was processed for histological examination, applying hematoxylin and eosin (H&E) stain to the tissue sections (Figure 3a–e). The tissue of the animals treated with the vehicle only, temozolomide only or mifepristone only showed hypercellularity, mitosis, pleomorphism and necrosis (Figure 3b–d). These characteristics were exhibited to a lesser extent in the rats administered mifepristone/temozolomide (Figure 3e), but with a hypo- versus hypercellular lesion.

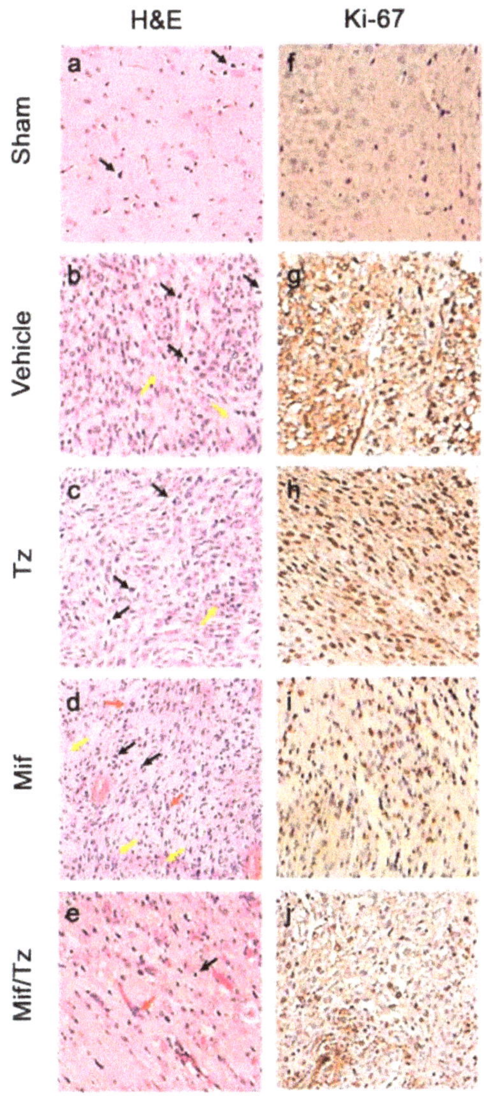

Figure 3. Haematoxylin and eosin (H&E) staining and immunohistochemical analysis of glioma tissue. Sections of tumor tissue were stained with haematoxylin and eosin (**a–e**): Mitosis (black arrows), nuclear pleomorphism (red arrows) and necrosis (yellow arrows). Immunostaining of Ki-67 (**f–j**), showing few cell nuclei positive to this protein in the tumors of the Mif/Tz group compared to those of the individual treatments (Tz or Mif). The images are representative of three animals per treatment. Magnification 40×.

Immunohistochemical staining of brain tumor tissue was employed to detect the presence of Ki-67 (Figure 3f–j), a protein expressed by proliferating cells. Compared to the groups receiving the vehicle only, temozolomide alone or mifepristone alone, rats given mifepristone/temozolomide displayed a lower proportion of Ki-67 positive cells in tumors. Indeed, the stained sections from the latter group were similar to those from the sham animals, indicating few cells undergoing proliferation. For animals administered mifepristone only, there was slightly diminished proliferation compared to those in the control and temozolomide only groups. In the mifepristone/temozolomide group, a close correlation can be observed between a lower level of Ki-67 and the significant reduction in tumor growth illustrated in the PET/CT images.

2.4. Expression of Apoptotic Proteins

The combination treatment also affected the expression of apoptotic proteins in tumor cells (Figure 4). Compared to the control group, animals with temozolomide only or mifepristone only displayed a slight increase in the Bax/Bcl-2 ratio, but no change in the expression of cleaved caspase 3. In animals given mifepristone/temozolomide, contrarily, a significant increase was found in the Bax/Bcl-2 ratio and the expression of cleaved caspase 3, revealing a synergy between these two drugs.

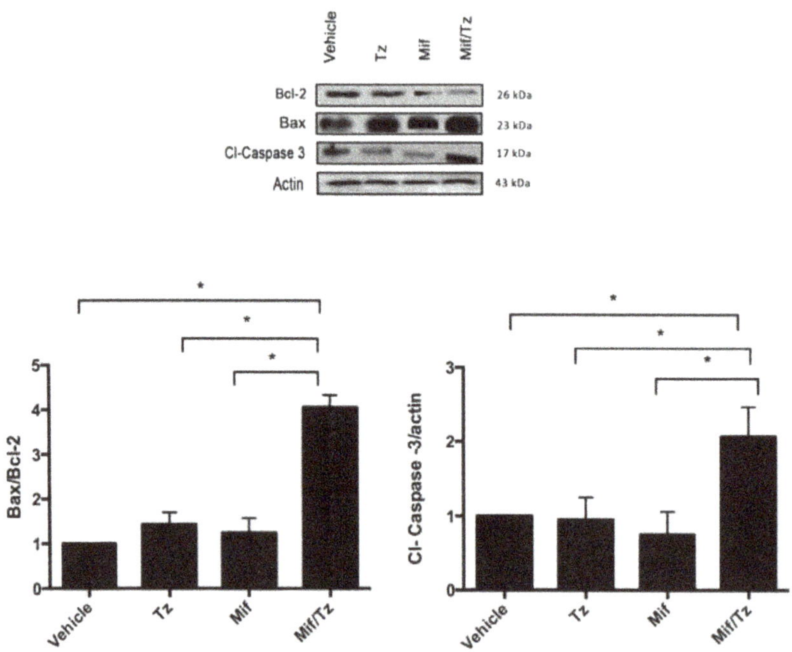

Figure 4. Expression of apoptotic proteins. Representative western blot results and densitometric analysis of Bcl-2, Bax and Cl-Caspase 3. Data are expressed as the mean ± SEM from three independent experiments. * Indicates a significant difference ($p < 0.05$) between Mif/Tz and the other groups with tumor cell implantation (given Tz, Mif or the vehicle).

2.5. Expression of MGMT

MGMT, an enzyme involved in DNA repair, plays a significant role in the resistance of glioma tumors to temozolomide. Western blot data and band intensity showed the protein expression of MGMT to be strikingly downregulated in the mifepristone/temozolomide group and slightly decreased in the mifepristone only group compared to the animals administered temozolomide only (Figure 5).

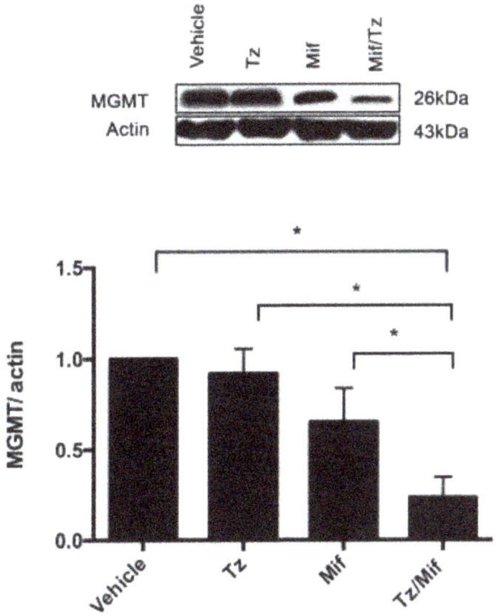

Figure 5. Expression of MGMT (O6-methylguanine-DNA-methyltransferase). Representative western blot results and densitometric analysis of the MGMT protein. Data are expressed as the mean ± SEM of three independent experiments. * A significant difference ($p < 0.05$) between Mif/Tz and the other groups with tumor cell implantation (given Tz, Mif or the vehicle).

3. Discussion

One of the major problems in the treatment of GBM is the drug resistance of tumors and the repopulation carried out by cells that escape chemotherapy. The standard treatment for this disease, consisting of surgery followed by radiation therapy plus chemotherapy with temozolomide, yields a median survival of only 1–2 years. Therefore, it is essential to find new strategies for treating GBM in order to lengthen patient survival and avoid recurrence.

The current study evaluated whether the combination of mifepristone and temozolomide could give a better outcome than temozolomide only applied to glioma tumors in rats. Mifepristone is known to act as an antagonist of progestins, glucocorticoids and androgens by blocking progesterone receptor (PR), glucocorticoid receptor (GR) and androgen receptor (AR), respectively [17,18]. However, the role of this antihormonal agent as a chemo-sensitizer in GBM has scarcely been explored. Although we previously suggested the potential of mifepristone as a chemo-radio-sensitizer for the standard treatment of a glioma tumor based on a subcutaneous xenograft model of GBM [15], the present intracerebral model more closely resembles the native niche of such a tumor in the central nervous system.

A significant difference in tumor growth was observed between the animals administered mifepristone/temozolomide and those given the vehicle only, temozolomide only or mifepristone only. Compared to the latter three groups, the animals submitted to the combination treatment exhibited significantly reduced tumor growth and greater survival time.

The level of Ki-67 was herein scrutinized due to its expression by proliferating cells in all phases of the active cell cycle and its absence in resting cells (G_0). Ki-67 was abundant in the tumor cells of the groups given temozolomide only, mifepristone only or the vehicle only, but was scarcely detected in tumor cells of the animals receiving mifepristone/temozolomide. Hence, the mifepristone/ temozolomide combination blocked the proliferation of glioma cells.

The PET/CT scans, which measured ^{18}F-FLT uptakes, were in agreement with the Ki-67 results, revealing a lack of tumor cell growth with the combined treatment. ^{18}F-FLT displays minimal uptake in normal brain tissue because the expression level of thymidine kinase-1 is very limited in neurons and glia. Thus, ^{18}F-FLT/PET may be advantageous for examining tumor recurrence [19,20].

Mifepristone is known to diminish the proliferation of cancer cells of reproductive and non-reproductive origin [21] and to inhibit the in vitro growth of cancer cells derived from the nervous system, breast, prostate, and ovary. It also gives rise to G1-S blockage of the cell cycle through inhibition of cdk2 activity in human ovarian cancer cells [22].

Whereas progesterone is known to generate the proliferation of tumor cells, the progesterone/mifepristone combination has proven to block the proliferative effect. This combination also significantly decreases the tumor area compared to progesterone only administered to rats [23]. Experimental studies have indicated that progesterone is capable of stimulating the infiltration and migration of astrocytes to the rat cortex, exerting its effects through two mechanisms. The first implies an interaction with nuclear PR (nPR), while the second requires the participation of membrane receptors (mPR) [24]. PR expression exists in several types of brain tumors, including meningiomas and gliomas, and increases with histological malignancy [25]. Moreover, mPRs are expressed in GBM cell lines such as U251 and U87 [26]. Upon evaluating the expression of PRs in C6 glioma cells, Su et al. observed no expression of classical nuclear PRs but identified significant levels of mPRs [27].

Although mifepristone plus temozolomide herein diminished tumor growth, its mechanism of action as a chemo-sensitizing agent is unknown. The promotion of apoptosis by mifepristone has been documented in distinct tumor types. Li et al. demonstrated that this drug produces an antiproliferative effect on human SGC-79901 gastric adenocarcinoma cells by downregulating Bcl-XL expression and upregulating caspase 3 activity [28]. Moreover, mifepristone inhibits cell growth by arresting cell cycle progression at the S phase, triggers apoptosis by activating caspase-3, and modulates the genes involved in apoptosis (including BCL2/BAX and FAS/FASLG) in Ishikawa cells [29]. According to Gonzalez-Agüero et al., progesterone significantly increases the growth of U373 and D54 human astrocytoma cells, an outcome blocked by mifepristone in both cases. Additionally, mifepristone administered without progesterone reduced the growth of these two cell lines [12], evidencing a possible antiproliferative effect on gliomas.

Mifepristone/temozolomide treatment presently exerted a strong downregulation of the expression of the antiapoptotic protein Bcl-2 and an upregulation of the expression of two apoptotic proteins (Bax and cl-caspase-3) (Figure 4). This suggests that mifepristone/temozolomide induces apoptosis in glioblastoma tumors in vivo, which could be relevant for improving the effectiveness of standard therapy for GBM.

On the other hand, the epigenetic silencing of MGMT was shown to enhance the response of GBM patients to chemotherapy based on alkylating agents, leading to longer mean survival. The median survival of patients treated with temozolomide is greater when the promoter of MGMT is hypermethylated [30,31]. Whereas most patients are resistant to temozolomide, epigenetic inactivation of MGMT, occurring in approximately 40% of patients, is associated with a better response. [32]. Therefore, the depletion of MGMT is important for increasing tumor sensitivity to temozolomide.

It is still unknown whether mifepristone epigenetically inhibits MGMT. However, some previously described mechanisms could partially account for the current results. Different nuclear transcription factors, including SP1 (protein 1-binding), AP-1 (activator protein), NF-κB (nuclear factor for the polypeptide gene enhancer in B cells) and HIF-1α (hypoxia inducible factor-1α), can also activate the transcription of the MGMT gene [33]. Some of the aforementioned transcription factors may participate in MGMT gene regulation by mifepristone. It has recently been documented that MGMT expression is associated with SP1 expression in glioma cell lines. Since knockdown of SP1 strongly reduced MGMT protein expression, SP1 is one of the main factors regulating MGMT [34]. MGMT expression also depends on p53, a transcription factor that sequesters SP1 and prevents its binding to MGMT [34]. Additionally, there is evidence of mifepristone increasing apoptosis due to p53 activation in several cancer cell lines [35,36]. Hence, the present findings could be partially explained by some of these mechanisms. Further studies are needed to clearly define the mechanisms responsible for the inhibition of MGMT by mifepristone.

Glucocorticoids are frequently used in glioblastoma therapy to address edema. According to some authors, they are involved in eliciting the expression of the MGMT gene, meaning that they could contribute to an elevated MGMT protein level. Biswas et al. and Ueda et al. detected an upregulation of MGMT in glioblastoma cell lines during glucocorticoid treatment [37,38]. Furthermore, Horiguchi et al. found that dexamethasone acts as a positive regulator of hepatic MGMT expression, an effect reversed by concomitant administration of its antagonist mifepristone [39]. The present results are the first, to our knowledge, in relation to the influence of mifepristone on MGMT expression when administered alone. The current mifepristone only treatment was able to decrease the levels of MGMT compared to the group given the vehicle only and temozolomide only. This could account for the enhancement of sensitivity of tumor cells to temozolomide, observed in the group receiving the mifepristone/temozolomide combination (Figure 5).

A correlation existed between a higher MGMT expression and greater resistance of the tumor to treatment (the latter reflected in the lower survival of the animals). Whereas the rats administered temozolomide only did not survive more than 24 days, those with the mifepristone/temozolomide combination showed a remarkable increase in mean survival. About 75% of the animals receiving the latter treatment survived to the end of the experiment (50 days).

A possible mechanism of mifepristone for improving the efficacy of temozolomide is proposed (Figure 6). Drug resistance to temozolomide is associated with DNA damage repair, which impedes apoptosis. Among the numerous mechanisms utilized by glioma cells to resist temozolomide-induced DNA damage, the avoidance of apoptosis is probably one of the most frequently studied. It has been reported that glioma tumors have elevated levels of the anti-apoptotic proteins Bcl-2 and reduced levels of the apoptotic protein Bax, giving rise to a predisposed anti-apoptotic state correlated with resistance to chemotherapy.

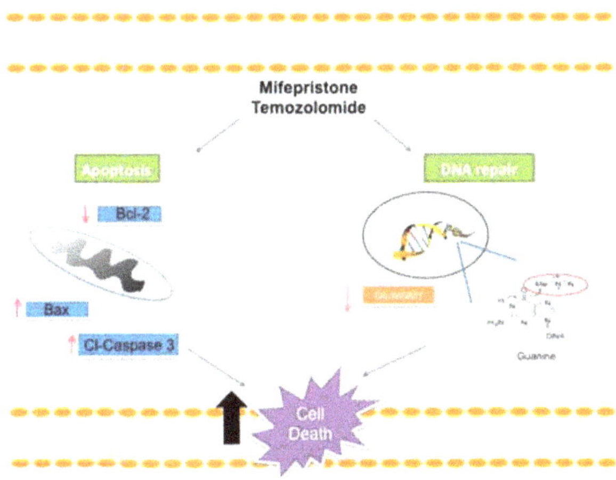

Figure 6. Schematic portrayal of the possible mechanisms of treatment with mifepristone/temozolomide. This combination decreased the level of anti-apoptotic protein Bcl-2 and increased the levels of pro-apoptotic proteins Bax and cl-caspase-3, leading to greater cellular apoptosis. As described elsewhere, temozolomide releases a compound, O^6-guanine, that causes DNA damage and cell death. However, there are different mechanisms of repair or avoidance employed by tumor cells. For example, resistance to temozolomide treatment is reportedly related to the presence of the MGMT enzyme, which removes the methyl group from O^6-guanine and thus restores the cellular replication of tumor cells. A lower expression of MGMT was herein detected when administering mifepristone/temozolomide versus temozolomide alone. The combination treatment perhaps allows methylation to occur in the purine bases of DNA at the O^6 position of guanine without removing the methyl groups, and consequently contributes to an increase in the effect of temozolomide.

Consequently, the inhibition of Bcl-2 through mifepristone treatment, fostering higher levels of Bax and Cl-Caspase, may be an effective strategy for overcoming the resistance of tumors to apoptosis when administering temozolomide. Although new small molecules have been investigated for the inhibition of the Bcl-2 family of proteins in numerous cancer types, their application to GBM has been scarcely explored.

Temozolomide generates numerous DNA adducts (e.g., O^6-methylguanine) that are regarded as very cytotoxic. Contrarily, MGMT removes alkylating adducts from the O^6 position of guanine and thus protects glioma cells from this cytotoxicity. A decrease in the levels of MGMT by mifepristone would impede the repair of temozolomide-induced damage, leading to an eventual accumulation of adducts, a reduction of replication and the triggering of cell death.

Mifepristone has been widely studied as an abortifacient and contraceptive drug. According to recent evidence and the current findings, this drug seems to have great potential for cancer treatment. It offers palliative benefits for a wide variety of human cancer types, including patients with glioblastoma (demonstrating that it successfully crosses the blood-brain barrier) [40,41]. Mifepristone has only a mild adverse effect, even when taken daily for up to 13 years [42]. Therefore, long-term administration of mifepristone may be feasible and clinically well-tolerated in combination with temozolomide to treat patients suffering from a glioma. We propose a Phase I clinical trial to test this combination in the near future.

4. Materials and Methods

4.1. Drugs and Reagents

Mifepristone and temozolomide were obtained from Sigma Chemical Co. (St. Louis, MO, USA). Dulbecco's modified Eagle's medium (DMEM), fetal calf serum (FCS), ethylenediaminetetracetic acid (EDTA) and SDS were purchased from Gibco, BRL (Grand Island, NY, USA). High-quality water for the solutions was processed with a Milli-Q Reagent Water System (Continental Water Systems; El Paso, TX, USA). Temozolomide was prepared in DMSO at a final concentration of 4% and mifepristone was reconstituted in polyethylenglycol and saline solution. All standard solutions were stored at −20 °C until use.

4.2. Animals

Male Wistar rats, (200–230 g) were supplied by the Instituto Nacional de Ciencias Médicas y Nutrición Salvador Zubiran (INCMNSZ), Mexico City, Mexico. All procedures for the care and handling of the animals were approved by the Ethics Committee of the "Instituto Nacional de Cancerología" (INCan, Mexico City, Mexico), and were in accordance with Mexican Federal Regulations for Animal Experimentation and Care (NOM-062-ZOO-1999, Ministry of Agriculture, Mexico).

4.3. Cell Culture

The glioma C6 cell line was acquired from American Type Culture Collection (Rockville, MD, USA). It was routinely maintained as a monolayer in DMEM, supplemented with 5% fetal bovine serum and incubated at 37 °C in 5% CO_2 atmosphere at high humidity. Cells were harvested with 1mM EDTA.

4.4. Tumor Cell Implantation

Posterior to anaesthetization with a combination of tiletamine hydrochloride (10 mg/kg) and acepromazine maleate (0.4 mg/kg) administered s.c., each animal was placed in a stereotactic device for surgery. After fastening the head in the frame, a midline incision was made and bregma was identified. The skull was then drilled at the coordinates of 2.0 mm right from bregma and 6 mm deep (hippocampus), in accordance with the Paxinos and Watson atlas [43].

C6 cells were harvested, washed three times and diluted in DMEM to a concentration of 7.5×10^5 in a volume of 3 μL. Employing an infusion pump and a 27-gauge needle, these cells were slowly implanted at a depth of 6 mm from the dura mater. The injection was made over a 6-min period. Upon closing the scalp, the rat was returned to the animal facility. The rats of the sham group were submitted only to the surgical procedure, without the implantation of tumor cells. Animals were subsequently weighed 3 times/week during the study.

4.5. Treatments

At two weeks post-implantation of tumor cells, the animals were divided into four groups (n = 6–8): (A) temozolomide only (5 mg/kg/day i.p), (B) mifepristone only (10 mg/kg/day s.c.), (C) mifepristone plus temozolomide (the same doses), and (D) control animals (vehicle only). A fifth group consisted of sham animals (inoculated only with DMEM, in the absence of glioma cells). The drug treatments were administered in three cycles of five consecutive days over a period of three weeks. The antitumor activity of temozolomide is schedule-dependent, with multiple administrations being more effective than a single treatment. In clinical practice, the recommended dose of temozolomide is 75–200 mg/m^2 given orally for 5 consecutive days every 28-day cycle (5/28 d) [44,45].

4.6. Tumor Growth Assessed by Molecular Imaging

Brain tumor growth was measured with a microPET/CT scanner (Albira ARS, Oncovision Valencia, Spain). PET/CT images were acquired at 2 weeks post-implantation (day 0 of drug treatment) and subsequently on a weekly basis up to the time of euthanization. 300 μCi of 18 F-FLT was injected into the caudal vein of rats under O_2/isoflurane anesthesia (1–3% isoflurane in 100% oxygen). The images were acquired at 40–60 min post-injection.

4.7. Determination of Body Weight and Overall Survival

Tumor growth was also evaluated by monitoring weight loss during treatments and recording the global survival of animals throughout the study, which lasted 50 days.

4.8. Histology and Immunohistochemistry

At the end of the experiment, all rats were euthanized and perfused with saline solution followed by 4% paraformaldehyde. Brains were removed and immersed in 4% paraformaldehyde for 2 weeks. The brain tissue was embedded in paraffin and sliced into sections (2 mm thick) on the coronal plane. The tissue slices were stained using the H&E method, and cell proliferation examined according to the level of the Ki-67 protein. A section of rat spleen was employed as a positive control.

4.9. Western Blot Analysis

Tissue samples were homogenized with a lysis buffer containing protease inhibitors. Proteins were obtained by centrifugation at 10,000 g and 4 °C, separated on 10% polyacrylamide gel and quantified by bicinchoninic acid assay (BCA). Colored markers (Bio-Rad, Hercules, CA, USA) were included to establish size. Proteins were then electrophoretically transferred from the gel onto PVDF membranes (Amersham, Buckinghamshire, UK), which were blocked with 5% non-fat dry milk at room temperature for 2 h. Membranes were incubated overnight at 4 °C with antibodies against MGMT (sc-166528, 1:500), Bcl-2 (sc-7382, 1:1000), Bax (sc-20067, 1:1000), actin (sc-69879, 1:1000; Santa Cruz Biotechnology, Dallas, TX, USA) and cleaved caspase 3 (9661, 1:1000; Cell Signaling Technology, MAB230, R&D Systems, Minneapolis, MN, USA). After washing, the membranes were incubated with anti-mouse or anti-rabbit secondary antibodies, IgG-HRP (1:1500 and 1:3000, respectively; Santa Cruz Biotechnology) for 1 h.

Blots were developed with a chemiluminescent substrate (ECL Prime). Chemiluminescent bands were developed on Kodak autoradiography film in a darkroom and their densities were measured on Image Studio software (version, 5.2 Li-cor, Lincoln, NE, USA). In each figure, representative blot images were selected from the same gel.

4.10. Statistical Analysis

Values are expressed as the mean ± SEM (standard error of the mean). Statistical analysis was performed with one-way analysis of variance (ANOVA) on SPSS Base 20.0 software (SPSS Inc., Chicago, IL, USA). Differences were scrutinized with multiple comparisons between groups. When necessary, the comparison of means was Bonferroni adjusted. In all cases, a value of $p < 0.05$ was considered significant.

5. Conclusions

The present results indicate that mifepristone was capable of enhancing the inhibitory effect of temozolomide on the proliferation of glioma tumors in an orthotopic rat model of glioblastoma. Hence, this antihormonal drug could possibly be beneficial as a sensitizer for temozolomide in standard therapy for GBM. Since standard GBM treatment is of limited effectiveness, evidenced by low patient survival, the current findings are encouraging and suggest the importance of a Phase I clinical study to further explore the combination of mifepristone and temozolomide.

Author Contributions: M.L.-M. participated in the experimental procedures for tumor cell implantation, in data processing and the analysis of the results. J.S.-B. took part in experimental procedures, while M.R.-P. and L.A.M. carried out the evaluation of tumor growth by molecular imaging. N.U.-U. contributed to histology and immunohistochemistry. A.M.S. and M.R.-D. designed the Western blot experiments. P.G.-L. planned and supervised the study, coordinated all activities, and wrote the manuscript. All authors read and approved the final version of the manuscript.

Funding: This work was partially financed by CONACYT (México) through grant CB-258823.

Acknowledgments: Monserrat Llaguno Munive, a doctoral student of the Programa de Doctorado en Ciencias Biomédicas, Universidad Nacional Autónoma de México (UNAM), is grateful for having received a fellowship 330137 from CONACYT. We thank Bruce Allan Larsen for proofreading the manuscript.

Conflicts of Interest: The authors declare no conflict of interest.

References

1. Johnson, D.R.; O'Neill, B.P. Glioblastoma survival in the United States before and during the temozolomide era. *J. Neurooncol.* **2012**, *107*, 359–364. [CrossRef]
2. Cardoso, F.L.; Brites, D.; Brito, M.A. Looking at the blood-brain barrier: Molecular anatomy and possible investigation approaches. *Brain Res. Rev.* **2010**, *64*, 328–363. [CrossRef]
3. van Tellingen, O.; Yetkin-Arik, B.; de Gooijer, M.C.; Wesseling, P.; Wurdinger, T.; de Vries, H.E. Overcoming the blood-brain tumor barrier for effective glioblastoma treatment. *Drug Resist. Updat.* **2015**, *19*, 1–12. [CrossRef] [PubMed]
4. Ramirez, Y.P.; Weatherbee, J.L.; Wheelhouse, R.T.; Ross, A.H. Glioblastoma multiforme therapy and mechanisms of resistance. *Pharmaceuticals* **2013**, *6*, 1475–1506. [CrossRef]
5. Denny, B.J.; Wheelhouse, R.T.; Stevens, M.F.; Tsang, L.L.; Slack, J.A. NMR and molecular modeling investigation of the mechanism of activation of the antitumor drug temozolomide and its interaction with DNA. *Biochemistry* **1994**, *33*, 9045–9051. [CrossRef] [PubMed]
6. Kitange, G.J.; Carlson, B.L.; Mladek, A.C.; Decker, P.A.; Schroeder, M.A.; Wu, W.; Grogan, P.T.; Giannini, C.; Ballman, K.V.; Buckner, J.C.; et al. Evaluation of MGMT promoter methylation status and correlation with temozolomide response in orthotopic glioblastoma xenograft model. *J. Neurooncol.* **2009**, *92*, 23–31. [CrossRef] [PubMed]
7. Weller, M.; Stupp, R.; Reifenberger, G.; Brandes, A.A.; van den Bent, M.J.; Wick, W.; Hegi, M.E. MGMT promoter methylation in malignant gliomas: Ready for personalized medicine? *Nat. Rev. Neurol.* **2010**, *6*, 39–51. [CrossRef] [PubMed]
8. Lefranc, F.; Rynkowski, M.; DeWitte, O.; Kiss, R. Present and potential future adjuvant issues in high-grade astrocytic glioma treatment. *Adv. Tech. Stand Neurosurg.* **2009**, *34*, 3–35.

9. Belhadj, Z.; Zhan, C.; Ying, M.; Wei, X.; Xie, C.; Yan, Z.; Lu, W. Multifunctional targeted liposomal drug delivery for efficient glioblastoma treatment. *Oncotarget* **2017**, *8*, 66889–66900. [CrossRef] [PubMed]
10. Darmon, I.; Morisse, M.C.; Coutte, A.; Blonski, M.; Le Rhun, E.; Taillandier, L.; Roufai, D.B.; Desenclos, C.; Trudel, S.; Faivre, J.C.; et al. Temozolomide and Bevacizumab Induction before Chemoradiotherapy in Patients with Bulky Glioblastoma and/or with Severe Neurological Impairment. *J. Cancer* **2017**, *8*, 1417–1424. [CrossRef]
11. Toms, S.A.; Tapinos, N. Recent Advances in the Treatment of Gliomas—Comprehensive Brain Tumor Center. *R. I. Med. J.* **2017**, *100*, 43–46.
12. Gonzalez-Aguero, G.; Gutierrez, A.A.; Gonzalez-Espinosa, D.; Solano, J.D.; Morales, R.; Gonzalez-Arenas, A.; Cabrera-Munoz, E.; Camacho-Arroyo, I. Progesterone effects on cell growth of U373 and D54 human astrocytoma cell lines. *Endocrine* **2007**, *32*, 129–135. [CrossRef] [PubMed]
13. Pina-Medina, A.G.; Hansberg-Pastor, V.; Gonzalez-Arenas, A.; Cerbon, M.; Camacho-Arroyo, I. Progesterone promotes cell migration, invasion and cofilin activation in human astrocytoma cells. *Steroids* **2016**, *105*, 19–25. [CrossRef] [PubMed]
14. Segovia-Mendoza, M.; Jurado, R.; Mir, R.; Medina, L.A.; Prado-Garcia, H.; Garcia-Lopez, P. Antihormonal agents as a strategy to improve the effect of chemo-radiation in cervical cancer: In vitro and in vivo study. *BMC Cancer* **2015**, *15*, 21. [CrossRef] [PubMed]
15. Llaguno-Munive, M.; Medina, L.A.; Jurado, R.; Romero-Pina, M.; Garcia-Lopez, P. Mifepristone improves chemo-radiation response in glioblastoma xenografts. *Cancer Cell Int.* **2013**, *13*, e29. [CrossRef] [PubMed]
16. Khawli, L.A.; Prabhu, S. Drug delivery across the blood-brain barrier. *Mol. Pharm.* **2013**, *10*, 1471–1472. [CrossRef] [PubMed]
17. Im, A.; Appleman, L.J. Mifepristone: Pharmacology and clinical impact in reproductive medicine, endocrinology and oncology. *Expert Opin. Pharmacother.* **2010**, *11*, 481–488. [CrossRef]
18. Spitz, I.M.; Bardin, C.W. Mifepristone (RU 486)—A modulator of progestin and glucocorticoid action. *N. Engl. J. Med.* **1993**, *329*, 404–412.
19. Schwarzenberg, J.; Czernin, J.; Cloughesy, T.F.; Ellingson, B.M.; Pope, W.B.; Geist, C.; Dahlbom, M.; Silverman, D.H.; Satyamurthy, N.; Phelps, M.E.; et al. 3′-deoxy-3′-18F-fluorothymidine PET and MRI for early survival predictions in patients with recurrent malignant glioma treated with bevacizumab. *J. Nucl. Med.* **2012**, *53*, 29–36. [CrossRef]
20. Moonshi, S.S.; Bejot, R.; Atcha, Z.; Vijayaragavan, V.; Bhakoo, K.K.; Goggi, J.L. A comparison of PET imaging agents for the assessment of therapy efficacy in a rodent model of glioma. *Am. J. Nucl. Med. Mol. Imaging* **2013**, *3*, 397–407.
21. Tieszen, C.R.; Goyeneche, A.A.; Brandhagen, B.N.; Ortbahn, C.T.; Telleria, C.M. Antiprogestin mifepristone inhibits the growth of cancer cells of reproductive and non-reproductive origin regardless of progesterone receptor expression. *BMC Cancer* **2011**, *11*, e207. [CrossRef] [PubMed]
22. Goyeneche, A.A.; Caron, R.W.; Telleria, C.M. Mifepristone inhibits ovarian cancer cell growth in vitro and in vivo. *Clin. Cancer Res.* **2007**, *13*, 3370–3379. [CrossRef] [PubMed]
23. German-Castelan, L.; Manjarrez-Marmolejo, J.; Gonzalez-Arenas, A.; Gonzalez-Moran, M.G.; Camacho-Arroyo, I. Progesterone induces the growth and infiltration of human astrocytoma cells implanted in the cerebral cortex of the rat. *Biomed. Res. Int.* **2014**, *2014*, e393174. [CrossRef] [PubMed]
24. Cabrera-Munoz, E.; Gonzalez-Arenas, A.; Saqui-Salces, M.; Camacho, J.; Larrea, F.; Garcia-Becerra, R.; Camacho-Arroyo, I. Regulation of progesterone receptor isoforms content in human astrocytoma cell lines. *J. Steroid Biochem. Mol. Biol.* **2009**, *113*, 80–84. [CrossRef] [PubMed]
25. Tavares, C.B.; Gomes-Braga, F.d.C.S.A.; Costa-Silva, D.R.; Escórcio-Dourado, C.S.; Borges, U.S.; Conde, A.M., Jr.; Barros-Oliveira, M.d.C.; Sousa, E.B.; Barros, L.d.R.; Martins, L.M.; et al. Expression of estrogen and progesterone receptors in astrocytomas: A literature review. *Clinics* **2016**, *71*, 481–486. [CrossRef]
26. Valadez-Cosmes, P.; German-Castelan, L.; Gonzalez-Arenas, A.; Velasco-Velazquez, M.A.; Hansberg-Pastor, V.; Camacho-Arroyo, I. Expression and hormonal regulation of membrane progesterone receptors in human astrocytoma cells. *J. Steroid Biochem. Mol. Biol.* **2015**, *154*, 176–185. [CrossRef] [PubMed]
27. Su, C.; Cunningham, R.L.; Rybalchenko, N.; Singh, M. Progesterone increases the release of brain-derived neurotrophic factor from glia via progesterone receptor membrane component 1 (Pgrmc1)-dependent ERK5 signaling. *Endocrinology* **2012**, *153*, 4389–4400. [CrossRef] [PubMed]

28. Li, D.Q.; Wang, Z.B.; Bai, J.; Zhao, J.; Wang, Y.; Hu, K.; Du, Y.H. Effects of mifepristone on invasive and metastatic potential of human gastric adenocarcinoma cell line MKN-45 in vitro and in vivo. *World J. Gastroenterol.* **2004**, *10*, 1726–1729. [CrossRef]
29. Li, A.; Felix, J.C.; Minoo, P.; Amezcua, C.A.; Jain, J.K. Effect of mifepristone on proliferation and apoptosis of Ishikawa endometrial adenocarcinoma cells. *Fertil. Steril.* **2005**, *84*, 202–211. [CrossRef]
30. Hegi, M.E.; Diserens, A.C.; Gorlia, T.; Hamou, M.F.; de Tribolet, N.; Weller, M.; Kros, J.M.; Hainfellner, J.A.; Mason, W.; Mariani, L.; et al. MGMT gene silencing and benefit from temozolomide in glioblastoma. *N. Engl. J. Med.* **2005**, *352*, 997–1003. [CrossRef]
31. Binabaj, M.M.; Bahrami, A.; ShahidSales, S.; Joodi, M.; Joudi Mashhad, M.; Hassanian, S.M.; Anvari, K.; Avan, A. The prognostic value of MGMT promoter methylation in glioblastoma: A meta-analysis of clinical trials. *J. Cell. Physiol.* **2018**, *233*, 378–386. [CrossRef] [PubMed]
32. Paz, M.F.; Yaya-Tur, R.; Rojas-Marcos, I.; Reynes, G.; Pollan, M.; Aguirre-Cruz, L.; Garcia-Lopez, J.L.; Piquer, J.; Safont, M.J.; Balana, C.; et al. CpG island hypermethylation of the DNA repair enzyme methyltransferase predicts response to temozolomide in primary gliomas. *Clin. Cancer Res.* **2004**, *10*, 4933–4938. [CrossRef] [PubMed]
33. Cabrini, G.; Fabbri, E.; Lo Nigro, C.; Dechecchi, M.C.; Gambari, R. Regulation of expression of O6-methylguanine-DNA methyltransferase and the treatment of glioblastoma (Review). *Int. J. Oncol.* **2015**, *47*, 417–428. [CrossRef] [PubMed]
34. Aasland, D.; Reich, T.R.; Tomicic, M.T.; Switzeny, O.J.; Kaina, B.; Christmann, M. Repair gene O (6)-methylguanine-DNA methyltransferase is controlled by SP1 and up-regulated by glucocorticoids, but not by temozolomide and radiation. *J. Neurochem.* **2018**, *144*, 139–151. [CrossRef] [PubMed]
35. Navo, M.A.; Smith, J.A.; Gaikwad, A.; Burke, T.; Brown, J.; Ramondetta, L.M. In vitro evaluation of the growth inhibition and apoptosis effect of mifepristone (RU486) in human Ishikawa and HEC1A endometrial cancer cell lines. *Cancer Chemother Pharmacol.* **2008**, *62*, 483–489. [CrossRef] [PubMed]
36. Li, C.; Ye, H. Mifepristone sensitizing cisplatin for cervical adenocarcinoma HeLa cell sensitivity to chemotherapy and its mechanism. *Eur. J. Gynaecol. Oncol.* **2013**, *34*, 142–147.
37. Biswas, T.; Ramana, C.V.; Srinivasan, G.; Boldogh, I.; Hazra, T.K.; Chen, Z.; Tano, K.; Thompson, E.B.; Mitra, S. Activation of human O6-methylguanine-DNA methyltransferase gene by glucocorticoid hormone. *Oncogene* **1999**, *18*, 525–532. [CrossRef]
38. Ueda, S.; Mineta, T.; Nakahara, Y.; Okamoto, H.; Shiraishi, T.; Tabuchi, K. Induction of the DNA repair gene O6-methylguanine-DNA methyltransferase by dexamethasone in glioblastomas. *J. Neurosurg.* **2004**, *101*, 659–663. [CrossRef]
39. Horiguchi, M.; Kim, J.; Matsunaga, N.; Kaji, H.; Egawa, T.; Makino, K.; Koyanagi, S.; Ohdo, S. Glucocorticoid-dependent expression of O (6)-methylguanine-DNA methyltransferase gene modulates dacarbazine-induced hepatotoxicity in mice. *J. Pharmacol. Exp. Ther.* **2010**, *333*, 782–787. [CrossRef]
40. Check, J.H.; Wilson, C.; Cohen, R.; Sarumi, M. Evidence that Mifepristone, a progesterone receptor antagonist, can cross the blood brain barrier and provide palliative benefits for glioblastoma multiforme grade IV. *Anticancer Res.* **2014**, *34*, 2385–2388. [CrossRef]
41. Check, J.H.; Check, D.; Cohen, R.; Sarumi, M. Mifepristone causing complete remission of rapidly advancing leukemia with measurement of progesterone-induced blocking factor. *Anticancer Res.* **2014**, *34*, 2413–2416. [PubMed]
42. Grunberg, S.M.; Weiss, M.H.; Russell, C.A.; Spitz, I.M.; Ahmadi, J.; Sadun, A.; Sitruk-Ware, R. Long-term administration of mifepristone (RU486): Clinical tolerance during extended treatment of meningioma. *Cancer Investig.* **2006**, *24*, 727–733. [CrossRef] [PubMed]
43. Paxinos, G.; Watson, C. *The Rat Brain in Stereotaxic Coordinates*, 4th ed.; Academic Press: New York, NY, USA, 1998.

44. Lee, S.Y. Temozolomide resistance in glioblastoma multiforme. *Genes Dis.* **2016**, *3*, 198–210. [CrossRef] [PubMed]
45. Nagane, M. Dose-dense temozolomide: Is it still promising? *Neurol. Med. Chir.* **2015**, *55*, 38–49. [CrossRef] [PubMed]

cancers

MDPI

Article

Intratumoural Heterogeneity Underlies Distinct Therapy Responses and Treatment Resistance in Glioblastoma

Seçkin Akgül [1,2,†], Ann-Marie Patch [3,†], Rochelle C.J. D'Souza [1], Pamela Mukhopadhyay [3], Katia Nones [3], Sarah Kempe [3], Stephen H. Kazakoff [3], Rosalind L. Jeffree [4], Brett W. Stringer [1], John V. Pearson [3], Nicola Waddell [3,‡] and Bryan W. Day [1,5,*,‡]

1 Cell and Molecular Biology Department, QIMR Berghofer Medical Research Institute, Brisbane 4006, QLD, Australia; Seckin.Akgul@qimrberghofer.edu.au (S.A.); Rochelle.D'Souza@qimrberghofer.edu.au (R.C.J.D.); brett.w.stringer@gmail.com (B.W.S.)

2 School of Medicine, Griffith University, Gold Coast 4215, QLD, Australia

3 Genetics and Computational Biology Department, QIMR Berghofer Medical Research Institute, Brisbane 4006, QLD, Australia; Ann-Marie.Patch@qimrberghofer.edu.au (A.-M.P.); Pamela.Mukhopadhyay@qimrberghofer.edu.au (P.M.); Katia.Nones@qimrberghofer.edu.au (K.N.); Sarah.Kempe@qimrberghofer.edu.au (S.K.); Stephen.Kazakoff@qimrberghofer.edu.au (S.H.K.); John.Pearson@qimrberghofer.edu.au (J.V.P); Nic.Waddell@qimrberghofer.edu.au (N.W.)

4 Department of Neurosurgery, Royal Brisbane and Women's Hospital, Brisbane 4006, QLD, Australia; Lindy.Jeffree@health.qld.gov.au

5 School of Biomedical Sciences, Faculty of Health, Queensland University of Technology, Brisbane 4059, QLD, Australia

* Correspondence: Bryan.Day@qimrberghofer.edu.au; Tel.: +61-738-453-885

† These authors contributed equally to the work.

‡ Co-senior authors.

Received: 24 December 2018; Accepted: 2 February 2019; Published: 6 February 2019

Abstract: Glioblastomas are the most common and lethal neoplasms of the central nervous system. Neighbouring glioma cells maintain extreme degrees of genetic and phenotypic variation that form intratumoural heterogeneity. This genetic diversity allows the most adaptive tumour clones to develop treatment resistance, ultimately leading to disease recurrence. We aimed to model this phenomenon and test the effectiveness of several targeted therapeutic interventions to overcome therapy resistance. Heterogeneous tumour masses were first deconstructed into single tumour cells, which were expanded independently as single-cell clones. Single nucleotide polymorphism arrays, whole-genome and RNA sequencing, and CpG methylation analysis validated the unique molecular profile of each tumour clone, which displayed distinct pathologic features, including cell morphology, growth rate, and resistance to temozolomide and ionizing radiation. We also identified variable sensitivities to AURK, CDK, and EGFR inhibitors which were consistent with the heterogeneous molecular alterations that each clone harboured. These targeted therapies effectively eliminated the temozolomide- and/or irradiation-resistant clones and also parental polyclonal cells. Our findings indicate that polyclonal tumours create a dynamic environment that consists of diverse tumour elements and treatment responses. Designing targeted therapies based on a range of molecular profiles can be a more effective strategy to eradicate treatment resistance, recurrence, and metastasis.

Keywords: glioblastoma; intratumoural heterogeneity; tumour resistance; personalised therapy; targeted therapy; combination therapy; drug screens

1. Introduction

Glioblastoma remains an aggressive disease despite decades of comprehensive research. The overall and progression-free survival has not changed since the introduction of concomitant and adjuvant radiotherapy and chemotherapy in the form of temozolomide (TMZ) in 2005 [1,2]. This poor outcome is partially linked to its original name, "glioblastoma multiforme", as glioblastoma displays extreme degrees of genetic and phenotypic variation. This intertumoural heterogeneity has been evidenced by the characterisation of numerous tumour samples that showed distinct and unique molecular profiles, including gene mutations, chromosomal alterations, copy number variations, epigenetic alterations, and protein or RNA expression signatures [3–7]. Thus, the classification of glioblastoma into different subgroups based on their molecular profiles has been suggested to simplify the complexity of the disease, and also facilitate personalised treatment strategies [8]. However, recent studies have identified another phenomenon: intratumoural heterogeneity, which refers to individual cells/clones within a single tumour mass that are associated with different molecular characteristics [9]. Intratumoural heterogeneity poses additional challenges, as current pathological and molecular diagnostics are based only on a small region of the tumour, which may not be representative of the tumour mass as a whole [9].

Initial attempts to better understand glioblastoma intratumoural heterogeneity was undertaken on spatially distinct regions of a single tumour, and analysing these distinct tumour fragments revealed both unique and common copy number alterations in different fragments [10]. Subsequent studies extended these findings and reported a wide spectrum of transcriptional programs, cell states (differentiation vs. stemness), and proliferation capacity between individual tumour cells/clones isolated from the same tumour mass [11,12]. Clonal diversity was further demonstrated by the discovery of a mosaic and mutually exclusive amplification of three receptor tyrosine kinase (RTK) genes in intermingled subpopulations of glioblastoma cells. The heterogeneous but co-existing distribution of these *EGFR*, *MET*, and *PDGFRA* gene amplifications suggests that glioblastoma may undergo a dynamic evolution during tumour progression that creates diversity within a single mass [13]. Importantly, the involvement of multiple kinases in tumour development raises the question of whether therapies or a combination of therapies targeting multiple oncogenic signals are needed to eradicate the whole tumour mass.

Intratumoural heterogeneity arises by the continuous acquisition of molecular alterations during tumour progression. As tumour growth proceeds, individual cells and clones persistently compete for nutrient, oxygen and space within the tumour microenvironment. In this selective environment, clones evolve and acquire alterations that enable them to survive and proliferate, essentially becoming dominant subclones while others either perish or remain quiescent [14]. Current treatment, including chemotherapy and radiotherapy, also provides strong selective pressures which trigger clonal evolution responses. Although treatment induces death in a significant proportion of the tumour, surviving cells acquire new alterations, becoming resistant to therapy and enabling tumour recurrence [14,15]. In support of this notion, it has been found that the mutation rate (mutation per megabase) in low-grade gliomas increased from (0.2–4.5) to (31.9–90.9) when they relapse as secondary glioblastomas. Importantly, >98.7% of these alterations have been associated with TMZ treatment and did not exist in the pre-treatment primary tumours. Thousands of de novo mutations and novel oncogenic signatures observed in these TMZ-resistant clones suggest that tumours branch out into new molecular profiles and evolve into even more malignant states after treatment [16].

It is therefore imperative to capture and recapitulate the ever-fluctuating intratumoural heterogeneity in order to fully comprehend the complex biology of glioblastoma. Furthermore, designing and testing rationalised therapeutic interventions in consideration of this phenomenon has important clinical implications. Here, we show that individual tumour clones have a wide range of genetic and biological features which ultimately determine their response to several clinically relevant compounds. Our results shed further light on the complexity and heterogeneity present within glioblastoma tumours and highlight that, despite this diversity, both treatment-resistant and sensitive clones can be effectively targeted. These findings may help to inform future clinical trial development to overcome tumour heterogeneity to improve clinical outcomes for glioblastoma sufferers.

2. Results

2.1. Single-Cell Clonal Model Development to Assess Intratumoural Heterogeneity in Glioblastoma

To model intratumoural heterogeneity, we developed a three-step approach (Figure 1A). Firstly, we prepared a polyclonal primary cell line from patient-derived tumour tissue. We then deconstructed this polyclonal cell line into individual cells and established single-cell clones grown under serum-free conditions. The passage number of the clones was kept to a minimum to reduce culture induced alterations. Next, we undertook a number of genomics analyses, including Single nucleotide polymorphism (SNP) arrays, RNA sequencing, and whole genome sequencing (WGS), allowing us to profile each clone in great detail. Secondly, we analysed the biological response of the clones to the clinical standards of care by treatment with TMZ and ionizing radiation (IR). This allowed us to identify a number of treatment sensitive and resistant clones. Lastly, we used our detailed knowledge of the clones to guide our treatment decisions to rationally target and eliminate resistant tumour cell populations. These three steps were achieved in a structured workflow (Figure 1B).

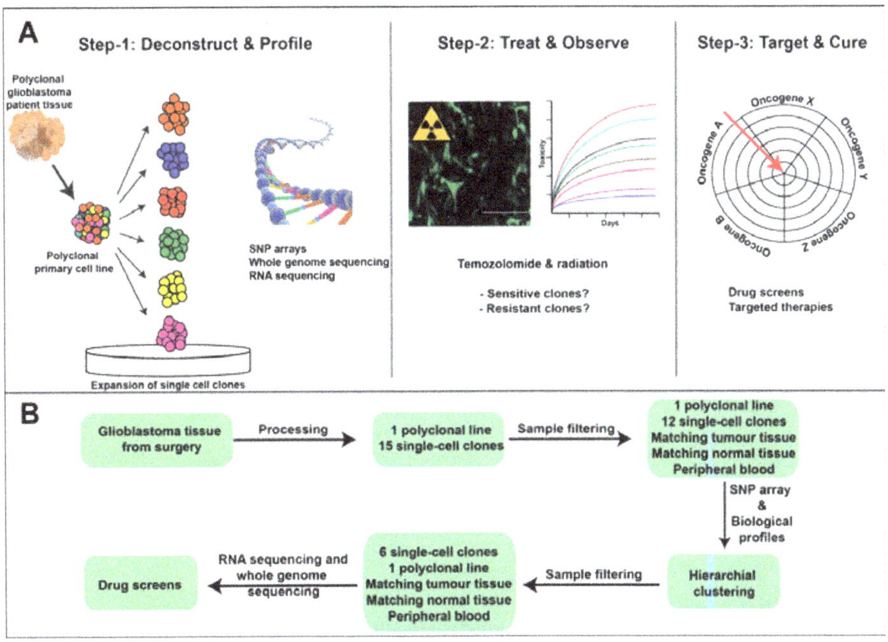

Figure 1. Modelling tumour heterogeneity. (**A**) Schematic representation of the three main steps used to investigate intratumoural heterogeneity in glioblastoma. Step 1, deconstructing a polyclonal tumour into single-cells that were expanded clonally prior to comprehensive genomics analyses. Step 2, assessment of individual clones' responses to the current standards of care, including temozolomide (TMZ) and IR. Step 3, drug screen development and rationale target validation. (**B**) A schema of the workflow used to undertake this study.

2.2. Single-Cell Clones Exhibit Unique Molecular Relationships with a Spectrum of Growth Rates

The copy number events in each sample, which were assessed by SNP array, were used to elucidate the clonal relationships between samples. Genotype analysis confirmed a high degree of similarity between all samples, with a concordance of $\geq$0.954 from pairwise comparisons of >400K polymorphic positions. The most variable cross sample LogR ratio values (the total normalised signal intensity of the two alleles), which indicate the copy number status of each array probe, were analysed using unsupervised hierarchical clustering to generate a relationship diagram (Figure 2A). The clustering indicated two major groups with all bulk tumour and matched normal samples in the first and cell lines in the second that contained three further identifiable subgroups of tumour clones (Figure 2A, yellow boxes).

The proliferation rate of each tumour clone was assessed for 12 single-cell clones and the polyclonal line to determine whether the observed molecular clonal SNP array variances correlated with mitogenic potential (Figure 2B). Interestingly, we observed a wide spectrum of growth rates among the tumour clones. For instance, while clone-F reached confluency in approximately 5 days, clone-B did not reach 50% confluency even after 8 days when the assay was ended. Other clones fell between these two extremes, which led to three main proliferation categories: high, medium, and low (Figure 2B). Using this relatively straight-forward two-parameter approach, using the SNP array and proliferative potential, we were able to separate tumour clones from each other.

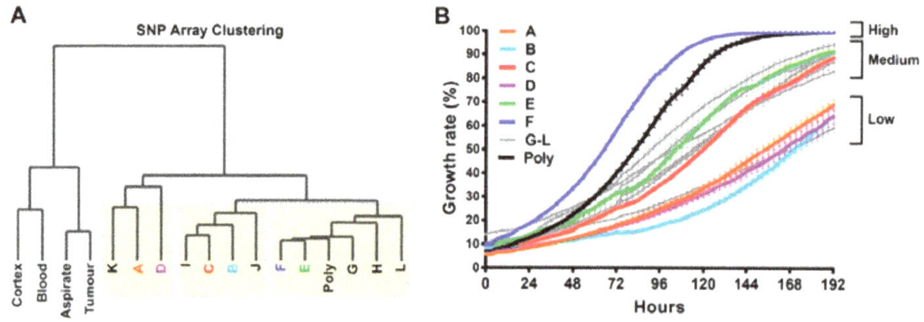

Figure 2. Single-cell clones exhibit unique molecular relationships and growth rates. (**A**) single nucleotide polymorphism (SNP) array-based dendrogram of 12 single-cell clones (A–L), parental polyclonal line (Poly), normal cortex tissue (Cortex), peripheral blood (Blood), aspirated tumour tissue (Aspirate), and surgically resected tumour tissue (Tumour). Three subgroups were found among the tumour clones, which are indicated with yellow boxes. (**B**) Growth rates of the 12 single-cell clones (A–L) and polyclonal cell line were recorded every 3 h for a period of 8 days using an IncuCyte instrument. The cell growth was categorised as high, medium or low. Growth curves for each single-cell clone are labelled with a different colour, except for the grey curves, which represent the clones that were not subjected to further comprehensive genomics analyses.

2.3. Comprehensive Genomic Analyses Identify Relationships between Single-Cell Clones

Based on the SNP array clustering and growth characteristics, we selected six clones (clones A–F) to further model intratumoural heterogeneity. These six clones were distributed within the three main arms of the SNP array cluster and covered the spectrum of observed growth rates (Figure 2A,B). We performed comprehensive genomic analyses for these selected clones, including whole genome sequencing, RNA sequencing, and CpG methylation profiling (Figure 3).

Genomic analysis focusing on the somatic differences between the clones revealed heterogeneity across all data types. Somatic DNA changes were catalogued for each sample, including copy number, structural variants (SVs), single nucleotide variants (SNVs) and small insertions/deletions (indels). Comparative analysis of the SNVs indicated that 3490 out of 8696 (40.14%, Supplementary Table S1 and Figure S1A–C) were shared by all tumour and clonal samples. These shared variants are likely to have occurred early on in the development of this cancer and are present in most cells as trunk variants. Genomic heterogeneity between the cell line clones was apparent as a further 3492 of 8696 (40.16%) SNVs were identified to be unique to one sample, with clone-A containing the most (1345) unique SNVs (Supplementary Table S1).

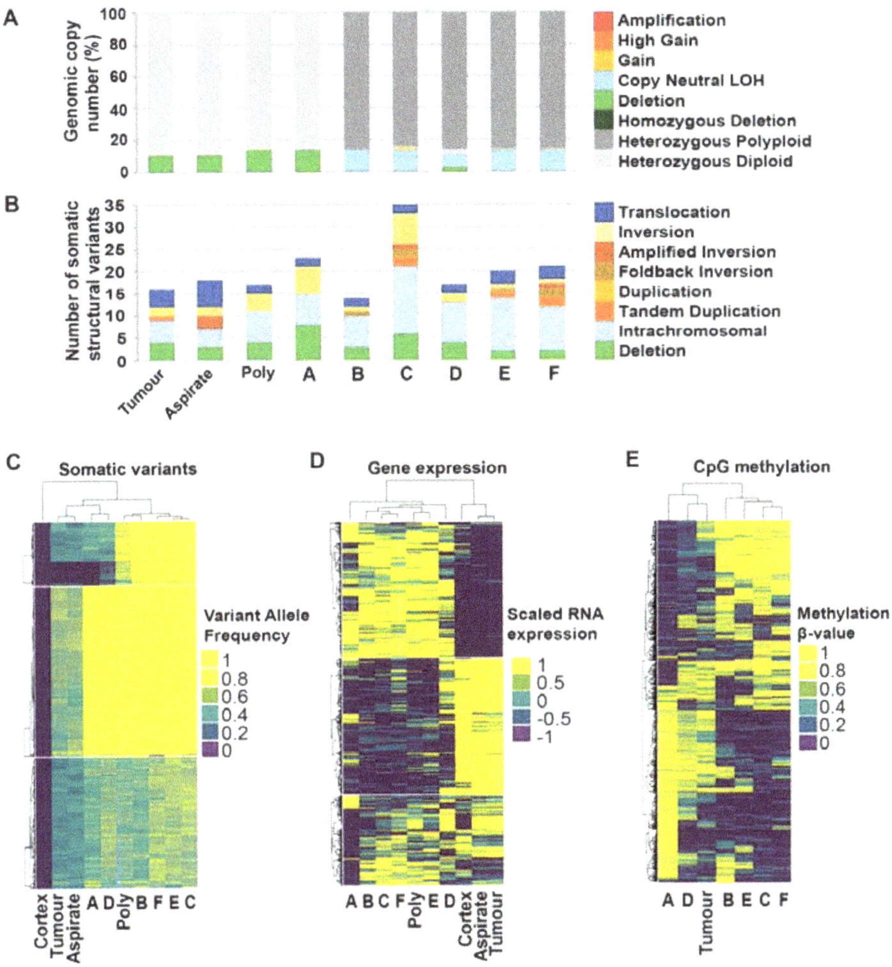

Figure 3. Comprehensive genomic analyses of single-cell clones. (**A**) Genomic proportion of copy number alteration. For clonal samples B–F, >84% of their genome is between copy number 3 and 4 (ranging from 84.82% to 85.9%, shown as dark grey bars) as compared to ≤3% for the other samples. This global whole genome duplication typically affected both alleles, preserving heterozygosity in these samples. (**B**) Number of somatic structural variants identified per sample with the type indicated by colours. (**C**) Hierarchical clustering of somatic substitution variant allele frequencies across tumour and clonal samples. The absence of a variant in the normal control cortex sample is demonstrated by a variant allele frequency of zero (dark blue). (**D**) Hierarchical clustering of Log2-normalised and gene-scaled RNA seq gene expression of genes with most variable expression as identified as contributing to the first principle component. Positive values indicate a sample with the highest expression and negative values with the lowest expression. (**E**) Hierarchical clustering of 1000 most variable β-values from DNA methylation sequencing data. B-value of 1 indicates completely methylated and 0 unmethylated CpGs.

The copy number data indicated that five of six clonal samples (B–F) had undergone whole genome duplication (WGD). For these samples, >84% of their genome was copy number 3 and 4 (range 84.82% to 85.9%) as compared to ≤3% for the other samples (Figure 3A and Supplementary Table S2). The WGD event increased copies of both alleles, and therefore it is difficult to assign a functional consequence at a gene level to such a global event. Therefore, a WGD correction was applied for these samples that revealed that all samples were relatively copy-number quiet, with <20% of the genome judged to be copy-number aberrant (i.e., not diploid or polyploidy heterozygous, Figure 3A).

Large structural rearrangements (SVs) were detectable in all samples at an average of 20.11 SVs (range 14 to 35 SVs) (Figure 3B, Supplementary Table S3 and Figure S1D–F). Of the 83 SVs detected, just 10 SVs (12.05%) were identified as early trunk events and were shared by all the samples. The majority 65 of 83 SVs (75.31%) were unique to individual samples. Clone-C harboured the highest number of somatic structural variants (35 SVs) with more than double the number of events in the bulk tumour sample (16 SVs). The predicted consequences for all rearrangements indicated that 43 genes were likely to have been disrupted, resulting in loss of function for those genes, none of which affected known cancer driver genes. There were three inter-gene fusion events identified, but none included known cancer driver genes or targets for current therapies (Supplementary Table S3).

In order to investigate the molecular similarity and relationships between the tumour and clonal samples, unsupervised hierarchical clustering of somatic substitution variant allele frequency (Supplementary Table S1), gene expression analysis (Supplementary Table S4), and CpG β-value methylation analysis (Supplementary Table S5) was carried out. All three analyses produced a consistent overall indication of the relationships between the samples. The clustering of the variant allele frequency of 8696 somatic substitutions grouped the bulk tumour samples with clones A and D and the remaining clones together (Figure 3C). Gene expression profiling revealed a similar pattern with clones A and D, which were most different from the other clones and the polyclonal cell line sample (Figure 3D). The clustering of the CpG methylation β-values from just the cell line and tumour sample grouped clones A and D together with the bulk tumour, separate from the other clones (Figure 3E).

Clones A and D had a reduced growth rate as compared to the other clones (Figure 2A,B). Therefore, we sought to identify somatic variants in genes associated with cell population proliferation (using gene ontology term GO:0008283) that were particular to clones A and D to explain the differences in growth rate (Supplementary Tables S3 and S6). Potentially interesting variants such as a missense variant in the G protein-coupled receptor *SMO* (chr7:128850945 G>A) and an SV affecting a putative tumour suppressor *CHD5* (predicted loss of function) were identified. However, the in silico prediction of the effect of the variants on the resulting proteins remains unclear, and further functional analysis would be required to assess the impact of these somatic variants on the growth rate of the clonal cell lines.

RNAseq data was used to classify samples into glioblastoma molecular subtypes [4,17,18]. Tumour and aspirate samples were more similar to the classical subtype, whereas all the tumour cultured clones, including the parental polyclonal line, displayed a mesenchymal phenotype (Figure S2A). This is consistent with previous observations suggesting that culture conditions can influence subtype identification [18–20]. Clones D and F, despite being of a predominantly mesenchymal subtype, also displayed strong elements associated with a classical subtype signature. These data suggested that clones of multiple subtypes were isolated from the single tumour mass.

RNAseq data was also used to analyse the expression of known glioblastoma-associated stem cell genes. Tumour clones, in general, showed higher stem cell gene expression compared to the parental tumour, aspirate and normal cortex tissue (Figure S2B). This was most likely a cause of the serum-free culture conditions used, which select for a more de-differentiated stem cell-like population [21]. We were unable to find a correlation between stem cell marker expression and the proliferation of the clones or response to TMZ and/or IR. These results imply that a more complex network between genes and signalling pathways controls key biological parameters, including proliferation and drug resistance.

2.4. Single-Cell Clones Respond Differently to the Current Standards of Care

We next sought to determine the responses of single-cell clones to the current standards of care used in the clinical management of glioblastoma, including TMZ and IR. To this end, we cultured single-cell clones and the polyclonal line in the presence of increasing TMZ concentrations. We detected a wide spectrum of cell viability in TMZ-treated tumour clones, particularly at lower concentrations. Interestingly, most of the clones were more resistant than the parental polyclonal line to TMZ treatment (Figure 4A). The clones also displayed distinct responses to IR, and encouragingly the polyclonal line seemed to respond approximately in the middle of IR response curves (Figure 4B). As expected, we observed better responses when TMZ and IR were combined; however, these responses were still heterogeneous (Figure 4C).

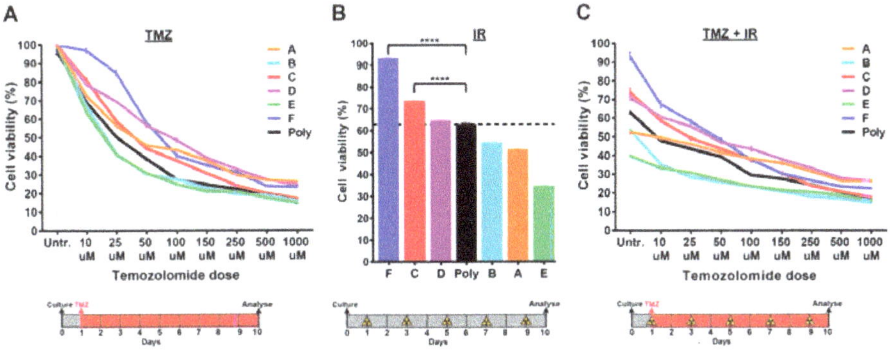

Figure 4. Single-cell clones respond differently to TMZ and/or IR treatments. (**A**) Single-cell clones were grown in the presence of varying TMZ dosages for 10 days. (**B**) A total of 10 Gy IR (2 Gy every other day) was applied to single-cell clones which were analysed at the end of the 10-day period. (**C**) The combination therapy of 10 Gy IR and varying dosages of TMZ on single-cell clones. Cell viability was measured by the MTT cell proliferation assay at day 10. Treatment and analysis protocols are shown under each graph. The unpaired t-test was used for statistical comparisons. **** = p-value < 0.0001.

Taken together, these results highlight the level of heterogeneity present at both the molecular and biological level, particularly with respect to response to the current standards of care. Of particular interest were clones which appear to be the most treatment-resistant and could potentially be responsible for tumour recurrence.

2.5. Genomic Analyses Can Be Used to Select Rationalised Therapies

To identify key oncogenic signals that could be targeted in a rationalised manner, we further investigated the somatic mutation, copy number status, and gene expression of known glioblastoma-associated candidate genes and related oncogenic pathways [3].

An analysis of somatic mutations revealed a homozygous transcript damaging *PTEN* splice site variant (chr10-89685268-A-G, Supplementary Table S6) reported as pathogenic in ClinVar and detectable in all tumour samples. The *PTEN* mutation was biallelic due to somatic loss or the copy-neutral loss of heterozygosity (LOH) of the loci.

Chromosome 10 (chr10) was consistently copy-number aberrant, present at copy number 1 or copy-neutral LOH across all tumour samples, indicating an early loss of the normal allele in the development of this tumour. A similar loss was consistently identified for chr13 and chr22 (Figure S3A) with chr16 altered in most cell line samples. Genes that were affected by copy number events included homozygous loss of *CDKN2A* (chr9) and broad LOH affecting *RB1* (chr13). In addition, the focal high gain of the *MYC* locus (chr8) with at least 6 copies was identified in clones C and F and amplification of *EGFR* (chr7) was identified at over 35 copies in the bulk tumour samples but none of the derived clonal samples (Figures 5A and S3B).

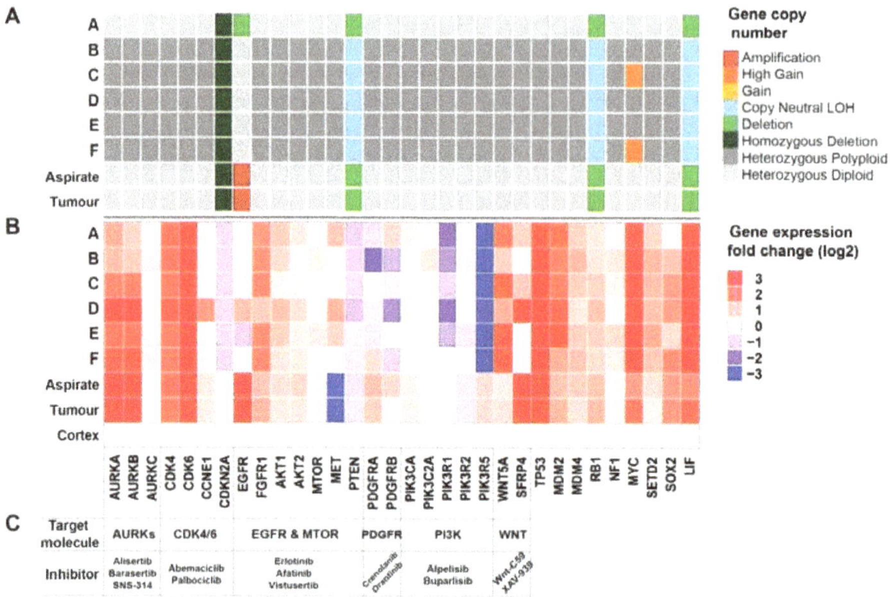

Figure 5. Comprehensive genomics analyses identified targetable oncogenes. (**A**) Gene copy number analysis of tumour clones based on whole genome sequencing. (**B**) Gene expression fold change across the tumour clones along with the aspirated tumour tissue (Aspirate), and surgically resected tumour tissue (Tumour). Log2 expression values were normalised to normal cortex tissue (Cortex). (**C**) Target molecules and their inhibitors are identified based on gene copy number and expression changes.

Gene expression levels measured relative to the adjacent normal cortex sample revealed numerous oncogenic driver genes with >2-fold higher expression in the tumour and clone samples, including genes encoding *AURKA/B*, *CDK4/6*, *FGFR1*, *AKT1/2*, *MTOR*, *MET*, *PIK3C2A*, *WNT5A*, *SFRP4*, and *MYC* (Figures 5B and S3C).

2.6. Drug Screens Identify Unique Sensitivities of Distinct Tumour Cells

Based on the genomic analyses, we created a drug panel which included inhibitors for AURKs, CDK4/6, EGFR, mTOR, PDGFR, PI3K, and WNT (Figure 5C). Except for mTOR, each of these candidate genes was targeted by two or more compounds in consideration of different drug efficacies. We started our screen with a constant drug concentration (5 μM) across the panel (Figure 6A). Tumour clones responded differently to each compound, and all drugs increased their effect with time (3 days versus 6 days). Overall, 8/14 compounds were effective for at least one clone and Buparlisib (BKM120), one of the most extensively tested PI3K inhibitors, was the most effective compound in our drug screen, as cell viability decreased dramatically with this treatment for all the clones (Figure 6A). Similar to Buparlisib, the mTOR inhibitor Vistusertib was effective in reducing the cell viability for most of the clones (Figure 6A). These findings are consistent with the previous studies showing that the PI3K/AKT/mTOR axis is very critical for glioma development [22]. We also observed distinct responses to certain drugs. For instance, the CDK4/6 inhibitor Abemaciclib was profoundly effective on clone-B and clone-C as shown by the almost 90% toxicity in these particular clones. The PDGFR inhibitor Crenolanib could also target these two clones specifically. We also detected unique drug resistances evidenced by the high viability of clone-A and clone-D in the presence of AURK inhibitors Alisertib and Barasertib. Furthermore, only clone-A showed noticeable resistance to EGFR inhibitor Afatinib. None of the Wnt inhibitors showed anti-tumorigenic effects on the six clones. Similar but less pronounced results were found when these drugs were used at a lower dose of 1 μM (Figure S4).

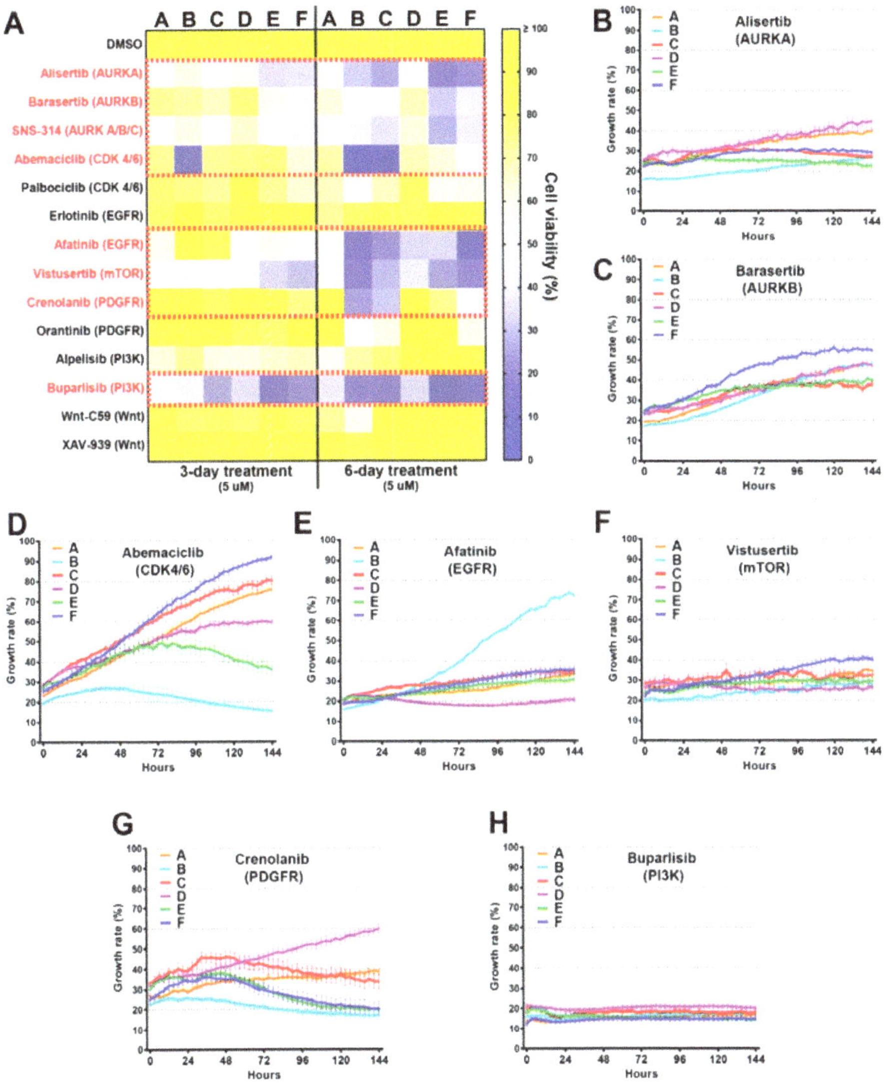

Figure 6. Drug screens identify unique sensitivities of tumour cells. (**A**) Cells were treated with 14 different compounds at 5 µM concentration and cell viability was tested 3 or 6 days after the treatment using MTT assay. Red dashed squares indicate the drugs that are most effective. (**B–H**) Effective seven drugs were tested at 5 µM concentration and the cell growth was examined using an IncuCyte instrument, which automatically took the pictures of the cells every 3 h for a period of 6 days.

To understand the real-time impact of the effective drugs on the growth of individual clones, we used a complimentary approach with IncuCyte. Accordingly, cell images were captured every 3 h for a period of 6 days after which growth rates were quantitated (Figure 6B–H). These results were mostly in concordance with our previous findings and suggested that tumour clones have shared and distinct sensitivities to specific compounds. For instance, clone-B was identified as a resistant clone to the EGFR inhibitor Afatinib; however, this clone showed the most sensitivity to the CDK4/6 inhibitor Abemaciclib (Figure 6E,D). Consistent with our previous findings, Vistusertib and Buparlisib were very effective in reducing growth rate for all tumour clones (Figure 6F,H). More importantly, we were able to target the IR and/or TMZ-resistant clones (e.g., clones C, D, and F,) effectively with Alisertib, Afatinib, and Vistusertib (Figure 6B,E,F).

Once effective drugs were identified, we tested whether these compounds could inhibit tumour cell growth at lower doses either as single agents or in combination. We chose to test these agents using a chimeric model, where all six clones were combined in equal parts prior to analysis. Interestingly, four of the compounds, Alisertib, Barasertib, SNS-314, and Abemaciclib significantly reduced cell growth when applied at 250 nM (Figure 7A). When the concentration was doubled to 500 nM, Afatinib also proved effective (Figure 7B). We then asked whether there could be a synergistic or additive effect of combining the most effective drugs. To this end, we designed combination therapies using a mixture of two (250 nM of each drug) or four (125 nM of each drug) drugs. Compared to the single-regimen treatment, none of the two-drug treatments showed a significant additive effect. However, the four-drug combination was more effective than both single or two-drug regimens (Figure 7C). Overall, these results suggested that TMZ and/or IR-resistant tumour cells can be targeted using specific combination therapies.

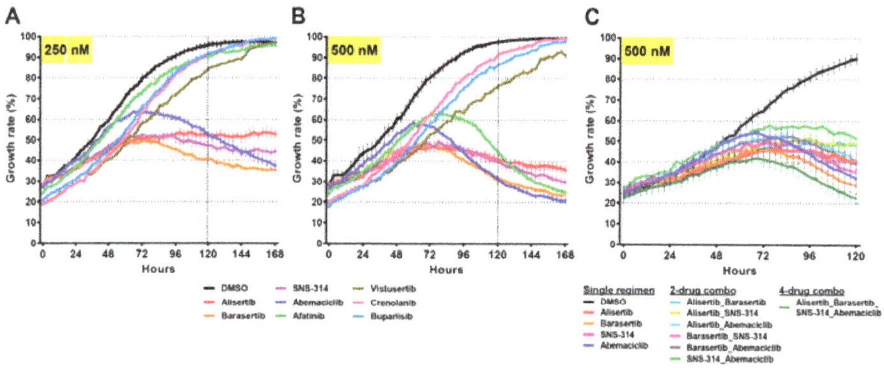

Figure 7. Combination therapies inhibit tumour growth. The 8 most effective drugs were tested at (**A**) 250 nM concentration on a mixed-tumour clone and (**B**) 500 nM concentration on mixed-tumour clones. (**C**) Combination therapies were designed using two drugs (2-drug combo, 250 nM each) or four drugs (4-drug combo, 125 nM each) in comparison to single drugs (single regimen) at a total of 500 nM concentration. Cell growth rates were examined using an IncuCyte instrument. Vertical dashed lines in A and B indicates day 5.

3. Discussion

It is well established that several cancers are characterised by multiple genomic, transcriptomic, and epigenetic alterations that accumulate during disease progression. These alterations lead to deregulation in various genes and networks, allowing tumours to become more malignant and resistant to therapy [23,24]. However, recent studies have also demonstrated that the reversal of only one or a few oncogenic pathways could cease tumour progression or even eliminate the tumour mass as a whole [25]. For instance, several clinical investigations in chronic myelogenous leukaemia, melanoma, and non-small-cell lung cancer suggested that identifying the unique sensitivities of tumour cells is an effective strategy in the treatment of cancer [24]. This phenomenon, described as oncogene addiction, is a characteristic whereby tumour cells become dependent on a single oncogenic pathway in order to survive [26,27]. Therefore, the identification of the unique 'Achilles' heel' of some cancer types is vital in determining the most beneficial therapy to improve patient outcomes [27].

Unfortunately, glioblastoma patients have benefited minimally from such targeted therapies. For instance, targeting the PI3K/AKT pathway alone or in combination with radiotherapy led to either varying or short-lived responses [28]. As a result, many glioblastoma patients relapse with a form of disease resistant to current therapies. This could in part be due to a lack of patient stratification to identify those patients who could benefit from particular therapies and also varying drug responses as a result of intratumoural heterogeneity [14,15].

We demonstrated intratumoural heterogeneity in glioblastoma using three genomic datasets and by analysing biological responses, including growth and response to the current standards of care. Although this heterogeneity between tumour clones was evident, these events were not found to affect the coding sequences of known cancer driver genes. Therefore, the clonal variation in the expression of the key candidate genes is likely due to genetic and/or epigenetic variants in regulatory regions. Further exploration of the non-coding variation may reveal underlying mechanisms of expression heterogeneity in these samples. We also detected whole genome doubling (WGD) in most of the tumour clones. The timing of this event is not completely clear. It is possible that cells from clones B, C, E, and F may have undergone WGD but were not detected in bulk tumours due to their low sub-clonal proportion. It is also possible that the WGD arose in cell culturing. Nevertheless, WGD in cancer has a suggested role in increasing tolerance to chromosome instability and genomic evolution [29], which may accelerate intratumoral heterogeneity.

Intratumoural heterogeneity suggests that despite the common presence of well-characterised oncogenic signals that can be targeted, some tumour cells may not necessarily show addiction to these oncogenic pathways [9]. This phenomenon is exemplified in our study, where almost all the clones analysed showed sensitivity to the EGFR inhibitor Afatinib; however, clone-B, which was slow-growing, was also highly resistant to Afatinib therapy. This finding may provide some explanation for the limited efficacy of Afatinib as a single agent in unselected patients with recurrent glioblastoma [30]. Similarly, only a small percentage (~17%) of patients benefited from CDK4/6 inhibitor Abemaciclib in another clinical trial [31]. Interestingly, we identified tumour clones (e.g., clone-C and clone-F) that showed significant resistance despite the notable sensitivity of clone-B and clone-E to Abemaciclib treatment. It is, therefore, of interest to determine whether the patients who did not benefit from Afatinib or Abemaciclib harboured tumour cells that have similar profiles to clone-B, or clone-C and clone-F, respectively. It is also worth noting that all tumour clones in our study displayed sensitivities to the PI3K inhibitor Buparlisib and mTOR inhibitor Vistusertib. This could have happened due to molecular similarities that we detected among tumour clones and also between tumour clones and the parental polyclonal line. Given that all of these cell lines originated from a single patient, we anticipated some degree of similarity in genetic profile and drug response. However, our cell culture methodology could have excluded other tumour clones that have a completely different molecular profile than the parental polyclonal line and show resistance to Buparlisib and/or mTOR inhibitors. These results highlight the importance of the careful examination of all the molecular profiles constituting a single mass in order to identify potentially resistant cells residing within the tumour. To achieve durable responses, rationalised combination therapies may also be required. Our four-drug combination therapy supported this notion, as we have detected a positive effect when four different cell cycle inhibitors are used simultaneously at relatively lower doses. Alternatively, during clinical progression, consecutive administration of targeted therapies that match the changing tumour dynamics could prove effective.

Taken together, our study not only illustrates the molecular and biological diversity of glioblastoma cells but also provides evidence that this information can be utilised to maximise treatment outcomes. Thus, our experimental design and methodology can serve as a template for rationalisation of clinical applications for both glioblastoma and other cancers in the future.

4. Materials and Methods

4.1. Primary Cell Culture

We have developed a characterised glioblastoma patient-derived cell line resource (Q-Cell) [32], in which glioblastoma lines are maintained as glioma neural stem cell (GNS) cultures [21] or as neurosphere cultures using StemPro NSC SFM (Invitrogen, cat. A1050901, Carlsbad, CA, USA) or KnockOut™ DMEM (Gibco cat. 10829018, Thermo Fisher Scientific, Waltham, MA, USA) as per the manufacturer's guidelines. Characterisation data is freely available from: https://www.qimrberghofer.edu.au/q-cell/. Clonal populations were derived from the HW1 glioblastoma patient-derived model. All tissue was collected following ethical approved by the Royal Brisbane Hospital and QIMR Berghofer Human Research Ethics Committees. Ethical approval number: P3420, HREC/17/QRBW/577 Novel Therapies for Brain Cancer. In order to maintain pluripotency, KnockOut™ DMEM (Gibco cat. 10829018) media was supplemented with GlutaMAX™ Supplement (Gibco cat. 35050061), StemPro™ Neural Supplement (Gibco cat. A1050801), Recombinant Human EGF (Gibco cat. PHG0314), Recombinant Human FGFb (Gibco cat. PHG0024), and Penicillin/Streptomycin (Gibco cat. 15140122). Cells were cultured on flasks coated with Basement Membrane Matrigel®Matrix (Corning cat. 354234). Passaging the cells was done by detaching the cells from the flask surface using Accutase®solution (Sigma-Aldrich cat. A6964, St. Louis, MO, USA).

Single-cell clones were prepared by culturing approximately 60 cells in each 96-well plate. A total of six 96-well plates were used, and the presence of a single cell was examined under a light microscope every week in order to make sure that two cells did not grow simultaneously in a single well. 15 single-cell clones were established from the parental polyclonal cell line. Of these 16 cell lines in total (15 × clones and 1 × polyclonal), 12 single-cell clones and the polyclonal line reached sufficient confluency that allowed us to extract DNA and RNA from them. These samples were subjected to SNP arrays in order to initially determine the genetic variance between individual clones. Furthermore, RNA sequencing and whole genome sequencing was performed for 6 of the single-cell clones and the polyclonal line. In all analyses, DNA and RNA extracted from the original tumour tissue, normal brain tissue (cerebral cortex), and the patient's peripheral blood were included.

4.2. DNA and RNA Extraction

Cell pellets of early-passage lines were frozen immediately after collection and kept at −80 °C followed by DNA and RNA extraction using the AllPrep DNA/RNA Mini Kit (Qiagen cat. 80204, Hilden, Germany). DNA and RNA samples were quantified using Qubit Fluorometric Quantification instrument, and RNA integrity was assessed using RNA 6000 Nano Eukaryote Total RNA kit (Agilent Technologies cat. 5067-1511, Santa Clara, CA, USA) and Bioanalyzer 2100 instrument (Thermo Fisher Scientific, Waltham, MA, USA). RNA Integrity Number (RIN) $\geq$7 was set as minimum threshold.

4.3. SNP Array, DNA/RNA Sequencing, CpG Methylation Analysis

4.3.1. SNP Arrays

Tumour and matched normal DNA was assayed with the Omni 2.5-8, V1.1 Illumina BeadChips as per the manufacturer's instructions (Illumina, San Diego, CA, USA), scanned on an iScan instrument (Illumina), and the data were processed using the Genotyping module (v.1.9.4) in GenomeStudio v.2011.1 (Illumina) to produce LogR ratio and B allele frequency values. The analyses of approximately >400K known commonly polymorphic positions were assessed for the concordance of genotype calls across the samples. The measure of hybridization intensity represented by the LogR ratio was used to profile copy number differences across samples. Probes that harboured difference across the samples was selected where the magnitude of the difference between the minimum and maximum LogR ratio vales was >0.3 and <1.5 capturing >700K probes. Euclidean distance and complete clustering methods were used in the generation of heatmaps and dendrogram.

4.3.2. Whole Genome Sequencing

Sequence libraries were generated from 500 ng DNA using the TruSeq DNA PCR-free (350 bp insert) kit (Illumina) from tumour and matched normal samples. Whole-genome paired-end sequencing reads of 150 bp were generated using an X-Ten instrument (Illumina) at Macrogen

(Geumcheon-gu, Seoul, Korea). Sequence reads were trimmed using Cutadapt (version 1.11) and aligned to GRCh37 using BWA-MEM (Li, 2013. version 0.7.12). Duplicate alignments were marked with Picard (version 1.129, http://picard.sourceforge.net) and BAM files were coordinate-sorted using Samtools [33]. Mean coverage was determined using qCoverage (http://sourceforge.net/projects/adamajava). Bulk tumour samples and the polyclonal cell line were sequenced to an average read depth of 95.15X. The single cell clones and matched control were sequenced to an average read depth of 67.57X. All sequencing data generated are available from the European Genome-phenome Archive accession number EGAS00001003438.

4.3.3. Somatic Variant Calling

Single nucleotide substitution variants were detected using a dual calling strategy using qSNP [34] and the GATK HaplotypeCaller [35]. Short insertion and deletions (1–50 bp) were called with the GATK Haplotype caller. Somatic calls from both tools and indels underwent further filtering to generate a high confidence call set: (1) a minimum coverage depth of 8 reads for control data and 12 reads for tumour data; (2) at least 5 variant-supporting reads with individual start positions and for which the variant was not within the first or last 5 bases, was supported by reads in both sequencing directions, and was more than 5 bp from a mono-nucleotide run of 7 or more bases in length; and (3) somatic variants had <3% variant evidence identified in the control sample. Variants were annotated with the Ensembl v75 gene feature information and transcript or protein consequences using SnpEff (version 4.2). Variant allele frequencies of substitutions as used in the sample clustering were generated from a pileup analysis of all high confidence SNV variants identified in at least one sample. Allele frequencies were used with dist (Euclidean) and hclust R functions for hierarchical clustering. Structural variants were determined using qSV as previously described [36]. Structural variant breakpoints and potential consequence of the SV, including potential in-frame gene fusions, was determined by annotation against Ensembl v75 known genes using in-house scripts. The copy number was determined using sequencing data and the tool ascatNGS [37].

4.3.4. Whole Transcriptome Sequencing

Sequence libraries were generated from 1 ug intact RNA using the TruSeq stranded mRNA kit (Illumin) from tumour and matched normal cortex samples. Transcriptome paired-end sequencing reads of 100 bp were generated using a Hiseq 2500 instrument (Illumina) at Macrogen (Geumcheon-gu, Seoul, Korea) to a depth of 100 million reads per sample. Adapter sequences were removed using Cutadapt (version 1.11) and aligned using STAR (version 2.5.2a) to the GRCh37 assembly and Ensembl gene model (release 75). Quality control metrics were computed using RNA-SeQC (version 1.1.8) and gene expression was estimated using RSEM (version 1.2.30). Further analysis was carried out in R using the edgeR package [38] for normalization of the expression data used a library size correction for gene count by million reads mapped and the trimmed mean of M-values (TMM) method [39] to facilitate cross sample comparisons. The prcomp function was used to identify genes with variable expression, essentially those contributing to the first and second principle components and clustering performed with Pearson correlation distance and hclust. Heatmaps were generated using a gene-wise centre scaled Log2 values of normalised expression. Log2 fold change of gene expression as compared to the normal cortex control samples was generated as Log2 (normalised tumour expression +1/normalised cortex expression +1).

4.3.5. Capture Methylation Sequencing

CpG region capture and bisulphite converted 100 bp paired end sequencing was carried out on 500 ng DNA using the Illumina Tru-seq Methyl Capture Epic Library Prep kit (cat. FC-151-1003) on the Illumina HiSeq-4000. Adaptors sequences were removed using Cutadapt (version 1.11), aligned to GRCh37 using bwa-meth (version 0.2.0) and duplicate alignments were marked using Picard (version 1.129). Methylation at 865,859 known CpGs from the Illumina Epic assay were counted using the qbasepileup tool (https://github.com/AdamaJava/adamajava), using non-duplicate reads with MAPQ quality over 10 that passed vendor quality testing and with a minimum of 10 reads. CpG positions reported as having poor accuracy on the epic array [40] were removed. Methylation value was scored as a β value ratio, ranging from 0 (fully unmethylated) to 1 (fully methylated). Quantile normalisation was carried out on beta values using the R package preprocessCore to control for technical variation and aid comparison across different samples. Further filtering removed any CpG position where two or more samples had low coverage and replaced a low coverage position from a single sample with the median value of the other CpG methylation. The top 1000 most variable probes across the clone samples were selected and used with dist (Euclidean) and hclust (complete) R functions for hierarchical clustering.

4.4. Growth Analysis Using IncuCyte

Cells were cultured in Matrigel®Matrix-coated 96-well plates (5000 cells/well) using 200 μL stem cell media. Plates were placed into the IncuCyte instrument approximately 24 h later. IncuCyte (Essen BioScience, Ann Arbor, MI, USA) is an instrument that takes the images of the cells every 3 h for a period of 6 or 8 days. Cell growth is graphed using the instrument's software (v6.1.7601.65536, Essen BioScience), which calculates the expansion of the cells based on the images.

4.5. Irradiation and/or Temozolomide Treatment, MTT Cell Viability Assay

Cells were cultured in Matrigel®Matrix-coated 96-well plates (5000 cells/well) using 100 μL stem cell media. Approximately 24 h later, various concentrations of TMZ (10 μM–1000 μM) were prepared and delivered in another 100 μL of stem cell media. Control samples or IR-only samples received only DMSO delivered in 100 μL stem cell media. Two Gy of irradiation were applied to IR-only or TMZ + IR samples every other day for a total of 10 Gy. Cell viability was measured used CellTiter 96®Non-Radioactive Cell Proliferation Assay (Promega cat. G4000, Madison, WI, USA) at day 10. Each reading is normalised to DMSO-treated controls.

4.6. Drug Screens

Compounds were purchased from Selleck Chemicals: Alisertib (MLN8237) (cat. S1133), Barasertib (AZD1152-HQPA) (cat. S1147), SNS-314 Mesylate (cat. S1154), Abemaciclib (LY2835219) (cat. S7158), Palbociclib (PD-0332991) HCl (cat. S1116), Erlotinib HCl (OSI-744) (cat. S1023), Afatinib (BIBW2992) (cat. S1011), Vistusertib (AZD2014), Crenolanib (CP-868596) (cat. S2730), Orantinib (TSU-68, SU6668) (cat. S1470), Alpelisib (BYL719) (cat. S2814), Buparlisib (BKM120, NVP-BKM120) (cat. S2247), Wnt-C59 (C59) (cat. S7037), XAV-939 (cat. S1180). Cells were cultured in Matrigel®Matrix-coated 96-well plates (5,000 cells/well) using 100 μL stem cell media. Approximately 24 h later, drugs with various concentrations (100 nM, 250 nM, 500 nM, 1 μM, and 5 μM) were prepared in DMSO and delivered in another 100 μL of stem cell media. Control samples received DMSO only. Cell viability was measured used CellTiter 96®Non-Radioactive Cell Proliferation Assay (Promega cat. G4000) at day 3 and day 6. Each reading is normalised to DMSO-treated controls. Growth rates in the presence of the drugs were monitored and graphed using IncuCyte instrument for a period of 6 days.

5. Conclusions

Our study establishes that the molecular and biological profiling of glioblastoma tumour cells constituting a single tumour mass has vital importance in determining the most effective treatments. Thus, our experimental set-up and methodology can serve as a template for the rationalisation of clinical applications for both glioblastoma and other cancers in the near future.

Supplementary Materials: The following are available online at http://www.mdpi.com/2072-6694/11/2/190/s1, Figure S1: Shared and unique somatic variants across tumour samples, Figure S2: Molecular subtypes and stem cell characteristics of tumour clones, Figure S3: Comprehensive genomics analyses of tumour samples, Figure S4: Drug screens identify unique sensitivities of tumour cells, Table S1: Variant allele frequencies (VAF) of genome wide somatic single nucleotide variants (SNVs), Table S2: Genomic copy number variants by count of bases and genomic proportion, Table S3: Somatic structural variant breakpoint pairs identified across all samples, Table S4: Most variable, normalised RNA expression values as identified from principle component analysis, Table S5: Beta value for top 1000 most variable positions of CpG methylation, Table S6: Somatic single nucleotide, small insertions and deletions identified in coding regions across all samples in MAF format.

Author Contributions: Conceptualization, S.A.; A.-M.P.; N.W. and B.W.D.; Methodology, S.A.; A.-M.P.; R.C.J.D.; P.M.; K.N.; S.H.K.; B.W.S.; J.V.P.; N.W. and B.W.D.; Formal Analysis and Investigation, S.A.; A.-M.P.; R.C.J.D.; P.M.; S.K.; N.W. and B.W.D.; Resources, R.L.J.; Data Curation, S.A.; A.-M.P.; P.M.; N.W. and B.W.D.; Writing-Original Draft, S.A.; A.-M.P.; N.W. and B.W.D.; Writing—Review & Editing, S.A.; A.-M.P.; R.C.J.D.; S.K.; B.W.S.; N.W. and B.W.D.; Supervision, N.W. and B.W.D. Project Administration, B.W.D. Funding Acquisition, N.W. and B.W.D.

Funding: This research was funded by the generous donation support from the Sid Faithfull Group and Cure Brain Cancer Foundation. N.W. is funded by an NHMRC research fellowship (1139071). R.C.J.D. is supported by an early career grant from Cancer Australia.

Acknowledgments: We acknowledge the QIMR Berghofer core facilities for their assistance and expertise in assisting us to complete this study. We thank Fiona M Smith, Zara C Bruce, and Kathleen S Ensbey for research support, and Conrad Leonard for submission of the sequence data to a public repository. We thank Roel Verhaak and Qianghu Wang for molecular subtype analysis.

Conflicts of Interest: The authors declare no conflict of interest.

References

1. Noroxe, D.S.; Poulsen, H.S.; Lassen, U. Hallmarks of glioblastoma: A systematic review. *ESMO Open* **2016**, *1*, e000144. [CrossRef] [PubMed]
2. Stupp, R.; Mason, W.P.; van den Bent, M.J.; Weller, M.; Fisher, B.; Taphoorn, M.J.; Belanger, K.; Brandes, A.A.; Marosi, C.; Bogdahn, U.; et al. Radiotherapy plus concomitant and adjuvant temozolomide for glioblastoma. *N. Engl. J. Med.* **2005**, *352*, 987–996. [CrossRef] [PubMed]
3. Brennan, C.W.; Verhaak, R.G.; McKenna, A.; Campos, B.; Noushmehr, H.; Salama, S.R.; Zheng, S.; Chakravarty, D.; Sanborn, J.Z.; Berman, S.H.; et al. The somatic genomic landscape of glioblastoma. *Cell* **2013**, *155*, 462–477. [CrossRef] [PubMed]
4. Verhaak, R.G.; Hoadley, K.A.; Purdom, E.; Wang, V.; Qi, Y.; Wilkerson, M.D.; Miller, C.R.; Ding, L.; Golub, T.; Mesirov, J.P.; et al. Integrated genomic analysis identifies clinically relevant subtypes of glioblastoma characterized by abnormalities in PDGFRA, IDH1, EGFR, and NF1. *Cancer Cell* **2010**, *17*, 98–110. [CrossRef] [PubMed]
5. Noushmehr, H.; Weisenberger, D.J.; Diefes, K.; Phillips, H.S.; Pujara, K.; Berman, B.P.; Pan, F.; Pelloski, C.E.; Sulman, E.P.; Bhat, K.P.; et al. Identification of a CpG island methylator phenotype that defines a distinct subgroup of glioma. *Cancer Cell* **2010**, *17*, 510–522. [CrossRef] [PubMed]
6. Sturm, D.; Witt, H.; Hovestadt, V.; Khuong-Quang, D.A.; Jones, D.T.; Konermann, C.; Pfaff, E.; Tonjes, M.; Sill, M.; Bender, S.; et al. Hotspot mutations in H3F3A and IDH1 define distinct epigenetic and biological subgroups of glioblastoma. *Cancer Cell* **2012**, *22*, 425–437. [CrossRef] [PubMed]
7. Phillips, H.S.; Kharbanda, S.; Chen, R.; Forrest, W.F.; Soriano, R.H.; Wu, T.D.; Misra, A.; Nigro, J.M.; Colman, H.; Soroceanu, L.; et al. Molecular subclasses of high-grade glioma predict prognosis, delineate a pattern of disease progression, and resemble stages in neurogenesis. *Cancer Cell* **2006**, *9*, 157–173. [CrossRef]
8. Parsons, D.W.; Jones, S.; Zhang, X.; Lin, J.C.; Leary, R.J.; Angenendt, P.; Mankoo, P.; Carter, H.; Siu, I.M.; Gallia, G.L.; et al. An integrated genomic analysis of human glioblastoma multiforme. *Science* **2008**, *321*, 1807–1812. [CrossRef]
9. Almendro, V.; Marusyk, A.; Polyak, K. Cellular heterogeneity and molecular evolution in cancer. *Annu. Rev. Pathol.* **2013**, *8*, 277–302. [CrossRef]
10. Sottoriva, A.; Spiteri, I.; Piccirillo, S.G.; Touloumis, A.; Collins, V.P.; Marioni, J.C.; Curtis, C.; Watts, C.; Tavare, S. Intratumor heterogeneity in human glioblastoma reflects cancer evolutionary dynamics. *Proc. Natl. Acad. Sci. USA* **2013**, *110*, 4009–4014. [CrossRef]
11. Patel, A.P.; Tirosh, I.; Trombetta, J.J.; Shalek, A.K.; Gillespie, S.M.; Wakimoto, H.; Cahill, D.P.; Nahed, B.V.; Curry, W.T.; Martuza, R.L.; et al. Single-cell RNA-seq highlights intratumoral heterogeneity in primary glioblastoma. *Science* **2014**, *344*, 1396–1401. [CrossRef] [PubMed]

12. Meyer, M.; Reimand, J.; Lan, X.; Head, R.; Zhu, X.; Kushida, M.; Bayani, J.; Pressey, J.C.; Lionel, A.C.; Clarke, I.D.; et al. Single cell-derived clonal analysis of human glioblastoma links functional and genomic heterogeneity. *Proc. Natl. Acad. Sci. USA* **2015**, *112*, 851–856. [CrossRef] [PubMed]
13. Snuderl, M.; Fazlollahi, L.; Le, L.P.; Nitta, M.; Zhelyazkova, B.H.; Davidson, C.J.; Akhavanfard, S.; Cahill, D.P.; Aldape, K.D.; Betensky, R.A.; et al. Mosaic amplification of multiple receptor tyrosine kinase genes in glioblastoma. *Cancer Cell* **2011**, *20*, 810–817. [CrossRef] [PubMed]
14. Greaves, M.; Maley, C.C. Clonal evolution in cancer. *Nature* **2012**, *481*, 306–313. [CrossRef] [PubMed]
15. Burrell, R.A.; McGranahan, N.; Bartek, J.; Swanton, C. The causes and consequences of genetic heterogeneity in cancer evolution. *Nature* **2013**, *501*, 338–345. [CrossRef] [PubMed]
16. Johnson, B.E.; Mazor, T.; Hong, C.; Barnes, M.; Aihara, K.; McLean, C.Y.; Fouse, S.D.; Yamamoto, S.; Ueda, H.; Tatsuno, K.; et al. Mutational analysis reveals the origin and therapy-driven evolution of recurrent glioma. *Science* **2014**, *343*, 189–193. [CrossRef] [PubMed]
17. Brennan, C.; Momota, H.; Hambardzumyan, D.; Ozawa, T.; Tandon, A.; Pedraza, A.; Holland, E. Glioblastoma subclasses can be defined by activity among signal transduction pathways and associated genomic alterations. *PLoS ONE* **2009**, *4*, e7752. [CrossRef] [PubMed]
18. Wang, Q.; Hu, B.; Hu, X.; Kim, H.; Squatrito, M.; Scarpace, L.; de Carvalho, A.C.; Lyu, S.; Li, P.; Li, Y.; et al. Tumor Evolution of Glioma-Intrinsic Gene Expression Subtypes Associates with Immunological Changes in the Microenvironment. *Cancer Cell* **2017**, *32*, 42–56.e46. [CrossRef] [PubMed]
19. Xie, Y.; Bergstrom, T.; Jiang, Y.; Johansson, P.; Marinescu, V.D.; Lindberg, N.; Segerman, A.; Wicher, G.; Niklasson, M.; Baskaran, S.; et al. The Human Glioblastoma Cell Culture Resource: Validated Cell Models Representing All Molecular Subtypes. *EBioMedicine* **2015**, *2*, 1351–1363. [CrossRef]
20. Bhat, K.P.L.; Balasubramaniyan, V.; Vaillant, B.; Ezhilarasan, R.; Hummelink, K.; Hollingsworth, F.; Wani, K.; Heathcock, L.; James, J.D.; Goodman, L.D.; et al. Mesenchymal Differentiation Mediated by NF-κB Promotes Radiation Resistance in Glioblastoma. *Cancer Cell* **2013**, *24*, 331–346. [CrossRef]
21. Pollard, S.M.; Yoshikawa, K.; Clarke, I.D.; Danovi, D.; Stricker, S.; Russell, R.; Bayani, J.; Head, R.; Lee, M.; Bernstein, M.; et al. Glioma stem cell lines expanded in adherent culture have tumor-specific phenotypes and are suitable for chemical and genetic screens. *Cell Stem Cell* **2009**, *4*, 568–580. [CrossRef] [PubMed]
22. Akgul, S.; Li, Y.; Zheng, S.; Kool, M.; Treisman, D.M.; Li, C.; Wang, Y.; Grobner, S.; Ikenoue, T.; Shen, Y.; et al. Opposing Tumor-Promoting and -Suppressive Functions of Rictor/mTORC2 Signaling in Adult Glioma and Pediatric SHH Medulloblastoma. *Cell Rep.* **2018**, *24*, 463–478.e5. [CrossRef] [PubMed]
23. Cloughesy, T.F.; Cavenee, W.K.; Mischel, P.S. Glioblastoma: From molecular pathology to targeted treatment. *Annu. Rev. Pathol.* **2014**, *9*, 1–25. [CrossRef] [PubMed]
24. Huse, J.T.; Holland, E.; DeAngelis, L.M. Glioblastoma: Molecular analysis and clinical implications. *Annu. Rev. Med.* **2013**, *64*, 59–70. [CrossRef]
25. Weinstein, I.B.; Joe, A.K. Mechanisms of disease: Oncogene addiction—A rationale for molecular targeting in cancer therapy. *Nat. Clin. Pract. Oncol.* **2006**, *3*, 448–457. [CrossRef] [PubMed]
26. Sharma, S.V.; Settleman, J. Oncogene addiction: Setting the stage for molecularly targeted cancer therapy. *Genes Dev.* **2007**, *21*, 3214–3231. [CrossRef] [PubMed]
27. Weinstein, I.B. Cancer. Addiction to oncogenes—The Achilles heal of cancer. *Science* **2002**, *297*, 63–64. [CrossRef] [PubMed]
28. Wen, P.Y.; Lee, E.Q.; Reardon, D.A.; Ligon, K.L.; Alfred Yung, W.K. Current clinical development of PI3K pathway inhibitors in glioblastoma. *Neuro Oncol.* **2012**, *14*, 819–829. [CrossRef] [PubMed]
29. Dewhurst, S.M.; McGranahan, N.; Burrell, R.A.; Rowan, A.J.; Gronroos, E.; Endesfelder, D.; Joshi, T.; Mouradov, D.; Gibbs, P.; Ward, R.L.; et al. Tolerance of whole-genome doubling propagates chromosomal instability and accelerates cancer genome evolution. *Cancer Discov.* **2014**, *4*, 175–185. [CrossRef] [PubMed]
30. Reardon, D.A.; Nabors, L.B.; Mason, W.P.; Perry, J.R.; Shapiro, W.; Kavan, P.; Mathieu, D.; Phuphanich, S.; Cseh, A.; Fu, Y.; et al. Phase I/randomized phase II study of afatinib, an irreversible ErbB family blocker, with or without protracted temozolomide in adults with recurrent glioblastoma. *Neuro Oncol.* **2015**, *17*, 430–439. [CrossRef] [PubMed]
31. Patnaik, A.; Rosen, L.S.; Tolaney, S.M.; Tolcher, A.W.; Goldman, J.W.; Gandhi, L.; Papadopoulos, K.P.; Beeram, M.; Rasco, D.W.; Hilton, J.F.; et al. Efficacy and Safety of Abemaciclib, an Inhibitor of CDK4 and CDK6, for Patients with Breast Cancer, Non-Small Cell Lung Cancer, and Other Solid Tumors. *Cancer Discov.* **2016**, *6*, 740–753. [CrossRef] [PubMed]

32. Day, B.W.; Stringer, B.W.; Wilson, J.; Jeffree, R.L.; Jamieson, P.R.; Ensbey, K.S.; Bruce, Z.C.; Inglis, P.; Allan, S.; Winter, C.; et al. Glioma surgical aspirate: A viable source of tumor tissue for experimental research. *Cancers* **2013**, *5*, 357–371. [CrossRef] [PubMed]
33. Li, H.; Handsaker, B.; Wysoker, A.; Fennell, T.; Ruan, J.; Homer, N.; Marth, G.; Abecasis, G.; Durbin, R.; 1000 Genome Project Data Processing Subgroup. The Sequence Alignment/Map format and SAMtools. *Bioinformatics* **2009**, *25*, 2078–2079. [CrossRef] [PubMed]
34. Kassahn, K.S.; Holmes, O.; Nones, K.; Patch, A.M.; Miller, D.K.; Christ, A.N.; Harliwong, I.; Bruxner, T.J.; Xu, Q.; Anderson, M.; et al. Somatic point mutation calling in low cellularity tumors. *PLoS ONE* **2013**, *8*, e74380. [CrossRef] [PubMed]
35. McKenna, A.; Hanna, M.; Banks, E.; Sivachenko, A.; Cibulskis, K.; Kernytsky, A.; Garimella, K.; Altshuler, D.; Gabriel, S.; Daly, M.; et al. The Genome Analysis Toolkit: A MapReduce framework for analyzing next-generation DNA sequencing data. *Genome Res.* **2010**, *20*, 1297–1303. [CrossRef] [PubMed]
36. Patch, A.M.; Christie, E.L.; Etemadmoghadam, D.; Garsed, D.W.; George, J.; Fereday, S.; Nones, K.; Cowin, P.; Alsop, K.; Bailey, P.J.; et al. Whole-genome characterization of chemoresistant ovarian cancer. *Nature* **2015**, *521*, 489–494. [CrossRef] [PubMed]
37. Raine, K.M.; Van Loo, P.; Wedge, D.C.; Jones, D.; Menzies, A.; Butler, A.P.; Teague, J.W.; Tarpey, P.; Nik-Zainal, S.; Campbell, P.J. ascatNgs: Identifying Somatically Acquired Copy-Number Alterations from Whole-Genome Sequencing Data. *Curr. Protoc. Bioinform.* **2016**, *56*, 15.9.1–15.9.17. [CrossRef] [PubMed]
38. Robinson, M.D.; McCarthy, D.J.; Smyth, G.K. edgeR: A Bioconductor package for differential expression analysis of digital gene expression data. *Bioinformatics* **2010**, *26*, 139–140. [CrossRef] [PubMed]
39. Robinson, M.D.; Oshlack, A. A scaling normalization method for differential expression analysis of RNA-seq data. *Genome Biol.* **2010**, *11*, R25. [CrossRef] [PubMed]
40. McCartney, D.L.; Walker, R.M.; Morris, S.W.; McIntosh, A.M.; Porteous, D.J.; Evans, K.L. Identification of polymorphic and off-target probe binding sites on the Illumina Infinium MethylationEPIC BeadChip. *Genom. Data* **2016**, *9*, 22–24. [CrossRef] [PubMed]

cancers

MDPI

Article

NMDA Receptor-Mediated Signaling Pathways Enhance Radiation Resistance, Survival and Migration in Glioblastoma Cells—A Potential Target for Adjuvant Radiotherapy

Adriana Müller-Längle [1], Henrik Lutz [1], Stephanie Hehlgans [2], Franz Rödel [2], Kerstin Rau [1] and Bodo Laube [1,*]

1 Neurophysiology and Neurosensory Systems, Technische Universität Darmstadt, Schnittspahnstrasse 3, 64287 Darmstadt, Germany; adriana1986@gmx.de (A.M.-L.); lutz@bio.tu-darmstadt.de (H.L.); kiki.rau@googlemail.com (K.R.)

2 Department of Radiotherapy and Oncology, Goethe-Universität Frankfurt, Theodor-Stern-Kai 7, 60590 Frankfurt am Main, Germany; stephanie.hehlgans@kgu.de (S.H.); Franz.Roedel@kgu.de (F.R.)

* Correspondence: laube@bio.tu-darmstadt.de; Tel.: +49-6151-16-20970

Received: 29 December 2018; Accepted: 4 April 2019; Published: 9 April 2019

Abstract: Glioblastoma is one of the most aggressive malignant brain tumors, with a survival time less than 15 months and characterized by a high radioresistance and the property of infiltrating the brain. Recent data indicate that the malignancy of glioblastomas depends on glutamatergic signaling via ionotropic glutamate receptors. In this study we revealed functional expression of Ca^{2+}-permeable NMDARs in three glioblastoma cell lines. Therefore, we investigated the impact of this receptor on cell survival, migration and DNA double-strand break (DSB) repair in the presence of both, glutamate and NMDAR antagonists, and after clinically relevant doses of ionizing radiation. Our results indicate that treatment with NMDAR antagonists slowed the growth and migration of glutamate-releasing LN229 cells, suggesting that activation of NMDARs facilitate tumor expansion. Furthermore, we found that DSB-repair upon radiation was more effective in the presence of glutamate. In contrast, antagonizing the NMDAR or the Ca^{2+}-dependent transcription factor CREB impaired DSB-repair similarly and resulted in a radiosensitizing effect in LN229 and U-87MG cells, indicating a common link between NMDAR signaling and CREB activity in glioblastoma. Since the FDA-approved NMDAR antagonists memantine and ifenprodil showed differential radiosensitizing effects, these compounds may constitute novel optimizations for therapeutic interventions in glioblastoma.

Keywords: ionotropic glutamate receptors; DNA repair; CREB inhibitor; NMDAR subunit GluN2B; radiotherapy; LN229; U-87MG; memantine; ifenprodil; sulfasalazine

1. Introduction

Glioblastoma (WHO grade IV) is one of the most common and aggressive malignant primary brain tumor in humans [1]. Despite recent incremental advances in surgical approaches, radiotherapy (RT, i.e., subsequent irradiation with single doses of 2 Gy daily, cumulatively max. 60 Gy), and chemotherapy (i.e., temozolomide), the median survival of patients with glioblastoma is only about 15 months [2,3]. Thus, the treatment of glioblastoma remains one of the most challenging task in clinical oncology. The so far limited success of the current multimodal therapy is likely due to the diffuse nature of the tumor and the prominent resistance to ionizing radiation (IR) with a high degree of recurrent invasive growth of the glioblastoma cells after the treatment [4–7]. In view of this and the outstanding role of RT for the prognosis of patients with glioblastoma, radiosensitizing substances addressing both, the high inherent migratory and invasive behavior and the intrinsic DNA double-strand break (DSB) repair capacity upon IR of glioblastoma cells may become increasingly important to improve therapy and patient survival.

Accordingly, with regard to a putative improvement of the efficacy of adjuvant RT, effects of the tumor microenvironment on glioma proliferation, invasion and survival should be considered [8]. One striking feature of glioblastoma is that glioma cells release excitotoxic levels of glutamate (Glu) into the extracellular space at the tumor margin which is a 100-fold excess as compared to normal levels (>100 μM) [9,10]. Consequently, the prolonged high levels of Glu lead to neuronal cell death in the surrounding nervous tissue, which offers more space for tumor growth [10] and induce seizures within the peritumoral border in the patients [11–13]. This active Glu release by glioma cells is mainly mediated by the system x_c^- cystine/glutamate exchanger [11,14]. Interestingly, pharmacological inhibition of this system by the FDA-approved drug sulfasalazine (SAS) was shown to slow tumor growth and to extend survival of tumor-bearing animals [15] and has therefore been considered for a treatment of both, gliomas and associated epilepsy [16]. In the brain, Glu is the major excitatory neurotransmitter by activating post-synaptic neuronal glutamate receptors upon pre-synaptic release and is crucially involved in normal brain function, including cognition, memory formation and learning [17]. Physiological Glu concentrations have also been shown to regulate proliferation, migration, and survival of neuronal progenitor cells and immature neurons during brain development [18].

Glu action is mediated by metabotropic glutamate receptors (mGluRs) and ionotropic glutamate receptors (iGluRs). The iGluRs are divided into three families: *N*-methyl-D-aspartate (NMDA), α-amino-3-hydroxy-5-methyl-4-isoxazolepropionic acid (AMPA) and kainate receptors [17]. It was originally considered that glutamatergic signaling is limited to neuronal cells in nervous systems. However, over the past decade, growing evidence occurred that iGluRs are also expressed in non-neuronal tissue [19], where they also play an important role in the proliferation, apoptosis, and migration of cells by interfering with their intracellular signaling pathways [20]. The ability of uncontrollable propagation and migration characterizes neoplastic cells as well; therefore, Glu has been suggested as a potential growth factor in tumor development [21]. Consistent with this notion, experimental data have implicated an important role of iGluRs in malignant tumor invasion and progression [22,23]. Furthermore, glioblastoma cells abundantly express AMPARs [22–25] promoting survival, invasion and migration and enhancing their malignant phenotype [10]. However, increased Glu concentrations can also mediate necrosis in malignant gliomas [26]. Thus, the multitude of paracrine and autocrine effects of Glu in glioma biology have supported the rationale for a pharmacological targeting of AMPARs in the treatment of glioblastoma [10]. Indeed, iGluR antagonists have been shown to increase the lifespan of human xenografts in mice [27]. In contrast, the role of Ca^{2+}-permeable NMDARs in glioblastoma cells is enigmatic and their pathophysiological significance is largely unknown. For example, several publications state that NMDARs are non-functional in glioblastoma cells [25,28]. However, other reports indicate a functional role of NMDAR-signaling pathways in glioblastoma progression [23,29,30] and in a xenograft study [31]. So far five distinct subunits are known to generate Glu-gated NMDARs (GluN1 and GluN2A-D; overview in [17]). Four subunits assemble to a heteromeric complex [32], which in the case of the GluN1/GluN2-NMDAR subtypes are gated by simultaneous binding of Glu and glycine [33]. Especially the GluN2A and B subunits have been implicated in neuronal survival and death [34]. Under physiological conditions in the brain, NMDARs are crucial for brain development and neuronal plasticity involved in learning and memory formation [35]. Activation of NMDARs is translated to the nucleus of neurons by signaling-cascades leading to the phosphorylation of the cAMP-responsive element binding protein (CREB) and activation of genes promoting survival [34]. Interestingly, a putative role of NMDARs in tumor progression has been recently re-vitalized by several reviews highlighting the potency of targeting NMDARs in tumor treatment [29,36–38].

In the present work, we analyzed the potency of NMDAR-mediated signaling as a promising therapy to combat glioblastoma with a focus on two pathognomonic features of their biology, i.e., cell migration/invasion and, associated with that, the apparent intrinsic resistance of glioblastoma cells to RT. We found in the glioblastoma cell lines LN229 and U-87MG that (i) Glu improved DNA double-strand break (DSB) repair and that (ii) blocking NMDAR-mediated glutamatergic signaling resulted in a decreased cell migration, survival and a sensitization to IR. In addition we could show that Glu activates the CREB pathway and that DSB repair is inhibited by a CREB antagonist. Furthermore, NMDAR inhibitors showed a radiosensitizing effect in LN229 and U-87MG glioblastoma cells. Thus, particularly the FDA-approved drugs memantine and the GluN2B subunit-selective NMDAR antagonist ifenprodil seem to be promising as adjuvant therapy besides RT.

2. Results

2.1. Functional and Immunohistochemical Characterization of NMDARs in LN229 Glioblastoma Cells

To analyze whether Ca^{2+}-permeable ionotropic glutamate receptors (iGluRs) are functionally expressed in glioblastoma cells, we measured intracellular calcium concentrations ($(Ca^{2+})_i$) in LN229 cells upon glutamate (Glu) application. In ~50% of the cells tested, a robust transient increase in the relative intracellular fluorescence intensity in their cell bodies was obtained after application of Glu (Figure 1A). Interestingly, a prominent $(Ca^{2+})_i$ rise was consistently seen in the cell nucleus (Figure 1B) and in the extensions of the responsive cells (Figure 1C) indicating both, a Glu-mediated generation of nuclear Ca^{2+}-transients and a spatial expression of Ca^{2+}-permeable iGluRs at cell extensions. To test which iGluRs are functionally expressed, we performed patch-clamp recordings upon application of Glu and the iGluR-subtype specific agonists AMPA, NMDA and kainate. Consistent with our imaging results, ~40% of the cells tested showed a robust Glu-mediated inward current that recovered after washout (Figure 1D). Application of NMDA, AMPA or kainate revealed a more differential contribution of the Glu-mediated inward currents with the highest maximal inducible currents (I_{max}) generated by NMDA and AMPA (Figure 1D). Since it has been postulated that NMDARs are not functionally expressed in glioblastoma cell lines (see [25,28,37]), we further tested three additional human glioblastoma cell lines (U-87MG, T98G, LN428) for their ability to respond to NMDA application. Two cell lines (U-87MG and T98G) revealed similar NMDA-inducible currents as obtained in LN229 cells, indicating that functional NMDARs are more commonly expressed in glioblastoma cells. To further verify the subunit composition of the expressed NMDARs and to determine their cellular localization, we labeled LN229 cells with antibodies targeting the GluN1, GluN2A and GluN2B NMDAR subunits. As shown by immunohistochemistry, membrane-associated localization of all three NMDAR subunits was evident (Figure 1E). However, the GluN2B subunit exhibited the most ubiquitous expression with an extensive punctuate pattern at long processes which is indicative of NMDAR clustering at the edge of lamellipodia (Figure 1E). Thus, our results demonstrate in three glioblastoma cell lines (LN229, U-87MG and T98G) functional expression of NMDARs. In addition, NMDARs in LN229 cells are composed of GluN1, GluN2A and GluN2B subunits with an enrichment of GluN2B-containing NMDARs at cellular extensions.

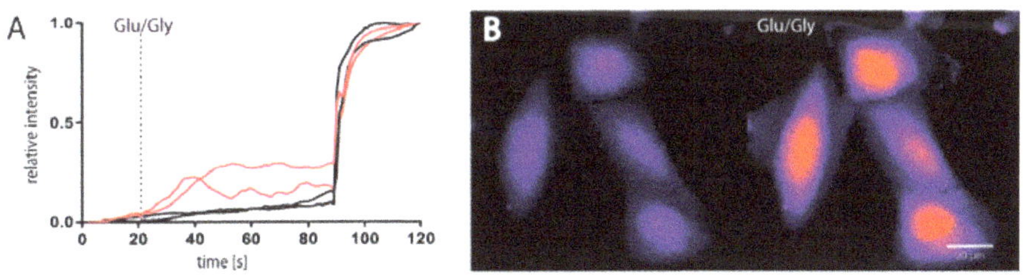

Figure 1. *Cont.*

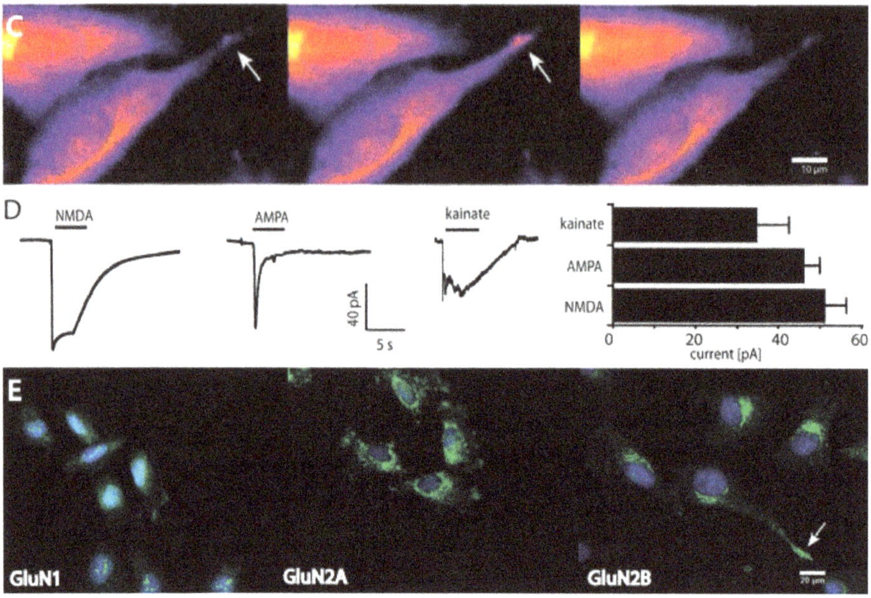

Figure 1. LN229 glioblastoma cells express functional NMDARs. (**A**) Ca^{2+}-imaging of LN229 cells labeled with Fluo-4 AM displayed in ~50% of the cells a significant increase in relative fluorescence after glutamate/glycine (100 µM Glu/10 µM Gly) treatment (20 s). As an internal control, 10 µM ionomycin was given after 90 seconds. The integrated density of two non-responding (black) and two responding (red) cells are displayed. (**B**) Representative image of nuclear Ca^{2+}-signals in four cells responding to Glu/Gly application (20 seconds before and after application). Scale bar 20 µm. (**C**) Images of Ca^{2+}-signal oscillations in cell extensions (white arrows) 1, 4 and 10 seconds after Glu-application. Scale bar 10 µm. (**D**) Agonist-mediated traces and maximal inducible currents (I_{max}) of kainate (100 µM), AMPA (100 µM) and NMDA (100 µM) analyzed by patch-clamp measurements (n = 22). Values of the I_{max} represent means ± SEM. (**E**) Immunostaining for GluN1, GluN2A and GluN2B NMDAR subunits with anti-GluN1, anti-GluN2A and anti-GluN2B (FITC, green) merged with DAPI (blue). GluN2B subunit expression in cell lamellipodia is indicated by white arrow. Scale bar 20 µm.

2.2. NMDAR-Activation is Crucial for LN229 Cell Viability, Migration and Survival

Our results so far indicate that functional NMDARs are expressed in three glioblastoma cell lines. Since it has been shown that iGluRs can induce Glu-mediated excitotoxicity in glioblastoma cells [26,39], we tested whether high concentrations of Glu may affect viability and survival of LN229 cells. We therefore incubated these cells with high concentrations of Glu and in the presence of the specific AMPAR antagonist Gyki-52466, the NMDAR antagonist MK801 and the GluN2B-subunit selective NMDAR antagonist ifenprodil and analyzed the morphology of the cells. Both, application of Glu and the AMPAR antagonist Gyki-52488 didn't show any obvious morphological changes compared to the untreated cells whereas antagonizing NMDARs by MK801 or ifenprodil induced a reduction in cell number, a more rounded cell appearance and a reduced number of lamellipodia (Figure 2A). Notably, ifenprodil resulted in the most pronounced effect with a prominent retraction of cellular extensions (Figure 2A). Next, we measured cell viability by a colorimetric MTT assay in the absence and presence of the antagonists. Similar to the morphological effects, both, MK801 and ifenprodil significantly reduced cell viability compared to control with a significant higher effect seen with ifenprodil (Figure 2B) whereas cell viability was not affected upon Glu treatment. Based on these findings and the fact that glioblastoma cells can secrete high amounts of Glu [11], we speculated whether Glu-release by LN229 cells may cause constitutive active NMDARs in an autocrine manner. Therefore we analyzed the time-course of Glu-concentrations in the supernatant of LN229 cells (Figure 2C). The Glu-assay revealed that LN229 cells release Glu concentrations sufficient to saturate NMDARs already after 1h in culture which could be significantly blocked by the system x_c^- cystine/glutamate transporter specific antagonist SAS (Figure 2C). Thus, our results indicate that i) LN229 cells release Glu in a high amount into the media and that ii) the changes in cell morphology and viability after blockage of NMDARs are likely due to a high concentration of endogenously released Glu.

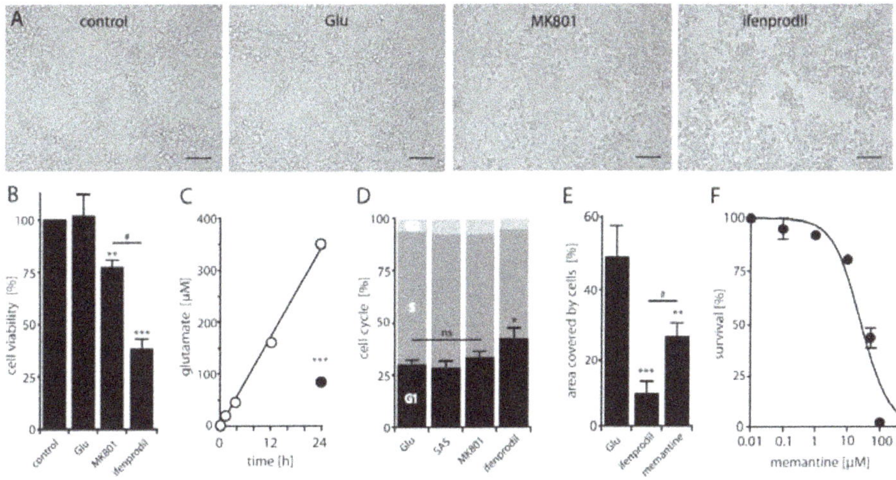

Figure 2. Effects of glutamate on LN229 cell viability, cell cycle distribution and migration. (**A**) Light microscope images of untreated cells (control), and in the presence of Glu (1 mM), MK801 (10 µM) and ifenprodil (25 µM). Scale bar 100 µm. (**B**) Cell viability measured after 48h by the MTT assay after Glu (1 mM), MK801 (10 µM) and ifenprodil (25 µM) treatment compared to control (untreated cells). The experiments were repeated at least five times performed in octuplet and values represent means ± SEM (one-way ANOVA followed by Bonferroni's post-hoc test, ** $p < 0.01$, *** $p < 0.001$, # $p < 0.05$). (**C**) Increase in extracellular Glu concentrations of 3.5×10^5 seeded cells at indicated time points (white circles) and after treatment with sulfasalazine (SAS, 250 µM, black circle) revealed a release of ~7.8 µg/mL Glu/h. Data are expressed as means ± SEM of three independent experiments performed in triplicate. Asterisks indicate a significant difference between treated and untreated cells as determined by Student's *t*-test (*** $p < 0.001$). (**D**) Cell cycle distribution after 24h in the presence of Glu (1mM), MK801 (10 µM) or ifenprodil (25 µM) compared to SAS-treated cells (250 µM) (n = 4; one-way ANOVA followed by Bonferroni's post-hoc test, * $p < 0.05$). (**E**) Cells were seeded for 48 h into two wells of an ibidi culture-insert for wound healing assays in the presence of ifenprodil (25 µM) and memantine (50 µM) compared to cells treated with Glu (1 mM). Data are expressed as means ± SEM of three independent experiments performed in triplicate. Asterisks indicate a significant difference between Glu-treated and antagonist-treated cells as determined by one-way ANOVA followed by Bonferroni's post-hoc test, ** $p < 0.01$, *** $p < 0.001$, # $p < 0.05$). (**F**) Colony formation of cells treated with memantine revealed an LD_{50} value of 26 ± 11 µM. Data represent the means ± SEM (n = 3).

To test whether activation of NMDARs may influence the cell cycle progression of LN229 cells, we performed a cell cycle analysis after treatment with Glu and in the presence of SAS, MK801 or ifenprodil. Neither Glu nor diminishing Glu-release or blocking NMDARs by MK801 revealed differences in cell cycle distribution after 24h whereas treatment with ifenprodil resulted in a slightly increased cell population in G1 (Figure 2D) which unlikely contributes to the reduced cell viability seen in the MTT assay (see Figure 2B). However, since GluN2B-subunit containing NMDARs are expressed in lamellipodia (see Figure 1E) and MK801 slowed the growth of gliomas in situ [31], we wondered whether NMDAR antagonists influence cell migration. Therefore, LN229 cells were exposed to ifenprodil or memantine and the migration rate estimated for 48h. The antagonist-treated cells showed a significant stagnation of cell migration (Figure 2E), especially for the ifenprodil treated cells (Figure 2E). Based on this result we next examined the effect of memantine on cell survival by a clonogenic survival assay. Figure 2F shows a dose-dependent decrease in clonogenic survival for memantine normalized to untreated controls with an LD_{50} value of 26 ± 11 μM. A similar result was obtained with MK801 with an LD_{50} value of 0.9 ± 1.1 μM. Thus, our results revealed that treatment of the Glu-secreting LN229 cells with NMDAR antagonists can slow the growth and migration of cells, suggesting that activation of NMDARs in glioblastomas by ambient Glu may facilitate tumor expansion in vivo.

2.3. Antagonizing NMDARs Increases LN229 Radiosensitivity and Impairs Radiation-Induced DNA Double-Strand Break Repair

To evaluate the impact of NMDARs on the DNA repair capacity in glioblastoma cells, we used a well-established DSB-marker, the Ser139 phosphorylated histone H2AX (γH2AX) to stain for γH2AX in S/G2–phase LN229 cells. As shown in Figure 3A, adding Glu resulted in a pronounced decrease in γH2AX foci 4h after a 2 Gy exposure as compared to mock-irradiated control cells. To further elucidate the impact of Glu on DSB repair, we analyzed the relative levels of γH2AX in control and Glu treated cells upon IR by western blotting. Consistent with the decrease in the number of γH2AX foci, we found that Glu induced a significant decrease in the amount of γH2AX protein at 4h post IR (Figure 3B), suggesting that the presence of Glu results in a more effective DSB repair. In a next step we counted γH2AX foci in a kinetic covering 8h post IR in the presence of Glu and upon adding MK801. A pronounced increase of γH2AX foci was observed 0.5, 2, 4 and 8 h after 2 Gy radiation in MK801 treated cells, which was highly significant at 4h (Figure 3C). We next tested the impact of the Ca^{2+}-chelator BAPTA-AM on γH2AX foci detection. Again, a highly significant increased number of γH2AX foci in BAPTA-AM-treated cells became evident at 4h after radiation (Figure 3C). Encouraged by these results and to test whether NMDAR antagonists are able to increase the sensitivity of LN229 cells to IR, we performed a clonogenic survival assay with cells treated with ifenprodil and memantine upon IR.

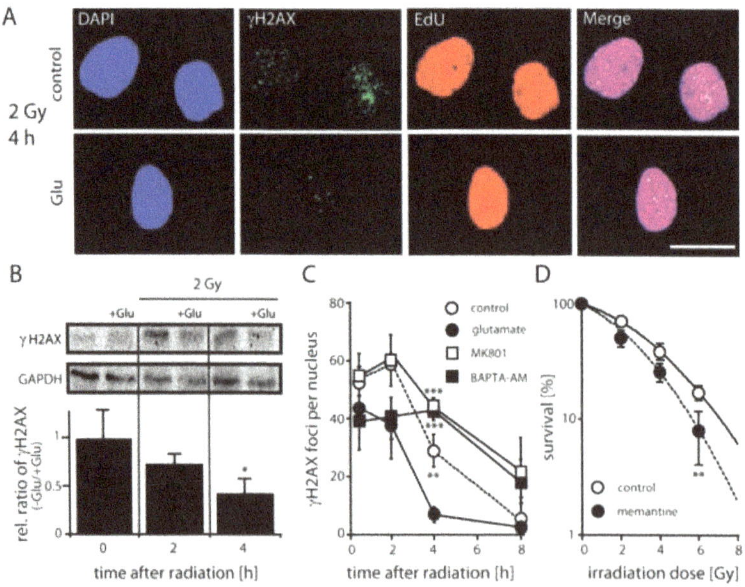

Figure 3. Effect of glutamate on DSB repair and clonogenic survival of LN229 cells.

(**A**) Immunofluorescence of γH2AX, EdU and DAPI from Glu-treated and untreated cells 4 h after irradiation (2 Gy). Note that addition of Glu (1 mM) resulted in a significant decrease in DSB levels in EdU-positive cells compared to mock-irradiated control cells from 29 ± 5 to 7 ± 2 foci ($p < 0.01$; n = 3; Student's *t*-test). Scale bar 20 μM. (**B**) γH2AX protein levels and their relative ratio (normalized to GAPDH expression) in the absence and presence of Glu (1 mM) before and after IR with a dose of 2 Gy detected by western blotting at the indicated time points (* $p < 0.05$; n = 3; Student's t-test). (**C**) Repair kinetics of γH2AX foci in S/G2 phase cells after IR (2 Gy) under control conditions, in the presence of Glu (1 mM), MK801 (10 μM) and BAPTA-AM (3 μM). The difference between Glu-treated/untreated or MK801 and BAPTA-AM treated cells at the time point of 4 h is determined by one-way ANOVA followed by Bonferroni's post-hoc test, ** $p < 0.05$, *** $p < 0.001$. (**D**) Colony formation of cells treated with memantine (50 μM) upon IR. The calculated radiation-induced cytotoxicity enhancement factor at 2 Gy is 1.4. Data are given as means ± SD of three independent experiments. The significant difference between treated and untreated cells is determined by Student's *t*-test (** $p < 0.01$).

Both ifenprodil and memantine significantly reduced the 10% clonogenic survival rates in comparison to mock-treated controls (Figure 3D), resulting in a calculated radiation-induced cytotoxicity enhancement factor at 2 Gy of 1.81 and 1.4 for ifenprodil and memantine, respectively. Thus, our data indicate that antagonizing NMDARs by memantine or by the GluN2B-specific compound ifenprodil sensitize LN229 cells to radiation, likely by decreasing DSB repair capacity.

2.4. NMDARs Increase DSB Repair Capacity and Clonogenic Survival in LN229 and U-87MG Cells by Activation of the Transcription Factor CREB

To understand mechanistically how NMDAR activation results in an altered DSB repair, we measured relative levels of the phosphorylated form of CREB at Ser133 (pCREB) in control and Glu-stimulated LN229 cells because CREB has been suggested to be involved in the resistance to IR [40]. We found that Glu induced a significant increase in the amount of pCREB protein after 30 min. (Figure 4A). To see whether activation of CREB through NMDARs can affect the expression of a downstream marker, we analyzed the induction of the brain derived neurotrophic factor (BDNF) gene in the absence of Glu and upon Glu treatment [34]. We found that Glu (1 mM) enhanced significantly the release of BDNF in LN229 cells (1.9 ± 0.15-fold, $p < 0.05$, n = 3, Student's t-test) compared to the situation in the absence of Glu (SAS, 250 μM). The Glu-mediated increase in BDNF release could be efficiently blocked by MK801 (10 μM) and by the CREB specific inhibitor KG-501 [41] (25 μM). These data indicate that Glu can induce transcriptional-active CREB phosphorylation in LN229 cells via the activation of NMDARs. To further analyze the impact of NMDAR-mediated CREB activation on DSB repair, we analyzed the relative levels of γH2AX in MK801 and KG-501 treated cells upon IR in the presence of Glu (Figure 4B). We found that both antagonists induced a significant increase in the amount of γH2AX protein after 30 min irradiation (Figure 4B), suggesting that an effective DNA repair of radiation-induced DSBs in LN229 cells may depend on both, the activation of NMDARs and CREB.

To further examine a possible involvement of CREB in Glu-induced radioresistance, we next analyzed DNA repair and clonogenic survival of LN229 and U-87MG cells in the presence of Glu and the CREB inhibitor KG-501 plus/minus memantine. We found an increased number of γH2AX foci after single KG-501 or memantine treatment in non-S-phase LN229 and U-87MG cells at 4 h after irradiation with 2 Gy, while 0.5 h after irradiation foci numbers were slightly decreased in LN229 and not changed in U-87MG cells upon inhibitor treatment (Figure 4C,D). Thus, our data revealed a significant increase of γH2AX foci in both, KG-501 and memantine treated cells after 4h compared to those treated with Glu alone. Double treatment with KG-501 and memantine only slightly increased the number of foci at 4 h after irradiation in comparison to single treatments, showing sub-additive effects of combined CREB and NMDAR inhibition (Figure 4C,D). Similarly, radiation survival of Glu-treated LN229 (Figure 4E) or U-87MG cells (Figure 4F) was only slightly further decreased upon combined treatment with KG-501 and memantine, highlighting a direct involvement of CREB in Glu-induced radiation resistance of glioblastoma cells and indicating that CREB is a key mediator in NMDAR-mediated DNA repair.

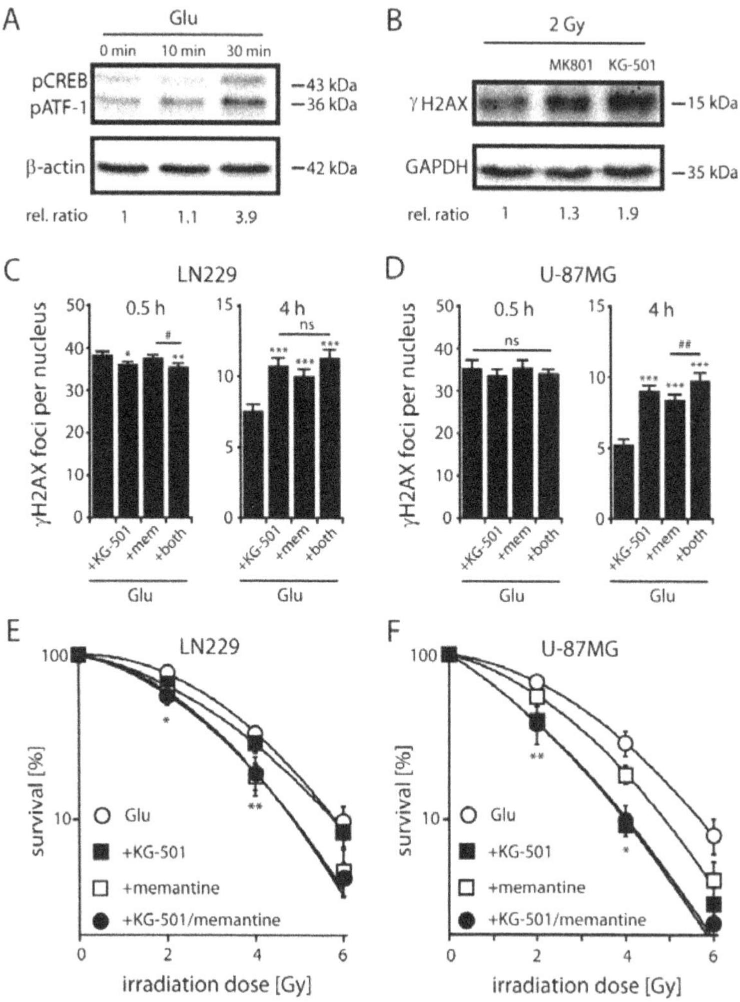

Figure 4. Effects of glutamate-induced pCREB on DNA repair and clonogenic survival in glioblastoma cells. (**A**) Relative pCREB levels in LN229 cells detected by western blotting (normalized to β-actin) at the indicated time points after treatment with Glu (1 mM). Note that the pCREB antibody is also detecting pATF-1, a protein closely related in structure and function to CREB (relative ratio at 30 min: 3.7 ± 1.1, $p < 0.05$; n = 3; Student's t-test). (**B**) γH2AX protein levels and their relative ratio (normalized to GAPDH expression) in the presence of Glu (1 mM) and MK801 (10 μM) or KG-501 (25 μM) 4h after IR with a dose of 2 Gy detected by western blotting. Relative ratio for MK801 treatment is 1.4 ± 0.3 (ns; n = 3; Student's t-test) and for KG-501 1.8 ± 0.4 ($p < 0.05$; n = 3; Student's t-test). (**C**) γH2AX foci detection in EdU-negative non-S-phase LN229 cells at 0.5 h and 4 h following 2 Gy exposure after treatment with Glu (1 mM) and with Glu in the presence of KG-501 (25 μM), memantine (mem, 100 μM) and both inhibitors. Data represent means ± SEM of γH2AX foci of three independent experiments. Data were analyzed by one-way ANOVA followed by Bonferroni´s post-hoc test. * indicates significant difference between Glu-treated control and Glu in the presence of the antagonists, # indicates significant

difference between the different antagonists ($^{*/\#}$ $p < 0.05$, ** $p < 0.01$, *** $p < 0.001$). (**D**) γH2AX foci detection in EdU-negative non-S-phase U-87MG cells at 0.5 h and 4 h following 2 Gy exposure after treatment with Glu (1 mM) and with Glu in the presence of KG-501 (25 μM), memantine (mem, 100 μM) and both inhibitors. Data represent means ± SEM of γH2AX foci of three independent experiments. Data analyses see Figure legend (**C**) ($^{\#\#}$ $p < 0.01$ *** $p < 0.001$). (**E**) Clonogenic survival of LN229 cells treated with 1 mM Glu in the absence and presence of KG-501 (1 μM), memantine (25 μM) and both inhibitors. Diagram shows fitted data. Memantine significantly reduces the survival starting at 2 Gy compared to Glu-treated cells (n = 3, each experiment was performed as triplet; error bars show SD; Student´s t-test, * $p < 0.05$, ** $p < 0.01$). (**F**) Clonogenic survival of U-87MG cells treated with 1 mM Glu in the absence and presence of KG-501 (1 μM), memantine (25 μM) and KG-501 and memantine. Both, memantine and KG-501 significantly reduce the survival starting at 2 Gy compared to Glu-treated cells. Note that co-application of KG-501 and memantine shows no significant further effect (n = 2, each experiment was performed as triplet; error bars show SD; student´s t-test, * $p < 0.05$, ** $p < 0.01$).

3. Discussion

Here, we present the first evidence that functional expression of Ca^{2+}-permeable NMDARs in human glioblastoma cell lines is correlated with a radiation response. Our results show that glutamate improves upon activation of NMDARs DSB repair whereas inhibition by NMDAR antagonists impairs this process. Furthermore, we could identify a decisive role of NMDAR activity and subsequent downstream CREB activation in DSB repair since inhibition of CREB by KG-501 showed similar effects compared to NMDAR antagonists. Hence, our work demonstrates that NMDAR and CREB activation play an important role in IR-induced cellular damage in glioblastoma cells. In addition, treatment with the GluN2B subunit-specific NMDAR antagonist ifenprodil showed a decreased cell survival and cell migration resulting in a more pronounced radiosensitizing effect as compared to MK801, highlighting the clinical potential of GluN2B subunit-specific NMDAR-antagonists and CREB-mediated downstream signaling in optimizing RT in glioblastoma treatment.

It has been shown that glioblastomas with a high Glu release have a distinct growth advantage [14,31]. Our results show that LN229 glioblastoma cells release glutamate into the media in an amount sufficient to activate NMDARs capable to induce NMDAR-mediated excitatory toxic effects [39,42]. Since we could not detect any Glu-induced excitotoxic effects in LN229 cells, we assume that this may reflect an increased capacity of LN229 cells to buffer intracellular Ca^{2+} allowing to survive prolonged Ca^{2+} influx. In contrast, we found a decrease in the viability of the LN229 cells when blocking glutamate release by SAS and by NMDAR antagonists, indicating that Glu induce NMDAR-mediated cell survival instead of cell death. Furthermore, when LN229 cells were cultured in the presence of SAS or NMDAR antagonists, the number of lamellipodia decreased, which might be suitable for tumor cell migration and invasion. In favor of our findings, suppression of Glu levels by SAS in orthotopic glioblastoma mice models prolonged survival and suppressed tumor growth in vitro and in vivo [15,24,43]. The potential of inhibiting the cystine/glutamate-mediated autocrine glutamate effect in glioblastoma treatment is also highlighted by a phase I clinical trial in newly diagnosed glioblasoma patients [10].

Concerning the target structures of an autocrine effect of Glu in glioblastoma biology, it has been shown that Glu stimulates the growth, migration and invasion of glioblastoma cells through activation of the AMPAR-subtype of iGluRs [22] and that iGluR antagonists increase the lifespan of human glioma xenografted in mice [27]. Other publications have shown a role of NMDAR-signaling pathways in tumor progression [23,29] and that NMDAR-antagonists have antitumoral effects when used in various xenograft tumors [44]. Here we show by Ca^{2+}-imaging and patch-clamp recording that three glioblastoma cell lines (LN229, U-87MG and T98G) express functional NMDARs and that treatment of LN229 cells by blocking NMDAR-mediated glutamatergic signaling results in i) a decreased cell migration and survival and ii) a sensitization to IR. Thus, our results show that at least some glioblastoma cells express functional NMDARs and that stimulation with Glu elevates intracellular Ca^{2+} concentrations. Deregulated Ca^{2+} signaling is a prominent feature of pathological states, including those defined as "hallmark of cancer" [45]. Since NMDAR antagonists have been shown to inhibit migration of various types of tumor cells, including glioblastoma cells ([23,31]; this study), memantine constitutes a promising antagonist against "oncogenic NMDARs" [31], which is currently investigated in a phase I trial for post-radiation therapy in glioblastoma [46]. Interestingly, we further could show by colony formation assays and viability tests that blocking GluN2B-containing NMDARs by ifenprodil suppresses tumor cell survival more efficiently compared to the broad NMDAR antagonists MK801 and memantine. Based on these results we assume differential effects of NMDAR subunit-specific antagonists, indicating that subunit composition and/or membrane localization of the NMDAR subunits contribute different to glioblastoma malignancy. Indeed, we found a specific localization of the GluN2B-subunit in lamellipodia and a pronounced effect of the GluN2B-specific antagonist ifenprodil on migration of LN229 cells. Due to the high invasiveness of glioblastoma cells, our data are consistent with the idea of a specific role of GluN2B-NMDAR-signaling pathways at the invasive front of glioblastoma progression. This assumption is supported by the finding that grade 4 glioblastoma patients with high expression levels of GluN2B in the tumor have a worse prognosis [29] and underpin GluN2B-specific antagonists as an interesting therapeutic approach for treating brain tumors.

IR-induced DSBs are the primary mechanism of tumor cell death [47]. Remarkably, we found a prominent role of the NMDAR and CREB in the modulation of the radiation response of LN229 and U-87MG cells. The effect of NMDARs and CREB on the repair of IR-induced DSBs is reflected by our finding that both, antagonizing NMDAR-mediated activation of CREB and CREB-itself increase the persistence of γH2AX foci after radiation in a non-additive manner. Thus, our results show that activation of NMDARs and the subsequent phosphorylation of CREB improve the repair of IR-induced DSBs and indicate that NMDAR-mediated downstream signaling via CREB plays a critical role in the protection of IR-induced cell damage in glioblastomas. In line with our findings, it has been shown that CREB is overexpressed in Acute Myeloid Leukemia (AML) cells and is associated with a poor prognosis [48]. The authors assume that CREB overexpression leads to chemotherapy resistance due to an increased DSB repair activity in AML cells. Expression of dominant negative CREB has also been demonstrated to reduce resistance to UV radiation in melanoma cells [49] and the tumorigenic potential in nude mice [50]. Remarkably, Amorino et al., 2003 described a direct link between CREB function and DSB repair [51]. Interestingly, the mechanism of radiosensitization included a reduction of the proliferating cell nuclear antigen (PCNA), a protein involved in the repair of IR-induced DNA damage [51]. However, CREB-activity increases also the release of BDNF in neoplastic cells) [34,52]. Interestingly, BDNF is thought to contribute to DNA damage repair (overview in [52]) and promotes survival and migration in C6 glioma cells [53]. Furthermore, Yano and colleagues (1998) have demonstrated in neuroblastoma cells that Ca^{2+}-entry through NMDAR phosphorylates a serine/threonine kinase Akt, which activates CREB and results in the release of a variety of anti-apoptotic signals facilitating cell survival [54]. Thus, several data suggest an important role of CREB for cellular response to IR and the regulation of a presumptive transcriptional program that mediates tumor growth and invasion. However, how NMDAR-mediated activation of CREB affects tumor cell survival and the response to IR remains elusive.

In conclusion, our work demonstrates that NMDARs and CREB play a role in the protection of IR-induced cell damage. CREB is phosphorylated in response to NMDAR-activation and the sensitivity of glioblastoma cells to IR is enhanced by the inhibition of both, NMDAR- and CREB-activity. Thus, further studies are needed to investigate NMDAR and CREB antagonism for enhancing tumor radiotherapy effects and to understand downstream effectors in glioblasoma physiology.

4. Materials and Methods

4.1. Cell Culture

Human glioblastoma cell lines LN229 ($IDH1^{wt}$), U-87MG ($IDH1^{wt}$; $P53^{wt}$), T98G and LN428 were obtained from the American Type Culture Collection and were cultured in Dulbecco's Modified

Eagle's Medium (LN229, DMEM, Sigma-Aldrich, Munich, Germany) or Minimum Essential Medium (U-87MG, T98G and LN428, MEM, Sigma-Aldrich) supplemented with 10% fetal bovine serum (FBS, Sigma–Aldrich), and 1% penicillin/streptomycin (Sigma-Aldrich). The cultures were kept at 37°C in a humidified atmosphere of 95% air and 5% CO_2.

4.2. Chemical Treatment and X-Irradiation

LN229 and U-87MG were plated in cell culture dishes or well plates 24-48h before treatment. Chemicals were dissolved in ddH_2O or DMSO (max. 0.2%). All antagonists (MK801, memantine, ifenprodil, BAPTA-AM (Tocris, Cologne, Germany) and KG-501 (Sigma-Aldrich) were applied in presence of the agonist Glu or NMDA (Sigma-Aldrich) and glycine (Roth, Karlsruhe, Germany) (if not stated otherwise) and maintained during the whole assay. All chemicals were added prior to irradiation. X-ray irradiation was performed at 90 kV and 19 mA with an aluminum filter with single doses of 2, 4 and 6 Gy using an x-ray tube equipped with a tungsten-anode (Philips, Amsterdam, Netherlands) as described previously [55]. X-ray treatment was performed using a power of 19 mA, 90 kV voltage and 30 cm distance to the IR source, which makes an applied dose of 1.96 Gy/min, established by Ficke-dosimetry. To all radiated samples equal control samples were performed and placed in the deactivated x-ray tube for the same time.

4.3. Measurement of Extracellular Glutamate and BDNF Concentrations

Extracellular Glu levels were detected using the Glu assay kit (KA1670, Abnova, Taipeh, Taiwan) according to the manufacturer's protocol where glutamate is converted to 2-oxoglutarate via glutamate dehydrogenase reducing NAD^+ to NADH. The subsequent oxidation of NADH drives a reaction catalysed by diaphorase, resulting in the conversion of *p*-INT to a formazan product, the absorption of which is measured spectrophotometrically. LN229 were grown to confluence in 60 mm dishes. Conditioned culture supernatant (250 μL) was collected after 2 h, 6 h and 24 h and analysed immediately. Collected media was transferred to a 96-well plate where glutamate dehydrogenase NAD^+, and formazan were mixed. Absorbance of OD 565 nm was measured at time 0 and time 30 min using an Infinite M200 microplate reader (Tecan, Männedorf, Switzerland). The Glu exchange inhibitor sulfasalazine (SAS; 250 μM) was added to block release of Glu. The Glu concentration in the experimental medium was calculated via linear regression analysis (GraphPad Prism 7.0, GraphPad Software, San Diego, CA, USA) using defined Glu concentrations as standard. Measurement of extracellular BDNF concentrations: 2×10^4 LN229 cells/well were seeded into 24 well plates and treated with sulfasalazine (SAS; 250 μM), 1 mM Glu, 20μM MK801 and 25 μM KG-501 directly after. The next day the supernatants of the cells were collected and centrifuged at 16200 g for 5 min. to remove remaining cells and debris. The BDNF concentration was determined with a BDNF ELISA Kit (ab99978, Abcam) following the manufacturer instructions. The absorbance at 450 nm was measured with the Tecan Infinite M 200 microplate reader.

4.4. 3-(4,5-Methylthiazol-2-yl)-2,5-Diphenyl-Tetrazolium Bromide (MTT) Assay

LN229 cells were plated at a density of 10^3 cells in 96-well plates. 24 h after plating, cells were treated with memantine or MK801 and/or irradiation and after additional 48 h DMEM was supplemented with 10 μL of 3-(4,5-dimethylthiazol-2-yl)-2,5-diphenletrazolium bromide (MTT) reagent (5 mg/mL in PBS) to each well and incubated for 1 h at 37 °C. After discarding the MTT solution 150 μL isopropanol/0.04 N HCl was added and the plate was incubated till the formazan crystals were dissolved. The reaction product was quantified by measuring the absorbance at 570 nm with reference of 630 nm using the Tecan Infinite M 200 microplate reader.

4.5. Clonogenic Survival Assay

Clonogenic cell survival of LN229 and U-87MG cells treated with memantine, KG-501 or MK801 was analyzed by means of standard colony formation assay. Briefly, cells were trypsinized and plated at a constant cell density (500 cells per well in 6-well plate), and incubated for 7 days in the presence of memantine, KG-501 or MK801 with different concentrations. After 7 days of colony growth, cultures were fixed and stained with methylene-blue solution. The number of colonies formed with more than 50 cells were then determined and the plating efficiency (PE) and survival fractions (SF) were calculated. Calculation of SF was performed using the number of colonies of treated cells divided by that for the control cells seeded -(PE/100), taking into consideration the individual PE. Clonogenic cell survival of cells treated with memantine and KG-501 was analyzed after radiation (2, 4, and 6 Gy). The number of seeded cells was increased with irradiation dose (LN229: 400 cells/well form 0 Gy and 2 Gy, 800 cells/well for 4 Gy and 1600 cells/well at 6 Gy. U-87MG: 800 cells/well for 0 Gy and 2 Gy, 1600 cells/well for 4Gy and 3000 cells/well for 6 Gy). The cells were allowed to attach for 3 h and then irradiated in a X-ray tube as descripted above. Colonies were allowed to form for 8 (LN229) and 11 (U-87MG) days, fixated with 70% ethanol and stained with 0.1% crystal violet in 25% ethanol. The colonies with more than 50 cells were manually counted. The X-ray dose-survival curves were fitted by GraphPad Prism software version 4.0, to the linear quadratic equation, surviving fraction (SF) = $\exp(-\alpha D - \beta D^2)$, where D is the X-ray dose. The radiation sensitizing enhancement ratio (SER) by treatment was used to evaluate the drug-radiation interaction and calculated at a dose of 2 Gy using the following formula: SER = $(SF2_{control})/(SF2_{antagonist})$. SER >1 suggests a radio-sensitizing effect. Each point on the survival curves represents the mean surviving fraction from at least two independent experiments performed in triplicate.

4.6. Western Blot Analysis

Samples were homogenized in 100 µL of whole-cell lysis buffer (150 M NaCl, 0.5% dodecylmaltosid, 50 mMTris, pH 7.5, 0.5% Triton X-100) mixed with complete protease inhibitor cocktail (Roche Diagnostics, Mannheim, Germany), and then lysed on ice. Equal amounts of protein samples (100 µg) were subjected to 14% SDS-PAGE and transferred to polyvinylidenedifluoride membrane (Amersham, Freiburg, Germany). Primary antibodies for γH2AX (Ser139) (JWB 301, Millipore, Schwalbach, Germany) and pCREB (Ser133) (Cell Signaling, Frankfurt, Germany) were used at 1:1000 dilutions. Secondary peroxidase- conjugated antibodies (Chemicon, Temecula, CA, USA and Santa Cruz, Heidelberg, Germany) were used at 1:10000, and the ECL Western Blotting Substrate (Pierce Thermo Fisher Scientific, Darmstadt, Germany) was used for visualizing the antibody-bound protein. To confirm equal protein loading, membranes were subsequently re-probed with anti-GABDH or anti-β-actin antibodies (Santa Cruz Biotechnology, Heidelberg, Germany).

4.7. Immunofluorescence Staining

LN229 cells (4×10^4 cells per channel) were grown in µ-slides $VI^{0,4}$ (Ibidi) over night, fixed with 4% PFA, permeabilized with 0.1% Triton X-100, blocked with 0.5% BSA/5% goat serum and incubated over night at 4 °C with following antibodies: anti-GluN1 (1:100, D65B7, Cell Signaling, Frankfurt, Germany), anti-GluN2A (1:100, N327A/38, Abcam), anti-GluN2B (1:200, S59-20, Stress Marq, Cadboro Bay, Victoria, Canada). After incubation with Alexa 488 labeled secondary antibodies (1:400, Abcam), the samples were stained with Hoechst 33342 and analyzed on an inverted epifluorescence microscope (Zeiss, Oberkochen, Germany).

4.8. γH2AX/EdU Double-Staining

LN229 and U-87MG cells grown on glass coverslips were treated with 10 µM EdU for 1 h to discriminate between S/G2- and G1-phase cells and irradiated with a dose of 2 Gy. Cells were fixed with 4% paraformaldehyde (PFA) for 10 minutes and permeabilized in 0.5% Triton X-100. Staining was performed with mouse-α-γH2AX antibody at 1:1000 (JWB 301, Millipore) and Click-it EdUAlexa Fluor 594 kit (Life Technologies, Darmstadt, Germany). Nuclei were counterstained with 4′,6-diamidino-2-phenylindole (DAPI) solution (Invitrogen, Karlsruhe, Germany) and coverslips were mounted with Mowiol (Roth). Images were taken using an AxioImager Z1 microscope and Axiovision

4.6. Micromanager software (Zeiss). For γH2AX foci quantification, 150–200 nuclei were evaluated for each data point from three independent experiments.

4.9. Fluorescence-Activated Cell Cycle Analysis

LN229 cells were grown in μ-slides $VI^{0,4}$ (Ibidi) over night with/without glutamate/MK801 and treated with 10 μM EdU for 30 min. The cells were fixed with 4% PFA, permeabilized with 0.1% Triton X-100 and labeled with Alexa 594 azide using the Click-iT EdU Imaging Kit (Invitrogen). The samples were stained with Hoechst 33342 and imaged on an inverted epifluorescence microscope (Zeiss) and analyzed with micro manager. For this the integrated density of the Hoechst signal of single nuclei was plotted against the mean EdU signal and G1/S/G2 phases were gated manually. Mean + SEM of at least three independent experiments were calculated and graphed. Analysis was performed on a FACScan (Becton Dickinson, Heidelberg, Germany) and data were analyzed using the ModFit LT 3.2 software (Verity Software House, Topsham, ME, USA).

4.10. Migration Assay

For measuring the change in the cell-covered area, over time as a parameter for migration rate we used culture inserts (Ibidi) consisting of two chambers. 2×10^4 LN229 cells were seeded into each chamber in DMEM with 10% FBS. After 24 h the culture inserts were removed and quantification of migration was done by taking six pictures of the gap after 6, 24 and 48 h and calculating the area of the gap with ImageJ/FIJI software (http://fiji.sc/Fiji). Each experiment was done in triplicate.

4.11. Electrophysiology

Patch clamp recordings of ligand-gated whole cell currents were performed at −70 mV with the Port-a-Patch System (Nanion, Munich, Germany). Cells were trypsinized before recording and resuspended in external solution. Cells were sealed in solution containing (in mM) 80 NaCl, 3 KCl, 10 $MgCl_2$, 35 $CaCl_2$, 10 HEPES /NaOH, pH 7.4. For recordings a buffer with (in mM) 4 KCl, 140 NaCl, 2 $CaCl_2$, 5 D-Glucose, 10 HEPES /NaOH, pH 7.4 was used as an external bath solution. The intracellular solution contained (in mM) 50 KCl, 10 NaCl, 60 K-Fluoride, 10 EGTA and 10 HEPES/KOH, pH 7.2. The sealing process and the access into the whole-cell mode were achieved with the help of PatchControl Software (Nanion). Solution exchange was executed by a rapid perfusion-system (Nanion). Currents were recorded with an EPC9 amplifier under the control of the Patchmaster Software (both from Heka Electronic, Lambrecht, Germany) as described previously [56]. Data were analyzed with Patchmaster and Fitmaster software (Heka Electronic).

4.12. Calcium Imaging

LN229 cells were seeded into 8-well μ-slides (Ibidi) and treated with the calcium dye Fluo-4 AM (Thermo Fisher) dissolved in DMSO loaded for 30 min. to a final concentration of 2 μM. The cells were washed once with the imaging-buffer (140 mM NaCl; 2.8 mM KCl; 1.8 mM $CaCl_2$; 10 mM HEPES; 20 mM Glucose; 10 μM EDTA; pH 7.2) and imaged in a recording chamber on a epifluorescence microscope (Zeiss) at a rate of 1 Hz for 120 seconds. Alterations in fluorescence as a function of time were measured at a single wavelength. Glutamate and glycine were applied to the cells to an end-concentration of 1 mM and 100 μM respectively after 20 seconds and 10 μM ionomycin was given after 90 seconds. The image sequences were analyzed for visual inspection and processed by using *ImageJ* /FIJI software (http://fiji.sc/Fiji) by subtracting the mean background of the integrated density of every cell and then normalize all values to the fist and the highest value of every cell. The data were plotted as relative fluorescence intensity scale versus time.

4.13. Data Analysis

Experimental data are presented as mean ± SEM from three or more independent experiments (if not indicated otherwise). Levels of significance were calculated using the Student´s unpaired t-test or one-way analyses of variance (ANOVA followed by Bonferroni´s post-hoc test) (GraphPadPrism).

5. Conclusions

In conclusion, our findings in LN229 and U-87MG cells support a new approach for the therapy of brain tumors based on antagonizing NMDAR-mediated signaling pathways upon radiation resulting in (i) an enhanced radiosensitivity by suppressing DSB repair capacity and (ii) a decreasing tumor cell survival and migration.

Author Contributions: A.M.-L., H.L., S.H., F.R., K.R. and B.L. conceived and designed the experiments; A.M.-L., H.L., S.H. and K.R. performed the experiments; A.M.-L., H.L., S.H. and K.R. analyzed the data; K.R. and B.L. wrote the paper.

Funding: This research was funded by the German Federal Ministry of Education and Research (BMBF; NeuroRad, 02NUK034B; GREWISalpha, 02NUK050D) and the German Research Society (DFG; GRK1657).

Acknowledgments: The authors gratefully acknowledge the excellent technical assistance of Katrin Merk and Gabriele Wenz and thank Markus Löbrich (TU Darmstadt, Germany) for excellent support. The authors acknowledge support by the German Research Foundation and the Open Access Publishing Fund of Technische Universität Darmstadt.

Conflicts of Interest: The authors declare no conflict of interest. The funders had no role in the design of the study, in the collection, analyses, or interpretation of data, in the writing of the manuscript, and in the decision to publish the results.

References

1. Kleihues, P.; Louis, D.N.; Scheithauer, B.W.; Rorke, L.B.; Reifenberger, G.; Burger, P.C.; Cavenee, W.K. The WHO classification of tumors of the nervous system. *J. Neuropathol. Exp. Neurol.* **2002**, *61*, 215–225. [CrossRef]
2. Wen, P.Y.; Kesari, S. Malignant gliomas in adults. *N. Engl. J. Med.* **2008**, *359*, 492–507. [CrossRef]
3. Stupp, R.; Hegi, M.E.; Mason, W.P.; van den Bent, M.J.; Taphoorn, M.J.; Janzer, R.C.; Ludwin, S.; Allgeier, A.; Fisher, B.; Belanger, K.; et al. Effects of radiotherapy with concomitant and adjuvant temozolomide versus radiotherapy alone on survival in glioblastoma in a randomised phase III study: 5-year analysis of the EORTC-NCIC trial. *Lancet Oncol.* **2009**, *10*, 459–466. [CrossRef]
4. Wild-Bode, C.; Weller, M.; Rimner, A.; Dichgans, J.; Wick, W. Sublethal irradiation promotes migration and invasiveness of glioma cells: Implications for radiotherapy of human glioblastoma. *Cancer Res.* **2001**, *61*, 2744–2750. [PubMed]
5. Stupp, R.; Mason, W.P.; Van Den Bent, M.J.; Weller, M.; Fisher, B.; Taphoorn, M.J.; Belanger, K.; Brandes, A.A.; Marosi, C.; Bogdahn, Y.; et al. Radiotherapy plus concomitant and adjuvant temozolomide for glioblastoma. *N. Engl. J. Med.* **2005**, *352*, 987–996. [CrossRef] [PubMed]
6. Bao, S.; Wu, Q.; McLendon, R.E.; Hao, Y.; Shi, Q.; Hjelmeland, A.B.; Dewhirst, M.W.; Bigner, D.D.; Rich, J.N. Glioma stem cells promote radioresistance by preferential activation of the DNA damage response. *Nature* **2006**, *444*, 756–760. [CrossRef]
7. Atkins, R.J.; Ng, W.; Stylli, S.S.; Hovens, C.M.; Kaye, A.H. Repair mechanisms help glioblastoma resist treatment. *J. Clin. Neurosci.* **2015**, *22*, 14–20. [CrossRef] [PubMed]
8. Manini, I.; Caponnetto, F.; Bartolini, A.; Ius, T.; Mariuzzi, L.; Di Loreto, C.; Beltrami, A.P.; Cesselli, D. Role of Microenvironment in Glioma Invasion: What We Learned from In Vitro Models. *Int. J. Mol. Sci.* **2018**, *19*, 147. [CrossRef] [PubMed]
9. Marcus, H.J.; Carpenter, K.L.; Price, S.J.; Hutchinson, P.J. In vivo assessment of high-grade glioma biochemistry using microdialysis: A study of energy-related molecules, growth factors and cytokines. *J. Neurooncol.* **2010**, *97*, 11–23. [CrossRef] [PubMed]
10. De Groot, J.; Sontheimer, H. Glutamate and the biology of gliomas. *Glia* **2011**, *59*, 1181–1189. [CrossRef] [PubMed]

11. Ye, Z.C.; Rothstein, J.D.; Sontheimer, H. Compromised glutamate transport in human glioma cells: Reduction-mislocalization of sodium-dependent glutamate transporters and enhanced activity of cystine-glutamate exchange. *J. Neurosci.* **1999**, *19*, 10767–10777. [CrossRef]
12. Buckingham, S.C.; Campbell, S.L.; Haas, B.R.; Montana, V.; Robel, S.; Ogunrinu, T.; Sontheimer, H. Glutamate release by primary brain tumors induces epileptic activity. *Nat. Med.* **2011**, *17*, 1269–1274. [CrossRef] [PubMed]
13. Simon, M.; von Lehe, M. Glioma-related seizures: Glutamate is the key. *Nat. Med.* **2011**, *17*, 1190. [CrossRef] [PubMed]
14. Lo, M.; Wang, Y.Z.; Gout, P.W. The x(c)-cystine/glutamate antiporter: A potential target for therapy of cancer and other diseases. *J. Cell. Physiol.* **2008**, *215*, 593–602. [CrossRef] [PubMed]
15. Chung, W.J.; Lyons, S.A.; Nelson, G.M.; Hamza, H.; Gladson, C.L.; Gillespie, G.Y.; Sontheimer, H. Inhibition of cystine uptake disrupts the growth of primary brain tumors. *J. Neurosci.* **2005**, *25*, 7101–7110. [CrossRef]
16. Huberfeld, G.; Vecht, C.J. Seizures and gliomas—towards a single therapeutic approach. *Nat. Rev. Neurol.* **2016**, *12*, 204–216. [CrossRef]
17. Dingledine, R.; Borges, K.; Bowie, D.; Traynelis, S.F. The glutamate receptor ion channels. *Pharmacol. Rev.* **1999**, *51*, 7–61. [PubMed]
18. Ikonomidou, C.; Bosch, F.; Miksa, M.; Bittigau, P.; Vöckler, J.; Dikranian, K.; Tenkova, T.I.; Stefovska, V.; Turski, L.; Olney, J.W. Blockade of NMDA receptors and apoptotic neurodegeneration in the developing brain. *Science* **1999**, *283*, 70–74. [CrossRef]
19. Prickett, T.D.; Samuels, Y. Molecular pathways: Dysregulated glutamatergic signaling pathways in cancer. *Clin. Cancer Res.* **2012**, *18*, 4240–4246. [CrossRef]
20. Bozic, M.; Valdivielso, J.M. The potential of targeting NMDA receptors outside the CNS. *Expert Opin. Ther. Targets* **2015**, *19*, 399–413. [CrossRef]
21. Ribeiro, M.P.; Custodio, J.B.; Santos, A.E. Ionotropic glutamate receptor antagonists and cancer therapy: Time to think out of the box? *Cancer Chemother. Pharmacol.* **2017**, *79*, 219–225. [CrossRef] [PubMed]
22. Ishiuchi, S.; Tsuzuki, K.; Yoshida, Y.; Yamada, N.; Hagimura, N.; Okado, H.; Miwa, A.; Kurihara, H.; Nakazato, Y.; Tamura, M.; et al. Blockage of Ca(2+)-permeable AMPA receptors suppresses migration and induces apoptosis in human glioblastoma cells. *Nat. Med.* **2002**, *8*, 971–978. [CrossRef]
23. Rzeski, W.; Turski, L.; Ikonomidou, C. Glutamate antagonists limit tumor growth. *Proc. Natl. Acad. Sci. USA* **2001**, *98*, 6372–6377. [CrossRef]
24. de Groot, J.F.; Piao, Y.; Lu, L.; Fuller, G.N.; Yung, W.A. Knockdown of GluR1 expression by RNA interference inhibits glioma proliferation. *J. Neurooncol.* **2008**, *88*, 121–133. [CrossRef] [PubMed]
25. Stepulak, A.; Luksch, H.; Gebhardt, C.; Uckermann, O.; Marzahn, J.; Sifringer, M.; Rzeski, W.; Staufner, C.; Brocke, K.S.; Turski, L.; et al. Expression of glutamate receptor subunits in human cancers. *Histochem. Cell Biol.* **2009**, *132*, 435–445. [CrossRef]
26. Noch, E.; Khalili, K. Molecular mechanisms of necrosis in glioblastoma: The role of glutamate excitotoxicity. *Cancer Biol. Ther.* **2009**, *8*, 1791–1797. [CrossRef]
27. Goudar, R.K.; Keir, S.T.; Bigner, D.D.; Friedman, H.S. NMDA and AMPA glutamate receptor antagonists in the treatment of human malignant glioma xenografts. *Exp. Mol. Ther.* **2004**, *64*, 41.
28. Lyons, S.A.; Chung, W.J.; Weaver, A.K.; Ogunrinu, T.; Sontheimer, H. Autocrine glutamate signaling promotes glioma cell invasion. *Cancer Res.* **2007**, *67*, 9463–9471. [CrossRef]
29. Li, L.; Hanahan, D. Hijacking the neuronal NMDAR signaling circuit to promote tumor growth and invasion. *Cell* **2013**, *153*, 86–100. [CrossRef] [PubMed]
30. Ramaswamy, P.; Devi, N.A.; Fathima, K.H.; Nanjaiah, N. Activation of NMDA receptor of glutamate influences MMP-2 activity and proliferation of glioma cells. *Neurol. Sci.* **2014**, *35*, 823–829. [CrossRef]
31. Takano, T.; Lin, J.H.C.; Arcuino, G.; Gao, Q.; Yang, J.; Nedergaard, M. Glutamate release promotes growth of malignant gliomas. *Nat. Med.* **2001**, *7*, 1010–1015. [CrossRef]
32. Laube, B.; Kuhse, J.; Betz, H. Evidence for a tetrameric structure of recombinant NMDA receptors. *J. Neurosci.* **1998**, *18*, 2954–2961. [CrossRef] [PubMed]
33. Laube, B.; Hirai, H.; Sturgess, M.; Betz, H.; Kuhse, J. Molecular determinants of agonist discrimination by NMDA receptor subunits: Analysis of the glutamate binding site on the NR2B subunit. *Neuron* **1997**, *18*, 493–503. [CrossRef]

34. Hardingham, G.E.; Bading, H. Synaptic versus extrasynaptic NMDA receptor signalling: Implications for neurodegenerative disorders. *Nat. Rev. Neurosci.* **2010**, *11*, 682–696. [CrossRef] [PubMed]
35. Cull-Candy, S.; Brickley, S.; Farrant, M. NMDA receptor subunits: Diversity, development and disease. *Curr. Opin. Neurobiol.* **2001**, *11*, 327–335. [CrossRef]
36. Deutsch, S.I.; Tang, A.H.; Burket, J.A.; Benson, A.D. NMDA receptors on the surface of cancer cells: Target for chemotherapy? *Biomed. Pharmacother.* **2014**, *68*, 493–496. [CrossRef] [PubMed]
37. Stepulak, A.; Rola, R.; Polberg, K.; Ikonomidou, C. Glutamate and its receptors in cancer. *J. Neural Transm. (Vienna)* **2014**, *121*, 933–944. [CrossRef] [PubMed]
38. Mehrotra, A.; Koiri, K.R. N-Methyl-D-Aspartate (NMDA) Receptors: Therapeutic Target against Cancer. *Int. J. Immunother. Cancer Res.* **2015**, *1*, 17.
39. Van Vuurden, D.G.; Yazdani, M.; Bosma, I.; Broekhuizen, A.J.; Postma, T.J.; Heimans, J.J.; van der Valk, P.; Aronica, E.; Tannous, B.A.; Würdinger, T.; et al. Attenuated AMPA receptor expression allows glioblastoma cell survival in glutamate-rich environment. *PLoS ONE* **2009**, *4*, e5953. [CrossRef]
40. D'Auria, F.; Centurione, L.; Centurione, M.A.; Angelini, A.; Di Pietro, R. Regulation of Cancer Cell Responsiveness to Ionizing Radiation Treatment by Cyclic AMP Response Element Binding Nuclear Transcription Factor. *Front. Oncol.* **2017**, *7*, 76. [CrossRef]
41. Best, J.L.; Amezcua, C.A.; Mayr, B.; Flechner, L.; Murawsky, C.M.; Emerson, B.; Zor, T.; Gardner, K.H.; Montminy, M. Identification of small-molecule antagonists that inhibit an activator: Coactivator interaction. *Proc. Natl. Acad. Sci. USA* **2004**, *101*, 17622–17627. [CrossRef] [PubMed]
42. Choi, D.W. Glutamate neurotoxicity and diseases of the nervous system. *Neuron* **1988**, *1*, 623–634. [CrossRef]
43. Savaskan, N.E.; Heckel, A.; Hahnen, E.; Engelhorn, T.; Doerfler, A.; Ganslandt, O.; Nimsky, C.; Buchfelder, M.; Eyüpogl, I.Y. Small interfering RNA-mediated xCT silencing in gliomas inhibits neurodegeneration and alleviates brain edema. *Nat. Med.* **2008**, *14*, 629–632. [CrossRef] [PubMed]
44. North, W.G.; Gao, G.; Memoli, V.A.; Pang, R.H.; Lynch, L. Breast cancer expresses functional NMDA receptors. *Breast Cancer Res. Treat.* **2010**, *122*, 307–314. [CrossRef] [PubMed]
45. Hanahan, D.; Weinberg, R.A. The hallmarks of cancer. *Cell* **2000**, *100*, 57–70. [CrossRef]
46. Lefranc, F.; Le Rhun, E.; Kiss, R.; Weller, M. Glioblastoma quo vadis: Will migration and invasiveness reemerge as therapeutic targets? *Cancer Treat. Rev.* **2018**, *68*, 145–154. [CrossRef]
47. Hartwell, L.H.; Kastan, M.B. Cell cycle control and cancer. *Science* **1994**, *266*, 1821–1828. [CrossRef]
48. Cho, E.C.; Mitton, B.; Sakamoto, K.M. CREB and leukemogenesis. *Crit. Rev. Oncog.* **2011**, *16*, 37–46. [CrossRef]
49. Yang, Y.M.; Dolan, L.R.; Ronai, Z. Expression of dominant negative CREB reduces resistance to radiation of human melanoma cells. *Oncogene* **1996**, *12*, 2223–2233.
50. Xie, S.; Price, J.E.; Luca, M.; Jean, D.; Ronai, Z.; Bar-Eli, M. Dominant-negative CREB inhibits tumor growth and metastasis of human melanoma cells. *Oncogene* **1997**, *15*, 2069–2075. [CrossRef]
51. Amorino, G.P.; Mikkelsen, R.B.; Valerie, K.; Schmidt-Ullrich, R.K. Dominant-negative cAMP-responsive element-binding protein inhibits proliferating cell nuclear antigen and DNA repair, leading to increased cellular radiosensitivity. *J. Biol. Chem.* **2003**, *278*, 29394–29399. [CrossRef]
52. Schmidt, R.H.; Nickerson, J.M.; Boatright, J.H. Exercise as Gene Therapy: BDNF and DNA Damage Repair. *Asia Pac. J. Ophthalmol.* **2016**, *5*, 309–311. [CrossRef]
53. Xiong, J.; Zhou, L.; Lim, Y.; Yang, M.; Zhu, Y.H.; Li, Z.W.; Zhou, F.H.; Xiao, Z.C.; Zhou, X.F. Mature BDNF promotes the growth of glioma cells in vitro. *Oncol. Rep.* **2013**, *30*, 2719–2724. [CrossRef]
54. Yano, S.; Tokumitsu, H.; Soderling, T.R. Calcium promotes cell survival through CaM-K kinase activation of the protein-kinase-B pathway. *Nature* **1998**, *396*, 584–587. [CrossRef]
55. Lutz, H.; Nguyen, T.; Joswig, J.; Rau, K.; Laube, B. NMDA receptor signaling mediates cFos expression via Top2β-induced DSBs in Glioblastoma cells. *Cancers* **2019**, *11*, 306. [CrossRef]
56. Laube, B.; Kuhse, J.; Betz, H. Kinetic and mutational analysis of Zn^{2+} modulation of recombinant human inhibitory glycine receptors. *J. Physiol.* **2000**, *522*, 215–230. [CrossRef]

cancers

MDPI

Article

Radiosensitization and a Less Aggressive Phenotype of Human Malignant Glioma Cells Expressing Isocitrate Dehydrogenase 1 (IDH1) Mutant Protein: Dissecting the Mechanisms

Jacqueline Kessler [1,†,*], Tim Hohmann [2,†], Antje Güttler [1], Marina Petrenko [1], Christian Ostheimer [1], Urszula Hohmann [2], Matthias Bache [1], Faramarz Dehghani [2] and Dirk Vordermark [1]

1 Department of Radiotherapy, Faculty of Medicine, Martin Luther University Halle-Wittenberg, Ernst-Grube-Str. 40, 06097 Halle (Saale), Germany; antje.guettler@uk-halle.de (A.G.); marina.petrenko@uk-halle.de (M.P.); christian.ostheimer@uk-halle.de (C.O.); matthias.bache@uk-halle.de (M.B.); dirk.vordermark@uk-halle.de (D.V.)

2 Department of Anatomy and Cell Biology, Faculty of Medicine, Martin Luther University Halle-Wittenberg, Große Steinstraße 52, 06108 Halle (Saale), Germany; tim.hohmann@medizin.uni-halle.de (T.H.); urszula.grabiec@medizin.uni-halle.de (U.H.); faramarz.dehghani@medizin.uni-halle.de (F.D.)

* Correspondence: jacqueline.kessler@uk-halle.de; Tel.: +49-345-557-7428; Fax: +49-345-557-5803

† These authors contributed equally to this work.

Received: 29 March 2019; Accepted: 11 June 2019; Published: 25 June 2019

Abstract: The presence of an isocitrate dehydrogenase 1 (IDH1) mutation is associated with a less aggressive phenotype, increased sensitivity to radiation, and increased overall survival in patients with diffuse glioma. Based on in vitro experimentations in malignant glioma cell lines, the consequences on cellular processes of $IDH1^{R132H}$ expression were analyzed. The results revealed that $IDH1^{R132H}$ expression enhanced the radiation induced accumulation of residual γH2AX foci and decreased the amount of glutathione (GSH) independent of the oxygen status. In addition, expression of the mutant IDH1 caused a significant increase of cell stiffness and induced an altered organization of the cytoskeleton, which has been shown to reinforce cell stiffness. Furthermore, $IDH1^{R132H}$ expression decreased the expression of vimentin, an important component of the cytoskeleton and regulator of the cell stiffness. The results emphasize the important role of mutant IDH1 in treatment of patients with diffuse gliomas especially in response to radiation. Hence, detection of the genetic status of IDH1 before therapy massively expands the utility of immunohistochemistry to accurately distinguish patients with a less aggressive and radiosensitive IDH1-mutant diffuse glioma suitable for radiotherapy from those with a more aggressive IDH1-wildtype diffuse glioma who might benefit from an individually intensified therapy comprising radiotherapy and alternative medical treatments.

Keywords: isocitrate dehydrogenase 1; IDH1; $IDH1^{R132H}$; glioma; glioblastoma; cell stiffness; atomic force microscopy

1. Introduction

Gliomas, primary tumors of the central nervous system (CNS) are relatively rare and form a heterogeneous group of neoplasms [1]. For multiple decades the histology of gliomas was the gold standard for classification and for assessment of prognosis or therapeutic management. Molecular characterization was mainly provided as supplementary information within these histologically defined categories [2]. Due to the increasing knowledge of molecular alterations in tumors of the CNS the revised fourth edition of the World Health Organization (WHO) Classification of CNS tumors (published in 2016) included molecular surrogates, which greatly expanded the utility of immunohistochemistry for providing diagnostic, prognostic, and predictive aid in the workup of gliomas [2–5]. Nowadays, classification of gliomas encompasses two principle subgroups: diffuse gliomas, displaying an extensive infiltration in the CNS parenchyma and the nondiffuse gliomas, showing a more circumscribed growth pattern.

Immunohistochemically gliomas are categorized according to their histologically equivalent normal cell type. Diffuse gliomas are historically allocated to diffuse astrocytomas (with glioblastoma as its most frequent and most malignant representative), oligodendrogliomas, or to tumors with a mixed astrocytic and oligodendroglial phenotype (oligoastrocytomas). The group of nondiffuse gliomas now includes pilocytic astrocytoma, subependymal giant cell astrocytoma (SEGA), pleomorphic xanthoastrocytoma (PXA), and anaplastic PXA as distinct entities [2–5]. Diffuse gliomas are the most common subtype of primary brain tumors, especially in adult patients [4,6]. Within the subgroups, based on the presence/absence of marked mitotic activity, necrosis and florid microvascular proliferation diffuse gliomas are graded as WHO grade II (low-grade), III (anaplastic), or IV (glioblastoma) [4,6]. Thereby, glioblastoma can occur as the result of progression from lower grade diffuse gliomas or can arise de novo. Histologically, both primary (de novo) and secondary glioblastomas seem identical [7].

Over the last decade the understanding of glioma tumorigenesis was substantially increased due to the discovery of mutations involving the genes encoding isocitrate dehydrogenase 1 and 2 (IDH1/IDH2) enzymes. Using whole-genome sequencing and mutational analysis, Parson and colleagues identified recurrent mutations in IDH1 and IDH2 at high frequencies in WHO grade II and III astrocytomas, oligodendrogliomas, oligoastrocytomas, as well as in glioblastomas [8,9]. In these cases IDH mutations seem to predispose a particular path for oncogenic progression resulting in an increased progression-free and overall survival of affected patients, irrespective of tumor malignancy [8,10,11]. In several studies the IDH1 mutation has proven to be a powerful prognostic factor in diffuse gliomas, irrespective of tumor grade and histology [12–14]. Mutations in the IDH genes are generally heterozygous missense substitutions, which remarkably occur in a mutually-exclusive manner affecting only the active sites of the enzymes [8,15]. IDH1 mutations always appear in the arginine residue at codon 132 resulting in a substitution of histidine for arginine (R132H) in over 90% of all IDH1 mutations [8,16].

Very early on, IDH1 mutations have successfully been linked to prognostic information. Besides, it was evident that the clinical outcome for tumors with identical histology was different for IDH1-wildtype and IDH1-mutant diffuse gliomas [2,8,9,15,17,18]. Furthermore, it has been shown that many histologically identified as WHO grade II and especially WHO grade III IDH1-wildtype diffuse gliomas in adults display molecular characteristics and behaviors of a glioblastoma [13,19–21]. These fundamental observations drive toward a molecular classification and have led to the decision to include IDH mutation as a crucial marker for the classification of diffuse gliomas. Importantly, in this histological–molecular classification the genetic characteristics of IDH-mutant glioma can actually override the histological diagnosis, which also leads to an reclassification of glioblastoma [2]. Based on the genetic status of IDH1, glioblastomas are now divided into IDH1-wildtype glioblastoma and IDH-mutant glioblastoma, whereas the latter one largely overlaps with secondary glioblastoma in older classifications [2].

Despite the central role of an IDH1 mutation in the current classification of diffuse gliomas, it appears reasonable to further focus on cellular functions and therapeutic effects that are influenced by an IDH1 mutation. The care of patients with diffuse gliomas is challenging. Especially high-grade gliomas (grade III/IV), with glioblastoma as its most frequent and most malignant representative, are aggressive, invasive, and exhibit intratumoral hypoxia [22–24]. Irrespective of the multimodal therapy options available, comprising surgery, radiotherapy and chemotherapy, high-grade gliomas remain lethal diseases with dismal prognosis [25,26]. Due to their intratumoral heterogeneity and various mutual signatures (e.g., IDH mutation/1p19q co-deletion status, MGMT promoter methylation status, TERT promoter mutations) the chances for establishing a universal standard treatment of diffuse gliomas are limited [27]. In this context, the identification, but also dissection of the mechanism of molecular markers represents a useful tool for understanding cancer biology and based on this for the development of tailored therapeutic options. Furthermore, such biomarkers permit a subclassification of diffuse gliomas, making it possible to differentiate between patients with higher risk for toxicity and those who may benefit from a particular treatment i.e. in some cases, molecular markers are able to guide treatment decisions [28].

Hence, various studies highlighted the physiological roles of IDH enzymes as well as the biochemical and cellular consequences of an altered genetic status of IDH genes. IDH enzymes catalyze the decarboxylation of isocitrate to α-ketoglutarate (α-KG). Thereby, IDH1 is involved in a variety of cellular processes, including glutamine metabolism, glucose sensing and lipid metabolism, synthesis of N-acetylated amino acids, and regulation of the cellular redox status via GSH [29–31]. Mutations in the active sites of the IDH1 enzyme cause a distinctly decreased enzyme activity to isocitrate and result in a neomorphic enzyme function, which catalyzes the NADPH (nicotinamide adenine dinucleotide phosphate hydrogen)-dependent reduction of α-KG to the 2-hydroxyglutarate (2-HG) enantiomer, D-2-hydroxyglutarate (D-2-HG). In turn, this leads to D-2-HG accumulation and lowering α-KG as well as NADPH levels [32–35]. However, NADPH is necessary for the regeneration of reduced GSH which functions as the main antioxidant in mammalian cells. Low levels of cytoplasmic NADPH have been linked to elevated oxidative stress through impaired reduction of GSH [36]. In general, oxidative stress is increased by irradiation and chemotherapy leading to the hypothesis that IDH1 mutations induce an enhanced response to therapy and may contribute to the prolonged survival of patients harboring the mutation [37]. Due to the fact that almost all patients with a malignant glioma receive a single treatment or a combined therapy it is difficult to specify whether IDH1 mutation is associated with a less aggressive phenotype or directly linked to increased sensitivity to therapy. Thus, different studies or clinical trials have been focused on the effect of expression of mutated IDH1 on cellular behavior and response to therapy since the first mutations of IDH1 were discovered in 2008 [17,38–40].

In previous studies and by using transduced malignant glioma cell lines U-251MG (glioblastoma) U-343MG (anaplastic astrocytoma) and LN-229 (glioblastoma) we showed that gene expression of mutated IDH1 ($IDH1^{R132H}$) resulted in elevated radiosensitivity [41–43]. Furthermore, gene expression of $IDH1^{R132H}$ caused a reduced aggressiveness based on slightly decreased cell proliferation and plating efficiency, altered growth properties in 3D spheroid culture and significantly reduced cell migration in these glioma cell lines [41]. In addition, the effect of gene expression of mutated IDH1 on the radiosensitivity and cellular behavior was independent of the oxygen concentrations [41].

Based on the complex role of an IDH1 mutation in progression, aggressive biological behavior and response to therapy of malignant diffuse gliomas, it appears reasonable to dissect the molecular mechanisms underlying the less aggressive phenotype and increased sensitivity to radiation of IDH1R132H-gliomas.

2. Results

2.1. Expression of $IDH1^{R132H}$ Enhanced the Radiation Induced Accumulation of Residual γH2AX Foci

In eukaryotic cells, DNA double-strand breaks (DSBs) occur frequently from endogenous cellular processes or are caused by exogenous sources such as ionizing radiation. In response to the introduction of DNA DSBs, the minor histone H2A variant is rapidly phosphorylated on Ser-139 to produce γH2AX [44,45]. Based on the direct correlation between the number of DSBs and γH2AX foci, quantitation of γH2AX foci formation was applied as a marker of DNA damage and repair [44,46]. Foci which persist for longer than 24 h, so-called residual γH2AX foci, indicate unrepaired or misrepaired DSBs. These unsuccessfully repaired DSBs are generally assumed to play a major role in radiation-induced cell death [47]. Previous studies have demonstrated a linear relationship between radiation dose and the number of γ-H2AX foci over a limited dose range (between 0.001 and 2 Gy) [48–51]. Therefore, analyzing the induction of γH2AX foci allows for the indirect evaluation an influence of IDH1^{R132H} gene expression on the effect of radiation in U-251MG, U-343MG, and LN-229 cells.

For this purpose, untreated cells and cells stably transduced with empty vector pLVX, pLVX IDH1^{wt}, or pLVX IDH1^{R132H} were irradiated with a single dose of 0, 2, and 4 Gy under normoxia (21% O_2) and hypoxia (< 0.1% O_2), respectively.

Induction of DNA DSBs and the associated cellular repair capacity was investigated by visualization (Figure 1, representative images of U-251MG; Figures A1 and A2, representative images of U-343MG and LN-229) and quantitation (Figures 2 and A3) of the residual γH2AX foci 24 h after radiation. Treatment with the empty vector or gene expression of IDH1^{wt} did not affect the number of γH2AX foci compared to the respective untreated U-251MG, U-343MG, and LN-229 cells in normoxia and hypoxia, respectively.

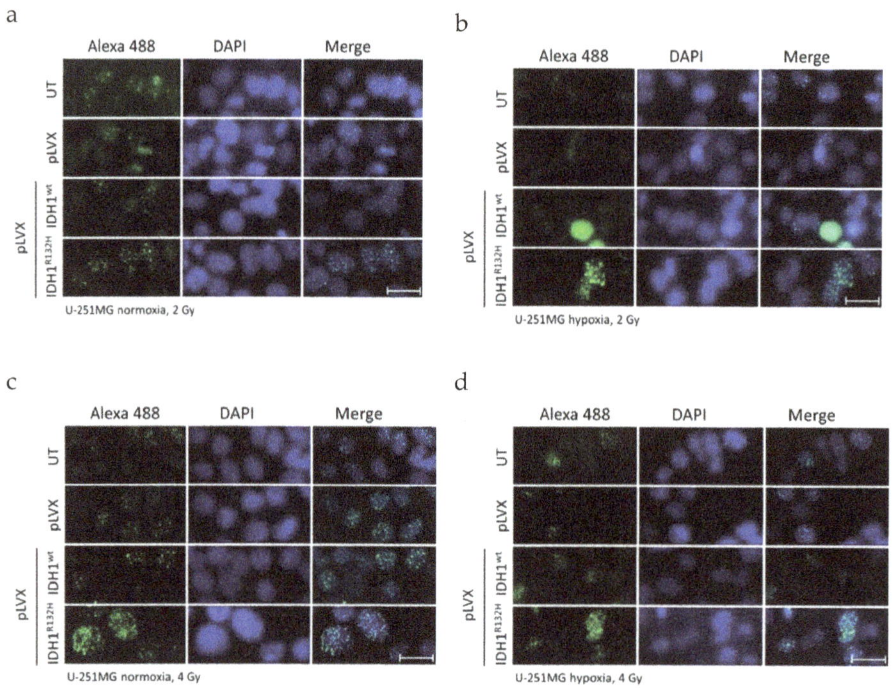

Figure 1. Effect of isocitrate dehydrogenase 1 (IDH1)^{R132H} gene expression on the accumulation of residual γH2AX foci after radiation in U-251MG cells. Representative immunofluorescence images of γH2AX of U-251MG cells 24 h after irradiation with 2 (**a**,**b**) and 4 Gy (**c**,**d**) under normoxia (**a**,**c**) and hypoxia (**b**,**d**). Green: γH2AX foci; blue: Cell nuclei (DAPI). n = 3 independent experiments; scale bar = 25 μm. Normoxia (21% O_2), hypoxia (< 0.1% O_2). UT: untreated, pLVX: cells stably transduced with empty vector, pLVX IDH1^{wt}: IDH1^{wt}-expressing cells, pLVX IDH1^{R132H}: IDH1^{R132H}-expressing cells.

After irradiation with 0, 2, and 4 Gy the average number of γH2AX foci per cell increased in a dose dependent manner in U-251MG, U-343MG, and LN-229 cells under normoxic and hypoxic conditions (Figure 2). Furthermore, in hypoxia γH2AX foci accumulation was decreased irrespective of the dose level in comparison to normoxic conditions in the investigated cell lines (Figure 2). Under hypoxic conditions, in untreated, empty vector and IDH1^{wt} cells, the γH2AX foci formation was up to 2.5-fold lower in U-251MG, up to 1.9-fold lower in U-343MG and up to 1.4-fold lower in LN-229 cells compared to the respective cells under normoxic conditions (Figure 2).

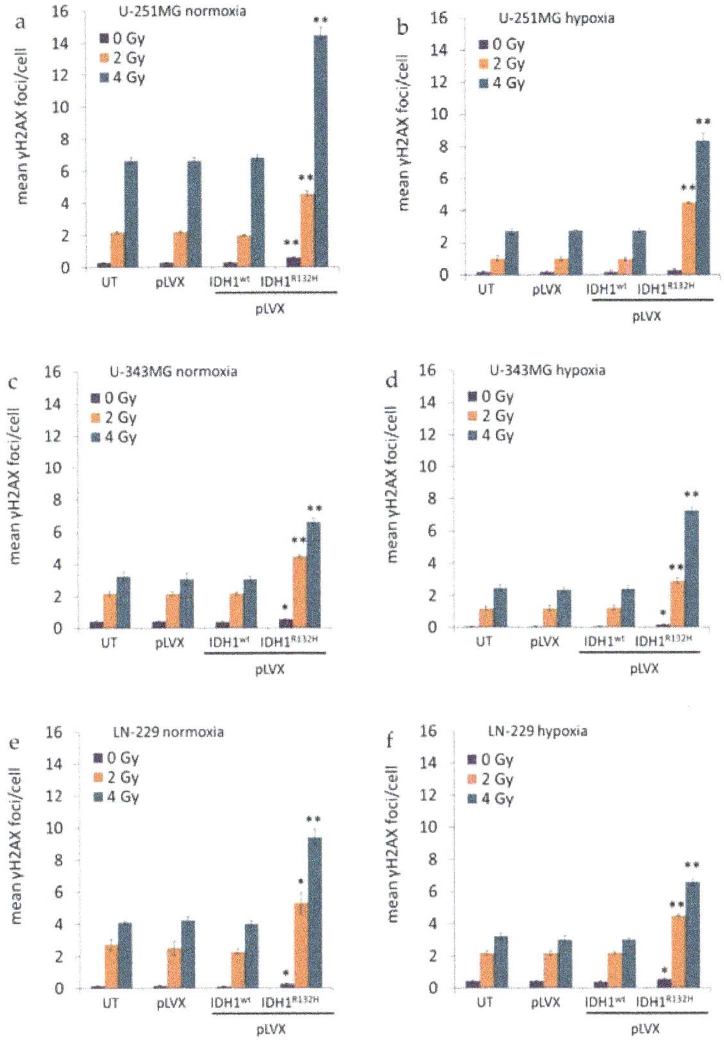

Figure 2. Effect of IDH1^{R132H} gene expression on the number of residual γH2AX foci after radiation in glioma cells. Mean γH2AX foci per cell nuclei of U-251MG (**a**,**b**), U-343MG (**c**,**d**), and LN-229 (**e**,**f**) cells. DNA damage was analyzed by γH2AX staining at 24 h after irradiation with 0, 2, and 4 Gy. γH2AX foci were counted manually in the nuclei of 300–400 untreated, empty vector, IDH1^{wt} or IDH1^{R132H} cells under normoxia (21% O_2) and hypoxia (< 0.1% O_2), respectively. Bars represent the mean values of three independent experiments. Error bars indicate standard deviations (±SD). UT: untreated, pLVX: cells stably transduced with empty vector, pLVX IDH1^{wt}: IDH1^{wt}-expressing cells, pLVX IDH1^{R132H}: IDH1^{R132H}-expressing cells; * $p < 0.05$ and ** $p < 0.01$ (compared to the respective IDH1^{wt} cells in normoxia or hypoxia).

In normoxia, the non-irradiated cells gene expression of IDH1^{R132H} increased the number of γH2AX foci by 2.1-fold ($p < 0.01$) from 0.28 foci/nucleus to 0.58 foci/nucleus in U-251MG, by 1.4-fold ($p < 0.05$) from 0.38 foci/nucleus to 0.54 foci/nucleus in U-343MG cells and by 2.5-fold ($p < 0.05$) from 0.1 foci/nucleus to 0.25 foci/nucleus in LN-229 cells compared to the respective IDH1^{wt} cells (Figure 2, purple bar). Furthermore, in normoxia, after irradiation at 2 Gy gene expression of IDH1^{R132H} increased the number of γH2AX foci by 2.3-fold ($p < 0.01$) from 2 foci/nucleus to 4.6 foci/nucleus in U-251MG, by 2.0-fold ($p < 0.01$) from 2.2 foci/nucleus to 4.5 foci/nucleus in U-343MG cells and by 2.3-fold ($p < 0.05$) from 2.3 foci/nucleus to 5.3 foci/nucleus in LN-229 cells compared to the respective $IDH1^{wt}$ cells (Figure 2, orange bar). In addition, after irradiation with 4 Gy $IDH1^{R132H}$ cells showed an increase of γH2AX foci formation by 2.1-fold ($p < 0.01$) from 6.8 foci/nucleus to 14.5 foci/nucleus in U-251MG, by 2.1-fold ($p < 0.01$) from 3.1 foci/nucleus to 6.6 foci/nucleus in U-343MG cells and by 2.4-fold ($p < 0.01$) from 4.0 foci/nucleus to 9.4 foci/nucleus in LN-229 cells in normoxia (Figure 2, blue bar).

Under hypoxic conditions, in the gene expression of $IDH1^{R132H}$ increased the number of γH2AX foci by 1.7-fold (not significant) from 0.17 foci/nucleus to 0.29 foci/nucleus in U-251MG, by 3.2-fold ($p < 0.05$) from 0.05 foci/nucleus to 0.16 foci/nucleus in U-343MG cells and by 1.4-fold ($p < 0.05$) from 0.38 foci/nucleus to 0.54 foci/nucleus in LN-229 cells compared to the respective $IDH1^{wt}$ cells (Figure 2, purple bar). In addition, under hypoxic conditions, when cells were irradiated at 2 Gy, the gene expression of $IDH1^{R132H}$ increased the number of γH2AX foci by 4.5-fold ($p < 0.01$) from 1.0 foci/nucleus to 4.5 foci/nucleus in U-251MG, by 2.4-fold ($p < 0.01$) from 1.2 foci/nucleus to 2.9 foci/nucleus in U-343MG cells and by 2.0-fold ($p < 0.01$) from 2.2 foci/nucleus to 4.5 foci/nucleus in LN-229 cells compared to the respective $IDH1^{wt}$ cells (Figure 2, orange bar). Furthermore, in hypoxia after irradiation at 4 Gy gene expression of $IDH1^{R132H}$ increased the γH2AX foci formation about 3.0-fold ($p < 0.01$) from 2.8 foci/nucleus to 8.4 foci/nucleus in U-251MG, 3.0-fold ($p < 0.01$) from 2.4 foci/nucleus to 7.3 foci/nucleus in U-343MG cells and 2.2-fold ($p < 0.01$) from 3.0 foci/nucleus to 6.6 foci/nucleus in LN-229 cells compared to the $IDH1^{wt}$ cells, respectively (Figure 2, blue bar).

Further, the fraction of cells in dependence of the number of residual γH2AX foci per nucleus was evaluated (Figure A3). In untreated, empty vector and $IDH1^{wt}$ cells a higher percentage of cells with low amount of foci per nucleus was observed (Figure A3). In contrast, $IDH1^{R132H}$-expressing cells showed an increased percentage of cells with high number of residual γH2AX foci per nucleus in normoxia and hypoxia (Figure A3).

2.2. Expression of $IDH1^{R132H}$ Decreased the Amount of GSH

IDH1 is involved in a variety of cellular processes, including the glutamine metabolism and regulation of the cellular redox status via GSH [29–31]. Based on the decreased enzyme activity of $IDH1^{R132H}$ and the neomorphic enzyme function, which lowers α-KG as well as NADPH levels, the GSH/GSSG ratio was measured. Different incubation times (1 h, 6 h, 24 h, and 48 h) after irradiation were analyzed in pilot experiments (data not shown). In accordance to pilot experiments and γH2AX assay, GSH/GSSG ratio was measured 24 h after irradiation with 0 or 5 Gy in an untreated, empty vector, $IDH1^{wt}$ and $IDH1^{R132H}$ cells of U-251MG, U-343MG, and LN-229 cells [41].

For evaluation of the data total GSH levels (sum of reduced GSH and oxidized GSSG) were set as 100%. Neither the treatment with the empty vector nor the gene expression of $IDH1^{wt}$ affected the amount of reduced GSH compared to the untreated U-251MG, U-343MG, and LN-229 cells, respectively (Figure 3). On the contrary, the gene expression of $IDH1^{R132H}$ resulted in a decreased amount of reduced GSH by 47.7% ± 4.2 ($p < 0.01$) in U-251MG, by 42.0% ± 1.04 ($p < 0.01$) in U-343MG and by 43.5% ± 2.8 ($p < 0.01$) in LN-229 cells compared to the respective non-irradiated $IDH1^{wt}$ cells (Figure 3, blue bars). After irradiation with 5 Gy in untreated, empty vector and $IDH1^{wt}$ cells, the level of reduced GSH was up to 10% lower in U-251MG, up to 15% lower in U-343MG and up to 21% lower in LN-229 cells compared to non-irradiated cells, respectively (Figure 3, blue vs. red bars). Furthermore, expression of $IDH1^{R132H}$ and irradiation with 5 Gy show an additive inhibitory effect on the amount of reduced GSH resulting in a decreased level of GSH by 79.5% ± 4.1 ($p < 0.01$) in U-251MG, by 65.8% ± 4.2 ($p < 0.01$) in U-343MG and by 70.7% ± 5.2 ($p < 0.01$) in LN-229 cells as compared to the irradiated (5 Gy) $IDH1^{wt}$ cells, respectively (Figure 3, red bars).

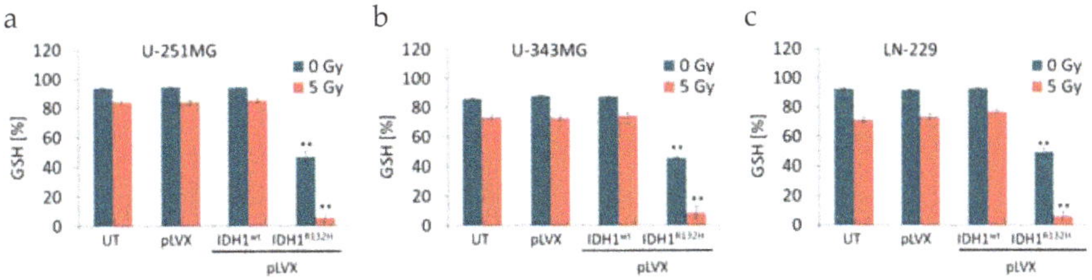

Figure 3. Effect of IDH1^{R132H} expression on the level of glutathione (GSH) in glioma cells. GSH level in stably transduced U-251MG (**a**), U-343MG (**b**), and LN-229 (**c**) cells after irradiation with 0 or 5 Gy. Total GSH levels (sum of reduced GSH and oxidized GSSG) were set as 100%. Bars represent the mean values of four independent experiments. Error bars indicate standard deviations (±SD). UT: untreated, pLVX: cells stably transduced with empty vector, pLVX IDH1^{wt}: IDH1^{wt}-expressing cells, pLVX IDH1^{R132H}: IDH1^{R132H}-expressing cells. ** $p < 0.01$ (compared to the respective IDH1^{wt} cells).

2.3. *Expression of IDH1^{R132H} Caused a Significant Increase of Cell Stiffness*

In a previous study we showed that gene expression of mutated IDH1 causes a reduced aggressiveness based on slightly decreased cell proliferation and plating efficiency, altered growth properties in 3D spheroid culture and significantly reduced cell migration [41]. In order to spread and form metastases, aggressive cancer cells with high metastatic potential might benefit from their softness and flexibility [52–54]. Based on these analyses we used AFM technology to quantify the relationship between several cell-specific parameters with a network analytical approach to compile the composite parameter "stiffness" (see Supplementary Methods 5) [55].

For AFM indentation measurements, single rounded cells were indented using a tip-less AFM cantilever (see Supplementary Methods 5). AFM measurements revealed that gene expression of IDH1^{wt} slightly increased the cell stiffness of glioma cell lines compared to untreated cells and empty vector cells, respectively (Figure 4). In addition, IDH1^{R132H} caused a considerable increase of cell stiffness of U-251MG ($p = 0.1$), U-343MG ($p < 0.01$), and LN-229 ($p < 0.01$) cells compared to untreated cells, empty vector cells (pLVX) and IDH1^{wt} cells (Figure 4).

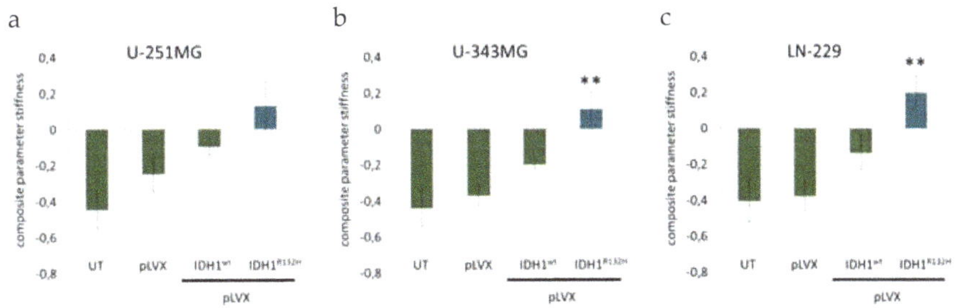

Figure 4. Effect of IDH1^{R132H} gene expression on the stiffness of glioma cells. Influence of IDH1^{R132H} on cell stiffness of U-251MG (**a**), U-343MG (**b**), and LN-229 (**c**) cells was investigated using AFM technology. Bars represent the mean values of 20 single cells per measurement. Error bars indicate standard deviations (±SD). UT: untreated cells, pLVX: cells stably transduced with empty vector, pLVX IDH1^{wt}: IDH1^{wt}-expressing cells, pLVX IDH1^{R132H}: IDH1^{R132H}-expressing cells. ** $p < 0.01$ (compared to the respective IDH1^{wt} cells).

2.4. *Expression of IDH1^{R132H} Induced an Altered Organization of the Cytoskeleton*

The cytoskeleton is a meshwork of a variety of biopolymers (e.g., actin, microtubules, intermediate filaments), which are essential for dynamic functions and mechanical stability. Actin filaments assemble into diverse protrusive and contractile structures to provide force for a number of vital cellular processes including cell adhesion, morphogenesis and mechanotransduction [56]. Therefore, the actin stress fibers can be divided into at least four different categories: dorsal and ventral stress fibers, transverse arcs and the perinuclear actin cap (schematic Figure 5a) [56,57].

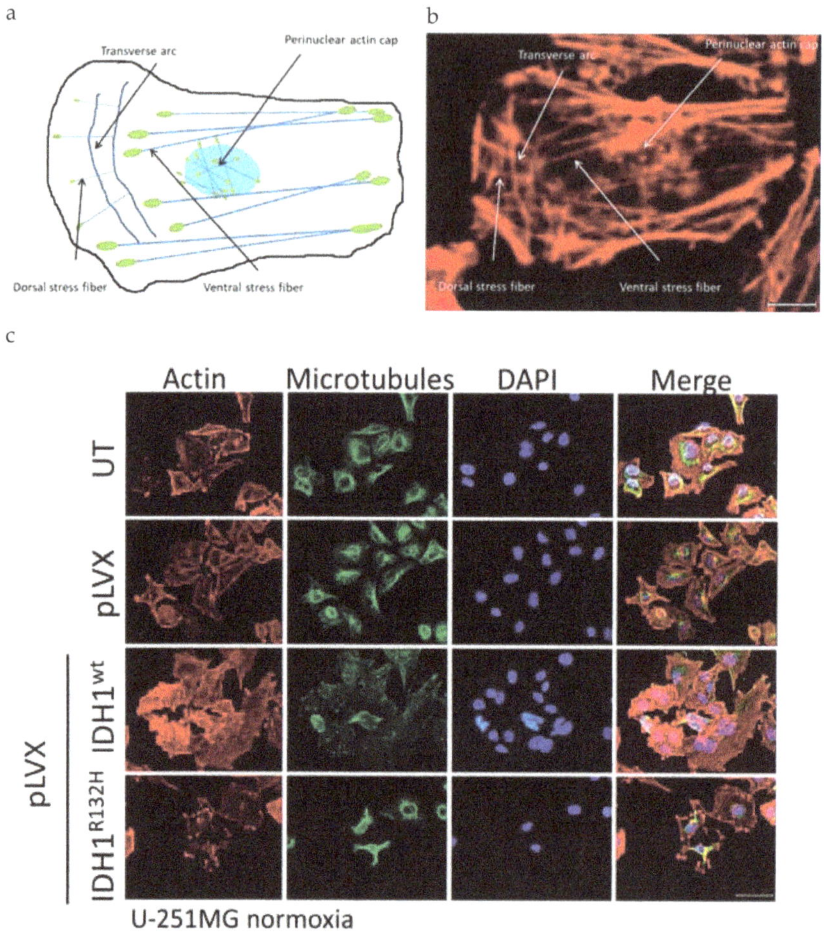

Figure 5. Organization of the cytoskeleton of transduced U-251MG glioma cells stably expressing IDH1^{wt} or IDH1^{R312H} protein. (**a**) Schematic representation of types of actin stress fibers. Four categories of actin stress fibers are observed: dorsal and ventral stress fibers, transverse arcs and the perinuclear actin cap. Blue: actin fibers, and green: focal adhesion. (**b**) Enlarged representative image as an example of different categories of actin stress fibers for untreated cells, empty vector cells (pLVX) and IDH1^{wt} cells; enlarged part of pLVX cells, scale bar = 10 μm (**c**) Representative immunofluorescence staining of actin stress fibers and microtubules in U-251MG cells using phalloidin-TRITC and anti-tubulin antibody. Immunofluorescence staining was performed 24 h after seeding. Cell nuclei were counterstained with DAPI. N = 3 independent experiments were performed; scale bar = 50 μm. UT: untreated, pLVX: cells stably transduced with empty vector, pLVX IDH1^{wt}: IDH1^{wt}-expressing cells, pLVX IDH1^{R132H}: IDH1^{R132H}-expressing cells.

Hence, immunofluorescence was applied to analyze the organization of actin stress fibers and microtubules in untreated cells, empty vector cells, and IDH1^{wt}- or IDH1^{R132H}-positive U-251MG, U-343MG, and LN-229 cells (Figure 5, Figure A4, Figure A5). Analysis of the stained cells showed no effect on the organization of microtubule fibers in the investigated cell lines. In addition, immunofluorescence staining indicated the presence of the four different categories of actin stress fibers in untreated cells, empty vector cells (pLVX) and IDH1^{wt} cells (enlarged and representative, Figure 5b). On the contrary, expression of IDH1^{R132H} induced an altered organization of the actin cytoskeleton of malignant glioma cells (Figure 5c, Figure A4, Figure A5). IDH1^{R132H}-positive cells displayed changes in spatial distribution of actin stress fibers with fibers located in the cell periphery and thicker fibers, which have been shown to reinforce cell stiffness (Figure 5c, Figure A4, Figure A5). Furthermore, expression of IDH1^{R132H} seems to induce a smaller surface area compared to the untreated cells, empty vector or IDH1^{wt}-positive cells but of LN-229 cells only (Figures A5 and 6b).

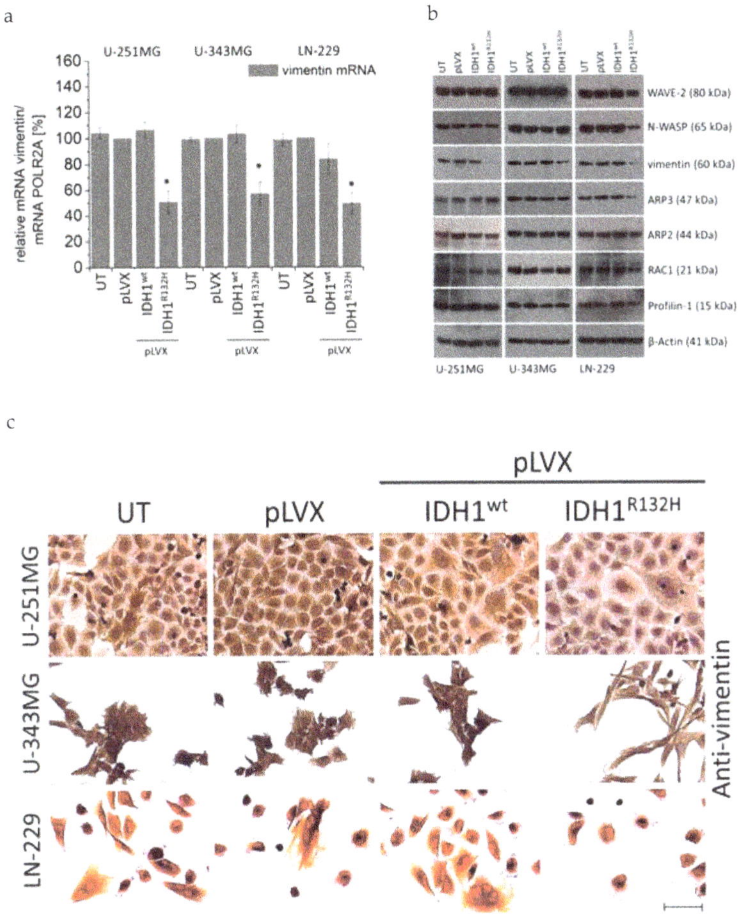

Figure 6. Effect of IDH1^{wt} or IDH1^{R132H} on the expression of actin-influencing proteins in glioma cells. (**a**) mRNA expression levels (qPCR) of vimentin in U-251MG, U-343MG, and LN-229 cells. Bars represent the mean values, relative to the control (empty vector pLVX set as 100%), of three independent experiments. Error bars indicate standard deviations (±SD). (**b**) Representative western blots for actin-influencing proteins of stably transduced U-251MG, U-343MG, and LN-229 cells using an Actin Nucleation and Polymerization Antibody Sampler Kit (cell signaling). n = 3 independent experiments. (**c**) Representative immunohistochemical staining of vimentin of U-251MG, U-343MG, and LN-229 cells. Cell nuclei were counterstained with 20% hematoxylin. n = 3 independent experiments; scale bar = 75 μm. UT: untreated, pLVX: cells stably transduced with empty vector, pLVX IDH1^{wt}: IDH1^{wt}-expressing cells, pLVX IDH1^{R132H}: IDH1^{R132H}-expressing cells.

2.5. Expression of IDH1^{R132H} Decreased the Expression of Vimentin

To determine the effect of an expression of IDH1 mutant protein on genes which are involved in the organization of the actin cytoskeleton (actin-regulating proteins, actin-binding proteins, actin structure proteins, and intermediate filaments) qPCR, western blot analyses and immunohistochemical staining were carried out. Analyses by qPCR included actin-regulating proteins: RAC1 (Ras-related C3 botulinum toxin substrate 1), RHOA (Ras homolog gene family, member A), RHOB (Ras homolog gene family, member B), RHOC (Ras homolog gene family, member C), CDC42 (Cell division control protein 42 homolog); actin-binding proteins: Cofilin-1, Profilin-1, Fascin, Filamin, Ezrin, Moesin, alpha-actinin-1, alpha-actinin-4; actin-structure proteins: Myo1B (Myosin1B), Myo9B (Myosin9B), Myo10B (Myosin10B), Myo18A (Myosin18A), and intermediate filaments: vimentin, desmin and GFAP (glial fibrillary acidic protein).

The qPCR experiments revealed a decreased mRNA level of the intermediate filament vimentin in IDH1^{R132H}-positive U-251MG (by 49.5%, $p < 0.05$), U-343MG (by 43.0%, $p < 0.05$), and LN-229 (by 50.7%, $p < 0.05$) cells (Figure 6a), whereas the other investigated genes (i.e., actin-regulating proteins see Figure A6; actin-binding proteins see Figures A7 and A8; actin-structure proteins see Figure A9 and intermediate filaments see Figure A10) showed no significant alteration in their mRNA expression level in U-251MG and U-343MG cells. Furthermore, in LN-229 cells expression of IDH1^{R132H} induced a reduced mRNA level of actin-regulating proteins RAC1 (by 33.7%, $p < 0.01$), CDC42 (by 45.1%, $p < 0.05$) and of the actin-binding protein Profilin-1 (by 46.3%, $p < 0.01$), whereas no influence on the other investigated genes (RHOA, RHOB, RHOC, see Figure A6; Cofilin-1, Fascin, Filamin, Ezrin, Moesin, alpha-actinin-1, alpha-actinin-4 see Figure A7, Figure A8; Myo1B, Myo9B, Myo10B, Myo18A see Figure A9; intermediate filaments: desmin and GFAP was observed (Figure A10).

In addition, western blot analyses using an Actin Nucleation and Polymerization Antibody Sampler Kit were applied to investigate the expression of actin-influencing proteins in stably transduced U-251MG, U-343MG, and LN-229 cells (Figure 6b). The Antibody Sampler Kit includes the actin-regulating protein RAC1 (GTPase) and its targets Wiskott–Aldrich syndrome protein (N-WASP) and WASP-family verprolin-homologous protein (WAVE-2). WAVE-2, N-WASP are scaffolds that link upstream signals to the activation of the ARP2/3 complex leading to a burst of actin polymerization [58]. When activated, the Arp2/3 complex contributes the actin branched junction and thus cross-links the polymerizing actin filaments [59]. Moreover, a further protein, namely Profilin-1, was analyzed. Profilin-1 has been shown to bind ATP-bound G-actin and promotes filament elongation [60].

Western blot analyses revealed a decreased protein level of the intermediate filament vimentin in IDH1^{R132H}-positive U-251MG, U-343MG, and LN-229 cells, whereas no influence on the other investigated proteins WAVE-2, N-WASP, ARP2/3, RAC1, and Profilin-1 was observed in U-251MG and U-343MG cells (Figure 6b). In accordance to the qPCR, western blot analyses confirmed the reduced expression of RAC1 and its targets WAVE-2, N-WASP in IDH1^{R132H}-positive LN-229 cells (Figure 6b). Consequently, in these cells a decreased protein levels of Profilin-1 and ARP2/3, targets of WAVE-2 and N-WASP, were detected by western blot analyses (Figure 6b).

Consistent with these findings immunohistochemical staining showed a reduction of vimentin expression in IDH1^{R132H}-positive U-251MG, U-343MG, and LN-229 cells (Figure 6c).

3. Discussion

Diffuse gliomas, the most common type of primary brain tumors, are often diagnosed in aged adults [4,6]. On a molecular level, diffuse gliomas are characterized by a variety of genetic and epigenetic alterations. Especially the molecular marker IDH1^{R132H} has been identified as prognostic marker with major roles in tumorigenesis and the response to therapy [18,61,62]. These observations paved the way to a genotype-driven classification by including the IDH mutation as a decisive marker for glioma classification. Nowadays, these tumors are classified according to the 2016 WHO system by both histologic and molecular characteristics as IDH-mutant or IDH-wildtype astrocytomas; IDH-mutant and 1p19q-codeleted oligodendrogliomas; and IDH-mutant or IDH-wildtype glioblastomas [5]. Glioblastoma

(WHO grade IV) as the most frequent and most malignant tumor in this group, i.e., is characterized by an aggressive invasiveness and a dismal prognosis with a median survival time ranging from 6 to 15 months under the current standard treatment regime, consisting of maximal surgical resection, whenever possible, followed by radiation and chemotherapy [6,25,63–66]. The heterogeneity of diffuse gliomas with respect to clinical presentation, pathology, genetic profile, and the poor response to treatment, reduces the chances for a universal treatment, particularly due to the fact that subtypes have different responses to therapy, which therefore may result in both over- or undertreatment of these tumors [3,4,38].

Hence, characterization of molecular tumor markers provides an opportunity to gain a deeper understanding of progression, aggressiveness and radioresistance of diffuse gliomas. In addition to the benefit in diagnosis, investigation of the IDH1 mutation represents a useful tool for the understanding of tumorigenesis and resistance to therapy of malignant diffuse gliomas. Functional studies support the development of personalized therapies as one of the main research objectives in the next years [66–69]. Based on a previous study, where the gene expression of mutated IDH1 caused an elevated radiosensitivity and a reduced aggressive biological behavior of different malignant glioma cell lines, this study dissects molecular mechanisms, underlying the influence of the most frequent IDH mutation ($IDH1^{R132H}$) in the investigated transduced malignant glioma cells [41].

In clinical studies it is difficult to specifically correlate the better prognosis and prolonged overall survival of patients with IDH-mutant gliomas to the cellular behavior itself or the improved response to therapy treatment, since patients with gliomas receive standard therapy according to guidelines on management of gliomas in any case. In a previous study, the effect of IDH1 mutant protein on radiobiological behavior was carried out with U-251MG, U-343MG, and LN-229 malignant glioma cell lines. In these cells the gene expression of $IDH1^{R132H}$ resulted in a decreased plating efficiency and enhanced induction of apoptosis under normoxia (21% O_2) and hypoxia ($< 0.1\%$ O_2) [26].

In the present work, gene expression of the mutant IDH1 effectively enhanced the radiation induced accumulation of γH2AX foci in U-251MG, U-343MG, and LN-229 malignant glioma cells irrespective of the oxygen conditions. Based on the direct correlation between the number of γH2AX foci and DSBs, quantitation of γH2AX foci formation was applied as a marker of DNA damage and repair [44,46]. Residual γH2AX foci indicate unrepaired or misrepaired DSBs, which have been shown to play a major role in radiation-induced cell death [47]. Therefore, analyzing the induction of γH2AX foci allows for the indirect evaluation an influence of $IDH1^{R132H}$ gene expression on the effect of radiation in U-251MG, U-343MG, and LN-229 cells. In addition, expression of $IDH1^{R132H}$ decreased the level of reduced GSH, further expression of $IDH1^{R132H}$ and irradiation with 5 Gy showed an additive inhibitory effect on the amount of reduced GSH. Under physiological conditions, IDH enzymes regulate a number of cellular functions and play an essential role in cellular protection as well as response to energetic and oxidative stress [29,31,36,70,71]. In addition, metabolic studies have shown that IDH activity is responsible for 65% of NADPH production capacity in the human brain. [72]. The mutated IDH1 enzyme exhibits a strongly decreased enzyme activity (loss of function) to isocitrate and NADP+ and gains an abnormal NADPH-dependent catalytic activity [33–36]. Notably, the NADPH production capacity is reduced in glioblastoma by 38% when IDH1 is mutated [36]. NADPH is required by glutathione reductase to recycle oxidized glutathione (GSSH) to reduced GSH, the major cellular ROS scavenger [33,36]. Additionally, α-KG itself functions as an antioxidant [73,74]. Thereby, it has been suggested that glioma cells expressing mutant IDH1 have a diminished antioxidant capacity and therefore may experience a subsequent loss of cytoprotection under conditions of oxidative stress [36,75–77]. Under these circumstances low NADPH levels might sensitize malignant gliomas for oxidative stress, amplifying the response to radiotherapy and thereby may account for the prolonged survival of patients harboring the mutations. Thus, it is possible that the enhanced radiosensitivity of U-251MG, U-343MG, and LN-229 cells is caused by the $IDH1^{R132H}$ induced reduction of the cytoprotection against oxidative stress. Due to the key role of NADPH in the cellular antioxidation systems it has been supposed that IDH1/2 mutations may increase intracellular reactive oxygen species (ROS) by the decrease of intracellular NADPH levels [75,76,78]. Insufficient control of intracellular

ROS has been associated with cellular senescence and apoptosis [79–81]. Consistent with these findings the gene expression of IDH1^{R132H} decreased the level of reduced GSH and consequently increased the number of residual γH2AX foci per nucleus in the non-irradiated glioma cell lines. Moreover, these effects on reduced GSH and the number of residual γH2AX foci were significantly reinforced by radiation. In accordance with these findings, IDH1 silencing of U87, A172, and U138 glioblastoma cell lines reduced levels of NADPH, deoxynucleotides and glutathione and increased their sensitivity to radiation-induced senescence [82]. In addition, a more recent study demonstrated that, even when NADPH is limiting, IDH1 mutants continue to synthesize 2-HG at the expense of other NADPH-requiring pathways that are essential for cell viability [83]. In addition, it was suggested that instead of trying to reduce 2-HG synthesis in patients, consumption of NADPH by the mutant IDH1 can be used as a metabolic weakness to sensitize tumor cells to ionizing radiation [83].

Among IDH-mutant gliomas, the very recent WHO classification system distinguishes three grades (II-IV) based on histopathologic features [5]. It has been widely reported in literature that with increasing malignancy gliomas exhibit intratumoral hypoxia, which has been associated with poor response to radio- or chemotherapy [7,84–87]. In the present work, gene expression of IDH1^{R132H} attenuated the hypoxia induced radioresistance of malignant glioma cells U-251MG, U-343MG and LN-229. Based on the intratumoral hypoxia of high-grade gliomas, these observations have important implications for the clinical consequences of an IDH1 mutation in these tumors. Since IDH1 is an enzyme of the tricarboxylic acid (TCA) cycle and plays a major role in energy and oxygen metabolism, it is crucial to understand how hypoxia alters the phenotype of IDH-mutant glioma cells as compared to its wildtype counterpart [88]. As already described, IDH1 catalyzes the reductive carboxylation of α-KG to isocitrate which is essential for citrate synthesis under hypoxic conditions. Likewise, cells grown under hypoxia rely almost exclusively on the reductive carboxylation of glutamine-derived α-KG for de novo lipogenesis [29,30]. The mutated IDH1 enzyme is not sufficient to catalyze the reductive carboxylation of α-KG to isocitrate, suggesting that this metabolic alteration contributes to the reduced aggressiveness of glioma cells expressing IDH1^{R132H} [89].

Based on the less aggressive phenotype of IDH1^{R132H}-expressing U-251MG, U-343MG, and LN-229 cells, including a decreased cell viability, proliferation, an altered growth in 3D culture and a reduced cell migration, we analyzed cell mechanics by atomic force microscopy, the organization of the cytoskeleton as well as factors responsible for migration, stiffness and deformability of cancer cells [41]. Investigation of cell mechanics by AFM technology revealed that IDH1^{R132H} caused a considerable increase of cell stiffness compared to untreated cells, empty vector cells and IDH1^{wt} cells in all three cell lines. Cell spreading has been shown to correlate with changes of important cell functions including DNA synthesis, differentiation, cell migration, and cell stiffness [90,91]. Using AFM technology living metastatic cancer cells extracted from the pleural fluids of patients with suspected lung, breast, and pancreas cancer were more than 70% to 80% softer compared to benign mesothelial cells taken from the body cavities [52]. This study also proved that cells of different cancer types exhibit a common stiffness, whereas the stiffness of benign mesothelial cells showed a log-normal distribution. In other words, the distribution of the stiffness of tumor cells was over five times narrower than the corresponding distribution for benign mesothelial cells. In addition, nanomechanical analysis correlated well with immunohistochemical testing generally used for detecting cancer [52]. A recent study analyzed the glioma stiffness via magnetic resonance elastography (MRE). MRE demonstrated that gliomas were not only softer than normal brain, but the degree of softening was directly correlated with tumor grade and IDH1 mutation status. This means that tumors with an IDH1 mutation were significantly stiffer than those with wild-type IDH1 [92].

Based on these and other nanomechanical studies it has been suggested that cancer cells with high metastatic potential might benefit from their altered mechanical properties. Increased softness and flexibility supports spreading, as glioma cells migrate through the healthy brain tissue, mostly along anatomical structures, such as nerve fiber tracts making it necessary for cells to effectively move along the interface of different cellular structures [65–68]. Furthermore, it has also been shown that cellular stiffness of ovarian cancer cell lines and primary cells derived from ascites of patients with advanced stage ovarian cancer is inversely proportional to migration and invasion [93]. This is consistent to a study by Watanabe and coworkers who revealed that highly motile melanoma cells (B16-F10) exhibit low cell stiffness while low motile and metastatic melanoma cells (B16-F1) cells are characterized by high cell stiffness [91]. Thus, the migration activity of cancer cells seems to correlate inversely with their cellular stiffness. These findings are in accordance with own observations where gene expression of IDH1^{R132H} caused a reduced proliferation, an altered growth in 3D culture, a reduced cell migration activity and an increased cellular stiffness of U-251MG, U-343MG, and LN-229 cells. Indeed, it has already been demonstrated that gene expression of IDH1^{R132H} decreased the proliferation and migration of U-87MG (glioblastoma, grade IV) cells in normoxia [94]. In turn, this would suggest a reduced invasiveness of the IDH1^{R132H}-expressing cells and may explain the better prognosis of patients with IDH1-mutant gliomas. Different studies support the relationship between the stiffness and the invasiveness of cancer cells. However, the underlying mechanism which drives cancer cells to softer mechanical characteristics and thereby to a creep deformability is not fully understood.

In the present work, immunofluorescence was applied to analyze the organization of actin stress fibers and microtubules in untreated cells, empty vector cells, and IDH1^{wt}- or IDH1^{R132H}-positive cells. Immunofluorescence staining indicated the presence of the four different categories of actin stress fibers in untreated cells, empty vector cells (pLVX) and IDH1^{wt} cells. On the contrary, expression of IDH1^{R132H} induced an altered organization of the cytoskeleton of glioma cells. IDH1^{R132H}-positive cells displayed changes in spatial distribution of actin stress fibers with fibers located in the cell periphery and thicker fibers, which have been shown to reinforce cell stiffness.

Furthermore, investigation by qPCR analyses, western blot analyses and immunohistochemical staining of genes involved in the organization (actin-regulating proteins, actin-binding proteins, actin structure proteins, and intermediate filaments) of the actin cytoskeleton revealed a decreased expression of the intermediate filament vimentin in IDH1^{R132H}-positive U-251MG, U-343MG, and LN-229 cells. In addition, expression of IDH1^{R132H} induced a smaller surface area and a reduced expression of RAC1 and CDC42 as well as their targets WAVE-2 and N-WASP in LN-229 cells. Consequently, in these cells a decreased protein levels of Profilin-1 and ARP2/3, targets of WAVE-2 and N-WASP, were detected by western blot analyses. CDC42 and RAC1, members of the Rho family of small G proteins, have been shown to induce actin polymerization by the Arp2/3 complex through binding to and activation of their effector proteins, N-WASP and WAVE-2, respectively [95–97]. The actin cytoskeleton plays a major role in the formation and function of the lamellipodia. In lamellipodia and filopodia, actin filaments are highly dynamic and directly pushes the plasma membrane forward [98]. Based on this it is likely that the reduced RAC/CDC42–WASP/WAVE–ARP2/3 signaling pathway induce a smaller surface area of LN-229 cells. Therefore, further work is required to elucidate the factors responsible for decreased surface area of LN-229 glioma cells.

Our results on cell stiffness are consistent with observations in several studies that have already shown that stiffness, deformation, and cell motility are regulated by different cellular processes. These studies have shown that actomyosin contractility, gene expression of the mesenchymal stiffness of various tumor cells [93,99–104]. Vimentin, one of the three major groups of cytoskeletal filaments, i.e., actin filaments, microtubules, and intermediate filaments has been shown to maintain mechanical cellular integrity and regulate cell stiffness [63,64,102,103]. Furthermore, vimentin was observed in a wide range of cancer types and seems to correlate with tumor aggressiveness and poor prognosis [105]. In our investigations, all three investigated cell lines showed reduced vimentin expression, an altered actin organization and increased cell stiffness. In contrast, only the cell line LN-229 showed an additional reduction of the actin-regulating proteins RAC1, CDC42 and a smaller cell surface. Therefore, it is possible that the change in vimentin expression caused the altered actin organization. Findings from different studies suggested that vimentin is indeed capable to interact and at least partly regulate focal adhesion formation and thus actin organization [106–108]. Jui and colleges reported a bidirectional interplay between vimentin intermediate filaments and contractile actin stress fibers. This study showed that specific actin stress fiber structures, transverse arcs, interacted with vimentin intermediate filaments and promoted their retrograde flow. Consequently, these transverse (myosin-II-containing) arcs were important for perinuclear localization of the vimentin network in cells. Further the vimentin network reciprocally restricted retrograde movement of arcs and hence controlled the width of flat lamellipodia at the leading edge of the cell [109]. In astrocytes it could furthermore be demonstrated that vimentin was necessary to maintain their polarization, indicating a possible important role for vimentin in explaining the effects of IDH1 mutations on glioma behavior [83].

In IDH mutated gliomas, elevated D-2-HG levels were induced by the gain-of-function of the mutant IDH1 protein. It is suspected that altered chromatin modifications and associated profound changes in the epigenetic status of these cells led to dysregulated gene expression [110–112]. The mechanism could also be a reason for the altered expression of vimentin. In the literature, clinical observations by Qi and coworkers suggest that prolonged survival of patients with IDH mutated gliomas is primarily linked to a less aggressive biological behavior assessed on the basis of preferred areas for tumorigenesis and magnetic resonance imaging (MRI) characteristics [113]. Further work will be needed to investigate possible epigenetic effects of an IDH1 mutation on genes which are involved in the organization of the actin cytoskeleton IDH1^{R132H}-expressing glioma cells.

In summary, the expression of IDH1^{R132H} enhanced the radiation induced accumulation of residual γH2AX foci, decreased the amount of reduced GSH and caused a significant increase of cell stiffness of different high-grade glioma cell lines. Furthermore, the expression of IDH1^{R132H} induced an altered organization of the actin cytoskeleton, which is supposed to be an effect of the reduced expression of the intermediate filament vimentin. The enhanced radiosensitivity and increased aggressive biological behavior of IDH1^{R132H}-positive malignant glioma cells are consistent with the clinical observation of a less aggressive tumor and better clinical outcome of gliomas of all WHO grades. Hence, the improved prognosis and the prolonged overall survival of glioma patients harboring the IDH1 mutation seem to be an interaction between two factors, namely the increased sensitivity to therapy and the reduced aggressiveness of these tumors. The results emphasize the important role of mutant IDH1 in treatment of patients with gliomas especially in response to radiation.

4. Materials and Methods

4.1. Generation of Constructs and Cell Culture Conditions

Cell culture conditions of the human malignant glioma cells U-251MG or LN-229 (both derived from glioblastomas, grade IV) and U-343MG cells (originate from an anaplastic astrocytoma, grade III) [42,43] were performed as previously described in detail [41]. The establishment of stable cells overexpressing IDH1^{wt} or IDH1^{R132H} was carried out as described in detail in the supplements (Supplementary Methods 1). Cell line authentication was achieved by genetic profiling using polymorphic short tandem repeat (STR) loci.

4.2. Hypoxia and Irradiation

Hypoxia ($< 0.1\%$ O_2) was achieved with an Anaerocult® A mini gas generation system (Merck, Darmstadt, Germany). The gas generator system is a special incubation bag that creates an anaerobic atmosphere through the presence of an Anaerocult® A mini-bag containing specific components (kieselguhr, iron powder, citric acid, and sodium carbonate) that are activated by water and rapidly bind oxygen (less than 0.1% residual oxygen in the bag after 1 h). Twenty-four hours after the cells were seeded in cell numbers suitable for the experiments (depending on the area, cell density approx. 50%), the flasks containing the untreated cells or cells stably transduced (with empty vector pLVX, IDH1^{wt}-, or IDH1^{R132H}-expressing cells) were transferred to the Anaerocult® A mini systems. After activating the Anaerocult® A mini bags by wetting them with 8 mL Aqua bidest, the bags were closed with Anaeroclips® and placed in a humidification incubator. The cells were incubated for another 24 h under hypoxic conditions. The detection of hypoxia was controlled by Anaerotest® strips on the covers of the flasks according to the manufacturer's protocol.

Irradiation was carried out on logarithmically growing cultures with 6 MV photons and adequate bolus material on a SIEMENS ONCOR (Erlangen, Germany) linear accelerator at a dose rate of 2 Gy/min.

4.3. *Quantitative Real-Time PCR, Western Blot Analysis, Immunofluorescence and Immunohistochemical Staining*

For real-time PCR (qPCR) analysis total RNA was extracted using TRIzol reagent (Thermo Scientific, Schwerte, Germany) as recommended by the manufacturer. DNA digestion was included by using 30 Kunitz units of RNase-free DNase in 80 μL RDD buffer (both Qiagen). RNA concentration was measured by a NanoDrop® ND-1000 Spectrophotometer. cDNA was synthesized from 1 μg of RNA using RevertAid H-Minus first-strand cDNA synthesis kit (Thermo Scientific) following the manufacturer's instructions. qPCR was performed in triplicate on a real-time PCR cycler (Rotor-Gene 6000; Qiagen, Hilden, Germany) by using Maxima SYBR Green/ROX qPCR Master Mix (Thermo Scientific). The PCR reaction conditions and a summary of all primer sequences are depicted in Table S1 of the Supplementary Material (Supplementary Methods 2; Table S1).

Protein expression was analyzed via western blot and immunofluorescence or immunohistochemical staining. The steps of protein isolation, western blot analysis and immunostaining procedures are described in detail in the Supplementary Materials (Supplementary Methods 3 and Supplementary Methods 4). The antibodies used are listed in Table S2 of the Supplement (Supplementary Methods 4; Table S2).

4.4. *Quantification of Phospho-Histone H2AX Foci Formation*

For quantification of phospho-histone H2AX (γH2AX) foci formation, untreated cells, empty vector cells and IDH1^{wt}- or IDH1^{R132H}-expressing cells (1×10^5) were seeded in 8-well chamber slides (Thermo Scientific). After 24 h at 37 °C in normoxia, chamber slides were either kept in normoxia or incubated in Anaerocult® A mini gas generator system (< 0.1% O_2) under hypoxia. After 24 h the cells were then irradiated with 0, 2, and 4 Gy and following further incubation in normoxia or hypoxia at 37 °C for 24 h, γH2AX was analyzed via immunofluorescence staining. Details of the immunofluorescence staining and the antibodies used are described in the supplements. Quantification of γH2AX foci formation was carried out using an AxioVert 200M microscope (Carl Zeiss, Jena, Germany). The foci were counted manually in the nuclei of 300–400 untreated, empty vector, IDH1^{wt} or IDH1^{R132H} cells under normoxia and hypoxia, respectively.

4.5. *GSH/GSSG Ratio*

For measuring the GSH/GSSG ratio, cells were trypsinized, plated in 96-well plates at different cell densities ranging from 3000–5000 cells/well depending on the cell line. After 24 h the cells were irradiated with 0 or 5 Gy and GSH/GSSG Ratio was measured using GSH/GSSG-Glo™ Assay (Promega) via a luminescence signal by GENios™ plate reader (Tecan, Crailsheim, Germany) 1 h, 2 h, and 24 h after irradiation following manufactures instructions.

4.6. *Atomic Force Microscopy*

The influence of IDH1^{R132H} mutant protein expression on cell stiffness was investigated by atomic force microscopy (AFM). AFM indentation is a useful tool to analyze the mechanical properties of living cells in physiological environment at the nanoscale. For characterization of mechanical properties of single glioma cells we used two parameters namely the indentation depth (physical deformation of the cell) and the Young´s modulus [114]. These two parameters were combined to a cluster parameter called "generalized stiffness", as described previously [55]. This parameter was used as it was demonstrated to show a strong correlation with the invasive properties of single malignant glioma cell lines [54].

Detailed information on AFM measurements can be found in Supplementary Materials (Supplementary Methods 5; Figure S1 and Figure S2).

4.7. Statistical Analysis

Data were analyzed by unpaired two-tailed t-tests using GraphPad Prism 6 (San Diego, CA, USA). All tests were performed using 95% confidence intervals (a p-value < 0.05 was considered significant). Results are presented as means ±SD of n = 3–4 independent experiments where * represents $p < 0.05$ and ** represents $p < 0.01$.

5. Conclusions

The enhanced radiosensitivity and increased cellular stiffness of $IDH1^{R132H}$-positive glioma cells are consistent with the clinical observation of a less aggressive tumor and a prolonged survival of diffuse glioma patients harboring the IDH1 mutation. The results improve the understanding of the molecular basis of diffuse gliomas and highlight the central role of mutant IDH1 in treatment of patients with these tumors especially in response to radiation. In conclusion, detection of the genetic status of IDH1 before therapy might guide treatment decisions for patients with a less aggressive and radiosensitive IDH1-mutant diffuse glioma who should receive radiotherapy and patients with a more aggressive IDH1-wildtype diffuse gliomas who could benefit from an individually intensified therapy comprising radiotherapy and alternative medical treatment.

Supplementary Materials: The following are available online at http://www.mdpi.com/2072-6694/11/6/889/s1, Supplementary Methods 1: Generation of constructs and stable overexpression of $IDH1^{wt}$ and $IDH1^{R132H}$ in glioma cell lines, Supplementary Methods 2: Real-time PCR and generation of plasmid standards, Table S1. Primers for qPCR, Supplementary Methods 3: Protein isolation and western blot analysis, Supplementary Methods 4: immunofluorescence or immunohistochemical staining, Table S2: Antibodies of western blot analysis, immunofluorescence and immunohistochemical staining, Supplementary Methods 5: Atomic force microscopy, Figure S1: Atomic force microscopy, Figure S2: Measurement principle of atomic force microscopy.

Author Contributions: Conceptualization, J.K., T.H., F.D and D.V.; Methodology, J.K., A.G., M.P.; Atomic force microscopy, T.H. and U.H.; software, T.H.; validation, J.K., T.H. and A.G.; formal analysis and investigation, J.K. and T.H.; writing—original draft preparation, J.K.; writing—review and editing, T.H., O.C., M.B., U.H.; supervision, M.B. and F.D.; project administration, D.V.; funding acquisition, D.V.

Funding: This work was supported by the Wilhelm-Roux-Program (FKZ: VF 02/39).

Acknowledgments: We would like to thank our colleagues from the Department of Radiotherapy and the Department of Anatomy and Cell Biology for their contribution to this study and their continuous support. We would also like to thank Gabriele Thomas for her excellent technical assistance.

Conflicts of Interest: The authors declare that they have no competing interests.

Appendix A

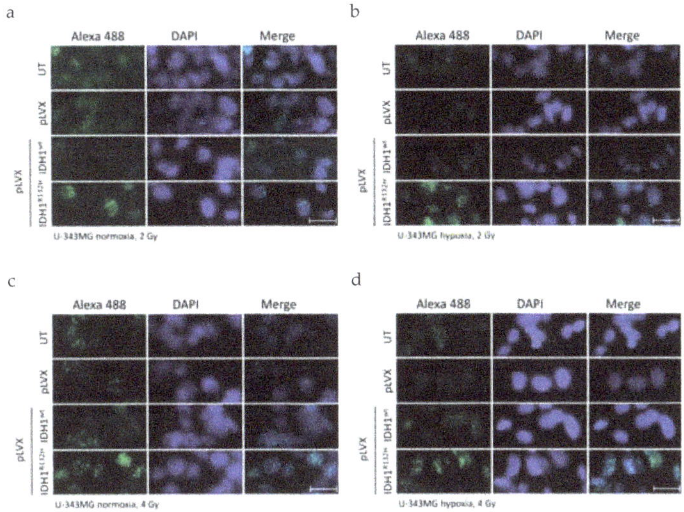

Figure A1. Effect of $IDH1^{R132H}$ gene expression on the accumulation of residual γH2AX foci after radiation in U-343MG cells. Representative immunofluorescence images of γH2AX of U-343MG cells 24 h after irradiation with 2 (**a**,**b**) and 4 Gy (c, d) under normoxia (**a**,**c**) and hypoxia (**b**,**d**). Green: γH2AX foci; Blue: Cell nuclei (DAPI). n = 3 independent experiments; scale bar = 25 μm. Normoxia (21% O_2), hypoxia (< 0.1% O_2). UT: untreated, pLVX: cells stably transduced with empty vector, pLVX $IDH1^{wt}$: $IDH1^{wt}$-expressing cells, pLVX $IDH1^{R132H}$: $IDH1^{R132H}$-expressing cells.

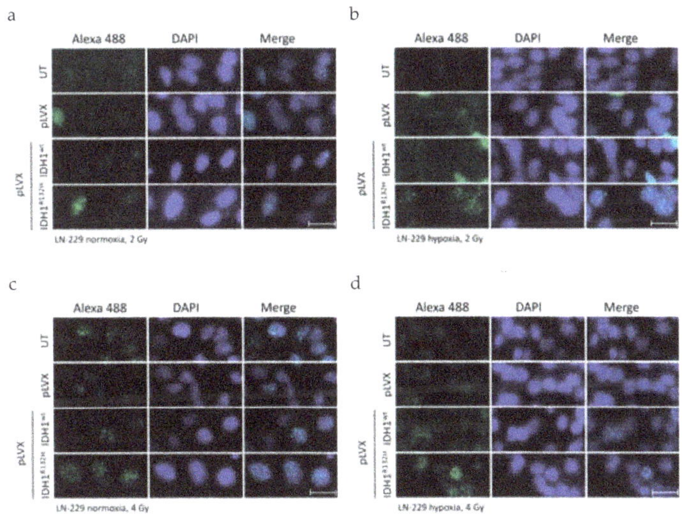

Figure A2. Effect of $IDH1^{R132H}$ gene expression on the accumulation of residual γH2AX foci after radiation in LN-229 cells. Representative immunofluorescence images of γH2AX of LN-229 cells 24 h after irradiation with 2 (**a**,**b**) and 4 Gy (**c**,**d**) under normoxia (**a**,**c**) and hypoxia (**b**,**d**). Green: γH2AX foci; Blue: Cell nuclei (DAPI). n = 3 independent experiments; scale bar = 25 μm. Normoxia (21% O_2), hypoxia (< 0.1% O_2). UT: untreated, pLVX: cells stably transduced with empty vector, pLVX $IDH1^{wt}$: $IDH1^{wt}$-expressing cells, pLVX $IDH1^{R132H}$: $IDH1^{R132H}$-expressing cells.

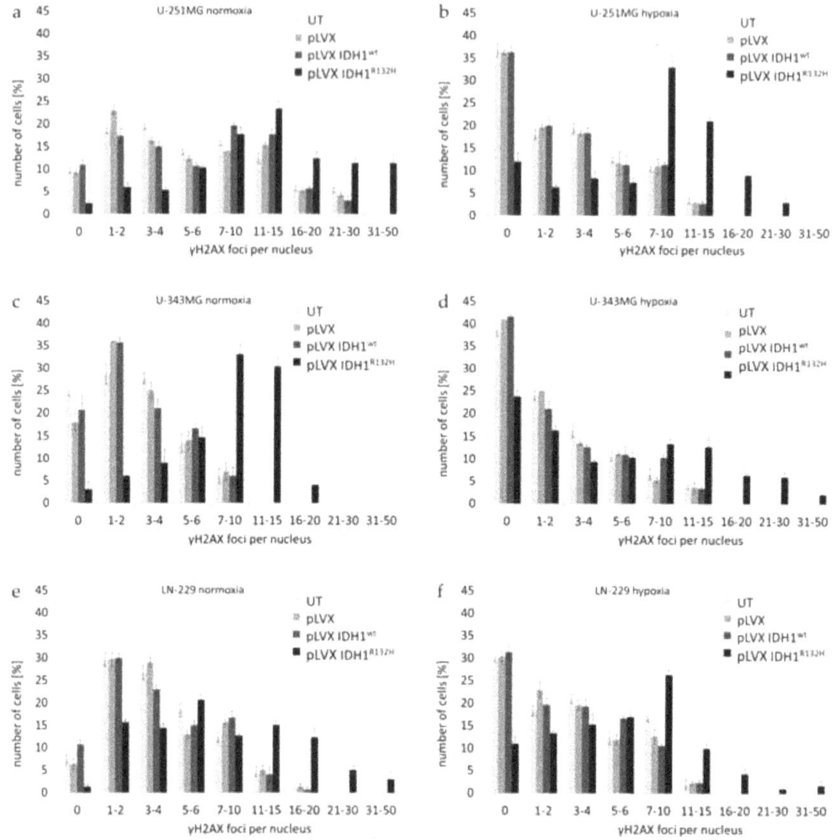

Figure A3. Effect of $IDH1^{R132H}$ gene expression on the distribution of residual γH2AX foci after radiation in glioma cells. Distribution of UT, pLVX, $IDH1^{wt}$-expressing and $IDH1^{R132H}$-expressing cells depending on the number of foci. In U-251MG (**a**,**b**), U-343MG (**c**,**d**), and LN-229 (**e**,**f**) cells residual γH2AX foci were quantified manually as foci per nucleus of 300–400 untreated, empty vector, $IDH1^{wt}$ or $IDH1^{R132H}$ cells under normoxia (21% O_2) and hypoxia (< 0.1% O_2). γH2AX staining was achieved 24 h after radiation with 4 Gy. Bars represent the mean values of three independent experiments. Error bars indicate standard deviations (±SD). UT: untreated, pLVX: cells stably transduced with empty vector, pLVX $IDH1^{wt}$: $IDH1^{wt}$-expressing cells, pLVX $IDH1^{R132H}$: $IDH1^{R132H}$-expressing cells.

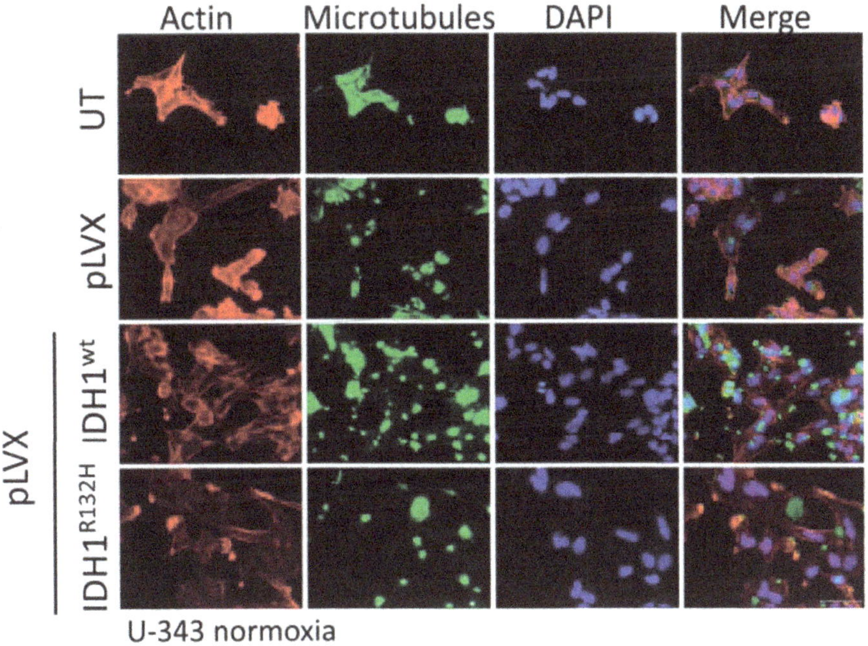

Figure A4. Immunofluorescence staining of transduced U-343MG glioma cells stably expressed $IDH1^{wt}$ or $IDH1^{R312H}$ protein. Representative immunofluorescence staining of actin stress fibers and microtubules in U-343MG cells using phalloidin-TRITC and anti-tubulin antibody. Immunofluorescence staining was achieved 24 h after seeding. Cell nuclei were counterstained with DAPI. n = 3 independent experiments; scale bar = 50 μm. UT: untreated, pLVX: cells stably transduced with empty vector, pLVX $IDH1^{wt}$: $IDH1^{wt}$-expressing cells, pLVX $IDH1^{R132H}$: $IDH1^{R132H}$-expressing cells.

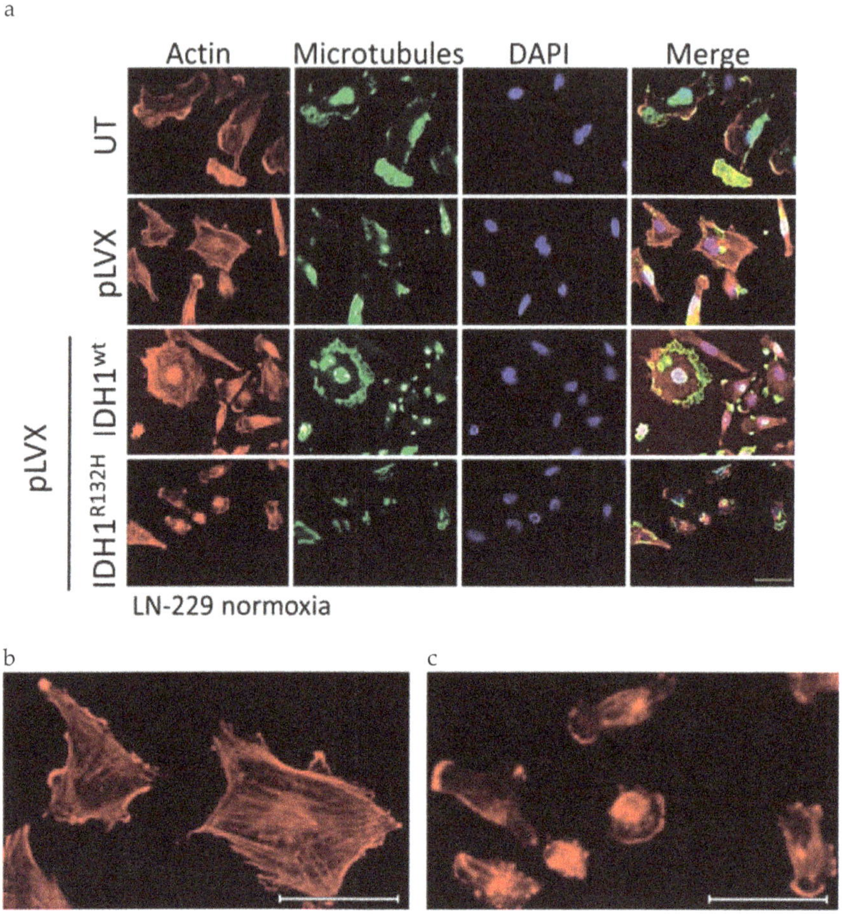

Figure A5. Immunofluorescence staining of transduced LN-229 glioma cells stably expressed IDH1^{wt} or IDH1^{R312H} protein. (**a**) Representative immunofluorescence staining of actin stress fibers and microtubules in LN-229 cells using phalloidin-TRITC and anti-tubulin antibody. Immunofluorescence staining was achieved 24 h after seeding. Cell nuclei were counterstained with DAPI. n = 3 independent experiments; scale bar = 50 µm. (**b**) Enlarged representative immunofluorescence staining as a representative example for untreated cells, empty vector cells (pLVX) and IDH1^{wt} cells; the enlarged part is taken from pLVX cells; scale bar = 50 µm. (**c**) Enlarged representative immunofluorescence staining as a representative example for untreated cells, empty vector cells (pLVX) and IDH1^{R132H} cells; the enlarged part is taken from IDH1^{R132H} cells; scale bar = 50 µm. UT: untreated, pLVX: cells stably transduced with empty vector, pLVX IDH1^{wt}: IDH1^{wt}-expressing cells, pLVX IDH1^{R132H}: IDH1^{R132H}-expressing cells.

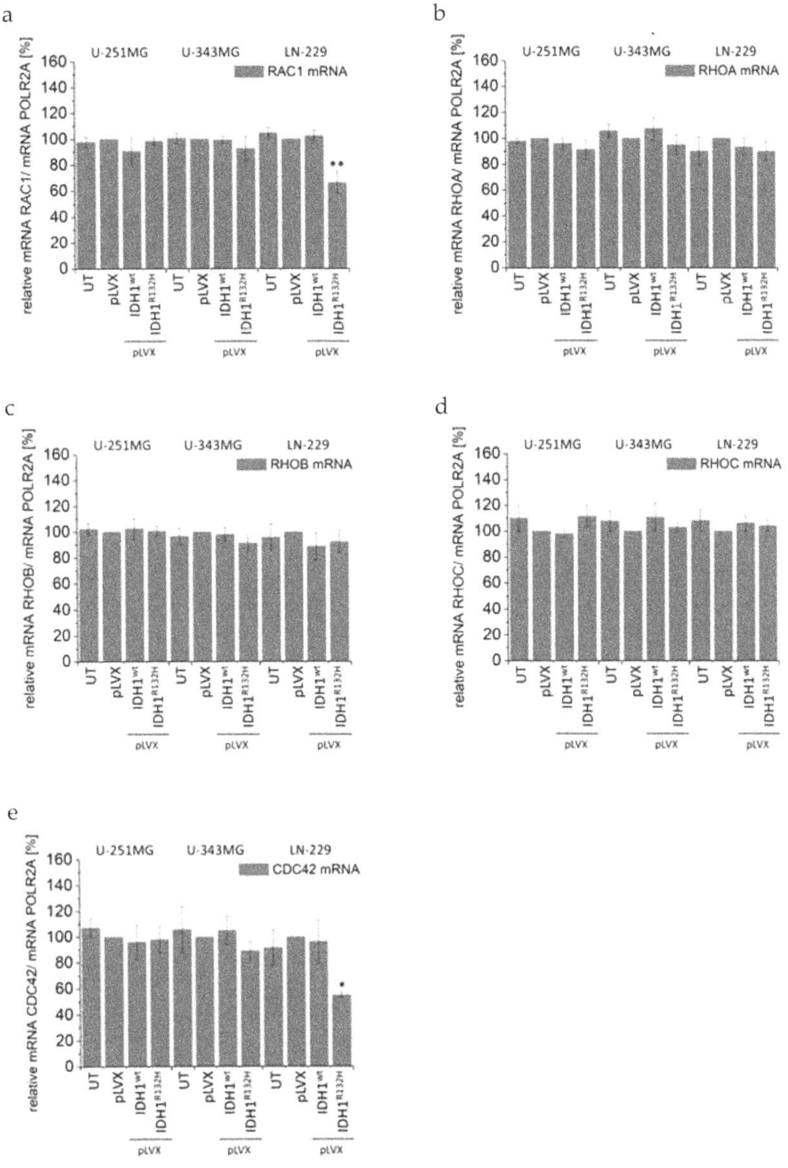

Figure A6. Effect of IDH1^{wt} or IDH1^{R132H} on mRNA expression of actin-regulating proteins. mRNA expression levels (qPCR) of RAC1 (**a**), RHOA (**b**), RHOB (**c**), RHOC (**d**), and CDC42 (**e**) in stably transduced U-251MG, U-343MG, and LN-229 cells. Bars represent the mean values, relative to the control (empty vector pLVX set as 100%), of three independent experiments. Error bars indicate standard deviations (±SD). UT: untreated cells, pLVX: cells stably transduced with empty vector, pLVX IDH1^{wt}: IDH1^{wt}-expressing cells, pLVX IDH1^{R132H}: IDH1^{R132H}-expressing cells. * $p < 0.05$ and ** $p < 0.01$ (compared to the respective IDH1^{wt} cells).

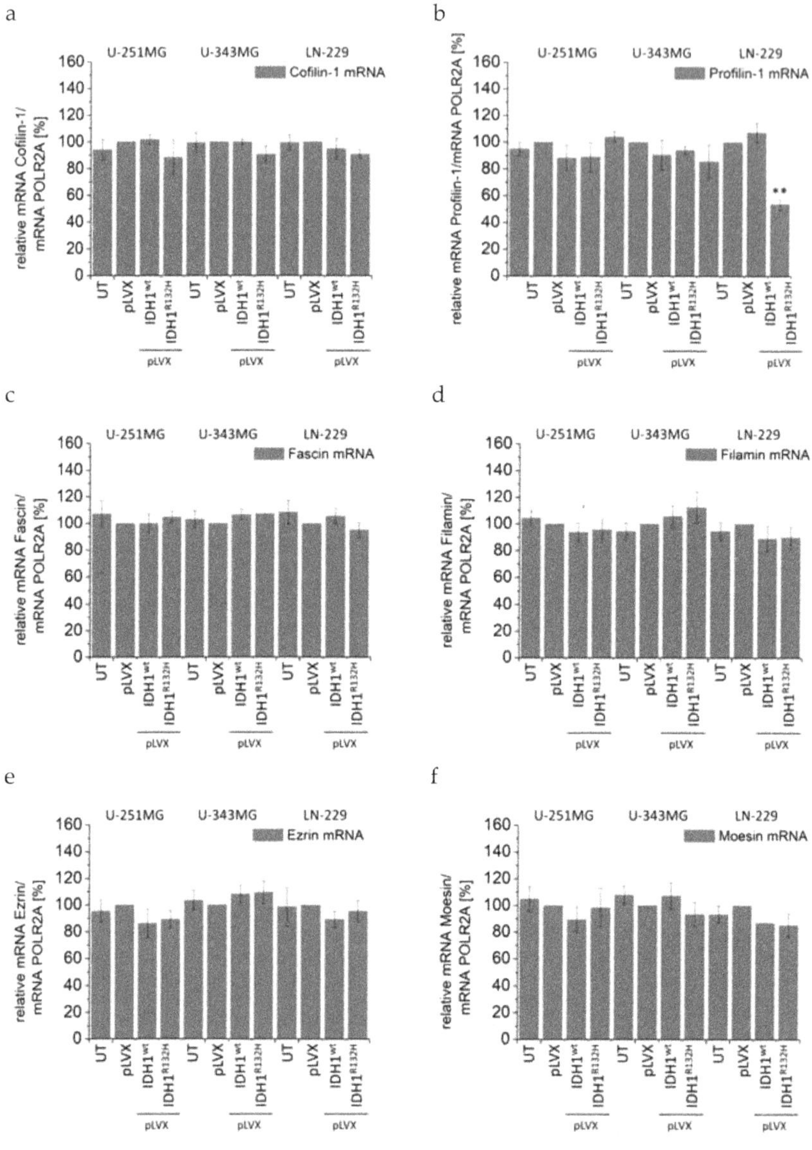

Figure A7. Effect of IDH1^{wt} or IDH1^{R132H} on mRNA expression of actin-binding proteins. mRNA expression levels (qPCR) of Cofilin-1 (**a**), Profilin-1 (**b**), Fascin (**c**), Filamin (**d**), Ezrin (**e**), and Moesin (**f**) in stably transduced U-251MG, U-343MG, and LN-229 cells. Bars represent the mean values, relative to the control (empty vector pLVX set as 100%), of three independent experiments. Error bars indicate standard deviations (±SD). UT: untreated cells, pLVX: cells stably transduced with empty vector, pLVX IDH1^{wt}: IDH1^{wt}-expressing cells, pLVX IDH1^{R132H}: IDH1^{R132H}-expressing cells. ** $p < 0.01$ (compared to the respective IDH1^{wt} cells).

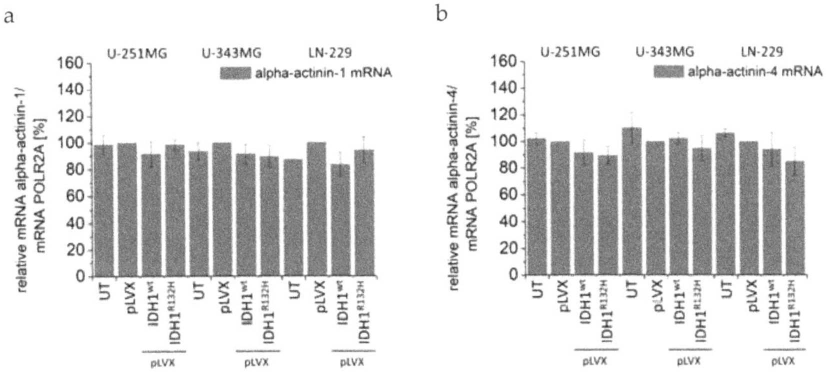

Figure A8. Effect of IDH1^{wt} or IDH1^{R132H} on mRNA expression of actin-binding proteins. mRNA expression levels (qPCR) of alpha-actinin-1 (**a**) and alpha-actinin-4 (**b**) in stably transduced U-251MG, U-343MG, and LN-229 cells. Bars represent the mean values, relative to the control (empty vector pLVX set as 100%), of three independent experiments. Error bars indicate standard deviations (±SD). UT: untreated cells, pLVX: cells stably transduced with empty vector, pLVX IDH1^{wt}: IDH1^{wt}-expressing cells, pLVX IDH1^{R132H}: IDH1^{R132H}-expressing cells.

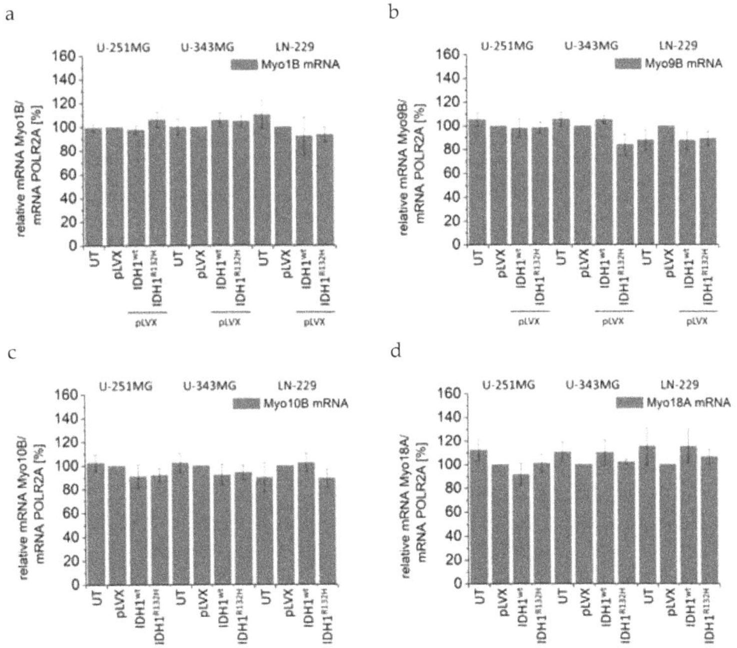

Figure A9. Effect of IDH1^{wt} or IDH1^{R132H} on mRNA expression of actin-structure proteins. mRNA expression levels (qPCR) of Myo1B (**a**), Myo9B (**b**), Myo10B (**c**), and Myo18A (**d**) in stably transduced U-251MG, U-343MG, and LN-229 cells. Bars represent the mean values, relative to the control (empty vector pLVX set as 100%), of three independent experiments. Error bars indicate standard deviations (±SD). UT: untreated cells, pLVX: cells stably transduced with empty vector, pLVX IDH1^{wt}: IDH1^{wt}-expressing cells, pLVX IDH1^{R132H}: IDH1^{R132H}-expressing cells.

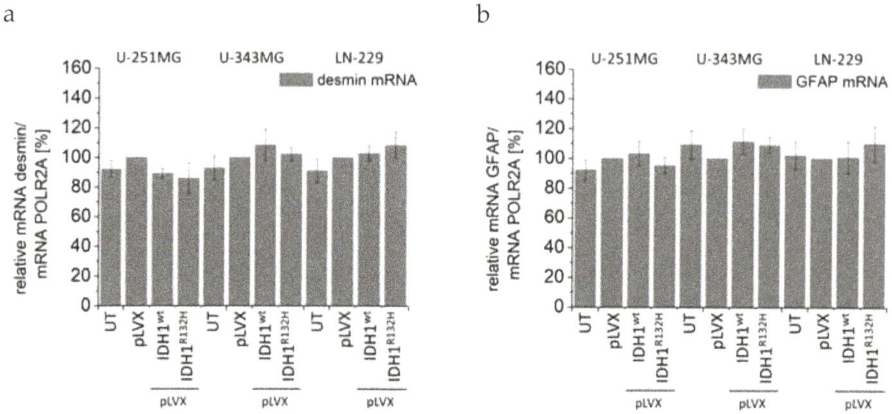

Figure A10. Effect of IDH1^{wt} or IDH1^{R132H} on mRNA expression of intermediate filaments. mRNA expression levels (qPCR) of desmin (**a**) and GFAP (**b**) in stably transduced U-251MG, U-343MG, and LN-229 cells. Bars represent the mean values, relative to the control (empty vector pLVX set as 100%), of three independent experiments. Error bars indicate standard deviations (±SD). UT: untreated cells, pLVX: cells stably transduced with empty vector, pLVX IDH1^{wt}: IDH1^{wt}-expressing cells, pLVX IDH1^{R132H}: IDH1^{R132H}-expressing cells.

References

1. Ostrom, Q.T.; Gittleman, H.; Liao, P.; Rouse, C.; Chen, Y.; Dowling, J.; Wolinsky, Y.; Kruchko, C.; Barnholtz-Sloan, J. CBTRUS Statistical Report: Primary Brain and Central Nervous System Tumors Diagnosed in the United States in 2007–2011. *Neuro Oncol.* **2014**, *16*, iv1–iv63. [CrossRef] [PubMed]
2. Wesseling, P.; Capper, D. WHO 2016 Classification of gliomas. *Neuropathol. Appl. Neurobiol.* **2018**, *44*, 139–150. [CrossRef] [PubMed]
3. Perry, A.; Wesseling, P. Chapter 5—Histologic classification of gliomas. In *Handbook of Clinical Neurology*; Berger, M.S., Weller, M., Eds.; Elsevier: Amsterdam, The Netherlands, 2016; Volume 134, pp. 71–95. [CrossRef]
4. Louis, D.N.; Perry, A.; Reifenberger, G.; von Deimling, A.; Figarella-Branger, D.; Cavenee, W.K.; Ohgaki, H.; Wiestler, O.D.; Kleihues, P.; Ellison, D.W.; et al. The 2016 World Health Organization Classification of Tumors of the Central Nervous System: A summary. *Acta Neuropathol.* **2016**, *131*, 803–820. [CrossRef]
5. Louis, D.; Ohgaki, H.; Wiestler, O.; Cavenee, W. *WHO Classification of Tumours of the Central Nervous System*; IARC Press: Lyon, France, 2016.
6. Louis, D.N.; Ohgaki, H.; Wiestler, O.D.; Cavenee, W.K.; Burger, P.C.; Jouvet, A.; Scheithauer, B.W.; Kleihues, P. The 2007 WHO Classification of Tumours of the Central Nervous System. *Acta Neuropathol.* **2007**, *114*, 97–109. [CrossRef] [PubMed]
7. Monteiro, A.R.; Hill, R.; Pilkington, G.J.; Madureira, P.A. The Role of Hypoxia in Glioblastoma Invasion. *Cells* **2017**, *6*, 45. [CrossRef] [PubMed]
8. Yan, H.; Parsons, D.W.; Jin, G.; McLendon, R.; Rasheed, B.A.; Yuan, W.; Kos, I.; Batinic-Haberle, I.; Jones, S.; Riggins, G.J.; et al. IDH1 and IDH2 Mutations in Gliomas. *N. Engl. J. Med.* **2009**, *360*, 765–773. [CrossRef] [PubMed]
9. Parsons, D.W.; Jones, S.; Zhang, X.; Lin, J.C.-H.; Leary, R.J.; Angenendt, P.; Mankoo, P.; Carter, H.; Siu, I.-M.; Gallia, G.L.; et al. An integrated genomic analysis of human glioblastoma multiforme. *Science* **2008**, *321*, 1807–1812. [CrossRef]
10. Bleeker, F.E.; Lamba, S.; Leenstra, S.; Troost, D.; Hulsebos, T.; Vandertop, W.P.; Frattini, M.; Molinari, F.; Knowles, M.; Cerrato, A.; et al. IDH1 mutations at residue p.R132 (IDH1R132) occur frequently in high-grade gliomas but not in other solid tumors. *Hum. Mutat.* **2009**, *30*, 7–11. [CrossRef]
11. Tabatabai, G.; Stupp, R.; van den Bent, M.J.; Hegi, M.E.; Tonn, J.C.; Wick, W.; Weller, M. Molecular diagnostics of gliomas: The clinical perspective. *Acta Neuropathol.* **2010**, *120*, 585–592. [CrossRef]

12. Tateishi, K.; Wakimoto, H.; Cahill, D.P. IDH1 Mutation and World Health Organization 2016 Diagnostic Criteria for Adult Diffuse Gliomas: Advances in Surgical Strategy. *Neurosurgery* **2017**, *64*, 134–138. [CrossRef]
13. Reuss, D.E.; Kratz, A.; Sahm, F.; Capper, D.; Schrimpf, D.; Koelsche, C.; Hovestadt, V.; Bewerunge-Hudler, M.; Jones, D.T.W.; Schittenhelm, J.; et al. Adult IDH wild type astrocytomas biologically and clinically resolve into other tumor entities. *Acta Neuropathol.* **2015**, *130*, 407–417. [CrossRef]
14. Olar, A.; Wani, K.M.; Alfaro-Munoz, K.D.; Heathcock, L.E.; van Thuijl, H.F.; Gilbert, M.R.; Armstrong, T.S.; Sulman, E.P.; Cahill, D.P.; Vera-Bolanos, E.; et al. IDH mutation status and role of WHO grade and mitotic index in overall survival in grade II-III diffuse gliomas. *Acta Neuropathol.* **2015**, *129*, 585–596. [CrossRef] [PubMed]
15. Reitman, Z.J.; Yan, H. Isocitrate Dehydrogenase 1 and 2 Mutations in Cancer: Alterations at a Crossroads of Cellular Metabolism. *J. Natl. Cancer Inst.* **2010**, *102*, 932–941. [CrossRef] [PubMed]
16. Hartmann, C.; Meyer, J.; Balss, J.; Capper, D.; Mueller, W.; Christians, A.; Felsberg, J.; Wolter, M.; Mawrin, C.; Wick, W.; et al. Type and frequency of IDH1 and IDH2 mutations are related to astrocytic and oligodendroglial differentiation and age: A study of 1010 diffuse gliomas. *Acta Neuropathol.* **2009**, *118*, 469–474. [CrossRef] [PubMed]
17. Dubbink, H.J.; Taal, W.; van Marion, R.; Kros, J.M.; van Heuvel, I.; Bromberg, J.E.; Zonnenberg, B.A.; Zonnenberg, C.B.L.; Postma, T.J.; Gijtenbeek, J.M.M.; et al. IDH1 mutations in low-grade astrocytomas predict survival but not response to temozolomide. *Neurology* **2009**, *73*, 1792–1795. [CrossRef] [PubMed]
18. Sanson, M.; Marie, Y.; Paris, S.; Idbaih, A.; Laffaire, J.; Ducray, F.; Hallani, S.E.; Boisselier, B.; Mokhtari, K.; Hoang-Xuan, K.; et al. Isocitrate Dehydrogenase 1 Codon 132 Mutation Is an Important Prognostic Biomarker in Gliomas. *JCO* **2009**, *27*, 4150–4154. [CrossRef] [PubMed]
19. Cancer Genome Atlas Research Network. Comprehensive, Integrative Genomic Analysis of Diffuse Lower-Grade Gliomas. *N. Engl. J. Med.* **2015**, *372*, 2481–2498. [CrossRef]
20. Preusser, M.; Haberler, C.; Hainfellner, J.A. Malignant glioma: Neuropathology and neurobiology. *Wien. Med. Wochenschr.* **2006**, *156*, 332–337. [CrossRef]
21. Huse, J.T.; Holland, E.C. Targeting brain cancer: Advances in the molecular pathology of malignant glioma and medulloblastoma. *Nat. Rev. Cancer* **2010**, *10*, 319–331. [CrossRef]
22. Vaupel, P.; Kelleher, D.K.; Höckel, M. Oxygen status of malignant tumors: Pathogenesis of hypoxia and significance for tumor therapy. *Semin. Oncol.* **2001**, *28*, 29–35. [CrossRef]
23. Höckel, M.; Vaupel, P. Biological consequences of tumor hypoxia. *Semin. Oncol.* **2001**, *28*, 36–41. [CrossRef]
24. Cuddapah, V.A.; Robel, S.; Watkins, S.; Sontheimer, H. A neurocentric perspective on glioma invasion. *Nat. Rev. Neurosci.* **2014**, *15*, 455–465. [CrossRef] [PubMed]
25. Stupp, R.; Mason, W.P.; van den Bent, M.J.; Weller, M.; Fisher, B.; Taphoorn, M.J.B.; Belanger, K.; Brandes, A.A.; Marosi, C.; Bogdahn, U.; et al. Radiotherapy plus Concomitant and Adjuvant Temozolomide for Glioblastoma. *N. Engl. J. Med.* **2005**, *352*, 987–996. [CrossRef]
26. Stupp, R.; Hegi, M.E.; Mason, W.P.; van den Bent, M.J.; Taphoorn, M.J.; Janzer, R.C.; Ludwin, S.K.; Allgeier, A.; Fisher, B.; Belanger, K.; et al. Effects of radiotherapy with concomitant and adjuvant temozolomide versus radiotherapy alone on survival in glioblastoma in a randomised phase III study: 5-year analysis of the EORTC-NCIC trial. *Lancet Oncol.* **2009**, *10*, 459–466. [CrossRef]
27. Jovčevska, I.; Kočevar, N.; Komel, R. Glioma and glioblastoma—How much do we (not) know? *Mol. Clin. Oncol.* **2013**, *1*, 935–941. [CrossRef] [PubMed]
28. De Groot, J.F. High-grade gliomas. *Continuum* **2015**, *21*, 332–344. [CrossRef] [PubMed]
29. Metallo, C.M.; Gameiro, P.A.; Bell, E.L.; Mattaini, K.R.; Yang, J.; Hiller, K.; Jewell, C.M.; Johnson, Z.R.; Irvine, D.J.; Guarente, L.; et al. Reductive glutamine metabolism by IDH1 mediates lipogenesis under hypoxia. *Nature* **2011**, *481*, 380–384. [CrossRef] [PubMed]
30. Wise, D.R.; Ward, P.S.; Shay, J.E.S.; Cross, J.R.; Gruber, J.J.; Sachdeva, U.M.; Platt, J.M.; DeMatteo, R.G.; Simon, M.C.; Thompson, C.B.; et al. Hypoxia promotes isocitrate dehydrogenase-dependent carboxylation of α-ketoglutarate to citrate to support cell growth and viability. *Proc. Natl. Acad. Sci. USA* **2011**, *108*, 19611–19616. [CrossRef] [PubMed]
31. Reitman, Z.J.; Jin, G.; Karoly, E.D.; Spasojevic, I.; Yang, J.; Kinzler, K.W.; He, Y.; Bigner, B.D.; Vogelstein, B.; Yan, H.; et al. Profiling the effects of isocitrate dehydrogenase 1 and 2 mutations on the cellular metabolome. *Proc. Natl. Acad. Sci. USA* **2011**, *108*, 3270–3275. [CrossRef] [PubMed]

32. Zhao, S.; Lin, Y.; Xu, W.; Jiang, W.; Zha, Z.; Wang, P.; Yu, W.; Li, Z.; Gong, L.; Peng, Y.; et al. Glioma-Derived Mutations in IDH1 Dominantly Inhibit IDH1 Catalytic Activity and Induce HIF-1α. *Science* **2009**, *324*, 261–265. [CrossRef]
33. Dang, L.; White, D.W.; Gross, S.; Bennett, B.D.; Bittinger, M.A.; Driggers, E.M.; Fantin, V.R.; Jang, H.G.; Jin, S.; Keenan, M.C.; et al. Cancer-associated IDH1 mutations produce 2-hydroxyglutarate. *Nature* **2010**, *465*, 966. [CrossRef] [PubMed]
34. Gross, S.; Cairns, R.A.; Minden, M.D.; Driggers, E.M.; Bittinger, M.A.; Jang, H.G.; Sasaki, M.; Jin, S.; Schenkein, D.P.; Su, S.M.; et al. Cancer-associated metabolite 2-hydroxyglutarate accumulates in acute myelogenous leukemia with isocitrate dehydrogenase 1 and 2 mutations. *J. Exp. Med.* **2010**, *207*, 339–344. [CrossRef] [PubMed]
35. Ward, P.S.; Patel, J.; Wise, D.R.; Abdel-Wahab, O.; Bennett, B.D.; Coller, H.A.; Cross, J.R.; Fantin, V.R.; Hedvat, C.V.; Perl, A.E.; et al. The common feature of leukemia-associated IDH1 and IDH2 mutations is a neomorphic enzyme activity converting alpha-ketoglutarate to 2-hydroxyglutarate. *Cancer Cell* **2010**, *17*, 225–234. [CrossRef] [PubMed]
36. Bleeker, F.E.; Atai, N.A.; Lamba, S.; Jonker, A.; Rijkeboer, D.; Bosch, K.S.; Tigchelaar, W.; Troost, D.; Vandertop, W.P.; Bardelli, A.; et al. The prognostic IDH1(R132) mutation is associated with reduced NADP+-dependent IDH activity in glioblastoma. *Acta Neuropathol.* **2010**, *119*, 487–494. [CrossRef] [PubMed]
37. Ozben, T. Oxidative stress and apoptosis: Impact on cancer therapy. *J. Pharm. Sci.* **2007**, *96*, 2181–2196. [CrossRef] [PubMed]
38. Van den Bent, M.J.; Dubbink, H.J.; Marie, Y.; Brandes, A.A.; Taphoorn, M.J.B.; Wesseling, P.; Frenay, M.; Tijssen, C.C.; Lacombe, D.; Idbaih, A.; et al. IDH1 and IDH2 Mutations Are Prognostic but not Predictive for Outcome in Anaplastic Oligodendroglial Tumors: A Report of the European Organization for Research and Treatment of Cancer Brain Tumor Group. *Clin. Cancer Res.* **2010**, *16*, 1597–1604. [CrossRef] [PubMed]
39. Lu, Y.; Kwintkiewicz, J.; Liu, Y.; Tech, K.; Frady, L.N.; Su, Y.-T.; Bautista, W.; Moon, S.I.; MacDonald, J.; Ewend, M.G.; et al. Chemosensitivity of IDH1-mutated gliomas due to an impairment in PARP1-mediated DNA repair. *Cancer Res.* **2017**, *77*, 1709–1718. [CrossRef] [PubMed]
40. Li, K.; Ouyang, L.; He, M.; Luo, M.; Cai, W.; Tu, Y.; Pi, R.; Liu, A. IDH1 R132H mutation regulates glioma chemosensitivity through Nrf2 pathway. *Oncotarget* **2017**, *8*, 28865–28879. [CrossRef] [PubMed]
41. Kessler, J.; Güttler, A.; Wichmann, H.; Rot, S.; Kappler, M.; Bache, M.; Vordermark, D. IDH1(R132H) mutation causes a less aggressive phenotype and radiosensitizes human malignant glioma cells independent of the oxygenation status. *Radiother. Oncol.* **2015**, *116*, 381–387. [CrossRef] [PubMed]
42. Westermark, B.; Pontén, J.; Hugosson, R. Determinants for the establishment of permanent tissue culture lines from human gliomas. *Acta Pathol. Microbiol. Scand. A* **1973**, *81*, 791–805. [CrossRef]
43. Ishii, N.; Maier, D.; Merlo, A.; Tada, M.; Sawamura, Y.; Diserens, A.C.; Van Meir, E.G. Frequent co-alterations of TP53, p16/CDKN2A, p14ARF, PTEN tumor suppressor genes in human glioma cell lines. *Brain Pathol.* **1999**, *9*, 469–479. [CrossRef] [PubMed]
44. Paull, T.T.; Rogakou, E.P.; Yamazaki, V.; Kirchgessner, C.U.; Gellert, M.; Bonner, W.M. A critical role for histone H2AX in recruitment of repair factors to nuclear foci after DNA damage. *Curr. Biol.* **2000**, *10*, 886–895. [CrossRef]
45. Rogakou, E.P.; Boon, C.; Redon, C.; Bonner, W.M. Megabase Chromatin Domains Involved in DNA Double-Strand Breaks in Vivo. *J. Cell Biol.* **1999**, *146*, 905–916. [CrossRef] [PubMed]
46. Sedelnikova, O.A.; Rogakou, E.P.; Panyutin, I.G.; Bonner, W.M. Quantitative Detection of 125IdU-Induced DNA Double-Strand Breaks with γ-H2AX Antibody. *Radiat. Res.* **2002**, *158*, 486–492. [CrossRef]
47. Frankenberg-Schwager, M. Induction, repair and biological relevance of radiation-induced DNA lesions in eukaryotic cells. *Radiat. Environ. Biophys.* **1990**, *29*, 273–292. [CrossRef] [PubMed]
48. Rothkamm, K.; Löbrich, M. Evidence for a lack of DNA double-strand break repair in human cells exposed to very low x-ray doses. *Proc. Natl. Acad. Sci. USA* **2003**, *100*, 5057–5062. [CrossRef] [PubMed]
49. Pilch, D.R.; Sedelnikova, O.A.; Redon, C.; Celeste, A.; Nussenzweig, A.; Bonner, W.M. Characteristics of γ-H2AX foci at DNA double-strand breaks sites. *Biochem. Cell Biol.* **2003**, *81*, 123–129. [CrossRef]
50. Rothkamm, K.; Horn, S. γ-H2AX as protein biomarker for radiation exposure. *Ann. Ist. Super. Sanita* **2009**, *45*, 265–271.
51. Banáth, J.P.; Klokov, D.; MacPhail, S.H.; Banuelos, C.A.; Olive, P.L. Residual γH2AX foci as an indication of lethal DNA lesions. *BMC Cancer* **2010**, *10*, 4. [CrossRef]

52. Cross, S.E.; Jin, Y.-S.; Rao, J.; Gimzewski, J.K. Nanomechanical analysis of cells from cancer patients. *Nat. Nano* **2007**, *2*, 780–783. [CrossRef]
53. Cross, S.E.; Jin, Y.-S.; Tondre, J.; Wong, R.; Rao, J.; Gimzewski, J.K. AFM-based analysis of human metastatic cancer cells. *Nanotechnology* **2008**, *19*, 384003. [CrossRef] [PubMed]
54. Hayashi, K.; Iwata, M. Stiffness of cancer cells measured with an AFM indentation method. *J. Mech. Behav. Biomed. Mater.* **2015**, *49*, 105–111. [CrossRef] [PubMed]
55. Hohmann, T.; Grabiec, U.; Ghadban, C.; Feese, K.; Dehghani, F. The influence of biomechanical properties and cannabinoids on tumor invasion. *Cell Adhes. Migr.* **2016**, *11*, 54–67. [CrossRef] [PubMed]
56. Tojkander, S.; Gateva, G.; Lappalainen, P. Actin stress fibers—Assembly, dynamics and biological roles. *J. Cell Sci.* **2012**, *125*, 1855–1864. [CrossRef] [PubMed]
57. Burridge, K.; Guilluy, C. Focal adhesions, stress fibers and mechanical tension. *Exp. Cell Res.* **2016**, *343*, 14–20. [CrossRef]
58. Takenawa, T.; Suetsugu, S. The WASP-WAVE protein network: Connecting the membrane to the cytoskeleton. *Nat. Rev. Mol. Cell Biol.* **2007**, *8*, 37–48. [CrossRef] [PubMed]
59. Molinie, N.; Gautreau, A. The Arp2/3 Regulatory System and Its Deregulation in Cancer. *Physiol. Rev.* **2017**, *98*, 215–238. [CrossRef] [PubMed]
60. Rotty, J.D.; Wu, C.; Haynes, E.M.; Suarez, C.; Winkelman, J.D.; Johnson, H.E.; Haugh, J.M.; Kovar, D.R.; Bear, J.E. Profilin-1 serves as a gatekeeper for actin assembly by Arp2/3-dependent and -independent pathways. *Dev. Cell* **2015**, *32*, 54–67. [CrossRef] [PubMed]
61. Ohgaki, H.; Kleihues, P. The Definition of Primary and Secondary Glioblastoma. *Clin. Cancer Res.* **2013**, *19*, 764–772. [CrossRef]
62. Hartmann, C.; Hentschel, B.; Tatagiba, M.; Schramm, J.; Schnell, O.; Seidel, C.; Stein, R.; Reifenberger, R.; Pietsch, T.; von Deimling, A.; et al. Molecular markers in low-grade gliomas: Predictive or prognostic? *Clin. Cancer Res.* **2011**, *17*, 4588–4599. [CrossRef]
63. Stupp, R.; Tonn, J.-C.; Brada, M.; Pentheroudakis, G.; ESMO Guidelines Working Group. High-grade malignant glioma: ESMO Clinical Practice Guidelines for diagnosis, treatment and follow-up. *Ann. Oncol.* **2010**, *21*, v190–v193. [CrossRef] [PubMed]
64. Stupp, R.; Brada, M.; van den Bent, M.J.; Tonn, J.-C.; Pentheroudakis, G. High-grade glioma: ESMO Clinical Practice Guidelines for diagnosis, treatment and follow-up. *Ann. Oncol.* **2014**, *25*, iii93–iii101. [CrossRef] [PubMed]
65. Wen, P.Y.; Kesari, S. Malignant gliomas in adults. *N. Engl. J. Med.* **2008**, *359*, 492–507. [CrossRef] [PubMed]
66. Chen, J.; McKay, R.M.; Parada, L.F. Malignant Glioma: Lessons from Genomics, Mouse Models, and Stem Cells. *Cell* **2012**, *149*, 36–47. [CrossRef] [PubMed]
67. Hilf, N.; Kuttruff-Coqui, S.; Frenzel, K.; Bukur, V.; Stevanović, S.; Gouttefangeas, C.; Platten, M.; Tabatabai, G.; Dutoit, V.; van der Burg, S.H.; et al. Actively personalized vaccination trial for newly diagnosed glioblastoma. *Nature* **2019**, *565*, 240–245. [CrossRef] [PubMed]
68. Silvestris, D.A.; Picardi, E.; Cesarini, V.; Fosso, B.; Mangraviti, N.; Massimi, L.; Martini, M.; Pesole, G.; Locatelli, F.; Gallo, A.; et al. Dynamic inosinome profiles reveal novel patient stratification and gender-specific differences in glioblastoma. *Genome Biol.* **2019**, *20*, 33. [CrossRef] [PubMed]
69. Rajaraman, S.; Canjuga, D.; Ghosh, M.; Codrea, M.C.; Sieger, R.; Wedekink, F.; Tatagiba, M.; Koch, M.; Lauer, U.M.; Nahnsen, S.; et al. Measles Virus-Based Treatments Trigger a Pro-inflammatory Cascade and a Distinctive Immunopeptidome in Glioblastoma. *Mol. Ther. Oncolytics* **2019**, *12*, 147–161. [CrossRef] [PubMed]
70. Ronnebaum, S.M.; Ilkayeva, O.; Burgess, S.C.; Joseph, J.W.; Lu, D.; Stevens, R.D.; Becker, T.C.; Sherry, A.D.; Newgard, C.B.; Jensen, M.V.; et al. A Pyruvate Cycling Pathway Involving Cytosolic NADP-dependent Isocitrate Dehydrogenase Regulates Glucose-stimulated Insulin Secretion. *J. Biol. Chem.* **2006**, *281*, 30593–30602. [CrossRef]
71. Filipp, F.V.; Scott, D.A.; Ronai, Z.A.; Osterman, A.L.; Smith, J.W. Reverse TCA cycle flux through isocitrate dehydrogenases 1 and 2 is required for lipogenesis in hypoxic melanoma cells. *Pigment Cell Melanoma Res.* **2012**, *25*, 375–383. [CrossRef]
72. Atai, N.A.; Renkema-Mills, N.A.; Bosman, J.; Schmidt, N.; Rijkeboer, D.; Tigchelaar, W.; Bosch, K.S.; Troost, D.; Jonker, A.; Bleeker, F.E.; et al. Differential Activity of NADPH-Producing Dehydrogenases Renders Rodents Unsuitable Models to Study IDH1R132 Mutation Effects in Human Glioblastoma. *J. Histochem. Cytochem.* **2011**, *59*, 489–503. [CrossRef]

73. Lee, S.M.; Koh, H.-J.; Park, D.-C.; Song, B.J.; Huh, T.-L.; Park, J.-W. Cytosolic NADP+-dependent isocitrate dehydrogenase status modulates oxidative damage to cells. *Free Radic. Biol. Med.* **2002**, *32*, 1185–1196. [CrossRef]
74. Mailloux, R.J.; Bériault, R.; Lemire, J.; Singh, R.; Chénier, D.R.; Hamel, R.D.; Appanna, D. The tricarboxylic acid cycle, an ancient metabolic network with a novel twist. *PLoS ONE* **2007**, *2*, e690. [CrossRef] [PubMed]
75. Molenaar, R.J.; Botman, D.; Smits, M.A.; Hira, V.V.; van Lith, S.A.; Stap, J.; Henneman, P.; Khurshed, M.; Lenting, K.; Mul, A.N.; et al. Radioprotection of IDH1-Mutated Cancer Cells by the IDH1-Mutant Inhibitor AGI-5198. *Cancer Res.* **2015**, *75*, 4790–4802. [CrossRef] [PubMed]
76. Molenaar, R.J.; Radivoyevitch, T.; Maciejewski, J.P.; van Noorden, C.J.F.; Bleeker, F.E. The driver and passenger effects of isocitrate dehydrogenase 1 and 2 mutations in oncogenesis and survival prolongation. *Biochim. Biophys. Acta (BBA) Rev. Cancer* **2014**, *1846*, 326–341. [CrossRef] [PubMed]
77. Labussiere, M.; Sanson, M.; Idbaih, A.; Delattre, J.-Y. IDH1 gene mutations: A new paradigm in glioma prognosis and therapy? *Oncologist* **2010**, *15*, 196–199. [CrossRef] [PubMed]
78. Sasaki, M.; Knobbe, C.B.; Itsumi, M.; Elia, A.J.; Harris, I.S.; Chio, I.I.C.; Cairns, R.A.; McCracken, S.; Wakeham, A.; Haight, J. D-2-hydroxyglutarate produced by mutant IDH1 perturbs collagen maturation and basement membrane function. *Genes Dev.* **2012**, *26*, 2038–2049. [CrossRef] [PubMed]
79. Ying, W. NAD+/NADH and NADP+/NADPH in cellular functions and cell death: Regulation and biological consequences. *Antioxid. Redox Signal.* **2008**, *10*, 179–206. [CrossRef]
80. Finkel, T.; Holbrook, N.J. Oxidants, oxidative stress and the biology of ageing. *Nature* **2000**, *408*, 239–247. [CrossRef]
81. Lee, J.-J.; Kim, B.C.; Park, M.-J.; Lee, Y.-S.; Kim, Y.-N.; Lee, B.L.; Lee, J.S. PTEN status switches cell fate between premature senescence and apoptosis in glioma exposed to ionizing radiation. *Cell Death Differ.* **2011**, *18*, 666–677. [CrossRef]
82. Wahl, D.R.; Dresser, J.; Wilder-Romans, K.; Parsels, J.D.; Zhao, S.G.; Davis, M.; Zhao, L.; Kachman, M.; Wernisch, S.; Burant, C.F. Glioblastoma Therapy Can Be Augmented by Targeting IDH1-Mediated NADPH Biosynthesis. *Cancer Res.* **2017**, *77*, 960–970. [CrossRef]
83. Gelman, S.J.; Naser, F.; Mahieu, N.G.; McKenzie, L.D.; Dunn, G.P.; Chheda, M.G.; Patti, G.J. Consumption of NADPH for 2-HG Synthesis Increases Pentose Phosphate Pathway Flux and Sensitizes Cells to Oxidative Stress. *Cell Rep.* **2018**, *22*, 512–522. [CrossRef] [PubMed]
84. Gray, L.H.; Conger, A.D.; Ebert, M.; Hornsey, S.; Scott, O.C. The concentration of oxygen dissolved in tissues at the time of irradiation as a factor in radiotherapy. *Br. J. Radiol.* **1953**, *26*, 638–648. [CrossRef] [PubMed]
85. Collingridge, D.R.; Piepmeier, J.M.; Rockwell, S.; Knisely, J.P. Polarographic measurements of oxygen tension in human glioma and surrounding peritumoural brain tissue. *Radiother. Oncol.* **1999**, *53*, 127–131. [CrossRef]
86. Nordsmark, M.; Bentzen, S.M.; Rudat, V.; Brizel, D.; Lartigau, E.; Stadler, P.; Becker, A.; Adam, M.; Molls, M.; Dunst, J. Prognostic value of tumor oxygenation in 397 head and neck tumors after primary radiation therapy. An international multi-center study. *Radiother. Oncol.* **2005**, *77*, 18–24. [CrossRef] [PubMed]
87. Combs, S.E.; Bruckner, T.; Mizoe, J.-E.; Kamada, T.; Tsujii, H.; Kieser, M.; Debus, J. Comparison of carbon ion radiotherapy to photon radiation alone or in combination with temozolomide in patients with high-grade gliomas: Explorative hypothesis-generating retrospective analysis. *Radiother. Oncol.* **2013**, *108*, 132–135. [CrossRef]
88. Parker, S.J.; Metallo, C.M. Metabolic consequences of oncogenic IDH mutations. *Pharmacol. Ther.* **2015**, *152*, 54–62. [CrossRef]
89. Leonardi, R.; Subramanian, C.; Jackowski, S.; Rock, C.O. Cancer-associated Isocitrate Dehydrogenase Mutations Inactivate NADPH-dependent Reductive Carboxylation. *J. Biol. Chem.* **2012**, *287*, 14615–14620. [CrossRef]
90. Bhadriraju, K.; Hansen, L.K. Extracellular Matrix- and Cytoskeleton-Dependent Changes in Cell Shape and Stiffness. *Exp. Cell Res.* **2002**, *278*, 92–100. [CrossRef]
91. Watanabe, T.; Kuramochi, H.; Takahashi, A.; Imai, K.; Katsuta, N.; Nakayama, T.; Fujiki, H.; Suganuma, M. Higher cell stiffness indicating lower metastatic potential in B16 melanoma cell variants and in (-)-epigallocatechin gallate-treated cells. *J. Cancer Res. Clin. Oncol.* **2012**, *138*, 859–866. [CrossRef]
92. Pepin, K.M.; McGee, K.P.; Arani, A.; Lake, D.S.; Glaser, K.J.; Manduca, A.; Parney, I.F.; Ehman, R.L.; Huston III, J. Magnetic resonance elastography analysis of glioma stiffness and IDH1 mutation status. *AJNR Am. J. Neuroradiol.* **2018**, *39*, 31–36. [CrossRef]

93. Swaminathan, V.; Mythreye, K.; O'Brien, E.T.; Berchuck, A.; Blobe, G.C.; Superfine, R. Mechanical stiffness grades metastatic potential in patient tumor cells and in cancer cell lines. *Cancer Res.* **2011**, *71*, 5075–5080. [CrossRef] [PubMed]
94. Li, S.; Chou, A.P.; Chen, W.; Chen, R.; Deng, Y.; Phillips, H.S.; Selfridge, J.; Zurayk, M.; Lou, J.J.; Everson, R.G.; et al. Overexpression of isocitrate dehydrogenase mutant proteins renders glioma cells more sensitive to radiation. *Neuro-Oncology* **2013**, *15*, 57–68. [CrossRef] [PubMed]
95. Spiering, D.; Hodgson, L. Dynamics of the Rho-family small GTPases in actin regulation and motility. *Cell Adhes. Migr.* **2011**, *5*, 170–180. [CrossRef] [PubMed]
96. Bonfim-Melo, A.; Ferreira, É.R.; Mortara, R.A. Rac1/WAVE2 and Cdc42/N-WASP Participation in Actin-Dependent Host Cell Invasion by Extracellular Amastigotes of *Trypanosoma cruzi*. *Front. Microbiol.* **2018**, *9*. [CrossRef] [PubMed]
97. Chen, B.; Chou, H.-T.; Brautigam, C.A.; Xing, W.; Yang, S.; Henry, L.; Doolittle, L.K.; Walz, T.; Rosen, M.K. Rac1 GTPase activates the WAVE regulatory complex through two distinct binding sites. *eLife* **2017**, *6*. [CrossRef]
98. Ridley, A.J. Life at the leading edge. *Cell* **2011**, *145*, 1012–1022. [CrossRef]
99. Reichl, E.M.; Ren, Y.; Morphew, M.K.; Delannoy, M.; Effler, J.C.; Girard, K.D.; Divi, S.; Iglesias, P.A.; Kuo, S.C.; Robinson, D.N.; et al. Interactions between myosin and actin crosslinkers control cytokinesis contractility dynamics and mechanics. *Curr. Biol.* **2008**, *18*, 471–480. [CrossRef]
100. Stamenović, D.; Coughlin, M.F. The role of prestress and architecture of the cytoskeleton and deformability of cytoskeletal filaments in mechanics of adherent cells: A quantitative analysis. *J. Ther. Biol.* **1999**, *201*, 63–74. [CrossRef]
101. Cui, D.; Ren, J.; Shi, J.; Feng, L.; Wang, K.; Zeng, T.; Yin, Y.; Gao, L. R132H mutation in IDH1 gene reduces proliferation, cell survival and invasion of human glioma by downregulating Wnt/β-catenin signaling. *Int. J. Biochem. Cell. Biol.* **2016**, *73*, 72–81. [CrossRef]
102. Mendez, M.G.; Restle, D.; Janmey, P.A. Vimentin Enhances Cell Elastic Behavior and Protects against Compressive Stress. *Biophys. J.* **2014**, *107*, 314–323. [CrossRef]
103. Ofek, G.; Wiltz, D.C.; Athanasiou, K.A. Contribution of the Cytoskeleton to the Compressive Properties and Recovery Behavior of Single Cells. *Biophys. J.* **2009**, *97*, 1873–1882. [CrossRef] [PubMed]
104. Pascalis, C.D.; Pérez-González, C.; Seetharaman, S.; Boëda, B.; Vianay, B.; Burute, M.; Leduc, C.; Borghi, N.; Trepat, X.; Etienne-Manneville, S.; et al. Intermediate filaments control collective migration by restricting traction forces and sustaining cell-cell contacts. *J. Cell Biol.* **2018**, *217*, 3031–3044. [CrossRef] [PubMed]
105. Yamashita, N.; Tokunaga, E.; Kitao, H.; Hisamatsu, Y.; Taketani, K.; Akiyoshi, S.; Okada, S.; Aishima, S.; Morita, M.; Maehara, Y. Vimentin as a poor prognostic factor for triple-negative breast cancer. *J. Cancer Res. Clin. Oncol.* **2013**, *139*, 739–746. [CrossRef] [PubMed]
106. Burgstaller, G.; Gregor, M.; Winter, L.; Wiche, G. Keeping the Vimentin Network under Control: Cell–Matrix Adhesion–associated Plectin 1f Affects Cell Shape and Polarity of Fibroblasts. *Mol. Biol. Cell* **2010**, *21*, 3362–3375. [CrossRef] [PubMed]
107. Lynch, C.D.; Lazar, A.M.; Iskratsch, T.; Zhang, X.; Sheetz, M.P. Endoplasmic spreading requires coalescence of vimentin intermediate filaments at force-bearing adhesions. *Mol. Biol. Cell* **2013**, *24*, 21–30. [CrossRef] [PubMed]
108. Gregor, M.; Osmanagic-Myers, S.; Burgstaller, G.; Wolfram, M.; Fischer, I.; Walko, G.; Resch, G.P.; Jörgl, A.; Herrmann, H.; Wiche, G. Mechanosensing through focal adhesion-anchored intermediate filaments. *FASEB J.* **2014**, *28*, 715–729. [CrossRef] [PubMed]
109. Jiu, Y.; Lehtimäki, J.; Tojkander, S.; Cheng, F.; Jäälinoja, H.; Liu, X.; Varjosalo, M.; Eriksson, J.E.; Lappalainen, P. Bidirectional Interplay between Vimentin Intermediate Filaments and Contractile Actin Stress Fibers. *Cell Rep.* **2015**, *11*, 1511–1518. [CrossRef]
110. Richarson, A.D.; Scott, D.A.; Zagnitko, O.; Aza-Blanc, P.; Chang, C.-C.; Russler-Germain, D.A.; Reproducibility Project: Cancer Biology. Registered report: IDH mutation impairs histone demethylation and results in a block to cell differentiation. *eLife* **2016**, *5*, e10860. [CrossRef]
111. Xu, W.; Yang, H.; Liu, Y.; Yang, Y.; Wang, P.; Kim, S.-H.; Ito, S.; Yang, C.; Wang, P.; Xiao, M.T.; et al. Oncometabolite 2-hydroxyglutarate is a competitive inhibitor of α-ketoglutarate-dependent dioxygenases. *Cancer Cell* **2011**, *19*, 17–30. [CrossRef]

112. Rohle, D.; Popovici-Muller, J.; Palaskas, N.; Turcan, S.; Grommes, C.; Campos, C.; Tsoi, J.; Clark, O.; Oldrini, B.; Komisopoulou, E.; et al. An Inhibitor of Mutant IDH1 Delays Growth and Promotes Differentiation of Glioma Cells. *Science* **2013**, *340*, 626–630. [CrossRef]
113. Qi, S.; Yu, L.; Li, H.; Ou, Y.; Qiu, X.; Ding, Y.; Han, H.; Zhang, X. Isocitrate dehydrogenase mutation is associated with tumor location and magnetic resonance imaging characteristics in astrocytic neoplasms. *Oncol. Lett.* **2014**, *7*, 1895–1902. [CrossRef] [PubMed]
114. Derjagin, B.W.; Muller, V.M.; Toporov, Y.P. Effect of Contact Deformations on the Adhesion of Particles. *J. Cell Sci.* **1975**, *53*, 314–326. [CrossRef]

cancers

MDPI

Article

Targeting Glioblastoma Stem Cells with 2-Deoxy-D-Glucose (2-DG) Potentiates Radiation-Induced Unfolded Protein Response (UPR)

Sumedh S. Shah [1,2], Gregor A. Rodriguez [1], Alexis Musick [3], Winston M. Walters [1], Nicolas de Cordoba [1], Eric Barbarite [1], Megan M. Marlow [1], Brian Marples [1,4,5], Jeffrey S. Prince [3], Ricardo J. Komotar [1,2,5], Steven Vanni [2] and Regina M. Graham [1,2,5,*]

1 University of Miami Brain Tumor Initiative, University of Miami Miller School of Medicine, Miami, FL 33136, USA; sumedhss@med.miami.edu (S.S.S.); gregor.rodriguez@med.miami.edu (G.A.R.); WWalters@med.miami.edu (W.M.W.); nicolas.a.decordoba@email.shc.edu (N.d.C.); eric_barbarite@meei.harvard.edu (E.B.); mmarlow135@yahoo.com (M.M.M.); brian.marples@med.miami.edu (B.M.); RKomotar@med.miami.edu (R.J.K.)

2 Department of Neurological Surgery, University of Miami Miller School of Medicine, Miami, FL 33136, USA; SVanni@med.miami.edu

3 Dauer Electron Microscopy Laboratory, Department of Biology, University of Miami, Coral Gables, FL 33146, USA; am722@duke.edu (A.M.); jeffprince@miami.edu (J.S.P.)

4 Department of Radiation Oncology, University of Miami Miller School of Medicine, Miami, FL 33136, USA

5 Sylvester Comprehensive Cancer Center, University of Miami Health System, Miami, FL 33136, USA

* Correspondence: RGraham@med.miami.edu

Received: 1 January 2019; Accepted: 29 January 2019; Published: 31 January 2019

Abstract: Glioblastoma (GBM) is the most common and aggressive primary brain tumor in adults, and despite optimized treatment options, median survival remains dismal. Contemporary evidence suggests disease recurrence results from expansion of a robustly radioresistant subset of GBM progenitor cells, termed GBM stem cells (GSCs). In this study, we utilized transmission electron microscopy to uncover ultrastructural effects on patient-derived GSC lines exposed to supratherapeutic radiotherapy levels. Elevated autophagosome formation and increased endoplasmic reticulum (ER) internal diameter, a surrogate for ER stress and activation of unfolded protein response (UPR), was uncovered. These observations were confirmed via protein expression through Western blot. Upon interrogating genomic data from an open-access GBM patient database, overexpression of UPR-related chaperone protein genes was inversely correlated with patient survival. This indicated controlled UPR may play a role in promoting radioresistance. To determine if potentiating UPR further can induce apoptosis, we exposed GSCs to radiation with an ER stress-inducing drug, 2-deoxy-D-glucose (2-DG), and found dose-dependent decreases in viability and increased apoptotic marker expression. Taken together, our results indicate GSC radioresistance is, in part, achieved by overexpression and overactivation of ER stress-related pathways, and this effect can be overcome via potentiation of UPR, leading to loss of GSC viability.

Keywords: cancer stem cells; ER stress; glioblastoma multiforme; radiation; unfolded protein response; autophagy

1. Introduction

Glioblastoma (GBM) is the most common and aggressive form of primary brain cancer, and despite optimized therapy consisting of surgery, chemotherapy, and radiation, its expected median survival remains under two years [1]. Informed by techniques used to enrich and characterize neural stem cells, several groups have demonstrated GBM, along with other primary brain tumors, contains self-renewing, tumorigenic cells referred to as tumor initiating cells, GBM stem-like cells, or GBM stem cells (GSCs) [2–5]. GSCs are implicated in tumor recurrence through mechanisms including expression of drug efflux pumps and ability to withstand radiotherapy, and thus, characterizing resistance mechanisms may uncover potential therapeutic targets and augment current cancer therapy [6].

Within the tumor microenvironment, a multitude of metabolic stresses (i.e., hypoxia, nutrient depletion, etc.) results in selective pressure on GSCs [7–9]. Activation of pro-survival pathways designed to mitigate the effect of these stressors can inadvertently impart a survival advantage to cells exposed to cytotoxic therapy. One such survival mechanism is stimulation of the endoplasmic reticulum (ER) stress pathway, commonly referred to as the unfolded protein response (UPR) [10–13]. Unfolded proteins accumulate in the ER secondary to energy depletion and/or environmental stresses, leading to the activation of PERK, IRE1, and ATF6—transducers of UPR that reestablish ER homeostasis via ER-associated protein degradation, translation of protein folding chaperones, and attenuation of global protein synthesis [14]. Activation of the ER stress pathway can also induce autophagy, which can help strengthen resistance to cellular stresses [15]. Ultimately, if ER homeostasis cannot be achieved, then UPR navigates the cell towards apoptosis.

One of the many critical downstream effects of UPR is the upregulation of glucose related proteins 78 (GRP78) and 94 (GRP94), multifunctional members of the heat shock protein family [16]. GRP78 acts to regulate PERK, IRE1, and ATF6 activation, while both GRP78 and GRP94 are protein-folding chaperones [17,18]. Elevated baseline GRP78 expression is also related to increased malignancy and radiotherapy resistance in multiple cancers [19–22]. In GBM, Dadey and colleagues showed that radiation potently activates UPR and subsequently upregulates GRP78, and antibody targeting of GRP78 in non-stem GBM cells enhanced efficacy of radiation [23,24]. Targeting UPR using the ER-stress potentiating drug, celecoxib, also sensitized GBM cells to radiotherapy [25].

Despite emerging evidence that ER-stress potentiation may be a viable therapeutic option in cancer treatment, exploitation of this pathway in cancer stem cells—notable for their robust radioresistance—remains understudied. In this present study, we exposed patient-derived GSC lines to radiotherapy and the ER-stress inducing agent, 2-deoxy-D-glucose (2-DG), to test our hypothesis that potentiating radiation-induced ER-stress in the GSC population leads to decreased cell viability. Our confirmation of this hypothesis reveals a novel understanding of a pathway GSCs utilize to evade radiation cytotoxicity and suggests a possible therapeutic target to circumvent this resistance.

2. Results

2.1. GSCs Exhibit Robust Radioresistance

To characterize GSC radioresistance, we exposed three established GSC lines—Glio9, Glio11, and Glio14—to increasing levels of ionized radiation (one-time dose: 0 Gy, 2 Gy, 4 Gy, 8 Gy, 16 Gy, and 20 Gy) and determined the cell viability after 48 h using a trypan blue exclusion assay. We observed a negligible decrease in cell viability despite administering supratherapeutic levels of radiation (Figure 1A). Utilizing light microscopy, we qualitatively analyzed GSC morphology 48 h post irradiation with 8 Gy and found no ultrastructural aberrations secondary to radiation treatment (Figure 1B). GSCs grown in normal culture conditions maintain their phenotype as spheroid clusters of cells, termed neurospheres, and a morphology change towards a monolayer pattern indicates loss of stemness [2]. Therefore, we found radiation did not negatively affect either stem phenotype or GSC viability.

To confirm that radiation induced cellular damage in our GSC lines, we examined H2AX phosphorylation 1h post radiation exposure. As a variant member of the H2A histone family, H2AX phosphorylation is the first step in recruiting and localizing DNA repair proteins, so it is often used as a marker for double-stranded breaks [26]. We found radiation induced DNA damage in treated cell lines; however, no identifiable loss of cell viability or form follows (Figure 1C). These data suggest that our patient-derived GSC lines are resistant to supratherapeutic levels of radiation despite accumulation of DNA insults.

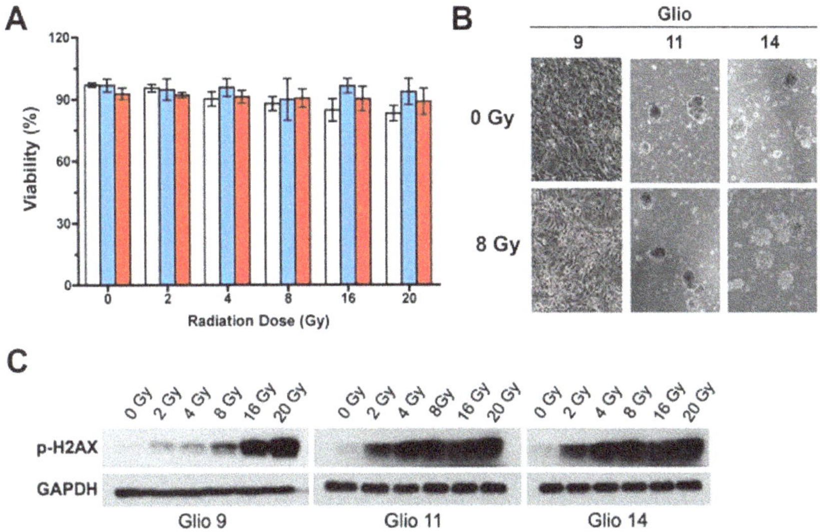

Figure 1. Glioblastoma stem cells (GSCs) exhibit robust radioresistance. (**A**) Viability as determined by trypan blue exclusion assay of Glio9 (white), Glio11 (blue), and Glio14 (red) treated with increasing doses of radiation and then analyzed 72 h after exposure. Results are representative of at least three experiments and displayed as mean ± SEM. (**B**) Light microscopy images of GSC neurosphere phenotype with and without radiation treatment. Magnification: 10×. (**C**) Western blot of p-H2AX, a marker for double strand breaks, in GSCs exposed to increasing radiation.

2.2. Radiation Exposure Induces Autophagosome Formation and Activates UPR in GSCs

After observing GSCs are highly radioresistant despite accumulation of significant DNA damage, we studied post-radiation cellular modifications using transmission electron microscopy (TEM) to elucidate potential survival strategies GSCs employ to circumvent radiation-mediated cell death. GSCs were treated with a single dose of 8 Gy radiation and were prepared for TEM after 48 h to allow sufficient time to capture any morphological changes. Compared to NT, radiotherapy treated lines exhibited increased ER luminal diameter (blue arrows) and higher number of autophagic vesicles (red asterisk) (Figure 2A,B). Dilation of the ER has been used as a marker for ER stress in cancer [27,28], thus our observations confirmed radiation induced ER stress. Table 1 shows ER diameter in microns and autophagic vesicles per cell between cell lines and treatment cohorts. Of note, while all GSCs lines exhibited increased stress responses, only Glio11 did not present with significantly higher autophagosome formation in NT versus radiotherapy treated cells. Consistent with these findings, inhibition of autophagy with chloroquine prior to radiotherapy decreased cell number compared to radiotherapy alone in both Glio9 and Glio14 but not in Glio11 suggesting that autophagy does not play a major role in mediating radioresistance in Glio11 cells (Figure S1).

After observing morphological changes using TEM, we performed western blot analysis (Figure 2C) for markers of UPR (GRP79, GRP94, and CCAAT-enhancer-binding protein homologous protein (CHOP)) and autophagy (LC3, Beclin1, and p62) at early (1 h) and late (48 h) timepoints after exposure to increasing doses of radiation. By 1 h post exposure, we see a dose dependent activation of stress factors like the GRPs; however, CHOP activation, a potent mediator of UPR-associated apoptosis did not follow identifiable trends. For autophagy-related protein products, we observed a dose-dependent increase in all targets probed. At 48 h, most effects seen at 1h plateaued (as in the case of Beclin1, p62, GRP94) or began returning to NT baseline (with LC3, GRP78, CHOP). Taken together, our results show that radiation rapidly induces stress adaptive mechanisms, such as UPR and autophagy, and these effects can persist 48 h after single dose.

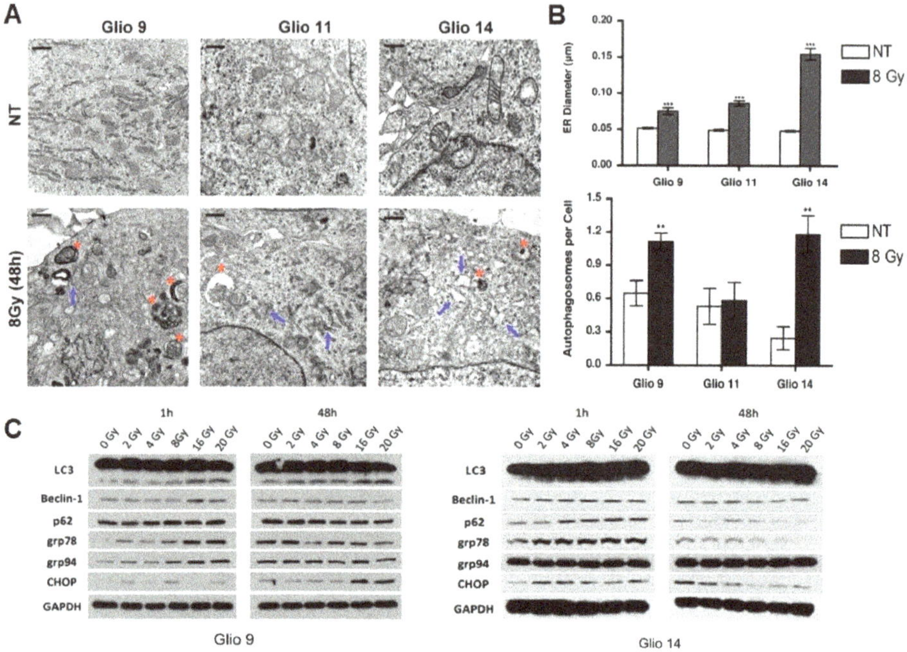

Figure 2. Radiation induces autophagosome formation and activation of unfolded protein response (UPR) in GSCs. (**A**) Transmission electron microscopy (TEM) analysis of Glio9, Glio11, and Glio14 comparing ultrastructural responses in non-treated lines and cells exposed to 8 Gy of radiation. Endoplasmic reticulum (ER, blue arrows) and autophagosomes (red asterisks) are highlighted. Five images from two independent studies were analyzed. Scale bar: 0.5 µm. (**B**) Quantitative analysis of ER luminal diameter, as a surrogate marker for ER stress, and autophagic vesicles per cell using TEM images. Results are representative averages of at least five images and are displayed as mean ± SEM. ** $p < 0.01$, *** $p < 0.001$. Mann-Whitney test. (**C**) Western blot analysis for ER stress markers (GRP78, GRP94, CHOP) and autophagy markers (LC3, Beclin-1, p62) in Glio9 and Glio14 at 1 h and 48 h post radiation exposure to increasing doses. See also Figure S1.

Table 1. Measurements of ER diameter (microns) and autophagic vesicles per cell in Glioblastoma stem cell (GSCs) treated with 8 Gy radiation. Mann-Whitney test.

Parameter	NT	Rad (8 Gy)	p-Value
ER Diameter (µm)			
Glio9	0.055 ± 0.002 µm	0.075 ± 0.005 µm	$p < 0.001$
Glio11	0.049 ± 0.002 µm	0.086 ± 0.003 µm	$p < 0.0001$
Glio14	0.048 ± 0.001 µm	0.154 ± 0.008 µm	$p < 0.0001$
AV per Cell			
Glio9	0.65 ± 0.11	1.11 ± 0.08	$p < 0.01$
Glio11	0.53 ± 0.16	0.58 ± 0.17	ns
Glio14	0.25 ± 0.10	1.18 ± 0.17	$p < 0.01$

Abbreviations: AV, autophagic vesicles; ER, endoplasmic reticulum; Gy, gray; ns, not significant; NT, non-treated; Rad, radiation; µm, microns.

2.3. Upregulation of UPR Genes in Human GBM Specimen Correlates with Reduced Patient Survival

Overexpression of the UPR genes that encode for GRP78 and GRP94 have been linked to radioresistance and in multiple cancer types, including breast, gastric, and pancreatic cancers [29–31]. We interrogated the TCGA database via the open-access analysis platform, GlioVis, to determine if upregulation of GRP78 and GRP94 is observed in GBM patients compared to non-tumor controls and if higher expression is clinically relevant to patient survival. Genomic data from the Human Genome U133 (HG-U133) array was deciphered. Comparisons were between the 75th percentile of expression vs. the 25th percentile (high vs. low expression). We found that GBMs overall exhibit increased GRP78 and GRP94 expression compared to non-tumor controls (Figure 3A).

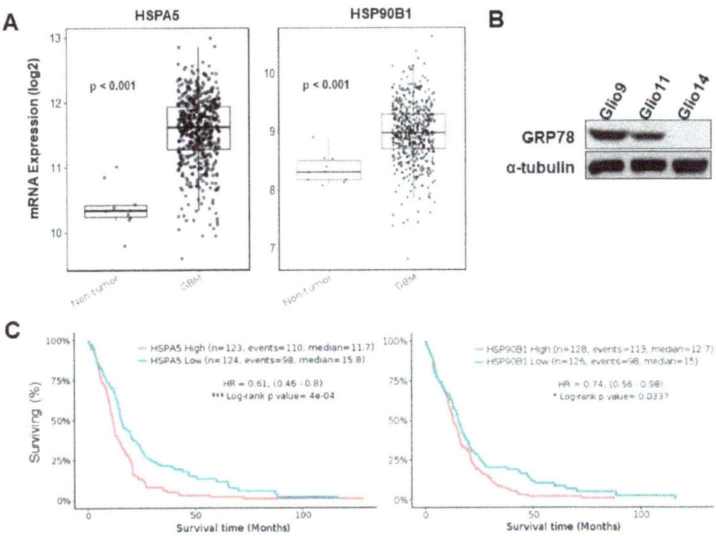

Figure 3. Upregulation of UPR genes in human GBM specimen correlates with reduced patient survival. (**A**) Comparison of non-tumor (n = 10) and GBM sample (n = 528) for mRNA expression of ER stress genes *HSPA5* and *HSP90B1*. Pairwise t-test. (**B**) Western blot analysis for baseline levels of GRP78 in Glio9, Glio11, and Glio14. (**C**) Kaplan-Meier plot of survival using the GlioVis portal comparing low expression (25th percentile) and high expression (75th percentile) of *HSPA5* and *HSP90B1*. Results reported as hazard ratio (95% confidence interval). * $p < 0.05$, *** $p < 0.001$. Log-rank test. Events = number of patients who died. See also Figure S2.

mRNA Log2 expression comparisons between non-tumor control and GBM specimen, respectively, were as follows: *HSPA5*, 10.38 ± 0.01 vs. 11.56 ± 0.02, $p < 0.001$; *HSP90B1*, 8.36 ± 0.08 vs. 8.98 ± 0.02, $p < 0.001$. From Western blots of our three patient samples, we noted heterogenous expression of GRP78; Glio9 displayed the highest level of baseline GRP78, followed by Glio11, and then Glio14 (Figure 3B). Should GRP78 expression be related to therapy resistance, we predicted that Glio9 would exhibit the most resistance to ER stress inducing stimuli. Interestingly, Glio9 was derived from a patient with a recurrent tumor. Finally, we found that higher vs. lower expression is correlated with significant differences in patient survival for both GRP78 and GRP94 (Figure 3C); Hazard ratio 0.61 (0.46–0.8) and 0.74 (0.56–0.98), respectively). Median survival time for high vs. low GRP78 and 94 expressions were 11.7 vs. 15.8 months, $p < 0.0001$, and 12.7 vs. 15.0 months, $p = 0.0337$, respectively.

We then decided to determine if differential autophagy gene expressions independently correlated with patient survival and found that the genes encoding for LC3, Beclin1, p62, ATG5, ATG7, and ATG12 were not correlated with patient survival (Figure S2). Additionally, the differences in non-tumor vs. GBM mRNA expression of the autophagy products were not as pronounced as those of UPR-related genes. These results indicate that upregulation of UPR genes, but not of autophagy genes, is clinically relevant and related to worse patient outcome.

2.4. 2-DG Induces ER Stress in GSCs in a Dose-Dependent Manner

While UPR attempts to reestablish ER homeostasis by managing accumulation of unfolded proteins, the outcome if homeostasis cannot be achieved is cell death [14]. We hypothesized that the addition of a supplementary ER stress source could potentiate the effects of radiation-induced UPR and force GSCs towards apoptosis. We, therefore, looked to the glucose analog, 2-DG, to induce ER stress in GSCs with the intention of combining it with radiation.

2-DG induces ER stress in multiple carcinoma types and has been used in combination with metformin to inhibit invasiveness and stem phenotype of GBM [32,33]. In a dose-escalation trial, we treated all our GSC lines with increasing concentrations of 2-DG (0.0, 0.1, 0.25, 0.5, 1.0, 1.5, and 2.0 mM 2-DG) and determined viability via MTS assay. A recent Phase I trial studying 2-DG combination with docetaxel in patients with advanced solid tumors found that peak plasma levels of 2-DG reach a median of 116 μg/mL, or 0.7 mM [34]. Therefore, utilizing concentrations from 0.0–2.0 mM is clinically relevant. Our viability assay revealed a modest, but significant, dose-dependent decrease in GSC viability with increasing 2-DG dose (Figure 4A). IC_{50} were determined using the calculated regression trendline for each cell line: Glio9, 1.38 ± 0.134 mM; Glio11, 1.77 ± 0.130 mM; Glio14, 1.46 ± 0.366 mM.

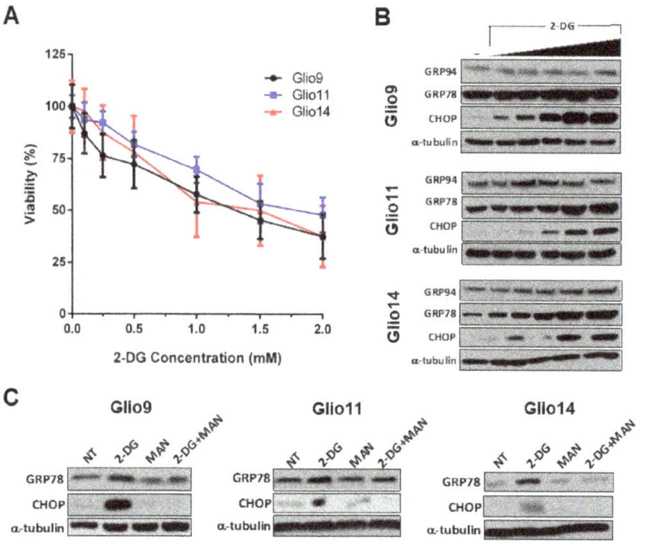

Figure 4. 2-Deoxy-D-glucose (2-DG) induces ER stress in GSCs in a dose-dependent manner. (**A**) CellTiter 96® AQ_{ueous} One Solution Cell Proliferation (Promega, Madison, WI, USA) assay of Glio9, Glio11, and Glio14 treated with increasing doses of 2-DG for 72 h. Viability is reported as mean ± SEM. Data represent average values of at least three independent experiments. (**B**) Western blot analysis for ER stress markers (GRP78, GRP94, and CHOP) in GSC lines treated with 0.0, 0.25, 0.5, 1.0, 2.0, and 4.0 mM of 2-DG for 24 h. (**C**) Western blot analysis of GSCs treated with 2-DG and/or mannose to determine mechanism of rescue from ER stress.

To confirm that 2-DG induced dose-dependent ER stress and activated UPR, we used western blot analysis of GRP78, GRP94, and CHOP in GSC lines treated with 0.0, 0.25, 0.5, 1.0, 2.0, and 4.0 mM 2-DG for 24 h. Similar to the effects of radiation, we see that UPR-associated products are upregulated in the presence of 2-DG; however, CHOP, a potent activator of UPR-mediated apoptosis, is also upregulated—a finding not seen in radiation-induced UPR (Figure 4B). Nonetheless, the viability of cells treated with achievable plasma 2-DG concentrations (0.7 mM) only is modest.

We next sought to elucidate the mechanism of ER stress induction via 2-DG in GSCs. While 2-DG is a glycolytic inhibitor [35], Kovács et al. found in endothelial cells that 2-DG interferes with N-linked glycosylation and cotreatment with 1 mM mannose can rescue cells [36]. GSC treatment using 2 mM 2-DG upregulates GRP78 and induces apoptosis via CHOP, but, similarly to the results of Kovács and colleagues, cotreatment with 1 mM mannose reverses these effects (Figure 4C). Taken together, we observed 2-DG induces UPR and apoptosis at clinically achievable concentrations, and that this ER stress is, in part, mediated by N-linked glycosylation since cotreatment with mannose prevents increased activation of UPR.

2.5. Combination Radiotherapy and 2-DG Increases ER Dilation in GSCs

Given our results that 2-DG induces ER stress in GSCs, we revisited TEM analysis to determine if combination therapy with radiation and 2-DG can induce greater ER dilation, and therefore ER stress, than either condition independently. Glio14 was treated with a single dose of 8 Gy radiation and incubated with either 0.5 mM or 2.0 mM 2-DG for 24 h, after which electron micrographs were taken and analyzed for ER dilation using ImageJ (Figure 5A). Qualitatively, we observed increased ER diameter in the dual-treated conditions than in radiation or 2-DG alone, 0.5 mM and 2.0 mM. When we quantitatively analyzed the data, we found statistically significant increases in ER diameter when dual treatment was compared to single therapy (Figure 5B). Refer to Table 2 for average diameter in microns with SEM. With this result, we see that ER stress can be potentiated by combination therapy of two ER stress inducing stimuli, and that there is an identifiable morphological consequence in GSCs.

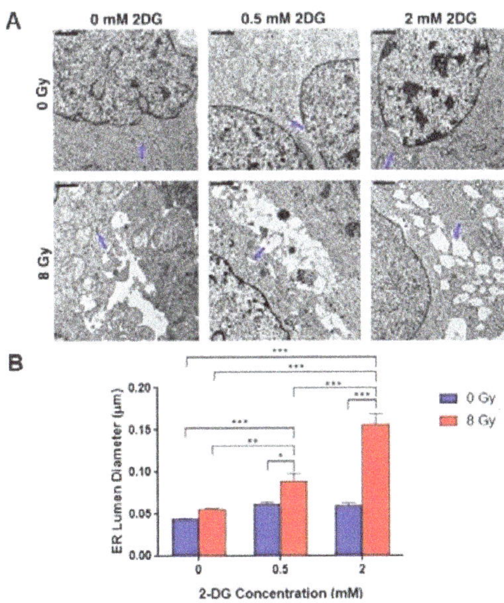

Figure 5. Combination radiotherapy and 2-DG increases ER dilation in GSCs. (**A**) Transmission electron microscopy (TEM) analysis of GSCs (Glio14 shown) comparing ultrastructural responses in non-treated lines and cells exposed to 8 Gy of radiation and/or increasing concentrations of 2-DG. Endoplasmic reticulum (ER) highlighted with blue arrows. Scale bar: 0.5 μm. (**B**) Quantitative analysis of ER luminal diameter, as a surrogate marker for ER stress, using TEM images (Glio14 shown). Results are representative averages of at least five images from two independent studies and are displayed as mean ± SEM. * $p < 0.05$, ** $p < 0.01$, *** $p < 0.001$. Mann-Whitney test.

Table 2. ER diameter measurements in Glio14 treated with increasing doses of 2-DG and 8 Gy radiation.

2-DG Dose	NT	Rad (8 Gy)
0.0 mM	0.045 ± 0.008 μm	0.056 ± 0.001 μm
0.5 mM	0.062 ± 0.0023 μm	0.089 ± 0.01 μm
2.0 mM	0.060 ± 0.0026 μm	0.157 ± 0.012 μm

Abbreviations: 2-DG; 2-deoxyglucose; Gy, gray; NT, non-treated; Rad, radiation; μm, micron.

2.6. Radiotherapy and 2-DG Cotreatment Potentiates UPR and Leads to Dose-Dependent Loss of Cell Viability and Apoptosis

To confirm our TEM observations that combination of multiple ER stress inducing stimuli led to increased ER stress, we first performed an MTS assay to determine viability of GSCs treated with radiation (single dose 8 Gy) and increasing concentrations of 2-DG. We found a significant decrease in viability between 2-DG treated cells versus 2-DG plus radiation for each concentration tested in all cell lines (Figure 6A). The IC_{50} of 2-DG for each cell line was also lower with 2-DG plus radiation compared to 2-DG alone: Glio9 IC_{50}, 0.86 ± 0.147 mM vs. 1.38 ± 0.134 mM, $p < 0.01$; Glio11 IC_{50}, 1.03 ± 0.206 mM vs. 1.77 ± 0.130 mM, $p < 0.01$; Glio14 IC_{50}, 0.57 ± 0.159 mM vs. 1.46 ± 0.366 mM, $p = 0.028$. We then performed Western blot analysis (Figure 6B) to see if combination therapy upregulated UPR associated proteins, GRP78 and CHOP, more than 2-DG (0.5 mM and 2 mM) or radiotherapy alone. A short timepoint (8 h) and long timepoint (48 h) were chosen to determine if UPR activation persists. Not only is GRP78 upregulation highest with 2 mM 2-DG plus 8 Gy radiation at the short timepoint, but we see an increased CHOP upregulation as well, indicating that cells are entering apoptosis. The only cell line that showed undiscernible CHOP activity with combination therapy compared to single therapy was Glio9, indicating that potentiating ER stress can effectively induce apoptosis in cells with lower baseline GRP78 levels but possibly not in cells with higher baseline GRP78 levels (refer to Figure 3B). At 48 post radiation, GRP78 remained upregulated, but a reduction in CHOP expression was seen in Glio11 and Glio14 when compared to 2-DG only treatment. An LDH assay performed confirmed that Glio9 does not experience apoptosis because of ER stress therapy, but that Glio11 and Glio14 do (Figure 6C). To further characterize activation of apoptotic pathways in 2-DG plus radiation treated cells, we performed Western blot for cleaved PARP and upregulation of Caspase 7 (Figure 6D). As expected, Glio9 did not experience either increased cleaved PARP or Caspase 7 (despite strong full PARP expression (Figure 6E), indicating that PARP is present, but is not cleaved to activate apoptosis.

These results suggest that combination therapy with multiple ER stress inducing stimuli activates, maintains activation of, and potentiates UPR in GSCs. We see an additive effect of 2-DG on radiotherapy. While this combination strategy is effective in reducing cell viability in GSCs, cells with higher baseline GRP78 expression like Glio9 do not respond with cell death when exposed to multiple ER stressors. Ultimately, there is a potential role in ER stress inducing combination therapy in well-selected patients who are found to have normal UPR-associated gene expression since conventional radiotherapy, and some chemotherapies, induce ER stress.

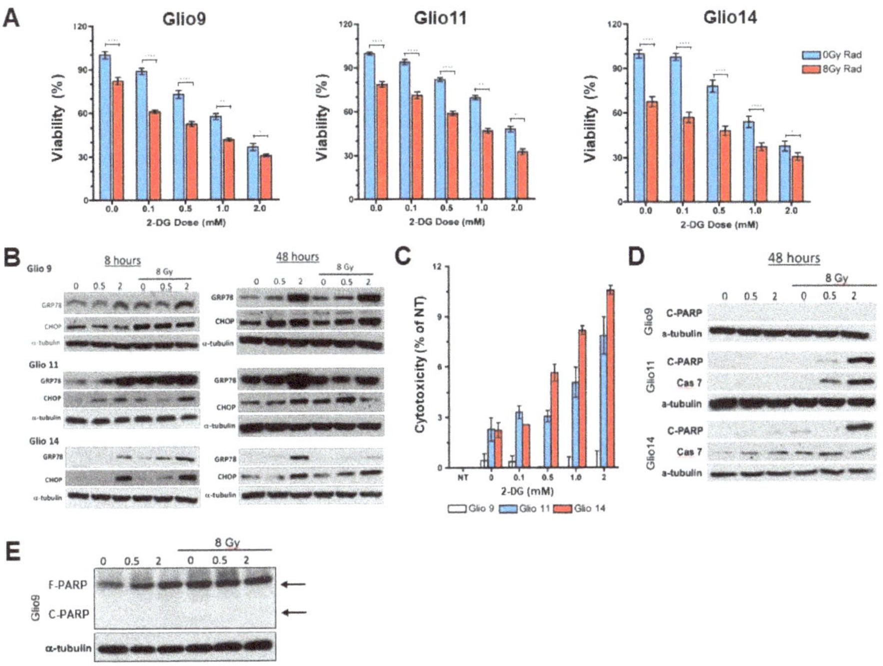

Figure 6. Radiotherapy and 2-DG cotreatment potentiates UPR and leads to dose-dependent loss of cell viability and apoptosis. (**A**) Proliferation assay of Glio9, Glio11, and Glio14 treated with increasing doses of 2-DG with or without 8 Gy radiation for 72 h. Viability is reported as mean ± SEM. Data represent average values of at least three independent experiments. * $p < 0.05$, ** $p < 0.01$, *** $p < 0.001$. Mann-Whitney Test. (**B**) Western blot analysis for ER stress markers GRP78 and CCAAT-enhancer-binding-protein homologous protein (CHOP) after treatment using increasing doses of 2-DG with or without 8 Gy radiation. Cells were analyzed at early (8 h) and late (48 h) time points. (**C**) Lactate dehydrogenase assay to determine cytotoxicity of Glio9 (white), Glio11 (blue), and Glio14 (red) treated with 2-DG. Cytotoxicity is reported as mean ± SEM. Data represent average values of at least three independent experiments. (**D**) Western blot analysis of apoptosis markers, cleaved-PARP and caspase-7, after 48 h treatment of GSCs using 2-DG with or without 8 Gy radiation. (**E**) Western blot of full-PARP and cleaved-PARP in Glio9 to confirm presence of full PARP and lack of PARP cleavage despite combination therapy.

3. Discussion

Despite advances in GBM therapy, overall patient survival remains dismal. A growing body of literature has focused on the role of GSCs in tumor formation, progression, and recurrence given their robust resistance to conventional chemotherapy and radiotherapy; thus, understanding mechanisms of resistance and developing cell specific therapies against GSCs are crucial in eliminating GBM [37,38]. Here, we demonstrated that radioresistance in GSCs is, in part, driven by increased expression of UPR-associated proteins, GRP78 and GRP94. However, by pharmacologically potentiating radiation-induced ER stress with 2-DG, GSC viability can be negatively affected through conversion of UPR towards CHOP-mediated apoptosis (Figure 7).

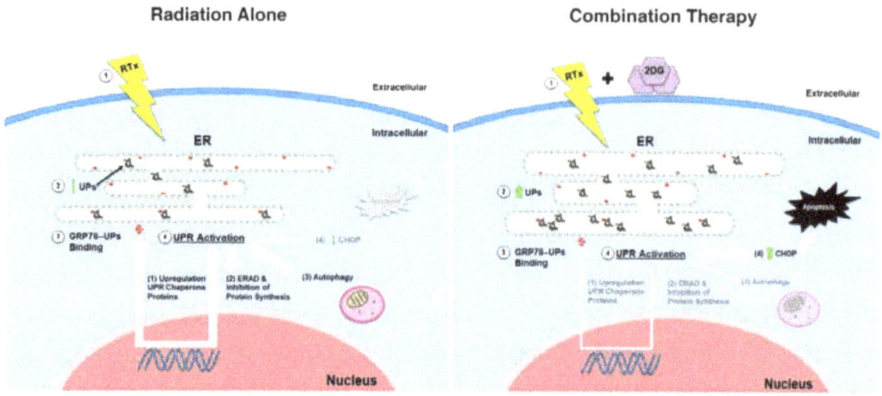

Figure 7. Graphic depiction of intracellular unfolded protein (UPR) activation induced by radiation versus radiation plus 2-deoxy-D-glucose (2-DG) on GSCs. **Left panel**: (1) Radiation induces accumulation of (2) unfolded proteins (UPs) in the endoplasmic reticulum (ER). This leads to release of the folding chaperone protein, GRP78, from PERK, IRE1, and ATF6 found on the ER membrane and (3) binding of GRP78 to UPs. GRP78 release activates PERK, IRE1, and ATF6 and (4) promotes UPR. Effects of activation include increased expression of chaperone proteins, ER associated protein degradation (ERAD), and autophagy, but limited apoptosis via CHOP upregulation. **Right panel**: (1) Radiation and 2-DG induce increased levels of ER stress and (2) higher levels of UPs. GRP78 binding to UPs and UPR activation follows (3,4); however, combination therapy induces higher ER stress than radiation alone, which prompt GSCs towards CHOP-associated apoptosis. Abbreviations: 2-DG, 2-deoxy-d-glucose; ER, endoplasmic reticulum; ERAD, ER-associated degradation; RTx, radiation; UP, unfolded proteins; UPR, unfolded protein response.

Adaptive mechanisms against the prolonged metabolic stresses in tumor microenvironment, constitutive activation of UPR, and overexpression of ER chaperone proteins like GRP78, have been studied in multiple cancer types as being linked to tumor processes such as invasion, resistance to environmental stressors, angiogenesis, oncogenic signaling, and resistance to traditional therapy [17,39]. However, should ER stress be too severe, the pro-survival function of UPR switches towards a cytotoxic pathway, leading to cell death. In this regard, targeting ER stress has emerged as a valuable mark [40].

Autophagy can be induced dependent or independent of UPR, and while we observed increased autophagosome formation in Glio9 and Glio14, we did not find increased vesicle formation in Glio11 (Figure 2A,B). When we utilized radiation and the autophagic inhibitor, chloroquine, we only observed loss of cell viability in Glio9 and Glio14. Recently, Ye et al., used chloroquine to radiosensitize glioma initiating cells (GICs); however, our results suggest that this modality is not effective in GBM cells that do not upregulate autophagy as a resistance mechanism, as is the case in Glio11 [41]. In a Phase I/II clinical trial conducted by Rosenfeld and colleagues, patients with newly diagnosed GBM were given hydroxychloroquine, a derivative of chloroquine, in addition to temozolomide and radiation. They concluded that despite hydroxychloroquine doses of 600 mg/day, consistent autophagy inhibition was not achieved and no significant improvement in overall survival was observed [42]. These findings in combination with our own suggest that GSCs are variable in their response to radiation and therapeutic use of autophagic inhibition may not be effective for all GBM patients.

Investigation into the feasibility of targeting ER stress as adjuvant to radiation therapy has proved successful. Dadey et al. concluded that induction of ER stress signaling by radiation in GBM contributes to the adaptive mechanisms encountered during radiotherapy, and that the PERK/ATF4 axis plays a critical role in maintaining viability of irradiated GBM cells [43]. Furthermore, it was also shown that antibody targeting of GRP78 attenuated cell proliferation, colony formation, PI3K/Akt/mTOR signaling, and enhanced cell death, indicating a therapeutic potential to targeting both mediators of UPR and UPR itself [24]. Pharmacological induction of ER stress has also been linked to favorable affects against GBM. Combination of 2-DG, also a potent activator of ER stress, and metformin suppressed neurosphere formation, expression of stem-cell related gene products, and invasive capacity [33]. In another study, the cyclooxygenase inhibitor, celecoxib, was utilized to enhance radiosensitivity in GBM cells through induction of ER stress [25].

Despite the expanding literature studying ER stress as a therapeutic target in GBM, the role of potentiating UPR in GBM stem cells to force cells towards CHOP-mediated cell death is still relatively novel. To our knowledge, this report represents a unique investigation seeking to characterize UPR as a radioresistance mechanism in specifically GSCs, correlate upregulation of UPR genes to overall patient survival, and propose a treatment strategy of potentiating ER stress to combat GSCs. We found evidence in The Cancer Genome Atlas (TCGA) that upregulation of UPR-associated genes is clinically relevant: patients with high expression of both *HSPA5* and *HSP90B1* exhibit decreased survival time when compared to lower expression levels (Figure 3). This data supplemented our microscopic analysis after we found a robust UPR response post radiation exposure with increased ER luminal diameter and increased autophagic vesicles—a response confirmed via Western blot. Of note, while increased autophagic vesicles were counted, autophagy related genes did not seem to be associated with survival differences in patients (Figure S1). Low levels of radiation did not induce higher CHOP expression, indicating that apoptosis may not be achieved in GSCs treated with clinical doses of radiation. We then opted to potentiate ER stress using a known ER stress inducing drug, 2-DG. In a recent phase I dose-escalation trial of 2-DG in advanced solid tumors, 2-DG was safely implemented at 63 mg/kg/day with plasma levels ~0.7 mM [34]. Combination therapy with 2-DG and radiation significantly reduced the IC_{50} of GSCs when compared to 2-DG alone, and combination therapy was responsible for increased CHOP-induced apoptosis in Glio11 and Glio14. However, this effect was not seen in Glio9, as confirmed by an LDH assay (Figure 6C).

An explanation for this lies in the variable expression of UPR protein expression at baseline (Figure 3B), with Glio9 having the highest expression, followed by Glio11, and then Glio14. While the genetic and phenotypic heterogeneity of GBM is well characterized, this result was particularly interesting since Glio9 was derived from a patient with recurrent GBM who had previously undergone the Stupp protocol (combination temozolomide with fractionated radiation) [1], as opposed to Glio11 and Glio14, which came from treatment-naïve patients. This suggests that higher levels of GRP78 expression confer a higher tolerance for ER stress inducing stimuli. As comparison, Glio14 had the lowest baseline GRP78 expression but had the highest evidence of combination therapy cytotoxicity per LDH assay. Despite a difference in cell death between cell lines, overall, GSC viability is negatively affected after treatment with combination therapy.

These results are relevant and novel for a few reasons. We showed that baseline levels of GRP78 are important in conferring resistance to radiation with or without ER stress inducing concurrent therapy. The fact that our highest GRP78 expressing cell line comes from a patient with recurrent disease also implies that GSCs with prior radiation exposure can maintain radioresistance. As such there may be a role for histologically examining GBM samples for UPR-associated protein expression prior to initiation of radiotherapy in primary or recurrent tumors, though this nascent idea would benefit from further clinically-oriented study. Our results also suggested that despite GRP78 expression, radiation plus 2-DG reduced viability of GSCs when compared to treatment with radiation alone. These results echo those achieved by several groups who used various ER stress-inducing stimuli to radiosensitize tumor cells [25,44–46]. Without radiation combination, the IC_{50} of GSCs exposed to 2-DG alone ranged between 1.38–1.77 mM—higher than what seems to be achievable in plasma. Combination therapy reduced the IC_{50} of all GSC lines collectively by roughly half (range, 0.57–1.03 mM).

Recently, 2-DG and radiotherapy was safely used in clinical trial against GBM in India, and trial results from over 100 patients revealed a modest survival advantage with improvement in patient quality of life [47,48]. Given intertumor variability though, future investigation would need to identify ways to increase the cytotoxicity of radiation and determine which patients would benefit most from this combination therapy. Our results suggest that examining ER-stress pathway protein expression could eventually be used for patient selection and that exploiting UPR-related apoptosis could enhance the degree of radiosensitization in GBM by targeting GSCs. Ultimately, further study is warranted via larger, prospectively-designed trials looking at multimodal potentiation of ER stress in patients selected based on UPR gene expression on histology.

4. Materials and Methods

4.1. Tissue Culture and Reagents

Collection of patient-derived GBM samples for use in biomedical research was approved by the Institutional Review Board (6 June 2007) at the University of Miami in accordance to the Declaration of Helsinki. Prior to harvesting of GBM specimen, informed consent was obtained from all patients by the attending Neurosurgeon, a clinical research coordinator, or a trained pathology assistant.

With Institutional Review Board (IRB) at the University of Miami approval and patient written informed consent, three GBM tumor samples were harvested at the time of surgical resection. Samples, named Glio9, Glio11, and Glio14, originated from a patient with recurrent disease (Glio9) and from treatment naïve patients (Glio11 and Glio14). Stem-like cell lines were generated as previously described [49,50]. Briefly, tumors were mechanically and enzymatically dissociated, red blood cells were lysed (Red Cell Lysis Buffer, Sigma-Aldrich, St. Louis, MO, USA), and then the single cell fraction was passed through a 40 μm filter to remove undigested particulates and tumor connective tissue. Cells were then plated in stem cell media: a 3:1 ratio of Dulbecco's Modified Eagle Medium (DMEM) and F12 (Gibco, Carlsbad, CA, USA) supplemented with 2% Gem21 NeuroPlex Serum Free Supplement without Vitamin A (Gemini Bioscience, Sacramento, CA, USA), 20 ng/mL of human epidermal growth factor and human basic fibroblast growth factor, and 1% penicillin and streptomycin. Neurospheres were grown in a humidified incubator kept at 37 °C and 5% CO_2. Neurospheres were passaged once spheroid diameter reached ~100 μm using 1 mL Accutase (StemCell Technologies, Cambridge, MA, USA) with a 5 min incubation at 37 °C. All cell lines were routinely tested for mycoplasma using LookOut mycoplasma PCR detection kit (Sigma Aldrich) according to manufacturer's instructions. Confirmation of the stem cell phenotypes of Glio9, Glio11, and Glio14 was performed via immunofluorescence staining for various stem markers and has been previously published [50].

The compounds used to potentiate ER stress and rescue cells, 2-deoxy-D-glucose (2-DG) and mannose, respectively, were obtained from Sigma-Aldrich and were reconstituted in phosphate buffered solution to generate stock concentrations. Autophagy inhibitor chloroquine was also obtained from Sigma-Aldrich and reconstituted in water to generate stock solution.

4.2. Radiation Source

Cells were irradiated using the RS-2000 Biological Irradiator (RadSource, Buford, GA, USA) at various dosages per equipment specifications. For dose-dependent radiation effects on GSC viability without addition of 2-DG, 2 Gray (Gy) to 20 Gy was administered, and 48 h post treatment, trypan blue was used to differentiate alive versus dead cells. For all transmission electron microscopy experiments, combination-therapy viability experiments, and western blot experiments, a single-time dose of 8 Gy was utilized.

4.3. Viability Assay

Viability was determined using the CellTiter 96® AQueous One Solution Cell Proliferation (MTS) Assay (Promega, Madison, WI, USA). Neurospheres were dissociated, and single cells were seeded into 96-well plates at a density of 1.0×10^4 cells per well in 100 μL of a modified neurosphere culture media containing 5% FBS and subsequently treated [50]. Following treatment, MTS reagent (20 μL per 100 μL media) was added to each well and incubated for 1–4 h. Optical density was measured at 490 nm using a Synergy HT plate reader (BioTek, Winooski, VT, USA). To determine viability effects of radiation ± 2-DG, GSCs were treated with or without 8 Gy radiation and increasing doses of 2-DG (0–2 mM) for 72 h. Data is represented as the average of 3 separate experiments in which viability is calculated as percent of non-treated (NT). The IC_{50}, or the concentration of drug at which 50% of cells was non-viable, was calculated from a composite of at least three experiments.

4.4. Lactate Dehydrogenase Assay

Cytotoxicity was determined by measuring the lactate dehydrogenase (LDH) activity released from damaged cells using the Cytotoxicity Detection Kit obtained from Roche Applied Science (Mannheim, Germany) as per manufacturer's instructions. Percent cytotoxicity was calculated using the following formula, in which the low control (minimal LDH release) was untreated cells and the high control (maximal release) was lysed cells. GSCs were treated with or without 8 Gy radiation and increasing doses of 2-DG (0–2 mM) and LDH release determined 48 h later:

$$Cytotoxicity\ \% = \frac{\text{experimental value} - \text{low control}}{\text{high control} - \text{low control}}$$

4.5. Transmission Electron Microscopy (TEM)

Prior to treatment, the neurospheres were first passed through a 40 μm filter (Falcon, Thermo Fisher Scientific, Waltham, MA, USA) to remove any single cells or debris then all the neurospheres greater than 40 μm were passed through a 100 μm to obtain a relative homogeneous population of neuropheres (ranging from 40–100 μm). Following treatment, the neurospheres were fixed in neutral buffered 2.5% glutaraldehyde at 25 °C. The specimens were post-fixed in 1% osmium tetroxide (OsO_4) for 10 min, dehydrated using a graded ethanol series, en bloc stained with 2% uranyl acetate in 50% ethanol for 30 min, and embedded in Spurr's epoxy resin. Semi-thin (1 μm) and ultra-thin (<90 nm) sections were cut using a Diatome 3-millimeter diamond knife on the EM UC6 ultramicrotome (Leica Microsystems Inc., Buffalo Groove, IL, USA). Semi-thin sections were stained with toluidine blue and examined using a BX60 light microscope (Olympus, Waltham, MA, USA) equipped with a digital camera (Olympus DP71). All ultra-thin sections were stained using lead citrate to be viewed under TEM with a 1400 EM (Jeol, Peabody, MA, USA) at 80 kV.

ER luminal diameter was analyzed using ImageJ software (Version 1.52e, NIH, Bethesda, MD, USA) on five electron micrographs taken at 10,000× magnification from two independent studies, while autophagy was determined by manual count. Autophagic vesicle were defined as a membrane bound vesicle containing either intact or degraded cytoplasmic material or organelles [51].

4.6. Western Blot

Protein extraction and western blot analysis were performed as previously described [52]. Anti-Beclin-1, anti-cleaved caspase 7, anti-CHOP, anti-GRP78, anti-phospho-H2AX, anti-LC3, anti-PARP, and anti-p62 were obtained from Cell Signaling Technology (Danvers, MA, USA). Anti-GRP94 was obtained from Enzo Life Sciences (Farmingdale, NY, USA). Anti-α-tubulin was obtained from Abcam (Eugene, OR, USA).

4.7. The Cancer Genome Atlas (TCGA) Data Analysis

Genomic data on GBM patients from TCGA was analyzed using an open-access brain tumor database, GlioVis (gliovis.bioinfo.cnio.es) [53]. A total of 528 GBM patient samples and 10 non-tumor patient samples were included. Data was limited to primary tumors with any 0-6-methylguanine-DNA methyltransferase (MGMT) status. Clinical survival data for high versus low expression—75th percentile versus (vs.) 25th percentile—of UPR genes (*HSPA5* (encodes GRP78) and *HSP90B1* (encodes

GRP94)) and autophagy genes (*MAP1LC3B* (encodes LC3), *BECN1* (encodes Beclin1), *SQSTM1* (encodes p62), *ATG5* (encodes ATG5), *ATG7* (encodes ATG7), and *ATG12* (encodes ATG12) was reported.

4.8. Statistical Analysis

Significance was determined via Mann-Whitney test for all pairwise comparisons of different treatments that were tested, ER luminal diameters, and autophagosomes per cell calculations using GraphPad Prism software (Version 6.0, La Jolla, CA, USA). GlioVis data was analyzed using a pairwise *t*-test. The results are presented as mean ± standard error mean (SEM). Significance was set at $p < 0.05$.

5. Conclusions

Our results indicate GSC radioresistance is, in part, achieved by overexpression and overactivation of ER stress-related pathways. Additionally, high expression levels of UPR-related genes, but not autophagy-related genes, are correlated with unfavorable patient survival. However, GSCs exhibit variable expression of UPR-related proteins like GRP78 and GRP94, indicating heterogeneity to UPR activation. Loss of GSC viability can be achieved via potentiation of UPR using multiple ER-stressors—such as radiation and the glycolytic inhibitor, 2-DG—particularly when baseline GRP78 levels are low. There may be a clinical role in pathological classification of GBM based on UPR-related proteins to determine which patients would benefit from ER stress inducing therapies; however, further investigation would first be required.

Supplementary Materials: The following are available online at http://www.mdpi.com/2072-6694/11/2/159/s1, Figure S1: GSCs treated with increasing doses of the autophagic inhibitor, chloroquine, and exposed to 8 Gy radiation, Figure S2: Kaplan-Meier curves displaying survival data from the GlioVis portal regarding autophagy related genes.

Author Contributions: Conceptualization; S.S.S., B.M., and R.M.G. Methodology; S.S.S., W.M.W., J.S.P., and R.M.G. Validation; S.S.S., W.M.W., and R.M.G. Formal analysis; S.S.S. and R.M.G. Investigation; S.S.S., G.A.R., A.M., W.M.W., N.d.C., E.B., and R.M.G. Resources; B.M., J.S.P., R.J.K., S.V., and R.M.G. Data curation; R.M.G. Writing—original draft preparation; S.S.S. and R.M.G. Writing-review and editing; S.S.S., G.A.R., A.M., W.M.W., N.d.C., E.B., M.M.M., B.M., R.J.K., S.V., and R.M.G. Visualization; S.S.S., G.A.R., A.M., E.B., M.M.M., and R.M.G. Supervision; B.M., J.S.P., R.J.K., and R.M.G. Project administration; S.S.S., J.S.P., and R.M.G. Funding acquisition, J.S.P., R.J.K., S.V., and R.M.G.

Funding: This research was funded by the Mystic Force Foundation and the University of Miami Brain Tumor Initiative (UMBTI).

Acknowledgments: We would like to thank our laboratory volunteers Frederic A. Vallejo, Wanda Gonzalez, Denis Ortega Ioni, Anthony Sanchez, Delaney Cooper, and Nicholas Pontillo for their contributions to our research efforts.

Conflicts of Interest: The authors declare no conflicts of interest.

References

1. Stupp, R.; Mason, W.P.; van den Bent, M.J.; Weller, M.; Fisher, B.; Taphoorn, M.J.B.; Belanger, K.; Brandes, A.A.; Marosi, C.; Bogdahn, U.; et al. Radiotherapy plus concomitant and adjuvant temozolomide for glioblastoma. *N. Engl. J. Med.* **2005**, *352*, 987–996. [CrossRef] [PubMed]
2. Ignatova, T.N.; Kukekov, V.G.; Laywell, E.D.; Suslov, O.N.; Vrionis, F.D.; Steindler, D.A. Human cortical glial tumors contain neural stem-like cells expressing astroglial and neuronal markers in vitro. *Glia* **2002**, *39*, 193–206. [CrossRef] [PubMed]
3. Hemmati, H.D.; Nakano, I.; Lazareff, J.A.; Masterman-Smith, M.; Geschwind, D.H.; Bronner-Fraser, M.; Kornblum, H.I. Cancerous stem cells can arise from pediatric brain tumors. *Proc. Natl. Acad. Sci. USA* **2003**, *100*, 15178–15183. [CrossRef]
4. Singh, K.; Clarke, I.D.; Terasaki, M.; Bonn, V.E.; Hawkins, C.; Squire, J.; Dirks, P.B. Identification of a cancer stem cell in human brain tumors. *Cancer Res.* **2003**, *63*, 5821–5828. [PubMed]
5. Sundar, S.J.; Hsieh, J.K.; Manjila, S.; Lathia, J.D.; Sloan, A. The role of cancer stem cells in glioblastoma. *Neurosurg. Focus* **2014**, *37*, E6. [CrossRef] [PubMed]

6. Lathia, J.D.; Mack, S.C.; Mulkearns-Hubert, E.E.; Valentim, C.L.; Rich, J.N. Cancer stem cells in glioblastoma. *Genes Dev.* **2015**, *29*, 1203–1217. [CrossRef]
7. Hui, L.; Chen, Y. Tumor microenvironment: Sanctuary of the devil. *Cancer Lett.* **2015**, *368*, 7–13. [CrossRef]
8. Mbenukui, F.; Johann, D.J. Cancer and the tumor microenvironment: A review of an essential relationship. *Cancer Chemother. Pharmacol.* **2009**, *63*, 571–582. [CrossRef]
9. Wang, M.; Zhao, J.; Zhang, L.; Wei, F.; Lian, Y.; Wu, Y.; Gong, Z.; Zhang, S.; Zhou, J.; Cao, K.; et al. Role of tumor microenvironment in tumorigenesis. *J. Cancer* **2017**, *8*, 761–773. [CrossRef]
10. Kaufman, R.J.; Scheuner, D.; Schroder, M.; Shen, X.; Lee, K.; Liu, C.Y.; Arnold, S.M. The unfolded protein response in nutrient sensing and differentiation. *Nat. Rev. Mol. Cell Biol.* **2002**, *3*, 411–421. [CrossRef]
11. Mokarram, P.; Albokashy, M.; Zarghooni, M.; Moosavi, M.A.; Sepehri, Z.; Chen, Q.M.; Hudecki, A.; Sargazi, A.; Alizadeh, J.; Moghadam, A.R.; et al. New frontiers in the treatment of colorectal cancer: Autophagy and the unfolded protein response as promising targets. *Autophagy* **2017**, *13*, 781–819. [CrossRef] [PubMed]
12. Ojha, R.; Amaravadi, R.K. Targeting the unfolded protein response in cancer. *Pharmacol. Res.* **2017**, *120*, 258–266. [CrossRef] [PubMed]
13. Epple, L.M.; Dodd, R.D.; Merz, A.L.; Dechkovskaia, A.M.; Herring, M.; Winston, B.M.; Lencioni, A.M.; Russell, R.L.; Madsen, H.; Nega, M.; et al. Induction of the unfolded protein response drives enhanced metabolism and chemoresistance in glioma cells. *PLoS ONE* **2013**, *8*, e73267. [CrossRef] [PubMed]
14. Lui, C.Y.; Kaufman, R.J. The unfolded protein response. *J. Cell Sci.* **2003**, *116*, 1861–1862.
15. Zhang, X.Y.; Zhang, T.T.; Song, D.D.; Zhou, J.; Han, R.; Qin, Z.H.; Sheng, R. Endoplasmic reticulum chaperone GRP78 is involved in autophagy activation induced by ischemic preconditioning in neural cells. *Mol. Brain* **2015**, *8*, e20. [CrossRef] [PubMed]
16. Lee, A.S. Glucose-regulated protein in cancer: Molecular mechanisms and therapeutic potential. *Nat. Rev. Cancer* **2014**, *14*, 263–276. [CrossRef] [PubMed]
17. Luo, B.; Lee, A.S. The critical roles in endoplasmic reticulum chaperone and unfolded protein response in tumorigenesis and anticancer therapies. *Oncogene* **2013**, *32*, 805–818. [CrossRef] [PubMed]
18. Marzec, M.; Eletto, D.; Argon, Y. GRP94: An HSP90-like protein specialized for protein folding and quality control in the endoplasmic reticulum. *Biochim. Biophys. Act* **2012**, *1823*, 774–787. [CrossRef] [PubMed]
19. Sun, C.; Han, C.; Jiang, Y.; Han, N.; Zhang, M.; Li, G.; Qiao, Q. Inhibition of GRP78 abrogates radioresistance in oropharyngeal carcinoma cells after EGFR inhibition by cetuximab. *PLoS ONE* **2017**, *12*, e0188932. [CrossRef] [PubMed]
20. Dong, D.; Ni, M.; Li, J.; Xiong, S.; Ye, W.; Virrey, J.J.; Mao, C.; Ye, R.; Wang, M.; Pen, L.; et al. Critical role of the stress chaperone GRP78/BiP in tumor proliferation, survival, and tumor angiogenesis in transgene-induced mammary tumor development. *Cancer Res.* **2008**, *68*, 498–505. [CrossRef] [PubMed]
21. Kwon, D.; Koh, J.; Kim, S.; Go, H.; Min, H.S.; Kim, Y.A.; Kim, D.K.; Jeon, Y.K.; Chung, D.H. Overexpression of endoplasmic reticulum stress-related proteins, XBP1s and GRP78, predicts poor prognosis in pulmonary adenocarcinoma. *Lung Cancer* **2018**, *122*, 131–137. [CrossRef] [PubMed]
22. Xi, J.; Chen, Y.; Huang, S.; Cui, F.; Wang, X. Suppression of GRP78 sensitizes human colorectal cancer cells to oxaliplatin by downregulation of CD24. *Oncol. Lett.* **2018**, *15*, 9861–9867. [CrossRef] [PubMed]
23. Dadey, D.Y.; Kapoor, V.; Khudanyan, A.; Urano, F.; Kim, A.H.; Thotala, D.; Hallahan, D.E. The ATF6 pathway of the ER stress response contributes to enhanced viability in glioblastoma. *Oncotarget* **2016**, *7*, 2080–2092. [CrossRef] [PubMed]
24. Dadey, D.Y.A.; Kapoor, V.; Hoye, K.; Khudanyan, A.; Collins, A.; Thotala, D.; Hallahan, D.E. Antibody targeting GRP78 enhances the efficacy of radiation therapy in human glioblastoma and non-small cell lung cancer cell lines and tumor models. *Clin. Cancer Res.* **2017**, *23*, 2556–2564. [CrossRef] [PubMed]
25. Suzuki, K.; Gerelchuluun, A.; Hong, Z.; Sun, L.; Zenkoh, J.; Moritake, T.; Tsuboi, K. Celecoxib enhances radiosensitivity of hypoxic glioblastoma cells through endoplasmic reticulum stress. *Neuro Oncol.* **2013**, *15*, 1186–1199. [CrossRef] [PubMed]
26. Kuo, L.J.; Yang, L.X. Gamma-H2AX—A novel biomarker for DNA double-strand breaks. *In Vivo* **2008**, *22*, 305–309. [PubMed]
27. Oslowsli, C.M.; Urano, F. Measuring ER stress and the unfolded protein response using mammalian tissue culture system. *Methods Enzymol.* **2011**, *490*, 71–92.

28. Wang, J.; Takeuchi, T.; Tanaka, S.; Kubo, S.K.; Kayo, T.; Lu, D.; Takata, K.; Koizumi, A.; Izumi, T. A mutation in the insulin 2 gene induces diabetes with severe pancreatic beta-cell dysfunction in the Mody mouse. *J. Clin. Investig.* **1999**, *103*, 27–37. [CrossRef]
29. Niu, Z.; Wang, M.; Zhou, L.; Yao, L.; Liao, Q.; Zhao, Y. Elevated GRP78 expression is associated with poor prognosis in patients with pancreatic cancer. *Sci. Rep.* **2015**, *5*, 16067. [CrossRef]
30. Nami, B.; Ghasemi-Dizgah, A.; Vaseghi, A. Overexpression of molecular chaperons GRP78 and GRP94 in CD44(hi)/CD24(lo) breast cancer stem cells. *Bioimpacts* **2016**, *6*, 105–110. [CrossRef]
31. Zhang, J.; Jiang, Y.; Jia, Z.; Li, Q.; Gong, W.; Wang, L.; Wei, D.; Yao, J.; Fang, S.; Xie, K. Association of elevated GRP78 expression with increased lymph node metastasis and poor prognosis in patients with gastric cancer. *Clin. Exp. Metastasis* **2006**, *23*, 401–410. [CrossRef] [PubMed]
32. Mert, I.; Chhina, J.; Allo, G.; Dai, J.; Seward, S.; Carey, M.S.; Llaurado, M.; Giri, S.; Rattan, R.; Munkarah, A.R. Synergistic effect of MEK inhibitor and metformin combination in low grade serous ovarian cancer. *Gynecol. Oncol.* **2017**, *146*, 319–326. [CrossRef] [PubMed]
33. Kim, E.H.; Lee, J.H.; Oh, Y.; Koh, I.; Shim, J.K.; Park, J.; Choi, J.; Yun, M.; Jeon, J.Y.; Huh, Y.M.; et al. Inhibition of glioblastoma tumorspheres by combined treatment with 2-deoxyglucose and metformin. *Neuro Oncol.* **2017**, *19*, 197–207. [CrossRef]
34. Raez, L.E.; Papadopoulos, K.; Ricart, A.D.; Chiorean, E.G.; Dipaola, R.S.; Stein, M.N.; Rocha Lima, C.M.; Schlesselman, J.J.; Tolba, K.; Langmuir, V.K.; et al. A phase I dose-escalation trial of 2-deoxy-D-glucose alone or combined with docetaxel in patients with advanced solid tumors. *Cancer Chemother. Pharmacol.* **2013**, *71*, 523–530. [CrossRef] [PubMed]
35. Xi, H.; Barredo, J.C.; Merchan, J.R.; Chiorean, E.G.; Dipaola, R.S.; Stein, M.N.; Rocha Lima, C.M.; Schlesselman, J.J.; Tolba, K.; Langmuir, V.K.; et al. Endoplasmic reticulum stress induced by 2-deoxyglucose but not glucose starvation activates AMPK through CaMKKβ leading to autophagy. *Biochem. Pharmacol.* **2013**, *85*, 1463–1477. [CrossRef] [PubMed]
36. Kovács, K.; Decatur, C.; Toro, M.; Pham, D.G.; Liu, H.; Jing, Y.; Murray, T.G.; Lampidis, T.J.; Merchan, J.R. 2-deoxy-glucose downregulates endothelial AKT and ARK via interverence with N-linked glycosylation, induction of endoplasmic reticulum stress and GSK3β activation. *Mol. Cancer Ther.* **2016**, *15*, 264–275. [CrossRef] [PubMed]
37. Chen, J.; Li, Y.; Yu, T.S.; McKay, R.M.; Burns, D.K.; Kernie, S.G.; Parada, L.F. A restricted cell population propagates glioblastoma growth after chemotherapy. *Nature* **2012**, *488*, 522–526. [CrossRef] [PubMed]
38. Flores, D.G.; Ledur, P.F.; Abujamra, A.L.; Brunetto, A.L.; Schwartsmann, G.; Lenz, G.; Roesler, R. Cancer stem cells and the biology of brain tumors. *Curr. Stem Cell Res. Ther.* **2009**, *4*, 306–313. [CrossRef]
39. Garg, A.D.; Maes, H.; van Vliet, A.R.; Agostinis, P. Targeting the hallmarks of cancer with therapy-induced endoplasmic reticulum (ER) stress. *Mol. Cell. Oncol.* **2014**, *2*, e975089. [CrossRef]
40. Verfaille, T.; Garg, A.D.; Agostinis, P. Targeting ER stress induced apoptosis and inflammation in cancer. *Cancer Lett.* **2013**, *332*, 249–264. [CrossRef]
41. Ye, H.; Chen, M.; Cao, F.; Huang, H.; Zhan, R.; Zheng, X. Chloroquine, an autophagy inhibitor, potentiates the radiosensitivity of glioma initiating cells by inhibiting autophagy and activating apoptosis. *BMC Neurol.* **2016**, *16*, 178. [CrossRef] [PubMed]
42. Rosenfeld, M.R.; Ye, X.; Supko, J.G.; Desideri, S.; Grossman, S.A.; Brem, S.; Mikkelson, T.; Wang, D.; Chang, Y.C.; Hu, J.; et al. A phase I/II trial of hydroxychloroquine in conjunction with radiation therapy and concurrent and adjuvant temozolomide in patients with newly diagnosed glioblastoma multiforme. *Autophagy* **2014**, *10*, 1359–1368. [CrossRef] [PubMed]
43. Dadey, D.Y.A.; Kapoor, V.; Khudanyan, A.; Thotala, D.; Hallahan, D.E. PERK regulates glioblastoma sensitivity to ER stress although promoting radiation resistance. *Mol. Cancer Res.* **2018**, *16*, 1447–1453. [CrossRef] [PubMed]
44. Pang, X.L.; He, G.; Liu, Y.B.; Wang, Y.; Zhang, B. Endoplasmic reticulum stress sensitizes human esophageal cancer cell to radiation. *World J. Gastroenterol.* **2013**, *19*, 1736–1748. [CrossRef] [PubMed]
45. Yasui, H.; Takeuchi, R.; Nagane, M.; Meike, S.; Nakamura, Y.; Yamamori, T.; Ikenaka, Y.; Kon, Y.; Murotani, H.; Oishi, M.; et al. Radiosensitization of tumor cells through endoplasmic reticulum stress induced by PEGylated nanogel containing cold nanoparticles. *Cancer Lett.* **2014**, *347*, 151–158. [CrossRef] [PubMed]

46. Gong, C.; Yang, Z.; Zhang, L.; Wang, Y.; Gong, W.; Liu, Y. Quercetin suppresses DNA double-strand break repair and enhances the radiosensitivity of human ovarian cancer cells via p53-dependent endoplasmic reticulum stress pathway. *OncoTargets Ther.* **2017**, *11*, 11–27. [CrossRef] [PubMed]
47. Dwarakanath, B.S.; Singh, D.; Banerji, A.K.; Sarin, R.; Venkataramana, N.K.; Jalali, R.; Vishwanath, P.N.; Mohanti, B.K.; Tripathi, R.P.; Kalia, V.K.; et al. Clinical studies for improving radiotherapy with 2-deoxy-D-glucose: Present status and future prospects. *J. Cancer Res. Ther.* **2009**, *5* (Suppl. 1), S21–S26. [CrossRef] [PubMed]
48. Singh, D.; Banerji, A.K.; Dwarakanath, B.S.; Tripathi, R.P.; Gupta, J.P.; Mathew, T.L.; Ravindranath, T.; Jain, V. Optimizing cancer radiotherapy with 2-deoxy-d-glucose dose escalation studies in patients with glioblastoma multiforme. *Strahlenther. Onkol.* **2005**, *181*, 507–514. [CrossRef] [PubMed]
49. Pastori, C.; Daniel, M.; Penas, C.; Volmar, C.H.; Johnstone, A.L.; Brothers, S.P.; Graham, R.M.; Allen, B.; Sarkaria, J.N.; Komotar, R.J.; et al. BET bromodomain proteins are required for glioblastoma cell proliferation. *Epigenetics* **2014**, *9*, 611–620. [CrossRef]
50. Gersey, Z.C.; Rodriguez, G.A.; Barbarite, E.; Sanchez, A.; Walters, W.M.; Ohaeto, K.C.; Komotar, R.J.; Graham, R.M. Curcumin decreases malignant characteristics of glioblastoma stem cells via induction of reactive oxygen species. *BMC Cancer* **2017**, *17*, 99. [CrossRef]
51. Ylä-Anttila, P.; Vihinen, H.; Jokitalo, E.; Eskelinen, E.L. Monitoring autophagy by electron microscopy in mammalian cells. *Methods Enzymol.* **2009**, *452*, 143–164. [PubMed]
52. Graham, R.M.; Hernandez, F.; Puerta, N.; De Angulo, G.; Webster, K.A.; Vanni, S. Resveratrol augments ER stress and the cytotoxic effects of glycolytic inhibition in neuroblastoma by downregulating Akt in a mechanism independent of SIRT1. *Exp. Mol. Med.* **2016**, *48*, e210. [CrossRef] [PubMed]
53. Bowman, R.L.; Wang, Q.; Carro, A.; Verhaak, R.G.; Squatrito, M. GlioVis data portal for visualization and analysis of brain tumor expression datasets. *Neuro Oncol.* **2017**, *19*, 139–141. [CrossRef] [PubMed]

cancers

MDPI

Article

Arsenic Trioxide and (−)-Gossypol Synergistically Target Glioma Stem-Like Cells via Inhibition of Hedgehog and Notch Signaling

Benedikt Linder [1,*], Andrej Wehle [1], Stephanie Hehlgans [2], Florian Bonn [3], Ivan Dikic [3,4], Franz Rödel [2], Volker Seifert [5] and Donat Kögel [1,6]

1 Experimental Neurosurgery, Department of Neurosurgery, Neuroscience Center, Goethe University Hospital, 60528 Frankfurt am Main, Germany; Andrej.Wehle93@gmx.net (A.W.); koegel@em.uni-frankfurt.de (D.K.)
2 Radiotherapy and Oncology, Goethe University Hospital, 60590 Frankfurt am Main, Germany; Stephanie.Hehlgans@kgu.de (S.H.); Franz.Roedel@kgu.de (F.R.)
3 Institute of Biochemistry II, Faculty of Medicine, Goethe University Hospital, 60590 Frankfurt am Main, Germany; bonn@med.uni-frankfurt.de (F.B.); dikic@biochem2.uni-frankfurt.de (I.D.)
4 Buchmann Institute for Molecular Life Sciences, Goethe University, 60438 Frankfurt am Main, Germany
5 Department of Neurosurgery, Goethe University Hospital, 60528 Frankfurt am Main, Germany; V.seifert@em.uni-frankfurt.de
6 German Cancer Consortium (DKTK), Partner Site Frankfurt, 60590 Frankfurt am Main, Germany
* Correspondence: linder@med.uni-frankfurt.de; Tel.: +49-69-6301-84051

Received: 28 December 2018; Accepted: 7 March 2019; Published: 12 March 2019

Abstract: Glioblastoma is one of the deadliest malignancies and is virtually incurable. Accumulating evidence indicates that a small population of cells with a stem-like phenotype is the major culprit of tumor recurrence. Enhanced DNA repair capacity and expression of stemness marker genes are the main characteristics of these cells. Elimination of this population might delay or prevent tumor recurrence following radiochemotherapy. The aim of this study was to analyze whether interference with the Hedgehog signaling (Hh) pathway or combined Hh/Notch blockade using small-molecule inhibitors can efficiently target these cancer stem cells and sensitize them to therapy. Using tumor sphere lines and primary patient-derived glioma cultures we demonstrate that the Hh pathway inhibitor GANT61 (GANT) and the arsenic trioxide (ATO)-mediated Hh/Notch inhibition are capable to synergistically induce cell death in combination with the natural anticancer agent (−)-Gossypol (Gos). Only ATO in combination with Gos also strongly decreased stemness marker expression and prevented sphere formation and recovery. These synergistic effects were associated with distinct proteomic changes indicating diminished DNA repair and markedly reduced stemness. Finally, using an organotypic brain slice transplantation model, we show that combined ATO/Gos treatment elicits strong growth inhibition or even complete elimination of tumors. Collectively, our data show for the first time that ATO and Gos, two drugs that can be used in the clinic, represent a promising targeted therapy approach for the synergistic elimination of glioma stem-like cells.

Keywords: cancer stem cells; glioblastoma; Hedgehog; Notch; DNA damage

1. Introduction

Glioblastoma (GBM, WHO grade IV astrocytoma) is the most common and most aggressive primary brain tumor in adults [1]. Due to its diffuse infiltrative growth, surgical resection is virtually impossible [2]. Thus, even with the current standard-of-care treatment consisting of temozolomide (TMZ)-based radiochemotherapy [3], the average survival barely exceeds one year and the 5-year-survival rate is below 5%. Another key characteristic of GBM is its high resistance to induction of apoptosis, which is in part mediated by overexpression of anti-apoptotic proteins and an enhanced DNA repair (reviewed in [4]).

Many cancers including GBM exhibit a hierarchical organization of cells including a population with a stem-like phenotype (GSCs: glioma stem cells) that can replenish the tumor after treatment and are thought to be responsible for disease recurrence (reviewed in [5]) and initiation [6]. GSCs harbor expression of stemness-related marker genes and phenotypical traits similar to normal stem cells (reviewed in [5]) including an unlimited regenerative potential and the ability to divide asymmetrically. It was further shown that GSCs are highly resistant to DNA damaging agents and ionizing radiation and have an increased DNA repair capacity, in part caused by enhanced expression of Checkpoint kinase 1 (CHK1) and CHK2 [7].

The Hh and Notch signaling pathways are developmentally important signaling pathways involved in body patterning, brain development and maintenance of stem cells [8]. During the last decade it became apparent that Hh signaling plays a crucial role for tumor formation and progression of many cancers including GBM ([9] and reviewed in [10]). Notch signaling is also involved in brain development and regulates the expression of a wide variety of target genes that increase cell proliferation and survival (reviewed in [11]). Recent research also revealed a central role in self-renewal and therapy resistance of GSCs ([12] and reviewed in [13]), indicating that Notch represents a potential therapeutic target in these cells.

Arsenic trioxide (ATO) was initially used in Traditional Chinese Medicine to treat various diseases, including cancer. Research based on these findings resulted in the application of ATO as a clinical drug for treatment of acute promyelocytic leukemia (APL). Further, many studies have demonstrated potent Notch-inhibitory and antitumoral effects of ATO in various other cancer entities [14] including glioma [15]. In addition, ATO has been shown to effectively inhibit Hh signaling at the level of GLI in multiple cancers including medulloblastoma [16].

Based on the findings that both pathways regulate key aspects of GSCs, we aimed at investigating the potential therapeutic effects of Hh and Notch inhibition in regard to stem cell differentiation and sensitivity to cell death induced by (−)-gossypol (AT-101, Gos), a natural compound that efficiently induces cell death in apoptosis-resistant, differentiated GBM cells [17]. Recent research indicates that Gos also targets cancer and cancer stem cells (CSCs) in colorectal [18] and prostate cancer [19], respectively. We hypothesized that an enforced differentiation of GSCs might render them more vulnerable to cell death induction, and that a combination of Hh and/or Notch inhibition consequently would synergistically increase the effects of Gos. We further aimed to scrutinize whether single pathway inhibition is sufficient in driving cell death or whether multiple intertwined (i.e., Hh and Notch) pathways need to be targeted. To this end, we employed the well-known Hh pathway inhibitor/GLI-antagonist 61 (GANT61, GANT; [20]) and ATO in combination with Gos.

Here, we demonstrate that ATO-mediated Hh/Notch inhibition together with the natural anticancer agent Gos, rather than combining the Hh-inhibitor GANT and Gos efficiently blocks GSC growth, induces cell death and diminishes the self-renewal capacity of tumor sphere lines. These therapeutic effects are accompanied by distinct proteomic changes suggesting alterations in cell movement and cell cycle progression. They are also associated with major alterations in the expression of (neuronal) development genes (e.g., oligodendrocyte transcription factor 2 (OLIG2), sex determining region Y (SRY)- box 2 (SOX2) and SOX9) and DNA damage repair genes (e.g., BRCA1 Associated ATM Activator 1 (*BRAT1*), CHK1, CHK2), suggesting that this drug combination can efficiently enforce GSC differentiation and sensitization to therapy. Finally, this treatment was applied to tumors transplanted onto adult organotypic brain slices where ATO and ATO/Gos decreases tumor size and even lead to the complete elimination of some tumors.

2. Results

2.1. Hh/Notch Inhibition in Combination with Gos Synergistically Induces Cell Death of GSCs

The general aim of this study was to identify novel approaches to selectively target GSCs using established small molecule inhibitors to inhibit either the Hh signaling pathway alone or in combination with the Notch pathway. To this end, we applied GANT or ATO to target the Hh or Hh/Notch signaling pathways, respectively, in combination with the cell death inducer Gos. Our hypothesis was that the enforced differentiation mediated by either Hh or Hh/Notch inhibition would increase the vulnerability of GSCs towards cell death stimulation with Gos, resulting in synergistic action of the two drugs. To test this hypothesis, we initially performed MTT assays (Figure 1) and calculated the combination indices (CIs) according to Chou et al. [21] (CI < 1: synergism; CI = 1: additive; CI > 1: antagonism). Morphologically, a reduction of sphere size and apparent disintegration of the spheres after single agent treatment with GANT, ATO or Gos was detectable, indicated by an increase of single cells (black arrows) and less tight appearance of the spheres (white arrows). Combination treatment further enhanced these effects (Figure 1a).

MTT assays with the tumor sphere line GS-5 (Figure 1b,c) showed that single agent treatment with GANT, ATO or Gos dose-dependently reduced the viability and combination treatments synergistically enhanced these effects (CI < 1). Similar findings were also made with the GANT/Gos and ATO/Gos combinations in GS-1 cells (Figure S1a,b), and with the GANT/Gos, but not ATO/Gos combination in GS-8 cells (Figure S1c,d), although GANT single agent treatment had no significant effects in these cells.

The decreases in viability were affirmed by increases in cell death as shown by FACS-based Annexin V/Propidium iodide (PI) double stainings (Figure 1d–f). Again the combination treatments were more effective than either single treatment. Similar findings were also made in two other GS-lines (GS-3 and GS-8, Figure S2a–d) and a GS-line with a restricted stem-like (progenitor-like) phenotype (GS-1, Figure S2e,f).

Next, we analyzed the expression of *GLI1*, *PTCH1*, *BCL2* and *CCND1*(Cyclin D1) as surrogate markers for Hh pathway activity as well as *HES5* and *HEY1* for Notch signaling in GS-5 (Figure 1g) and the primary culture 17/02 (Figure 1h). Despite the fact that we applied GANT at 2.5 μM, a concentration that exhibits robust inhibitory activity of Hh signaling in the Gli-responsive cell line Shh light II [22] (Figure S3), it had little effect on any of the analyzed target genes, although a small tendency towards *PTCH1* and *CCND1* inhibition was apparent. Gos alone strongly reduced *PTCH1* and *HES5* expression. *HES5* expression was also reduced after GANT + Gos treatment. ATO and ATO + Gos reduced the expression of all markers, except *HEY1* in 17/02, whereas the combination exerted greater inhibitory effects. Similar findings were also observed for GS-8 and a second primary culture, 17/01. Notably, 17/01 appeared to be insensitive towards Hh-inhibition and only showed minor inhibition of the Notch-targets. Curiously, we observed that Gos increased the expression of *GLI1* in GS-5, GS-8 and 17/02, while simultaneously decreasing *PTCH1*-expression.

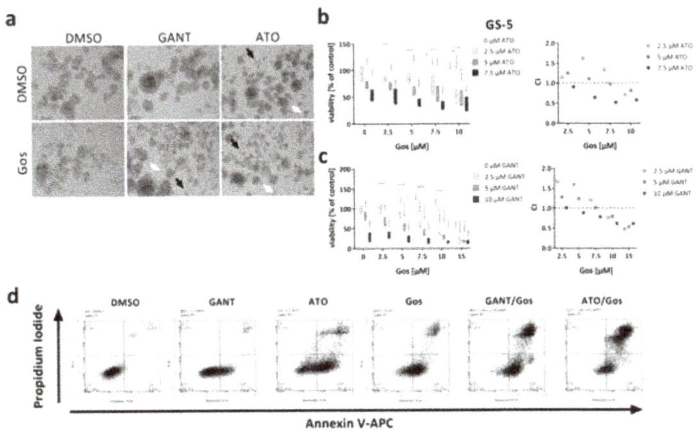

Figure 1. *Cont.*

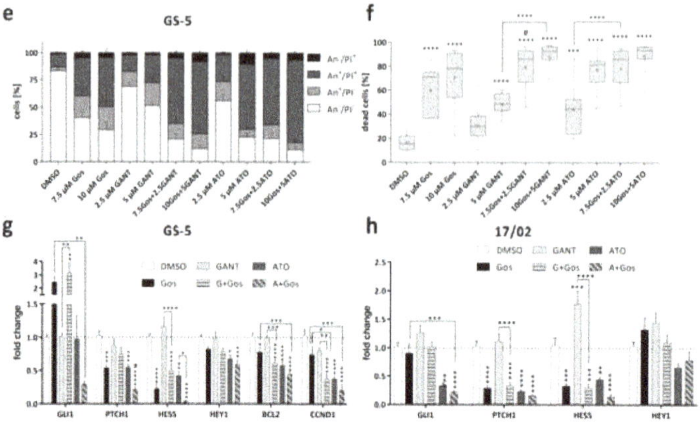

Figure 1. Synergistic inhibition of viability and induction of cell death of GS-5 and 17/02 GSCs via inhibition of Hh and Notch signaling. (**a**) Morphological appearance of GS-5 24 h after treatment with solvent (DMSO), 5 μM GANT, 5 μM ATO or 7.5 μM Gos alone or in combination. Note that after ATO (upper right) single treatment more single cells (black arrows) and disintegrating spheres (white arrows) can be detected which is further enhanced by ATO/Gos combination treatment and also after GANT/Gos; scale bar: 100 μm. (**b**,**c**, left side) Box-Plots (Tukey) of MTT assay of GS-5 after treatment for 24 h with increasing concentrations of (**b**) ATO or (**c**) GANT in combination with Gos. (**b**,**c**, right side) The CI was calculated from the data obtained according to Chou et al. [21] using the non-constant ratio setting (CI > 1: antagonism; CI = 1: additive; CI < 1: synergism). The CI value is given as a single value calculated from the summary of all experiments. (**d**) Representative Dot-Plots of GS-5 after treatment for 24 h with solvent (DMSO), 2.5 μM GANT, 2.5 μM ATO or 7.5 μM Gos alone or in combination. (**e**) Stacked bar chart of GS-5 after treatment for 24 h with the drugs and concentrations as indicated. (**f**) Box-Plots (Tukey) of the percentages of dead cells (100%—An-/PI-) after treatment. The lines in the Box-Plots represents the median, the plus-symbol the mean. (**g**,**h**) Bar graph of Taqman-based gene expression analysis of (**g**) GS-5 and (**h**) 17/02 after treatment with 5 μM Gos, 2.5 μM GANT or ATO or the combination of GANT or ATO with Gos (G + Gos; A + Gos). The MTT assays were performed at least three times in 6 biological replicates. All other experiments were performed at least three times in three biological replicates. Error bars are SEM. * $p < 0.05$; ** $p < 0.01$; *** $p < 0.001$; **** $p < 0.0001$ against solvent or as indicated; ° $p < 0.05$; °° $p < 0.01$; °°° $p < 0.001$; °°°° $p < 0.0001$ against GANT or ATO single treatment; # $p < 0.05$ against both single treatments.

2.2. Proteomic Analysis Reveals Global Changes Related to Impaired Cell Movement, DNA Repair and Stemness Properties after ATO/Gos Treatment

To analyze the underlying processes mediating the observed synergistic antitumoral effects in more detail, we next performed unbiased proteomic analyses using label-free quantification of GS-5 cells (Figure 2). For our proteomics approach we chose lower drug concentrations that did not result in massive cell death (2.5 μM GANT/ATO each and 5 μM Gos). Using this approach, we obtained reproducible quantitative data for 5008 proteins. Compared to solvent-treated cells 255, 47 and one protein(s) differed significantly after single treatment with Gos (Figure 2a), ATO and GANT (Figure S4), respectively. The combination treatments markedly elevated the amount of significant changes found to 424 and 648 after GANT/Gos (Figure 2c) and ATO/Gos (Figure 2e) treatment, respectively, thereby further providing evidence for the observed drug synergism. The five most increased and decreased proteins for each condition are depicted in the Volcano Plots (Figure 2a,c,e) along with the known stemness markers OLIG2, SOX2 and SOX9. The decreased proteins include proteins with known functions for glioma stemness and migration/invasion like CD9 [23] (Gos, ATO/Gos), Ephrin receptor A2 (EPHA2 [24]; Gos, GANT/Gos) and mesenchymal (i.e., more aggressive) differentiation like FRAS1 related extracellular matrix protein 2 (FREM2 [25]; ATO/Gos). We validated selected genes among the top five hits via qRT-PCR (Figure S4c–e). A list displaying all significantly changed proteins is provided in the Supplements (Supplement File 2).

To analyze for Gene Ontology (GO) terms (Figure 2b,d,f) that are enriched among the changed proteins we used the STRING-platform, (string-db.org; ver. 10.5; accessed Feb. 2018; [26]) with a consecutive analysis using Revigo (revigo.irb.hr; accessed Sep. 2018; [27]). Treatment with Gos (Figure 2b) caused a decrease in proteins related to (neuronal) development and movement, while proteins from the translation machinery were overrepresented among the increased proteins. The combined treatment of GANT and Gos (Figure 2d) enriched for cell cycle- and movement-related proteins among the decreased proteins. Among the increased hits, proteins involved in the processes "response to oxidative stress", "cellular respiration" and "response to unfolded protein" are overrepresented.

Combined ATO and Gos treatment (Figure 2f) caused an even stronger enrichment of (neuronal) development proteins among the significantly decreased proteins. This group includes the known GSC markers OLIG2, SOX2 and SOX9. In addition, movement and cell cycle proteins are found to a large portion in decreased amounts. Similar to the single treatment with Gos the increased proteins were often assigned to translation and again "response to oxidative stress" and "cellular respiration". A list displaying all enriched GO:BP terms is provided as a supplemental file (Supplement File 3).

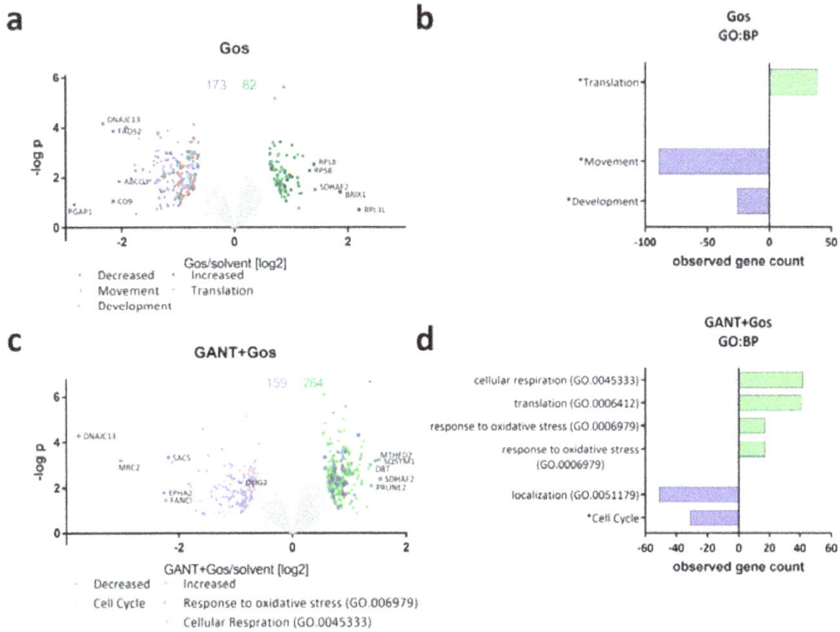

Figure 2. *Cont.*

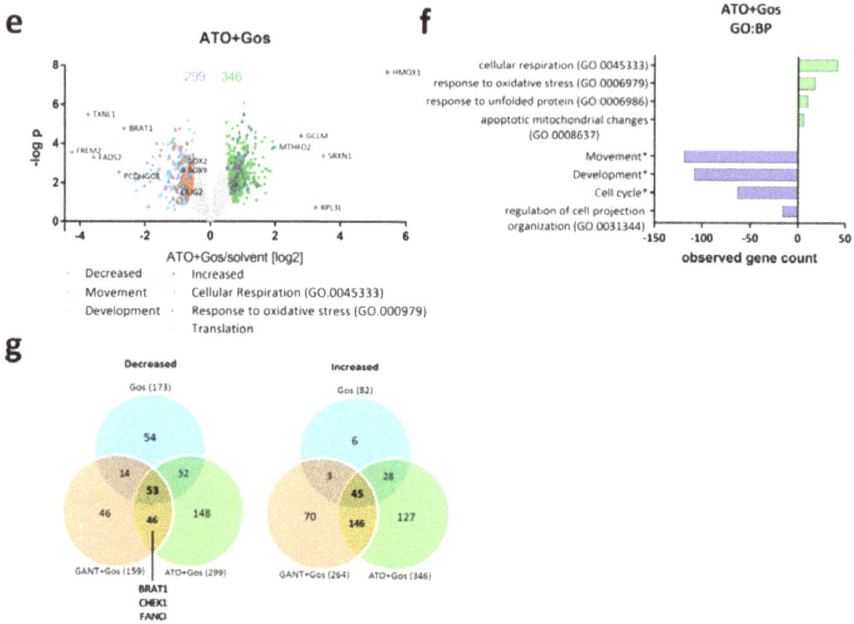

Figure 2. GANT and ATO-treatment in combination with Gos causes global proteomic changes affecting differentiation, cell cycle and proliferation. (**a**,**c**,**e**) Volcano Plots showing the protein ratios (in log2) as a function of the –log p-values of label-free quantification proteomics data of GS-5 after treatment for 24 h with (**a**) 5 μM Gos alone or in combination with (**c**) 2.5 μM GANT or (**e**) 2.5 μM ATO. 5008 proteins were quantified, the amount of significantly changed proteins is written above the Plot. (**b**,**d**,**f**) Bar Chart after bioinformatic analysis using the STRING-platform (string-db.com, [26]) and Revigo (revigo.irb.hr, [27]) for enriched Gene Ontology biological processes (GO:BP) and manual clustering (marked with an *) after treatment with (**b**) Gos alone, (**d**) GANT + Gos or (**f**) ATO + Gos. (**g**) Venn diagram depicting the overlap between the significantly (left side) decreased and (right side) increased proteins. Note that the x-axis in (**b**,**d**,**f**) depicts the number of proteins that are significantly decreased (negative values) and increased (positive values).

Next we analyzed the degree of overlap between the three groups (Figure 2g, Supplement File 4). Here, 53 and 45 proteins are significantly decreased and increased following all three treatments, respectively. 46 and 146 proteins are decreased and increased only by the combination treatments (GANT/Gos, ATO/Gos), but not by Gos alone. Notably, proteins decreased only after both combination treatments include proteins involved in the DNA damage response (DDR) like BRAT1, Fanconi Anemia Complementation Group I (FANCI) and CHEK1, indicating that the cells might be more prone to DNA damage. GO-analysis (Supplement File 5) of the proteins regulated by both combination treatments revealed that only the GO:BP term mitotic cell cycle is significantly enriched among the decreased proteins. Some terms were assigned to translation, cellular respiration, and response to oxidative stress among the increased proteins.

2.3. *ATO and Gos Treatment Reduces Expression of Stemness Markers*

A key hallmark of GSCs is the expression of stemness-related genes like OLIG2, SOX2 or SOX9 with a concomitant lack of known differentiation genes like GFAP. Indeed, our proteomic analysis showed that OLIG2 abundance decreases upon GANT/Gos and ATO/Gos treatment, while ATO/Gos also significantly decreased SOX2 and SOX9 protein levels and increased GFAP.

Western Blot (Figure 3a,b) analyses confirmed that all treatments except GANT alone reduce OLIG2 and SOX9 expression in GS-5 cells (Figure 3a) with ATO and the ATO/Gos combination being most effective. Additionally, SOX2 was only decreased after ATO treatment. ATO also strongly reduced OLIG2, SOX2 and SOX9 mRNA expression which was further enhanced by Gos (Figure 3c). In 17/02 especially ATO and ATO/Gos reduce the protein levels (Figure 3b) of the three markers, while all treatments except GANT reduced the expression of the corresponding mRNAs (Figure 3d). In GS-8 (Figure S5a,c) we obtained similar data compared to 17/02, although the mRNA expression was only changed after ATO or ATO/Gos treatment and 17/01 (Figure S5b,d) was mainly reacting to ATO/Gos treatment. In order to analyze expression changes on a single cell level we performed immunofluorescence microscopy of GS-5 and could show that all treatments except GANT can reduce the amount of marker positive cells with the combination treatment being the most effective.

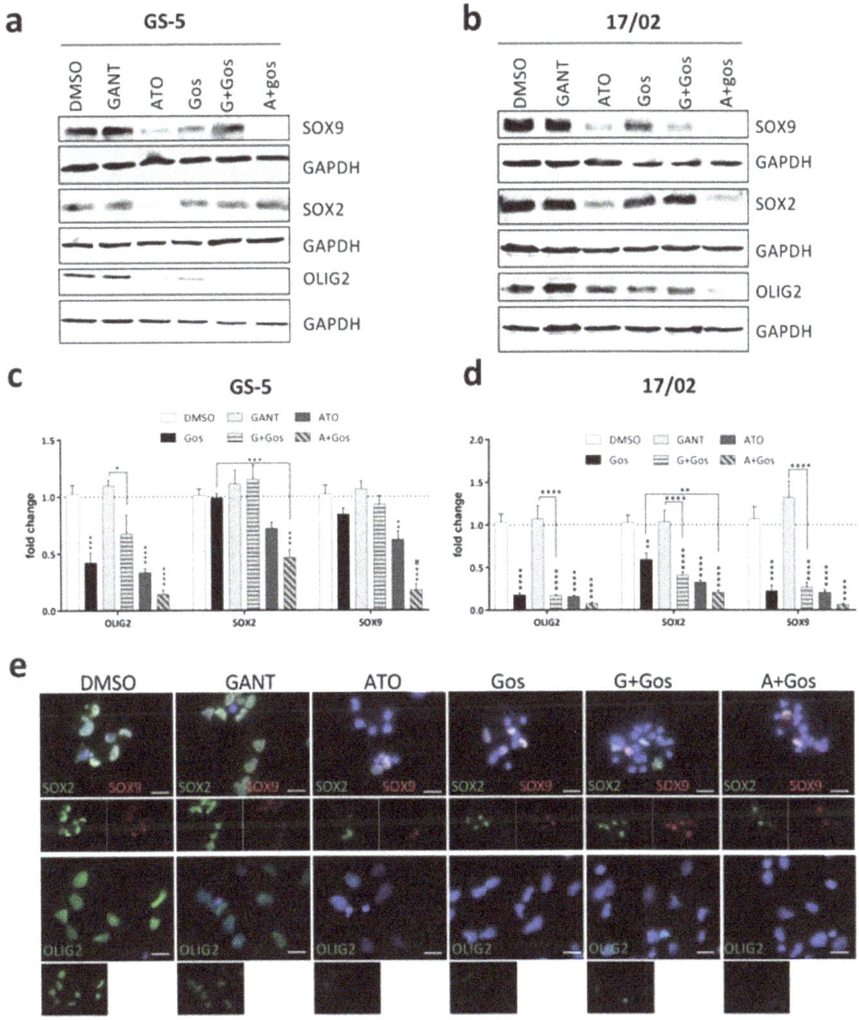

Figure 3. *Cont.*

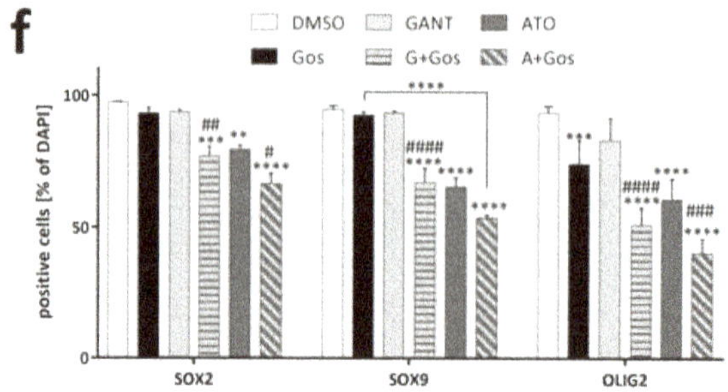

Figure 3. ATO, but not GANT in combination with Gos reduces GSC marker and Hh and Notch pathway activity. (**a**,**b**) Western Blot analysis of (**a**) GS-5 and (**b**) 17/02 after treatment for 24 h with solvent (DMSO), 7.5 μM Gos, 5 μM GANT, 5 μM ATO or a combination of Gos with GANT (GANT + Gos) or ATO (ATO + Gos). (**c**,**d**) Bar graph of Taqman-based gene expression analysis of GS-5 cells after treatment with 5 μM Gos, 2.5 μM GANT or ATO or the combination of GANT or ATO with Gos (G+Gos; A+Gos). (**e**) Representative images of immunofluorescent microscopy of GS-5 72h after treatment as in (**c**) of the stemness markers SOX2 (green) and SOX9 (red) (upper row) and OLIG2 (green; lower row); scale bar: 20 μm. (**f**) Summarized quantification of three independent experiments; at least 100 cells per condition and experiment were counted. All experiments were performed at least 3 times; gene expression data was further performed in 3 biological replicates. * $p < 0.05$; ** $p < 0.01$; *** $p < 0.001$; **** $p < 0.0001$. # $p < 0.05$; ## $p < 0.01$; ### $p < 0.001$; #### $p < 0.0001$. against both single treatments One-way ANOVA followed by Tukey Post-Hoc-Test (GraphPad Prism 7).

2.4. ATO and Gos Treatment Induces DNA Damage Via Downregulation of DDR Genes

A key hallmark of GSC is their treatment resistance towards conventional chemotherapy by enhanced DNA repair, which is in part facilitated by overexpression of CHK1 and CHK2 [7]. Interestingly, CHK1 was significantly decreased according to our proteomic data. This finding prompted us to analyze additional key targets involved in the DNA damage response (DDR) including *CHK1*, *CHK2*, *RAD51* and Survivin (*BIRC5*) (Figure 4a). Survivin is a known downstream target of both Hh [28] and Notch signaling [29] and has been linked to DNA damage [30] and radiation resistance of GBM [31]. Accordingly, we observed that GANT/Gos and ATO/Gos significantly decreased *CHEK1*, *CHEK2*, RAD51 Recombinase (*RAD51*) and *BIRC5* (Survivin) expression, while ATO/Gos also decreased *BRAT1* and Ataxia Telangiectasia Mutated (*ATM*) expression. Similary, CHK1, CHK2 and Survivin protein levels were also decreased upon treatment, most pronouncedly after ATO and ATO/Gos treatment. This was also accompanied by loss of CHK1 phosphorylation (Figure 4b), while phosphorylated CHK2 could not be detected. Since functional loss of CHK1 and CHK2, two main effectors of the DDR might lead to accumulating DNA damage, finally resulting in cell death, we next analyzed DNA damage induction after drug treatment by immunofluorescent microscopy of TP53BP1- and γH2AX-positive foci (Figure 4c–e).

All single treatments significantly increased the number of TP53BP1- (Figure 4c) and γH2AFX-positive foci (Figure 4d) in GS-5, which could even be increased using the combination treatment. Of note, the increase in γH2AFX foci did not reach statistical significance for Gos and GANT alone.

Strikingly, the amount of TP53BP1-positive foci of the combination treatment is significantly higher than either single treatment, indicative of synergism. As a visual control for DNA damage/foci induction the cells were also treated with Etoposide (Figure 4e), a known inducer of DNA damage. Similar findings were also observed in GS-3 (Figure S6a,b), while GS-8 only showed detectable induction of DNA damage after ATO and ATO/Gos treatment.

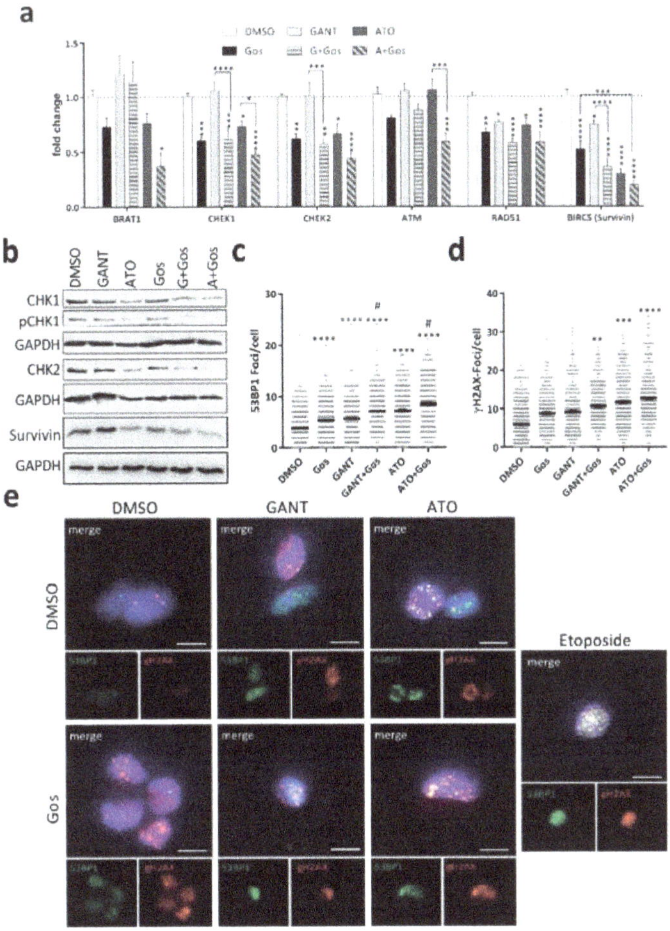

Figure 4. GANT or ATO in combination with Gos synergistically induces DNA damage in GSC via downregulation of CHK1, CHK2 and Survivin. (**a**) Taqman-based gene expression analysis of GS-5 after treatment with 5 µM Gos, 2.5 µM GANT or ATO or the combination of GANT or ATO with Gos (G + Gos; A + Gos). (**b**) Western Blot analysis of GS-5 after treatment for 24 h with solvent (DMSO), 7.5 µM Gos, 5 µM GANT, 5 µM ATO or a combination of Gos with GANT (GANT + Gos) or ATO (ATO + Gos). (**c**,**d**) Dot-Plots of (**c**) TP53BP1 (green)- or (**d**) γH2AFX (red)-positive foci per nucleus of GS-5 24 h after treatment with 5 µM Gos, 3 µM GANT, 2.5 µM ATO or a combination of GANT and Gos (GANT + Gos) or ATO and Gos (ATO + Gos). Each point represents the number of foci per nucleus. Representative pictures are presented in (**e**); scale bar: 20 µm. The experiments were performed at least 3 times in triplicates. For each replicate at least 20 nuclei were counted. Statistics in (**c**,**d**) were performed using the mean value of each replicate. * $p < 0.05$; ** $p < 0.01$; *** $p < 0.001$; **** $p < 0.0001$ against solvent; # $p < 0.05$ against both single treatments. One-way ANOVA followed by Tukey Post-Hoc-Test (GraphPad Prism 7).

2.5. Effects of ATO and Gos on Sphere Forming Capacity and Stem-Cell Frequency of GSCs

Another key hallmark of GSCs is the ability to form new spheres from single cells in vitro [32]. Furthermore, our proteomic analyses clearly showed that multiple GO-terms related to neuronal differentiation and development are enriched among the decreased proteins following ATO/Gos treatment. In order to functionally test whether the treatment indeed reduces stemness properties, we performed limiting dilution assays (LDA; Figure 5a,b) and extended sphere formation assays (SFA [33]; Figure 5c–f) (for a schematic overview see Figure 5c).

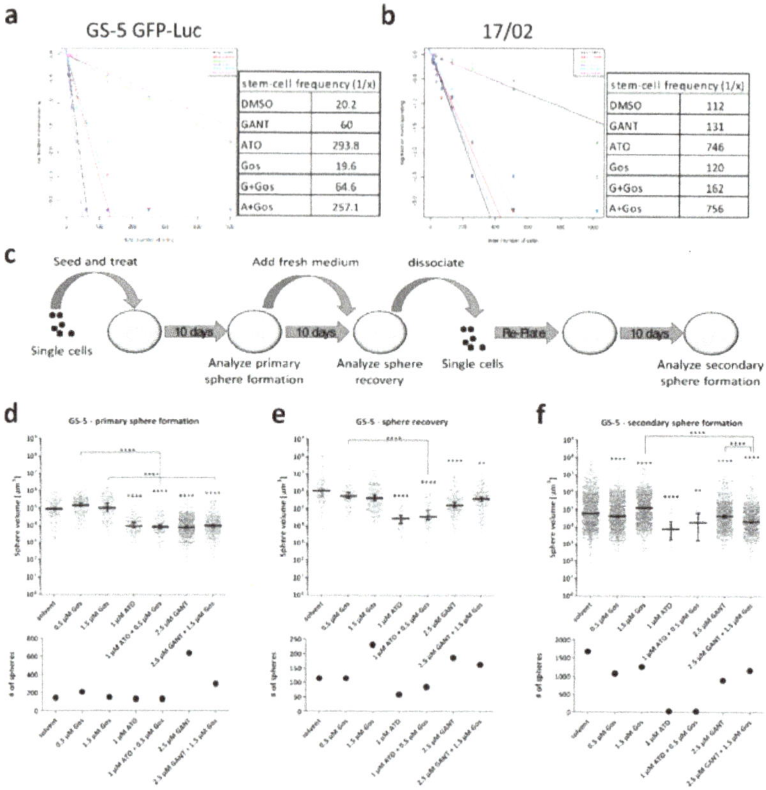

Figure 5. ATO and Gos treatment reduces stemness properties of GS-5/GS-5 GFP-Luc and 17/02 in vitro. (**a**,**b**) Extreme limiting dilution analysis [34] of (**a**) GS-5 GFP-Luc and (**b**) 17/02 7 days after seeding and treatment with 1 μM ATO, 2.5 μM GANT61 or 0.5 μM Gos alone or in combinaton of increasing cell numbers (4 to 500 cells for GS-5 and 8 to 1024 for 17/02). The stem cell frequency calculated by ELDA software [34] is depicted next to the graph. (**c**) Scheme of the extended SFA. (**d**–**f**) Dot-Plots displaying the (upper row) sphere volume and (lower row) total number of spheres of GS-5 spheres after measurement of primary sphere formation, sphere recovery and secondary sphere formation. Lines in (**d**–**f**) are the median ± 95% confidence intervals. Limiting dilution assays were performed three times in 12 biological replicates. Sphere formation assays were performed twice in triplicates and three vision fields were analyzed for each biological replicate. * $p < 0.05$; ** $p < 0.01$; *** $p < 0.001$; **** $p < 0.0001$. One-way ANOVA followed by Tukey Post-Hoc-Test (GraphPad Prism 7).

The LDA revealed that treatment with ATO and ATO/Gos treatment reduced the stem-cell fraction about 15- and 7-fold in GS-5 and 17/02, respectively, whereas GANT or GANT/Gos treatment only reduced the stem-cell fraction 3-fold in GS-5 and only slightly in 17/02. In contrast, Gos did not change the ratio of sphere-forming vs total cells. GS-8 reacted similar to GS-5 (Figure S7).

Next, we performed the extended SFA (Figure 5d–f). We determined primary sphere formation (Figure 5d), sphere recovery (Figure 5e) and secondary sphere formation (Figure 5f) of GS-5 cells after treatment with lower doses of the drugs (1 μM ATO, 2.5 μM GANT and 0.5 μM Gos). Treatment with Gos did not change primary sphere formation or sphere recovery. GANT or GANT/Gos led to smaller spheres, but did not reduce sphere number. In fact, we observed more primary spheres after GANT and GANT/Gos treatment. In comparison, ATO treatment was far more effective as it resulted in significantly smaller and fewer spheres, even in the secondary sphere formation assay. Similar findings (no to low effects of Gos, moderate effects of GANT, strong effects of ATO) were also observed in GS-3 (Figure S8a–c) and GS-8 (Figure S8d–f). Here we could show that ATO/Gos strongly decreases the stem-cell frequency and disturbs sphere formation and recovery.

2.6. *ATO and ATO/Gos Treatment Inhibits Tumor Growth in Adult Organotypic Slice Cultures*

To finally test the proposed therapeutic approach in a more complex and pathophysiological relevant model, we used adult organotypic brain slices (OTC), transplanted GFP-Luc-positive GS-5 (Figure 6), GS-8 (Figure S9) or CellTracker CM Dil-labeled 17/02 (Figure S9) cells onto these OTCs and analyzed tumor growth under treatment over time. We observed that GANT or Gos single agent treatment of GS-5 tumors partially affected the growth of the tumors, but did not cause any reduction of tumor sizes over time. The GANT/Gos treatment resulted in a fast (within 6 days) decrease of tumor sizes, and these effects remained stable until the end of the experiment (28 days). In contrast, treatment with ATO, and even more profoundly with ATO/Gos, continuously decreased the tumor sizes until only small tumor cell populations remained, while other tumors disappeared entirely (Figure 6a, white arrows).

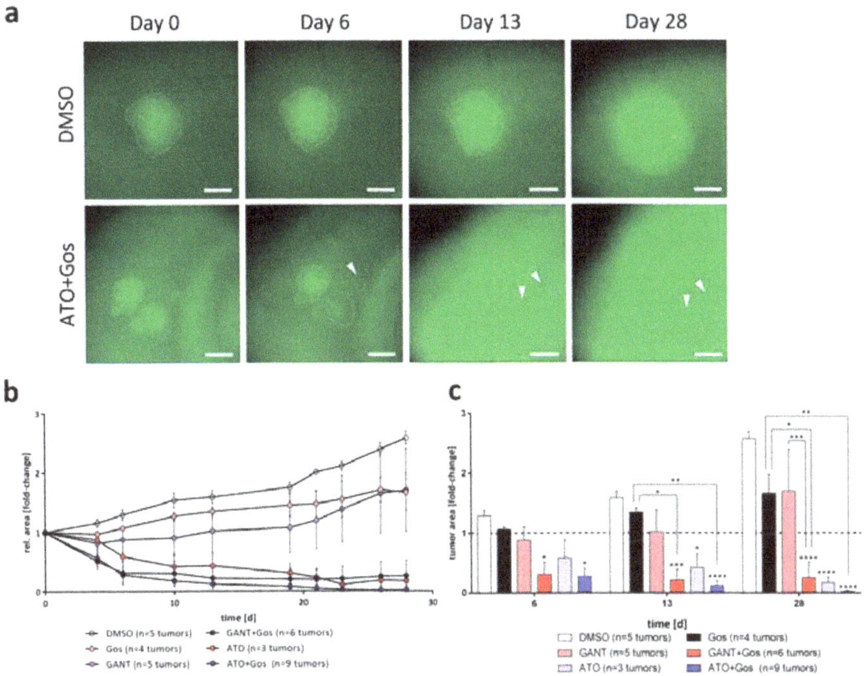

Figure 6. ATO, ATO/Gos and GANT/Gos decrease tumor size in adult OTC transplantation experiments. (**a**) Representative images of tumors of GS-5 GFP-Luc grown on adult OTCs after treatment with solvent (DMSO) or ATO/Gos. Within 6 days some tumors treated with ATO/Gos show complete elimination (white arrows), while others decrease in size; scale bar: 500 µm. (**b**) Growth kinetic of tumors over time after treatment with solvent (DMSO, white), 2.5 µM GANT (pink), 2.5 µM ATO (light blue), 5 µM Gos (black), GANT/Gos (red) or ATO/Gos (blue). (**c**) Bar graph for selected time point displaying the mean (+ SEM) tumor size. * $p < 0.05$; ** $p < 0.01$; *** $p < 0.001$; **** $p < 0.0001$ compared to DMSO for each time-point. One-way ANOVA followed by Tukey Post-Hoc-Test (GraphPad Prism 7).

The GS-8 tumors were more sensitive towards ATO and ATO/Gos, while only showing a growth retardation using GANT and GANT/Gos. Finally, we observed that the patient-derived primary culture 17/02 shows growth retardation after treatment with ATO, Gos and GANT/Gos, and tumor sizes remained almost at baseline after ATO/Gos treatment. It should be emphasized that the primary culture 17/02 was derived from a heavily pre-treated patient (see materials and methods), likely associated with an aggressive phenotype and high therapy resistance of the corresponding tumor. Taken together we can show that ATO and the ATO/Gos drug combination can effectively reduce tumor growth or even eliminate tumors in the absence of observable detrimental effects in the surrounding brain tissue in an organotypic environment.

3. Discussion

Cancer recurrence usually results in more aggressive and treatment-resistant disease and finally in the death of the patient. Since diffusely infiltrating GBMs cannot be completely resected and a fraction of residual cancer cells are thought to exhibit or obtain a stem-like [35] phenotype, disease recurrence is virtually unavoidable. Previous research showed that when GSCs are enforced to differentiate through cultivation in fetal calf serum (FCS) containing media, they are more susceptible to conventional chemotherapy [36]. Hence, a proposed strategy for eliminating GSCs is to enforce their differentiation. Thus, we inquired whether small molecule-mediated inhibition of the Hh signaling pathway, which has been described as a major driver of GSCs in several studies [9,37], is sufficient for depleting this cell population in combination with the natural anticancer agent AT-101/Gossypol (Gos) [17], or if additional inhibition of other pathways implicated in therapy resistance of GBM, in particular Notch-signaling [13,33], does so more efficiently.

Here, we show for the first time that although the Hh inhibitor GANT and cell-death inducer Gos synergistically induce cell death, only combined Hh/Notch inhibition using ATO in combination with Gos is able to efficiently target multiple GSC lines and two primary GBM cultures in vitro and ex vivo. Other groups have described Hh inhibition alone to be sufficient for targeting GSCs [9,37]. However, we did not observe a robust depletion of GSC-properties by the Hh inhibitor GANT in our lines. Nonetheless, we regularly observed minor effects of single agent treatment with GANT, including a reduction of cell viability, induction of cell death and DNA damage, as well as depletion of sphere forming cells, as described by other groups using genetic models of Hh inhibition [9,37,38]. How could these rather moderate effects of GANT in our experiments be explained? Notably, one major aim of this study was to assess the synergy of combination treatments, which prompted us to use lower concentrations of all drugs when used as single agent treatments. Although Hh/GLI-signaling has been described to be essential for GBM [39], so far only one study reported the use of GANT [40] using FCS-cultured GBM cells. In the study by Li et al. [40] relatively high concentrations (up to 10 μM) and treatment periods (up to 72 h) were applied. Considering that GANT61 requires a hydrolysis step to yield the biologically relevant form of the drug [41], it is also possible that this step is disturbed or can even be reversed in glioma stem-like cells. This proposition should be analyzed in future studies. However, considering that ATO alone already depleted the GSC population in this and other studies [12] quite effectively, while reducing the expression of known Hh and Notch target genes, we conclude that the ATO/Gos combination is superior to GANT/Gos and that ATO is the more potent, although less specific, Hh-pathway inhibitor. In any case, our data supports our main hypothesis, i.e., that multiple intertwined signaling events need to be targeted to effectively eliminate GSCs.

Our proteomic data suggested the disturbance of processes related to cell movement, highlighted by decreased levels of several proteins involved in glioma migration, such as CD9 [23] and EPHB3 [42] after treatment with Gos or ATO/Gos, as well as EPHA2 after treatment with GANT/Gos. More prominently, we also found processes related to the cell cycle/DDR and to neuronal development to be disturbed. These changes indicate a reduced capability of tumor invasiveness/migration. Additionally, the proteomic analysis revealed an increase of many proteins involved in cellular respiration (GO.0045333) and the response to oxidative stress (GO.006979) after

GANT/Gos and ATO/Gos treatment. In line with these observations, Gos [43], ATO [44] and GANT61 [45] have been described to impair mitochondrial function and to induce ROS-formation in various settings.

Based on our proteomic data, we next analyzed whether the expression of key mRNAs/proteins of the DNA damage response may be altered following drug treatment. Indeed, we could show a pronounced reduction of CHK1/*CHEK1*, CHK2/*CHEK2* and *RAD51* expression after both GANT/Gos and ATO/Gos, and additionally of *ATM*, *BRAT1* and *FANCI* after ATO/Gos combination treatment. The facts that (1) an enhanced DNA repair capacity is an established hallmark of GSCs [7] and (2) we observed increased DNA damage in three GSC lines with the concurrent reduction in DDR proteins/genes lend further support to the notion that the ATO/Gos combination efficiently depletes stem-like cells by reversing their phenotype and increasing their susceptibility for cell death stimulation. Notably, a central role of Hh (reviewed in [46]) and Notch signaling [47] in regulating the DDR has been proposed. Accordingly, inhibition of Notch1 increased the radiosensitivity of TALL-1 T-cell acute lymphoblastic leukaemia cells in vitro [47] and inhibition of Hh signaling in U87 glioma cells [48] increased their radiosensitivity. Interestingly, Gos has been shown to induce ROS-mediated DNA damage in cancers such as gastric cancer [49], where it also disturbed self-renewal of CD133-positive cells. Furthermore, Gos induced DNA damage and cell death in prostate cancer cell lines, stem-like cells and in vivo xenografts [19]. Our findings further imply that the ATO/Gos treatment sensitizes GSCs to chemo- and/or radiation therapy. Thus, future research should also be directed towards analyzing the combination of ATO/Gos with ionizing radiation, preferably in an in vivo setting.

Most importantly, we detected reduced expression of the stemness marker OLIG2, SOX2 or SOX9, after ATO/Gos treatment in two GS lines and two primary patient-derived cultures. The GANT/Gos combination did not result in such pronounced effects compared to ATO/Gos. These data suggest that dual inhibition of Hh and Notch signaling is superior to Hh inhibition alone, indicating that the stem-like phenotype is maintained via input from multiple signaling pathways. Our data also support the notion that ATO enforces differentiation of GSCs, while Gos seems to be dispensable for limiting stemness as it did not reduce stem-cell frequency or sphere formation in the phenotypic assays. Counterintuitively, our proteomic data suggested that Gos reduced developmental processes, whereas ATO did not. This apparent discrepancy may in part be explained by the lower concentrations of ATO used for the proteomic screen (i.e., 2.5 μM) in comparison to the western blot analyses (i.e., 5 μM), although the lower concentration used for the proteomic screen was also applied for the qRT-PCR analyses, where ATO clearly reduced stem marker expression. At the applied concentration (5 μM), Gos reduced proteins related to stemness in the proteomic dataset which could be confirmed in our qRT-PCR and Western Blot analyses. In line with these observations Gos is capable to reduce stem-cell markers in stem-like prostate cancer cells [19]. However, these proteomic changes appear to be insufficient to evoke a phenotypic response, since we observed neither changes in stem-cell frequency nor in sphere formation could be observed. From these data we conclude that ATO is the drug primarily serving to de-differentiate GSCs following combined ATO/Gos treatment, while Gos enhances these effects.

Aside from the effects on differentiation, ATO is also known to regulate a variety of unrelated processes that likely contribute to the depletion of GSCs. Accordingly, Sun et al. showed that ATO induces ROS formation in rat glioma cells thereby inducing apoptosis [50]. In line with these findings, we observed increased DNA damage after ATO or Gos treatment, and even more pronounced after the combination treatment, likely caused by excessive ROS formation [51]. This is also reflected in our proteomic analysis showing increased expression of proteins of the GO-BP-term "response to oxidative stress" (GO.0006979). Conclusively, the most increased protein after ATO and ATO+Gos treatment is heme oxygenase 1 (HMOX-1). Of note, HMOX-1 is believed to mediate ATO-induced mitochondrial damage in rat astrocytes [52]. In fact, we could recently show that HMOX-1 is induced after Gos-mediated mitochondrial dysfunction in non-GSC glioma cells and that this is accompanied by induction of BNIP3 and Nix (BNIP3L) [53], leading to an autophagic type of cell death. Strikingly, a similar mechanism (BNIP3/BNIP3L-mediated autophagic cell death) has also been proposed for ATO in non-GSC glioma cells [54], indicating that Gos and ATO might target overlapping pathways to promote cell demise of GSCs.

Following this line of thougt, we also observed an increase of proteins of the GO-BP term "response to unfolded protein" (GO.0006986). Interestingly, these alterations may in part reflect the binding properties of ATO to target proteins via thiol-adduction of cysteine-residues. For example, this has been shown in acute promyelocytic leukemia (APL), where ATO is already used in the clinic, and binds to PML and the PML-RARα fusion protein that is found in almost all APL-cells [55]. This binding is thought to disturb the correct protein folding and function, thereby inducing target degradation via induction of the unfolded-protein response [56,57]. ATO was additionally shown to bind to thioredoxins and glutathione, proteins that are involved in cellular respiration and the antioxidant defense, respectively (reviewed in [58]), which may further contribute to ROS formation and cell death.

Notch signaling was shown to enhance Hh signaling in neural stem cells by modulating the ciliary trafficking of the canonical Hh pathway component Smoothened (Smo) by regulating the presence of Patched [59]. Thus, the blockade of Notch through ATO might target Hh signaling by multiple mechanisms: either directly at the level of Gli and/or more upstream by preventing ciliary entry of Smo. Accordingly, we detected an enrichment of the GO-term "regulation of cell projection organization" (GO.0034344) after treatment with ATO/Gos, indicative of reduced ciliary formation. In analogy to our findings, it was shown that combined Hh and Notch inhibition using the archetypical Smo inhibitor cyclopamine in combination with the gamma-secretase inhibitor GSI-1 induced cell death in CD133-enriched glioma cells and increased the cytotoxicity of TMZ [60]. Strikingly, Gos has also been shown to indirectly inhibit Notch (and Wnt) signaling via targeting of the translational activator and repressor Musashi-1 in colon carcinoma cell lines and nude mice xenografts [18]. Similarly, it has been shown using transgenic mice that Notch and Hedgehog signaling regulate a chemotherapy-resistant and possibly stem-like subpopulation of cells in prostate cancer that can be re-sensitized by targeted therapy [61]. Based on our findings, we propose that combined Hh/Notch inhibition strongly reduces the stemlike phenotype of GS cells, which is further evidenced by the ~15-fold reduction of stem-cell frequency in GS-5 and GS-8 and 7~-fold reduction in 17/02. Our ex vivo experiments provide further support for the possible clinical potential of the ATO/Gos combination in limiting tumor growth in an organotypic environment. Moreover, (1) ATO is already used in the clinic to treat APL, (2) is tested in 5 clinical trials for glioma treatment (clinicaltrial.gov; NCT00095771; NCT00720564; NCT00045565; NCT00185861 and NCT00275067) and (3) is known to cross the blood-brain-barrier [62,63]. Gos has already been tested in a clinical trial with measurable, but not significant effects [64], which might be due to its low penetration of the blood brain barrier [65]. Future research should therefore address if the ATO/Gos combination can be applied to murine models and/or if Gos can be substituted with other cell-death inducing drugs that can better permeate the blood-brain barrier.

In summary, it seems likely that ATO evokes a multifactorial cellular response consisting of stemness-specific (i.e., targeting the expression of stemness marker genes and inhibition of driver pathways like Hh and Notch) and general cell death-inducing (i.e., enhanced DNA damage through ROS-induction) mechanisms that are further enhanced and/or complemented by Gos (e.g., induction of mitochondrial damage), finally leading to the effective depletion of GSCs. Importantly, we could show that ATO (± Gos) leads to an almost complete abrogation of sphere forming capacity, indicating grossly reduced self-renewal capacity of GSCs. In conclusion, the combination of ATO and Gos has the potential to specifically target brain tumor stem cell populations and should be further investigated, especially in regard to the multitude of cancer hallmarks that are addressed by this treatment.

4. Materials and Methods

4.1. Cells and Cell Culture

The glioma stem-like cells (GS) with a restricted phenotype (GSr) GS-1 and with the full stem-like phenotype (GSf) GS-3, GS-5 and GS-8 were a kind gift from Kathrin Lamszus (UKE Hamburg, Germany) and have been described previously [66]. GS-5 and GS-8 were transduced with a construct for the stable expression of GFP/Luciferase as described previously [67]. The primary GS-lines were prepared as described previously [68]. 17/01 is from a 51 year old female, 17/02 is from a recurrent tumor of a 60 year old male. The initial tumor was treated with percutaneous radiotherapy (60 Gy) followed by four cycles of adjuvant TMZ chemotherapy followed by further cycles of TMZ therapy with reduced dosages. The recurrent tumor was first treated with fractionated irradiation (36 Gy) and subtotally resected. From this resection, the line was established. Both tumors were classified as GBM (IDH1-neg.). Tumor samples were obtained after patients gave informed consent. The University Cancer Center (UCT; Universitäres Centrum für Tumorkrankheiten Frankfurt) of the University Hospital Frankfurt provided the biomaterial after approval of the local ethics committee (SNO-12-2016). An STR profile was generated for all GS-cell lines using the multiplex STR system for human identification Power Plex 21® from Promega (Mannheim, Germany) performed by Genolytic GmbH (Leipzig, Germany) in 2018, except GS-8 GFP-Luc. All cells display a different genotype and were positively compared to DNA from the original tumor or from very-low passage cells (kindly provided by K. Lamszus (Department of Neurological Surgery, UKE Hamburg, Germany) except GS-5 GFP-Luc, which is derived from GS-5 and harbors the same genotype as GS-5. All GS-cells were kept in Neurobasal A Medium (Gibco, Darmstadt, Germany) containing B27 Supplement (Gibco), 100 U/mL Penicillin 100 µg/mL Streptomycin (Gibco), 1× Glutamax (Gibco), 1× B27 Supplement (Gibco) and 20 ng/mL epidermal growth factor (EGF, Peprotech, Hamburg, Germany) and fibroblast growth factor (FGF, Peprotech). HEK-293-T, HEK293-Shh and Shh light II were cultured as described [22]. The GS-cells were dissociated using Accutase (Sigma-Aldrich, Taufkirchen, Germany); all other cells using Trypsin/EDTA (Gibco) to create a single cell suspension prior to seeding. All cells are tested monthly for mycoplasma using the PCR Mycoplasma Test Kit II (AppliChem, Darmstadt, Germany) according to the manufacturer's instructions.

4.2. Compounds

Arsenic trioxide (As_2O_3, ATO; Sigma-Aldrich) was solved in 1 M NaOH, diluted with PBS (Gibco) to 0.5 M and solved at 80 °C while stirring. The solution was than sterile filtrated and diluted to 1 mM for long-term storage. An intermediate dilution of 15 mM was used for all consecutive dilutions. AT-101 ((−)-Gossypol, Bio-Techne GmbH, Wiesbaden–Nordenstadt, Germany), GANT61 (Sigma-Aldrich) and etoposide (Enzo Life Sciences, Lörrach, Germany) were diluted in DMSO (Carl Roth GmbH, Karlsruhe, Germany).

4.3. SDS-PAGE and Western Blot

Western Blotting was carried out as described previously [69]. After blocking with 5% BSA/TBS-Tween 20 (TBS-T) or 5% Milk/TBS-T the primary antibodies were incubated overnight in 5% BSA/TBS-T at 4 °C, while secondary goat anti-mouse, goat anti-rabbit or donkey anti-goat antibodies (dilution 1:10,000, LI-COR Biosciences, Bad Homburg, Germany) were incubated at room temperature for 1 h. Detection was achieved using a LI-COR Odyssey reader (LI-COR Biosciences).

4.4. Immunofluorescence Microscopy

For immunofluorescent microscopy of GS-5 10,000 to 12,000 cells were seeded on Laminin-coated 8-well chamber slides (Falcon, Corning, Amsterdam, NY, USA). Laminin-coating (10 µg/mL, sigma, L2020) was performed at 4 °C overnight. One day after seeding the cells were treated as indicated and fixed with 4% paraformaldehyde for 20 min at RT. The slides were washed with TBS-Tween

(0.1%; TBS-T), blocked with 4% BSA in TBS with 0.3% Triton X-100 for 1 h at RT and primary antibody incubation occurred at 4 °C overnight. Hereafter, the slides were washed at least three times with TBS-T and secondary antibody was diluted 1:500 in TBS-T and incubated for 1 h at RT. After an additional wash with TBS-T the slides were mounted with DAPI containing Immunoselect antifading mounting medium (Dianova, Hamburg, Germany) or Fluoroshield with DAPI (Thermo Fisher, Frankfurt, Germany). Images were acquired with an Eclipse TS100 inverted fluorescence microscope (Nikon, Düsseldorf, Germany) operated by NIS Elements AR software (version 3.22, Nikon) or an AxioImager Z1 (Carl Zeiss, Jena, Germany).

4.5. Antibodies

The following antibodies and dilutions were used: Sox9 (#ab185966, abcam, Cambridge, UK) 1:5000 for western blot (WB); 1:500 for immunofluorescence (IF); Sox2 (#MAB2016, R&D Systems, Wiesbaden, Germany) 1:1000 (WB) and 1:250 (IF); Olig2 (#AF2418, R&D Systems) 1:10,000 (WB) and 1:5000 (IF); GAPDH (#CB1001, Calbiochem, Darmstadt, Germany) 1:20,000; CHK1 (#2360, Cell Signaling Technologies (CST), Frankfurt am Main, Germany) 1:1000; phosphoCHK1 (CST #2348), 1:1000; CHK2 (CST #2662S) 1:1000; Survivin (R&D #AF886) 1:1000; TP53BP1 (NB #100-304, Novus Biologicals, Wiesbaden, Germany) 1:1000; γH2AFXSer139 (#05-636, clone JBW301, Merck Millipore, Darmstadt, Germany) 1:1000; F(ab′)2-Goat anti-Rabbit IgG (H + L) Cross-Adsorbed Secondary Antibody, Alexa Fluor 488 (A-11070, Thermo Fisher) 1:500; F(ab′)2-Goat anti-mouse IgG (H + L) Cross-Adsorbed Secondary Antibody, Alexa Fluor 594 (A-11020, Thermo Fisher) 1:500; F(ab′)2-donkey anti-goat IgG (H + L) Cross-Adsorbed Secondary Antibody, Alexa Fluor 488 (A-11055, Thermo Fisher) 1:500; donkey anti-goat IgG (sc2042, Santa Cruz, Dallas, TX, USA) 1:10,000.

4.6. Cell-Based Assays

MTT (3-(4,5-Dimethylthiazol-2-yl)-2,5-diphenyltetrazolium bromide) assay was basically performed as described previously [70] and measured on a Tecan Genios or Spark (Tecan, Grödig, Austria) plate reader at 560 nm. MTT (Sigma-Aldrich) was solved in sterile PBS at 5 mg/mL. The combination index (CI) [21] was calculated as described previously [71] using CompuSyn (combosyn.com) software [21]. Cell death was measured using Annexin-V-APC (BD Biosciences, Heidelberg, Germany) and propidium iodide (PI, Merck) as described previously [69] on a BD Accuri C6 (BD Biosciences). To measure the sphere forming ability an extended sphere formation assay (SFA) based on Gilbert et al. [33] was performed. Briefly, 6000 single cells were seeded in 6 well plates in the presence of the compound indicated and measured after 10, 20 and 30 days (for details see Figure 5C). Limiting dilution assay (LDA) was performed as described [72] by seeding serial dilutions of 4 to 500 cells/well for GS-5 GFP-Luc and GS-8 and 8 to 1024 cells/well for 17/02 and analyzed using extreme limiting dilution analysis (ELDA) software [34]. For DNA damage analyses the cells were seeded on laminin-coated (10 µg/mL, Sigma-Aldrich) 8-well-chamber-slides (Corning, Wiesbaden, Germany) and processed as described previously [73]. For GANT-functionality tests using Shh light II Sonic hedgehog conditioned medium (Shh-CM) or control medium (CoM) was prepared as described [71]. For the reporter assay 10,000 Shh light II cells were seeded in lumitrac 600 96-well plates (Greiner bio-one, Frickenhausen, Germany) in culture medium. The next day the medium was replaced with CoM or Shh-CM and after 24 h the cells were treated as indicated in Shh-CM for 48 h. Hereafter a dual-luciferase assay (Promega, Mannheim, Germany) was performed per the manufacturer's instructions.

4.7. Proteomics

For mass-spectrometry-based proteomics the cells were seeded in 6-well plates at 300,000 cells per well in triplicates and treated for 24 h. Sample-preparation was performed as described recently [74] with minor modifications. In brief, a cell lysate was digested in a urea buffer with LysC and Trypsin after reduction and alkylation of disulfide bonds with DTT and iodoacedamide. Tryptic peptides were analysed on a QExactive HF after reversed phase separation on an easy nLC 1200. Data were analysed with MaxQuant and differentially expressed proteins were identified by MaxLFQ quantification [75] applying an FDR of 0.05 and an S0 of 0.2 as cut-offs. The mass spectrometry proteomics data have been deposited to the ProteomeXchange Consortium via the PRIDE [76] partner repository with the dataset identifier PXD009249.

4.8. Taqman-Based qRT-PCR

RNA-Isolation was conducted using the RNeasy Mini Kit (Qiagen, Hilden, Germany) in combination with the QiaShredder Kit or the ExtractMe Total RNA Kit (Blirt S.A., Gdansk, Poland) according to the manufacturer's instruction. cDNA-Synthesis was achieved using the SuperScript III System (Life Technologies, Darmstadt, Germany) according to the manufacturers instruction using 100 U SuperScript per sample. The quantitative Real-Time PCR (qRT-PCR) was performed using 20× Taqman Probes (Applied Biosystems, Darmstadt, Germany) and 2× Fast-Start Universal Probe Master Mix (Roche, Mannheim, Germany) on a StepOne Plus System (Applied Biosystems) using the standard setting. The target gene expression values were normalized to the reference gene TATA-box binding protein (TBP). A list of all FAM-MGB probes is provided as a Supplemental file (Supplement File 1).

4.9. Adult Organotypic Slice Cultures and Ex Vivo Tumor Growth assay

Adult, organotypic brain slices (OTC) from 8–12 week old C56BL/6JOlaHsd mice were prepared as recently described [77]. Briefly, the brains were embedded in 2% low-melting agarose (Carl Roth) and cut on a Vibratome VT1000 (Leica, Wetzlar, Germany) in 250 µm thick transverse sections and cultured on Millicell cell culture inserts (PICMORG50, Merck) in 6-well plates containing OTC-specific, FCS-free medium (DMEM/F12, 1× B27, 1× N2-supplement, 1% P/S; all from Gibco). For the ex vivo tumor growth assay 2000 GFP-Luc positive GS-5 or GS-8 or CellTracker CM Dil-labeled (Molecular Probes, Thermo Fisher, dilution: 1:1000) 17/02 were grown in u-shaped 96-well plates 4–6 days prior to OTC preparation in order to generate large spheres. One day after OTC-preparation multiple spheres were spotted onto the OTCs by careful pipetting. Starting from the next day after spotting (hereafter defined as Day 1) the OTCs were treated as indicated three times per week by exchanging the medium. Tumor growth was monitored using a Nikon SMZ25 stereomicroscope equipped with a P2-SHR Plan Apo 2× objective operated by NIS elements software (version 4.30.02). The tumor area was measured and processed using FIJI [78]. To generate growth curves each tumor was normalized to the size of day 1. All animal experiments comply with the ARRIVE guidelines and were performed in accordance with the German animal protection law authorized by the regional administrative council (Regierungspräsidium Darmstadt, Germany) to the department of medicine (University Hospital Frankfurt).

4.10. Statistics

All statistical analyses applying One-ANOVA followed by Tukey Post-Hoc-Tests were performed using GraphPad Prism 7 (GraphPad Software, La Jolla, CA, USA) basically as described previously [71].

5. Conclusions

Glioblastoma recurrence is in part mediated by so-called glioblastoma stem-like cells (GSCs) that are characterized by expression of stemness marker genes, high Hedgehog and Notch pathway activity and treatment resistance through enhanced DNA repair. Here, we show for the first time that arsenic trioxide-mediated Hedgehog/Notch inhibition in combination with the natural anticancer agent (−)-Gossypol synergistically targets GSCs in part by interfering with DNA double strand break repair by reducing CHEK1 and CHEK2 expression. The combination of ATO and Gos was also proven successful in ex vivo models where it could strongly reduce tumor growth and even eliminate some tumors.

Supplementary Materials: The following are available online at http://www.mdpi.com/2072-6694/11/3/350/s1, Figure S1: Synergistic inhibition of viability of glioma stem-like cells with a restricted and full stem-like phenotype via inhibition of Hh and Notch signaling, Figure S2: Synergistic induction of cell death after combined treatment with GANT or ATO with Gos, Figure S3: Functionality test of GANT61 using Shh light II, Figure S4: Proteomic changes after GANT61 and ATO treatment and Taqman-based validation of proteomic hits, Figure S5: ATO, but not GANT in combination with Gos reduces GSC marker proteins and Hh and Notch pathway activity in GS-8 and 17/01, Figure S6: GANT or ATO in combination with Gos synergistically induce DNA damage in GSC, Figure S7: ATO treatment reduces stemness properties of GS-8 in vitro, Figure S8: ATO and Gos treatment reduces stemness properties of GS-3 and GS-8 in vitro, Figure S9: ATO and ATO/Gos decrease tumor size in adult OTC transplantation experiments of GS-8 GFP-Luc and 17/02, Supplemental File 1: List of all used TAQMAN-Probes, A list displaying all used TAQMAN-probes (Applied Biosystems) including the official gene name and the Assay ID, Supplemental File 2: Significantly changed proteins of the proteomic data set of GS-5, A list displaying the significantly decreased and increased proteins after treatment with 2.5 μM GANT, 2.5 μM ATO, 5 μM Gos or a combination of GANT and Gos or ATO and Gos. The list shows the protein names, the official gene name the –log *p* value and the log2 difference against solvent treated cells, Supplemental File 3: Enriched GO-terms of the significantly changed proteins of the proteomic data set, A list displaying the GO pathway ID, the pathway description, the observed gene count, false discovery rate and the gene names of the proteins after bioinformatics processing of the proteomic data set (Supplement File 2) using the string platform (string-db.org), Supplemental File 4: Comparison of the protein lists of GS-5 after treatment with Gos or GANT + Gos or ATO + Gos, A list displaying the gene names of the proteins that are only changed after treatment with Gos, GANT + Gos or ATO + Gos or are changed after two treatment (Gos AND GANT + Gos; Gos AND ATO/Gos; GANT/Gos AND ATO/Gos) or after all three treatments, Supplemental File 5: Enriched GO-terms of the overlapping proteins after Gos + GANT/Gos + ATO/Gos treatment and after GANT/Gos and ATO/Gos, A list displaying the GO pathway ID, the pathway description, the observed gene count, false discovery rate and the gene names of the proteins after bioinformatics processing of the proteomic data set (Supplement File 2) using the string platform (string-db.org) of the proteins that are regulated after both GANT/Gos and ATO/Gos treatment or after Gos and GANT/Gos and ATO/Gos treatment.

Author Contributions: Conceptualization, B.L. and D.K.; Data curation, B.L. and F.B.; Formal analysis, B.L., A.W. and F.B.; Funding acquisition, I.D. and D.K.; Investigation, B.L., A.W., S.H. and F.B.; Methodology, B.L., A.W., S.H. and F.B.; Project administration, B.L. and D.K.; Resources, S.H., F.B., I.D., F.R., V.S. and D.K.; Supervision, B.L. and D.K.; Validation, B.L.; Visualization, B.L., A.W. and S.H.; Writing—original draft, B.L., S.H., F.B., F.R. and D.K.; Writing—review & editing, B.L. and D.K.

Funding: This study was supported by the Deutsche Forschungsgemeinschaft (SFB 1177 on selective autophagy) to D.K. and I.D.

Acknowledgments: We would like to thank Gabriele Köpf and Hildegard König for excellent technical assistance. We are also grateful to Michael Meister (Institute for Experimental Cancer Research in Pediatrics, University Hospital, Frankfurt, Germany), who provided us with the initial stock of ATO, Cecile Maire and Kathrin Lamszus (Department of Neurological Surgery, UKE Hamburg, Germany) for helpful insights in GS cell culture specifics and providing of DNA and low-passage cells for STR-genotyping. Additional thank is directed to Patrick Harter, Stefan Liebner and Stefan Momma (Edinger Institute, University Hospital Frankfurt, Germany) for help with primary tumor cultures, the SMZ500 stereomicroscope and with OTC preparation, respectively. We also like to thank Bushra Rais and Stephan Müller from the lab of Evelyn Ullrich (Pediatric stem cell transplantation and Immunology, University Hosptal Frankfurt, Germany) for performing the GFP-Luc transductions of GS cells.

Conflicts of Interest: The authors declare no conflict of interest.

References

1. Louis, D.N.; Perry, A.; Reifenberger, G.; von Deimling, A.; Figarella-Branger, D.; Cavenee, W.K.; Ohgaki, H.; Wiestler, O.D.; Kleihues, P.; Ellison, D.W. The 2016 World Health Organization Classification of Tumors of the Central Nervous System: A summary. *Acta Neuropathol.* **2016**, *131*, 803–820. [CrossRef] [PubMed]
2. Becker, K.P.; Yu, J. Status quo-standard-of-care medical and radiation therapy for glioblastoma. *Cancer J.* **2012**, *18*, 12–19. [CrossRef] [PubMed]
3. Stupp, R.; Hegi, M.E.; Gilbert, M.R.; Chakravarti, A. Chemoradiotherapy in malignant glioma: Standard of care and future directions. *J. Clin. Oncol.* **2007**, *25*, 4127–4136. [CrossRef]
4. Signore, M.; Ricci-Vitiani, L.; De Maria, R. Targeting apoptosis pathways in cancer stem cells. *Cancer Lett.* **2013**, *332*, 374–382. [CrossRef]
5. Bradshaw, A.; Wickremsekera, A.; Tan, S.T.; Peng, L.; Davis, P.F.; Itinteang, T. Cancer Stem Cell Hierarchy in Glioblastoma Multiforme. *Front. Surg.* **2016**, *3*, 21. [CrossRef] [PubMed]

6. Tirosh, I.; Venteicher, A.S.; Hebert, C.; Escalante, L.E.; Patel, A.P.; Yizhak, K.; Fisher, J.M.; Rodman, C.; Mount, C.; Filbin, M.G.; et al. Single-cell RNA-seq supports a developmental hierarchy in human oligodendroglioma. *Nature* **2016**, *539*, 309–313. [CrossRef]
7. Bao, S.; Wu, Q.; McLendon, R.E.; Hao, Y.; Shi, Q.; Hjelmeland, A.B.; Dewhirst, M.W.; Bigner, D.D.; Rich, J.N. Glioma stem cells promote radioresistance by preferential activation of the DNA damage response. *Nature* **2006**, *444*, 756–760. [CrossRef] [PubMed]
8. Ferent, J.; Traiffort, E. Hedgehog: Multiple Paths for Multiple Roles in Shaping the Brain and Spinal Cord. *Neuroscientist* **2015**, *21*, 356–371. [CrossRef]
9. Clement, V.; Sanchez, P.; de Tribolet, N.; Radovanovic, I.; Ruiz i Altaba, A. HEDGEHOG-GLI1 signaling regulates human glioma growth, cancer stem cell self-renewal, and tumorigenicity. *Curr. Biol.* **2007**, *17*, 165–172. [CrossRef]
10. Briscoe, J.; Therond, P.P. The mechanisms of Hedgehog signalling and its roles in development and disease. *Nat. Rev. Mol. Cell Biol.* **2013**, *14*, 416–429. [CrossRef]
11. Bray, S.J.; Gomez-Lamarca, M. Notch after cleavage. *Curr. Opin. Cell Biol.* **2017**, *51*, 103–109. [CrossRef] [PubMed]
12. Zhen, Y.; Zhao, S.; Li, Q.; Li, Y.; Kawamoto, K. Arsenic trioxide-mediated Notch pathway inhibition depletes the cancer stem-like cell population in gliomas. *Cancer Lett.* **2010**, *292*, 64–72. [CrossRef]
13. Stockhausen, M.T.; Kristoffersen, K.; Poulsen, H.S. The functional role of Notch signaling in human gliomas. *Neuro Oncol.* **2010**, *12*, 199–211. [CrossRef] [PubMed]
14. Beauchamp, E.M.; Ringer, L.; Bulut, G.; Sajwan, K.P.; Hall, M.D.; Lee, Y.C.; Peaceman, D.; Ozdemirli, M.; Rodriguez, O.; Macdonald, T.J.; et al. Arsenic trioxide inhibits human cancer cell growth and tumor development in mice by blocking Hedgehog/GLI pathway. *J. Clin. Investig.* **2011**, *121*, 148–160. [CrossRef] [PubMed]
15. Ding, D.; Lim, K.S.; Eberhart, C.G. Arsenic trioxide inhibits Hedgehog, Notch and stem cell properties in glioblastoma neurospheres. *Acta Neuropathol. Commun.* **2014**, *2*, 31. [CrossRef] [PubMed]
16. Kim, J.; Aftab, B.T.; Tang, J.Y.; Kim, D.; Lee, A.H.; Rezaee, M.; Chen, B.; King, E.M.; Borodovsky, A.; Riggins, G.J.; et al. Itraconazole and arsenic trioxide inhibit Hedgehog pathway activation and tumor growth associated with acquired resistance to smoothened antagonists. *Cancer Cell* **2013**, *23*, 23–34. [CrossRef] [PubMed]
17. Voss, V.; Senft, C.; Lang, V.; Ronellenfitsch, M.W.; Steinbach, J.P.; Seifert, V.; Kogel, D. The pan-Bcl-2 inhibitor (−)-gossypol triggers autophagic cell death in malignant glioma. *Mol. Cancer Res.* **2010**, *8*, 1002–1016. [CrossRef] [PubMed]
18. Lan, L.; Appelman, C.; Smith, A.R.; Yu, J.; Larsen, S.; Marquez, R.T.; Liu, H.; Wu, X.; Gao, P.; Roy, A.; et al. Natural product (−)-gossypol inhibits colon cancer cell growth by targeting RNA-binding protein Musashi-1. *Mol. Oncol.* **2015**, *9*, 1406–1420. [CrossRef]
19. Volate, S.R.; Kawasaki, B.T.; Hurt, E.M.; Milner, J.A.; Kim, Y.S.; White, J.; Farrar, W.L. Gossypol induces apoptosis by activating p53 in prostate cancer cells and prostate tumor-initiating cells. *Mol. Cancer Ther.* **2010**, *9*, 461–470. [CrossRef]
20. Lauth, M.; Bergstrom, A.; Shimokawa, T.; Toftgard, R. Inhibition of GLI-mediated transcription and tumor cell growth by small-molecule antagonists. *Proc. Natl. Acad. Sci. USA* **2007**, *104*, 8455–8460. [CrossRef] [PubMed]
21. Chou, T.C. Theoretical basis, experimental design, and computerized simulation of synergism and antagonism in drug combination studies. *Pharmacol. Rev.* **2006**, *58*, 621–681. [CrossRef]
22. Chen, J.K.; Taipale, J.; Young, K.E.; Maiti, T.; Beachy, P.A. Small molecule modulation of Smoothened activity. *Proc. Natl. Acad. Sci. USA* **2002**, *99*, 14071–14076. [CrossRef] [PubMed]
23. Podergajs, N.; Motaln, H.; Rajcevic, U.; Verbovsek, U.; Korsic, M.; Obad, N.; Espedal, H.; Vittori, M.; Herold-Mende, C.; Miletic, H.; et al. Transmembrane protein CD9 is glioblastoma biomarker, relevant for maintenance of glioblastoma stem cells. *Oncotarget* **2016**, *7*, 593–609. [CrossRef] [PubMed]
24. Chen, H.; Yuan, B.; Zheng, Z.; Liu, Z.; Wang, S.; Liu, Y. A novel vaccine containing EphA2 epitope and LIGHT plasmid induces robust cellular immunity against glioma U251 cells. *Cell. Immunol.* **2011**, *272*, 102–106. [CrossRef] [PubMed]
25. Nagaishi, M.; Kim, Y.H.; Mittelbronn, M.; Giangaspero, F.; Paulus, W.; Brokinkel, B.; Vital, A.; Tanaka, Y.; Nakazato, Y.; Legras-Lachuer, C.; et al. Amplification of the STOML3, FREM2, and LHFP genes is associated with mesenchymal differentiation in gliosarcoma. *Am. J. Pathol.* **2012**, *180*, 1816–1823. [CrossRef] [PubMed]

26. Szklarczyk, D.; Morris, J.H.; Cook, H.; Kuhn, M.; Wyder, S.; Simonovic, M.; Santos, A.; Doncheva, N.T.; Roth, A.; Bork, P.; et al. The STRING database in 2017: Quality-controlled protein-protein association networks, made broadly accessible. *Nucleic Acids Res.* **2017**, *45*, D362–D368. [CrossRef] [PubMed]
27. Supek, F.; Bosnjak, M.; Skunca, N.; Smuc, T. REVIGO summarizes and visualizes long lists of gene ontology terms. *PLoS ONE* **2011**, *6*, e21800. [CrossRef] [PubMed]
28. Vlckova, K.; Ondrusova, L.; Vachtenheim, J.; Reda, J.; Dundr, P.; Zadinova, M.; Zakova, P.; Pouckova, P. Survivin, a novel target of the Hedgehog/GLI signaling pathway in human tumor cells. *Cell Death Dis.* **2016**, *7*, e2048. [CrossRef] [PubMed]
29. Chen, Y.; Li, D.; Liu, H.; Xu, H.; Zheng, H.; Qian, F.; Li, W.; Zhao, C.; Wang, Z.; Wang, X. Notch-1 signaling facilitates survivin expression in human non-small cell lung cancer cells. *Cancer Biol. Ther.* **2011**, *11*, 14–21. [CrossRef] [PubMed]
30. Capalbo, G.; Dittmann, K.; Weiss, C.; Reichert, S.; Hausmann, E.; Rodel, C.; Rodel, F. Radiation-induced survivin nuclear accumulation is linked to DNA damage repair. *Int. J. Radiat. Oncol. Biol. Phys.* **2010**, *77*, 226–234. [CrossRef] [PubMed]
31. Chakravarti, A.; Zhai, G.G.; Zhang, M.; Malhotra, R.; Latham, D.E.; Delaney, M.A.; Robe, P.; Nestler, U.; Song, Q.; Loeffler, J. Survivin enhances radiation resistance in primary human glioblastoma cells via caspase-independent mechanisms. *Oncogene* **2004**, *23*, 7494–7506. [CrossRef] [PubMed]
32. Ludwig, K.; Kornblum, H.I. Molecular markers in glioma. *J. Neurooncol.* **2017**, *134*, 505–512. [CrossRef] [PubMed]
33. Gilbert, C.A.; Daou, M.C.; Moser, R.P.; Ross, A.H. Gamma-secretase inhibitors enhance temozolomide treatment of human gliomas by inhibiting neurosphere repopulation and xenograft recurrence. *Cancer Res.* **2010**, *70*, 6870–6879. [CrossRef] [PubMed]
34. Hu, Y.; Smyth, G.K. ELDA: Extreme limiting dilution analysis for comparing depleted and enriched populations in stem cell and other assays. *J. Immunol. Methods* **2009**, *347*, 70–78. [CrossRef] [PubMed]
35. Lan, X.; Jorg, D.J.; Cavalli, F.M.G.; Richards, L.M.; Nguyen, L.V.; Vanner, R.J.; Guilhamon, P.; Lee, L.; Kushida, M.M.; Pellacani, D.; et al. Fate mapping of human glioblastoma reveals an invariant stem cell hierarchy. *Nature* **2017**, *549*, 227–232. [CrossRef] [PubMed]
36. Chen, J.; McKay, R.M.; Parada, L.F. Malignant glioma: Lessons from genomics, mouse models, and stem cells. *Cell* **2012**, *149*, 36–47. [CrossRef] [PubMed]
37. Uchida, H.; Arita, K.; Yunoue, S.; Yonezawa, H.; Shinsato, Y.; Kawano, H.; Hirano, H.; Hanaya, R.; Tokimura, H. Role of sonic hedgehog signaling in migration of cell lines established from CD133-positive malignant glioma cells. *J. Neurooncol.* **2011**, *104*, 697–704. [CrossRef]
38. Takezaki, T.; Hide, T.; Takanaga, H.; Nakamura, H.; Kuratsu, J.; Kondo, T. Essential role of the Hedgehog signaling pathway in human glioma-initiating cells. *Cancer Sci.* **2011**, *102*, 1306–1312. [CrossRef]
39. Santoni, M.; Burattini, L.; Nabissi, M.; Morelli, M.B.; Berardi, R.; Santoni, G.; Cascinu, S. Essential role of Gli proteins in glioblastoma multiforme. *Curr. Protein Pept. Sci.* **2013**, *14*, 133–140. [CrossRef]
40. Li, J.; Cai, J.; Zhao, S.; Yao, K.; Sun, Y.; Li, Y.; Chen, L.; Li, R.; Zhai, X.; Zhang, J.; et al. GANT61, a GLI inhibitor, sensitizes glioma cells to the temozolomide treatment. *J. Exp. Clin. Cancer Res.* **2016**, *35*, 184. [CrossRef]
41. Calcaterra, A.; Iovine, V.; Botta, B.; Quaglio, D.; D'Acquarica, I.; Ciogli, A.; Iazzetti, A.; Alfonsi, R.; Lospinoso Severini, L.; Infante, P.; et al. Chemical, computational and functional insights into the chemical stability of the Hedgehog pathway inhibitor GANT61. *J. Enzyme Inhib. Med. Chem.* **2018**, *33*, 349–358. [CrossRef] [PubMed]
42. Nakada, M.; Hayashi, Y.; Hamada, J. Role of Eph/ephrin tyrosine kinase in malignant glioma. *Neuro Oncol.* **2011**, *13*, 1163–1170. [CrossRef] [PubMed]
43. Ko, C.H.; Shen, S.C.; Yang, L.Y.; Lin, C.W.; Chen, Y.C. Gossypol reduction of tumor growth through ROS-dependent mitochondria pathway in human colorectal carcinoma cells. *Int. J. Cancer* **2007**, *121*, 1670–1679. [CrossRef] [PubMed]
44. Woo, S.H.; Park, I.C.; Park, M.J.; Lee, H.C.; Lee, S.J.; Chun, Y.J.; Lee, S.H.; Hong, S.I.; Rhee, C.H. Arsenic trioxide induces apoptosis through a reactive oxygen species-dependent pathway and loss of mitochondrial membrane potential in HeLa cells. *Int. J. Oncol.* **2002**, *21*, 57–63. [CrossRef] [PubMed]
45. Lim, C.B.; Prele, C.M.; Baltic, S.; Arthur, P.G.; Creaney, J.; Watkins, D.N.; Thompson, P.J.; Mutsaers, S.E. Mitochondria-derived reactive oxygen species drive GANT61-induced mesothelioma cell apoptosis. *Oncotarget* **2015**, *6*, 1519–1530. [CrossRef] [PubMed]

46. Palle, K.; Mani, C.; Tripathi, K.; Athar, M. Aberrant GLI1 Activation in DNA Damage Response, Carcinogenesis and Chemoresistance. *Cancers* **2015**, *7*, 2330–2351. [CrossRef]
47. Vermezovic, J.; Adamowicz, M.; Santarpia, L.; Rustighi, A.; Forcato, M.; Lucano, C.; Massimiliano, L.; Costanzo, V.; Bicciato, S.; Del Sal, G.; et al. Notch is a direct negative regulator of the DNA-damage response. *Nat. Struct. Mol. Biol.* **2015**, *22*, 417–424. [CrossRef] [PubMed]
48. Chiang, M.F.; Chen, H.H.; Chi, C.W.; Sze, C.I.; Hsu, M.L.; Shieh, H.R.; Lin, C.P.; Tsai, J.T.; Chen, Y.J. Modulation of Sonic hedgehog signaling and WW domain containing oxidoreductase WOX1 expression enhances radiosensitivity of human glioblastoma cells. *Exp. Biol. Med.* **2015**, *240*, 392–399. [CrossRef]
49. Wei, X.; Duan, W.; Li, Y.; Zhang, S.; Xin, X.; Sun, L.; Gao, M.; Li, Q.; Wang, D. AT101 exerts a synergetic efficacy in gastric cancer patients with 5-FU based treatment through promoting apoptosis and autophagy. *Oncotarget* **2016**, *7*, 34430–34441. [CrossRef]
50. Sun, Y.; Wang, C.; Wang, L.; Dai, Z.; Yang, K. Arsenic trioxide induces apoptosis and the formation of reactive oxygen species in rat glioma cells. *Cell. Mol. Biol. Lett.* **2018**, *23*, 13. [CrossRef]
51. Cadet, J.; Wagner, J.R. DNA base damage by reactive oxygen species, oxidizing agents, and UV radiation. *Cold Spring Harb. Perspect. Biol.* **2013**, *5*, a012559. [CrossRef] [PubMed]
52. Schipper, H.M. Heme oxygenase-1: Role in brain aging and neurodegeneration. *Exp. Gerontol.* **2000**, *35*, 821–830. [CrossRef]
53. Meyer, N.; Zielke, S.; Michaelis, J.B.; Linder, B.; Warnsmann, V.; Rakel, S.; Osiewacz, H.D.; Fulda, S.; Mittelbronn, M.; Munch, C.; et al. AT 101 induces early mitochondrial dysfunction and HMOX1 (heme oxygenase 1) to trigger mitophagic cell death in glioma cells. *Autophagy* **2018**, *14*, 1693–1709. [CrossRef]
54. Kanzawa, T.; Zhang, L.; Xiao, L.; Germano, I.M.; Kondo, Y.; Kondo, S. Arsenic trioxide induces autophagic cell death in malignant glioma cells by upregulation of mitochondrial cell death protein BNIP3. *Oncogene* **2005**, *24*, 980–991. [CrossRef] [PubMed]
55. Zhang, X.W.; Yan, X.J.; Zhou, Z.R.; Yang, F.F.; Wu, Z.Y.; Sun, H.B.; Liang, W.X.; Song, A.X.; Lallemand-Breitenbach, V.; Jeanne, M.; et al. Arsenic trioxide controls the fate of the PML-RARalpha oncoprotein by directly binding PML. *Science* **2010**, *328*, 240–243. [CrossRef] [PubMed]
56. Srivastava, R.K.; Li, C.; Chaudhary, S.C.; Ballestas, M.E.; Elmets, C.A.; Robbins, D.J.; Matalon, S.; Deshane, J.S.; Afaq, F.; Bickers, D.R.; et al. Unfolded protein response (UPR) signaling regulates arsenic trioxide-mediated macrophage innate immune function disruption. *Toxicol. Appl. Pharmacol.* **2013**, *272*, 879–887. [CrossRef]
57. Weng, C.Y.; Chiou, S.Y.; Wang, L.; Kou, M.C.; Wang, Y.J.; Wu, M.J. Arsenic trioxide induces unfolded protein response in vascular endothelial cells. *Arch. Toxicol.* **2014**, *88*, 213–226. [CrossRef]
58. Shen, S.; Li, X.F.; Cullen, W.R.; Weinfeld, M.; Le, X.C. Arsenic binding to proteins. *Chem. Rev.* **2013**, *113*, 7769–7792. [CrossRef] [PubMed]
59. Kong, J.H.; Yang, L.; Dessaud, E.; Chuang, K.; Moore, D.M.; Rohatgi, R.; Briscoe, J.; Novitch, B.G. Notch activity modulates the responsiveness of neural progenitors to sonic hedgehog signaling. *Dev. Cell* **2015**, *33*, 373–387. [CrossRef] [PubMed]
60. Ulasov, I.V.; Nandi, S.; Dey, M.; Sonabend, A.M.; Lesniak, M.S. Inhibition of Sonic Hedgehog and Notch Pathways Enhances Sensitivity of CD133(+) Glioma Stem Cells to Temozolomide Therapy. *Mol. Med.* **2011**, *17*, 103–112. [CrossRef]
61. Domingo-Domenech, J.; Vidal, S.J.; Rodriguez-Bravo, V.; Castillo-Martin, M.; Quinn, S.A.; Rodriguez-Barrueco, R.; Bonal, D.M.; Charytonowicz, E.; Gladoun, N.; de la Iglesia-Vicente, J.; et al. Suppression of acquired docetaxel resistance in prostate cancer through depletion of notch- and hedgehog-dependent tumor-initiating cells. *Cancer Cell* **2012**, *22*, 373–388. [CrossRef] [PubMed]
62. Zhou, W.; Cheng, L.; Shi, Y.; Ke, S.Q.; Huang, Z.; Fang, X.; Chu, C.W.; Xie, Q.; Bian, X.W.; Rich, J.N.; et al. Arsenic trioxide disrupts glioma stem cells via promoting PML degradation to inhibit tumor growth. *Oncotarget* **2015**, *6*, 37300–37315. [CrossRef] [PubMed]
63. Kiguchi, T.; Yoshino, Y.; Yuan, B.; Yoshizawa, S.; Kitahara, T.; Akahane, D.; Gotoh, M.; Kaise, T.; Toyoda, H.; Ohyashiki, K. Speciation of arsenic trioxide penetrates into cerebrospinal fluid in patients with acute promyelocytic leukemia. *Leuk. Res.* **2010**, *34*, 403–405. [CrossRef] [PubMed]
64. Bushunow, P.; Reidenberg, M.M.; Wasenko, J.; Winfield, J.; Lorenzo, B.; Lemke, S.; Himpler, B.; Corona, R.; Coyle, T. Gossypol treatment of recurrent adult malignant gliomas. *J. Neurooncol.* **1999**, *43*, 79–86. [CrossRef] [PubMed]
65. Kalla, N.R.; Sud, S. Distribution of gossypol. *Acta Eur. Fertil.* **1990**, *21*, 77–80. [PubMed]

66. Gunther, H.S.; Schmidt, N.O.; Phillips, H.S.; Kemming, D.; Kharbanda, S.; Soriano, R.; Modrusan, Z.; Meissner, H.; Westphal, M.; Lamszus, K. Glioblastoma-derived stem cell-enriched cultures form distinct subgroups according to molecular and phenotypic criteria. *Oncogene* **2008**, *27*, 2897–2909. [CrossRef] [PubMed]
67. Wagner, J.; Pfannenstiel, V.; Waldmann, A.; Bergs, J.W.J.; Brill, B.; Huenecke, S.; Klingebiel, T.; Rodel, F.; Buchholz, C.J.; Wels, W.S.; et al. A Two-Phase Expansion Protocol Combining Interleukin (IL)-15 and IL-21 Improves Natural Killer Cell Proliferation and Cytotoxicity against Rhabdomyosarcoma. *Front. Immunol.* **2017**, *8*, 676. [CrossRef] [PubMed]
68. Lemke, D.; Weiler, M.; Blaes, J.; Wiestler, B.; Jestaedt, L.; Klein, A.C.; Low, S.; Eisele, G.; Radlwimmer, B.; Capper, D.; et al. Primary glioblastoma cultures: Can profiling of stem cell markers predict radiotherapy sensitivity? *J. Neurochem.* **2014**, *131*, 251–264. [CrossRef] [PubMed]
69. Antonietti, P.; Linder, B.; Hehlgans, S.; Mildenberger, I.C.; Burger, M.C.; Fulda, S.; Steinbach, J.P.; Gessler, F.; Rodel, F.; Mittelbronn, M.; et al. Interference with the HSF1/HSP70/BAG3 Pathway Primes Glioma Cells to Matrix Detachment and BH3 Mimetic-Induced Apoptosis. *Mol. Cancer Ther.* **2017**, *16*, 156–168. [CrossRef] [PubMed]
70. Gessler, F.; Voss, V.; Dutzmann, S.; Seifert, V.; Gerlach, R.; Kogel, D. Inhibition of tissue factor/protease-activated receptor-2 signaling limits proliferation, migration and invasion of malignant glioma cells. *Neuroscience* **2010**, *165*, 1312–1322. [CrossRef]
71. Linder, B.; Weber, S.; Dittmann, K.; Adamski, J.; Hahn, H.; Uhmann, A. A Functional and Putative Physiological Role of Calcitriol in Patched1/Smoothened Interaction. *J. Biol. Chem.* **2015**, *290*, 19614–19628. [CrossRef]
72. Rahman, M.; Reyner, K.; Deleyrolle, L.; Millette, S.; Azari, H.; Day, B.W.; Stringer, B.W.; Boyd, A.W.; Johns, T.G.; Blot, V.; et al. Neurosphere and adherent culture conditions are equivalent for malignant glioma stem cell lines. *Anat. Cell Biol.* **2015**, *48*, 25–35. [CrossRef] [PubMed]
73. Hehlgans, S.; Storch, K.; Lange, I.; Cordes, N. The novel HDAC inhibitor NDACI054 sensitizes human cancer cells to radiotherapy. *Radiother. Oncol.* **2013**, *109*, 126–132. [CrossRef]
74. Das, C.K.; Linder, B.; Bonn, F.; Rothweiler, F.; Dikic, I.; Michaelis, M.; Cinatl, J.; Mandal, M.; Kogel, D. BAG3 Overexpression and Cytoprotective Autophagy Mediate Apoptosis Resistance in Chemoresistant Breast Cancer Cells. *Neoplasia* **2018**, *20*, 263–279. [CrossRef] [PubMed]
75. Cox, J.; Hein, M.Y.; Luber, C.A.; Paron, I.; Nagaraj, N.; Mann, M. Accurate proteome-wide label-free quantification by delayed normalization and maximal peptide ratio extraction, termed MaxLFQ. *Mol. Cell. Proteom.* **2014**, *13*, 2513–2526. [CrossRef] [PubMed]
76. Vizcaino, J.A.; Csordas, A.; Del-Toro, N.; Dianes, J.A.; Griss, J.; Lavidas, I.; Mayer, G.; Perez-Riverol, Y.; Reisinger, F.; Ternent, T.; et al. 2016 update of the PRIDE database and its related tools. *Nucleic Acids Res.* **2016**, *44*, 11033. [CrossRef] [PubMed]
77. Remy, J.; Linder, B.; Weirauch, U.; Konovalova, J.; Marschalek, R.; Aigner, A.; Kogel, D. Inhibition of PIM1 blocks the autophagic flux to sensitize glioblastoma cells to ABT-737-induced apoptosis. *Biochim. Biophys. Acta Mol. Cell Res.* **2018**. [CrossRef] [PubMed]
78. Schindelin, J.; Arganda-Carreras, I.; Frise, E.; Kaynig, V.; Longair, M.; Pietzsch, T.; Preibisch, S.; Rueden, C.; Saalfeld, S.; Schmid, B.; et al. Fiji: An open-source platform for biological-image analysis. *Nat. Methods* **2012**, *9*, 676–682. [CrossRef] [PubMed]

cancers

MDPI

Article

Novel *Curcumin Inspired* Bis-Chalcone Promotes Endoplasmic Reticulum Stress and Glioblastoma Neurosphere Cell Death

Lorenzo Sansalone [1], Eduardo A. Veliz [1], Nadia G. Myrthil [2], Vasileios Stathias [2,3,4,5], Winston Walters [2], Ingrid I. Torrens [2], Stephan C. Schürer [3,5], Steven Vanni [6], Roger M. Leblanc [1] and Regina M. Graham [2,3,6,*]

1 Department of Chemistry, University of Miami, 1301 Memorial Drive, Coral Gables, FL 33146, USA; lo.sansalone@gmail.com (L.S.); eav10@miami.edu (E.A.V.); rml@miami.edu (R.M.L.)
2 University of Miami Brain Tumor Initiative, Department of Neurosurgery, University of Miami Miller School of Medicine, 1095 NW 14th Terrace, Miami, FL 33136, USA; ngm17@med.miami.edu (N.G.M.); v.stathias@med.miami.edu (V.S.); wwalters@med.miami.edu (W.W.); iit4@miami.edu (I.I.T.)
3 Sylvester Comprehensive Cancer Center, University of Miami Miller School of Medicine, 1475 NW 12th Ave, Miami, FL 33136, USA; sschuerer@med.miami.edu
4 Center for Therapeutic Innovation, Department of Psychiatry and Behavioral Sciences, University of Miami Miller School of Medicine, 1120 NW 14th St, Miami, FL 33136, USA
5 Department of Molecular and Cellular Pharmacology, Center for Computational Science, University of Miami Miller School of Medicine, 1120 NW 14th St, Miami, FL 33136, USA
6 Department of Neurosurgery, University of Miami Miller School of Medicine, 1095 NW 14th Terrace, Miami, FL 33136, USA; Svanni@med.miami.edu
* Correspondence: rgraham@med.miami.edu; Tel.: +(305)-321-4972

Received: 28 January 2019; Accepted: 7 March 2019; Published: 13 March 2019

Abstract: Glioblastoma (GBM) has a dismal prognosis and successful elimination of GBM stem cells (GSCs) is a high-priority as these cells are responsible for tumor regrowth following therapy and ultimately patient relapse. Natural products and their derivatives continue to be a source for the development of effective anticancer drugs and have been shown to effectively target pathways necessary for cancer stem cell self-renewal and proliferation. We generated a series of curcumin inspired bis-chalcones and examined their effect in multiple patient-derived GSC lines. Of the 19 compounds synthesized, four analogs robustly induced GSC death in six separate GSC lines, with a half maximal inhibitory concentration (IC50) ranging from 2.7–5.8 μM and significantly reduced GSC neurosphere formation at sub-cytotoxic levels. Structural analysis indicated that the presence of a methoxy group at position 3 of the lateral phenylic appendages was important for activity. Pathway and drug connectivity analysis of gene expression changes in response to treatment with the most active bis-chalcone 4j (the 3,4,5 trimethoxy substituted analog) suggested that the mechanism of action was the induction of endoplasmic reticulum (ER) stress and unfolded protein response (UPR) mediated cell death. This was confirmed by Western blot analysis in which 4j induced robust increases in CHOP, p-jun and caspase 12. The UPR is believed to play a significant role in GBM pathogenesis and resistance to therapy and as such represents a promising therapeutic target.

Keywords: cancer stem cell; endoplasmic reticulum stress; glioblastoma multiforme; unfolded protein response; curcumin; bis-chalcones; brain tumor; glioblastoma stem cell

1. Introduction

Glioblastoma (GBM) is one of the worst diagnoses that a person can receive, with an average survival time of approximately 12–15 months [1]. It is the most common and malignant form of primary brain cancer and is currently treated with surgery to remove the tumor followed by radiotherapy with concurrent chemotherapy. It is a highly invasive cancer, invading into the normal tissue surrounding the tumor, making a total resection virtually impossible. Despite an enormous effort to identify the genetic and epigenetic alterations and develop molecularly targeted therapies, the outcome has not significantly improved in over fifteen years since the introduction of the Stupp protocol, which is the addition of the DNA alkylating agent temozolomide to the standard of care, affording patients an additional 2–3 months of survival [2]. Ultimately however, the tumor recurs and the patient succumbs to the disease. Treatment failure has been attributed to the highly invasive nature of these tumors as well as the presence of treatment resistant glioblastoma stem-like cancer cell also known as tumor initiating cells or glioblastoma stem cells (GSCs) [3].

The stem cell theory of carcinogenesis postulates that a small proportion of tumor cells are responsible for driving tumor growth, by giving rise to additional cancer stem cells as well as more differentiated progeny, contributing to tumor heterogeneity. The role of GSCs has been elegantly illustrated in cell lineage studies, which demonstrated the tumor recurrence following treatment was in fact due to GSCs in GBM mouse models [3]. Therefore, for long-term cures, treatments must successfully target GSCs. Multiple signaling pathways have been implicated in GSC self-renewal and therapy-resistance [4]; however, effective clinical therapies remain elusive.

It is now becoming more evident that the adaptive response known as the unfolded protein response (UPR) plays a major role in the development and progression of several cancers including GBM [5–8]. GBM is a fast growing tumor, often outgrowing its blood supply resulting in hypoxia, nutrient deprivation and acidosis. In addition to these extrinsic stresses, GBM cells are subject to intrinsic stresses including oncogene pathway activation and an increased demand for protein synthesis and folding. The endoplasmic reticulum (ER) plays a critical role in protein folding and secretion; and cellular stress can perturb the normal protein homeostasis leading to the accumulation of unfolded or misfolded proteins. In response to this stress, cells initiate the unfolded protein response (UPR), an adaptive response that aims to restore ER proteostasis. The UPR consists of three signal transduction pathways initiated by ER resident proteins, inositol requiring enzyme 1 (IRE1), double-stranded RNA-activated protein kinase (PKR)–like ER kinase (PERK), and activating transcription factor 6 (ATF6) and activation of these pathways is the attempt to mitigate, and reverse ER stress by promotion of protein degradation, decrease in global protein synthesis and upregulation of the expression of specific UPR downstream genes such as molecular chaperones and foldases to aid in protein folding. However, if the ER stress cannot be resolved, the UPR switches from adaptation and cell survival to the induction of apoptosis and cell death. Specifically, the convergence of the aforementioned pathways is actuated by the multi-functional transcription factor CCAAT-enhancer-binding protein homologous protein (CHOP), which plays a key role in ER stress induced cell death [9]. Evidence is accumulating and suggesting that the UPR plays an important role in GBM growth and progression, and supports tumor cell survival response to radiotherapy and chemotherapy and as such represents a promising therapeutic target [9].

Natural products and their derivatives have long provided a diverse source of new medicinal leads, especially in the development of anti-cancer drugs [10]. One of the most studied natural products is curcumin, which is a bioactive component of the popular Indian spice turmeric. Curcumin has shown anti-cancer properties for multiple cancers including GBM [11–13]. Specifically, curcumin has been shown to downregulate many cellular pathways critical for cancer stem cell self-renewal [14]. We previously demonstrated that curcumin caused GSC death by inducing reactive oxygen species (ROS) and downregulation of STAT3 activity [15]. However, a major obstacle to curcumin therapy is its poor bioavailability. Even at high oral doses (8 g/day), curcumin peak plasma levels are below 2 μM [16]. Attempts to increase plasma levels resulted in the development of different formulations including, theracurmin®, curcumin within N-trimethyl chitosan coated solid lipid nanoparticles or nano-emulsions encapsulating curcumin which demonstrated to increase peak plasma levels up to 12.6 μM in rodent models [17–19]. An alternate approach is to design curcumin structural analogs to optimize specific chemotherapeutic properties. Specifically, chalcones, both synthetic and natural, have demonstrated anti-glioma effects by multiple mechanisms. [20–22]. Recently, symmetric bis-chalcones were demonstrated to be potent inhibitors of the breast cancer resistance protein (BCRP/ABCG2); however, the anti-cancer effects of bis-chalcones have not been investigated in GBM [23]. Here, we generated a series of *curcumin inspired* bis-chalcone derivatives and examined their effect on GBM stem cells (GSCs). Patient derived GSCs have been shown to recapitulate the original tumor upon transplantation into mice confirming their reliability as an in vitro model system. [24].

2. Results

2.1. Bis-Chalcone Synthesis

The synthesis of bis-chalcones 4a–4s is outlined in the following reaction scheme (Figure 1). The bis-chalcones were prepared by a base-catalyzed Claisen–Schmidt condensation between 2,6-diacetylpyridine (1 equivalent) and the appropriate aryl aldehyde (2.1 equivalents) using either method a or b. Bis-chalcones 4a, 4d, 4f [25], 4g [26] 4l [27] and 4p [28] were previously cited in the literature. More detailed description of the synthesis along with the spectral data for each compound can be found in the experimental section of the Supplemental Materials.

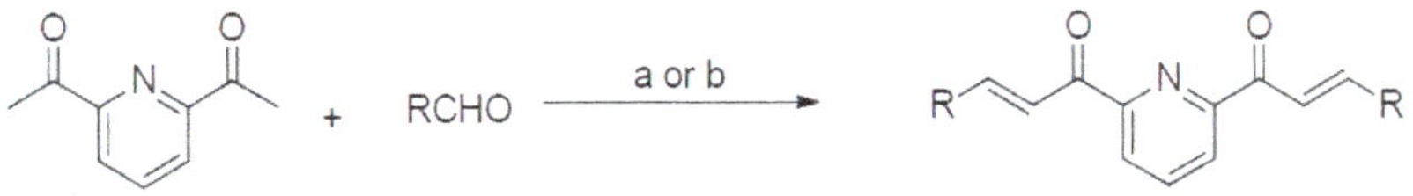

4a R = Ph Ph = Phenyl
4b R = 2-OMe-Ph
4c R = 3-OMe-Ph
4d R = 4-OMe-Ph
4e R = 2,3-DimethoxyPh
4f R = 2,4-DimethoxyPh
4g R = 3,4-DimethoxyPh
4h R = 3,5-DimethoxyPh
4i R = 3,4-(OCH_2O)
4j R = 3,4,5-TrimethoxyPh
4k R = 4-MOM-Ph
4l R = 4-OH-Ph
4m R = 3-OMe-4-MOM-Ph → c
4n R = 4-OH-3-OMe-Ph
4o R = 4-OBn-3-OMe-Ph
4p R = 4-DimethylaminoPh
4q R = 4-Dimethylaminocinnamyl
4r R = 2,3,4-TrimethoxyPh
4s R = 2,4,5-TrimethoxyPh
MOM = OCH_2OCH_3

Figure 1. Reaction scheme for the synthesis of bis-chalcones[a]. [a]Reagents and conditions: (a) 20%NaOH, MeOH, RT; (b) cat. Piperidine, MeOH, ref lux; (c) Trifluoroacetic acid/conc. HCl, Dichloromethane.

2.2. Bis-Chalcones Reduce Viability in GSCs

We previously found curcumin induced GSC death with an approximate IC_{50} of 25 μM. To determine if these bis-chalcones were more cytotoxic than curcumin, GSC lines Glio3, Glio9 and Glio38 were treated with increasing concentrations of each analog and viability was determined 72 h later by 3-(4,5-dimethylthiazol-2-yl)-5-(3-carboxymethoxyphenyl)-2-(4-sulfophenyl)-2H- tetrazolium)

MTS assay. The percent viable cells for concentrations of 0.1 μM, 1 μM and 10 μM are shown in Figure 2. Interestingly, 10 μM of 4a and 4e (Figure 2a) induced robust cell death in Glio9, to approximately 6% and 45% of non-treated cells respectively, but only slightly reduced cell viability in the remaining two cell lines. On the other hand, 4r (Figure 2d) significantly reduced viability in all cell lines, although not all below 50% viability (approximately 20%–62% compared to non-treated controls). Morphological examination of Glio3 (62% viability) suggested that 4r might promote GSC differentiation as well as cell death as indicated by the loss of neurospheres and the corresponding increase in a more differentiated phenotype (Supplementary Figure S1). Bis-chalcone 4g (Figure 2b) reduced viability by more than 50% in Glio38 but was less effective in Glio3 and Glio9. At a concentration of 10 μM, bis-chalcones 4c (Figure 2a: blue), 4h and 4j (Figure 2b: orange and red), 4m and 4n (Figure 2c: dark blue and green) reduced cell viability below 50% compared to non-treated controls (100% viability) in all three GSC lines; Glio3, Glio9 and Glio38 (arrows).

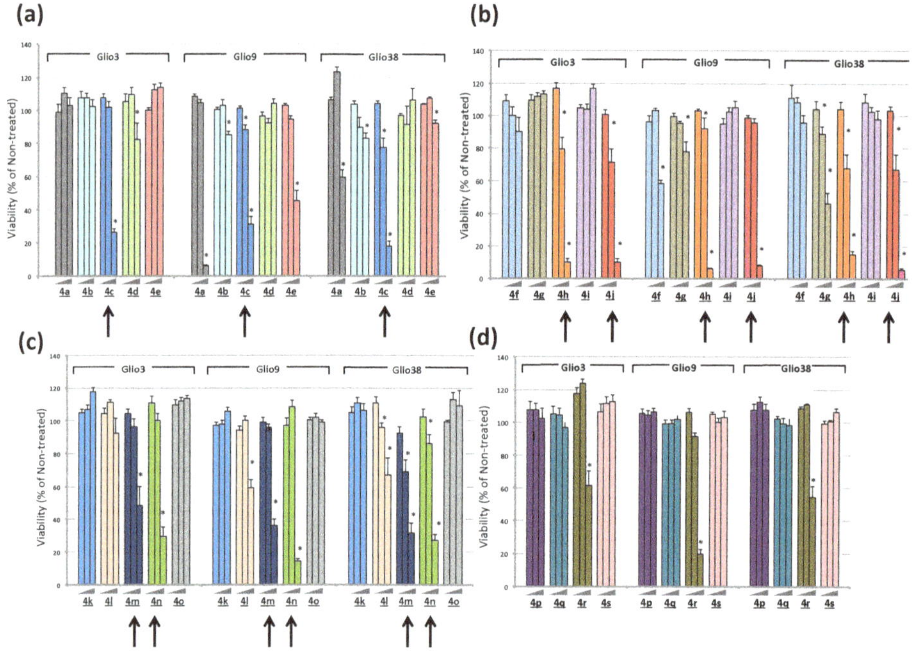

Figure 2. Bis-chalcones reduce GSC viability. GSC lines Glio3, Glio9 and Glio38 were treated with 0.1 μM, 1 μM, or 10 μM of each bis-chalcone analog and viability determined by MTS assay. The data is presented as percent viability compared to non-treated controls. * $p < 0.05$, compared to non-treated controls. Arrows indicate bis-chalcones that reduced viability over 50% at the 10 μM in all three GSC lines. (**a**) bis-chalcones 4a–4e; (**b**) bis-chalcones 4f–4j; (**c**) bis-chalcones 4k–4o; (**d**) bis-chalcones 4p–4s.

2.3. Bis-Chalcones 4c, 4h, 4j and 4n Substantially Reduce Viability in Six GSC Lines

Since we are interested in finding an analog that is substantially more potent than curcumin and demonstrates efficacy across multiple GSC lines, we chose to continue further analysis only with the analogs in which upon treatment with 10 μM decreased the viability over 50% in all three cell lines compared to non-treated controls (arrows, Figure 2). To confirm the GSC cytotoxicity of bis-chalcones 4c, 4h, 4j, 4m and 4n, we treated three additional GSC lines, Glio4, Glio11, and Glio14, with increasing concentrations of each analog and determined cell viability. Similar to previous results, 4c, 4h, 4j and 4n induced robust cell death in the three additional GSC lines. The structures and IC50 for these analogs are shown in Figure 3a,b, respectively. Previously, we determined that the IC50s for curcumin were as follows: Glio3 25.5 ± 2.7 μM, Glio4 39.5 ± 5.4 μM, Glio9 22.5 ± 1.7 μM, Glio11 20.3 ± 3.7 μM, and Glio14 13.9 ± 5.0 μM [15]. Overall, on average, these analogs were approximately 5–10 fold more cytotoxic than curcumin, with 4j, the 2,6-di-[3-(3,4,5-trimethoxyphenyl)propenoyl]pyridine, having the lowest IC50. (Glio3 3.37 ± 0.9 μM, Glio4 2.32 ± 0.6 μM, Glio9 2.51 ± 0.4 μM, Glio11 2.73 ± 0.5 μM, and Glio14 2.52 ± 0.8 μM, Glio38 2.58 ± 0.9). Interestingly, 4m induced robust loss of cell viability in two additional cell lines (Glio11 and Glio14) but to a much lesser extent in Glio4. This is consistent with our previous data demonstrating that, among the five GSC lines examined, Glio4 was the most resistant to curcumin. Percent viability and the IC50 of 4m for Glio4, Glio11 and Glio14 are shown in Supplementary Figure S2a,b, respectively. Taken together, these data suggest that the genetic or epigenetic differences between the various GSC lines may regulate the susceptibility to the bis-chalcones.

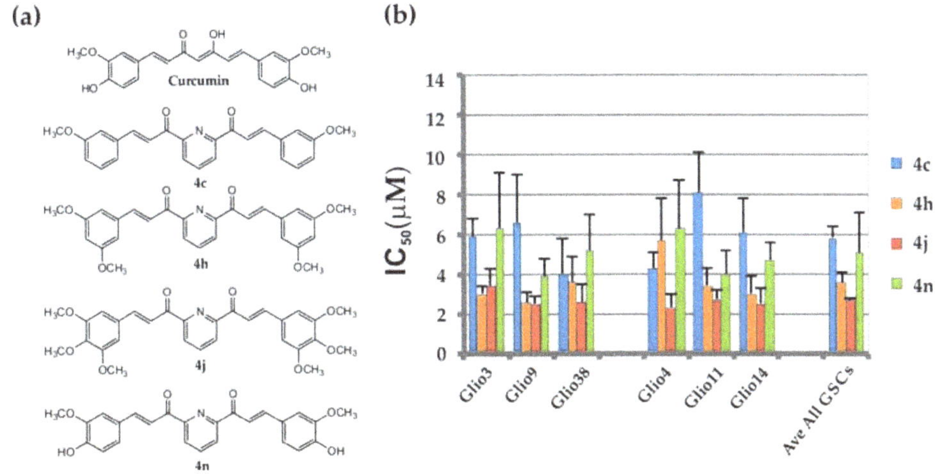

Figure 3. Bis-chalcones 4c, 4h, 4j, 4n induces robust cell death in 6 GSC lines. (**a**) structures of 4c, 4h, 4j, and 4n; (**b**) IC50 of each bis-chalcone for Glio3, Glio4, Glio9, Glio11, Glio14 and Glio38.

2.4. Bis-Chalcones Reduce Neurosphere Formation at Sub Cytotoxic Levels

We previously demonstrated that 2.5 μM curcumin, a 10-fold lower dose than the average IC50, could significantly interfere with GSC neurosphere formation, suggesting an inhibition of GSC self-renewal properties [15]. To determine the effect of the bis-chalcones on neurosphere formation, GSC lines, Glio3 and Glio38, were dissociated and 50–100 single cells/well were plated into 96-well plates, treated with each analog at 100 nM, 250 nM and 500 nM concentrations and the number of neurospheres counted 14 days later. Sub cytotoxic levels of each bis-chalcone significantly reduced neurosphere formation in both cell lines (Figure 4). Consistent with the viability results, bis-chalcone 4j was the most effective, virtually eliminating sphere formation at concentrations as low as 250 nM.

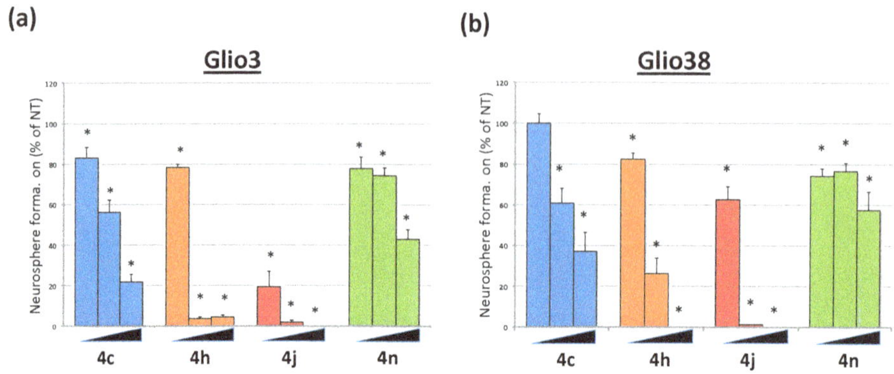

Figure 4. Bis-chalcones 4c, 4h, 4j, 4n significantly reduce neurosphere formation. GSCs Glio3 and Glio38 were plated as single cells in 96-well plates and treated with increasing concentration of 4c, 4h, 4j, 4n (100 nM, 250 nM or 500 nM) and the number of neurospheres counted on day 14. Data is presented as percent neurospheres relative to non-treated controls. (**a**) Glio3; (**b**) Glio38. * $p < 0.05$.

2.5. *Bis-Chalcone 4j Does not Substantially Reduce p-Stat Activity*

We and others have shown that curcumin induces anti-cancer effects via STAT3 inhibition [15,29]. To determine if the mechanism of the most active bis-chalcone, 4j, is similar, we investigated the levels of STAT3 phosphorylation at tyrosine 705 (p-STAT3) and total STAT3 in Glio3 and Glio38 in response to either 5 µM bis-chalcone 4j as well as 4c and 4n. However, unlike curcumin, the bis-chalcones did not substantially reduce p-STAT3 levels suggesting an alternate mechanism of action. Bis-chalcones did, however, induce caspase activity, indicating activation of an apoptotic pathway (Supplemental Figure S3).

2.6. *Bis-Chalcone 4j Induces Gene Expression Changes Consistent with ER Stress and UPR*

To more extensively study the mechanism of action of 4j, we proceeded with profiling its transcriptional impact on Glioblastoma cells. For this, the GSC lines Glio9, Glio11, Glio14 and Glio38 were treated with 2.5 µM bis-chalcone 4j or, for comparison, bis-chalcone 4n and curcumin for 24 h and mRNA were collected for gene expression profiling using the L1000 Platform [30].

2.6.1. Bis-Chalcones 4j Induces a Greater Transcriptional Impact Compared to 4n or Curcumin

We then calculated the Transcriptional Activity Score (TAS) for each treatment and compared the results among the different conditions. TAS is a quality metric developed for the L1000 gene expression data that quantifies the strength and reproducibility of the transcriptional changes induced by a perturbation [30]. As shown in Figure 5a, bis-chalcone 4j elicited the strongest transcriptional response among the four GSC lines tested (Average TAS = 0.388). Moreover, curcumin exhibited the lowest transcriptional response among the three compounds with an average TAS of 0.285. By comparing our results to CLUE (https://clue.io), an external L1000 database, we observed that curcumin exhibits a similar weak transcriptional response in non-glioblastoma cancer cell lines (TAS = 0.21).

2.6.2. Bis-Chalcone 4j Induces a Transcriptional Signature Consistent with ER Stress

To identify genes that differentially regulated across all four glioblastoma cell lines, we calculated the Transcriptional Consensus Signature for each compound, as previously described (Table S1 in expression profiling data section of the Supplemental Materials) [31]. The upregulated and downregulated genes of each compound were then used to perform a functional enrichment analysis using the online annotation tool DAVID [32]. As shown in Figure 5b, the transcriptional signature of

4j is enriched in terms related to ER stress response, including "chaperone", "stress response" and "response to unfolded protein". No obvious discernable pathway was evident for bis-chalcone 4n or curcumin at the concentration tested.

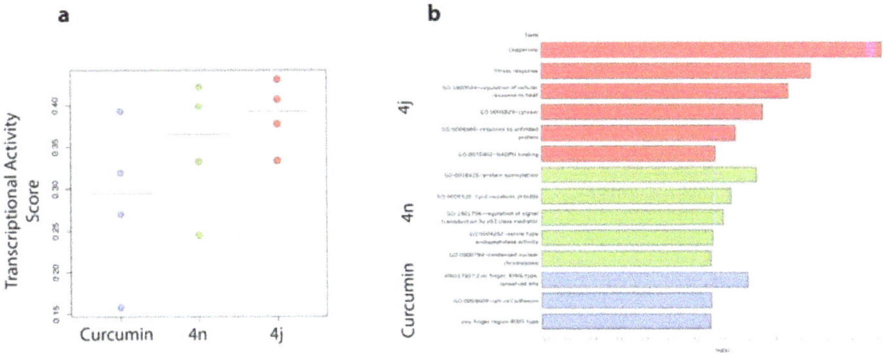

Figure 5. Bis-chalcone 4j elicits a high transcriptional response. (**a**) GSCs Glio9, Glio11, Glio14 and Glio38 were treated with 2.5 μM bis-chalcone 4j, 4n or curcumin for 24 h; RNA was collected and processed by the L1000 Platform. The Transcriptional Activity Score (TAS) for each condition was calculated from the resulted L1000 gene expression data; (**b**) Bis-chalcone 4j induces a transcriptional signature consistent with ER stress. For each compound, a Transcriptional Consensus Signature was calculated from the L1000 gene expression profiling data and subjected to functional enrichment analysis using DAVID.

2.6.3. Bis-Chalcone 4j Induce Similar Transcriptional Responses only in the Neurosphere Cell Lines

The Transcriptional Consensus Signature was generated by aggregating across all four GSC lines; however, we also wanted to evaluate whether there were cell-specific transcriptional responses to the 4j, 4n, and curcumin treatments. For this, we created cell-specific gene expression signatures that would be indicative of a cell's transcriptional response to a treatment. This signature was created by computing the median gene expression across all biological replicates for a particular GSC line–compound pair. We then performed the same functional enrichment analysis as above (filtered genes with |z-score| >= 1). As shown in Figure 6a, we observed that three (Glio11, Glio14 and Glio38) out of the four cell lines treated with bis-chalcone 4j were enriched in stress response terms similarly to Figure 5b; however, Glio9 followed a different transcriptional response. Interestingly, Glio11, Glio14 and Glio38 were derived from treatment naïve tumors and grow as neurospheres, whereas Glio9 was derived from a recurrent tumor and grows adherently. The functional enrichment analysis for bis-chalcone 4n and curcumin can be found in the expression profiling data of Supplementary Materials.

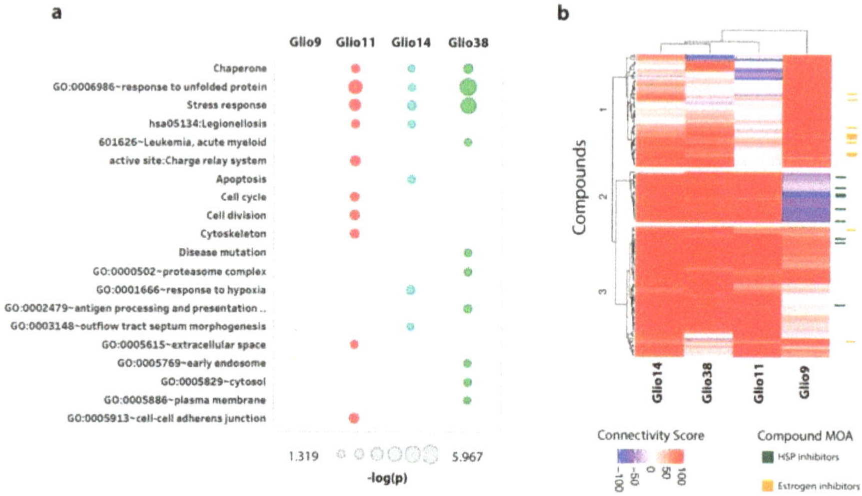

Figure 6. Bis-chalcone 4j induce similar transcriptional responses only in the neurosphere cell lines. (**a**) The individual gene expression signatures for each condition were subjected to functional enrichment analysis using DAVID and the significant (p-value < 0.05) enrichment terms were ranked based on the numbers of cells that were enriched in them; (**b**) Drug connectivity analysis of 4j indicates ER perturbation of neurosphere cells. Drug connectivity scores were determined by comparing the 4j, 4n and Curcumin transcriptional response to 2911 compounds, part of the CLUE Touchstone dataset. Hierarchical clustering was performed on the 50 most connected compounds for each condition. Glio11, Glio14 and Glio38 cluster together and are highly connected to heat shock protein HSP inhibitors (green). Glio9 was highly connected to estrogen inhibitors (orange).

2.6.4. Drug Connectivity Analysis of 4j Supports ER Perturbation of Neurosphere Cells

We further validated the difference in the transcriptional response of Glio9, Glio11, Glio14 and Glio38 after bis-chalcone 4j treatment by evaluating their connectivity to other compounds in the CLUE reference dataset (Touchstone), a dataset consisting of cellular signatures representing systematic perturbations of small-molecule and genetic perturbations. The cell-specific gene expression signatures were used as input in the CLUE tool and for each signature we extracted the 50 most connected/similar compounds (out of a total of 2911 reference CLUE compounds). We then plotted the corresponding Connectivity Scores in Figure 6b. The differential response to treatments was more prominent in the case of 4j, where we can see that the Glio14, Glio38 and Glio11 signatures are highly connected to compounds in Cluster 2 and 3 and the Glio9 signature is highly connected to compounds in Cluster 1. Moreover, by examining the mechanisms of action of the compounds in each Cluster, we noticed that the Glio14, Glio38 and Glio11 signatures were highly connected to heat shock protein (HSP) inhibitors (Cluster 2), while the Glio9 signature was highly connected to estrogen inhibitors (Cluster 1). Drug connectivity analysis for bis-chalcone 4n and curcumin can be found in the expression profiling data of the Supplementary Materials.

2.7. Bis-Chalcone 4j Induces Robust Expression of CHOP and Promotes JNK and Caspase 12 Activity

The above analysis indicates that 4j cytotoxicity may be mediated through the ER stress induced UPR response. The UPR is an adaptive mechanism initiated in response to ER stress. To confirm that bis-chalcone 4j was inducing the UPR, we examined UPR markers in the GSCs that grow as neurospheres (glio3, 14 and 38) by Western blot analysis. Consistent with the transcriptional response and drug connectivity data, 4j induced robust increase in the protein levels of UPR markers including CHOP, p-jun at serine 73 (indicative of the stress activated kinase c-Jun N-terminal kinase (JNK) activity) and caspase 12 only in the neurosphere cell lines. No obvious increase in glucose related protein 78 (GRP78), a molecular chaperone important for mediating cell adaptation and survival in response to ER stress, was observed (Figure 7a,b). The other analogs examined either failed to induce protein expression or did so at a much lower level compared to 4j, indicating that 4c, 4h, 4n induce GSC death, at least in part, by an alternate mechanism. All analogs however induced Poly (ADP-ribose) polymerase (PARP) cleavage suggestive of apoptotic cell death. As expected, 4j treatment did not result in an increase in CHOP or caspase 12 in Glio9. Bis-chalcone 4j did, however, induce a small increase in p-jun compared to non-treated controls; however, this was much less than that observed in the neurosphere cells or what was observed in response to curcumin (Figure 7c).

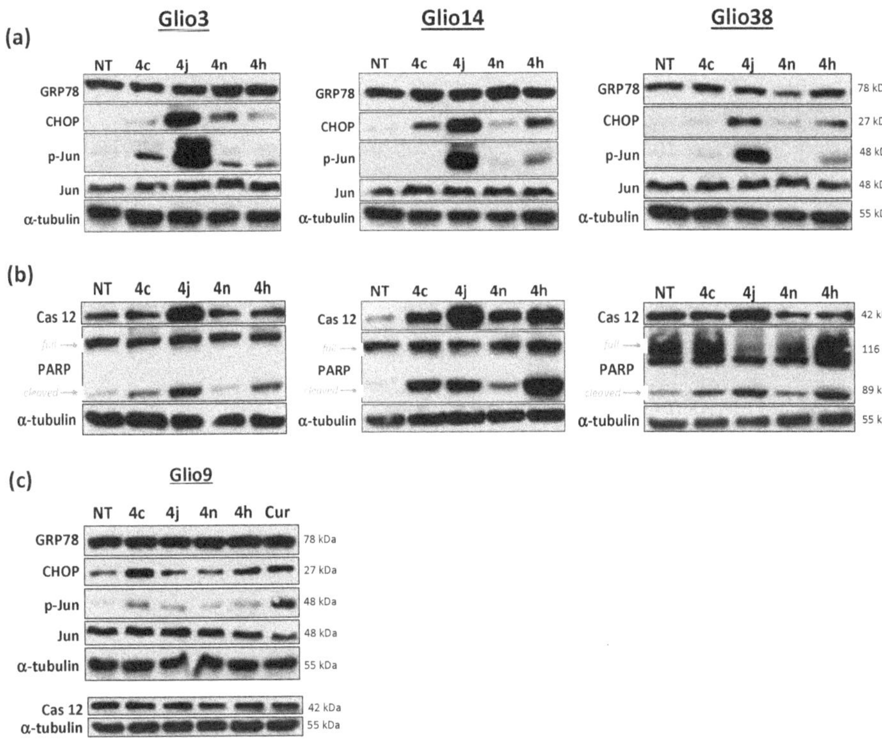

Figure 7. Western blot analysis indicates that 4j induces robust ER stress and UPR in neurosphere cells. Glio3, Glio9, Glio14 and Glio38 were treated with 5 μM of bis-chalcones 4c, 4h, 4j, and 4n for 8 or 24 h and subjected to Western blot analysis. Levels of GRP78, CHOP, p-jun, jun, were examined at 8 h. Levels of caspase 12 and cleaved PARP were examined at 24 h. α-tubulin serves as a loading control. (**a**) levels of ER stress/UPR markers at eight hour in the neurosphere cell lines; (**b**) levels of apoptosis indicators at 24 h in the neurosphere cell lines; (**c**) levels of ER stress/UPR markers in the adherent Glio9 cell line.

2.8. Bis-Chalcone 4j Demonstrates Reduced Toxicity to Non-Cancer Stem Cells

To determine the cytotoxicity of bis-chalcone 4j in a non-cancer stem cell line, we treated human mesenchymal stem cells (MSCs) with increasing concentrations of 4j (0.1–10 μM) and examined viability at 72 h as previously described. Compared to GSCs, the cytotoxicity induced by treatment was substantially lower, with an estimated IC50 of 13.1 μM (Supplemental Figure S4), suggesting a selective targeted effect of 4j towards GSCs.

3. Discussion

Despite an aggressive treatment regimen, the prognosis for GBM patients remains dismal. This poor outcome has in part been attributed to the presence of a small number of treatment resistant cells that are responsible for tumor recurrence and patient relapse. Cancer stem cells and cancer cells, in general, have an innate ability to adapt to extracellular stress (hypoxia, acidosis, nutrient and oxygen deprivation, etc.) and intracellular stress (ROS, oncogenic signaling pathways, etc.). Cellular stresses can result in a loss of protein homeostasis in the ER, prompting the activation of the UPR that consists of three different parallel signaling pathways initiated by ER resident transmembrane proteins, PERK, IRE1) and ATF6. Activation of these pathways aims to restore ER protein homeostasis by attenuating protein synthesis, upregulating the expression of specific protein chaperones to aid in protein folding and to stimulate ER associated protein degradation. However, if protein homeostasis cannot be restored, the UPR initiates apoptosis via the activation of the transcription factor CHOP. CHOP regulates the expression of both pro-survival and pro-cell death BCL-2 family members [33]. In addition, IRE1 mediated JNK activation has been shown to play an important role in ER stress mediated cell death by promoting mitochondrial-mediated cell death [34,35]. Lastly, ER localized Caspase 12 has been shown to mediate ER stress cell death as caspase-12-deficient mice were resistant to ER stress-induced apoptosis [36].

Of the 19 bis-chalcones synthesized and tested, only four (4c, 4g, 4j and 4n) significantly reduced GSC viability across a panel of six genetically distinct GBM patient-derived cancer stem cell lines with IC50s in the very low micromolar range (2.3–5.8 μM), considerably less than the previously reported IC50 for curcumin of approximately 25 Mm [15]. Our results indicate that a methoxy group ($-OCH_3$) at the 3 position of the phenylic side appendages is important for inducing GSC death. In fact, at 10 μM, bis-chalcones 4c, 4e, 4g, 4h, 4j, 4m, 4n, and 4r reduced GSC viability to below 50% of non-treated controls in at least one cell line. Similarly, the methoxy group on the curcumin scaffold has been shown to be important for its biological activities [37,38]. Although bis-chalcone 4o possesses a methoxy group at the 3 position, it appears that its activity is compromised by the lipophilic benzyloxy moiety at position 4. Furthermore, it has been shown that the introduction of additional methoxy groups on the aromatic rings enhanced the anticancer effect on multiple cancer cell lines [39]. In analogy, we found that bis-chalcones containing multiple methoxy groups on the phenyl rings could be very cytotoxic. The most effective bis-chalcone was the 3,4,5 trimethoxy substituted analog (4j); however, the 2,4,5 trimethoxy substituted analog (4s), lacking substitution at position 3 was ineffective.

Bis-chalcone 4j induced robust cell death in all GSC lines with IC50 ranging from 2.3 μM (Glio4) to 2.7 μM (Glio11). Furthermore, 4j significantly reduced neurosphere formation at concentrations as low as 100 nM and virtually eliminating neurosphere formation at 250 nM suggesting modulation of GSC self-renewal properties. Pathway analysis of expression profiles generated using L1000 assay indicated that 4j treatment induced a stress response consistent with ER stress/UPR in the GSCs that grow as neurospheres but not in the GSC that grows adherently. Furthermore, drug connectivity analysis using the Clue compound database indicated that the signatures were highly connected to HSP inhibitors for the neurosphere cell lines, which is consistent with the pathway analysis, as HSP inhibition is associated with ER stress and UPR [40]. Consistent with these findings, Western blot analysis demonstrated that 4j induced robust induction of CHOP, p-jun (indicative of JNK activity) and caspase 12.

A putative mechanism by which bis-chalcone 4j is inducing ER stress in the neurospheres cells may be by directly increasing the number of misfolded proteins in the ER. Like curcumin, 4j is electrophilic. In fact, the presence of the pyridine ring between the two carbonyls makes the dienones more electrophilic, i.e., more reactive. Electrophiles (electron deficient) target electron rich nucleophiles such as side chains of the amino acids cysteine, histidine and lysine. One of the major modifications occurring in the ER is the formation of disulfide bonds, which stabilizes the newly formed protein [41]. It is possible that the reactive bis-chalcone is inducing ER stress by disrupting the formation of disulfide bonds between the thiol groups of the cysteine residues by forming a carbon-sulfur bond (Michael adduct formation). This disruption leads to an accumulation of unfolded proteins and, thus, ER stress similar to that of arylating quinones [42]. Additionally, it has been shown that GBM cells display a higher antioxidant capacity compared to normal cells, particularly higher level of Glutathione Reductase (GSR) and Glutathione (GSH) were detected [43]. Such elevated levels have been linked to the well-known GBM resistance to standard treatment with Temozolomide. It is then conceivable that the subtle chemical reactivity of our bis-chalcones could be targeting these antioxidants species (GSR, GSH), which would, in turn, lead to an increase in reactive oxygen species and, ultimately, ER stress [44].

However, the induction of ER stress/UPR was observed only in the neurosphere cell lines and not the adherent cell line, Glio9. Drug connectivity analysis indicated that 4j might induce cell death in Glio9 similar to that of estrogen inhibition. One potential connection of the UPR to estrogen signaling is the IRE1/XBP1 signaling pathway. Activated IRE1 cleaves XBP1 mRNA leading to the translation of a highly active transcription factor. The estrogen receptor antagonist, fulvestrant, was demonstrated to downregulate the IRE1/XBP1 signaling pathway in prolactinoma cells [45]. Conversely, Minchenko et al. recently demonstrated that IRE1 inhibition modulates the expression of genes encoding estrogen related proteins in glioma cells [46]. Recent data supports a role for IRE1/XBP1 in glioblastoma development and progression and as such a promising therapeutic target [47,48]. Glio9 was derived from a recurrent tumor, previously treated with both chemotherapy and radiation; therefore, it is not inconceivable that these cells would respond differently compared to cells generated from naive tumors. Regardless, our bis-chalcones, in particular 4j, induced robust cell death in all GSC lines examined. Moreover, our analogs obey Lipinski's rule of 5, an important "benchmark test" in drug development, suggesting the *likely* oral bioavailability of such compounds [49]. Our encouraging results support 4j as a potential therapeutic lead for the development of a novel drug for the treatment of this deadly disease.

4. Materials and Methods

4.1. Chemistry

All reagents were obtained from Sigma-Aldrich (St. Louis, MO, USA) and were used directly without further purification. ^{1}H- and ^{13}C-NMR spectra were recorded at 500 and 125 MHz on Bruker. The spectra were referenced to the residual protonated solvents. Abbreviations such as *s*, *d*, *t*, *m*, *br*, and *dd* used in the description denote *singlet*, *doublet*, *triplet*, *multiplet*, *broad*, and *double doublet*, respectively. The chemical shifts and coupling constants were reported in parts per million (ppm) and hertz (Hz), respectively. High-resolution mass spectra were obtained on Bruker micrOTO-Q II mass spectrometer (Bruker, Billerica, MA, USA). The NMR and mass spectrometry data for the synthesized compounds are provided in the supplemental document. All intermediate and final products were monitored by thin layer chromatography (TLC) on 250 µm silica plates. Where applicable, the compounds were recrystallized from the proper solvent or purified by flash column chromatography on silica gel (200–300 mesh) with ethyl acetate/hexanes (1:1) as eluant.

The synthesis of bis-chalcones 4a–4s is outlined in Figure 1. The bis-chalcones were prepared by the Claisen–Schmidt condensation between 2,6-diacetylpyridine (1 equivalent) and the appropriate aryl aldehyde (2.1 equivalents) using either method a or b as shown. The reaction monitored by TLC. Upon completion, the reaction mixture was diluted with water and the solid formed was collected by vacuum filtration. The bis-chalcone was either purified by flash column chromatography or recrystallization. All the compounds were characterized by NMR (Nuclear Magnetic Resonance) Analysis and EI-HRMS (Electrospray Ionization-High Resolution Mass Spectrometry, Bruker, Billerica, MA, USA) Analysis. MOM-protected derivatives of 4-hydroxybenzaldehyde and vanillin were synthesized according to literature procedures [50]. More detailed description of the synthesis for each compound can be found in the Supplemental Materials.

4.2. Cell Culture

GSC lines Glio3, Glio4, Glio9, Glio11 and Glio14 have been previously described [15]. With Institutional Review Board (IRB) approval (number 20060858), the Glio38 cell line was derived from a patient's resected tumor after receiving written consent. Briefly, tumor samples were physically and enzymatically digested and single cells were plated in DMEM/F12 3:1 supplemented with 20 ng/mL each of epidermal growth factor (EGF) and fibroblast growth factor (FGF), 2% Gem21 and 1% Penicillin/Streptomycin (P/S) to promote the growth of glioblastoma stem-like cells. Similar to the other neurosphere cell lines Glio3, Glio4, Glio11 and Glio14, Glio38 cells grew as neurospheres and expressed the putative GBM stem cell markers; cell surface proteins CD133 and A2B5, intermediate filament Nestin, RNA binding protein Musashi, oncogene BMi-1 and the transcriptional regulator Sox2 (Supplemental Figure S5). Glio9 was derived from a recurrent, post therapy, tumor and grows adherently and fails to express Sox2. [15] Human MSCs were obtained from Thermo Fisher Scientific (Waltham, MA, USA) and maintained in MEM supplemented with 20% fetal bovine serum (FBS) and 1% P/S. Our cell lines were routinely tested for mycoplasma using LookOut mycoplasma PCR detection kit (Sigma Aldrich, St. Louis, MO, USA) according to the manufacturer's instructions and maintained at 37 °C in a humidified 5% CO_2 incubator.

4.3. Drug Treatment

The bis-chalcones were dissolved in dimethyl sulfoxide (DMSO) at a concentration of 10 mM, vortexed and subsequently diluted 1:10 to obtain 1 mM and 0.1 mM stock concentrations. Viability was determined using the CellTiter 96®Aqueous One Solution Cell Proliferation Assay (MTS) assay (Promega Madison, WI, USA) as previously described [15]. Briefly, GSCs were seeded into 96-well plates using a modified neurosphere media containing 5% FBS at a density of 5000–10,000 cells per well, depending on the cell line. Cells were treated with increasing concentrations (0.1–10 μM) of each curcumin analog for 72 h. Media was aspirated and 100 μL of a 1:5 solution of MTS to cell culture media was added to each well and incubated for 1–4 h. Optical density was measured at 490 nm using a BoiTek Synergy HT plate reader (Biotek, Winooski, VT, USA). Viability of drug treated cells is expressed as the percent viable cells relative to non-treated cells (100% viability). Experiments were done in triplicate.

4.4. Neurosphere Forming Assay

To determine the effect of the curcumin analogs on stem cell activity, neurosphere assays were performed as previously described [15]. Briefly, single cells were seeded at 50–100 cells per well in a 96-well plate and treated with 100 nM, 250 nM or 500 nM bis-chalcones on day 0. Spheres greater than 50 microns were manually counted under microscopy on day 14. All experiments were done in triplicate.

4.5. RNA Analysis

To determine the molecular mechanism of bis-chalcone induced cell death, we exposed Glio11, Glio14 and Glio38 as neurospheres as well as the adherent Glio9 to 2.5 μM of bis-chalcones 4j, 4n or 2.5 μM curcumin (for comparison) and RNA was extracted 24 h later. Neurospheres were collected, spun down, washed with sterile PBS and RNA harvested using RNeasy Mini kit (Qiagen, Valencia, CA, USA) as per the manufacturer's instructions. For the adherent culture, Glio9, cells were collected using accutase (Gemini), spun down, washed and RNA isolated as described above. RNA concentration was determined using Nanodrop 2000 spectrophotometer (Thermo Scientific, Waltham, MA, USA) and subsequently aliquotted for gene expression profiling. For Glio9, Glio14 and Glio38, three biological replicates each with four technical replicates were analyzed. For Glio11, two biological replicates each with six technical replicates were analyzed. Transcriptional profiles were generated by the LINCS

Project, which utilizes a novel gene expression profiling method that measures the expression of 978 representative landmark transcripts [30].

4.5.1. Transcriptional Impact

Transcriptional Consensus Signatures (TCSs) for each compound were calculated as described previously [24] using the Level 4 population-normalized L1000 data. Briefly, the TCSs quantify the genes that are consistently over/under expressed in multiple cell lines after a compound treatment.

4.5.2. Functional Enrichment Analysis

Functional enrichment analysis was performed using the Functional Annotation Chart tool in David [25]. The median gene expression was calculated between biological replicates of the Level 5 L1000 data and genes with a |z-score| >= 1 were used as input in DAVID.

4.5.3. CLUE Analysis

Connectivity Scores between the CLUE Touchstone (Reference) perturbagens were calculated using https://clue.io/. Hierarchical clustering (complete linkage, Euclidean distance) was performed on the 50 highest connected compounds for each condition (cell line treated with drug).

4.6. Western Blot Assay

Our protocol for Western blot assays has been described previously [51]. Cells were treated with 5 µM bis-chalcones or curcumin for 8 or 24 h cells are lysed with a RIPA buffer (1% sodium deoxycholate, 0.1% Sodium dodecyl sulfate (SDS), 1% Triton X-100, 10 mM Tris pH 8 and 140 mM NaCl) supplemented with 250 units per ml Benzonase, 1 mM dithiothreitol and phosSTOP phosphatase inhibitor cocktail and a cOmplete protease inhibitor cocktail (both from Roche, Indianapolis, IN, USA). Protein concentration determined using a bicinchoninic acid (BCA) protein assay (Thermo Scientific, Waltham, MA, USA), and 20 µg of protein was loaded onto 8, 12 or 15% polyacrylamide gels (BioRad Hercules, CA, USA) gels for electrophoresis and subsequently transferred onto nitrocellulose membranes. Membranes are incubated overnight with primary antibodies, washed and incubated with HRP conjugated secondary antibodies for 1 hour. Bands were visualized using Super-Signal™ West Pico Chemiluminescent Substrate (Thermo Scientific Waltham, MA, USA). Anti-C/EBP homologous protein (CHOP), anti-78 kDa glucose-regulated protein (GRP78), anti-caspase 12, anti-phospho c-jun (Ser63), anti-c-jun, anti-signal transducer and activator of transcription 3 (STAT3), anti-phospho-STAT3 (Tyr705), anti-poly ADP ribose polymerase (PARP) and anti-cleaved caspase 3 were all obtained from Cell Signaling Technology (Danvers, MA, USA). Anti-α-tubulin was obtained from Abcam (Eugene, OR, USA).

4.7. Statistical Analysis

Significance was determined using Student's *t*-tests for all pairwise comparisons of the different treatments that were tested. The results are presented as the mean ± standard error mean (SEM). Significance was set at $p < 0.05$.

5. Conclusions

Despite advances in neuroimaging and neurosurgical techniques, and an abundance of research aimed at understanding and targeting cell-signaling pathways driving GBM pathogenesis, GBM remains one of the most lethal brain tumors. The UPR is an adaptive mechanism initiated to mitigate ER stress resulting from the tumor microenvironment, oncogene activation, rapid cell proliferation as well as anti-cancer therapies. Specifically, the UPR plays a role in temozolomide and radiotherapy resistance in GBM [52,53]. However, if protein homeostasis cannot be restored, the UPR induces cell death. Here, we discovered a novel bis-chalcone (4j) capable of "weaponizing" the UPR to promote

GBM stem cell death. Targeting the UPR is a novel strategy for treating this deadly disease and 4j is a promising lead compound for drug development.

Supplementary Materials: The following are available online at http://www.mdpi.com/2072-6694/11/3/357/s1, Figure S1: Bis-chalcone 4r promotes morphological changes consistent with cell differentiation. Glio3 cells were treated with 10 μM bis-chalcone 4r for 3 days and examined by light microscopy. Arrows indicate neurospheres in non-treated cells, Figure S2: Bis-chalcone 4m induced cell death in 3 additional GSC lines. (a) Percent viability of bis-chalcone 4m treated Glio4, Glio11, and Glio14 cell lines. * $p < 0.05$ compared to non-treated controls; (b) IC50 of 4m for each cell line tested, Figure S3: Bis-chalcones do not substantially reduce STAT3 activity. Glio3 and Glio38 were treated with 5 μM of 4c, 4j and 4n and harvested 8 h later. Levels of p-STAT3, STAT3, cleaved caspase 3 were examined by Western blot analysis, Figure S4: Bis-chalcone 4j is less toxic to human MSCs. MSCs were treated with 0.1 μM–μ10 M 4j and percent viability determined at 72 h by MTS assay. * $p < 0.05$, compared to non-treated controls, Figure S5: Glio3 Characterization. Bright field images indicate that Glio3 grows as neurospheres in defined media. Expression of putative stem cell markers A2B5, CD133, Nestin, Bmi1, Musashi and Sox2 were evaluated by immunocytochemistry, Table S1: Transcriptional Consensus Signature for each compound, bis-chalcone 4j, 4n and curcumin. Table S1, the functional enrichment analysis for bis-chalcone 4n and curcumin and the Drug connectivity analysis for bis-chalcone 4n and curcumin can be found in the expression profiling data of the supplementary materials. Detailed description of bis-chalcone synthesis along with the spectral data for each compound can be found in the experimental section of the supplemental materials.

Author Contributions: Conceptualization, L.S., S.V., R.M.L. and R.M.G.; Data curation, L.S., E.A.V., N.G.M., V.S. and R.M.G.; Formal analysis, E.A.V., V.S., I.I.T. and R.M.G.; Funding acquisition, S.V., R.M.L. and R.M.G.; Investigation, L.S., E.A.V., N.G.M., W.W., I.I.T. and R.M.G.; Methodology, L.S., E.A.V., V.S. and R.M.G.; Project administration, R.M.L. and R.M.G.; Resources, S.V.; Supervision, S.C.S. and R.M.L.; Validation, L.S., E.A.V., N.G.M., V.S., W.W., S.C.S. and R.M.G.; Visualization, E.A.V., V.S., R.M.L. and R.M.G.; Writing—original draft, R.M.G.; Writing—review and editing, L.S., E.A.V., V.S., S.V., R.M.L. and R.M.G.

Funding: This research was funded by the Mystic Force Foundation, National Science Foundation (NSF): grant 1809060, National Institute of Health (NIH): grant 1809060 and NIH: U54HL127624 (BD2K LINCS Data Coordination and Integration Center, DCIC).

Acknowledgments: We would like to thank the following volunteer students for their contributions to our research efforts: Anthony Sanchez, Denis Ioni Ortega, Wanda Gonzalez, Amelia Bahamonde, Katrina Kostenko, Frederic Vallejo and Nicolas De Cordoba.

Conflicts of Interest: The authors declare no conflict of interest.

References

1. Ostrom, Q.T.; Gittleman, H.; Truitt, G.; Boscia, A.; Kruchko, C.; Barnholtz-Sloan, J.S. CBTRUS Statistical Report: Primary Brain and Other Central Nervous System Tumors Diagnosed in the United States in 2011–2015. *Neuro-Oncology* **2018**, *20* (Suppl. 4), iv1–iv86. [CrossRef]
2. Stupp, R.; Mason, W.P.; van den Bent, M.J.; Weller, M.; Fisher, B.; Taphoorn, M.J.; Belanger, K.; Brandes, A.A.; Marosi, C.; Bogdahn, U.; et al. Radiotherapy plus concomitant and adjuvant temozolomide for glioblastoma. *N. Engl. J. Med.* **2005**, *352*, 987–996. [CrossRef]
3. Chen, J.; Li, Y.; Yu, T.S.; McKay, R.M.; Burns, D.K.; Kernie, S.G.; Parada, L.F. A restricted cell population propagates glioblastoma growth after chemotherapy. *Nature* **2012**, *488*, 522–526. [CrossRef] [PubMed]
4. Kalkan, R. Glioblastoma Stem Cells as a New Therapeutic Target for Glioblastoma. *Clin. Med. Insights Oncol.* **2015**, *9*, 95–103. [CrossRef]
5. Madden, E.; Logue, S.E.; Healy, S.J.; Manie, S.; Samali, A. The role of the unfolded protein response in cancer progression: From oncogenesis to chemoresistance. *Biol. Cell* **2019**, *111*, 1–17. [CrossRef] [PubMed]
6. Ciavattini, A.; Delli Carpini, G.; Serri, M.; Tozzi, A.; Leoni, F.; Di Loreto, E.; Saccucci, F. Unfolded protein response, a link between endometrioid ovarian carcinoma and endometriosis: A pilot study. *Oncol. Lett.* **2018**, *16*, 5449–5454. [CrossRef]
7. Obacz, J.; Avril, T.; Le Reste, P.J.; Urra, H.; Quillien, V.; Hetz, C.; Chevet, E. Endoplasmic reticulum proteostasis in glioblastoma-From molecular mechanisms to therapeutic perspectives. *Sci. Signal.* **2017**, *10*. [CrossRef]
8. Obacz, J.; Avril, T.; Rubio-Patino, C.; Bossowski, J.P.; Igbaria, A.; Ricci, J.E.; Chevet, E. Regulation of tumor-stroma interactions by the unfolded protein response. *FEBS J.* **2017**. [CrossRef]
9. Penaranda Fajardo, N.M.; Meijer, C.; Kruyt, F.A. The endoplasmic reticulum stress/unfolded protein response in gliomagenesis, tumor progression and as a therapeutic target in glioblastoma. *Biochem. Pharmacol.* **2016**, *118*, 1–8. [CrossRef]

10. Mann, J. Natural products in cancer chemotherapy: Past, present and future. *Nat. Rev. Cancer* **2002**, *2*, 143–148. [CrossRef] [PubMed]
11. Rodriguez, G.A.; Shah, A.H.; Gersey, Z.C.; Shah, S.S.; Bregy, A.; Komotar, R.J.; Graham, R.M. Investigating the therapeutic role and molecular biology of curcumin as a treatment for glioblastoma. *Ther. Adv. Med. Oncol.* **2016**, *8*, 248–260. [CrossRef] [PubMed]
12. Wang, Y.; Yu, J.; Cui, R.; Lin, J.; Ding, X. Curcumin in Treating Breast Cancer: A Review. *J. Lab. Autom.* **2016**, *21*, 723–731. [CrossRef] [PubMed]
13. Mehta, H.J.; Patel, V.; Sadikot, R.T. Curcumin and lung cancer—A review. *Targeted Oncol.* **2014**, *9*, 295–310. [CrossRef] [PubMed]
14. Li, Y.; Zhang, T. Targeting cancer stem cells by curcumin and clinical applications. *Cancer Lett.* **2014**, *346*, 197–205. [CrossRef] [PubMed]
15. Gersey, Z.C.; Rodriguez, G.A.; Barbarite, E.; Sanchez, A.; Walters, W.M.; Ohaeto, K.C.; Komotar, R.J.; Graham, R.M. Curcumin decreases malignant characteristics of glioblastoma stem cells via induction of reactive oxygen species. *BMC Cancer* **2017**, *17*, 99. [CrossRef]
16. Cheng, A.L.; Hsu, C.H.; Lin, J.K.; Hsu, M.M.; Ho, Y.F.; Shen, T.S.; Ko, J.Y.; Lin, J.T.; Lin, B.R.; Ming-Shiang, W.; et al. Phase I clinical trial of curcumin, a chemopreventive agent, in patients with high-risk or pre-malignant lesions. *Anticancer Res.* **2001**, *21*, 2895–2900.
17. Ramalingam, P.; Ko, Y.T. Enhanced oral delivery of curcumin from N-trimethyl chitosan surface-modified solid lipid nanoparticles: Pharmacokinetic and brain distribution evaluations. *Pharm. Res.* **2015**, *32*, 389–402. [CrossRef]
18. Zhongfa, L.; Chiu, M.; Wang, J.; Chen, W.; Yen, W.; Fan-Havard, P.; Yee, L.D.; Chan, K.K. Enhancement of curcumin oral absorption and pharmacokinetics of curcuminoids and curcumin metabolites in mice. *Cancer Chemother. Pharmacol.* **2012**, *69*, 679–689. [CrossRef]
19. Sasaki, H.; Sunagawa, Y.; Takahashi, K.; Imaizumi, A.; Fukuda, H.; Hashimoto, T.; Wada, H.; Katanasaka, Y.; Kakeya, H.; Fujita, M.; et al. Innovative preparation of curcumin for improved oral bioavailability. *Biol. Pharm. Bull.* **2011**, *34*, 660–665. [CrossRef]
20. Chen, P.H.; Chang, C.K.; Shih, C.M.; Cheng, C.H.; Lin, C.W.; Lee, C.C.; Liu, A.J.; Ho, K.H.; Chen, K.C. The miR-204-3p-targeted IGFBP2 pathway is involved in xanthohumol-induced glioma cell apoptotic death. *Neuropharmacology* **2016**, *110 (Pt A)*, 362–375. [CrossRef]
21. Champelovier, P.; Chauchet, X.; Hazane-Puch, F.; Vergnaud, S.; Garrel, C.; Laporte, F.; Boutonnat, J.; Boumendjel, A. Cellular and molecular mechanisms activating the cell death processes by chalcones: Critical structural effects. *Toxicol. In Vitro Int. J. Publ. Assoc. BIBRA* **2013**, *27*, 2305–2315. [CrossRef] [PubMed]
22. Robinson, M.W.; Overmeyer, J.H.; Young, A.M.; Erhardt, P.W.; Maltese, W.A. Synthesis and evaluation of indole-based chalcones as inducers of methuosis, a novel type of nonapoptotic cell death. *J. Med. Chem.* **2012**, *55*, 1940–1956. [CrossRef] [PubMed]
23. Winter, E.; Devantier Neuenfeldt, P.; Chiaradia-Delatorre, L.D.; Gauthier, C.; Yunes, R.A.; Nunes, R.J.; Creczynski-Pasa, T.B.; Di Pietro, A. Symmetric bis-chalcones as a new type of breast cancer resistance protein inhibitors with a mechanism different from that of chromones. *J. Med. Chem.* **2014**, *57*, 2930–2941. [CrossRef]
24. Lee, J.; Kotliarova, S.; Kotliarov, Y.; Li, A.; Su, Q.; Donin, N.M.; Pastorino, S.; Purow, B.W.; Christopher, N.; Zhang, W.; et al. Tumor stem cells derived from glioblastomas cultured in bFGF and EGF more closely mirror the phenotype and genotype of primary tumors than do serum-cultured cell lines. *Cancer Cell* **2006**, *9*, 391–403. [CrossRef] [PubMed]
25. Al-Omran, F.; Al-Awadi, N.; Edun, M. Corrigendum-Synthesis of New 2-Pyrazoline Derivatives from 2, 6-Dicinnamoylpyridine and 1, 3-Dicinnamoylbenzene. *J. Chem. Res.-Part S Synop.* **1994**, *333*, 168–169.
26. Reddy, D.B.; Seshamma, T.; Seenaiah, B.; Reddy, M.R. Synthesis and Biological Activity of Some New Bis (2-pyrazolin-3-yl) benzenes and-pyridines. *Indian J. Chem.* **1991**, *30B*, 46–51. [CrossRef]
27. Constable, E.C.; Figgemeier, E.; Hougen, I.A.; Housecroft, C.E.; Neuburger, M.; Schaffner, S.; Whall, L.A. Hairpin helicates: A missing link between double-helicates and trefoil knots. *Dalton Trans.* **2005**, 1168–1175. [CrossRef] [PubMed]
28. Tan, Y.; Zhang, Q.; Yu, J.; Zhao, X.; Tian, Y.; Cui, Y.; Hao, X.; Yang, Y.; Qian, G. Solvent effect on two-photon absorption (TPA) of three novel dyes with large TPA cross-section and red emission. *Dyes Pigments* **2013**, *97*, 58–64. [CrossRef]

29. Weissenberger, J.; Priester, M.; Bernreuther, C.; Rakel, S.; Glatzel, M.; Seifert, V.; Kogel, D. Dietary curcumin attenuates glioma growth in a syngeneic mouse model by inhibition of the JAK1,2/STAT3 signaling pathway. *Clin. Cancer Res. Off. J. Am. Assoc. Cancer Res.* **2010**, *16*, 5781–5795. [CrossRef]
30. Subramanian, A.; Narayan, R.; Corsello, S.M.; Peck, D.D.; Natoli, T.E.; Lu, X.; Gould, J.; Davis, J.F.; Tubelli, A.A.; Asiedu, J.K.; et al. A Next Generation Connectivity Map: L1000 Platform and the First 1,000,000 Profiles. *Cell* **2017**, *171*, 1437–1452e1417. [CrossRef]
31. Stathias, V.; Jermakowicz, A.M.; Maloof, M.E.; Forlin, M.; Walters, W.; Suter, R.K.; Durante, M.A.; Williams, S.L.; Harbour, J.W.; Volmar, C.H.; et al. Drug and disease signature integration identifies synergistic combinations in glioblastoma. *Nat. Commun.* **2018**, *9*, 5315. [CrossRef]
32. Huang da, W.; Sherman, B.T.; Lempicki, R.A. Systematic and integrative analysis of large gene lists using DAVID bioinformatics resources. *Nat. Protocols* **2009**, *4*, 44–57. [CrossRef]
33. Li, Y.; Guo, Y.; Tang, J.; Jiang, J.; Chen, Z. New insights into the roles of CHOP-induced apoptosis in ER stress. *Acta Biochim. Biophys. Sin.* **2014**, *46*, 629–640. [CrossRef] [PubMed]
34. Joo, H.; Lee, H.J.; Shin, E.A.; Kim, H.; Seo, K.H.; Baek, N.I.; Kim, B.; Kim, S.H. c-Jun N-terminal Kinase-Dependent Endoplasmic Reticulum Stress Pathway is Critically Involved in Arjunic Acid Induced Apoptosis in Non-Small Cell Lung Cancer Cells. *Phytother. Res.* **2016**, *30*, 596–603. [CrossRef] [PubMed]
35. Zheng, Q.Y.; Li, P.P.; Jin, F.S.; Yao, C.; Zhang, G.H.; Zang, T.; Ai, X. Ursolic acid induces ER stress response to activate ASK1-JNK signaling and induce apoptosis in human bladder cancer T24 cells. *Cell. Signal.* **2013**, *25*, 206–213. [CrossRef] [PubMed]
36. Nakagawa, T.; Zhu, H.; Morishima, N.; Li, E.; Xu, J.; Yankner, B.A.; Yuan, J. Caspase-12 mediates endoplasmic-reticulum-specific apoptosis and cytotoxicity by amyloid-beta. *Nature* **2000**, *403*, 98–103. [CrossRef] [PubMed]
37. Yang, H.; Du, Z.; Wang, W.; Song, M.; Sanidad, K.; Sukamtoh, E.; Zheng, J.; Tian, L.; Xiao, H.; Liu, Z.; et al. Structure-Activity Relationship of Curcumin: Role of the Methoxy Group in Anti-inflammatory and Anticolitis Effects of Curcumin. *J. Agric. Food Chem.* **2017**, *65*, 4509–4515. [CrossRef]
38. Indira Priyadarsini, K. Chemical and structural features influencing the biological activity of curcumin. *Curr. Pharm. Des.* **2013**, *19*, 2093–2100.
39. Zhou, D.; Ding, N.; Zhao, S.; Li, D.; Van Doren, J.; Qian, Y.; Wei, X.; Zheng, X. Synthesis and evaluation of curcumin-related compounds containing inden-2-one for their effects on human cancer cells. *Biol. Pharm. Bull.* **2014**, *37*, 1977–1981. [CrossRef]
40. Bi, K.; Nishihara, K.; Machleidt, T.; Hermanson, S.; Wang, J.; Sakamuru, S.; Huang, R.; Xia, M. Identification of known drugs targeting the endoplasmic reticulum stress response. *Anal. Bioanal. Chem.* **2015**, *407*, 5343–5351. [CrossRef]
41. Trivedi, M.V.; Laurence, J.S.; Siahaan, T.J. The role of thiols and disulfides on protein stability. *Curr. Protein Pept. Sci.* **2009**, *10*, 614–625. [CrossRef] [PubMed]
42. Wang, X.; Thomas, B.; Sachdeva, R.; Arterburn, L.; Frye, L.; Hatcher, P.G.; Cornwell, D.G.; Ma, J. Mechanism of arylating quinone toxicity involving Michael adduct formation and induction of endoplasmic reticulum stress. *Proc. Natl. Acad. Sci. USA* **2006**, *103*, 3604–3609. [CrossRef] [PubMed]
43. Zhu, Z.; Du, S.; Du, Y.; Ren, J.; Ying, G.; Yan, Z. Glutathione reductase mediates drug resistance in glioblastoma cells by regulating redox homeostasis. *J. Neurochem.* **2018**, *144*, 93–104. [CrossRef] [PubMed]
44. Jackson, P.A.; Widen, J.C.; Harki, D.A.; Brummond, K.M. Covalent Modifiers: A Chemical Perspective on the Reactivity of alpha,beta-Unsaturated Carbonyls with Thiols via Hetero-Michael Addition Reactions. *J. Med. Chem.* **2017**, *60*, 839–885. [CrossRef] [PubMed]
45. Wang, C.; Bai, M.; Wang, X.; Tan, C.; Zhang, D.; Chang, L.; Li, G.; Xie, L.; Su, J.; Wen, Y. Estrogen receptor antagonist fulvestrant inhibits proliferation and promotes apoptosis of prolactinoma cells by regulating the IRE1/XBP1 signaling pathway. *Mol. Med. Rep.* **2018**, *18*, 4037–4041. [CrossRef] [PubMed]
46. Minchenko, D.O.; Riabovol, O.O.; Ratushna, O.O.; Minchenko, O.H. Hypoxic regulation of the expression of genes encoded estrogen related proteins in U87 glioma cells: Effect of IRE1 inhibition. *Endocr. Regul.* **2017**, *51*, 8–19. [PubMed]
47. Lhomond, S.; Avril, T.; Dejeans, N.; Voutetakis, K.; Doultsinos, D.; McMahon, M.; Pineau, R.; Obacz, J.; Papadodima, O.; Jouan, F.; et al. Dual IRE1 RNase functions dictate glioblastoma development. *EMBO Mol. Med.* **2018**, *10*, e7929. [CrossRef] [PubMed]

48. Jabouille, A.; Delugin, M.; Pineau, R.; Dubrac, A.; Soulet, F.; Lhomond, S.; Pallares-Lupon, N.; Prats, H.; Bikfalvi, A.; Chevet, E.; et al. Glioblastoma invasion and cooption depend on IRE1alpha endoribonuclease activity. *Oncotarget* **2015**, *6*, 24922–24934. [CrossRef]
49. Lipinski, C.A.; Lombardo, F.; Dominy, B.W.; Feeney, P.J. Experimental and computational approaches to estimate solubility and permeability in drug discovery and development settings. *Adv. Drug Deliv. Rev.* **2001**, *46*, 3–26. [CrossRef]
50. Wuts, P.G.; Greene, T.W. *Greene's Protective Groups in Organic Synthesis*; John Wiley & Sons: Hoboken, NJ, USA, 2006.
51. Graham, R.M.; Hernandez, F.; Puerta, N.; De Angulo, G.; Webster, K.A.; Vanni, S. Resveratrol augments ER stress and the cytotoxic effects of glycolytic inhibition in neuroblastoma by downregulating Akt in a mechanism independent of SIRT1. *Exp. Mol. Med.* **2016**, *48*, e210. [CrossRef]
52. Hombach-Klonisch, S.; Mehrpour, M.; Shojaei, S.; Harlos, C.; Pitz, M.; Hamai, A.; Siemianowicz, K.; Likus, W.; Wiechec, E.; Toyota, B.D.; et al. Glioblastoma and chemoresistance to alkylating agents: Involvement of apoptosis, autophagy, and unfolded protein response. *Pharmacol. Ther.* **2018**, *184*, 13–41. [CrossRef] [PubMed]
53. Dadey, D.Y.; Kapoor, V.; Khudanyan, A.; Urano, F.; Kim, A.H.; Thotala, D.; Hallahan, D.E. The ATF6 pathway of the ER stress response contributes to enhanced viability in glioblastoma. *Oncotarget* **2016**, *7*, 2080–2092. [CrossRef] [PubMed]

cancers

MDPI

Article

Reduction of Human Glioblastoma Spheroids Using Cold Atmospheric Plasma: The Combined Effect of Short- and Long-Lived Reactive Species

Angela Privat-Maldonado [1,2,*], Yury Gorbanev [1], Sylvia Dewilde [3], Evelien Smits [2] and Annemie Bogaerts [1]

1 PLASMANT, Chemistry Department, University of Antwerp, 2610 Antwerp, Belgium; yury.gorbanev@uantwerpen.be (Y.G.); annemie.bogaerts@uantwerpen.be (A.B.)
2 Solid Tumor Immunology Group, Center for Oncological Research, University of Antwerp, 2610 Antwerp, Belgium; evelien.smits@uantwerpen.be
3 Protein Chemistry, Proteomics and Epigenetic Signaling, Department of Biomedical Sciences, University of Antwerp, 2610 Antwerp, Belgium; sylvia.dewilde@uantwerpen.be
* Correspondence: angela.privatmaldonado@uantwerpen.be; Tel.: +32-3265-25-76

Received: 20 September 2018; Accepted: 19 October 2018; Published: 23 October 2018

Abstract: Cold atmospheric plasma (CAP) is a promising technology against multiple types of cancer. However, the current findings on the effect of CAP on two-dimensional glioblastoma cultures do not consider the role of the tumour microenvironment. The aim of this study was to determine the ability of CAP to reduce and control glioblastoma spheroid tumours in vitro. Three-dimensional glioblastoma spheroid tumours (U87-Red, U251-Red) were consecutively treated directly and indirectly with a CAP using dry He, He + 5% H_2O or He + 20% H_2O. The cytotoxicity and spheroid shrinkage were monitored using live imaging. The reactive oxygen and nitrogen species produced in phosphate buffered saline (PBS) were measured by electron paramagnetic resonance (EPR) and colourimetry. Cell migration was also assessed. Our results demonstrate that consecutive CAP treatments (He + 20% H_2O) substantially shrank U87-Red spheroids and to a lesser degree, U251-Red spheroids. The cytotoxic effect was due to the short- and long-lived species delivered by CAP: they inhibited spheroid growth, reduced cell migration and decreased proliferation in CAP-treated spheroids. Direct treatments were more effective than indirect treatments, suggesting the importance of CAP-generated, short-lived species for the growth inhibition and cell cytotoxicity of solid glioblastoma tumours. We concluded that CAP treatment can effectively reduce glioblastoma tumour size and restrict cell migration, thus demonstrating the potential of CAP therapies for glioblastoma.

Keywords: cancer; cold atmospheric plasma (CAP); spheroid shrinkage; cytotoxicity; tumour reduction; glioblastoma; short-lived reactive species; cell migration; proliferation

1. Introduction

Glioblastoma multiforme (GBM) is a highly aggressive neoplasia and the most common aggressive tumour on the central nervous system in the adult population [1]. Current therapies against GBM include maximal surgical resection, radiation and chemotherapy. However, tumour recurrence is highly common, which contributes to the poor survival rates after diagnosis (median survival of 15 months; five-year survival of less than 5%) [2,3]. Despite significant improvements in cancer treatment, GBM remains an incurable neoplasia.

Cold atmospheric plasmas (CAPs) are currently being explored due to their anti-cancer properties [4,5]. The chemical components produced by CAPs, identified as the active agents responsible for the biological effects, have been extensively studied [6,7]. The reactive oxygen and nitrogen species (RONS) generated and delivered by plasma cause functional and structural damage to lipids in cell membranes [8], oxidize proteins [9], induce DNA breaks [10], and promote cell death by different mechanisms [11,12]. In contrast, the physical components produced by plasma such as UV photons and electromagnetic fields on their own have a negligible cellular impact [13]. Furthermore, it has been suggested that CAP is selective towards cancer cells due to the increased presence of aquaporins and the reduced amount of cholesterol in their cell membranes [14,15], which facilitates the access of RONS into the cell. Cancer cells are also less effective in removing extracellular H_2O_2 compared to normal cells [16] (one of the main reactive species produced by CAP), adding to the selectivity of CAP treatments.

The cytotoxic effect of CAPs in glioblastoma has been mainly demonstrated using the conventional two-dimensional (2D) monolayer cell model. Previous studies on glioblastoma cells have shown that CAP reduces the cell viability and induces apoptosis [17]. Even more, CAP treatment reduces clonogenicity and induces cell-cycle arrest in glioblastoma cells resistant to temozolomide, a chemotherapeutic agent used to treat glioblastoma patients [18]. The biomedical effects of CAPs are achieved not only by direct treatment, but also by applying plasma-treated liquids, rich in long-lived species [19,20]. These studies carried out using the 2D monolayer model, although informative, do not reproduce the biological microenvironment of tumours. Recent findings indicate that the tumour microenvironment plays a key role in the response to treatment, regulating tumour progression and metastatic processes [21]. Thus, it is possible that the therapies developed using 2D systems would not meet the requirements to achieve the desired response in three-dimensional (3D) tumours. To date, only few studies have used more complex models such as in vitro 3D culture systems to assess the effect of CAP on different types of cancer [22–24] and one study has tested CAP on glioblastoma tumours in mice [25].

The COST plasma jet was developed as a standard reference plasma jet [26,27]. Our group has previously reported the anticancer capacity of this CAP on 2D glioblastoma cultures [17]. In the present study, we used an in vitro 3D tumour spheroid model that allows cells to behave in a manner that is closer to natural conditions. The model is more relevant than 2D cultures as it enables the in vitro spheroid to develop biophysical properties characteristic of solid tumours (nutrients and oxygen gradients, catabolite accumulation) and the synthesis and assembly of extracellular matrix (ECM) proteins. In this environment, the cells in the spheroid conformation also present proliferation gradients and gene expression profiles closer to the clinical and in vivo models [28,29]. Using this model, we investigated the ability of CAP to reduce tumour size in glioblastoma spheroids in vitro. Both biological and chemical assays were performed to assess the effect of CAP treatments in U87-Red and U251-Red glioblastoma spheroids. We found that both short- and long-lived species were required to exert an efficient inhibitory effect. Our approach thus highlights the need of plasma-generated short-lived RONS to reduce tumour size and control tumour growth, and emphasizes the importance of the tumour microenvironment for the development of more effective plasma therapies.

2. Results

2.1. Identifying an Effective Plasma Treatment

The purpose of our study was to determine the effect of plasma treatment on a 3D tumour spheroid in vitro. We used red-fluorescent cells to facilitate the identification of living cells. The spheroid core (red, proliferative and viable part of the spheroid) was monitored to determine variations in its area relating to growth inhibition or spheroid shrinkage upon CAP treatment. The total spheroid area was measured to assess changes in spheroid size due to cell death and destruction of the spheroid architecture. Cell death in glioblastoma spheroids was monitored using the Cytotox Green reagent, a DNA-binding dye that produces a fluorescent signal in cells with a damaged cell membrane.

We first assessed the cytotoxic effect of a single direct plasma treatment on glioblastoma spheroids in phosphate buffered saline (PBS) using the COST jet with dry He, He + 5% H_2O or He + 20% H_2O. No cytotoxic effect or inhibition of spheroid growth was achieved with the plasma generated using dry He or He + 5% H_2O (not shown). However, we observed a delay in spheroid growth in U87-Red and U251-Red, as well as cell death in U87-Red spheroids treated with the COST jet He + 20% H_2O (Supplementary Information, Figure S1). Admittedly, this effect was not enough to inhibit spheroid growth or to decrease the spheroid core area. Thus, we administered an additional plasma treatment at 24 h.

The effective outcome was achieved after the 2× direct treatment by He + 20% H_2O plasma. Under this condition, the plasma not only inhibited spheroid growth, but also reduced the spheroid core area. This effect was maintained seven days post-treatment in U87-Red spheroids ($p \leq 0.0001$; Figures 1a and 2). In contrast, spheroids 2×-treated with He + 5%H_2O and dry He plasma reduced their spheroid core area, but the spheroid growth resumed after 72 h in both cell lines (Figure 1a). He + 5%H_2O and dry He plasma-treated U251-Red spheroids yielded the same core area as untreated controls ($p \leq 0.0001$; Figure 1a), evidencing a lower response to plasma treatment. The analysis of the total spheroid area evidenced similar values for all the conditions tested in both cell lines, except for He + 20% H_2O in U87-Red (Figure 1b). CAP treatment was cytotoxic only when H_2O was added to the feed gas ($p \leq 0.0001$; Figure 1c), which suggests the participation of reactive oxygen species (ROS) in the overall biological effect. The least effective treatment was with dry He, where only a reduction in spheroid growth but no cytotoxic effect was observed (similar levels to untreated PBS control, Figure 1). This could suggest a modulatory rather than a cytotoxic effect of dry He plasma on glioblastoma cells. We concluded that the 2× He + 20% H_2O plasma treatment effectively inhibited spheroid growth and had a cytotoxic effect in U87-Red and U251-Red glioblastoma spheroids.

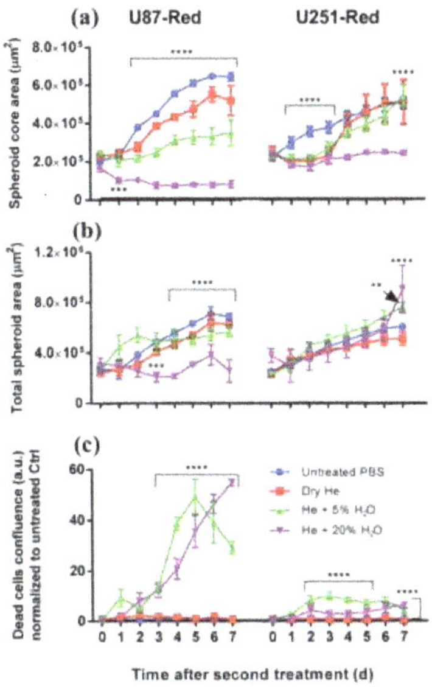

Figure 1. Consecutive plasma treatment inhibited growth in glioblastoma spheroids. (**a**) The area of the spheroid core (viable cells) was reduced by 2× direct plasma treatment generated with dry He, He + 5% H_2O or He + 20% H_2O vapour saturation. The most effective inhibition of growth in both U87-Red and U251-Red was achieved using plasma generated with He + 20% H_2O ($p \leq 0.0001$). (**b**) Similar values of the total spheroid area (comprising living spheroid core and dead cells) were obtained for all the conditions in both cell lines, except for U87-Red, He + 20% H_2O, where the total spheroid area was smaller. (**c**) The addition of H_2O to the gas feed resulted in increased cytotoxicity for both cell lines. Cell death is expressed as the ratio of Cytotox Green$^+$ cells in treated spheroids/untreated controls at each time point (a.u.). Dead cells confluence = confluence percentage of the image area occupied by dead cells. Results corresponding to U87-Red (**left**) and U251-Red spheroids (**right**). Data representative of two independent experiments, 4–6 spheroids per condition. Mean ± SD; ** = $p \leq 0.01$; **** = $p \leq 0.0001$.

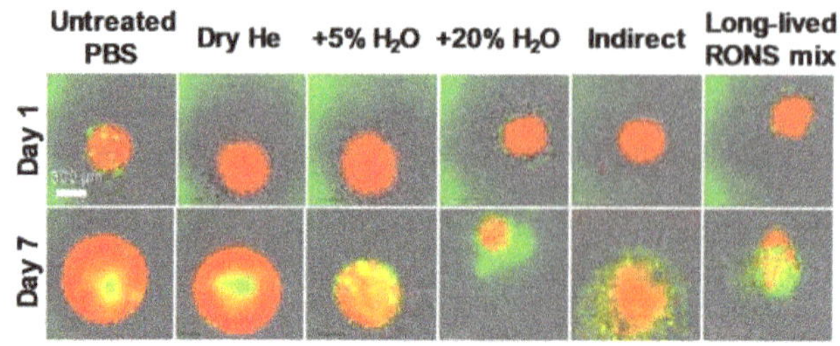

Figure 2. Spheroid growth/shrinkage upon the different treatments applied. Representative images of spheroid growth/shrinkage and cytotoxic effect in U87-Red spheroids on day 1 and 7 days after exposure to the different treatments tested. Living cells in red, Cytotox Green$^+$ cells in green. Scale bar = 300 μm.

2.2. RONS Present in Plasma-Treated PBS (pPBS)

To determine which RONS could be responsible for the biological outcome described above, we measured the short- and long-lived species present in pPBS by electron paramagnetic resonance (EPR) and colourimetry. H_2O_2 was found in pPBS in high concentrations. The concentration was dependent on the H_2O vapour saturation of the feed gas: a higher percent of H_2O saturation resulted in more H_2O_2 production (Figure 3a), in agreement with our previous report for this plasma jet [27]. The short-lived species cannot be detected directly and requires the use of spin traps (which react with radicals to form more persistent nitroxide radical adducts) [30]. We used 5,5-dimethyl-1-pyrroline *N*-oxide (DMPO) to detect •OH, and 5-(diethoxyphosphoryl)-5-methyl-1-pyrroline *N*-oxide (DEPMPO) for $O_2•^-$/•OOH radicals (Figures S2 and S3) [30–32]. DMPO-OH and DEPMPO-OOH adducts were detected in the μM range. Their concentration increased with the increasing H_2O vapour content of the feed gas (Figure 3b), again in agreement with the previously reported data [27].

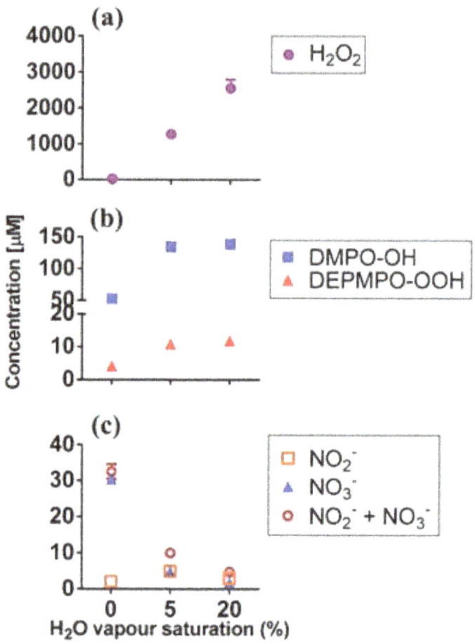

Figure 3. Short- and long-lived RONS present in plasma-treated PBS. (**a**) H_2O_2; (**b**) DMPO-OH and DEPMPO-OOH adducts; (**c**) NO_2^-, NO_3^- and NO_2^- + NO_3^- were detected in plasma-treated PBS by EPR and colourimetry (see Mat. and Met.). Experiments were carried out in 200 μL pPBS, using dry He, He + 5% or 20% H_2O vapour saturation. Data represent mean values ($n \geq 2$).

We note that the concentration of the spin adducts did not correspond to the total amount of the radical induced by CAP in PBS. However, the changes in their concentration under different plasma conditions corresponded to the changes in the radical amount [31,33]. Additionally, the nitroxide moiety of spin adducts was prone to decay in plasma-treated liquids [34]. The direct comparison of different conditions is possible if the concentration of a spin adduct does not decay or reach a flat plateau within the experimental (plasma exposure) time frame [33,35]. We demonstrate here that the DMPO-OH concentration increased near-linearly within the experimental time frame (up to 4 min), thus confirming the applicability of the method (Figure S4).

The spin trap 2,2,6,6-tetramethylpiperidine (TEMP) was used to detect the oxygen species produced by plasma, such as O, $O_2(a^1\Delta g)$, and O_3 [31,33]. The use of NaN_3, an $O_2(a^1\Delta g)$ scavenger, allowed to determine the specific contribution of $O_2(a^1\Delta g)$ to the formation of 2,2,6,6-tetramethylpiperidine 1-oxyl (TEMPO), a stable and EPR-detectable nitroxide radical, via the oxidation of TEMP. However, we only detected minor amounts of the formed TEMPO (below 9 μM) compared to the impurities in untreated samples (Figure S5), and they did not change with varied H_2O saturation of the feed gas or introduction of NaN_3 (not shown). Previously, we have demonstrated a very efficient plasma-induced degradation of TEMPO derivatives [34]. Hence, we attribute the observed effect to two factors: (1) low concentrations of $O/O_2(a^1\Delta g)/O_3$ produced by plasma in the presence of water vapour [31]; and (2) rapid degradation of the formed TEMPO by •OH radicals which are present in high concentrations (see Figure 3b).

We also attempted to detect •NO radicals using N-methyl-D-glucamine dithiocarbamate (MGD)-iron(II) complex ($(MGD)_2Fe^{2+}$) and 2-phenyl-4,4,5,5-tetramethylimidazoline 1-oxyl 3-oxide (PTIO) spin traps, as described in Section 4.8 (see Mat. and Met.). Under all plasma conditions, no •NO was detected. These results agreed with the low concentrations of other measured (long-lived) reactive nitrogen species (RNS) induced by CAP in pPBS (Figure 3c).

Thus, we can tentatively ascribe the biological effects of CAP to the following species: •OH, O_2•$^-$/•OOH, H_2O_2, NO_2^-/NO_3^-. We acknowledge that other biologically-relevant, plasma-induced species such as ClO^- [36], $ONOO^-$ [37,38] and $OONOO^-$ [39] could be present in pPBS, but these were not monitored in this study. However, the two latter species are likely to be present in low concentrations (if any), like all other detected RNS, and ClO^- is formed in large quantities, mostly in the presence of oxygen species such as atomic oxygen [36].

2.3. Determining the Role of the Short- and Long-Lived Species in Plasma on Spheroid Viability

To determine whether the biological effect observed was due only to the long-lived species or to the combination of short- and long-lived species produced by CAP, we assessed the effect of: (1) 2× direct plasma treatment; (2) 2× indirect plasma treatment (2× pPBS); and (3) 2× RONS mix in PBS (with the same concentrations of the long-lived species H_2O_2 and NO_2^- as in the pPBS used) on glioblastoma spheroids. NO_2^- can undergo further oxidation in the liquid to yield NO_3^- [40]. Thus, we used NO_2^- in the concentration that corresponded to the combined amount of NO_2^- and NO_3^-. We acknowledge that the actual concentration of NO_2^- induced in PBS by CAP may have been lower.

The 2× treatments with either pPBS or RONS mix were less effective than the 2× direct treatment in inhibiting spheroid growth in both glioblastoma cell lines, especially in U87-Red ($p \leq 0.0001$; Figures 2 and 4a). The direct treatment induced spheroid shrinkage in U87-Red that was maintained for the whole duration of the experiment, whereas spheroid growth resumed 72 h after in the spheroids exposed to 2× pPBS and 2× RONS mix. Although the 2× direct plasma treatment was not able to completely inhibit U251-Red spheroid growth (Figure 4a), a statistically significant difference was observed between this treatment and the other two conditions tested here ($p \leq 0.001$ at t = 168 h). Thus, we conclude that the 2× direct plasma treatment was more efficient than the 2× pPBS and RONS mix in inhibiting spheroid growth.

In both cell lines, spheroids exposed to 2× RONS mix showed an increased total spheroid area compared to the untreated control (Figure 4b). In U87-Red, this was also observed in 2× pPBS-treated spheroids. This effect could be explained by the cell detachment of dead cells from the spheroid core that add to the total spheroid area (Figure 4c). However, we observed that the high cytotoxicity of the 2× direct treatment (Figure 4c) did not correlate with an increase in the total spheroid area (Figure 4b). Thus, it appears that the 2× direct treatment was able not only to inhibit spheroid growth and kill cancer cells as shown here, but also to restrict the cell detachment of dead cells and spheroid tissue from the spheroid core. Although we acknowledge the participation of the long-lived species present in pPBS and in the RONS mix, our results indicate that the short-lived species delivered by plasma during the 2× direct treatment are required to fully inhibit spheroid growth in U87-Red and to delay it in U251-Red.

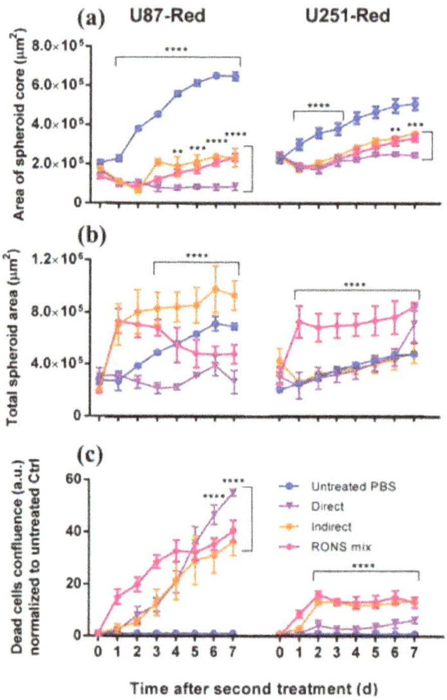

Figure 4. Short- and long-lived species delivered by plasma are needed to inhibit spheroid growth. (**a**) A statistically significant difference was observed in the area of the spheroid core (viable cells) in spheroids exposed to 2× direct plasma treatment compared to those treated with pPBS (indirect) and the RONS mix (U87-Red, $p \leq 0.0001$; U251-Red, $p \leq 0.001$). (**b**) U87-Red spheroids treated with pPBS and the RONS mix showed higher total spheroid area than spheroids treated 2× directly with plasma ($p \leq 0.0001$). In U251-Red spheroids, higher total spheroid area was observed only in spheroids treated with the RONS mix. (**c**) The amount of cell death induced in spheroids by pPBS and RONS mix were comparable. Cell death is expressed as the ratio of Cytotox Green$^+$ cells in treated spheroids/untreated controls at each time point (a.u.). Dead cells confluence = confluence percentage of the image area occupied by dead cells. Results corresponding to U87-Red (**left**) and U251-Red (**right**) spheroids. Data representative of two independent experiments, 4–6 spheroids per condition. Mean ± SD; ** = $p \leq 0.01$; *** = $p \leq 0.001$; **** = $p \leq 0.0001$.

2.4. *The 2× Direct Plasma Treatments Inhibited Cell Migration*

Glioblastomas are characterized by their high migratory nature, which allows them to metastasize into surrounding brain tissues [1]. We investigated the ability of our highly cytotoxic direct plasma treatment (2× plasma treatment, He + 20% H_2O) to inhibit the migration of the cells that survive the treatment. Our results demonstrated that the 2× direct plasma treatment reduced cell migration (Figure 5a,b). The 2× direct treatment had a significant effect on both cell lines, with little or no cell migration 24 h post-treatment ($p \leq 0.0001$) and a hindered migration in U87-Red cells 48 h post-treatment ($p \leq 0.0001$; Figure 5a,b). This effect was maintained 72 h post-treatment only in U251-Red spheroids ($p \leq 0.0001$). Lower cell migration inhibition was observed in 2× pPBS- and RONS mix-treated spheroids at all time points (except after 48 h for U87-Red and 72 h for both cell lines). We observed morphological changes in the spheroids upon plasma treatment, specifically the shrinkage of the spheroid core and presence of dead cells around it (Figure 5c). Whereas control spheroids presented a homogeneous migration pattern with cells spreading in all directions around the spheroid core, the pattern was asymmetrical in plasma-treated spheroids with a reduced number of migratory cells. This could be due to the amount of dead cells in the outer layer of the spheroid. The differences observed between the migration dynamics of both cell lines could also indicate different survival strategies.

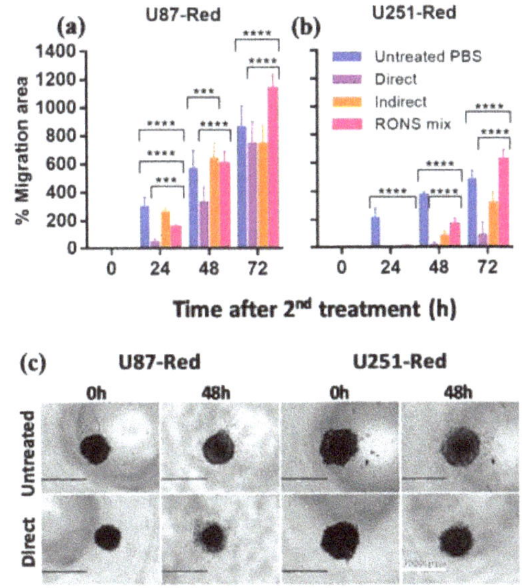

Figure 5. The migratory ability of U87-Red and U251-Red cells is affected by the 2× direct plasma treatment. The 2× direct plasma treatment inhibited cell migration for (**a**) up to 48 h in U87-Red and (**b**) up to 72 h in U251-Red. 2× pPBS and RONS mix had a significant but less inhibitory effect, reaching similar migration areas than the untreated PBS controls after 48 h (U87-Red) and 72 h (U251-Red). (**c**) Representative images of U87-Red and U251-Red cell migration (0 and 48 h after consecutive treatments). % Migration area = 100*(area T_x-T_0/area T_0). Data representative of two independent experiments, 4–6 spheroids per condition. Mean ± SD; *** = $p \leq 0.001$; **** = $p \leq 0.0001$. Scale bar = 1000 μm.

Based on these results, we concluded that the 2× direct plasma treatment effectively inhibited the migratory ability of U87-Red and U252-Red spheroids and this treatment was more efficient than the pPBS and RONS mix. Thus, the combination of short-lived species delivered by CAP is needed to effectively reduce cell migration in glioblastoma spheroids.

2.5. ki67 Expression Was Reduced by 2× Direct Plasma Treatment

To further characterize the inhibitory ability of our plasma treatment, the expression of the proliferation marker ki67 in spheroids was assessed by immunohistochemistry. The percentage of ki67$^+$

cells was reduced in U87-Red spheroids exposed to the 2× direct plasma treatment ($p < 0.05$; Figure 6a). Although no statistically significant difference was found between untreated and 2× directly-treated U251-Red, a trend to a reduced ki67 expression in 2× direct plasma-treated cells was observed (Figure 6b). The pPBS and RONS mix treatments produced similar %ki67⁺ cells comparable to the untreated controls in both cell lines. In addition to the reduced ki67 expression, the CAP-treated spheroids presented a more fragile structure, as shown in Figure 6c. Altogether, our results suggest that our 2× direct plasma treatment reduced ki67 expression in glioblastoma spheroids. This reduction correlates with the inhibition of growth observed in the 2× plasma-treated spheroids, as described in Sections 2.1 and 2.3.

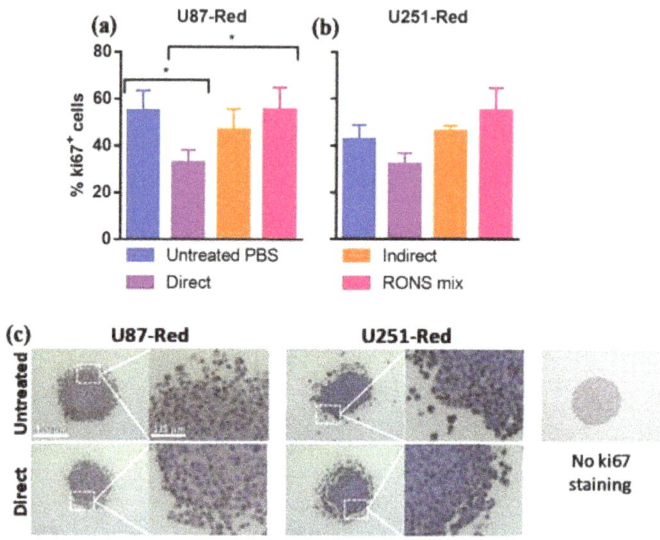

Figure 6. The 2× direct plasma treatments reduced the expression of the proliferative marker ki67 in glioblastoma spheroids. The 2× direct plasma treatment decreased the % ki67⁺ cells in (**a**) U87-Red spheroids ($p < 0.05$) and was lower than in those untreated or treated with pPBS and RONS mix ($p < 0.05$). (**b**) The same trend was observed in U251-Red spheroids ($p > 0.05$). The staining was scored using IHC Profiler in ImageJ. (**c**) Representative images of ki67 staining of U87-Red and U251-Red spheroids exposed to untreated PBS or He + 20% H_2O direct treatment. Enlarged image demonstrates variable ki67 staining. No ki67 staining control on the right. Data representative of two independent experiments, 4–6 spheroids per condition. Mean ± SD; * = $p < 0.05$. Scale bar = 500 μm.

3. Discussion

Although significant preclinical work has been done in the field of medical plasmas for cancer, there are still unknowns regarding the efficacy of these treatments in natural conditions. A major concern is the use of models that do not consider the tumour microenvironment, such as the widely used 2D cell culture model. This comes at the cost of developing therapies tailored for a 2D model that might not be able to control tumour cell growth in natural conditions due to physiological and structural differences [41]. To address this problem, plasma treatment in 3D glioblastoma spheroids was explored in this study for the first time. The model resembled the nutrient and oxygen gradients and architecture of solid tumours and provided a more accurate representation of treatment outcomes under 3D conditions.

In this study, we demonstrate the importance of short-lived species delivered by CAP for the reduction of 3D glioblastoma spheroids. Interestingly, a single plasma treatment that would induce cell cytotoxicity in the 2D culture model was not sufficient to inhibit spheroid growth or to induce spheroid shrinkage, thus demonstrating the importance of using 3D culture models. However, the 2× direct plasma treatment administered 24 h after the initial treatment generated with 20% vapour saturation elicited a long-lasting cytotoxic and inhibitory effect in U87-Red spheroids and in U251-Red spheroids, though to a lesser degree. Whereas the total spheroid area and spheroid core area of untreated spheroids were of similar magnitude (no cell death, no spheroid damage), the increased total spheroid area of plasma-treated spheroids could be due to other factors. It is possible that the combination of cell death and oxidation of ECM caused by RONS damaged the spheroid architecture, resulting in an increased total spheroid area. In turn, the high cytotoxic effect could at least partially be explained by the high concentration of H_2O_2 produced in the PBS and its diffusion to the intracellular compartment aided by aquaporins [42]. However, we observed that only the 2× direct plasma treatment induced complete growth inhibition in U87-Red spheroids and reduced spheroid growth in U251-Red, while 2× pPBS and RONS mix treatments failed to induce such effects. We tentatively attribute this difference to the short-lived reactive species, most likely ROS such as •OH and $O_2•^-$ radicals (as RNS were only formed in very low concentrations), present only during the direct plasma treatment. Both the cytotoxic effect and the concentration of ROS in our plasma treatment increased with the increase of H_2O vapour content in the feed gas, confirming this hypothesis.

As cell migration is an important feature of malignant, metastatic cancers, we assessed the ability of glioblastoma spheroids to migrate upon plasma treatment. Cells that survived the 2× direct plasma treatment presented reduced migratory activity compared to those exposed to the pPBS and RONS mix. This is in agreement with the findings on human metastatic breast cancer cells, where plasma treatment significantly inhibited migration and invasion of BrCa cells and fibroblasts [43,44]. Besides oxidation, plasma can also reduce the expression of integrin, an important cell surface protein that participates in cell motility and adhesion [43]. This effect is independent of the cell cycle phase, as plasma affects cells in all stages of the cell cycle [45]. Thus, this effect could be due to the destruction of ECM components and the oxidation of cell adhesion molecules on the cell surface [12]. Further studies on the expression of adhesion molecules and ECM oxidation in glioblastoma spheroids upon CAP treatment are needed to better understand the effect of plasma on cell migration.

The brain tissue is particularly susceptible to oxidative stress due to its high requirement for oxygen and its poor defence mechanisms against oxidative stress [46]. The difference in the response to treatment observed between both glioblastoma cell lines studied here could be due to different sensitivities to oxidative stress as a consequence of specific mutations [17], to different proliferative rates or metabolic activity [47]. For example, p53 (p53) and p16 (encoded by CDKN2A) are tumour suppressor proteins that control the cell cycle and are associated with cancer development [48]. It has been reported that mutations in p53 and CDKN2A impair their corresponding protein function, favouring the accumulation of DNA damage and making cells more sensitive to oxidative damage [49,50]. We have previously confirmed that the p53 status of both cell lines remained the same after cell passage (data not shown). U87-Red (p53 wild type, CDKN2A^{mut}) was more sensitive to plasma treatment than U251-Red (p53^{mut}, CDKN2A^{mut}). Although p53^{mut} cells are reported to be more sensitive to plasma [51], this was not observed in our study. Thus, the mechanism of action of plasma might be independent of p53 [52]. Mutations in other genes such as CDKN2A could contribute to the sensitivity to plasma [17]. The combination of multiple mutations present in cancer cell lines could provide one explanation for the different responses to CAP treatment obtained here. In addition, the difference in protein expression between both cell lines could also explain the differences in the treatment outcome, as it has been demonstrated that the U87 and U251 cell lines differ in their proliferation, migration, and invasion abilities [53].

Supporting the growth inhibition exhibited in the 2× direct plasma-treated spheroids, we found a reduction in the expression of the proliferative marker ki67 in U87-Red and to a lesser extent in U251-Red. This marker is present in cells in G_1, S, G_2 and mitosis and absent in quiescent cells [54]. We demonstrated that the 2× direct plasma treatment, unlike pPBS or RONS mix, decreased the percentage of ki67$^+$ cells in spheroids of both cell lines. Ki67$^+$ cells were mainly located in the outer layer of the spheroids, corresponding to the proliferative fraction of the tumour. Our results correlated with previous findings reporting a decreased ki67 expression in the pancreatic tumours of mice treated with plasma-treated medium [55] and in 3D colorectal spheroids directly exposed to a plasma jet [24]. It is worth noticing that a reduction in ki67 expression was observed in spheroids immediately after the application of the second CAP treatment. Due to technical reasons, it was not possible to collect the spheroids for immunohistochemistry at later stages without disturbing their overall integrity, as plasma reduced the spheroid size and affected the structure of the spheroid (as observed in Figures 2 and 6c). We speculate that ki67 expression in CAP-treated spheroids could further decrease over time, correlating with the sustained inhibition of growth observed here. The anti-proliferative effect of plasma, combined with the cytotoxic effect displayed, contributed to the reduction and control of the glioblastoma spheroids described here. Thus, we concluded that the 2× direct plasma treatment, rich in short- and long-lived species, can decrease ki67 expression and reduce the proliferative ability of cancer cells.

The results obtained in the present work could explain the findings of Chen et al. in athymic nude mice [25]. In that study, CAP treatment reduced the size of glioblastoma tumours in immunocompromised mice. This could have been achieved by the combination of effects described in our work: damage to the tumour microenvironment, the induction of cell death, the reduction of cell proliferation and the inhibition of cell migration. Our findings correlate with the results obtained by Chen et al., as the direct CAP treatments effectively reduced the glioblastoma spheroid size. In addition, the findings support the need of CAP-generated short-lived species for more effective therapies against glioblastoma. Our findings support the benefits of using the 3D spheroid model for CAP research, as spheroids can provide more biological information on CAP treatment in a 3D structures, which are closer to natural conditions. The use of this technology reduces animal testing and allows high throughput analyses, bridging the gap between in vitro and in vivo models for anticancer CAP treatments.

To date, the 3D spheroid model has been used only in three studies to investigate the effect of plasma treatments on HTC116 colorectal and head and neck FaDu cancer cells [22,23]. While the most recent article indicates that four consecutive treatments with plasma-activated medium were needed to inhibit FaDu spheroid growth [23], our therapy was able to achieve tumour reduction with only two applications, as it does not rely solely on the presence of long-lived species such as H_2O_2 for its efficacy. In principle, liquid solutions with CAP-induced long-lived species may not require plasma for their generation: we were able to generate a RONS mix with commercially available chemicals. However, plasma is required to efficiently deliver short-lived species, which substantially contribute to the effects of CAP, as we demonstrate here.

This is the first report of plasma inducing sustained tumour shrinkage and growth inhibition in glioblastoma spheroids. Also for the first time, we demonstrated the direct correlation between the in-liquid identified plasma RONS and the biological effect of CAP on 3D spheroid models. We showed that the 2× plasma treatment with the COST jet caused the inhibition of both spheroid growth and cell migration, and ultimately a reduction in ki67 expression in glioblastoma tumour spheroids. Our results demonstrated the importance of short-lived species delivered by plasma in the overall inhibitory effect, as both contribute to the elimination of 3D glioblastoma tumours. Furthermore, the different responses to treatment observed between the cell lines studied here emphasize the need to use relevant models to investigate the response to treatment under more representative conditions. Our findings regarding glioblastoma spheroids support the potential of CAP as a tool against cancer. Further studies on 3D spheroids or more complex 3D in vitro models using co-cultures of glioblastoma cells and astrocytes could bring light to the effect of CAP on the ECM and the safety of CAP application on benign cells. Likewise, the CAP treatments described here could be applied to spheroids generated from primary cells or complex organoids that more closely resemble the architecture of solid tumours, as this is an indispensable step towards the development of more effective CAP therapies.

4. Materials and Methods

4.1. Cell Lines and Reagents

Human GBM cell lines U87 and U251 were obtained from the Cell Line Service GmbH and kindly provided by Dr. Nicolas Goffart (Université de Liège, ULG), respectively. Cells were grown in Dulbecco's Modified Eagle's medium (DMEM) supplemented with 10% foetal bovine serum (FBS, Gibco, Fisher Scientific, Merelbeke, Belgium) with 2 mM L-Glutamine (Life Technologies, Eggenstein, Germany) and maintained in a tissue culture incubator at 37 °C with 5% CO_2. Cells were transduced with the Nuclight Red Lentivirus reagent (Essen Biosciences, Ann Arbor, MI, USA) using their standard transduction protocol. The transduced cells referred here to as U87-Red and U251-Red, expressed nuclear mKate2 red fluorescent protein.

4.2. Generation of Spheroids

Cell suspensions were prepared at 5×10^4 cells/mL density in culture medium supplemented with 0.24% methylcellulose to enhance spheroid formation. Cells were seeded in ultra-low attachment (ULA) 96-well plates (round bottom, Corning Costar, Corning, NY, USA) at a concentration of 5000 cells/well in 100 µL of DMEM and centrifuged for 10 min at 200× g. Methylcellulose was prepared as previously described [56]. Spheroids were formed for 3 days at 37 °C in a 5% CO_2 humidified atmosphere.

4.3. COST Jet Plasma Setup

The COST plasma jet used in this work is described elsewhere [17,26,27]. The plasma was sustained at 250 VRMS and an operating frequency of 13.56 MHz. It was operated with a feed gas of He with 0%, 5% and 20% H_2O vapour admixture achieved using the split He flow by passing part of it through an H_2O-filled Drechsel flask. For the detailed setup description, see Supplementary Information (SI), Figure S6. H_2O content is expressed as the percentage of the relative saturation of the feed gas at 21–22 °C, room temperature (RT) during the experiments.

4.4. Treatments

Before treatment, 3-day-old spheroids were washed once with PBS to remove the culture medium. The treatment regimen consisted of a single or two consecutive plasma treatments of 3 min each administered 24 h apart (referred to as 2× plasma treatment). Direct (spheroids in 200 µL of PBS in ULA plate) or indirect treatments (pPBS: 200 µL of PBS in ULA plate exposed to CAP, then transferred to spheroids) were performed with the different feed gas admixtures described above. Spheroids in 200 µL of untreated PBS were used as negative controls. A RONS mix solution of PBS containing the same concentrations of H_2O_2 and NO_2^- detected in He + 20% H_2O plasma-treated PBS was used to assess the role of the long- and short-lived species. Spheroids were incubated for 90 min with the corresponding treatment, after which it was replaced with warm, fresh medium prepared according to the downstream experiment.

4.5. Cytotoxicity Assay

The Cytotox Green reagent (10 nM, Essen Biosciences) was used for real-time quantification of cell death. Plates were incubated in the IncuCyte Live-Cell Analysis System (Sartorius, Ann Arbor, MI, USA) and images were collected every 24 h for 7 days. Data were analysed with the IncuCyte ZOOM version 2016B (Essen BioScience) to collect the following metrics: (1) spheroid core area, corresponding to the proliferative region only (red live cells; calculated as the mean fluorescent object area); (2) total spheroid area, corresponding to the combined viable, proliferative core and the Cytotox Green$^+$ region (total spheroid region measured from phase contrast images); and (3) the amount of Cytotox Green$^+$ in treated spheroids (confluence percentage of the image area occupied by green objects) normalized to the untreated control at each time point, representing the cytotoxic effect of the treatment.

4.6. Cell Migration Assay

A 2D migration assay was done in 96-well, flat-bottomed plates coated with 0.1% (*v*/*v*) gelatin [57]. Unbound gelatin was rinsed twice with PBS and wells were blocked with 1% (*w*/*v*) BSA in PBS for 1 h (RT). Spheroids were gently transferred to a migration plate containing 200 μL of DMEM supplemented with 2% FBS and 2 mM L-Glutamine. Spheroids were allowed to adhere to the plate for 30 min before imaging. Images were collected using the EVOS XL Core Cell Imaging System (Life Technologies). Migration was scored using ImageJ. To calculate the migration, the following formula was used: %Migration area = 100*(area T_x-T_0/area T_0), where T_0 corresponds to the migrated area at t = 0 h and T_x corresponds to the migrated area at each time point assessed.

4.7. Immunohistochemical Analysis

Spheroids were collected immediately after the second treatment and fixed with 4% paraformaldehyde for 24 h at 4 °C. Fixed spheroids were transferred to a 4% agarose mould as described before [58]. The agarose pads were paraffin-embedded. Sections of 5 μm were cut, deparaffinised and rehydrated prior to staining. Antigen retrieval was performed with citrate buffer (10 mM, pH 6), at 96 °C for 20 min. Sections were permeabilised in 0.1% Tween-20 and blocked with 3% H_2O_2 in PBS (10 min, RT). The primary antibody incubation was 40 min at RT (1/75 dilution; Mouse Anti-Human Ki-67 Antigen, Clone MIB-1, Agilent, Santa Clara, CA, USA), followed by incubation with the secondary antibody (30 min at RT; Envision Flex HRP). Diaminobenzidine was used to visualize positive staining and haematoxylin to counterstain. Sections were imaged with Zeiss AxioImager Z1 microscope (Carl Zeiss, Göttingen, Germany) equipped with an AxioCam MR ver.3.0. The number of ki67$^+$ and haematoxylin$^+$ cells was counted using ImageJ and the plugin IHC Profiler [59]. %ki67$^+$ cells = (number ki67$^+$ cells/number haematoxylin$^+$ cells)*100.

4.8. Electron Paramagnetic Resonance (EPR) Spectroscopy

The presence of hydroxyl radical (•OH), ozone (O_3)/atomic oxygen (O), singlet oxygen (O2(a1Δg)) and superoxide radical anions (O_2•$^-$) was assessed by EPR using spin trapping, as described elsewhere [14,31]. PBS solutions of spin traps 2,2,6,6-tetramethylpiperidine (TEMP, ≥99%, Sigma Aldrich, Overijse, Belgium), 5,5-dimethyl-1-pyrroline *N*-oxide (DMPO, ≥98%, Sigma Aldrich) and 5-(diethoxyphosphoryl)-5-methyl-1-pyrroline *N*-oxide (DEPMPO, ≥99%, Santa Cruz Biotechnology, Heidelberg, Germany) were freshly prepared at 0.1 M concentration prior to use. The $O_2(a^1\Delta g)$ scavenger sodium azide (NaN_3, ≥99%, Sigma Aldrich) at a 0.1 M concentration was added to a solution of TEMP to distinguish between the amount of 2,2,6,6-tertramethylpiperidine 1-oxyl (TEMPO) formed from O_3/O and that from $O_2(a^1\Delta g)$ [6,30]. In a typical experiment, 200 μL of PBS containing spin traps were exposed to plasma for 1–4 min.

The •NO radical was monitored using spin traps 2-phenyl-4,4,5,5-tetramethylimidazoline 1-oxyl 3-oxide (PTIO; ≥98%, Enzo Life Sciences, Antwerp, Belgium) and *N*-methyl-D-glucamine dithiocarbamate (MGD)-iron(II) complex $(MGD)_2Fe^{2+}$ [30,34,35]. In the first case, 0.2 mM solution of PTIO was exposed to plasma, after which it was immediately analysed by EPR to assess the formation of 2-phenyl-4,4,5,5-tetramethylimidazoline 1-oxyl (PTI). To prepare $(MGD)_2Fe^{2+}$, 100 μL of 4 mM solution of $FeSO_4$•$7H_2O$ (≥99%, Sigma Aldrich, Overijse, Belgium) were mixed with 100 μL of 20 mM solution of MGD sodium monohydrate (≥98%, Enzo Life Sciences). To minimize the oxidation of Fe^{2+} to Fe^{3+} and the loss of the paramagnetic nature of the complex, both solutions were prepared in PBS that was degassed with Ar (99.998%, Praxair, Schoten, Belgium) for 2 min, and 50 μL of 20 mM solution of $Na_2S_2O_3$ (99%, Sigma Aldrich) were added after the exposure [34,60].

EPR measurements were carried out on a Magnettech MiniScope MS 200 spectrometer with the following parameters: frequency 9.4 GHz, power 5 dBm (3.16 mW) or 15 dBm (31.6 mW), modulation frequency 100 kHz, modulation amplitude 0.1 mT, sweep time 40 s, time constant 0.1, sweep width 15 mT, number of scans 3. For the measurements, the analysed samples were contained in 50 μL glass capillaries (Hirschmann, Eberstadt, Germany). The concentrations reported were obtained via double integration (SpectrumViewer ver. 2.6.3 [61]) of the respective simulated spectra (NIH P.E.S.T. WinSIM software ver. 0.96 [62]) of the formed nitroxide radical adducts (Figure S2). For simulations, the hyperfine coupling constants were obtained from the available literature [63]. The calibration of the EPR signal (double integral intensity as a function of a radical concentration) was carried out using a stable radical 4-hydroxy-2,2,6,6-tetramethylpiperidine 1-oxyl (96%, Sigma Aldrich), as previously described [27].

4.9. Colourimetric Assays

Colourimetric assays were used to quantify H_2O_2, nitrite (NO_2^-) and nitrate (NO_3^-) in pPBS in the absence of spheroids. The potassium titanium (IV) oxalate method was used to measure H_2O_2 as described elsewhere [40,64]. Immediately after plasma treatment, 50 μL of 80 mM NaN_3 solution, 200 μL of pPBS and 50 μL of 0.1 M $K_2TiO(C_2O_4)_2 \bullet 2H_2O$ (≥98% Ti basis, Sigma Aldrich) were mixed and analyzed by UV-Vis absorption using a spectrophotometer (Thermo Fischer Genesys 6) at 400 nm. The concentrations of NO_2^- and NO_3^- were measured using the Nitrate/Nitrite Colourimetric Assay Kit (Cayman Chemical, Ann Arbor, Michigan, USA) [65]. For NO_2^- detection, 100 μL of pPBS were mixed with 50 μL of Griess Reagent 1 and 50 μL of Griess Reagent 2, incubated for 6 min. Samples were measured in a 96 flat-bottom well plate using the BIO-RAD iMark microplate reader at 540 nm. For NO_3^- detection, 80 μL of pPBS were mixed with 10 μL of nitrate reductase and 10 μL of nitrate reductase cofactor and incubated for 1h at RT. This way, all NO_3^- is converted into NO_2^-, which was measured. The amount of NO_3^- was calculated as the difference of the two values obtained.

The calibration was performed using solutions of H_2O_2 (30 wt%, Fisher Scientific S.P.R.L., Brussels, Belgium) and $NaNO_2$ (provided in the kit). Background absorbance and PBS evaporation were accounted for in the final concentration values.

4.10. Statistical Analysis

One- (ki67 staining) and two-way ANOVA (cytotoxicity and migration assays) followed by Tukey's multiple comparison test were performed using Prism v.6.01 (GraphPad Software, La Jolla, CA, USA). Statistical significance was set at $p < 0.05$.

5. Conclusions

We conclude that CAP treatment can effectively reduce 3D glioblastoma spheroid growth, cell migration and cell proliferation. Our results indicate the importance of the CAP-generated short-lived species for the growth inhibition and cell cytotoxicity of solid glioblastoma tumours, as they are necessary to achieve a sustained reduction of 3D glioblastoma spheroids in vitro. These results demonstrate the potential of CAP therapies for cancer treatment.

Supplementary Materials: The following are available online at http://www.mdpi.com/2072-6694/10/11/394/s1, Figure S1: Cytotoxic effect of single plasma treatment in U87-Red and U251-Red spheroids; Figure S2: Calculation of the concentration of spin radical adduct present in the sample using DMPO; Figure S3: Representative experimental and simulated EPR spectra of the radical adducts of DEPMPO formed in pPBS; Figure S4: DMPO-OH adduct formation as a function of plasma treatment time; Figure S5: Representative experimental and simulated EPR spectra of TEMPO: impurities in an untreated PBS sample and TEMPO measured in pPBS; Figure S6: Schematic of the COST plasma jet.

Author Contributions: Conceptualization, A.P.-M., Y.G. and A.B.; Formal analysis, A.P.-M.; Funding acquisition, Y.G., E.S. and A.B.; Investigation, A.P.-M. and Y.G.; Methodology, A.P.-M.; Resources, S.D., E.S. and A.B.; Supervision, S.D., E.S. and A.B.; Validation, A.P.-M.; Visualization, A.P.-M.; Writing–original draft, A.P.-M.; Writing–review & editing, A.P.-M., Y.G., S.D., E.S. and A.B.

Funding: This work was supported by the Olivia Hendrickx Research Fund and by Methusalem Funding. Y.G. thanks the Horizon2020 European Marie Skladowska-Curie Individual Fellowship programme (LTPAM 743151) for funding his work.

Acknowledgments: The authors thank Paul Cos (Department of Pharmaceutical Sciences, University of Antwerp) for providing EPR equipment and Christophe Hermans for his help with the immunohistochemical experiments.

Conflicts of Interest: The authors declare no conflict of interest.

References

1. *Glioblastoma [Internet]*; de Vleeschouwer, S. (Ed.) Codon Publications: Brisbane, Australia, 2017.
2. Koshy, M.; Villano, J.L.; Dolecek, T.A.; Howard, A.; Mahmood, U.; Chmura, S.J.; Weichselbaum, R.R.; McCarthy, B.J. Improved survival time trends for glioblastoma using the SEER 17 population-based registries. *J. Neurooncol.* **2012**, *107*, 207–212. [CrossRef] [PubMed]
3. Ostrom, Q.T.; Gittleman, H.; Xu, J.; Kromer, C.; Wolinsky, Y.; Kruchko, C.; Barnholtz-Sloan, J.S. CBTRUS statistical report: Primary brain and other central nervous system tumors diagnosed in the United States in 2009–2013. *Neuro. Oncol.* **2016**, *18*, v1–v75. [CrossRef] [PubMed]
4. Yan, D.; Sherman, J.H.; Keidar, M. Cold atmospheric plasma, a novel promising anti-cancer treatment modality. *Oncotarget* **2017**, *8*, 15977–15995. [CrossRef] [PubMed]
5. Weltmann, K.D.; von Woedtke, T. Plasma medicine—Current state of research and medical application. *Plasma. Phys. Contr. F.* **2017**, *59*. [CrossRef]
6. Lu, X.; Naidis, G.V.; Laroussi, M.; Reuter, S.; Graves, D.B.; Ostrikov, K. Reactive species in non-equilibrium atmospheric-pressure plasmas: Generation, transport, and biological effects. *Phys. Rep.* **2016**, *630*, 1–84. [CrossRef]
7. Graves, D.B. Reactive species from cold atmospheric plasma: Implications for cancer therapy. *Plasma Process. Polym.* **2014**, *11*, 1120–1127. [CrossRef]
8. Furuta, R.; Kurake, N.; Ishikawa, K.; Takeda, K.; Hashizume, H.; Tanaka, H.; Kondo, H.; Sekine, M.; Hori, M. Intracellular responses to reactive oxygen and nitrogen species, and lipid peroxidation in apoptotic cells cultivated in plasma-activated medium. *Plasma Process. Polym.* **2017**, *14*. [CrossRef]
9. De Backer, J.; Razzokov, J.; Hammerschmid, D.; Mensch, C.; Hafideddine, Z.; Kumar, N.; van Raemdonck, G.; Yusupov, M.; van Doorslaer, S.; Johannessen, C.; et al. The effect of reactive oxygen and nitrogen species on the structure of cytoglobin: A potential tumor suppressor. *Redox Biol.* **2018**, *19*. [CrossRef] [PubMed]
10. Hirst, A.M.; Simms, M.S.; Mann, V.M.; Maitland, N.J.; O'Connell, D.; Frame, F.M. Low-temperature plasma treatment induces DNA damage leading to necrotic cell death in primary prostate epithelial cells. *Br. J. Cancer* **2015**, *112*, 1536–1545. [CrossRef] [PubMed]
11. Lin, A.G.; Xiang, B.; Merlino, D.J.; Baybutt, T.R.; Sahu, J.; Fridman, A.; Snook, A.E.; Miller, V. Non-thermal plasma induces immunogenic cell death in vivo in murine CT26 colorectal tumors. *OncoImmunology* **2018**, *7*, 1–13. [CrossRef] [PubMed]
12. Dezest, M.; Chavatte, L.; Bourdens, M.; Quinton, D.; Camus, M.; Garrigues, L.; Descargues, P.; Arbault, S.; Burlet-Schiltz, O.; Casteilla, L.; et al. Mechanistic insights into the impact of cold atmospheric pressure plasma on human epithelial cell lines. *Sci. Rep.* **2017**, *7*, 41163. [CrossRef] [PubMed]
13. Graves, D.B. Low temperature plasma biomedicine: A tutorial review. *Phys. Plasmas.* **2014**, *21*. [CrossRef]
14. Yan, D.Y.; Talbot, A.; Nourmohammadi, N.; Sherman, J.H.; Cheng, X.Q.; Keidar, M. Toward understanding the selective anticancer capacity of cold atmospheric plasma-A model based on aquaporins. *Biointerphases* **2015**, *10*. [CrossRef] [PubMed]
15. Van der Paal, J.; Neyts, E.C.; Verlackt, C.C.W.; Bogaerts, A. Effect of lipid peroxidation on membrane permeability of cancer and normal cells subjected to oxidative stress. *Chem. Sci.* **2016**, *7*, 489–498. [CrossRef] [PubMed]
16. Doskey, C.M.; Buranasudja, V.; Wagner, B.A.; Wilkes, J.G.; Du, J.; Cullen, J.J.; Buettner, G.R. Tumor cells have decreased ability to metabolize H_2O_2: Implications for pharmacological ascorbate in cancer therapy. *Redox Biol.* **2016**, *10*, 274–284. [CrossRef] [PubMed]
17. Vermeylen, S.; De Waele, J.; Vanuytsel, S.; De Backer, J.; Van der Paal, J.; Ramakers, M.; Leyssens, K.; Marcq, E.; Van Audenaerde, J.; Smits, E.L.J.; et al. Cold atmospheric plasma treatment of melanoma and glioblastoma cancer cells. *Plasma Process. Polym.* **2016**, *13*, 1195–1205. [CrossRef]

18. Koritzer, J.; Boxhammer, V.; Schafer, A.; Shimizu, T.; Klampfl, T.G.; Li, Y.F.; Welz, C.; Schwenk-Zieger, S.; Morfill, G.E.; Zimmermann, J.L.; et al. Restoration of sensitivity in chemo-resistant glioma cells by cold atmospheric plasma. *PLoS ONE* **2013**, *8*, e64498. [CrossRef] [PubMed]
19. Tanaka, H.; Mizuno, M.; Ishikawa, K.; Takeda, K.; Kondo, H.; Sekine, M.; Hashizume, H.; Nakamura, K.; Kajiyama, H.; Okazaki, Y.; et al. Similarities and differences in the cellular responses between plasma-activated medium-treated glioblastomas and plasma-activated Ringer's lactate solution-treated glioblastomas. *Clin. Plasma. Med.* **2018**, *9*, 42–43. [CrossRef]
20. Kurake, N.; Tanaka, H.; Ishikawa, K.; Kondo, T.; Sekine, M.; Nakamura, K.; Kajiyama, H.; Kikkawa, F.; Mizuno, M.; Hori, M. Cell survival of glioblastoma grown in medium containing hydrogen peroxide and/or nitrite, or in plasma-activated medium. *Arch. Biochem. Biophys.* **2016**, *605*, 102–108. [CrossRef] [PubMed]
21. Son, B.; Lee, S.; Youn, H.; Kim, E.; Kim, W.; Youn, B. The role of tumor microenvironment in therapeutic resistance. *Oncotarget* **2017**, *8*, 3933–3945. [CrossRef] [PubMed]
22. Judee, F.; Fongia, C.; Ducommun, B.; Yousfi, M.; Lobjois, V.; Merbahi, N. Short and long time effects of low temperature Plasma Activated Media on 3D multicellular tumor spheroids. *Sci. Rep.* **2016**, *6*, 21421. [CrossRef] [PubMed]
23. Merbahi, N.; Chauvin, J.; Vicendo, P.; Judee, F. Effects of plasma activated medium on head and neck FaDu cancerous cells: Comparison of 3D and 2D response. *Anticancer Agents Med. Chem.* **2017**, *17*, 1–8.
24. Plewa, J.-M.; Yousfi, M.; Frongia, C.; Eichwald, O.; Ducommun, B.; Merbahi, N.; Lobjois, V. Low-temperature plasma-induced antiproliferative effects on multi-cellular tumor spheroids. *New J. Phys.* **2014**, *16*. [CrossRef]
25. Chen, Z.; Simonyan, H.; Cheng, X.; Gjika, E.; Lin, L.; Canady, J.; Sherman, J.H.; Young, C.; Keidar, M. A novel micro cold atmospheric plasma device for glioblastoma both in vitro and in vivo. *Cancers* **2017**, *9*, 61. [CrossRef] [PubMed]
26. Golda, J.; Held, J.; Redeker, B.; Konkowski, M.; Beijer, P.; Sobota, A.; Kroesen, G.; Braithwaite, N.S.; Reuter, S.; Turner, M.; et al. Concepts and characteristics of the 'COST reference microplasma jet'. *J. Phys. D Appl. Phys.* **2016**, *49*. [CrossRef]
27. Gorbanev, Y.; Verlackt, C.C.W.; Tinck, S.; Tuenter, E.; Foubert, K.; Cos, P.; Bogaerts, A. Combining experimental and modelling approaches to study the sources of reactive species induced in water by the COST RF plasma jet. *Phys. Chem. Chem. Phys.* **2018**, *20*, 2797–2808. [CrossRef] [PubMed]
28. Katt, M.E.; Placone, A.L.; Wong, A.D.; Xu, Z.S.; Searson, P.C. In vitro tumor models: Advantages, disadvantages, variables, and selecting the right platform. *Front. Bioeng. Biotechnol.* **2016**, *4*, 12. [CrossRef] [PubMed]
29. Achilli, T.M.; Meyer, J.; Morgan, J.R. Advances in the formation, use and understanding of multi-cellular spheroids. *Expert Opin. Biol. Ther.* **2012**, *12*, 1347–1360. [CrossRef] [PubMed]
30. Takamatsu, T.; Uehara, K.; Sasaki, Y.; Miyahara, H.; Matsumura, Y.; Iwasawa, A.; Ito, N.; Azuma, T.; Kohno, M.; Okino, A. Investigation of reactive species using various gas plasmas. *RSC Adv.* **2014**, *4*, 39901–39905. [CrossRef]
31. Gorbanev, Y.; O'Connell, D.; Chechik, V. Non-thermal plasma in contact with water: The origin of species. *Chem. Eur. J.* **2016**, *22*, 3496–3505. [CrossRef] [PubMed]
32. Gorbanev, Y.; Soriano, R.; O'Connell, D.; Chechik, V. An atmospheric pressure plasma setup to investigate the reactive species formation. *J. Vis. Exp.* **2016**, *2016*. [CrossRef] [PubMed]
33. Elg, D.T.; Yang, I.W.; Graves, D.B. Production of TEMPO by O atoms in atmospheric pressure non-thermal plasma–liquid interactions. *J. Phys. D Appl. Phys.* **2017**, *50*, 475201. [CrossRef]
34. Gorbanev, Y.; Stehling, N.; O'Connell, D.; Chechik, V. Reactions of nitroxide radicals in aqueous solutions exposed to non-thermal plasma: Limitations of spin trapping of the plasma induced species. *Plasma. Sources Sci. Technol.* **2016**, *25*. [CrossRef]
35. Chauvin, J.; Judée, F.; Yousfi, M.; Vicendo, P.; Merbahi, N. Analysis of reactive oxygen and nitrogen species generated in three liquid media by low temperature helium plasma jet. *Sci. Rep.* **2017**, *7*, 4562. [CrossRef] [PubMed]
36. Kondeti, V.S.S.K.; Phan, C.Q.; Wende, K.; Jablonowski, H.; Gangal, U.; Granick, J.L.; Hunter, R.C.; Bruggeman, P.J. Long-lived and short-lived reactive species produced by a cold atmospheric pressure plasma jet for the inactivation of *Pseudomonas aeruginosa* and *Staphylococcus aureus*. *Free Radic. Biol. Med.* **2018**, *124*, 275–287. [CrossRef] [PubMed]

37. Lukes, P.; Dolezalova, E.; Sisrova, I.; Clupek, M. Aqueous-phase chemistry and bactericidal effects from an air discharge plasma in contact with water: Evidence for the formation of peroxynitrite through a pseudo-second-order post-discharge reaction of H_2O_2 and HNO_2. *Plasma Sources Sci. Technol.* **2014**, *23*. [CrossRef]
38. Wende, K.; Williams, P.; Dalluge, J.; Van Gaens, W.; Aboubakr, H.; Bischof, J.; von Woedtke, T.; Goyal, S.M.; Weltmann, K.D.; Bogaerts, A.; et al. Identification of the biologically active liquid chemistry induced by a nonthermal atmospheric pressure plasma jet. *Biointerphases* **2015**, *10*, 29518–103307. [CrossRef] [PubMed]
39. Ikawa, S.; Tani, A.; Nakashima, Y.; Kitano, K. Physicochemical properties of bactericidal plasma-treated water. *J. Phys. D Appl. Phys.* **2016**, *49*, 425401. [CrossRef]
40. Van Boxem, W.; Van der Paal, J.; Gorbanev, Y.; Vanuytsel, S.; Smits, E.; Dewilde, S.; Bogaerts, A. Anti-cancer capacity of plasma-treated PBS: Effect of chemical composition on cancer cell cytotoxicity. *Sci. Rep.* **2017**, *7*. [CrossRef] [PubMed]
41. Ravi, M.; Paramesh, V.; Kaviya, S.R.; Anuradha, E.; Solomon, F.D.P. 3D Cell culture systems: Advantages and applications. *J. Cell. Physiol.* **2015**, *230*, 16–26. [CrossRef] [PubMed]
42. Yan, D.Y.; Xiao, H.J.; Zhu, W.; Nourmohammadi, N.; Zhang, L.G.; Bian, K.; Keidar, M. The role of aquaporins in the anti-glioblastoma capacity of the cold plasma-stimulated medium. *J. Phys. D Appl. Phys.* **2017**, *50*. [CrossRef]
43. Wang, M.; Holmes, B.; Cheng, X.; Zhu, W.; Keidar, M.; Zhang, L.G. Cold atmospheric plasma for selectively ablating metastatic breast cancer cells. *PLoS ONE* **2013**, *8*, e73741. [CrossRef] [PubMed]
44. Volotskova, O.; Shashurin, A.; Stepp, M.A.; Pal-Ghosh, S.; Keidar, M. Plasma-controlled cell migration: Localization of cold plasma-cell interaction region. *Plasma Med.* **2011**, *1*, 85–92. [CrossRef]
45. Volotskova, O.; Hawley, T.S.; Stepp, M.A.; Keidar, M. Targeting the cancer cell cycle by cold atmospheric plasma. *Sci. Rep.* **2012**, *2*. [CrossRef] [PubMed]
46. Salazar-Ramiro, A.; Ramirez-Ortega, D.; de la Cruz, V.P.; Hernandez-Pedro, N.Y.; Gonzalez-Esquivel, D.F.; Sotelo, J.; Pineda, B. Role of redox status in development of glioblastoma. *Front. Immunol.* **2016**, *7*. [CrossRef] [PubMed]
47. Naciri, M.; Dowling, D.; Al-Rubeai, M. Differential sensitivity of mammalian cell lines to non-thermal atmospheric plasma. *Plasma Process Polym.* **2014**, *11*, 391–400. [CrossRef]
48. Otto, T.; Sicinski, P. Cell cycle proteins as promising targets in cancer therapy. *Nat. Rev. Cancer* **2017**, *17*, 93. [CrossRef] [PubMed]
49. Jenkins, N.C.; Liu, T.; Cassidy, P.; Leachman, S.A.; Boucher, K.M.; Goodson, A.G.; Samadashwily, G.; Grossman, D. The p16(INK4A) tumor suppressor regulates cellular oxidative stress. *Oncogene* **2011**, *30*, 265–274. [CrossRef] [PubMed]
50. Sablina, A.A.; Budanov, A.V.; Ilyinskaya, G.V.; Agapova, L.S.; Kravchenko, J.E.; Chumakov, P.M. The antioxidant function of the p53 tumor suppressor. *Nat. Med.* **2005**, *11*, 1306–1313. [CrossRef] [PubMed]
51. Ma, Y.; Ha, C.S.; Hwang, S.W.; Lee, H.J.; Kim, G.C.; Lee, K.-W.; Song, K. Non-thermal atmospheric pressure plasma preferentially induces apoptosis in p53-mutated cancer cells by activating ROS stress-response pathways. *PLoS ONE* **2014**, *9*, e91947. [CrossRef] [PubMed]
52. Guerrero-Preston, R.; Ogawa, T.; Uemura, M.; Shumulinsky, G.; Valle, B.L.; Pirini, F.; Ravi, R.; Sidransky, D.; Keidar, M.; Trink, B. Cold atmospheric plasma treatment selectively targets head and neck squamous cell carcinoma cells. *Int. J. Mol. Med.* **2014**, *34*, 941–946. [CrossRef] [PubMed]
53. Li, H.; Lei, B.; Xiang, W.; Wang, H.; Feng, W.; Liu, Y.; Qi, S. Differences in protein expression between the U251 and U87 cell lines. *Turk. Neurosurg.* **2017**, *27*, 894–903. [PubMed]
54. Yang, C.; Zhang, J.; Ding, M.; Xu, K.; Li, L.; Mao, L.; Zheng, J. Ki67 targeted strategies for cancer therapy. *Clin. Transl. Oncol.* **2018**, *20*, 570–575. [CrossRef] [PubMed]
55. Liedtke, K.R.; Bekeschus, S.; Kaeding, A.; Hackbarth, C.; Kuehn, J.-P.; Heidecke, C.-D.; von Bernstorff, W.; von Woedtke, T.; Partecke, L.I. Non-thermal plasma-treated solution demonstrates antitumor activity against pancreatic cancer cells in vitro and in vivo. *Sci. Rep.* **2017**, *7*, 8319. [CrossRef] [PubMed]
56. Longati, P.; Jia, X.H.; Eimer, J.; Wagman, A.; Witt, M.R.; Rehnmark, S.; Verbeke, C.; Toftgard, R.; Lohr, M.; Heuchel, R.L. 3D pancreatic carcinoma spheroids induce a matrix-rich, chemoresistant phenotype offering a better model for drug testing. *BMC Cancer* **2013**, *13*. [CrossRef] [PubMed]

57. Vinci, M.; Box, C.; Zimmermann, M.; Eccles, S.A. Tumor spheroid-based migration assays for evaluation of therapeutic agents. In *Target identification and Validation in Drug Discovery: Methods and Protocols*; Moll, J., Colombo, R., Eds.; Springer: New York, NY, USA, 2013; Volume 986, pp. 253–266.
58. Ivanov, D.P.; Grabowska, A.M. Spheroid arrays for high-throughput single-cell analysis of spatial patterns and biomarker expression in 3D. *Sci. Rep.* **2017**, *7*. [CrossRef] [PubMed]
59. Varghese, F.; Bukhari, A.B.; Malhotra, R.; De, A. IHC Profiler: An open source plugin for the quantitative evaluation and automated scoring of immunohistochemistry images of human tissue samples. *PLoS ONE* **2014**, *9*. [CrossRef] [PubMed]
60. Nedeianu, S.; Pali, T. EPR spectroscopy of common nitric oxide–spin trap complexes. *Cell Mol. Biol. Lett.* **2002**, *7*, 142–143. [PubMed]
61. Timmerman, E. Spectrum Viewer 2.6.3. Available online: http://www.phys.tue.nl/people/etimmerman/specview/ (accessed on 12 February 2018).
62. Duling, D.R. Simulation of Multiple Isotropic Spin Trap EPR Spectra. *J Magn Reson B.* **1994**, *104*, 105–110. [CrossRef] [PubMed]
63. Database, N.I.S.T. Available online: http://tools.niehs.nih.gov/stdb/index.cfm (accessed on 1 March 2018).
64. Privat-Maldonado, A.; Gorbanev, Y.; O'Connell, D.; Vann, R.; Chechik, V.; Woude, M.W.v.d. Non-target biomolecules alter macromolecular changes induced by bactericidal low-temperature plasma. *IEEE Trans. Radiat. Plasma Med. Sci.* **2017**, *2*, 121–128. [CrossRef]
65. Gibson, A.R.; McCarthy, H.O.; Ali, A.A.; O'Connell, D.; Graham, W.G. Interactions of a non-thermal atmospheric pressure plasma effluent with PC-3 prostate cancer cells. *Plasma Process Polym.* **2014**, *11*, 1142–1149. [CrossRef]

cancers

MDPI

Article

Synergistic Effects of Bortezomib-OV Therapy and Anti-Invasive Strategies in Glioblastoma: A Mathematical Model

Yangjin Kim [1], Junho Lee [1], Donggu Lee [1] and Hans G. Othmer [2,*]

1 Department of Mathematics, Konkuk University, Seoul 05029, Korea; ahyouhappy@gmail.com (Y.K.); juneho2222@gmail.com (J.L.); donggu9211@gmail.com (D.L.)

2 School of Mathematics, University of Minnesota, Minneapolis, MN 55455, USA

* Correspondence: othmer@math.umn.edu; Tel.: +1-(612)-624-8325; Fax: +1-(612)-626-2017

Received: 31 December 2018; Accepted: 6 February 2019; Published: 13 February 2019

Abstract: It is well-known that the tumor microenvironment (TME) plays an important role in the regulation of tumor growth and the efficacy of anti-tumor therapies. Recent studies have demonstrated the potential of combination therapies, using oncolytic viruses (OVs) in conjunction with proteosome inhibitors for the treatment of glioblastoma, but the role of the TME in such therapies has not been studied. In this paper, we develop a mathematical model for combination therapies based on the proteosome inhibitor bortezomib and the oncolytic herpes simplex virus (oHSV), with the goal of understanding their roles in bortezomib-induced endoplasmic reticulum (ER) stress, and how the balance between apoptosis and necroptosis is affected by the treatment protocol. We show that the TME plays a significant role in anti-tumor efficacy in OV combination therapy, and illustrate the effect of different spatial patterns of OV injection. The results illustrate a possible phenotypic switch within tumor populations in a given microenvironment, and suggest new anti-invasion therapies.

Keywords: glioblastoma; oncolytic virus; mathematical model; bortezomib; ER stress; apoptosis; CSPG

1. Introduction

Glioblastoma multiforme (GBM) is one of the deadliest human cancers, with mean survival times of <15 month because of its aggressive growth and rapid, widespread invasion of the brain [1]. Because it spreads rapidly, a significant number of cancer deaths following resection of a primary tumor are due to secondary tumors that stem from tumor cells that have migrated to other parts of the brain [1,2]. Thus there are two major problems involved; (i) the detection and local treatment of a primary tumor via resection, chemo/radiotherapy, or other means, and (ii) coping with the diffuse, dispersed tumor cells. To date, most therapies focus on the first problem, which is itself complex because of unintended effects of therapies [3], but to increase the mean survival time significantly, the second problem must be confronted.

Since cancer cells are usually highly stressed due to their rapid growth or conditions in the tumor microenvironment (TME), they may contain an excess of un- or misfolded proteins in the endoplasmic reticulum (ER), which leads to ER stress. NF-κB, a regulator of genes that control cell proliferation, is usually sequestered in the cytoplasm and maintained in an inactive state by its inhibitor, IκB. Numerous stimuli, including ER stress, enhance the degradation of IκB, thereby activating NFκB. Upon activation NFκB translocates to the nucleus and activates a wide variety of genes, including *Bcl-2*, which in turn down-regulates the apoptosis gene *Bax* (cf. Figure 1A) [4–8]. In normal tissue this leads to production of IκB, and the feedback loop maintains homeostasis in the tissue [9]. However, a constitutively active NF-κB pathway fosters resistance to anti-cancer therapy [10–12] in cancer cells [13,14] and in tumor-associated macrophages [15]. Drugs that activate *Bax* show promise as anti-tumor treatments by inducing apoptosis [5,16]. Necroptosis is a caspase-independent, massive cell death program mediated by high expression levels of reactive oxygen species (ROS) induced by activity of RIP1, a receptor-interacting-protein kinase [17]. Apart from upregulation of *Bcl2L12*, low levels of caspase-8 and upregulated RIP1 were associated with necroptosis in glioma cells [18]. RIP1, a serine/threonine kinase, contains a key death domain and is a major activator of necroptosis, but is not required for activation of NFκB and the apoptotic signaling pathway [17,19].

Proteasomes are protein complexes that regulate degradation and recycling of proteins that control cellular functions such as signal transduction, differentiation and cell death, and their Inhibition leads to the accumulation of various regulatory proteins, including the pro-apoptotic protein p53, in the ER [7]. Thus one therapeutic approach is to use agents such as bortezomib (BTZ), a peptide-based inhibitor of proteosomes, to interfere with protein turnover, and thereby to stimulate immunogenic apoptosis (cf. Figure 1B) [20,21]. BTZ induces apoptosis by inhibiting the phosphorylation of IκB protein, thus inhibiting NF-κB activity [22,23]. While it is used as a single agent or in combination with other radio-/chemo-agents for many cancers, including multiple myeloma, ovarian cancer, and head and neck cancers, different combination strategies are being developed for better efficacy [24–27] in these and other tumor types.

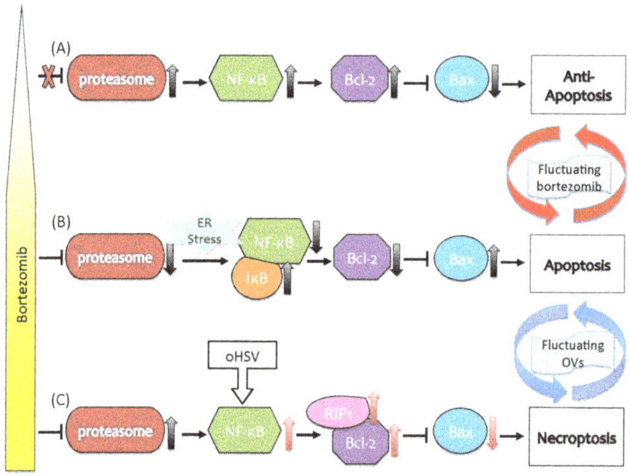

Figure 1. A schematic, non-mechanistic, diagram of the role of bortezomib (BTZ) and onocolytic viruses (OVs) in the regulation of cell death pathways and the effects of fluctuating BTZ and OVs [20,21]. OV activities determine cell apoptosis or necroptosis in response to BTZ (yellow triangle-bar on the left). (**A**) A low BTZ level promotes proteasome activities and leads to upregulation of NFκB/Bcl-2 and downregulation of Bax, thereby suppressing apoptosis. (**B**) A high BTZ level in the absence of OVs suppresses proteasome activities by activating IκB, which in turn inhibits the NFκB/Bcl-2 complex and up-regulates Bax, leading to apoptosis. (**C**) In the presence of OVs, the BTZ-induced endoplasmic reticulum (ER) stress pathway is modified: the NFκB/Bcl-2 complex is upregulated and the Bax level is downregulated. This also results in upregulation of RIP1, leading to necroptosis. Red arrows on the right indicate the switching behavior between an anti-apoptosis mode in (**A**) and the apoptosis state in (**B**) in response to fluctuating BTZ. Blue arrows on the right indicate the switching behavior between a apoptosis mode in (**B**) and the necroptosis state in (**C**) in response to fluctuating OVs.

One promising combination therapy for GBM utilizes an oncolytic virus (OV), specifically, the herpes simplex virus HSV-1, together with BTZ. OV therapy utilizes viruses that are genetically engineered to efficiently target, infect and replicate in cancer cells, with minimal damage to normal cells in surrounding tissue. In the combination BTZ-oHSV approach, HSV-1 exploits the host proteasome [28,29], leading to modified signaling pathways. When cells are infected by viruses, RIP1 may form a complex with RIP3 that mediates necroptosis [30], but RIP1 may have pro-tumor effects as well under different treatment protocols [31,32]. BTZ also increases the expression of HSP60 and HSP90, which leads to expression of receptors that activate natural killer cells, and by promoting breakdown of the cell membrane, necroptosis can induce local changes in the TME that enhance the anti-tumor activity of natural killer cells [33]. In recent experimental studies, Yoo et al. [20,21] demonstrated that the induction of the unfolded protein response (UPR) following BTZ treatment not only improved oHSV replication, but synergistically increased the cancer cell death rate in vitro and in vivo through necroptosis (cf. Figure 1C). They also found that RIP1 and Jun N-terminal kinase (JNK) levels were up-regulated in synergistic cell death of OV-infected cells after BTZ treatment [21]. These findings demonstrated that the synergistic interaction between oHSV and BTZ increases overall therapeutic efficacy. OV therapy is currently being evaluated for its efficacy and safety in many pre-clinical and clinical trials for various cancers [34,35] including gliomas [36–41]. While these trials have shown promising results [42], the overall survival data have to be evaluated, and development of efficient, optimized OV treatment strategies remains to be done.

While the BTZ-oHSV combination shows significant positive synergistic effects for local treatment of GBM and leads to localized changes in the TME, it does not address the problem of eliminating the dispersed glioma cells. Chondroitin sulfate proteoglycans (CSPGs) are a major component of the Extracellular matrix (ECM) in brain tissue, and play an important role in regulation of glioma invasion and in the spread of OVs [43]. On the one hand, CSPGs block movement of OVs throughout the tumor and reduce their effectiveness, but on the other hand, other ECM components such as chondroitin sulfate-glycosaminoglycans (CS-GAGs) are reported to block glioma invasion by forming strong adhesion between glioma cells and ECM components [44–46]. Kim et al. [47] recently developed a mathematical model of oncolytic virus spread with Chondroitinase ATP-binding cassette (Chase-ABC), a bacterial enzyme that can remove CS-GAGs. This and other work shows that the TME plays a complex role in tumor treatment, and treatment for one effect may enhance other effects.

The model developed herein addresses the two issues mentioned earlier—the local treatment of a tumor, and ways of mitigating the effect of rapid tumor cell dispersal throughout the brain. To address the first issue we define a system of differential equations that describe the evolution of the intracellular variables shown in Figure 1 under specified protocols of BTZ and oHSV treatment. This local component involves the concentration of IκB, NFκB, Bax, and RIP1, and we use this to study the regulation of anti-apoptosis, apoptosis, and necroptosis in response to injections of OVs and BTZ, as well as the anti-tumor efficacy of a BTZ-OV combination therapy, at the tumor level. We then incorporate the local model into a multi-scale model to describe the spatial distribution of uninfected, infected, and dead cancer cells, the density of virus particles, and the concentration of BTZ, This multi-scale model is used to investigate the role of a heterogeneous tumor microenvironment in the regulation of the combination therapy. In silico experiments were performed to investigate the effect of CSPGs on control of invasive tumor cells during the combination therapy, and possible anti-invasion strategies in the complex tumor microenvironment.

2. Materials and Methods

2.1. The Intracellular Network

Anti-apoptosis, apoptosis, and necroptosis of tumor cells are mediated by a complex intracellular signaling network of NFκB, its inhibitor Iκ, proteasomes, Bcl-2, and Bax, amongst others, and at present the detailed interactions amongst these components are not known. Therefore we focused on a minimal network of four components shown in Figure 2B that incorporated the primary known interactions, and we described these interactions phenomenologically rather than mechanistically. The best-characterized interactions of the components are between IκB and NFκB. As stated earlier, IκB inhibits NFκB, but it has also been shown that the homolog of NFκB not only activates, but also inhibits the homolog of IκB via the action of a micro-RNA [48,49]. The levels of BTZ and the presence or absence of OVs were treated parametrically in the model for the local dynamics, but the production and diffusion of BTZ were incorporated in the spatially-distributed model later.

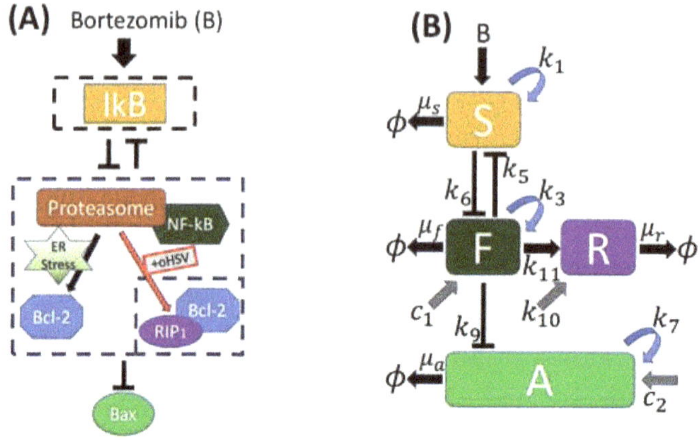

Figure 2. (**A**) A simplified modular model of the network of the IκB-NFκB-Bcl2-Bax-RIP system for anti-apoptosis, apoptosis, and necroptosis [8,20,21,50,51]. The network encodes the mutual antagonism between IκB (top box) and the NFκB-Bcl2 complex (center box), and inhibition of bax by the NFκB-Bcl2 complex. Up- and down-regulation of these modules affects the pro- and anti-apoptosis signaling pathways in response to low and high bortezomib levels B. Introduction of oncolytic herpes simplex virus (oHSVs) (small box in the lower right of the center panel) leads to changes in the expression levels of these modules, enhancing the necroptotic pathway. (**B**) A schematic of the interactions amongst the primary variables.

The four primary variables in the local model are IκB, NFκB, Bax, and RIP1, which are denoted as S, F, A and R. The governing equations of S, F, A and R are

$$\frac{dS}{dt} = k_{SB}\frac{B}{k_{12} + k_{13}[oHSV]} + \frac{k_1 k_2^2}{k_2^2 + k_5 F^2} - \mu_s S, \tag{1}$$

$$\frac{dF}{dt} = c_1 + \frac{k_3 k_4^2}{k_4^2 + k_6 S^2} - \mu_f F, \tag{2}$$

$$\frac{dA}{dt} = c_2 + \frac{k_7 k_8^2}{k_8^2 + k_9 F^2} - \mu_a A. \tag{3}$$

$$\frac{dR}{dt} = k_{10} + k_{11}[oHSV]F - \mu_r R. \tag{4}$$

Here B encodes the level of BTZ, which serves as a surrogate for the signaling pathways from BTZ to IκB, [oHSV] is a two-state switch for the oncolytic virus, with [oHSV] = $\frac{v}{k+v}$ where v is the OV density, as introduced below, and k is the Hill type parameter, giving [oHSV] = 0 (1) below (above) a threshold of virus. When $B = 0$ the first two equations reflect the fact that IκB and NFκB repress one another. In the first equation k_2 and k_5 denote the coefficients in the Hill function that models the inhibition—here assumed to be quadratic—while k_1 encodes the strength of this inhibition on IκB. Similarly, c_1, c_2, k_{10} encode inputs to the NFκB-Bcl-2 complex, Bax, and RIP1, pathways, resp., k_3 and k_7 are the autocatalytic enhancement parameters for NFκB-Bcl-2 complex and Bax, resp., k_4 and k_6 (k_8 and k_9) denote the coefficients in the Hill function that models inhibition of the NFκB-Bcl-2 complex, (Bax), resp. Finally, μ_s, μ_f, μ_a and μ_r represent the decay rates of the species.

It should be noted that the equations for IκB and NFκB are decoupled from the equations for Bax and RIP1. Since the interaction between S and F is mutual repression, the parameters could easily be chosen to produce a region of bistability, which defines a switch, for a suitable range of parameters in the Hill functions. However, as we show later, the parameters used here do not produce a switch, but rather a rapid transition from a high to low value of NFκB as B is varied.

The third equation describes the evolution of Bax, which in our formulation is inhibited by NFκB via the second term in Equation (3). This again is a simplification of the network from NFκB to Bax, as suggested in Figure 1A, but reflects the fact that NFκB↑ leads to Bax↓. The last equation governs RIP1, and reflects the fact that RIP1 is upregulated in response to combination treatment with BTZ and oHSV, leading to necroptotic cell death [21]. In the absence of treatment, glioma cells maintain upregulated Bcl-2 and develop resistance to cytotoxic and pro-apoptotic agents [17,52]. BTZ induces apoptosis via ER stress in the absence of oHSV but, in the presence of OVs, the combination OVs+BTZ induces increased ROS and JNK activity, which leads to the critical transition from either apoptotic or anti-apoptotic pathways to necroptosis [21]. Therefore, the term $k_{11}[oHSV]F$ is activated only in the presence of sufficient oHSV, and the strength of the effect depends on the level of NFκB.

The values of all parameters used in these equations are given in Table 1.

Table 1. Model parameters for the intracellular dynamics.

Par	Description	Parameter Values	Ref
k_{SB}	Bortezomib signaling scaling factor	1.0×10^{-1} h^{-1}μM	Estimated
k_{12}	Inhibition strength of bortezomib by oHSV	$= B^* = 2.6 \times 10^{-2}$ μM	Estimated
k_{13}	Inhibition strength of bortezomib by oHSV	7.8×10^{-1} μM	Estimated
c_1	Signaling strength of the NFκB-Bcl2 complex	3.64×10^{-2} h^{-1}μM	Estimated
c_2	Signaling strength of Bax	3.43×10^{-4} h^{-1}μM	Estimated
k_{10}	Signaling strength of RIP1	5.2×10^{-1} h^{-1}μM	Estimated
k_{11}	Activation rate of RIP1 in the presence of OVs	1.35 h^{-1}	Estimated
k_1	Autocatalytic enhancement rate of IκB	2.08×10^{-1} h^{-1}μM	Estimated
k_3	Autocatalytic enhancement rate of the NFκB-Bcl2 complex	6.91×10^{-1} h^{-1}μM	Estimated
k_7	Autocatalytic enhancement rate of Bax	1.155×10^{-2} h^{-1}μM	Estimated
k_5	Inhibition strength of IκB by the NFκB-Bcl2 complex	8.8 (μM)$^{-2}$	Estimated
k_6	Inhibition strength of the NFκB-Bcl2 complex by IκB	400 (μM)$^{-2}$	Estimated
k_9	Inhibition strength of Bax by the NFκB-Bcl2 complex	4.0 (μM)$^{-2}$	Estimated
k_2	Hill-type parameter	1.0	Estimated
k_4	Hill-type parameter	1.0	Estimated
k_8	Hill-type parameter	1.0	Estimated
μ_s	Decay rate of IκB	1.0397 h^{-1}	[53–55]
μ_f	Decay rate of NFκB-Bcl2 complex	3.151×10^{-1} h^{-1}	[53,54]
μ_a	Decay rate of Bax	2.17×10^{-2} h^{-1}	[53,54,56]
μ_r	Decay rate of RIP1	1.444×10^{-1} h^{-1}	[53,54,57]
k	Hill type parameter of oHSV switching	$0.01\, v^*$	Estimated

2.2. The Spatially-Distributed Component of the Model

In the second component of modeling we expanded the local model to describe the evolution of various states of glioma cells and other factors in both space and time. To do this we added partial differential equations that describe the spatial distributions of three states of glioma cells—those that are uninfected by virus, infected, and dead. The densities of these are denoted x, y and n, resp., and we also describe the density of virus particles (v), and the concentration of BTZ (B). The postulated interactions of these components are shown in Figure 3. The spatial distributions are restricted to a planar surface, and thus all variables are functions of time and two spatial variables.

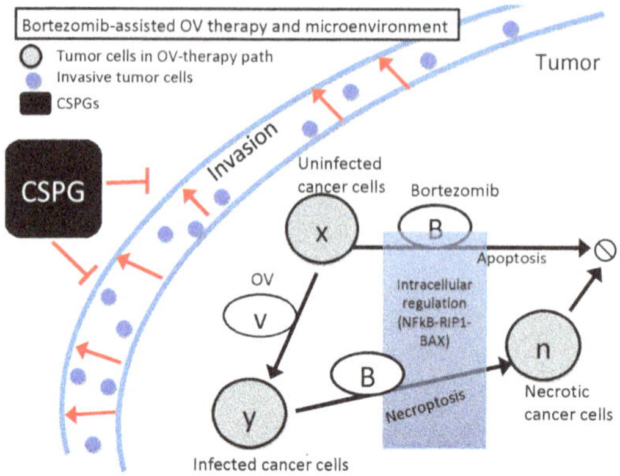

Figure 3. A regulatory network involving OV therapy (gray circles) and the tumor microenvironment (TME) in BTZ-assisted OV therapy. Arrows indicate induction and activation. Hammerheads indicate inhibition.

The generic form of the evolution equation for all spatially-distributed components is

$$\frac{\partial w}{\partial t} = -\nabla \cdot J_w + P_w, \tag{5}$$

where w is one of x, y, n, v or B, J_w is the flux of that species, P_w is the birth/death (or production/destruction) rate of that species, and ∇ is the divergence operator in two dimensions. We suppose that all components are restricted to a closed bounded domain in the plane and impose the no-flux condition $n \cdot J_w = 0$ on the boundary, where n is normal to the boundary.

We assume that the fluxes of mobile species (all but dead cells n) are due to Fickian diffusion, and thus the flux of any species other than n is given by

$$J_w = -D_w \nabla w. \tag{6}$$

The birth/death of the production/destruction rate of each species is the sum of all processes that lead to creation or destruction of that species [20,47,51,58].

For the uninfected cells these processes are proliferation, which we assume follows a logistic growth law, apoptosis, necroptosis, and viral infection. Therefore

$$P_x = \lambda x(1 - x/x_0) - \beta x v - \beta_1 x B I_{apop} - \beta_3 x v I_{Necrop}, \tag{7}$$

where λ is the proliferation rate of uninfected glioma cells whose carrying capacity is x_0, β is the infection rate in the absence of BTZ, β_1 is the BTZ-induced apoptosis of unifected cells, β_3 is the BTZ-induced necroptotic cell death rate in the presence of OVs. Further, $I_{apop}(\cdot)$ and $I_{necroptosis}(\cdot)$ are the indicator or characteristic functions of the apoptotic and necrotic regions in F, A, R space. These functions are either one or zero, depending on whether F, A and R are in specified ranges defined in the following section. Thus, the governing equation for x is

$$\frac{\partial x}{\partial t} = \nabla \cdot (D_1 \nabla x) + \lambda x(1 - x/x_0) - \beta x v - \beta_1 x B I_{apop} - \beta_3 x v I_{Necrop}. \tag{8}$$

We assume that (i) dying cells do not move, (ii) infected tumor cells become dying cells at a rate δ, (iii) dying cells are cleared from the system at a rate μ, and (iv) infected cells diffuse at a rate D_2. From this it follows that the governing equations for densities of uninfected (y) and dead cells (n) are as follows:

$$\frac{\partial y}{\partial t} = \nabla \cdot (D_2 \nabla y) + \beta x v - \delta y + \beta_3 x v I_{Necrop} \tag{9}$$

$$\frac{\partial n}{\partial t} = \delta y - \mu n. \tag{10}$$

The oncolytic virus is replication-competent, and BTZ enhances viral replication by a factor proportional to B [20]. We denote by b the number of viral particles released after OV-mediated lysis of infected cancer cells. Hence, the equation for v is the following:

$$\frac{\partial v}{\partial t} = \nabla \cdot (D_v \nabla v) + b\delta y(1 + \alpha_1 B) - \gamma v. \tag{11}$$

BTZ is supplied to the glioma via direct injection and diffusion through the brain tissue. We took into account the consumption from the internalization of BTZ in tumor cells and natural decay at rate μ_B. Hence, the governing equation is

$$\frac{\partial B}{\partial t} = \nabla \cdot (D_B \nabla B) + I_B - (\mu_1 x + \mu_2 y)\frac{B}{k_B + B} - \mu_B B, \tag{12}$$

where D_B is the diffusion coefficient of BTZ, I_B is the effective injection rate of BTZ, μ_1, μ_2 are consumption rate of BTZ by uninfected and infected tumor cells, respectively, and k_B is the Hill type coefficient.

The list of parameters and their values are given in Table 2 for the distributed variables. In the foregoing Equations (1)–(12) are stated in terms of dimensional quantities. These are cast into dimensionless form for computational purposes in the Supplementary Information file, and the basis for the parameter estimations is given there. The simulations were done using the alternating direction implicit method and the non-linear solver, nksol, for algebraic systems. The equations were solved on a regular uniform spatial grid ($h_x = 0.01$, $h_y = 0.01$) using an adaptive time-stepping method [59].

Table 2. Model parameters for the distributed variables. ‡ Dimensionless values.

Par	Description	Parameter Value	Ref.
	Diffusion coefficients/Random motility (mm^2/h)		
D_1	random motility of uninfected glioma cells	3.6×10^{-6}	[51]
D_2	Random motility of infected glioma cells	3.6×10^{-9}	Estimated
D_v	random motility of virus (PBS)	3.89×10^{-2}	[51,60]
D_B	diffusion coefficient of bortezomib	2.5×10^{-2}	[51]
	Production/remodeling rates		
λ	proliferation rate of tumor cells	4.2×10^{-1} 1/h	[58], Estimated
x_0	Carrying capacity of uninfected tumor cells	$= x^*$	[47,51,58]
β	Virus infection rate	2.43×10^{-11} mm^3/(h·virus)	[51,58]
b	Burst size of infected cells	‡ 1.1364×10^1	[58], Estimated
α_1	bortezomib-induced viral replication rate	10^{11} mm^3/g	[51]
I_B	bortezomib supply rate	1.8×10^{-12} g/(mm$^3\cdot$h)	[51]
	Inhibition/degradation/decay rates		
β_1	Bortezomib-induced apoptosis rate of tumor cells	8.0×10^8 mm^3/(g·h)	Estimated
β_3	Necroptosis rate of tumor cells	1.37×10^3 mm^3/(g·h)	Estimated
δ	infected cell lysis rate	8.2×10^{-2} h^{-1}	[51,58]
μ	Removal rate of dead cells	1.04×10^{-1} h^{-1}	[51,58], Estimated
γ	clearance rate of viruses	1.8×10^{-3} h^{-1}	[51,58]
μ_1	consumption rate of bortezomib by uninfected tumor cells	2.075×10^{-9} h^{-1}	[51]
μ_2	consumption rate of bortezomib by infected tumor cells	2.075×10^{-9} h^{-1}	[51]
k_B	Hill-type parameter	$= B^*$	[51]
μ_B	decay rate of bortezomib	3.47×10^{-2} h^{-1}	[61,62]

Table 2. *Cont.*

Par	Description	Parameter Value	Ref.
	Reference values of main variables		
S^*	IκB concentration	0.05 μM	[63,64]
F^*	Concentration of the NFκB-Bcl2 complex	0.5 μM	[63–66]
A^*	Bax concentration	0.1 μM	[67]
R^*	RIP1 concentration	5.0 μM	[68]
x^*	Uninfected cell density	10^6 cells/mm^3	[51,58,69]
y^*	Infected cell density	$= x^*$	[51,58,69]
n^*	Dead cell density	$= x^*$	[51,58,69]
v^*	Virus concentration	2.2×10^8 virus/mm^3	[51,58,69]
B^*	Bortezomib concentration	1.0×10^{-11} g/mm^3	[20,21,51]

3. Results

3.1. Intracellular Dynamics

We recall that low levels of NFκB and Bcl-2 lead to up-regulated Bax and induce apoptosis, while over-expression of NFκB and Bcl-2 leads to down-regulation of Bax and up-regulation of RIP1, which induces necroptosis (Figure 1). As a first step toward understanding the effects of BTZ and oHSV on cell death in the full model, we analyzed how the BTZ level (B) affects the levels of the key effectors of cell death (F, A, R) in the intracellular model. When the core IκB-NFκB-Bax-RIP1 system (1)–(4) is at a steady state, we can solve the algebraic equations for the steady-state values of F, A, R as a function of the extracellular BTZ level (B), and we denoted the resulting values by F^s, A^s, R^s. Figure 4A shows the graphs of $F = F(B)$ (blue), $A = A(B)$ (red), $R = R(B)$ (green) in the absence (without circle markers) and presence (with circle markers) of OVs. The response curves of NFκB and Bax show that Bax inherits the IκB-NFκB mutual antagonism, with a crossover at $B \sim 0.5$. We define thresholds for the variables at the expression level of NFκB at crossover as th_F = 1.7 for NFκB, th_A = 1.7 for Bax, and th_R =1.7 for RIP1, and we use these to define the anti-apoptotic ($\mathbb{T}_t$), apoptotic ($\mathbb{T}_a$), and necroptotic ($\mathbb{T}_n$) regions as

$$\mathbb{T}_t = \{(F, A, R) \in \mathbb{R}^3 : F > th_F,\ A < th_A,\ R < th_R\}, \tag{13}$$
$$\mathbb{T}_a = \{(F, A, R) \in \mathbb{R}^3 : F < th_F,\ A > th_A,\ R < th_R\}, \tag{14}$$
$$\mathbb{T}_n = \{(F, A, R) \in \mathbb{R}^3 : F > th_F,\ A < th_A,\ R > th_R\}, \tag{15}$$

respectively (see Figure 4B).

In the absence of OV therapy, and under low levels of BTZ, the system of Equations (1)–(4) exhibits low Bax levels, high F levels, and low A and R levels, and the cells are in the anti-apoptotic state $\mathbb{T}_t$. Under these conditions the cancer cells would continue to grow as B increases until it reaches the crossover point (~0.5). In the vicinity of this point the Bax level rises rapidly, which leads to down-regulated NFκB, and the cells are characterized as being in the apoptotic state. For an illustration of the dynamics, Figure 4C,D show the dynamics of the core control system in response to low ($B = 0.0$; Figure 4C) and high ($B = 1.0$; Figure 4D) BTZ levels. Starting from various initial conditions (($S, F, A, R)(0)$ = (0, 0, 2, 1), (0, 1, 5, 3), (0, 2, 5, 4), (0, 5, 5, 5), (0, 5, 3, 5), (0, 6, 0, 0).) the dynamics lead to either the $\mathbb{T}_t$, $\mathbb{T}_a$ or $\mathbb{T}_n$ state. For a low BTZ level, the system converges to the unique steady state ((F^s, A^s, R^s) (~(4.64,0.41,0.72)) in the anti-apoptotic region where Bax (A) and RIP1 (R) expressions are low, but NFκB (F) activity is high. On the other hand, for high BTZ levels ($B > 0.5$), there is only one stable steady state ((F^s, A^s, R^s) ~(0.42, 4.89, 0.72)) in the apoptotic region $\mathbb{T}_a$ where NFκB and RIP1 expressions are low but Bax activity is high. However, the presence of oHSV leads to upregulation of RIP1 and NFκB and downregulation of Bax ((F^s, A^s, R^s) (~(4.57, 0.42, 5.0)) in response to the high BTZ level (Figure 4E). This dichotomous behavior of the NFκB and Bax modules via ER stress in response to high and low BTZ levels are well known experimentally [13,20,21,30,70], but have not been replicated with a mathematical model heretofore.

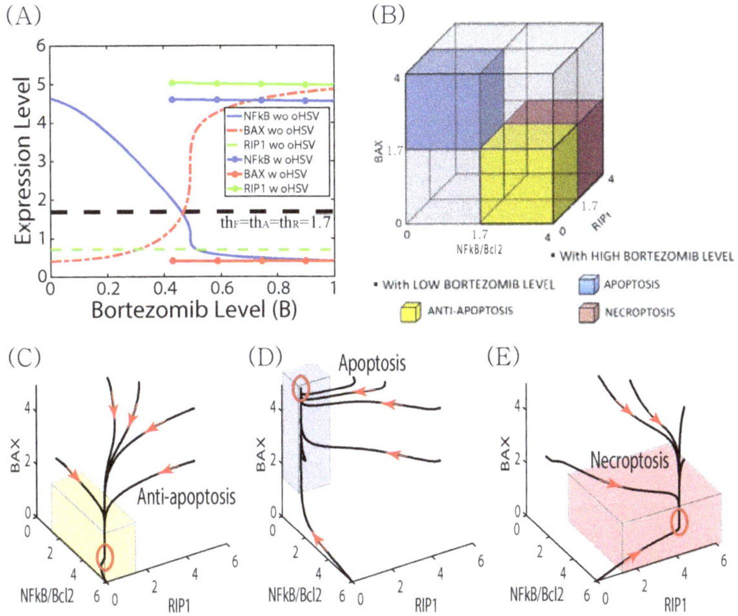

Figure 4. The dependence of the intracellular variables (IκB-NFκB-Bax-RIP1) on BTZ, and characterization of the cell death program (anti-apoptosis, apoptosis, and necroptosis). (**A**) BTZ levels produce an effective on-off switch for control of NFκB, Bax, and RIP1 levels, and activate the cell death program: anti-apoptosis, apoptosis, or necroptosis. (**B**) Characterization of cell death in the NFκB-RIP1-Bax state space, where the labeled domains are defined in Equations (13)–(15). (**C**–**E**) Dynamics of the core control system in response to low ($B = 0.0$ in (**C**)) and high ($G = 1.0$ in (**D**,**E**)) BTZ levels in the absence (**C**,**D**) and presence (**E**) of OVs. The stable steady states in each subframe lie within the red circles.

One can predict the qualitative responses of the solution components to time-varying BTZ levels in the absence of OVs from Figure 4, and these are shown explicitly for a periodic variation in Figure 5. The BTZ level is defined as $B(t) = 0.25 \times cos(\pi t/250) + 0.45)$, and starting at high BTZ and Bax, the trajectory of Bax follows the upper branch of the red loop for decreasing BTZ until BTZ drops to $\sim$0.2, whereupon it follows the increasing part of the cycle. The blue solid curve and blue arrows represent NFκB and its flow, where NFκB decreases along the upper branch as BTZ increases, while it increases along the lower branch as BTZ decreases. Figure 5C shows the solutions $(F(t), A(t), R(t))$ as a function of their position in the BTZ cycle. Fluctuating BTZ induces the up- or down-regulation of NFκB, leading to periodic transitions between $\mathbb{T}_t$ (the white region) and $\mathbb{T}_a$ (the pink region). The corresponding solutions are shown in the $F-A-R$ space (Figure 5D).

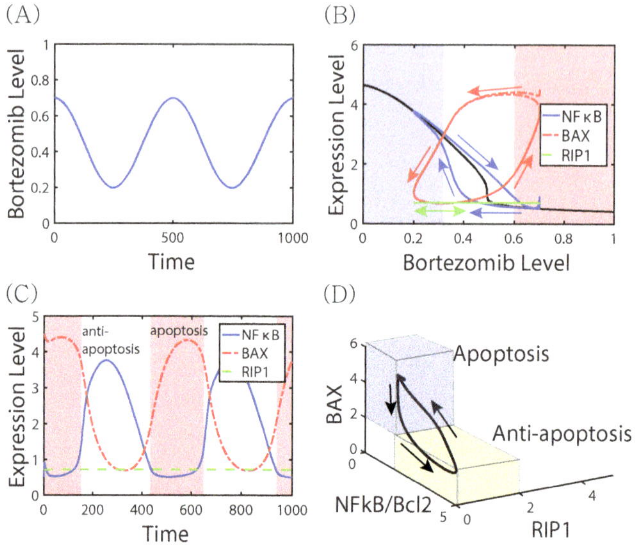

Figure 5. The effect of fluctuating BTZ on the transition between $\mathbb{T}_t$ and $\mathbb{T}_a$ in the absence of OVs. (**A**) A time-dependent BTZ level ($B(t) = 0.25 \times cos(\pi t/250) + 0.45$) was assigned for a periodic injection of BTZ in tumor microenvironment. (**B**) Trajectories of solutions ($(F(t), B(t))$ and $(A(t), B(t))$) in response to BTZ in (**A**). The black curve represents the upper and lower branches of steady states ($F - B$ bifurcation loop in Figure 4A). Red and blue arrows = solution flow of Bax and NFκB-Bcl2, respectively. Fluctuating BTZ levels induce transitions between anti-apoptotic and apoptotic status. (**C**) Time courses of concentrations of intracellular variables (NFκB-Bcl2 (blue solid), Bax (red dashed), and RIP1 (green dashed)) in response to periodic B injection in (**A**). (**D**) Trajectories of solutions corresponding to (**C**) in the $F - A - R$ space. Initial conditions: $B(0) = 0.7, S(0) = 0, F(0) = 0.5, A(0) = 4.5, R(0) = 0.7$.

3.2. Spatial Effects

Next, we compared experimental results with simulation results for various treatments, using the full model that incorporates spatial variations of components. Figure 6 shows the time course of the tumor volume for wild type (PBS), BTZ treatment, oHSV treatment, and BTZ+oHSV treatment with basic parameter values of BTZ injection (Tumor volumes are calculated based on the calculated tumor diameter ($D = 2r$, r = radius, volume (V)= $(4\pi/3) \times (D/2)^3$) in simulation (Figure 6A) [47] and on tumor length (L) and width (W) using the formula $V = 0.5\ LW^2$ for subcutaneous studies in the experiments (Figure 6B,C) [20]). One sees that when BTZ is combined with oHSV therapy, the tumor size is significantly reduced compared with other treatment protocols [20]. Of course, the killing rate of a tumor in the presence of BTZ may depend on various cell lines and tumor microenvironments in experiments. For example, mice implanted with CAL27 head and neck cancer cells (Figure 6B) and U251T3 glioma cells (Figure 6C) show slightly different time curves of tumor growth [20]. However, the overall anti-tumor efficacy of the combined therapy BTZ+oHSV was evident in these experiments. Results from the mathematical model shown in Figure 6A are in good agreement with experimental data, in particular with the growth pattern of CAL27 head and neck cancer cell lines (Figure 6B) in [20]. Moreover, the overall synergistic effect of bortezomib on tumor growth is similar to that in U251T3 glioma cells (Figure 6C). In general, subcutaneous head and neck tumors and orthotopic glioma models present different growth and invasion patterns, and the location of the tumor might influence the experimental data. In the subcutaneous tumors, the tumor usually does not grow invasive as in the brain, but is surrounded by a capsule. The details of the growth patterns can be fitted to experimental data using different parameters in the mathematical model (cf. Figure S7 in supplementary information Appendix in [51].) The mathematical model developed here presents the general framework of tumor growth in response to the combination therapy OVs+BTZ so that these results can be used to explore the various tumor dynamics in the presence of CSPGs and other microenvironments in brain tissue.

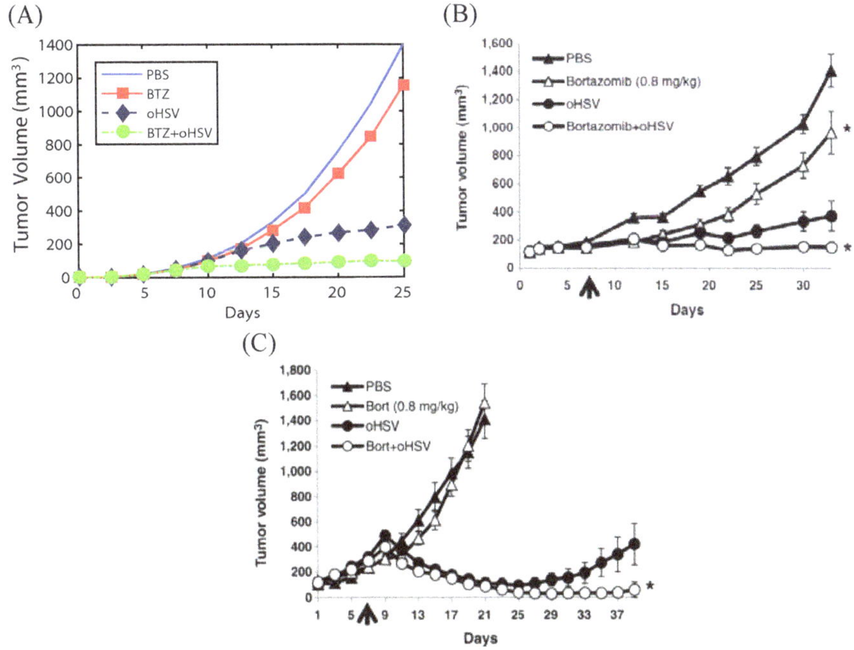

Figure 6. (**A**) Simulation results of tumor volume from simulations for PBS (blue solid), BTZ treatment (red solid square), oHSV treatment (black dotted diamond), and combination treatment (oHSV+BTZ; green dotted diamond). (**B**,**C**) Experimental results on tumor growth in response to the random motility of virus (PBS), BTZ treatment, oHSV treatment and combination therapy oHSVs+BTZ [20]: Athymic nude mice were subcutaneously implanted with CAL27 head and neck cancer cells (**B**) and U251T3 glioma cells (**C**). PBS or BTZ (0.8 mg/kg) treatments were administered through intraperitoneal injection twice a week. Following one week of BTZ treatment, mice were injected intratumorally with HBSS or oHSV (1×10^5 pfu in (**B**) and 5×10^4 pfu of 34.5ENVE in (**C**)). Figure 6B,C and legend are taken from [20].

In Figure 7, we show the effect on cell death of altering the BTZ dosage in the absence and presence of OVs. In the presence of OVs, tumor volume was decreased as the BTZ level was increased compared to the control case ($B(-)$; $I_B = 0.0$), leading to higher anti-tumor efficacy (blue; Figure 7A). BTZ also enhanced virus replication (blue; Figure 7D) and infected cell population (blue; Figure 7C), leading to higher necrosis and higher OV-enhanced cell death. The enhanced OV activities in the presence of BTZ is in good agreement with experimental observations in previous studies [20,21]. However, the anti-tumoral effect of the apoptotic pathways via BTZ alone was minimal. For example, ER stress and apoptosis (signaling pathways in Figure 7E–H) were insignificant, leading to little change in anti-tumor efficacy (yellow, Figure 7A). This relatively low response for BTZ treatment was also observed in the experiments [20,21]. These results confirm the synergistic effect of BTZ and OV treatment on anti-tumor efficacy.

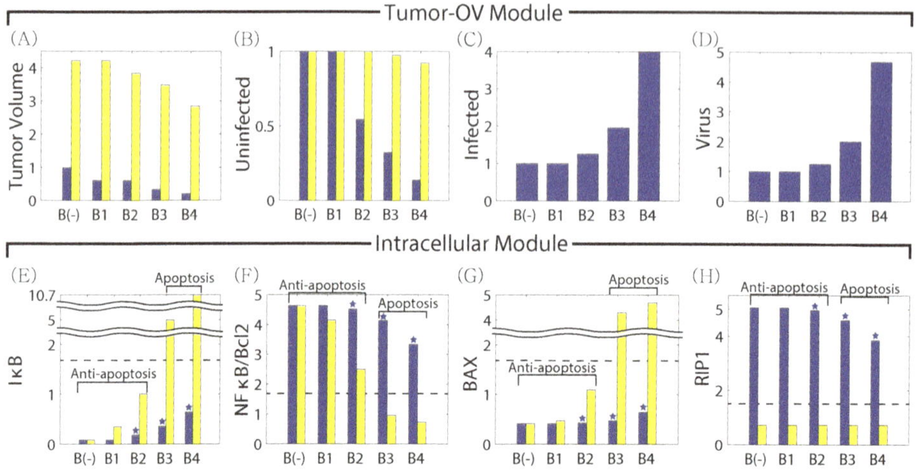

Figure 7. Effect of BTZ on anti-tumor efficacy in the absence (yellow bar) and presence (blue bar) of OVs. (**A**) Normalized tumor volume at day 20 for various BTZ supply rates ($I_B = 0(B(-))$, $3.3 \times 10^{-2}(B1), 7.7 \times 10^{-2}(B2), 1.8 \times 10^{-1}(B3^*), 3.3 \times 10^{-1}(B4)$). (**B–D**) Relative populations of uninfected cancer cells (**B**), infected cancer cells (**C**), and oHSV (**D**). (**E–H**) levels of intracellular variables, IκB (**E**), NFκB (**F**), Bax (**G**), and RIP1 (**H**).

In Figure 8, we show the effects of three different strategies for killing cancer cells. When a bolus of OVs were injected on the periphery of the growing tumor (asterisks in the lower panel of Figure 8A), the growing tumor cells were infected and killed. However, some of the tumor cells avoided the OV attacks and regrew outside the infected areas (upper panel in Figure 8A). When more OVs were injected in the center of the tumor core in addition to the periphery, tumor cells were infected in the central region of the tumor and more cells die (Figure 8B). When BTZ was added at the same injection points, OVs were amplified (Figure 8F) and more tumor cells were infected (Figure 8E), increasing the anti-tumor efficacy (Figure 8G). However, in all cases, there was a large growing ring of tumor cells that survive treatment. It is known that the location of injection sites of OVs for multifocal glioma [71–73] and anti-tumor immunity [74–77] may influence the anti-tumor efficacy and killing of invasive glioma cells.

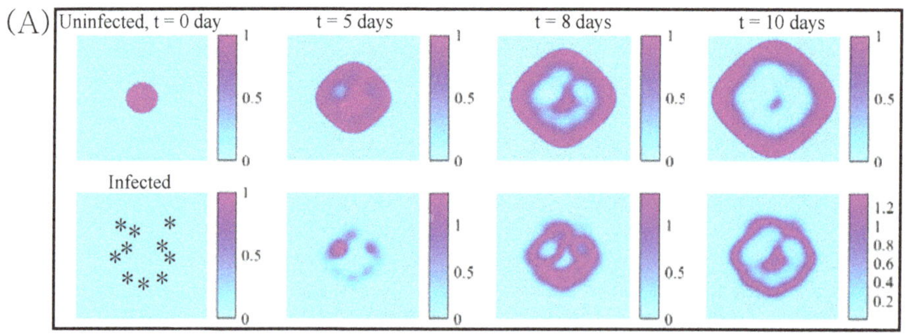

Figure 8. *Cont.*

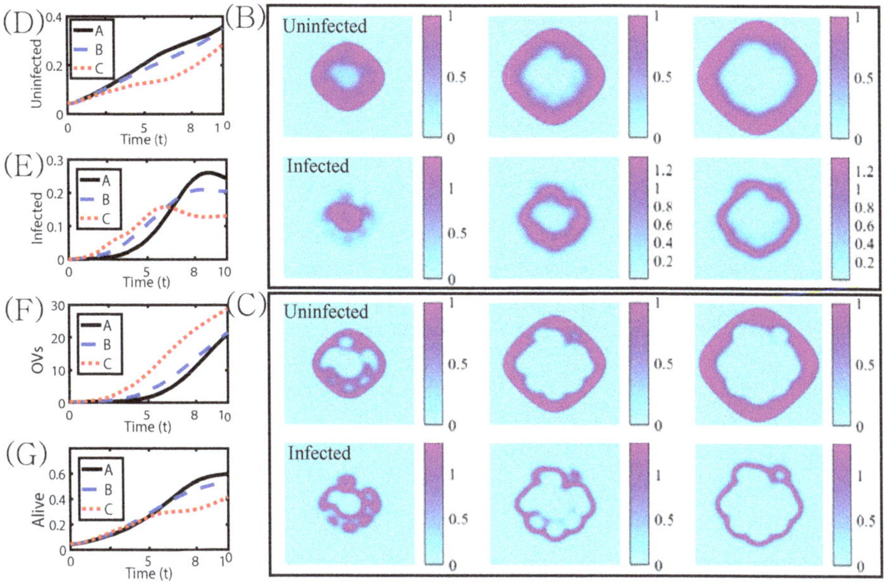

Figure 8. The dynamics of tumor growth and cell death under the combination therapy. (**A**) Profiles of solutions of uninfected (x) and infected (y) tumor cells at time $t = 0, 5, 8, 10$ days in response to OV injections on the periphery of the growing tumor. (**B**,**C**) Profiles of solutions in response to OV injections on the periphery and center of the tumor with OV only (**B**) and combination therapy OVs+BTZ; (**C**) at the times shown in the header for (**A**); (**D**–**F**) time courses of uninfected tumor populations (**D**), infected tumor population (**E**), and OVs (**F**) in three cases of (**A**–**C**). (**G**) Time courses of the populations of alive (ininfected+infected) tumor cells in three cases of (**A**–**C**).

Next we investigated how the TME influences the effect of the combination therapy. Figure 9A–C shows the spatial profiles of uninfected and infected tumor densities in response to OV injection on the periphery, OV injection on the periphery and center of the tumor, and the combination therapy OVs+BTZ. In all cases the diffusion coefficients of diffusible variables were 100-fold less in the upper half Ω_+ of the computational domain $\Omega = [0,1] \times [0,1]$ than in the lower half Ω_-. The injection sites were marked in asterisks in the lower t = 0 panel in Figure 9A. While the combination therapy resulted in the enhanced anti-tumor efficacy compared to OV therapies, tumor growth (Figure 9G) and OV spread (Figure 9E,F) are slower in the upper domain in all cases. A challenging microenvironment such as gray matter or dense ECM structure in the brain has been shown to prevent tumor invasion [78,79] and limit viral spread in the tissue [43,80]. Therefore, ECM-degrading substances such as Relaxin/decorin [81,82] and Chase-ABC [43] have been suggested to improve the anti-tumor efficacy [83–88]. MRI histological analysis showed the presence of preferred migratory paths of glioma cells and different diffuse growth patterns in complex brain microenvironment [89]. This undoubtedly involves chemotaxis of glioma cells, but this has not been included in the present model. Nonetheless, our simulation results illustrate how the tumor microenvironment can affect the OVs+BTZ combination therapy.

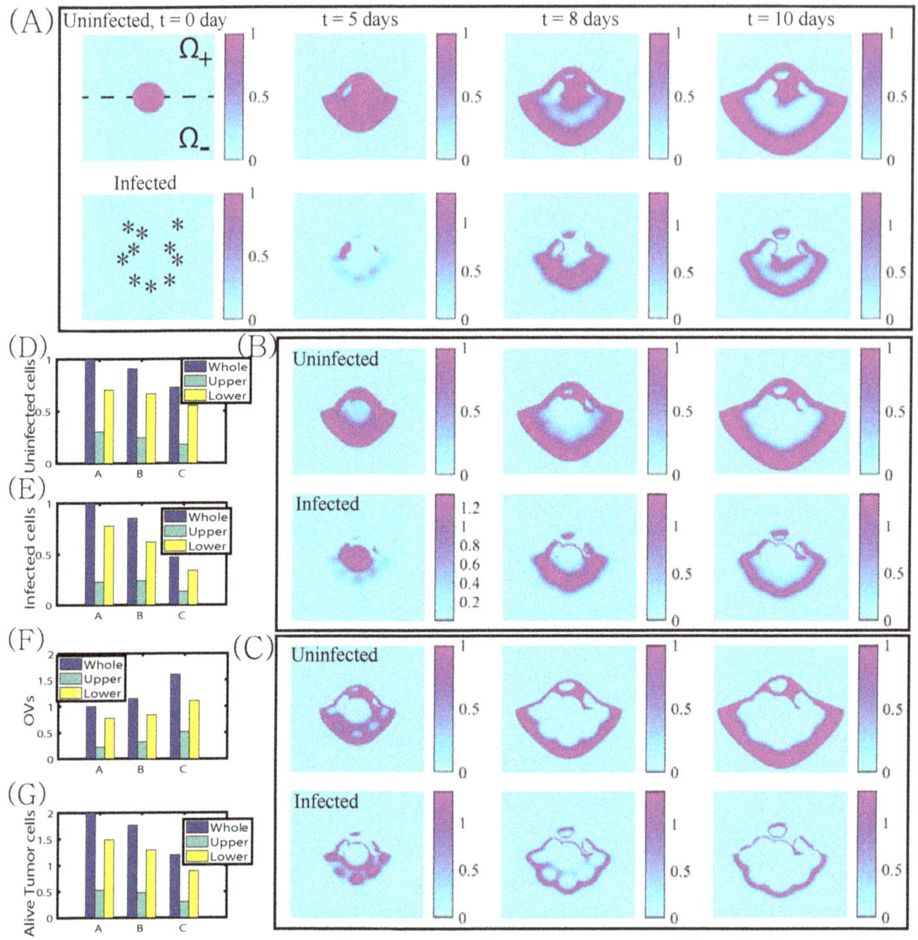

Figure 9. Dynamics of tumor growth and cell killing under combination therapy in different TMEs. In Ω_+ the diffusivities of diffusible variables are reduced 100-fold. (**A**) Profiles of solutions of uninfected (x) and infected (y) tumor cells at time $t = 0, 5, 8, 10$ *day* in response to OV injections on the periphery of the growing tumor. (**B**,**C**) Profiles of solutions in response to OV injections alone (**B**) and combination therapy OVs+BTZ; (**C**) on the periphery and center of the tumor at the corresponding time. (**D**–**G**) Populations of uninfected tumor cells (**D**), infected tumor cells (**E**), OVs (**F**), and alive tumor cells (**G**) at final time. * Whole population (blue; Ω), population in the upper domain (green; Ω_+), and population in the lower domain (yellow; Ω_-).

3.3. Effect of CSPGs on Glioma Invasion and Anti-Invasion Strategies

CSPGs, which are major parts of the ECM in the brain, can form heterogenous structures such as a ring with partial openings [90] or heterogenous patches [91], and it is known that CSPGs characterize the invasive and non-invasive phenotypes of tumors [44,92]. In the following computations the four quadrants in the computational domain are marked Q1-Q4—see Figure 10A1—and the spatial distributions of CSPGs surrounding the tumor are shown in Figure 10C (closed) and Figure 10D (open). Figure 10A,B show the time courses of uninfected tumor density at $t = 0$ (A1, B1), 5 (A2, B2), 8 (A3, B3), 10 days (A4, B4) in response to OVs+BTZ injections in the presence of a CSPG ring surrounding the tumor without (Figure 10A), and with (Figure 10B) an open section in the first quadrant. Both OVs and BTZ were injected at the six spots (black disks in Figure 10C) in the interior region. Figure 10E,F show the time courses of tumor populations within the whole area (black solid), inside (blue dashed), and outside (red dotted) the CSPG surrounding for the closed (Figure 10E) and open (Figure 10F) cases, respectively.

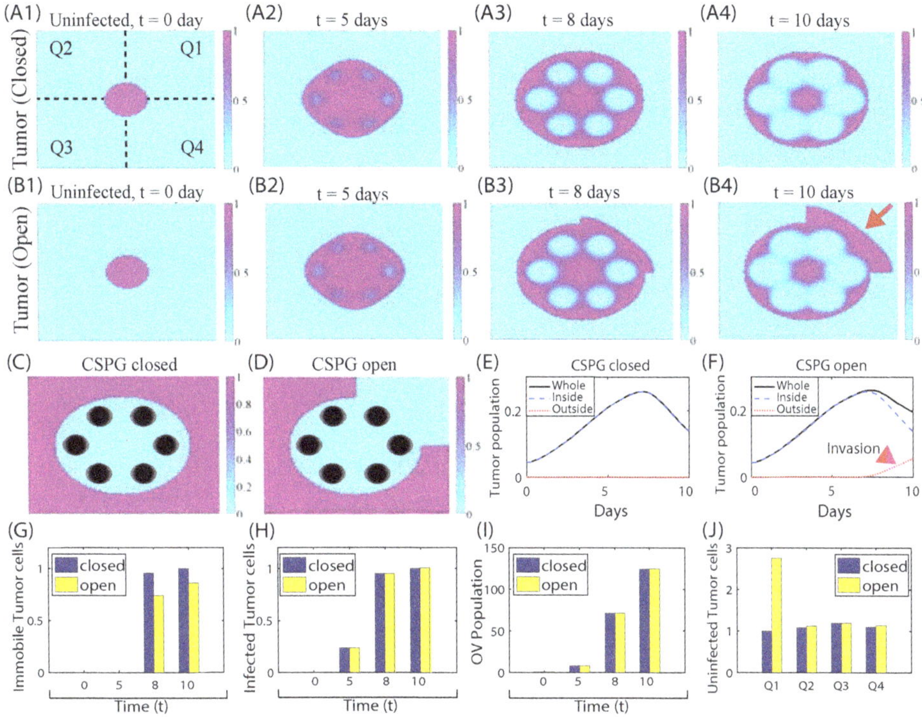

Figure 10. The role of the TME in the regulation of glioma invasion. (**A**) Profiles of uninfected tumor cells (x) in response to OVs and BTZ in the presence of CSPG surrounding the tumor at $t = 0, 5, 8,$ 10 days. OVs (v) and BTZ (B) were injected in the interior of CSPG bands marked in black circles in (**C**,**D**). (**B**) Profiles of uninfected tumor cells (x) under the same conditions, but with a partial gap in Q1. One sees that some of surviving tumor cells invade the brain tissue in Q_1. (**C**,**D**) The spatial distribution of CSPG in the closed (**C**) and open cases (**D**) corresponding to (**A**,**B**), respectively. Spatial locations of injection sites for both OVs and BTZ are marked in black filled circles in the interior. (**E**,**F**) The time courses of the tumor index (the normalized tumor cell populations); total tumor population (black solid), tumor population inside the CSPG ring structure (blue dashed), and invasive tumor cells outside the ring structure (red dotted). (**G**) Immobile tumor cells near the CSPG boundary in closed and open cases at $t = 0, 5, 8, 10$ days. (**H**,**I**) Infected tumor population (**H**) and OV (**I**) population in closed and open cases. (**J**) Uninfected tumor populations in the four quadrants at the final time in the closed (blue) and open (yellow) cases.

In both cases, uninfected tumor cells initially invaded the nearby surrounding tissue after treatment with OVs and BTZ. In the closed case, cell invasion was blocked at the CSPG barrier in all quadrants at around $t = 8$ days. The blockage was incorporated in the model by setting the diffusion coefficient of the cells to zero at the boundary of the magenta-colored region in Figure 10C,D. In the presence of the open sector, some uninfected tumor cells in the Q1 region invade the brain tissue through the open gap in the CSPG barrier (red arrow in Figure 10B4) (see also the increasing population (red arrowhead) of invasive cells in Figure 10F). Figure 10G shows the normalized populations of uninfected tumor cells that are strongly adhered to the CSPG boundary at $t = 0, 5, 8, 10$ days in the open and closed regions. Figure 10H,I show the normalized populations of infected tumor cells and OVs at the corresponding time points for open (yellow) and closed (blue) cases, respectively. No significant differences in OV infection activities are observed between the open and closed cases. The relative amounts of uninfected tumor cells at 20 days in the four quadrants are shown in more detail in the open (yellow) and closed cases (blue) in Figure 10J. In the open case, the total number of cells was higher and more are present in Q1, while the numbers in Q1-Q4 were roughly equal in the closed case. These results show the critical role of the CSPG distribution in the regulation of glioma invasion, even in the presence of OV therapy. This also suggests the possibility of blocking aggressive glioma invasion by rearrangement or manipulation of CSPG ECM through LAR-CSGAG interaction on the periphery of a growing tumor [44,92].

In Figure 11 we investigated anti-invasion strategies in the presence of a CSPG band with a partial gap. When OVs were injected in the invasion area (Ω_i in Figure 11A1) in addition to the interior of the CSPG band, some of the surviving invasive cells were killed by these OVs (Figure 11A). However, OVs alone were not enough to completely block the invasion of the glioma cells. When BTZ was added at the same invasive site, the anti-invasion efficacy is increased Figure 11B. Therefore, strategic injection of OVs+BTZ may be effective in controlling the invasive multi-focal gliomas [93]. Figure 11C–E show populations of uninfected tumor cells (C), infected tumor cells (D), and OVs (E), respectively, in the invasion region (Ω_i) of Q1 at $t = 8, 10$ days. Higher rates of OV replication and infection compared to the control lead to increased killing of invasive tumor cells in the invasive area. Figure 11F shows populations of invasive tumor cells in Ω_i at the final time t=10 days for control (without OV injection; blue) and various levels of OVs (left column) and OVs+BTZ (right column) on the invasion sites.

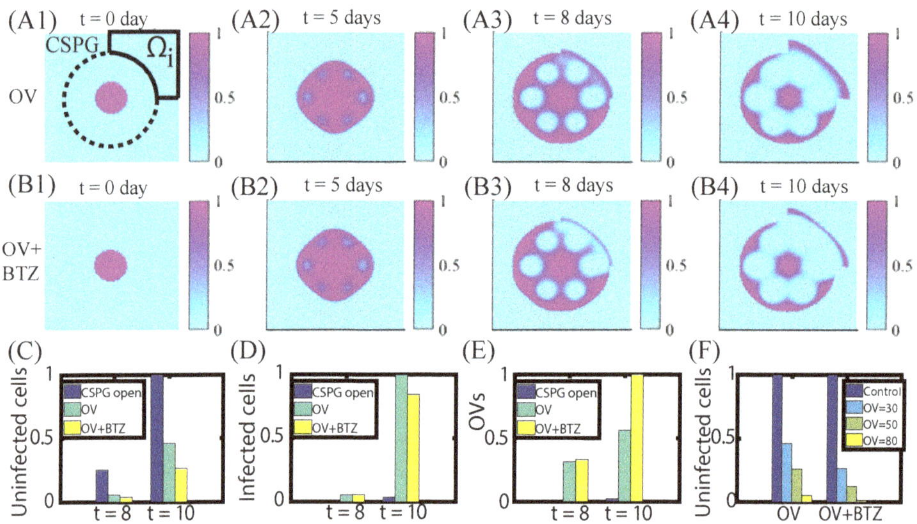

Figure 11. Anti-invasion strategies. (**A**,**B**) The distributions of uninfected tumor cells at time $t = 0, 5, 8,$ 10 days in response to injection of OVs (**A1–A4**), or OVs+BTZ (**B1–B4**), in the invasion area (Ω_i) and OV-BTZ treatment in the interior. (**C–E**) Populations of uninfected tumor cells (**C**), infected tumor cells (**D**), and OVs (**E**), respectively, in the invasion region (Ω_i) of the Q1 area at $t = 8, 10$ days in response to no treatment (blue), OV treatment (green), and combination treatment OVs+BTZ (yellow) in the Ω_i area. (**F**) Populations of invasive tumor cells in Ω_i at final time for control (without OV injection; blue) and various levels of OVs (left column) and OVs+BTZ (right column) on the invasion sites.

3.4. Anti-Invasion Strategies with OV Therapy: Localization

Finally, we investigated the effect of surgical resection when there is a gap in the surrounding CSPGs. Figure 12 shows different patterns of invasion as a function of the diffusivity of cells,

OVs and BTZ in response to resection of the tumor at the center followed by combination therapy. The following combinations of D_1 and D_v were used: $(D_1/10, 10D_v)$ in Figure 12A, $(D_1^*, 10D_v^*)$ in Figure 12B, $(10D_1^*, 10D_v^*)$ in Figure 12C, $(D_1^*/10, D_v^*)$ in Figure 12D, (D_1^*, D_v^*) in Figure 12E, $(10D_1^*, D_v^*)$ in Figure 12F, $(D_1^*/10, D_v^*/10)$ in Figure 12G, $(D_1^*, D_v^*/10)$ in Figure 12H, $(10D_1^*, D_v^*/10)$ in Figure 12I. Here, $(D_1^*$ and $D_v^*)$ are the reference diffusion coefficients. The model predicts that different TME conditions such as tissue composition—which affect D_1 and D_v—may induce different spatial invasion patterns. When both D_1 and D_v are small, the dynamics are largely controlled by localized infection of OVs and slow migration of tumor cells, leading to limited invasion and tumor cell killing. As D_1 is increased (from the left panel to the right panel), the increasing diffusion rate of uninfected cells leads to highly invasive phenotypes through the open gap in Q1 (Figure 12G → Figure 12H → Figure 12I). On the other hand, as D_v is increased, the dynamics lead to quick spread of OVs in the area, killing more tumor cells (Figure 12G → Figure 12D → Figure 12A). Interestingly, high values of D_1 and D_v lead to the effective eradication of tumor cells in the enclosed area but effective tumor invasion through the gap in Q1, thereby increasing the potential for tumor recurrence at other sites even after surgery and combination therapy (Figure 12C). This type of glioma recurrence in other parts of brain after surgery and OV therapy has been observed in clinical trials by Market et al. [72].

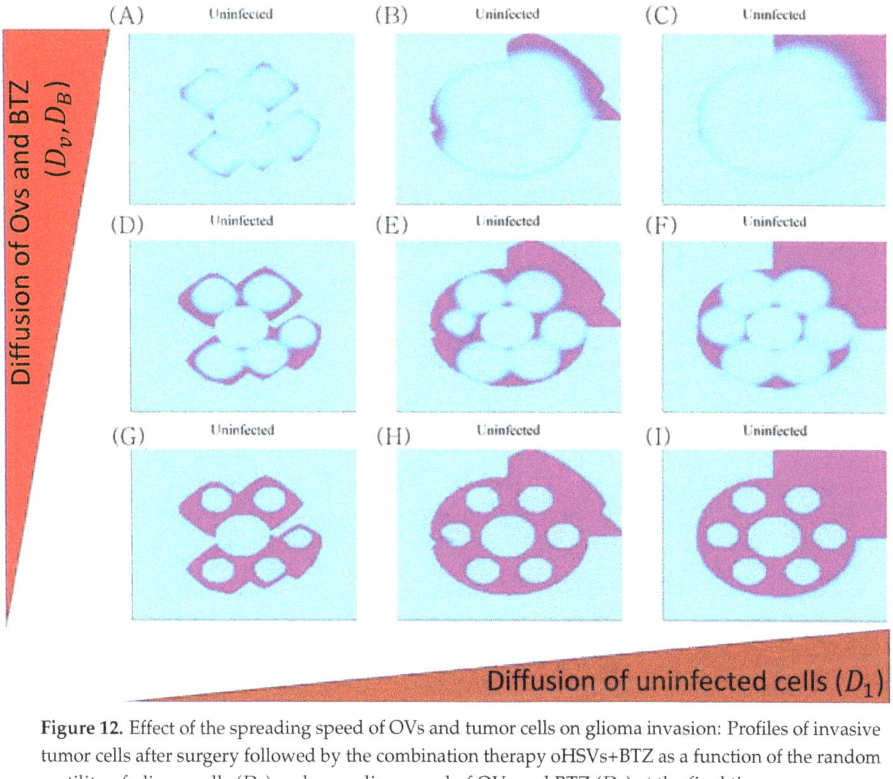

Figure 12. Effect of the spreading speed of OVs and tumor cells on glioma invasion: Profiles of invasive tumor cells after surgery followed by the combination therapy oHSVs+BTZ as a function of the random motility of glioma cells (D_1) and spreading speed of OVs and BTZ (D_v) at the final time.

4. Discussion

Cellular apoptosis is a typical target of a significant number of anti-cancer chemotherapeutic drugs such as doxorubicin, cisplatin and BTZ. However, patients often develop resistance to such drugs, leading to reduced clinical outcomes. Interestingly, HSV-1 can hijack the usual cellular pathways of ER stress and apoptosis in response to BTZ to override this response [94,95] and induce necroptotic cell death independently of both autophagic cell death and/or apoptosis [21]. While apoptotic cell death involves an immunologically silent death [96,97], necrotic cells release cytokines, resulting in a robust inflammatory response and long term immune response [98–101]. Interestingly, necroptotic cell death is in the downstream pathway of virus replication and does not affect oHSV replication in vitro [21] but the associated inflammation can also induce a mechanism for pathogen clearance [102]. The combination treatment can induce secretion of several cytokines such as IL1a, in vitro and in vivo [21], which may lead to a significant and long-term anti-tumor immune response. BTZ may also sensitize tumor cells to NK cell- and/or TNF-related apoptosis-inducing ligand (TRAIL)-mediated killing using death receptors, such as DR5 [103,104]. While NK cell treatment in addition to OV injection increases the anti-tumor efficacy, depletion of endogenous NK cells was also shown to enhance the anti-tumor efficacy in the OVs+BTZ combination therapy [51]. Other immune cells such as tumor-associated macrophages (TAMs) [105,106] also play a significant role in regulation of tumor growth and invasion by exchanging signaling molecules such as transforming growth factor-β (TGF-β) and colony stimulating factor 1 (CSF-1), adding complexities in optimizing the OV therapy [107]. Since OV therapy may be able to skew the balance between M2 and M1 activation of TAMs toward the M1 phenotype, thereby activating the anti-tumor immune responses, it may be worthwhile trying to manipulate OVs so as to 'fully educate' the existing immune cells [107]. In recent in vitro experiments, TAMs were reported to be able to increase the oncolysis of attenuated measles and mumps virus [108], opening these possibilities. These results illustrate the nonlinear complex aspects of the innate immune system in the TME in regard to regulation of OV therapy [107].

Despite the localized synergistic effects of combination therapies on cancer cell killing in glioma [20,21], its anti-tumor efficacy is not high in the TME, where the heterogenous distribution of extracellular matrix components such as CSPGs [43,47,109], and other structures such as white matter, as well as the host immune system, [51,105,110] play critical roles in the regulation of cell movement [83]. CSPGs, one of the major components of brain ECM, has been shown to inhibit glioma invasion by forming strong tumor-ECM adhesion [44–46] and a band of surrounding astrocytes repelled by the dense CSPGs in the tumor center [92]. In the context of OV therapy, CSPGs were shown to inhibit OV spread within the growing glioma [43] and Chase-ABC, a degrading enzyme, was suggested to improve OV spread throughout the tumor core [43,47,111]. Our work shows that a heterogeneous spatial distribution of CSPGs in the brain can affect the anti-tumor efficacy, in particular on invasive glioma cells. Localization of invasive glioma cells to the periphery of the resection site by injecting chemoattractants on the tumor site could improve anti-cancer therapy, as suggested in [92,112,113]. In this regard, recent advances in obtaining vector fields of GBM cell movements may also provide a useful guide for future modeling of tumour growth prior to surgical treatment [114].

In this study we focused on developing BTZ-dependent therapeutic strategies for glioblastoma in the context of OV therapies, but our mathematical model can be applied to other types of cancers. For instance, therapeutic effects of BTZ were studied in colon [115,116], prostate [117,118], breast [119,120], lung [121,122], melanoma [123,124], ovarian [20,125,126], myeloma [25,127,128], head and neck [20] cancers, EBV-associated lymphomas [129], hepatocellular carcinoma [130], and glioma [20,21]. Various types of OVs have been studied in combination with BTZ, including reovirus [131], HSV-1 [20,21,132], adenovirus [133] and VSV [134]. For instance, HSV-1 and reovirus have shown synergistic effects [135] in breast cancer [136] and pancreatic cancer [131]. A combination BTZ+adenovirus treatment led to caspase-dependent apoptosis and suppression of the antiviral immune responses in a hepatocellular carcinoma model [133]. Interestingly, the combination BTZ+OV treatment was found to inhibit VSV replication, showing less than additive cell killing rates in vitro, but additive anti-tumor activity in vivo in myeloma cells [134]. Recently, Yoo et al. [20,21] observed that the combination BTZ+OV therapy can induce the synergistic anti-tumor effects in ovarian and head and neck cancers, as well as in glioma cells and malignant peripheral nerve sheath tumor cells. Other groups have also shown that BTZ therapy can sensitize multiple myeloma therapies to ReoV infection by up-regulating the expression of the viral receptor JAM-A [137–139]. Therefore, our mathematical model can provide a framework of investigating the anti-tumor efficacy and developing therapeutic strategies of these cancers by taking into account the organ-specific tumor microenvironmental factors.

In our modeling framework, we can further investigate the effect of the combination BTZ+OV therapy on tumor growth and develop anti-invasion strategies in a complex tumor microenvironment. For instance, the model can be used to predict the anti-tumor responses when some of components in the BTZ-affected signaling network are either promoted or inhibited by anti-cancer drugs, or when tumor ECM components such as CSPGs are modified or manipulated in the brain. Since the location of OV administration is a major determinant of the fundamental characteristics of initial host responses against these armed OVs [140], our mathematical model can be adapted to optimize the injection location of OVs and BTZ in a given tumor location in order to maximize synergistic anti-tumor effects and minimize the negative anti-viral effects from the host defense system. The mathematical model can be also used to optimize the schedule and amount of BTZ and OVs, maximizing the anti-tumor effect and minimizing the administrative costs in the clinic [141–143]. However, there exist several disadvantages of the current mathematical approach. For example, the intracellular signaling networks and their regulation in cancerous cells are treated at a cell population level, not at the individual cell level. Therefore, it is quite challenging to localize and control the core IκB-NFκB-Bcl2-Bax-RIP1 signaling network at the individual cell level, especially for highly invasive cells in the tumor microenvironment. In addition, the heterogeneity in the cancer cell population plays an important role in cell motility, cellular invasion, resistance to drugs, and recurrence of the tumor cells. However, it is difficult to take such heterogeneity into account in the current modeling framework. A multi-scale hybrid modeling framework can better represent the cellular process at the individual cell level. For instance, in this work, we did not take into account the detailed biochemical reactions of tumor-ECM adhesion processes. More detailed [CSPG]-[CSGAG] receptor binding kinetics at an individual glioma cell site were modeled in [92]. A useful anti-invasion strategy may emerge by using a hybrid mathematical model [112,113,144–149] where important cellular features of highly invasive glioma cells such as cell-ECM adhesion, signaling networks [141,150–153], cell movement in a dense tissue [154], and interaction with other cells in the TME [105,155,156] can be taken into account. A review of eukaryotic cell motility detailing some of the modeling difficulties appears in [157]. Finally, one of the major components of the core control system in our model, the RIP1-RIP3 axis, is not always intact in all glioblastoma cells. Therefore, the current model may not be applicable in such cases, and we plan to modify the model in order to take into account alternative RIP1-independent cell death mechanisms in a future mathematical model.

Even though BTZ, like other agents, has a low blood-brain barrier (BBB) permeability for treating neurological disorders and spinal muscular atrophy [158–162], recent advances in the delivery method of BTZ improved anti-tumor efficacy by bypassing the BBB, making BTZ a more affordable, effective treatment for GBM [163]. Pre-treatment with an ATP-binding cassette (ABC) transporter inhibitor may also overcome the low biodistribution of BTZs in the CNS by preventing BTZ's efflux at the BBB [159]. Efficient use of BTZ and combination therapy at different locations in the brain for multi-focal glioma could also improve overall eradication of cancerous cells. For instance, BTZ alone may be strategically applied to kill tumor cells via ER stress in a location with limited access to OVs and the combination therapy OVs+BTZ can be adapted for massive cell killing in more accessible locations, such as the core of a growing tumor.

As it was clearly shown by recent Nobel prize awards, immune therapy and normalization of the immune system greatly enhanced understanding of the role of the immune system as well as the potential of immune therapy. However, recent work by Kim et al. [51] illustrated the dual role of immune systems in the regulation of tumor dynamics in response to a triple combination therapy—OVs+BTZ+NK cells. Furthermore, many immune factors can play a role in regulating killing, replication, and spread of OVs. For instance, IFNα/β can limit virus replication [42,140,164–166]. In addition, innate immune cells such as NK cells and macrophages can block spread of OVs throughout the tumor [51,140]. Intravenous, intra-aterial, or intratumoral administration of various types of OVs (vaccinia virus [167], Newcastle disease virus [168], measles virus [169], and adenovirus [170,171]) can be eliminated by the circulating humoral defense agents [172,173]. Intracellular defense mechanism can also limit OV replication [174–180]. These factors would certainly influence how OV dynamics and anti-tumor effects of therapeutic drugs in the TME are modeled, and some of these will be taken into account in future work. A better understanding of immune-tumor interactions in the TME would certainly lead to development of new therapeutic strategies in OV therapy.

Recurrence of tumors after surgical resection is a major contributing factor to the low survival rate of GBM and conventional treatments such as chemotherapy and radiotherapy are not effective in treating infiltrating glioma cells because of changes in characteristics in signaling pathways such as the apoptotic programs, and in drug resistance [93]. A new innovative strategy in addition to development of OVs, effectively targeting invasive glioma cells, is needed to eradicate cancerous cells and prevent recurrence.

Supplementary Materials: The following are available online at http://www.mdpi.com/2072-6694/11/2/215/s1, Figure S1: (A) A simplied model of the network of IκB-NFκB-Bcl2-Bax-RIP system for anti-apoptosis, apoptosis, and necroptosis of glioma cells [1–5]. (B) Schematic components of IκB, proteasome-NFκB -Bcl2 complex, BAX, and RIP1 are represented by 'S', 'F', 'A', and 'R', respectively. Table S1: Reference value used in the model.

Author Contributions: Design of models: Y.K.; writing the manuscript: Y.K., H.G.O.; analysis of the model: Y.K., J.L.; numerical simulations: J.L., D.L. discussion: Y.K., H.G.O.

Funding: This paper was supported by Konkuk University in 2016: 2016-A019-0137 (Yangjin Kim)†.

Acknowledgments: We thank professors Balveen Kaur and Ji Young Yoo at University of Texas Health Science Center at Houston for helpful discussions on the topic.

Conflicts of Interest: The authors declare no conflict of interest.

References

1. Dupont, C.; Vermandel, M.; Leroy, H.A.; Quidet, M.; Lecomte, F.; Delhem, N.; Mordon, S.; Reyns, N. INtraoperative photoDYnamic Therapy for GliOblastomas: Study Protocol for a Phase I Clinical Trial. *Neurosurgery* **2018**. [CrossRef] [PubMed]
2. Stupp, R.; Taillibert, S.; Kanner, A.; Read, W.; Steinberg, D.; Lhermitte, B.; Toms, S.; Idbaih, A.; Ahluwalia, M.S.; Fink, K.; et al. Effect of Tumor-Treating Fields Plus Maintenance Temozolomide vs Maintenance Temozolomide Alone on Survival in Patients With Glioblastoma: A Randomized Clinical Trial. *JAMA* **2017**, *318*, 2306–2316. [CrossRef] [PubMed]
3. Labi, V.; Erlacher, M. How cell death shapes cancer. *Cell Death Dis.* **2016**, *6*, e1675. [CrossRef] [PubMed]
4. Campbell, K.J.; Tait, S.W.G. Targeting BCL-2 regulated apoptosis in cancer. *Open Biol.* **2018**, *8*, 180002. [CrossRef] [PubMed]
5. Westphal, D.; Kluck, R.M.; Dewson, G. Building blocks of the apoptotic pore: How Bax and Bak are activated and oligomerize during apoptosis. *Cell Death Differ.* **2014**, *21*, 196–205. [CrossRef] [PubMed]
6. Mitsiades, N.; Mitsiades, C.S.; Poulaki, V.; Chauhan, D.; Richardson, P.G.; Hideshima, T.; Munshi, N.; Treon, S.P.; Anderson, K.C. Biologic sequelae of nuclear factor-kappaB blockade in multiple myeloma: Therapeutic applications. *Blood* **2002**, *29*, 4079–4086. [CrossRef]
7. Richardson, P.G.; Hideshima, T.; Anderson, K.C. Bortezomib (PS-341): A novel, first-in-class proteasome inhibitor for the treatment of multiple myeloma and other cancers. *Cancer Control* **2003**, *10*, 361–369. [CrossRef] [PubMed]
8. Ashkenazi, A.; Salvesen, G. Regulated cell death: Signaling and mechanisms. *Annu. Rev. Cell Dev. Biol.* **2014**, *30*, 337–356. [CrossRef] [PubMed]
9. Brown, K.; Park, S.; Kanno, T.; Franzoso, G.; Siebenlist, U. Mutual regulation of the transcriptional activator NF-kappa B and its inhibitor, I kappa B-alpha. *Proc. Natl. Acad. Sci. USA* **1993**, *90*, 2532–2536. [CrossRef] [PubMed]
10. Fan, P.; Tyagi, A.K.; Agboke, F.A.; Mathur, R.; Pokharel, N.; Jordan, V.C. Modulation of nuclear factor-kappa B activation by the endoplasmic reticulum stress sensor PERK to mediate estrogen-induced apoptosis in breast cancer cells. *Cell Death Discov.* **2018**, *4*, 15. [CrossRef] [PubMed]

11. Ma, M.H.; Parker, K.M.; Manyak, S.; Altamirano, C.V.; Wu, Z.Q.; Borad, M.J.; Berenson, J.R. Proteasome inhibitor PS-341 markedly enhances sensitivity of multiple myeloma cells to chemotherapeutic agents and overcomes chemoresistance through inhibition of the NF-kappaB pathway. *Blood* **2001**, *98*, 473a.
12. Berenson, J.R.; Ma, H.M.; Vescio, R. The role of nuclear factor-kappaB in the biology and treatment of multiple myeloma. *Semin Oncol.* **2001**, *28*, 626–633. [CrossRef]
13. Masilamani, A.P.; Ferrarese, R.; Kling, E.; Thudi, N.K.; Kim, H.; Scholtens, D.M.; Dai, F.; Hadler, M.; Unterkircher, T.; Platania, L.; et al. KLF6 depletion promotes NF-κB signaling in glioblastoma. *Oncogene* **2017**, *36*, 3562–3575. [CrossRef] [PubMed]
14. Karin, M. Nuclear factor-kappaB in cancer development and progression. *Nature* **2006**, *441*, 431–436. [CrossRef] [PubMed]
15. Achyut, B.R.; Angara, K.; Jain, M.; Borin, T.F.; Rashid, M.H.; Iskander, A.S.M.; Ara, R.; Kolhe, R.; Howard, S.; Venugopal, N.; et al. Canonical NFkB signaling in myeloid cells is required for the glioblastoma growth. *Sci. Rep.* **2017**, *7*, 13754. [CrossRef] [PubMed]
16. Strobel, T.; Tai, Y.T.; Korsmeyer, S.; Cannistra, S.A. BAD partly reverses paclitaxel resistance in human ovarian cancer cells. *Oncogene* **1998**, *17*, 2419–2427. [CrossRef]
17. Wojton, J.; Meisen, W.; Kaur, B. How to train glioma cells to die: Molecular challenges in cell death. *J. Neurooncol.* **2016**, *126*, 377–384. [CrossRef]
18. Melo-Lima, S.; Lopes, M.C.; Mollinedo, F. Necroptosis is associated with low procaspase-8 and active RIPK1 and -3 in human glioma cells. *Oncoscience* **2014**, *1*, 649–664. [CrossRef]
19. Rosenfeld, M.; Ye, X.; Supko, J.; Desideri, S.; Grossman, S.; Brem, S.; Mikkelson, T.; Wang, D.; Chang, Y.; Hu, J.; et al. A phase I/II trial of hydroxychloroquine in conjunction with radiation therapy and concurrent and adjuvant temozolomide in patients with newly diagnosed glioblastoma multiforme. *Autophagy* **2014**, *10*, 1359–1368. [CrossRef]
20. Yoo, J.Y.; Hurwitz, B.S.; Bolyard, C.; Yu, J.; Zhang, J.; Selvendiran, K.; Rath, K.S.; He, S.; Bailey, Z.; Eaves, D.; et al. Bortezomib-Induced Unfolded Protein Response Increases Oncolytic HSV-1 Replication Resulting in Synergistic Antitumor Effects. *Clin. Cancer Res.* **2014**, *20*, 3787–3798. [CrossRef]
21. Yoo, J.Y.; Jaime-Ramirez, A.C.; Bolyard, C.; Dai, H.; Nallanagulagari, T.; Wojton, J.; Hurwitz, B.; Relation, T.; Yu, J.G.; Lee, T.; et al. Bortezomib treatment sensitizes oncolytic HSV-1 treated tumors to NK cell immunotherapy. *Clin. Cancer Res.* **2016**, *22*, 5265–5276. [CrossRef] [PubMed]
22. Perkins, N.D. The diverse and complex roles of NF-κB subunits in cancer. *Nat. Rev. Cancer* **2012**, *12*, 121–132. [CrossRef] [PubMed]
23. Yin, D.; Zhou, H.; Kumagai, T.; Liu, G.; Ong, J.M.; Black, K.L.; Koeffler, H.P. Proteasome inhibitor PS-341 causes cell growth arrest and apoptosis in human glioblastoma multiforme (GBM). *Oncogene* **2005**, *24*, 344–354. [CrossRef]
24. Manasanch, E.E.; Shah, J.J.; Lee, H.C.; Weber, D.M.; Thomas, S.K.; Amini, B.; Feng, L.; Berkova, Z.; Hildebrandt, M.; Orlowski, R.Z. Bortezomib, lenalidomide, and dexamethasone with panobinostat for front-line treatment of patients with multiple myeloma who are eligible for transplantation: A phase 1 trial. *Lancet Haematol.* **2018**, *5*, e628–e640. [CrossRef]
25. de la Fuente, F.; Duran, M.S.; Alvarez, M.A.; Sanroman, I.L.; Dios, A.M.; Tamayo, R.R.; Garcia, R.; Gonzalez, M.S.; Prieto, E.; Barez, A.; et al. Subcutaneous bortezomib in newly diagnosed patients with multiple myeloma nontransplant eligible: Retrospective evaluation. *Semin. Hematol.* **2018**, *55*, 189–196. [CrossRef] [PubMed]
26. Dudek, A.Z.; Lesniewski-Kmak, K.; Shehadeh, N.J.; Pandey, O.N.; Franklin, M.; Kratzke, R.A.; Greeno, E.W.; Kumar, P. Phase I study of bortezomib and cetuximab in patients with solid tumours expressing epidermal growth factor receptor. *Br. J. Cancer* **2009**, *100*, 1379–1384. [CrossRef] [PubMed]
27. Gilbert, J.; Lee, J.W.; Argiris, A.; Haigentz, M.; Feldman, L.E.; Jang, M.; Arun, P.; Waes, C.V.; Forastiere, A.A. Phase II 2-arm trial of the proteasome inhibitor, PS-341 (bortezomib) in combination with irinotecan or PS-341 alone followed by the addition of irinotecan at time of progression in patients with locally recurrent or metastatic squamous cell carcinoma of the head and neck (E1304): A trial of the Eastern Cooperative Oncology Group. *Head Neck* **2013**, *35*, 942–948. [PubMed]
28. Burch, A.D.; Weller, S.K. Nuclear sequestration of cellular chaperone and proteasomal machinery during herpes simplex virus type 1 infection. *J. Virol.* **2004**, *78*, 7175–7185. [CrossRef] [PubMed]

29. Gross, S.; Catez, F.; Masumoto, H.; Lomonte, P. Centromere architecture breakdown induced by the viral E3 ubiquitin ligase ICP0 protein of herpes simplex virus type 1. *PLoS ONE* **2012**, *7*, e44227. [CrossRef] [PubMed]
30. Christofferson, D.E.; Yuan, J. Necroptosis as an alternative form of programmed cell death. *Curr. Opin. Cell Biol.* **2010**, *22*, 263–268. [CrossRef] [PubMed]
31. de Almagro, M.C.; Vucic, D. Necroptosis: Pathway diversity and characteristics. In *Seminars in Cell and Developmental Biology*; Elsevier: Amsterdam, The Netherlands, 2015; Volume 39, pp. 56–62.
32. Das, A.; McDonald, D.G.; Dixon-Mah, Y.N.; Jacqmin, D.J.; Samant, V.N.; Vandergrift, W.A.; Lindhorst, S.M.; Cachia, D.; Varma, A.K.; Vanek, K.N.; et al. RIP1 and RIP3 complex regulates radiation-induced programmed necrosis in glioblastoma. *Tumor Biol.* **2016**, *37*, 7525–7534. [CrossRef] [PubMed]
33. Suryadevara, C.M.; Riccione, K.A.; Sampson, J.H. Immunotherapy gone viral: Bortezomib and oHSV enhance antitumor NK cell activity. *Clin. Cancer Res.* **2016**, *22*, 5164–5166. [CrossRef] [PubMed]
34. Lundstrom, K. New frontiers in oncolytic viruses: Optimizing and selecting for virus strains with improved efficacy. *Biologics* **2018**, *12*, 43–60. [CrossRef] [PubMed]
35. Taguchi, S.; Fukuhara, H.; Homma, Y.; Todo, T. Current status of clinical trials assessing oncolytic virus therapy for urological cancers. *Int. J. Urol.* **2017**, *24*, 342–351. [CrossRef] [PubMed]
36. Paraskevakou, G.; Allen, C.; Nakamura, T.; Zollman, P.; James, C.D.; Peng, K.W.; Schroeder, M.; Russell, S.J.; Galanis, E. Epidermal growth factor receptor (EGFR)-retargeted measles virus strains effectively target EGFR- or EGFRvIII expressing gliomas. *Mol. Ther.* **2007**, *15*, 677–686. [CrossRef] [PubMed]
37. Crommentuijn, M.H.; Kantar, R.; Noske, D.P.; Vandertop, W.P.; Badr, C.E.; Wurdinger, T.; Maguire, C.A.; Tannous, B.A. Systemically administered AAV9-sTRAIL combats invasive glioblastoma in a patient-derived orthotopic xenograft model. *Mol. Ther. Oncol.* **2016**, *3*, 16017. [CrossRef] [PubMed]
38. Ylosmaki, E.; Martikainen, M.; Hinkkanen, A.; Saksela, K. Attenuation of Semliki Forest virus neurovirulence by microRNA-mediated detargeting. *J. Virol.* **2013**, *87*, 335–344. [CrossRef] [PubMed]
39. Chen, B.; Timiryasova, T.M.; Haghighat, P.; Andres, M.L.; Kajioka, E.H.; Dutta-Roy, R.; Gridley, D.S.; Fodor, I. Low-dose vaccinia virus-mediated cytokine gene therapy of glioma. *J. Immunother.* **2001**, *24*, 46–57. [CrossRef] [PubMed]
40. Jha, B.K.; Dong, B.; Nguyen, C.T.; Polyakova, I.; Silverman, R.H. Suppression of antiviral innate immunity by sunitinib enhances oncolytic virotherapy. *Mol. Ther.* **2013**, *21*, 1749–1757. [CrossRef] [PubMed]
41. Zhen, Z.; Yang, K.; Ye, L.; You, Z.; Chen, R.; Liu, Y. Decorin gene upregulation mediated by an adeno-associated virus vector increases intratumoral uptake of nab-paclitaxel in neuroblastoma via inhibition of stabilin-1. *Investig. New Drugs* **2017**, *35*, 566–575. [CrossRef] [PubMed]
42. de Graaf, J.; de Vor, L.; Fouchier, R.; van den Hoogen, B. Armed oncolytic viruses: A kick-start for anti-tumor immunity. *Cytokine Growth Factor Rev.* **2018**, *41*, 28–39. [CrossRef] [PubMed]
43. Dmitrieva, N.; Yu, L.; Viapiano, M.; Cripe, T.P.; Chiocca, E.A.; Glorioso, J.C.; Kaur, B. Choindroitinase ABC I-mediated enhancement of oncolytic virus spread and antitumor efficacy. *Clin. Cancer Res.* **2011**, *17*, 1362–1372. [CrossRef] [PubMed]
44. Silver, D.J.; Siebzehnrubl, F.A.; Schildts, M.J.; Yachnis, A.T.; Smith, G.M.; Smith, A.A.; Scheffler, B.; Reynolds, B.A.; Silver, J.; Steindler, D.A. Chondroitin sulfate proteoglycans potently inhibit invasion and serve as a central organizer of the brain tumor microenvironment. *J. Neurosci.* **2013**, *33*, 15603–15617. [CrossRef] [PubMed]
45. Silver, D.J.; Silver, J. Contributions of chondroitin sulfate proteoglycans to neurodevelopment, injury, and cancer. *Curr. Opin. Neurobiol.* **2014**, *27*, 171–178. [CrossRef] [PubMed]
46. Kundu, S.; Forsberg-Nilsson, K. Glycosaminoglycans and Glioma Invasion. *Eur. Assoc. NeuroOncol. Mag.* **2014**, *4*, 75–80.
47. Kim, Y.; Lee, H.G.; Dmitrieva, N.; Kim, J.; Kaur, B.; Friedman, A. Choindroitinase ABC I-mediated enhancement of oncolytic virus spread and anti-tumor efficacy: A mathematical model. *PLoS ONE* **2014**, *9*, e102499. [CrossRef] [PubMed]
48. Zuo, H.; Yuan, J.; Chen, Y.; Li, S.; Su, Z.; Wei, E.; Li, C.; Weng, S.; Xu, X.; He, J. A microRNA-mediated positive feedback regulatory loop of the NF-κB pathway in Litopenaeus vannamei. *J. Immunol.* **2016**, *196*, 3842–3853. [CrossRef]
49. Kaur, B.; Yoo, J.Y. (University of Texas Health Science Center at Houston, Houston, TX, USA). Private communication, 2018.

50. Najafov, A.; Chen, H.; Yuan, J. Necroptosis and Cancer. *Trends Cancer* **2017**, *3*, 294–301. [CrossRef]
51. Kim, Y.; Yoo, J.Y.; Lee, T.J.; Liu, J.; Yu, J.; Caligiuri, M.A.; Kaur, B.; Friedman, A. Complex role of NK cells in regulation of oncolytic virus-bortezomib therapy. *Proc. Natl. Acad. Sci. USA* **2018**, *115*, 4927–4932. [CrossRef]
52. Qiu, B.; Wang, Y.; Tao, J.; Wang, Y. Expression and correlation of Bcl-2 with pathological grades in human glioma stem cells. *Oncol. Rep.* **2012**, *28*, 155–160.
53. Bergqvist, S.; Ghosh, G.; Komives, E.A. The IkBa/NF-kB complex has two hot spots, one at either end of the interface. *Protein Sci.* **2008**, *17*, 2051–2058. [CrossRef] [PubMed]
54. Mathes, E.; O'Dea, E.L.; Hoffmann, A.; Ghosh, G. NF-kB dictates the degradation pathway of IkBa. *EMBO J.* **2008**, *27*, 1357–1367. [CrossRef] [PubMed]
55. Krappmann, D.; Scheidereit, C. Regulation of NF-kappa B activity by I kappa B alpha and I kappa B beta stability. *Immunobiology* **1997**, *198*, 3–13. [CrossRef]
56. Xin, M.; Deng, X. Nicotine Inactivation of the Proapoptotic Function of Bax through Phosphorylation. *J. Biol. Chem.* **2005**, *280*, 10781–10789. [CrossRef]
57. Wang, Q.; Chen, W.; Xu, X.; Li, B.; He, W.; Padilla, M.T.; Jang, J.H.; Nyunoya, T.; Amin, S.; Wang, X.; Lin, Y. RIP1 potentiates BPDE-induced transformation in human bronchial epithelial cells through catalase-mediated suppression of excessive reactive oxygen species. *Carcinogenesis* **2013**, *34*, 2119–2128. [CrossRef] [PubMed]
58. Friedman, A.; Tian, J.P.; Fulci, G.; Chiocca, E.A.; Wang, J. Glioma virotherapy: Effects of innate immune suppression and increased viral replication capacity. *Cancer Res.* **2006**, *66*, 2314–2319. [CrossRef] [PubMed]
59. Othmer, H.G.; Xie, M. *Implicit and Explicit Methods for Excitable Systems*; Technical Report; University of Minesota: Minneapolis, MN, USA, 2006.
60. Mok, W.; Stylianopoulos, T.; Boucher, Y.; Jain, R.K. Mathematical modeling of herpes simplex virus distribution in solid tumors: Implications for cancer gene therapy. *Clin. Cancer Res.* **2009**, *15*, 2352–2360. [CrossRef]
61. Leveque, D.; Carvalho, M.C.M.; MALOISEL, F. Clinical Pharmacokinetics of Bortezomib. *In Vivo* **2007**, *21*, 273–278. [PubMed]
62. Kane, R.C.; Bross, P.F.; Farrell, A.T.; Pazdur, R. The mean elimination half-life of bortezomib after the first dose ranged from 9-15 hours at doses ranging from 1.45–2.00 mg/m^2 in patients with advanced malignancies. *Oncologist* **2003**, *8*, 508–513. [CrossRef]
63. Lipniacki, T.; Paszek, P.; Brasier, A.; Luxon, B.; Kimmel, M. Mathematical model of NF-kappaB regulatory module. *J. Theor. Biol.* **2004**, *228*, 195–215. [CrossRef]
64. Lee, E.; Boone, D.; Chai, S.; Libby, S.; Chien, M.; Lodolce, J.; Ma, A. Failure to regulate TNF-induced NF-kappaB and cell death responses in A20-deficient mice. *Science* **2000**, *289*, 2350–2354. [CrossRef] [PubMed]
65. Mothes, J.; Busse, D.; Kofahl, B.; Wolf, J. Sources of dynamic variability in NF-kB signal transduction: A mechanistic model. *Bioessays* **2015**, *37*, 452–462. [CrossRef] [PubMed]
66. Xue, X.; Xia, W.; Wenzhong, H. A modeled dynamic regulatory network of NF-kappaB and IL-6 mediated by miRNA. *Biosystems* **2013**, *114*, 214–218. [CrossRef] [PubMed]
67. Kirkland, R.; Saavedra, G.; Cummings, B.; Franklin, J. Bax regulates production of superoxide in both apoptotic and nonapoptotic neurons: Role of caspases. *J. Neurosci.* **2010**, *30*, 16114–16127. [CrossRef] [PubMed]
68. Li, J.; McQuade, T.; Siemer, A.; Napetschnig, J.; Moriwaki, K.; Hsiao, Y.; Damko, E.; Moquin, D.; Walz, T.; McDermott, A.; et al. The RIP1/RIP3 necrosome forms a functional amyloid signaling complex required for programmed necrosis. *Cell* **2012**, *150*, 339–350. [CrossRef] [PubMed]
69. ODonoghue, J.A.; Bardies, M.; Wheldon, T.E. Relationships between tumor size and curability for uniformly targeted therapy with beta-emitting radionuclides. *J. Nucl. Med.* **1995**, *36*, 1902–1909.
70. Feinman, R.; Gangurde, P.; Miller, S. Proteasome inhibitor PS-341 inhibits constitutive NF-kappaB activation and bypasses the anti-apoptotic bcl-2 signal in human multiple myeloma cells. *Blood* **2001**, *98*, 640a.
71. Ozduman, K.; Wollmann, G.; Piepmeier, J.M.; van den Pol, A.N. Systemic vesicular stomatitis virus selectively destroys multifocal glioma and metastatic carcinoma in brain. *J. Neurosci.* **2008**, *28*, 1882–1893. [CrossRef]

72. Markert, J.M.; Liechty, P.G.; Wang, W.; Gaston, S.; Braz, E.; Karrasch, M.; Nabors, L.B.; Markiewicz, M.; Lakeman, A.D.; Palmer, C.A.; et al. Phase Ib trial of mutant herpes simplex virus G207 inoculated pre-and post-tumor resection for recurrent GBM. *Mol. Ther.* **2009**, *17*, 199–207. [CrossRef]
73. Wollmann, G.; Ozduman, K.; van den Pol, A.N. Oncolytic Virus Therapy of Glioblastoma Multiforme—Concepts and Candidates. *Cancer J.* **2012**, *18*, 69–81. [CrossRef]
74. Jiang, H.; Rivera-Molina, Y.; Gomez-Manzano, C.; Clise-Dwyer, K.; Bover, L.; Vence, L.; Yuan, Y.; Lang, F.; Toniatti, C.; Hossain, M.; et al. Oncolytic Adenovirus and Tumor-Targeting Immune Modulatory Therapy Improve Autologous Cancer Vaccination. *Cancer Res.* **2017**, *77*, 3894–3907. [CrossRef]
75. Ali, S.; King, G.; Curtin, J.; Candolfi, M.; Xiong, W.; Liu, C.; Puntel, M.; Cheng, Q.; Prieto, J.; Ribas, A.; et al. Combined immunostimulation and conditional cytotoxic gene therapy provide long-term survival in a large glioma model. *Cancer Res.* **2005**, *65*, 7194–7204. [CrossRef] [PubMed]
76. King, G.; Kroeger, K.; Bresee, C.; Candolfi, M.; Liu, C.; Manalo, C.; Muhammad, A.; Pechnick, R.; Lowenstein, P.; Castro, M. Flt3L in Combination With HSV1-TK-mediated Gene Therapy Reverses Brain Tumor-induced Behavioral Deficits. *Mol. Ther.* **2008**, *16*, 682–690. [CrossRef] [PubMed]
77. King, G.; Muhammad, A.; Curtin, J.; Barcia, C.; Puntel, M.; Liu, C.; Honig, S.; Candolfi, M.; Mondkar, S.; Lowenstein, P.; Castro, M. Flt3L and TK gene therapy eradicate multifocal glioma in a syngeneic glioblastoma model. *Neuro-Oncology* **2008**, *10*, 19–31. [CrossRef]
78. Wesseling, P.; Kros, J.M.; Jeuken, J.W.M. The pathological diagnosis of diffuse gliomas: Towards a smart synthesis of microscopic and molecular information in a multidisciplinary context. *Diagn. Histopathol.* **2011**, *17*, 486–494. [CrossRef]
79. Beadle, C.; Assanah, M.C.; Monzo, P.; Vallee, R.; Rosenfield, S.S.; Canoll, P. The role of myosin II in glioma invasion of the brain. *Mol. Biol. Cell* **2008**, *19*, 3357–3368. [CrossRef] [PubMed]
80. Yaacov, B.; Lazar, I.; Tayeb, S.; Frank, S.; Izhar, U.; Lotem, M.; Perlman, R.; Ben-Yehuda, D.; Zakay-Rones, Z.; Panet, A. Extracellular matrix constituents interfere with Newcastle disease virus spread in solid tissue and diminish its potential oncolytic activity. *J. Gen. Virol.* **2012**, *93*, 1664–1672. [CrossRef] [PubMed]
81. Kim, J.H.; Lee, Y.S.; Kim, H.; Huang, J.H.; Yoon, A.R.; Yun, C.O. Relaxin expression from tumor targeting adenoviruses and its intra tumoral spread, apoptosis induction, and efficacy. *Gene Ther.* **2006**, *98*, 1482–1493.
82. Choi, I.; Lee, Y.; Yoo, J.; Yoon, A.; Kim, H.; Kim, D.; Seidler, D.; Kim, J.; Yun, C. Effect of decorin on overcoming the extracellular matrix barrier for oncolytic virotherapy. *Gene Ther.* **2010**, *17*, 190–201. [CrossRef]
83. Wojton, J.; Kaur, B. Impact of tumor microenvironment on oncolytic viral therapy. *Cytokine Growth Factor Rev.* **2010**, *21*, 127–134. [CrossRef]
84. Ganesh, S.; Edick, M.G.; Idamakanti, N.; Abramova, M.; Vanroey, M.; Robinson, M.; Yun, C.O.; Jooss, K. Relaxin-expressing, fiber chimeric oncolytic adenovirus prolongs survival of tumor-bearing mice. *Cancer Res.* **2007**, *67*, 4399–4407. [CrossRef] [PubMed]
85. Ganesh, S.; Gonzalez-Edick, M.; Gibbons, D.; Roey, M.V.; Jooss, K. Intratumoral coadministration of hyaluronidase enzyme and oncolytic adenoviruses enhances virus potency in metastatic tumor models. *Clin. Cancer Res.* **2008**, *14*, 3933–3941. [CrossRef] [PubMed]
86. Kuriyama, N.; Kuriyama, H.; Julin, C.M.; Lamborn, K.; Israel, M.A. Pretreatment with protease is a useful experimental strategy for enhancing adenovirus-mediated cancer gene therapy. *Hum. Gene Ther.* **2000**, *11*, 2219–2230. [CrossRef] [PubMed]
87. McKee, T.D.; Grandi, P.; Mok, W.; Alexandrakis, G.; Insin, N.; Zimmer, J.P.; Bawendi, M.G.; Boucher, Y.; Breakefield, X.O.; Jain, R.K. Degradation of fibrillar collagen in a human melanoma xenograft improves the efficacy of an oncolytic herpes simplex virus vector. *Cancer Res.* **2006**, *66*, 2509–2513. [CrossRef]
88. Cheng, J.; Sauthoff, H.; Huang, Y.; Kutler, D.I.; Bajwa, S.; Rom, W.N.; Hay, J.G. Human matrix metalloproteinase-8 gene delivery increases the oncolytic activity of a replicating adenovirus. *Mol. Ther.* **2007**, *15*, 1982–1990. [CrossRef] [PubMed]
89. Mughal, A.A.; Zhang, L.; Fayzullin, A.; Server, A.; Li, Y.; Wu, Y.; Glass, R.; Meling, T.; Langmoen, I.A.; Leergaard, T.B.; et al. Patterns of Invasive Growth in Malignant Gliomas-The Hippocampus Emerges as an Invasion-Spared Brain Region. *Neoplasia* **2018**, *20*, 643–656. [CrossRef] [PubMed]
90. Andrews, E.M.; Richards, R.J.; Yin, F.Q.; Viapiano, M.S.; Jakemana, L.B. Alterations in chondroitin sulfate proteoglycan expression occur both at and far from the site of spinal contusion injury. *Exp. Neurol.* **2012**, *235*, 174–187. [CrossRef]

91. Hayashi, N.; Tatsumi, K.; Okuda, H.; Yoshikawa, M.; Ishizaka, S.; Miyata, S.; Manabe, T.; Wanaka, A. DACS, novel matrix structure composed of chondroitin sulfate proteoglycan in the brain. *Biochem. Biophys. Res. Commun.* **2007**, *364*, 410–415. [CrossRef]
92. Kim, Y.; Kang, H.; Powathil, G.; Kim, H.; Trucu, D.; Lee, W.; Lawler, S.; Chaplain, M. Role of extracellular matrix and microenvironment in regulation of tumor growth and LAR-mediated invasion in glioblastoma. *PLoS ONE* **2018**, *13*, e0204865. [CrossRef]
93. Lun, X.; Senger, D.L.; Alain, T.; Oprea, A.; Parato, K.; Stojdl, D.; Lichty, B.; Power, A.; Johnston, R.N.; Hamilton, M.; et al. Effects of intravenously administered recombinant vesicular stomatitis virus (VSV(deltaM51)) on multifocal and invasive gliomas. *J. Natl. Cancer Inst.* **2006**, *98*, 1546–1557. [CrossRef]
94. Mulvey, M.; Arias, C.; Mohr, I. Maintenance of endoplasmic reticulum (ER) homeostasis in herpes simplex virus type 1-infected cells through the association of a viral glycoprotein with PERK, a cellular ER stress sensor. *J. Virol.* **2007**, *81*, 3377–3390. [CrossRef]
95. Jerome, K.R.; Fox, R.; Chen, Z.; Sears, A.E.; Lee, H.Y.; Corey, L. Herpes simplex virus inhibits apoptosis through the action of two genes, Us5 and Us3. *J. Virol.* **1999**, *73*, 8950–8957.
96. Nagata, S.; Hanayama, R.; Kawane, K. Autoimmunity and the clearance of dead cells. *Cell* **2010**, *140*, 619–630. [CrossRef]
97. Lauber, K.; Blumenthal, S.G.; Waibel, M.; Wesselborg, S. Clearance of apoptotic cells: Getting rid of the corpses. *Mol. Cell* **2004**, *14*, 277–287. [CrossRef]
98. Fadok, V.A.; Bratton, D.L.; Rose, D.M.; Pearson, A.; Ezekewitz, R.A.; Henson, P.M. A receptor for phosphatidylserine-specific clearance of apoptotic cells. *Nature* **2000**, *405*, 85–90. [CrossRef] [PubMed]
99. Golstein, P.; Kroemer, G. Cell death by necrosis: Towards a molecular definition. *Trends Biochem. Sci.* **2007**, *32*, 37–43. [CrossRef] [PubMed]
100. Zong, W.X.; Thompson, C.B. Necrotic death as a cell fate. *Genes Dev.* **2006**, *20*, 1–15. [CrossRef]
101. Pasparakis, M.; Vandenabeele, P. Necroptosis and its role in inflammation. *Nature* **2015**, *517*, 311–320. [CrossRef] [PubMed]
102. Kaczmarek, A.; Vandenabeele, P.; Krysko, D.V. Necroptosis: The release of damage-associated molecular patterns and its physiological relevance. *Immunity* **2013**, *38*, 209–223. [CrossRef] [PubMed]
103. Lundqvist, A.; Abrams, S.I.; Schrump, D.S.; Alvarez, G.; Suffredini, D.; Berg, M.; Childs, R. Bortezomib and depsipeptide sensitize tumors to tumor necrosis factorrelated apoptosis-inducing ligand: A novel method to potentiate natural killer cell tumor cytotoxicity. *Cancer Res.* **2006**, *66*, 7317–7325. [CrossRef]
104. Hallett, W.H.; Ames, E.; Motarjemi, M.; Barao, I.; Shanker, A.; Tamang, D.L.; Sayers, T.J.; Hudig, D.; Murphy, W.J. Sensitization of tumor cells to NK cell-mediated killing by proteasome inhibition. *J. Immunol.* **2008**, *180*, 163–170. [CrossRef] [PubMed]
105. Kim, Y.; Jeon, H.; Othmer, H.G. The role of the tumor microenvironment in glioblastoma: A mathematical model. *IEEE Trans. Biomed. Eng.* **2017**, *64*, 519–527. [CrossRef] [PubMed]
106. Wesolowska, A.; Kwiatkowska, A.; Slomnicki, L.; Dembinski, M.; Master, A.; Sliwa, M.; Franciszkiewicz, K.; Chouaib, S.; Kaminska, B. Microglia-derived TGF-beta as an important regulator of glioblastoma invasion—An inhibition of TGF-beta-dependent effects by shRNA against human TGF-beta type II receptor. *Oncogene* **2008**, *27*, 918–930. [CrossRef] [PubMed]
107. Marelli, G.; Howells, A.; Lemoine, N.R.; Wang, Y. Oncolytic Viral Therapy and the Immune System: A Double-Edged Sword Against Cancer. *Front. Immunol.* **2018**, *9*, 866. [CrossRef] [PubMed]
108. Tan, D.Q.; Zhang, L.; Ohba, K.; Ye, M.; Ichiyama, K.; Yamamoto, N. Macrophage response to oncolytic paramyxoviruses potentiates virus-mediated tumor cell killing. *Eur. J. Immunol.* **2016**, *46*, 919–928. [CrossRef] [PubMed]
109. Lee, H.G.; Kim, Y.; Kim, J. Mathematical model and its fast numerical method for the tumor growth. *Math. Biosci. Eng.* **2015**, *12*, 1173–1187. [CrossRef] [PubMed]
110. Batich, K.; Reap, E.; Archer, G.; Sanchez-Perez, L.; Nair, S.; Schmittling, R.; Norberg, P.; Xie, W.; Herndon, J.; Healy, P.; et al. Long-term Survival in Glioblastoma with Cytomegalovirus pp65-Targeted Vaccination. *Clin. Cancer Res.* **2017**, *23*, 1898–1909. [CrossRef]
111. Lee, H.G.; Kim, Y. The role of the microenvironment in regulation of CSPG-driven invasive and non-invasive tumor growth in glioblastoma. *Jpn. J. Ind. Appl. Math.* **2016**, *32*, 771–805. [CrossRef]
112. Kim, Y. Regulation of cell proliferation and migration in glioblastoma: New therapeutic approach. *Front. Mol. Cell. Oncol.* **2013**, *3*, 53. [CrossRef]

113. Kim, Y.; Powathil, G.; Kang, H.; Trucu, D.; Kim, H.; Lawler, S.; Chaplain, M. Strategies of eradicating glioma cells: A multi-scale mathematical model with miR-451-AMPK-mTOR control. *PLoS ONE* **2015**, *10*, e0114370. [CrossRef]
114. Esmaeili, M.; Stensjoen, A.; Berntsen, E.; Solheim, O.; Reinertsen, I. The Direction of Tumour Growth in Glioblastoma Patients. *Sci. Rep.* **2018**, *8*, 1199. [CrossRef] [PubMed]
115. Wang, S.; Wang, L.; Zhou, Z.; Deng, Q.; Li, L.; Zhang, M.; Liu, L.; Li, Y. Leucovorin Enhances the Anti-cancer Effect of Bortezomib in Colorectal Cancer Cells. *Sci. Rep.* **2017**, *7*, 682. [CrossRef] [PubMed]
116. Cusack, J.C., Jr.; Liu, R.; Houston, M.; Abendroth, K.; Elliott, P.; Adams, J.; Baldwin, A., Jr. Enhanced chemosensitivity to CPT-11 with proteasome inhibitor PS-341: Implications for systemic nuclear factor-kappaB inhibition. *Cancer Res.* **2001**, *61*, 3535–3540. [PubMed]
117. Modernelli, A.; Naponelli, V.; Giovanna-Troglio, M.; Bonacini, M.; Ramazzina, I.; Bettuzzi, S.; Rizzi, F. EGCG antagonizes bortezomib cytotoxicity in prostate cancer cells by an autophagic mechanism. *Sci. Rep.* **2015**, *5*, 15270. [CrossRef]
118. Befani, C.; Vlachostergios, P.; Hatzidaki, E.; Patrikidou, A.; Bonanou, S.; Simos, G.; Papandreou, C.; Liakos, P. Bortezomib represses HIF-1α protein expression and nuclear accumulation by inhibiting both PI3K/Akt/TOR and MAPK pathways in prostate cancer cells. *J. Mol. Med.* **2012**, *90*, 45–54. [CrossRef] [PubMed]
119. Periyasamy-Thandavan, S.; Jackson, W.; Samaddar, J.; Erickson, B.; Barrett, J.; Raney, L.; Gopal, E.; Ganapathy, V.; Hill, W.; Bhalla, K.; et al. Bortezomib blocks the catabolic process of autophagy via a cathepsin-dependent mechanism, affects endoplasmic reticulum stress and induces caspase-dependent cell death in antiestrogen-sensitive and resistant ER+ breast cancer cells. *Autophagy* **2010**, *6*, 19–35. [CrossRef] [PubMed]
120. Jones, M.; Liu, J.; Barthel, T.; Hussain, S.; Lovria, E.; Cheng, D.; Schoonmaker, J.; Mulay, S.; Ayers, D.; Bouxsein, M.; et al. A proteasome inhibitor, bortezomib, inhibits breast cancer growth and reduces osteolysis by downregulating metastatic genes. *Clin. Cancer Res.* **2010**, *16*, 4978–4989. [CrossRef] [PubMed]
121. Ando, M.; Hoyos, V.; Yagyu, S.; Tao, W.; Ramos, C.; Dotti, G.; Brenner, M.; Bouchier-Hayes, L. Bortezomib sensitizes non-small cell lung cancer to mesenchymal stromal cell-delivered inducible caspase-9-mediated cytotoxicity. *Cancer Gene Ther.* **2014**, *21*, 472–482. [CrossRef]
122. Davies, A.; Lara, P.L., Jr.; Mack, P.; Gandara, D. Incorporating bortezomib into the treatment of lung cancer. *Clin Cancer Res.* **2007**, *13*, s4647–s4651. [CrossRef]
123. Selimovic, D.; Porzig, B.; El-Khattouti, A.; Badura, H.; Ahmad, M.; Ghanjati, F.; Santourlidis, S.; Haikel, Y.; Hassan, M. Bortezomib/proteasome inhibitor triggers both apoptosis and autophagy-dependent pathways in melanoma cells. *Cell. Signal.* **2013**, *25*, 308–318. [CrossRef] [PubMed]
124. Poklepovic, A.; Youssefian, L.; Winning, M.; Birdsell, C.; Crosby, N.; Ramakrishnan, V.; Ernstoff, M.; Roberts, J. Phase I trial of bortezomib and dacarbazine in melanoma and soft tissue sarcoma. *Investig. New Drugs* **2013**, *31*, 937–942. [CrossRef]
125. Kao, C.; Chao, A.; Tsai, C.; Chuang, W.; Huang, W.; Chen, G.; Lin, C.; Wang, T.; Wang, H.; Lai, C. Bortezomib enhances cancer cell death by blocking the autophagic flux through stimulating ERK phosphorylation. *Cell Death Dis.* **2014**, *5*, e1510. [CrossRef] [PubMed]
126. Kao, C.; Chao, A.; Tsai, C.; Lin, C.; Chuang, W.; Chen, H.; Yen, T.; Wang, T.; Lai, C.; Wang, H. Phosphorylation of signal transducer and activator of transcription 1 reduces bortezomib-mediated apoptosis in cancer cells. *Cell Death Dis.* **2013**, *4*, e512. [CrossRef] [PubMed]
127. Zaal, E.; Wu, W.; Jansen, G.; Zweegman, S.; Cloos, J.; Berkers, C. Bortezomib resistance in multiple myeloma is associated with increased serine synthesis. *Cancer Metab.* **2017**, *5*, 7. [CrossRef] [PubMed]
128. Zi, F.; He, J.; Li, Y.; Wu, C.; Wu, W.; Yang, Y.; Wang, L.; He, D.; Yang, L.; Zhao, Y.; et al. Fibroblast activation protein protects bortezomib-induced apoptosis in multiple myeloma cells through β-catenin signaling pathway. *Cancer Biol. Ther.* **2014**, *15*, 1413–1422. [CrossRef] [PubMed]
129. Zou, P.; Kawada, J.; Pesnicak, L.; Cohen, J.I. Bortezomib induces apoptosis of Epstein-Barr virus (EBV)-transformed B cells and prolongs survival of mice inoculated with EBV-transformed B cells. *J. Virol.* **2007**, *81*, 10029–10036. [CrossRef] [PubMed]
130. Hui, B.; Shi, Y.H.; Ding, Z.B.; Zhou, J.; Gu, C.Y.; Peng, Y.F.; Yang, H.; Liu, W.R.; Shi, G.M.; Fan, J. Proteasome inhibitor interacts synergistically with autophagy inhibitor to suppress proliferation and induce apoptosis in hepatocellular carcinoma. *Cancer* **2012**, *118*, 5560–5571. [CrossRef]

131. Carew, J.; Espitia, C.; Zhao, W.; Kelly, K.; Coffey, M.; Freeman, J.; Nawrocki, S. Reolysin is a novel reovirus-based agent that induces endoplasmic reticular stress-mediated apoptosis in pancreatic cancer. *Cell Death Dis.* **2013**, *4*, e728. [CrossRef]
132. Michaud, M.; Martins, I.; Sukkurwala, A.; Adjemian, S.; Ma, Y.; Pellegatti, P.; Shen, S.; Kepp, O.; Scoazec, M.; Mignot, G.; et al. Autophagy-dependent anticancer immune responses induced by chemotherapeutic agents in mice. *Oncolytic Virother.* **2011**, *334*, 1573–1577. [CrossRef]
133. Boozari, B.; Mundt, B.; Woller, N.; Strüver, N.; Gürlevik, E.; Schache, P.; Kloos, A.; Knocke, S.; Manns, M.P.; Wirth, T.C.; et al. Antitumoural immunity by virus-mediated immunogenic apoptosis inhibits metastatic growth of hepatocellular carcinoma. *Gut* **2010**, *59*, 1416–1426. [CrossRef]
134. Yarde, D.N.; Nace, R.A.; Russell, S.J. Oncolytic vesicular stomatitis virus and bortezomib are antagonistic against myeloma cells in vitro but have additive anti-myeloma activity in vivo. *Exp. Hematol.* **2013**, *41*, 1038–1049. [CrossRef]
135. Simpson, G.; Relph, K.; Harrington, K.; Melcher, A.; Pandha, H. Cancer immunotherapy via combining oncolytic virotherapy with chemotherapy: Recent advances. *Oncolytic Virother.* **2016**, *5*, 1–13. [PubMed]
136. Milani, M.; Rzymski, T.; Mellor, H.; Pike, L.; Bottini, A.; Generali, D.; Harris, A. The role of ATF4 stabilization and autophagy in resistance of breast cancer cells treated with bortezomib. *Cancer Res.* **2009**, *69*, 4415–4423. [CrossRef] [PubMed]
137. Bartee, E. Potential of oncolytic viruses in the treatment of multiple myeloma. *Oncolytic Virother.* **2018**, *7*, 1–12. [CrossRef] [PubMed]
138. Kelly, K.; Espitia, C.; Zhao, W.; Wendlandt, E.; Tricot, G.; Zhan, F.; Carew, J.; Nawrocki, S. Junctional adhesion molecule-A is overexpressed in advanced multiple myeloma and determines response to oncolytic reovirus. *Oncotarget* **2015**, *6*, 41275–41289. [CrossRef] [PubMed]
139. Stiff, A.; Caserta, E.; Sborov, D.; Nuovo, G.; Mo, X.; Schlotter, S.; Canella, A.; Smith, E.; Badway, J.; Old, M.; et al. Histone deacetylase inhibitors enhance the therapeutic potential of reovirus in multiple myeloma. *Mol. Cancer Ther.* **2016**, *15*, 830–841. [CrossRef] [PubMed]
140. Chiocca, E.; Rabkin, S. Oncolytic viruses and their application to cancer immunotherapy. *Cancer Immunol. Res.* **2014**, *2*, 295–300. [CrossRef]
141. Reyes, A.A.V.L.; Jung, E.; Kim, Y. Optimal control strategies of eradicating invisible glioblastoma cells after conventional surgery. *J. R. Soc. Interface* **2015**, *12*, 20141392.
142. Kim, Y.; Lee, D.; Lee, J.; Lee, S.; Lawler, S. Role of tumor-associated neutrophils in regulation of tumor growth in lung cancer development: A mathematical model. *PLoS ONE* **2019**, *14*, e0211041. [CrossRef]
143. Schattler, H.; Kim, Y.; Ledzewicz, U.; los Reyes V, A.A.; Jung, E. On the control of cell migration and proliferation in glioblastoma. In Proceedings of the 52nd IEEE Conference on Decision and Control, Florence, Italy, 10–13 December 2013; pp. 1810–1815.
144. Kim, Y.; Stolarska, M.; Othmer, H.G. A hybrid model for tumor spheroid growth in vitro I: Theoretical development and early results. *Math. Models Methods Appl. Sci.* **2007**, *17*, 1773–1798. [CrossRef]
145. Stolarska, M.; Kim, Y.; Othmer, H.G. Multiscale Models of Cell and Tissue Dynamics. *Phil. Trans. R. Soc. A* **2009**, *367*, 3525–3553. [CrossRef] [PubMed]
146. Kim, Y.; Stolarska, M.; Othmer, H.G. The role of the microenvironment in tumor growth and invasion. *Prog. Biophys. Mol. Biol.* **2011**, *106*, 353–379. [CrossRef] [PubMed]
147. Kim, Y.; Roh, S. A hybrid model for cell proliferation and migration in glioblastoma. *Discret. Contin. Dyn. Syst. B* **2013**, *18*, 969–1015. [CrossRef]
148. Kim, Y.; Othmer, H.G. A hybrid model of tumor-stromal interactions in breast cancer. *Bull. Math. Biol.* **2013**, *75*, 1304–1350. [CrossRef] [PubMed]
149. Kim, Y.; Othmer, H.G. Hybrid models of cell and tissue dynamics in tumor growth. *Math. Biosci. Eng.* **2015**, *12*, 1141–1156. [PubMed]
150. Aguda, B.D.; Kim, Y.; Kim, H.S.; Friedman, A.; Fine, H. Qualitative network modeling of the MYC-p53 control system of cell proliferation and differentiation. *Biophys. J.* **2011**, *101*, 2082–2091. [CrossRef] [PubMed]
151. Kim, Y.; Roh, S.; Lawler, S.; Friedman, A. miR451 and AMPK/MARK mutual antagonism in glioma cells migration and proliferation. *PLoS ONE* **2011**, *6*, e28293. [CrossRef] [PubMed]
152. Aguda, B.D.; Kim, Y.; Hunter, M.G.; Friedman, A.; Marsh, C.B. MicroRNA Regulation of a Cancer Network: Consequences of the Feedback Loops Involving miR-17-92, E2F, and Myc. *Proc. Natl. Acad. Sci. USA* **2008**, *105*, 19678–19683. [CrossRef]

153. Kim, Y.; Kang, H.; Lawler, S. The role of miR-451-AMPK signaling pathways in regulation of cell migration and proliferation in glioblastoma. In *Mathematical Models of Tumor-Immune System Dynamics*; Kim, P., Eladdadi, A., Mallet, D., Eds.; Springer: New York, NY, USA, 2014; pp. 125–155.
154. Lee, W.; Lim, S.; Kim, Y. The role of myosin II in glioma invasion: A mathematical model. *PLoS ONE* **2017**, *12*, e0171312. [CrossRef]
155. Kim, Y.; Wallace, J.; Li, F.; Ostrowski, M.; Friedman, A. Transformed epithelial cells and fibroblasts/myofibroblasts interaction in breast tumor: A mathematical model and experiments. *J. Math. Biol.* **2010**, *61*, 401–421. [CrossRef]
156. Kim, Y.; Friedman, A. Interaction of tumor with its microenvironment: A Mathematical Model. *Bull. Math. Biol.* **2010**, *72*, 1029–1068. [CrossRef] [PubMed]
157. Othmer, H. Eukaryotic cell dynamics from crawlers to swimmers. *WIREs Comput. Mol. Sci.* **2019**, *9*, e1376. [CrossRef]
158. Gozzetti, A.; Cerase, A. Novel agents in CNS myeloma treatment. *Cent. Nerv. Syst. Agents Med. Chem.* **2014**, *14*, 23–27. [CrossRef] [PubMed]
159. Foran, E.; Kwon, D.; Nofziger, J.; Arnold, E.; Hall, M.; Fischbeck, K.; Burnett, B. CNS uptake of bortezomib is enhanced by P-glycoprotein inhibition: Implications for spinal muscular atrophy. *Neurobiol. Dis.* **2016**, *88*, 118–124. [CrossRef] [PubMed]
160. Nakamura, T.; Tanaka, K.; Matsunobu, T.; Okada, T.; Nakatani, F.; Sakimura, R.; Hanada, M.; Iwamoto, Y. The mechanism of cross-resistance to proteasome inhibitor bortezomib and overcoming resistance in Ewing's family tumor cells. *Int. J. Oncol.* **2007**, *31*, 803–811. [CrossRef] [PubMed]
161. Rumpold, H.; Salvador, C.; Wolf, A.; Tilg, H.; Gastl, G.; Wolf, D. Knockdown of Pgp resensitizes leukemic cells to proteasome inhibitors. *Biochem. Biophys. Res. Commun.* **2007**, *361*, 549–554. [CrossRef] [PubMed]
162. Lu, S.; Chen, Z.; Yang, J.; Chen, L.; Zhou, H.; Xu, X.; Li, J.; Han, F.; Wang, J. The effects of proteasome inhibitor bortezomib on a P-gp positive leukemia cell line K562/A02. *Int. J. Lab. Hematol.* **2010**, *32*, e123–e131. [CrossRef]
163. Wang, W.; Cho, H.; Rosenstein-Sisson, R.; Ramos, N.; Price, R.; Hurth, K.; Schonthal, A.; Hofman, F.; Chen, T. Intratumoral delivery of bortezomib: Impact on survival in an intracranial glioma tumor model. *J. Neurosurg* **2018**, *128*, 695–700. [CrossRef]
164. Marvin, S.; Huerta, C.; Sharp, B.; Freiden, P.; Cline, T.; Schultz-Cherry, S. Type I Interferon Response Limits Astrovirus Replication and Protects against Increased BarrierPermeability In Vitro and In Vivo. *J. Virol.* **2015**, *90*, 1988–1996. [CrossRef]
165. Westcott, M.; Liu, J.; Rajani, K.; D'Agostino, R., Jr.; Lyles, D.; Porosnicu, M. Interferon Beta and Interferon Alpha 2a Differentially Protect Head and Neck Cancer Cells from Vesicular Stomatitis Virus-Induced Oncolysis. *J. Virol.* **2015**, *89*, 7944–7954. [CrossRef]
166. Buijs, P.; van Nieuwkoop, S.; Vaes, V.; Fouchier, R.; van Eijck, C.; Hoogen, B.V.D. Recombinant immunomodulating lentogenic or mesogenic oncolytic newcastle disease virus for treatment of pancreatic adenocarcinoma. *Viruses* **2015**, *7*, 2980–2998. [CrossRef] [PubMed]
167. Magge, D.; Guo, Z.; O'Malley, M.; Francis, L.; Ravindranathan, R.; Bartlett, D. Inhibitors of C5 complement enhance vaccinia virus oncolysis. *Cancer Gene Ther.* **2013**, *20*, 342–350. [CrossRef] [PubMed]
168. Biswas, M.; Johnson, J.; Kumar, S.; Parks, G.; Elankumarana, S. Incorporation of host complement regulatory proteins into Newcastle disease virus enhances complement evasion. *J. Virol.* **2012**, *86*, 12708–12716. [CrossRef] [PubMed]
169. Russell, S.; Peng, K. Measles virus for cancer therapy. *Curr. Top. Microbiol. Immunol.* **2009**, *330*, 213–241. [PubMed]
170. Tomita, K.; Sakurai, F.; Tachibana, M.; Mizuguchi, H. Correlation between adenovirus-neutralizing antibody titer and adenovirus vector-mediated transduction efficiency following intratumoral injection. *Anticancer Res.* **2012**, *32*, 1145–1152. [PubMed]
171. Raki, M.; Sarkioja, M.; Escutenaire, S.; Kangasniemi, L.; Haavisto, E.; Kanerva, A.; Cerullo, V.; Joensuu, T.; Oksanen, M.; Pesonen, S.; et al. Switching the fiber knob of oncolytic adenoviruses to avoid neutralizing antibodies in human cancer patients. *J. Gene Med.* **2011**, *13*, 253–261. [CrossRef] [PubMed]
172. Wakimoto, H.; Fulci, G.; Tyminski, E.; Chiocca, E. Altered expression of antiviral cytokine mRNAs associated with cyclophosphamideas enhancement of viral oncolysis. *Gene Ther.* **2004**, *11*, 214–223. [CrossRef]

173. Kueberuwa, G.; Cawood, R.; Seymour, L. Blood compatibility of enveloped viruses. *Curr. Opin. Mol. Ther.* **2010**, *12*, 412–420.
174. Yebdri, F.B.; Grevenynghe, J.V.; Tang, V.; Goulet, M.; Wu, J.H.; Stojdl, D.; Hiscott, J.; Lin, R. Triptolide-mediated inhibition of interferon signaling enhances vesicular stomatitis virus-based oncolysis. *Mol. Ther.* **2013**, *21*, 2043–2053. [CrossRef]
175. Alvarez-Breckenridge, C.; Yu, J.; Price, R.; Wei, M.; Wang, Y.; Nowicki, M.; Ha, Y.; Bergin, S.; Hwang, C.; Fernandez, S.; et al. The histone deacetylase inhibitor valproic acid lessens NK cell action against oncolytic virus-infected glioblastoma cells by inhibition of STAT5/T-BET signaling and generation of gamma interferon. *J. Virol.* **2012**, *86*, 4566–4577. [CrossRef]
176. Otsuki, A.; Patel, A.; Kasai, K.; Suzuki, M.; Kurozumi, K.; Chiocca, E.; Saeki, Y. Histone deacetylase inhibitors augment antitumor efficacy of herpes-based oncolytic viruses. *Mol. Ther.* **2008**, *16*, 1546–1555. [CrossRef] [PubMed]
177. Liu, Y.; Suksanpaisan, L.; Steele, M.; Russell, S.; Peng, K. Induction of antiviral genes by the tumor microenvironment confers resistance to virotherapy. *Sci. Rep.* **2013**, *3*, 2375. [CrossRef] [PubMed]
178. Okemoto, K.; Kasai, K.; Wagner, B.; Haseley, A.; Meisen, H.; Bolyard, C.; Mo, X.; Wehr, A.; Lehman, A.; Fernandez, S.; et al. DNA Demethylating Agents Synergize with Oncolytic HSV1 against Malignant Gliomas. *Clin Cancer Res.* **2013**, *9*, 5952–5959. [CrossRef] [PubMed]
179. Okemoto, K.; Wagner, B.; Meisen, H.; Haseley, A.; Kaur, B.; Chiocca, E. STAT3 activation promotes oncolytic HSV1 replication in glioma cells. *PLoS ONE* **2013**, *8*, e71932. [CrossRef] [PubMed]
180. Berchtold, S.; Lampe, J.; Weiland, T.; Smirnow, I.; Schleicher, S.; Handgretinger, R.; Kopp, H.; Reiser, J.; Stubenrauch, F.; Mayer, N.; et al. Innate immune defense defines susceptibility of sarcoma cells to measles vaccine virus-based oncolysis. *J. Virol.* **2013**, *87*, 3484–3501. [CrossRef] [PubMed]

cancers

MDPI

Article

Epilepsy Associates with Decreased HIF-1α/STAT5b Signaling in Glioblastoma

Sharon Berendsen [1], Wim G. M. Spliet [2], Marjolein Geurts [1], Wim Van Hecke [2], Tatjana Seute [1], Tom J. Snijders [1], Vincent Bours [3], Erica H. Bell [4], Arnab Chakravarti [4] and Pierre A. Robe [1,3,4,*]

1 Department of Neurology and Neurosurgery, Brain Center Rudolf Magnus, University Medical Center of Utrecht, Heidelberglaan 100, 3584 CX Utrecht, The Netherlands; s.berendsen-2@umcutrecht.nl (S.B.); M.Geurts-2@umcutrecht.nl (M.G.); t.seute@umcutrecht.nl (T.S.); T.J.Snijders@umcutrecht.nl (T.J.S.)
2 Department of Pathology, University Medical Center of Utrecht, Heidelberglaan 100, 3584 CX Utrecht, The Netherlands; W.G.M.Spliet@umcutrecht.nl (W.G.M.S.); w.vanhecke@umcutrecht.nl (W.V.H.)
3 Department of Human Genetics, GIGA Research Center, University of Liège, Avenue de l Hôpital, 1, 4000 Liège, Belgium; vbours@uliege.be
4 Department of Radiation Oncology, The Ohio State University Comprehensive Cancer Center—Arthur G. James Cancer Hospital, The Ohio State University, West 10th Avenue, Columbus, OH 43210, USA; Erica.Bell@osumc.edu (E.H.B.); Arnab.Chakravarti@osumc.edu (A.C.)
* Correspondence: p.robe@umcutrecht.nl; Tel.: +31-88-7557059

Received: 9 November 2018; Accepted: 28 December 2018; Published: 4 January 2019

Abstract: Epilepsy at presentation is an independent favorable prognostic factor in glioblastoma (GBM). In this study, we analyze the oncologic signaling pathways that associate with epilepsy in human GBMs, and that can underlie this prognostic effect. Following ethical approval and patient consent, fresh frozen GBM tissue was obtained from 76 patient surgeries. Hospital records were screened for the presence of seizures at presentation of the disease. mRNA and miRNA expression-based and gene set enrichment analyses were performed on these tissues, to uncover candidate oncologic pathways that associate with epilepsy. We performed qPCR experiments and immunohistochemistry on tissue microarrays containing 286 GBMs to further explore the association of these candidate pathways and of markers of mesenchymal transformation (NF-κB, CEBP-β, STAT3, STAT5b, VEGFA, SRF) with epilepsy. Gene sets involved in hypoxia/HIF-1α, STAT5, CEBP-β and epithelial-mesenchymal transformation signaling were significantly downregulated in epileptogenic GBMs. On confirmatory protein expression analyses, epileptogenic tumors were characterized by a significant downregulation of phospho-STAT5b, a target of HIF-1α. Epilepsy status did not associate with molecular subclassification or miRNA expression patterns of the tumors. Epileptogenic GBMs correlate with decreased hypoxia/ HIF-1α/STAT5b signaling compared to glioblastomas that do not present with epilepsy.

Keywords: glioblastoma; epilepsy; translational research; GSEA; tissue microarrays; hypoxia; HIF-1α; STAT5b; mesenchymal transformation

---

1. Introduction

Glioblastoma (GBM) is the most malignant primary brain tumor with a dismal prognosis. Median patient survival in those who can undergo aggressive therapy is 15–20 months from diagnosis [1]. Genetic and epigenetic (e.g., DNA methylation, histone acetylation) events condition the deregulation of key signaling pathways in glioblastoma and hence, its growth, invasion and therapeutic resistance [2]. Specific molecular characteristics provide distinct molecular glioblastoma subtypes, of which the mesenchymal subtype has been linked to a more aggressive and invasive tumor phenotype [3] and is driven by alterations in master transcription factors such as STAT3, CEBP-β and NF-κB [4,5].

The disease presents itself with epileptic seizures in 30–40% of the cases [6,7], and we previously showed that epilepsy at presentation is an independent favorable prognostic factor for overall survival in glioblastoma patients [7]. This prognostic effect could not be explained by treatment with specific anti-epileptic drugs [7,8], a smaller tumor volume at presentation, due to an earlier detection of the tumor, or IDH1 mutations [7]. The mechanisms underlying this prognostic effect in glioblastoma thus remain to be elucidated.

Glioma-associated epilepsy is (in part) mediated by tumor specific biological changes. Most studies have however been performed in low grade gliomas (LGGs), since these tumors more often present with epilepsy compared to GBMs [9].

Tumor-associated epilepsy results, at least in part, from a local neuronal excitation/inhibition disbalance. For instance, the release of D-2-hydroxyglutarate (D2HG), a substance structurally similar to glutamate, in the tumor microenvironment was associated with epilepsy [10]. NR2B, a predominantly extrasynaptic NMDA glutamate receptor, is highly phosphorylated in peritumoral mouse brain tissue, and increases Ca^{2+} influx in the cells, leading to a self-activating circle and overexcitation of neurons [11,12]. Another proposed mechanism of tumor-associated seizures is mediated by the glutamate release pathway cystine/glutamate transporter System Xc−(SXC), which expression is elevated in a subset of glioblastoma tissues, and in peritumoral tissues. Higher glutamate concentrations and lower glutamine synthetase expression in the tumors and peritumoral tissues were also linked to glioma-related seizures [13–15].

Some studies have also linked glioma-associated epilepsy with *IDH1* mutation [16–18]. We have shown that such mutations cannot solely explain the favorable prognosis of epileptogenic glioblastomas [7]. A lower expression of *OLIG2*, linked to the proneural GBM subtype [3], was also related to an increased risk of tumor-associated seizures [19]. How this could relate to the prognosis is however unknown.

The mechanisms underlying the prognostic effect of epilepsy at presentation on overall survival in glioblastoma patients have thus not yet been investigated. Therefore, in this study, we focus on the oncogenic signaling pathways that associate with epilepsy in human glioblastomas, in search for the mechanisms that underlie this prognostic effect.

2. Results

Fresh-frozen tissue from 76 'de novo' GBM patients was included in this study. Baseline characteristics are shown in Table S1. Of these patients, 30 presented with epilepsy and 46 had different symptoms at presentation of the disease. mRNA expression analysis was performed for 73 patients, after quality control and removal of three outliers. miRNA expression analysis was performed for 72 patients, as four samples were removed due to insufficient RNA quality. Molecular classification could be assigned to 66/76 samples. TMA's included archival fresh-frozen paraffin-embedded (FFPE) tissue from 286 consecutive GBM patients. Details on this cohort are shown in Table S2.

2.1. Epileptogenic GBMs Show Downregulation of HIF1a/STAT5b Signaling

None of the individual mRNAs or miRNAs reached a significant association value after correction for multiple testing (Figure 1A,B). Exploratory gene set enrichment analyses (GSEA) were performed with use of the mRNA expression data and showed significant downregulation of 218 gene sets in the epilepsy subgroup compared to GBMs that did not present with epilepsy, with use of the Broad Institute MySig (MSigDB) libraries of curated gene sets C2 collection (curated gene sets, $p < 0.05$, false discovery rate (FDR) < 0.25). The top results are shown in Table 1, and full table is available in the supplementary material (Table S3). The majority of the associated gene sets are involved in hypoxia and HIF-1α signaling. Additionally, a gene set containing STAT5 targets ($p < 0.0001$, FDR = 0.06) and gene sets involved in CEBP-β, STAT3 signaling and epithelial to mesenchymal transition (EMT) signaling ($p < 0.05$, FDR < 0.25) were downregulated in the epilepsy subgroup compared to patients without epilepsy.

GSEA with the MSigDB C3 collection containing motif gene sets showed, among others, downregulation of a gene set involved in CEBP-β signaling ($p < 0.05$, FDR < 0.25, Table 1) and SRF signaling ($p < 0.05$, FDR < 0.25). Analysis with the MSigDB C7 gene set showed downregulation of multiple gene sets containing genes responsive to LPS stimulation in macrophages ($p < 0.05$, FDR < 0.25, Table 1).

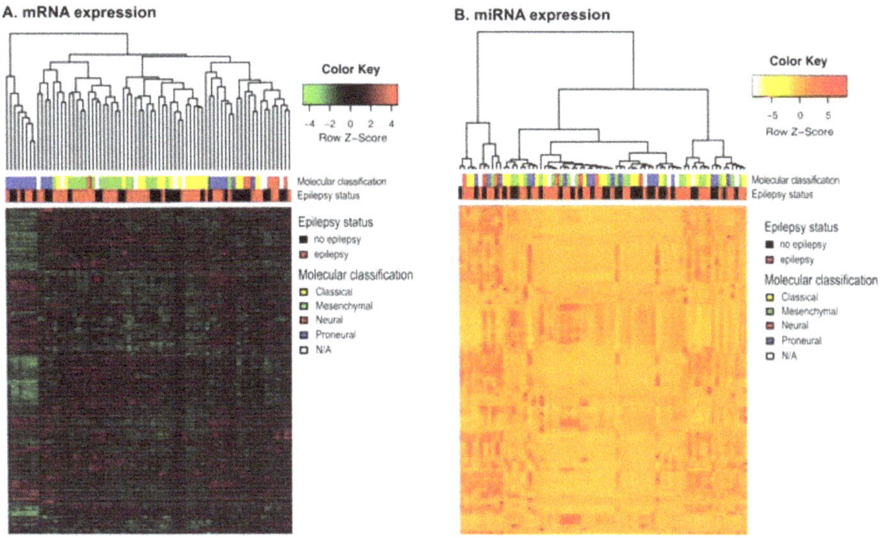

Figure 1. Lack of correlation between GBM-associated epilepsy and gene expression data. (**A**) Heatmap showing gene expression patterns of the 1000 RNA microarray probes with the highest standard deviation. Relative downregulation is shown as green, relative upregulation is shown as red, see color key. No differentially expressed genes were observed after correction for multiple testing (BH adjusted $p < 0.05$). (**B**) MiRNA expression patterns of the 100 probes with the highest standard deviation. 67 samples from our institute were included in this analysis. Relative miRNA downregulation is shown as white/yellow, and relative miRNA upregulation is shown as red, see color key. No differentially expressed miRNAs were observed after correction for multiple testing (Benjamini Hochberg adjusted $p < 0.05$).

Table 1. Gene set enrichment analysis with MSigDB collections. Gene sets are significantly downregulated in the epilepsy group compared to the patients without epilepsy. Cutoff values for significance were $p < 0.05$ and FDR < 0.25. Analyses were performed with MSigDB collections C1–C7. A subset of significant C2 results were displayed in this table. See supplementary Table S1 for full results with the C2 collection. NES: normalized enrichment score, FDR: false-discovery rate.

Gene Sets	NES	*p*-Value	FDR
C2 collection—curated gene sets			
ELVIDGE_HIF1A_TARGETS_DN	−2.355	<0.0001	<0.0001
ELVIDGE_HYPOXIA_BY_DMOG_UP	−2.283	<0.0001	0.0007
ELVIDGE_HIF1A_AND_HIF2A_TARGETS_DN	−2.301	<0.0001	0.0007
ELVIDGE_HYPOXIA_UP	−2.259	<0.0001	0.0009
LEONARD_HYPOXIA	−2.285	<0.0001	0.001
FARDIN_HYPOXIA_11	−2.198	<0.0001	0.007
PID_HIF1_TFPATHWAY	−2.149	<0.0001	0.02
GROSS_HIF1A_TARGETS_DN	−2.136	<0.0001	0.02
GROSS_HYPOXIA_VIA_ELK3_AND_HIF1A_UP	−2.116	<0.0001	0.03

Table 1. *Cont.*

Gene Sets	NES	*p*-Value	FDR
C3 collection—transcription factor targets			
V$ROAZ_01	−1.69	0.03	0.11
V$SRF_01	−1.71	0.02	0.13
CCAWWNAAGG_V$SRF_Q4	−1.71	0.01	0.16
TTGCWCAAY_V$CEBPB_02	−1.73	0.004	0.18
GGGNRMNNYCAT_UNKNOWN	−1.62	0.006	0.20
KRCTCNNNNMANAGC_UNKNOWN	−1.77	0.013	0.25
C7 collection—immunologic signatures			
GSE14769_UNSTIM_VS_40MIN_LPS_BMDM_DN	−1.90	0.006	0.13
GSE37416_CTRL_VS_12H_F_TULARENSIS_LVS_NEUTROPHIL_DN	−1.90	0.006	0.15
GSE14769_UNSTIM_VS_80MIN_LPS_BMDM_DN	−1.91	0.008	0.17
GSE14769_UNSTIM_VS_60MIN_LPS_BMDM_DN	−1.91	0.006	0.21
GSE37416_CTRL_VS_3H_F_TULARENSIS_LVS_NEUTROPHIL_DN	−1.93	0.004	0.24

qPCR experiments in a subset of fresh-frozen GBM samples (epilepsy $n = 5$, no epilepsy $n = 16$) did not show significant differential expression of *HIF1a*, *VEGF* or *SRF* (Figure 2A). However, on a protein level, we observed a significant decrease in protein expression of nuclear phosphorylated STAT5b in the epilepsy subgroup (Mann-Whitney *U* test, $p = 0.004$, Figure 2B). The JAK/STAT5b pathway has been shown to closely correlate with hypoxia and HIF-1α signaling in multiple cancers [20–23]. VEGFA and SRF protein expression did not differ between the two groups (Figure 2B and Figure S1).

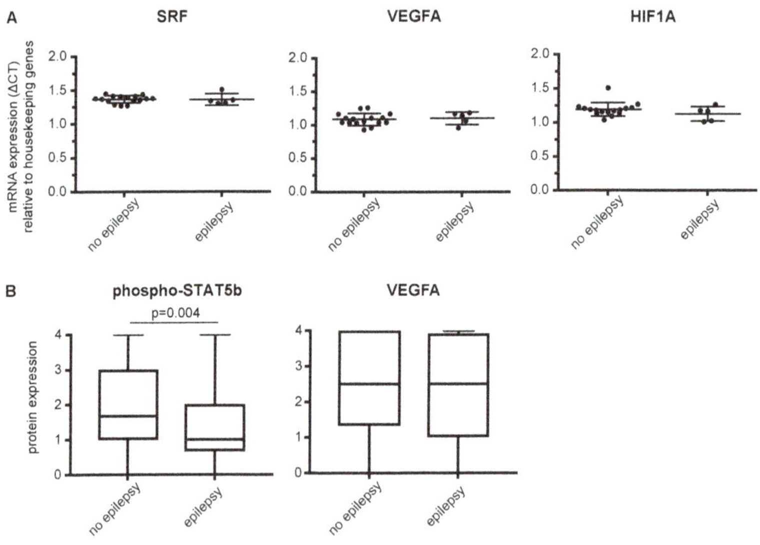

Figure 2. Differential activation of STAT5b between epileptogenic and non-epileptogenic GBMs. (**A**) qPCR experiments showing mRNA expression (ΔCT) of SRF, VEGFA and HIF1A in GBM patients with epilepsy ($n = 5$) and without epilepsy ($n = 16$). Graphs show mean ±SD and expression values per patient. Expression values were normalized to the average of 3 housekeeping genes (ACTB, GAPDH, GUSB). There was no significant difference in mRNA expression between epilepsy and non-epilepsy samples (independent samples *t*-test, SRF: $p = 0.9$, VEGFA: $p = 0.76$, HIF1A: $p = 0.23$). (**B**) Protein expression of phosphorylated STAT5b and VEGFA on GBM samples included on a tissue microarray. STAT5b expression was significantly lower in patients that presented with epilepsy compared to GBM patients with other presenting symptoms ($n = 265$, Mann Whitney *U* test, $p = 0.004$). VEGFA expression did not differ between the groups ($n = 275$, Mann Whitney *U* test, $p = 0.43$).

2.2. Epilepsy Does Not Correlate with a Mesenchymal Signature in GBMs

As described above, GSEA showed downregulation of, among others, gene sets involved in EMT, STAT3 and CEBP-β signaling in the epileptogenic GBMs. Based on these results, we further investigated the association of epileptogenic GBMs with mesenchymal transformation.

In the subgroup of patients without epilepsy a relatively larger percentage of tumors was classified as the mesenchymal subtype compared to the epilepsy group (40% vs. 25%), but this difference did not reach statistical significance (Chi-square test, $p = 0.48$, Figure 3A). In line with these results, no significant difference was observed in protein expression of phosphorylated NF-κB p65, phosphorylated STAT3 and CEBP-β (Figure 3B), which are master transcriptional regulators of the mesenchymal gene signature in GBM.

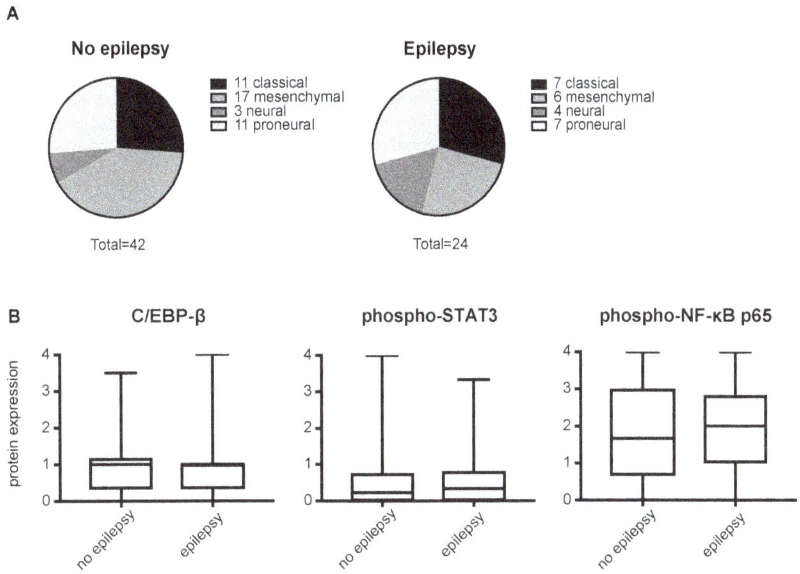

Figure 3. Molecular subclassification and mesenchymal activation in epileptogenic GBM. (**A**) Distribution of molecular GBM subtypes in subgroup of patients with epilepsy (n = 24) and without epilepsy (n = 42), based on mRNA expression profile. There was no significant difference in subtype distribution (Chi-square test, p = 0.48). (**B**) Protein expression of CEBP-β, phosphorylated STAT3 and phosphorylated NF-κB p65 on GBM tissues included on a tissue microarray. Mann Whitney U test did not show significant differences in protein expression between the epilepsy and non-epilepsy subgroups (CEBP-β: n = 266, p = 0.81, phosphorylated STAT3: n = 272, p = 0.68, phosphorylated NF-κB p65: n = 278, p = 0.56). Boxes represent median and quartiles, whiskers show data range.

3. Discussion

In our previous work, we observed that glioblastoma patients with epilepsy at the time of diagnosis have a significantly longer overall survival compared to patients that present with other symptoms. This prognostic effect was independent of anti-epileptic treatment and other important clinical characteristics [7].

Others have demonstrated that specific tumor biological effects are associated with epilepsy in glioma patients, such as neuronal excitation/inhibition disbalance and expression changes in the glutamate pathway [10–15]. Also, *IDH1* mutation was linked to the risk of tumor-associated epilepsy and seizure prognosis [16,17]. In our previous study, however, we found no correlation between epilepsy and IDH1 mutation in primary glioblastoma patients [7].

NF-κB, CEBP-β and STAT proteins, including STAT3 and −5, are transcription factors that mediate a wide range of cellular cytokine responses in physiological and disease processses [24–26]. Previous studies have identified STAT3 and CEBP-β as principal regulators of the mesenchymal gene expression signature in glioblastoma [4,27]. High expression of these mesenchymal markers has been associated with worse survival of GBM patients [27]. Additionally, NF-κB has been shown to drive mesenchymal transformation in glioma stem cells by induction of master transcription factors STAT3, CEBP-β and TAZ, which can subsequently contribute to resistance to radiation [5] and chemotherapy [28]. In our study, GSEA showed a downregulation of gene sets involved in epithelial to mesenchymal transformation (C2) and CEBP-β signaling (C3), suggesting a possible link between epileptogenic GBMs and molecular subclassification or mesenchymal transformation. Subsequent analyses could however not confirm this relationship. Although the epilepsy subgroup contained fewer tumors with the mesenchymal subtype, this was not a statistically significant difference. We also did not observe significant differences in protein expression of the key mesenchymal transcription factors NF-κB p65, STAT3 and CEBP-β.

Hypoxia-inducible factors, such as HIF-1α, contribute to the cell's response to hypoxia [29]. HIF-1α is a major player in the oncologic signaling in GBM. In hypoxic conditions, a mesenchymal shift mediated by HIF-1α is induced in glioblastoma cells [30,31]. Several of the key regulators of transcription involved in mesenchymal transformation (e.g., NF-κB, CEBP-β) upregulate under hypoxic conditions in cancer cells [27,32], and hypoxia activates the JAK2/STAT5b pathway in several types of cancer [20–23]. Also, crosstalk between HIF-1α and STAT3 was reported, in different tumor types, including glioblastoma [33,34]. Interest in STAT5 signaling in glioblastoma is growing, as multiple studies have shown that STAT5b drives proliferation and invasion in glioma [35–38]. This may be, in part mediated by the oncogenic EGFRvIII variant [39–41].

In our study, we observed downregulation of multiple gene sets involved in HIF-1α and hypoxia signaling in the epileptogenic GBMs. Additionally, STAT5 target genes seemed downregulated in this group. Interestingly, nuclear phosphorylated STAT5b protein expression was downregulated in the epileptogenic GBMs. Activated STAT5 proteins translocate to the nucleus [26], and serine 730 phosphorylation induces its intrinsic transcriptional activity [42]. We did not observe differential RNA or protein expression of HIF-1α and its target VEGFA between the epileptogenic GBMs and tumors that did not cause epilepsy. These results indicate that epilepsy in GBM patients more specifically correlates to decreased hypoxia/HIF-1α/STAT5b signaling. Interestingly, STAT5b protein expression associated with glioblastoma patient survival in our population of patients [43].

Besides its role in tumor growth and cancerogenesis, HIF-1α/STAT5b signaling could also be directly related to the epileptogenicity of the tumors, by altering the buffering and networking properties of the glial network. STAT5-null mice have indeed undetectable levels of connexin 32 [44], while connexin 43 (Cx43) mediates HIF-1α in astrocytes [45]. High expression of connexin 43 has in turn been linked to both glioma-related [46] and epilepsy due to mesiotemporal sclerosis [47], while this latter presents reduced levels of connexin 32. On the other hand, STAT5 is a negative regulator of xCT Expression and System Xc-Activity [48]. The exploration of these potential correlations is beyond the scope of the present work but represents an interesting venue for future research. With respect to tumor epileptogenicity as well, our GSEA also showed decreased expression of gene sets involved in serum response factor (SRF) signaling in the epileptogenic tumors. SRF depletion has been associated with increased seizure frequency in a mouse model [49,50]. Interestingly, SRF has also been linked to tumor invasion, proliferation, metastasis and resistance to therapy in different cancer types [51–53]. Its role in glioblastoma remains to be explored. In a subset of patients however, we did not observe differential SRF RNA expression between the epilepsy and non-epilepsy subgroups by qRT-PCR. Likewise, SRF protein expression was evaluated with two different antibodies on our tissue microarrays and did not associate with epileptogenicity. Hence, we could not confirm the association of SRF signaling with epileptogenic GBMs.

Our study presents some limitations. Not all results from the gene set enrichment analyses could be confirmed by other RNA or protein expression-based experiments. This discrepancy confirms the exploratory role of GSEA. The number of fresh-frozen tumor tissues available for confirmatory qRT-PCR experiments was somewhat limited. On the other hand, we included a large cohort of patients in the TMA analyses for our confirmatory protein expression analyses. Yet, we cannot rule out that intratumoral tissue heterogeneity could alter these analyses as well. Fresh-frozen tumor specimens were all collected by an experienced neuro-oncological neurosurgeon (P.R.) from the vital tumor tissue during surgery. Due to ethical reasons, we were not able to include peritumoral brain tissue from the patients. Archival FFPE tissue was used for protein expression analyses on tissue microarrays and multiple tumor tissue regions per patient were included. To minimize this bias however, a combined protein expression score was computed per patient.

4. Materials and Methods

4.1. Ethics Statement

This study was conducted following approval by the local ethical committee, and institutional review board (protocols 09-420, 16-229, 16-348).

4.2. Clinical data and Tumor Tissues

Following ethical approval and written patient consent, fresh frozen glioblastoma tissue was prospectively obtained from 76 patients at first surgery for their disease between 2010–2015. Hospital records were screened for the presence of seizures at diagnosis.

4.3. mRNA Expression Analysis

Seventy-six fresh-frozen surgical samples of de novo GBMs were prospectively collected between 2010 and 2015. RNA was extracted with the Nucleospin® TriPrep (Macherey-Nagel, Düren, Germany) and the QIASymphony RNA (Qiagen, Venlo, The Netherlands) kits according to the manufacturers' instructions. Affymetrix HG U133 plus 2.0 arrays were prepared and scanned according to the manufacturer's protocol and as reported previously [54]. Quality control and differential gene expression analyses were performed with R (v3.2.2). Based on principal component analysis (PCA) plots, 3 outliers were removed. Robust Multi-array Average (RMA) normalization was applied. Batch correction was performed with the 'sva' package. Differential expression was analyzed with the 'limma' package and heatmaps were created with the 'heatmap3' package (R).

Exploratory Gene Set Enrichment Analyses (GSEA) were performed after RMA-normalization [55] and batch correction, with the Partek® Genomics suite platform (v6.6) (Partek, St. Louis, MO, USA). Analyses were performed with the Broad Institute MySig libraries of curated gene sets C1—C7 version 5.0 [56], 1000 permutations and default additional parameters. An FDR threshold of 0.25 was applied as recommended [55].

4.4. miRNA Expression Analysis

RNA was isolated from the 76 fresh-frozen surgical samples of GBM patients with the MiRNeasy Micro Kit (Qiagen). Expression profiling of 800 miRNA probes was performed with the nCounter® Human v2 miRNA Expression Assay (NanoString, Seattle, WA, USA) at The Ohio State University Nucleic Acid Core Facility. 250 ng RNA was used per sample and conditions were set according to the manufacturer's instructions. RNA quality was insufficient for 4 samples. Data were processed with the Partek® Genomics suite platform (v6.6) by geometric mean normalization, average background subtraction and normalization to housekeeping genes. miRNA expression levels in GBMs in patients with and without seizures were analyzed with the 'limma' and 'heatmap3' package (R).

4.5. Class Prediction

Molecular subclassification (proneural, neural, classical, mesenchymal) was predicted by hierarchical clustering [3]. Microarray normalization, data filtering and analysis of inter-array homogeneity were performed as reported previously [3,57]. Affymetrix HG U133 plus 2.0 probesets were matched to 840 genes originally published for the classification of GBMs (http://tcga-data.nci.nih.gov/docs/publications/gbm_exp/). Relative gene expression values were calculated. Genes were then excluded for a median absolute deviation below 0.5 [3]. After filtering, 768 genes were used for the class prediction. The hierarchical clustering of samples was performed with cluster 3 software [58] with the agglomerative average linkage for the structure and 1 minus the Pearson's correlation for the distance metric [3].

4.6. qPCR Analyses

RNA for qPCR was available from the RNA extraction for miRNA expression analysis for 5 GBM patients with epilepsy and 16 patients without epilepsy. RNA quality was checked by measuring of A260/280 and A260/230 values with Nanodrop. qPCR expression analyses were performed for *HIF1A, SRF, VEGFA*. Expression values were normalized to the average of 3 housekeeping genes (*ACTB, GAPDH, GUSB*).

4.7. Tissue Microarrays and Immunohistochemistry

Archival FFPE GBM tumor tissues from a consecutive cohort of 286 patients treated in the UMCU between 2009 and 2013 were included on tissue microarrays (TMA's) as described previously [7]. Immunohistochemistry was performed, as described previously [7], with antibodies against STAT5b (phospho S731, Rabbit polyclonal, Abcam, Cambridge, UK), VEGF (Rabbit polyclonal, ThermoScientific, Waltham, MA USA), anti-NF-κB p65 (phospho S276, Rabbit polyclonal, Abcam); anti-STAT3 (phospho Y705) (Rabbit monoclonal, Cell Signaling, Leiden, The Netherlands); anti-CEBP-β (Mouse monoclonal, Abcam); anti-SRF (Rabbit polyclonal, Abcam, and rabbit polyclonal, Sigma-Aldrich, St. Louis, MO, USA).

Protein expression evaluation was performed with blinding for the clinical data, and was supervised by a senior neuropathologist. The percentage of nuclear and/or cytoplasmatic staining was scored as: 0, negative; 1, 1–25% positive cells; 2, 26–50% positive cells; 3, 51–75% positive cells and 4, 76–100% positive cells. A mean expression score was computed per patient. Due to insufficient tissue quality on TMA, a variable number of tissues per staining could not be evaluated (n = 8–21).

4.8. Statistical Analyses

Expression analyses of individual mRNAs and miRNAs were performed with the 'limma' package in R (v3.2.2). To control for inflation of type I error by multiple testing, p-values were adjusted by default Benjamini-Hochberg procedure. Adjusted p-values < 0.05 were considered significant.

GSEA was performed with the Partek® Genomics suite platform (v6.6). A recommended cutoff of $p < 0.05$ and false-discovery rate (FDR) < 0.25 was applied. All other analyses were performed with SPSS (v25, IBM, Armonk, NY, USA). A p-value < 0.05 was considered significant. Differential distributions of molecular subtypes across epilepsy and non-epilepsy patients were tested with a Chi-square test. Differences in RNA expression between patients with and without epilepsy were analyzed by independent samples t-test. Data distribution of protein expression was evaluated graphically and with a Kolmogorov-Smirnov test and differential expression was subsequently analyzed with a Mann Whitney U test.

5. Conclusions

A reduced activity of the hypoxia/HIF-1α/STAT5b signaling pathway is associated with epileptogenicity in glioblastomas. For the first time to our knowledge, these results provide biological insight in the favorable prognosis of tumor-associated epilepsy.

Supplementary Materials: The following are available online at http://www.mdpi.com/2072-6694/11/1/41/s1, Figure S1: SRF protein expression does not correlate to GBM-associated epilepsy, Table S1: Baseline table fresh-frozen tissues, Table S2: Baseline table TMA cohort, Table S3: GSEA results MSigDB C2 collection.

Author Contributions: Conceptualization, S.B., A.C. and P.A.R.; Data curation, S.B., M.G., E.H.B. and P.A.R.; Formal analysis, S.B., M.G., T.J.S., E.H.B. and P.A.R.; Funding acquisition, T.S., V.B., A.C. and P.A.R.; Investigation, S.B., M.G. and E.H.B.; Methodology, S.B., W.G.M.S., M.G., W.V.H., T.J.S., E.H.B. and P.A.R.; Project administration, A.C. and P.A.R.; Resources, W.G.M.S., W.V.H., T.S., T.J.S., V.B., A.C. and P.A.R.; Supervision, T.J.S., E.H.B., A.C. and P.A.R.; Visualization, S.B., T.J.S. and P.A.R.; Writing—original draft, S.B., T.J.S. and P.A.R.; Writing—review & editing, S.B., W.G.M.S., M.G., W.V.H., T.S., T.J.S., V.B., E.H.B., A.C. and P.A.R.

Funding: This work was supported by the following grants: PNC-029-006 of the Belgian Ministry of Health, grants FRSM 3.4.562.12, Televie 7.4.564.11.F and 1.5.162.10 of the FNRS of Belgium, FBC 2010.14 of the Belgian Foundation against Cancer, and an unrestricted grant from the T&P Bohnenn fund for Neuro-Oncological Research to T.S. and P.R. In addition, this work was supported by National Cancer Institute R01CA169368, R01CA108633, R01CA188228, R01CA188500, 1RC2CA148190 and The Ohio State University Comprehensive Cancer Center Award (all to A.C.). We also thank The Ohio State University (OSU) Comprehensive Cancer Center Genomics Shared Resource supported in part by grant P30 CA016058, National Cancer Institute, Bethesda, MD.

Acknowledgments: The authors would like to acknowledge C.H. Poulet for his help with the molecular classification of the tumors.

Conflicts of Interest: The authors declare no conflict of interest.

Abbreviations

ACTB	Actin β
CEBP-β	CCAAT/Enhancer Binding Protein β
D2HG	(D)-2-hydroxyglutarate
EGFRvIII	Epidermal growth factor receptor variant III
GAPDH	Glyceraldehyde 3-phosphate dehydrogenase
GUSB	β-glucuronidase
HIF-1α	Hypoxia-inducible factor-1α
IDH1	Isocitrate dehydrogenase 1
JAK	Janus kinase
LPS	Lipopolysaccharide
NF-kB	Nuclear Factor binding near the κ light chain gene in B cells
NMDA	N-methyl-D-aspartate
NR2B	N-methyl D-aspartate receptor subtype 2B
OLIG2	Oligodendrocyte transcription factor
SRF	Serum response factor
STAT3	Signal transducer of activation 3
STAT5	Signal transducer of activation 5
TAZ	Tafazzin
VEGF	Vascular endothelial growth factor

References

1. Stupp, R.; Taillibert, S.; Kanner, A.A.; Kesari, S.; Steinberg, D.M.; Toms, S.A. Maintenance Therapy With Tumor-Treating Fields Plus Temozolomide vs. Temozolomide Alone for Glioblastoma: A Randomized Clinical Trial. *JAMA* **2015**, *314*, 2535–2543. [CrossRef] [PubMed]
2. Brennan, C.W.; Verhaak, R.G.; McKenna, A.; Campos, B.; Noushmehr, H.; Salama, S.R. The somatic genomic landscape of glioblastoma. *Cell* **2013**, *155*, 462–477. [CrossRef] [PubMed]

3. Verhaak, R.G.; Hoadley, K.A.; Purdom, E.; Wang, V.; Qi, Y.; Wilkerson, M.D. Integrated genomic analysis identifies clinically relevant subtypes of glioblastoma characterized by abnormalities in PDGFRA, IDH1, EGFR, and NF1. *Cancer Cell* **2010**, *17*, 98–110. [CrossRef] [PubMed]
4. Carro, M.S.; Lim, W.K.; Alvarez, M.J.; Bollo, R.J.; Zhao, X.; Snyder, E.Y. The transcriptional network for mesenchymal transformation of brain tumours. *Nature* **2010**, *463*, 318–325. [CrossRef] [PubMed]
5. Bhat, K.P.; Balasubramaniyan, V.; Vaillant, B.; Ezhilarasan, R.; Hummelink, K.; Hollingsworth, F. Mesenchymal differentiation mediated by NF-kappaB promotes radiation resistance in glioblastoma. *Cancer Cell* **2013**, *24*, 331–346. [CrossRef] [PubMed]
6. Kerkhof, M.; Dielemans, J.C.; van Breemen, M.S.; Zwinkels, H.; Walchenbach, R.; Taphoorn, M.J.; Vecht, C.J. Effect of valproic acid on seizure control and on survival in patients with glioblastoma multiforme. *Neuro-Oncology* **2013**, *15*, 961–967. [CrossRef] [PubMed]
7. Berendsen, S.; Varkila, M.; Kroonen, J.; Seute, T.; Snijders, T.J.; Kauw, F. Prognostic relevance of epilepsy at presentation in glioblastoma patients. *Neuro-Oncology* **2016**, *18*, 700–706. [CrossRef] [PubMed]
8. Happold, C.; Gorlia, T.; Chinot, O.; Gilbert, M.R.; Nabors, L.B.; Wick, W. Does Valproic Acid or Levetiracetam Improve Survival in Glioblastoma? A Pooled Analysis of Prospective Clinical Trials in Newly Diagnosed Glioblastoma. *J. Clin. Oncol.* **2016**, *34*, 731–739. [CrossRef] [PubMed]
9. van Breemen, M.S.; Wilms, E.B.; Vecht, C.J. Epilepsy in patients with brain tumours: Epidemiology, mechanisms, and management. *Lancet Neurol.* **2007**, *6*, 421–430. [CrossRef]
10. Dang, L.; White, D.W.; Gross, S.; Bennett, B.D.; Bittinger, M.A.; Driggers, E.M. Cancer-associated IDH1 mutations produce 2-hydroxyglutarate. *Nature* **2009**, *462*, 739–744. [CrossRef] [PubMed]
11. Gao, X.; Wang, H.; Cai, S.; Saadatzadeh, M.R.; Hanenberg, H.; Pollok, K.E. Phosphorylation of NMDA 2B at S1303 in human glioma peritumoral tissue: Implications for glioma epileptogenesis. *Neurosurg. Focus* **2014**, *37*, E17. [CrossRef] [PubMed]
12. Gao, X.; Wang, H.; Pollok, K.E.; Chen, J.; Cohen-Gadol, A.A. Activation of death-associated protein kinase in human peritumoral tissue: A potential therapeutic target. *J. Clin. Neurosci.* **2015**, *22*, 1655–1660. [CrossRef] [PubMed]
13. Yuen, T.I.; Morokoff, A.P.; Bjorksten, A.; D'abaco, G.; Paradiso, L.; Finch, S. Glutamate is associated with a higher risk of seizures in patients with gliomas. *Neurology* **2012**, *79*, 883–889. [CrossRef] [PubMed]
14. Robert, S.M.; Buckingham, S.C.; Campbell, S.L.; Robel, S.; Holt, K.T.; Ogunrinu-Babarinde, T. SLC7A11 expression is associated with seizures and predicts poor survival in patients with malignant glioma. *Sci. Transl. Med.* **2015**, *7*, 289ra86. [CrossRef] [PubMed]
15. Rosati, A. Epilepsy in glioblastoma multiforme: Correlation with glutamine synthetase levels. *J. Neurooncol.* **2009**, *93*, 319–324. [CrossRef] [PubMed]
16. Neal, A. IDH1 and IDH2 mutations in postoperative diffuse glioma-associated epilepsy. *Epilepsy Behav.* **2018**, *78*, 30–36. [CrossRef]
17. Li, Y. IDH1 mutation is associated with a higher preoperative seizure incidence in low-grade glioma: A systematic review and meta-analysis. *Seizure* **2018**, *55*, 76–82. [CrossRef]
18. Chen, H. Mutant IDH1 and seizures in patients with glioma. *Neurology* **2017**, *88*, 1805–1813. [CrossRef]
19. Lee, J.W.; Norden, A.D.; Ligon, K.L.; Golby, A.J.; Beroukhim, R.; Quackenbush, J. Tumor associated seizures in glioblastomas are influenced by survival gene expression in a region-specific manner: A gene expression imaging study. *Epilepsy Res.* **2014**, *108*, 843–852. [CrossRef]
20. Pak, S.H. Hypoxia upregulates Hsp90alpha expression via STAT5b in cancer cells. *Int. J. Oncol.* **2012**, *41*, 161–168.
21. Joung, Y.H. Hypoxia activates the IGF-1 expression through STAT5b in human HepG2 cells. *Biochem. Biophys. Res. Commun.* **2007**, *358*, 733–738. [CrossRef] [PubMed]
22. Joung, Y.H.; Lim, E.J.; Lee, M.Y.; Park, J.H.; Ye, S.K.; Park, E.U. Hypoxia activates the cyclin D1 promoter via the Jak2/STAT5b pathway in breast cancer cells. *Exp. Mol. Med.* **2005**, *37*, 353–364. [CrossRef] [PubMed]
23. Lee, M.Y. Phosphorylation and activation of STAT proteins by hypoxia in breast cancer cells. *Breast* **2006**, *15*, 187–195. [CrossRef] [PubMed]
24. Zhang, Q.; Lenardo, M.J.; Baltimore, D. 30 Years of NF-kappaB: A Blossoming of Relevance to Human Pathobiology. *Cell* **2017**, *168*, 37–57. [CrossRef]
25. van der Krieken, S.E. CCAAT/enhancer binding protein beta in relation to ER stress, inflammation, and metabolic disturbances. *Biomed. Res. Int.* **2015**, *2015*, 324815. [CrossRef]

26. Villarino, A.V. Mechanisms of Jak/STAT signaling in immunity and disease. *J. Immunol.* **2015**, *194*, 21–27. [CrossRef] [PubMed]
27. Cooper, L.A.; Gutman, D.A.; Chisolm, C.; Appin, C.; Kong, J.; Rong, Y. The tumor microenvironment strongly impacts master transcriptional regulators and gene expression class of glioblastoma. *Am. J. Pathol.* **2012**, *180*, 2108–2119. [CrossRef] [PubMed]
28. Bredel, M. Tumor necrosis factor-alpha-induced protein 3 as a putative regulator of nuclear factor-kappaB-mediated resistance to O6-alkylating agents in human glioblastomas. *J. Clin. Oncol.* **2006**, *24*, 274–287. [CrossRef]
29. Carmeliet, P.; Dor, Y.; Herbert, J.M.; Fukumura, D.; Brusselmans, K.; Dewerchin, M. Role of HIF-1alpha in hypoxia-mediated apoptosis, cell proliferation and tumour angiogenesis. *Nature* **1998**, *394*, 485–490. [CrossRef]
30. Joseph, J.V.; Conroy, S.; Pavlov, K.; Sontakke, P.; Tomar, T.; Eggens-Meijer, E. Hypoxia enhances migration and invasion in glioblastoma by promoting a mesenchymal shift mediated by the HIF1alpha-ZEB1 axis. *Cancer Lett.* **2015**, *359*, 107–116. [CrossRef]
31. Talasila, K.M. The angiogenic switch leads to a metabolic shift in human glioblastoma. *Neuro-Oncology* **2017**, *19*, 383–393. [CrossRef]
32. Murat, A. Modulation of angiogenic and inflammatory response in glioblastoma by hypoxia. *PLoS ONE* **2009**, *4*, e5947. [CrossRef]
33. Jung, J.E.; Lee, H.G.; Cho, I.H.; Chung, D.H.; Yoon, S.H.; Yang, Y.M. STAT3 is a potential modulator of HIF-1-mediated VEGF expression in human renal carcinoma cells. *FASEB J.* **2005**, *19*, 1296–1298. [CrossRef] [PubMed]
34. Ganguly, D. The critical role that STAT3 plays in glioma-initiating cells: STAT3 addiction in glioma. *Oncotarget* **2018**, *9*, 22095–22112. [CrossRef]
35. Alkharusi, A. Stimulation of prolactin receptor induces STAT-5 phosphorylation and cellular invasion in glioblastoma multiforme. *Oncotarget* **2016**, *7*, 79572–79583. [CrossRef] [PubMed]
36. Gressot, L.V.; Doucette, T.A.; Yang, Y.; Fuller, G.N.; Heimberger, A.B.; Bögler, O. Signal transducer and activator of transcription 5b drives malignant progression in a PDGFB-dependent proneural glioma model by suppressing apoptosis. *Int. J. Cancer* **2015**, *136*, 2047–2054. [CrossRef] [PubMed]
37. Liang, Q.C.; Xiong, H.; Zhao, Z.W.; Jia, D.; Li, W.X.; Qin, H.Z. Inhibition of transcription factor STAT5b suppresses proliferation, induces G1 cell cycle arrest and reduces tumor cell invasion in human glioblastoma multiforme cells. *Cancer Lett.* **2009**, *273*, 164–171. [CrossRef] [PubMed]
38. Cao, S.; Wang, C.; Zheng, Q.; Qiao, Y.; Xu, K.; Jiang, T.; Wu, A. STAT5 regulates glioma cell invasion by pathways dependent and independent of STAT5 DNA binding. *Neurosci. Lett.* **2011**, *487*, 228–233. [CrossRef] [PubMed]
39. Fan, Q.W. EGFR phosphorylates tumor-derived EGFRvIII driving STAT3/5 and progression in glioblastoma. *Cancer Cell* **2013**, *24*, 438–449. [CrossRef] [PubMed]
40. Roos, A. EGFRvIII-Stat5 Signaling Enhances Glioblastoma Cell Migration and Survival. *Mol. Cancer Res.* **2018**, *16*, 1185–1195. [CrossRef]
41. Latha, K.; Li, M.; Chumbalkar, V.; Gururaj, A.; Hwang, Y.; Dakeng, S. Nuclear EGFRvIII-STAT5b complex contributes to glioblastoma cell survival by direct activation of the Bcl-XL promoter. *Int. J. Cancer* **2013**, *132*, 509–520. [CrossRef] [PubMed]
42. Park, S.H.; Yamashita, H.; Rui, H.; Waxman, D.J. Serine phosphorylation of GH-activated signal transducer and activator of transcription 5a (STAT5a) and STAT5b: Impact on STAT5 transcriptional activity. *Mol. Endocrinol.* **2001**, *15*, 2157–2171. [CrossRef] [PubMed]
43. Robe, P.A. (University Medical Center Utrecht, Utrecht, The Netherlands). Personal communication, 2018.
44. Miyoshi, K.; Shillingford, J.M.; Smith, G.H.; Grimm, S.L.; Wagner, K.U.; Oka, T. Signal transducer and activator of transcription (Stat) 5 controls the proliferation and differentiation of mammary alveolar epithelium. *J. Cell Biol.* **2001**, *155*, 531–542. [CrossRef] [PubMed]
45. Valle-Casuso, J.C. HIF-1 and c-Src mediate increased glucose uptake induced by endothelin-1 and connexin43 in astrocytes. *PLoS ONE* **2012**, *7*, e32448. [CrossRef]
46. Dong, H. Complex role of connexin 43 in astrocytic tumors and possible promotion of gliomaassociated epileptic discharge (Review). *Mol. Med. Rep.* **2017**, *16*, 7890–7900. [CrossRef]

47. Collignon, F. Altered expression of connexin subtypes in mesial temporal lobe epilepsy in humans. *J. Neurosurg.* **2006**, *105*, 77–87. [CrossRef]
48. Linher-Melville, K. Chronic Inhibition of STAT3/STAT5 in Treatment-Resistant Human Breast Cancer Cell Subtypes: Convergence on the ROS/SUMO Pathway and Its Effects on xCT Expression and System xc- Activity. *PLoS ONE* **2016**, *11*, e0161202. [CrossRef] [PubMed]
49. Lösing, P.; Niturad, C.E.; Harrer, M.; zu Reckendorf, C.M.; Schatz, T.; Sinske, D. SRF modulates seizure occurrence, activity induced gene transcription and hippocampal circuit reorganization in the mouse pilocarpine epilepsy model. *Mol. Brain* **2017**, *10*, 30. [CrossRef]
50. Kuzniewska, B. Adult Deletion of SRF Increases Epileptogenesis and Decreases Activity-Induced Gene Expression. *Mol. Neurobiol.* **2016**, *53*, 1478–1493. [CrossRef]
51. Ohrnberger, S. Dysregulated serum response factor triggers formation of hepatocellular carcinoma. *Hepatology* **2015**, *61*, 979–989. [CrossRef]
52. Qiao, J. SRF promotes gastric cancer metastasis through stromal fibroblasts in an SDF1-CXCR4-dependent manner. *Oncotarget* **2016**, *7*, 46088–46099. [CrossRef] [PubMed]
53. Lundon, D.J.; Boland, A.; Prencipe, M.; Hurley, G.; O'Neill, A.; Kay, E. The prognostic utility of the transcription factor SRF in docetaxel-resistant prostate cancer: In-vitro discovery and in-vivo validation. *BMC Cancer* **2017**, *17*, 163. [CrossRef] [PubMed]
54. Turkheimer, F.E.; Roncaroli, F.; Hennuy, B.; Herens, C.; Nguyen, M.; Martin, D. Chromosomal patterns of gene expression from microarray data: Methodology, validation and clinical relevance in gliomas. *BMC Bioinform.* **2006**, *7*, 526. [CrossRef] [PubMed]
55. Subramanian, A.; Tamayo, P.; Mootha, V.K.; Mukherjee, S.; Ebert, B.L.; Gillette, M.A. Gene set enrichment analysis: A knowledge-based approach for interpreting genome-wide expression profiles. *Proc. Natl. Acad. Sci. USA* **2005**, *102*, 15545–15550. [CrossRef] [PubMed]
56. Reich, M. GenePattern 2.0. *Nat. Genet.* **2006**, *38*, 500–501. [CrossRef] [PubMed]
57. Wislet-Gendebien, S.; Poulet, C.; Neirinckx, V.; Hennuy, B.; Swingland, J.T.; Laudet, E. In vivo tumorigenesis was observed after injection of in vitro expanded neural crest stem cells isolated from adult bone marrow. *PLoS ONE* **2012**, *7*, e46425. [CrossRef] [PubMed]
58. Eisen, M.B.; Spellman, P.T.; Brown, P.O.; Botstein, D. Cluster analysis and display of genome-wide expression patterns. *Proc. Natl. Acad. Sci. USA* **1998**, *95*, 14863–14868. [CrossRef] [PubMed]

cancers

MDPI

Article

Macrovascular Networks on Contrast-Enhanced Magnetic Resonance Imaging Improves Survival Prediction in Newly Diagnosed Glioblastoma

Josep Puig [1,2,*], Carles Biarnés [2], Pepus Daunis-i-Estadella [3], Gerard Blasco [2], Alfredo Gimeno [2], Marco Essig [1], Carme Balaña [4], Angel Alberich-Bayarri [5], Ana Jimenez-Pastor [5], Eduardo Camacho [5], Santiago Thio-Henestrosa [3], Jaume Capellades [6], Javier Sanchez-Gonzalez [7], Marian Navas-Martí [2], Blanca Domenech-Ximenos [2], Sonia Del Barco [8], Montserrat Puigdemont [9], Carlos Leiva-Salinas [10], Max Wintermark [11], Kambiz Nael [12], Rajan Jain [13] and Salvador Pedraza [2]

1 Department of Radiology, University of Manitoba, Winnipeg, MB R3T 2N2, Canada; messig@exchange.hsc.mb.ca
2 Research Unit of Diagnostic Imaging Institute (IDI), Department of Radiology (Girona Biomedical Research Institute) IDIBGI, Hospital Universitari Dr Josep Trueta, 17007 Girona, Spain; carlesbiarnes90@gmail.com (C.B.); gbs.blasco@gmail.com (G.B.); fredigimeno1989@gmail.com (A.G.); marianmarti33@gmail.com (M.N.-M.); bl.domenech@gmail.com (B.D.-X.); sapedraza@gmail.com (S.P.)
3 Department of Computer Science, Applied Mathematics and Statistics, University of Girona, 17003 Girona, Spain; pepus@imae.udg.edu (P.D.-i.-E.); santiago.thio@udg.edu (S.-T.H.)
4 Medical Oncology, Institut Catala Oncologia (ICO), Applied Research Group in Oncology (B-ARGO), IGTP, Badalona, 08916 Barcelona, Spain; cbalana@iconcologia.net
5 QUIBIM SL, Quantitative Imaging Biomarkers in Medicine, 46026 Valencia, Spain; angel@quibim.com (A.A.-B.); anajimenez@quibim.com (A.J.-P.); educamacho@quibim.com (E.C.)
6 Department of Radiology, Hospital del Mar, 08003 Barcelona, Spain; jaumecapellades@gmail.com
7 Philips Healthcare Ibérica, 28050 Madrid, Spain; javier.sanchez.gonzalez@philips.com
8 Medical Oncology, Institut Catala Oncologia (ICO), 17007 Girona, Spain; Sdelbarco@iconcologia.net
9 Hospital Cancer Registry, ICO, Hospital Universitari Dr Josep Trueta, 17007 Girona, Spain; mpuigdemont@iconcologia.net
10 Department of Radiology, University of Missouri, Columbia, MO 65212, USA; carlosleivasalinas@gmail.com
11 Department of Radiology, Neuroradiology Division, Stanford University, Stanford, CA 94304, USA; mwinterm@stanford.edu
12 Department of Radiology, Icahn School of Medicine at Mount Sinai, New York, NY 10029, USA; kambiznael@gmail.com
13 Departments of Radiology and Neurosurgery, New York University School of Medicine, New York, NY 10016, USA; Rajan.Jain@nyumc.org
* Correspondence: jpuig@hsc.mb.ca; Tel.: +1-204-510-3126

Received: 27 October 2018; Accepted: 20 December 2018; Published: 14 January 2019

Abstract: A higher degree of angiogenesis is associated with shortened survival in glioblastoma. Feasible morphometric parameters for analyzing vascular networks in brain tumors in clinical practice are lacking. We investigated whether the macrovascular network classified by the number of vessel-like structures (nVS) visible on three-dimensional T1-weighted contrast–enhanced (3D-T1CE) magnetic resonance imaging (MRI) could improve survival prediction models for newly diagnosed glioblastoma based on clinical and other imaging features. Ninety-seven consecutive patients (62 men; mean age, 58 ± 15 years) with histologically proven glioblastoma underwent 1.5T-MRI, including anatomical, diffusion-weighted, dynamic susceptibility contrast perfusion, and 3D-T1CE sequences after 0.1 mmol/kg gadobutrol. We assessed nVS related to the tumor on 1-mm isovoxel 3D-T1CE images, and relative cerebral blood volume, relative cerebral flow volume (rCBF), delay mean time, and apparent diffusion coefficient in volumes of interest for contrast-enhancing lesion (CEL), non-CEL, and contralateral normal-appearing white matter. We also assessed Visually Accessible Rembrandt

Images scoring system features. We used ROC curves to determine the cutoff for nVS and univariate and multivariate cox proportional hazards regression for overall survival. Prognostic factors were evaluated by Kaplan-Meier survival and ROC analyses. Lesions with nVS > 5 were classified as having highly developed macrovascular network; 58 (60.4%) tumors had highly developed macrovascular network. Patients with highly developed macrovascular network were older, had higher $\text{volume}_{\text{CEL}}$, increased rCBF_{CEL}, and poor survival; nVS correlated negatively with survival ($r = -0.286$; $p = 0.008$). On multivariate analysis, standard treatment, age at diagnosis, and macrovascular network best predicted survival at 1 year (AUC 0.901, 83.3% sensitivity, 93.3% specificity, 96.2% PPV, 73.7% NPV). Contrast-enhanced MRI macrovascular network improves survival prediction in newly diagnosed glioblastoma.

Keywords: glioblastoma; magnetic resonance imaging; angiogenesis; biomarker; survival

---

1. Introduction

Glioblastoma is the most common angiogenic malignant astrocytic tumor. Despite therapeutic advances, the prognosis of glioblastoma is dismal, with median overall survival of 16 months [1]. Angiogenesis, a key step in tumor progression, is among the most important prognostic factors in glioblastoma and correlates with worse survival [2–8]. However, morphological vascular parameters cannot easily differentiate between pre-existing brain vessels incorporated into tumors and neoangiogenesis [8], though tumor vascularity and leakiness measured with perfusion imaging have been shown to indirectly correlate with different stages of angiogenesis with increasing glioma grade [9]. Thus, it is more useful to assess vascularity by considering the tumor's "vascular network", which comprises of both pre-existing vessels incorporated into the tumor and microvessels arising from neoangiogenesis [10,11]. However, no morphometric parameter has been validated as a biomarker of vascularity. A biomarker that enabled tumor grading based on angiogenic sub-patterns could help improve diagnosis and prognosis, and would facilitate the translation of antiangiogenic therapy from the experimental stage into clinical practice. In addition to classic angiogenesis seen at histology as evenly distributed capillary-like microvascular sprouting, immunohistochemistry for CD34 reveals unevenly distributed bizarre vascular formations (glomeruloid vascular formations, vascular garlands, and vascular clusters), which are considered a histological hallmark of glioblastoma [3,5]. These formations are considered a late, secondary development that is insufficient to save tumoral tissue from hypoxia-mediated death. Birner et al. [3] found that a predominantly bizarre vascular pattern was associated with worse survival than a predominantly classic pattern. Moreover, histological microvessel density is higher in the classic vascular pattern than in the predominantly bizarre pattern.

However, various factors impede the use of histopathological or molecular biomarkers to assess glioblastoma angiogenesis in clinical practice. Continuous monitoring is difficult because sampling is invasive, does not represent the heterogeneity within tumors, requires long processing times as well as complex storage and testing techniques, and yields poor interobserver agreement [12]. By contrast, imaging biomarkers for patient selection and therapeutic monitoring are not limited by invasive sampling and can assess the heterogeneity of the whole tumor. Novel magnetic resonance imaging (MRI) biomarkers of angiogenesis of newly diagnosed glioblastomas have aroused great interest. The degree of vascularity of glioblastomas has been evaluated directly by digital subtraction angiography [13–16] and magnetic resonance angiography (MRA) [17], as well as indirectly by dynamic susceptibility contrast perfusion MRI (DSC-MRI) [9,15,16,18–25] and CT perfusion [26]. DSC-MRI shows that glioblastomas have regions of significantly elevated relative cerebral blood volume (rCBV) or relative cerebral blood flow (rCBF) consistent with their increased vascularity [18–22]. However, some newly diagnosed glioblastomas show low vascularity [14–17]. Wetzel et al. [16] documented the absence of vascularity in 30% of 231 gliomas, and a recent pilot study using gadofosveset blood-pool contrast agent found that half of glioblastomas had decreased vascularity on high-resolution MRA [17]; both these studies found that increased vascularity predicted worse outcome [16,17].

Numerous studies suggest that pretreatment MRI features may be prognostic indicators of survival in patients with glioblastoma [16,17,27–33]. Features included in survival models include the degree of necrosis, the degree of enhancement, multifocality, satellite lesions, volume of contrast-enhancing lesion (CEL), volume of non-CEL, and extent of edema. Additionally, a few studies have suggested that including imaging features based on Visually AcceSAble Rembrandt Images (VASARI), a controlled vocabulary system developed to standardize the grading of visual MRI features in gliomas [34], could enhance the predictive power of survival models based on clinical features [27,31]. Other studies have suggested that the extent of resection, age at diagnosis, and Karnofsky Performance Scale (KPS) score can determine survival in glioblastoma [32,35,36].

Using MRI biomarkers to assess baseline risk at diagnosis can help researchers stratify patients into risk groups and help clinicians make treatment decisions. This study aimed to assess whether including measures of the macrovascular network of newly diagnosed glioblastoma in routine clinical contrast-enhanced MRI protocols could improve the predictive power of survival models based on clinical and other imaging features. We counted vessel-like structures related to the CEL or non-CEL components on three-dimensional contrast-enhanced spin-echo T1-weighted imaging (3D-T1CE) to assess the macrovascular network. We found that the macrovascular network on 3D-T1CE, together with age at diagnosis and standard treatment, best predicted survival with AUC = 0.901 ($p < 0.001$), yielding 83.3% sensitivity, 93.3% specificity, 96.2% positive predictive value, and 73.7% negative predictive value.

2. Results

2.1. Determination of the Cutoff for Number of Vessel-Like Structures in Glioblastoma

The best cutoff for nVS related to the CEL or non-CEL components to discriminate between highly developed macrovascular network and less developed macrovascular network in newly diagnosed glioblastomas on 3D-T1CE (Figure 1), was 5, yielding 100% sensitivity, 97.7% specificity, 98.1% positive predictive value, and 100% negative predictive values. Considering glioblastomas with nVS > 5 to have highly developed macrovascular network, 53 (54.6%) tumors had highly developed macrovascular network.

Interobserver agreement for macrovascular network was excellent ($\kappa = 0.85$).

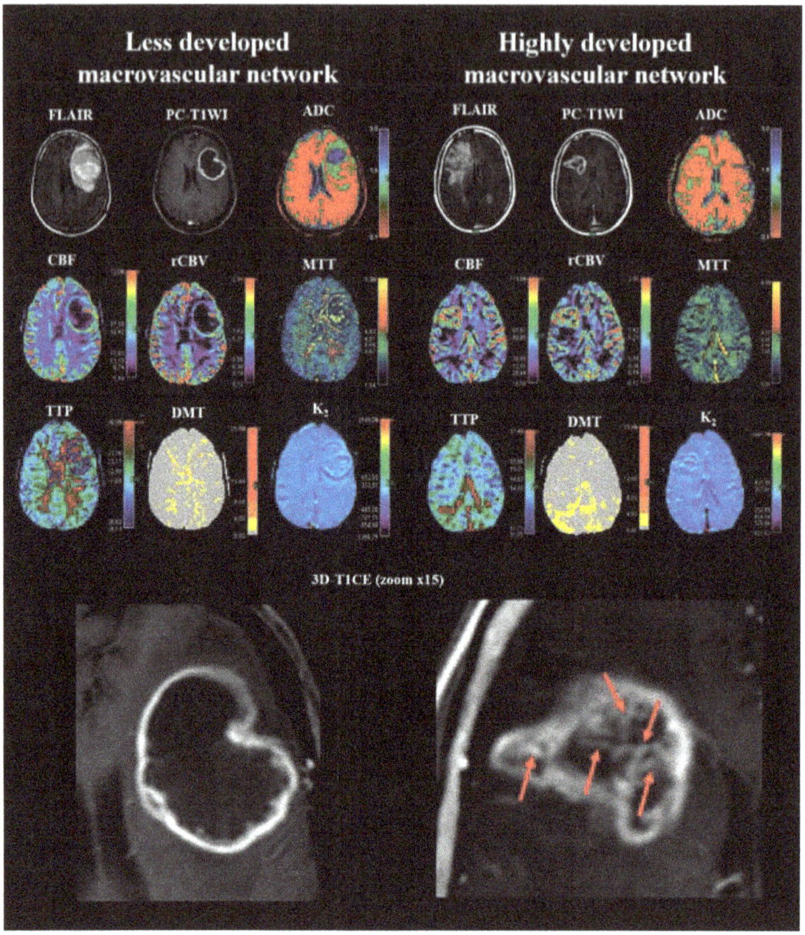

Figure 1. Macrovascular network on 3D-T1 contrast enhancement MRI. Panel of 2 cases illustrating MRI characteristics for less developed and highly developed macrovascular networks. The glioblastoma with a highly developed macrovascular network (red arrows) shows an unevenly distributed bizarre large-vessel pattern. ADC indicates apparent diffusion coefficient; DMT, delay mean time; FLAIR, fluid-attenuated inversion recovery; K_2, microvascular permeability; MTT, mean transit time; rCBF, relative cerebral blood flow; rCBV, relative cerebral blood volume; PC-T1WI, postcontrast T1 weighted image; TTP, time to peak.

2.2. Patient Characteristics

Table 1 summarizes the patients' data. All 97 patients (66 male, mean age 58 ± 15 years) died during the observation period. The median KPS score at diagnosis was 80.0 (IQR, 70.0–80.0).

Patients with glioblastomas with highly developed macrovascular network were older, had higher mean volume$_{CEL}$ and higher rCBF$_{CEL}$, and tended toward lower KPS scores (median, 85.76% vs. 90.20% in patients with glioblastomas with less developed macrovascular network, $p = 0.063$) (Table 1).

Standard treatment was administered in 64 (65.98%) patients, 53 with highly developed macrovascular network and 44 with less developed macrovascular network. None of the patients underwent antiangiogenic therapy.

Table 1. Patient characteristics.

Characteristic	Overall (n = 97)	Less Developed Macrovascular Network (n = 44)	Highly Developed Macrovascular Network (n = 53)	p-Value
Male (%)	62 (63.9%)	66 (68.2%)	58 (60.4%)	0.426
Age at diagnosis (years)	58 (15)	54 (15)	61 (12)	0.026
Karnofsky score	87.45 (18.23)	90.20 (7.65)	85.76 (14.02)	0.063
$Volume_{CEL}$ (cm^3)	20.4 (19.12)	13.55 (10.52)	26.28 (22.71)	0.001
$Volume_{NCEL}$ (cm^3)	50.67 (32.64)	44.91 (27.6)	55.62 (36.02)	0.260
Volume of necrosis (cm^3)	20.37 (21.84)	21.52 (28.2)	19.41 (14.86)	0.310
$rCBF_{CEL}$	16.08 (4.92)	14.31 (4.06)	18.52 (5.04)	0.001
$rCBF_{NCEL}$	16.45 (4.34)	15.5 (4.05)	17.75 (4.49)	0.118
$rCBV_{CEL}$	1.76 (0.93)	1.66 (1.09)	1.83 (0.81)	0.181
$rCBV_{NCEL}$	2.08 (1.18)	1.88 (0.86)	2.22 (1.37)	0.361
MTT_{CEL} (s)	5.74 (1.93)	5.71 (1.68)	5.77 (2.12)	0.938
MTT_{NCEL} (s)	6.06 (2.23)	6.14 (2.57)	6 (2)	0.898
TTP_{CEL} (s)	25.99 (8.67)	25.86 (9.26)	26.09 (8.38)	0.667
TTP_{NCEL} (s)	25.85 (7.12)	24.67 (5.67)	26.64 (7.96)	0.296
DMT_{CEL} (s)	−0.26 (1.21)	−0.38 (1.35)	−0.18 (1.11)	0.573
DMT_{NCEL} (s)	−0.12 (0.57)	−0.16 (0.51)	−0.1 (0.62)	0.725
Microvascular permeability, K_{2CEL}	−56.48 (57.21)	−51.75 (67.45)	−59.9 (49.48)	0.274
Microvascular permeability, K_{2NCEL}	−53.82 (61.23)	−46.25 (83.83)	−59.31 (38.29)	0.099
ADC_{CEL} ($mm^2\ s^{-1} \times 10^{-3}$)	0.30 (0.07)	0.30 (0.07)	0.31 (0.08)	0.952
ADC_{NCEL} ($mm^2\ s^{-1} \times 10^{-3}$)	0.44 (0.03)	0.44 (0.03)	0.44 (0.03)	0.944
Vessel-like structures (n)	9.56 (8.11)	2.75 (1.94)	15.21 (6.83)	<0.001
Treatment				0.280
Surgery + RT + TMZ (%)	64 (65.98%)	33 (34.01%)	31 (31.96%)	
Surgery + RT (%)	11 (11.34%)	5 (5.14%)	6 (6.19%)	
RT + TMZ (%)	14 (14.42%)	3 (3.08%)	11 (11.34%)	
TMZ (%)	7 (7.22%)	3 (3.08%)	4 (4.11%)	
Palliative (%)	1 (1.03%)	0 (0%)	1 (1.03%)	

Note—Unless otherwise specified, data are means with standard deviations in parentheses. Significant differences between less developed macrovascular network and highly developed macrovascular network cohorts are highlighted in bold. ADC indicates apparent diffusion coefficient; CEL, contrast-enhancing lesion; DMT, delay mean time; MTT, mean transit time; NCEL, non-CEL; rCBF, relative cerebral blood flow; rCBV, relative cerebral blood volume; RT, radiotherapy; TMZ, temozolamide; TTP, time to peak.2.1. Sub-types of macrovascular networks on contrast-enhanced MRI.

Ependymal invasion was more common in lesions with highly developed macrovascular network (55% vs. 38%, $p = 0.013$). Interobserver agreement for macrovascular network was excellent ($\kappa = 0.85$). Interobserver agreements for the 13 features from the VASARI features were good to excellent; the highest agreement was for midline cross ($\kappa = 0.957$) and the lowest for eloquent area involvement ($\kappa = 0.778$).

2.3. Survival Analysis According the Treatment Received and Macrovascular Network

Figure 2 shows the Kaplan-Meier survival curves according to the macrovascular network and treatment received. In patients who received standard treatment, the survival was significantly longer (Figure 2A). For the subgroup of patients who received standard treatment, nVS was also negatively correlated with survival ($r = -0.347$; $p = 0.016$). Median survival rates for patients with less developed macrovascular network and patients with highly developed macrovascular network were 11.67 months (95% CI, 4.51–18.05) and 7.80 months (95% CI, 3.48–13.21), respectively (Figure 2B). Median survival rates for patients with less developed macrovascular network increased and patients with highly developed macrovascular network decreased receiving standard treatment were 15.9 months (95% CI, 11.24–20.70) and 10.26 months (95% CI, 8.03–18.26), respectively. When treatment was other, median survival for patients with less developed macrovascular network and patients with highly developed macrovascular network was 3.80 months (95%CI, 2.59–8.03) and 3.30 months (95% CI, 1.75–5.69), respectively (Figure 2C). For overall tumors, nVS was negatively correlated with survival ($r = -0.286$; $p = 0.008$) (Figure 2D).

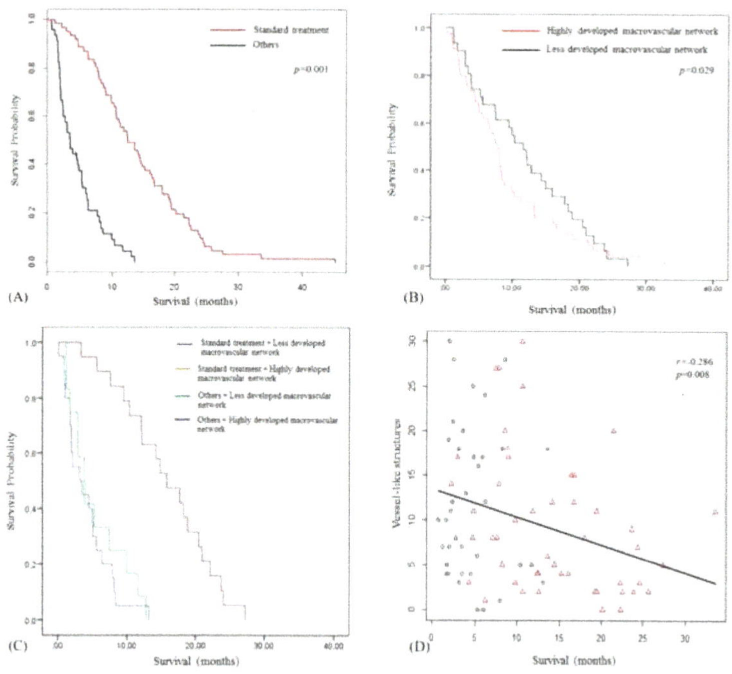

Figure 2. Survival analysis. Kaplan-Meier survival curves comparing survival rates according to treatment received (**A**) between macrovascular network subtypes on contrast-enhanced MRI (**B**), and 4 branches of groups combining treatment received and macrovascular network on contrast-enhanced MRI (**C**). Patients with glioblastomas with less developed macrovascular networks that received standard treatment had better survival. The number of visible vessels significantly correlates with survival (**D**).

Univariate Cox proportional hazards regression analysis showed that age at diagnosis, macrovascular network, delay mean time at CEL (DMT_{CEL}), DMT_{NCEL}, apparent diffusion coefficient at NCEL (ADC_{NCEL}), ependymal invasion, thickness of CEL margin, and the treatment received are the variables that significantly discriminate the survival time (Table 2).

Table 2. Associations between variables and survival time.

Variable	Overall (n = 97)	Less than 1 Year (n = 56)	More than 1 Year (n = 29)	p-Value
Male (%)	62 (63.9%)	23 (79.3%)	33 (58.9%)	0.060
Age at diagnosis (years)	57.75 (14.43)	62.07 (12.66)	50 (15.11)	<0.001
Karnofsky score	87.45 (18.23)	82.52 (13.15)	91.03 (12.66)	0.084
Highly/less developed macrovascular network	44/53	16/40	18/11	0.002
CEL (cm^3)	20.4 (19.12)	22.41 (21.16)	19.36 (17.04)	0.421
Non-CEL (cm^3)	50.67 (32.64)	48.33 (27.55)	50.73 (40.82)	1.000
Necrosis (cm^3)	20.37 (21.84)	17.99 (15.45)	23.1 (22.48)	0.647
$rCBF_{CEL}$	16.08 (4.92)	15.95 (4.74)	15.59 (5.39)	0.512
$rCBF_{NCEL}$	16.45 (4.34)	16.69 (3.46)	15.33 (5.94)	0.463
$rCBV_{CEL}$	1.27 (0.73)	1.42 (0.88)	1.05 (0.36)	0.342
$rCBV_{NCEL}$	1.54 (0.95)	1.64 (1.09)	1.42 (0.73)	0.763
MTT_{CEL} (s)	5.74 (1.93)	5.86 (1.79)	5.59 (2.29)	0.424
MTT_{NCEL} (s)	6.06 (2.23)	5.79 (1.91)	6.31 (2.79)	0.485
TTP_{CEL} (s)	25.99 (8.67)	24.55 (5.91)	28.78 (13.08)	0.590
TTP_{NCEL} (s)	25.85 (7.12)	25.15 (6.13)	27.02 (9.75)	0.808
DMT_{CEL} (s)	−0.26 (1.21)	−0.58 (1.2)	0.46 (1.09)	0.006
DMT_{NCEL} (s)	−0.12 (0.57)	−0.26 (0.58)	0.18 (0.47)	0.010
Microvascular permeability, $K_{2\ CEL}$	−56.48 (57.21)	−64.46 (68.17)	−52.38 (26.57)	0.730
Microvascular permeability, $K_{2\ NCEL}$	−53.82 (61.23)	−58.03 (68.31)	−48.27 (55.32)	0.730
ADC_{CEL} ($mm^2\ s^{-1} \times 10^{-3}$)	0.3 (0.07)	0.3 (0.06)	0.31 (0.08)	0.831
ADC_{NCEL} ($mm^2\ s^{-1} \times 10^{-3}$)	0.44 (0.03)	0.43 (0.03)	0.45 (0.02)	0.041
Vessel-like structures (n)	9.56 (8.11)	12.46 (8.62)	6.59 (5.48)	0.002
Treatment				0.003
Standard treatment (%)	64 (65.98%)	28 (28.87%)	27 (27.84%)	
Surgery + RT (%)	11 (11.34%)	8 (8.25%)	1 (1.03%)	
RT + TMZ (%)	14 (14.43%)	12 (12.36%)	1 (1.03%)	
TMZ (%)	7 (7.22%)	7 (7.22%)	0 (0%)	
Palliative (%)	1 (1.03%)	1 (1.03%)	0 (0%)	

Note—Data are represented as means (standard deviations). Significant associations are highlighted in bold. ADC indicates apparent diffusion coefficient; CEL, contrast-enhancing lesion; DMT, delay mean time; MTT, mean transit time; NCEL, non-CEL; rCBF, relative cerebral blood flow; rCBV, relative cerebral blood volume; RT, radiotherapy; TMZ, temozolamide; TTP, time to peak.

The cutoff values are reported in Table 3. Standard treatment was the best predictor of survival (hazard ratio: 0.163, 95% CI, 0.092–0.288; p = 0.001) with a sensitivity, specificity, positive predictive value and negative predictive value of 62.5%, 93.1%, 94.6%, 56.2%, respectively (AUC = 0.778) (Table 3). However, in the best multivariate model selection hazard ratios, the most important combined factors were age at diagnosis, standard treatment and macrovascular network with a sensitivity, specificity, positive predictive value and negative predictive value of 83.3%, 93.3%, 96.2%, 73.7%, respectively (AUC = 0.901, p < 0.001).

Table 3. Survival prediction: Summary of class performance.

Variable	Area under Curve	Cutoff	Sensitivity	Specificity	Positive Predictive Value	Negative Predictive Value	Hazard Ratio (95% CI)	*p*-Value	Likelihood *p* (Multivariate)
Univariate Analysis									
Age at diagnosis	0.737	59.73	0.679	0.724	0.826	0.538	1.042 (1.022,1.063)	<0.001	
DMT_{NCEL}	0.697	−0.50	0.267	1.000	1.000	0.405	0.444 (0.232,0.852)	0.015	
Vessel-like structures	0.709	6.94	0.696	0.621	0.780	0.514	1.029 (0.998,1.061)	0.033	
Highly developed macrovascular network	0.667	Present	0.714	0.621	0.784	0.529	1.254 (0.788,1.998)	0.029	
Standard treatment	0.778	Present	0.625	0.931	0.946	0.562	0.163 (0.092,0.288)	<0.001	
Bivariate Analysis									
Age at diagnosis DMT_{NCEL}	0.859	58 −0.48	0.867	0.733	0.867	0.733	1.042 (1.014–1.071) 0.560 (0.284–1.105)	0.002 0.095	<0.001
Age at diagnosis Standard treatment	0.850	54.8 2.0	0.714	0.897	0.930	0.619	1.026 (1.005–1.048) 0.213 (0.117–0.388)	<0.001 0.015	<0.001
Vessel-like structures Standard treatment	0.864	5 Present	0.768	0.897	0.935	0.667	1.017 (0.987–1.048) 0.170 (0.096–0.301)	0.044 <0.001	<0.001
Highly developed macrovascular network Standard treatment	0.850	-	0.625	0.931	0.946	0.562	1.265 (0.792–2.019) 0.163 (0.092–0.288)	0.032 <0.001	<0.001
Trivariate Analysis *									
Age at diagnosis Standard treatment Highly developed macrovascular network	0.901	-	0.833	0.933	0.962	0.737	0.604 (0.459–0.796) 0.163 (0.090–0.297) 1.481 (0.909–2.414)	<0.001 <0.001 0.045	<0.001

* This trivariate analysis provided the highest performance of the model. Other combinations of variables with lower performance included age at diagnosis, standard treatment and vessel-like structures with a sensitivity, specificity, positive predictive value and negative predictive value of 0.801, 0.908, 0.925, 0.716, respectively (AUC = 0.863, $p = 0.02$), and age at diagnosis, standard treatment and DMT_{NCE} with a sensitivity, specificity, positive predictive value and negative predictive value of 0.793, 0.891, 0.901, 0.705, respectively (AUC = 0.824, $p = 0.03$).

3. Discussion

This study using a routine clinical MRI protocol found that the macrovascular network of newly diagnosed glioblastoma could be classified based on nVS on standard postcontrast sequences and that macrovascular network is a prognostic biomarker of survival. Patients with highly developed macrovascular network had worse survival. Interestingly, the clinical and imaging characteristics of patients with the two subtypes of macrovascular network were very similar, the only difference being that glioblastomas with highly developed macrovascular network had higher $volume_{CEL}$ and $rCBF_{CEL}$ than those with less developed macrovascular network. These findings are consistent with the well-recognized observations that glioblastomas are highly vascularized tumors with highly heterogeneous angioarchitectures and that solid tumor growth depends on vascularity [3,4,11]. Although the vascular network initially develops by incorporating existing host vessels, solid tumors probably cannot grow more than 1 mm^3 unless they synthesize their own new vessels [37].

Our results allow us to speculate that MRI might offer two approaches to assessing glioblastomas' vascularity in the pretreatment stage: the macrovascular network and the microvascular network. In fact, DSC-MRI perfusion evaluates the dynamic process of the first-pass effect in the short period of time during which the contrast agent enters the brain parenchyma, providing physiological information regarding vascularity: rCBV indirectly reflects vascularity and is useful for grading gliomas, as well as a predictive and prognostic tool [38–42]. In all patients, compared with contralateral normal-appearing WM, DSC-MRI perfusion indices and permeability values were increased in both CEL and non-CEL, supporting the idea that newly diagnosed glioblastomas have a highly developed vascular network. Recently, Jia et al. [26] demonstrated that imaging permeability parameters correlated positively with microvascular density in glioblastomas. However, the possible role of microvascular density as a biomarker of survival is still controversial. Some studies have defended its value as predictor of glioma growth and survival [2,43,44], and another found that low-grade gliomas with higher microvascular density had a higher risk of malignant transformation and were associated with shorter survival [45]. By contrast, other studies found no significant correlation with survival [2,4]. These discrepancies may be related to the type of angiogenesis that is predominant in the tumor. Birner et al. [3] found that a prominent classical capillary pattern of angiogenesis with fewer bizarre glomeruloid vessels was an independent predictor of longer survival in patients with glioblastoma. In tumors with the bizarre pattern, cell growth seems to outpace neovascularization, so their progression seems less dependent on adequate vascular networks. Glioblastomas with a predominantly bizarre pattern are unlikely to benefit from antiangiogenic therapy. By contrast, in glioblastomas with a classic capillary-like pattern, cell growth seems to parallel neovascularization; the evenly distributed pattern of delicate vessels would make it easier for chemotherapeutic drugs to reach tumor cells, so patients are more likely to respond better to antiangiogenic therapy.

We assessed the vascular network in two ways. The first, DSC-MRI perfusion indices, clearly showed that all the glioblastomas in our patients had developed a microvascular network. The second, visually counting vessel-like structures related to the CEL or non-CEL on 3D-T1CE, detected an increased presence of large-diameter vessels (cutoff, nVS = 5) in only 60% of glioblastomas. The multivariate analysis found that nVS had a greater impact than other quantitative or qualitative imaging variables when added to age and standard treatment to predict survival in the multivariate analysis. Age and standard treatment were independent prognostic factors (AUC = 0.850), and adding macrovascular network increased the predictive power of the model (AUC = 0.901). Since glioblastomas need the vascular network to provide nutrients and oxygen for metabolism and removal of waste products, the nVS could reflect differences in vascular growth, remodeling related to blood flow and metabolic demands (glioblastomas with nVS > 5 had higher $rCBF_{CEL}$ and higher $volume_{CEL}$). Although ependymal invasion was more frequent in glioblastomas with highly developed macrovascular network, adding ependymal invasion in multivariate analysis did not increase the predictive power of the model. Nevertheless, recent studies show that glioblastomas with ependymal invasion are more frequently associated with multifocal lesions and tumor recurrence than those without and that this variable can also serve as an independent prognostic factor for poor survival [35,46–48]. We focused on variables available in the pretreatment stage because being able to predict survival at this stage can help clinicians determine the timing, mode, and aggressiveness of treatment; by contrast, variables related to surgery or postoperative treatments cannot be assessed until later in the course of disease and are not useful in initial treatment planning.

Glioblastoma stem cells stimulate tumor angiogenesis by expressing elevated levels of vascular endothelial growth factor (VEGF), and high levels of VEGF correlate with worse prognosis [49]. Bevacizumab (Avastin; Genentech, San Francisco, CA, USA), a monoclonal antibody that inhibits VEGF, results in tumor shrinkage in 55% of patients with recurrent glioblastomas, palliating neurological symptoms and increasing progression-free survival [50]. However, bevacizumab does not benefit all patients. Bevacizumab combined with temozolomide and radiotherapy failed to improve survival in some patients with newly diagnosed glioblastoma [51,52], and bevacizumab combined with lomustine did not increase overall survival over lomustine alone in patients with recurrent glioblastoma [53]. Therefore, there is an urgent need for new biomarkers to predict the response to antiangiogenic treatment so responders can be selected before initiating treatment. Along these lines, rCBV and rCBF maps have been used to predict survival in patients under antiangiogenic therapy [54,55]. Our results suggest that macrovascular network would be a helpful biomarker to identify high-risk patients who would likely benefit from more aggressive and experimental treatments, and promises to be helpful in predicting survival under antiangiogenic therapy and in guiding subsequent treatment.

These results are a step toward validating the results of our recent pilot study [17], where we described these patterns of vascularity on gadofosveset trisodium high-resolution MRA in a small sample of 35 newly diagnosed glioblastomas [17]. Gadofosveset trisodium (MS-325; in that study, Vasovist®; Bayer Schering Pharma AG, Berlin, Germany; now commercially available as Ablavar®, Lantheus Medical Imaging, North Bilerica, MA, USA) is a blood-pool contrast agent that binds reversibly to albumin, resulting in much higher blood relaxivity and better contrast-to-noise ratio of the vessels than non-protein binding extracellular contrast agents, such as gadobutrol. Gadofosveset trisodium has a half-life about 16 h with optimal enhancement and diagnostic information at 6 h after administration [56]. Nevertheless, gadofosveset trisodium is indicated for MRA only to evaluate aortoiliac occlusive disease, and it is not included in clinical protocols to study brain tumors. Thus, in the present study, we used intravenous gadobutrol (Gadovist®; Bayer Schering Pharma, Berlin, Germany), an extracellular contrast agent whose high relaxivity and high concentration yields the highest T1 shortening time and excellent image enhancement. Extensive routine clinical experience has demonstrated that gadobutrol is effective and well tolerated [57].

Various limitations of this study merit comment. This was a single-center study, with a modest sample size. Thus, further studies with larger populations are necessary to determine the usefulness of nVS in predicting survival in patients with glioblastoma. We quantified the number of vessels without postprocessing. This approach is easy and fast, but also subjective; nevertheless, the interobserver agreement was excellent. Although our cutoff of nVS = 5 discriminated well between highly vascularized and less vascularized glioblastomas well, further studies are needed to explore the performance of other possible cutoffs. Moreover, interactions between factors such as the field strength of equipment, type of gadolinium-based contrast agents, and MRI acquisition protocols present challenges to establishing macrovascular network as a biomarker for predicting survival, because differences across institutions impede the replication of single-center study results [38].

Glioblastoma's macrovascular network phenotype may result from unusual levels or combinations of angiogenic factors that lead to unbalanced angiogenesis. Bennett et al. [58] reported that circulating endothelial cell counts were increased in patients with glioblastoma, and that preoperative circulating progenitor cell counts appeared to be proportional to tumor vascularity, as they correlated strongly with $rCBV_{CEL}$. Further studies correlating with histological, molecular techniques, and angiogenic gene expression might help validate different glioblastoma subtypes according to their macrovascular network. Likewise, macrovascular network would be a MRI biomarker signature that may identify distinct GBM phenotypes associated with highly significant survival differences and specific molecular pathways. It seems imaging could detect different angioarchitectural patterns of glioblastoma (not included in the present analysis) (Figures 1 and 3) in a manner similar to histopathology [3,10]. In low-grade gliomas, rounder vessel contour and microvascular branching may reflect increased intraluminal pressure due to disorderly vasculature [59]. This qualitative approach should be followed by quantitative morphometric analyses. Thus, future studies could also assess other vessel-related parameters, such as size, shape, density, distribution, and branching patterns to better express the glioblastoma's complex angioarchitecture [60,61]. High- and ultra-high field MRI might enable better analysis and quantification of vessel morphology [62,63]. Computer-assisted fractal analysis is potentially useful for quantifying different vascular networks and identifying different vascular patterns [64]. Standard pathology tests are part of clinical practice and essential criteria for most clinical trials, so a direct comparison between the imaging vascular networks and histological assessments (e.g., blood-vessel immunohistochemistry for blood vessels) must be performed in the field to adopt an additional method to complement or replace the histological analysis

4. Material and Methods

4.1. Study Data

We prospectively recruited consecutive patients aged > 18 years with newly diagnosed, histopathologically confirmed glioblastoma. We excluded patients with a history of other malignancies. Between April 2014 and November 2016, we considered 115 patients; 18 patients were excluded because histology ruled out glioblastoma. Therefore, 97 patients (62 men; age, 58 ± 15 years) were included in the analysis. Our institution's ethics committee approved the study protocol, and all patients provided written informed consent. This Ethic Committee does not provide a code because in that institution clinical trials have a code only. This is not a clinical trial, this an academic study. The date when we obtained the approval was on 7 March, 2014. Patients were managed according to published guidelines [42], recommending surgical resection followed by radiotherapy combined with concomitant and adjuvant temozolomide chemotherapy as standard treatment. Survival was defined as the interval between the date of initial glioblastoma diagnosis and the date of either death from any cause or the last follow-up date on which the patient was known to be alive.

4.2. MRI Protocol

Patients received no corticosteroids before MRI. Patients underwent MRI on a standard clinical 1.5-T system (Intera, Philips Healthcare, Best, The Netherlands) with an eight-channel head coil. The protocol included axial spin-echo T1-weighted imaging, fluid-attenuated inversion recovery imaging (FLAIR), single-shot echo-planar diffusion-weighted imaging (DWI), and first-pass perfusion DSC-MRI with gadobutrol (Gadovist; Bayer Schering Pharma, Berlin, Germany), and 3D-T1CE. Parameters for axial T1-weighted spin-echo imaging were repetition time (TR) 536 ms, echo time (TE) 15 ms; flip angle, 90°; matrix, 256 × 192; section thickness, 5 mm; intersection gap, 1 mm; field of view (FOV), 230 × 180 mm; and brain coverage, 120 mm. Parameters for FLAIR were TR, 7569 ms; TE, 115 ms; inversion delay, 2200 ms; flip angle, 90°; matrix, 256 × 192; section thickness, 5 mm; intersection gap, 1 mm; (FOV), 230 × 180 mm; and brain coverage, 120 mm. For DWI, parameters were TR, 3758 ms; TE, 99 ms ($b = 0$ and 1000 s/mm^2); 20 sections; section thickness, 5 mm; intersection gap, 1 mm; FOV, 230 × 230 mm; and matrix, 192 × 128; two signals were acquired in three orthogonal directions and combined into a trace image. Apparent diffusion coefficient (ADC) maps were calculated on a voxel-by-voxel basis. For perfusion DSC-MRI, multislice T2* single-shot echo-planar images were acquired before, during, and after rapid administration of a contrast bolus (sixteen 7-mm sections without gap; matrix, 128 × 128; FOV, 230 × 230 mm; TR, 1800 ms; TE, 25 ms; flip angle, 90°). The TR of each multi-shot block was 17 ms, and the acquisition time for each dynamic volume was 1.8 s. TE was 17 ms and the flip angle 7°. Each perfusion series consisted of 50 dynamic acquisitions with temporal resolution set to 1.8 s during the first pass of a standard dose (0.1 mmol/kg bolus of gadobutrol injected at 5 mL/s with a power injector, followed by a 20-mL bolus of saline at the same rate). To reduce the effect of contrast leakage on CBV calculations, a 5 mL bolus of gadobutrol was administered 5 min prior to DSC-MRI perfusion acquisition, at a rate of 1 mL/s, followed by a 15 mL saline flush. Parameters for isovoxel 3D-T1CE images were TR/TE, 536/15 ms; flip angle, 90°; matrix, 256 × 192; and section thickness, 1 mm.

4.3. Quantitative Image Analysis

Digital imaging and communications in medicine files were transferred to an external computing station for processing. Using Olea Sphere V.3.0 software (Olea Medical; La Ciotat, France), we applied a semi-automatic region-growing segmentation algorithm on selected seeds to delimit volumes of interest (VOI) of CEL, necrosis, and non-CEL. Non-CEL was defined as the hyperintense area on FLAIR surrounding the CEL that was associated with mass effect and architectural distortion, including blurring of the gray matter/white matter (WM) interface [43]. We also obtained the VOI in normal-appearing WM of the centrum semiovale of the hemisphere contralateral to the tumor from three contiguous slices manually. To create DSC-MRI perfusion parametric maps with contrast-leakage correction, we applied a fully automated deconvolution analysis of the tissue concentration-versus-time curve with arterial input function following the technique outlined by Boxerman et al. [43]. All VOIs were co-registered onto the DSC-MRI parametric maps (Figure 3). The following parameters were calculated: rCBF, rCBV, mean transit time (MTT), delay mean time (DMT), time when the residue function reaches its maximum (Tmax), time to peak (TTP), and ADC.

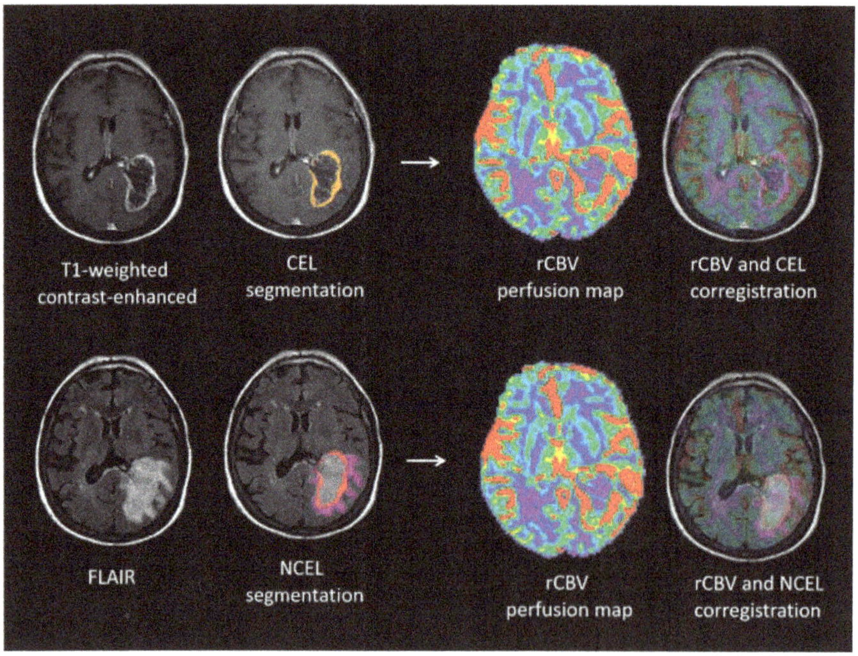

Figure 3. Lesion segmentation and co-registration MRI analysis. Regions of interest for contrast-enhancing lesion and non-contrast enhancing lesion in a mass consistent with glioblastoma involving the left temporal and parietal lobes. DSC-MRI perfusion color maps demonstrating the areas of increased relative cerebral blood volume (red) within the tumor. Postcontrast imaging shows an increased macrovascular network consisting of an evenly distributed, more linear vessel pattern within the necrotic component of the lesion (upper left).

4.4. Qualitative Image Analysis

For each patient, two neuroradiologists (J.P. and S.P., with 13 and 25 years' experience, respectively) blinded to clinical data independently reviewed axial T1-weighted images, 3D-T1CE, and axial FLAIR images on a local picture archiving and communication system (Starviewer, Gilab, University of Girona; Girona, Spain), scoring 13 features according to VASARI (tumor location; side of lesion center; eloquent area involvement; multifocality; satellite lesion; enhancement quality; thickness of CEL margin; deep WM invasion; midline cross; hemorrhage; ependymal invasion; pial invasion; and mass effect). Eloquent area involvement was defined as the presence of tumor in the cortex or immediate subcortical WM of speech-motor or speech-receptive brain areas [45]. Multifocality was defined as the presence of at least one region of tumor not contiguous with the dominant lesion, outside the region of signal abnormality surrounding the dominant mass. A satellite lesion was defined as an area of enhancement within the region of signal abnormality surrounding the dominant lesion, but not contiguous in any part with the dominant lesion mass. Deep WM invasion was defined as the presence of CEL or non-CEL in the internal capsule, corpus callosum, or brain stem. Ependymal involvement was defined as CEL in contact with the lining of the ventricles. Interobserver agreement for the MRI features was assessed, and discordant interpretations were resolved by consensus. To assess the macrovascular network, readers counted vessel-like structures related to the CEL or non-CEL components on 3D-T1CE (Figure 1), analyzing source images and multiplanar reconstructions together with subvolume maximum-intensity projection slabs. To achieve optimal views of the vessels of interest, readers were free to manipulate these images by changing subvolume position and thickness interactively in real time. Before the final analysis, readers reached a consensus on discordant classifications. Assessing the macrovascular network took less than four min per patient.

4.5. Statistical Analysis

Continuous variables are expressed as means and standard deviations or medians and ranges; categorical variables are expressed as frequencies and percentages. We used Student's *t*-tests to determine differences in continuous variables when normal distribution was able to be assumed or Wilcoxon test when normality was not able to be assumed, and chi-square tests to determine differences in categorical variables with respect to the macrovascular network and survival. To calculate the cutoff point for nVS to discriminate between glioblastomas with high nVS and those with low nVS (i.e., those with a highly developed macrovascular network vs. those with a less developed macrovascular network), we used receiver operating characteristic (ROC) curves. To determine independent predictors of overall survival, we used univariate Cox proportional hazards regression, selecting variables with p-values < 0.05 to generate prognostic models and calculating hazard ratios with corresponding 95% confidence intervals. Then, we developed multivariate Cox proportional hazards models from clinical and MRI findings including age at diagnosis, macrovascular network, DMT_{CEL}, DMT_{NCEL}, nVS, ependymal invasion, thickness of CEL, and treatment (dichotomized "standard treatment" or "others"), to adjust for the influence of these prognostic factors. We also used ROC analysis to determine the optimal cutoff for these variables for predicting survival. We defined the optimal cutoff value as the value that maximizes the sum of sensitivity and specificity. Survival curves were calculated by the Kaplan-Meier method, using the variables most significant in differentiating between survival for more than one year versus less than one year. We also combined these variables to achieve the best predictive capability, using the log-rank test to analyze overall differences. To assess interobserver agreement for the imaging features, we used the kappa consistency test, considering $k > 0.81$ excellent agreement, $0.61 < k < 0.80$ good agreement, and $k < 0.60$ poor agreement. Statistical analyses were performed with R version 3.4.2 (R Foundation for Statistical Computing, Vienna, Austria) and IBM SPSS version 23.0.0.0 (IBM Corp. Armonk, NY, USA). Significance was set at $p < 0.05$ for all tests.

5. Conclusions

In conclusion, the macrovascular network of newly diagnosed glioblastoma on contrast-enhanced MRI promises to be an easily assessable biomarker of survival. Cross-validation studies in other populations are necessary to test the generalizability of our findings, to expand our understanding of the pathophysiology of glioblastoma's vascular network, and to determine whether this approach can effectively improve subpopulation selection before treatment. Clinical trials need to clarify the possible influence of vascular patterns on the effects of antiangiogenic therapies. In addition to helping clinicians plan patients' management, the macrovascular network, together with age and standard treatment may also be useful for stratifying glioblastoma patients for enrollment in clinical trials.

Author Contributions: G.B. and J.S.-G. contributed to the design of the study together with J.P., C.B. (Carles Biarnés) and S.D.B. C.B. (Carles Biarnés), M.N.-M., A.J.-P., E.C., A.G., B.D.-X. and A.A.-B. analyzed imaging quantitative data as well. S.P. and J.P. did the qualitative interpretation of the studies. M.P. provided the registry of the patients and demographic and clinical data. P.D.-i.-E. and S.T.-H. did the statistical analysis. M.E. R.J., J.C., C.B. (Carme Balaña), C.L.-S., M.W. and K.N. supervised the project, and wrote the manuscript together with J.P. All authors were involved in the final version of the manuscript.

Funding: This research was funded by "Fundació La Marató TV3" (665/C/2013) (http://www.ccma.cat/tv3/marato/projectes-financats/2012/231/). No grant number is applicable.

Acknowledgments: The radiological platform was supported by the citizens' initiative "Associació Montgat en Acció" (G66504572).

Conflicts of Interest: The authors have no conflict of interest.

References

1. Stupp, R.; Hegi, M.E.; Mason, W.P.; van den Bent, M.J.; Taphoorn, M.J.; Janzer, R.C.; Ludwin, S.K.; Allgeier, A.; Fisher, B.; Belanger, K.; et al. Effects of radiotherapy with concomitant and adjuvant temozolomide versus radiotherapy alone on survival in glioblastoma in a randomised phase III study: 5-year analysis of the EORTC-NCIC trial. *Lancet Oncol.* **2009**, *10*, 459–466. [CrossRef]
2. Leon, S.P.; Folkerth, R.D.; Black, P.M. Microvessel density is a prognostic indicator for patients with astroglial brain tumors. *Cancer* **1996**, *77*, 362–372. [CrossRef]
3. Birner, P.; Piribauer, M.; Fischer, I.; Gatterbauer, B.; Marosi, C.; Ambros, P.F.; Ambros, I.M.; Bredel, M.; Oberhuber, G.; Rössler, K.; et al. Vascular patterns in glioblastoma influence clinical outcome and associate with variable expression of angiogenic proteins: Evidence for distinct angiogenic subtypes. *Brain Pathol.* **2003**, *13*, 133–143. [CrossRef] [PubMed]
4. Folkerth, R.D. Descriptive analysis and quantification of angiogenesis in human brain tumors. *J. Neurooncol.* **2000**, *50*, 165–172. [CrossRef] [PubMed]
5. Wesseling, P.; van der Laak, J.A.; Link, M.; Teepen, H.L.; Ruiter, D.J. Quantitative analysis of microvascular changes in diffuse astrocytic neoplasms with increasing grade of malignancy. *Hum. Pathol.* **1998**, *29*, 352–358. [CrossRef]
6. Lund, E.L.; Spang-Thomsen, M.; Skovgaard-Poulsen, H.; Kristjansen, P.E. Tumor angiogenesis—A new therapeutic target in gliomas. *Acta Neurol. Scand.* **1998**, *97*, 52–56. [CrossRef]
7. Zhong, H.; De Marzo, A.M.; Laughner, E.; Lim, M.; Hilton, D.; Zagzag, D.; Buechler, P.; Isaacs, W.B.; Semenza, G.L.; Simons, J.W. Overexpression of hypoxia-inducable factor 1a in common human cancers and their metastases. *Cancer Res.* **1999**, *59*, 5830–5835. [PubMed]
8. Argyriou, A.A.; Giannopoulou, E.; Kalofonos, H.P. Angiogenesis and anti-angiogenic molecularly targeted therapies in malignant gliomas. *Oncology* **2009**, *77*, 1–11. [CrossRef] [PubMed]
9. Jain, R.; Griffith, B.; Alotaibi, F.; Zagzag, D.; Fine, H.; Golfinos, J.; Schultz, L. Glioma angiogenesis and perfusion imaging: Understanding relationship between tumor blood volume and leakiness with increasing glioma grade. *AJNR Am. J. Neuroradiol.* **2015**, *36*, 2030–2035. [CrossRef]
10. Sharma, S.; Sharma, M.C.; Sarkar, C. Morphology of angiogenesis in human cancer: A conceptual overview, histoprognostic perspective and significance of neoangiogenesis. *Histopathology* **2005**, *46*, 481–489. [CrossRef]
11. Di Ieva, A. Angioarchitectural morphometrics of brain tumors: Are there any potential histopathological biomarkers? *Microvasc. Res.* **2010**, *80*, 522–533. [CrossRef] [PubMed]

12. Preusser, M.; Heinzl, H.; Gelpi, E.; Schonegger, K.; Haberler, C.; Birner, P.; Marosi, C.; Hegi, M.; Gorlia, T.; Hainfellner, J.A.; European Organization for Research and Treatment of Cancer Brain Tumor Group. Histopathologic assessment of hot-spot microvessel density and vascular patterns in glioblastoma: Poor observer agreement limits clinical utility as prognostic factors: A translational research project of the European Organization for Research and Treatment of Cancer Brain Tumor Group. *Cancer* **2006**, *107*, 162–170. [PubMed]
13. Russell, S.M.; Elliott, R.; Forshaw, D.; Golfinos, J.G.; Nelson, P.K.; Kelly, P.J. Glioma vascularity correlates with reduced patient survival and increased malignancy. *Surg. Neurol.* **2009**, *72*, 242–246. [CrossRef] [PubMed]
14. Kadota, T.; Nakagawa, H.; Kuroda, C. Malignant glioma. Evaluation with 3D time-of-flight MR angiography. *Acta Radiol.* **1998**, *39*, 227–232. [PubMed]
15. Sugahara, T.; Korogi, Y.; Kochi, M.; Ikushima, I.; Hirai, T.; Okuda, T.; Shigematsu, Y.; Liang, L.; Ge, Y.; Ushio, Y.; et al. Correlation of MR imaging determined cerebral blood volume maps with histologic and angiographic determination of vascularity of gliomas. *AJR Am. J. Roentgenol.* **1998**, *171*, 1479–1486. [CrossRef] [PubMed]
16. Wetzel, S.G.; Cha, S.; Law, M.; Johnson, G.; Golfinos, J.; Lee, P.; Nelson, P.K. Preoperative assessment of intracranial tumors with perfusion MR and a volumetric interpolated examination: A comparative study with DSA. *AJNR Am. J. Neuroradiol.* **2002**, *23*, 1767–1774.
17. Puig, J.; Blasco, G.; Daunis-i-Estadella, J.; Alberich-Bayarri, A.; Essig, M.; Jain, R.; Remollo, S.; Hernández, D.; Puigdemont, M.; Sánchez-González, J.; et al. High-resolution blood-pool-contrast-enhanced MR angiography in glioblastoma: Tumor-associated neovascularization as a biomarker for patient survival. A. preliminary study. *Neuroradiology* **2016**, *58*, 17–26. [CrossRef]
18. Aronen, H.; Gazit, I.; Louis, D.; Buchbinder, B.R.; Pardo, F.S.; Weisskoff, R.M.; Harsh, G.R.; Cosgrove, G.R.; Halpern, E.F.; Hochberg, F.H.; et al. Cerebral blood volume maps of gliomas: Comparison with tumor grade and histologic findings. *Radiology* **1994**, *191*, 41–51. [CrossRef]
19. Le Bas, J.F.; Kremer, S.; Grand, S.; Rémy, C.; Le Duc, G.; Tropres, I.; Fonchy, E.; Peoc'h, M.; Szabo de Edelenyi, F.; Estève, F.; et al. NMR perfusion imaging: Applications to the study of brain tumor angiogenesis. *Bull. Acad. Natl. Med.* **2000**, *184*, 557–567.
20. Hirai, T.; Murakami, R.; Nakamura, H.; Kitajima, M.; Fukuoka, H.; Sasao, A.; Akter, M.; Hayashida, Y.; Toya, R.; Oya, N.; et al. Prognostic value of perfusion MR imaging of high-grade astrocytomas: Long-term follow-up study. *AJNR Am. J. Neuroradiol.* **2008**, *29*, 1505–1510. [CrossRef]
21. Jain, R.; Narang, J.; Griffith, B.; Bagher-Ebadian, H.; Scarpace, L.; Mikkelsen, T.; Littenberg, B.; Schultz, L.R. Prognostic vascular imaging biomarkers in high-grade gliomas: Tumor permeability as an adjunct to blood volume estimates. *Acad. Radiol.* **2013**, *20*, 478–485. [CrossRef] [PubMed]
22. Jain, R.; Gutierrez, J.; Narang, J.; Scarpace, L.; Schultz, L.; Lemke, N.; Patel, S.C.; Mikkelsen, T.; Rock, J. In vivo correlation of tumor blood volume and permeability with histological and molecular angiogenic markers in gliomas. *Am. J. Neuroradiol.* **2011**, *32*, 388–394. [CrossRef]
23. Mills, S.J.; Patankar, T.A.; Haroon, H.A.; Balériaux, D.; Swindell, R.; Jackson, A. Do cerebral blood volume and contrast transfer coefficient predict prognosis in human glioma? *AJNR Am. J. Neuroradiol.* **2006**, *27*, 853–858. [PubMed]
24. Catalaa, I.; Henry, R.; Dillon, W.P.; Graves, E.E.; McKnight, T.R.; Lu, Y.; Vigneron, D.B.; Nelson, S.J. Perfusion, diffusion and spectroscopy values in newly diagnosed cerebral gliomas. *NMR Biomed.* **2006**, *19*, 463–475. [CrossRef] [PubMed]
25. Jafari-Khouzani, K.; Emblem, K.E.; Kalpathy-Cramer, J.; Bjørnerud, A.; Vangel, M.G.; Gerstner, E.R.; Schmainda, K.M.; Paynabar, K.; Wu, O.; Wen, P.Y.; et al. Repeatability of Cerebral Perfusion Using Dynamic Susceptibility Contrast MRI in Glioblastoma Patients. *Transl. Oncol.* **2015**, *8*, 137–146. [CrossRef] [PubMed]
26. Jia, Z.Z.; Shi, W.; Shi, J.L.; Shen, D.D.; Gu, H.M.; Zhou, X.J. Comparison between perfusion computed tomography and dynamic contrast-enhanced magnetic resonance imaging in assessing glioblastoma microvasculature. *Eur. J. Radiol.* **2017**, *87*, 120–124. [CrossRef] [PubMed]
27. Hammoud, M.A.; Sawaya, R.; Shi, W.; Thall, P.F.; Leeds, N.E. Prognostic significance of preoperative MRI scans in glioblastoma multiforme. *J. Neurooncol.* **1996**, *27*, 65–73. [CrossRef]
28. Zinn, P.O.; Sathyan, P.; Mahajan, B.; Bruyere, J.; Hegi, M.; Majumder, S.; Colen, R.R. A novel volume-age-KPS (VAK) glioblastoma classification identifies a prognostic cognate microRNAgene signature. *PLoS ONE* **2012**, *7*, e41522. [CrossRef]

29. Gutman, D.A.; Cooper, L.A.; Hwang, S.N.; Holder, C.A.; Gao, J.; Aurora, T.D.; Dunn, W.D., Jr.; Scarpace, L.; Mikkelsen, T.; Jain, R.; et al. MR imaging predictors of molecular profile and survival: Multi-institutional study of the TCGA glioblastoma data set. *Radiology* **2013**, *267*, 560–569. [CrossRef]
30. Park, J.K.; Hodges, T.; Arko, L.; Shen, M.; Dello Iacono, D.; McNabb, A.; Olsen Bailey, N.; Kreisl, T.N.; Iwamoto, F.M.; Sul, J.; et al. Scale to predict survival after surgery for recurrent glioblastoma multiforme. *J. Clin. Oncol.* **2010**, *28*, 3838–3843. [CrossRef]
31. Lacroix, M.; Abi-Said, D.; Fourney, D.R.; Gokaslan, Z.L.; Shi, W.; DeMonte, F.; Lang, F.F.; McCutcheon, I.E.; Hassenbusch, S.J.; Holland, E.; et al. A multivariate analysis of 416 patients with glioblastoma multiforme: Prognosis, extent of resection, and survival. *J. Neurosurg.* **2001**, *95*, 190–198. [CrossRef] [PubMed]
32. Pope, W.B.; Sayre, J.; Perlina, A.; Villablanca, J.P.; Mischel, P.S.; Cloughesy, T.F. MR imaging correlates of survival in patients with high-grade gliomas. *AJNR Am. J. Neuroradiol.* **2005**, *26*, 2466–2474. [PubMed]
33. Mazurowski, M.A.; Desjardins, A.; Malof, J.M. Imaging descriptors improve the predictive power of survival models for glioblastoma patients. *Neuro Oncol.* **2013**, *15*, 1389–1394. [CrossRef] [PubMed]
34. Wiki for the VASARI Feature Set the National Cancer Institute Web Site. Available online: https://wiki.cancerimagingarchive.net/display/Public/VASARI+Research+Project (accessed on 2 September 2018).
35. Scott, J.G.; Bauchet, L.; Fraum, T.J.; Nayak, L.; Cooper, A.R.; Chao, S.T.; Suh, J.H.; Vogelbaum, M.A.; Peereboom, D.M.; Zouaoui, S.; et al. Recursive partitioning analysis of prognostic factors for glioblastoma patients aged 70 years or older. *Cancer* **2012**, *118*, 5595–5600. [CrossRef] [PubMed]
36. Oh, S.W.; Jee, T.K.; Kong, D.S.; Nam, D.H.; Lee, J.I.; Seol, H.J. Outcome of conventional treatment and prognostic factor in elderly glioblastoma patients. *Acta Neurochir.* **2014**, *156*, 641–651. [CrossRef] [PubMed]
37. Denekamp, J. Angiogenesis, neovascular proliferation and vascular pathophysiology as targets for cancer therapy. *Br. J. Radiol.* **1993**, *66*, 181–196. [CrossRef] [PubMed]
38. Zhang, J.; Liu, H.; Tong, H.; Wang, S.; Yang, Y.; Liu, G.; Zhang, W. Clinical Applications of Contrast-Enhanced Perfusion MRI Techniques in Gliomas: Recent Advances and Current Challenges. *Contrast Media Mol. Imaging* **2017**, *2017*, 7064120. [CrossRef]
39. Jain, R.; Poisson, L.; Narang, J.; Gutman, D.; Scarpace, L.; Hwang, S.N.; Holder, C.; Wintermark, M.; Colen, R.R.; Kirby, J.; et al. Genomic mapping and survival prediction in glioblastoma: Molecular subclassification strengthened by hemodynamic imaging biomarkers. *Radiology* **2013**, *267*, 212–220. [CrossRef]
40. Jain, R.; Poisson, L.M.; Gutman, D.; Scarpace, L.; Hwang, S.N.; Holder, C.A.; Wintermark, M.; Rao, A.; Colen, R.R.; Kirby, J.; et al. Outcome prediction in patients with glioblastoma by using imaging, clinical, and genomic biomarkers: Focus on the nonenhancing component of the tumor. *Radiology* **2014**, *272*, 484–493. [CrossRef]
41. Margiewicz, S.; Cordova, C.; Chi, A.S.; Jain, R. State of the Art Treatment and Surveillance Imaging of Glioblastomas. *Semin. Roentgenol.* **2018**, *53*, 23–36. [CrossRef]
42. Griffith, B.; Jain, R. Perfusion Imaging in Neuro-Oncology: Basic Techniques and Clinical Applications. *Magn Reson. Imaging Clin. N. Am.* **2016**, *24*, 765–779. [CrossRef] [PubMed]
43. Weidner, N. The importance of tumour angiogenesis. The evidence continues to grow. *Am. J. Clin. Pathol.* **2004**, *122*, 675–677. [CrossRef] [PubMed]
44. Folkerth, R.D. Histologic measures of angiogenesis in human primary brain tumors. *Cancer Treat Res.* **2004**, *117*, 79–95. [PubMed]
45. Abdulrauf, S.I.; Evardsen, K.; Ho, K.L.; Yang, X.Y.; Rock, J.P.; Rosenblum, M.L. Vascular endothelial growth factor expression and vascular density as prognostic markers of survival in patients with low-grade astrocytoma. *J. Neurosurg.* **1998**, *88*, 513–520. [CrossRef] [PubMed]
46. Jafri, N.F.; Clarke, J.L.; Weinberg, V.; Barani, I.J.; Cha, S. Relationship of glioblastoma multiforme to the subventricular zone is associated with survival. *Neuro Oncol.* **2013**, *15*, 91–96. [CrossRef] [PubMed]
47. Chaichana, K.; Parker, S.; Olivi, A.; Quinones-Hinojosa, A. A proposed classification system that projects outcomes based on preoperative variables for adult patients with glioblastoma multiforme. *J. Neurosurg.* **2010**, *112*, 997–1004. [CrossRef]
48. Adeberg, S.; Bostel, T.; Konig, L.; Welzel, T.; Debus, J.; Combs, S.E. A comparison of long-term survivors and short-term survivors with glioblastoma, subventricular zone involvement: A predictive factor for survival? *Radiat Oncol.* **2014**, *9*, 95. [CrossRef]

49. Takano, S. Glioblastoma angiogenesis: VEGF resistance solutions and new strategies based on molecular mechanisms of tumor vessel formation. *Brain Tumor Pathol.* **2012**, *29*, 73–86. [CrossRef]
50. Friedman, H.S.; Prados, M.D.; Wen, P.Y.; Mikkelsen, T.; Schiff, D.; Abrey, L.E.; Yung, W.K.; Paleologos, N.; Nicholas, M.K.; Jensen, R.; et al. Bevacizumab alone and in combination with irinotecan in recurrent glioblastoma. *J. Clin. Oncol.* **2009**, *27*, 4733–4740. [CrossRef]
51. Gilbert, M.R.; Dignam, J.J.; Armstrong, T.S.; Wefel, J.S.; Blumenthal, D.T.; Vogelbaum, M.A.; Colman, H.; Chakravarti, A.; Pugh, S.; Won, M.; et al. A randomized trial of bevacizumab for newly diagnosed glioblastoma. *N. Engl. J. Med.* **2014**, *370*, 699–708. [CrossRef]
52. Chinot, O.L.; Wick, W.; Mason, W.; Henriksson, R.; Saran, F.; Nishikawa, R.; Carpentier, A.F.; Hoang-Xuan, K.; Kavan, P.; Cernea, D.; et al. Bevacizumab plus radiotherapy-temozolomide for newly diagnosed glioblastoma. *N. Engl. J. Med.* **2014**, *370*, 709–722. [CrossRef] [PubMed]
53. Wick, W.; Gorlia, T.; Bendszus, M.; Taphoorn, M.; Sahm, F.; Harting, I.; Brandes, A.A.; Taal, W.; Domont, J.; Idbaih, A.; et al. Lomustine and Bevacizumab in Progressive Glioblastoma. *N. Engl. J. Med.* **2017**, *377*, 1954–1963. [CrossRef] [PubMed]
54. Sawlani, R.N.; Raizer, J.; Horowitz, S.W.; Shin, W.; Grimm, S.A.; Chandler, J.P.; Levy, R.; Getch, C.; Carroll, T.J. Glioblastoma: A method for predicting response to antiangiogenic chemotherapy by using MR perfusion imaging—Pilot study. *Radiology* **2010**, *255*, 622–628. [CrossRef]
55. Kickingereder, P.; Wiestler, B.; Burth, S.; Wick, A.; Nowosielski, M.; Heiland, S.; Schlemmer, H.P.; Wick, W.; Bendszus, M.; Radbruch, A. Relative cerebral blood volume is a potential predictive imaging biomarker of bevacizumab efficacy in recurrent glioblastoma. *Neuro Oncol.* **2015**, *17*, 1139–1147. [CrossRef] [PubMed]
56. Puig, J.; Blasco, G.; Essig, M.; Daunis-i-Estadella, J.; Laguillo, G.; Quiles, A.M.; Remollo, S.; Bergmann, K.; Joly, C.; Bernado, L.; et al. Albumin-binding MR blood pool contrast agent improves diagnostic performance in human brain tumour: Comparison of two contrast agents for glioblastoma. *Eur. Radiol.* **2013**, *23*, 1093–1101. [CrossRef]
57. Scott, L.J. Gadobutrol: A review of its use for contrast-enhanced magnetic resonance imaging in adults and children. *Clin. Drug Investig.* **2013**, *33*, 303–314. [CrossRef]
58. Bennett, I.E.; Guo, H.; Kountouri, N.; D'abaco, G.M.; Hovens, C.M.; Moffat, B.A.; Desmond, P.; Drummond, K.; Kaye, A.H.; Morokoff, A.P. Preoperative biomarkers of tumour vascularity are elevated in patients with glioblastoma multiforme. *J. Clin. Neurosci.* **2015**, *22*, 1802–1808. [CrossRef]
59. Korkolopoulou, P.; Patsouris, E.; Kavantzas, N.; Konstantinidou, A.E.; Christodoulou, P.; Thomas-Tsagli, E.; Pananikolaou, A.; Eftychiadis, C.; Pavlopoulos, P.M.; Angelidakis, D.; et al. Prognostic implications of microvessel morphometry in diffuse astrocytic neoplasms. *Neuropathol. Appl. Neurobiol.* **2002**, *28*, 57–66. [CrossRef]
60. Bullitt, E.; Gerig, G.; Pizer, S.M.; Lin, W.; Aylward, S.R. Measuring tortuosity of the intracerebral vasculature from MRA images. *IEEE Trans. Med. Imaging* **2003**, *22*, 1163–1171. [CrossRef]
61. Jain, R.K. Normalizing tumor vasculature with antiangiogenic therapy: A new paradigm for combination therapy. *Nat. Med.* **2001**, *7*, 987–989. [CrossRef]
62. Wardlaw, G.; Wong, R.; Noseworthy, M.D. Identification of intratumour low frequency microvascular components via BOLD signal fractal dimension mapping. *Phys. Med.* **2008**, *24*, 87–91. [CrossRef] [PubMed]
63. Bullitt, E.; Reardon, D.A.; Smith, J.K. A review of micro- and macrovascular analysis in the assessment of tumor-associated vasculature as visualized by MR. *Neuroimage* **2007**, *37*, 116–119. [CrossRef] [PubMed]
64. Di Ieva, A. Fractal analysis of microvascular networks in malignant brain tumors. *Clin. Neuropathol.* **2012**, *31*, 342–351. [CrossRef] [PubMed]

cancers

MDPI

Article

Location-Dependent Patient Outcome and Recurrence Patterns in IDH1-Wildtype Glioblastoma

Christine Jungk [1,*], Rolf Warta [1], Andreas Mock [1,2], Sara Friauf [1], Bettina Hug [3], David Capper [4,5], Amir Abdollahi [6], Jürgen Debus [6], Martin Bendszus [3], Andreas von Deimling [4], Andreas Unterberg [1] and Christel Herold-Mende [1]

1 Division of Experimental Neurosurgery, Department of Neurosurgery, University Hospital Heidelberg, INF 400, D-69120 Heidelberg, Germany; rolf.warta@med.uni-heidelberg.de (R.W.); andreas.mock@med.uni-heidelberg.de (A.M.); sara.friauf@med.uni-heidelberg.de (S.F.); andreas.unterberg@med.uni-heidelberg.de (A.U.); h.mende@med.uni-heidelberg.de (C.H.-M.)

2 Department of Medical Oncology, National Center for Tumor Diseases, INF 460, D-69120 Heidelberg, Germany

3 Department of Neuroradiology, Clinic of Neurology, University Hospital Heidelberg, INF 400, D-69120 Heidelberg, Germany; bettina.hug@muerle.de (B.H.); martin.bendszus@med.uni-heidelberg.de (M.B.)

4 Department of Neuropathology, Institute of Pathology, University of Heidelberg, German Cancer Consortium, CCU Neuropathology, German Cancer Research Center, INF 224, D-69120 Heidelberg, Germany; david.capper@charite.de (D.C.); andreas.vondeimling@med.uni-heidelberg.de (A.v.D.)

5 Department of Neuropathology, Charité-Universitätsmedizin, Charitéplatz 1, D-10117 Berlin, Germany

6 Department of Radiation Oncology, Heidelberg University Hospital, Molecular and Translational Radiation Oncology, National Center for Tumor Diseases, German Cancer Research Center, D-69120 Heidelberg, Germany; amir.abdollahi@med.uni-heidelberg.de (A.A.); juergen.debus@med.uni-heidelberg.de (J.D.)

* Correspondence: christine.jungk@med.uni-heidelberg.de

Received: 29 December 2018; Accepted: 15 January 2019; Published: 21 January 2019

Abstract: Recent studies suggest that glioblastomas (GBMs) contacting the subventricular zone (SVZ) as the main adult neurogenic niche confer a dismal prognosis but disregard the unique molecular and prognostic phenotype associated with isocitrate dehydrogenase 1 (IDH1) mutations. We therefore examined location-dependent prognostic factors, growth, and recurrence patterns in a consecutive cohort of 285 IDH1-wildtype GBMs. Based on pre-operative contrast-enhanced MRI, patients were allotted to four location-dependent groups with (SVZ+; groups I, II) and without (SVZ−; groups III, IV) SVZ involvement or with (cortex+; groups I, III) and without (cortex−; groups II, IV) cortical involvement and compared for demographic, treatment, imaging, and survival data at first diagnosis and recurrence. SVZ involvement was associated with lower Karnofsky performance score ($p < 0.001$), lower frequency of complete resections at first diagnosis ($p < 0.0001$), and lower non-surgical treatment intensity at recurrence ($p < 0.001$). Multivariate survival analysis employing a Cox proportional hazards model identified SVZ involvement as an independent prognosticator of inferior overall survival ($p < 0.001$) and survival after relapse ($p = 0.041$). In contrast, multifocal growth at first diagnosis ($p = 0.031$) and recurrence ($p < 0.001$), as well as distant recurrences ($p < 0.0001$), was more frequent in cortex+ GBMs. These findings offer the prospect for location-tailored prognostication and treatment based on factors assessable on pre-operative MRI.

Keywords: Glioblastoma; isocitrate dehydrogenase 1 (IDH1)-wildtype; subventricular zone; survival; multifocal growth; distant recurrence

1. Introduction

Glioblastoma (GBM) World Health Organization (WHO) grade IV, the most common and lethal primary brain tumor, is a clinically, radiographically, and molecularly heterogeneous disease. Standard therapy comprising maximal safe resection, irradiation, and temozolomide (TMZ)-based chemotherapy confers a median survival of less than 15 months [1], with high inter-individual variability and only 3–5% of patients experiencing long-term survival (LTS) of more than three years [2]. Understanding the ontogeny of GBM, and mechanisms leading to its heterogeneity, will help to tailor individualized treatments based on personalized prognostication and improve patient outcomes.

Mounting evidence suggests that glioma ontogeny is linked to a subpopulation of neural stem (NSC) and progenitor (NPC) cells persisting in neurogenic niches throughout adulthood, in particular NSC-like astrocytes in the subventricular zone (SVZ) lining the walls of the lateral ventricles [3,4]. In animal models, inactivation of tumor suppressor genes (TP53, NF1, PTEN) or activation of oncogenes (Akt, Ras) resulted in the formation of astrocytomas through malignant transformation of NSCs [5–8]. In humans, the SVZ was identified as a reservoir of malignant precursor clones by employing a novel approach of intraoperative fluorescence-guided multiple sampling of GBMs and their adjacent fluorescent SVZ [9]. Accordingly, targeting the ipsilateral SVZ by postoperative irradiation has been shown to confer improved progression-free (PFS) and overall (OS) survival [10]. Ultimately, several studies have reported that the proximity of GBM to the SVZ is associated with inferior patient outcomes [11–16]. However, none of these studies considered isocitrate dehydrogenase 1 (IDH1) mutation status for survival analysis, although IDH1-mutant (IDH1-mut) gliomas comprise a molecularly distinct GBM subgroup with favorable patient outcomes [17,18]. This has prompted the 2016 WHO classification of CNS tumors to discriminate between IDH1-wildtype (IDH1-wt) and IDH1-mut GBMs [19]. Moreover, conflicting data exist regarding the impact of SVZ involvement on growth and recurrence patterns of GBM. Initially, Lim et al. reported a series of 53 GBMs assigned to four groups dependent on their spatial relationship to the SVZ and the cortex [20]. Group I GBMs (contacting SVZ and cortex) were most frequently multifocal at the first diagnosis and recurred in a remote location, while group IV GBMs (neither contacting SVZ nor cortex) were always solitary lesions with tumor recurrence exclusively adjacent to the resection cavity [20]. Accordingly, Adeberg et al. reported that distant and multifocal progression was more common in GBMs contacting the SVZ [21]. In contrast, Kappadakunnel et al. found the highest rate of multifocal disease in group III GBMs (involving cortex but not SVZ), while the presence of distant tumor recurrence was independent of tumor location [12]. Intriguingly, in a MRI analysis of 49 GBM patients, no location-dependent recurrence pattern was observed [22]. Finally, computerized simulation of glioma growth provided evidence that GBMs involving the SVZ do not necessarily originate in the periventricular region, but also in the white matter with centrifugal growth and ultimate contact to the SVZ, dependent on increasing tumor size [23]. To shed light into these conflicting data, possibly owing to small sample sizes and the inherent molecular bias conferred by IDH1 mutations, we sought to analyze a large cohort of 285 patients with newly diagnosed IDH1-wt GBM with respect to location-dependent survival, growth, and recurrence patterns.

2. Results

2.1. Location-Dependent IDH1 Mutation Status and Treatment-Inherent Differences

Altogether, 302 consecutive newly diagnosed GBM patients were allocated to one of the four location groups depicted in Figure 1A. Group I consisted of contrast-enhancing lesions (CEL) contacting the SVZ and infiltrating the cortex, group II of CELs contacting the SVZ only, group III of CELs contacting the cortex only, and group IV of CELs residing in the subcortical white matter, neither contacting SVZ nor cortex. Location-dependent molecular and clinical aspects, growth and recurrence patterns and patient outcome were investigated in each of the following comparisons: SVZ+ (groups I, II) vs. SVZ− (groups III, IV); cortex+ (groups I, III) vs. cortex− (groups II, IV); and group II ("pure"

SVZ involvement) vs. group III ("pure" cortical involvement). IDH1 mutations were identified in 17/302 patients (5.6%), leaving 285 IDH1-wt patients for further comparison. Noteworthy, IDH1-mut tumors were more common among GBMs without SVZ involvement (SVZ+ vs. SVZ−: $p = 0.04$). Demographic, radiographic, treatment-related and outcome data of the IDH1-wt GBM cohort are summarized in Table 1. Radiographic classification assigned the majority of tumors to group III (cortical involvement only; 37%) and group I (SVZ and cortical involvement; 34%) with a balanced distribution between SVZ+ (52%) and SVZ− (48%) patients (Table 1, Figure 1B). No location-dependent age difference was observed. Karnofsky performance score (KPS) was significantly lower in patients with SVZ involvement (SVZ+ vs. SVZ−: $p < 0.001$; II vs. III: $p = 0.002$). At first diagnosis, gross total resection (GTR) was achieved significantly more often in patients without SVZ involvement (group III: 47%; group IV: 45%; SVZ+ vs. SVZ−: $p < 0.0001$; II vs. III: $p = 0.003$), while intensified postoperative treatment was independent of tumor location. At relapse, MRI was available in 187/285 patients (66%) of whom 163 patients (87%) received any kind of salvage therapy. Neither the number of re-resections, nor the extent of resection (EOR), showed location-dependent differences; however, the number of non-surgical interventions ("non-surgical treatment intensity") was significantly higher among SVZ− GBM patients ($p < 0.001$).

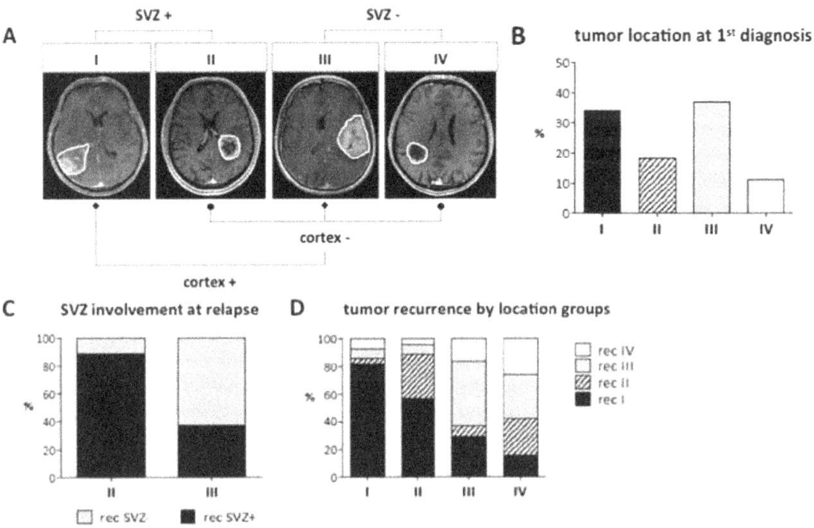

Figure 1. As exemplified in (**A**), 285 IDH1-wt GBM patients were allocated to four different location groups based on the CEL's contact to the SVZ and/or the cortex on pre-operative MRI. Groups were further pooled into GBMs with (SVZ+) and without (SVZ−) SVZ involvement or with (cortex+) and without (cortex−) cortical involvement. (**B**) At 1st diagnosis, most GBMs were allocated to groups III (37%) and I (34%). (**C**) SVZ involvement increased from 1st diagnosis to recurrence, as depicted for "pure" SVZ (group II) and "pure" cortical (group III) GBMs. Group III tumors now recurred as SVZ+ GBM in 38.5% of cases while only 11% of group II tumors shed their SVZ contact and recurred as SVZ− GBM. (**D**) No predilection was found for each of the four groups regarding location of recurrent tumor. Most group I and III GBMs recurred at the same location, while recurrences of group IV GBMs were distributed throughout all groups.

Table 1. Patient Characteristics of the IDH1-Wildtype Cohort.

Patients; *n* (%)	Group I	Group II	Group III	Group IV	Total	All	SVZ +/−	cortex +/−	II vs. III
	96 (34)	**53 (18)**	**105 (37)**	**31 (11)**	**285 (100)**	*p*-value			
Age (years); median (range)	67 (20–84)	60 (30–81)	61 (36–87)	64 (38–78)	64 (20–87)	0.095	0.404	0.218	0.443
Sex; n (male/female)	58/38	34/19	67/38	19/12	178/107	0.953	0.808	1.0	1.0
KPS pre-op; median (range)	85 (20–100)	80 (30–100)	90 (40–100)	90 (50–100)	90 (20–100)	**0.002**	**<0.001**	0.092	**0.002**
Survival Data									
Death; *n* (%)	93 (97)	49 (91)	95 (90)	25 (81)	272 (95)	**0.032**	**0.031**	0.152	0.775
OS (months); median (range)	10 (0–69)	8 (0–83)	14 (0–99)	18 (0–68)	12 (0–99)	**<0.001**	**<0.0001**	0.701	**0.018**
PFS (months); median (range)	4.5 (0–57)	3 (0–43)	4 (0–90)	5 (0–57)	4 (0–90)	0.197	0.189	0.978	0.271
Survival after relapse (months); median (range)	5 (0–49)	6 (0–43)	9 (0–78)	11.5 (0–63)	8 (0–78)	**0.041**	**0.032**	0.286	0.91
LTS (> 36 months); *n* (%)	7 (7)	4 (8)	11 (10)	5 (16)	27 (9)	0.478	0.229	0.66	0.775
STS (> 6 < 10 months); *n* (%)	21 (22)	11 (20)	16 (15)	4 (13)	52 (18)	0.512	0.167	1.0	0.381
Molecular Data (*n* = 285)									
MGMT meth; *n* (%) - Yes - No - N/A	**96** (100) 32 (33) 24 (25) 40 (42)	**53** (100) 14 (26) 16 (31) 23 (43)	**105** (100) 20 (19) 42 (40) 43 (41)	**31** (100) 9 (29) 12 (39) 10 (32)	**285** (100) 75 (26) 94 (33) 116 (41)	0.203	**0.045**	0.944	0.394
Radiographic Characteristics at 1st Diagnosis (*n* = 285)									
Multifocal growth; *n* (%) - CE - FLAIR	**96** (100) 14 (15) 12 (13)	**53** (100) 4 (8) 3 (6)	**105** (100) 18 (17) 19 (18)	**31** (100) 2 (6) 2 (6)	**285** (100) 38 (13) 36 (13)	 0.235 0.100	 0.602 0.212	 0.056 **0.031**	 0.146 **0.049**
Radiographic Characteristics at Recurrence (*n* = 187)									
Imaging available; *n* (%) - Yes - No - Alive & no recurrence	**96** (100) 57 (59) 39 (41) 0 (0)	**53** (100) 28 (53) 25 (47) 0 (0)	**105** (100) 83 (79) 20 (19) 2 (2)	**31** (100) 19 (61) 11 (35) 1 (3)	**285** (100) 187 (66) 95 (33) 3 (1)				
Location at recurrence; *n* (%) - Group I (rec) - Group II (rec) - Group III (rec) - Group IV (rec)	**57** (100) 47 (82) 2 (3.5) 4 (7) 4 (7)	**28** (100) 16 (57) 9 (32) 2 (7) 1 (4)	**83** (100) 24 (29) 8 (9.5) 38 (46) 13 (15.5)	**19** (100) 3 (16) 5 (26) 6 (32) 5 (26)	**187** (100) 90 (48) 24 (13) 50 (27) 23 (12)	**<0.0001**	**<0.0001**	**0.0003**	**<0.0001**

Table 1. *Cont.*

Patients; n (%)	Group I	Group II	Group III	Group IV	Total	All	SVZ +/−	cortex +/−	II vs. III
	96 (34)	**53 (18)**	**105 (37)**	**31 (11)**	**285 (100)**	*p*-value			
Recurrence pattern; *n* (%)	**57** (100)	**28** (100)	**83** (100)	**19** (100)	**187** (100)	**0.0001**	0.078		
- Local	41 (72)	27 (96)	51 (61)	19 (100)	138 (74)			**<0.0001**	**0.0002**
- Distant	3 (5)	0 (0)	8 (10)	0 (0)	11 (6)				
- Local & Distant	13 (23)	1 (4)	24 (29)	0 (0)	38 (20)				
Multifocal growth; *n* (%)	**57** (100)	**28** (100)	**83** (100)	**19** (100)	**187** (100)				
- Multifocal CE (rec)	18 (32)	2 (7)	28 (34)	2 (11)	50 (27)	**0.013**	0.409	**0.001**	**0.006**
- Multifocal FLAIR (rec)	7 (12)	2 (7)	18 (22)	2 (11)	29 (15.5)	0.193	0.106	0.164	0.096
Treatment at 1st Diagnosis (*n* = 285)									
EOR; *n* (%)	**96** (100)	**53** (100)	**105** (100)	**31** (100)	**285** (100)				
- GTR	19 (20)	10 (19)	49 (47)	14 (45)	92 (32)				
- Partial	62 (64.5)	30 (57)	42 (40)	15 (48)	149 (52)	**0.0005**	**<0.0001**	0.556	**0.003**
- Biopsy	1 (1)	6 (11)	3 (3)	0 (0)	10 (4)				
- Unknown	14 (14.5)	7 (13)	11 (10)	2 (7)	34 (12)				
Adjuvant therapy; *n* (%)	**96** (100)	**53** (100)	**105** (100)	**31** (100)	**285** (100)				
- RT	77 (80)	44 (83)	92 (88)	27 (87)	240 (84)				
- TMZ concomitant	46 (48)	34 (64)	63 (60)	19 (61)	162 (57)				
- Stupp	22 (23)	12 (23)	35 (33)	10 (32)	79 (28)	0.289	0.064	0.773	0.199
- Clinical trial	29 (30)	17 (32)	40 (38)	8 (26)	94 (33)	0.509	0.451	0.492	0.488
Treatment at Recurrence (*n* = 187)									
Salvage-therapy; *n* (%)	**57** (100)	**28** (100)	**83** (100)	**19** (100)	**187** (100)				
- Treatment received	50 (88)	23 (82)	73 (88)	17 (89)	163 (87)				
- No treatment received	30	23	19	4	76	**0.008**	**0.004**	**0.037**	**0.002**
- Lost to follow-up	17	8	14	9	48				
- Alive & no recurrence	0	0	2	1	3				
Re-resection; *n* (%)	**10** (20)	**5** (22)	**24** (33)	**6** (35)	**45** (28)				
- GTR	4 (8)	2 (9)	15 (21)	3 (18)	24 (15)				
- Partial	5 (10)	2 (9)	5 (7)	1 (6)	13 (8)	0.24	0.194	0.45	0.538
- Unknown	1 (2)	0	1 (1)	2 (12)	4 (2)				
Non-surgical therapies; *n* (%)	**57** (100)	**28** (100)	**83** (100)	**19** (100)	**187** (100)				
- 0 (re-resection only)	2 (4)	1 (4)	1 (1)	0	4 (2)	**0.013**	**<0.001**	0.789	0.209
- 1	34 (68)	13 (57)	29 (40)	5 (29)	81 (50)				
- 2–5	14 (28)	9 (39)	43 (59)	12 (71)	78 (48)				

KPS. Karnofsky Performance Score; OS: overall survival; PFS: progression-free survival; LTS: long-term survivor; STS: short-term survivor; MGMT: O^6-methylguanin-DNA-methyltransferase; CE: contrast-enhancing; FLAIR: fluid-attenuated inversion recovery; EOR: extent or resection, GTR: gross total resection; RT: radiotherapy; TMZ: temozolomide; and Stupp: Stupp protocol including >3 cycles TMZ.

2.2. Location-Dependent Growth and Recurrence

Next, conflicting data in the recent literature on location-dependent GBM growth and recurrence patterns prompted us to investigate these features in our large, molecularly homogeneous dataset. Multifocal disease noncontiguous with the primary CEL was evaluated both on contrast-enhancing (CE) T1-weighted (T1-w) and fluid-attenuated inversion recovery (FLAIR) images. At first diagnosis, multifocal disease was detected mainly in GBMs with cortical involvement (group III: 17% (CE), 18% (FLAIR); group I: 15% (CE), 13% (FLAIR)) with a significant increase in multifocal FLAIR lesions in cortex+ vs. cortex− ($p = 0.031$) or group III vs. II GBMs ($p = 0.049$), but not in SVZ+ vs. SVZ− GBMs (Figure 2A,B). Longitudinal analysis of all patients with MRI available at tumor relapse ($n = 187$) revealed an increase in patients with multifocal disease compared to the first diagnosis, again with a significant difference in cortex+ vs. cortex− (CE: $p = 0.001$; FLAIR: $p = 0.164$) or group III vs. II GBMs (CE: $p = 0.006$; FLAIR: $p = 0.096$), but this time particularly in CELs (group III: 34% (CE), 22% (FLAIR); group I: 32% (CE), 12% (FLAIR)) (Figure 2C,D). Thus, multifocal disease was primarily found in cortical GBMs and increased with tumor relapse.

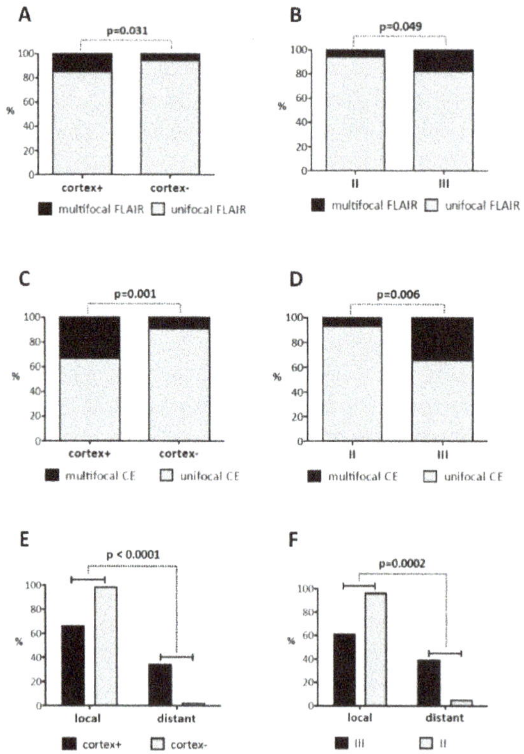

Figure 2. Multifocal growth and distant recurrences were hallmarks of cortical IDH1-wt GBMs. At 1st diagnosis ($n = 285$), multifocal FLAIR lesions were significantly increased in cortex+ vs. cortex− (**A**) and group III vs. group II GBMs (**B**), but not in SVZ+ vs. SVZ− GBMs. At recurrence ($n = 187$), multifocal CELs were significantly enhanced in cortex+ (**C**) and group III GBMs (**D**). Local and distant recurrence patterns differed markedly within location-specific groups. Distant tumor growth remote from the initial resection site was significantly more frequent in GBMs with cortical involvement ((**E**) cortex+ vs. cortex−; (**F**) group II vs. group III).

While most GBMs were evenly allocated to groups III (37%) and I (34%) at first diagnosis, group I GBMs represented the majority of tumors at relapse (48%; Table 1). Accordingly, SVZ involvement was increased to 61% of all recurrent GBMs compared to 52% at first diagnosis (Table 1). When comparing the two most unambiguous location groups II (SVZ involvement only) and III (cortical involvement only), group III GBMs now were found to extend towards the SVZ and recur as SVZ+ GBMs in 38.5% of cases while only 11% of group II GBMs shed their SVZ contact and recurred as SVZ− GBMs (Figure 1C). No predilection was found for each of the four groups regarding the location of recurrent tumor, although most group I and III GBMs recurred at the same location (group I: 82% rec group I; group III: 46% rec group III; Table 1, Figure 1D). In contrast, recurrences of group IV GBMs (neither SVZ nor cortical involvement) were observed throughout all groups (Table 1, Figure 1D). Of 187 patients with recurrent GBM, the vast majority (94%; $n = 176$) presented with local tumor recurrence adjacent to the primary resection site, among those 38 patients (20%) with concurrent distant tumor growth. Tumor growth exclusively remote from the primary resection site occurred in only 11 patients (6%) (Table 1). Local and distant recurrence patterns differed markedly within location-specific groups ($p = 0.0001$). Distant tumor growth was most commonly observed in group III (30%) and I (28%) GBMs resulting in a significant increase in distant tumor recurrences in GBMs with cortical involvement (cortex+ vs. cortex−: $p < 0.0001$; II vs. III: $p = 0.0002$) (Figure 2E,F).

In summary, tumor location had a distinct impact on tumor growth and recurrence: GBMs with cortical involvement were prone to grow at multifocal sites and to recur distant from the primary resection site.

2.3. Location-Dependent Patient Outcome

Previous studies reported that GBMs with SVZ involvement confer inferior survival [11–16] but disregarded the IDH1 mutation status as a major prognostic confounder. Since we found IDH1-mut tumors to be more frequent among SVZ− GBMs, we analyzed location-dependent survival exclusively in IDH1-wt GBMs. To this end, we compared SVZ+ vs. SVZ− GBMs as well as group II vs. III GBMs, the latter being the most unambiguous separation between GBMs with SVZ and cortical involvement. In univariate analysis of the complete IDH1-wt cohort ($n = 285$), both OS and survival after relapse differed significantly with longest median survival observed in group IV and shortest median survival in group II patients (OS: $p < 0.001$; survival after relapse: $p = 0.041$ (Table 1, Table S2, Figure 3A,B). Noteworthy, PFS was comparable between groups ($p = 0.197$). In general, SVZ involvement conferred inferior OS ($p < 0.0001$) and survival after relapse ($p = 0.023$) (Figure 3C,D). There was no location-dependent preponderance of STS or LTS (Table 1). Since cortical GBMs showed a tendency to extend towards the SVZ upon tumor relapse, we included SVZ involvement at recurrence (SVZ+ rec) into our prognostic model and found this also to be negatively associated with OS and survival after relapse (OS: $p < 0.001$; survival after relapse: $p < 0.001$; Table S2). Beside tumor location, well-known demographic and clinical parameters were identified as prognostic factors for OS (age at 1st diagnosis, pre-operative KPS, GTR and intensified treatment at the first diagnosis and recurrence), PFS (age, GTR and intensified treatment at the first diagnosis), and survival after relapse (age at the first diagnosis; GTR and intensified treatment at 1st diagnosis and recurrence) (Table S2). Interestingly, multifocal disease at the first diagnosis also predicted inferior OS (CEL: $p = 0.003$; FLAIR lesions: $p = 0.028$), PFS (CEL: $p = 0.034$; FLAIR lesions: $p = 0.002$) and survival after relapse (CEL: $p = 0.005$) (Table S2). The important prognostic impact of SVZ involvement was confirmed by multivariate analysis in which SVZ involvement at the first diagnosis was identified as an independent prognostic factor of inferior OS ($p = 0.008$) and SVZ involvement at the recurrence of inferior survival after relapse ($p = 0.015$) (Table 2).

Previous outcome studies analyzed group I and II tumors together. However, group I GBMs are voluminous tumors extending from the SVZ throughout the white matter to the cortex and may as well originate from the cortex spreading towards the SVZ. Consequently, we performed a survival analysis, exclusively comparing group II (SVZ involvement only) and III (cortical involvement only) patients. In line with our previous findings, OS was significantly shorter for "pure" SVZ patients ($p = 0.028$) (Figure 3E). Likewise, SVZ involvement at recurrence was associated with inferior survival after relapse ($p = 0.045$; Table S3, Figure 3F). Consistent with the complete cohort, age, EOR, intensified treatment as well as multifocal disease on CE and FLAIR images at 1st diagnosis were predictive of OS and PFS, while EOR and intensified treatment at recurrence were predictive of survival after relapse (Table S3). In multivariate analysis, SVZ involvement was confirmed as an independent prognostic factor for OS ($p = 0.007$), while tumor location no longer impacted survival after relapse (Table 3).

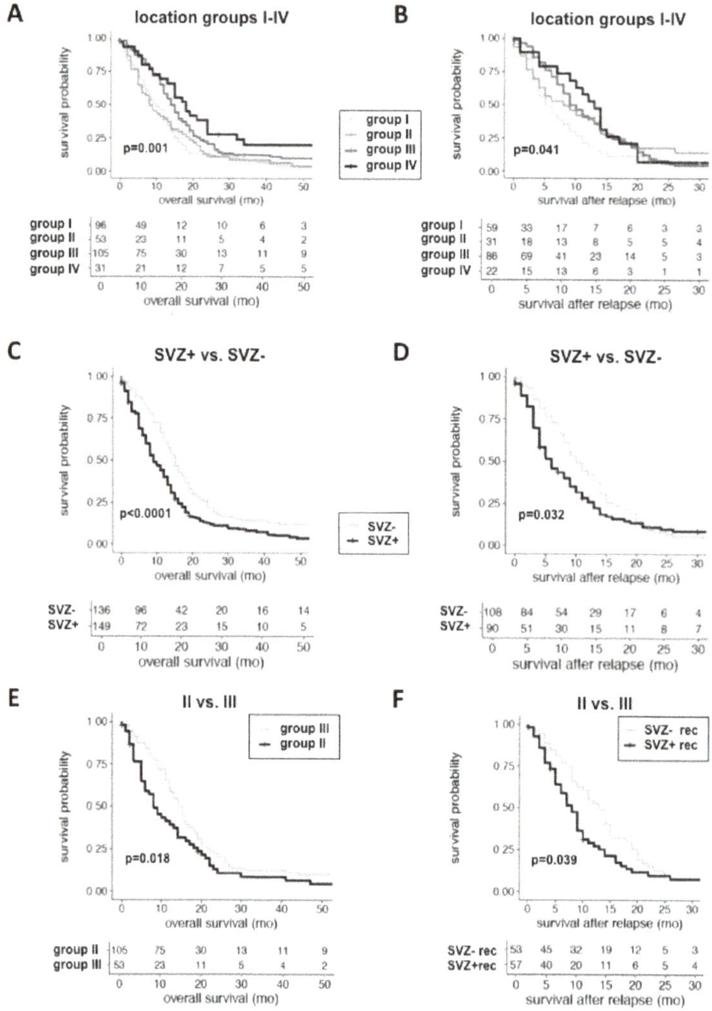

Figure 3. Kaplan-Meier plots depicting the negative prognostic impact of SVZ involvement in 285 IDH1-wt GBM patients; numbers at risk are given. OS and survival after relapse, but not PFS differed significantly when comparing all location groups (**A**,**B**), SVZ+ vs. SVZ− GBMs (**C**,**D**) and group II vs. group III GBMs ($n = 155$) (**E**,**F**). Shortest median OS and survival after relapse were observed in group I (10 and 5 months) and group II (8 and 6 months) patients. For comparison of survival after relapse in group II vs. group III GBMs, SVZ involvement at recurrence (SVZ+/− rec) was valued.

Table 2. Multivariate Analysis of Overall Survival (n = 253), Progression-free Survival (n = 253), and Survival after Relapse (n = 150) for the Complete IDH1-Wildtype Cohort.

Clinical and Radiographic Factors	p-Value	HR	95% CI
Overall Survival			
SVZ+ (1st diagnosis)	**0.008** **	1.434	1.099–1.872
Age (above median)	**0.036** *	1.343	1.02–1.77
KPS pre-operative	0.110	0.993	0.985–1.002
EOR: STR (1st diagnosis)	**<0.0001** ***	1.923	1.423–2.599
Intensified Treatment (1st diagnosis)	**<0.0001** ***	0.302	0.221–0.412
Multifocal disease CE (1st diagnosis)	**0.022** *	1.56	1.067–2.280
Progression-Free Survival			
SVZ+ (1st diagnosis)	0.529	0.918	0.703–1.199
EOR: STR (1st diagnosis)	**<0.0001** ***	1.811	1.348–2.433
Intensified Treatment (1st diagnosis)	**<0.0001** ***	0.431	0.322–0.579
Multifocal disease FLAIR (1st diagnosis)	**0.013** *	1.614	1.108–2.350
Survival after Relapse			
SVZ+ (at relapse)	**0.015** *	1.575	1.092–2.273
Cortex+ (at relapse)	**<0.001** ***	2.069	1.355–3.157
Treatment Intensity (at relapse)	**<0.001** ***	0.768	0.67–0.88
Multifocal disease FLAIR (at relapse)	0.097	1.538	0.925–2.558

HR: Hazard ratio; CI: confidence interval; significance levels: * $p < 0.05$; ** $p < 0.01$; *** $p < 0.001$. SVZ+ (1st diagnosis) [vs. SVZ−]; age (median splitted; 64 years); KPS pre-operative (numeric variable); EOR: STR (subtotal resection) [vs. GTR]; intensified treatment at 1st diagnosis (Stupp protocol including > 3 cycles TMZ); multifocal disease CE; multifocal disease FLAIR; SVZ+ (at relapse) [vs. SVZ−]; cortex+ (at relapse) [vs. cortex−]; treatment intensity at relapse (continuous variable); and multifocal disease FLAIR (at relapse).

Table 3. Multivariate Analysis of Overall Survival (n = 140), Progression-free Survival (n = 140) and Survival after Relapse (n = 59) for Group II versus Group III IDH1-Wildtype GBM.

Clinical and Radiographic Factors	p-Value	HR	95% CI
Overall Survival			
Location group II (1st diagnosis)	**0.007** *	1.725	1.164–2.557
Age (above median)	**<0.0001** ***	2.531	1.690–3.788
KPS pre-operative	0.139	1.009	0.997–1.021
EOR: STR (1st diagnosis)	**<0.001** ***	2.15	1.452–3.184
Intensified Treatment (1st diagnosis)	**<0.0001** ***	0.271	0.179–0.412
Progression-Free Survival			
Location group II (1st diagnosis)	0.432	0.848	0.563–1.278
EOR: STR (1st diagnosis)	**0.003** **	1.829	1.233–2.712
Intensified Treatment (1st diagnosis)	**<0.0001** ***	0.406	0.270–0.611
Multifocal disease FLAIR (1stdiagnosis)	0.104	1.519	0.918–2.514
Survival after Relapse			
Location group II (at relapse)	0.148	0.623	0.328–1.183
Treatment Intensity (at relapse)	**0.002** **	0.702	0.560–0.879
Multifocal disease FLAIR (at relapse)	**0.005** **	2.966	1.399–6.291

HR: Hazard ratio; CI: confidence interval; significance levels: * $p < 0.05$; ** $p < 0.01$; *** $p < 0.001$. Location group II (1st diagnosis) [vs. group III]; age (median splitted; 64 years); KPS pre-operative (numeric variable); EOR: STR (subtotal resection) [vs. GTR]; intensified treatment at 1st diagnosis (Stupp protocol including > 3 cycles TMZ); multifocal disease FLAIR; Location group II (at relapse) [vs. group III]; treatment intensity at relapse (continuous variable); and multifocal disease FLAIR (at relapse).

In conclusion, SVZ involvement both at 1st diagnosis and recurrence needs to be considered as an important prognostic factor for OS and survival after relapse in IDH1-wt GBM.

3. Discussion

GBM is characterized by high molecular heterogeneity affecting clinical and radiographic presentation, treatment response, and survival. The underlying mechanisms are still poorly understood. Therefore, we interrogated if tumor location, particularly proximity to the SVZ, contributes to heterogeneous growth, recurrence patterns and patient outcome [16,20]. It has consistently been shown that IDH1 mutations entail a molecularly and prognostically distinct GBM subtype [17,18]. Interestingly, in this study, IDH1-mut tumors were significantly enriched in SVZ− GBMs. This is in line with an immunohistochemical study that detected IDH1 mutations more frequently among group III GBMs, admitted that newly diagnosed and secondary GBMs were analyzed together [24]. Our finding contributes to the hypothesis of location-specific molecular signatures, but also stresses the need to stratify any location-dependent survival analysis for IDH1 mutation status. To eliminate this molecular bias, we only analyzed IDH1-wt GBM patients (n = 285) and found remarkable location-specific differences. First, growth and recurrence patterns were dependent on cortical involvement. Multifocal growth, both at first diagnosis (FLAIR lesions) and recurrence (CEL), was significantly enhanced in cortex+ and group III GBMs. Additionally, distant recurrences were observed more often in these tumors. While SVZ involvement was increased from newly diagnosed to recurrent tumors, it did not affect growth and recurrence patterns. In contrast, SVZ involvement at the first diagnosis and recurrence was found to be an independent prognostic factor for inferior OS (SVZ+ GBMs, group II GBMs) and survival after relapse (SVZ+ GBMs), while cortical involvement did not impact on survival.

Our finding that cortical involvement predicts tumor growth and recurrence resolves the ambiguity of previous studies. Lim et al. reported that group I tumors were most often multifocal at first diagnosis [20], whereas Kappadakunnel et al. found the highest rate of multifocal disease, both at first diagnosis and recurrence, in group III tumors [12]. This might account to small sample sizes but also to the fact that Kappadakunnel et al. [12] only analyzed CELs, while Lim et al. [20] considered, but did not discriminate for, CE and FLAIR lesions. In our study, multifocal growth was comparably high in group I and III GBMs (CE: 15% and 17%; FLAIR: 13% and 18%) in contrast to only 6% of group II and IV GBMs and supports the data by Kappadakunnel et al. [12]. Indeed, although Lim et al. concluded that multifocality is characteristic of SVZ involvement, they encountered multifocal growth more often in tumors with (group I: 56%, group III: 29%) than without (group II: 11%, group IV: 0%) cortical involvement. In this context, it is impossible to compare the finding by Adeberg et al. that multifocal progression was more common in SVZ+ GBMs since this study discriminated only by SVZ involvement, but not by location-specific groups [21]. Noteworthy, group I tumors extend from the SVZ to the cortex and therefore can be assigned both to SVZ+ and cortex+ GBMs although their true spatial origin remains unknown. In fact, a computerized simulation model of GBM growth suggested that a tumor invading both cortex and SVZ more likely originates from the subcortical white matter than from the SVZ and that SVZ involvement is rather a matter of increasing tumor size [23]. This is in line with our own findings that group I GBMs represented the majority of tumors at relapse as opposed to group III GBMs at first diagnosis and that SVZ involvement increased from first diagnosis to recurrence. To overcome this potential bias for our radiographic and survival analysis, "pure" cortical (group III) and "pure" SVZ (group II) tumors were compared separately, still analyzing a significant number of patients (n = 158). Consistent with the complete cohort, we found a significant increase of multifocal lesions in "pure" cortical GBMs. The incidence of multifocal lesions in GBM is reported to range between 10–15% [25], but can be as high as 35% [26,27], and is considered exceedingly aggressive with significantly worse outcome compared to unifocal lesions [25]. This is supported by our multivariate survival analysis in which multifocal disease at first diagnosis independently predicted shorter PFS and OS (complete cohort), while multifocal disease at tumor relapse predicted shorter survival after relapse (group II vs. III). There is still uncertainty whether multifocal lesions arise from the same precursor cell. Two recent publications applying extensive genomic analyses obtained conflicting results, either describing that multiple lesions are derived from different clones with heterogeneous drug response [28] or reporting that multiple lesions are of monoclonal origin and share an unexpected high frequency of genetic alterations in core regulatory pathways (RTK/PI3K, p53, RB), which may account for their highly invasive phenotype [29]. In this regard, our finding that multifocal growth is location-dependent may indicate that these tumors are derived from different location-specific cells of origin with varying migratory potential.

GBM is characterized by its inevitable recurrence. In more than 80 %, tumors recur adjacent to the initial resection site [30,31], as confirmed by our radiographic analysis. In 20% of all cases, tumors recurred both locally and remote from the initial resection site while exclusive distant tumor recurrence was infrequent (6%). However, our analysis adds a remarkable location-specific picture since distant tumor recurrence was more frequently encountered in GBMs with cortical involvement (cortex+ GBMs, group III GBMs). In this respect, previous location studies reported ambiguous findings: Two small-sized studies did not observe any location-dependent recurrence pattern [12,22], while others reported that distant recurrence was more common in GBMs with SVZ involvement, but did not discriminate between all four location-specific groups [21]. Thus far, O^6-methylguanin-DNA-methyltransferase (MGMT) promoter methylation [32], extensive resections [33], and large tumor volumes [34] were identified as risk factors for distant tumor recurrence. Our study identified cortical involvement (i.e., groups I (the largest tumors) and III (the highest frequency of GTR)) as another risk factor. From a molecular point of view, distant recurrence may reflect a highly invasive phenotype attributed to a distinct (possibly location-dependent) cell of origin, but experimental evidence is sparse. A recent longitudinal analysis found distant GBM recurrences to have a low rate of retention of the primary tumor's driver mutations, indicating a divergent, rather than a clonal, evolution [35]. This finding is critical for guiding targeted therapies at relapse and advocates repeat surgery, particularly in distant recurrences. For prevention of tumor relapse, it is worth elucidating molecular drivers of distant recurrences for which cortical GBMs, based on our findings, seem to be the ideal workhorse.

Despite its impact on tumor growth and recurrence, cortical involvement was not associated with survival. In contrast, SVZ involvement at first diagnosis and recurrence was predictive of inferior OS and survival after relapse. The negative prognostic impact of SVZ involvement has been described before [16], but our study provides robust survival data from a large cohort that strictly excluded confounding IDH1 mutations [17,18]. Moreover, we applied multivariate survival analysis in which SVZ involvement at first diagnosis was identified as an independent negative prognosticator of OS, and for the first time, SVZ involvement at tumor relapse as an independent negative prognosticator of survival after relapse. Moreover, to exclude a potential bias by group I GBMs that may arise from the cortex, the subcortical white matter or the SVZ, we separately analyzed "pure" SVZ (i.e., group II) and "pure" cortical (i.e., group III) tumors, a comparison that has been disregarded by previous outcome studies. Importantly, SVZ involvement at the first diagnosis remained an independent negative prognosticator of OS. Moreover, SVZ involvement at tumor recurrence conferred significantly shorter survival after relapse (SVZ+ rec: 8 months; SVZ− rec: 13 months; $p = 0.045$), but did not translate into an independent prognostic factor in multivariate analysis. The prognostic significance of SVZ involvement can be best illustrated by comparing our survival data to the most recent "historic control": In the control arm of a multicenter trial on tumor-treating fields in newly diagnosed GBM [36], patients receiving standard radio-chemotherapy experienced a median PFS and OS of 4 and 16 months, respectively. In our study, analyzing patients with radio-chemotherapy only ($n = 79$), SVZ involvement conferred comparable outcomes (PFS = 6 months; OS = 18 months), while median PFS and OS were meaningfully prolonged to nine and 24 months in SVZ− GBM patients (Figure 4A,B). It has been hypothesized that the poor prognosis associated with SVZ involvement results from an impaired clinical condition and distinct therapeutic challenges encountered with this specific tumor location. Indeed, well-known prognostic factors differed significantly between SVZ+

and SVZ− GBMs since pre-operative KPS, the rate of GTR at the first diagnosis and non-surgical treatment intensity at recurrence were significantly lower in SVZ+ GBMs. Noteworthy, PFS and OS were comparable in patients undergoing GTR regardless of SVZ involvement (SVZ+ GBM: PFS = 6.5 months; OS = 15.5 months; SVZ− GBM: PFS = 6 months; OS = 16.5 months; and Figure 4C,D), underlining the need to strive for maximum safe tumor resection even in tumors involving the lateral ventricles. Nevertheless, all these confounders were included into the multivariate model in which SVZ involvement was confirmed as a robust negative prognostic factor. Admittedly, MGMT promoter methylation status was not considered for multivariate analysis because of missing data in 41% of patients (Table 1) but OS was comparable for patients with (15 months) and without (14 months) methylated MGMT promoter ($p = 0.255$). Therefore, it is reasonable that SVZ involvement does not merely influence survival by the accumulation of negative clinical prognostic factors, but also by its inherent tumor biology. As preliminary evidence, we were recently able to identify molecular markers that were differentially expressed in SVZ+ GBM and conferred a prognostic impact [37].

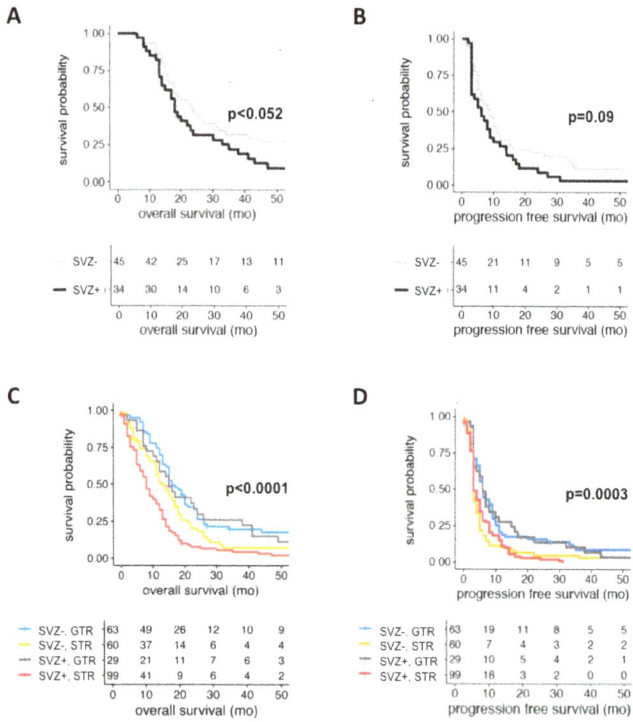

Figure 4. When analyzing IDH1-wt GBM patients with intensified adjuvant treatment (i.e., completion of concomitant radio-chemotherapy and ≥ 3 cycles of TMZ-based chemotherapy) only ($n = 79$), absence of radiographic SVZ involvement conferred a non-significant, but meaningful prolongation of OS (24 vs. 18 months; **A**) and PFS (9 vs. 6 months; **B**) that also compared favorable to the outcomes of the most recent "historic control" derived from a multicenter trial on tumor-treating fields in newly diagnosed GBM [36] (OS = 16 months; PFS = 4 months; data not shown). (**C**,**D**) Kaplan-Meier plots depicting the prognostic significance of SVZ involvement in 285 IDH1-wt GBM patients on OS (**C**) and PFS (**D**) when stratified for EOR. Noteworthy, the negative prognostic impact of SVZ involvement can be, in part, resolved by GTR since OS and PFS were comparable in patients undergoing GTR regardless of SVZ involvement (OS: SVZ+GBM 15.5 months vs. SVZ−GBM = 16.5 months; PFS: SVZ+GBM 6.5 months vs. SVZ−GBM 6 months). A tabular overview of numbers at risk is given below each Kaplan-Meier plot.

4. Materials and Methods

4.1. Patient Cohort

Our institutional database was searched retrospectively for all patients treated for newly diagnosed GBM at the Department of Neurosurgery (University Hospital Heidelberg, Germany) from 2004 to 2011 for whom demographic, treatment-related and outcome data were available and preoperative MRI was accessible on the Picture Archiving and Communication System (PACS). In accordance with the Declaration of Helsinki and the research proposals approved by the Institutional Review Board at Heidelberg Medical Faculty, informed consent was obtained in all cases. (Ethical code: S-005/2003, permission date: 31 March 2003). Histological diagnosis was confirmed by neuropathological review. IDH1 mutation and MGMT promoter methylation status was evaluated as described [2,38,39]. 302 consecutive patients were identified, 285 of those (94.4%) lacking IDH1 mutations. Except for patients with biopsies, EOR was determined for each patient on MRI scans taken within 72 hours post surgery and was valued "complete" (gross total resection = GTR) if no residual contrast enhancement was detected; otherwise, EOR was classified as "subtotal" or "unknown" if no postoperative MRI was available. Adjuvant treatments comprised radiotherapy, concomitant and/or stand-alone TMZ-based chemotherapy as well as treatment within clinical trials. Intensified adjuvant treatment ("Stupp regimen") was defined as completion of ≥ 3 cycles of TMZ after concomitant radio-chemotherapy. Definition of tumor progression/recurrence was based on the Response Assessment in Neuro-Oncology (RANO) criteria [40] with salvage therapies based on interdisciplinary decision. Salvage treatments comprised re-resection, re-irradiation, cytotoxic (TMZ, nitrosoureas, carboplatin), antiangiogenic (bevacizumab), or targeted therapies. Primary outcome measures were OS, PFS, and survival after relapse. OS was defined as the time from the first histologic diagnosis until death or last follow-up and PFS as the time from first histologic diagnosis to radiographic signs of progression/recurrence or death. Survival after relapse was defined as the time from radiographic signs of progression/recurrence until death. Patients still alive in June 2016 were censored. LTS were defined by an OS of > 36 months [2], while STS lived between six and 10 months after diagnosis.

4.2. Radiographic Analysis

For every patient, preoperative and follow-up MRI with standard sequences comprising T1-w, FLAIR and post contrast three-dimensional MPRAGE T1-w images in at least the axial plane with coronal and sagittal reformations were acquired on 1.5 or 3.0 Tesla scanners. As proposed [20], radiographic classification of GBMs according to their vicinity to the SVZ was performed on immediately preoperative CE T1-w MRI by two independent reviewers (CJ, BH). SVZ involvement was valued if the CEL contacted the lining of the ventricle. Group I consisted of tumors contacting the SVZ and infiltrating the cortex, group II of tumors contacting the SVZ only, group III of tumors contacting the cortex only and group IV of tumors residing in the subcortical white matter, neither contacting SVZ nor cortex (Figure 1A). Multifocal disease noncontiguous with the primary CEL was evaluated both on CE T1-w and FLAIR images at 1st diagnosis and at recurrence. At recurrence, distant tumor growth was defined as a new CEL remote from the initial resection cavity. Patients with infratentorial tumor location were precluded from analysis. For comparison of location-dependent outcome, growth and recurrence patterns, patients were further categorized into SVZ+ (groups I, II) versus SVZ− (groups III, IV) GBMs and cortex+ (groups I, III) versus cortex− (groups II, IV) GBMs (Figure 1A). Moreover, group II (involving the SVZ only) and group III (involving the cortex only) tumors were compared separately since this discriminates best between GBMs with and without SVZ involvement.

4.3. Statistical Analysis

GraphPad Prism version 6.0c was used for statistical analysis of clinico-pathological patient data as well as growth and recurrence patterns. Survival was analyzed in a Cox proportional hazards model in the *R* package "survival". Covariate inclusion was defined by stepwise forward selection conducted by the stepAIC algorithm in the *R* package "MASS" (covariates listed in Table S1). Only cases with all covariates available were considered for multivariate analysis. Significance threshold was set at $p < 0.05$.

5. Conclusions

In this large-scale analysis of IDH1-wt glioblastomas, tumor location was associated with a distinct growth and recurrence pattern and patient outcome, significantly contributing to the heterogeneous nature of this disease. While the underlying molecular factors still need to be identified, these findings may help to tailor location-dependent treatment strategies and allow for individualized prognostication based on factors easily assessable on pre-operative MRI. SVZ involvement, both at the first diagnosis and tumor relapse, proved to be a robust prognostic factor that should be incorporated in future outcome studies.

Supplementary Materials: The following are available online at http://www.mdpi.com/2072-6694/11/1/122/s1, Table S1: Covariates included into the multivariate model, Table S2: Univariate Analysis of Survival Endpoints in the Complete IDH1-Wildtype Cohort ($n = 285$), Table S3: Univariate Analysis of Survival Endpoints in Group II and Group III IDH1-Wildtype GBM ($n = 158$).

Author Contributions: Conceptualization, C.J.; methodology, C.J., R.W., A.M.; software, A.A., R.W., A.M.; validation, C.J., R.W., A.M.; formal analysis, C.J., R.W., A.M.; investigation, C.J., S.F., B.H., D.C., A.v.D.; resources, A.A., J.D., M.B., A.v.D., A.U., C.H.-M.; data curation, C.J., R.W., A.M., C.H.-M.; writing—original draft preparation, C.J.; writing—review and editing, R.W., A.M., S.F., B.H., D.C., A.A., J.D., M.B., A.U., C.H.-M..; visualization, C.J..; supervision, C.J., C.H.-M.; project administration, C.J., C.H.-M.; funding acquisition, C.H.-M.

Funding: This study has been funded by a grant from the Anni-Hofmann-Foundation.

Acknowledgments: We thank Mandy Barthel, Anja Metzner and Daniela Zito for review of patient data. Also, we acknowledge financial support by Deutsche Forschungsgemeinschaft within the funding programme Open Access Publishing, by the Baden-Württemberg Ministry of Science, Research and the Arts and by Ruprecht-Karls-Universität Heidelberg.

Conflicts of Interest: The authors declare no conflict of interest.

Abbreviations

GBM	glioblastoma
WHO	World Health Organization
TMZ	temozolomide
LTS	long-term survivor
NSC	neural stem cell
NPC	neural progenitor cell
SVZ	subventricular zone
PFS	progression-free survival
OS	overall survival
IDH1	isocitrate dehydrogenase 1
CEL	contrast-enhancing lesion
KPS	Karnofsky performance score
GTR	gross total resection
EOR	extent of resection
STS	short-term survivor
MGMT	O^6-methylguanin-DNA-methyltransferase
CE	contrast-enhancing
T1-w	T1-weighted
FLAIR	fluid-attenuated inversion recovery
RT	radiotherapy
HR	hazard ratio
CI	confidence intervall
PACS	Picture Archiving and Communication System
RANO	Response Assessment in Neuro-Oncology
MPRAGE	magnetization-prepared rapid gradient-echo

References

1. Stupp, R.; Mason, W.P.; Van Den Bent, M.J.; Weller, M.; Fisher, B.; Taphoorn, M.J.; Belanger, K.; Brandes, A.A.; Marosi, C.; Bogdahn, U.; et al. Radiotherapy plus concomitant and adjuvant temozolomide for glioblastoma. *N. Engl. J. Med.* **2005**, *352*, 987–996. [CrossRef] [PubMed]
2. Krex, D.; Klink, B.; Hartmann, C.; von Deimling, A.; Pietsch, T.; Simon, M.; Sabel, M.; Steinbach, J.P.; Heese, O.; Reifenberger, G.; et al. Long-term survival with glioblastoma multiforme. *Brain J. Neurol.* **2007**, *130*, 2596–2606. [CrossRef] [PubMed]
3. Sanai, N.; Tramontin, A.D.; Quinones-Hinojosa, A.; Barbaro, N.M.; Gupta, N.; Kunwar, S.; Lawton, M.T.; McDermott, M.W.; Parsa, A.T.; Verdugo, J.M.; et al. Unique astrocyte ribbon in adult human brain contains neural stem cells but lacks chain migration. *Nature* **2004**, *427*, 740–744. [CrossRef] [PubMed]
4. Sanai, N.; Alvarez-Buylla, A.; Berger, M.S. Neural stem cells and the origin of gliomas. *N. Engl. J. Med.* **2005**, *353*, 811–822. [CrossRef] [PubMed]
5. Llaguno, S.A.; Chen, J.; Kwon, C.H.; Jackson, E.L.; Li, Y.; Burns, D.K.; Alvarez-Buylla, A.; Parada, L.F. Malignant astrocytomas originate from neural stem/progenitor cells in a somatic tumor suppressor mouse model. *Cancer Cell* **2009**, *15*, 45–56. [CrossRef] [PubMed]
6. Kwon, C.H.; Zhao, D.; Chen, J.; Alcantara, S.; Li, Y.; Burns, D.K.; Mason, R.P.; Eva, Y.H.; Wu, H.; Parada, L.F. Pten haploinsufficiency accelerates formation of high-grade astrocytomas. *Cancer Res.* **2008**, *68*, 3286–3294. [CrossRef] [PubMed]
7. Zhu, Y.; Guignard, F.; Zhao, D.; Liu, L.; Burns, D.K.; Mason, R.P.; Messing, A.; Parada, L.F. Early inactivation of p53 tumor suppressor gene cooperating with NF1 loss induces malignant astrocytoma. *Cancer Cell* **2005**, *8*, 119–130. [CrossRef]
8. Holland, E.C.; Celestino, J.; Dai, C.; Schaefer, L.; Sawaya, R.E.; Fuller, G.N. Combined activation of Ras and Akt in neural progenitors induces glioblastoma formation in mice. *Nat. Genet.* **2000**, *25*, 55–57. [CrossRef]
9. Piccirillo, S.G.; Dietz, S.; Madhu, B.; Griffiths, J.; Price, S.J.; Collins, V.P.; Watts, C. Fluorescence-guided surgical sampling of glioblastoma identifies phenotypically distinct tumour-initiating cell populations in the tumour mass and margin. *Br. J. Cancer* **2012**, *107*, 462–468. [CrossRef]
10. Gupta, T.; Nair, V.; Jalali, R. Stem cell niche irradiation in glioblastoma: Providing a ray of hope? *CNS Oncol.* **2014**, *3*, 367–376. [CrossRef]
11. Chaichana, K.L.; McGirt, M.J.; Frazier, J.; Attenello, F.; Guerrero-Cazares, H.; Quinones-Hinojosa, A. Relationship of glioblastoma multiforme to the lateral ventricles predicts survival following tumor resection. *J. Neurooncol.* **2008**, *89*, 219–224. [CrossRef] [PubMed]
12. Kappadakunnel, M.; Eskin, A.; Dong, J.; Nelson, S.F.; Mischel, P.S.; Liau, L.M.; Ngheimphu, P.; Lai, A.; Cloughesy, T.F.; Goldin, J.; et al. Stem cell associated gene expression in glioblastoma multiforme: Relationship to survival and the subventricular zone. *J. Neurooncol.* **2010**, *96*, 359–367. [CrossRef] [PubMed]
13. Jafri, N.F.; Clarke, J.L.; Weinberg, V.; Barani, I.J.; Cha, S. Relationship of glioblastoma multiforme to the subventricular zone is associated with survival. *Neuro-Oncol.* **2013**, *15*, 91–96. [CrossRef] [PubMed]
14. Young, G.S.; Macklin, E.A.; Setayesh, K.; Lawson, J.D.; Wen, P.Y.; Norden, A.D.; Drappatz, J.; Kesari, S. Longitudinal MRI evidence for decreased survival among periventricular glioblastoma. *J. Neurooncol.* **2011**, *104*, 261–269. [CrossRef] [PubMed]
15. Adeberg, S.; Bostel, T.; König, L.; Welzel, T.; Debus, J.; Combs, S.E. A comparison of long-term survivors and short-term survivors with glioblastoma, subventricular zone involvement: A predictive factor for survival? *Radiat. Oncol. Lond. Engl.* **2014**, *9*, 95. [CrossRef] [PubMed]
16. Mistry, A.M.; Hale, A.T.; Chambless, L.B.; Weaver, K.D.; Thompson, R.C.; Ihrie, R.A. Influence of glioblastoma contact with the lateral ventricle on survival: A meta-analysis. *J. Neurooncol.* **2017**, *131*, 125–133. [CrossRef] [PubMed]
17. Noushmehr, H.; Weisenberger, D.J.; Diefes, K.; Phillips, H.S.; Pujara, K.; Berman, B.P.; Pan, F.; Pelloski, C.E.; Sulman, E.P.; Bhat, K.P.; et al. Identification of a CpG island methylator phenotype that defines a distinct subgroup of glioma. *Cancer Cell* **2010**, *17*, 510–522. [CrossRef]
18. Turcan, S.; Rohle, D.; Goenka, A.; Walsh, L.A.; Fang, F.; Yilmaz, E.; Campos, C.; Fabius, A.W.; Lu, C.; Ward, P.S.; et al. IDH1 mutation is sufficient to establish the glioma hypermethylator phenotype. *Nature* **2012**, *483*, 479–483. [CrossRef]

19. Louis, D.N.; Perry, A.; Reifenberger, G.; Von Deimling, A.; Figarella-Branger, D.; Cavenee, W.K.; Ohgaki, H.; Wiestler, O.D.; Kleihues, P.; Ellison, D.W. The 2016 World Health Organization Classification of Tumors of the Central Nervous System: A summary. *Acta Neuropathol.* **2016**, *131*, 803–820. [CrossRef]
20. Lim, D.A.; Cha, S.; Mayo, M.C.; Chen, M.H.; Keles, E.; VandenBerg, S.; Berger, M.S. Relationship of glioblastoma multiforme to neural stem cell regions predicts invasive and multifocal tumor phenotype. *Neuro-Oncol.* **2007**, *9*, 424–429. [CrossRef]
21. Adeberg, S.; König, L.; Bostel, T.; Harrabi, S.; Welzel, T.; Debus, J.; Combs, S.E. Glioblastoma recurrence patterns after radiation therapy with regard to the subventricular zone. *Int. J. Radiat. Oncol. Biol. Phys.* **2014**, *90*, 886–893. [CrossRef] [PubMed]
22. Kimura, M.; Lee, Y.; Miller, R.; Castillo, M. Glioblastoma multiforme: Relationship to subventricular zone and recurrence. *Neuroradiol. J.* **2013**, *26*, 542–547. [CrossRef] [PubMed]
23. Bohman, L.E.; Swanson, K.R.; Moore, J.L.; Rockne, R.; Mandigo, C.; Hankinson, T.; Assanah, M.; Canoll, P.; Bruce, J.N. Magnetic resonance imaging characteristics of glioblastoma multiforme: Implications for understanding glioma ontogeny. *Neurosurgery* **2010**, *67*, 1319–1327. [CrossRef] [PubMed]
24. Batista, K.M.; Vega, I.F.; de Eulate-Beramendi, S.A.; Morales, J.C.; Kurbanov, A.; Asnel, D.; Meilan, A.; Astudillo, A. Prognostic significance of the markers IDH1 and YKL40 related to the subventricular zone. *Folia Neuropathol.* **2015**, *53*, 52–59. [CrossRef]
25. Patil, C.G.; Yi, A.; Elramsisy, A.; Hu, J.; Mukherjee, D.; Irvin, D.K.; John, S.Y.; Bannykh, S.I.; Black, K.L.; Nuño, M. Prognosis of patients with multifocal glioblastoma: A case-control study. *J. Neurosurg.* **2012**, *117*, 705–711. [CrossRef]
26. Thomas, R.P.; Xu, L.W.; Lober, R.M.; Li, G.; Nagpal, S. The incidence and significance of multiple lesions in glioblastoma. *J. Neurooncol.* **2013**, *112*, 91–97. [CrossRef] [PubMed]
27. Lasocki, A.; Gaillard, F.; Tacey, M.; Drummond, K.; Stuckey, S. Multifocal and multicentric glioblastoma: Improved characterisation with FLAIR imaging and prognostic implications. *J. Clin. Neurosci. Off. J. Neurosurg. Soc. Australas.* **2016**, *31*, 92–98. [CrossRef]
28. Lee, J.K.; Wang, J.; Sa, J.K.; Ladewig, E.; Lee, H.O.; Lee, I.H.; Kang, H.J.; Rosenbloom, D.S.; Camara, P.G.; Liu, Z.; et al. Spatiotemporal genomic architecture informs precision oncology in glioblastoma. *Nat. Genet.* **2017**, *49*, 594–599. [CrossRef] [PubMed]
29. Abou-El-Ardat, K.; Seifert, M.; Becker, K.; Eisenreich, S.; Lehmann, M.; Hackmann, K.; Rump, A.; Meijer, G.; Carvalho, B.; Temme, A.; et al. Comprehensive molecular characterization of multifocal glioblastoma proves its monoclonal origin and reveals novel insights into clonal evolution and heterogeneity of glioblastomas. *Neuro-Oncol.* **2017**, *19*, 546–557. [CrossRef] [PubMed]
30. Chamberlain, M.C. Radiographic patterns of relapse in glioblastoma. *J. Neurooncol.* **2011**, *101*, 319–323. [CrossRef] [PubMed]
31. Petrecca, K.; Guiot, M.-C.; Panet-Raymond, V.; Souhami, L. Failure pattern following complete resection plus radiotherapy and temozolomide is at the resection margin in patients with glioblastoma. *J. Neurooncol.* **2013**, *111*, 19–23. [CrossRef] [PubMed]
32. Brandes, A.A.; Tosoni, A.; Franceschi, E.; Sotti, G.; Frezza, G.; Amista, P.; Morandi, L.; Spagnolli, F.; Ermani, M. Recurrence pattern after temozolomide concomitant with and adjuvant to radiotherapy in newly diagnosed patients with glioblastoma: Correlation with MGMT promoter methylation status. *J. Clin. Oncol. Off. J. Am. Soc. Clin. Oncol.* **2009**, *27*, 1275–1279. [CrossRef]
33. De Bonis, P.; Anile, C.; Pompucci, A.; Fiorentino, A.; Balducci, M.; Chiesa, S.; Lauriola, L.; Maira, G.; Mangiola, A. The influence of surgery on recurrence pattern of glioblastoma. *Clin. Neurol. Neurosurg.* **2013**, *115*, 37–43. [CrossRef]
34. Tejada, S.; Díez-Valle, R.; Aldave, G.; Marigil, M.; de Gallego, J.; Domínguez, P.D. Factors associated with a higher rate of distant failure after primary treatment for glioblastoma. *J. Neurooncol.* **2014**, *116*, 169–175. [CrossRef]
35. Kim, J.; Lee, I.H.; Cho, H.J.; Park, C.K.; Jung, Y.S.; Kim, Y.; Nam, S.H.; Kim, B.S.; Johnson, M.D.; Kong, D.S.; et al. Spatiotemporal Evolution of the Primary Glioblastoma Genome. *Cancer Cell* **2015**, *28*, 318–328. [CrossRef]
36. Stupp, R.; Taillibert, S.; Kanner, A.; Read, W.; Steinberg, D.M.; Lhermitte, B.; Toms, S.; Idbaih, A.; Ahluwalia, M.S.; Fink, K.; et al. Effect of Tumor-Treating Fields Plus Maintenance Temozolomide vs Maintenance Temozolomide Alone on Survival in Patients with Glioblastoma: A Randomized Clinical Trial. *JAMA* **2017**, *318*, 2306–2316. [CrossRef]

37. Jungk, C.; Mock, A.; Exner, J.; Geisenberger, C.; Warta, R.; Capper, D.; Abdollahi, A.; Friauf, S.; Lahrmann, B.; Grabe, N.; et al. Spatial transcriptome analysis reveals Notch pathway-associated prognostic markers in IDH1 wild-type glioblastoma involving the subventricular zone. *BMC Med.* **2016**, *14*, 170. [CrossRef] [PubMed]
38. Capper, D.; Weißert, S.; Balss, J.; Habel, A.; Meyer, J.; Jäger, D.; Ackermann, U.; Tessmer, C.; Korshunov, A.; Zentgraf, H.; et al. Characterization of R132H mutation-specific IDH1 antibody binding in brain tumors. *Brain Pathol.* **2010**, *20*, 245–254. [CrossRef]
39. Hartmann, C.; Meyer, J.; Balss, J.; Capper, D.; Mueller, W.; Christians, A.; Felsberg, J.; Wolter, M.; Mawrin, C.; Wick, W.; et al. Type and frequency of IDH1 and IDH2 mutations are related to astrocytic and oligodendroglial differentiation and age: A study of 1,010 diffuse gliomas. *Acta Neuropathol.* **2009**, *118*, 469–474. [CrossRef]
40. Wen, P.Y.; Macdonald, D.R.; Reardon, D.A.; Cloughesy, T.F.; Sorensen, A.G.; Galanis, E.; DeGroot, J.; Wick, W.; Gilbert, M.R.; Lassman, A.B.; et al. Updated response assessment criteria for high-grade gliomas: Response assessment in neuro-oncology working group. *J. Clin. Oncol.* **2010**, *28*, 1963–1972. [CrossRef] [PubMed]

cancers

MDPI

Article

Prognostic Gene Discovery in Glioblastoma Patients using Deep Learning

Kelvin K. Wong [1,2,3,*], Robert Rostomily [4] and Stephen T. C. Wong [1,3,5,6]

1 Department of Systems Medicine and Bioengineering, Houston Methodist, Houston, TX 77030, USA; stwong@houstonmethodist.org
2 Department of Neurological Surgery, Weill Cornell Medicine, New York, NY 10065, USA
3 Department of Radiology, Weill Cornell Medicine, New York, NY 10065, USA
4 Department of Neurosurgery, Houston Methodist Neurological Institute, Houston, TX 77030, USA; rrostomily@houstonmethodist.org
5 Department of Neuroscience, Weill Cornell Medicine, New York, NY 10065, USA
6 Department of Pathology and Laboratory Medicine, Weill Cornell Medicine, New York, NY 10065, USA
* Correspondence: KWong@houstonmethodist.org; Tel.: +1-713-441-8694

Received: 14 November 2018; Accepted: 24 December 2018; Published: 8 January 2019

Abstract: This study aims to discover genes with prognostic potential for glioblastoma (GBM) patients' survival in a patient group that has gone through standard of care treatments including surgeries and chemotherapies, using tumor gene expression at initial diagnosis before treatment. The Cancer Genome Atlas (TCGA) GBM gene expression data are used as inputs to build a deep multilayer perceptron network to predict patient survival risk using partial likelihood as loss function. Genes that are important to the model are identified by the input permutation method. Univariate and multivariate Cox survival models are used to assess the predictive value of deep learned features in addition to clinical, mutation, and methylation factors. The prediction performance of the deep learning method was compared to other machine learning methods including the ridge, adaptive Lasso, and elastic net Cox regression models. Twenty-seven deep-learned features are extracted through deep learning to predict overall survival. The top 10 ranked genes with the highest impact on these features are related to glioblastoma stem cells, stem cell niche environment, and treatment resistance mechanisms, including *POSTN*, *TNR*, *BCAN*, *GAD1*, *TMSB15B*, *SCG3*, *PLA2G2A*, *NNMT*, *CHI3L1* and *ELAVL4*.

Keywords: deep learning; discovery; glioblastoma; glioblastoma stem cells; survival prediction

1. Introduction

Deep learning [1–7] has been used to learn prognostic subtypes of glioblastoma using pan-cancer gene expression data from The Cancer Genome Atlas (TCGA) [8], predict drug synergy based on cancer cell gene expression data [9], and predict survival based on multi-omics integrated data in liver cancer [10] etc. [11–13]. It is important to explain the model in a meaningful way to understand the deep learning model and its limitations. A typical deep learning model involves millions of parameters, which makes it a difficult task to understand. We propose to use feature importance ranking within the deep learning model. While feature importance ranking is popular with machine learning [14], its use within the deep learning model is rare, especially in cancer genomics where a model usually includes thousands of features. In this paper, we expand the permutation feature importance techniques to deep learning. Our goal is to study the inside of a trained deep learning model to discover prognostic gene features in glioblastoma.

Conventional machine learning approaches have been used to determine the gene expression that are prognostic to glioblastoma patient survival [15–17]. Glioblastoma gene expression regression modeling using the least absolute shrinkage and selection operator (Lasso) flavor strategy performed better when cancer pathway genes were used as input variables compared to whole genome input [16]. However, these types of penalized regression methods often require dropping a large number of genes in order to fit the survival outcome, hindering biological pathway interpretation and introducing random bias toward the selected gene factors. Deep learning offers the capacity to model a large number of differentially expressed genes, is less susceptible to multicollinearity problem, and generalizes better. Deep learning based on transcriptome data has only recently been used to determine the primary effects of gene features that are prognostic to survival of glioblastoma (GBM) or other cancer types [8,18]. However, few studies work on what was learned in these deep learning models.

Using differentially expressed genes from the GBM-specific TCGA database as inputs, we tested the hypothesis that deep learning can model the relationship between specific genes and the corresponding protein effect to predict patient survival prognosis. Like most of the deep learning models, our model learned a set of features at the last hidden layer, which in this case linearly modulates the survival risk of patients. We hypothesize these features contains the key factors that determine patients' overall survival for those who have gone through standard of care therapy, from surgery to chemotherapy, without undergoing targeted therapy.

2. Results

2.1. Deep Learning Model

We trained and optimized the deep learning model to generate a network architecture consisting of an input layer feeding to two hidden layers (82 nodes then 27 nodes) and connected to one single output node to predict patient survival prognosis.

The validation concordance index is 0.69, the corresponding training concordance index is 0.73, and the validation concordance index of each sample partition is evaluated to be with a mean $\pm$ 1 SD concordance index of 0.70 $\pm$ 0.07. The out-of-sample testing concordance is 0.63 and is within the uncertainty of the validation concordance. The 95% confidence intervals of all testing, validation, and training concordance indexes do not include 0.5.

Important genes that contribute to the overall deep learned model were identified according to the input permutation method. A frequency analysis of genes occurring in the last hidden layer is listed in Table 1. The top 10 ranked genes are either known to be associated with glioblastoma survival, glioblastoma cancer cell migration, or glioblastoma cancer stem cells, or are known to be related to other types of cancer with mechanistic significance. The top 10 genes are *TNR*, *GAD1*, *TMSB15B*, *POSTN*, *SCG3*, *PLA2G2A*, *NNMT*, *CHI3L1*, and *ELAVL4*. The ranked gene importance at each network node is available in Supplementary Table S1. The entire frequency analysis table is available in Supplementary Table S2.

Table 1. Frequency analysis of important genes in the 27 deep-learned network nodes at the top hidden layer. Only the top 100 frequently occurring genes are listed for brevity.

Gene	Frequency	Gene	Frequency	Gene	Frequency	Gene	Frequency
TNR	17	*DPYSL4*	10	*MEG3*	9	*GRB10*	8
GAD1	16	*EGFR*	10	*NES*	9	*KDELR3*	8
TMSB15B	15	*F13A1*	10	*NPTX2*	9	*KIF1A*	8
POSTN	15	*FBN2*	10	*NRXN1*	9	*LSAMP*	8
SCG3	15	*NEFM*	10	*NTSR2*	9	*LYPD1*	8
PLA2G2A	14	*PTGDS*	10	*PEG3*	9	*MMP9*	8
NNMT	13	*RAB6B*	10	*PROM1*	9	*MYT1L*	8
CHI3L1	13	*RAPGEF4*	10	*SH3GL3*	9	*NMNAT2*	8
ELAVL4	13	*RUNDC3A*	10	*SOX11*	9	*NNAT*	8
TF	13	*SERPINA3*	10	*SPOCK1*	9	*NOL4*	8
UGT8	13	*SH3GL2*	10	*TMEM35*	9	*NSG1*	8
AQP1	12	*SNAP25*	10	*C4B*	8	*PLBD1*	8

Table 1. *Cont.*

Gene	Frequency	Gene	Frequency	Gene	Frequency	Gene	Frequency
COL6A3	12	*TCEAL2*	10	*SLC16A3*	8	*RGS1*	8
ERBB3	12	*TIMP4*	10	*SOD2*	8	*RGS17*	8
KCNQ2	12	*LOC101060835*	9	*AIM1*	8	*RGS4*	8
LTF	12	*ADAM22*	9	*ANXA1*	8	*RTN1*	8
MEOX2	12	*BCAN*	9	*APOD*	8	*S100A2*	8
PCDH9	12	*C1orf61*	9	*ATP2B2*	8	*SLC17A7*	8
STMN2	12	*DDX25*	9	*ATP6V1G2*	8	*SRD5A1*	8
FCGR2B	11	*ETNPPL*	9	*CFI*	8	*STC1*	8
FGFR3	11	*FAM107A*	9	*DSP*	8	*STEAP3*	8
SLC1A2	11	*GABRB1*	9	*ENPP2*	8	*STK32B*	8
CA10	10	*GDF15*	9	*FCGBP*	8	*TAC1*	8
CXCL14	10	*GNAO1*	9	*FUT9*	8	*VSNL1*	8
CXorf57	10	*LGI1*	9	*FZD6*	8	*WIF1*	8

In the deep-learned model, no proportional hazard assumption is used. Since the last hidden layer network node outputs are combined linearly with a weight vector to predict patient survival prognosis, the weight vector determines the relative risk from each of the 27 network nodes. Putting these network nodes in a Cox proportional hazard model showed a model concordance of 0.71, with 14 out of 27 network nodes statistically significant using Wald test at $p < 0.05$; only network node 10 is stratified for all network nodes to satisfy the proportional hazard assumption at $p < 0.05$ (Table 2).

Table 2. The prognostic value of network node outputs at the top hidden layer are evaluated using Cox proportional hazard model. The hazard ratios (HR) and 95% confidence intervals (95% CI) are listed with corresponding *p*-value. Thirteen network nodes are statistically significant in overall survival prognosis.

Cox Model with Deep Learning Features	HR (95% CI)	*p*-Value
Network Node 0	1.26 (0.98–1.62)	0.0718
Network Node 1	1.13 (0.94–1.36)	0.1996
Network Node 2	1.03 (0.81–1.32)	0.7931
Network Node 3	1.15 (0.94–1.41)	0.1681
Network Node 4	0.73 (0.59–0.89)	0.0022
Network Node 5	0.95 (0.77–1.16)	0.5935
Network Node 6	1.13 (0.88–1.44)	0.341
Network Node 7	1.19 (0.97–1.46)	0.0929
Network Node 8	1.71 (1.40–2.08)	<0.0001
Network Node 9	1.02 (0.81–1.29)	0.8505
Network Node 10		
≥1.6	0.45 (0.25–0.81)	0.0076
<1.6	1	
Network Node 11	0.80 (0.66–0.96)	0.0197
Network Node 12	1.36 (1.11–1.68)	0.0034
Network Node 13	0.93 (0.77–1.14)	0.4994
Network Node 14	1.12 (0.88–1.42)	0.3495
Network Node 15	0.86 (0.67–1.10)	0.2324
Network Node 16	0.57 (0.45–0.72)	<0.0001
Network Node 17	1.35 (1.09–1.67)	0.0056
Network Node 18	0.80 (0.64–1.00)	0.0478
Network Node 19	0.78 (0.64–0.95)	0.0132
Network Node 20	0.91 (0.73–1.15)	0.4437
Network Node 21	1.34 (1.10–1.62)	0.0035
Network Node 22	1.16 (0.95–1.42)	0.1575
Network Node 23	0.77 (0.62–0.97)	0.0281
Network Node 24	1.87 (1.54–2.27)	<0.0001
Network Node 25	1.41 (1.10–1.80)	0.0063
Network Node 26	0.80 (0.65–1.00)	0.0476
Overall Model		<0.0001

2.2. Deep Learning Model Performance Comparison with Penalized Cox Regression Models

Using the training dataset, the k-fold cross validation training resulted in the ridge, adaptive Lasso, and elastic net Cox regression models. These models perform consistently well with concordance indexes at 0.70, 0.69, and 0.70, respectively and are comparable to the deep learning model in performance. In the testing dataset, however, ridge, adaptive Lasso, and elastic net Cox regression models resulted in concordance indexes of 0.58, 0.56, and 0.56, with the 95% confidence intervals of the model concordance index including 0.5. In another word, these models performed close to random and failed to predict survival in the testing dataset, which could be due to the challenge of prediction in a small dataset with 49 patients. This is in sharp contrast to the deep learning model which still performs well in the small testing dataset.

2.3. Network Node Parameters Improved the Baseline Cox Proportional Hazard Model

A baseline Cox proportional hazard survival model constructed with clinical covariates, including age, gender, Karnofsky Performance Status (KPS), tumor subtypes, and therapy options performed at a concordance value of 0.68. Age, gender, KPS, chemoradiation, and proneural subtype all reached statistical significance using Wald test at $p < 0.05$, with no parameters violating the proportional hazard assumption at $p < 0.05$. Since most patients with the Glioma CpG island methylator phenotype (G-CIMP) mutation also harbor the *IDH1* heterozygous Arg132-to-His (R132H) point mutation, the multicollinearity between these two variables (correlation coefficient = 0.845) led to large standard error on the parameter estimation, affecting statistical significance.

After adding all deep-learned network nodes, the baseline Cox model (Table 3) concordance index increased from 0.68 to 0.76 with age, KPS, chemoradiation, and proneural subtype statistically significant (Wald test, $p < 0.05$); four risk-increasing network nodes (8, 17, 24, and 25) and four risk-decreasing nodes (4, 10, 16, and 23) were also statistically significant (Wald test, $p < 0.05$), with all covariates satisfying the proportional hazard assumption.

Table 3. Combined multivariate Cox proportional hazard model including clinical covariates and deep learning network node covariates to predict overall survival (p-value < 0.0001). The hazard ratios (HR) and 95% confidence intervals (95% CI) are listed with corresponding p-value.

Cox Model with Clinical Covariates and Deep Learning Features	HR (95% CI)	p-Value
Age		
≥54 years old	1.50 (1.10–2.03)	0.0098
<54 years old	1	
Gender		
Male	1.25 (0.92–1.68)	0.1542
Female	1	
KPS		
≥60	0.35 (0.17–0.72)	0.0042
<60	1	
Therapy		
Chemoradiation	0.27 (0.12–0.62)	0.0018
Chemotherapy	1.06 (0.35–3.17)	0.9193
Radiation	0.51 (0.22–1.17)	0.1122
Subtype		
Proneural	1.70 (1.01–2.87)	0.0464
Classical	1.26 (0.79–2.01)	0.3311
Mesenchymal	1.41 (0.89–2.25)	0.1462
MGMT Methylated	1.18 (0.85–1.62)	0.3181
G-CIMP Methylated	1.03 (0.35–3.06)	0.9553

Table 3. *Cont.*

Cox Model with Clinical Covariates and Deep Learning Features	HR (95% CI)	*p*-Value
R132C/R132G/R132H Mutation	1.08 (0.35–3.31)	0.8986
Network Node 0	1.09 (0.80–1.48)	0.5888
Network Node 1	1.10 (0.87–1.39)	0.4348
Network Node 2	1.15 (0.86–1.55)	0.3407
Network Node 3	1.11 (0.86–1.44)	0.4264
Network Node 4	0.77 (0.60–0.99)	0.0387
Network Node 5	0.85 (0.64–1.12)	0.2558
Network Node 6	1.12 (0.80–1.55)	0.514
Network Node 7	1.12 (0.86–1.44)	0.4041
Network Node 8	1.73 (1.36–2.21)	<0.0001
Network Node 9	1.07 (0.81–1.42)	0.6384
Network Node 10		
≥1.6	0.44 (0.20–0.95)	0.0363
Network Node 11	0.86 (0.67–1.10)	0.2336
Network Node 12	1.27 (0.99–1.65)	0.0645
Network Node 13	1.04 (0.82–1.32)	0.7484
Network Node 14	1.10 (0.82–1.48)	0.5313
Network Node 15	0.79 (0.57–1.11)	0.1711
Network Node 16	0.64 (0.48–0.86)	0.0029
Network Node 17	1.55 (1.16–2.07)	0.0027
Network Node 18	0.79 (0.60–1.05)	0.1049
Network Node 19	0.86 (0.68–1.08)	0.2001
Network Node 20	0.74 (0.54–1.00)	0.0528
Network Node 21		
≥1.6	0.96 (0.55–1.69)	0.8937
Network Node 22	1.22 (0.94–1.58)	0.1327
Network Node 23	0.75 (0.57–1.00)	0.048
Network Node 24	1.66 (1.30–2.12)	<0.0001
Network Node 25	1.49 (1.10–2.01)	0.0101
Network Node 26	0.83 (0.64–1.07)	0.1555
Overall Model		<0.0001

2.4. *Prognostic Significance Validation of Gene Set with External Data*

The chance of a gene deemed important to a small number of network nodes is much higher than it is to a large number of nodes, as shown in Figure 1. We used a probability threshold of $p < 0.01$ and selected a set of genes that are important in at least 10 out of 27 network nodes. These important genes form a 39-gene signature and they are listed in Table 4. Using this gene signature, statistical significance (log-rank test, $p < 1.5 \times 10^{-6}$) was achieved in seven glioblastoma studies and three low-grade glioma studies in separating the low-risk and high-risk groups with mean $\pm$ 1 SD concordance index of 0.79 $\pm$ 0.09. The Kaplan–Meier survival curves of the low-risk and high-risk groups in these studies are shown in Figure 2. There are no notable pathway enrichment in the 39-gene signature.

In the discovery process, two particular genes stand out with high hazard ratio (HR) and concordance index (CI) in univariate Cox model, *AQP1* (HR = 3.3, $p < 0.05$, CI = 0.711) and *MEOX2* (HR=2.5, $p < 0.05$, CI = 0.675), both of which are lesser known to be associated with survival in gliomas. Both genes are statistically significant and independently prognostic to survival in the multivariate Cox model (*MEOX2* HR = 2.2, *AQP1* HR = 1.55, both $p < 0.05$, CI = 0.796) in the TCGA GBMLGG dataset.

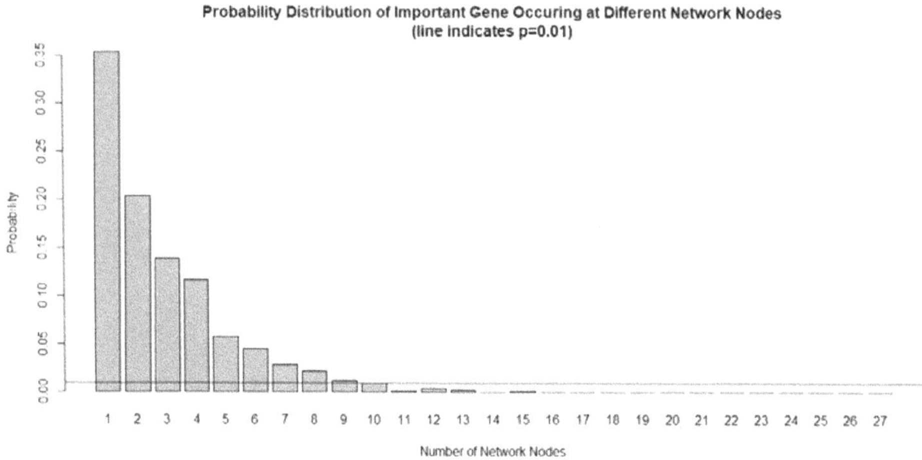

Figure 1. Probability distribution of important genes occurring at different network nodes. It is very rare for an important gene to occur in many nodes that are prognostic to survival. Using a threshold of $p < 0.01$, only genes that occurred at least 10 out of 27 network nodes meet the criteria and are included in the 39-gene signature.

Table 4. Gene list of the 39-gene signature selected based on $p < 0.01$ occurrence at the deep-learned network nodes at the top hidden layer.

39-Gene Signature			
TNR	*UGT8*	*FGFR3*	*PTGDS*
GAD1	*AQP1*	*SLC1A2*	*RAB6B*
TMSB15B	*COL6A3*	*CA10*	*RAPGEF4*
POSTN	*ERBB3*	*CXCL14*	*RUNDC3A*
SCG3	*KCNQ2*	*CXorf57*	*SERPINA3*
PLA2G2A	*LTF*	*DPYSL4*	*SH3GL2*
NNMT	*MEOX2*	*EGFR*	*SNAP25*
CHI3L1	*PCDH9*	*F13A1*	*TCEAL2*
ELAVL4	*STMN2*	*FBN2*	*TIMP4*
TF	*FCGR2B*	*NEFM*	

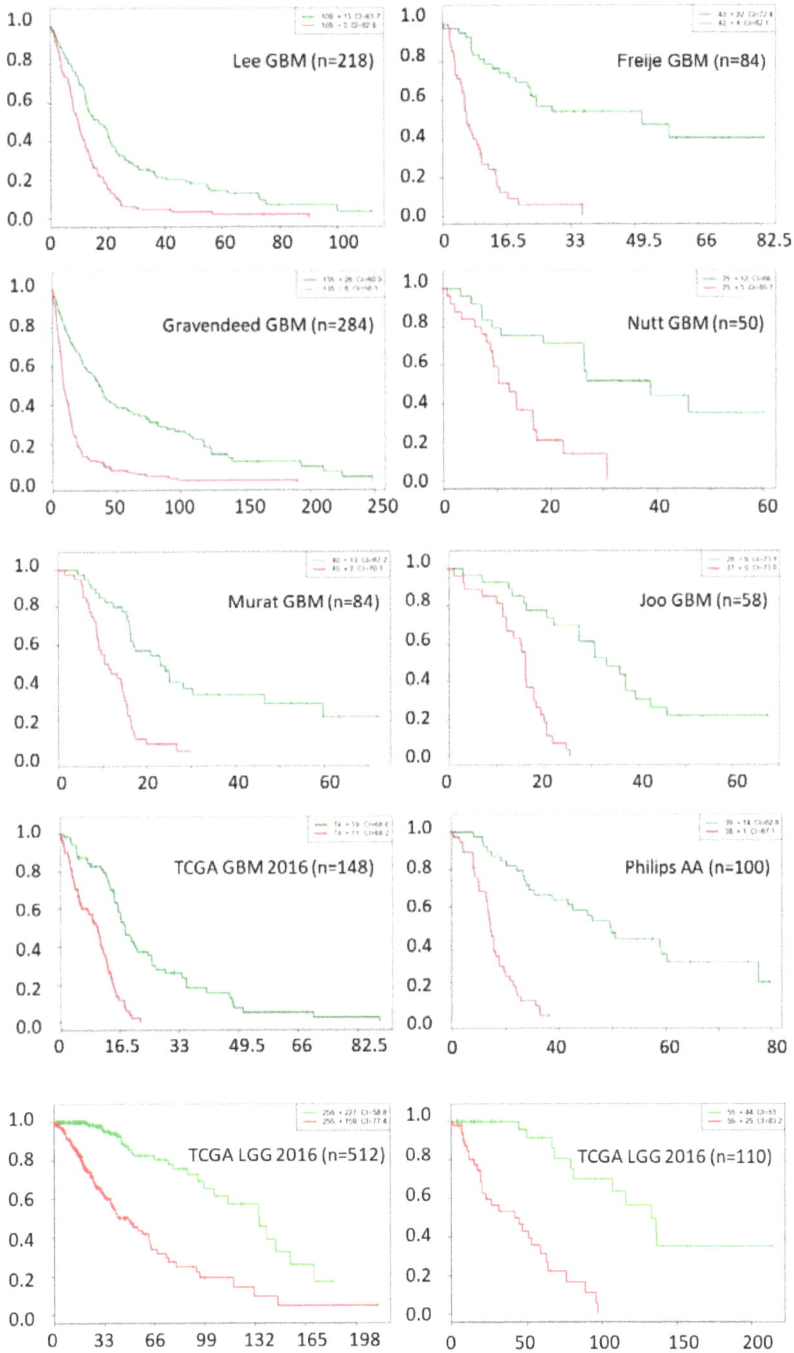

Figure 2. Kaplan–Meier survival fraction versus survival time (months) of the low-risk (green color) and high-risk (red color) groups are well separated using top 39 genes across nine different datasets, including data from seven glioblastoma and three low-grade glioma studies.

3. Discussion

Prior reports of deep learning model in cancer research [8,18,19] were derived from multiple cancer types to increase sample numbers, but fall short in identifying genes or primary mutations of mechanistic interest. A recent deep feedforward network studies used gene expression feature selection and outcome classification in TCGA breast cancer data as well as in TCGA kidney renal cell carcinoma [20]. In that study, only importance features to the model output were discovered and a number of biologically relevant features were found.

In this paper, we expand the input permutation method for feature importance ranking in deep learning network. Input permutation method is a useful technique for feature importance ranking in machine learning [14] and is broadly applicable to various models. It is usually used to rank feature importance at the model output. Expanding this method to rank features at any hidden layer within a deep learning model opened up many possibilities. It solidifies our understanding of the model and helps in explaining the deep learning model, which is usually considered a black box. In the proposed GBM model, we found that the last hidden layer contains important features that are far more biologically relevant than those obtained from the output layer.

Compared with gene-signature from previously identified GBM molecular subtypes (classical, proneural, and mesenchymal) [17,21,22], our 39-gene signature has a small overlaps with the 840-gene signatures in classical (8 out of 210), proneural (6 out of 210), and mesenchymal (5 out of 210) subtypes, or 12 genes overlap among all subtypes. The other 27 prognostic genes are likely due to shared biological mechanism(s) among tumor subtypes that are crucial to patient survival. The improvement in model concordance from 0.68 to 0.76 with the addition of deep learned network parameters confirmed the prognostic value of these additional genes in additional to known tumor subtype. It is also quite remarkable that the 39-gene signature learned from one GBM dataset is able to stratify patients in several other low grade glioma datasets as well as other GBM datasets. In addition, two lesser known genes, *AQP1* and *MEOX2*, are discovered to be prognostic to gliomas patients overall survival through the deep learning approach.

Our deep learning features revealed many genes of interest to glioblastoma stem cells mechanism. For example, both glutamate decarboxylase 1 (*GAD1*) and Chitinase 3 Like 1 (*CHI3L1/YKL-40*) have been recently identified as targets of Notch inhibitors (alpha secretase and gamma secretase inhibitors) in treating glioblastoma stem cells. Notch inhibitors work via Notch binding to YKL-40 and leukemia inhibitory factor (LIF) promotors and increased survival in a GBM stem cell orthotopic mouse model [23]. Epigenetic upregulation of glutamate decarboxylase 1 (*GAD1*) has been shown to program the aggressive features of cancer cell metabolism in brain metastatic microenvironment [24]. Chitinase 3 Like 1 (*CHI3L1/YKL-40*) was also reported previously to be prognostic to glioma patient survival [25] and involved in the angiogenesis, radioresistance, and progression of glioblastoma in vivo [26]. Periostin (*POSTN*) has been shown to impact GBM stem cell tumorigenicity and GBM patient survival [27]. *POSTN* is secreted by glioblastoma stem cells to recruit tumor-associated macrophages in order to promote malignant growth [28] and regulate tumor resistance to anti-angiogenic therapy [29]. Nicotinamide N-methyltransferase (*NNMT*) was also reported to regulate mesenchymal glioblastoma stem cell maintenance by depletion of methionine and shift tumor towards a mesenchymal phenotype and accelerated tumor growth [30]. *NNMT* was reported to be a prognostic marker for glioblastoma [31], inhibiting tumor suppressor protein phosphatase 2 (PP2A) at the epigenome and proteome level and concomitantly activates prosurvival serine/threonine kinases. Receptor tyrosine-protein kinase ErbB-3 (*ERBB3*) is known to mediate glioblastoma cancer stem-like cell resistance to EGFR inhibition [32].

Brevican (*BCAN*) which is known to bind to tenascin-R (*TNR*) with high affinity [33], is highly expressed in gliomas, initiating cells' extracellular niche in human GBM tumors and is expressed by glioma initiating cells in vitro [34]. Though *BCAN* knockdown does not affect glioblastoma initiating cell viability in vitro [34], it promotes glioma cell adhesion and migration in vitro [35] and the knockdown of the gene inhibits both cell motility in vitro and tumorigenicity in vivo [35]. Tenascin-C (*TNC*) and tenascin-R (*TNR*) are two of the three members of the tenascin family of extracellular matrix glycoproteins. TNC (which occurred in four networks) has been shown to promote glioblastoma invasion [36] and is heavily involved in pro-angiogenic and anti-angiogenic signaling in glioblastoma [37], as well as having an impact on survival [38]. A strong *TNR* expression is linked to non-invasive brain tumor (pilocytic astrocytomas) whereas a weak expression is detected in glioblastoma [39]. The exact role of *TNR*, particularly in glioblastoma stem cells extracellular niche, is a subject worth exploring.

On the other hand, ELAV-like RNA binding protein 4 (*ELAVL4*) has been shown to modulate radiation sensitivity in vitro in non-small cell lung cancer [40]. Secretogranin III (*SCG3*) has been shown to be involved in anti-angiogenesis in diabetic retinopathy [41]. Secreted phospholipase A2 group IIA (*PLA2G2A*) was shown to induce phosphorylation of the *EGFR* to induce proliferation through a PKC-dependent pathway in human astrocytoma in vitro [42]. Interestingly, the thymosin beta 15B (*TMSB15B*) is involved in epidermal growth factor-induced migration of prostate cancer cells [43].

Deep learning models that are fully connected and have high dimensional inputs are notoriously difficult to train due to their large number of variables. In our case, the number of variables, about 300,000, is much larger than the number of cases ($n = 492$). Our model is able to generalize on out-of-sample patient cases and has comparable performance in validation concordance index across different data sample splits. Due to the limitation of a single gene chip platform used in this study, its out-of-sample performance on another chip platform may need evaluation.

There are other limitations on the performance of deep learning based on differentially expressed genes. For instance, with 2-fold change cutoff used in this study, we may be artificially removing genes that are prognostic to survival. In addition, the relatively small number of patient samples in this study limits the depth of the model.

Finally, deep learning without explicit biological knowledge or network architecture constraints is not expected to learn biological structure within the data, so care must be taken with biological interpretation. In our case, we identify the prognostic genes by identifying genes with disproportional impact to last hidden layer and the occurrence frequency. Lastly, using deep learning to extract prognostic differential expressed genes for survival prediction provides a flexible way to combine gene expression data with other clinical covariates such as age, KPS, therapy options, and tumor subtypes to enable better patient survival stratification.

4. Materials and Methods

4.1. Gene Expression Data Analysis

The Cancer Genome Atlas (TCGA) is a publicly repository with patient-derived clinical, imaging, and genomic data that has been deidentified and contains no linkage to patient identifiers, no institutional review board or Health Insurance Portability and Accountability Act approval was required for our study. Microarray data from untreated glioblastoma patients ($n = 492$) were retrieved from TCGA. Gene expression level 1 data from the Affymetrix Human Genome U133A platform were used. The data were processed by software script using R (version 3.2, https://www.R-project.org/, Vienna, Austria), affy [44] and affycoretools [45] packages, and quality control was conducted using affyQCReport package. The probe level data were normalized by gcrma [46] package to control for batch variations, and the probe level expressions are compared to the normal brain tissue group ($n = 10$). Statistically significant gene probe changes were selected with a threshold of $p < 0.01$ with at least 2-fold biological change adjusted for multiple comparisons using the beta uniform mixture model [47]. Genes with a Pearson correlation coefficient higher than 0.8 were represented by one gene to reduce the strong intrinsic correlation. The number of gene probes was reduced to 3581 and used as model inputs.

Clinical data such as age, gender, *MGMT* methylation, G-CIMP, *IDH1/2* mutation, cancer subtype, and therapy information were retrieved from a TCGA publication [22].

4.2. Deep Learning Model

The deep learning model was built using Tensorflow 1.3 and Python 3.6 platforms using the deep learning survival modeling framework [48]. Gene expression dataset is randomly partitioned into ten equal partitions, with the first as the testing set, the second as the validation set, and the remaining as training sets. Stratified sampling was used to preserve the survival time distribution among each data partitions to fully capture the heterogeneity from multiple sites. The testing set contained samples from 11 sites whereas the validation set contained samples from 12 sites. The network structure consists of an input layer, one/two hidden layers with rectifier linear unit (Relu) functions, and an output layer with a single node corresponding to the survival prognosis of each patient. The partial likelihood function was used as a loss function and an L_2 penalty was applied on all network weights to prevent overfitting, [49] retaining the advantage of interpretation like Cox's proportional hazard. Batch-normalization [50] was used in each layer to improve learning stability.

Network structures, including number of hidden layers and number of hidden nodes, are varied to arrive at different models with a maximum of two hidden layers. Hyperparameter tuning on hidden layer(s), nodes and learning rate used concordance index as performance criteria in both the training and validation datasets. A maximum of 1000 epochs were allowed for computation convergence. The optimum hidden layers and nodes were determined by the maximum concordance index in the validation dataset; parameters were stored as a model for testing.

The gene expression dataset was randomly split to training, validation, and testing datasets with ratios of 80%, 10%, and 10%, respectively, while preserving the distribution of survival time to maximize the available datasets for training. The performance variability of the deep learning model was tested by rotating each sample partition as a validation dataset while using the rest for training. After the validation statistics were evaluated, its performance was tested on an out-of-sample data set that was never used in the modeling process.

4.3. Deep Learning Model Performance Comparison with Penalized Cox Regression Models

A comparison was made between deep learning model and penalized Cox regression models, including ridge, adaptive Lasso [51], and elastic net [52] using the glmnet package [53] (version 2.0.12).

Cox regression method assumes a semi-parametric hazard form of:

$$h_i(t) = h_0(t)e^{x_i^T\beta}$$

where $h_i(t)$ is the hazard for patient I at time t, $h_0(t)$ is the baseline hazard, and β is a fixed length vector of length p. In penalized Cox regression, models are fitted by maximizing the penalized partial log-likelihood function. The penalized partial log-likelihood function is given by:

$$\prod_{i=1}^{m} \frac{e^{x_i^T\beta}}{\sum_{j\in R_i} e^{x_i^T\beta}} - \sum_{j=1}^{p} p_{\alpha,\lambda}(|\beta_j|)$$

where $p_{\alpha,\lambda}(|\cdot|)$ is the penalty function with tuning parameters λ and α.

For ridge regression, the penalty function takes the form:

$$p_{\alpha,\lambda}(|\beta_j|) = \lambda\beta_j^2$$

For adaptive Lasso regression, the penalty function takes the form:

$$p_{\alpha,\lambda}(|\beta_j|) = \lambda w_j|\beta_j|$$

where $w_j = 1/\beta_{j0}$. β_{j0} the initial estimated of β_j, which in this case is estimated by ridge regression.

For elastic net, the penalty function takes the form:

$$p_{\alpha,\lambda}(|\beta_j|) = \lambda\left(\alpha|\beta_j| + (1-\alpha)\frac{1}{2}\beta_j^2\right)$$

where $\alpha \in (0,1]$.

Model performances were evaluated using concordance index and the same survival time stratified training/validation/testing datasets are used as in the deep learning model for a fair comparison. A 9-fold cross validation was used. A minimum lambda model was chosen as the lamda.1se models are not numerically stable. Feature importance of Lasso, adaptive Lasso, and elastic net methods was identified by the absolute amplitude of regression coefficients.

4.4. Impact of Deep Learning Network Features on Baseline Survival Model

The hidden nodes' outputs are Relu functions that are positive or zero, functioning like an on/off switch that allows effects from a previous level of interacting genes to pass through. Whether these network signatures provide distinct or complementary prognostic value to the baseline survival model was evaluated using Cox proportional hazard model.

A baseline Cox model was constructed from clinical covariates, tumor subtypes, therapy options and genetic mutation/methylation status known to be associated with patient survival. Clinical and genetic mutation/methylation covariates include age, Karnofsky performance scale, gender, *IDH1/2* mutation status, MGMT, G-CIMP methylation status, tumor subtypes (proneural, mesenchymal and classical) as well as chemotherapy/radiation/chemoradiation therapy options. To evaluate the additional prognostic value of deep learning network nodes, they were added to the baseline Cox proportional hazard model. The contribution of these deep learning features in improving survival prediction over the baseline model was evaluated using the concordance index.

4.5. Identifying Important Genes in Deep Learning Model

To identify the genes important to survival, we permuted the input genes one gene at a time to break the correlation between the input gene and the output risk [54]. An important gene that contributes significantly to the overall model when permuted across the patient group will impose a significant change to the predicted patient survival, whereas an unimportant gene will not. The process was repeated five times and the average change in patient risk was used. The high impact genes were identified as those that affect predicted patient survival outside the 95% confidence interval of the average change due to single gene permutation.

4.6. Prognostic Significance Validation of Gene Set with External Data

To compare the prognostic significance of our gene set in predicting survival in glioblastoma patients, we evaluated it in seven glioblastoma and three low-grade glioma studies (Table 5) using Cox proportional analysis with SurvExpress platform [55]. The samples were split by the median of the prognostic index to designate low-risk and high-risk groups. The top ranked 39 genes, which correspond to $p < 0.01$ or equivalently any gene occurring at least 10 times in the 27 network nodes, were chosen as gene biomarkers. The gene biomarkers were evaluated for survival difference between the low-risk and high-risk groups using log-rank test. The prognostic index is the linear component of the exponential function in the Cox model.

Table 5. List of glioblastoma studies used in survival prognosis validation using the gene set discovered by deep learning.

Study Datasets	Samples	Source
Lee Nelson Glioblastoma GSE13041 GPL96	218	Lee [56]
Freije Nelson Glioblastoma GSE4412 GPL96	85	Freije [57]
Gravendeed French Glioblastoma GSE16011	284	Gravendeel [58]
Nutt Louis Glioblastoma BROAD	50	Nutt [59]
Murat Hegi Glioblastoma GSE7696	84	Murat [60]
Joo Kim Jin Kim Seol Nam Glioblastoma GSE42669	58	Joo [61]
Philips Aldape Astrocytoma GSE4271 GPL96	100	Phillips [62]
Brain Low Grade Glioma TCGA 2016	110	TCGA
GBM-TCGA June 2016	148	TCGA
LGG-TCGA-Low Grade Gliomas June 2016	512	TCGA

5. Conclusions

In conclusion, we discovered that deep learning survival prediction model learned genes that are strongly related to glioblastoma stem cells and/or treatment resistant genes which may be useful to inform patient therapy. Compared with traditional Cox proportional hazard survival models, deep learning networks provide non-redundant prognostic covariates to patient survival even in the presence of strong clinical predictors. Using this approach, we identified many specific genes that are potential biomarkers or therapeutic targets.

Supplementary Materials: The following are available online at http://www.mdpi.com/2072-6694/11/1/53/s1, Table S1: Ranked gene importance in prognostic network nodes, Table S2: Frequency analysis of genes occuring in 27 prognostic network nodes.

Author Contributions: Conceptualization, K.K.W., S.T.C.W.; Methodology, K.K.W., S.T.C.W.; Formal analysis, K.K.W.; Writing—original draft preparation, K.K.W.; Writing—review and editing, S.T.C.W, R.R.; Supervision, K.K.W., S.T.C.W.; Project administration, K.K.W., S.T.C.W.; Funding acquisition, K.K.W., S.T.C.W.

Funding: This research was funded by Ting Tsung and Wei Fong Chao Foundation, John S Dunn Research Foundation, NIH U01 CA188388 and NIH R01 NS091251.

Acknowledgments: The results published here are in whole or part based upon data generated by the TCGA Research Network: http://cancergenome.nih.gov/. The authors sincerely acknowledge the helpful discussions with Nan Xiang and Hong Zhao, both from the Systems Medicine and Bioengineering Department, Houston Methodist Research Institute.

Conflicts of Interest: The authors declare no conflict of interest.

References

1. Ripley, B.D. *Pattern Recognition and Neural Networks*; Cambridge University Press: Cambridge, UK, 1996.
2. Bishop, C.M. *Neural Networks for Pattern Recognition*; Oxford University Press: Oxford, UK, 1995.
3. Cheng, B.; Titterington, D.M. Neural Networks: A Review from a Statistical Perspective. *Stat. Sci.* **1994**, *9*, 2–30. [CrossRef]
4. Kuan, C.M.; White, H. Artificial Neural Networks: An Econometric Perspective. *Econom. Rev.* **1994**, *13*, 1–91. [CrossRef]
5. Ripley, B.D. *Statistical Aspects of Neural Networks*; Chapman & Hall: Boca Raton, FL, USA, 1993.
6. Schmidhuber, J. Deep learning in neural networks: An overview. *Neural Netw.* **2015**, *61*, 85–117. [CrossRef]
7. Cherkassky, V.; Friedman, J.H.; Wechsler, H. *Statistics to Neural Networks: Theory and Pattern Recognition Applications*; Springer: Berlin, Germany, 1994.
8. Young, J.D.; Cai, C.; Lu, X. Unsupervised deep learning reveals prognostically relevant subtypes of glioblastoma. *BMC Bioinform.* **2017**, *18*, 381. [CrossRef]
9. Preuer, K.; Lewis, R.P.I.; Hochreiter, S.; Bender, A.; Bulusu, K.C.; Klambauer, G. DeepSynergy: Predicting anti-cancer drug synergy with Deep Learning. *Bioinformatics* **2017**. [CrossRef] [PubMed]

10. Chaudhary, K.; Poirion, O.B.; Lu, L.; Garmire, L.X. Deep Learning based multi-omics integration robustly predicts survival in liver cancer. *Clin. Cancer Res.* **2017**. [CrossRef] [PubMed]
11. Putin, E.; Mamoshina, P.; Aliper, A.; Korzinkin, M.; Moskalev, A.; Kolosov, A.; Ostrovskiy, A.; Cantor, C.; Vijg, J.; Zhavoronkov, A. Deep biomarkers of human aging: Application of deep neural networks to biomarker development. *Aging* **2016**, *8*, 1021–1033. [CrossRef] [PubMed]
12. Angermueller, C.; Lee, H.J.; Reik, W.; Stegle, O. DeepCpG: Accurate prediction of single-cell DNA methylation states using deep learning. *Genome Biol.* **2017**, *18*, 67. [CrossRef] [PubMed]
13. Tan, J.; Ung, M.; Cheng, C.; Greene, C.S. Unsupervised feature construction and knowledge extraction from genome-wide assays of breast cancer with denoising autoencoders. *Pac. Symp. Biocomput.* **2015**, 132–143. [CrossRef]
14. Breiman, L. Random Forests. *Mach. Learn.* **2001**, *45*, 5–32. [CrossRef]
15. Kim, Y.W.; Koul, D.; Kim, S.H.; Lucio-Eterovic, A.K.; Freire, P.R.; Yao, J.; Wang, J.; Almeida, J.S.; Aldape, K.; Yung, W.K. Identification of prognostic gene signatures of glioblastoma: A study based on TCGA data analysis. *Neuro-Oncology* **2013**, *15*, 829–839. [CrossRef] [PubMed]
16. Kim, H.; Bredel, M. Feature selection and survival modeling in The Cancer Genome Atlas. *Int. J. Nanomed.* **2013**, *8* (Suppl. 1), 57–62. [CrossRef]
17. Verhaak, R.G.; Hoadley, K.A.; Purdom, E.; Wang, V.; Qi, Y.; Wilkerson, M.D.; Miller, C.R.; Ding, L.; Golub, T.; Mesirov, J.P.; et al. Integrated genomic analysis identifies clinically relevant subtypes of glioblastoma characterized by abnormalities in PDGFRA, IDH1, EGFR, and NF1. *Cancer Cell* **2010**, *17*, 98–110. [CrossRef] [PubMed]
18. Yousefi, S.; Amrollahi, F.; Amgad, M.; Dong, C.; Lewis, J.E.; Song, C.; Gutman, D.A.; Halani, S.H.; Velazquez Vega, J.E.; Brat, D.J.; et al. Predicting clinical outcomes from large scale cancer genomic profiles with deep survival models. *Sci. Rep.* **2017**, *7*, 11707. [CrossRef] [PubMed]
19. Martinez-Ledesma, E.; Verhaak, R.G.; Trevino, V. Identification of a multi-cancer gene expression biomarker for cancer clinical outcomes using a network-based algorithm. *Sci. Rep.* **2015**, *5*, 11966. [CrossRef]
20. Kong, Y.; Yu, T. A graph-embedded deep feedforward network for disease outcome classification and feature selection using gene expression data. *Bioinformatics* **2018**, *34*, 3727–3737. [CrossRef]
21. Wang, Q.; Hu, B.; Hu, X.; Kim, H.; Squatrito, M.; Scarpace, L.; deCarvalho, A.C.; Lyu, S.; Li, P.; Li, Y.; et al. Tumor Evolution of Glioma-Intrinsic Gene Expression Subtypes Associates with Immunological Changes in the Microenvironment. *Cancer Cell* **2017**, *32*, 42–56. [CrossRef]
22. Brennan, C.W.; Verhaak, R.G.; McKenna, A.; Campos, B.; Noushmehr, H.; Salama, S.R.; Zheng, S.; Chakravarty, D.; Sanborn, J.Z.; Berman, S.H.; et al. The somatic genomic landscape of glioblastoma. *Cell* **2013**, *155*, 462–477. [CrossRef] [PubMed]
23. Floyd, D.H.; Kefas, B.; Seleverstov, O.; Mykhaylyk, O.; Dominguez, C.; Comeau, L.; Plank, C.; Purow, B. Alpha-secretase inhibition reduces human glioblastoma stem cell growth in vitro and in vivo by inhibiting Notch. *Neuro-Oncology* **2012**, *14*, 1215–1226. [CrossRef]
24. Schnepp, P.M.; Lee, D.D.; Guldner, I.H.; O'Tighearnaigh, T.K.; Howe, E.N.; Palakurthi, B.; Eckert, K.E.; Toni, T.A.; Ashfeld, B.L.; Zhang, S. GAD1 Upregulation Programs Aggressive Features of Cancer Cell Metabolism in the Brain Metastatic Microenvironment. *Cancer Res.* **2017**, *77*, 2844–2856. [CrossRef]
25. Steponaitis, G.; Skiriute, D.; Kazlauskas, A.; Golubickaite, I.; Stakaitis, R.; Tamasauskas, A.; Vaitkiene, P. High CHI3L1 expression is associated with glioma patient survival. *Diagn. Pathol.* **2016**, *11*, 42. [CrossRef] [PubMed]
26. Francescone, R.A.; Scully, S.; Faibish, M.; Taylor, S.L.; Oh, D.; Moral, L.; Yan, W.; Bentley, B.; Shao, R. Role of YKL-40 in the angiogenesis, radioresistance, and progression of glioblastoma. *J. Biol. Chem.* **2011**, *286*, 15332–15343. [CrossRef] [PubMed]
27. Mikheev, A.M.; Mikheeva, S.A.; Trister, A.D.; Tokita, M.J.; Emerson, S.N.; Parada, C.A.; Born, D.E.; Carnemolla, B.; Frankel, S.; Kim, D.H.; et al. Periostin is a novel therapeutic target that predicts and regulates glioma malignancy. *Neuro-Oncology* **2015**, *17*, 372–382. [CrossRef] [PubMed]
28. Zhou, W.; Ke, S.Q.; Huang, Z.; Flavahan, W.; Fang, X.; Paul, J.; Wu, L.; Sloan, A.E.; McLendon, R.E.; Li, X.; et al. Periostin secreted by glioblastoma stem cells recruits M2 tumour-associated macrophages and promotes malignant growth. *Nat. Cell Biol.* **2015**, *17*, 170–182. [CrossRef] [PubMed]

29. Park, S.Y.; Piao, Y.; Jeong, K.J.; Dong, J.; de Groot, J.F. Periostin (POSTN) Regulates Tumor Resistance to Antiangiogenic Therapy in Glioma Models. *Mol. Cancer Ther.* **2016**, *15*, 2187–2197. [CrossRef]
30. Jung, J.; Kim, L.J.Y.; Wang, X.; Wu, Q.; Sanvoranart, T.; Hubert, C.G.; Prager, B.C.; Wallace, L.C.; Jin, X.; Mack, S.C.; et al. Nicotinamide metabolism regulates glioblastoma stem cell maintenance. *JCI Insight* **2017**, *2*. [CrossRef]
31. Palanichamy, K.; Kanji, S.; Gordon, N.; Thirumoorthy, K.; Jacob, J.R.; Litzenberg, K.T.; Patel, D.; Chakravarti, A. NNMT Silencing Activates Tumor Suppressor PP2A, Inactivates Oncogenic STKs, and Inhibits Tumor Forming Ability. *Clin. Cancer Res.* **2017**, *23*, 2325–2334. [CrossRef]
32. Clark, P.A.; Iida, M.; Treisman, D.M.; Kalluri, H.; Ezhilan, S.; Zorniak, M.; Wheeler, D.L.; Kuo, J.S. Activation of multiple ERBB family receptors mediates glioblastoma cancer stem-like cell resistance to EGFR-targeted inhibition. *Neoplasia* **2012**, *14*, 420–428. [CrossRef]
33. Anlar, B.; Gunel-Ozcan, A. Tenascin-R: Role in the central nervous system. *Int. J. Biochem. Cell Biol.* **2012**, *44*, 1385–1389. [CrossRef]
34. Dwyer, C.A.; Bi, W.L.; Viapiano, M.S.; Matthews, R.T. Brevican knockdown reduces late-stage glioma tumor aggressiveness. *J. Neurooncol.* **2014**, *120*, 63–72. [CrossRef]
35. Lu, R.; Wu, C.; Guo, L.; Liu, Y.; Mo, W.; Wang, H.; Ding, J.; Wong, E.T.; Yu, M. The role of brevican in glioma: Promoting tumor cell motility in vitro and in vivo. *BMC Cancer* **2012**, *12*, 607. [CrossRef] [PubMed]
36. Xia, S.; Lal, B.; Tung, B.; Wang, S.; Goodwin, C.R.; Laterra, J. Tumor microenvironment tenascin-C promotes glioblastoma invasion and negatively regulates tumor proliferation. *Neuro-Oncology* **2016**, *18*, 507–517. [CrossRef] [PubMed]
37. Rupp, T.; Langlois, B.; Koczorowska, M.M.; Radwanska, A.; Sun, Z.; Hussenet, T.; Lefebvre, O.; Murdamoothoo, D.; Arnold, C.; Klein, A.; et al. Tenascin-C Orchestrates Glioblastoma Angiogenesis by Modulation of Pro- and Anti-angiogenic Signaling. *Cell Rep.* **2016**, *17*, 2607–2619. [CrossRef] [PubMed]
38. Midwood, K.S.; Hussenet, T.; Langlois, B.; Orend, G. Advances in tenascin-C biology. *Cell. Mol. Life Sci.* **2011**, *68*, 3175–3199. [CrossRef] [PubMed]
39. El Ayachi, I.; Baeza, N.; Fernandez, C.; Colin, C.; Scavarda, D.; Pesheva, P.; Figarella-Branger, D. KIAA0510, the 3′-untranslated region of the tenascin-R gene, and tenascin-R are overexpressed in pilocytic astrocytomas. *Neuropathol. Appl. Neurobiol.* **2010**, *36*, 399–410. [CrossRef] [PubMed]
40. Choi, K.J.; Lee, J.H.; Kim, K.S.; Kang, S.; Lee, Y.S.; Bae, S. Identification of ELAVL4 as a modulator of radiation sensitivity in A549 non-small cell lung cancer cells. *Oncol. Rep.* **2011**, *26*, 55–63. [CrossRef] [PubMed]
41. LeBlanc, M.E.; Wang, W.; Chen, X.; Caberoy, N.B.; Guo, F.; Shen, C.; Ji, Y.; Tian, H.; Wang, H.; Chen, R.; et al. Secretogranin III as a disease-associated ligand for antiangiogenic therapy of diabetic retinopathy. *J. Exp. Med.* **2017**, *214*, 1029–1047. [CrossRef]
42. Hernandez, M.; Martin, R.; Garcia-Cubillas, M.D.; Maeso-Hernandez, P.; Nieto, M.L. Secreted PLA2 induces proliferation in astrocytoma through the EGF receptor: Another inflammation-cancer link. *Neuro-Oncology* **2010**, *12*, 1014–1023. [CrossRef]
43. Banyard, J.; Barrows, C.; Zetter, B.R. Differential regulation of human thymosin beta 15 isoforms by transforming growth factor beta 1. *Genes Chromosomes Cancer* **2009**, *48*, 502–509. [CrossRef]
44. Gautier, L.; Cope, L.; Bolstad, B.M.; Irizarry, R.A. Analysis of Affymetrix GeneChip data at the probe level. *Bioinformatics* **2004**, *20*, 307–315. [CrossRef]
45. MacDonald, J.W. Affycoretools: Functions Useful for Those Doing Repetitive Analyses with Affymetrix GeneChips. Available online: https://bioconductor.org/packages/release/bioc/html/affycoretools.html (accessed on 23 September 2017).
46. Wu, Z.J.; Irizarry, R.A.; Gentleman, R.; Martinez-Murillo, F.; Spencer, F. A model-based background adjustment for oligonucleotide expression arrays. *J. Am. Stat. Assoc.* **2004**, *99*, 909–917. [CrossRef]
47. Pounds, S.; Morris, S.W. Estimating the occurrence of false positives and false negatives in microarray studies by approximating and partitioning the empirical distribution of p-values. *Bioinformatics* **2003**, *19*, 1236–1242. [CrossRef] [PubMed]
48. Hallam, A. TensorFlow-Survival-Analysis. Available online: https://github.com/alexhallam/TensorFlow-Survival-Analysis (accessed on 23 September 2017).
49. Faraggi, D.; Simon, R. A neural network model for survival data. *Stat. Med.* **1995**, *14*, 73–82. [CrossRef] [PubMed]

50. Ioffe, S.; Szegedy, C. Batch Normalization: Accelerating Deep Network Training by Reducing Internal Covariate Shift. In Proceedings of the International Conference on Machine Learning, Lille, France, 6–11 July 2015; pp. 448–456.
51. Wang, L.; Shen, J.; Thall, P.F. A Modified Adaptive Lasso for Identifying Interactions in the Cox Model with the Heredity Constraint. *Stat. Probab. Lett.* **2014**, *93*, 126–133. [CrossRef] [PubMed]
52. Suchting, R.; Hebert, E.T.; Ma, P.; Kendzor, D.E.; Businelle, M.S. Using Elastic Net Penalized Cox Proportional Hazards Regression to Identify Predictors of Imminent Smoking Lapse. *Nicotine Tob. Res.* **2017**. [CrossRef] [PubMed]
53. Simon, N.; Friedman, J.; Hastie, T.; Tibshirani, R. Regularization Paths for Cox's Proportional Hazards Model via Coordinate Descent. *J. Stat. Softw.* **2011**, *39*, 1–13. [CrossRef] [PubMed]
54. Guam, X.; Olden, J.D. A new R2-based metric to shed greater insight on variable importance in artificial neural networks. *Ecol. Model.* **2015**, *313*, 307–313. [CrossRef]
55. Aguirre-Gamboa, R.; Gomez-Rueda, H.; Martinez-Ledesma, E.; Martinez-Torteya, A.; Chacolla-Huaringa, R.; Rodriguez-Barrientos, A.; Tamez-Pena, J.G.; Trevino, V. SurvExpress: An online biomarker validation tool and database for cancer gene expression data using survival analysis. *PLoS ONE* **2013**, *8*, e74250. [CrossRef]
56. Lee, Y.; Scheck, A.C.; Cloughesy, T.F.; Lai, A.; Dong, J.; Farooqi, H.K.; Liau, L.M.; Horvath, S.; Mischel, P.S.; Nelson, S.F. Gene expression analysis of glioblastomas identifies the major molecular basis for the prognostic benefit of younger age. *BMC Med. Genom.* **2008**, *1*, 52. [CrossRef]
57. Freije, W.A.; Castro-Vargas, F.E.; Fang, Z.; Horvath, S.; Cloughesy, T.; Liau, L.M.; Mischel, P.S.; Nelson, S.F. Gene expression profiling of gliomas strongly predicts survival. *Cancer Res.* **2004**, *64*, 6503–6510. [CrossRef]
58. Gravendeel, L.A.; Kouwenhoven, M.C.; Gevaert, O.; de Rooi, J.J.; Stubbs, A.P.; Duijm, J.E.; Daemen, A.; Bleeker, F.E.; Bralten, L.B.; Kloosterhof, N.K.; et al. Intrinsic gene expression profiles of gliomas are a better predictor of survival than histology. *Cancer Res.* **2009**, *69*, 9065–9072. [CrossRef] [PubMed]
59. Nutt, C.L.; Mani, D.R.; Betensky, R.A.; Tamayo, P.; Cairncross, J.G.; Ladd, C.; Pohl, U.; Hartmann, C.; McLaughlin, M.E.; Batchelor, T.T.; et al. Gene expression-based classification of malignant gliomas correlates better with survival than histological classification. *Cancer Res.* **2003**, *63*, 1602–1607. [PubMed]
60. Murat, A.; Migliavacca, E.; Gorlia, T.; Lambiv, W.L.; Shay, T.; Hamou, M.F.; de Tribolet, N.; Regli, L.; Wick, W.; Kouwenhoven, M.C.; et al. Stem cell-related "self-renewal" signature and high epidermal growth factor receptor expression associated with resistance to concomitant chemoradiotherapy in glioblastoma. *J. Clin. Oncol.* **2008**, *26*, 3015–3024. [CrossRef] [PubMed]
61. Joo, K.M.; Kim, J.; Jin, J.; Kim, M.; Seol, H.J.; Muradov, J.; Yang, H.; Choi, Y.L.; Park, W.Y.; Kong, D.S.; et al. Patient-specific orthotopic glioblastoma xenograft models recapitulate the histopathology and biology of human glioblastomas in situ. *Cell Rep.* **2013**, *3*, 260–273. [CrossRef] [PubMed]
62. Phillips, H.S.; Kharbanda, S.; Chen, R.; Forrest, W.F.; Soriano, R.H.; Wu, T.D.; Misra, A.; Nigro, J.M.; Colman, H.; Soroceanu, L.; et al. Molecular subclasses of high-grade glioma predict prognosis, delineate a pattern of disease progression, and resemble stages in neurogenesis. *Cancer Cell* **2006**, *9*, 157–173. [CrossRef] [PubMed]

cancers

MDPI

Article

Mining-Guided Machine Learning Analyses Revealed the Latest Trends in Neuro-Oncology

Taijun Hana, Shota Tanaka *, Takahide Nejo, Satoshi Takahashi, Yosuke Kitagawa, Tsukasa Koike, Masashi Nomura, Shunsaku Takayanagi and Nobuhito Saito

Department of Neurosurgery, Graduate School of Medicine, The University of Tokyo, 7-3-1 Hongo, Bunkyo-ku, Tokyo 113-8655, Japan; thana-tky@umin.ac.jp (T.H.); tnejo-tky@umin.ac.jp (T.N.); takahashi-satoshi0719@g.ecc.u-tokyo.ac.jp (S.T.); yokitagawa-tky@umin.ac.jp (Y.K.); tkoike-ham@umin.ac.jp (T.K.); nomura-m@umin.ac.jp (M.N.); takayanagi-nsu@umin.ac.jp (S.T.); nsaito-tky@umin.net (N.S.)

* Correspondence: tanakas-tky@umin.ac.jp; Tel.: +81-3-5800-8853

Received: 25 December 2018; Accepted: 30 January 2019; Published: 3 February 2019

Abstract: In conducting medical research, a system which can objectively predict the future trends of the given research field is awaited. This study aims to establish a novel and versatile algorithm that predicts the latest trends in neuro-oncology. Seventy-nine neuro-oncological research fields were selected with computational sorting methods such as text-mining analyses. Thirty journals that represent the recent trends in neuro-oncology were also selected. As a novel concept, the annual impact (AI) of each year was calculated for each journal and field (number of articles published in the journal × impact factor of the journal). The AI index (AII) for the year was defined as the sum of the AIs of the 30 journals. The AII trends of the 79 fields from 2008 to 2017 were subjected to machine learning predicting analyses. The accuracy of the predictions was validated using actual past data. With this algorithm, the latest trends in neuro-oncology were predicted. As a result, the linear prediction model achieved relatively good accuracy. The predicted hottest fields in recent neuro-oncology included some interesting emerging fields such as microenvironment and anti-mitosis. This algorithm may be an effective and versatile tool for prediction of future trends in a particular medical field.

Keywords: impact factor; machine learning; neuro-oncology; regression analysis; trend prediction; text-mining

1. Introduction

In science, new research fields and methods are continuously evolving. The design and conduct of cutting-edge research mandate a keen perception of the current trends and future directions of the field. However, proper assessment of a large number of related literatures and information in a timely fashion would not necessarily be easy. A systematic review would be one solution; it prioritizes an overwhelmingly large number of research studies by their level of importance to retrieve the key information in a specific research field [1,2]. However, it will never be free from bias, given the nature of a study which involves manual data gathering. For that reason, a systematic review will not always be used for the decision-making process of researchers. Computer data mining is superior in coverage and objectivity for trend analysis in a specific research field. Attempts have already been made in various medical fields to predict future trends by machine learning analyses. In such studies, regression formulas are created from the obtained data to do so, but the prediction accuracy is yet to be satisfactory [3–5].

The field of neuro-oncology has undergone rapid progress in recent years. Treatment with tumor treating (TT) [6], glioma classification based on the tumor's genomic abnormalities [7], and epigenetic dysregulation during the process of tumor initiation and progression [8] are some of the emerging fields. For any physician and researcher, appreciation of 'hot topics' in neuro-oncology is key to successful, future collaborative work.

We developed a novel algorithm, named Mining Integrated Navigation and Estimation Research Via Articles (MINERVA), which makes use of text mining and machine learning of a large amount of data extracted from the literature (Figure 1). In this study, the MINERVA algorithm was used to predict 20 neuro-oncology fields that are expected to make great leaps in 2019.

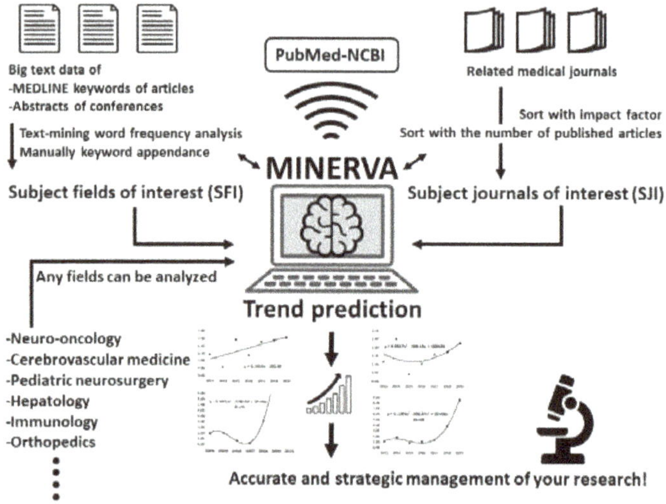

Figure 1. Conceptual diagram of Mining Integrated Navigation and Estimation Research Via Articles (MINERVA) algorithm. Based on the input data of subject fields of interest (SFI) and subject journals of interest (SJI), MINERVA will analyze the trends using PubMed database. By switching the input keywords, MINERVA can analyze the trends of any medical field.

2. Results

2.1. Data Collection for Analysis

2.1.1. Subject Fields of Interest

To analyze the trends of several research fields, the fields to be analyzed were first determined. Using the PubMed database (https://www.ncbi.nlm.nih.gov/pubmed/), all brain tumor-related articles published in 2017 were identified. Text-mining word frequency analysis was performed on text data of these articles and abstracts of neuro-oncological conferences. As a result, 79 keywords that seemed to represent recent trends in neuro-oncology were extracted as subject fields of interest (SFI) (Table 1).

2.1.2. Subject Journals of Interest

Thirty subject journals of interest (SJI) (Table 1) that can be representative of trends in neuro-oncology were determined. They included 15 highest impact factor (IF) journals among 1935 journals that published at least one neuro-oncological article in 2017 and 15 journals with the largest number of published articles in 2017. Impact factors were obtained from the 2017 Journal Citation Reports [9].

Table 1. Seventy-nine keywords of SFI and 30 journals of SJI.

SFI		SJI
1p/19q	lymphoma	*Cancer cell*
2-HG	medulloblastoma	*Cancer research*
acute myeloid leukemia	melanoma	*Cell*
anaplastic astrocytoma	meningioma	*Cell stem cell*
anaplastic oligodendroglioma	metabolism	*Child's nervous system*
angiogenesis and invasion	metastasis	*Immunity*
anti-mitotic	methylation	*International journal of radiation oncology, biology, physics*
ATRX	methyltransferase	*JAMA*
bevacizumab	MGMT	*Journal of clinical neuroscience*
biomarker	microenvironment	*Journal of clinical oncology*
BRAF	midline	*Journal of neuro-oncology*
castleman	MRI	*Journal of neurosurgery*
cell biology	neuro-imaging	*Lancet*
cell signal	neurotoxicity	*Lancet oncology*
chemotherapy	next generation sequencing	*Medicine*
cholangiocarcinoma	oligodendroglioma	*Molecular neurobiology*
complications	p53	*Nature*
craniopharyngioma	palliative care	*Nature genetics*
diffuse astrocytoma	PD-1	*Nature medicine*
diffuse midline glioma	pediatric	*Nature methods*
DIPG	pilocytic astrocytoma	*Nature reviews cancer*
drug resistance	PNET	*Neuro-oncology*
EGFR	progression	*Neurosurgery*
ependymoma	quality of life	*Oncology letters*
epidemiology	radio	*Oncotarget*
epigenetics	recurrent	*PloS one*
epithelioid	schwannoma	*Science*
genetics	single cell	*Scientific reports*
germ cell tumor	spinal	*The New England journal of medicine*
glioblastoma	STAT	*World neurosurgery*
glioneuronal	stem cell	30 journals of interest
H3K9	targeted therapy	
hemangioblastoma	temozolomide	
histone	TET2	
IDH	thalamic	
immunology	translational	
inhibitor	TT	
K27M	tumor models	
leptomeningeal	WHO	
low-grade glioma		
79 fields of interest		

Abbreviations: 2-HG = 2-hydroxyglutarate. ATRX = alpha-thalassemia/mental retardation syndrome, nondeletion type, x-linked. BRAF = B-Raf. DIPG = diffuse intrinsic pontine glioma. EGFR = epidermal growth factor receptor. IDH = isocitrate dehydrogenase. MGMT = O6-methylguanine DNA methyltransferase. MRI = magnetic resonance imaging. PD-1 = programmed death-1. PNET = primitive neuroectodermal tumor. STAT = signal transducer and activator of transcription. TET2 = ten-eleven translocation 2. TT = tumor treating. WHO = world health organization.

2.1.3. Annual Impact and Annual Impact Index

A system to objectively and sequentially evaluate the frequency and importance of SFI within the SJI is necessary. Therefore, the novel concepts of annual impact (AI) and annual impact index (AII) were employed in the current study. The AI of a certain subject field in a certain subject journal of a certain year was calculated according to the following formula:

$$\mathrm{AI} = \mathrm{N} \times \mathrm{IF},$$

where N is the number of articles related to the subject field published in the subject journal in the year. The AII value of a certain subject field of a certain year was calculated according to the following formula:

$$\mathrm{AII} = \mathrm{AI1} + \mathrm{AI2} + \mathrm{AI3} + \cdots + \mathrm{AI28} + \mathrm{AI29} + \mathrm{AI30},$$

where AI1 to AI30 are the AIs of each subject journal in that year. There were 30 journals in SJI, so numbers from 1 to 30 were assigned. In other words, the AII value of a certain subject field of a certain year is the sum of the AIs of each of the 30 journals for the year. Data of the number of articles that matched each condition were collected using PubMed. For example, the number of articles on "pilocytic astrocytoma" published in *Nature* in 2015 was obtained by entering the following command in the PubMed search window, and the number was 6.

((pilocytic astrocytoma) AND "Nature" [Journal]) AND ("2015" [Date-Publication]: "2015" [Date-Publication]).

Since the IF of *Nature* is 40.137, the AI of "pilocytic astrocytoma" in *Nature* in 2015 was calculated as follows.

$$\mathrm{AI} = \mathrm{N} \times \mathrm{IF} = 6 \times 40.137 = 240.822$$

Besides *Nature*, 25 journals contained articles on "pilocytic astrocytoma" in 2015. For example, there were two articles in the *Journal of the American Medical Association* (AI = 2 × 44.405 = 88.81) and one in *Science* (AI = 1 × 37.205 = 37.205). Thus, the AII of "pilocytic astrocytoma" in 2015 was 4068.629, as calculated by aggregating the 25 AIs according to the above mathematical rule.

2.2. Trend Predictions of Neuro-Oncology

2.2.1. Mathematical Analyses

A total of 79 SFI fields were analyzed using a comparison of the rate of change of AII (Δ-AII). In other words, Δ-AII is one year's AII change rate compared with the previous year. The Δ-AIIs of all 79 SFI fields were tracked for 10 years from 2008 to 2017. Three types of mathematical regression analyses based on the least-square method (linear prediction, quadratic polynomial, and cubic polynomial) were performed to predict trends in neuro-oncology in 2019. Regarding the prediction of the top 20 subject fields, we tried to predict the rankings for 1–3 years after collection of the data and also examined the accuracy rate of each.

2.2.2. Accuracy of the Predictions by Multiple Regression Analyses

Regression analytic graphs are illustrated in Figure 2a, with "single cell" and "IDH" as examples. Predictions did not necessarily match between linear prediction, quadratic polynomial and cubic polynomial methods. Relatively accurate prediction was possible with the linear prediction method. With an accuracy of 70.6%, we were able to predict the Δ-AII of the following year of the data collection range within an error of 1.0-fold (Table 2). This means that the latest trends of about 56 out of 79 fields can be predicted with high accuracy. In addition, we were able to predict the top 20 fields with a high Δ-AII of 2 years after the data collection period with an accuracy of 38.3% (binomial test, $p = 0.018$), using linear prediction. Figure 3a shows the details of prediction. This method can constantly achieve a high-match rate of predictions. Interestingly, this method predicted a same field as highest expected field every year, "PD-1". Although "PD-1" had never actually been ranked as highest, it had ranked in top 20 fields every year. "Cholangiocarcinoma", which seems to have little relation to brain tumor, had been ranked sometimes. However, cholangiocarcinoma is known to have isocitrate dehydrogenase (IDH) mutation, and it had been frequently discussed with gliomas. On the other hand, the accuracy of quadratic polynomial with the same condition was 30.0% (binomial test, $p = 0.241$) (Figure 3b). This method achieved a high match rate in particular years, but as a whole was not such an accurate method. The results obtained with other analytical methods are shown in Table 2. Regarding the accuracy of B, C, and D, given that the coincidence match rate was 25.3% (20/79 = 0.253), most analytical methods tended to have higher predictive accuracy than by sheer coincidence, and some methods had statistical significance. Accuracy D of linear prediction was not statistically significant, but the p-value was 0.06. As a result of comprehensive judgment of these accuracy data, we concluded that linear prediction can be the most reliable prediction method.

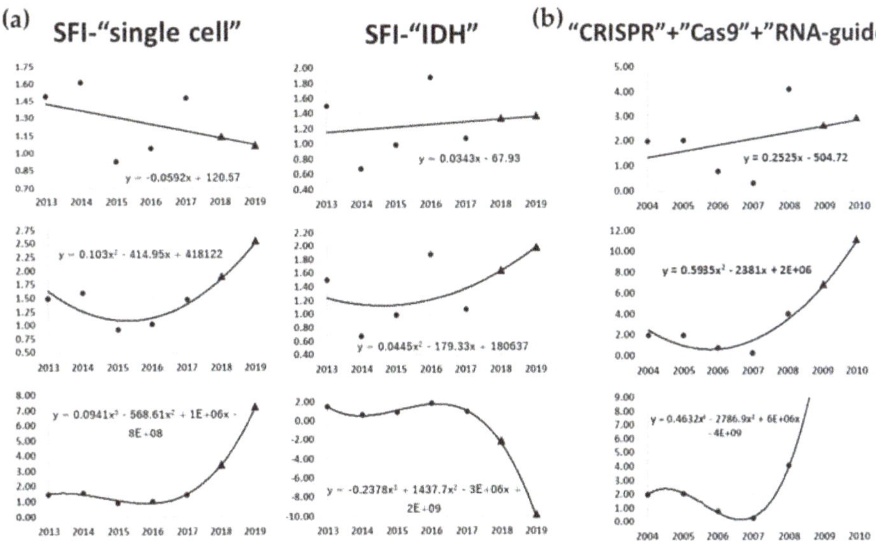

Figure 2. (**a**) Regression analytic graphs are exemplified. Left side shows the result of "single cell", and the right shows "IDH". The vertical axis is the fold value of Δ-AII. The upper row is the result of linear prediction, the middle is quadratic polynomial and the lower is cubic polynomial. The circles are actual measured values and the triangles are predicted values. (**b**) Regression analytic predictions of "CRISPR", "Cas 9" and "RNA-guided". The vertical axis is the fold value of Δ-AII. The analyzed AII is the sum of these three keywords' AII. Other arrangements are the same as Figure 2a.

Table 2. Accuracy of the predictions by multiple regression analyses.

Regression Analyses	Accuracy A	Accuracy B	Accuracy C	Accuracy D
Linear prediction	70.6%	25.0% *	38.3% **	37.5% *
Quadratic polynomial	54.4%	15.0% *	30.0% *	45.0% ****
Cubic polynomial	34.8%	28.8% *	36.7% ***	20.0% *

Accuracy A: The accuracy of prediction of the Δ-AII of the following year of data collection range within an error of 1.0-fold. B: The accuracy of prediction of the top 20 high Δ-AII fields of the following year of data collection range. C: The accuracy of prediction of the top 20 high Δ-AII fields of 2 years after data collection range. D: The accuracy of prediction of the top 20 high Δ-AII fields of 3 years after data collection range. Binominal test: * $p > 0.05$, ** $p = 0.018$, *** $p = 0.034$, **** $p = 0.005$.

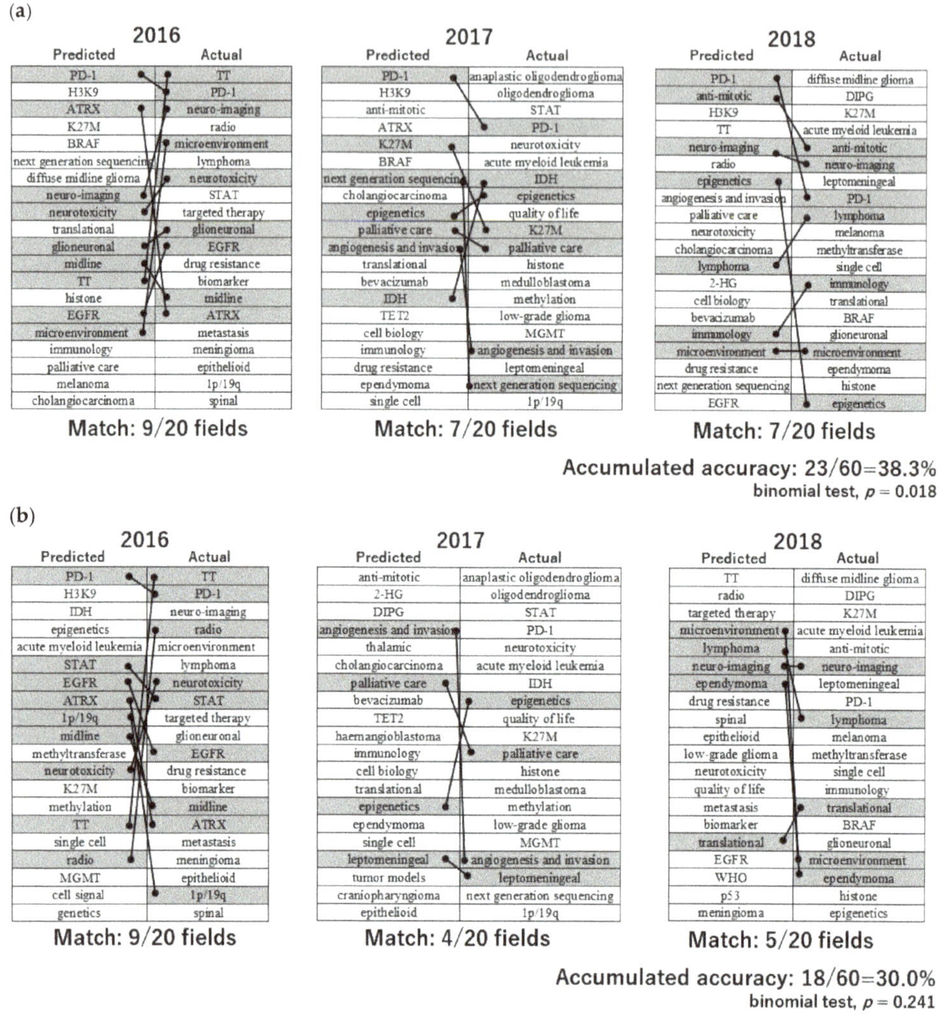

Figure 3. (**a**) The details of prediction with linear prediction using 5 years. The AII dataset to predict the top 20 fields with high Δ-AII of 2 years after the data collection period. Predicted top 20 fields with high Δ-AII are enumerated in left columns in order of expected Δ-AII scores. Right columns are the year's actual top 20 fields with high Δ-AII in order of actual Δ-AII scores. Fields matching the prediction are connected by lines. The accumulated accuracy means Accuracy C of linear prediction of Table 2. (**b**) The details of prediction with quadratic polynomial. Other analytical conditions are the same as Figure 3a. Fields matching the prediction are connected by lines. The accumulated accuracy means accuracy C of quadratic polynomial of Table 2.

2.2.3. The Top 20 Hottest Fields of Neuro-Oncology in 2019

Based on the policy above, the top 20 hottest fields of neuro-oncology in 2019 were predicted using the linear prediction method and using the data of Δ-AII for the period of 2013–2017 (Table 3). The predicted rank was ordered by predicted Δ-AII scores. In the predicted fields, there are many emerging fields such as "microenvironment" and "anti-mitotic", as well as some important fields already widely recognized, such as "immunology" and "epigenetics". "Epigenetics", "microenvironment", "neurotoxicity", "palliative care", "anti-mitotic", "angiogenesis and invasion", "EGFR", "lymphoma", "TT" and "translational" had been repeatedly ranked in the top 20 fields in 2016–2018 with this method, in both predicted and actual fields (Figure 3a). Those four fields at the beginning, "epigenetics", "microenvironment", "neurotoxicity", and "palliative care", appeared a total of four times, which was the most frequent. This may reinforce the hopeful prospectivity of those fields in another form different from the predicted rank of 2019.

Table 3. Top 20 hottest fields of neuro-oncology in 2019.

Predicted Rank	Fields
1	anti-mitotic
2	anaplastic oligodendroglioma
3	oligodendroglioma
4	TT
5	STAT
6	neurotoxicity
7	angiogenesis and invasion
8	radio
9	translational
10	cell biology
11	quality of life
12	palliative care
13	immunology
14	low-grade glioma
15	microenvironment
16	epigenetics
17	WHO
18	lymphoma
19	EGFR
20	biomarker

2.2.4. Prediction of Milestone Discoveries in Other Fields

The current study focused on neuro-oncology, but the MINERVA algorithm can potentially be used to predict trends in any field by changing the input keywords. Meanwhile, the discovery of the CRISPR-Cas9 gene editing system in 2012 was a recent milestone in the field of medical biology [10]. We retrospectively assessed the ability of the MINERVA algorithm to predict its discovery in advance with the use of only three keywords for SFI: "CRISPR" "Cas 9" and "RNA-guided". Then the following 15 journals dealing with cell engineering and genetics, or general science with the highest IFs were selected for SJI: *Nature Reviews—Molecular Cell Biology*, *Nature Biotechnology*, *Nature Reviews—Genetics*, *Nature*, *Science*, *Cell*, *Nature Genetics*, *Nature Reviews—Microbiology*, *Cell—Stem Cell*, *Nature Cell Biology*, *Cell Research*, *Trends in Cell Biology*, *Cell Host & Microbe*, *Annual Review of Cell and Developmental Biology*, and *Molecular Cell*. The AII data since 2000 were collected. As a result, a marked increase in the sum of the three keywords' Δ-AII was predicted in 2008 by all three analytic models (Figure 2b). This means the possibility that an important discovery in the fields related to "CRISPR", "Cas 9" and "RNA-guided" after 2009 could have been predicted as early as 2008. Individual keywords "CRISPR", "Cas 9" and "RNA-guided" had been broadly used in the fields of archaea and genetics. However, by analyzing these three keywords together, it can be related to the trend of the CRISPR-Cas9 gene editing system. This prediction result indicates the good versatility of the MINERVA algorithm.

3. Discussion

In clinical and basic research, it is important to grasp the current status of the research field to predict future trends. This type of evaluation is often performed by the individual's impression based on notable journals, conferences, and lectures. Scientific reporters have many similar personal impressions and tend to have subjective bias in their evaluations. The current study suggested that the MINERVA algorithm can be used to grasp current trends and predict at least some future trends in a specific research field. Notably, the MINERVA algorithm is characterized by its wide applicability; it can analyze any medical field simply by changing the selected keywords of SFI and SJI, whether it is another central nervous system disorder such as cerebrovascular disease or basic science, exemplified by our results regarding the CRISPR-Cas9 gene editing system.

Among the 20 fields predicted by the MINERVA algorithm, there were several fields of particular interest. Regarding the "anti-mitotic" field, the AII value was 25.353-fold greater than the value of the previous year in 2014. Likewise, the AII value of "TT" in 2015 also increased by 5.412-fold, so it would be reasonable to understand that the "anti-mitotic" field was paid attention to as the mechanism of the TT field [6]. Actually, an article about NovoTTF (Novocure, St. Heliar, Jersey) was responsible for the rapid increase in the "anti-mitotic" field in 2014 [11]. "Immunology" and "microenvironment" were ranked. Recently, many high-IF articles related to microenvironment and glioma have been published [12–15]. Tumor microenvironment is a concept related to micro cellular signals, receptors, structures, angiogenesis, molecules, etc. In particular, immunosurveillance systems are deeply related to tumor microenvironment, and many articles have recently been published which focus on immunology and microenvironment [13,15,16]. Considering that the word "microenvironment" appeared 146 times in the official abstracts of the Annual Meeting of the Society for Neuro-Oncology (SNO-2017, San Francisco, CA, USA) [17], we can see how this field has recently gained attention of neuro-oncologists. With respect to the field "epigenetics," the discovery of mutation to the IDH1 and IDH2 genes in gliomas [18] has determined the prosperity of subsequent glioma and epigenetic studies. Even now, after 10 years have passed, important articles on gliomas and methylation as well as other epigenetic systems continue to be published [19–21]. It strongly suggests the importance of the field. The field "lymphoma" was a relatively unexpected result because there were relatively few titles in SNO-2017 pertaining to this subject. However, in 2015, seven articles including research on new drugs were published in high-IF journals [22–28], which seems to have contributed to its popularity.

As regards limitations to the current study, even though multiple regression analyses were performed, the prediction accuracy was not perfect. The prediction results of linear prediction, quadratic polynomial, and cubic polynomial are not always concordant (Figure 2a,b). In order to improve accuracy, advanced machine learning by artificial intelligence, such as deep learning, should be considered in future studies. In keyword selection, SFI and SJI are selected as objectively and extensively as possible, but it would be desirable to conduct analyses with a much larger number of keywords in SFI and SJI. In addition, compiling similar words in the current study is a work involving human intervention, which may have contained some degree of subjectivity. Natural language processing may achieve more objectivity in this process.

4. Material and Methods

4.1. Choosing the Subject Fields of Interest

With all brain tumor-related articles published in 2017, the MEDLINE information of all the articles was extracted as text data. Only Other Terms (OTs), but not medical subject headings (MeSHs), were subjected to word frequency analysis, because MsSHs are basically broad terms, so it was difficult to choose keywords that only appear in a specific field of brain tumors, whereas OTs include many keywords specific to the field of neuro-oncology, such as signal transducer and activator of transcription (STAT), IDH, and TT. For the extracted OT text data, text-mining word frequency analysis was performed using a publicly available program (https://textmining.userlocal.jp/) to select frequently used keywords. Next, 28 abstract categories of SNO-2017 were registered as 28 keywords [17]. Word frequency analysis of the English presentation titles of the SNO-2017 and the English symposium presentation titles of the Annual Meeting of the Japan Society of Brain Tumor Pathology (JSBTP-2017, Utsunomiya, Tochigi, Japan) was also performed and frequently cited keywords were extracted.

In addition to these objectively extracted keywords, some keywords of brain tumor names were added manually to investigate the trends. Many synonyms and similar words that appeared as keywords in each data source were checked manually and then integrated into one keyword. We also manually excluded relatively vague keywords (e.g., survival, malignancy, etc.) that were judged as not suitable for tracking a specific research field of neuro-oncology. According to this method, 79 keywords that seemed to represent recent trends in neuro-oncology were extracted as SFI.

4.2. Data Collection with PubMed

Python 3.7.0 (https://www.python.org) was used to collect a large amount of journal data with PubMed. In addition, regarding the searching keywords, such as "genetics", "single cell", etc., which may lead to the identification of many other articles not related to neuro-oncology, the search was conducted by adding the term "brain tumor" as a keyword. By doing so, the directionality of the keywords was increased. About the possibility of retrieving the same article with multiple keywords, even if one article was found to be a related article of more than one keyword, we considered that it was not particularly problematic and did not affect the trend analysis of each subject field.

4.3. Mathematical Background of Δ-AII

The rule that Δ-AII is used for the analysis was adopted because inequality arises simply by comparing the AII values themselves. Each keyword has a difference in the "breadth" the word covers, and the "advantage" caused by the research scale to date. Such inequities need to be corrected. For example, the AII value of "epigenetics" in 2012 and 2013 were 15.760 and 88.141, respectively, and the Δ-AII for the period of 2012–2013 was calculated as 88.141/15.760 = 5.593. On the other hand, the Δ-AII of 2012–2013 of "glioblastoma" was similarly calculated as 3972.712/3298.539 = 1.204. Although the AII value was much higher for "glioblastoma", "epigenetics" was judged as having more momentum than "glioblastoma" over this two-year period. In addition, when the AII value of a certain subject field is zero, it is not possible to calculate the rate of change. In such a case, 1.081 was substituted for zero. This number is the same score of one article of SJI's lowest IF journal, "Child's nervous system".

4.4. Accuracy Validation of the Predictions

We validated the accuracy of the predictions by the three regression analyses using actual past data. Specifically, the Δ-AII data over a period of 5 years (year 1–5) were analyzed to predict the Δ-AII of the following year (year 6), which was then compared with the actual Δ-AII of year 6. We also rearranged the SFI fields in order of the predicted Δ-AII to determine the top 20 subject fields for each year. Similarly, the accuracy rate was calculated using the actual past data. Based on these analyses, we determined which regression method was most reliable for prediction.

4.5. Statistical Analyses

All statistical analyses were performed using JMP Pro 13 software (SAS Institute Inc., Cary, NC, USA) and R 3.4.1 [29]. The level of significance was set at a p-value of 0.05.

5. Conclusions

The current study highlights some evolving fields in clinical practice as well as research in neuro-oncology. In the upcoming few years, molecular biological approaches will become far more popular in neuro-oncological research. The fields of tumor microenvironment and epigenetics, both of which are closely related to many research fields like tumor immunology, will be more extensively discussed in the context of pathological mechanisms. In clinical practice, fine-grained treatments that emphasize the patient's quality of life would be desirable, reflecting the current social situations.

New therapeutic drugs, devices and techniques are constantly appearing. Clinicians and researchers are encouraged to remain keen on grasping the latest trends.

The MINERVA algorithm may provide useful data of current and future trends in the field of neuro-oncology. Moreover, it could potentially be applied to any medical research field by changing keywords. However, the accuracy of future prediction has room for improvement. A larger-scale system that takes into consideration more complex factors, such as deep learning by artificial intelligence, might be one future direction for better prediction.

Author Contributions: Conceptualization, T.H.; Methodology, T.H.; Formal Analysis, T.H.; Investigation, T.H.; Data Curation, T.H., S.T.(Shota Tanaka), T.N., S.T.(Satoshi Takahashi), Y.K., T.K., M.N. and S.T. (Shunsaku Takayanagi); Writing—Original Draft Preparation, T.H.; Supervision, S.T.(Shota Tanaka) and N.S.; Project Administration, N.S.

Funding: This research received no external funding.

Acknowledgments: We sincerely thank Mr. Yamato Nagata for his technical support of writing and editing the Python codes.

Conflicts of Interest: The authors declare no conflict of interest.

References

1. Ganau, L.; Prisco, L.; Ligarotti, G.K.I.; Ambu, R.; Ganau, M. Understanding the Pathological Basis of Neurological Diseases Through Diagnostic Platforms Based on Innovations in Biomedical Engineering: New Concepts and Theranostics Perspectives. *Medicines* **2018**, *5*. [CrossRef] [PubMed]
2. Ganau, M.; Paris, M.; Syrmos, N.; Ganau, L.; Ligarotti, G.K.I.; Moghaddamjou, A.; Prisco, L.; Ambu, R.; Chibbaro, S. How Nanotechnology and Biomedical Engineering Are Supporting the Identification of Predictive Biomarkers in Neuro-Oncology. *Medicines* **2018**, *5*. [CrossRef] [PubMed]
3. Cook, B.L.; Progovac, A.M.; Chen, P.; Mullin, B.; Hou, S.; Baca-Garcia, E. Novel Use of Natural Language Processing (NLP) to Predict Suicidal Ideation and Psychiatric Symptoms in a Text-Based Mental Health Intervention in Madrid. *Comput. Math. Methods Med.* **2016**, *2016*, 8708434. [CrossRef] [PubMed]
4. Saberi Hosnijeh, F.; Kavousi, M.; Boer, C.G.; Uitterlinden, A.G.; Hofman, A.; Reijman, M.; Oei, E.H.G.; Bierma-Zeinstra, S.M.; van Meurs, J.B.J. Development of a prediction model for future risk of radiographic hip osteoarthritis. *Osteoarthritis Cartilage* **2018**, *26*, 540–546. [CrossRef] [PubMed]
5. Seely, A.J. Prediction Is Difficult, Especially About Future Unexpected Deterioration. *Crit. Care Med.* **2016**, *44*, 1781–1783. [CrossRef] [PubMed]
6. Stupp, R.; Taillibert, S.; Kanner, A.; Read, W.; Steinberg, D.; Lhermitte, B.; Toms, S.; Idbaih, A.; Ahluwalia, M.S.; Fink, K.; et al. Effect of Tumor-Treating Fields Plus Maintenance Temozolomide vs Maintenance Temozolomide Alone on Survival in Patients with Glioblastoma: A Randomized Clinical Trial. *JAMA* **2017**, *318*, 2306–2316. [CrossRef] [PubMed]
7. Suzuki, H.; Aoki, K.; Chiba, K.; Sato, Y.; Shiozawa, Y.; Shiraishi, Y.; Shimamura, T.; Niida, A.; Motomura, K.; Ohka, F.; et al. Mutational landscape and clonal architecture in grade II and III gliomas. *Nat. Genet.* **2015**, *47*, 458–468. [CrossRef]
8. Sturm, D.; Witt, H.; Hovestadt, V.; Khuong-Quang, D.A.; Jones, D.T.; Konermann, C.; Pfaff, E.; Tonjes, M.; Sill, M.; Bender, S.; et al. Hotspot mutations in H3F3A and IDH1 define distinct epigenetic and biological subgroups of glioblastoma. *Cancer Cell* **2012**, *22*, 425–437. [CrossRef]
9. Journal Citation Reports. Available online: http://ipscience-help.thomsonreuters.com/incitesLiveJCR/JCRGroup/jcrOverview.html (accessed on 1 December 2017).
10. Jinek, M.; Chylinski, K.; Fonfara, I.; Hauer, M.; Doudna, J.A.; Charpentier, E. A programmable dual-RNA-guided DNA endonuclease in adaptive bacterial immunity. *Science* **2012**, *337*, 816–821. [CrossRef]
11. Lacouture, M.E.; Davis, M.E.; Elzinga, G.; Butowski, N.; Tran, D.; Villano, J.L.; DiMeglio, L.; Davies, A.M.; Wong, E.T. Characterization and management of dermatologic adverse events with the NovoTTF-100A System, a novel anti-mitotic electric field device for the treatment of recurrent glioblastoma. *Semin. Oncol.* **2014**, *41*, S1–S14. [CrossRef]

12. Miller, T.E.; Liau, B.B.; Wallace, L.C.; Morton, A.R.; Xie, Q.; Dixit, D.; Factor, D.C.; Kim, L.J.Y.; Morrow, J.J.; Wu, Q.; et al. Transcription elongation factors represent in vivo cancer dependencies in glioblastoma. *Nature* **2017**, *547*, 355–359. [CrossRef] [PubMed]
13. Wang, Q.; Hu, B.; Hu, X.; Kim, H.; Squatrito, M.; Scarpace, L.; deCarvalho, A.C.; Lyu, S.; Li, P.; Li, Y.; et al. Tumor Evolution of Glioma-Intrinsic Gene Expression Subtypes Associates with Immunological Changes in the Microenvironment. *Cancer Cell* **2017**, *32*, 42–56. [CrossRef] [PubMed]
14. Venteicher, A.S.; Tirosh, I.; Hebert, C.; Yizhak, K.; Neftel, C.; Filbin, M.G.; Hovestadt, V.; Escalante, L.E.; Shaw, M.L.; Rodman, C.; et al. Decoupling genetics, lineages, and microenvironment in IDH-mutant gliomas by single-cell RNA-seq. *Science* **2017**, *355*. [CrossRef] [PubMed]
15. O'Rourke, D.M.; Nasrallah, M.P.; Desai, A.; Melenhorst, J.J.; Mansfield, K.; Morrissette, J.J.D.; Martinez-Lage, M.; Brem, S.; Maloney, E.; Shen, A.; et al. A single dose of peripherally infused EGFRvIII-directed CAR T cells mediates antigen loss and induces adaptive resistance in patients with recurrent glioblastoma. *Sci. Transl. Med.* **2017**, *9*. [CrossRef] [PubMed]
16. Felthun, J.; Reddy, R.; McDonald, K.L. How immunotherapies are targeting the glioblastoma immune environment. *J. Clin. Neurosci.* **2018**, *47*, 20–27. [CrossRef] [PubMed]
17. de Groot, J. *Abstracts from the 22nd Annual Scientific Meeting and Education Day of the Society for Neuro-Oncology. The 22nd annual scientific meeting of the Society for Neuro-Oncology (SNO), San Francisco, CA, USA, November 16–19, 2017*; Oxford University Press: London, UK, 2017.
18. Parsons, D.W.; Jones, S.; Zhang, X.; Lin, J.C.; Leary, R.J.; Angenendt, P.; Mankoo, P.; Carter, H.; Siu, I.M.; Gallia, G.L.; et al. An integrated genomic analysis of human glioblastoma multiforme. *Science* **2008**, *321*, 1807–1812. [CrossRef] [PubMed]
19. de Souza, C.F.; Sabedot, T.S.; Malta, T.M.; Stetson, L.; Morozova, O.; Sokolov, A.; Laird, P.W.; Wiznerowicz, M.; Iavarone, A.; Snyder, J.; et al. A Distinct DNA Methylation Shift in a Subset of Glioma CpG Island Methylator Phenotypes during Tumor Recurrence. *Cell Rep.* **2018**, *23*, 637–651. [CrossRef]
20. Cui, Q.; Shi, H.; Ye, P.; Li, L.; Qu, Q.; Sun, G.; Sun, G.; Lu, Z.; Huang, Y.; Yang, C.G.; et al. m(6)A RNA Methylation Regulates the Self-Renewal and Tumorigenesis of Glioblastoma Stem Cells. *Cell Rep.* **2017**, *18*, 2622–2634. [CrossRef]
21. Mazor, T.; Pankov, A.; Johnson, B.E.; Hong, C.; Hamilton, E.G.; Bell, R.J.A.; Smirnov, I.V.; Reis, G.F.; Phillips, J.J.; Barnes, M.J.; et al. DNA Methylation and Somatic Mutations Converge on the Cell Cycle and Define Similar Evolutionary Histories in Brain Tumors. *Cancer Cell* **2015**, *28*, 307–317. [CrossRef]
22. Conklin, H.M.; Ogg, R.J.; Ashford, J.M.; Scoggins, M.A.; Zou, P.; Clark, K.N.; Martin-Elbahesh, K.; Hardy, K.K.; Merchant, T.E.; Jeha, S.; et al. Computerized Cognitive Training for Amelioration of Cognitive Late Effects Among Childhood Cancer Survivors: A Randomized Controlled Trial. *J. Clin. Oncol.* **2015**, *33*, 3894–3902. [CrossRef]
23. Costa, D.B.; Shaw, A.T.; Ou, S.H.; Solomon, B.J.; Riely, G.J.; Ahn, M.J.; Zhou, C.; Shreeve, S.M.; Selaru, P.; Polli, A.; et al. Clinical Experience with Crizotinib in Patients with Advanced ALK-Rearranged Non-Small-Cell Lung Cancer and Brain Metastases. *J. Clin. Oncol.* **2015**, *33*, 1881–1888. [CrossRef] [PubMed]
24. Hoang-Xuan, K.; Bessell, E.; Bromberg, J.; Hottinger, A.F.; Preusser, M.; Ruda, R.; Schlegel, U.; Siegal, T.; Soussain, C.; Abacioglu, U.; et al. Diagnosis and treatment of primary CNS lymphoma in immunocompetent patients: Guidelines from the European Association for Neuro-Oncology. *Lancet Oncol.* **2015**, *16*, e322–e332. [CrossRef]
25. Sakurai, H.; Sugimoto, K.J.; Shimada, A.; Imai, H.; Wakabayashi, M.; Sekiguchi, Y.; Ota, Y.; Izutsu, K.; Takeuchi, K.; Komatsu, N.; et al. Primary CNS CCND1/MYC-Positive Double-Hit B-Cell Lymphoma: A Case Report and Review of the Literature. *J. Clin. Oncol.* **2015**, *33*, e79–e83. [CrossRef] [PubMed]
26. Zhang, I.; Zaorsky, N.G.; Palmer, J.D.; Mehra, R.; Lu, B. Targeting brain metastases in ALK-rearranged non-small-cell lung cancer. *Lancet Oncol.* **2015**, *16*, e510–e521. [CrossRef]

27. Zou, H.Y.; Friboulet, L.; Kodack, D.P.; Engstrom, L.D.; Li, Q.; West, M.; Tang, R.W.; Wang, H.; Tsaparikos, K.; Wang, J.; et al. PF-06463922, an ALK/ROS1 Inhibitor, Overcomes Resistance to First and Second Generation ALK Inhibitors in Preclinical Models. *Cancer Cell* **2015**, *28*, 70–81. [CrossRef] [PubMed]
28. Johung, K.L.; Yeh, N.; Desai, N.B.; Williams, T.M.; Lautenschlaeger, T.; Arvold, N.D.; Ning, M.S.; Attia, A.; Lovly, C.M.; Goldberg, S.; et al. Extended Survival and Prognostic Factors for Patients With ALK-Rearranged Non-Small-Cell Lung Cancer and Brain Metastasis. *J. Clin. Oncol.* **2016**, *34*, 123–129. [CrossRef] [PubMed]
29. R Core Team. *R: A Language and Environment for Statistical Computing*; R Foundation for Statistical Computing: Vienna, Austria, 2017. Available online: https://www.R-project.org/ (accessed on 1 August 2017).

cancers

MDPI

Review

Role of Notch Signaling Pathway in Glioblastoma Pathogenesis

Riccardo Bazzoni [1,2,3] **and Angela Bentivegna** [3,4,*]

1 Stem Cell Research Laboratory, Section of Hematology, Department of Medicine, University of Verona, Pz.le Scuro 10, 37134 Verona, Italy; riccardo.bazzoni@univr.it
2 Program in Clinical and Experimental Biomedical Sciences, University of Verona, 37134 Verona, Italy
3 NeuroMi, Milan Center for Neuroscience, Department of Neurology and Neuroscience, San Gerardo Hospital, University of Milano-Bicocca, 20900 Monza, Italy
4 School of Medicine and Surgery, University of Milano-Bicocca, via Cadore 48, 20900 Monza, Italy
* Correspondence: angela.bentivegna@unimib.it

Received: 30 December 2018; Accepted: 25 February 2019; Published: 1 March 2019

Abstract: Notch signaling is an evolutionarily conserved pathway that regulates important biological processes, such as cell proliferation, apoptosis, migration, self-renewal, and differentiation. In mammals, Notch signaling is composed of four receptors (Notch1–4) and five ligands (Dll1-3–4, Jagged1–2) that mainly contribute to the development and maintenance of the central nervous system (CNS). Neural stem cells (NSCs) are the starting point for neurogenesis and other neurological functions, representing an essential aspect for the homeostasis of the CNS. Therefore, genetic and functional alterations to NSCs can lead to the development of brain tumors, including glioblastoma. Glioblastoma remains an incurable disease, and the reason for the failure of current therapies and tumor relapse is the presence of a small subpopulation of tumor cells known as glioma stem cells (GSCs), characterized by their stem cell-like properties and aggressive phenotype. Growing evidence reveals that Notch signaling is highly active in GSCs, where it suppresses differentiation and maintains stem-like properties, contributing to Glioblastoma tumorigenesis and conventional-treatment resistance. In this review, we try to give a comprehensive view of the contribution of Notch signaling to Glioblastoma and its possible implication as a target for new therapeutic approaches.

Keywords: glioblastoma; GSCs; Notch signaling pathway; new therapeutic approaches

1. Introduction

Glioblastoma is the most common and fatal type of primary brain tumor [1]. It comprises 70% of all gliomas and is classified as a Grade IV astrocytoma (World Health Organization classification) [2]. The current standard of care for Glioblastoma patients includes maximal safe resection, followed by concurrent radiotherapy (RT) and chemotherapy with temozolomide (TMZ), followed by adjuvant TMZ [3]. Despite these aggressive therapies, the median survival time is currently 14.6 months [4], with a final mortality rate of close to 100% [5]. The failure of current therapies is due to the coexistence of heterogeneous tumor cell populations with different grades of differentiation [6,7] and the presence of a small subset of tumor cells that display stem cell-like properties, which are responsible for tumor relapse after conventional treatments [8,9]. This population, known as glioma stem cells (GSCs), exhibits an enhanced self-renewal capacity, compromised differentiation, and in vivo tumorigenicity, besides being radio- and chemoresistant [10,11]. Recent studies have demonstrated that Notch signaling is highly active in GSCs, contributing to suppressing differentiation and maintaining stem cell-like properties. Therefore, this review aims to discuss the role of Notch signaling in Glioblastoma pathogenesis and in the development of potential therapeutic strategies.

2. Notch Signaling Pathway Overview

Notch signaling is an evolutionarily conserved pathway that plays a critical role in various cellular and developmental processes, including cell proliferation, apoptosis, stem cell maintenance, cell fate decision, and tissue homeostasis [12–14]. Notch functions as a cytoplasmic receptor and, in mammals, there are four homologous proteins known as Notch1, Notch2, Notch3, and Notch4, which can bind two ligands families: Delta-like (Dll1-3 and -4) and Jagged (Jagged1 and -2) [15]. Both the receptors and ligands are single-pass transmembrane proteins (Figure 1). The interaction between Notch and its ligands can occur in two ways: In trans, when they are present on neighboring cells, or in cis, when the receptor and ligand are present on the same cell [16,17]. In the first case, binding leads to pathway activation, while in cis form interaction inhibits the signaling cascade [18].

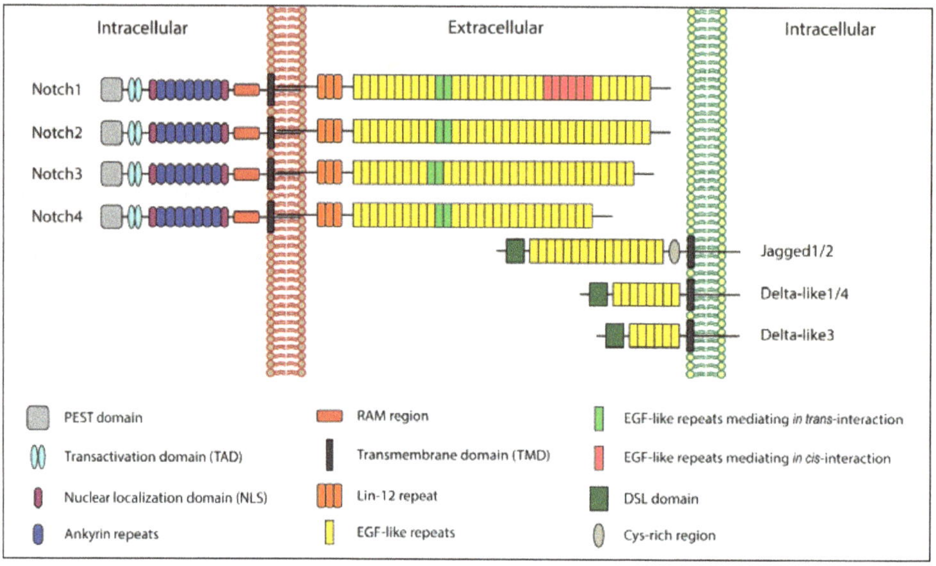

Figure 1. Structure of human Notch receptors and ligands.

Structurally, the Delta and Jagged ligands are very similar: They lack an intracellular domain, while their extracellular portion is constituted by 6–16 EGF-like repeats [19] and a distal Cys-rich region called the Delta/Serrate/Lag-2 (DSL) domain [20]. The DSL region is responsible for interaction with Notch receptors [21]. Moreover, Jagged1 and -2 possess a Cys-rich region closer to the membrane [21]. Unlike Delta and Jagged, the extracellular domain of Notch receptors is organized in 29–36 EGF-like repeats [19,21] and exists in a Cys-rich region called Lin-12 that noncovalently binds the extracellular Notch with the membrane-tethered intracellular Notch [18]. The Notch intracellular domain (NICD) is composed of an RBPJK associate molecule (RAM) region, repeated structural motifs named ankyrin repeats (which mediate the interaction between NICD and CBF1, a transcriptional activator), a transactivation domain (TAD), and a PEST domain (involved in Notch degradation) [21].

The life of Notch receptors begins in the endoplasmic reticulum (ER), where they are synthesized as inactive single-peptide precursors (Figure 2).

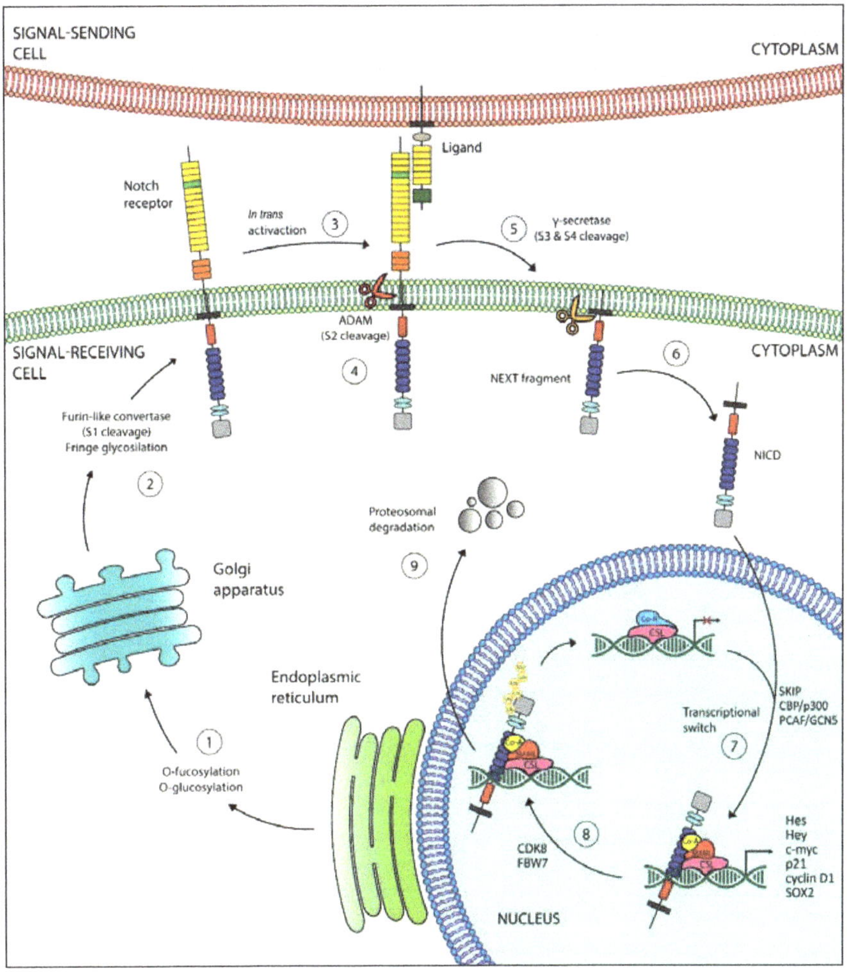

Figure 2. Schematic representation of the Notch signaling pathway. Once synthesized in the endoplasmic reticulum (①), the inactive single peptide precursor moves to the Golgi where it is cleaved by a furin-like convertase (S1 cleavage) (②) and translocates into the cell membrane. The binding with a Notch ligand (③) induces the second cleavage (S2) by a member of the disintegrin and metalloproteinases (ADAM) family (④), resulting in the formation of a membrane-tethered Notch truncated (NEXT) fragment, which is further processed in two sites (S3 and S4) by a presenilin-dependent γ-secretase complex (⑤), generating the Notch intracellular domain (NICD), the active form of the Notch receptor (⑥). The NICD can now enter into the nucleus, where it exerts its transcriptional activity (⑦). The ubiquitination of the NICD (⑧) leads to its proteasome degradation (⑨).

Once in the trans-Golgi, a furin-like convertase proteolytically cleaves the inactive precursor [22]. This first cleavage site (S1) allows the mature Notch receptor to form, which is able to translocate in the cytoplasmic membrane and interact with its ligands [22]. The in-trans activation triggers a conformational change in the Notch receptor that exposes the second cleavage site (S2) to a member of the disintegrin and metalloproteinases (ADAM) family (ADAM10 or ADAM17) [23,24]. The result is a membrane-tethered Notch-truncated (NEXT) fragment [24], which is further processed into two sites (S3 and S4) by a presenilin-dependent γ-secretase complex, constituted by presenilin 1

(PSEN1) or PSEN2, nicastrin, presenilin enhancer 2 (PEN2), and anterior pharynx-defective 1 (APH1) [25,26]. These cleavages release the active form of the NICD into the cytoplasm. The NICD can then translocate into the nucleus [27], where it binds several DNA-binding proteins of the CSL family (RBPJK/CBF1/KBPF2 in mammals). The newly formed complex switches from a transcriptional-repressor state to a transcriptional-activator one, recruiting the transcriptional coactivator mastermind-like protein (MAML), the Ski-interacting Protein (SKIP), and histone acetyltransferases CBP/p300 and PCAF/GCN5 [28–30]. All these events allow the transcription of transcriptional repressors such as the Hairy Enhancer of Split (Hes) family of proteins and HES-related proteins (Hey) [31,32], two families responsible for lineage-commitment decisions. Other Notch target genes include p21/Waf1, cyclin D1 and -3, c-Myc, HER2, Notch-regulated ankyrin repeat protein (NRAR), NF-κB, pre-Tα, IGF1-R, survivin, Snail homolog 2 (SLUG), SOX2, and PAX5 [33–36]. The Notch-mediated transcriptional activation ends with the degradation of the NICD. The mechanism consists of the phosphorylation of a degron within the PEST domain of NICD, mediated by cyclin-dependent kinase 8 (CDK8) and targeted for proteasome-mediated degradation by E3 ubiquitin ligases SEL10 (also known as FBW7) [18,37].

In addition to canonical signaling activation, several groups have identified a novel way to activate the Notch pathway through a group of different, unrelated proteins that lack the DSL domain [38,39]. These proteins can be membrane-integral (for example, the Delta/Notch-like epidermal growth factor-related receptor—DNER), a glycosylphosphatidylinositol (GPI)-linked membrane (for example NB3/Contactin6), or secreted proteins (for example, MAGP1 and -2) [40]. Notably, Notch can non-canonically exert its biological functions either in a ligand-dependent or -independent way [41].

3. Notch Signaling in Brain Development

The Notch signaling pathway plays a fundamental role in central nervous system (CNS) development, from embryonic stages to the adult brain [41]. Although Notch signaling members show differential expression patterns throughout the brain (Table 1), they are master regulators of neurogenic niches—specialized microenvironments that are able to modulate the properties of stem cells, such as their cell number, self-renewal, and fate decision, in order to avoid the depletion of the NSC pool [42,43]. Being the starting point for neurogenesis, NSCs are extremely important [44]. In the adult brain, there are two major sites for neurogenesis: The subventricular zone (SVZ) of the lateral ventricle [43,45] and the dentate gyrus (DG) in the hippocampus [46].

Table 1. Expression pattern of Notch receptors and ligands in the adult brain.

Notch Signaling Members	Expression Pattern	References
Notch1	Neurons, astrocytes, precursor cells, ependymal cells, endothelium	[47–50]
Notch2	Precursor cells	[47,51]
Notch3	Precursor cells	[51]
Notch4	Endothelium	[52]
Delta/Notch-like epidermal growth factor-related receptor (DNER)	Neurons	[53]
Dll1	Intermediate neural progenitors, neurons	[49,51,54,55]
Dll3	Intermediate neural progenitors	[54]
Dll4	Endothelium	[56]
Jagged1	Precursor cells, intermediate neural progenitors, neurons	[48,49,54,57–59]
Jagged2	Neurons	[49,52]

It was demonstrated that low Notch levels induce the proliferation of NSCs, followed by an exit from the cell cycle and differentiation into neurons [60]. On the contrary, high Notch levels lead to growth arrest, or even induce a quiescent state [61,62]. The cell-fate regulation of quiescent NSCs results under RBPJK activity, probably through Notch2 and Notch3, but not Notch1 [63,64]. Instead, Notch1 seems to be crucial for the active proliferation of the NSC pool, which is selectively lost during aging [43,46]. Additionally, the ablation or loss of RBPJK function shows progenitor-cell depletion in the postnatal SVZ and DG, and reduced neurogenesis [46,65]. Thus, it is plausible to conclude that Notch signaling is instrumental for the regenerative capacity of a mature brain. Concerning neuronal migration, in the cerebral cortex, reelin-DAB1 signaling avoids NICD degradation, promoting Notch signaling in order to alter neuron morphology and, therefore, perturbing their migration [66]. This interaction between reelin-DAB1 and NICD was also reported in a study by Sibbe et al., where they showed that this mechanism is essential for the correct development of the radial glial scaffold [67].

To explicate their functions, neurons require glial cells which surround and insulate them, providing physical support. Like neurons, glial cells differentiate from NSCs [68]. During neurogenesis, Notch receptors are ubiquitously expressed, whereas Delta ligands appear transiently on the surface of newly differentiating neurons. NSCs, when exposed to a Delta signal, tend to resist prevailing neurogenic signals, activate Notch signaling and ultimately differentiate into glial cells. Therefore, the resistance to neurogenic signals is dependent on the activation of Notch by Delta [69,70]. Recently, several studies have suggested a new model in which Notch signaling prevents all types of NSC differentiation, rather than inhibiting neural differentiation. According to this view, NSCs will choose their fate according to whichever instructive factors prevail when they are released from the influence of Delta [71]. Indeed, Notch signaling seems to have an instructive role in gliogenesis, promoting the differentiation of many glial subtypes, with the exception of oligodendrocytes. The results of in vivo studies are consistent with the notion that Notch signaling plays an instructive role in gliogenesis through their conventional basic helix-loop-helix (bHLH) targets [72,73]. Moreover, if Notch activation promotes glial differentiation, it would be expected that some Notch receptors downstream elements (for example transcriptional targets [74]) would also influence gliogenesis.

Finally, although the role of Notch signaling in brain function is complicated and controversial, all this evidence points out the essential role of Notch in several aspects of the CNS.

4. Notch Pathway Deregulation in Brain Tumors and Brain CSCs

Deregulation of the Notch pathway has been associated with the development of a wide range of diseases through both germline and somatic mutations. Concerning the latter events, Notch signaling mutations lead to cancer and malignancy, of which the best-known example is T-cell acute lymphoblastic leukemia [75,76]. Over the years, aberrant Notch signaling was also found in several solid tumors, including brain tumors, although its components are rarely mutated [76,77].

Brain tumors are a heterogeneous group of neoplasms. Among them, the most common are gliomas and medulloblastomas [2]. Several studies have reported the abnormal expression of various Notch components in brain tumors. For instance, a higher expression of ASCL1, Dll1, Notch1, -3, -4, and Hey1 correlates with a higher glioma grade and a worse prognosis [78–81], indicating that more activated Notch signaling promotes a more undifferentiated and aggressive tumor phenotype.

In the last twenty years, the cancer stem cell (CSC) theory has said that CSCs are responsible for tumor development and growth, as well as being the reason for cancer metastasis and relapse [33,82–85]. Brain CSCs (bCSCs), or brain-tumor initiating cells (bTICs), share similarities with NSCs [86]: They are able to grow as floating aggregates (called neurospheres) in a serum-free medium [87,88], express stem cell markers (such as nestin, GFAP, β-III tubulin, and CD133) [87,89], and can differentiate into all three neural lineages [87,89,90]. Since Notch signaling is a key regulator of the NSC state, it is easy to suppose its involvement in the maintenance of the NSC tumor counterpart. For instance, intracellular modulators such as Numb4 and Numb4Δ7 regulate the expression of stem cell markers in bCSCs, despite functioning like an inhibitor and activator for Notch signaling, respectively. This assumes that Notch mediators can alter bCSC differentiation independently from Notch inhibition itself [91]. Furthermore, NICD-overexpressing bCSCs induce tumor formation in nude mice [92].

bCSCs are generally more resistant to treatment under hypoxic conditions. In support of this, the depletion of HIF-1α alters the proliferation of glioma-derived bCSCs through blocking the interaction of HIF-1α and NICD [93]. This interaction was confirmed by another group, which showed that HIF-1α displaced HIF-2α (a Notch inhibitor) from the NICD under hypoxic conditions [94]. From these data, it emerges that active Notch signaling is necessary for maintaining stem cell features and the tumorigenic potential of bCSCs, and sets the stage for promising Notch-based therapies.

5. Notch Signaling and Glioblastoma

5.1. Expression Pattern of Notch Signaling in Glioblastoma

mRNA and protein levels of Notch1, Notch4, Dll1, Dll4, Jagged1, CBF1, Hey1, Hey2, and Hes1 are higher in brain tumor cells than normal brain cells, correlating with an elevated expression of VEGF and pAKT, and reduced levels of PTEN [79,95–98]. In particular, Notch1 expression is higher in the survival of >1 year patients than <1 year [99], whereas Notch1 overexpression is associated with low overall survival (OS) [100], suggesting a controversial role of Notch1 in gliomagenesis. Moreover, Notch1 is more expressed in peritumor-tissue GSCs compared to tumor-core GSCs [101]. Notch1 and Notch4 levels correlate with those of GFAP and vimentin, respectively. Notch4 expression increases with higher-grade and primary tumors [102]. Notch2 expression levels in Glioblastoma tissue correlate with stemness genes (*nestin, SOX2*), astrocyte fate genes (*vimentin* and *GFAP*), and anti-apoptotic proteins (BCL6 and BCL-W), but are inversely correlated with Olig2, CNP, and PLP1 (oligodendrocyte fate) and pro-apoptotic proteins (BAX and BCLAF1) [102,103]. The overexpression of Hey1, which is associated with survival and tumor grade, might be due to the impairment of Notch and E2F signaling; it was demonstrated that its overexpression in NSCs triggers neurosphere formation and contributes to Glioblastoma proliferation [95]. On the contrary, several groups reported a weak expression of Notch1, Notch2, MAML1, and p300 in Glioblastoma [102,104,105]. Intriguingly, the impairment of Notch signaling in secondary Glioblastoma, in which Hes1 expression is almost absent, is associated with the overexpression of ASCL1. On the other hand, the activation of Notch signaling in primary Glioblastoma is associated with low levels of ASCL1, suggesting that Notch inhibition via ASCL1 upregulation might be responsible for a potential progression into secondary Glioblastomas [78].

Correlation between Glioblastoma molecular subtypes and Notch expression was also demonstrated. Concerning the mesenchymal subtype (the most aggressive one), Notch-related genes are the most highly enriched in high p-STAT3 patients, suggesting a synergy between Notch and STAT3 signaling [106]. Verhaak et al. reported that Notch signaling is highly expressed in the classic subtype [107]. The expression levels of Dll3 and Hey2 are low in proneural Glioblastomas, while the expression level of Notch1 is high [81,105,107,108], although Cooper et al. reported reduced levels of Dll3 and Hey2 [108]. Always concerning the proneural subtype, the majority of Glioblastomas with the *IDH* mutation have a proneural gene expression pattern, even if only 30% of proneural Glioblastomas have the mutation [109]. Spino et al. reported that *IDH*-mutant gliomas (mostly low grade) have high and homogenous Dll3 expression, whereas approximately half of *IDH*-wild type Glioblastomas had either no expression or only scattered cells expressing Dll3. Regardless, Dll3 expression, if present in a *IDH*-wild type Glioblastoma, is generally restricted to non-mesenchymal subtypes [110]. Notably, Jungk et al. provided a link between Notch expression and tumor location in the proneural subtype. They observed that, in *IDH*-wild-type tissue near to the SVZ, the overexpression of Hes4 and Dll3 predicts inferior OS [111].

Finally, even the non-canonical Notch pathway plays a role in gliomagenesis. Huber et al. found that Deltex1 (DTX1) levels were higher in Glioblastoma compared to the normal brain, inducing several pathways involved in glioma aggressiveness such as RTK/PI3K/PKB and MAPK/ERK signaling, and the anti-apoptotic protein Mcl-1 [112].

5.2. Epigenetic Regulation of Notch Signaling in Glioblastoma

A peculiar characteristic of epigenetic alterations is their reversibility, making them a promising therapeutic target to explore in order to reset the abnormal cancer epigenome. To date, we do not know much about the epigenetic regulation of Notch signaling in Glioblastoma. Tsung et al. showed how the methylation status of the transcription factor *HEY1* contributes to Glioblastoma pathogenesis [113]. They found low levels of methylation on CpG islands within the *HEY1* promoter across Glioblastoma specimens when compared to a healthy brain, resulting in Hey1 overexpression [113]. In support of this, treatment with sodium butyrate (NaB), a histone deacetylase (HDAC) inhibitor, on 4910 and 5310 xenograft cell lines induced Glioblastoma cell apoptosis, decreased Hey1 expression, and increased DNMT1 levels. Moreover, the knockdown of *HEY1* reduced cell invasion, migration, and proliferation [113]. Sun et al. highlighted the role of the Delta/Notch-like epidermal growth factor-related receptor (DNER), which regulates cerebellar development and neurodevelopmental interactions between Purkinje cells and Bergmann glia which express Notch via a Deltex-dependent mechanism [114]. HDAC inhibition is able to activate the DNER/Deltex signaling pathway in Glioblastoma-derived neurospheres, resulting in cell differentiation and neurosphere-growth inhibition [114]. However, due to lack of sufficient evidence relating to the epigenetic regulation of the Notch signaling pathway in Glioblastoma, to date there are no epigenetic Notch biomarkers for cancer diagnosis.

5.3. Role of miRNAs in Notch-Dependent Gliomagenesis

MicroRNAs (miRNAs or miRs) are small (20–22 nucleotides), non-coding RNA molecules that can play a gene-regulatory role by pairing to the mRNAs of protein-coding genes to direct the inhibition of their translation or induce their destabilization and degradation. By regulating gene expression and therefore various cell processes, like proliferation and apoptosis, their alterations are often associated with the pathogenesis of several cancers. Starting from a network topological analysis of the Glioblastoma Notch regulatory network, Sun et al. pointed out 32 miRNAs that might be involved in the Notch pathway, and six of them (miR-9, miR-34a, miR-92b, miR-124, miR-137, and miR-219-5p) might play a key role [115]. Among the Notch-related miRNAs involved in gliomagenesis (Figure 3). The miR-34 family is the most studied. It is downregulated in Glioblastoma tissue compared to normal brain tissue and is more expressed in wild-type *p53* Glioblastomas than mutant *p53* Glioblastomas [116,117].

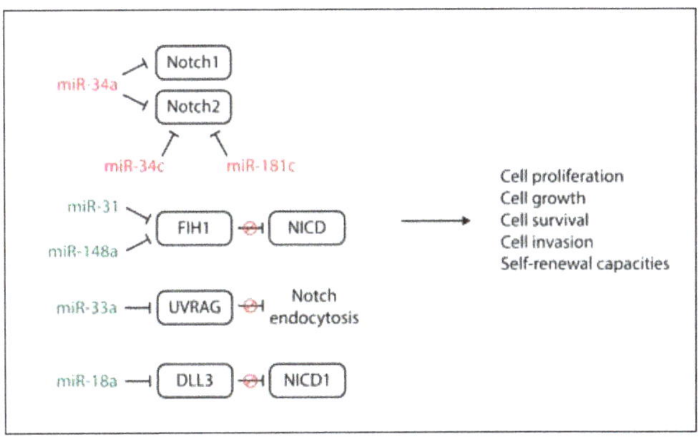

Figure 3. Functional effects of Notch-regulated miRNAs in glioblastoma. Red miRNAs are downregulated while the green ones are upregulated in Glioblastoma cells.

miR-34a and miR-34a-5p function as tumor-suppressive miRNAs, inhibiting cell proliferation, cell-cycle progression, and cell invasion by targeting Notch1, Notch2, c-Met, CDK6, and EGFR [116,117].

Di Bari et al. reported that miR-34a-5p expression levels are inversely correlated to Notch1 and Notch2 expression, and its function is restored by the activation of M2 acetylcholine muscarinic receptors, which in turn downregulate Notch1 and consequently cell proliferation [117]. Wu et al. showed that lower levels of miR-34c-3p and miR-34c-5p correlate with a higher glioma grade. The overexpression of both miRNAs strongly inhibits glioma invasion and miR-34c-3p but not miR-34c-5p, promotes S-phase arrest, increases cell apoptosis, and reduces Notch2 expression [118]. Notch2 is a target of another tumor-suppressive miRNA, miR-181c, which reduces cell proliferation, cell invasion, and self-renewal capacities through Notch2 downregulation. Unfortunately, miR-181c is commonly downregulated in Glioblastoma, especially in the mesenchymal subtype, suggesting a potential relationship between miR-181c and the malignant behavior of Glioblastoma [119]. Among the miRNAs associated with shorter survival in Glioblastoma, Wong et al. discovered miR-148a and miR-31 [120]. miR-148a is frequently upregulated in Glioblastoma and correlated with hypoxia-induced and extracellular-matrix genes, while high levels of miR-31 are appreciated only in a small group of Glioblastomas and are associated with proliferation and immune-response genes. A common target of both miRNAs is factor-inhibiting HIF-1 (FIH1), which mediates their effects on tumor growth, counteracting HIF-1α and the NICD. In particular, HIF-1α is able to stabilize the NICD in order to expand and maintain GSCs. The inhibition of miR-148a and miR-31 in Glioblastoma mouse models prolongs animal survival, depletes the stem cell pool, suppresses tumor growth, and normalizes tumor vasculature [120]. With regard to GSC plasticity, miR-18a is a key player in controlling the switch between the self-renewing and non-self-renewing states [121]. By downregulating Dll3 and strengthening Notch1 signaling, miR-18a induces the expression of *SHH* and *GLI-1* via ERK, maintaining the self-renewal and stemness abilities of GSCs [121]. The last investigated Notch-related miRNA in Glioblastoma is miR-33a, which promotes GSC growth and is responsible for their self-renewal abilities. This occurs because, among miR-33a target genes, there is a UV-radiation resistance-associated gene (UVRAG) that negatively regulates the Notch signaling pathway through the repression of Notch endocytosis. Moreover, an inverse correlation between the expression levels of miR-33a and UVRAG exists in Glioblastoma, and patients with a higher expression of miR-33a are characterized by poor prognosis and shorter survival [122].

5.4. Tumorigenesis and Other Tumor Aspects Driven by Notch Pathway

The overexpression of Notch1 promotes AKT activation, which in turn induces the nuclear localization of β-catenin and NF-κB, together with the Notch-mediated overexpression of Snail, Zeb1, and vimentin, promotes cell invasion and migration (Figure 4) [98]. The crosstalk between Notch and AKT is also mediated by collapsine response mediator protein-5 (CRMP5), preventing Notch degradation and favoring Glioblastoma proliferation [123]. In addition, CMRP5^{high} Glioblastoma has elevated Hey1 expression compared to CMRP5^{low} Glioblastoma, suggesting CMRP5 as an indicator of poor survival [123]. Two transient receptor cation channels, TRPM7 and TRPC6, are linked to Notch pathway-stimulating proliferation, invasion, and the migration of glioma cells [124,125]. In particular, TRPM7 levels correlate with those of Notch1, Jagged1, Hey2, and survivin [124]. Neurosphere cultures with high Notch1 levels show a more infiltrative phenotype when compared to Notch1^{low} cultures [126]; furthermore, the suppression of cell migration, tumor invasion, and angiogenesis can be achieved by targeting the urokinase-type plasminogen activator/urokinase-type plasminogen activator receptor (uPA/uPAR) system in order to inhibit Notch-signaling-induced AKT, NF-κB, and ERK pathways [127]. The oncogenic role of Notch1 can also be due to the infection of human cytomegalovirus, which upregulates Notch1, ATF5 (an anti-apoptotic protein already highly expressed in Glioblastoma), and stem cell markers CD133, nestin, SOX2, OCT4, KLF4, and BMI-1 [128]. Finally, Wang et al. reported that silencing Notch1 reduced GSC proliferation and oncogenicity in vitro and in vivo [129].

Besides Notch1, even Notch2 overexpression enhances cell migration in an RBPJK-dependent manner [130]. Notch2 deregulation might also promote NSC transformation and gliomagenesis, preventing neuronal lineage [103]. The stem cell phonotype is also supported and maintained by stanniocalcin-1 (STC1) and LMO2. STC1, a secretory glycoprotein, is highly expressed in glioma spheres; it is able to bind Notch1 and activate the Notch1-SOX2 signaling pathway, therefore supporting the stemness and tumorigenicity of GSCs [131]. Transcription factor LMO2, besides being inversely correlated with shorter survival, promotes the endothelial-like conversion of GSCs through the activation of VE-cadherin [132]. Jeon et al. demonstrated that inhibitor of differentiation 4 (ID4) increases Jagged1 expression, followed by Notch1 activation to drive astrocytes into a neural stem-like cell state and to increase cyclin E to produce a hyperproliferative state [133]. The downregulation of E3 ubiquitin ligase TRIM3 and the high levels of Musashi found in Glioblastoma promote the growth, survival, and self-renewal of stem cells and differentiation through the Numb/Notch pathway [134].

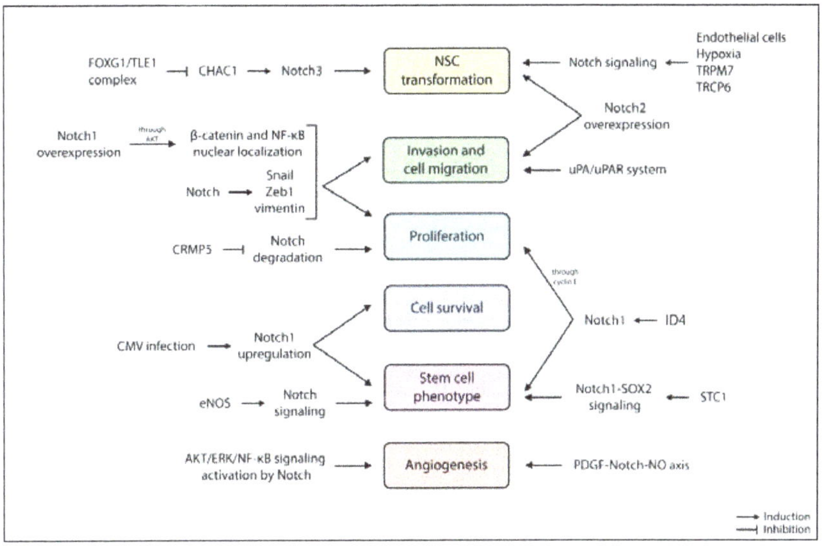

Figure 4. Other Notch-related tumorigenic aspects in the pathogenesis of Glioblastoma.

The tumor microenvironment also contributes to tumorigenesis. Endothelial cells function as a CSC niche by providing Notch ligands to Notch receptors on GSCs. The absence of Notch ligands on endothelial cells can reduce the $CD133^{+}$ glioma subpopulation in vitro, inhibiting neurosphere propagation [135]. Moreover, differentiated cells within the tumor express higher levels of Dll1 compared to GSCs, contributing to Notch signaling activation in GSCs [135]. Similarly, silenced *DLL1* in GSCs decreases stem cell markers and impairs self-renewal ability [136]. On the contrary, mesenchymal stem cells (MSCs) have a tumor-suppressor effect on GSCs. Indeed, paracrine signals from MSCs sensitize NCH421k and NCH644 cell lines to TMZ, probably turning them toward more differentiated cell types, downregulating Notch1 and SOX2 and upregulating vimentin and GFAP [137].

Under hypoxic conditions, GSCs increases the expression of several Notch genes (*NOTCH1, NOTCH3, DLL1, JAGGED1, JAGGED2, HES1, HEY1,* and *HEY2*), and hypoxia-related genes (*HIF-1α, VEGF, LOX,* and *HIG2*) [125,138,139]. The hypoxia-Notch gene subset might hold a prognostic implication, as the overexpression of the hypoxia-Notch axis is associated with poor survival [138]. Notch1 activation under hypoxic conditions also induces the expression of transient receptor TRCP6, which has emerged as a critical player in Glioblastoma aggressiveness, promoting NFAT activity, a crucial factor for glioma proliferation [125]. Han et al. showed that Notch1 inhibition in Glioblastoma xenografts reduces the hypoxic fraction and delays tumor growth, further supporting the crosstalk between Notch signaling and hypoxia, suggesting a potential mechanism whereby Notch1 downregulation radiosensitizes Glioblastoma cells [100]. On the contrary, Bayin et al. demonstrated that the NICD is strongly expressed in perivascular tumor regions and not in the hypoxic zone. However, the authors highlighted an intratumoral GSC heterogeneity with a divergent activation of Notch signaling, which coexist in tumors, but populates distinct niches and accordingly organizes their metabolisms [140]. Charles et al. showed that endothelial nitric oxide (eNOS) maintains the GSC phenotype in perivascular niches, activating the Notch pathway via paracrine signaling and promoting in vivo tumorigenicity [141]. In another report, the convergence of the PDGF-Notch-NO axis simultaneously drove perivascular promotion of the GSC phenotype and angiogenesis [142]. Jubb et al. reported the presence of Dll4 and Jagged1 in Glioblastoma vasculature [143]. These data define Glioblastoma subsets that might be sensitive ($Dll4^+$/$Jagged1^+$) or resistant ($Dll4^+$/$Jagged1^-$) to bevacizumab, a humanized monoclonal antibody against VEGF [143]. It was also reported that the Wnt signaling pathway represses Notch signaling in a Glioblastoma hypoxic microenvironment and supports the ablation of $CD133^+$ subpopulations [144], which can populate tumors regardless of local vascularity and selectively utilize anaerobic glycolysis to expand in hypoxia [140]. Moreover, Notch1-stimulated GSCs induce highly vascularized tumors in vivo, with the production of several angiogenesis-related factors and the expression of pericyte cell markers [92].

Finally, the FOXG1/TLE1 complex cooperates with Notch signaling to promote gliomagenesis by directly repressing CHAC1 expression, a negative regulator of Notch3. Dali et al. also identified DNER as an additional potential transcription-repression target of FOXG1/TLE1: DNER inhibits Glioblastoma-derived neurosphere growth and promotes their differentiation, opposite to the effect of FOXG1 and TLE1 [145].

5.5. Therapeutic Approaches against Notch Signaling in Glioblastoma

To date, several classes of Notch inhibitors have been developed (Table 2). The most employed in cancer are γ-secretase inhibitors (GSIs), which prevent the release of the active NICD from the receptor by the γ-secretase complex, while α-secretase inhibitors (ASIs), which inhibit members of the ADAM family by preventing the second cleavage (S2) of the Notch receptor, are less used. Finally, to overcome GSI/ASI resistance, new therapeutic approaches that directly or indirectly target Notch signaling have been developed.

5.5.1. γ-Secretase Inhibitors

DAPT (GSI-IX), the most known and used GSI, amplifies the effect of radiation, and reduces GSC proliferation and the number of endothelial cells disrupting the perivascular niche [146]. The suppression of cell proliferation and induction of apoptosis is mediated by the decrease of NF-κB (p65) expression through Notch inhibition [105]. Glioblastoma neurospheres with high Notch activity are more sensitive to DAPT and DAPT-treated cultures and show a more differentiated state and low Hes5 levels [126,146,147]. The combined treatment of DAPT with Iressa, an EGFR inhibitor, reduces VEGF expression and secretion when compared to single treatments. Unfortunately, this combined treatment is not sufficient to fully block endothelial cells from sprouting, probably due to other angiogenesis factors. Moreover, by blocking EGFR signaling, Hes1 levels decreased, suggesting that EGFR signaling stimulates Notch pathway activity [148]. Like DAPT, LLNle and L-685,458 are also able to kill GSCs, inhibiting NICD generation and inducing proteasome inhibition, proteolytic stress, and mitotic arrest [149]. DAPT and L-685,458 strongly reduce neurosphere formation in Glioblastoma cell lines, although DAPT and L-685,458 work to a much lesser extent than LLNle [149,150]. GSI RO4929097, in combination with radiation and TMZ, decreases the expression of CD133, SOX2, and nestin (inducing neural and astrocytic differentiation), has an anti-proliferative effect (reducing 3D spheroid growth), and increases the survival of the orthotopic Glioblastoma mouse model [151,152]. The triple combination is more effective than radio- and chemotherapy, or GSI alone [152]. Notably, Saito et al. showed that GSCs sensitive to DAPT, RO4929097, and BMS-708163, another GSI, have a gene signature enriched in proneural genes such as *OLIG2*, *SOX2*, *ERBB3*, *HDAC2*, *TGFB3*, *DLL3*, *CHIL3I*, and *NKX2-2* [151].

Table 2. Therapeutic approaches against Notch signaling in Glioblastoma.

Class of Inhibitors	Molecules	Biological Effects							
		Decreased Cell Growth	Anti-Proliferative Activity	Pro-Apoptotic Activity	Reduced Neurosphere Formation	Reduced Colony Formation	Decreased Tumor Size In Vivo	Prolonged Animal Survival	Others
γ-Secretase inhibitor (GSI)	DAPT (GSI-IX)		✓	✓	✓				• Amplifies the effects of radiation • Induces differentiation • In combination with Iressa, reduces VEGF
	LLNle			✓					
	L-685,458			✓	✓				
	RO4929097		✓		✓			✓	• In combination with RT and TMZ, reduces CD133, SOX2, and nestin expression
	BMS-708163								
	MRK003		✓	✓			✓		• Sensitizes cells to TMZ • Reduces onco-metabolite levels • In combination with GSI-XVIII, reduces CD133, nestin, OLIG2, and BMI-1 expression, and neurosphere formation
	GSI-XVIII			✓			✓		
	GSI-X	✓							• No effects on cell-cycle distribution, apoptosis, and cell invasion
α-Secretase inhibitor (ASI)	INCB3619	✓					✓	✓	• In combination with DAPT, represses *HES1* and *HEY1* expression
Others	Arsenic trioxide (ATO)			✓		✓			• Enhances radiation-induced killing of Glioblastoma cells • Decreases Notch1, Hes1, nestin, and CD133 levels
	Tipifarnib			✓					• Sensitizes GSCs to GSIs • Combined treatment with RO4929097 sensitizes cells to radiation and reduces in vivo tumor growth
	Honokiol			✓					• Combined treatment with o6-BG increases TMZ sensitivity and suppresses Notch3 and Hes1 mRNA levels
	mAb428.2			✓			✓	✓	• Blocks ADAM17 activation • Enhancement of inflammatory macrophage infiltration • Reduces in vivo vascularization

Table 2. *Cont.*

Class of Inhibitors	Molecules	Biological Effects							
		Decreased Cell Growth	Anti-Proliferative Activity	Pro-Apoptotic Activity	Reduced Neurosphere Formation	Reduced Colony Formation	Decreased Tumor Size In Vivo	Prolonged Animal Survival	Others
Others	ZR30						✓	✓	• Prevents MMP2 activation • Limits tumor invasion • Blocks EGFR/Notch/AKT signaling
	dnMAML peptide		✓	✓					• Reduces Hes1 and Hey3 expression
	Retinoic acid	✓			✓	✓			• Downregulates the Hes and Hey family • Induces neural differentiation
	aPKC-PSP			✓					• Reduces Notch1 levels

MRK003 reduces cell growth in vitro and in vivo and sensitizes cell lines and neurospheres to radiation and TMZ [153]. It was demonstrated that MRK003 has a strong therapeutic potential in CD44^{high}/CD133^{low} GICs [150]. Intriguingly, Kahlert et al., through the metabolomic analysis of MRK003-treated Glioblastoma neurospheres, found reduced levels of intracellular glutamate, glutaminase (known to promote cancer cell proliferation), phosphocoline (which is elevated in fast-dividing glioma cells and in malignant high-grade brain tumors), and glycine (involved in the survival regulation of hypoxic glioma cells) [154]. However, MRK003 and GSI-XVII reduce the expression of stemness markers, such as CD133, nestin, BMI-1, and OLIG2, concomitant with the reduction of neurosphere formation and clonogenicity ability in vitro and in vivo. Moreover, both GSIs increase apoptosis, reducing AKT and STAT3 phosphorylation [150,155], and GSCs pretreated with MRK003 and GSI-XVIII show reduced tumor formation in vivo. In order to avoid GSI gastrointestinal toxicity, Fan et al. administered inhibitors through the brain implantation of drug-impregnated polymer beads that effectively blocked tumor growth and significantly prolonged animal survival [155].

Finally, GSI-X significantly impairs c-CSC cell growth compared with p-CSC pools, with no effects observed in cell-cycle distribution, apoptosis, and cell-invasion assays [156].

In conclusion, GSIs, being pan-Notch inhibitors, cause intestinal toxicity through the goblet cell metaplasia of the small intestinal epithelium. To limit toxicity, a preclinical study using antibodies for specific Notch receptor has shown that inhibition of the Notch1 receptor alone induces mild goblet cell metaplasia, whereas the inhibition of Notch2 receptor alone can eliminate this effect [157]

5.5.2. α-Secretase Inhibitors

In the literature, only one ASI has been used in Glioblastoma. Floyd et al. found that INCB3619 decreases cell growth and tumor size and prolongs the survival of a Glioblastoma animal model [158]. Moreover, combined treatment with DAPT represses *HES1* and *HEY1* expression, as well as *LIF* and *YKL-40* levels, two new key players in Glioblastoma pathogenesis [158,159].

5.5.3. Other Molecules

Besides ASIs and GSIs, other molecules have been employed to block Notch signaling in Glioblastoma with the aim of overcoming side effects and resistance to GSIs.

Arsenic trioxide (ATO) is an inorganic compound that was approved by the Food and Drug Administration (FDA) in 2000 for the treatment of acute promyelocytic leukemia (APL) because of its strong anti-growth APL-derived stem cells [160]. With this rationale, Zhen et al. discovered that ATO reduces colony formation and nestin expression, induces apoptosis, and enhances the radiation-induced killing of Glioblastoma cells by decreasing Notch1 and Hes1 protein levels [161]. These data were confirmed by other studies, which further demonstrated that ATO treatment decreases CD133 expression and induces apoptosis through the repression of phosphorylation of AKT and STAT3 through the Notch pathway [162,163].

Tipifarnib, a farnesyltransferase inhibitor, sensitizes GSCs to GSIs, whereas the combined treatment with RO4929097 makes tumor cells more sensitive to radiation, resulting in significantly reduced tumor growth and improved survival in animal models. Intriguingly, non-stem GSCs are resistant to treatment, suggesting that combined treatment selectively targets the CSC pool [164].

Honokiol, a natural extract from different *Magnolia* species, can readily cross the blood-brain barrier and shows pro-apoptotic activity in Glioblastoma [165]. Combined treatment with O^6-benzylguanine (O6-BG), an MGMT inhibitor, increases TMZ sensitivity and suppresses Notch3 and Hes1 mRNA levels with pro-apoptotic effects on GSCs [166].

Fibulin-3, an extracellular matrix glycoprotein, is highly expressed in Glioblastoma and functions as an autocrine/paracrine activator of NF-κB and the Notch pathway, promoting tumor invasion, angiogenesis, and drug resistance, besides being a marker of poor prognosis [167]. Nandhu et al. developed an antibody against fibulin-3 called mAb428.2, which is able to prevent the fibulin-3-activation of ADAM17, resulting in cell apoptosis, enhancement of inflammatory macrophage infiltration, reduced tumor growth and vascularization, and extended survival in Glioblastoma mouse models [167]. Always with regard to fibulin-3, Li et al. developed ZR30, an in vitro synthetized protein, based on the human fibulin-3 protein variant (ETSP), lacking the N-terminal signal peptide (responsible for extracellular exportation). ZR30 prevents MMP2 activation, thus limiting tumor invasion, blocking EGFR/Notch/AKT signaling, exerting an anti-tumor effect on different Glioblastoma cell subpopulations in a mouse model miming intratumor heterogeneity [168].

A dominant negative MAML peptide was developed in order to inhibit Notch signaling, and it was demonstrated that it has anti-proliferative and pro-apoptosis effects, decreasing Hes1 and Hey3 expression [153,169]. Hes and Hey family members are also downregulated by retinoic acid, which can inhibit sphere and colony formation, promote cell-growth arrest, and induce neural differentiation, both in vitro and in vivo [170].

Protein kinase C iota (PRKCi) is in close proximity with Notch1. In a study by Phillips et al., it was found that inhibiting PRKCi by aPKC-PSP results in lower Notch1 levels and increased GSC death [171].

5.6. Therapeutic Resistance to Notch Inhibitors

The elevated heterogeneity of Glioblastoma tissue is why the therapeutic approach fails. Concerning Notch-targeting strategies (Figure 5), it was shown that GSCs express higher levels of *RBPJK* compared to non-GSCs, and *RBPJK* knockdown reduced tumor propagation in vitro and in vivo [172]. Notably, RBPJK could regulate a different transcription program than Notch, binding CDK9 and therefore affecting Pol-II transcription elongation. Targeting CDK9, or even c-Myc, an upstream regulator of RBPJK, decreased the propagation of GSCs and prolonged survival in an orthotopic mouse model [172]. Another possible resistance mechanism is the overactivation of the Hedgehog pathway under Notch inhibition, which in turn increases GLI-1 expression via the inhibition of Hes1. Indeed, targeting both Notch and Hedgehog increases apoptosis, inhibiting cell growth and colony-forming ability more dramatically compared to monotherapy [173]. The upregulation of Dll4-Notch signaling might have a possible role in chemoresistance and be related to a classical vascular pattern, tumor edema, and MGMT-methylated promoter [97].

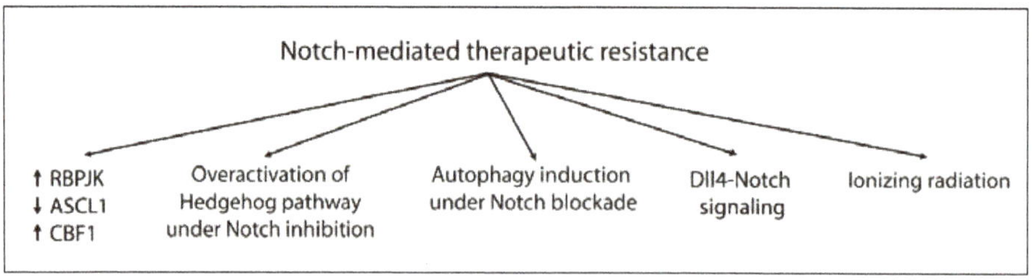

Figure 5. Potential mechanisms involved in Notch-mediated therapeutic resistance.

High CBF1 levels are also implicated in drug resistance. Maciaczyk et al. found that targeting CBF1 impairs glioma invasion, suppressing Zeb1, an activator of the epithelial-mesenchymal transition program, and sensitizes cells to drugs other than GSIs. Unfortunately, it was reported that a CBF1 blockade can activate Notch target genes [174].

It was also reported that ionizing radiation induces Notch pathway activation, resulting in GSC expansion, whereas Notch inhibition sensitizes GSCs, but not non-GSCs, to radiotherapy [175]. Similarly, CD133^{high} GSCs were resistant to MRK003 treatment due the high level of drug-resistance genes, such as *BCRP1* [176]. It was also demonstrated that numerous patient-derived GSCs, responding to Notch inhibition, express high levels of ASCL1, a proneural transcription factor involved in normal neurogenesis. ASCL1^{high} GSCs exhibit a latent capacity for terminal neuronal differentiation in response to Notch signaling inhibition, whereas ASCL1^{low} GSCs do not [177].

Finally, Nastumeda et al. found that MRK003 promotes autophagy in GSCs, as revealed by the higher expression of LC3B-II/LC3B-I autophagy markers after treatment, resulting in chemoresistance induction. Fortunately, this can be avoided by treating cells with autophagy inhibitors together with GSIs [178].

5.7. Glioblastoma Clinical Studies and Notch Inhibitors

In the last decade, several clinical trials have been conducted to evaluate the dose-limiting toxicity (DLT) and tumor-suppressive effects of Notch inhibitors in Glioblastoma patients. Despite different clinical studies being conducted, only two molecules were tested: RO4929097, a γ-secretase inhibitor, and CB-103, a novel first-in-class orally active small-molecule inhibitor of the Notch transcriptional activation complex in the nucleus [179].

Targeting recurrent or progressive Glioblastoma patients with GSI RO4929097 (NCT01122901; https://clinicaltrials.gov/ct2/show/NCT01122901) showed that the six-month and two-year progression-free survival was 1.7 months, while OS (two years) was 6.7 months. Moreover, according to the response assessment in the neuro-oncology (RANO) criteria, the majority of enrolled patients presented disease progression, one complete remission, and three occasions of stable disease. The lack of activity by RO4929097 is probably due to its auto-induction metabolism, resulting in an increased CYP3A4 activity and a marked reduction of steady-state drug levels [180].

Since the standard of care for glioblastoma patients is radiotherapy followed by chemotherapy, a phase 0/I was conducted to study the effect of chemo-radiotherapy in combination with RO4929097 in newly diagnosed malignant gliomas (NCT01119599; https://clinicaltrials.gov/ct2/show/NCT01119599). The combined regimen was well-tolerated and no dose-limiting toxicities were observed. The study showed a significant decrease in proliferation and NICD expression by tumor cells and blood vessels, with a concomitant reduction of the $CD133^+$ GSC subpopulation. Investigators also reported alterations in the angiogenesis pathways under treatment, although the tumor was able to adapt to perturbations in the microenvironment, switching to a Notch-independent form of angiogenesis, highlighting the necessity of concomitantly targeting multiple signaling pathways in Glioblastoma [181]. Angiogenesis is the issue of another study (NCT01189240; https://clinicaltrials.gov/ct2/show/NCT01189240), in which the combined treatment of RO4929097 with bevacizumab on patients with progressive or recurrent malignant glioma was evaluated. The drug combination was well-tolerated, and two (both Glioblastoma) of 12 evaluable subjects showed a radiographic response with complete or partial remission, although bevacizumab has often been associated with an invasive phenotype [182].

Finally, a new clinical trial (phase I/IIA) began in January 2018 (NCT03422679; https://clinicaltrials.gov/ct2/show/record/NCT03422679). In this study, CB-103 was employed to treat several advanced or metastatic solid tumors and haematological malignancies, including Glioblastoma. The current primary outcomes are to evaluate DLT and antitumor efficacy. The study is ongoing.

6. Conclusions

Glioblastoma is the most aggressive types of brain tumor, with a final mortality rate close to 100%. Conventional therapies have failed to improve patient survival due to a small subpopulation of cancer cells known as GSCs that exhibit an enhanced self-renewal capacity, compromised differentiation, in vivo tumorigenicity, and resistance to radio- and chemotherapy. Blocking certain pathways can be useful to induce GSC terminal differentiation and apoptosis in order to reduce their survival. Among them, the Notch signaling pathway is one of the most important. The dysregulation of Notch signaling can take place at different levels, involving genetic, epigenetic, miRNA, and protein regulation in order to maintain cells in an undifferentiated state. Therefore, the Notch pathway could represent an attractive target for treatment in order to induce cell differentiation and kill both undifferentiated and differentiated tumor cells. The most employed and effective inhibitors in Glioblastoma are GSIs, which inhibit the activity of the γ-secretase complex, resulting in no formation of the active Notch form. Even if Notch-targeted monotherapy seems to be promising, a better accomplishment is to combine Notch inhibitors with other molecules or chemotherapeutic agents like bevacizumab and TMZ, respectively. This is because Notch signaling is at the center of a signaling network, which also includes pathways such as PI3K/AKT, NF-κB, STAT3, Hedgehog, and Wnt/β-catenin, which are involved in cell differentiation, cell growth, and survival. Although effective, GSIs present many several side effects, like acute gastrointestinal toxicity. Thus, new compounds or other administration routes should be developed to avoid side effects. To further complicate the possibility of achieving satisfactory results in patient treatment, no specific Notch-related biomarkers are available for Notch-targeted treatment selection and/or treatment response. This situation suggests that patients might be exposed to an ineffective therapeutic regimen, with consequent side effects and higher costs for health institutions.

In conclusion, the Notch pathway is definitely involved in Glioblastoma tumorigenesis, even if its role seems to be quite controversial. However, Notch-signaling inhibition represents a potential therapeutic approach to kill tumor cells, especially in combination with other molecules. Further investigations are needed to better understand the molecular mechanisms of Notch signaling in Glioblastoma pathogenesis, and how they can be overcome to develop an effective therapeutic approach against Notch to improve patient survival.

Author Contributions: R.B. wrote the manuscript and prepared the figures and tables. A.B. critically revised the manuscript. Both authors approved the submitted version of the manuscript and agreed to be personally accountable for their own contribution.

Funding: This research received no external funding.

Conflicts of Interest: The authors declare no conflict of interest.

References

1. Seymour, T.; Nowak, A.; Kakulas, F. Targeting Aggressive Cancer Stem Cells in Glioblastoma. *Front. Oncol.* **2015**, *5*, 159. [CrossRef] [PubMed]
2. Louis, D.N.; Perry, A.; Reifenberger, G.; von Deimling, A.; Figarella-Branger, D.; Cavenee, W.K.; Ohgaki, H.; Wiestler, O.D.; Kleihues, P.; Ellison, D.W. The 2016 World Health Organization Classification of Tumors of the Central Nervous System: A summary. *Acta Neuropathol.* **2016**, *131*, 803–820. [CrossRef] [PubMed]
3. Clarke, J.; Butowski, N.; Chang, S. Recent advances in therapy for glioblastoma. *Arch. Neurol.* **2010**, *67*, 279–283. [CrossRef] [PubMed]
4. Wilson, T.A.; Karajannis, M.A.; Harter, D.H. Glioblastoma multiforme: State of the art and future therapeutics. *Surg. Neurol. Int.* **2014**, *5*, 64. [PubMed]
5. Roy, S.; Lahiri, D.; Maji, T.; Biswas, J. Recurrent Glioblastoma: Where we stand. *South Asian J. Cancer* **2015**, *4*, 163–173. [CrossRef] [PubMed]
6. Baronchelli, S.; Bentivegna, A.; Redaelli, S.; Riva, G.; Butta, V.; Paoletta, L.; Isimbaldi, G.; Miozzo, M.; Tabano, S.; Daga, A.; et al. Delineating the cytogenomic and epigenomic landscapes of glioma stem cell lines. *PLoS ONE* **2013**, *8*, e57462. [CrossRef] [PubMed]
7. Sottoriva, A.; Spiteri, I.; Piccirillo, S.G.; Touloumis, A.; Collins, V.P.; Marioni, J.C.; Curtis, C.; Watts, C.; Tavare, S. Intratumor heterogeneity in human glioblastoma reflects cancer evolutionary dynamics. *Proc. Natl. Acad. Sci. USA* **2013**, *110*, 4009–4014. [CrossRef] [PubMed]
8. Vescovi, A.L.; Galli, R.; Reynolds, B.A. Brain tumour stem cells. *Nat. Rev. Cancer* **2006**, *6*, 425–436. [CrossRef] [PubMed]
9. Alifieris, C.; Trafalis, D.T. Glioblastoma multiforme: Pathogenesis and treatment. *Pharmacol. Ther.* **2015**, *152*, 63–82. [CrossRef] [PubMed]
10. Bradshaw, A.; Wickremsekera, A.; Tan, S.T.; Peng, L.; Davis, P.F.; Itinteang, T. Cancer Stem Cell Hierarchy in Glioblastoma Multiforme. *Front. Surg.* **2016**, *3*, 21. [CrossRef] [PubMed]
11. Ahmed, A.U.; Auffinger, B.; Lesniak, M.S. Understanding glioma stem cells: Rationale, clinical relevance and therapeutic strategies. *Expert Rev. Neurother.* **2013**, *13*, 545–555. [CrossRef] [PubMed]
12. Guruharsha, K.G.; Kankel, M.W.; Artavanis-Tsakonas, S. The Notch signalling system: Recent insights into the complexity of a conserved pathway. *Nat. Rev. Genet.* **2012**, *13*, 654–666. [CrossRef] [PubMed]

13. Andersson, E.R.; Sandberg, R.; Lendahl, U. Notch signaling: Simplicity in design, versatility in function. *Development* **2011**, *138*, 3593–3612. [CrossRef] [PubMed]
14. Yang, X.; Klein, R.; Tian, X.; Cheng, H.T.; Kopan, R.; Shen, J. Notch activation induces apoptosis in neural progenitor cells through a p53-dependent pathway. *Dev. Biol.* **2004**, *269*, 81–94. [CrossRef] [PubMed]
15. Hori, K.; Sen, A.; Artavanis-Tsakonas, S. Notch signaling at a glance. *J. Cell Sci.* **2013**, *126 Pt 10*, 2135–2140. [CrossRef]
16. Del Alamo, D.; Rouault, H.; Schweisguth, F. Mechanism and significance of cis-inhibition in Notch signalling. *Curr. Biol.* **2011**, *21*, R40–R47. [CrossRef] [PubMed]
17. LeBon, L.; Lee, T.V.; Sprinzak, D.; Jafar-Nejad, H.; Elowitz, M.B. Fringe proteins modulate Notch-ligand cis and trans interactions to specify signaling states. *Elife* **2014**, *3*, e02950. [CrossRef] [PubMed]
18. Kopan, R.; Ilagan, M.X. The canonical Notch signaling pathway: Unfolding the activation mechanism. *Cell* **2009**, *137*, 216–233. [CrossRef] [PubMed]
19. Wang, M.M. Notch signaling and Notch signaling modifiers. *Int. J. Biochem. Cell Biol.* **2011**, *43*, 1550–1562. [CrossRef] [PubMed]
20. Chillakuri, C.R.; Sheppard, D.; Lea, S.M.; Handford, P.A. Notch receptor–ligand binding and activation: Insights from molecular studies. *Semin. Cell Dev. Biol.* **2012**, *23*, 421–428. [CrossRef] [PubMed]
21. Fiuza, U.M.; Arias, A.M. Cell and molecular biology of Notch. *J. Endocrinol.* **2007**, *194*, 459–474. [CrossRef] [PubMed]
22. Logeat, F.; Bessia, C.; Brou, C.; LeBail, O.; Jarriault, S.; Seidah, N.G.; Israël, A. The Notch1 receptor is cleaved constitutively by a furin-like convertase. *Proc. Natl. Acad. Sci. USA* **1998**, *95*, 8108–8112. [CrossRef] [PubMed]
23. Zolkiewska, A. ADAM proteases: Ligand processing and modulation of the Notch pathway. *Cell. Mol. Life Sci.* **2008**, *65*, 2056–2068. [CrossRef] [PubMed]
24. Mumm, J.S.; Schroeter, E.H.; Saxena, M.T.; Griesemer, A.; Tian, X.; Pan, D.J.; Ray, W.J.; Kopan, R. A ligand-induced extracellular cleavage regulates gamma-secretase-like proteolytic activation of Notch1. *Mol. Cell* **2000**, *5*, 197–206. [CrossRef]
25. Fortini, M.E. Gamma-secretase-mediated proteolysis in cell-surface-receptor signalling. *Nat. Rev. Mol. Cell Biol.* **2002**, *3*, 673–684. [CrossRef] [PubMed]
26. Zanotti, S.; Canalis, E. Notch Signaling and the Skeleton. *Endocr. Rev.* **2016**, *37*, 223–253. [CrossRef] [PubMed]
27. De Strooper, B.; Annaert, W.; Cupers, P.; Saftig, P.; Craessaerts, K.; Mumm, J.S.; Schroeter, E.H.; Schrijvers, V.; Wolfe, M.S.; Ray, W.J.; et al. A presenilin-1-dependent gamma-secretase-like protease mediates release of Notch intracellular domain. *Nature* **1999**, *398*, 518–522. [CrossRef] [PubMed]
28. Wallberg, A.E.; Pedersen, K.; Lendahl, U.; Roeder, R.G. p300 and PCAF act cooperatively to mediate transcriptional activation from chromatin templates by notch intracellular domains in vitro. *Mol. Cell. Biol.* **2002**, *22*, 7812–7819. [CrossRef] [PubMed]
29. Kurooka, H.; Honjo, T. Functional Interaction between the Mouse Notch1 Intracellular Region and Histone Acetyltransferases PCAF and GCN5. *J. Biol. Chem.* **2000**, *275*, 17211–17220. [CrossRef] [PubMed]
30. Wu, L.; Griffin, J.D. Modulation of Notch signaling by mastermind-like (MAML) transcriptional co-activators and their involvement in tumorigenesis. *Semin. Cancer Biol.* **2004**, *14*, 348–356. [CrossRef] [PubMed]
31. Jarriault, S.; Brou, C.; Logeat, F.; Schroeter, E.H.; Kopan, R.; Israel, A. Signalling downstream of activated mammalian Notch. *Nature* **1995**, *377*, 355–358. [CrossRef] [PubMed]
32. Iso, T.; Sartorelli, V.; Chung, G.; Shichinohe, T.; Kedes, L.; Hamamori, Y. HERP, a new primary target of Notch regulated by ligand binding. *Mol. Cell. Biol.* **2001**, *21*, 6071–6079. [CrossRef] [PubMed]
33. Capaccione, K.M.; Pine, S.R. The Notch signaling pathway as a mediator of tumor survival. *Carcinogenesis* **2013**, *34*, 1420–1430. [CrossRef] [PubMed]
34. Teodorczyk, M.; Schmidt, M.H.H. Notching on Cancer's Door: Notch Signaling in Brain Tumors. *Front. Oncol.* **2014**, *4*, 341.
35. Yin, L.; Velazquez, O.C.; Liu, Z.J. Notch signaling: Emerging molecular targets for cancer therapy. *Biochem. Pharmacol.* **2010**, *80*, 690–701. [CrossRef] [PubMed]
36. Borggrefe, T.; Oswald, F. The Notch signaling pathway: Transcriptional regulation at Notch target genes. *Cell. Mol. Life Sci.* **2009**, *66*, 1631–1646. [CrossRef] [PubMed]
37. Kopan, R. Notch signaling. *Cold Spring Harb. Perspect. Biol.* **2012**, *4*, a011213. [CrossRef] [PubMed]
38. Andersen, P.; Uosaki, H.; Shenje, L.T.; Kwon, C. Non-canonical Notch signaling: Emerging role and mechanism. *Trends Cell Biol.* **2012**, *22*, 257–265. [CrossRef] [PubMed]

39. Sanalkumar, R.; Dhanesh, S.B.; James, J. Non-canonical activation of Notch signaling/target genes in vertebrates. *Cell. Mol. Life Sci.* **2010**, *67*, 2957–2968. [CrossRef] [PubMed]
40. D'Souza, B.; Meloty-Kapella, L.; Weinmaster, G. Canonical and Non-Canonical Notch Ligands. *Curr. Top. Dev. Biol.* **2010**, *92*, 73–129. [PubMed]
41. Mathieu, P.; Adami, P.V.; Morelli, L. Notch signaling in the pathologic adult brain. *Biomol. Concepts* **2013**, *4*, 465–476. [CrossRef] [PubMed]
42. Imayoshi, I.; Sakamoto, M.; Yamaguchi, M.; Mori, K.; Kageyama, R. Essential roles of Notch signaling in maintenance of neural stem cells in developing and adult brains. *J. Neurosci.* **2010**, *30*, 3489–3498. [CrossRef] [PubMed]
43. Basak, O.; Giachino, C.; Fiorini, E.; MacDonald, H.R.; Taylor, V. Neurogenic Subventricular Zone Stem/Progenitor Cells Are Notch1-Dependent in Their Active But Not Quiescent State. *J. Neurosci.* **2012**, *32*, 5654–5666. [CrossRef] [PubMed]
44. Taupin, P. Adult neural stem cells, neurogenic niches, and cellular therapy. *Stem Cell Rev.* **2006**, *2*, 213–219. [CrossRef] [PubMed]
45. Nyfeler, Y.; Kirch, R.D.; Mantei, N.; Leone, D.P.; Radtke, F.; Suter, U.; Taylor, V. Jagged1 signals in the postnatal subventricular zone are required for neural stem cell self-renewal. *EMBO J.* **2005**, *24*, 3504–3515. [CrossRef] [PubMed]
46. Lugert, S.; Basak, O.; Knuckles, P.; Haussler, U.; Fabel, K.; Gotz, M.; Haas, C.A.; Kempermann, G.; Taylor, V.; Giachino, C. Quiescent and active hippocampal neural stem cells with distinct morphologies respond selectively to physiological and pathological stimuli and aging. *Cell Stem Cell* **2010**, *6*, 445–456. [CrossRef] [PubMed]
47. Sestan, N.; Artavanis-Tsakonas, S.; Rakic, P. Contact-dependent inhibition of cortical neurite growth mediated by notch signaling. *Science* **1999**, *286*, 741–746. [CrossRef] [PubMed]
48. Breunig, J.J.; Silbereis, J.; Vaccarino, F.M.; Sestan, N.; Rakic, P. Notch regulates cell fate and dendrite morphology of newborn neurons in the postnatal dentate gyrus. *Proc. Natl. Acad. Sci. USA* **2007**, *104*, 20558–20563. [CrossRef] [PubMed]
49. Stump, G.; Durrer, A.; Klein, A.L.; Lutolf, S.; Suter, U.; Taylor, V. Notch1 and its ligands Delta-like and Jagged are expressed and active in distinct cell populations in the postnatal mouse brain. *Mech. Dev.* **2002**, *114*, 153–159. [CrossRef]
50. Carlen, M.; Meletis, K.; Goritz, C.; Darsalia, V.; Evergren, E.; Tanigaki, K.; Amendola, M.; Barnabe-Heider, F.; Yeung, M.S.; Naldini, L.; et al. Forebrain ependymal cells are Notch-dependent and generate neuroblasts and astrocytes after stroke. *Nat. Neurosci.* **2009**, *12*, 259–267. [CrossRef] [PubMed]
51. Irvin, D.K.; Zurcher, S.D.; Nguyen, T.; Weinmaster, G.; Kornblum, H.I. Expression patterns of Notch1, Notch2, and Notch3 suggest multiple functional roles for the Notch-DSL signaling system during brain development. *J. Comp. Neurol.* **2001**, *436*, 167–181. [CrossRef] [PubMed]
52. Murphy, P.A.; Lam, M.T.; Wu, X.; Kim, T.N.; Vartanian, S.M.; Bollen, A.W.; Carlson, T.R.; Wang, R.A. Endothelial Notch4 signaling induces hallmarks of brain arteriovenous malformations in mice. *Proc. Natl. Acad. Sci. USA* **2008**, *105*, 10901–10906. [CrossRef] [PubMed]
53. Kurisu, J.; Fukuda, T.; Yokoyama, S.; Hirano, T.; Kengaku, M. Polarized targeting of DNER into dendritic plasma membrane in hippocampal neurons depends on endocytosis. *J. Neurochem.* **2010**, *113*, 1598–1610. [CrossRef] [PubMed]
54. Irvin, D.K.; Nakano, I.; Paucar, A.; Kornblum, H.I. Patterns of Jagged1, Jagged2, Delta-like 1 and Delta-like 3 expression during late embryonic and postnatal brain development suggest multiple functional roles in progenitors and differentiated cells. *J. Neurosci. Res.* **2004**, *75*, 330–343. [CrossRef] [PubMed]
55. Yoon, K.J.; Koo, B.K.; Im, S.K.; Jeong, H.W.; Ghim, J.; Kwon, M.C.; Moon, J.S.; Miyata, T.; Kong, Y.Y. Mind bomb 1-expressing intermediate progenitors generate notch signaling to maintain radial glial cells. *Neuron* **2008**, *58*, 519–531. [CrossRef] [PubMed]
56. Shutter, J.R.; Scully, S.; Fan, W.; Richards, W.G.; Kitajewski, J.; Deblandre, G.A.; Kintner, C.R.; Stark, K.L. Dll4, a novel Notch ligand expressed in arterial endothelium. *Genes Dev.* **2000**, *14*, 1313–1318. [PubMed]
57. Alberi, L.; Liu, S.; Wang, Y.; Badie, R.; Smith-Hicks, C.; Wu, J.; Pierfelice, T.J.; Abazyan, B.; Mattson, M.P.; Kuhl, D.; et al. Activity-induced Notch signaling in neurons requires Arc/Arg3.1 and is essential for synaptic plasticity in hippocampal networks. *Neuron* **2011**, *69*, 437–444. [CrossRef] [PubMed]

58. Lavado, A.; Lagutin, O.V.; Chow, L.M.L.; Baker, S.J.; Oliver, G. Prox1 Is Required for Granule Cell Maturation and Intermediate Progenitor Maintenance During Brain Neurogenesis. *PLoS Biol.* **2010**, *8*, e1000460. [CrossRef] [PubMed]
59. Miller, A.C.; Lyons, E.L.; Herman, T.G. cis-Inhibition of Notch by endogenous Delta biases the outcome of lateral inhibition. *Curr. Biol.* **2009**, *19*, 1378–1383. [CrossRef] [PubMed]
60. Borghese, L.; Dolezalova, D.; Opitz, T.; Haupt, S.; Leinhaas, A.; Steinfarz, B.; Koch, P.; Edenhofer, F.; Hampl, A.; Brustle, O. Inhibition of notch signaling in human embryonic stem cell-derived neural stem cells delays G1/S phase transition and accelerates neuronal differentiation in vitro and in vivo. *Stem Cells* **2010**, *28*, 955–964. [CrossRef] [PubMed]
61. Guentchev, M.; Ronald, D.G.M. Notch controls proliferation and differentiation of stem cells in a dose-dependent manner. *Eur. J. Neurosci.* **2006**, *23*, 2289–2296. [CrossRef] [PubMed]
62. Boareto, M.; Iber, D.; Taylor, V. Differential interactions between Notch and ID factors control neurogenesis by modulating Hes factor autoregulation. *Development* **2017**, *144*, 3465–3474. [CrossRef] [PubMed]
63. Engler, A.; Rolando, C.; Giachino, C.; Saotome, I.; Erni, A.; Brien, C.; Zhang, R.; Zimber-Strobl, U.; Radtke, F.; Artavanis-Tsakonas, S.; et al. Notch2 Signaling Maintains NSC Quiescence in the Murine Ventricular-Subventricular Zone. *Cell Rep.* **2018**, *22*, 992–1002. [CrossRef] [PubMed]
64. Than-Trong, E.; Ortica-Gatti, S.; Mella, S.; Nepal, C.; Alunni, A.; Bally-Cuif, L. Neural stem cell quiescence and stemness are molecularly distinct outputs of the Notch3 signalling cascade in the vertebrate adult brain. *Development* **2018**, *145*, dev161034. [CrossRef] [PubMed]
65. Faigle, R.; Song, H. Signaling mechanisms regulating adult neural stem cells and neurogenesis. *Biochim. Biophys. Acta* **2013**, *1830*, 2435–2448. [CrossRef] [PubMed]
66. Hashimoto-Torii, K.; Torii, M.; Sarkisian, M.R.; Bartley, C.M.; Shen, J.; Radtke, F.; Gridley, T.; Šestan, N.; Rakic, P. Interaction between Reelin and Notch signaling regulates neuronal migration in the cerebral cortex. *Neuron* **2008**, *60*, 273–284. [CrossRef] [PubMed]
67. Sibbe, M.; Forster, E.; Basak, O.; Taylor, V.; Frotscher, M. Reelin and Notch1 cooperate in the development of the dentate gyrus. *J. Neurosci.* **2009**, *29*, 8578–8585. [CrossRef] [PubMed]
68. Lewis, J. Neurogenic genes and vertebrate neurogenesis. *Curr. Opin. Neurobiol.* **1996**, *6*, 3–10. [CrossRef]
69. Henrique, D.; Hirsinger, E.; Adam, J.; Le Roux, I.; Pourquie, O.; Ish-Horowicz, D.; Lewis, J. Maintenance of neuroepithelial progenitor cells by Delta-Notch signalling in the embryonic chick retina. *Curr. Biol.* **1997**, *7*, 661–670. [CrossRef]
70. Dorsky, R.I.; Chang, W.S.; Rapaport, D.H.; Harris, W.A. Regulation of neuronal diversity in the Xenopus retina by Delta signalling. *Nature* **1997**, *385*, 67–70. [CrossRef] [PubMed]
71. Lowell, S. Notch signalling: You make me feel so glial. *Curr. Biol.* **2000**, *10*, R595–R597. [CrossRef]
72. Hojo, M.; Ohtsuka, T.; Hashimoto, N.; Gradwohl, G.; Guillemot, F.; Kageyama, R. Glial cell fate specification modulated by the bHLH gene Hes5 in mouse retina. *Development* **2000**, *127*, 2515–2522. [PubMed]
73. Satow, T.; Bae, S.-K.; Inoue, T.; Inoue, C.; Miyoshi, G.; Tomita, K.; Bessho, Y.; Hashimoto, N.; Kageyama, R. The Basic Helix-Loop-Helix Gene hesr2 Promotes Gliogenesis in Mouse Retina. *J. Neurosci.* **2001**, *21*, 1265–1273. [CrossRef] [PubMed]
74. Patten, B.A.; Sardi, S.P.; Koirala, S.; Nakafuku, M.; Corfas, G. Notch1 signaling regulates radial glia differentiation through multiple transcriptional mechanisms. *J. Neurosci.* **2006**, *26*, 3102–3108. [CrossRef] [PubMed]
75. Penton, A.L.; Leonard, L.D.; Spinner, N.B. Notch signaling in human development and disease. *Semin. Cell Dev. Biol.* **2012**, *23*, 450–457. [CrossRef] [PubMed]
76. Takebe, N.; Nguyen, D.; Yang, S.X. Targeting notch signaling pathway in cancer: Clinical development advances and challenges. *Pharmacol. Ther.* **2014**, *141*, 140–149. [CrossRef] [PubMed]
77. Ranganathan, P.; Weaver, K.L.; Capobianco, A.J. Notch signalling in solid tumours: A little bit of everything but not all the time. *Nat. Rev. Cancer* **2011**, *11*, 338–351. [CrossRef] [PubMed]
78. Somasundaram, K.; Reddy, S.P.; Vinnakota, K.; Britto, R.; Subbarayan, M.; Nambiar, S.; Hebbar, A.; Samuel, C.; Shetty, M.; Sreepathi, H.K.; et al. Upregulation of ASCL1 and inhibition of Notch signaling pathway characterize progressive astrocytoma. *Oncogene* **2005**, *24*, 7073–7083. [CrossRef] [PubMed]
79. Kanamori, M.; Kawaguchi, T.; Nigro, J.M.; Feuerstein, B.G.; Berger, M.S.; Miele, L.; Pieper, R.O. Contribution of Notch signaling activation to human glioblastoma multiforme. *J. Neurosurg.* **2007**, *106*, 417–427. [CrossRef] [PubMed]

80. Zhang, X.P.; Zheng, G.; Zou, L.; Liu, H.L.; Hou, L.H.; Zhou, P.; Yin, D.D.; Zheng, Q.J.; Liang, L.; Zhang, S.Z.; et al. Notch activation promotes cell proliferation and the formation of neural stem cell-like colonies in human glioma cells. *Mol. Cell. Biochem.* **2008**, *307*, 101–108. [CrossRef] [PubMed]
81. Phillips, H.S.; Kharbanda, S.; Chen, R.; Forrest, W.F.; Soriano, R.H.; Wu, T.D.; Misra, A.; Nigro, J.M.; Colman, H.; Soroceanu, L.; et al. Molecular subclasses of high-grade glioma predict prognosis, delineate a pattern of disease progression, and resemble stages in neurogenesis. *Cancer Cell* **2006**, *9*, 157–173. [CrossRef] [PubMed]
82. Jing, Y.; Han, Z.; Zhang, S.; Liu, Y.; Wei, L. Epithelial-Mesenchymal Transition in tumor microenvironment. *Cell Biosci.* **2011**, *1*, 29. [CrossRef] [PubMed]
83. Adams, J.M.; Strasser, A. Is Tumor Growth Sustained by Rare Cancer Stem Cells or Dominant Clones? *Cancer Res.* **2008**, *68*, 4018–4021. [CrossRef] [PubMed]
84. Venkatesh, V.; Nataraj, R.; Thangaraj, G.S.; Karthikeyan, M.; Gnanasekaran, A.; Kaginelli, S.B.; Kuppanna, G.; Kallappa, C.G.; Basalingappa, K.M. Targeting Notch signalling pathway of cancer stem cells. *Stem Cell Investig.* **2018**, *5*, 5. [CrossRef] [PubMed]
85. Dragu, D.L.; Necula, L.G.; Bleotu, C.; Diaconu, C.C.; Chivu-Economescu, M. Therapies targeting cancer stem cells: Current trends and future challenges. *World J. Stem Cells* **2015**, *7*, 1185–1201. [PubMed]
86. Dirks, P.B. Brain tumor stem cells: The cancer stem cell hypothesis writ large. *Mol. Oncol.* **2010**, *4*, 420–430. [CrossRef] [PubMed]
87. Yuan, X.; Curtin, J.; Xiong, Y.; Liu, G.; Waschsmann-Hogiu, S.; Farkas, D.L.; Black, K.L.; Yu, J.S. Isolation of cancer stem cells from adult glioblastoma multiforme. *Oncogene* **2004**, *23*, 9392–9400. [CrossRef] [PubMed]
88. Nakamura, Y.; Sakakibara, S.I.; Miyata, T.; Ogawa, M.; Shimazaki, T.; Weiss, S.; Kageyama, R.; Okano, H. The bHLH Gene Hes1 as a Repressor of the Neuronal Commitment of CNS Stem Cells. *J. Neurosci.* **2000**, *20*, 283–293. [CrossRef] [PubMed]
89. Gunther, H.S.; Schmidt, N.O.; Phillips, H.S.; Kemming, D.; Kharbanda, S.; Soriano, R.; Modrusan, Z.; Meissner, H.; Westphal, M.; Lamszus, K. Glioblastoma-derived stem cell-enriched cultures form distinct subgroups according to molecular and phenotypic criteria. *Oncogene* **2008**, *27*, 2897–2909. [CrossRef] [PubMed]
90. Ignatova, T.N.; Xing, H.; Kim, T.M.; Jung, Y.; Huang, W.; Yang Hong, W.; Song, S.; Park Peter, J.; Carroll Rona, S.; Johnson Mark, D. Human cortical glial tumors contain neural stem-like cells expressing astroglial and neuronal markers in vitro. *Glia* **2002**, *39*, 193–206. [CrossRef] [PubMed]
91. Jiang, X.; Xing, H.; Kim, T.M.; Jung, Y.; Huang, W.; Yang Hong, W.; Song, S.; Park Peter, J.; Carroll Rona, S.; Johnson Mark, D. Numb Regulates Glioma Stem Cell Fate and Growth by Altering Epidermal Growth Factor Receptor and Skp1-Cullin-F-Box Ubiquitin Ligase Activity. *Stem Cells* **2012**, *30*, 1313–1326. [CrossRef] [PubMed]
92. Guichet, P.O.; Guelfi, S.; Teigell, M.; Hoppe, L.; Bakalara, N.; Bauchet, L.; Duffau, H.; Lamszus, K.; Rothhut, B.; Hugnot, J.P. Notch1 Stimulation Induces a Vascularization Switch With Pericyte-Like Cell Differentiation of Glioblastoma Stem Cells. *Stem Cells* **2014**, *33*, 21–34. [CrossRef] [PubMed]
93. Qiang, L.; Wu, T.; Zhang, H.W.; Lu, N.; Hu, R.; Wang, Y.J.; Zhao, L.; Chen, F.H.; Wang, X.T.; You, Q.D.; et al. HIF-1α is critical for hypoxia-mediated maintenance of glioblastoma stem cells by activating Notch signaling pathway. *Cell Death Differ.* **2011**, *19*, 284. [CrossRef] [PubMed]
94. Hu, Y.-Y.; Fu, L.-A.; Li, S.-Z.; Chen, Y.; Li, J.-C.; Han, J.; Liang, L.; Li, L.; Ji, C.-C.; Zheng, M.-H.; et al. Hif-1α and Hif-2α differentially regulate Notch signaling through competitive interaction with the intracellular domain of Notch receptors in glioma stem cells. *Cancer Lett.* **2014**, *349*, 67–76. [CrossRef] [PubMed]
95. Hulleman, E.; Quarto, M.; Vernell, R.; Masserdotti, G.; Colli, E.; Kros, J.M.; Levi, D.; Gaetani, P.; Tunici, P.; Finocchiaro, G.; et al. A role for the transcription factor HEY1 in glioblastoma. *J. Cell. Mol. Med.* **2009**, *13*, 136–146. [CrossRef] [PubMed]
96. Brennan, C.; Momota, H.; Hambardzumyan, D.; Ozawa, T.; Tandon, A.; Pedraza, A.; Holland, E. Glioblastoma subclasses can be defined by activity among signal transduction pathways and associated genomic alterations. *PLoS ONE* **2009**, *4*, e7752. [CrossRef] [PubMed]
97. El Hindy, N.; Keyvani, K.; Pagenstecher, A.; Dammann, P.; Sandalcioglu, I.E.; Sure, U.; Zhu, Y. Implications of Dll4-Notch signaling activation in primary glioblastoma multiforme. *Neuro Oncol.* **2013**, *15*, 1366–1378. [CrossRef] [PubMed]

98. Zhang, X.; Chen, T.; Zhang, J.; Mao, Q.; Li, S.; Xiong, W.; Qiu, Y.; Xie, Q.; Ge, J. Notch1 promotes glioma cell migration and invasion by stimulating β-catenin and NF-κB signaling via AKT activation. *Cancer Sci.* **2011**, *103*, 181–190. [CrossRef] [PubMed]
99. Xing, Z.Y.; Sun, L.G.; Guo, W.J. Elevated expression of Notch-1 and EGFR induced apoptosis in glioblastoma multiforme patients. *Clin. Neurol. Neurosurg.* **2015**, *131*, 54–58. [CrossRef] [PubMed]
100. Han, N.; Hu, G.; Shi, L.; Long, G.; Yang, L.; Xi, Q.; Guo, Q.; Wang, J.; Dong, Z.; Zhang, M. Notch1 ablation radiosensitizes glioblastoma cells. *Oncotarget* **2017**, *8*, 88059–88068. [CrossRef] [PubMed]
101. Biswas, S.; Rao, C.M. Epigenetics in cancer: Fundamentals and Beyond. *Pharmacol. Ther.* **2017**, *173*, 118–134. [CrossRef] [PubMed]
102. Dell'albani, P.; Rodolico, M.; Pellitteri, R.; Tricarichi, E.; Torrisi, S.A.; D'Antoni, S.; Zappia, M.; Albanese, V.; Caltabiano, R.; Platania, N.; et al. Differential patterns of NOTCH1-4 receptor expression are markers of glioma cell differentiation. *Neuro Oncol.* **2014**, *16*, 204–216. [CrossRef] [PubMed]
103. Tchorz, J.S.; Tome, M.; Cloetta, D.; Sivasankaran, B.; Grzmil, M.; Huber, R.M.; Rutz-Schatzmann, F.; Kirchhoff, F.; Schaeren-Wiemers, N.; Gassmann, M.; et al. Constitutive Notch2 signaling in neural stem cells promotes tumorigenic features and astroglial lineage entry. *Cell Death Dis.* **2012**, *3*, e325. [CrossRef] [PubMed]
104. Margareto, J.; Larrarte, E.; Leis, O.; Carrasco, A.; Lafuente, J.V.; Idoate, M.A. Gene expression profiling of human gliomas reveals differences between GBM and LGA related to energy metabolism and notch signaling pathways. *J. Mol. Neurosci.* **2007**, *32*, 53–63. [CrossRef] [PubMed]
105. Hai, L.; Zhang, C.; Li, T.; Zhou, X.; Liu, B.; Li, S.; Zhu, M.; Lin, Y.; Yu, S.; Zhang, K.; et al. Notch1 is a prognostic factor that is distinctly activated in the classical and proneural subtype of glioblastoma and that promotes glioma cell survival via the NF-κB(p65) pathway. *Cell Death Dis.* **2018**, *9*, 158. [CrossRef] [PubMed]
106. Cheng, W.; Zhang, C.; Ren, X.; Jiang, Y.; Han, S.; Liu, Y.; Cai, J.; Li, M.; Wang, K.; Liu, Y.; et al. Bioinformatic analyses reveal a distinct Notch activation induced by STAT3 phosphorylation in the mesenchymal subtype of glioblastoma. *J. Neurosurg.* **2017**, *126*, 249–259. [CrossRef] [PubMed]
107. Verhaak, R.G.W.; Hoadley, K.A.; Purdom, E.; Wang, V.; Qi, Y.; Wilkerson, M.D.; Miller, C.R.; Ding, L.; Golub, T.; Mesirov, J.P.; et al. An integrated genomic analysis identifies clinically relevant subtypes of glioblastoma characterized by abnormalities in PDGFRA, IDH1, EGFR and NF1. *Cancer Cell* **2010**, *17*, 98. [CrossRef] [PubMed]
108. Cooper, L.A.; Gutman, D.A.; Long, Q.; Johnson, B.A.; Cholleti, S.R.; Kurc, T.; Saltz, J.H.; Brat, D.J.; Moreno, C.S. The proneural molecular signature is enriched in oligodendrogliomas and predicts improved survival among diffuse gliomas. *PLoS ONE* **2010**, *5*, e12548. [CrossRef] [PubMed]
109. Ohgaki, H.; Kleihues, P. The definition of primary and secondary glioblastoma. *Clin. Cancer Res.* **2013**, *19*, 764–772. [CrossRef] [PubMed]
110. Spino, M.; Kurz, S.C.; Chiriboga, L.; Serrano, J.; Zeck, B.; Sen, N.; Patel, S.; Shen, G.; Vasudevaraja, V.; Tsirigos, A.; et al. Cell Surface Notch Ligand DLL3 is a Therapeutic Target in Isocitrate Dehydrogenase-mutant Glioma. *Clin. Cancer Res.* **2018**. [CrossRef] [PubMed]
111. Jungk, C.; Mock, A.; Exner, J.; Geisenberger, C.; Warta, R.; Capper, D.; Abdollahi, A.; Friauf, S.; Lahrmann, B.; Grabe, N.; et al. Spatial transcriptome analysis reveals Notch pathway-associated prognostic markers in IDH1 wild-type glioblastoma involving the subventricular zone. *BMC Med.* **2016**, *14*, 170. [CrossRef] [PubMed]
112. Huber, R.M.; Rajski, M.; Sivasankaran, B.; Moncayo, G.; Hemmings, B.A.; Merlo, A. Deltex-1 activates mitotic signaling and proliferation and increases the clonogenic and invasive potential of U373 and LN18 glioblastoma cells and correlates with patient survival. *PLoS ONE* **2013**, *8*, e57793. [CrossRef] [PubMed]
113. Tsung, A.J.; Guda, M.R.; Asuthkar, S.; Labak, C.M.; Purvis, I.J.; Lu, Y.; Jain, N.; Bach, S.E.; Prasad, D.V.R.; Velpula, K.K. Methylation regulates HEY1 expression in glioblastoma. *Oncotarget* **2017**, *8*, 44398–44409. [CrossRef] [PubMed]
114. Sun, P.; Xia, S.; Lal, B.; Eberhart, C.G.; Quinones-Hinojosa, A.; Maciaczyk, J.; Matsui, W.; Dimeco, F.; Piccirillo, S.M.; Vescovi, A.L.; et al. DNER, an epigenetically modulated gene, regulates glioblastoma-derived neurosphere cell differentiation and tumor propagation. *Stem Cells* **2009**, *27*, 1473–1486. [CrossRef] [PubMed]
115. Sun, J.; Gong, X.; Purow, B.; Zhao, Z. Uncovering MicroRNA and Transcription Factor Mediated Regulatory Networks in Glioblastoma. *PLoS Comput. Biol.* **2012**, *8*, e1002488. [CrossRef] [PubMed]

116. Li, Y.; Guessous, F.; Zhang, Y.; DiPierro, C.; Kefas, B.; Johnson, E.; Marcinkiewicz, L.; Jiang, J.; Yang, Y.; Schmittgen, T.D.; et al. microRNA-34a inhibits glioblastoma growth by targeting multiple oncogenes. *Cancer Res.* **2009**, *69*, 7569–7576. [CrossRef] [PubMed]

117. Di Bari, M.; Bevilacqua, V.; De Jaco, A.; Laneve, P.; Piovesana, R.; Trobiani, L.; Talora, C.; Caffarelli, E.; Tata, A.M. Mir-34a-5p Mediates Cross-Talk between M2 Muscarinic Receptors and Notch-1/EGFR Pathways in U87MG Glioblastoma Cells: Implication in Cell Proliferation. *Int. J. Mol. Sci.* **2018**, *19*, 1631. [CrossRef] [PubMed]

118. Wu, Z.; Wu, Y.; Tian, Y.; Sun, X.; Liu, J.; Ren, H.; Liang, C.; Song, L.; Hu, H.; Wang, L.; et al. Differential effects of miR-34c-3p and miR-34c-5p on the proliferation, apoptosis and invasion of glioma cells. *Oncol. Lett.* **2013**, *6*, 1447–1452. [CrossRef] [PubMed]

119. Ruan, J.; Lou, S.; Dai, Q.; Mao, D.; Ji, J.; Sun, X. Tumor suppressor miR-181c attenuates proliferation, invasion, and self-renewal abilities in glioblastoma. *Neuroreport* **2015**, *26*, 66–73. [CrossRef] [PubMed]

120. Wong, H.A.; Fatimy, R.E.; Onodera, C.; Wei, Z.; Yi, M.; Mohan, A.; Gowrisankaran, S.; Karmali, P.; Marcusson, E.; Wakimoto, H.; et al. The Cancer Genome Atlas Analysis Predicts MicroRNA for Targeting Cancer Growth and Vascularization in Glioblastoma. *Mol. Ther.* **2015**, *23*, 1234–1247. [CrossRef] [PubMed]

121. Turchi, L.; Debruyne, D.N.; Almairac, F.; Virolle, V.; Fareh, M.; Neirijnck, Y.; Burel-Vandenbos, F.; Paquis, P.; Junier, M.P.; Van Obberghen-Schilling, E.; et al. Tumorigenic potential of miR-18A* in glioma initiating cells requires NOTCH-1 signaling. *Stem Cells* **2013**, *31*, 1252–1265. [CrossRef] [PubMed]

122. Wang, H.; Sun, T.; Hu, J.; Zhang, R.; Rao, Y.; Wang, S.; Chen, R.; McLendon, R.E.; Friedman, A.H.; Keir, S.T.; et al. miR-33a promotes glioma-initiating cell self-renewal via PKA and NOTCH pathways. *J. Clin. Investig.* **2014**, *124*, 4489–4502. [CrossRef] [PubMed]

123. Moutal, A.; Honnorat, J.; Massoma, P.; Desormeaux, P.; Bertrand, C.; Malleval, C.; Watrin, C.; Chounlamountri, N.; Mayeur, M.E.; Besancon, R.; et al. CRMP5 Controls Glioblastoma Cell Proliferation and Survival through Notch-Dependent Signaling. *Cancer Res.* **2015**, *75*, 3519–3528. [CrossRef] [PubMed]

124. Liu, M.; Inoue, K.; Leng, T.; Guo, S.; Xiong, Z.G. TRPM7 channels regulate glioma stem cell through STAT3 and Notch signaling pathways. *Cell Signal.* **2014**, *26*, 2773–2781. [CrossRef] [PubMed]

125. Chigurupati, S.; Venkataraman, R.; Barrera, D.; Naganathan, A.; Madan, M.; Paul, L.; Pattisapu, J.V.; Kyriazis, G.A.; Sugaya, K.; Bushnev, S.; et al. Receptor channel TRPC6 is a key mediator of Notch-driven glioblastoma growth and invasiveness. *Cancer Res.* **2010**, *70*, 418–427. [CrossRef] [PubMed]

126. Kristoffersen, K.; Nedergaard, M.K.; Villingshoj, M.; Borup, R.; Broholm, H.; Kjaer, A.; Poulsen, H.S.; Stockhausen, M.T. Inhibition of Notch signaling alters the phenotype of orthotopic tumors formed from glioblastoma multiforme neurosphere cells but does not hamper intracranial tumor growth regardless of endogene Notch pathway signature. *Cancer Biol. Ther.* **2014**, *15*, 862–877. [CrossRef] [PubMed]

127. Raghu, H.; Gondi, C.S.; Dinh, D.H.; Gujrati, M.; Rao, J.S. Specific knockdown of uPA/uPAR attenuates invasion in glioblastoma cells and xenografts by inhibition of cleavage and trafficking of Notch -1 receptor. *Mol. Cancer* **2011**, *10*, 130. [CrossRef] [PubMed]

128. Wang, X.; Hu, M.; Xing, F.; Wang, M.; Wang, B.; Qian, D. Human cytomegalovirus infection promotes the stemness of U251 glioma cells. *J. Med. Virol.* **2017**, *89*, 878–886. [CrossRef] [PubMed]

129. Wang, J.; Wang, C.; Meng, Q.; Li, S.; Sun, X.; Bo, Y.; Yao, W. siRNA targeting Notch-1 decreases glioma stem cell proliferation and tumor growth. *Mol. Biol. Rep.* **2012**, *39*, 2497–2503. [CrossRef] [PubMed]

130. Sivasankaran, B.; Degen, M.; Ghaffari, A.; Hegi, M.E.; Hamou, M.F.; Ionescu, M.C.; Zweifel, C.; Tolnay, M.; Wasner, M.; Mergenthaler, S.; et al. Tenascin-C is a novel RBPJkappa-induced target gene for Notch signaling in gliomas. *Cancer Res.* **2009**, *69*, 458–465. [CrossRef] [PubMed]

131. Li, Y.; He, Z.C.; Zhang, X.N.; Liu, Q.; Chen, C.; Zhu, Z.; Chen, Q.; Shi, Y.; Yao, X.H.; Cui, Y.H.; et al. Stanniocalcin-1 augments stem-like traits of glioblastoma cells through binding and activating NOTCH1. *Cancer Lett.* **2018**, *416*, 66–74. [CrossRef] [PubMed]

132. Kim, S.H.; Kim, E.J.; Hitomi, M.; Oh, S.Y.; Jin, X.; Jeon, H.M.; Beck, S.; Jin, X.; Kim, J.K.; Park, C.G.; et al. The LIM-only transcription factor LMO2 determines tumorigenic and angiogenic traits in glioma stem cells. *Cell Death Differ.* **2015**, *22*, 1517–1525. [CrossRef] [PubMed]

133. Jeon, H.M.; Jin, X.; Lee, J.S.; Oh, S.Y.; Sohn, Y.W.; Park, H.J.; Joo, K.M.; Park, W.Y.; Nam, D.H.; DePinho, R.A.; et al. Inhibitor of differentiation 4 drives brain tumor-initiating cell genesis through cyclin E and notch signaling. *Genes Dev.* **2008**, *22*, 2028–2033. [CrossRef] [PubMed]

134. Chen, G.; Kong, J.; Tucker-Burden, C.; Anand, M.; Rong, Y.; Rahman, F.; Moreno, C.S.; Van Meir, E.G.; Hadjipanayis, C.G.; Brat, D.J. Human Brat ortholog TRIM3 is a tumor suppressor that regulates asymmetric cell division in glioblastoma. *Cancer Res.* **2014**, *74*, 4536–4548. [CrossRef] [PubMed]
135. Zhu, T.S.; Costello, M.A.; Talsma, C.E.; Flack, C.G.; Crowley, J.G.; Hamm, L.L.; He, X.; Hervey-Jumper, S.L.; Heth, J.A.; Muraszko, K.M.; et al. Endothelial cells create a stem cell niche in glioblastoma by providing NOTCH ligands that nurture self-renewal of cancer stem-like cells. *Cancer Res.* **2011**, *71*, 6061–6072. [CrossRef] [PubMed]
136. Zhang, C.; Hai, L.; Zhu, M.; Yu, S.; Li, T.; Lin, Y.; Liu, B.; Zhou, X.; Chen, L.; Zhao, P.; et al. Actin cytoskeleton regulator Arp2/3 complex is required for DLL1 activating Notch1 signaling to maintain the stem cell phenotype of glioma initiating cells. *Oncotarget* **2017**, *8*, 33353–33364. [CrossRef] [PubMed]
137. Kolosa, K.; Motaln, H.; Herold-Mende, C.; Korsic, M.; Lah, T.T. Paracrine effects of mesenchymal stem cells induce senescence and differentiation of glioblastoma stem-like cells. *Cell Transplant.* **2015**, *24*, 631–644. [CrossRef] [PubMed]
138. Irshad, K.; Mohapatra, S.K.; Srivastava, C.; Garg, H.; Mishra, S.; Dikshit, B.; Sarkar, C.; Gupta, D.; Chandra, P.S.; Chattopadhyay, P.; et al. A combined gene signature of hypoxia and notch pathway in human glioblastoma and its prognostic relevance. *PLoS ONE* **2015**, *10*, e0118201. [CrossRef] [PubMed]
139. Bar, E.E.; Lin, A.; Mahairaki, V.; Matsui, W.; Eberhart, C.G. Hypoxia increases the expression of stem-cell markers and promotes clonogenicity in glioblastoma neurospheres. *Am. J. Pathol.* **2010**, *177*, 1491–1502. [CrossRef] [PubMed]
140. Bayin, N.S.; Frenster, J.D.; Sen, R.; Si, S.; Modrek, A.S.; Galifianakis, N.; Dolgalev, I.; Ortenzi, V.; Illa-Bochaca, I.; Khahera, A.; et al. Notch signaling regulates metabolic heterogeneity in glioblastoma stem cells. *Oncotarget* **2017**, *8*, 64932–64953. [CrossRef] [PubMed]
141. Charles, N.; Ozawa, T.; Squatrito, M.; Bleau, A.M.; Brennan, C.W.; Hambardzumyan, D.; Holland, E.C. Perivascular nitric oxide activates notch signaling and promotes stem-like character in PDGF-induced glioma cells. *Cell Stem Cell* **2010**, *6*, 141–152. [CrossRef] [PubMed]
142. Jeon, H.-M.; Kim, S.-H.; Jin, X.; Park, J.B.; Kim, S.H.; Joshi, K.; Nakano, I.; Kim, H. Crosstalk between Glioma-Initiating Cells and Endothelial Cells Drives Tumor Progression. *Cancer Res.* **2014**, *74*, 4482–4492. [CrossRef] [PubMed]
143. Jubb, A.M.; Browning, L.; Campo, L.; Turley, H.; Steers, G.; Thurston, G.; Harris, A.L.; Ansorge, O. Expression of vascular Notch ligands Delta-like 4 and Jagged-1 in glioblastoma. *Histopathology* **2012**, *60*, 740–747. [CrossRef] [PubMed]
144. Rampazzo, E.; Persano, L.; Pistollato, F.; Moro, E.; Frasson, C.; Porazzi, P.; Della Puppa, A.; Bresolin, S.; Battilana, G.; Indraccolo, S.; et al. Wnt activation promotes neuronal differentiation of glioblastoma. *Cell Death Dis.* **2013**, *4*, e500. [CrossRef] [PubMed]
145. Dali, R.; Verginelli, F.; Pramatarova, A.; Sladek, R.; Stifani, S. Characterization of a FOXG1:TLE1 transcriptional network in glioblastoma initiating cells. *Mol. Oncol.* **2018**, *12*, 775–787. [CrossRef] [PubMed]
146. Hovinga, K.E.; Shimizu, F.; Wang, R.; Panagiotakos, G.; Van Der Heijden, M.; Moayedpardazi, H.; Correia, A.S.; Soulet, D.; Major, T.; Menon, J.; et al. Inhibition of notch signaling in glioblastoma targets cancer stem cells via an endothelial cell intermediate. *Stem Cells* **2010**, *28*, 1019–1029. [CrossRef] [PubMed]
147. Kristoffersen, K.; Villingshoj, M.; Poulsen, H.S.; Stockhausen, M.T. Level of Notch activation determines the effect on growth and stem cell-like features in glioblastoma multiforme neurosphere cultures. *Cancer Biol. Ther.* **2013**, *14*, 625–637. [CrossRef] [PubMed]
148. Staberg, M.; Michaelsen, S.R.; Olsen, L.S.; Nedergaard, M.K.; Villingshoj, M.; Stockhausen, M.T.; Hamerlik, P.; Poulsen, H.S. Combined EGFR- and notch inhibition display additive inhibitory effect on glioblastoma cell viability and glioblastoma-induced endothelial cell sprouting in vitro. *Cancer Cell Int.* **2016**, *16*, 34. [CrossRef] [PubMed]
149. Monticone, M.; Biollo, E.; Fabiano, A.; Fabbi, M.; Daga, A.; Romeo, F.; Maffei, M.; Melotti, A.; Giaretti, W.; Corte, G.; et al. z-Leucinyl-leucinyl-norleucinal induces apoptosis of human glioblastoma tumor-initiating cells by proteasome inhibition and mitotic arrest response. *Mol. Cancer Res.* **2009**, *7*, 1822–1834. [CrossRef] [PubMed]
150. Tanaka, S.; Nakada, M.; Yamada, D.; Nakano, I.; Todo, T.; Ino, Y.; Hoshii, T.; Tadokoro, Y.; Ohta, K.; Ali, M.A.; et al. Strong therapeutic potential of gamma-secretase inhibitor MRK003 for CD44-high and CD133-low glioblastoma initiating cells. *J. Neurooncol.* **2015**, *121*, 239–250. [CrossRef] [PubMed]

151. Saito, N.; Fu, J.; Zheng, S.; Yao, J.; Wang, S.; Liu, D.D.; Yuan, Y.; Sulman, E.P.; Lang, F.F.; Colman, H.; et al. A high Notch pathway activation predicts response to gamma secretase inhibitors in proneural subtype of glioma tumor-initiating cells. *Stem Cells* **2014**, *32*, 301–312. [CrossRef] [PubMed]
152. Yahyanejad, S.; King, H.; Iglesias, V.S.; Granton, P.V.; Barbeau, L.M.; van Hoof, S.J.; Groot, A.J.; Habets, R.; Prickaerts, J.; Chalmers, A.J.; et al. NOTCH blockade combined with radiation therapy and temozolomide prolongs survival of orthotopic glioblastoma. *Oncotarget* **2016**, *7*, 41251–41264. [CrossRef] [PubMed]
153. Chen, J.; Kesari, S.; Rooney, C.; Strack, P.R.; Chen, J.; Shen, H.; Wu, L.; Griffin, J.D. Inhibition of notch signaling blocks growth of glioblastoma cell lines and tumor neurospheres. *Genes Cancer* **2010**, *1*, 822–835. [CrossRef] [PubMed]
154. Kahlert, U.D.; Cheng, M.; Koch, K.; Marchionni, L.; Fan, X.; Raabe, E.H.; Maciaczyk, J.; Glunde, K.; Eberhart, C.G. Alterations in cellular metabolome after pharmacological inhibition of Notch in glioblastoma cells. *Int. J. Cancer* **2016**, *138*, 1246–1255. [CrossRef] [PubMed]
155. Fan, X.; Khaki, L.; Zhu, T.S.; Soules, M.E.; Talsma, C.E.; Gul, N.; Koh, C.; Zhang, J.; Li, Y.M.; Maciaczyk, J.; et al. NOTCH pathway blockade depletes CD133-positive glioblastoma cells and inhibits growth of tumor neurospheres and xenografts. *Stem Cells* **2010**, *28*, 5–16. [CrossRef] [PubMed]
156. Cenciarelli, C.; Marei, H.E.; Zonfrillo, M.; Pierimarchi, P.; Paldino, E.; Casalbore, P.; Felsani, A.; Vescovi, A.L.; Maira, G.; Mangiola, A. PDGF receptor alpha inhibition induces apoptosis in glioblastoma cancer stem cells refractory to anti-Notch and anti-EGFR treatment. *Mol. Cancer* **2014**, *13*, 247. [CrossRef] [PubMed]
157. Wu, Y.; Cain-Hom, C.; Choy, L.; Hagenbeek, T.J.; de Leon, G.P.; Chen, Y.; Finkle, D.; Venook, R.; Wu, X.; Ridgway, J.; et al. Therapeutic antibody targeting of individual Notch receptors. *Nature* **2010**, *464*, 1052–1057. [CrossRef] [PubMed]
158. Floyd, D.H.; Kefas, B.; Seleverstov, O.; Mykhaylyk, O.; Dominguez, C.; Comeau, L.; Plank, C.; Purow, B. Alpha-secretase inhibition reduces human glioblastoma stem cell growth in vitro and in vivo by inhibiting Notch. *Neuro Oncol.* **2012**, *14*, 1215–1226. [CrossRef] [PubMed]
159. Faibish, M.; Francescone, R.; Bentley, B.; Yan, W.; Shao, R. A YKL-40-neutralizing antibody blocks tumor angiogenesis and progression: A potential therapeutic agent in cancers. *Mol. Cancer Ther.* **2011**, *10*, 742–751. [CrossRef] [PubMed]
160. Alimoghaddam, K. A review of arsenic trioxide and acute promyelocytic leukemia. *Int. J. Hematol. Oncol. Stem Cell Res.* **2014**, *8*, 44–54. [PubMed]
161. Zhen, Y.; Zhao, S.; Li, Q.; Li, Y.; Kawamoto, K. Arsenic trioxide-mediated Notch pathway inhibition depletes the cancer stem-like cell population in gliomas. *Cancer Lett.* **2010**, *292*, 64–72. [CrossRef] [PubMed]
162. Wu, J.; Ji, Z.; Liu, H.; Liu, Y.; Han, D.; Shi, C.; Shi, C.; Wang, C.; Yang, G.; Chen, X.; et al. Arsenic trioxide depletes cancer stem-like cells and inhibits repopulation of neurosphere derived from glioblastoma by downregulation of Notch pathway. *Toxicol. Lett.* **2013**, *220*, 61–69. [CrossRef] [PubMed]
163. Ding, D.; Lim, K.S.; Eberhart, C.G. Arsenic trioxide inhibits Hedgehog, Notch and stem cell properties in glioblastoma neurospheres. *Acta Neuropathol. Commun.* **2014**, *2*, 31. [CrossRef] [PubMed]
164. Ma, Y.; Cheng, Z.; Liu, J.; Torre-Healy, L.; Lathia, J.D.; Nakano, I.; Guo, Y.; Thompson, R.C.; Freeman, M.L.; Wang, J. Inhibition of Farnesyltransferase Potentiates NOTCH-Targeted Therapy against Glioblastoma Stem Cells. *Stem Cell Rep.* **2017**, *9*, 1948–1960. [CrossRef] [PubMed]
165. Chang, K.H.; Yan, M.D.; Yao, C.J.; Lin, P.C.; Lai, G.M. Honokiol-induced apoptosis and autophagy in glioblastoma multiforme cells. *Oncol. Lett.* **2013**, *6*, 1435–1438. [CrossRef] [PubMed]
166. Lai, I.C.; Shih, P.H.; Yao, C.J.; Yeh, C.T.; Wang-Peng, J.; Lui, T.N.; Chuang, S.E.; Hu, T.S.; Lai, T.Y.; Lai, G.M. Elimination of cancer stem-like cells and potentiation of temozolomide sensitivity by Honokiol in glioblastoma multiforme cells. *PLoS ONE* **2015**, *10*, e0114830. [CrossRef] [PubMed]
167. Nandhu, M.S.; Behera, P.; Bhaskaran, V.; Longo, S.L.; Barrera-Arenas, L.M.; Sengupta, S.; Rodriguez-Gil, D.J.; Chiocca, E.A.; Viapiano, M.S. Development of a Function-Blocking Antibody Against Fibulin-3 as a Targeted Reagent for Glioblastoma. *Clin. Cancer Res.* **2018**, *24*, 821–833. [CrossRef] [PubMed]
168. Li, Y.; Hu, Y.; Liu, C.; Wang, Q.; Han, X.; Han, Y.; Xie, X.S.; Chen, X.H.; Li, X.; Siegel, E.R.; et al. Human fibulin-3 protein variant expresses anti-cancer effects in the malignant glioma extracellular compartment in intracranial xenograft models. *Oncotarget* **2017**, *8*, 106311–106323. [CrossRef] [PubMed]
169. Opacak-Bernardi, T.; Ryu, J.S.; Raucher, D. Effects of cell penetrating Notch inhibitory peptide conjugated to elastin-like polypeptide on glioblastoma cells. *J. Drug Target.* **2017**, *25*, 523–531. [CrossRef] [PubMed]

170. Ying, M.; Wang, S.; Sang, Y.; Sun, P.; Lal, B.; Goodwin, C.R.; Guerrero-Cazares, H.; Quinones-Hinojosa, A.; Laterra, J.; Xia, S. Regulation of glioblastoma stem cells by retinoic acid: Role for Notch pathway inhibition. *Oncogene* **2011**, *30*, 3454–3467. [CrossRef] [PubMed]
171. Phillips, E.; Lang, V.; Bohlen, J.; Bethke, F.; Puccio, L.; Tichy, D.; Herold-Mende, C.; Hielscher, T.; Lichter, P.; Goidts, V. Targeting atypical protein kinase C iota reduces viability in glioblastoma stem-like cells via a notch signaling mechanism. *Int. J. Cancer* **2016**, *139*, 1776–1787. [CrossRef] [PubMed]
172. Xie, Q.; Wu, Q.; Kim, L.; Miller, T.E.; Liau, B.B.; Mack, S.C.; Yang, K.; Factor, D.C.; Fang, X.; Huang, Z.; et al. RBPJ maintains brain tumor–initiating cells through CDK9-mediated transcriptional elongation. *J. Clin. Investig.* **2016**, *126*, 2757–2772. [CrossRef] [PubMed]
173. Schreck, K.C.; Taylor, P.; Marchionni, L.; Gopalakrishnan, V.; Bar, E.E.; Gaiano, N.; Eberhart, C.G. The Notch target Hes1 directly modulates Gli1 expression and Hedgehog signaling: A potential mechanism of therapeutic resistance. *Clin. Cancer Res.* **2010**, *16*, 6060–6070. [CrossRef] [PubMed]
174. Maciaczyk, D.; Picard, D.; Zhao, L.; Koch, K.; Herrera-Rios, D.; Li, G.; Marquardt, V.; Pauck, D.; Hoerbelt, T.; Zhang, W.; et al. CBF1 is clinically prognostic and serves as a target to block cellular invasion and chemoresistance of EMT-like glioblastoma cells. *Br. J. Cancer* **2017**, *117*, 102–112. [CrossRef] [PubMed]
175. Wang, J.; Wakeman, T.P.; Lathia, J.D.; Hjelmeland, A.B.; Wang, X.F.; White, R.R.; Rich, J.N.; Sullenger, B.A. Notch promotes radioresistance of glioma stem cells. *Stem Cells* **2010**, *28*, 17–28. [CrossRef] [PubMed]
176. Liu, G.; Yuan, X.; Zeng, Z.; Tunici, P.; Ng, H.; Abdulkadir, I.R.; Lu, L.; Irvin, D.; Black, K.L.; Yu, J.S. Analysis of gene expression and chemoresistance of CD133+ cancer stem cells in glioblastoma. *Mol. Cancer* **2006**, *5*, 67. [CrossRef] [PubMed]
177. Park, N.I.; Guilhamon, P.; Desai, K.; McAdam, R.F.; Langille, E.; O'Connor, M.; Lan, X.; Whetstone, H.; Coutinho, F.J.; Vanner, R.J.; et al. ASCL1 Reorganizes Chromatin to Direct Neuronal Fate and Suppress Tumorigenicity of Glioblastoma Stem Cells. *Cell Stem Cell* **2017**, *21*, 209–224.e7. [CrossRef] [PubMed]
178. Natsumeda, M.; Maitani, K.; Liu, Y.; Miyahara, H.; Kaur, H.; Chu, Q.; Zhang, H.; Kahlert, U.D.; Eberhart, C.G. Targeting Notch Signaling and Autophagy Increases Cytotoxicity in Glioblastoma Neurospheres. *Brain Pathol.* **2016**, *26*, 713–723. [CrossRef] [PubMed]
179. Weber, D.; Lehal, R.; Frismantas, V.; Bourquin, J.; Bauer, M.; Murone, M.; Radtke, F. Pharmacological Activity of CB-103 in Haematological Malignancies—An Oral Pan-Notch Inhibitor with a Novel Mode of Actcion. *Hematol. Oncol.* **2017**, *35*, 46. [CrossRef]
180. Strosberg, J.R.; Yeatman, T.; Weber, J.; Coppola, D.; Schell, M.J.; Han, G.; Almhanna, K.; Kim, R.; Valone, T.; Jump, H.; et al. A phase II study of RO4929097 in metastatic colorectal cancer. *Eur. J. Cancer* **2012**, *48*, 997–1003. [CrossRef] [PubMed]
181. Xu, R.; Shimizu, F.; Hovinga, K.; Beal, K.; Karimi, S.; Droms, L.; Peck, K.K.; Gutin, P.; Iorgulescu, J.B.; Kaley, T.; et al. Molecular and Clinical Effects of Notch Inhibition in Glioma Patients: A Phase 0/I Trial. *Clin. Cancer Res.* **2016**, *22*, 4786–4796. [CrossRef] [PubMed]
182. Pan, E.; Supko, J.G.; Kaley, T.J.; Butowski, N.A.; Cloughesy, T.; Jung, J.; Desideri, S.; Grossman, S.; Ye, X.; Park, D.M. Phase I study of RO4929097 with bevacizumab in patients with recurrent malignant glioma. *J. Neurooncol.* **2016**, *130*, 571–579. [CrossRef] [PubMed]

cancers

MDPI

Review

The Role of SVZ Stem Cells in Glioblastoma

Christine Altmann [1], Stefanie Keller [1] and Mirko H. H. Schmidt [1,2,3,*]

[1] Institute for Microscopic Anatomy and Neurobiology, University Medical Center of the Johannes Gutenberg University, 55131 Mainz, Germany; Christine.Altmann@unimedizin-mainz.de (C.A.); Stefanie.Keller@unimedizin-mainz.de (S.K.)

[2] German Cancer Consortium (DKTK), partner site Frankfurt/Mainz, 60590 Frankfurt/55131 Mainz, Germany

[3] German Cancer Research Center (DKFZ), 69120 Heidelberg, Germany

* Correspondence: Mirko.Schmidt@unimedizin-mainz.de

Received: 27 February 2019; Accepted: 26 March 2019; Published: 29 March 2019

Abstract: As most common primary brain cancer, glioblastoma is also the most aggressive and malignant form of cancer in the adult central nervous system. Glioblastomas are genetic and transcriptional heterogeneous tumors, which in spite of intensive research are poorly understood. Over the years conventional therapies failed to affect a cure, resulting in low survival rates of affected patients. To improve the clinical outcome, an important approach is to identify the cells of origin. One potential source for these are neural stem cells (NSCs) located in the subventricular zone, which is one of two niches in the adult nervous system where NSCs with the capacity of self-renewal and proliferation reside. These cells normally give rise to neuronal as well as glial progenitor cells. This review summarizes current findings about links between NSCs and cancer stem cells in glioblastoma and discusses current therapeutic approaches, which arise as a result of identifying the cell of origin in glioblastoma.

Keywords: glioblastoma; subventricular zone; neural stem cells; neurogenesis; brain tumor stem cells; therapy

---

1. Introduction

Glioblastoma (GB) is the most frequent form of brain tumor in adults and is associated with a poor prognosis and a short median patient survival [1]. Limited therapeutic options, combined with a poor response to currently used therapies, increased the pressure to discover new genetic, epigenetic and molecular pathways involved in GB to create new therapies. One of the most significant questions in GB research is aiming at the hierarchical organization and the cell of origin. Conventional theories state that cancer arises from an accumulation of somatic mutations, resulting in uncontrolled proliferation as well as selective growth advantage [2,3]. Most commonly, cancer occurs in epithelial tissues [4]. Whether a tumor originates from a differentiated cell, which regains the ability to proliferate, or whether it originates from a stem cell, which already has the capacity to proliferate, is not fully resolved, and depends on the tissue and the tumor itself. The existence of brain tumor propagating cells (BTPCs) and their molecular, genetic, and epigenetic footprint could open new ways of therapeutic approaches. In the last years, diverse tumors could be retraced to mutations in stem cells [4] and various studies have suggested that NSCs might be the cells of origin of GB, including mutated astrocyte-like NSCs from the SVZ [5–8]. Recent studies reported from clinics and mouse models that glioblastoma arise from migration of mutated astrocyte-like NSCs from the SVZ [8].

2. Glioblastoma

2.1. General Facts

Glioma is an umbrella term, compromising around 30 percent of all brain tumors that are thought to grow from intrinsic glia cells. As an umbrella term glioma consolidates different types of tumors including ependymoma, astrocytoma, and oligodendroglioma, which vary in their symptoms, aggressiveness, malignancy, and treatment strategy.

Glioblastoma multiforme (GB) belongs to the category of astrocytoma, is the most common and most aggressive of all malignant glial tumor in adults [1], and is less common in children [9]. Based on the World Health Organization classification, GB is the most malignant form of glioma and is classified as a grade IV tumor (ICD-O 9440/3) [10]. GB can be divided into primary (arising de novo) or secondary (developed from a pre-existing tumor) intrinsic brain tumor, however, 90% of all GB are primary [9]. Specific mutations in the gene of isocitrate dehydrogenase (IDH) 1/2 are characteristic for secondary glioblastomas, which are more frequent in younger patients. High invasiveness of GB is recorded, with tumor cells mainly spreading into distinct brain regions, whereas metastasis into other organs is infrequent [1].

Diagnosis of GB comes with a poor prognosis with high morbidity and mortality [1]. The median survival of patients diagnosed with GB and treated with the common medication is only 12 to 15 months [1]. GB can occur in each age group; however, most of the patients are between 45–75 years old. Gliomas are mainly located in the cerebral cortex of adult brains, with 40% in the frontal lobe, followed by the temporal lobe (29%), the parietal lobe (14%), the occipital lobe (3%) and 14 % of gliomas are positioned in deeper brain structures [11].

2.2. Genetic Alterations

GB features a complex pathogenesis that involves mutations and alterations of several key cellular pathways, associated with cell proliferation, angiogenesis, migration, and survival [9]. However, the lack of effective therapies increases the importance to understand pathogenesis in detail. Cellular signaling pathways involved in GB are reviewed in [9,12].

The most common mutations in GB are found in p53 (85.3–87%) [13,14], the epidermal growth factor receptor (EGFR) (45–57%) [13–15], the platelet-derived growth factor receptor (PDGFR) (60%) [16,17] the mouse double minute homolog 2 (10–15%) (MDM2) [18], the phosphatase and tensin homolog (PTEN) gene (20–34%) [19,20], the RTK/Ras/PI3K signaling pathway (86–89.6%) [13,14] and in the pRB signaling pathway (77–78.9%) [13,16].

2.3. Conventional Therapy

The current therapy of GB is limited and inefficient and focuses on surgical resection of as much of the tumorigenic tissue as possible with subsequent radiation- and chemotherapy, hereby mostly using oral alkylating agent temozolomide [21]. However, this therapy strategy is insufficient and no adequate cure for GB was described, yet. One major disadvantage of GB is the tissue-distribution pattern of the tumors resulting from dispersion of the tumors cells in the neighboring brain tissue [22,23]. This characteristic hinders complete surgical resection, increases recurrence rate and thereby reduces healing abilities.

Despite ongoing research, survival of GB patients could not be increased in the last years. However, there are rare case reports about long-term survival and zero recurrence of single patients with GB, which are explained by young age, as well as aggressive and complete surgical removal [24,25]. These cases, as rarely as they are, rise the hope that GB can be cured and show that more research effort and innovative treatment approaches are desperately needed to better understand the tumor development and progression.

3. Neural Stem Cell Niches in the Adult Brain

Neural stem cells (NSCs), a subpopulation of astroglial cells, are self-renewing cells with the capacity to differentiate into multiple neural cell types like neurons and glial cells (astrocytes and oligodendrocytes) (reviewed in [26]). During development, NSCs are obligatory for the formation of the nervous system. They are most active in this period; however, since 1992 it is described that NSCs can also be found in the adult brain. Here, small populations of NSCs are located in specific stem cell niches that divide occasionally to generate differentiated cells including neurons (neurogenesis) and glial cells (gliogenesis) [27,28].

3.1. Adult Neurogenesis

NSCs have a relative slow division rate and generate progenitor cells, which are able to differentiate into one of the three major cell lineages of the brain: neurons, astrocytes and oligodendrocytes [26,29,30]. The two neurogenic niches in the adult rodent brain, which contain NSCs and produce new neural cells, are the subventricular zone (SVZ) of the lateral ventricle and the dentate gyrus of the hippocampus [31,32]. While adult neurogenesis in the SVZ of humans is widely accepted [33,34], it is still under discussion in the hippocampus [33,35–37]. Recently, Sorrells et al. showed that neurogenesis in the dentate gyrus declines sharply in children and that in the adult brain they could not detect any new neurons [36]. However, shortly after this publication, Boldrini et al. published a manuscript stating that neurogenesis still persists throughout life [37]. Thus, it is still controversial if the potential to produce new neurons still exists in the adult brain; but many studies show that it is drastically decreased as compared to embryonic stages [5,33,38,39]. During the last decades adult neurogenesis was also described in other mammalian brain regions, including the hypothalamus, the cortex, the striatum, and the amygdala [40–43]. Furthermore, an increase of neurogenesis in the adult brain was reported after injury and in disease [44–48].

The impact and consequence of adult neurogenesis in humans is largely unknown and can only be estimated. However, many studies suggest that the generation of new neurons in the adult brain could be important for learning and memory, degradation as well as regeneration processes underlying aging, injury and diseases, including dementia and cancer [49–54].

3.2. Stem Cells in the Subventricular Zone

The SVZ is the largest neurogenic niche in the adult mammalian brain. It is located at the border of both lateral cerebral ventricles, fitting in between the ependyma and the parenchyma of the striatum. The NSCs in the SVZ are found in an astrocytic ribbon in the sub-ependymal zone. They are surrounded by ependymal cells, vascular endothelial cells, astrocytes, and oligodendrocytes [5], which are important to support the stem cells and to control the proliferation rate [55,56]. Next to the SVZ, the brain parenchyma is located, which is mostly composed of differentiated neurons and glia cells.

In adult rodents, the SVZ contains four major cell types: ependymal cells, NSCs, fast proliferation precursors and neuroblasts. The ependymal cells form a monolayer-border to the ventricle. This layer is followed by the other three cell types, which are not arranged in layers and keep close contact to the ependymal layer. Astrocyte-like NSCs have an apical cilium, which extends into the ventricle lumen and might influence cell proliferation and differentiation [57–59]. These astrocyte-like NSCs (type-B cells) occasionally give rise to multipotent intermediate progenitors (type-C cells), which correspond to transit-amplifying cells that further divide to generate neuroblasts (type-A cells) (reviewed in detail in [60]). In rodents, these neuroblasts migrate in chains along the rostral migratory stream (RMS) to the cortical layer of the olfactory bulb (OB) [61,62], where they differentiate into interneurons and integrate into the pre-existing neural circuitry [61–64].

NSCs in the adult human SVZ were first described in 2004 by Sanai et al. [5], followed by several groups, which identified and isolated human NSCs from the anterior SVZ (Figure 1) [65–67]. In the human SVZ, there are some differences in the anatomical organization and in the behavior of NSCs compared to rodents:

(I) The human SVZ features a hypocellular gap between the ependymal layer and the neurogenic astrocytes, which mainly contains ependymal and astrocytic expansions [5,38,68]. This layer is missing in rodents.

(II) The SVZ in humans can be separated into four distinct layers, whereas the rodent SVZ is not separated into layers. The first layer (ependymal layer, layer I) lines the ventricle and is composed of ependymal cells that possess several apical microvilli and basal expansions [38]. This layer is followed by the hypocellular gap forming the second layer (hypocellular layer, layer II) which is characterized by the absence of cell bodies and has an unknown function. As it mainly contains astrocytic and ependymal processes, one theory would be that these are forming connections in layer II [60,69,70]. The third layer (astrocytic ribbon, layer III) mainly contains the astrocyte-like NSCs and neuroblasts [5], whereas the forth layer, the transitional zone (layer IV) mainly contains myelinated axons and oligodendrocytes [38].

(III) In the human SVZ, multipotent transit-amplifying cells (in rodents: type-C cells) are missing. Only NSCs and neuroblasts can be found in the human SVZ.

(IV) The migration of neural progenitor cells from the SVZ along the RMS to the olfactory bulb is still under discussion and the migration of non-neuronal progenitor cells is at least reduced [5,33,34,67,71], whereas migration of these in rodents is verified. In the OB, mostly new oligodendrocytes from the SVZ are needed to maintain the myelin sheath of the neurons [72,73]. The fate of newborn neurons is unclear. It was shown by Ernst et al. in 2014 that in the adult human brain, new neurons from the SVZ are integrated not into the OB but in form of interneurons into the striatum [74]. Thus, the SVZ is still considered as an important pool of neuronal and glial progenitor cells in the adult mammalian brain, which provides an opportunity for neuroregenerative repair, learning, and memory [75–79]. However, this pool was also implicated in other conditions such as injury, neurodegeneration and cancer [80–84].

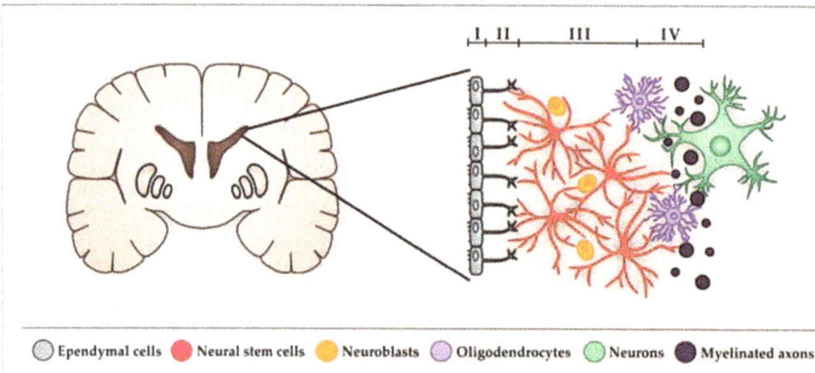

Figure 1. Schematic representation of the subventricular zone (SVZ) in a coronal view of the human brain and the adult human neurogenic niche in the sub-ependymal zone showing the cellular composition. The neurogenic niche in the SVZ can be divided into four layers (layer I–IV). The ependymal layer (layer I) separates the SVZ form the lumen of the ventricle by a thin single layer (monolayer) of ependymal cells (grey), which feature several apical microvilli and basal expansions. The ependyma is followed by the hypocellular layer (layer II), mainly containing astrocytic and ependymal processes. The astrocytic ribbon (layer III) primary contains astrocyte-like neural stem cells (NSCs) (red) and neuroblasts (orange), closely followed by the transitional zone (layer IV) which contains myelinated axons (black) and oligodendrocytes (violet). Thereafter the parenchyma begins, which mostly consists of neurons (green) and glia.

4. Cancer Stem Cells

Gradually, many studies have reported that tumor cells are not homogenous, but that a tumor consists of a variety of cell types forming a hierarchical organization containing slowly dividing cancer stem cells (CSCs) rapidly dividing precursor cells and non-dividing differentiated cells [85,86]. This cellular hierarchy in tumors was identified by genetic, molecular, epigenetic as well as behavioral variations of the cells. But what is the origin of these different tumor cells?

4.1. Cancer Stem Cell Theories

The transformation of a cell into a tumorigenic cell includes multiple mutations. There are two prominent theories about the origin of cancer cells. One property of at least some cancer cells is the ability to divide. However, do cancer cells acquire this ability or do they arise from a stem cell population that already possesses this ability?

The first theory about the origin of CSCs states that any body cell can become a cancer stem cell by mutation, meaning that already differentiated, somatic cells become tumorigenic. Therefore, an accumulation of mutations is needed in oncogenes (gain of function) or tumor suppressor genes (loss of function), which regulate cell growth, to transform somatic cells into CSCs [87–89]. These mutations occur through replication errors or DNA damage, combined with a missing or incorrect repair mechanism [90]. However, many critics on this conventional theory emerged in the last decades, because of the unlikelihood that various mutations occur in one mature (non-dividing) cell with a limited lifespan. These critics favor the second theory, which is called cancer stem cell theory. This theory is based on the self-renewal ability of stem cells or progenitor cells and states that CSCs arise through oncogenic mutation in stem cells. The remaining of this review will deal with this second cancer stem cell theory, including the influence of CSCs on glioblastoma, their origin in the SVZ as well as possible treatment therapies targeting CSCs.

The idea of stem cells derived CSCs was minted by studies using human leukemia cancer cells, which were transferred into immunodeficient mice [91,92]. When characterizing these cells, the authors found that the cells were quite heterogeneous and only a minor portion had the potential of producing leukemia in mice. This suggests that not all cancer cells but only the slowly dividing stem cells have the potential to reproduce the tumor itself [91,93]. Another study addressed breast cancer cells and described the heterogeneous phenotype of the cells. Only a limited number of cells in the tumor displayed tumorigenic potential which they identified by cell surface markers ($CD44^{+}$ $CD24^{-/low}$ $lineage^{-}$) [94]. Thus, targeting these cells by cancer therapy would be most promising.

In addition to their tumorigenic properties and extensive proliferative potential, CSCs share various qualities with normal stem cells: (I) The capacity of multipotency, meaning the ability to differentiate into multiple lineages, self-renewal, and the capacity to divide into either new stem cells or into differentiated cells [86]. (II) A low self-renewal rate and rare occurrence (only one in a million cells) [92,93,95,96]. (III) A strict control by their microenvironment to regulate the balance between proliferation and cell death [97,98]. (IV) The usage of similar signaling pathways [85].

4.2. Neural Brain Tumor Propagating Cells

The hypothesis of CSCs can also be extended to brain tumors, here referred to as brain tumor propagating cells (BTPCs), however, with some minor deviations. As discussed above, stem cells are scarce in the adult brain and can only grow in protective stem cells niches, including the hippocampus and the SVZ [31,32]. These NSCs already possess the ability to proliferate and thus they could transform more easily and rapidly into BTPCs than any other post-mitotic neural cell in the brain [99]. After certain variations, neural precursor cells could become BTPCs (Figure 2). However, other than their offspring, NSCs normally do not leave their neurogenic niches. One hypothesis would be that BTPCs originate from a mutation or deregulation that enables the NSCs to migrate and leave the niche. This exit and a subsequent dysregulation of the stem cell might result in unpredictable proliferation and thus tumorigenesis [100–103].

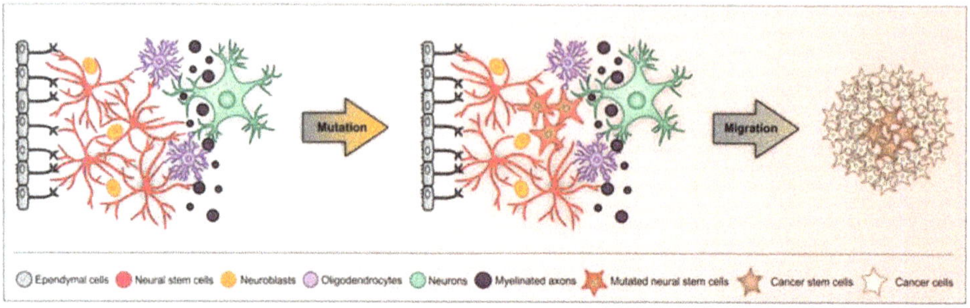

Figure 2. Schematic representation of the origin of brain tumor propagating cells (BTPCs) based on the cancer stem cell theory. In the SVZ of the adult brain, NSCs can be found in the sub-ependymal zone. There are two properties that NSCs need to achieve to become BTPCs. First, the cells need to become tumorigenic, including several mutations, resulting in mutated NSCs. Second these mutated NSCs have to be able to migrate over a long distance into other brain regions, where they form a tumor, which consists of BTPCs and more differentiated cancer cells which originate from the BTPCs themselves.

Like other stem cells, NSCs have the ability of self-renewal and by asymmetric cell division, another NSC and differentiated daughter cell is created. To ensure that the number of NSCs remains constant, proliferation is strictly regulated by intrinsic and extrinsic factors from proliferative cells themselves, neighboring cells, as well as the adjacent blood vessels (reviewed in [104]). Disruption/mutation of these regulatory mechanisms could result in creation of a BTPC. However, brain tumors are not found in the SVZ itself, but rather in the cerebral cortex of adult brains. Thus, besides various tumorigenic mutations, the mutated neural precursor cells need a signal to migrate into other brain regions.

Tumorigenic mutations in migratory SVZ neural precursor cells would be of advantage, because of their ability to migrate over long distances [61,62]. However, in the adult human brain, migration of neuroblasts to the OB is limited, which raises the question if these cells still have the ability to migrate [33,34,67]. Brain injuries, like ischemia, increase the number of neural precursors and some NSCs migrate into the lesion site to increase regeneration, suggesting that the NSCs themselves are able to migrate [105,106]. Furthermore, BTPCs could migrate along tumor microtubes, which have been described as ultra-long membrane protrusions combining BTPC niches scattered in the brain [107,108]. Those BTPC niches are similar to NSC niches with a specific protective microenvironment composed of specific cell types. Based on their location within the tumor and their composition, they are mostly described as peri-vascular, necrotic and hypoxic niches [101,109]. BTPCs in peri-vascular niches are closely located to endothelial cells and express vascular endothelial growth factor VEGF, a well-known growth factor which regulates angiogenesis [110]. Those endothelial cells include capillaries, venules, lymph vessels or arterioles, whereby in case of arterioles, the niche can be directly specified as peri-arteriolar niche [111]. In hypoxic niches, the level of hypoxia-inducible factor HIF is increased, which can increase pro-angiogenic growth factors [112,113]. Most recently, a combined concept was presented, comprising all BTPC niches to one integral hypoxic peri-arteriolar niche model (reviewed in [114]).

Over the years, various studies on mouse models of brain tumors suggested that the cell of origin for brain tumors are NSCs located in the neurogenic niches [6,115–118]. These NSCs further proliferate in other brain regions, forming a tumor, which consist of BTPCs and more differentiated cancer cells. In the following the cancer stem cell theory is discussed in the context of glioblastoma.

5. The Cell of Origin in Glioblastoma

Over the years of glioblastoma research, various theories emerged about the cell of origin, including tissue-specific stem cells like NSCs or committed precursor cells, like astrocyte precursor cells (APCs) and oligodendrocyte precursor cells (OPCs) [119,120].

5.1. NSCs as the Cells of Origin in Glioblastoma

Gradually, several groups identified that glioblastomas are organized hierarchically and contain tumorigenic cells, which could correspond to BTPCs [121–125] with the basic BTPC and NSC characteristics like multi-lineage potency, self-renewal capacity and the ability of tumor initiation and migration [122,123,126–128]. Another evidence for an involvement of undifferentiated BTPCs in GB was provided by discovering that the cells share molecular pathways with NSCs, e.g., the Notch receptor activation. Notch 1 along with its signaling partners is important for NSC maintenance [129] and glioma cell survival [130]. Furthermore, BTPCs and NSCs both express characteristic genes, which can be found in human tumor tissue [125], including Nestin [131–136], CD133 [133,137], Sox [132], Musashi1 [136,138], Olig1/2 [139,140], Ras [141], Akt [141], and GFAP [142,143]. However, these genes are not exclusively expressed in NSCs and BTPCs, but can be found in various other cell types, e.g., GFAP is expressed in astrocytes and radial glial cells. Also, the inactivation of certain tumor suppression genes is similar, e.g., PTEN, which increases proliferation of NSCs as well as of tumor cells [144].

One of the neurogenic niches, the SVZ, came into the focus of BTPC research. Already in the early 21st century, GB was linked to the NSCs of the SVZ. Ignatova et al. isolated cancer cells from human glioblastoma, took them in culture and discovered that these tumors contain neurosphere-forming cells [121]. Those BTPC niches Comparing the structural organization of these tumor-derived spheres to those derived from adult human SVZ highlighted a similar hierarchy, concerning the general organization and an outward gradient of differentiation [145]. This gradient in neurospheres is believed to be important for maintaining the stem cell pool. However, for a long time, data from mouse models could not provide a direct link between SVZ NSCs and GB BTPCs.

One novel factor involved in both GB and SVZ NSCs is the extracellular matrix protein epidermal growth factor-like protein 7 (EGFL7), which was previously described to be secreted by endothelial cells as well as NSCs in the SVZ [146–150]. Loss of EGFL7 increased proliferation of NSCs and decreased cell differentiation in the SVZ, mainly via Notch signaling [146,148]. Additionally, the interaction of Notch 1 and EGFR increased survival of GB cells [147] via promoting angiogenesis [15]. The influence of EGFL7 in angiogenesis [149] also plays an important role in GB [150]. EGFL7 is secreted by glioma blood vessels and increases angiogenesis in GB, suggesting inhibition of EGFL7 as a possible treatment strategy [150].

Analyses of the survival of patients with GB revealed that it is crucial if the tumor has a direct connection to the SVZ [151,152]. Patients bearing tumors which were in contact to the SVZ had a shortened survival period [151,153–155], but the size of these tumors was not altered [156] and thereby did not influence the survival time [152]. Potentially, a contact to the SVZ might provide the tumor with a pool of tumorigenic stem cells, therefore increasing the growth and invasiveness of the tumor. Proteomic analyses by Gollapalli et al. revealed significant alterations in acute phase proteins (e.g., proteins-hemopexin, alpha-1-antichymotrypsin), lipid carrying proteins (e.g., apolipoprotein A1), cytoskeletal proteins (e.g., brain acid soluble protein 1, thymosin beta 4), lipid binding proteins, chaperones, and regulating proteins between tumors with SVZ contact and without [157]. However, gene expression studies showed that there is no evidence that tumors with contact to the SVZ are more likely to be stem-cell derived than tumors without any contact to the SVZ [153]. Following this, two scenarios were suggested: [I] Tumors without contact are not derived from SVZ NSCs or [II] NSCs from the SVZ migrate into other brain regions and only then start to form a tumor. Various studies are pointing to the second scenario [8,158], although it could not be explained why the survival of the patients is altered.

The origin of the tumor can influence the malignance and survival of the patients. A study in mice showed the induction of GB originating in NSCs of the SVZ increases tumor development and resistance to cancer drugs, giving a hind to the aggressiveness and the malignance of glioblastoma [159]. Furthermore, chronic inflammation in the SVZ could facilitate and accelerate mutations in NSCs and thus contribute to the transformation into BTPCs (reviewed in [99]). The SVZ represents a unique inflammatory niche, which differs significantly from other brain regions, due to close contact to the cerebrospinal fluid (CSF).

Recently, Lee et al. showed direct genetic evidence that brain BTPCs arise from cells of the SVZ using human patients and glioblastoma mouse models [8]. The authors performed single-cell sequencing and laser microdissection on tumor free SVZ tissue and tumor tissue and found a clonal relationship of driver mutations between the SVZ and GB-derived tissue. This is the first proof that SVZ NSCs indeed are the cell of origin in glioblastoma. Furthermore, the authors could show that astrocyte-like NSCs leave the SVZ and migrate into distant brain regions to form gliomas. Restoration of homeobox protein EMX2 expression, which is normally decreased in cancer tissue, but is also expressed in NSCs of the SVZ, affects the cell cycle in glioblastoma, leading to a cell cycle arrest and cell death [160]. A recent study of the tumorigenic potential in different stages of neuronal progenitor cells showed that the tumorigenic potential decreases with an increasing lineage restriction, hinting that mutations likely occur in early stages of neurogenesis, e.g., in NSCs [161]. Taken together, these latest reports on the involvement of SVZ derived NSCs in GB are a huge step to discover new therapeutic targets for the treatment of GB patients.

5.2. Committed Precursor Cells as the Cell of Origin in Glioblastoma

Another theory about the cell of origin states that glioblastoma BTPCs arise from committed precursor cells, like APCs or OPCs. Proliferating multipotent NSCs can create committed neuronal precursor cells that may further divide and differentiate into mature neurons and committed glial precursor cells (GPCs) that can further differentiate into OPCs to generate mature oligodendrocytes and into APCs to generate mature astrocytes [162,163].

For a long time, astrocytes were believed to be the only proliferating cells in the adult brain [164] und thus, were subject in glioblastoma research, also because of high GFAP levels in glioma tissue [165]. However, these mature astrocytes would need to dedifferentiate to become tumorigenic, which is possible but unlikely [166,167]. Furthermore, GFAP is also expressed by radial glial cells [168] and NSCs of the adult SVZ [55].

Lindberg et al. and Hide et al. introduced OPCs as the cell of origin by specific mouse models to study OPCs in glioblastoma development, like the MADM-based lineage tracing model to mutate sporadically p53/Nf1 [120,169]. Hide et al. suggested that transformation of both OPCs and NSCs could lead to formation of BTPCs with tumorigenic properties [120]. Additionally, Liu et al. noted that it is important to analyze premalignant stages of tumors to identify the cell of origin, because the tumor cell could acquire plasticity and veil their origin [119]. Further, they demonstrated that OPCs, but not NSC or any other NSC-derived lineage, show aberrant growth prior to malignancy. In OPCs are some overlapping marker expressions, like PDGFRα and NG2, which are involved in development of OPCs and are altered in glioma [13,170,171]. OPCs might form a stem cell niche at the tumor border, increasing chemo-radioresistance and promoting recurrence [171].

Taken together, the current data suggests that BTPCs might develop from various stem or progenitor cells, which needs to be considered when developing treatment strategies. NSCs, ASCs, OPCs, and GPCs might all be the cell of origin, which lead to the development of GB. However, most recent studies using state of the art techniques clearly point to an involvement of NSCs in GB [8].

6. Brain Tumor Propagating Cells as Target for Glioblastoma Treatment

Traditional glioma treatments include chemotherapy, radiation, and surgery. However, these approaches are limited due to multiple factors like the development of therapy resistance coming with a decreasing sensitivity of tumor cells towards ionizing radiation, the unspecific targeting of both cancer and healthy brain cells, as well as the recurrence of the tumor after treatment. Recent advances suggest that the failure of therapies might be caused by the assumption that the tumor consists of a homogenous group of cells. In fact, the tumor is distinguishable into many subgroups of cells varying on genetics, epigenetics and phenotype as well as their capabilities to replicate and migrate. Thus, elimination of BTPCs is a promising alternative to specifically target the cell of origin of the tumor, which raises the possibility to prevent tumor proliferation, tumor cell differentiation and BTPC regeneration.

6.1. Complications in Glioblastoma Therapy

Due to specific BTPC characteristics, like slow cell division rate, self-renewal properties, high capacity for DNA repairing and high expression of drug transporters, the identification and targeting of this cell population represents a challenge to this day. Moreover, BTPCs are capable of developing resistance mechanisms in multiple ways complicating conventional drug efficacies. High expression of ATP-binding cassette drug transporters can impede cytotoxic agents to enter the cell [172], resulting in resistance to different chemotherapeutic drugs including the commonly used alkylating agent temozolomide [173,174] and increasing the risk of tumor recurrence after the treatment [175]. Besides the chemo-resistance, BTPCs are capable of developing a radio-resistance by an increase in the activation of the DNA repair machinery, which is promoted by the expression of stem cell marker CD133 [176]. This combined chemo- and radio-resistance hampers a successful treatment and therefore many patients require combinational therapeutic strategies to improve the survival.

Another way how BTPCs escape especially surgery is by forming stem cell niches and using ultra-long membrane protrusions, tumor microtubes, which can be found in various brain tumors and can be used as migration routes for cells located in BTPCs niches scattered in the brain. Disconnecting these tumor microtubes was suggested as a new target for treatment of GB [107,108].

The brain and especially brain tumors are always considered as extremely difficult for treatment, due to the blood–brain barrier (BBB). The BBB normally hinders harmful substances and toxins to enter the brain via different cellular and molecular components as well as divers transport systems. However, the location of the SVZ at the border to the lateral ventricle introduces a new aspect to the system, the CSF, which is secreted by the choroid plexus, forming the blood–cerebrospinal fluid barrier (CSFB). This barrier is functionally distinct and is not as tight as the BBB; mostly all non-cellular substances can enter the CSF [177]. It raises an opportunity for future drugs to enter the brain via the CSF and thereby increase the range of possible treatment molecules. This new approach was already successfully applied to medulloblastoma, another brain tumor type [178].

6.2. BTPCs as Target in Glioblstoma

Over the years, different approaches to target BTPCs were developed and tested (Figure 3), mostly in combination with traditional treatments and with more or less successful outcome [179,180]. There are three major therapeutic strategies for targeting BTPCs: (I) Targeting specific cell surface markers, signaling pathways or BTPCs microenvironment, (II) the induction of apoptosis or stem cell differentiation, as well as inhibition of autophagy and (III) the application of vaccines or epigenetic drugs.

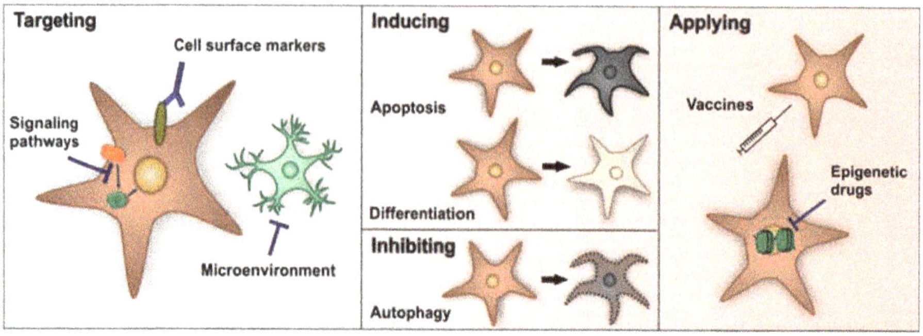

Figure 3. There are three therapeutic strategies to target brain tumor propagating cells (BTPCs). (I) Most common approaches include targeting specific cell surface markers, aberrantly up- or down-regulated signaling pathways or the BTPCs' microenvironment. (II) The induction of apoptosis or stem cell differentiation as well as the inhibition of autophagy, which aims to reduce the number of BTPCs. (III) Application of vaccines or epigenetic drugs, which are still in development.

6.2.1. Targeting Specific Cell Surface Markers of Signaling Pathways or the Microenvironment of BTPCs

The most prominent marker to target the cell surface with antibody-drug conjugates is CD133, which is expressed not only in brain tumors [124], but also in other cancer types like lung cancer [181] or colorectal cancer [182]. The expression of CD133, alone or together with proliferation marker Ki67, is correlated to a glioma patients' prognosis with poor clinical outcome [137,183]. However, only recent studies found a population of BTPCs that does not express CD133 and displays a lower proliferation rate compared to the CD133-positive cell population, with equal tumorigenic characteristics [184]. Thus, targeting CD133-positive cells alone could spare some proliferating BTPCs and thereby increase the reoccurrence rate, but targeting CD133-positive cells could be combined with other treatment strategies.

Targeting aberrant signaling pathways is a common approach in all forms of cancer. As for many cancer cell types, Notch signaling, the Wnt/β-catenin pathway and the PI3K/Akt cascade are among the pathways, which are significantly up-regulated and in the focus for targeting BTPCs in glioblastoma [185–187]. As an example, the upregulation of the Notch pathway in glioma plays a significant role in increasing tumorigenesis by promoting proliferation and cell survival. Evidence emerged that it also directly regulates maintenance and cell differentiation of BTPCs [130,188]. Application of γ-secretase inhibitors to impair Notch signaling led to decreased growth of glioma, but also induced both neuronal and astrocytic differentiation in the stem cell niches and thereby decreased the number of CD133-positive BTPCs [189,190].

Another way to target BTPCs is to inhibit or modify the signaling from their microenvironment that sustains the growth of the cells and is necessary for fundamental processes e.g., the regulation of hypoxia and angiogenesis. These processes are closely linked, with angiogenesis being regulated through hypoxia. The hypoxia-inducible factor (HIF) is regulating numerous genes, including vascular endothelial growth factor (VEGF) expression. The overexpression of HIF-1α and HIF-2α have been linked to poor prognosis in many cancer types [191]. HIF-2α-positive cells are in close contact to BTPCs, which indicates that BTPCs might be able to self-regulate their maintenance through communication with surrounding cells and support of angiogenesis [113]. Different agents to target HIF proteins are in development with some being tested in clinical trials [191] and it was also previously shown that bevacizumab, a VEGF neutralizing antibody, could specifically inhibit the proangiogenic effects of BTPCs [192]. A combinational therapy, targeting both hypoxic cancer cells and angiogenesis, may therefore represent a promising strategy in depleting BTPCs.

6.2.2. The Induction of Apoptosis, Autophagy or Stem Cell Differentiation

A further approach to diminish the number of BTPCs and to erase the tumors origin is the induction of apoptosis. Apoptosis includes a complex signaling network and the evasion of this system is crucial for the stem cell survival as well as tumor development. Agents to manipulate apoptotic pathways include the inhibition of tumor necrosis factor related apoptosis inducing ligand (TRAIL) and bortezomib, a proteasome inhibitor, which, given in combination, could trigger stem cell death in glioblastoma [193]. Also, both EGFR inhibitors, AG1478 and gefitinib, did not only decrease cell proliferation, but also induced apoptosis of BTPCs [194].

Autophagy, a process of degradation and recycling of cellular components, has also been studied in glioblastoma BTPCs, especially in the context of drug-therapy resistance [195]. Modulation of the autophagic process provides an opportunity to increase cell death and to interfere with the cell cycle in BTPCs. Impairment of the autophagic flux decreases the cell self-renewal capacity of BTPCs. Furthermore, it was reported that chemotherapy increases autophagy in cancer cells [196] and that combining cytotoxic drugs and autophagy inhibitors (e.g., chloroquine or quinacrine) increase sensitivity of BTPCs [197]. Combination of bevacizumab or temozolomide with autophagy inhibitor chloroquine increased efficiency of the chemotherapy and affected survival of BTPCs [198,199]. Another novel and promising approach is the autophagy inhibitor quinacrine, which is able to cross the blood-brain barrier and increases the responsiveness of BTPCs to temozolomide and thereby death of BTPCs [200].

Stem cell differentiation also displays a promising strategy to limit the number of cells with tumorigenic potential and to produce cells with higher sensitivity to radio- and chemotherapy. Among the anti-cancer drugs affecting cell differentiation are bone morphogenetic proteins (BMP), retinoic acid and histone deacetylase (HDAC) inhibitors [201]. The application of BMP has been shown to block glioma development and to reduce CD133-positive cells, probably by manipulating symmetric cell cycles that generate new stem cells [202].

6.2.3. The Application of Vaccines or Epigenetic Drugs

HDAC inhibitors are belonging to the category of epigenetic drugs and target reversible histone acetylation and methylation. HDAC inhibitor vorinostat is approved by the Food and Drug Administration [203] and is blocking HDAC SirT1 (silencing information regulator), which both increased apoptosis and cell differentiation of CD133-positive glioma cells [204].

Immune cell therapy has a lot of potential and receives increasing attention in the field of cancer treatments. The dendritic cell vaccine ICT-107 consists of six synthetic peptides derived from tumor-associated and on glioma BTPCs overexpressed antigens (AIM-2, MAGE1, TRP-2, gp100, HER2/neu, and IL-13Rα2). It is currently tested in phase III studies and it was already reported that ICT-107 treatment decreased CD133-positive cells and prolonged survival of the patients [205].

6.2.4. Radiation of the SVZ

Mutated tumor cells might reside in the SVZ and other BTPCs niches, forming a reservoir of BTPCs leading to increased tumor recurrence and decreased outcome of GB patients. An approach to overcome distribution of tumor cells as well as radio resistance would be to decrease the number of BTPCs in the SVZ by radiation. Studies in this field however report conflicting results, depending on the dose of radiation and the size of the resected area [128,206–215].

Besides controversy of the benefit of this treatment, side effects are difficult to estimate. Delivering high doses of radiation to healthy SVZ tissue could decrease quality of life by loss of memory. One recent study on a large group of patients could not find any improvement in survival [216]. This might indicate that BTPCs either do not reside in the SVZ or that there are other reservoirs of BTPCs (BTPC niches) outside the SVZ which lead to a regrow of the tumors.

All these therapeutical approaches are only a small selection of the broad range of agents under development and examination to target BTPCs in GB and how this, especially in combination with classical therapies, may bring benefits in erasing cancer stem cells for a successful treatment of GB patients.

7. Conclusions

Each year, 240,000 cases of glioma are recorded worldwide, with GB accounting for most of these cases. As the most aggressive and malignant form of primary brain cancer, GB leads to thousands of deaths per year. The current standard care of therapy consists of a combination of surgery, radiation, and chemotherapy but the therapeutical success remains poor. This may be due to invasive tumor cells infiltrating neighboring brain regions, intra-tumoral heterogeneity and the capacity of distinct tumor cell populations to develop therapy resistance mechanisms. In this review we summarized various studies, which show that major reasons for the almost inevitable tumor recurrence are BTPCs, which are often spared by conventional therapy. It is suggested that these cells originate from astrocytic-like NSCs in the SVZ with the ability to leave the niche and to migrate over long distances. Through division into either differentiated cells or new BTPCs, they can provide an infinite pool of tumor cells and, additionally, the cells' chemo- and radio-resistance in turn leads to resistance of recurrent GB to standard therapy. The identification of BTPCs and their origin raises hope to identify new molecular, epigenetic and genetic characteristics for the development of combination therapies to erase tumor cells as well as BTPCs. Current research focuses on multiple ways to target BTPCs including targeting specific cell surface markers, signaling pathways or the microenvironment, inducing either apoptosis or cell differentiation, inhibiting autophagy or applying vaccines or epigenetic drugs. Some of the agents are already investigated in clinical trials, showing promising outcomes. Further research is still needed to verify the latest results and technical advantages will help to discover specific GB BTPC treatment to terminate tumor stem cell proliferation and thereby increase the survival of GB patients. Taken together, this leads towards the development of a cure for this dreadful disease.

Author Contributions: C.A. and S.K. wrote and edited the manuscript and M.H.H.S. edited the manuscript.

Funding: S.K. is funded by the German Research Foundation (DFG) Collaborative Research Center 1292, project TP09.

Acknowledgments: We thank members of the Schmidt lab for all their support.

Conflicts of Interest: The authors declare no conflict of interest.

References

1. Wen, P.Y.; Kesari, S. Malignant gliomas in adults. *N. Engl. J. Med.* **2008**, *359*, 492–507. [CrossRef]
2. Greaves, M.; Maley, C.C. Clonal evolution in cancer. *Nature* **2012**, *481*, 306–313. [CrossRef]
3. Merlo, L.M.; Pepper, J.W.; Reid, B.J.; Maley, C.C. Cancer as an evolutionary and ecological process. *Nat. Rev. Cancer* **2006**, *6*, 924–935. [CrossRef] [PubMed]
4. Blanpain, C. Tracing the cellular origin of cancer. *Nat. Cell Biol.* **2013**, *15*, 126. [CrossRef] [PubMed]
5. Sanai, N.; Tramontin, A.D.; Quinones-Hinojosa, A.; Barbaro, N.M.; Gupta, N.; Kunwar, S.; Lawton, M.T.; McDermott, M.W.; Parsa, A.T.; Manuel-Garcia Verdugo, J.; et al. Unique astrocyte ribbon in adult human brain contains neural stem cells but lacks chain migration. *Nature* **2004**, *427*, 740–744. [CrossRef]
6. Alcantara Llaguno, S.; Chen, J.; Kwon, C.H.; Jackson, E.L.; Li, Y.; Burns, D.K.; Alvarez-Buylla, A.; Parada, L.F. Malignant astrocytomas originate from neural stem/progenitor cells in a somatic tumor suppressor mouse model. *Cancer Cell* **2009**, *15*, 45–56. [CrossRef] [PubMed]
7. Tomasetti, C.; Li, L.; Vogelstein, B. Stem cell divisions, somatic mutations, cancer etiology, and cancer prevention. *Science* **2017**, *355*, 1330–1334. [CrossRef]
8. Lee, J.H.; Lee, J.E.; Kahng, J.Y.; Kim, S.H.; Park, J.S.; Yoon, S.J.; Um, J.-Y.; Kim, W.K.; Lee, J.-K.; Park, J.; et al. Human glioblastoma arises from subventricular zone cells with low-level driver mutations. *Nature* **2018**, *560*, 243–247. [CrossRef] [PubMed]

9. Pearson, J.R.D.; Regad, T. Targeting cellular pathways in glioblastoma multiforme. *Signal Transduct. Target. Ther.* **2017**, *2*, 17040. [CrossRef] [PubMed]
10. Louis, D.N.; Ohgaki, H.; Wiestler, O.D.; Cavenee, W.K.; Burger, P.C.; Jouvet, A.; Scheithauer, B.W.; Kleihues, P. The 2007 WHO classification of tumours of the central nervous system. *Acta Neuropathol.* **2007**, *114*, 97–109. [CrossRef]
11. Larjavaara, S.; Mantyla, R.; Salminen, T.; Haapasalo, H.; Raitanen, J.; Jaaskelainen, J.; Auvinen, A. Incidence of gliomas by anatomic location. *Neuro Oncol.* **2007**, *9*, 319–325. [CrossRef]
12. Aldape, K.; Zadeh, G.; Mansouri, S.; Reifenberger, G.; von Deimling, A. Glioblastoma: Pathology, molecular mechanisms and markers. *Acta Neuropathol.* **2015**, *129*, 829–848. [CrossRef]
13. The Cancer Genome Atlas Research Network. Comprehensive genomic characterization defines human glioblastoma genes and core pathways. *Nature* **2008**, *455*, 1061–1068. [CrossRef]
14. Philpott, C.; Tovell, H.; Frayling, I.M.; Cooper, D.N.; Upadhyaya, M. The NF1 somatic mutational landscape in sporadic human cancers. *Hum. Genom.* **2017**, *11*, 13. [CrossRef]
15. Keller, S.; Schmidt, M.H.H. EGFR and EGFRvIII promote angiogenesis and cell invasion in glioblastoma: Combination therapies for an effective treatment. *Int. J. Mol. Sci.* **2017**, *18*, 1295. [CrossRef]
16. Brennan, C.W.; Verhaak, R.G.; McKenna, A.; Campos, B.; Noushmehr, H.; Salama, S.R.; Zheng, S.; Chakravarty, D.; Sanborn, J.Z.; Berman, S.H.; et al. The somatic genomic landscape of glioblastoma. *Cell* **2013**, *155*, 462–477. [CrossRef]
17. Ozawa, T.; Brennan, C.W.; Wang, L.; Squatrito, M.; Sasayama, T.; Nakada, M.; Huse, J.T.; Pedraza, A.; Utsuki, S.; Yasui, Y.; et al. PDGFRA gene rearrangements are frequent genetic events in PDGFRA-amplified glioblastomas. *Genes Dev.* **2010**, *24*, 2205–2218. [CrossRef]
18. Shapiro, W.R.; Green, S.B.; Burger, P.C.; Mahaley, M.S., Jr.; Selker, R.G.; VanGilder, J.C.; Robertson, J.T.; Ransohoff, J.; Mealey, J., Jr.; Strike, T.A.; et al. Randomized trial of three chemotherapy regimens and two radiotherapy regimens and two radiotherapy regimens in postoperative treatment of malignant glioma. Brain Tumor Cooperative Group Trial 8001. *J. Neurosurg.* **1989**, *71*, 1–9. [CrossRef]
19. Duerr, E.M.; Rollbrocker, B.; Hayashi, Y.; Peters, N.; Meyer-Puttlitz, B.; Louis, D.N.; Schramm, J.; Wiestler, O.D.; Parsons, R.; Eng, C.; et al. PTEN mutations in gliomas and glioneuronal tumors. *Oncogene* **1998**, *16*, 2259–2264. [CrossRef]
20. Smith, J.S.; Tachibana, I.; Passe, S.M.; Huntley, B.K.; Borell, T.J.; Iturria, N.; O'Fallon, J.R.; Schaefer, P.L.; Scheithauer, B.W.; James, C.D.; et al. PTEN mutation, EGFR amplification, and outcome in patients with anaplastic astrocytoma and glioblastoma multiforme. *J. Natl. Cancer Inst.* **2001**, *93*, 1246–1256. [CrossRef]
21. Stupp, R.; Mason, W.P.; van den Bent, M.J.; Weller, M.; Fisher, B.; Taphoorn, M.J.; Belanger, K.; Brandes, A.A.; Marosi, C.; Bogdahn, U.; et al. Radiotherapy plus concomitant and adjuvant temozolomide for glioblastoma. *N. Engl. J. Med.* **2005**, *352*, 987–996. [CrossRef]
22. Holland, E.C. Glioblastoma multiforme: The terminator. *Proc. Natl. Acad. Sci. USA* **2000**, *97*, 6242–6244. [CrossRef]
23. Vescovi, A.L.; Galli, R.; Reynolds, B.A. Brain tumour stem cells. *Nat. Rev. Cancer* **2006**, *6*, 425. [CrossRef] [PubMed]
24. Caruso, R.; Pesce, A.; Wierzbicki, V. A very rare case report of long-term survival: A patient operated on in 1994 of glioblastoma multiforme and currently in perfect health. *Int. J. Surg. Case Rep.* **2017**, *33*, 41–43. [CrossRef]
25. Kumar, A.; Deopujari, C.; Karmarkar, V. A case of glioblastoma multiforme with long term survival: Can we predict the outcome? *Turk. Neurosurg.* **2012**, *22*, 378–381. [CrossRef] [PubMed]
26. Joyner, A.L.; Zervas, M. Genetic inducible fate mapping in mouse: Establishing genetic lineages and defining genetic neuroanatomy in the nervous system. *Dev. Dyn. Off. Publ. Am. Assoc. Anat.* **2006**, *235*, 2376–2385. [CrossRef]
27. Reynolds, B.A.; Weiss, S. Generation of neurons and astrocytes from isolated cells of the adult mammalian central nervous system. *Science* **1992**, *255*, 1707–1710. [CrossRef]
28. Reynolds, B.A.; Weiss, S. Clonal and population analyses demonstrate that an EGF-responsive mammalian embryonic CNS precursor is a stem cell. *Dev. Biol.* **1996**, *175*, 1–13. [CrossRef]
29. Simitzi, C.; Karali, K.; Ranella, A.; Stratakis, E. Controlling the outgrowth and functions of neural stem cells: The effect of surface topography. *Chem. Phys. Chem.* **2018**, *19*, 1143–1163. [CrossRef]

30. Das, S.; Srikanth, M.; Kessler, J.A. Cancer stem cells and glioma. *Nat. Clin. Pract. Neurol.* **2008**, *4*, 427. [CrossRef] [PubMed]
31. Lois, C.; Alvarez-Buylla, A. Proliferating subventricular zone cells in the adult mammalian forebrain can differentiate into neurons and glia. *Proc. Natl. Acad. Sci. USA* **1993**, *90*, 2074–2077. [CrossRef]
32. Eriksson, P.S.; Perfilieva, E.; Bjork-Eriksson, T.; Alborn, A.M.; Nordborg, C.; Peterson, D.A.; Gage, F.H. Neurogenesis in the adult human hippocampus. *Nat. Med.* **1998**, *4*, 1313–1317. [CrossRef]
33. Sanai, N.; Nguyen, T.; Ihrie, R.A.; Mirzadeh, Z.; Tsai, H.H.; Wong, M.; Gupta, N.; Berger, M.S.; Huang, E.; Garcia-Verdugo, J.M.; et al. Corridors of migrating neurons in the human brain and their decline during infancy. *Nature* **2011**, *478*, 382–386. [CrossRef]
34. Wang, C.; Liu, F.; Liu, Y.Y.; Zhao, C.H.; You, Y.; Wang, L.; Zhang, J.; Wei, B.; Ma, T.; Zhang, Q.; et al. Identification and characterization of neuroblasts in the subventricular zone and rostral migratory stream of the adult human brain. *Cell Res.* **2011**, *21*, 1534–1550. [CrossRef]
35. Roy, N.S.; Wang, S.; Jiang, L.; Kang, J.; Benraiss, A.; Harrison-Restelli, C.; Fraser, R.A.; Couldwell, W.T.; Kawaguchi, A.; Okano, H.; et al. In vitro neurogenesis by progenitor cells isolated from the adult human hippocampus. *Nat. Med.* **2000**, *6*, 271–277. [CrossRef]
36. Sorrells, S.F.; Paredes, M.F.; Cebrian-Silla, A.; Sandoval, K.; Qi, D.; Kelley, K.W.; James, D.; Mayer, S.; Chang, J.; Auguste, K.I.; et al. Human hippocampal neurogenesis drops sharply in children to undetectable levels in adults. *Nature* **2018**, *555*, 377–381. [CrossRef]
37. Boldrini, M.; Fulmore, C.A.; Tartt, A.N.; Simeon, L.R.; Pavlova, I.; Poposka, V.; Rosoklija, G.B.; Stankov, A.; Arango, V.; Dwork, A.J.; et al. Human hippocampal neurogenesis persists throughout aging. *Cell Stem Cell* **2018**, *22*, 589–599. [CrossRef]
38. Quiñones-Hinojosa, A.; Sanai, N.; Soriano-Navarro, M.; Gonzalez-Perez, O.; Mirzadeh, Z.; Gil-Perotin, S.; Romero-Rodriguez, R.; Berger, M.S.; Garcia-Verdugo, J.M.; Alvarez-Buylla, A. Cellular composition and cytoarchitecture of the adult human subventricular zone: A niche of neural stem cells. *J. Comp. Neurol.* **2006**, *494*, 415–434. [CrossRef]
39. Guerrero-Cazares, H.; Gonzalez-Perez, O.; Soriano-Navarro, M.; Zamora-Berridi, G.; Garcia-Verdugo, J.M.; Quinones-Hinojosa, A. Cytoarchitecture of the lateral ganglionic eminence and rostral extension of the lateral ventricle in the human fetal brain. *J. Comp. Neurol.* **2011**, *519*, 1165–1180. [CrossRef]
40. Mitchell, B.D.; Emsley, J.G.; Magavi, S.S.; Arlotta, P.; Macklis, J.D. Constitutive and induced neurogenesis in the adult mammalian brain: Manipulation of endogenous precursors toward CNS repair. *Dev. Neurosci.* **2004**, *26*, 101–117. [CrossRef]
41. Gould, E. How widespread is adult neurogenesis in mammals? *Nat. Rev. Neurosci.* **2007**, *8*, 481–488. [CrossRef]
42. Fowler, C.D.; Liu, Y.; Wang, Z. Estrogen and adult neurogenesis in the amygdala and hypothalamus. *Brain Res. Rev.* **2008**, *57*, 342–351. [CrossRef]
43. Vessal, M.; Darian-Smith, C. Adult neurogenesis occurs in primate sensorimotor cortex following cervical dorsal rhizotomy. *J. Neurosci.* **2010**, *30*, 8613–8623. [CrossRef]
44. Parent, J.M. The role of seizure-induced neurogenesis in epileptogenesis and brain repair. *Epilepsy Res.* **2002**, *50*, 179–189. [CrossRef]
45. Parent, J.M.; Vexler, Z.S.; Gong, C.; Derugin, N.; Ferriero, D.M. Rat forebrain neurogenesis and striatal neuron replacement after focal stroke. *Ann. Neurol.* **2002**, *52*, 802–813. [CrossRef]
46. Hou, S.W.; Wang, Y.Q.; Xu, M.; Shen, D.H.; Wang, J.J.; Huang, F.; Yu, Z.; Sun, F.Y. Functional integration of newly generated neurons into striatum after cerebral ischemia in the adult rat brain. *Stroke* **2008**, *39*, 2837–2844. [CrossRef]
47. Danilov, A.I.; Covacu, R.; Moe, M.C.; Langmoen, I.A.; Johansson, C.B.; Olsson, T.; Brundin, L. Neurogenesis in the adult spinal cord in an experimental model of multiple sclerosis. *Eur. J. Neurosci.* **2006**, *23*, 394–400. [CrossRef]
48. Chi, L.; Ke, Y.; Luo, C.; Li, B.; Gozal, D.; Kalyanaraman, B.; Liu, R. Motor neuron degeneration promotes neural progenitor cell proliferation, migration, and neurogenesis in the spinal cords of amyotrophic lateral sclerosis mice. *Stem Cells* **2006**, *24*, 34–43. [CrossRef]
49. Drapeau, E.; Montaron, M.F.; Aguerre, S.; Abrous, D.N. Learning-induced survival of new neurons depends on the cognitive status of aged rats. *J. Neurosci.* **2007**, *27*, 6037–6044. [CrossRef]

50. Pereira, A.C.; Huddleston, D.E.; Brickman, A.M.; Sosunov, A.A.; Hen, R.; McKhann, G.M.; Sloan, R.; Gage, F.H.; Brown, T.R.; Small, S.A. An in vivo correlate of exercise-induced neurogenesis in the adult dentate gyrus. *Proc. Natl. Acad. Sci. USA* **2007**, *104*, 5638–5643. [CrossRef]
51. Saxe, M.D.; Battaglia, F.; Wang, J.W.; Malleret, G.; David, D.J.; Monckton, J.E.; Garcia, A.D.; Sofroniew, M.V.; Kandel, E.R.; Santarelli, L.; et al. Ablation of hippocampal neurogenesis impairs contextual fear conditioning and synaptic plasticity in the dentate gyrus. *Proc. Natl. Acad. Sci. USA* **2006**, *103*, 17501–17506. [CrossRef] [PubMed]
52. Aimone, J.B.; Wiles, J.; Gage, F.H. Potential role for adult neurogenesis in the encoding of time in new memories. *Nat. Neurosci.* **2006**, *9*, 723–727. [CrossRef] [PubMed]
53. Mu, Y.; Gage, F.H. Adult hippocampal neurogenesis and its role in Alzheimer's disease. *Mol. Neurodegener.* **2011**, *6*, 85. [CrossRef]
54. Hamilton, L.K.; Joppe, S.E.; Cochard, L.M.; Fernandes, K.J. Aging and neurogenesis in the adult forebrain: What we have learned and where we should go from here. *Eur. J. Neurosci.* **2013**, *37*, 1978–1986. [CrossRef] [PubMed]
55. Lim, D.A.; Alvarez-Buylla, A. Interaction between astrocytes and adult subventricular zone precursors stimulates neurogenesis. *Proc. Natl. Acad. Sci. USA* **1999**, *96*, 7526–7531. [CrossRef]
56. Lim, D.A.; Tramontin, A.D.; Trevejo, J.M.; Herrera, D.G.; Garcia-Verdugo, J.M.; Alvarez-Buylla, A. Noggin antagonizes BMP signaling to create a niche for adult neurogenesis. *Neuron* **2000**, *28*, 713–726. [CrossRef]
57. Tong, C.K.; Han, Y.-G.; Shah, J.K.; Obernier, K.; Guinto, C.D.; Alvarez-Buylla, A. Primary cilia are required in a unique subpopulation of neural progenitors. *Proc. Natl. Acad. Sci. USA* **2014**, *111*, 12438–12443. [CrossRef]
58. Ihrie, R.A.; Álvarez-Buylla, A. Lake-front property: A unique germinal niche by the lateral ventricles of the adult brain. *Neuron* **2011**, *70*, 674–686. [CrossRef] [PubMed]
59. Mirzadeh, Z.; Merkle, F.T.; Soriano-Navarro, M.; Garcia-Verdugo, J.M.; Alvarez-Buylla, A. Neural stem cells confer unique pinwheel architecture to the ventricular surface in neurogenic regions of the adult brain. *Cell Stem Cell* **2008**, *3*, 265–278. [CrossRef]
60. Fuentealba, L.C.; Obernier, K.; Alvarez-Buylla, A. Adult neural stem cells bridge their niche. *Cell Stem Cell* **2012**, *10*, 698–708. [CrossRef]
61. Kornack, D.R.; Rakic, P. The generation, migration, and differentiation of olfactory neurons in the adult primate brain. *Proc. Natl. Acad. Sci. USA* **2001**, *98*, 4752–4757. [CrossRef]
62. Pencea, V.; Bingaman, K.D.; Freedman, L.J.; Luskin, M.B. Neurogenesis in the subventricular zone and rostral migratory stream of the neonatal and adult primate forebrain. *Exp. Neurol.* **2001**, *172*, 1–16. [CrossRef]
63. Altman, J. Autoradiographic and histological studies of postnatal neurogenesis. IV. Cell proliferation and migration in the anterior forebrain, with special reference to persisting neurogenesis in the olfactory bulb. *J. Comp. Neurol.* **1969**, *137*, 433–457. [CrossRef]
64. Lois, C.; Alvarez-Buylla, A. Long-distance neuronal migration in the adult mammalian brain. *Science* **1994**, *264*, 1145–1148. [CrossRef]
65. Ayuso-Sacido, A.; Roy, N.S.; Schwartz, T.H.; Greenfield, J.P.; Boockvar, J.A. Long-term expansion of adult human brain subventricular zone precursors. *Neurosurgery* **2008**, *62*, 223–231. [CrossRef]
66. Roy, N.S.; Benraiss, A.; Wang, S.; Fraser, R.A.; Goodman, R.; Couldwell, W.T.; Nedergaard, M.; Kawaguchi, A.; Okano, H.; Goldman, S.A. Promoter-targeted selection and isolation of neural progenitor cells from the adult human ventricular zone. *J. Neurosci. Res.* **2000**, *59*, 321–331. [CrossRef]
67. Curtis, M.A.; Kam, M.; Nannmark, U.; Anderson, M.F.; Axell, M.Z.; Wikkelso, C.; Holtas, S.; van Roon-Mom, W.M.; Bjork-Eriksson, T.; Nordborg, C.; et al. Human neuroblasts migrate to the olfactory bulb via a lateral ventricular extension. *Science* **2007**, *315*, 1243–1249. [CrossRef] [PubMed]
68. Sawamoto, K.; Hirota, Y.; Alfaro-Cervello, C.; Soriano-Navarro, M.; He, X.; Hayakawa-Yano, Y.; Yamada, M.; Hikishima, K.; Tabata, H.; Iwanami, A.; et al. Cellular composition and organization of the subventricular zone and rostral migratory stream in the adult and neonatal common marmoset brain. *J. Comp. Neurol.* **2011**, *519*, 690–713. [CrossRef]
69. Haydon, P.G. Glia: Listening and talking to the synapse. *Nat. Rev. Neurosci.* **2001**, *2*, 185. [CrossRef]
70. Pascual, O.; Haydon, P.G. Synaptic inhibition mediated by glia. *Neuron* **2003**, *40*, 873–875. [CrossRef]
71. Kam, M.; Curtis, M.A.; McGlashan, S.R.; Connor, B.; Nannmark, U.; Faull, R.L. The cellular composition and morphological organization of the rostral migratory stream in the adult human brain. *J. Chem. Neuroanat.* **2009**, *37*, 196–205. [CrossRef]

72. Bergmann, O.; Liebl, J.; Bernard, S.; Alkass, K.; Yeung, M.S.; Steier, P.; Kutschera, W.; Johnson, L.; Landen, M.; Druid, H.; et al. The age of olfactory bulb neurons in humans. *Neuron* **2012**, *74*, 634–639. [CrossRef]
73. McKenzie, I.A.; Ohayon, D.; Li, H.; de Faria, J.P.; Emery, B.; Tohyama, K.; Richardson, W.D. Motor skill learning requires active central myelination. *Science* **2014**, *346*, 318–322. [CrossRef]
74. Ernst, A.; Alkass, K.; Bernard, S.; Salehpour, M.; Perl, S.; Tisdale, J.; Possnert, G.; Druid, H.; Frisen, J. Neurogenesis in the striatum of the adult human brain. *Cell* **2014**, *156*, 1072–1083. [CrossRef]
75. Serwanski, D.R.; Rasmussen, A.L.; Brunquell, C.B.; Perkins, S.S.; Nishiyama, A. Sequential contribution of parenchymal and neural stem cell-derived oligodendrocyte precursor cells toward remyelination. *Neuroglia* **2018**, *1*, 8. [CrossRef]
76. Faiz, M.; Sachewsky, N.; Gascon, S.; Bang, K.W.; Morshead, C.M.; Nagy, A. Adult neural stem cells from the subventricular zone give rise to reactive astrocytes in the cortex after stroke. *Cell Stem Cell* **2015**, *17*, 624–634. [CrossRef]
77. Xing, Y.L.; Roth, P.T.; Stratton, J.A.; Chuang, B.H.; Danne, J.; Ellis, S.L.; Ng, S.W.; Kilpatrick, T.J.; Merson, T.D. Adult neural precursor cells from the subventricular zone contribute significantly to oligodendrocyte regeneration and remyelination. *J. Neurosci.* **2014**, *34*, 14128–14146. [CrossRef]
78. Klingener, M.; Chavali, M.; Singh, J.; McMillan, N.; Coomes, A.; Dempsey, P.J.; Chen, E.I.; Aguirre, A. N-cadherin promotes recruitment and migration of neural progenitor cells from the SVZ neural stem cell niche into demyelinated lesions. *J. Neurosci.* **2014**, *34*, 9590–9606. [CrossRef]
79. Obernier, K.; Tong, C.K.; Alvarez-Buylla, A. Restricted nature of adult neural stem cells: Re-evaluation of their potential for brain repair. *Front. Neurosci.* **2014**, *8*, 162. [CrossRef]
80. Parent, J.M. Injury-induced neurogenesis in the adult mammalian brain. *Neurosci. A Rev. J. Bringing Neurobiol. Neurol. Psychiatry* **2003**, *9*, 261–272. [CrossRef]
81. Van den Berge, S.A.; van Strien, M.E.; Hol, E.M. Resident adult neural stem cells in Parkinson's disease—The brain's own repair system? *Eur. J. Pharmacol.* **2013**, *719*, 117–127. [CrossRef]
82. Nait-Oumesmar, B.; Picard-Riera, N.; Kerninon, C.; Baron-Van Evercooren, A. The role of SVZ-derived neural precursors in demyelinating diseases: From animal models to multiple sclerosis. *J. Neurol. Sci.* **2008**, *265*, 26–31. [CrossRef]
83. Fainstein, N.; Dan-Goor, N.; Ben-Hur, T. Resident brain neural precursor cells develop age-dependent loss of therapeutic functions in Alzheimer's mice. *Neurobiol. Aging* **2018**, *72*, 40–52. [CrossRef]
84. Chen, L.; Qiu, R.; Li, L.; He, D.; Lv, H.; Wu, X.; Gu, N. The role of exogenous neural stem cells transplantation in cerebral ischemic stroke. *J. Biomed. Nanotechnol.* **2014**, *10*, 3219–3230. [CrossRef]
85. Reya, T.; Morrison, S.J.; Clarke, M.F.; Weissman, I.L. Stem cells, cancer, and cancer stem cells. *Nature* **2001**, *414*, 105–111. [CrossRef]
86. Galderisi, U.; Cipollaro, M.; Giordano, A. Stem cells and brain cancer. *Cell Death Differ.* **2005**, *13*, 5. [CrossRef]
87. Rodenhuis, S. Ras and human tumors. *Semin. Cancer Biol.* **1992**, *3*, 241–247.
88. Alt, F.W.; Kellems, R.E.; Bertino, J.R.; Schimke, R.T. Selective multiplication of dihydrofolate reductase genes in methotrexate-resistant variants of cultured murine cells. *J. Biol. Chem.* **1978**, *253*, 1357–1370.
89. Croce, C.M. Oncogenes and cancer. *N. Engl. J. Med.* **2008**, *358*, 502–511. [CrossRef]
90. Martincorena, I.; Campbell, P.J. Somatic mutation in cancer and normal cells. *Science* **2015**, *349*, 1483–1489. [CrossRef]
91. Hope, K.J.; Jin, L.; Dick, J.E. Acute myeloid leukemia originates from a hierarchy of leukemic stem cell classes that differ in self-renewal capacity. *Nat. Immunol.* **2004**, *5*, 738–743. [CrossRef]
92. Lapidot, T.; Sirard, C.; Vormoor, J.; Murdoch, B.; Hoang, T.; Caceres-Cortes, J.; Minden, M.; Paterson, B.; Caligiuri, M.A.; Dick, J.E. A cell initiating human acute myeloid leukaemia after transplantation into SCID mice. *Nature* **1994**, *367*, 645–648. [CrossRef]
93. Bonnet, D.; Dick, J.E. Human acute myeloid leukemia is organized as a hierarchy that originates from a primitive hematopoietic cell. *Nat. Med.* **1997**, *3*, 730–737. [CrossRef]
94. Al-Hajj, M.; Wicha, M.S.; Benito-Hernandez, A.; Morrison, S.J.; Clarke, M.F. Prospective identification of tumorigenic breast cancer cells. *Proc. Natl. Acad. Sci. USA* **2003**, *100*, 3983–3988. [CrossRef]
95. Uckun, F.M.; Sather, H.; Reaman, G.; Shuster, J.; Land, V.; Trigg, M.; Gunther, R.; Chelstrom, L.; Bleyer, A.; Gaynon, P.; et al. Leukemic cell growth in SCID mice as a predictor of relapse in high-risk B-lineage acute lymphoblastic leukemia. *Blood* **1995**, *85*, 873–878.

96. Papaccio, F.; Paino, F.; Regad, T.; Papaccio, G.; Desiderio, V.; Tirino, V. Concise review: Cancer cells, cancer stem cells, and mesenchymal stem cells: Influence in cancer development. *Stem Cells Transl. Med.* **2017**, *6*, 2115–2125. [CrossRef]
97. Marusyk, A.; Polyak, K. Tumor heterogeneity: Causes and consequences. *Biochim. Biophys. Acta* **2010**, *1805*, 105–117. [CrossRef]
98. Weigelt, B.; Glas, A.M.; Wessels, L.F.; Witteveen, A.T.; Peterse, J.L.; van't Veer, L.J. Gene expression profiles of primary breast tumors maintained in distant metastases. *Proc. Natl. Acad. Sci. USA* **2003**, *100*, 15901–15905. [CrossRef]
99. Bardella, C.; Al-Shammari, A.R.; Soares, L.; Tomlinson, I.; O'Neill, E.; Szele, F.G. The role of inflammation in subventricular zone cancer. *Prog. Neurobiol.* **2018**, *170*, 37–52. [CrossRef]
100. Wodarz, A.; Gonzalez, C. Connecting cancer to the asymmetric division of stem cells. *Cell* **2006**, *124*, 1121–1123. [CrossRef]
101. Calabrese, C.; Poppleton, H.; Kocak, M.; Hogg, T.L.; Fuller, C.; Hamner, B.; Oh, E.Y.; Gaber, M.W.; Finklestein, D.; Allen, M.; et al. A perivascular niche for brain tumor stem cells. *Cancer Cell* **2007**, *11*, 69–82. [CrossRef]
102. Clarke, M.F.; Fuller, M. Stem cells and cancer: Two faces of eve. *Cell* **2006**, *124*, 1111–1115. [CrossRef]
103. Khalifa, J.; Tensaouti, F.; Lusque, A.; Plas, B.; Lotterie, J.A.; Benouaich-Amiel, A.; Uro-Coste, E.; Lubrano, V.; Cohen-Jonathan Moyal, E. Subventricular zones: New key targets for glioblastoma treatment. *Radiat. Oncol.* **2017**, *12*, 67. [CrossRef]
104. Homem, C.C.; Repic, M.; Knoblich, J.A. Proliferation control in neural stem and progenitor cells. *Nat. Rev. Neurosci.* **2015**, *16*, 647–659. [CrossRef]
105. Alagappan, D.; Lazzarino, D.A.; Felling, R.J.; Balan, M.; Kotenko, S.V.; Levison, S.W. Brain injury expands the numbers of neural stem cells and progenitors in the SVZ by enhancing their responsiveness to EGF. *Asn Neuro* **2009**, *1*, e00009. [CrossRef]
106. Capilla-Gonzalez, V.; Lavell, E.; Quinones-Hinojosa, A.; Guerrero-Cazares, H. Regulation of subventricular zone-derived cells migration in the adult brain. *Adv. Exp. Med. Biol.* **2015**, *853*, 1–21. [CrossRef]
107. Osswald, M.; Jung, E.; Sahm, F.; Solecki, G.; Venkataramani, V.; Blaes, J.; Weil, S.; Horstmann, H.; Wiestler, B.; Syed, M.; et al. Brain tumour cells interconnect to a functional and resistant network. *Nature* **2015**, *528*, 93–98. [CrossRef]
108. Weil, S.; Osswald, M.; Solecki, G.; Grosch, J.; Jung, E.; Lemke, D.; Ratliff, M.; Hänggi, D.; Wick, W.; Winkler, F. Tumor microtubes convey resistance to surgical lesions and chemotherapy in gliomas. *Neuro Oncol.* **2017**, *19*, 1316–1326. [CrossRef]
109. Seidel, S.; Garvalov, B.K.; Wirta, V.; von Stechow, L.; Schanzer, A.; Meletis, K.; Wolter, M.; Sommerlad, D.; Henze, A.T.; Nister, M.; et al. A hypoxic niche regulates glioblastoma stem cells through hypoxia inducible factor 2 alpha. *Brain A J. Neurol.* **2010**, *133*, 983–995. [CrossRef]
110. Sharma, A.; Shiras, A. Cancer stem cell-vascular endothelial cell interactions in glioblastoma. *Biochem. Biophys. Res. Commun.* **2016**, *473*, 688–692. [CrossRef]
111. Hira, V.V.; Ploegmakers, K.J.; Grevers, F.; Verbovsek, U.; Silvestre-Roig, C.; Aronica, E.; Tigchelaar, W.; Turnsek, T.L.; Molenaar, R.J.; Van Noorden, C.J. CD133+ and Nestin+ glioma stem-like cells reside around CD31+ arterioles in niches that express SDF-1alpha, CXCR4, osteopontin and cathepsin K. *J. Histochem. Cytochem.* **2015**, *63*, 481–493. [CrossRef]
112. Filatova, A.; Seidel, S.; Bogurcu, N.; Graf, S.; Garvalov, B.K.; Acker, T. Acidosis acts through HSP90 in a PHD/VHL-independent manner to promote HIF function and stem cell maintenance in glioma. *Cancer Res.* **2016**, *76*, 5845–5856. [CrossRef]
113. Li, Z.; Bao, S.; Wu, Q.; Wang, H.; Eyler, C.; Sathornsumetee, S.; Shi, Q.; Cao, Y.; Lathia, J.; McLendon, R.E.; et al. Hypoxia-inducible factors regulate tumorigenic capacity of glioma stem cells. *Cancer Cell* **2009**, *15*, 501–513. [CrossRef]
114. Aderetti, D.A.; Hira, V.V.V.; Molenaar, R.J.; van Noorden, C.J.F. The hypoxic peri-arteriolar glioma stem cell niche, an integrated concept of five types of niches in human glioblastoma. *Biochim. Biophys. Acta Rev. Cancer* **2018**, *1869*, 346–354. [CrossRef]
115. Yang, Z.-J.; Ellis, T.; Markant, S.L.; Read, T.-A.; Kessler, J.D.; Bourboulas, M.; Schüller, U.; Machold, R.; Fishell, G.; Rowitch, D.H.; et al. Medulloblastoma can be initiated by deletion of Patched in lineage-restricted progenitors or stem cells. *Cancer Cell* **2008**, *14*, 135–145. [CrossRef]

116. Wang, Y.; Yang, J.; Zheng, H.; Tomasek, G.J.; Zhang, P.; McKeever, P.E.; Lee, E.Y.; Zhu, Y. Expression of mutant p53 proteins implicates a lineage relationship between neural stem cells and malignant astrocytic glioma in a murine model. *Cancer Cell* **2009**, *15*, 514–526. [CrossRef]
117. Macas, J.; Ku, M.C.; Nern, C.; Xu, Y.; Buhler, H.; Remke, M.; Synowitz, M.; Franz, K.; Seifert, V.; Plate, K.H.; et al. Generation of neuronal progenitor cells in response to tumors in the human brain. *Stem Cells* **2014**, *32*, 244–257. [CrossRef]
118. Abou-Kheir, W.G.; Hynes, P.G.; Martin, P.L.; Pierce, R.; Kelly, K. Characterizing the contribution of stem/progenitor cells to tumorigenesis in the Pten-/-TP53-/- prostate cancer model. *Stem Cells* **2010**, *28*, 2129–2140. [CrossRef]
119. Liu, C.; Sage, J.C.; Miller, M.R.; Verhaak, R.G.; Hippenmeyer, S.; Vogel, H.; Foreman, O.; Bronson, R.T.; Nishiyama, A.; Luo, L.; et al. Mosaic analysis with double markers reveals tumor cell of origin in glioma. *Cell* **2011**, *146*, 209–221. [CrossRef]
120. Hide, T.; Takezaki, T.; Nakatani, Y.; Nakamura, H.; Kuratsu, J.-I.; Kondo, T. Combination of a Ptgs2 Inhibitor and an epidermal growth factor receptor-signaling inhibitor prevents tumorigenesis of oligodendrocyte lineage-derived glioma-initiating cells. *Stem Cells* **2011**, *29*, 590–599. [CrossRef]
121. Ignatova, T.N.; Kukekov, V.G.; Laywell, E.D.; Suslov, O.N.; Vrionis, F.D.; Steindler, D.A. Human cortical glial tumors contain neural stem-like cells expressing astroglial and neuronal markers in vitro. *Glia* **2002**, *39*, 193–206. [CrossRef]
122. Galli, R.; Binda, E.; Orfanelli, U.; Cipelletti, B.; Gritti, A.; De Vitis, S.; Fiocco, R.; Foroni, C.; Dimeco, F.; Vescovi, A. Isolation and characterization of tumorigenic, stem-like neural precursors from human glioblastoma. *Cancer Res.* **2004**, *64*, 7011–7021. [CrossRef]
123. Singh, S.K.; Hawkins, C.; Clarke, I.D.; Squire, J.A.; Bayani, J.; Hide, T.; Henkelman, R.M.; Cusimano, M.D.; Dirks, P.B. Identification of human brain tumour initiating cells. *Nature* **2004**, *432*, 396–401. [CrossRef]
124. Singh, S.K.; Clarke, I.D.; Terasaki, M.; Bonn, V.E.; Hawkins, C.; Squire, J.; Dirks, P.B. Identification of a cancer stem cell in human brain tumors. *Cancer Res.* **2003**, *63*, 5821–5828.
125. Hemmati, H.D.; Nakano, I.; Lazareff, J.A.; Masterman-Smith, M.; Geschwind, D.H.; Bronner-Fraser, M.; Kornblum, H.I. Cancerous stem cells can arise from pediatric brain tumors. *Proc. Natl. Acad. Sci. USA* **2003**, *100*, 15178–15183. [CrossRef]
126. Kroonen, J.; Nassen, J.; Boulanger, Y.G.; Provenzano, F.; Capraro, V.; Bours, V.; Martin, D.; Deprez, M.; Robe, P.; Rogister, B. Human glioblastoma-initiating cells invade specifically the subventricular zones and olfactory bulbs of mice after striatal injection. *Int. J. Cancer* **2011**, *129*, 574–585. [CrossRef]
127. Lee, J.; Kotliarova, S.; Kotliarov, Y.; Li, A.; Su, Q.; Donin, N.M.; Pastorino, S.; Purow, B.W.; Christopher, N.; Zhang, W.; et al. Tumor stem cells derived from glioblastomas cultured in bFGF and EGF more closely mirror the phenotype and genotype of primary tumors than do serum-cultured cell lines. *Cancer Cell* **2006**, *9*, 391–403. [CrossRef]
128. Gupta, T.; Nair, V.; Jalali, R. Stem cell niche irradiation in glioblastoma: Providing a ray of hope? *CNS Oncol.* **2014**, *3*, 367–376. [CrossRef]
129. Androutsellis-Theotokis, A.; Leker, R.R.; Soldner, F.; Hoeppner, D.J.; Ravin, R.; Poser, S.W.; Rueger, M.A.; Bae, S.K.; Kittappa, R.; McKay, R.D. Notch signalling regulates stem cell numbers in vitro and in vivo. *Nature* **2006**, *442*, 823–826. [CrossRef]
130. Purow, B.W.; Haque, R.M.; Noel, M.W.; Su, Q.; Burdick, M.J.; Lee, J.; Sundaresan, T.; Pastorino, S.; Park, J.K.; Mikolaenko, I.; et al. Expression of Notch-1 and its ligands, Delta-like-1 and Jagged-1, is critical for glioma cell survival and proliferation. *Cancer Res.* **2005**, *65*, 2353–2363. [CrossRef]
131. Neradil, J.; Veselska, R. Nestin as a marker of cancer stem cells. *Cancer Sci.* **2015**, *106*, 803–811. [CrossRef]
132. Feng, W.; Liu, S.; Zhu, R.; Li, B.; Zhu, Z.; Yang, J.; Song, C. SOX10 induced Nestin expression regulates cancer stem cell properties of TNBC cells. *Biochem. Biophys. Res. Commun.* **2017**, *485*, 522–528. [CrossRef]
133. Wu, B.; Sun, C.; Feng, F.; Ge, M.; Xia, L. Do relevant markers of cancer stem cells CD133 and nestin indicate a poor prognosis in glioma patients? A systematic review and meta-analysis. *J. Exp. Clin. Cancer Res.* **2015**, *34*, 44. [CrossRef]
134. Shih, A.H.; Holland, E.C. Notch signaling enhances nestin expression in gliomas. *Neoplasia* **2006**, *8*, 1072–1082. [CrossRef]
135. Dahlstrand, J.; Collins, V.P.; Lendahl, U. Expression of the class VI intermediate filament nestin in human central nervous system tumors. *Cancer Res.* **1992**, *52*, 5334–5341.

136. Strojnik, T.; Rosland, G.V.; Sakariassen, P.O.; Kavalar, R.; Lah, T. Neural stem cell markers, nestin and musashi proteins, in the progression of human glioma: Correlation of nestin with prognosis of patient survival. *Surg. Neurol.* **2007**, *68*, 133–143. [CrossRef]
137. Pallini, R.; Ricci-Vitiani, L.; Banna, G.L.; Signore, M.; Lombardi, D.; Todaro, M.; Stassi, G.; Martini, M.; Maira, G.; Larocca, L.M.; et al. Cancer stem cell analysis and clinical outcome in patients with glioblastoma multiforme. *Clin. Cancer Res.* **2008**, *14*, 8205–8212. [CrossRef]
138. Toda, M.; Iizuka, Y.; Yu, W.; Imai, T.; Ikeda, E.; Yoshida, K.; Kawase, T.; Kawakami, Y.; Okano, H.; Uyemura, K. Expression of the neural RNA-binding protein Musashi1 in human gliomas. *Glia* **2001**, *34*, 1–7. [CrossRef]
139. Bouvier, C.; Bartoli, C.; Aguirre-Cruz, L.; Virard, I.; Colin, C.; Fernandez, C.; Gouvernet, J.; Figarella-Branger, D. Shared oligodendrocyte lineage gene expression in gliomas and oligodendrocyte progenitor cells. *J. Neurosurg.* **2003**, *99*, 344–350. [CrossRef]
140. Riemenschneider, M.J.; Koy, T.H.; Reifenberger, G. Expression of oligodendrocyte lineage genes in oligodendroglial and astrocytic gliomas. *Acta Neuropathol.* **2004**, *107*, 277–282. [CrossRef]
141. Holland, E.C. A mouse model for glioma: Biology, pathology, and therapeutic opportunities. *Toxicol. Pathol.* **2000**, *28*, 171–177. [CrossRef]
142. Maunoury, R.; Courdi, A.; Vedrenne, C.; Constans, J.P. Immunocytochemical localization of the GFAP in heterotransplanted human gliomas (author's transl). *Neuro Chir.* **1978**, *24*, 221–226.
143. Kimura, T.; Budka, H.; Soler-Federsppiel, S. An immunocytochemical comparison of the glia-associated proteins glial fibrillary acidic protein (GFAP) and S-100 protein (S100P) in human brain tumors. *Clin. Neuropathol.* **1986**, *5*, 21–27.
144. Groszer, M.; Erickson, R.; Scripture-Adams, D.D.; Lesche, R.; Trumpp, A.; Zack, J.A.; Kornblum, H.I.; Liu, X.; Wu, H. Negative regulation of neural stem/progenitor cell proliferation by the Pten tumor suppressor gene in vivo. *Science* **2001**, *294*, 2186–2189. [CrossRef]
145. Vik-Mo, E.O.; Sandberg, C.; Joel, M.; Stangeland, B.; Watanabe, Y.; Mackay-Sim, A.; Moe, M.C.; Murrell, W.; Langmoen, I.A. A comparative study of the structural organization of spheres derived from the adult human subventricular zone and glioblastoma biopsies. *Exp. Cell Res.* **2011**, *317*, 1049–1059. [CrossRef]
146. Bicker, F.; Vasic, V.; Horta, G.; Ortega, F.; Nolte, H.; Kavyanifar, A.; Keller, S.; Stankovic, N.D.; Harter, P.N.; Benedito, R.; et al. Neurovascular EGFL7 regulates adult neurogenesis in the subventricular zone and thereby affects olfactory perception. *Nat. Commun.* **2017**, *8*, 15922. [CrossRef]
147. Fassl, A.; Tagscherer, K.E.; Richter, J.; Berriel Diaz, M.; Alcantara Llaguno, S.R.; Campos, B.; Kopitz, J.; Herold-Mende, C.; Herzig, S.; Schmidt, M.H.H.; et al. Notch1 signaling promotes survival of glioblastoma cells via EGFR-mediated induction of anti-apoptotic Mcl-1. *Oncogene* **2012**, *31*, 4698–4708. [CrossRef]
148. Schmidt, M.H.H.; Bicker, F.; Nikolic, I.; Meister, J.; Babuke, T.; Picuric, S.; Muller-Esterl, W.; Plate, K.H.; Dikic, I. Epidermal growth factor-like domain 7 (EGFL7) modulates Notch signalling and affects neural stem cell renewal. *Nat. Cell Biol.* **2009**, *11*, 873–880. [CrossRef]
149. Nikolić, I.; Stanković, N.D.; Bicker, F.; Meister, J.; Braun, H.; Awwad, K.; Baumgart, J.; Simon, K.; Thal, S.C.; Patra, C.; et al. EGFL7 ligates $\alpha_v\beta_3$ integrin to enhance vessel formation. *Blood* **2013**, *121*, 3041–3050. [CrossRef]
150. Dudvarski Stankovic, N.; Bicker, F.; Keller, S.; Jones, D.T.; Harter, P.N.; Kienzle, A.; Gillmann, C.; Arnold, P.; Golebiewska, A.; Keunen, O.; et al. EGFL7 enhances surface expression of integrin alpha5beta1 to promote angiogenesis in malignant brain tumors. *EMBO Mol. Med.* **2018**, *10*, e8420. [CrossRef]
151. Lim, D.A.; Cha, S.; Mayo, M.C.; Chen, M.H.; Keles, E.; VandenBerg, S.; Berger, M.S. Relationship of glioblastoma multiforme to neural stem cell regions predicts invasive and multifocal tumor phenotype. *Neuro Oncol.* **2007**, *9*, 424–429. [CrossRef]
152. Pope, W.B.; Sayre, J.; Perlina, A.; Villablanca, J.P.; Mischel, P.S.; Cloughesy, T.F. MR imaging correlates of survival in patients with high-grade gliomas. *Am. J. Neuroradiol.* **2005**, *26*, 2466–2474.
153. Kappadakunnel, M.; Eskin, A.; Dong, J.; Nelson, S.F.; Mischel, P.S.; Liau, L.M.; Ngheimphu, P.; Lai, A.; Cloughesy, T.F.; Goldin, J.; et al. Stem cell associated gene expression in glioblastoma multiforme: Relationship to survival and the subventricular zone. *J. Neuro Oncol.* **2010**, *96*, 359–367. [CrossRef]
154. Chaichana, K.L.; McGirt, M.J.; Frazier, J.; Attenello, F.; Guerrero-Cazares, H.; Quinones-Hinojosa, A. Relationship of glioblastoma multiforme to the lateral ventricles predicts survival following tumor resection. *J. Neurooncol.* **2008**, *89*, 219–224. [CrossRef]

155. Mistry, A.M.; Dewan, M.C.; White-Dzuro, G.A.; Brinson, P.R.; Weaver, K.D.; Thompson, R.C.; Ihrie, R.A.; Chambless, L.B. Decreased survival in glioblastomas is specific to contact with the ventricular-subventricular zone, not subgranular zone or corpus callosum. *J. Neurooncol.* **2017**, *132*, 341–349. [CrossRef]
156. Barami, K.; Sloan, A.E.; Rojiani, A.; Schell, M.J.; Staller, A.; Brem, S. Relationship of gliomas to the ventricular walls. *J. Clin. Neurosci.* **2009**, *16*, 195–201. [CrossRef]
157. Gollapalli, K.; Ghantasala, S.; Kumar, S.; Srivastava, R.; Rapole, S.; Moiyadi, A.; Epari, S.; Srivastava, S. Subventricular zone involvement in Glioblastoma—A proteomic evaluation and clinicoradiological correlation. *Sci. Rep.* **2017**, *7*, 1449. [CrossRef]
158. Zhu, Y.; Guignard, F.; Zhao, D.; Liu, L.; Burns, D.K.; Mason, R.P.; Messing, A.; Parada, L.F. Early inactivation of p53 tumor suppressor gene cooperating with NF1 loss induces malignant astrocytoma. *Cancer Cell* **2005**, *8*, 119–130. [CrossRef]
159. Jiang, Y.; Marinescu, V.D.; Xie, Y.; Jarvius, M.; Maturi, N.P.; Haglund, C.; Olofsson, S.; Lindberg, N.; Olofsson, T.; Leijonmarck, C.; et al. Glioblastoma cell malignancy and drug sensitivity are affected by the cell of origin. *Cell Rep.* **2017**, *18*, 977–990. [CrossRef]
160. Monnier, A.; Boniface, R.; Bouvet, R.; Etcheverry, A.; Aubry, M.; Avril, T.; Quillien, V.; Chevet, E.; Mosser, J. The expression of EMX2 lead to cell cycle arrest in glioblastoma cell line. *BMC Cancer* **2018**, *18*, 1213. [CrossRef]
161. Llaguno, S.A.; Sun, D.; Pedraza, A.M.; Vera, E.; Wang, Z.; Burns, D.K.; Parada, L.F. Cell-of-origin susceptibility to glioblastoma formation declines with neural lineage restriction. *Nat. Neurosci.* **2019**, *22*, 545–555. [CrossRef]
162. Rowitch, D.H.; Kriegstein, A.R. Developmental genetics of vertebrate glial—Cell specification. *Nature* **2010**, *468*, 214. [CrossRef]
163. Jiang, Y.; Uhrbom, L. On the origin of glioma. *Upsala J. Med. Sci.* **2012**, *117*, 113–121. [CrossRef]
164. Cavanagh, J.B. The proliferation of astrocytes around a needle wound in the rat brain. *J. Anat.* **1970**, *106*, 471–487.
165. Jones, T.R.; Bigner, S.H.; Schold, S.C., Jr.; Eng, L.F.; Bigner, D.D. Anaplastic human gliomas grown in athymic mice. Morphology and glial fibrillary acidic protein expression. *Am. J. Pathol.* **1981**, *105*, 316–327.
166. Sharif, A.; Legendre, P.; Prevot, V.; Allet, C.; Romao, L.; Studler, J.M.; Chneiweiss, H.; Junier, M.P. Transforming growth factor alpha promotes sequential conversion of mature astrocytes into neural progenitors and stem cells. *Oncogene* **2007**, *26*, 2695–2706. [CrossRef]
167. Dufour, C.; Cadusseau, J.; Varlet, P.; Surena, A.L.; de Faria, G.P.; Dias-Morais, A.; Auger, N.; Leonard, N.; Daudigeos, E.; Dantas-Barbosa, C.; et al. Astrocytes reverted to a neural progenitor-like state with transforming growth factor alpha are sensitized to cancerous transformation. *Stem Cells* **2009**, *27*, 2373–2382. [CrossRef]
168. Alves, J.A.; Barone, P.; Engelender, S.; Froes, M.M.; Menezes, J.R. Initial stages of radial glia astrocytic transformation in the early postnatal anterior subventricular zone. *J. Neurobiol.* **2002**, *52*, 251–265. [CrossRef]
169. Lindberg, N.; Kastemar, M.; Olofsson, T.; Smits, A.; Uhrbom, L. Oligodendrocyte progenitor cells can act as cell of origin for experimental glioma. *Oncogene* **2009**, *28*, 2266–2275. [CrossRef]
170. Shoshan, Y.; Nishiyama, A.; Chang, A.; Mork, S.; Barnett, G.H.; Cowell, J.K.; Trapp, B.D.; Staugaitis, S.M. Expression of oligodendrocyte progenitor cell antigens by gliomas: Implications for the histogenesis of brain tumors. *Proc. Natl. Acad. Sci. USA* **1999**, *96*, 10361–10366. [CrossRef]
171. Parsons, D.W.; Jones, S.; Zhang, X.; Lin, J.C.; Leary, R.J.; Angenendt, P.; Mankoo, P.; Carter, H.; Siu, I.M.; Gallia, G.L.; et al. An integrated genomic analysis of human glioblastoma multiforme. *Science* **2008**, *321*, 1807–1812. [CrossRef]
172. Dean, M.; Fojo, T.; Bates, S. Tumour stem cells and drug resistance. *Nat. Rev. Cancer* **2005**, *5*, 275. [CrossRef]
173. Liu, G.; Yuan, X.; Zeng, Z.; Tunici, P.; Ng, H.; Abdulkadir, I.R.; Lu, L.; Irvin, D.; Black, K.L.; John, S.Y. Analysis of gene expression and chemoresistance of CD133+ cancer stem cells in glioblastoma. *Mol. Cancer* **2006**, *5*, 67. [CrossRef]
174. Eramo, A.; Ricci-Vitiani, L.; Zeuner, A.; Pallini, R.; Lotti, F.; Sette, G.; Pilozzi, E.; Larocca, L.M.; Peschle, C.; De Maria, R. Chemotherapy resistance of glioblastoma stem cells. *Cell Death Differ.* **2006**, *13*, 1238. [CrossRef]
175. Chen, J.; Li, Y.; Yu, T.-S.; McKay, R.M.; Burns, D.K.; Kernie, S.G.; Parada, L.F. A restricted cell population propagates glioblastoma growth after chemotherapy. *Nature* **2012**, *488*, 522. [CrossRef]

176. Bao, S.; Wu, Q.; McLendon, R.E.; Hao, Y.; Shi, Q.; Hjelmeland, A.B.; Dewhirst, M.W.; Bigner, D.D.; Rich, J.N. Glioma stem cells promote radioresistance by preferential activation of the DNA damage response. *Nature* **2006**, *444*, 756. [CrossRef]
177. Pardridge, W.M. Drug transport in brain via the cerebrospinal fluid. *Fluids Barriers CNS* **2011**, *8*, 7. [CrossRef]
178. Glantz, M.; Kesari, S.; Recht, L.; Fleischhack, G.; Van Horn, A. Understanding the origins of gliomas and developing novel therapies: Cerebrospinal fluid and subventricular zone interplay. *Semin. Oncol.* **2009**, *36*, S17–S24. [CrossRef]
179. Dragu, D.L.; Necula, L.G.; Bleotu, C.; Diaconu, C.C.; Chivu-Economescu, M. Therapies targeting cancer stem cells: Current trends and future challenges. *World J. Stem Cells* **2015**, *7*, 1185.
180. Batlle, E.; Clevers, H. Cancer stem cells revisited. *Nat. Med.* **2017**, *23*, 1124. [CrossRef]
181. Eramo, A.; Lotti, F.; Sette, G.; Pilozzi, E.; Biffoni, M.; Di Virgilio, A.; Conticello, C.; Ruco, L.; Peschle, C.; De Maria, R. Identification and expansion of the tumorigenic lung cancer stem cell population. *Cell Death Differ.* **2008**, *15*, 504. [CrossRef]
182. Ricci-Vitiani, L.; Lombardi, D.G.; Pilozzi, E.; Biffoni, M.; Todaro, M.; Peschle, C.; De Maria, R. Identification and expansion of human colon-cancer-initiating cells. *Nature* **2007**, *445*, 111. [CrossRef]
183. Zeppernick, F.; Ahmadi, R.; Campos, B.; Dictus, C.; Helmke, B.M.; Becker, N.; Lichter, P.; Unterberg, A.; Radlwimmer, B.; Herold-Mende, C.C. Stem cell marker CD133 affects clinical outcome in glioma patients. *Clin. Cancer Res.* **2008**, *14*, 123–129. [CrossRef]
184. Beier, D.; Hau, P.; Proescholdt, M.; Lohmeier, A.; Wischhusen, J.; Oefner, P.J.; Aigner, L.; Brawanski, A.; Bogdahn, U.; Beier, C.P. CD133+ and CD133− glioblastoma-derived cancer stem cells show differential growth characteristics and molecular profiles. *Cancer Res.* **2007**, *67*, 4010–4015. [CrossRef]
185. Bleau, A.-M.; Hambardzumyan, D.; Ozawa, T.; Fomchenko, E.I.; Huse, J.T.; Brennan, C.W.; Holland, E.C. PTEN/PI3K/Akt pathway regulates the side population phenotype and ABCG2 activity in glioma tumor stem-like cells. *Cell Stem Cell* **2009**, *4*, 226–235. [CrossRef]
186. Wang, J.; Wakeman, T.P.; Lathia, J.D.; Hjelmeland, A.B.; Wang, X.F.; White, R.R.; Rich, J.N.; Sullenger, B.A. Notch promotes radioresistance of glioma stem cells. *Stem Cells* **2010**, *28*, 17–28. [CrossRef]
187. Rheinbay, E.; Suvà, M.L.; Gillespie, S.M.; Wakimoto, H.; Patel, A.P.; Shahid, M.; Oksuz, O.; Rabkin, S.D.; Martuza, R.L.; Rivera, M.N. An aberrant transcription factor network essential for Wnt signaling and stem cell maintenance in glioblastoma. *Cell Rep.* **2013**, *3*, 1567–1579. [CrossRef]
188. Stockhausen, M.-T.; Kristoffersen, K.; Poulsen, H.S. The functional role of Notch signaling in human gliomas. *Neuro Oncol.* **2009**, *12*, 199–211. [CrossRef]
189. Saito, N.; Fu, J.; Zheng, S.; Yao, J.; Wang, S.; Liu, D.D.; Yuan, Y.; Sulman, E.P.; Lang, F.F.; Colman, H. A high notch pathway activation predicts response to γ secretase inhibitors in proneural subtype of glioma tumor-initiating cells. *Stem Cells* **2014**, *32*, 301–312. [CrossRef]
190. Fan, X.; Khaki, L.; Zhu, T.S.; Soules, M.E.; Talsma, C.E.; Gul, N.; Koh, C.; Zhang, J.; Li, Y.M.; Maciaczyk, J. NOTCH pathway blockade depletes CD133-positive glioblastoma cells and inhibits growth of tumor neurospheres and xenografts. *Stem Cells* **2010**, *28*, 5–16. [CrossRef]
191. Semenza, G.L. Targeting HIF-1 for cancer therapy. *Nat. Rev. Cancer* **2003**, *3*, 721. [CrossRef]
192. Bao, S.; Wu, Q.; Sathornsumetee, S.; Hao, Y.; Li, Z.; Hjelmeland, A.B.; Shi, Q.; McLendon, R.E.; Bigner, D.D.; Rich, J.N. Stem cell–like glioma cells promote tumor angiogenesis through vascular endothelial growth factor. *Cancer Res.* **2006**, *66*, 7843–7848. [CrossRef]
193. Unterkircher, T.; Cristofanon, S.; Vellanki, S.H.K.; Nonnenmacher, L.; Karpel-Massler, G.; Wirtz, C.; Debatin, K.-M.; Fulda, S. Bortezomib primes glioblastoma including glioblastoma stem cells for TRAIL by increasing tBid stability and mitochondrial apoptosis. *Clin. Cancer Res.* **2011**, *7*, 4019–4030. [CrossRef]
194. Soeda, A.; Inagaki, A.; Oka, N.; Ikegame, Y.; Aoki, H.; Yoshimura, S.-I.; Nakashima, S.; Kunisada, T.; Iwama, T. Epidermal growth factor plays a crucial role in mitogenic regulation of human brain tumor stem cells. *J. Biol. Chem.* **2008**, *283*, 10958–10966. [CrossRef]
195. Nazio, F.; Bordi, M.; Cianfanelli, V.; Locatelli, F.; Cecconi, F. Autophagy and cancer stem cells: Molecular mechanisms and therapeutic applications. *Cell Death Differ.* **2019**, *26*, 690–702. [CrossRef]
196. Sui, X.; Chen, R.; Wang, Z.; Huang, Z.; Kong, N.; Zhang, M.; Han, W.; Lou, F.; Yang, J.; Zhang, Q.; et al. Autophagy and chemotherapy resistance: A promising therapeutic target for cancer treatment. *Cell Death Dis.* **2013**, *4*, e838. [CrossRef]

197. Sun, R.; Shen, S.; Zhang, Y.-J.; Xu, C.-F.; Cao, Z.-T.; Wen, L.-P.; Wang, J. Nanoparticle-facilitated autophagy inhibition promotes the efficacy of chemotherapeutics against breast cancer stem cells. *Biomaterials* **2016**, *103*, 44–55. [CrossRef]
198. Encouse, B.G.; Hee-Yeon, C.; Ardeshir, J.; Florence, M.H.; Stan, G.L.; Axel, H.S.; Thomas, C.C. Chloroquine enhances temozolomide cytotoxicity in malignant gliomas by blocking autophagy. *Neurosurg. Focus* **2014**, *37*, e12. [CrossRef]
199. Huang, H.; Song, J.; Liu, Z.; Pan, L.; Xu, G. Autophagy activation promotes bevacizumab resistance in glioblastoma by suppressing Akt/mTOR signaling pathway. *Oncol. Lett.* **2018**, *15*, 1487–1494. [CrossRef]
200. Buccarelli, M.; Marconi, M.; Pacioni, S.; De Pascalis, I.; D'Alessandris, Q.G.; Martini, M.; Ascione, B.; Malorni, W.; Larocca, L.M.; Pallini, R.; et al. Inhibition of autophagy increases susceptibility of glioblastoma stem cells to temozolomide by igniting ferroptosis. *Cell Death Dis.* **2018**, *9*, 841. [CrossRef]
201. Massard, C.; Deutsch, E.; Soria, J. Tumour stem cell-targeted treatment: Elimination or differentiation. *Ann. Oncol.* **2006**, *17*, 1620–1624. [CrossRef]
202. Piccirillo, S.; Reynolds, B.A.; Zanetti, N.; Lamorte, G.; Binda, E.; Broggi, G.; Brem, H.; Olivi, A.; Dimeco, F.; Vescovi, A.L. Bone morphogenetic proteins inhibit the tumorigenic potential of human brain tumour-initiating cells. *Nature* **2006**, *444*, 761. [CrossRef]
203. Bezecny, P. Histone deacetylase inhibitors in glioblastoma: Pre-clinical and clinical experience. *Med. Oncol.* **2014**, *31*, 985. [CrossRef]
204. Chang, C.-J.; Hsu, C.-C.; Yung, M.-C.; Chen, K.-Y.; Tzao, C.; Wu, W.-F.; Chou, H.-Y.; Lee, Y.-Y.; Lu, K.-H.; Chiou, S.-H. Enhanced radiosensitivity and radiation-induced apoptosis in glioma CD133-positive cells by knockdown of SirT1 expression. *Biochem. Biophys. Res. Commun.* **2009**, *380*, 236–242. [CrossRef]
205. Phuphanich, S.; Wheeler, C.J.; Rudnick, J.D.; Mazer, M.; Wang, H.; Nuno, M.A.; Richardson, J.E.; Fan, X.; Ji, J.; Chu, R.M. Phase I trial of a multi-epitope-pulsed dendritic cell vaccine for patients with newly diagnosed glioblastoma. *Cancer Immunol. Immunother.* **2013**, *62*, 125–135. [CrossRef]
206. Elicin, O.; Inac, E.; Uzel, E.K.; Karacam, S.; Uzel, O.E. Relationship between survival and increased radiation dose to subventricular zone in glioblastoma is controversial. *J. Neuro-Oncol.* **2014**, *118*, 413–419. [CrossRef]
207. Chua, M.; Kusumawidjaja, G.; Gan, P.; Tan, D.H.Y.; Chua, E.T.; Tham, C.K.; Wong, F.Y. Dose-escalated intensity modulated radiotherapy (IMRT) and increased radiation doses to subventricular zones (SVZ) in treatment outcomes of patients with glioblastoma multiforme (GBM). *J. Clin. Oncol.* **2014**, *32*, e13031. [CrossRef]
208. Malik, M.; Akram, K.S.; Joseph, D.; Valiyaveettil, D.; Ahmed, S.F. Prospective study of irradiation of potential stem cell niches in glioblastoma. *Int. J. Radiat. Oncol. Biol. Phys.* **2015**, *93*, S111. [CrossRef]
209. Nourallah, B.; Digpal, R.; Jena, R.; Watts, C. Irradiating the subventricular zone in glioblastoma patients: Is there a case for a clinical trial? *Clin. Oncol.* **2017**, *29*, 26–33. [CrossRef]
210. Adeberg, S.; Harrabi, S.B.; Bougatf, N.; Bernhardt, D.; Mohr, A.; Rieber, J.; Koelsche, C.; Rieken, S.; Debus, J. Do increased doses to stem-cell niches during radiation therapy improve glioblastoma survival? *Stem Cells Int.* **2016**, *2016*, 8793462. [CrossRef]
211. Gupta, T.; Nair, V.; Paul, S.N.; Kannan, S.; Moiyadi, A.; Epari, S.; Jalali, R. Can irradiation of potential cancer stem-cell niche in the subventricular zone influence survival in patients with newly diagnosed glioblastoma? *J. Neurooncol.* **2012**, *109*, 195–203. [CrossRef]
212. Evers, P.; Lee, P.P.; DeMarco, J.; Agazaryan, N.; Sayre, J.W.; Selch, M.; Pajonk, F. Irradiation of the potential cancer stem cell niches in the adult brain improves progression-free survival of patients with malignant glioma. *BMC Cancer* **2010**, *10*, 384. [CrossRef]
213. Lee, P.; Eppinga, W.; Lagerwaard, F.; Cloughesy, T.; Slotman, B.; Nghiemphu, P.L.; Wang, P.C.; Kupelian, P.; Agazaryan, N.; Demarco, J.; et al. Evaluation of high ipsilateral subventricular zone radiation therapy dose in glioblastoma: A pooled analysis. *Int. J. Radiat. Oncol. Biol. Phys.* **2013**, *86*, 609–615. [CrossRef]
214. Chen, L.; Guerrero-Cazares, H.; Ye, X.; Ford, E.; McNutt, T.; Kleinberg, L.; Lim, M.; Chaichana, K.; Quinones-Hinojosa, A.; Redmond, K. Increased subventricular zone radiation dose correlates with survival in glioblastoma patients after gross total resection. *Int. J. Radiat. Oncol. Biol. Phys.* **2013**, *86*, 616–622. [CrossRef]

215. Mathew, B.S.; Kaliyath, S.B.; Krishnan, J.; Bhasi, S. Impact of subventricular zone irradiation on outcome of patients with glioblastoma. *J. Cancer Res. Ther.* **2018**, *14*, 1202–1206. [CrossRef]
216. Murchison, S.C.; Wiksyk, B.; Gossman, S.; Jensen, B.; Sayers, D.; Lesperance, M.; Truong, P.T.; Alexander, A. Subventricular zone radiation dose and outcome for glioblastoma treated between 2006 and 2012. *Cureus* **2018**, *10*, e3618. [CrossRef]

cancers

MDPI

Review

Glioblastoma: Microenvironment and Niche Concept

Davide Schiffer [1], Laura Annovazzi [2], Cristina Casalone [3,*], Cristiano Corona [3] and Marta Mellai [4,5]

1 Professore Emerito di Neurologia, Università di Torino, Corso Bramante 88/90, 10126 Torino, Italy; davide.schiffer@unito.it
2 Ex Centro Ricerche/Fondazione Policlinico di Monza, Via P. Micca 29, 13100 Vercelli, Italy; lannov16@gmail.com
3 Istituto Zooprofilattico Sperimentale del Piemonte, Liguria e Valle d'Aosta, Via Bologna 148, 10154 Torino, Italy; cristiano.corona@izsto.it
4 Dipartimento di Scienze della Salute, Scuola di Medicina, Università del Piemonte Orientale "A. Avogadro", Corso Mazzini 18, 28100 Novara, Italy; martamel73@gmail.com
5 Fondazione Edo ed Elvo Tempia Valenta—Onlus, Via Malta 3, 13900 Biella, Italy
* Correspondence: cristina.casalone@izsto.it; Tel.: +39-011-2686341; Fax: +39-011-2686360

Received: 17 October 2018; Accepted: 7 December 2018; Published: 20 December 2018

Abstract: The niche concept was originally developed to describe the location of normal neural stem cells (NSCs) in the subependymal layer of the sub-ventricular zone. In this paper, its significance has been extended to the location of tumor stem cells in glioblastoma (GB) to discuss the relationship between GB stem cells (GSCs) and endothelial cells (ECs). Their interaction is basically conceived as responsible for tumor growth, invasion and recurrence. Niches are described as the points of utmost expression of the tumor microenvironment (TME), therefore including everything in the tumor except for tumor cells: NSCs, reactive astrocytes, ECs, glioma-associated microglia/macrophages (GAMs), myeloid cells, pericytes, fibroblasts, etc. and all intrinsic and extrinsic signaling pathways. Perivascular (PVNs), perinecrotic (PNNs) and invasive niches were described from the pathological point of view, highlighting the basic significance of the EC/tumor stem cell couple. PNN development was reinterpreted based on the concept that hyperproliferative areas of GB are composed of GSCs/progenitors. TME was depicted in its function as the main regulator of everything that happens in the tumor. A particular emphasis was given to GAMs, pericytes and reactive astrocytes as important elements affecting proliferation, growth, invasion and resistance to therapies of tumor cells.

Keywords: glioblastoma; microenvironment; niche; pericytes; reactive astrocytes

1. Introduction

Glioblastoma (GB) is the most aggressive primary brain tumor in adults accounting for >50% of the tumors of the brain. After surgery, radio- and chemotherapy, survival remains dismal and less than 15 months [1].

Three main properties of the tumor hampers its successful treatment: *(i)* The occurrence of GB stem cells (GSCs); *(ii)* the tumor heterogeneity; *(iii)* the microenvironment and the niches. All these features represent crucial points in the tumor therapy.

The hypothesis of a GSC origin of the tumor is based on the assumption that they represent a rare subset of cells within GB with significant expansion capacity and the ability to generate new tumors [2]. The rest of the tumor is composed of variously differentiated cells with limited progenitor capacity or terminally differentiated non-tumorigenic cells [3]. Therefore, cell heterogeneity and hierarchical organization of GB largely depends on its origin from stem cells or progenitors. Other possible origins of GB are: *(i)* From mature astrocytes that may acquire stemness properties through a dedifferentiation process [4,5]; *(ii)* from neuron glial antigen 2 (NG2) or chondroitin sulphate proteoglyacan 4 (CSPG4)-positive cells, mostly in tumors arising far from the ventricles or with (secondary) *Isocitrate Dehydrogenase* (*IDH*)-mutant GB [6]. Reactive astrocytes may contribute to glioma development, too [7,8]. In fact, they are derived from precursors with a stem-like phenotype [9]. Alternatively, GSCs may represent a sheer functional status [10], depending on the microenvironment regulation [11–14]. The location and generation of GSCs inside the tumor have long been discussed [3]. They may occur either throughout the tumor [15] or, most probably, in proximity of the central necrosis [16–18]. They can be found in the highly proliferative areas of GB close to central necrosis [12,19]. These areas are characterized by high cell and vessel density, high values of proliferation markers, high expression of hypoxia and by the occurrence of circumscribed necrosis; they are in contiguity with the infiltration edges of the tumor.

There is a general agreement that GSCs in the tumor reside in niches that are similar to those hosting normal neural stem cells (NSCs) in the subventricular zone (SVZ) [20]. In these niches, neuroblasts, quiescent NSCs and transit-amplifying cells (A, B and C cells, respectively) occur [21]. They are surrounded by ependymal cells projecting an apical process toward the ventricle. They also develop close to vessels, essential for the stemness maintenance [22]. The main function of the niche in the SVZ is thus to preserve stemness of NSCs [22,23].

The niche concept in malignant gliomas was originally developed to describe the sites where GSCs reside in the tumor and where the tumor microenvironment (TME) exerts its maximum influence. Therefore, for the definition of niche, two conditions must be respected: *(i)* That GSCs do occur and *(ii)* that they have direct contact with endothelial cells (ECs). These conditions are fully realized only in perivascular niches (PVNs) that develop in exchange vessels and not in larger transport vessels with a well-defined layer wall. However, in a broader sense, the term niche also includes perinecrotic niches (PNNs) that do contain GSCs, but not ECs, the occurrence of which precedes necrosis development, being not a reactive phenomenon. On the other hand, GSCs/progenitors in different differentiation stages populate solid proliferative areas of GB, apparently not associated with necrosis or vessels; their differentiation stage is regulated by the TME.

2. Pathology of Niches

2.1. Perivascular Niches

A census of the possible cell components, besides tumor cells, would include normal and reactive astrocytes, pericytes, glioma-associated microglia/macrophages (GAMs), myeloid cells, fibroblasts, and, obviously, GSCs and normal NSCs [24–27]. In their simplest form, PVNs are represented by capillaries or arterioles where ECs are in direct contact with stem cells [28] (Figure 1a). Larger vessels with defined layers, such as transport vessels, cannot function as niches, because they do not allow direct contact between GSCs and ECs. Therefore, not all areas containing vessels and tumor cells would deserve the name niche and are crucial for tumor growth, diffusion, and resistance to therapies [27–30]. The non-cellular component is given by intrinsic and extrinsic signaling pathways [25].

Niches are mainly found in infiltration and invasion areas of the tumor, where they are called invasive niches [29,31]. In invasion areas, tumor cells infiltrate normal tissue as single cells and grow along the basal lamina of vessels to form the so-called vessel co-option (Figure 1b,c), slipping between vessels and reactive astrocytes [32]; the detachment of their end-feet from vessels contributes to the brain-blood-barrier (BBB) disruption (Figure 1d–f) [33].

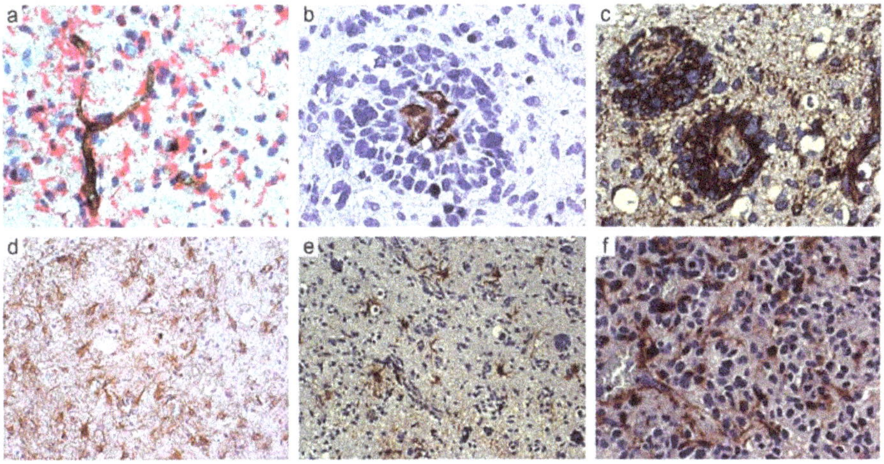

Figure 1. Glioblastoma, IDH wild type. (**a**) CD34+ endothelial cells of arterioles and capillaries in direct contact with Nestin+ tumor cells; double immunostaining with CD34 (DAB) and Nestin (*Fast Red*), original magnification (OM) ×400. (**b**) Vessel co-option. Sleeve of tumor cells around capillaries; CD34, DAB, OM ×400. (**c**) *Id*., with several Nestin+ tumor cells; DAB, OM ×200. (**d**) Mild infiltration. Reactive astrocytes on small vessels; GFAP, DAB, OM ×200; (**e**) More intense infiltration. Reactive astrocytes on vessels; GFAP, DAB, OM ×200; (**f**) High infiltration. Reactive astrocytes with end-feet on small vessels; GFAP, DAB, OM ×400. IDH, isocitrate dehydrogenase; DAB, 3,3′-*Diaminobenzidine*.

Reactive astrocytes produce angiopoietins 1 (Ang-1) and 2 (Ang-2) and vascular endothelial growth factor (VEGF) [34–36]. Nestin+ and Sox2+ tumor cells representing the neoplastic counterpart of normal progenitor cells. They trigger pericyte dissociation, matrix and basal lamina degradation, vessel dilation, leakiness and extracellular deposition of fibrin, to form the so-called "mother vessels" or chronic hyperplasia [37]. In the absence of inhibition from pericytes, ECs proliferate (Figure 2a,b); a switch from an avascular to a vascular state follows and sprouts are formed through EC proliferation (Figure 2c,d). BBB undergoes disruption with leak of macrophages from the vessels (Figure 2e–g). Pericytes, recruited by platelet-derived growth factor receptor β (PDGFRβ) [38], dissociate and ECs proliferate to form new channels that cover with an increased number of pericytes (Figure 2h,i). Whether they are venules or arterioles or neo-formed tumor vessels that do not correspond to any type of normal vessels is difficult to demonstrate. Hypoxia obviously occurs, as everywhere in GB. Glomeruli appear later, surrounded by macrophages and reactive astrocytes. In gliomas, glomeruli formation during angiogenesis takes place as in normal embryos, with the difference that, in the tumor, it is dysregulated and bumpy structures are built that do not contribute to the supply of nutrients and oxygen to the tumor [39].

Angiogenesis is not the only possibility the vasculature has to expand, since vasculogenesis and trans-differentiation of tumor cells into ECs may occur as well [35,40,41].

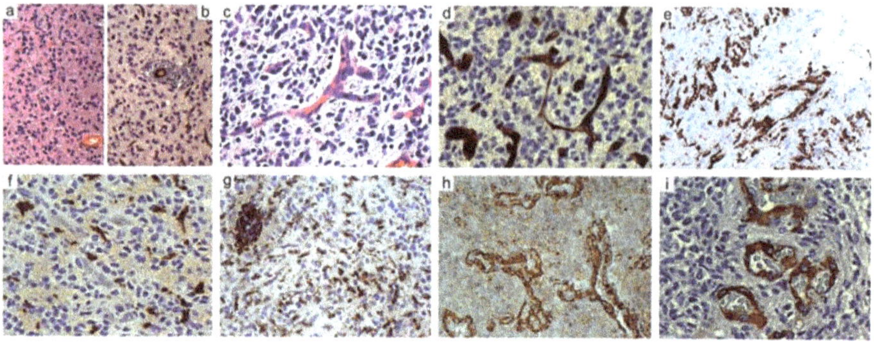

Figure 2. Glioblastoma, IDH wild type. (**a**) Mild infiltration with inital vessel increase; H&E, original magnification (OM) ×200. (**b**) *Id.*, CD34+ endothelial cells; DAB, OM ×200. (**c**) More advanced tumor infiltration, small vessel with endothelial proliferation and sprouts; H&E, OM ×400. (**d**) *Id.*, CD34+ endothelial cells; DAB, OM ×400. (**e**) Mild infiltration with perivascular macrophages; CD163, DAB, OM ×200. (**f**) Infiltration area with leaked perivascular macrophages; CD163, DAB, OM ×200. (**g**) Ramified microglia in tumor parenchyma and perivascular macrophages; Iba-1, DAB, OM ×200. (**h**) Proliferated tumor vessels with NG2/CSPG4+ endothelial cells and pericytes; DAB, OM ×200. (**i**) Glomeruli with α-SMA pericytes; DAB, OM ×200. IDH, isocitrate dehydrogenase; H&E, hematoxylin and eosin; DAB, DAB, 3,3′-*Diaminobenzidine*.

2.2. Perinecrotic Niches

Classically, circumscribed necrosis has been interpreted as due to a vessel occlusion or an intravascular thrombosis [42]. Perinecrotic pseudopalisades have been considered as due to tumor cells fleeing necrosis [43]. GSCs would be induced by hypoxia and hypoxia-inducible factor 1 (HIF-1) and 2 (HIF-2) [35]. Alternatively, or additionally, GSCs/progenitors are believed to regularly populate hyperproliferative areas of GB near central necrosis with high cell and small vessel density, several mitoses and a high Ki-67/MIB-1 labeling index [3]. These areas are regulated by the microenvironment and are recognizable because of their Nestin, Sox2 and CD133 positivity (Figure 3a–k). In these areas, circumscribed necrosis develops as the result of the imbalance between the high proliferation rate of tumor cells and the low one of ECs [44,45]. Indeed, circumscribed necroses are always found in avascular areas of hyperproliferative districts (Figure 3a), close to central necrosis of GB. The cell population expressing Nestin, Sox2 and other stemness markers, including CD133, remains to border the necrosis as remnants of GSCs/progenitors that populated the area and escaped necrosis [12,13] (Figure 3d,e,h–k).

A recent paper took into consideration all niche types described in the literature (perivascular, hypoxic, immune, extracellular matrix niches, etc.) concluding that they are not distinct from one another but they are parts of a single GSC niche, according to the hypoxic periarteriolar niche model [46,47] in which cathepsin K would play a functional role [48]. Roughly, they correspond to the one described as prototype of the PVN.

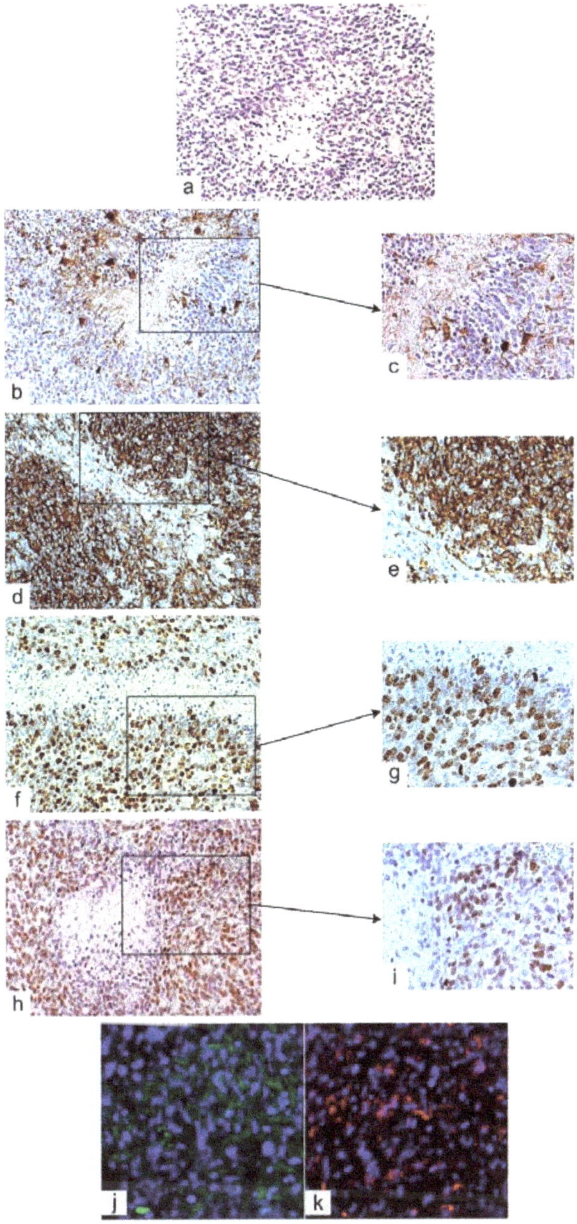

Figure 3. Glioblastoma, IDH wild type. (**a**) Circumscribed necrosis in a hyperproliferative area; H&E, original magnification (OM) ×200. The hyperproliferative zone bordering necrosis is almost GFAP-negative; DAB, OM ×200 (**b**) and ×400 (**c**). The same area is highly Nestin-positive; DAB, OM ×200 (**d**) and ×400 (**e**). The same area shows a high Ki-67/MIB-1 labeling index; DAB, OM ×200 (**f**) and ×400 (**g**). The same area is highly Sox2-positive; DAB, OM ×200 (**h**) and ×400 (**i**). The same area is positive for Musashi-1, cryostat section, immunofluorescence (green) (**j**) and highly CD133-positive, cryostat sections, immunofluorescence (red), both OM ×400 (**k**). IDH, isocitrate dehydrogenase; H&E, hematoxylin and eosin; DAB, DAB, 3,3′-*Diaminobenzidine* [49].

3. Tumor Microenvironment (TME)

TME represents the non-cancerous cells inside the tumor, including normal and reactive astrocytes, GSCs, fibroblasts, immune cells, microglia/macrophages, ECs and vascular pericytes. It also includes proteins and non-protein biomolecules (polysaccharides, hormones, nitric oxide (NO), etc.) produced by all cell types within the TME to support the tumor growth. TME can be mainly detected and demonstrated in niches, but is supposed to regulate everything in the tumor and in the tissue around the tumor or in the brain adjacent to the tumor. GSCs, for instance, mainly occur in niches, but they can also be found in proliferative areas of GB as well [13,50,51], even conceived as being of hypoxic origin. As a matter of fact, hypoxia occurs in the whole tumor, distributed with variable intensity. In necrotic foci, it assumes the typical phenotype of necrosis [52,53], but it may occur in a spot-like manner, with minor intensity and with a not yet modified phenotype. Therefore, besides central and circumscribed necrosis, isolated tumor cells may undergo death as phenotypic hypoxia translation because of individual tumor cell responses to a range of oxygen tension [29]. Other examples are available in the brain tumor pathology. Single apoptotic cells may occur in proliferative areas, due to both the intrinsic, transcriptional pathway of apoptosis associated with duplication, and to the extrinsic pathway associated with a phenotypically subliminal necrosis [3].

In PVNs, the most important signaling is supposed to occur between GSCs/progenitors and ECs [3] (Figure 4).

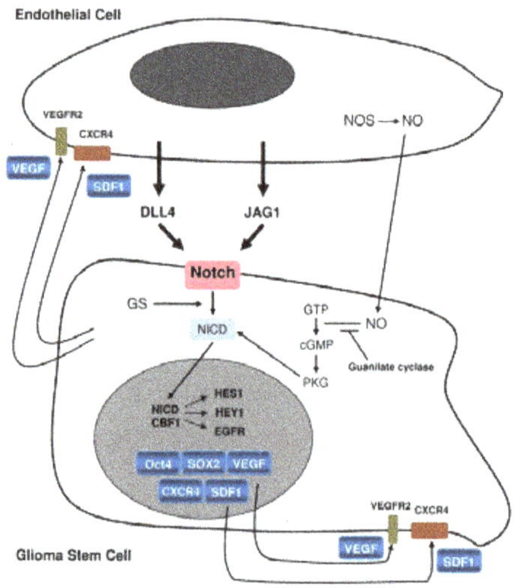

Figure 4. Relationship between a stem cell/progenitor and an endothelial cell [13].

The stemness status of GSCs/progenitors is maintained by ECs via pathways such as NO, cyclic guanosine monophosphate (cGMP) [46] and Notch activation [13,25,54]. Notch-1 and Notch-2 are expressed on GSCs whereas their ligands, Delta-like ligand 4 (DLL4) and Jagged 1 (JAG1), are expressed on the ECs [55]. GSCs/progenitors would promote EC proliferation, eliciting angiogenesis through VEGF, and hosting the bone marrow-derived endothelial precursor cells (EPCs) at the tumor. After activation through its ligands Notch also leads to the final activation of target genes such as *HES1* and *HEY1* [51,56]. This has been confirmed by the blockade of Notch by γ-secretase inhibition that reduces the expression of stemness antigens such as Nestin, CD133 and Bmi-1. It also inhibits in vitro human GB-derived neurosphere formation and xenografts [57] promoting their differentiation into blood vessels inserted into the pre-existing vasculature [27].

Another efficient factor is hypoxia, considered to be a hallmark of GB [53,58–61] that activates pro-angiogenic factors, such as Ang-1/2, transforming growth factor β (TGF-β), PDGF-BB/PDGFR and VEGF/VEGFR through HIF-1/2 [62]. Therefore, hypoxia triggers multiple signaling pathways that affect GSCs self-renewal, proliferation, cell invasion and survival [63]. In addition, it also influences therapeutic resistance of GB and enhances genetic instability of tumor cells. The low oxygen content in the tumor tissue attenuates the expression of *DNA Mismatch Repair (MMR)* genes and inhibits free radicals generated from radiation treatment thus impeding therapeutic efficacy. The *Multi-Drug Resistance Gene 1 (MDR1/ABCB1)* encoding for P-glycoprotein (P-gp) is activated in response to hypoxia [64]. A complicated series of spatially heterogeneous tissue events follows hypoxia in GB [including energetic metabolism (29)] promoting the malignant phenotype and tumor heterogeneity [65].

Finally, TME may control the regulation of the equilibrium between tumor stem and non-stem cells (Figure 5).

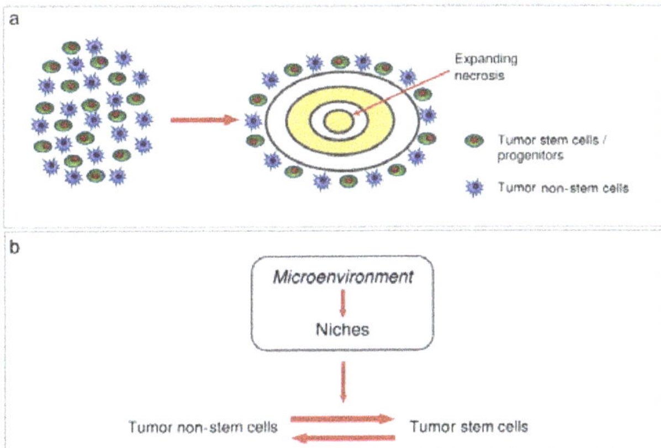

Figure 5. (**a**) Development of circumscribed necrosis. (**b**) Equilibrium between tumor non-stem cells and tumor stem cells [66].

4. Glioma-Associated Microglia/Macrophages (GAMs)—Inflammatory Microenvironment

GAMs cannot be interpreted using the same criteria as for macrophages in other pathological conditions. A fundamental distinction is made between: *(i)* Resident microglia-derived cells (i.e., the so-called reactive microglia with a typical histological appearance) and *(ii)* blood-borne macrophages. The described dichotomy is given with a certain degree of approximation as, often, the distinction between the two cell types is not so sharp. However, different types of myeloid cells occur. It is worth considering that both resident microglia and blood-borne monocytes derive from the yolk sac in different times during embryonic development, the former earlier and directly, and the latter later and through the bone marrow.

The great amount of GAMs, almost equal to the number of tumor cells, raises many still unanswered questions. For instance, GAMs are mainly identified as reactive ramified microglia in low-grade gliomas (LGGs) and as blood-borne monocytes in high-grade gliomas (HGGs) (Figure 2e–g). Since microglia/macrophages and other myeloid cells are strictly connected with the immunological features of gliomas, there is wide literature on the subject suggesting various therapeutic strategies. Basically, there is compelling evidence that GAMs favor tumor progression [67–79], but uncertainties concerning the M1/M2 polarization and the extent of phagocytosis still exist. Very likely, the dilemma whether they are "friends or foes" [80] has not been completely solved because of some demonstrations on the "good" role of GAMs. Scavenger receptors and phagocytosis seem to be completely lacking; however, Fc-γ receptor expression occurs in the solid tumor and, at a lesser extent, in the peritumoral tissue [81,82].

Monocytes [83], tumor-associated neutrophils (TANs) [84] and myeloid-derived suppressor cells (MDSCs) [85] are commonly found within the TME [70,86–88]. An intense signaling exchange takes place among MDSCs, ECs, macrophages, tumor cells and reactive cells [29]. In addition, the influence of chemokines and their receptors must be considered. The most studied signal axes include CXCL12 (SDF-1)-CXCR4, CXCL2-CXCR2, CCL2-CCR2, CX3CL1-CX3CR1, but the problem is far from being completely clarified. The first interaction occurs between macrophages and GSCs [83,89] with the latter activating M2 anti-inflammatory macrophages and, conversely, being maintained in stemness through CXCL12 (SDF-1)-CXCR4 axis. GSCs secrete periostin that recruits M2 tumor-associated macrophages (TAMs) and promotes glioma growth through intergin $\alpha_v\beta3$ [90]. Moreover, TGF-β, released from TAMs, induces matrix metalloproteinase 2 (MMP-2) and 9 (MMP-9) expression from the tumor to enhance GSC invasion [91–94]. On the other hand, TGF-β, shed from GSCs, promotes the polarization of microglia/macrophages into the M2 immunosuppressive phenotype enhancing the capacity of TAMs to inhibit T cell proliferation, thereby promoting tumor progression [25,95]. MDSCs mediate immune suppression and support glioma growth, also interacting with GSCs [96], mainly by immunosuppressing monocytes and other T cell populations [29].

GB can be classified into Proneural (PN), Neural (N), Mesenchymal (MES) and Classical (CL) subtypes, each with its own GSC content [97]. MES GSCs show a preferential activation of the Notch signaling pathway and PDGF receptor, whereas activation of the nuclear factor-κB (NF-κB) pathway and glycolysis-mediated metabolism pathway prevail in PN GSCs. Radiation therapy may induce in GSCs a cellular transformation resembling the epithelial-mesenchymal transition (EMT), called proneural-mesenchymal-transition (PMT) [98]. Triggering PMT GSCs are maintained, and in this step crucial is osteopontin (OPN) that, secreted by immune cells, promotes GSCs phenotype by activating CD44 [99]. A complicated mechanism involves PN and MES GB expression subtypes, PMT, CD44, tumor necrosis factor α (TNF-α), but how GSCs, ECs and TAMs interact has not yet been completely understood. Several studies on microglia/macrophages in gliomas focused on improving patient survival [70,100,101]; some of them, mainly in recent times, concerned the use of dendritic cells (DCs). DCs are granular lymphocytes with cell surface markers: major histocompatibility complex (MHC) class I molecules, MHC class II molecules and CD86, all of which can help to identify DCs from other myeloid lineage cells [102]. They recognize and bind antigens in their immature state and then migrate to lymphoid organs where they present processed peptides to T cells in the context of MHC I or II molecules [103,104] inducing tumor antigen-specific immune responses. Additionally, DCs display various features in the immune system that balance the complex system of inflammatory and inhibitory immune reactions in the TME [105]. Several studies have been designed with a therapeutic task [106–109].

5. Pericytes

There is a cross-talk among vascular pericytes and the other components of the TME, mainly ECs and GSCs. Their interactions during tumor angiogenesis have been widely discussed. Basically, aberrations in pericyte-EC signaling networks have been regarded as contributing to tumor angiogenesis [110]. Pericytes promote vascular maturation, express PDGFRβ, α-smooth muscle Actin (α-SMA), Desmin and NG2/CSPG4 (Figure 2h,i). Pericytes originate from mesoderm-derived mesenchymal stem cells (MSCs) or from neuroectoderm-derived neural crest cells. They are an essential element of the neurovascular unit and participate in the function of BBB. Their reciprocal signaling with ECs, mainly through PDGFRβ and CXCL12 (SDF-1)-CXCR4, TGF-β and Ang-1 has been widely discussed [111]. Pericytes may derive from GSCs undergoing mesenchymal differentiation and support vessel function and tumor growth. GSCs are recruited toward ECs via the CXCL12 (SDF-1)-CXCR4 axis and induced to become pericytes predominantly by TGF-β. Thus, GSCs contribute to vascular pericytes that may actively remodel PVNs [112].

NG2/CSPG4 promotes tumor growth as a component of both tumor and stromal cells; it is expressed by other cell types, mainly oligodendrocyte precursor cells (OPCs). In myeloid-specific and pericyte-specific NG2/CSPG4 null mice, a reduced growth of the tumor was observed. The loss of pericyte-EC interactions reduces the formation of endothelial junctions, assembly of the basal lamina and reduces macrophage recruitment [113]. MSCs injected into brain tumors in mouse models resulted in close associations with the tumor vasculature, also with up-regulation of the expression of pericyte markers [25]. Through the NG2/CSPG4 knockdown in pericytes by small interfering RNA (siRNA) transfection, 60% reduction of β1 integrin activation and 40% of FAK phosphorylation occur with a concomitant decrease of pericyte proliferation and migration [114]. The NG2/CSPG4 ectodomain, shed from pericytes due to a proteolytic cleavage, may recruit at a distance ECs to sites of angiogenesis and may activate β1 integrin on ECs.

In the neo-angiogenesis of GB, pericytes start increasing together with the disruption of BBB becoming a good marker of neo-vascularization [31].

6. Reactive Astrocytes

Reactive astrocytes are a constant phenomenon associated with gliomas [115]. They can surround the tumor or can be located inside. Outside the tumor, they can be found in early or in mature stages. In the first case, they are GFAP+ and Nestin+, regularly distributed with round cytoplasms and short processes, often in mitosis. In the second case, they are more regularly distributed, mainly GFAP+ and with several and long processes. Inside the tumor, they can be observed in continuity with the peritumoral gliosis or entrapped in the advancing tumor with a large, gemistocytic type cytoplasm. They are often distributed around vessels, or they may form areas with a dense GFAP+ fibrillary net. Reactive astrocytes can also be located in highly proliferative areas, around circumscribed necrosis. This means that they may persist for a long time inside the tumor, often in the form of round, GFAP+ cells. The tumor growth speed plays an important role in the reactive astrocyte morphology; in fact, slow growing tumors may include mature astrocytes. The distinction between reactive astrocytes and tumor cells is not easy [116]: the finding of a GFAP+ cell in mitosis does not rule out the possibility that it could be a reactive astrocyte. However, their histological aspect and distribution has been known for a long time and most GFAP positivity of cells in (primary) *IDH*-wild type GB, must be ascribed to entrapped reactive astrocytes [39]. From the functional point of view, peritumoral gliosis cannot be compared with gliosis in other pathological conditions and, for this reason, it must deserve a different interpretation. Notably, reactive astrocytes from tumor infiltration areas send end-feet to arterioles and capillaries, from which they are detached by infiltrating tumor cells, thus contributing to BBB disruption. In addition, reactive astrocytes play a major role in the TME.

There are recent and exhaustive reviews on the subject [117]. Currently, the general opinion is that reactive astrocytes favor invasion and progression of gliomas exerting a chemoprotection and an immune protection of tumor cells. Reactive astrocytes interact with glioma cells and facilitate the progression, aggression and survival of tumors by releasing different cytokines. This interaction is further promoted through ion channels and ion transporters that enhance the migratory capability and invasiveness of tumor cells by modifying H^+ and Ca^{2+} concentrations and stimulating cell volume changes [115].

Several mechanisms involved in the cross-talk between reactive astrocytes and gliomas favor their proliferation, invasion and resistance to radio- and chemotherapy:

(i) Expression of MMP-2, which favors infiltration and secretes CXCL12 (SDF-1) for proliferation and migration [118];

(ii) synergistic relationship with tumor cells concerning the p53 function between apoptosis and proliferation [119];

(iii) regulation through NF-κB activated by receptor activator of NF-κB ligand (RANKL) and lipopolysaccharides (LPS) that decreases IkBα [120,121];

(iv) the gap junction channel protein 43 (Cx43) that confers resistance to glioma cells and prevents apoptosis [122];

(v) indirect cross-talk via chemokines (interleukin 6, IL-6), TGF-β, insulin-like growth factor 1 (IGF-1), monocyte chemotactic protein 4 (MCP-4), interleukin 19 (IL-19), VEGF and leukemia inhibitory factor (LIF), promoting tumor cell invasion [123,124];

(vi) microRNAs [125,126], oncogene astrocyte elevated gene-1 (AEG-1), which is associated with poor survival of gliomas [127] and acts modulating PI3K/Akt, NF-kB, MMP-2 and MMP-9 [128,129] the inhibition of which induces apoptosis [130];

(vii) L-Glutamin [117].

In a murine glioma resection and recurrence model, surgical resection has been showed to alter the reactive astrocyte component of the peritumoral microenvironment and injured astrocytes to induce in vitro alterations of transcriptome and secretome that significantly influence tumor biology. This may be important for therapies [131].

All available speculations on the significance of reactive astrocytes are based on in vitro experiments or on the demonstration that certain pathways play a role in tumor progression. It is possible that these pathways belong to the tumor cells themselves. Moreover, there is no direct demonstration of a negative influence of reactive astrocytes on survival in human gliomas. The possibility that reactive astrogliosis opposes tumor invasion, without success, cannot be completely ruled out. Another unanswered question is whether entrapped reactive astrocytes in the advancing tumor may transform into tumor cells.

7. Conclusions

The great amount of contributions on radio- and chemotherapy did not substantially modify survival of GB patients. Studies on cell death-based treatments continue [132] and new approaches are suggested [133], but more recently, studies on immunity of GB have appeared in the literature and special attention is being paid to vaccines, cytokines, DCs, gene therapy and viruses [134–136]. This seems to be a possible path to advantageous novelties.

Funding: This work was supported by Fondazione Compagnia di San Paolo (Turin, Italy) (Grant N° 2016.AAI2705.U3302) and Fondazione Cassa di Risparmio di Vercelli (Vercelli, Italy).

Conflicts of Interest: The authors declare no conflict of interest.

References

1. Stupp, R.; Mason, W.P.; van den Bent, M.J.; Weller, M.; Fisher, B.; Taphoorn, M.J.; Belanger, K.; Brandes, A.A.; Marosi, C.; Bogdahn, U.; et al. Radiotherapy plus concomitant and adjuvant temozolomide for glioblastoma. *N. Engl. J. Med.* **2005**, *352*, 987–996. [CrossRef] [PubMed]
2. Holland, E.C. Progenitor cells and glioma formation. *Curr. Opin. Neurol.* **2001**, *14*, 683–688. [CrossRef] [PubMed]
3. Schiffer, D.; Mellai, M.; Annovazzi, L.; Casalone, C.; Cassoni, P. Tumor Microenvironment: Perivascular and Perinecrotic Niches in Gliomas. In *Molecular Considerations and Evolving Surgical Management Issues in the Treatment of Patients with a Brain Tumor*, 1st ed.; Morgan, L.R., Ed.; InTech: Rjieka, Croatia, 2015; pp. 49–82, ISBN 978-953-51-2031-5.
4. Sanai, N.; Alvarez-Buylla, A.; Berger, M.S. Neural stem cells and the origin of gliomas. *N. Engl. J. Med.* **2005**, *353*, 811–822. [CrossRef] [PubMed]
5. Assanah, M.; Lochhead, R.; Ogden, A.; Bruce, J.; Goldman, J.J.; Canoll, P. Glial progenitors in adult white matter are driven to form malignant gliomas by platelet-derived growth factor-expressing retroviruses. *J. Neurosci.* **2006**, *26*, 6781–6790. [CrossRef] [PubMed]
6. Siebzehnrubl, F.A.; Reynolds, B.A.; Vescovi, A.; Steindler, D.A.; Deleyrolle, L.P. The origins of glioma: E Pluribus Unum? *Glia* **2011**, *59*, 1135–1147. [CrossRef] [PubMed]
7. Dufour, C.; Cadusseau, J.; Varlet, P.; Surena, A.L.; de Faria, G.P.; Dias-Morais, A.; Auger, N.; Léonard, N.; Daudigeos, E.; Dantas-Barbosa, C.; et al. Astrocytes reverted to a neural progenitor-like state with transforming growth factor alpha are sensitized to cancerous transformation. *Stem Cells* **2009**, *27*, 2373–2382. [CrossRef] [PubMed]

8. Silver, D.J.; Steindler, D.A. Common astrocytic programs during brain development, injury and cancer. *Trends Neurosci.* **2009**, *32*, 303–311. [CrossRef]
9. Buffo, A.; Rite, I.; Tripathi, P.; Lepier, A.; Colak, D.; Horn, A.P.; Mori, T.; Götz, M. Origin and progeny of reactive gliosis: A source of multipotent cells in the injured brain. *Proc. Natl. Acad. Sci. USA* **2008**, *105*, 3581–3586. [CrossRef]
10. Vescovi, A.L.; Galli, R.; Reynolds, B.A. Brain tumour stem cells. *Nat. Rev. Cancer* **2006**, *6*, 425–436. [CrossRef]
11. Safa, A.R.; Saadatzadeh, M.R.; Cohen-Gadol, A.A.; Pollok, K.E.; Bijangi-Vishehsaraei, K. Glioblastoma stem cells (GSCs) epigenetic plasticity and interconversion between differentiated non-GSCs and GSCs. *Gene Dis.* **2015**, *2*, 152–163. [CrossRef]
12. Schiffer, D.; Mellai, M.; Annovazzi, L.; Caldera, V.; Piazzi, A.; Denysenko, T.; Melcarne, A. Stem cell niches in glioblastoma: A neuropathological view. *Biomed. Res. Int.* **2014**. [CrossRef] [PubMed]
13. Schiffer, D.; Annovazzi, L.; Mazzucco, M.; Mellai, M. The microenvironment in gliomas: Phenotypic expressions. *Cancers* **2015**, *7*, 2352–2359. [CrossRef]
14. Schiffer, D.; Annovazzi, L.; Mellai, M. A comprehensive view of tumor stem cells and their regulation by the microenvironment in glioblastoma. *Neurol. Sci.* **2017**, *38*, 527–529. [CrossRef] [PubMed]
15. Pallini, R.; Ricci-Vitiani, L.; Banna, G.L.; Signore, M.; Lombardi, D.; Todaro, M.; Stassi, G.; Martini, M.; Maira, G.; Larocca, L.M.; et al. Cancer stem cell analysis and clinical outcome in patients with glioblastoma multiforme. *Clin. Cancer Res.* **2008**, *14*, 8205–8212. [CrossRef]
16. Piccirillo, S.G.; Combi, R.; Cajola, L.; Patrizi, A.; Redaelli, S.; Bentivegna, A.; Baronchelli, S.; Maira, G.; Pollo, B.; Mangiola, A.; et al. Distinct pools of cancer stem-like cells coexist within human glioblastomas and display different tumorigenicity and independent genomic evolution. *Oncogene* **2009**, *28*, 1807–1811. [CrossRef] [PubMed]
17. Pistollato, F.; Abbadi, S.; Rampazzo, E.; Persano, L.; Della Puppa, A.; Frasson, C.; Sarto, E.; Scienza, R.; D'avella, D.; Basso, G.; et al. Intratumoral hypoxic gradient drives stem cells distribution and MGMT expression in glioblastoma. *Stem Cells* **2010**, *28*, 851–862. [CrossRef]
18. Persano, L.; Rampazzo, E.; Della Puppa, A.; Pistollato, F.; Basso, G. The three-layer concentric model of glioblastoma: Cancer stem cells, microenvironmental regulation, and therapeutic implications. *Sci. World J.* **2011**, *11*, 1829–1841. [CrossRef] [PubMed]
19. Valentini, M.C.; Mellai, M.; Annovazzi, L.; Melcarne, A.; Denysenko, T.; Cassoni, P.; Casalone, C.; Maurella, C.; Grifoni, S.; Fania, P.; et al. Comparison among conventional and advanced MRI, (18)F-FDG PET/CT, phenotype and genotype in glioblastoma. *Oncotarget* **2017**, *8*, 91636–91653. [CrossRef]
20. Palmer, T.D.; Willhoite, A.R.; Gage, F.H. Vascular niche for adult hippocampal neurogenesis. *J. Comp. Neurol.* **2000**, *425*, 479–494. [CrossRef]
21. Veeravagu, A.; Bababeygy, S.R.; Kalani, M.Y.; Hou, L.C.; Tse, V. The cancer stem cell-vascular niche complex in brain tumor formation. *Stem Cells Dev.* **2008**, *17*, 859–867. [CrossRef]
22. Mirzadeh, Z.; Merkle, F.T.; Soriano-Navarro, M.; Garcia-Verdugo, J.M.; Alvarez-Buylla, A. Neural stem cells confer unique pinwheel architecture to the ventricular surface in neurogenic regions of the adult brain. *Cell Stem Cell* **2008**, *3*, 265–278. [CrossRef] [PubMed]
23. Shen, Q.; Wang, Y.; Kokovay, E.; Lin, G.; Chuang, S.M.; Goderie, S.K.; Roysam, B.; Temple, S. Adult SVZ stem cells lie in a vascular niche: A quantitative analysis of niche cell-cell interactions. *Cell Stem Cell* **2008**, *3*, 289–300. [CrossRef] [PubMed]
24. Lorger, M. Tumor microenvironment in the brain. *Cancers* **2012**, *4*, 218–243. [CrossRef] [PubMed]
25. Charles, N.A.; Holland, E.C.; Gilbertson, R.; Glass, R.; Kettenmann, H. The brain tumor microenvironment. *Glia* **2012**, *60*, 502–514. [CrossRef] [PubMed]
26. Filatova, A.; Acker, T.; Garvalov, B.K. The cancer stem cell niche(s): The crosstalk between glioma stem cells and their microenvironment. *Biochim. Biophys. Acta* **2013**, *1830*, 2496–2508. [CrossRef] [PubMed]
27. Hambardzumyan, D.; Bergers, G. Glioblastoma: Defining Tumor Niches. *Trends Cancer* **2015**, *1*, 252–265. [CrossRef] [PubMed]
28. Calabrese, C.; Poppleton, H.; Kocak, M.; Hogg, T.L.; Fuller, C.; Hamner, B.; Oh, E.Y.; Gaber, M.W.; Finklestein, D.; Allen, M.; et al. A perivascular niche for brain tumor stem cells. *Cancer Cell* **2007**, *11*, 69–82. [CrossRef]
29. Ho, I.A.W.; Shim, W.S.N. Contribution of the microenvironmental niche to glioblastoma heterogeneity. *Biomed. Res. Int.* **2017**, *9634172*, 13. [CrossRef]

30. Lathia, J.D.; Mack, S.C.; Mulkearns-Hubert, E.E.; Valentim, C.L.; Rich, J.N. Cancer stem cells in glioblastoma. *Genes Dev.* **2015**, *29*, 1203–1217. [CrossRef]
31. Schiffer, D.; Mellai, M.; Bovio, E.; Bisogno, I.; Casalone, C.; Annovazzi, L. Glioblastoma niches: From the concept to the phenotypical reality. *Neurol. Sci.* **2018**, *39*, 1161–1168. [CrossRef]
32. Cuddapah, V.A.; Robel, S.; Watkins, S.; Sontheimer, H. A neurocentric perspective on glioma invasion. *Nat. Rev. Neurosci.* **2014**, *15*, 455–465. [CrossRef] [PubMed]
33. Watkins, S.; Robel, S.; Kimbrough, I.F.; Robert, S.M.; Ellis-Davies, G.; Sontheimer, H. Disruption of astrocyte-vascular coupling and the blood-brain barrier by invading glioma cells. *Nat. Commun.* **2014**, *5*, 4196. [CrossRef] [PubMed]
34. Holash, J.; Maisonpierre, P.C.; Compton, D.; Boland, P.; Alexander, C.R.; Zagzag, D.; Yancopoulos, G.D.; Wiegand, S.J. Vessel cooption, regression, and growth in tumors mediated by angiopoietins and VEGF. *Science* **1999**, *284*, 1994–1998. [CrossRef] [PubMed]
35. Hardee, M.E.; Zagzag, D. Mechanisms of glioma-associated neovascularization. *Am. J. Pathol.* **2012**, *181*, 1126–1141. [CrossRef] [PubMed]
36. Westphal, M.; Lamszus, K. The neurobiology of gliomas: From cell biology to the development of therapeutic approaches. *Nat. Rev. Neurosci.* **2011**, *12*, 495–508. [CrossRef] [PubMed]
37. Dvorak, H.F. Tumors: Wounds that do not heal-redux. *Cancer Immunol. Res.* **2015**, *3*, 1–11. [CrossRef] [PubMed]
38. Lindahl, P.; Johansson, B.R.; Levéen, P.; Betsholtz, C. Pericyte loss and microaneurysm formation in PDGF-B-deficient mice. *Science* **1997**, *277*, 242–245. [CrossRef] [PubMed]
39. Schiffer, D. *Brain Tumors. Biology, Pathology and Clinical References*, 2nd ed.; Springer: Berlin/Heidelberg, Germany; New York, NY, USA, 1997; pp. 1–695.
40. Du, R.; Lu, K.V.; Petritsch, C.; Liu, P.; Ganss, R.; Passegué, E.; Song, H.; Vandenberg, S.; Johnson, R.S.; Werb, Z.; et al. HIF1alpha induces the recruitment of bone marrow-derived vascular modulatory cells to regulate tumor angiogenesis and invasion. *Cancer Cell* **2008**, *13*, 206–220. [CrossRef]
41. Ricci-Vitiani, L.; Pallini, R.; Biffoni, M.; Todaro, M.; Invernici, G.; Cenci, T.; Maira, G.; Parati, E.A.; Stassi, G.; Larocca, L.M.; et al. Tumour vascularization via endothelial differentiation of glioblastoma stem-like cells. *Nature* **2010**, *468*, 824–828. [CrossRef]
42. Rong, Y.; Durden, D.L.; Van Meir, E.G.; Brat, D.J. 'Pseudopalisading' necrosis in glioblastoma: A familiar morphologic feature that links vascular pathology, hypoxia, and angiogenesis. *J. Neuropathol. Exp. Neurol.* **2006**, *65*, 529–539. [CrossRef]
43. Brat, D.J.; Castellano-Sanchez, A.A.; Hunter, S.B.; Pecot, M.; Cohen, C.; Hammond, E.H.; Devi, S.N.; Kaur, B.; Van Meir, E.G. Pseudopalisades in glioblastoma are hypoxic, express extracellular matrix proteases, and are formed by an actively migrating cell population. *Cancer Res.* **2004**, *64*, 920–927. [CrossRef] [PubMed]
44. Schiffer, D.; Chiò, A.; Giordana, M.T.; Mauro, A.; Migheli, A.; Vigliani, M.C. The vascular response to tumor infiltration in malignant gliomas. Morphometric and reconstruction study. *Acta Neuropathol.* **1989**, *77*, 369–378. [CrossRef] [PubMed]
45. Kargiotis, O.; Rao, J.S.; Kyritsis, A.P. Mechanisms of angiogenesis in gliomas. *J. Neurooncol.* **2006**, *78*, 281–293. [CrossRef] [PubMed]
46. Hira, V.V.V.; Wormer, J.R.; Kakar, H.; Breznik, B.; van der Swaan, B.; Hulsbos, R.; Tigchelaar, W.; Tonar, Z.; Khurshed, M.; Molenaar, R.J.; et al. Periarteriolar Glioblastoma Stem Cell Niches Express Bone Marrow Hematopoietic Stem Cell Niche Proteins. *J. Histochem. Cytochem.* **2018**, *66*, 155–173. [CrossRef] [PubMed]
47. Aderetti, D.A.; Hira, V.V.V.; Molenaar, R.J.; van Noorden, C.J.F. The hypoxic peri-arteriolar glioma stem cell niche, an integrated concept of five types of niches in human glioblastoma. *Biochim. Biophys. Acta* **2018**, *1869*, 346–354. [CrossRef] [PubMed]
48. Breznik, B.; Limbaeck Stokin, C.; Kos, J.; Khurshed, M.; Hira, V.V.V.; Bošnjak, R.; Lah, T.T.; Van Noorden, C.J.F. Cysteine cathepsins B, X and K expression in peri-arteriolar glioblastoma stem cell niches. *J. Mol. Histol.* **2018**, *49*, 481–497. [CrossRef] [PubMed]
49. Schiffer, D.; Annovazzi, L.; Mazzucco, M.; Mellai, M. The origin of circumscribed necroses and perinecrotic niches in glioblastoma multiforme: An additional hypothesis. *Integr. Cancer Sci. Ther.* **2015**, *2*, 75–78. [CrossRef]
50. Christensen, K.; Schrøder, H.D.; Kristensen, B.W. CD133 identifies perivascular niches in grade II-IV astrocytomas. *J. Neurooncol.* **2008**, *90*, 157–170. [CrossRef]

51. Seidel, S.; Garvalov, B.K.; Wirta, V.; von Stechow, L.; Schänzer, A.; Meletis, K.; Wolter, M.; Sommerlad, D.; Henze, A.T.; Nistér, M.; et al. A hypoxic niche regulates glioblastoma stem cells through hypoxia inducible factor 2 alpha. *Brain* **2010**, *133*, 983–995. [CrossRef]
52. Evans, S.M.; Judy, K.D.; Dunphy, I.; Jenkins, W.T.; Hwang, W.T.; Nelson, P.T.; Lustig, R.A.; Jenkins, K.; Magarelli, D.P.; Hahn, S.M.; et al. Hypoxia is important in the biology and aggression of human glial brain tumors. *Clin. Cancer Res.* **2004**, *10*, 8177–8184. [CrossRef]
53. Hanahan, D.; Weinberg, R.A. Hallmarks of cancer: The next generation. *Cell* **2011**, *144*, 646–674. [CrossRef] [PubMed]
54. Hambardzumyan, D.; Becher, O.J.; Rosenblum, M.K.; Pandolfi, P.P.; Manova-Odorova, K.; Holland, E.C. PI3K pathway regulates survival of cancer stem cells residing in the perivascular niche following radiation in medulloblastoma in vivo. *Genes Dev.* **2008**, *22*, 436–448. [CrossRef] [PubMed]
55. Li, L.; Neaves, W.B. Normal stem cells and cancer stem cells: The niche matters. *Cancer Res.* **2006**, *66*, 4553–4557. [CrossRef] [PubMed]
56. Bar, E.E.; Lin, A.; Mahairaki, V.; Matsui, W.; Eberhart, C.G. Hypoxia increases the expression of stem-cell markers and promotes clonogenicity in glioblastoma neurospheres. *Am. J. Pathol.* **2010**, *177*, 1491–1502. [CrossRef] [PubMed]
57. Fan, X.; Khaki, L.; Zhu, T.S.; Soules, M.E.; Talsma, C.E.; Gul, N.; Koh, C.; Zhang, J.; Li, Y.M.; Maciaczyk, J.; et al. NOTCH pathway blockade depletes CD133-positive glioblastoma cells and inhibits growth of tumor neurospheres and xenografts. *Stem Cells* **2010**, *28*, 5–16. [CrossRef] [PubMed]
58. Wong, E.T.; Brem, S. Antiangiogenesis treatment for glioblastoma multiforme: Challenges and opportunities. *J. Natl. Compr. Cancer Netw.* **2008**, *6*, 515–522. [CrossRef]
59. Fischer, U.; Radermacher, J.; Mayer, J.; Mehraein, Y.; Meese, E. Tumor hypoxia: Impact on gene amplification in glioblastoma. *Int. J. Oncol.* **2008**, *33*, 509–515. [CrossRef]
60. Irshad, K.; Mohapatra, S.K.; Srivastava, C.; Garg, H.; Mishra, S.; Dikshit, B.; Sarkar, C.; Gupta, D.; Chandra, P.S.; Chattopadhyay, P.; et al. A combined gene signature of hypoxia and Notch pathway in human glioblastoma and its prognostic relevance. *PLoS ONE* **2015**, *10*, e0118201. [CrossRef]
61. Gordan, J.D.; Bertout, J.A.; Hu, C.J.; Diehl, J.A.; Simon, M.C. HIF-2alpha promotes hypoxic cell proliferation by enhancing c-myc transcriptional activity. *Cancer Cell* **2007**, *11*, 335–347. [CrossRef]
62. Gordan, J.D.; Thompson, C.B.; Simon, M.C. HIF and c-Myc: Sibling rivals for control of cancer cell metabolism and proliferation. *Cancer Cell* **2007**, *12*, 108–113. [CrossRef]
63. Binello, E.; Germano, I.M. Targeting glioma stem cells: A novel framework for brain tumors. *Cancer Sci.* **2011**, *102*, 1958–1966. [CrossRef] [PubMed]
64. Yang, L.; Lin, C.; Wang, L.; Guo, H.; Wang, X. Hypoxia and hypoxia-inducible factors in glioblastoma multiforme progression and therapeutic implications. *Exp. Cell Res.* **2012**, *318*, 2417–2426. [CrossRef]
65. Jawhari, S.; Ratinaud, M.H.; Verdier, M. Glioblastoma, hypoxia and autophagy: A survival-prone 'ménage-à-trois'. *Cell Death Dis.* **2016**, *7*, e2434. [CrossRef] [PubMed]
66. Schiffer, D.; Mellai, M.; Corona, C.; Casalone, C.; Annovazzi, L. Glioblastoma: Equilibrium and Interconvesion between Tumor Non-Stem Cells and Tumor Stem Cells. *Biomed. J. Sci. Tech. Res.* **2018**, *8*, 1–6. [CrossRef]
67. Li, W.; Graeber, M.B. The molecular profile of microglia under the influence of glioma. *Neuro Oncol.* **2012**, *14*, 958–978. [CrossRef] [PubMed]
68. Gabrusiewicz, K.; Ellert-Miklaszewska, A.; Lipko, M.; Sielska, M.; Frankowska, M.; Kaminska, B. Characteristics of the alternative phenotype of microglia/macrophages and its modulation in experimental gliomas. *PLoS ONE* **2011**, *6*, e23902. [CrossRef] [PubMed]
69. Glass, R.; Synowitz, M. CNS macrophages and peripheral myeloid cells in brain tumours. *Acta Neuropathol.* **2014**, *128*, 347–362. [CrossRef] [PubMed]
70. Hambardzumyan, D.; Gutmann, D.H.; Kettenmann, H. The role of microglia and macrophages in glioma maintenance and progression. *Nat. Neurosci.* **2016**, *9*, 20–27. [CrossRef] [PubMed]
71. Brandenburg, S.; Müller, A.; Turkowski, K.; Radev, Y.T.; Rot, S.; Schmidt, C.; Bungert, A.D.; Acker, G.; Schorr, A.; Hippe, A.; et al. Resident microglia rather than peripheral macrophages promote vascularization in brain tumors and are source of alternative pro-angiogenic factors. *Acta Neuropathol.* **2016**, *131*, 365–378. [CrossRef] [PubMed]

72. Hussain, S.F.; Yang, D.; Suki, D.; Aldape, K.; Grimm, E.; Heimberger, A.B. The role of human glioma-infiltrating microglia/macrophages in mediating antitumor immune responses. *Neuro Oncol.* **2006**, *8*, 261–279. [CrossRef]
73. Parney, I.F.; Waldron, J.S.; Parsa, A.T. Flow cytometry and in vitro analysis of human glioma-associated macrophages. *J. Neurosurg.* **2009**, *110*, 572–582. [CrossRef] [PubMed]
74. Wei, J.; Gabrusiewicz, K.; Heimberger, A. The controversial role of microglia in malignant gliomas. *Clin. Dev. Immunol.* **2013**, 285246. [CrossRef] [PubMed]
75. Prinz, M.; Tay, T.L.; Wolf, Y.; Jung, S. Microglia: Unique and common features with other tissue macrophages. *Acta Neuropathol.* **2014**, *128*, 319–331. [CrossRef]
76. Müller, A.; Brandenburg, S.; Turkowski, K.; Müller, S.; Vajkoczy, P. Resident microglia, and not peripheral macrophages, are the main source of brain tumor mononuclear cells. *Int. J. Cancer* **2015**, *137*, 278–288. [CrossRef]
77. Zhu, W.; Carney, K.E.; Pigott, V.M.; Falgoust, M.N.; Clark, P.A.; Kuo, J.S.; Sun, D. Glioma-mediated microglial activation promotes glioma proliferation and migration: Roles of Na+/H+ exchanger isoform 1. *Carcinogenesis* **2016**, *37*, 839–851. [CrossRef]
78. Szulzewsky, F.; Arora, S.; de Witte, L.; Ulas, T.; Markovic, D.; Schultze, J.L.; Holland, E.C.; Synowitz, M.; Wolf, S.A.; Kettenmann, H. Human glioblastoma associated microglia/monocytes express a distinct RNA profile compared to human control and murine samples. *Glia* **2016**, *64*, 1416–1436. [CrossRef]
79. Placone, A.L.; Quiñones-Hinojosa, A.; Searson, P.C. The role of astrocytes in the progression of brain cancer: Complicating the picture of the tumor microenvironment. *Tumour Biol.* **2016**, *37*, 61–69. [CrossRef]
80. Kennedy, B.C.; Showers, C.R.; Anderson, D.E.; Anderson, L.; Canoll, P.; Bruce, J.N.; Anderson, R.C. Tumor-associated macrophages in glioma: Friend or foe? *J. Oncol.* **2013**, 486912. [CrossRef]
81. Morimura, T.; Neuchrist, C.; Kitz, K.; Budka, H.; Scheiner, O.; Kraft, D.; Lassmann, H. Monocyte subpopulations in human gliomas: Expression of Fc and complement receptors and correlation with tumor proliferation. *Acta Neuropathol.* **1990**, *80*, 287–294. [CrossRef]
82. Schiffer, D.; Mellai, M.; Bovio, E.; Annovazzi, L. The neuropathological basis to the functional role of microglia/macrophages in gliomas. *Neurol. Sci.* **2017**, *38*, 1571–1577. [CrossRef]
83. Feng, X.; Szulzewsky, F.; Yerevanian, A.; Chen, Z.; Heinzmann, D.; Rasmussen, R.D.; Alvarez-Garcia, V.; Kim, Y.; Wang, B.; Tamagno, I.; et al. Loss of CX3CR1 increases accumulation of inflammatory monocytes and promotes gliomagenesis. *Oncotarget* **2015**, *6*, 15077–15094. [CrossRef] [PubMed]
84. Liang, J.; Piao, Y.; Holmes, L.; Fuller, G.N.; Henry, V.; Tiao, N.; de Groot, J.F. Neutrophils promote the malignant glioma phenotype through S100A4. *Clin. Cancer Res.* **2014**, *20*, 187–198. [CrossRef] [PubMed]
85. Kohanbash, G.; Okada, H. Myeloid-derived suppressor cells (MDSCs) in gliomas and glioma-development. *Immunol. Investig.* **2012**, *41*, 658–679. [CrossRef] [PubMed]
86. Badie, B.; Schartner, J.M. Flow cytometric characterization of tumor-associated macrophages in experimental gliomas. *Neurosurgery* **2000**, *46*, 957–961. [PubMed]
87. Kettenmann, H.; Hanisch, U.K.; Noda, M.; Verkhratsky, A. Physiology of microglia. *Physiol. Rev.* **2011**, *91*, 461–553. [CrossRef] [PubMed]
88. Gabrusiewicz, B.; Rodriguez, J.; Wei, J.; Hashimoto, Y.; Healy, L.M.; Maiti, S.N.; Thomas, G.; Zhou, S.; Wang, Q.; Elakkad, A.; et al. Glioblastoma-infiltrated innate immune cells resemble M0 macrophage phenotype. *JCI Insight* **2016**, *1*, e85841. [CrossRef] [PubMed]
89. Yi, L.; Xiao, H.; Xu, M.; Ye, X.; Hu, J.; Li, F.; Li, M.; Luo, C.; Yu, S.; Bian, X.; et al. Glioma-initiating cells: A predominant role in microglia/macrophages tropism to glioma. *J. Neuroimmunol.* **2011**, *232*, 75–82. [CrossRef]
90. Zhou, W.; Ke, S.Q.; Huang, Z.; Flavahan, W.; Fang, X.; Paul, J.; Wu, L.; Sloan, A.E.; McLendon, R.E.; Li, X.; et al. Periostin secreted by glioblastoma stem cells recruits M2 tumour-associated macrophages and promotes malignant growth. *Nat. Cell Biol.* **2015**, *17*, 170–182. [CrossRef]
91. Ye, X.Z.; Xu, S.L.; Xin, Y.H.; Yu, S.C.; Ping, Y.F.; Chen, L.; Xiao, H.L.; Wang, B.; Yi, L.; Wang, Q.L.; et al. Tumor-associated microglia/macrophages enhance the invasion of glioma stem-like cells via TGF-β1 signaling pathway. *J. Immunol.* **2012**, *189*, 444–453. [CrossRef]
92. Markovic, D.S.; Glass, R.; Synowitz, M.; Rooijen, N.; Kettenmann, H. Microglia stimulate the invasiveness of glioma cells by increasing the activity of metalloprotease-2. *J. Neuropathol. Exp. Neurol.* **2005**, *64*, 754–762. [CrossRef]

93. Tartour, E.; Pere, H.; Maillere, B.; Terme, N.; Merillon, N.; Taieb, J.; Sandoval, F.; Quintin-Colonna, F.; Lacerda, K.; Karadimou, A.; et al. Angiogenesis and immunity: A bidirectional link potentially relevant for the monitoring of antiangiogenic therapy and the development of novel therapeutic combination with immunotherapy. *Cancer Metastasis Rev.* **2011**, *30*, 83–95. [CrossRef] [PubMed]
94. Hu, F.; Ku, M.C.; Markovic, D.; Dzaye, O.D.; Lehnardt, S.; Synowitz, M.; Wolf, S.A.; Kettenmann, H. Glioma-associated microglial MMP9 expression is upregulated by TLR2 signaling and sensitive to minocycline. *Int. J. Cancer* **2014**, *135*, 2569–2578. [CrossRef] [PubMed]
95. Wu, A.; Wei, J.; Kong, L.Y.; Wang, Y.; Priebe, W.; Qiao, W.; Sawaya, R.; Heimberger, A.B. Glioma cancer stem cells induce immunosuppressive macrophages/microglia. *Neuro Oncol.* **2010**, *12*, 1113–11125. [CrossRef]
96. Domenis, R.; Cesselli, D.; Toffoletto, B.; Bourkoula, E.; Caponnetto, F.; Manini, I.; Beltrami, A.P.; Ius, T.; Skrap, M.; Di Loreto, C.; et al. Systemic T cells immunosuppression of glioma stem cell-derived exosomes is mediated by monocytic myeloid-derived suppressor cells. *PLoS ONE* **2017**, *12*, e0169932. [CrossRef] [PubMed]
97. Verhaak, R.G.; Hoadley, K.A.; Purdom, E.; Wang, V.; Qi, Y.; Wilkerson, M.D.; Miller, C.R.; Ding, L.; Golub, T.; Mesirov, J.P.; et al. Integrated genomic analysis identifies clinically relevant subtypes of glioblastoma characterized by abnormalities in PDGFRA, IDH1, EGFR, and NF1. *Cancer Cell* **2010**, *17*, 98–110. [CrossRef] [PubMed]
98. Nakano, I. Proneural-mesenchymal transformation of glioma stem cells: Do therapies cause evolution of target in glioblastoma? *Future Oncol.* **2014**, *10*, 1527–1530. [CrossRef] [PubMed]
99. Pietras, A.; Katz, A.M.; Ekström, E.J.; Wee, B.; Halliday, J.J.; Pitter, K.L.; Werbeck, J.L.; Amankulor, N.M.; Huse, J.T.; Holland, E.C. Osteopontin-CD44 signaling in the glioma perivascular niche enhances cancer stem cell phenotypes and promotes aggressive tumor growth. *Cell Stem Cell.* **2014**, *14*, 357–369. [CrossRef]
100. da Fonseca, A.C.; Badie, B. Microglia and macrophages in malignant gliomas: Recent discoveries and implications for promising therapies. *Clin. Dev. Immunol.* **2013**, 264124. [CrossRef]
101. Poon, C.C.; Sarkar, S.; Ying, W.; Kelly, J.P. Glioblastoma-associated microglia and macrophages: Targets for therapy to improve prognosis. *Brain* **2017**, *140*, 1548–1560. [CrossRef]
102. Butterfield, L.H. Dendritic cells in cancer immunotherapy clinical trials: Are we making progress? *Front. Immunol.* **2013**, *4*, 454. [CrossRef]
103. Trombetta, E.S.; Mellman, I. Cell biology of antigen processing in vitro and in vivo. *Clin. Dev. Immunol.* **2005**, *23*, 975–1028. [CrossRef] [PubMed]
104. Cohn, L.; Delamarre, L. Dendritic cell-targeted vaccines. *Front. Immunol.* **2014**, *5*, 255. [CrossRef] [PubMed]
105. Mineharu, Y.; Castro, M.G.; Lowenstein, P.R.; Sakai, N.; Miyamoto, S. Dendritic Cell-based Immunotherapy for glioma: Multiple regimens and implications in clinical trials. *Neurol. Med. Chir* **2013**, *53*, 741–754. [CrossRef]
106. Yang, L.; Guo, G.; Niu, X.Y.; Liu, J. Dendritic cell-based immunotherapy treatment for glioblastoma multiforme. *Biomed. Res. Int.* **2015**, 717530 [CrossRef]
107. Nava, S.; Lisini, D.; Pogliani, S.; Dossena, M.; Bersano, A.; Pellegatta, S.; Parati, E.; Finocchiaro, G.; Frigerio, S. Safe and reproducible preparation of functional dendritic cells for immunotherapy in glioblastoma patients. *Stem Cells Transl. Med.* **2015**, *4*, 1164–1172. [CrossRef] [PubMed]
108. Schaller, T.H.; Sampson, J.H. Advances and challenges: Dendritic cell vaccination strategies for glioblastoma. *Expert Rev. Vaccines* **2017**, *16*, 27–36. [CrossRef] [PubMed]
109. Finocchiaro, G.; Pellegatta, S. Immunotherapy with dendritic cells loaded with glioblastoma stem cells: From preclinical to clinical studies. *Cancer Immunol. Immunother.* **2016**, *65*, 101–109. [CrossRef] [PubMed]
110. Raza, A.; Franklin, M.J.; Dudek, A.Z. Pericytes and vessel maturation during tumor angiogenesis and metastasis. *Am. J. Hematol.* **2010**, *85*, 7593–7598. [CrossRef]
111. Harrell, C.R.; Simovic Markovic, B.; Fellabaum, C.; Arsenijevic, A.; Djonov, V.; Volarevic, V. Molecular mechanisms underlying therapeutic potential of pericytes. *J. Biomed. Sci.* **2018**, *25*, 21. [CrossRef]
112. Cheng, L.; Huang, Z.; Zhou, W.; Wu, Q.; Donnola, S.; Liu, J.K.; Fang, X.; Sloan, A.E.; Mao, Y.; Lathia, J.D.; et al. Glioblastoma stem cells generate vascular pericytes to support vessel function and tumor growth. *Cell* **2013**, *153*, 139–152. [CrossRef]
113. Stallcup, W.B.; You, W.K.; Kucharova, K.; Cejudo-Martin, P.; Yotsumoto, F. NG2 proteoglycan-dependent contributions of pericytes and macrophages to brain tumor vascularization and progression. *Microcirculation* **2016**, *23*, 122–133. [CrossRef] [PubMed]

114. Stallcup, W.B. NG2 proteoglycan enhances brain tumor progression by promoting beta-1 Integrin activation in both Cis and Trans orientations. *Cancers* **2017**, *9*, 31. [CrossRef] [PubMed]
115. Guan, X.; Hasan, M.N.; Maniar, S.; Jia, W.; Sun, D. Reactive astrocytes in glioblastoma multiforme. *Mol. Neurobiol.* **2018**, *55*, 6927–6938. [CrossRef] [PubMed]
116. Rivera-Zengotita, M.; Yachnis, A.T. Gliosis Versus Glioma? Don't Grade Until You Know. *Adv. Anat. Pathol.* **2012**, *19*, 239–249. [CrossRef] [PubMed]
117. O'Brien, E.R.; Howarth, C.; Sibson, N.R. The role of astrocytes in CNS tumors: Pre-clinical models and novel imaging approaches. *Front. Cell Neurosci.* **2013**, *7*, 40. [CrossRef] [PubMed]
118. Barbero, S.; Bajetto, A.; Bonavia, R.; Porcile, C.; Piccioli, P.; Pirani, P.; Ravetti, J.L.; Zona, G.; Spaziante, R.; Florio, T.; et al. Expression of the chemokine receptor CXCR4 and its ligand stromal cell-derived factor 1 in human brain tumors and their involvement in glial proliferation in vitro. *Ann. N. Y. Acad. Sci.* **2002**, *973*, 60–69. [CrossRef] [PubMed]
119. Biasoli, D.; Sobrinho, M.F.; da Fonseca, A.C.C.; de Matos, G.G.; Romão, L.; de Moraes Maciel, R.; Rehen, S.K.; Moura-Neto, V.; Borges, H.L.; Lima, F.R.; et al. Glioblastoma cells inhibit astrocytic p53-expression favoring cancer malignancy. *Oncogenesis* **2014**, *3*, e123. [CrossRef]
120. Lin, Q.; Liu, Z.; Ling, F.; Xu, G. Astrocytes protect glioma cells from chemotherapy and upregulate survival genes via gap junctional communication. *Mol. Med. Rep.* **2016**, *13*, 1329–1335. [CrossRef] [PubMed]
121. Kim, J.K.; Jin, X.; Sohn, Y.W.; Jin, X.; Jeon, H.Y.; Kim, E.J.; Ham, S.W.; Jeon, H.M.; Chang, S.Y.; Oh, S.Y.; et al. Tumoral RANKL activates astrocytes that promote glioma cell invasion through cytokine signaling. *Cancer Lett.* **2014**, *353*, 194–200. [CrossRef]
122. Gielen, P.R.; Aftab, Q.; Ma, N.; Chen, V.C.; Hong, X.; Lozinsky, S.; Naus, C.C.; Sin, W.C. Connexin43 confers temozolomide resistance in human glioma cells by modulating the mitochondrial apoptosis pathway. *Neuropharmacology* **2013**, *75*, 539–548. [CrossRef]
123. Bharti, R.; Dey, G.; Mandal, M. Cancer development, chemoresistance, epithelial to mesenchymal transition and stem cells: A snapshot of IL-6 mediated involvement. *Cancer Lett.* **2016**, *375*, 51–61. [CrossRef] [PubMed]
124. Chen, W.; Xia, T.; Wang, D.; Huang, B.; Zhao, P.; Wang, J.; Qu, X.; Li, X. Human astrocytes secrete IL-6 to promote glioma migration and invasion through upregulation of cytomembrane MMP14. *Oncotarget* **2016**, *7*, 62425–62438. [CrossRef] [PubMed]
125. Hong, X.; Sin, W.C.; Harris, A.L.; Naus, C.C. Gap junctions modulate glioma invasion by direct transfer of microRNA. *Oncotarget* **2015**, *6*, 15566–15577. [CrossRef] [PubMed]
126. Tardito, S.; Oudin, A.; Ahmed, S.U.; Fack, F.; Keunen, O.; Zheng, L.; Miletic, H.; Sakariassen, P.Ø.; Weinstock, A.; Wagner, A.; et al. Glutamine synthetase activity fuels nucleotide biosynthesis and supports growth of glutamine-restricted glioblastoma. *Nat. Cell Biol.* **2015**, *17*, 1556–1568. [CrossRef] [PubMed]
127. Hu, B.; Emdad, L.; Bacolod, M.D.; Kegelman, T.P.; Shen, X.N.; Alzubi, M.A.; Das, S.K.; Sarkar, D.; Fisher, P.B. Astrocyte elevated gene-1 interacts with Akt isoform 2 to control glioma growth, survival, and pathogenesis. *Cancer Res.* **2014**, *4*, 7321–7332. [CrossRef] [PubMed]
128. Fisher, P.B. Activation of the nuclear factor kappaB pathway by astrocyte elevated gene-1: Implications for tumor progression and metastasis. *Cancer Res.* **2006**, *66*, 1509–1516.
129. Sarkar, D.; Park, E.S.; Emdad, L.; Lee, S.G.; Su, Z.Z.; Fisher, P.B. Molecular basis of nuclear factor-kappaB activation by astrocyte elevated gene-1. *Cancer Res.* **2008**, *8*, 1478–1484. [CrossRef] [PubMed]
130. Zou, M.; Duan, Y.; Wang, P.; Gao, R.; Chen, X.; Ou, Y.; Liang, M.; Wang, Z.; Yuan, Y.; Wang, L.; et al. DYT-40, a novel synthetic 2-styryl-5-nitroimidazole derivative, blocks malignant glioblastoma growth and invasion by inhibiting AEG-1 and NF-κB signaling pathways. *Sci. Rep.* **2016**, *6*, 27331. [CrossRef] [PubMed]
131. Okolie, O.; Bago, J.R.; Schmid, R.S.; Irvin, D.M.; Bash, E.; Miller, C.R.; Hingtgen, S.D. Reactive astrocytes potentiate tumor aggressiveness in a murine glioma resection and recurrence model. *Neuro Oncol.* **2016**, *18*, 1622–1633. [CrossRef] [PubMed]
132. Fulda, S. Cell death-based treatment of glioblastoma. *Cell Death Dis.* **2018**, *9*, 121. [CrossRef]
133. Siddharth, K.J.; Zuniga, R. High Grade Glioma—Standard Approach, Obstacles and Future Directions. In *Tumors of Central Nervous System. Primary and Secondary*, 1st ed.; Morgan, L.R., Ed.; InTech: Rjieka, Croatia, 2014; pp. 3–29. ISBN 978-953-51-1576-2.
134. Sikorski, C.W.; Lesniak, M.S. Immunotherapy for malignant glioma: Current approaches and future directions. *Neurol. Res.* **2005**, *27*, 703–716. [CrossRef] [PubMed]

135. Binder, D.C.; Davis, A.A.; Wainwright, D.A. Immunotherapy for cancer in the central nervous system: Current and future directions. *Oncoimmunology* **2015**, *5*, e1082027. [CrossRef] [PubMed]
136. Zloza, A.; Karolina Palucka, A.; Coussens, L.M.; Gotwals, P.J.; Headley, M.B.; Jaffee, E.M.; Lund, A.W.; Sharpe, A.H.; Sznol, M.; Wainwright, D.A.; et al. Workshop on challenges, insights, and future directions for mouse and humanized models in cancer immunology and immunotherapy: A report from the associated programs of the 2016 annual meeting for the Society for Immunotherapy of cancer. *J. Immunother. Cancer* **2017**, *5*, 77. [CrossRef] [PubMed]

cancers

MDPI

Review

Targeting the Post-Irradiation Tumor Microenvironment in Glioblastoma via Inhibition of CXCL12

Frank A. Giordano [1,*], Barbara Link [1], Martin Glas [2], Ulrich Herrlinger [3], Frederik Wenz [4], Viktor Umansky [5,6], J. Martin Brown [7] and Carsten Herskind [1]

1 Department of Radiation Oncology, Universitätsmedizin Mannheim, Medical Faculty Mannheim, Heidelberg University, 68167 Mannheim, Germany; barbara.link@medma.uni-heidelberg.de (B.L.); carsten.herskind@medma.uni-heidelberg.de (C.H.)
2 Division of Clinical Neurooncology, Department of Neurology and West German Cancer Center (WTZ), University Hospital Essen and German Cancer Consortium, Partner Site University Hospital Essen, University Duisburg-Essen, 45147 Essen, Germany; Martin.Glas@UK-Essen.de
3 Division of Clinical Neurooncology, Department of Neurology, University of Bonn Medical Center, 53105 Bonn, Germany; Ulrich.Herrlinger@ukbonn.de
4 CEO, University Medical Center Freiburg, 79110 Freiburg, Germany; frederik.wenz@uniklinik-freiburg.de
5 Skin Cancer Unit, German Cancer Research Center (DKFZ), 69120 Heidelberg, Germany; v.umansky@dkfz.de
6 Department of Dermatology, Venereology and Allergology, University Medical Center Mannheim, Ruprecht-Karl University of Heidelberg, 68167 Mannheim, Germany
7 Department of Neurology, Stanford University School of Medicine, Stanford, CA 94305, USA; mbrown@stanford.edu
* Correspondence: Frank.Giordano@umm.de; Tel.: +49-621-383-6020; Fax: +49-621-383-3493

Received: 7 January 2019; Accepted: 20 February 2019; Published: 26 February 2019

Abstract: Radiotherapy is a mainstay in glioblastoma therapy as it not only directly targets tumor cells but also depletes the tumor microvasculature. The resulting intra-tumoral hypoxia initiates a chain of events that ultimately leads to re-vascularization, immunosuppression and, ultimately, tumor-regrowth. The key component of this cascade is overexpression of the CXC-motive chemokine ligand 12 (CXCL12), formerly known as stromal-cell derived factor 1 (SDF-1). We here review the role of CXCL12 in recruitment of pro-vasculogenic and immunosuppressive cells and give an overview on future and current drugs that target this axis.

Keywords: glioblastoma; radiotherapy; CXCL12

1. Introduction

Glioblastoma (GB, WHO Grade IV astrocytoma) is the most common malignant and most aggressive primary brain tumor. The incidence generally increases with age, and the median age of diagnosis is 64 years [1]. The age-adjusted incidence in the U.S. is approximately 3 per 100,000 persons, and survival time of patients diagnosed with GB is usually between 12 and 24 months, with less than 5% living up to 5 years [2]. Focal neurological deficits, symptoms of increased intracranial pressure, epilepsy, and cognitive dysfunction are prominent symptoms which may arise in any stage of the disease [3,4]. Key prognostic factors for survival include general factors such as age, clinical performance status and the extent of resection.

The presence or absence of a promoter methylation of the O^6-methylguanine DNA methyltransferase (MGMT) has been shown to be of specific relevance for outcome as it may predict the response to chemotherapy and overall prognosis as a confirmed prognostic biomarker [5,6].

Treatment of GBM requires a multidisciplinary approach and, for more than a decade, patients with adequate performance scores and tumors amenable to resection undergo surgery and then combined external-beam radiotherapy (EBRT) and chemotherapy with the alkylating agent temozolomide followed by maintenance temozolomide, the 'Stupp protocol' [7]. Elderly or frail patients may, depending on molecular markers (MGMT), alternatively receive either radiotherapy alone, temozolomide alone, or short-course radiotherapy with or without temozolomide [8–10].

A recent randomized clinical phase III trial has shown survival benefits after treatment with a portable, non-invasive device that delivers low-intensity, intermediate-frequency, alternating electric fields to the brain and acts by reversing tumor growth by inhibiting cell division [11]. This therapy, commonly referred to as Tumor-Treating Fields (TTF), has evolved as an additional treatment modality on top of maintenance temozolomide chemotherapy, which is usually initiated after radiochemotherapy.

Despite their effectiveness in GB, all modalities exhibit characteristic adverse effects. Common complications from surgical resection are focal neurological deficits; radiotherapy induces vascular injury, radiation necrosis and gliosis and in patients with longer survival there is also a risk of long-term neurocognitive impairment [12]. Common adverse reactions to temozolomide are mostly limited to the many features of myelotoxicity (anemia, leukopenia, thrombocytopenia) but may also include more unspecific side effects such as nausea, skin rashes and liver toxicity [13–16].

The benefit of the current standard of care, surgical resection followed by radiotherapy and adjuvant chemotherapy, is modest. The Stupp regimen demonstrated a median survival of 14.6 months for treatment with radiotherapy plus temozolomide vs. 12.1 months with radiotherapy alone [7], which is still higher than an expected survival of approximately 7 months with best supportive care only [17]. Although the success of other 'classical' chemotherapies has been limited, the German CeTeG trial demonstrated that the addition of lomustine (CCNU) is beneficial in terms of overall survival in MGMT methylated patients [18]. Other approaches addressing classical cellular signaling pathways (epidermal growth factor receptor: EGFR; fibroblast growth factor receptor: FGFR; tyrosine-protein kinase c-Met: MET; platelet-derived growth factor receptor: PDGFR; phosphoinositide-3-kinase/protein kinase B/mammalian target of rapamycin: PI3K/AKT/mTOR; mitogen-activated protein kinase: MAPK) have failed for various reasons such as weak penetration of the blood brain barrier [19] or bypasses (e.g., resistance to EGFR therapy via insulin-like growth factor receptor (IGFR)-I signaling) and downregulation of pathways [20,21].

Almost all GB recur (mostly local) after first line treatment and, to date, no standard of care has been established for treating recurrent GB. Commonly applied treatment options are re-surgery (if applicable), re-irradiation, chemotherapy with CCNU or therapy with the angiogenesis inhibitor bevacizumab [22]. Almost all salvage options may be considered as palliative: (i) surgery may not tackle the complete extent of the (mostly dispersed) tumor, (ii) EBRT cannot be applied in the same intensity as in the first line due to the limited tolerance of brain tissue towards radiation and (iii) chemotherapy with temozolomide or CCNU is rendered ineffective by the repair enzyme MGMT.

2. Tumor Environment and Immunosuppression in the Brain

It has been controversially discussed for decades if and how the central nervous system (CNS) could be a subject of active immunosurveillance and vigorous immune responses [23]. However, the recent finding that T cells primed by antigen presenting cells in cervical lymph nodes could reach the brain via connecting lymphatic vessels [24] suggests that although the CNS is clearly an immunologically distinct site, its immune microenvironment offers opportunities to implement immunotherapy for brain tumors [25]. Nevertheless, GB is considered as a poorly immunogenic [26], "cold" tumor with only few tumor-infiltrating lymphocytes (TILs) that frequently express markers of exhaustion [27,28]. Furthermore, GB could suppress TIL functions through increased concentrations of extracellular potassium typical for hypoxic tumor areas [29]. In contrast, brain tumors were found to be infiltrated by large numbers of myeloid cells such as microglia, macrophages, and myeloid-derived suppressor cells (MDSCs) [30].

It has been demonstrated that MDSCs play a critical role in the development of an immunosuppressive tumor microenvironment [31–33] (Figure 1). This extremely heterogeneous population of myeloid cells was found to efficiently inhibit T-cell mediated anti-tumor reactivity through various mechanisms [31,32,34]. In mice, MDSCs express Gr1 and CD11b surface molecules and consist of two major subsets: polymorphonuclear $CD11b^{+}Ly6G^{+}Ly6C^{lo/-}$ (PMN-MDSCs) and monocytic $CD11b^{+}Ly6G^{-}Ly6C^{hi}$ (M-MDSCs) [31,32,35]. In humans, M-MDSCs are defined as $CD11b^{+}CD14^{+}CD15^{-}HLA\text{-}DR^{low/-}$ cells. Human PMN-MDSCs are characterized as $CD11b^{+}CD14^{-}CD15^{+}HLA^{-}DR^{-}$ or $CD11b^{+}CD14^{-}CD66b^{+}$ [36,37]. In addition, a subset of more immature human MDSCs characterized as Lin^{-} (including CD3, CD14, CD15, CD19, CD56) HLA-DR-$CD33^{+}$ cells were defined as early-stage MDSCs (eMDSCs) [35]. Recently, a lectin-type oxidized LDL receptor-1 (LOX-1) has been proposed as a new marker for human PMN-MDSCs to distinguish them from neutrophils [38].

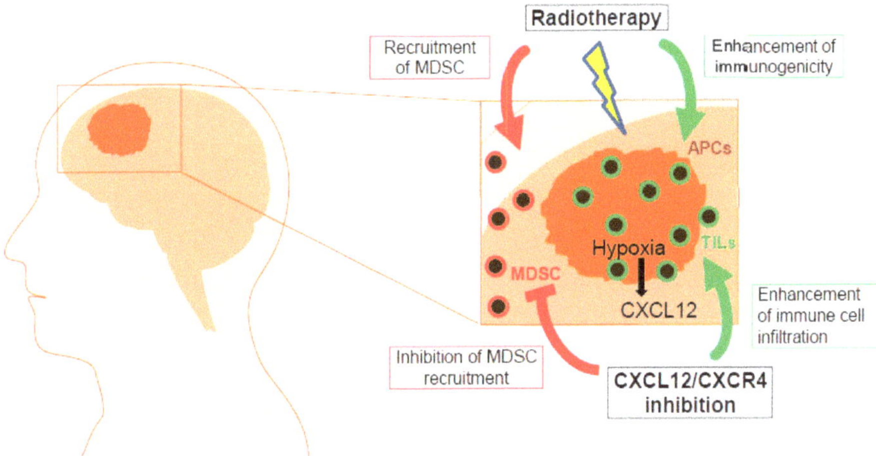

Figure 1. Components of the glioblastoma tumor microenvironment. APC: antigen-presenting cells; MDSC: myeloid-derived suppressor cells; TIL: tumor-infiltrating lymphocytes; CXCL12: C-X-C chemokine ligand 12; CXCR4: C-X-C chemokine receptor 4.

MDSCs derive from bone marrow hematopoietic precursors due to the altering of myelopoiesis induced by chronic inflammatory mediators such as granulocyte-macrophage colony-stimulating factor (GM-CSF), granulocyte colony-stimulating factor (G-CSF), macrophage colony-stimulating factor (M-CSF), stem cell factor (SCF), Vascular Endothelial Growth Factor (VEGF), Interleukin (IL) 6 and 1β [31,32,37,39]. The signaling mainly involves the signal transducer and activator of transcription 3 (STAT3), preventing MDSC differentiation and promoting their proliferation [40,41]. However, the induction of MDSC activity in tumor lesions is provided by pro-inflammatory molecules such as interferon (IFN)-γ, IL-1β, IL-4, Tumor necrosis factor (TNF)-α, toll-like receptor (TLR) ligands, prostaglandin (PGE) E2 and is mediated by STAT1, STAT6 and nuclear factor (NF)-κB transcription factors [31,40,42].

The migration of MDSCs into the tumor microenvironment is mediated by chemokines produced by tumor and host cells. It has been reported that the trafficking of M-MDSCs occurred via an interaction between chemokine (C-C motif) ligand CCL2 and its receptors CCR2, CCR4, and CCR5 [43]. Moreover, the targeting of CCL2/CCR2 axis with antibodies carlumab showed a modest activity as a single-agent therapeutic in patients with metastatic, castration-resistant prostate cancer [44]. On the other hand, the enhanced production of C-X-C motif receptor (CXCR) 2 ligands supported the migration of PMN-MDSCs to the tumor site [45]. Previous publications have indicated that CCL5 supported the tumor growth, invasion, and angiogenesis, as well as immune cell recruitment to the tumor microenvironment via the interaction with CCR5 [46].

In patients with glioblastoma, the frequency of circulating M-MDSCs and PMN-MDSCs was increased as compared to healthy donors [47]. In addition, most of tumor infiltrating MDSC represented PMN-MDSC characterized as $CD15^{+}HLA^{-}DR^{-}$ cells [47]. Another study revealed that that glioblastoma patients displayed higher frequencies of circulating MDSC than age-matched healthy donors and patients with other tumor entities [48]. Interestingly, the majority of these cells were $CD15^{+}$ $CD14^{-}$ PMN-MDSCs. Analyzing a cohort of 52 glioblastoma patients, it was reported a strong accumulation of both PMN- and M-MDSC as compared healthy donors [49]. Importantly, reduced MDSC frequency in the peripheral blood of newly diagnosed glioblastoma was found to be associated with their extended survival, providing a basis for developing strategies to target MDSC in glioblastoma [50].

3. The Glioblastoma Microenvironment Switches from "Angiogenesis" to "Vasculogenesis"

A hallmark of glioblastoma is an abnormal and dysfunctional vasculature which in combination with the glycolytic environment leads to hypoxia [51]. It is now considered that severely hypoxic zones contribute to invasion by causing so called 'pseudopalisades' of cells to migrate away from the central hypoxia [52,53]. Radiotherapy (RT) is one of the key treatment modalities of glioblastoma and has been reported to cause blood vessel damage that may lead to additional tumor cell kill with sufficiently high radiation doses per fraction [54]. Although radiation-induced loss of blood vessels is widely considered a positive effect of RT, it initiates a chain of events that counteract tumor cell killing [55]. Depletion of microvasculature results in weakly perfused or non-perfused areas in which hypoxia-surviving cells respond with overexpression of the otherwise constitutively but only moderately expressed transcription factor hypoxia induced factor 1 alpha (HIF-1α). HIF-1α then induces expression of a wide variety of genes such as VEGF and angiopoietin 2 (ANGPT2), which in turn counteract hypoxia by promoting angiogenesis [56,57]. Besides these rather "directly" acting factors, HIF-1α also induces overexpression of CXC-motive chemokine ligand 12 (CXCL12), formerly known as stromal-cell derived factor 1 (SDF-1). CXCL12 in turn attracts hematopoietic progenitor cells that carry the chemokine (C-X-C motif) receptor 4 (CXCR4; also known as CD184 or fusin).

4. The CXCL12/CXCR4/ACKR3 Axis in Inflammation and Vasculogenesis

Overexpression of CXCL12 and CXCR4 is associated with higher grade and poor prognosis of GBM [58,59]. A potentially important aspect of targeting CXCL12 is its ability to form a heterodimer with high mobility group B1 (HMGB1), a non-histone nuclear protein which acts as a damage associated molecular pattern (DAMP) molecule [60]. Thus, HMGB1 and other DAMP molecules (calreticulin, ATP, dsDNA) are released by damaged cells after irradiation with doses high enough to initiate an inflammatory response [61]. Heterodimerization of CXCL12 with HMGB1 is necessary for attracting monocytes to injured tissue, and migration of mouse embryo fibroblasts and leukocytes towards HMGB1 could be blocked by the CXCR4 inhibitor plerixafor (formerly known as AMD 3100) [62,63]. The redox state of HMGB1 determines which signal pathways are activated; thus, the reduced all-thiol form binds to CXCL12 in a heterocomplex which is more efficient in attracting lymphocytes than CXCL12 alone [60,64]. In relation to tumors, CXCR4 is present on a number of immune suppressor cells, including regulatory T cells (Treg) and myeloid-derived suppressor cells (MDSCs) [65].

The production of CXCL12 can be induced by prostaglandin E2 resulting in a strong enrichment of MDSCs in ovarian and gastric cancer microenvironment [66]. Furthermore, CXCL12–CXCR4 signaling pathways are also reported to be involved in MDSC trafficking in a breast tumor mouse model [67]. Moreover, CXCL12 produced by cancer associated fibroblasts can recruit MDSCs to the tumor microenvironment to exert tumor-promoting effects in mouse model of hepatic carcinoma [68] and estrogen receptor-negative breast cancer [69]. Therefore, CXCL12/CXCR4 interaction could be considered as an important driving force of MDSC recruitment into the tumor microenvironment. Thus, future immune-based strategies may be focused on combinations of different immune checkpoint inhibitors with substances targeting CXCL12/CXCR4 and reversal of local immunosuppression in the microenvironment, converting a 'cold' tumor into a 'hot' tumor [25,70].

Significantly, expression of CXCL12 and CXCR4 was observed in hypoxic regions but not in regions with proliferating cells [71–73]. Hypoxia induces expression of HIF-1α which activates transcription of VEGF and upregulates CXCL12, leading to recruitment of CXCR4$^+$ bone marrow cells and neovasculogenesis [74,75]. A major component of these bone marrow-derived cells for neovasculogenesis is monocytes, which in the tumors differentiate into tumor-associated macrophages (TAMs). These TAMs promote tumor invasiveness, immune suppression and formation of new blood vessels [76,77].

Irradiation of GBM leads to loss of vasculature and increasing hypoxia [78,79] (Figure 2). Doses of 5–15 Gy stimulate vasculogenesis by upregulating CXCL12, in part via stimulation of HIF-1α [74,75]. Since recurrence of glioblastoma after radiotherapy relies on vasculogenesis, targeting CXCL12/CXCR4 was suggested as a novel therapeutic rationale in combination with radiotherapy in these tumors [79,80]. In a mouse GBM model, inhibition of CXCR4 with plerixafor (formerly known as AMD3100) prevented recruitment of bone marrow-derived cells (mainly CD11b$^+$ cells), which together with irradiation (5x2Gy or 1x15Gy) prevented tumor recurrence [78], suggesting that these CD11b$^+$ monocytes or their differentiated progeny, TAMs, supply signals for the formation of new vessels. This was corroborated by the ability of the CXCL12 antagonist NOX-A12, to control autochthonous brain tumors in rats. NOX-A12 is a so-called 'Spiegelmer', a class of synthetic L-ribonucleic acids that are designed to bind to specific targets conceptually similar to antibodies [81]. In this rat brain tumor model, which is extremely resistant to anticancer therapy, and which results in tumors with a genetic diversity and aggressiveness comparable to human brain tumors [82], NOX-A12 in combination with 20 Gy single dose irradiation significantly reduced tumor burden and prolonged rat survival even if the rats were irradiated when the tumors were already established [83]. In another model, NOX-A12 also enhanced the effect of the anti-VEGF antibody bevacizumab [84].

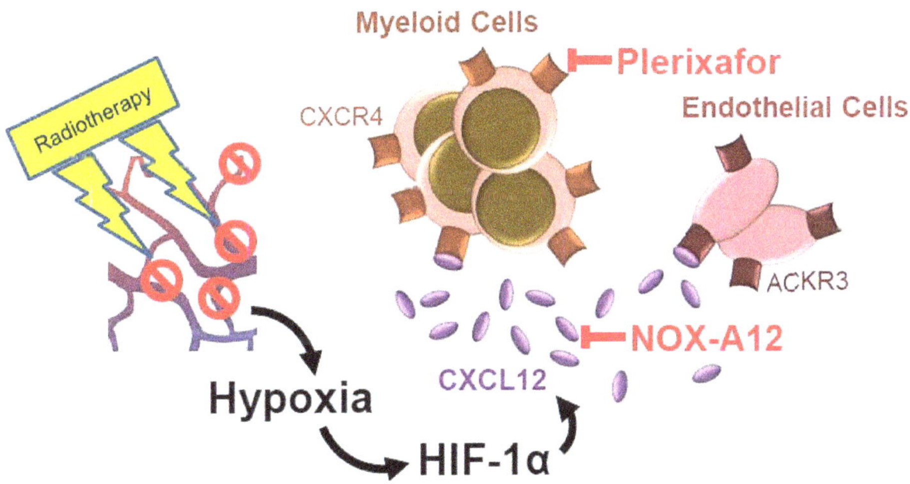

Figure 2. Radiotherapy increases hypoxia in glioblastoma by damaging the tumor vasculature. This leads to upregulation of CXCL12 via HIF-1α (hypoxia-induced factor-1α). Plerixafor (AMD3100) is representative of a number of different types of CXCR4 inhibitors. NOX-A12 binds to and inhibits CXCL12 directly.

CXCL12 was also shown to recruit endothelial progenitor cells from the bone marrow to the tumor microenvironment where they form a new, functional vessel architecture that re-nourishes the tumor and propagates its regrowth [55]. Although some studies have reported that such ACKR3+ endothelial progenitor cells are supplied by the bone marrow, evidence from other studies point towards recruitment of circulating progenitor cells [79]. In fact, it could be shown experimentally that all components of the CXCL12/CXCR4/ACKR3 axis are relevant for recurrence of glioblastoma after radiotherapy, as inhibition of ACKR3 alone in combination with irradiation was also effective in preclinical models [85]. Furthermore, CXCL12-dependent vascularization turned out to be the most prominent bypass mechanism of VEGF-targeted therapy [86].

5. Function of the CXCL12-CXCR4 Axis in Brain and Glioblastoma

CXCL12 belongs to the evolutionary conserved CXC chemokine family expressed especially in the immune system, central nervous systems, and vasculature, of higher vertebrates [87]. CXCL12 has multiple functions that are essential during development and is perinatally lethal in knockout mice. Embryos show defective B-cell lymphopoiesis and myelopoiesis, abnormal neuron migration, and lacking or defective vasculature in kidneys, heart, and skin [88–93]. CXCL12 is expressed by different cell types including stromal cells in the bone marrow, glial and neuronal cells in the nervous system, and endothelial cells in various organs (reviewed in reference [87]). Six different splice variants in humans (CXCL12α to ϕ) are transcriptionally regulated in different cell types. The functions of their products are further regulated by posttranslational modification, including proteolytic truncation, citrullination of Arg residues, and nitration of Tyr residues (reviewed in reference [94]).

CXCL12 is the only chemokine ligand binding to CXCR4, a heptahelical G-protein coupled receptor (GPCR) that is expressed on circulating leukocytes, hematopoietic progenitor and stem cells, as well as on endothelial, stromal and epithelial cells in various tissues [95]. Dissociation of G-protein subunits leads to intracellular signal transduction through MAP Kinase, PI3K/AKT, and phospholipase C pathways, inhibits cAMP signaling, and activates Ca^{2+} and K^+ ion channels [75,87]. A further receptor for CXCL12 is atypical chemokine receptor (ACKR) 3 (formerly known as CXCR7). ACKRs are heptahelical membrane receptors that do not activate G-protein and are expressed mainly on non-leukocyte cell types, including vascular endothelial cells [96]. ACKR3 has a higher affinity for CXCL12 than CXCR4 and can act as a scavenger by internalization. However, it is also able to signal to MAP Kinase and p38 via β-arrestin [87,94] and has been reported to activate MAP Kinase via G protein in astrocytes and glioma cells [97]. In addition to membrane receptors, CXCL12 can bind to glycosaminoglycans (GAG) such as heparin sulfate found in the extracellular matrix and heparin secreted by mast cells, which is important for setting up chemokine gradients required for directed cell migration [94,98]. A schematic overview of the CXCL12/CXCR4/ACKR3 signaling cascade is shown in Figure 3.

Homo-dimerization of CXCR4 and hetero-dimerization of CXCR4 with ACKR3 or other chemokine receptors determines the choice between multiple pathways leading to proliferation, migration, cell adhesion, or survival [75]. Thus, CXCL12 signaling via CXCR4 versus ACKR3 was shown to activate different transcriptional profiles, including two miRNAs [99].

An important function of the CXCL12/CXCR4/ACKR3 axis is regulation of migration and homing of stem and progenitor cells to their niches [75,100,101]. Thus, it regulates stem cells in the bone marrow as well as in the peripheral and central nervous system, and attracts lymphocytes and monocytes expressing the CXCR4 receptor [75]. In the brain, adult neural stem cells are found in vascular niches, and mesenchymal stem cells with potential for differentiating into astrocytes have been found in the perivascular niche [102]. Similarly, glioblastoma stem cells are found in perivascular niches [103] which have recently been reported to be hypoxic periarteriolar rather than pericapillary, and in which endothelial cells and smooth muscle cells provide signals required for maintenance [104]. CXCL12 is expressed by microvascular endothelial cells and stimulated proliferation of GBM progenitor cells but not differentiated tumor cells via CXCR4, whereas ACKR3 was not involved [105,106]. Furthermore, a CXCL12 gradient stimulates migration of GBM cells [107] by Ca^{2+}-dependent activation of BK (Big-Potassium) K^+ channels [108]. Notably, CXCL12-CXCR4 interactions target GBM stem cells to endothelial cells in the perivascular niche where they may be induced to differentiate into pericytes by transforming growth factor (TGF)-β1 [109].

Recent evidence supports the hypothesis human glioblastoma arises from neural stem cells in the subventricular zone (SVZ) [110] though, in a xenograft model, GBM cells have been found to migrate from the tumor to SVZ which was reported to constitute a CXCL12-dependent radioresistant niche [111]. However, the evidence for radioresistance is open to alternative interpretations and no direct evidence for an effect on cell survival was presented. Nevertheless, GBM stem cells are generally considered to show radioresistance mediated by intracellular cell signaling pathways and microenvironmental factors [112].

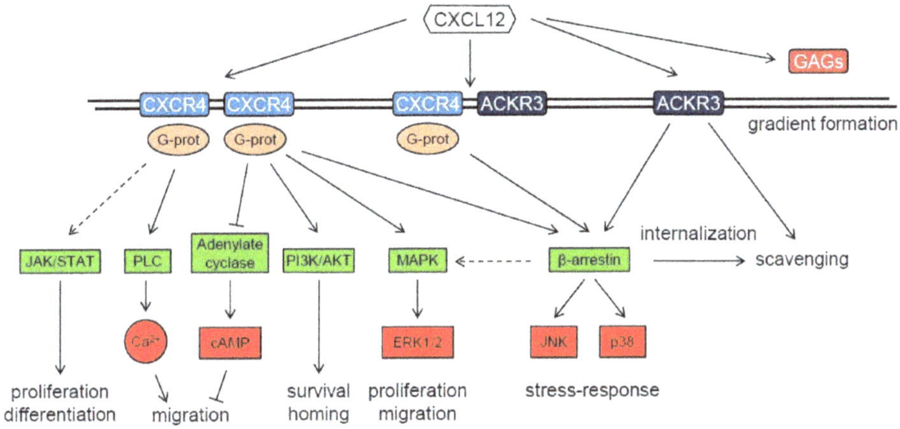

Figure 3. Schematic overview of the CXCL12-CXCR4-ACKR3 signaling cascade. Intermediate kinases have been omitted (for a detailed description, see reviews by Würth et al. [75] and Janssen et al. [94]). cAMP: cyclic adenosine-3′,5′-monophosphate; ERK1/2: extracellular signal-regulated kinase 1/2; GAG: glycosaminoglycans (heparin, heparan sulfate); G-prot: G-proteins; JAK/STAT: Janus kinase/signal transducers and activators of transcription; JNK: c-Jun N-terminal kinase; MAPK: mitogen-activated protein kinase; PI3K/AKT: phosphoinositide-3-kinase/Protein kinase B; PLC: phospholipase C.

6. CXCL12 in Extracranial Tumors

The CXCL12/CXCR4/ACKR3 axis is not only implicated in GB, but also in extracranial tumors and is involved in tumor progression, angiogenesis, metastasis, and survival as well as contributing to immunosuppressive networks within the tumor microenvironment [113]. In hematological malignancies, the cross-talk among lymphoma, myeloma and leukemia cells and their microenvironments is mediated via CXCL12 and its receptors [114] and interference with CXCL12 signaling not only directly inhibits leukemic cell proliferation, but also mobilizes leukemic-lymphoma cells from their niches improving the efficacy of conventional treatments [115].

In solid malignancies, CXCL12 was shown to synergize with vascular endothelial growth factor (VEGF) to promote tumor angiogenesis [116]. CXCL12 also promotes tumor cell proliferation and survival [117] as well as invasion and metastasis via CXCR4 [118,119]. Interestingly, CXCR4+ tumor cells may have stem-like properties, a high metastatic potential, and show radiation resistance [120,121]. Some tumors also express ACKR3, which can promote adhesion, invasion, survival and growth [122,123].

Targeting the CXCL12/CXCR4 axis attenuated non-small cell lung cancer growth and augmented the effects of chemotherapy and radiotherapy [124]. In pancreatic cancer, inhibition of the CXCL12/CXCR4 axis arrested cell growth and abrogated gemcitabine resistance [125] as well as reduced tumor growth in animal models by blocking CXCR4-dependent mast cell migration [126]. CXCL12/CXCR4 axis blockade in combination with sorafenib treatment was reported to inhibit hepatocellular cancer growth [127]. Blocking the CXCR4/CXCL12 axis also inhibited tumor growth and metastasis formation in breast cancer models [128,129]. In models of prostate cancer, inhibition of CXCR4-dependent vascularization delayed tumor growth [130]. In addition, inhibition of CXCL12/CXCR4 signaling by antibodies, peptide analogues, or small molecules has been found to reduce metastatic burden in various orthotopic and metastatic mouse xenograft models [131–133].

Recently, it was reported that T cells may be excluded from cancer cell nests by high CXCL12 levels. Consequently, blocking the CXCL12/CXCR4 axis improved immuno-oncological approaches such as checkpoint inhibition in models of pancreatic cancer [134] and hepatocellular cancer [135]. Clinically, patients with cervical cancer treated with standard-of-care radio-chemotherapy and the CXCR4 inhibitor plerixafor showed improved primary tumor response and reduced metastases [136].

7. Current CXCL12/CXCR4/ACKR3 Inhibitors

A number of antagonists and inhibitors targeting the CXCL12/CXCR4/ACKR3 axis are being developed in cancer indications, mostly in non-brain tumor related cancers and not in combination with radiotherapy (Table 1). Whereas three agents—plerixafor, NOX-A12 and USL311—have been or are reported to be tested in glioblastoma, only the first two are currently explored in combination with radiotherapy, thus targeting the angiogenic-vasculogenic 'switch' induced by radiotherapy. Plerixafor is a bicyclam molecule that antagonizes the CXCR4 receptor. It rapidly and reversibly mobilizes hematopoietic stem cells into the peripheral circulation and was first approved in combination with granulocyte-colony stimulating factor (G-CSF) to mobilize hematopoietic stem cells to the peripheral blood for collection and subsequent autologous transplantation in patients with hematological malignancies.

This approach is currently being tested in a Phase 1/2 study in adult patients with newly diagnosed glioblastoma assessing the impact of CXCR4 blockade with plerixafor combined with radiotherapy and temozolomide (NCT01977677). Interim data presented at ASCO 2018, with 29 patients enrolled, validate the concept of targeting the axis. This study showed that inhibition of the CXCL12/CXCR4 axis (i) is safe in brain tumor patients, (ii) leads to improved local control, (iii) may improve overall survival with an estimated median OS of 20.7 months, and (iv) shifts the pattern of recurrence towards an out-of-field pattern (58.8% out-of-field recurrences vs. 10% in a control group) [137]. This study therefore validates preclinical data by demonstrating that inhibition of the CXCL12/CXCR4 pathway improves local control of GBM following radiotherapy [78,83]. Another study with plerixafor was performed in recurrent high-grade glioma (HGG) patients in combination with VEGF inhibition but without concomitant radiotherapy. This approach tested whether blocking CXCR4 could break the resistance of the tumor to anti-angiogenic therapy. However, while confirming good safety and tolerability of the combination in these patients, the effect on the clinical outcome of these particularly hard-to-treat patients was limited [138].

Targeting the ligand, CXCL12, rather than its receptor, CXCR4, NOX-A12 is another inhibitor of the axis that is to be tested in brain tumors in combination with radiotherapy. The compound was previously tested in two phase I studies in healthy volunteers and two phase IIa studies, one in multiple myeloma patients [139], and one in patients with chronic lymphocytic leukemia. Another phase I/II study in patients with pancreatic and colorectal cancer in combination with PD-1 checkpoint inhibition is ongoing (Keynote-559; NCT03168139). The translational relevance of the preclinical data generated with NOX-A12 described earlier [83] is supported by clinical correlative studies showing that elevated expression of CXCL12 and its receptors, CXCR4 and ACKR3, is associated with higher tumor grade and invasion as well as decreased apoptosis in glioblastoma [140,141].

Table 1. Antagonists of the CXCL12/CXCR4/ACKR3 axis currently in clinical development for cancer.

Compound	Company	Type of Molecule	Target	Indications and Study Pase
Mozobil® (plerixafor, AMD3100)	Sanofi-Genzyme	Small molecule	CXCR4	Stem cell mobilization: approved Multiple myeloma: Phase 1/2 Glioblastoma: Phase 1/2 Advanced solid tumors (pancreas, ovarian, colorectal): Phase 1 AML/ALL: Phase 1
BL-8040	BioLineRx	Peptide	CXCR4	Stem cell mobilization: Phase 3 AML*: Phase 2 Metastatic pancreatic cancer*: Phase 2 T-ALL: Phase 2 Advanced gastric cancer*: Phase 1/2 Metastatic non-small cell lung cancer*: Phase 1/2
NOX-A12 Olaptesed pegol	NOXXON Pharma	L-RNA, PEGylated	CXCL12	Metastatic colorectal and pancreatic cancer: Phase 1/2* Glioblastoma: Phase 1/2
Burixafor TG-0054	TaiGen	Small molecule	CXCR4	Metastatic prostate cancer: Phase 1
LY-2510924	Eli Lilly	Peptide	CXCR4	AML: Phase 1 *Discontinued:* *Metastatic clear cell renal cell carcinoma combined with sunitinib (Phase 2)* *Extensive-disease small-cell lung cancer combined with carboplatin/etoposide (Phase 2)*
Balixafortide POL6326	Polyphor	Peptide	CXCR4	Metastatic breast cancer: Phase 1
Ulocuplumab BMS 936564	BMS	Monoclonal antibody	CXCR4	Waldenström's macroglobulinemia: Phase 1/2 AML: Phase 1/2 *Discontinued:* *Advanced solid tumors (pancreatic and lung cancer patients) combined with nivolumab (Phase 1/2)*
X4P-001	X4 Pharmaceuticals	Small molecule	CXCR4	Metastatic clear cell renal cell carcinoma*, **: Phase 1/2 Resectable melanoma*: Phase 1
USL311	Proximagen	Small molecule	CXCR4	Recurrent glioblastoma: Phase 2

*plus PD-1/PD-L1 inhibitor; **plus axitinib. ALL—acute lymphoblastic leukemia; AML—acute myeloid leukemia; CLL—chronic lymphocytic leukemia; MM—multiple myeloma.

8. Conclusions

The CXCL12/CXCR4 axis is a key regulator of the post-radiotherapy microenvironment in glioblastoma. Blocking of CXCL12 or CXCR4 could prevent the influx of tumor associated macrophages and myeloid suppressor cells which mediate re-vascularization and immunosuppression respectively. Initial results from an early clinical trial warrant further investigation of this therapeutic approach in glioblastoma.

Funding: This research was funded in part by the German Research Council (RTG2099 to VU) and the DKFZ-MOST Cooperation in Cancer Research (CA181 to VU).

Conflicts of Interest: F.A.G.: consultant and speaker for Carl Zeiss Meditec AG, NOXXON Pharma, MSD Sharp and Dohme, Roche Pharma AG, AstraZeneca, Bristol-Myers Squibb and Oncare, holds patents related with Carl Zeiss Meditec AG, co-founder and stakeholder of Implacit. Research grants from Guerbet SA, Carl Zeiss Medotec, Elekta AB and NOXXON Pharma. B.L.: none. M.G.: consulting and advisory role to Roche, Novartis, Daiichi Sankyo, Novocure. Travel support received from Medac, Novocure. Honoraria received from Novartis, Merck, Novocure, Medac, Kyowa Kirin, Bayer. U.H.: Consultant and/or speaker for Roche, Medac, Bristol-Myers Squibb, Novocure, Novartis, Daichii-Sankyo, Riemser, Noxxon. F.W.: advisor, consultant and/or speaker for Celgene GmbH, Roche Pharma AG, Eli Lilly and Company, Ipsen Pharma GmbH, receives travel and research grants from Carl Zeiss Meditec AG and Elekta AB, is on the Carl Zeiss Meditec AG speaker's bureau, holds patents related with Carl Zeiss Meditec AG. V.U.: consultant for NOXXON Pharma. Research grants from NOXXON Pharma and Apogenix. J.M.B.: None. C.H.: teaching the radiobiology of low-energy X-rays for Carl Zeiss Meditec AG; co-inventor of patent held by Carl Zeiss Meditec AG.

References

1. Agnihotri, S.; Burrell, K.E.; Wolf, A.; Jalali, S.; Hawkins, C.; Rutka, J.T.; Zadeh, G. Glioblastoma, a brief review of history, molecular genetics, animal models and novel therapeutic strategies. *Arch. Immunol. Ther. Exp.* **2013**, *61*, 25–41. [CrossRef] [PubMed]
2. Ostrom, Q.T.; Gittleman, H.; Farah, P.; Ondracek, A.; Chen, Y.; Wolinsky, Y.; Stroup, N.E.; Kruchko, C.; Barnholtz-Sloan, J.S. CBTRUS statistical report: Primary brain and central nervous system tumors diagnosed in the United States in 2006–2010. *Neuro-Oncology* **2013**, *15* (Suppl. 2), ii1–ii56. [CrossRef] [PubMed]
3. Behin, A.; Hoang-Xuan, K.; Carpentier, A.F.; Delattre, J.Y. Primary brain tumours in adults. *Lancet* **2003**, *361*, 323–331. [CrossRef]
4. Correa, D.D. Cognitive functions in brain tumor patients. *Hematol. Oncol. Clin. N. Am.* **2006**, *20*, 1363–1376. [CrossRef] [PubMed]
5. Kreth, F.W.; Thon, N.; Simon, M.; Westphal, M.; Schackert, G.; Nikkhah, G.; Hentschel, B.; Reifenberger, G.; Pietsch, T.; Weller, M.; et al. Gross total but not incomplete resection of glioblastoma prolongs survival in the era of radiochemotherapy. *Ann. Oncol.* **2013**, *24*, 3117–3123. [CrossRef] [PubMed]
6. Gittleman, H.; Lim, D.; Kattan, M.W.; Chakravarti, A.; Gilbert, M.R.; Lassman, A.B.; Lo, S.S.; Machtay, M.; Sloan, A.E.; Sulman, E.P.; et al. An independently validated nomogram for individualized estimation of survival among patients with newly diagnosed glioblastoma: NRG Oncology RTOG 0525 and 0825. *Neuro-Oncology* **2017**, *19*, 669–677. [CrossRef] [PubMed]
7. Stupp, R.; Mason, W.P.; van den Bent, M.J.; Weller, M.; Fisher, B.; Taphoorn, M.J.; Belanger, K.; Brandes, A.A.; Marosi, C.; Bogdahn, U.; et al. Radiotherapy plus concomitant and adjuvant temozolomide for glioblastoma. *N. Engl. J. Med.* **2005**, *352*, 987–996. [CrossRef] [PubMed]
8. Malmstrom, A.; Gronberg, B.H.; Marosi, C.; Stupp, R.; Frappaz, D.; Schultz, H.; Abacioglu, U.; Tavelin, B.; Lhermitte, B.; Hegi, M.E.; et al. Temozolomide versus standard 6-week radiotherapy versus hypofractionated radiotherapy in patients older than 60 years with glioblastoma: The Nordic randomised, phase 3 trial. *Lancet Oncol.* **2012**, *13*, 916–926. [CrossRef]
9. Wick, W.; Platten, M.; Meisner, C.; Felsberg, J.; Tabatabai, G.; Simon, M.; Nikkhah, G.; Papsdorf, K.; Steinbach, J.P.; Sabel, M.; et al. Temozolomide chemotherapy alone versus radiotherapy alone for malignant astrocytoma in the elderly: The NOA-08 randomised, phase 3 trial. *Lancet Oncol.* **2012**, *13*, 707–715. [CrossRef]
10. Sulman, E.P.; Ismaila, N.; Armstrong, T.S.; Tsien, C.; Batchelor, T.T.; Cloughesy, T.; Galanis, E.; Gilbert, M.; Gondi, V.; Lovely, M.; et al. Radiation Therapy for Glioblastoma: American Society of Clinical Oncology Clinical Practice Guideline Endorsement of the American Society for Radiation Oncology Guideline. *J. Clin. Oncol.* **2017**, *35*, 361–369. [CrossRef] [PubMed]
11. Stupp, R.; Taillibert, S.; Kanner, A.; Read, W.; Steinberg, D.; Lhermitte, B.; Toms, S.; Idbaih, A.; Ahluwalia, M.S.; Fink, K.; et al. Effect of Tumor-Treating Fields Plus Maintenance Temozolomide vs. Maintenance Temozolomide Alone on Survival in Patients with Glioblastoma: A Randomized Clinical Trial. *JAMA* **2017**, *318*, 2306–2316. [CrossRef] [PubMed]
12. Klein, M. Neurocognitive functioning in adult WHO grade II gliomas: Impact of old and new treatment modalities. *Neuro-Oncology* **2012**, *14* (Suppl. 4), iv17–iv24. [CrossRef] [PubMed]
13. Stupp, R.; Hegi, M.E.; Mason, W.P.; van den Bent, M.J.; Taphoorn, M.J.; Janzer, R.C.; Ludwin, S.K.; Allgeier, A.; Fisher, B.; Belanger, K.; et al. Effects of radiotherapy with concomitant and adjuvant temozolomide versus radiotherapy alone on survival in glioblastoma in a randomised phase III study: 5-year analysis of the EORTC-NCIC trial. *Lancet Oncol.* **2009**, *10*, 459–466. [CrossRef]
14. Bae, S.H.; Park, M.J.; Lee, M.M.; Kim, T.M.; Lee, S.H.; Cho, S.Y.; Kim, Y.H.; Kim, Y.J.; Park, C.K.; Kim, C.Y. Toxicity profile of temozolomide in the treatment of 300 malignant glioma patients in Korea. *J. Korean Med. Sci.* **2014**, *29*, 980–984. [CrossRef] [PubMed]
15. Sengupta, S.; Marrinan, J.; Frishman, C.; Sampath, P. Impact of temozolomide on immune response during malignant glioma chemotherapy. *Clin. Dev. Immunol.* **2012**, *2012*, 831090. [CrossRef] [PubMed]
16. Hilverda, K.; Bosma, I.; Heimans, J.J.; Postma, T.J.; Peter Vandertop, W.; Slotman, B.J.; Buter, J.; Reijneveld, J.C.; Klein, M. Cognitive functioning in glioblastoma patients during radiotherapy and temozolomide treatment: Initial findings. *J. Neurooncol.* **2010**, *97*, 89–94. [CrossRef] [PubMed]

17. Keime-Guibert, F.; Chinot, O.; Taillandier, L.; Cartalat-Carel, S.; Frenay, M.; Kantor, G.; Guillamo, J.S.; Jadaud, E.; Colin, P.; Bondiau, P.Y.; et al. Radiotherapy for glioblastoma in the elderly. *N. Engl. J. Med.* **2007**, *356*, 1527–1535. [CrossRef] [PubMed]
18. Herrlinger, U.; Tzaridis, T.; Mack, F.; Steinbach, J.; Schlegel, U.; Sabel, M.; Hau, P.; Kortman, R.-D.; Krex, D.; Grauer, O.; et al. ACTR-58. Phase III trial of CCNU/temozolomide (TMZ) combination therapy vs. standard TMZ therapy for newly diagnosed MGMT-methylated glioblastoma patients: THE CeTeg/NOA-09 trial. *Neuro-Oncology* **2017**, *19*, vi13–vi14. [CrossRef]
19. Mellinghoff, I.K.; Lassman, A.B.; Wen, P.Y. Signal transduction inhibitors and antiangiogenic therapies for malignant glioma. *Glia* **2011**, *59*, 1205–1212. [CrossRef] [PubMed]
20. Wang, M.; Maier, P.; Wenz, F.; Giordano, F.A.; Herskind, C. Mitogenic signalling in the absence of epidermal growth factor receptor activation in a human glioblastoma cell line. *J. Neurooncol.* **2013**, *115*, 323–331. [CrossRef] [PubMed]
21. Chakravarti, A.; Loeffler, J.S.; Dyson, N.J. Insulin-like growth factor receptor I mediates resistance to anti-epidermal growth factor receptor therapy in primary human glioblastoma cells through continued activation of phosphoinositide 3-kinase signaling. *Cancer Res.* **2002**, *62*, 200–207. [PubMed]
22. Wick, W.; Gorlia, T.; Bendszus, M.; Taphoorn, M.; Sahm, F.; Harting, I.; Brandes, A.A.; Taal, W.; Domont, J.; Idbaih, A.; et al. Lomustine and Bevacizumab in Progressive Glioblastoma. *N. Engl. J. Med.* **2017**, *377*, 1954–1963. [CrossRef] [PubMed]
23. Waksman, B.H.; Adams, R.D. Allergic neuritis: An experimental disease of rabbits induced by the injection of peripheral nervous tissue and adjuvants. *J. Exp. Med.* **1955**, *102*, 213–236. [CrossRef] [PubMed]
24. Louveau, A.; Smirnov, I.; Keyes, T.J.; Eccles, J.D.; Rouhani, S.J.; Peske, J.D.; Derecki, N.C.; Castle, D.; Mandell, J.W.; Lee, K.S.; et al. Structural and functional features of central nervous system lymphatic vessels. *Nature* **2015**, *523*, 337–341. [CrossRef] [PubMed]
25. Lim, M.; Xia, Y.; Bettegowda, C.; Weller, M. Current state of immunotherapy for glioblastoma. *Nat. Rev. Clin. Oncol.* **2018**, *15*, 422–442. [CrossRef] [PubMed]
26. Patel, M.A.; Pardoll, D.M. Concepts of immunotherapy for glioma. *J. Neurooncol.* **2015**, *123*, 323–330. [CrossRef] [PubMed]
27. Hao, C.; Parney, I.F.; Roa, W.H.; Turner, J.; Petruk, K.C.; Ramsay, D.A. Cytokine and cytokine receptor mRNA expression in human glioblastomas: Evidence of Th1, Th2 and Th3 cytokine dysregulation. *Acta Neuropathol.* **2002**, *103*, 171–178. [CrossRef] [PubMed]
28. Wherry, E.J. T cell exhaustion. *Nat. Immunol.* **2011**, *12*, 492–499. [CrossRef] [PubMed]
29. Eil, R.; Vodnala, S.K.; Clever, D.; Klebanoff, C.A.; Sukumar, M.; Pan, J.H.; Palmer, D.C.; Gros, A.; Yamamoto, T.N.; Patel, S.J.; et al. Ionic immune suppression within the tumour microenvironment limits T cell effector function. *Nature* **2016**, *537*, 539–543. [CrossRef] [PubMed]
30. Li, B.; Severson, E.; Pignon, J.C.; Zhao, H.; Li, T.; Novak, J.; Jiang, P.; Shen, H.; Aster, J.C.; Rodig, S.; et al. Comprehensive analyses of tumor immunity: Implications for cancer immunotherapy. *Genome Biol.* **2016**, *17*, 174. [CrossRef] [PubMed]
31. Gabrilovich, D.I.; Ostrand-Rosenberg, S.; Bronte, V. Coordinated regulation of myeloid cells by tumours. *Nat. Rev. Immunol.* **2012**, *12*, 253–268. [CrossRef] [PubMed]
32. Parker, K.H.; Beury, D.W.; Ostrand-Rosenberg, S. Myeloid-Derived Suppressor Cells: Critical Cells Driving Immune Suppression in the Tumor Microenvironment. *Adv. Cancer Res.* **2015**, *128*, 95–139. [CrossRef] [PubMed]
33. Veglia, F.; Perego, M.; Gabrilovich, D. Myeloid-derived suppressor cells coming of age. *Nat. Immunol.* **2018**, *19*, 108–119. [CrossRef] [PubMed]
34. Meirow, Y.; Kanterman, J.; Baniyash, M. Paving the Road to Tumor Development and Spreading: Myeloid-Derived Suppressor Cells are Ruling the Fate. *Front. Immunol.* **2015**, *6*, 523. [CrossRef] [PubMed]
35. Bronte, V.; Brandau, S.; Chen, S.H.; Colombo, M.P.; Frey, A.B.; Greten, T.F.; Mandruzzato, S.; Murray, P.J.; Ochoa, A.; Ostrand-Rosenberg, S.; et al. Recommendations for myeloid-derived suppressor cell nomenclature and characterization standards. *Nat. Commun.* **2016**, *7*, 12150. [CrossRef] [PubMed]
36. Filipazzi, P.; Huber, V.; Rivoltini, L. Phenotype, function and clinical implications of myeloid-derived suppressor cells in cancer patients. *Cancer Immunol. Immunother.* **2012**, *61*, 255–263. [CrossRef] [PubMed]

37. Fleming, V.; Hu, X.; Weber, R.; Nagibin, V.; Groth, C.; Altevogt, P.; Utikal, J.; Umansky, V. Targeting Myeloid-Derived Suppressor Cells to Bypass Tumor-Induced Immunosuppression. *Front. Immunol.* **2018**, *9*, 398. [CrossRef] [PubMed]
38. Condamine, T.; Dominguez, G.A.; Youn, J.I.; Kossenkov, A.V.; Mony, S.; Alicea-Torres, K.; Tcyganov, E.; Hashimoto, A.; Nefedova, Y.; Lin, C.; et al. Lectin-type oxidized LDL receptor-1 distinguishes population of human polymorphonuclear myeloid-derived suppressor cells in cancer patients. *Sci. Immunol.* **2016**, *1*. [CrossRef] [PubMed]
39. Kanterman, J.; Sade-Feldman, M.; Baniyash, M. New insights into chronic inflammation-induced immunosuppression. *Semin. Cancer Biol.* **2012**, *22*, 307–318. [CrossRef] [PubMed]
40. Condamine, T.; Gabrilovich, D.I. Molecular mechanisms regulating myeloid-derived suppressor cell differentiation and function. *Trends Immunol.* **2011**, *32*, 19–25. [CrossRef] [PubMed]
41. Yu, H.; Pardoll, D.; Jove, R. STATs in cancer inflammation and immunity: A leading role for STAT3. *Nat. Rev. Cancer* **2009**, *9*, 798–809. [CrossRef] [PubMed]
42. Sade-Feldman, M.; Kanterman, J.; Ish-Shalom, E.; Elnekave, M.; Horwitz, E.; Baniyash, M. Tumor necrosis factor-alpha blocks differentiation and enhances suppressive activity of immature myeloid cells during chronic inflammation. *Immunity* **2013**, *38*, 541–554. [CrossRef] [PubMed]
43. Lesokhin, A.M.; Hohl, T.M.; Kitano, S.; Cortez, C.; Hirschhorn-Cymerman, D.; Avogadri, F.; Rizzuto, G.A.; Lazarus, J.J.; Pamer, E.G.; Houghton, A.N.; et al. Monocytic CCR2(+) myeloid-derived suppressor cells promote immune escape by limiting activated CD8 T-cell infiltration into the tumor microenvironment. *Cancer Res.* **2012**, *72*, 876–886. [CrossRef] [PubMed]
44. Ugel, S.; De Sanctis, F.; Mandruzzato, S.; Bronte, V. Tumor-induced myeloid deviation: When myeloid-derived suppressor cells meet tumor-associated macrophages. *J. Clin. Investig.* **2015**, *125*, 3365–3376. [CrossRef] [PubMed]
45. Sawanobori, Y.; Ueha, S.; Kurachi, M.; Shimaoka, T.; Talmadge, J.E.; Abe, J.; Shono, Y.; Kitabatake, M.; Kakimi, K.; Mukaida, N.; et al. Chemokine-mediated rapid turnover of myeloid-derived suppressor cells in tumor-bearing mice. *Blood* **2008**, *111*, 5457–5466. [CrossRef] [PubMed]
46. Appay, V.; Rowland-Jones, S.L. RANTES: A versatile and controversial chemokine. *Trends Immunol.* **2001**, *22*, 83–87. [CrossRef]
47. Gielen, P.R.; Schulte, B.M.; Kers-Rebel, E.D.; Verrijp, K.; Petersen-Baltussen, H.M.; ter Laan, M.; Wesseling, P.; Adema, G.J. Increase in both CD14-positive and CD15-positive myeloid-derived suppressor cell subpopulations in the blood of patients with glioma but predominance of CD15-positive myeloid-derived suppressor cells in glioma tissue. *J. Neuropathol. Exp. Neurol.* **2015**, *74*, 390–400. [CrossRef] [PubMed]
48. Raychaudhuri, B.; Rayman, P.; Ireland, J.; Ko, J.; Rini, B.; Borden, E.C.; Garcia, J.; Vogelbaum, M.A.; Finke, J. Myeloid-derived suppressor cell accumulation and function in patients with newly diagnosed glioblastoma. *Neuro-Oncology* **2011**, *13*, 591–599. [CrossRef] [PubMed]
49. Dubinski, D.; Wolfer, J.; Hasselblatt, M.; Schneider-Hohendorf, T.; Bogdahn, U.; Stummer, W.; Wiendl, H.; Grauer, O.M. CD4+ T effector memory cell dysfunction is associated with the accumulation of granulocytic myeloid-derived suppressor cells in glioblastoma patients. *Neuro-Oncology* **2016**, *18*, 807–818. [CrossRef] [PubMed]
50. Alban, T.J.; Alvarado, A.G.; Sorensen, M.D.; Bayik, D.; Volovetz, J.; Serbinowski, E.; Mulkearns-Hubert, E.E.; Sinyuk, M.; Hale, J.S.; Onzi, G.R.; et al. Global immune fingerprinting in glioblastoma patient peripheral blood reveals immune-suppression signatures associated with prognosis. *JCI Insight* **2018**, *3*. [CrossRef] [PubMed]
51. Huang, W.J.; Chen, W.W.; Zhang, X. Glioblastoma multiforme: Effect of hypoxia and hypoxia inducible factors on therapeutic approaches. *Oncol. Lett.* **2016**, *12*, 2283–2288. [CrossRef] [PubMed]
52. Monteiro, A.R.; Hill, R.; Pilkington, G.J.; Madureira, P.A. The Role of Hypoxia in Glioblastoma Invasion. *Cells* **2017**, *6*, 45. [CrossRef] [PubMed]
53. Rong, Y.; Durden, D.L.; Van Meir, E.G.; Brat, D.J. 'Pseudopalisading' necrosis in glioblastoma: A familiar morphologic feature that links vascular pathology, hypoxia, and angiogenesis. *J. Neuropathol. Exp. Neurol.* **2006**, *65*, 529–539. [CrossRef] [PubMed]
54. Song, C.W.; Park, I.; Cho, L.C.; Yuan, J.; Dusenbery, K.E.; Griffin, R.J.; Levitt, S.H. Is indirect cell death involved in response of tumors to stereotactic radiosurgery and stereotactic body radiation therapy? *Int. J. Radiat. Oncol. Biol. Phys.* **2014**, *89*, 924–925. [CrossRef] [PubMed]

55. Greenfield, J.P.; Cobb, W.S.; Lyden, D. Resisting arrest: A switch from angiogenesis to vasculogenesis in recurrent malignant gliomas. *J. Clin. Investig.* **2010**, *120*, 663–667. [CrossRef] [PubMed]
56. Han, Z.B.; Ren, H.; Zhao, H.; Chi, Y.; Chen, K.; Zhou, B.; Liu, Y.J.; Zhang, L.; Xu, B.; Liu, B.; et al. Hypoxia-inducible factor (HIF)-1 alpha directly enhances the transcriptional activity of stem cell factor (SCF) in response to hypoxia and epidermal growth factor (EGF). *Carcinogenesis* **2008**, *29*, 1853–1861. [CrossRef] [PubMed]
57. Kelly, B.D.; Hackett, S.F.; Hirota, K.; Oshima, Y.; Cai, Z.; Berg-Dixon, S.; Rowan, A.; Yan, Z.; Campochiaro, P.A.; Semenza, G.L. Cell type-specific regulation of angiogenic growth factor gene expression and induction of angiogenesis in nonischemic tissue by a constitutively active form of hypoxia-inducible factor 1. *Circ. Res.* **2003**, *93*, 1074–1081. [CrossRef] [PubMed]
58. Bian, X.W.; Yang, S.X.; Chen, J.H.; Ping, Y.F.; Zhou, X.D.; Wang, Q.L.; Jiang, X.F.; Gong, W.; Xiao, H.L.; Du, L.L.; et al. Preferential expression of chemokine receptor CXCR4 by highly malignant human gliomas and its association with poor patient survival. *Neurosurgery* **2007**, *61*, 570–578; discussion 578–579. [CrossRef] [PubMed]
59. Salmaggi, A.; Gelati, M.; Pollo, B.; Marras, C.; Silvani, A.; Balestrini, M.R.; Eoli, M.; Fariselli, L.; Broggi, G.; Boiardi, A. CXCL12 expression is predictive of a shorter time to tumor progression in low-grade glioma: A single-institution study in 50 patients. *J. Neurooncol.* **2005**, *74*, 287–293. [CrossRef] [PubMed]
60. Venereau, E.; Schiraldi, M.; Uguccioni, M.; Bianchi, M.E. HMGB1 and leukocyte migration during trauma and sterile inflammation. *Mol. Immunol.* **2013**, *55*, 76–82. [CrossRef] [PubMed]
61. Herskind, C.; Wenz, F.; Giordano, F.A. Immunotherapy Combined with Large Fractions of Radiotherapy: Stereotactic Radiosurgery for Brain Metastases-Implications for Intraoperative Radiotherapy after Resection. *Front. Oncol.* **2017**, *7*, 147. [CrossRef] [PubMed]
62. Kew, R.R.; Penzo, M.; Habiel, D.M.; Marcu, K.B. The IKKalpha-dependent NF-kappaB p52/RelB noncanonical pathway is essential to sustain a CXCL12 autocrine loop in cells migrating in response to HMGB1. *J. Immunol.* **2012**, *188*, 2380–2386. [CrossRef] [PubMed]
63. Schiraldi, M.; Raucci, A.; Munoz, L.M.; Livoti, E.; Celona, B.; Venereau, E.; Apuzzo, T.; De Marchis, F.; Pedotti, M.; Bachi, A.; et al. HMGB1 promotes recruitment of inflammatory cells to damaged tissues by forming a complex with CXCL12 and signaling via CXCR4. *J. Exp. Med.* **2012**, *209*, 551–563. [CrossRef] [PubMed]
64. D'Agostino, G.; Cecchinato, V.; Uguccioni, M. Chemokine Heterocomplexes and Cancer: A Novel Chapter to Be Written in Tumor Immunity. *Front. Immunol.* **2018**, *9*, 2185. [CrossRef] [PubMed]
65. Susek, K.H.; Karvouni, M.; Alici, E.; Lundqvist, A. The Role of CXC Chemokine Receptors 1–4 on Immune Cells in the Tumor Microenvironment. *Front. Immunol.* **2018**, *9*, 2159. [CrossRef] [PubMed]
66. Kalinski, P. Regulation of immune responses by prostaglandin E2. *J. Immunol.* **2012**, *188*, 21–28. [CrossRef] [PubMed]
67. Nagarsheth, N.; Wicha, M.S.; Zou, W. Chemokines in the cancer microenvironment and their relevance in cancer immunotherapy. *Nat. Rev. Immunol.* **2017**, *17*, 559–572. [CrossRef] [PubMed]
68. Deng, Y.; Cheng, J.; Fu, B.; Liu, W.; Chen, G.; Zhang, Q.; Yang, Y. Hepatic carcinoma-associated fibroblasts enhance immune suppression by facilitating the generation of myeloid-derived suppressor cells. *Oncogene* **2017**, *36*, 1090–1101. [CrossRef] [PubMed]
69. Ouyang, L.; Chang, W.; Fang, B.; Qin, J.; Qu, X.; Cheng, F. Estrogen-induced SDF-1alpha production promotes the progression of ER-negative breast cancer via the accumulation of MDSCs in the tumor microenvironment. *Sci. Rep.* **2016**, *6*, 39541. [CrossRef] [PubMed]
70. Wu, A.; Cardarelli, P.; Oyasu, M.; Menezes, D.; Ponath, P.; Cogswell, J.; Maxwell, R.; Luksik, A.; Hung, A.; Kim, E.; et al. Abstract 1736: The combination of CXCR4 and checkpoint receptor inhibition improves survival in an orthotopic murine glioma model. *Cancer Res.* **2018**, *78*, 1736. [CrossRef]
71. Rempel, S.A.; Dudas, S.; Ge, S.; Gutierrez, J.A. Identification and localization of the cytokine SDF1 and its receptor, CXC chemokine receptor 4, to regions of necrosis and angiogenesis in human glioblastoma. *Clin. Cancer Res.* **2000**, *6*, 102–111. [PubMed]
72. Salmaggi, A.; Riva, M.; Silvani, A.; Merli, R.; Tomei, G.; Lorusso, L.; Russo, A.; Marchioni, E.; Imbesi, F.; Lombardia Neuro-oncology, G. A multicentre prospective collection of newly diagnosed glioblastoma patients in Lombardia, Italy. *Neurol. Sci.* **2005**, *26*, 227–234. [CrossRef] [PubMed]

73. Zagzag, D.; Esencay, M.; Mendez, O.; Yee, H.; Smirnova, I.; Huang, Y.; Chiriboga, L.; Lukyanov, E.; Liu, M.; Newcomb, E.W. Hypoxia- and vascular endothelial growth factor-induced stromal cell-derived factor-1alpha/CXCR4 expression in glioblastomas: One plausible explanation of Scherer's structures. *Am. J. Pathol.* **2008**, *173*, 545–560. [CrossRef] [PubMed]
74. Petit, I.; Jin, D.; Rafii, S. The SDF-1-CXCR4 signaling pathway: A molecular hub modulating neo-angiogenesis. *Trends Immunol.* **2007**, *28*, 299–307. [CrossRef] [PubMed]
75. Wurth, R.; Bajetto, A.; Harrison, J.K.; Barbieri, F.; Florio, T. CXCL12 modulation of CXCR4 and CXCR7 activity in human glioblastoma stem-like cells and regulation of the tumor microenvironment. *Front. Cell. Neurosci.* **2014**, *8*, 144. [CrossRef] [PubMed]
76. Wang, S.C.; Yu, C.F.; Hong, J.H.; Tsai, C.S.; Chiang, C.S. Radiation therapy-induced tumor invasiveness is associated with SDF-1-regulated macrophage mobilization and vasculogenesis. *PLoS ONE* **2013**, *8*, e69182. [CrossRef] [PubMed]
77. Brown, J.M.; Recht, L.; Strober, S. The Promise of Targeting Macrophages in Cancer Therapy. *Clin. Cancer Res.* **2017**, *23*, 3241–3250. [CrossRef] [PubMed]
78. Kioi, M.; Vogel, H.; Schultz, G.; Hoffman, R.M.; Harsh, G.R.; Brown, J.M. Inhibition of vasculogenesis, but not angiogenesis, prevents the recurrence of glioblastoma after irradiation in mice. *J. Clin. Investig.* **2010**, *120*, 694–705. [CrossRef] [PubMed]
79. Tseng, D.; Vasquez-Medrano, D.A.; Brown, J.M. Targeting SDF-1/CXCR4 to inhibit tumour vasculature for treatment of glioblastomas. *Br. J. Cancer* **2011**, *104*, 1805–1809. [CrossRef] [PubMed]
80. Brown, J.M. Vasculogenesis: A crucial player in the resistance of solid tumours to radiotherapy. *Br. J. Radiol.* **2014**, *87*, 20130686. [CrossRef] [PubMed]
81. Hoellenriegel, J.; Zboralski, D.; Maasch, C.; Rosin, N.Y.; Wierda, W.G.; Keating, M.J.; Kruschinski, A.; Burger, J.A. The Spiegelmer NOX-A12, a novel CXCL12 inhibitor, interferes with chronic lymphocytic leukemia cell motility and causes chemosensitization. *Blood* **2014**, *123*, 1032–1039. [CrossRef] [PubMed]
82. Jang, T.; Savarese, T.; Low, H.P.; Kim, S.; Vogel, H.; Lapointe, D.; Duong, T.; Litofsky, N.S.; Weimann, J.M.; Ross, A.H.; et al. Osteopontin expression in intratumoral astrocytes marks tumor progression in gliomas induced by prenatal exposure to *N*-ethyl-*N*-nitrosourea. *Am. J. Pathol.* **2006**, *168*, 1676–1685. [CrossRef] [PubMed]
83. Liu, S.C.; Alomran, R.; Chernikova, S.B.; Lartey, F.; Stafford, J.; Jang, T.; Merchant, M.; Zboralski, D.; Zollner, S.; Kruschinski, A.; et al. Blockade of SDF-1 after irradiation inhibits tumor recurrences of autochthonous brain tumors in rats. *Neuro-Oncology* **2014**, *16*, 21–28. [CrossRef] [PubMed]
84. Deng, L.; Stafford, J.H.; Liu, S.C.; Chernikova, S.B.; Merchant, M.; Recht, L.; Martin Brown, J. SDF-1 Blockade Enhances Anti-VEGF Therapy of Glioblastoma and Can Be Monitored by MRI. *Neoplasia* **2017**, *19*, 1–7. [CrossRef] [PubMed]
85. Walters, M.J.; Ebsworth, K.; Berahovich, R.D.; Penfold, M.E.; Liu, S.C.; Al Omran, R.; Kioi, M.; Chernikova, S.B.; Tseng, D.; Mulkearns-Hubert, E.E.; et al. Inhibition of CXCR7 extends survival following irradiation of brain tumours in mice and rats. *Br. J. Cancer* **2014**, *110*, 1179–1188. [CrossRef] [PubMed]
86. de Groot, J.F.; Lamborn, K.R.; Chang, S.M.; Gilbert, M.R.; Cloughesy, T.F.; Aldape, K.; Yao, J.; Jackson, E.F.; Lieberman, F.; Robins, H.I.; et al. Phase II study of aflibercept in recurrent malignant glioma: A North American Brain Tumor Consortium study. *J. Clin. Oncol.* **2011**, *29*, 2689–2695. [CrossRef] [PubMed]
87. Guyon, A. CXCL12 chemokine and its receptors as major players in the interactions between immune and nervous systems. *Front. Cell. Neurosci.* **2014**, *8*, 65. [CrossRef] [PubMed]
88. Cavallero, S.; Shen, H.; Yi, C.; Lien, C.L.; Kumar, S.R.; Sucov, H.M. CXCL12 Signaling Is Essential for Maturation of the Ventricular Coronary Endothelial Plexus and Establishment of Functional Coronary Circulation. *Dev. Cell* **2015**, *33*, 469–477. [CrossRef] [PubMed]
89. Ivins, S.; Chappell, J.; Vernay, B.; Suntharalingham, J.; Martineau, A.; Mohun, T.J.; Scambler, P.J. The CXCL12/CXCR4 Axis Plays a Critical Role in Coronary Artery Development. *Dev. Cell* **2015**, *33*, 455–468. [CrossRef] [PubMed]
90. Li, W.; Kohara, H.; Uchida, Y.; James, J.M.; Soneji, K.; Cronshaw, D.G.; Zou, Y.R.; Nagasawa, T.; Mukouyama, Y.S. Peripheral nerve-derived CXCL12 and VEGF-A regulate the patterning of arterial vessel branching in developing limb skin. *Dev. Cell* **2013**, *24*, 359–371. [CrossRef] [PubMed]

91. Ma, Q.; Jones, D.; Borghesani, P.R.; Segal, R.A.; Nagasawa, T.; Kishimoto, T.; Bronson, R.T.; Springer, T.A. Impaired B-lymphopoiesis, myelopoiesis, and derailed cerebellar neuron migration in CXCR4- and SDF-1-deficient mice. *Proc. Natl. Acad. Sci. USA* **1998**, *95*, 9448–9453. [CrossRef] [PubMed]
92. Nagasawa, T.; Hirota, S.; Tachibana, K.; Takakura, N.; Nishikawa, S.; Kitamura, Y.; Yoshida, N.; Kikutani, H.; Kishimoto, T. Defects of B-cell lymphopoiesis and bone-marrow myelopoiesis in mice lacking the CXC chemokine PBSF/SDF-1. *Nature* **1996**, *382*, 635–638. [CrossRef] [PubMed]
93. Takabatake, Y.; Sugiyama, T.; Kohara, H.; Matsusaka, T.; Kurihara, H.; Koni, P.A.; Nagasawa, Y.; Hamano, T.; Matsui, I.; Kawada, N.; et al. The CXCL12 (SDF-1)/CXCR4 axis is essential for the development of renal vasculature. *J. Am. Soc. Nephrol.* **2009**, *20*, 1714–1723. [CrossRef] [PubMed]
94. Janssens, R.; Struyf, S.; Proost, P. The unique structural and functional features of CXCL12. *Cell. Mol. Immunol.* **2018**, *15*, 299–311. [CrossRef] [PubMed]
95. Pawig, L.; Klasen, C.; Weber, C.; Bernhagen, J.; Noels, H. Diversity and Inter-Connections in the CXCR4 Chemokine Receptor/Ligand Family: Molecular Perspectives. *Front. Immunol.* **2015**, *6*, 429. [CrossRef] [PubMed]
96. Massara, M.; Bonavita, O.; Mantovani, A.; Locati, M.; Bonecchi, R. Atypical chemokine receptors in cancer: Friends or foes? *J. Leukoc. Biol.* **2016**, *99*, 927–933. [CrossRef] [PubMed]
97. Odemis, V.; Lipfert, J.; Kraft, R.; Hajek, P.; Abraham, G.; Hattermann, K.; Mentlein, R.; Engele, J. The presumed atypical chemokine receptor CXCR7 signals through G(i/o) proteins in primary rodent astrocytes and human glioma cells. *Glia* **2012**, *60*, 372–381. [CrossRef] [PubMed]
98. Handel, T.M.; Johnson, Z.; Crown, S.E.; Lau, E.K.; Proudfoot, A.E. Regulation of protein function by glycosaminoglycans—As exemplified by chemokines. *Annu. Rev. Biochem.* **2005**, *74*, 385–410. [CrossRef] [PubMed]
99. Heckmann, D.; Maier, P.; Laufs, S.; Li, L.; Sleeman, J.P.; Trunk, M.J.; Leupold, J.H.; Wenz, F.; Zeller, W.J.; Fruehauf, S.; et al. The disparate twins: A comparative study of CXCR4 and CXCR7 in SDF-1alpha-induced gene expression, invasion and chemosensitivity of colon cancer. *Clin. Cancer Res.* **2014**, *20*, 604–616. [CrossRef] [PubMed]
100. Kaplan, R.N.; Psaila, B.; Lyden, D. Niche-to-niche migration of bone-marrow-derived cells. *Trends Mol. Med.* **2007**, *13*, 72–81. [CrossRef] [PubMed]
101. Li, M.; Ransohoff, R.M. The roles of chemokine CXCL12 in embryonic and brain tumor angiogenesis. *Semin. Cancer Biol.* **2009**, *19*, 111–115. [CrossRef] [PubMed]
102. Ozen, I.; Boix, J.; Paul, G. Perivascular mesenchymal stem cells in the adult human brain: A future target for neuroregeneration? *Clin. Transl. Med.* **2012**, *1*, 30. [CrossRef] [PubMed]
103. Filatova, A.; Acker, T.; Garvalov, B.K. The cancer stem cell niche(s): The crosstalk between glioma stem cells and their microenvironment. *Biochim. Biophys. Acta* **2013**, *1830*, 2496–2508. [CrossRef] [PubMed]
104. Hira, V.V.V.; Aderetti, D.A.; van Noorden, C.J.F. Glioma Stem Cell Niches in Human Glioblastoma Are Periarteriolar. *J. Histochem. Cytochem.* **2018**, *66*, 349–358. [CrossRef] [PubMed]
105. Ehtesham, M.; Mapara, K.Y.; Stevenson, C.B.; Thompson, R.C. CXCR4 mediates the proliferation of glioblastoma progenitor cells. *Cancer Lett.* **2009**, *274*, 305–312. [CrossRef] [PubMed]
106. Rao, S.; Sengupta, R.; Choe, E.J.; Woerner, B.M.; Jackson, E.; Sun, T.; Leonard, J.; Piwnica-Worms, D.; Rubin, J.B. CXCL12 mediates trophic interactions between endothelial and tumor cells in glioblastoma. *PLoS ONE* **2012**, *7*, e33005. [CrossRef] [PubMed]
107. Sciaccaluga, M.; Fioretti, B.; Catacuzzeno, L.; Pagani, F.; Bertollini, C.; Rosito, M.; Catalano, M.; D'Alessandro, G.; Santoro, A.; Cantore, G.; et al. CXCL12-induced glioblastoma cell migration requires intermediate conductance Ca2+-activated K+ channel activity. *Am. J. Physiol. Cell Physiol.* **2010**, *299*, C175–C184. [CrossRef] [PubMed]
108. Edalat, L.; Stegen, B.; Klumpp, L.; Haehl, E.; Schilbach, K.; Lukowski, R.; Kuhnle, M.; Bernhardt, G.; Buschauer, A.; Zips, D.; et al. BK K+ channel blockade inhibits radiation-induced migration/brain infiltration of glioblastoma cells. *Oncotarget* **2016**, *7*, 14259–14278. [CrossRef] [PubMed]
109. Cheng, L.; Huang, Z.; Zhou, W.; Wu, Q.; Donnola, S.; Liu, J.K.; Fang, X.; Sloan, A.E.; Mao, Y.; Lathia, J.D.; et al. Glioblastoma stem cells generate vascular pericytes to support vessel function and tumor growth. *Cell* **2013**, *153*, 139–152. [CrossRef] [PubMed]

110. Lee, J.H.; Lee, J.E.; Kahng, J.Y.; Kim, S.H.; Park, J.S.; Yoon, S.J.; Um, J.Y.; Kim, W.K.; Lee, J.K.; Park, J.; et al. Human glioblastoma arises from subventricular zone cells with low-level driver mutations. *Nature* **2018**, *560*, 243–247. [CrossRef] [PubMed]
111. Goffart, N.; Lombard, A.; Lallemand, F.; Kroonen, J.; Nassen, J.; Di Valentin, E.; Berendsen, S.; Dedobbeleer, M.; Willems, E.; Robe, P.; et al. CXCL12 mediates glioblastoma resistance to radiotherapy in the subventricular zone. *Neuro-Oncology* **2016**. [CrossRef] [PubMed]
112. Liu, Y.; Shen, Y.; Sun, T.; Yang, W. Mechanisms regulating radiosensitivity of glioma stem cells. *Neoplasma* **2017**, *64*, 655–665. [CrossRef] [PubMed]
113. Teicher, B.A.; Fricker, S.P. CXCL12 (SDF-1)/CXCR4 pathway in cancer. *Clin. Cancer Res.* **2010**, *16*, 2927–2931. [CrossRef] [PubMed]
114. Peled, A.; Klein, S.; Beider, K.; Burger, J.A.; Abraham, M. Role of CXCL12 and CXCR4 in the pathogenesis of hematological malignancies. *Cytokine* **2018**, *109*, 11–16. [CrossRef] [PubMed]
115. Barbieri, F.; Bajetto, A.; Thellung, S.; Wurth, R.; Florio, T. Drug design strategies focusing on the CXCR4/CXCR7/CXCL12 pathway in leukemia and lymphoma. *Expert Opin. Drug Discov.* **2016**, *11*, 1093–1109. [CrossRef] [PubMed]
116. Kryczek, I.; Lange, A.; Mottram, P.; Alvarez, X.; Cheng, P.; Hogan, M.; Moons, L.; Wei, S.; Zou, L.; Machelon, V.; et al. CXCL12 and vascular endothelial growth factor synergistically induce neoangiogenesis in human ovarian cancers. *Cancer Res.* **2005**, *65*, 465–472. [PubMed]
117. Porcile, C.; Bajetto, A.; Barbieri, F.; Barbero, S.; Bonavia, R.; Biglieri, M.; Pirani, P.; Florio, T.; Schettini, G. Stromal cell-derived factor-1alpha (SDF-1alpha/CXCL12) stimulates ovarian cancer cell growth through the EGF receptor transactivation. *Exp. Cell Res.* **2005**, *308*, 241–253. [CrossRef] [PubMed]
118. Dillenburg-Pilla, P.; Patel, V.; Mikelis, C.M.; Zarate-Blades, C.R.; Doci, C.L.; Amornphimoltham, P.; Wang, Z.; Martin, D.; Leelahavanichkul, K.; Dorsam, R.T.; et al. SDF-1/CXCL12 induces directional cell migration and spontaneous metastasis via a CXCR4/Galphai/mTORC1 axis. *FASEB J.* **2015**, *29*, 1056–1068. [CrossRef] [PubMed]
119. Zeelenberg, I.S.; Ruuls-Van Stalle, L.; Roos, E. The chemokine receptor CXCR4 is required for outgrowth of colon carcinoma micrometastases. *Cancer Res.* **2003**, *63*, 3833–3839. [PubMed]
120. Jung, M.J.; Rho, J.K.; Kim, Y.M.; Jung, J.E.; Jin, Y.B.; Ko, Y.G.; Lee, J.S.; Lee, S.J.; Lee, J.C.; Park, M.J. Upregulation of CXCR4 is functionally crucial for maintenance of stemness in drug-resistant non-small cell lung cancer cells. *Oncogene* **2013**, *32*, 209–221. [CrossRef] [PubMed]
121. Cioffi, M.; D'Alterio, C.; Camerlingo, R.; Tirino, V.; Consales, C.; Riccio, A.; Ierano, C.; Cecere, S.C.; Losito, N.S.; Greggi, S.; et al. Identification of a distinct population of CD133(+)CXCR4(+) cancer stem cells in ovarian cancer. *Sci. Rep.* **2015**, *5*, 10357. [CrossRef] [PubMed]
122. Burns, J.M.; Summers, B.C.; Wang, Y.; Melikian, A.; Berahovich, R.; Miao, Z.; Penfold, M.E.; Sunshine, M.J.; Littman, D.R.; Kuo, C.J.; et al. A novel chemokine receptor for SDF-1 and I-TAC involved in cell survival, cell adhesion, and tumor development. *J. Exp. Med.* **2006**, *203*, 2201–2213. [CrossRef] [PubMed]
123. Miao, Z.; Luker, K.E.; Summers, B.C.; Berahovich, R.; Bhojani, M.S.; Rehemtulla, A.; Kleer, C.G.; Essner, J.J.; Nasevicius, A.; Luker, G.D.; et al. CXCR7 (RDC1) promotes breast and lung tumor growth in vivo and is expressed on tumor-associated vasculature. *Proc. Natl. Acad. Sci. USA* **2007**, *104*, 15735–15740. [CrossRef] [PubMed]
124. Fahham, D.; Weiss, I.D.; Abraham, M.; Beider, K.; Hanna, W.; Shlomai, Z.; Eizenberg, O.; Zamir, G.; Izhar, U.; Shapira, O.M.; et al. In vitro and in vivo therapeutic efficacy of CXCR4 antagonist BKT140 against human non-small cell lung cancer. *J. Thorac. Cardiovasc. Surg.* **2012**, *144*, 1167–1175. [CrossRef] [PubMed]
125. Singh, S.; Srivastava, S.K.; Bhardwaj, A.; Owen, L.B.; Singh, A.P. CXCL12-CXCR4 signalling axis confers gemcitabine resistance to pancreatic cancer cells: A novel target for therapy. *Br. J. Cancer* **2010**, *103*, 1671–1679. [CrossRef] [PubMed]
126. Ma, Y.; Hwang, R.F.; Logsdon, C.D.; Ullrich, S.E. Dynamic mast cell-stromal cell interactions promote growth of pancreatic cancer. *Cancer Res.* **2013**, *73*, 3927–3937. [CrossRef] [PubMed]
127. Chen, Y.; Huang, Y.; Reiberger, T.; Duyverman, A.M.; Huang, P.; Samuel, R.; Hiddingh, L.; Roberge, S.; Koppel, C.; Lauwers, G.Y.; et al. Differential effects of sorafenib on liver versus tumor fibrosis mediated by stromal-derived factor 1 alpha/C-X-C receptor type 4 axis and myeloid differentiation antigen-positive myeloid cell infiltration in mice. *Hepatology* **2014**, *59*, 1435–1447. [CrossRef] [PubMed]

128. Huang, E.H.; Singh, B.; Cristofanilli, M.; Gelovani, J.; Wei, C.; Vincent, L.; Cook, K.R.; Lucci, A. A CXCR4 antagonist CTCE-9908 inhibits primary tumor growth and metastasis of breast cancer. *J. Surg. Res.* **2009**, *155*, 231–236. [CrossRef] [PubMed]
129. Smith, M.C.; Luker, K.E.; Garbow, J.R.; Prior, J.L.; Jackson, E.; Piwnica-Worms, D.; Luker, G.D. CXCR4 regulates growth of both primary and metastatic breast cancer. *Cancer Res.* **2004**, *64*, 8604–8612. [CrossRef] [PubMed]
130. Darash-Yahana, M.; Pikarsky, E.; Abramovitch, R.; Zeira, E.; Pal, B.; Karplus, R.; Beider, K.; Avniel, S.; Kasem, S.; Galun, E.; et al. Role of high expression levels of CXCR4 in tumor growth, vascularization, and metastasis. *FASEB J.* **2004**, *18*, 1240–1242. [CrossRef] [PubMed]
131. Taichman, R.S.; Cooper, C.; Keller, E.T.; Pienta, K.J.; Taichman, N.S.; McCauley, L.K. Use of the stromal cell-derived factor-1/CXCR4 pathway in prostate cancer metastasis to bone. *Cancer Res.* **2002**, *62*, 1832–1837. [PubMed]
132. Porvasnik, S.; Sakamoto, N.; Kusmartsev, S.; Eruslanov, E.; Kim, W.J.; Cao, W.; Urbanek, C.; Wong, D.; Goodison, S.; Rosser, C.J. Effects of CXCR4 antagonist CTCE-9908 on prostate tumor growth. *Prostate* **2009**, *69*, 1460–1469. [CrossRef] [PubMed]
133. Sun, Y.X.; Schneider, A.; Jung, Y.; Wang, J.; Dai, J.; Wang, J.; Cook, K.; Osman, N.I.; Koh-Paige, A.J.; Shim, H.; et al. Skeletal localization and neutralization of the SDF-1(CXCL12)/CXCR4 axis blocks prostate cancer metastasis and growth in osseous sites in vivo. *J. Bone Miner. Res.* **2005**, *20*, 318–329. [CrossRef] [PubMed]
134. Feig, C.; Jones, J.O.; Kraman, M.; Wells, R.J.; Deonarine, A.; Chan, D.S.; Connell, C.M.; Roberts, E.W.; Zhao, Q.; Caballero, O.L.; et al. Targeting CXCL12 from FAP-expressing carcinoma-associated fibroblasts synergizes with anti-PD-L1 immunotherapy in pancreatic cancer. *Proc. Natl. Acad. Sci. USA* **2013**, *110*, 20212–20217. [CrossRef] [PubMed]
135. Chen, Y.; Ramjiawan, R.R.; Reiberger, T.; Ng, M.R.; Hato, T.; Huang, Y.; Ochiai, H.; Kitahara, S.; Unan, E.C.; Reddy, T.P.; et al. CXCR4 inhibition in tumor microenvironment facilitates anti-programmed death receptor-1 immunotherapy in sorafenib-treated hepatocellular carcinoma in mice. *Hepatology* **2015**, *61*, 1591–1602. [CrossRef] [PubMed]
136. Chaudary, N.; Pintilie, M.; Jelveh, S.; Lindsay, P.; Hill, R.P.; Milosevic, M. Plerixafor Improves Primary Tumor Response and Reduces Metastases in Cervical Cancer Treated with Radio-Chemotherapy. *Clin. Cancer Res.* **2017**, *23*, 1242–1249. [CrossRef] [PubMed]
137. Thomas, R.P.; Nagpal, S.; Iv, M.; Soltys, S.G.; Bertrand, S.; Pelpola, J.S.; Yang, J.; Ball, R.L.; Brown, M.; Recht, L.D. CXCR4 blockade at the end of irradiation to improve local control of glioblastoma (GBM). *J. Clin. Oncol.* **2018**, *36*. [CrossRef]
138. Lee, E.Q.; Duda, D.G.; Muzikansky, A.; Gerstner, E.R.; Kuhn, J.G.; Reardon, D.A.; Nayak, L.; Norden, A.D.; Doherty, L.; LaFrankie, D.; et al. Phase I and Biomarker Study of Plerixafor and Bevacizumab in Recurrent High-Grade Glioma. *Clin. Cancer Res.* **2018**, *24*, 4643–4649. [CrossRef] [PubMed]
139. Ludwig, H.; Weisel, K.; Petrucci, M.T.; Leleu, X.; Cafro, A.M.; Garderet, L.; Leitgeb, C.; Foa, R.; Greil, R.; Yakoub-Agha, I.; et al. Olaptesed pegol, an anti-CXCL12/SDF-1 Spiegelmer, alone and with bortezomib-dexamethasone in relapsed/refractory multiple myeloma: A Phase IIa Study. *Leukemia* **2017**, *31*, 997–1000. [CrossRef] [PubMed]
140. Hattermann, K.; Held-Feindt, J.; Lucius, R.; Muerkoster, S.S.; Penfold, M.E.; Schall, T.J.; Mentlein, R. The chemokine receptor CXCR7 is highly expressed in human glioma cells and mediates antiapoptotic effects. *Cancer Res.* **2010**, *70*, 3299–3308. [CrossRef] [PubMed]
141. Maderna, E.; Salmaggi, A.; Calatozzolo, C.; Limido, L.; Pollo, B. Nestin, PDGFRbeta, CXCL12 and VEGF in glioma patients: Different profiles of (pro-angiogenic) molecule expression are related with tumor grade and may provide prognostic information. *Cancer Biol. Ther.* **2007**, *6*, 1018–1024. [CrossRef] [PubMed]

cancers

MDPI

Review

EMT Regulation by Autophagy: A New Perspective in Glioblastoma Biology

Barbara Colella [1], Fiorella Faienza [2] and Sabrina Di Bartolomeo [1,*]

1 Department of Biosciences and Territory, University of Molise, 86090 Pesche (IS), Italy; b.colella@studenti.unimol.it

2 Department of Biology, University of Rome Tor Vergata, 00133 Rome, Italy; fiorella.faienza@uniroma2.it

* Correspondence: sabrina.dibartolomeo@unimol.it

Received: 30 January 2019; Accepted: 1 March 2019; Published: 6 March 2019

Abstract: Epithelial-to-mesenchymal transition (EMT) and its reverse process MET naturally occur during development and in tissue repair in vertebrates. EMT is also recognized as the crucial event by which cancer cells acquire an invasive phenotype through the activation of specific transcription factors and signalling pathways. Even though glial cells have a mesenchymal phenotype, an EMT-like process tends to exacerbate it during gliomagenesis and progression to more aggressive stages of the disease. Autophagy is an evolutionary conserved degradative process that cells use in order to maintain a proper homeostasis, and defects in autophagy have been associated to several pathologies including cancer. Besides modulating cell resistance or sensitivity to therapy, autophagy also affects the migration and invasion capabilities of tumor cells. Despite this evidence, few papers are present in literature about the involvement of autophagy in EMT-like processes in glioblastoma (GBM) so far. This review summarizes the current understanding of the interplay between autophagy and EMT in cancer, with special regard to GBM model. As the invasive behaviour is a hallmark of GBM aggressiveness, defining a new link between autophagy and EMT can open a novel scenario for targeting these processes in future therapeutical approaches.

Keywords: autophagy; epithelial-to-mesenchymal transition (EMT); glioblastoma (GBM); cadherins; Wnt/β-catenin signalling

1. Introduction

Glioblastoma (GBM) is the most malignant and frequent form of glioma. This brain tumor is derived from glial cells and is characterized by high proliferation rate and local dissemination. Despite the improvement of chemo- and radio-therapies obtained in last decades, GBM prognosis is very poor with a median survival time that rarely exceed 18 months. Notably, GBM is characterized by resistance to apoptosis and high invasiveness, driving the search for novel targets useful to design effective therapeutical strategies.

The Epithelial-to-Mesenchymal Transition (EMT) programme is considered to be crucial for the acquisition of an invasive phenotype in epithelial cancer cells. Although some differences have been outlined, glioma cells also undergo an EMT-like process, through the activation of specific transcription factors and signalling pathways. Considering the critical role played by EMT on GBM dissemination, resistance to apoptosis and maintenance of cancer stem cells staminality, a number of preclinical studies have been launched to target the process as therapeutic approach.

In addition, the role of autophagy in tumor onset, progression is acquiring increasing clinical inerest. For instance, it has been recently shown that, besides modulating cell resistance or sensitivity to therapy, autophagy can also modulate GBM invasion.

In this review, we examined the current understanding of the role of autophagy in regulating the EMT and EMT-like programmes and in directing GBM cells to a more or less invasive phenotype.

Moreover, we provided some clues to argue that blocking autophagy for therapeutic purposes requires careful consideration.

2. Epithelial-to-Mesenchymal Transition

EMT is a biological process by which epithelial polarized cells undergo various biochemical modifications that convert them in mesenchymal, isolated, and not-polarized cells. A lot of evidence suggests that EMT and its reverse process (mesenchymal-to-epithelial transition, MET) are crucial for tissue remodeling during development, wound repair and the initiation of cancer metastasis. In the early 1980s, Elizabeth Hay described epithelial-to-mesenchymal phenotype changes in the primitive streak of chick embryos [1]. Initially named as "epithelial to mesenchymal transformation", this process is now known as "epithelial-to-mesenchymal transition" to emphasize its transient nature and to distinguish it from the neoplastic transformation. To acquire a mesenchymal phenotype, epithelial cells undergo morphological and biochemical changes, reorganize their cytoskeleton, and activate a specific transcriptional programme. Indeed, epithelial cells are characterized by an apical-basal polarity, a polygonal shape and various kind of junctions. The latter tightly holds epithelial cells against each other and anchors them to the basement membrane, thus ensuring the structural integrity of epithelial sheets within the body tissues [2]. Conversely, mesenchymal cells exhibit a spindle-like morphology, do not have any polarity or connection with other and they are loosely anchored to the extracellular matrix (ECM) by focal adhesions.

Transcription factors (TFs) belonging to the SNAI family, named Snail and Slug, the zinc-finger E-box-binding homeobox (ZEB)1/2, and Twist1/2 are recognized to be the master regulators of EMT execution, since they promote the transcription of genes normally expressed in mesenchymal cells, such as N-cadherin, vimentin, and fibronectin. On the contrary, they suppress the expression of the epithelial markers E-cadherin, claudins, occludins, and cytokeratins. Loss of E-cadherin, in turn, promotes Wnt signaling and β-catenin accumulation in the nucleus, where it activates Tcf/LEF-dependent transcription of genes promoting proliferation and migration [3].

Cells undergoing EMT lose their apical–basolateral polarization and acquire a fibroblast-like morphology that allow them to degrade the underlying basement membrane and to migrate from the epithelial layer in which they originated [4]. Metalloproteases are also activated during EMT, and favor cell migration by degrading the membrane basement and the extracellular matrix components [2].

Furthermore, it has been shown that noncoding microRNAs play a role in EMT, by regulating the translation of EMT players, as extensively addressed in Abba et al. [5].

EMT can be classified in three different subtypes according to EMT meetings discussion: the so-called "type 1" EMT occurs during implantation, embryogenesis and organogenesis, the "type 2" EMT is associated to wound healing, tissue regeneration and organ fibrosis, and the "type 3" EMT characterizes neoplastic cells during metastatization. For a comprehensive overview of type 1 and type 2, refer to Kalluri et al., 2009 [4]. The activation of a "type 3" EMT programme (hereafter referred to as EMT) has been proposed to be pivotal for the acquisition of a malignant phenotype by cancer cells [6], as discussed below in detail.

2.1. Signals Stimulating Epithelial-to-Mesenchymal Transition in Cancer

Epithelial carcinoma cells, and typically those present at the external part of the tumoral mass, can acquire a mesenchymal phenotype upon specific stimuli. The microenvironment surrounding the primary tumor (TME) is characterized by inflammation, hypoxia, ECM components and tissue-specific soluble factors [7]. Notably, tumor cells recruit activated fibroblasts and immune cells that secrete cytokines, that, in turn, can activate the EMT programme. TGF-β is secreted by stromal fibroblasts, platelets and tumor cells themselves, and is considered the main EMT activator as it induces the expression of specific TFs in different cancer models [8–10].

Among cytokines, Tumor Necrosis Factor-α (TNFα) is crucial for EMT induction, and its effects are mediated through NFκB signaling pathway activation [11–13]. It has been also described that interleukins, particularly IL6, released by TME cells, can contribute to EMT stimulation. Other EMT-inducing signals originated from the tumor stroma are represented by growth factors such as HGF, EGF, PDGF that are able to activate EMT-specific TFs [14–17].

Furthermore, tumor microenvironment is characterized by hypoxia that promotes EMT via hypoxia-inducible-factor-1α (HIF1α) activation [18,19]. HIF1α stimulates inflammatory cytokines [20] and cooperates with Wnt/β-catenin signaling to enforce the EMT induction [21]. Moreover, during hypoxia, mitochondria increase the production of reactive oxygen species (ROS) that further contribute to EMT activation via both stimulating NFκB signaling [22], ECM regulation and cytoskeleton remodeling [23].

Once they have acquired a mesenchymal phenotype, cancer cells can dissociate from the primary mass, migrate and eventually enter the blood vessels by intravasation to initiate the metastatic process. Following extravasation and micrometastases formation, invading cells activate a MET programme and form macroscopic metastases resembling the epithelial features of the originating primary tumor, although the molecular mechanisms of MET have been less investigated if compared to those regulating EMT. However, both EMT and MET activation seem to be highly tissue-specific and strictly dependent on the local microenvironment encountered [2,6,24,25].

2.2. *Epithelial-to-Mesenchymal Transition in Glioblastoma Dissemination*

EMT has been mainly characterized in carcinoma models, and the role of EMT in glioma has only recently been investigated [26–28]. Candidate cells for originating gliomas (cells of origin) are mostly neural stem cells (NSCs), normally present in the adult brain, and oligodendrocyte precursor cells (OPCs) [29–32]. However, the involvement of more differentiated cells cannot be completely excluded [33]. As a consequence of the neurodevelopmental process, neural cells assume a mesenchymal phenotype, different from the epithelial one typical of the ectodermal cells they derive from. Therefore, gliomas do not undergo the classical EMT programme during tumorigenesis, and the terms "EMT-like" or "glial-to-mesenchymal transition (GMT)" have been proposed to indicate this peculiar process [34]. Although it is not clear whether cells of origin undergo or not significant modifications toward more mesenchymal features, glioma cells show a high plasticity in terms of EMT-like/MET-like conversion, likely mediated by epigenetic alterations induced by the tumor microenvironment [35,36]. Based on The Cancer Genome Atlas (TCGA) classification, GBMs can be differentiated into four genetic subtypes: Mesenchymal, Classical, Neural and Proneural [37]. Verhaak results suggest that Proneural GBM patients do not have a survival advantage from aggressive therapeutical treatments, unlike Classical and Mesenchymal GBM patients [37]. Role of EMT in each GBM genetic subtype has been investigated by Zarkoob et al. in 2013 [38]. A significant overlap between the genes that are up-regulated in the EMT signature and those that are up-regulated in each of the GBM subtypes exists, although, among all, the mesenchymal subtype has the highest number and expression levels of up-regulated genes [38]. Indeed, GBMs belonging to the mesenchymal subtype are characterized by an elevated invasive potential, poor clinical prognosis, and significantly shortened time to recurrence following initial treatment, compared to the other subtypes. The most commonly used glioma cell lines also present a predominant mesenchymal signature.

GBM cells in the invasive front, differently from those of the inner mass, commonly execute an invasion programme characterized by detachment from the mass, direct adhesion and degradation of ECM (lack of basement membrane) and widespread dissemination in the surrounding brain tissue. Remarkably, unlike other tumors, GBMs only rarely form metastases outside CNS, even though a hallmark of their aggressiveness is the infiltration and the diffuse growth in the surrounding parenchyma.

Large-scale genetic analyses have suggested that signaling networks activated during the physiological neural development are also employed by GBM cells to promote tumor growth and invasion [39–41]. In detail, pathways mediated by Wnt/β-catenin, TGF-/β, Tyrosine kinase receptors and SDF/CXCR4 have been involved in the activation of EMT-like related genes to promote GBM dissemination [26,27]. Kahlert et al. found that the Wnt/β-catenin pathway is predominantly activated within cells located at the invasive peritumoral front of patient specimens belonging to the mesenchymal subtype. Chiefly, it induces the expression of Zeb1, Twist1 and Slug, thus promoting the migratory capability of GBM cells in vitro [42].

Regarding TGF-β pathway, a number of evidence demonstrated its critical role for the promotion of invasive properties of glioma cells [43–45], although the molecular mechanisms involved need to be further investigated. Interestingly, TGF-β signaling is known to be crucial in the maintenance of the mesenchymal stem-like population in GBM [45,46]. ZEB1 seems to be the pivotal mesenchymal transcription factor activated by TGF-β signaling since, differently from Snail, Slug and Twist, it accumulates in the nucleus of GBM cells [47].

The Hepatocyte growth factor (HGF) binding to the tyrosine kinase receptor c-MET is another crucial event highly activated. c-MET is overexpressed within GSC populations [48] and in patient-derived GSCs belonging to the mesenchymal subtype [49]. Accordingly, elevated c-MET signaling enhances GSC migration by activating EMT TFs [49,50] and is associated with poor survival and increased tumor invasiveness in patients [51–53].

The majority of molecules involved in the classical EMT have also been shown to play also a role in the EMT-like process. An increased activity of the TFs that mainly orchestrate the typical EMT, such as SNAI proteins, ZEB1/2, and Twist, promotes the invasion of GBM cells [54–57]. For instance, SNAIL silencing reduces invasion, migration and proliferation in GBM cell lines [58,59] and overexpression of Slug correlates with GBM grade [56]. ZEB1 and ZEB2 expression is also correlated with invasive features and with survival of GBM patients; ZEB1 knock-down cells formed less invasive and more drug-sensitive masses than wild type cells when inoculated in mouse brain [54,60]. Moreover, Twist1 and Twist 2 expression, besides affecting stemness properties, has been associated to the invasive properties of GBM cells as it mediates the expression of crucial EMT-related genes such as metalloproteinase 2 (MMP2), Slug and HGF among others [57].

It is worth mentioning that the classical cadherin switch, which is widely accepted as an EMT hallmark in carcinomas, is a controversial matter in GBM. Differently from carcinomas, E-cadherin expression is almost absent in neural tissues, where its expression appears limited to GCSs cells and to a subset of highly aggressive GBM cells. Otherwise, N-cadherin is absent in epithelial tumors before the EMT execution, whereas is highly expressed in astrocytes, where it contributes to regulate cell polarity and migration and in GBM cells, that show a faster and less-directed movement to respect to astrocytes [54,61]. It was found that N-cadherin expression is inversely correlated with the invasive behaviour of GBM, and its ectopical expression reduces cell migration and restores polarity in GBM cells [62,63]. Notably, it has also been shown that differences in N-cadherin distribution rather than in its expression levels are responsible for different motility behaviours [64,65].

In addition, several studies showed that the treatment of primary GBMs with radiation therapy or with the anti-angiogenic agents Bevacizumab promotes the acquisition of a mesenchymal phenotype in recurrent tumors [34,66–68]. Indeed, glioma cells that have acquired radioresistant properties following treatment exhibit a gene expression profile enriched for genes involved in EMT-related processes [34,69,70], and the pathways promoting EMT result strongly upregulated in these cells, thereby resulting in an increased invasion capability [71,72]. An in vivo study by Halliday et al. demonstrated that proneural GBM cells rapidly shifted their gene expression pattern towards a mesenchymal phenotype in response to radiation therapy in a tumor-bearing mouse model [66]. As radiation is a universal component in the treatment of GBMs, this subtype shift poses an important clinical challenge, especially considering that cells shifted to a mesenchymal subtype display an increased radioresistance [73]. If this shift is due to changes in the microenvironment or to clonal selection of mutant therapy-resistant cells is controversial, but both the hypotheses seem to be possible.

3. Autophagy

The term "autophagy" was coined by the discoverer of lysosomes Christian de Duve and it means, in Greek language, "self-eating" [74–76]. From the first description of the process in 1960s, many studies described the process of self-degradation by a morphological point of view, until in 1993 a genetic screen led to isolation of some yeast autophagy-defective mutants and to identification of the so-called AuTophaGy-related (ATG) genes [77]. This seminal work allowed Yoshinori Oshumi to be awarded the Nobel Prize in Physiology and Medicine in 2016. Oshumi's screen identified 15 genes involved in autophagy regulation in yeast undergoing nutrient deprivation, but today, more than 40 genes have been described in yeast, many of them having orthologues in vertebrates [76]. Three main types of autophagy have been described: macroautophagy, microautophagy and Chaperone-Mediated Autophagy (CMA).

3.1. Mechanisms and Molecules

During macroautophagy (hereafter referred to as autophagy), double-membrane vesicles named autophagosomes form and engulf cytoplasmic material, including long-lived proteins and old or damaged organelles which are then delivered to lysosomes for degradation and recycling [78]. Always activated at basal level within the cell, autophagy can be modulated by several signals, mainly by nutrient signaling, growth factors, energy status, oxidative or ER stress and pathogen infection. The input from these upstream signals is integrated by the serine/threonine protein kinase mTOR (mechanistic or mammalian target of Rapamycin), which acts upstream of the ATG genes, thus controlling autophagy activation [79]. mTOR belongs to the phosphoinositide 3-kinase (PI3K)-related kinase family, and, in mammalian cells, works as the catalytic subunit of two multi-protein complexes known as mTORC1 and mTORC2 [80]. Under nutrient deprivation, mTOR is inhibited and Ulk1/Atg13-FIP200 complex initates and drives a massive autophagy activation [81]. Autophagosome formation requires the formation of a multi-protein complex, composed by class III PI3K, Beclin1 and p150, although other proteins such as UVRAG, Ambra1 and Atg14L are able to bind and regulate the complex [82–86]. Elongation and maturation of autophagosomes involve two ubiquitin-like conjugation systems, both requiring Atg7, which catalyze the covalent linkage of ATG5 to ATG12 and ATG16-like 1, and the attachment of phosphatidylethanolamine to proteins of the microtubule-associated protein 1 light chain 3 (MAP1LC3 or LC3) family [87].

Lipidated LC3 protein is then recruited to the autophagosome membrane that docks and fuse with lysosome, resulting in the formation of a single membrane vesicle named autophagolysosome or autolysosome. Lysosomal hydrolases, ultimately, degrade and recycle the content of autolysosomes.

Although originally suggested to be just a nonspecific and bulk degradation mechanism, autophagy is now recognized as a highly regulated process, enabling cells to sense and promptly respond to a plethora of stimuli, thereby conferring adaptation to the ever-changing environment. Nevertheless, even though a basal level of autophagy contributes to maintain the proper cell homeostasis both during embryogenesis and adulthood in physiological conditions, it is now ascertained that autophagy is deregulated in various human pathologies, including cancer [88–90].

3.2. Autophagy Role in Tumorigenesis

The observation in 1999 that the gene encoding Beclin1 is monoallelically deleted in a high percentage of human breast, ovarian and prostate cancers provided the first evidence of the involvement of the autophagic process in tumorigenesis [91]. Disruption of Beclin1 in mice results in an increased proliferation and in the spontaneous development of various malignancies, confirming Beclin1 as an haploinsufficient tumor suppressor gene [92,93]. In a similar way, mice lacking one copy of the Beclin1 interactor Ambra1 exhibit a higher incidence of spontaneous tumors than their wild type littermates, and cells depleted of the gene are characterized by a hyperproliferative phenotype [94]. Notably, *Ambra1* homozygous disruption in mouse leads to a strong hyperproliferation and lethal defects in the developing nervous system during embryogenesis [82]. Mice bearing systemic or tissue-specific deletion of Atg5 and Atg7 also develop tumoral masses a higher frequency than the wild type counterparts [95] and are more prone to develop cancers upon carcinogen-induced stimuli [96–98].

Several mechanisms have been proposed to explain the oncosuppressive functions of autophagy [90]. First of all, the autophagy-mediated clearance of proteins and organelles ensures the proper cellular homeostasis, avoiding the accumulation of genotoxic molecules, such as reactive oxygen species (ROS) produced by dysfunctional mitochondria, as well as aggregates of ubiquitinated proteins [99,100]. An intact autophagic machinery is also required to deal with cytotoxic stress and to maintain genome stabilization, although further investigation is required to underlie the mechanisms involved [101,102]. Moreover, autophagy counteracts the metabolic switch accompanying malignant transformation by eliminating old and damaged mitochondria, thus preserving the optimal bioenergetic needs and maintaining the physiological metabolic homeostasis [103,104].

Other potential mechanisms through which autophagy acts as an oncosuppressive process are linked to its role in the regulation of immune response [105], maintenance of the staminal niches [106], defens of the organism against pathogen infections and degradation of oncogenic proteins, like mutant (but not wild-type) TP53 [107].

On the other hand, it is well accepted that, in an established tumor, cancer cells use autophagy as a strategy to overcome microenvironmental stresses, including nutrient deprivation, hypoxia and drugs. Advanced tumors sometimes exhibit an increased autophagic flux and ex-vivo cell lines in which BECN1 or ATG5 have been down-regulated are virtually unable to survive within the metastatic niche [108]. Analogously, autophagy-defective tumoral cells appear more sensitive to pro-apoptotic stimuli than autophagy-proficient cells [109–112].

Due to this dual function, autophagy has been defined a 'Janus-faced' player in cancer progression [113]: in the early stages of tumorigenesis it plays onco-suppressive functions by limiting cell proliferation, DNA damage and tumor progression; on the contrary, when the tumor mass is established, it helps cells to counteract the stressful conditions characterizing the tumor microenvironment.

3.3. *Autophagy and Glioblastoma: Friends or Foes?*

It was demonstrated that high-grade gliomas exhibit lower expression of some autophagy related proteins with respect to low-grade ones, and that the progression of astrocytomas toward higher grades is accompanied by a decrease in autophagic proficiency. Pirtoli et al. observed that both BECN1 mRNA (encoding for Beclin1) and protein levels are lower in GBM tissue than in low-grade and healthy brain tissue [114]. Accordingly, following Karnofski classification, high Beclin1 levels have been positively correlated with patient survival and performance status, whereas low Beclin1 expression correlates with an increase of proliferation [114]. Similarly to Beclin1 expression, also LC3B II expression (index of autophagy activation) is low in high-grade astrocytomas, thus suggesting an impairment of the autophagic process in these tumors [115]. On the other hand, in 2012, through a proteomic screening, Galavotti et al. found that some genes involved in autophagy regulation are highly expressed in the GBM mesenchymal subtype [116]. Among these, the autophagy associated genes DRAM1 and SQSTM1 encoding for the key regulator p62 are highly expressed in Glioma stem cells (GSCs), and modulate their migration and invasion capabilities [116]. Although these studies suggest that autophagy may regulate gliomagenesis, a systematic and comprehensive investigation of autophagy role among the GBM subtypes is missing, but needed. Indeed, a different expression of autophagy regulators across GBM genetic groups could be responsible for a different susceptibility to autophagy modulation.

In addition to the growing evidences showing a direct involvement of autophagy-regulating genes in GBM progression, several autophagy-associated molecules are frequently altered in brain tumors. As an example, the tyrosine kinase EGF receptor is often amplified in gliomas, and suppresses autophagy through both kinase-dependent and -independent mechanisms [117]. PTEN, is commonly mutated in gliomas, and positively regulates autophagy by inhibiting the PI3K/Akt pathway, although PTEN and NF1 co-mutation determines an autophagy suppression through the hyperactivation of the PI3K/Akt signalling [118].Furthermore, the oncosuppressor p53, frequently mutated in gliomas, may have a dual role in autophagy regulation, as the nuclear protein is able to promote autophagy through the transcriptionally regulation of autophagy-related genes, whereas cytoplasmic p53 suppresses autophagy [119]. Further investigation are needed to define whether autophagy machinery may be considered as a novel useful prognostic and/or therapeutic marker of glial tumors.

4. Autophagy and Epithelial-to-Mesenchymal Transition

Autophagy was only recently connected to EMT. In the last years, some observations indicated that an intricate relationship exists between these two important processes in cancer [120]. According to its dual role in tumorigenesis, the effect of autophagy on EMT appears controversial and strictly dependent on the cellular type and on the stimulus employed for activating or inhibiting autophagy, as summarized in Table 2 and in Figure 1.

Literature data highlights that an autophagy stimulation of metastatization could be merely the consequence of its pro-survival activity against the apoptotic signals coming from changes in adhesion and cytoskeleton reorganization [108]. A number of compounds or microenvironmental conditions that are able to activate the EMT programme, can also induce an autophagic response in different types of cultured cancer and non-cancerous cells; autophagy inhibition in these models impairs EMT (Table 2).

However, emerging evidence also indicates that autophagy activation can induce a reversion of the EMT phenotype in different healthy and cancer models and that several anticancer compounds that induce autophagy also inhibit EMT [121–123] (Table 2). Selective degradation of EMT players seems to be a general mechanism by which autophagy can modulate EMT process [124]. Notably, in ATG3, ATG5, ATG9 and ATG12 KO mice, p62 accumulation accumulation determines stabilization of Twist1, which is normally degraded by both proteasome and autophagosomes [125]. This regulation can be crucial in those tumors characterized by p62 up-regulation, as observed so far in human squamous cell carcinoma and in melanoma (Table 2). Autophagy deficiency reduces the expression of epithelial markers and increases that of mesenchymal ones also in ATG7 KO hepatocytes [126] (Table 2).

Table 1. Autophagy and EMT.

Cell/Tissue	Autophagy Modulation	Effect on EMT on EMT	Mechanism	References
HCC cells	Induction by starvation	Induction	activation of TGFβ/Smad3 dependent and cAMP/PKA/CREB signalling	[108,127–129]
	Inhibition by ATG KD or CQ	Inhibition		
Colorectal cancer cells	Induction by mTOR inhibition or by ALS treatment	Inhibition	decreased activation of RhoA and Rac1	[122,130,131]
	Inhibition by BECN1 knockdown	Inhibition	unknown	
Ovarian cancer cells	Induction by Danu treatment	Inhibition	unknown	[123]
Non tumorigenic hepatocytes	Induction by starvation + TGF β_1	Inhibition	Snail degradation	[126]
	Inhibition by BECN1 or ATG7 KD	Induction	unknown	
NPC cells	Induction by Cisplatin	Induction	unknown	[132]
NSCL cells	Induction by TGFβ_1 treatment	Induction	unknown	[133]

Table 2. Autophagy and EMT.

Cell/Tissue	Autophagy Modulation	Effect on EMT on EMT	Mechanism	References
Lung adenocarcinoma cells	Induction by MSCs co-colture	Induction	Snail up-regulation	[134]
Endometrial cells	Induction by Hypoxia	Induction	unknown	[135]
Uroepithelial cells	Induction by DBP exposure or starvation	Induction	E-cadherin degradation or TGFβ1/Smad3 pathway activation	[136,137]
Kidney podocytes	Inhibition by V-ATPase	Inhibition	Reduction of p62 phosphorylation	[138]
MEFs, keratinocytes, melanoma cells	Inhibition by ATG KD	Induction	p62-mediated Twist stabilization	[139]
Breast	Activation by DEDD overexpression	Inhibition	Snail and Twist degradation	[140]
Gastric cancer cells and tissue	Inhibition by BECN1 KD	Inhibition	ROS-NFκB-HIF-1α pathway activation	[141]

3-MA, 3-methyladenine; ALS, Alisertib (Aurora kinase A inhibitor); ATG, autophagy related gene; Baf, bafilomycin; BECN1, Beclin1; cAMP, cyclic adenosine monophosphate; CQ, chloroquine; CREB, cAMP responsive element binding; Danu, Danusertib; DBP, n-butyl phthalate; DEDD, death effector domain containing; DRAM1, DNA damage-regulated autophagy modulator 1; EMT, epithelial to mesenchymal transition; HCC, hepatocarcinoma cells; HIF-1α, hypoxia-inducible factor 1; KD, knockdown; MEFs, mouse embryonic fibroblasts; MSC, mesenchymal stem cell; NF- κB, nuclear factor kappa beta; NPC, nasopharyngeal carcinoma; NSCL, Non-small cell lung; PI3KC3, phosphatidylinositol 3-kinase; PKA, protein kinase A; Rac1, Ras-related C3 botulinum toxin substrate 1; RhoA, Ras homolog gene family, member A; ROS, reactive oxygen species; Smad3, small mother against decapentaplegic3; SQSTM1, sequestosome 1; TGFβ, Transforming growth factor beta; V-H-ATPase, vacuolar-type H^+-adenosine triphosphatase.

Taken together, these observations show a complex crosstalk between autophagy and EMT processes. It is conceivable that, at the early stages of metastatization, autophagy acts as oncosuppressive signal, tending to inhibit the EMT programme mainly by destabilizing EMT crucial players. Later on, metastatic cells could require a sustained autophagy to survive to environmental and metabolic stressful conditions encountered [113].

4.1. *Autophagy Role on Glioblastoma Dissemination*

To date, only a few studies correlate autophagy to GBM cells capability to migrate and spread toward surrounding tissues, and, similarly to what observed in other tumor models, they highlight two opposite point of views, as detailed below, in Section 4.1.1, and as illustrated in Figure 1.

4.1.1. Autophagy Promotes Glioblastoma Dissemination

In 2012, Galavotti et al. showed that some autophagy players are up-regulated in GBM mesenchymal subtype, and that their modulation modifies the migration properties of GBM cells. They observed that GSCs require autophagy activation to migrate, and down-regulation of the autophagy-associated factors DRAM1, p62 and ATG7 limit their invasion capabilities potentially through the regulation of energy metabolism [116] (Figure 1A). In line with these observations, a study conducted by using an 3D organotypic model of GBM showed that ATG12 RNA silencing reduced cellular invasion, although no modifications of cellular migration capabilities was observed [142] (Figure 1A). More recently, another couple of studies have correlated autophagy activation induced by a combination of different stimuli with an enhanced mesenchymal phenotype in GBM cells through various mechanisms [143,144]. Lastly, inhibiting the late stages of autophagy by using Chloroquine, Liu et al. showed a potentiation of the effect of the multi kinase inhibitor Sorafenib in reducing cell proliferation and migration, through a further stimulation of apoptosis [145] (Figure 1A).

4.1.2. Autophagy Impairs Glioblastoma Dissemination

In spite of the previous mentioned evidence, we and other groups have recently demonstrated a direct effect of autophagy modulation on migration and invasion capabilities of GBM cells, as illustrated in Figure 1B. Autophagy induction by nutrient deprivation or by mTOR inhibition determines a reversion of EMT phenotype in immortalized and primary GBM cells [146–149]. In 2014, Palumbo et al. observed that ATG7 RNA silencing restored clonogenic ability of irradiated GBM cells with inactive EGFR and, conversely, that rapamycin-mediated autophagy further impaired their clonogenic and migration capabilities [148]. Later on, we demonstrated that autophagy induction in GBM immortalized and primary cells, obtained by nutrient starvation or by mTOR pharmacological inhibition, induced a drastic impairment of both migration and invasiveness. On the contrary, autophagy deficiency, obtained by silencing the autophagy master genes ATG5, ATG7 or BECN1, stimulated cell motility [146], similarly to what observed in highly metastatic breast, colon and hepatocellular cancer models [122,124,130]. We correlated the migration properties of the cells analyzed with a molecular shift from a mesenchymal to an epithelial-like phenotype (Figure 1B). Upon autophagy induction, in fact, we found a down-regulation of the EMT players Snail and Slug, likely due to a general impairment of protein synthesis mediated by mTOR inhibition, rather than to the autophagosome-mediated degradation [146]. As an outcome of SNAI down-regulation, the up-regulation of N- and R-cadherin mRNA and protein expression was observed, whereas no significant differences in other cadherin family members were found (Figure 1B). Remarkably, as above discussed, although a "cadherin shift" from the E- to the N- isoform is actually believed a hallmark of carcinoma cells undergoing EMT, the role of cadherins in non-epithelial tumors is much less documented and elucidated. N-cadherin overexpression or re-distribution has been associated with the recovery of the cell polarity and the inhibition of migration of GBM cells [62–65]. Notably, we have recently found that, upon autophagy induction, the Wnt signaling effector β-catenin relocalises within the cell and associates to N-cadherin in sub-membrane compartments to form epithelial-like cell-cell adhesion structures [147], thus contributing to a mesenchymal-to-epithelial-like transition of GBM cells (Figure 1B). Similarly to other tumors, Wnt pathway is constitutively active in GBM and its deregulation is likely responsible for initiation and/or progression of the disease [150–153]. β-catenin translocation to the nucleus characterizes the mesenchymal and invasive phenotype of tumoral cells as it promotes SNAIL and ZEB1 transcription and mediates EGFR pathway-induced EMT [154–156].

In summary, a complicated interplay between autophagy process and EMT/MET activation in GBM is recently emerging. Contradictory results could be explained by the different stimuli and different models employed. Notably, autophagy induction obtained through mTOR inhibition or nutrient deprivation always results in migration/invasion reduction in independent experiments and different GBM cell lines. Conversely, autophagy activation resulting by the employment of drugs like TMZ plus low glucose, or by TGF-β, or by AEG overexpression is associated to promotion of migration capabilities of GBM cells, maybe suggesting that different and complex mechanisms contribute to the migratory phenotype in those contexes.

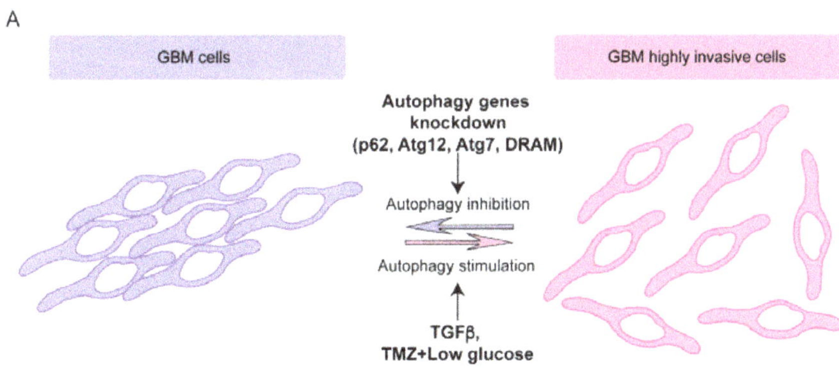

Figure 1. *Cont.*

B

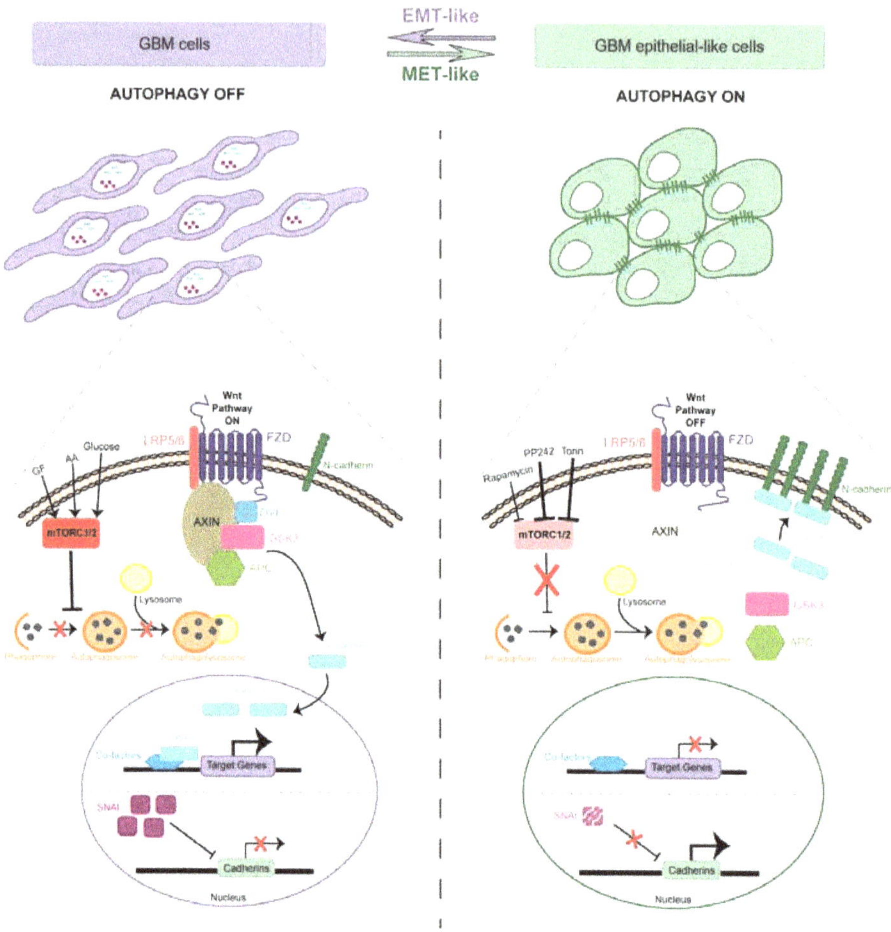

Figure 1. Effects of autophagy modulation on GBM migration/invasion capabilities from opposite point of views. (**A**) Autophagy induction promotes GBM cells invasiveness and viceversa. Upon different stimuli, autophagy is activated and a more invasive phenotype is observed in some models of GBM cells (right). Conversely, when autophagy is impaired by genetic knockdown of some autophagy-related genes, a less invasive phenotype is obtained (left). TGFβ, Tumor necrosis factor β; TMZ, Temozolomide. (**B**) Autophagy modulation promotes EMT/MET-like shifts in GBM cells. In nutrient-rich conditions, hyperactivation of the mTORC1/2 complexes impairs autophagy and Wnt pathway is active thus allowing β-catenin translocation to the nucleus where it promotes the transcription of pro-invasive molecules. In this condition, EMT players of the SNAI family express and repress cadherins expression. The genetic knockdown of autophagy related genes exacerbates the mesenchymal phenotype and enhances the cell migration capability. Upon autophagy induction, shown on the right, Dishvelled (DVL) is degraded and Wnt pathway inactivated leading to β-catenin accumulation into the cytosol. In autophagic cells, SNAI factors are down-regulated and, consequently, N-cadherin accumulates and binds β-catenin, thus promoting cell-cell adhesion.

Effects observed following autophagy inhibition also require attention, although Chloroquine effects on cells often depend on autophagy-independent mechanisms, as better illustrated in the next section. More controversial, although conducted in different model systems, are the results on GBM invasiveness obtained by inactivating ATG genes. In light of such contradictory results, further in-depth investigation is, for sure, required to better unravel the question.

4.2. Targeting Autophagy in Glioblastoma Therapy

In the last decade, the standard therapeutic regimen for GBM has been surgical resection, if feasible, followed by radiation therapy (IR) and temozolomide (TMZ)-based chemotherapy. Both TMZ and IR are able to induce an autophagic response in GBM cell lines, likely through DNA damage induction, but the outcome of the autophagic response is so far inconclusive (Figure 2). In 2004, Kanzawa et al. reported that TMZ treatment induced autophagy instead of apoptosis in GBM cells and that the co-treatment with a pharmacological inhibitor of the late stages of autophagy restored TMZ-induced cytotoxicity, although the opposite effect was obtained by blocking autophagy initiation [157]. In line with this observation, the combination between standard therapies and autophagy flux inhibitors Chloroquine (CQ) or its analog hydroxychloroquine (HCQ) is a promising therapeutic approach in GBM treatment, and combined therapies including these molecules are currently employed in ongoing phase III clinical trials (Figure 2). Indeed, CQ and its derivatives are, in fact, the only autophagy inhibitors already approved by the USA Food and Drug Administration due to their anti-inflammatory and anti-malarial properties. Similar to other tumoral settings, CQ-treated GBM patients exhibit a better median survival if compared to placebo-treated individuals [158]. However, despite the potential promising beneficial effect, a trial combining HCQ with TMZ and IR showed that the maximum tolerated HCQ dose was ineffective in inhibiting autophagy, suggesting that CQ effect could be due to autophagy-independent mechanisms [159]. Notably, it has been recently shown that, in addition to DNA intercalation properties and ROS production, CQ also induces a strong autophagy-independent disorganization of the Golgi apparatus and of the endo-lysosomal system in mice tissues, thus suggesting caution in interpreting results obtained with this drug [160,161].

In an apparently contradictory way, other therapeutic approaches and clinical trials aimed at inhibiting mTOR pathway, inducing in turn an autophagic response, have been launched. PI3K/Akt/mTOR pathway is often deregulated in human cancers including GBM, and is involved in cancer stem cell maintenance, thus inducing an uncontrolled proliferation [162,163]. It was observed that the co-treatment of TMZ and the mTOR inhibitor RAD001 (also known as everolimus) induced cell death [164] (Figure 2). RAD001 was shown to enhance the cytotoxic effect of an oncolytic adenovirus in a viral-mediated therapy by inducing an autophagy-dependent cell death [165,166].

mTOR inhibition also promotes TMZ-induced senescence in in vitro and in vivo models [167,168]. However, the link between autophagy and senescence needs to be further clarified in order to be exploited a potential therapeutic tool.

Very recently, it has been suggested that the failure of some clinical trials targeting PI3K and mTOR, could be due to the employment of Rapalogs (rapamycin and its analogs) which are known to inhibit mTORC1, but not mTORC2 [149]. In fact, a feedback mechanism activated by mTORC1 inhibition stimulates mitogenic pathways, thus compromising the rapalog efficacy on cell proliferation [169]. In order to overcome the emerged limitations, a second generation of mTOR inhibitors, (named ATP-competitive mTOR kinase inhibitors or TORKIs) have been developed and are revealing more efficacious than rapalogs in GBM treatment [170–172]. As an example, the novel TORKI PP242, able to target both mTORC1 and mTORC2, impairs cell proliferation and reduces stemness and invasiveness properties in a group of GBM cell lines carrying different genetic alterations [149] (Figure 2). In this respect, a putative difference in the "autophagy signature" among GBM subtypes, if found, could affect the response to the treatments.

Notably, in our model, we have obtained β-catenin relocalisation and migration impairment by both nutrient deprivation and by inhibiting mTOR complexes by means of Torin 1, an ATP-competitive mTOR inhibitor able to target both the mTOR complexes [147].

Finally, a number of compounds, such as arsenic trioxide, sodium selenite and cannabinoids (THC), used in combination with traditional therapy showed beneficial effects through the induction of an autophagic response, in some cases potentiating drug-induced cell death, some other inducing mitochondrial damage or ER stress [173–176] (Figure 2). In addition, autophagy-induced cell death seems also to be the mechanism by which some compounds overcome the apoptosis-resistance typical of the anoxic cells inside tumors. For instance, the class of small molecules able to bind the BH3 domain of the anti-apoptotic protein Bcl-2, known as BH3 mimetics, were found to induce an autophagy-dependent cell death in GBM [177,178]. If these compounds are also able to trigger a beneficial or detrimental effect on GBM invasiveness remains to be investigated.

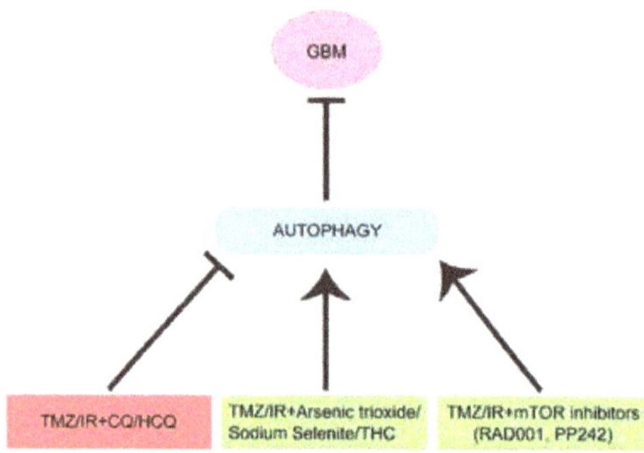

Figure 2. Targeting autophagy in GBM therapy. The effects of different therapeutic combinations on autophagy and the outcome on GBM progression in patients are shown. Chloroquine or hydroxy-chloroquine (CQ/HCQ) addition to standard protocols (TMZ/IR, temozolomide/radiation therapy) impairs the autophagy flux and sensitizes GBM to the treatment (pink box). Otherwise, several chemical compounds and mTOR inhibitors, that are able to stimulate autophagy, also increase the efficacy of the standard treatments (green boxes). THC, tetrahydrocannabinol.

5. Concluding Remarks

GBM is the most common and aggressive brain malignancy, and is characterized by a highly invasive behaviour, although the role of EMT in GBM dissemination has only been recently investigated. Although several molecules and signalling pathways mediating EMT in carcinomas play a role also in glioma invasion, further investigation will be necessary to better characterize EMT-like and its reverse MET-like processes occurring in GBM.

Targeting autophagy in GBM therapy is still a matter of debate; autophagy induction has been observed in GBM in response to radio- and temozolomide-based therapy and even though a number of clinical trials aimed at inhibiting autophagy execution, mainly by CQ, have been launched, others directed to inhibiting mTOR pathway, and thus activating autophagy, are ongoing.

Intriguingly, we observed that autophagy induction by nutrient starvation or by mTOR inhibition impairs migration and invasion of GBM cells, in line with other studies conducted on other cancer models. Further in-depth studies will be crucial to clearly dissect the autophagy role in GBM biology and to carefully evaluate autophagy modulation as therapeutic strategy to contrast GBM progression.

Funding: This work was supported by a grant from FFABR (Italian Ministry of University and Research) to SDB.

Conflicts of Interest: The authors declare no conflict of interest.

References

1. Hay, E.D. An Overview of Epithelio-Mesenchymal Transformation. *Acta Anat.* **1995**, *22*, 8–20. [CrossRef]
2. Radisky, D.C. Epithelial-mesenchymal transition. *J. Cell Sci.* **2005**, *2005*, 4325–4326. [CrossRef] [PubMed]

3. McCrea, P.D.; Gottardi, C.J. Beyond β-catenin: Prospects for a larger catenin network in the nucleus. *Nat. Rev. Mol. Cell Biol.* **2015**, *17*, 55. [CrossRef] [PubMed]
4. Kalluri, R.; Weinberg, R.A. The basics of epithelial-mesenchymal transition. *J. Clin. Investig.* **2009**, *120*, 1786. [CrossRef]
5. Abba, M.L.; Patil, N.; Leupold, J.H.; Allgayer, H. MicroRNA Regulation of Epithelial to Mesenchymal Transition. *J. Clin. Med.* **2016**, *5*, 8. [CrossRef] [PubMed]
6. Thiery, J.P. Epithelial-mesenchymal transitions in tumour progression. *Nat. Rev. Cancer* **2002**, *2*, 442–454. [CrossRef] [PubMed]
7. Jung, H.; Fattet, L.; Yang, J.; Diego, S.; Jolla, L.; Diego, S.; Jolla, L. Molecular Pathways: Linking Tumor Microenvironment to Epithelial–Mesenchymal Transition in Metastasis. *Clin. Cancer Res.* **2015**, *21*, 962–968. [CrossRef] [PubMed]
8. Shiota, M.; Zardan, A.; Takeuchi, A.; Kumano, M.; Beraldi, E.; Naito, S. The role of transforming growth factor-beta in primary brain tumors. *Cancer Res.* **2012**, 5261–5273. [CrossRef] [PubMed]
9. Brandl, M.; Seidler, B.; Haller, F.; Adamski, J.; Schmid, R.M.; Saur, D.; Brandl, M.; Seidler, B.; Haller, F.; Adamski, J.; et al. IKK-alpha controls canonical TGFbeta—SMAD signaling to regulate genes expressing SNAIL and SLUG during EMT in Panc1 cells. *J. Cell Sci.* **2013**. [CrossRef]
10. Vincent, T.; Neve, E.P.A.; Johnson, J.R.; Kukalev, A.; Rojo, F.; Albanell, J.; Pietras, K.; Virtanen, I.; Philipson, L.; Leopold, P.L.; et al. A SNAIL1—SMAD3/4 transcriptional repressor complex promotes TGF—β mediated epithelial—Mesenchymal transition. *Nat. Cell Biol.* **2009**, *11*, 943–950. [CrossRef] [PubMed]
11. Storci, G.; Sansone, P.; Mari, S.; Uva, G.D.; Tavolari, S.; Guarnieri, T.; Taffurelli, M.; Ceccarelli, C.; Santini, D.; Marcu, K.B.; et al. TNFalpha up-regulates SLUG via the NF-kappaB/HIF1alpha axis, which imparts breast cancer cells with a stem cell-like phenotype. *J. Cell Physiol.* **2010**, *225*, 682–691. [CrossRef] [PubMed]
12. Wu, J.; Chen, X.; Liu, X.; Huang, S.; He, C.; Chen, B.; Liu, Y. Autophagy regulates TGF—β 2—induced epithelial—mesenchymal transition in human retinal pigment epithelium cells. *Mol. Med. Rep.* **2018**, 3607–3614. [CrossRef] [PubMed]
13. Chua, H.L.; Clare, S.E.; Morimiya, A.; Badve, S.; Nakshatri, H. NF-kB represses E-cadherin expression and enhances epithelial to mesenchymal transition of mammary epithelial cells: Potential involvement of ZEB-1 and ZEB-2. *Oncogene* **2007**, 711–724. [CrossRef] [PubMed]
14. Kim, J.; Kong, J.; Chang, H.; Kim, H.; Kim, A. EGF induces epithelial-mesenchymal transition through phospho-Smad2/3-Snail signaling pathway in breast cancer cells. *Oncotarget* **2016**, *7*, 85021–85032. [CrossRef] [PubMed]
15. Liu, F.; Song, S.; Yi, S.; Zhang, M.; Li, J.; Yang, F.; Yin, H.; Yu, X.; Guan, C.; Liu, Y.; et al. HGF induces EMT in non-small-cell lung cancer through the hBVR pathway. *Eur. J. Pharmacol.* **2017**. [CrossRef] [PubMed]
16. Sylvester, P.W. Targeting met mediated epithelial-mesenchymal transition in the treatment of breast cancer. *Clin. Transl. Med.* **2014**, *3*, 30. [CrossRef] [PubMed]
17. Wu, Q.; Hou, X.; Xia, J.; Qian, X.; Miele, L.; Sarkar, F.H.; Wang, Z. Emerging roles of PDGF-D in EMT progression during tumorigenesis. *Cancer Treat. Rev.* **2013**, *39*, 640–646. [CrossRef] [PubMed]
18. Yang, M.; Wu, M.; Chiou, S.; Chen, P.; Chang, S.; Liu, C. Direct regulation of TWIST by HIF-1 α promotes metastasis. *Nat. Cell Biol.* **2008**, *10*. [CrossRef] [PubMed]
19. Imai, T.; Horiuchi, A.; Wang, C. Hypoxia Attenuates the Expression of E-Cadherin via Up-Regulation of SNAIL in Ovarian Carcinoma Cells. *Am. J. Pathol.* **2003**, *163*, 1437–1447. [CrossRef]
20. Kim, H.; Park, J.; Cho, Y.; Cho, C.; Kim, J.; Shin, H.; Hyun, D.; Jeong, S.; Chun, Y. Pathogenic role of HIF-1 α in prostate hyperplasia in the presence of chronic in fl ammation. *BBA Mol. Basis Dis.* **2013**, *1832*, 183–194. [CrossRef] [PubMed]
21. Cannito, S.; Novo, E.; Compagnone, A.; Busletta, C.; Zamara, E.; Paternostro, C.; Povero, D.; Bandino, A.; Bozzo, F.; Cravanzola, C.; et al. Redox mechanisms switch on hypoxia-dependent epithelial—Mesenchymal transition in cancer cells. *Carcinogenesis* **2008**, *29*, 2267–2278. [CrossRef] [PubMed]
22. Cichon, M.A.; Radisky, D.C. ROS-induced epithelial-mesenchymal transition in mammary epithelial cells is mediated by NF-κB-dependent activation of Snail. *Oncotarget* **2014**, *5*, 2827. [CrossRef] [PubMed]
23. Jiang, J.; Wang, K.; Chen, Y.; Chen, H.; Nice, E.C.; Huang, C. Redox regulation in tumor cell epithelial—mesenchymal transition: Molecular basis and therapeutic strategy. *Nat. Publ. Gr.* **2017**, *2*, 1–12. [CrossRef] [PubMed]

24. Bissel, M.J.; Radisky, D.C.; Rizki, A. The organizing principle: Microenvironmental influences in the normal and malignant breast. *Differentiation* **2002**, *70*, 537–546. [CrossRef] [PubMed]
25. Jechlinger, M.; Gr, S.; Beug, H. Mechanisms in Epithelial Plasticity and Metastasis: Insights From 3D Cultures and Expression Profiling. *J. Mammary Gland Biol. Neoplasia* **2003**, *7*, 415–432. [CrossRef]
26. Iser, I.C.; Pereira, M.B.; Lenz, G.; Wink, R. The Epithelial-to-Mesenchymal Transition-Like Process in Glioblastoma: An Updated Systematic Review and In Silico Investigation. *Med. Res. Rev.* **2017**, *37*, 271–313. [CrossRef] [PubMed]
27. Mehta, S.; Lo Cascio, C. Developmentally regulated signaling pathways in glioma invasion. *Cell. Mol. Life Sci.* **2018**, *75*, 385–402. [CrossRef] [PubMed]
28. Kubelt, C.; Hattermann, K.; Sebens, S.; Mehdorn, H.M.; Held-feindt, J. Epithelial-to-mesenchymal transition in paired human primary and recurrent glioblastomas. *Int. J. Oncol.* **2015**, *46*, 2515–2525. [CrossRef] [PubMed]
29. Lindberg, N.; Kastemar, M.; Olofsson, T.; Smits, A.; Uhrbom, L. Oligodendrocyte progenitor cells can act as cell of origin for experimental glioma. *Oncogene* **2009**, *28*, 2266–2275. [CrossRef] [PubMed]
30. Liu, C.; Sage, J.C.; Miller, M.R.; Verhaak, R.G.W.; Vogel, H.; Foreman, O.; Bronson, R.T.; Nishiyama, A. Mosaic Analysis with Double Markers (MADM) Reveals Tumor Cell-of-Origin in Glioma Chong. *Cell* **2011**, *146*, 209–221. [CrossRef] [PubMed]
31. Alcantara, L.; Chen, J.; Kwon, C.; Jackson, E.L.; Li, Y.; Burns, D.K.; Alvarez-buylla, A.; Parada, L.F. Malignant Astrocytomas Originate from Neural Stem/Progenitor Cells in a Somatic Tumor Suppressor Mouse Model. *Cancer Cell* **2009**, *15*, 45–56. [CrossRef] [PubMed]
32. Alcantara Llaguno, S.R.; Wang, Z.; Sun, D.; Chen, J.; Xu, J.; Kim, E.; Hatanpaa, K.J.; Raisanen, J.M.; Burns, D.K.; Johnson, J.; et al. Adult Lineage Restricted CNS Progenitors Specify Distinct Glioblastoma Subtypes. *Cancer Cell* **2015**, *28*, 429–440. [CrossRef] [PubMed]
33. Maher, E.A.; Furnari, F.B.; Bachoo, R.M.; Rowitch, D.H.; Louis, D.N.; Cavenee, W.K.; Depinho, R.A. Malignant glioma: Genetics and biology of a grave matter. *Genes Dev.* **2001**, *15*, 1311–1333. [CrossRef] [PubMed]
34. Mahabir, R.; Tanino, T.; Elmansuri, A.; Wang, L.; Kimura, T.; Itoh, T.; Ohba, Y.; Nishihara, H.; Shirato, H.; Tsuda, M.; et al. Sustained elevation of Snail promotes glial-mesenchymal transition after irradiation in malignant glioma. *Neuro Oncol.* **2014**, *16*, 671–685. [CrossRef] [PubMed]
35. Brabletz, T. To differentiate or not—Routes towards metastasis. *Nat. Rev. Cancer* **2012**, *12*, 425–436. [CrossRef] [PubMed]
36. Baysan, M.; Woolard, K.; Bozdag, S.; Riddick, G.; Kotliarova, S.; Cam, M.C.; Belova, G.I.; Ahn, S.; Zhang, W.; Song, H.; et al. Micro-Environment Causes Reversible Changes in DNA Methylation and mRNA Expression Profiles in Patient-Derived Glioma Stem Cells. *PLoS ONE* **2014**, *9*, 1–9. [CrossRef] [PubMed]
37. Verhaak, R.G.; Hoadley, K.A.; Purdom, E.; Wang, V.; Qi, Y.; Wilkerson, M.D.; Miller, C.R.; Ding, L.; Golub, T.; Mesirov, J.P.; et al. Integrated genomic analysis identifies clinically relevant subtypes of glioblastoma characterized by abnormalities in PDGFRA, IDH1, EGFR, and NF1. *Cancer Cell* **2010**. [CrossRef] [PubMed]
38. Zarkoob, H.; Taube, J.H.; Singh, S.K.; Mani, S.A.; Kohandel, M. Investigating the Link between Molecular Subtypes of Glioblastoma, Epithelial-Mesenchymal Transition, and CD133 Cell Surface Protein. *PLoS ONE* **2013**. [CrossRef] [PubMed]
39. Liau, B.B.; Sievers, C.; Donohue, L.K.; Gillespie, S.M.; William, A.; Miller, T.E.; Venteicher, A.S.; Herbert, C.H.; Carey, C.D.; Rodig, S.J.; et al. Adaptive chromatin remodeling drives glioblastoma stem cell plasticity and drug tolerance. *Cell Stem Cell* **2017**, *20*, 233–246. [CrossRef] [PubMed]
40. Patel, A.P.; Tirosh, I.; Trombetta, J.J.; Shalek, A.K.; Shawn, M.; Wakimoto, H.; Cahill, D.P.; Nahed, B.V.; Curry, W.T.; Martuza, R.L.; et al. Single-cell RNA-seq highlights intratumoral heterogeneity in primary glioblastoma. *Science* **2014**, *344*, 1396–1401. [CrossRef] [PubMed]
41. Meyer, M.; Reimand, J.; Lan, X.; Head, R.; Zhu, X.; Kushida, M.; Bayani, J. Single cell-derived clonal analysis of human glioblastoma links functional and genomic heterogeneity. *Proc. Natl. Acad. Sci. USA* **2015**, *112*. [CrossRef] [PubMed]
42. Kahlert, U.D.; Nikkhah, G.; Maciaczyk, J. Epithelial-to-mesenchymal (-like) transition as a relevant molecular event in malignant gliomas. *Cancer Lett.* **2013**, *331*, 131–138. [CrossRef] [PubMed]
43. Weller, M.; Wick, W.; Platten, M. Role of TGF-beta in Oncogenesis. *Microsc. Res. Tech.* **2001**, *353*, 72076.
44. Wick, W.; Platten, M.; Weller, M. Glioma cell invasion: Regulation of metalloproteinase activity by TGF-β. *J. Neurooncol.* **2001**, *53*, 177–185. [CrossRef] [PubMed]

45. Anido, J.; Saez-Borderias, A.; Gonzales-Junca, A.; Rodon, L.; Folch, G.; Carmona, M.A.; Prieto-Sanchez, R.M.; Barba, I.; Martınez, E.; Prudkin, L.; et al. TGF- beta Receptor Inhibitors Target Population in Human Glioblastoma. *Cancer Cell* **2010**, 655–668. [CrossRef] [PubMed]
46. Zhang, Y.; Chen, K.; Sloan, S.A.; Bennett, M.L.; Scholze, A.R.; Keeffe, S.O.; Phatnani, H.P.; Guarnieri, X.P.; Caneda, C.; Ruderisch, N.; et al. An RNA-Sequencing Transcriptome and Splicing Database of Glia, Neurons, and Vascular Cells of the Cerebral Cortex. *J. Neurosci.* **2014**, *34*, 11929–11947. [CrossRef] [PubMed]
47. Joseph, J.V.; Conroy, S.; Tomar, T.; Bhat, K.; Copray, S.; Walenkamp, A.M.E.; Boddeke, E.; Balasubramanyian, V.; Wagemakers, M.; Den Dunnen, W.F.A.; et al. TGF-β is an inducer of ZEB1-dependent mesenchymal transdifferentiation in glioblastoma that is associated with tumor invasion. *Cell Death Dis.* **2014**, *5*, e1443. [CrossRef] [PubMed]
48. Li, Y.; Li, A.; Glas, M.; Lal, B.; Ying, M.; Sang, Y.; Xia, S. c-Met signaling induces a reprogramming network and supports the glioblastoma stem-like phenotype. *Proc. Natl. Acad. Sci. USA* **2011**, *108*, 1–6. [CrossRef] [PubMed]
49. De Bacco, F.; Casanova, E.; Medico, E.; Pellegatta, S.; Orzan, F.; Albano, R.; Luraghi, P.; Reato, G.; Ambrosio, A.D.; Porrati, P.; et al. The MET Oncogene Is a Functional Marker of a Glioblastoma Stem Cell Subtype. *Cancer Res.* **2012**. [CrossRef] [PubMed]
50. Abounader, R.; Laterra, J. Scatter factor/hepatocyte growth factor in brain tumor growth and angiogenesis. *Neuro Oncol.* **2005**. [CrossRef] [PubMed]
51. Boccaccio, C.; Comoglio, P.M. The MET Oncogene in Glioblastoma Stem Cells: Implications as a Diagnostic Marker and a Therapeutic Target. *Cancer Res.* **2013**, 3193–3200. [CrossRef] [PubMed]
52. Kong, D.; Song, S.; Kim, D.; Joo, K.M. Prognostic Significance of c-Met Expression in Glioblastomas. *Cancer* **2009**. [CrossRef] [PubMed]
53. Nabeshima, K.; Shimao, Y.; Sato, S.; Kataoka, H.; Moriyama, T.; Kawano, H.; Wakisaka, S.; Koono, M. Expression of c-Met correlates with grade of malignancy in human astrocytic tumours: An immunohistochemical study. *Histopathology* **1997**, *31*, 436–443. [CrossRef] [PubMed]
54. Siebzehnrubl, F.A.; Silver, D.J.; Tugertimur, B.; Deleyrolle, L.P.; Siebzehnrubl, D.; Sarkisian, M.R.; Devers, K.G.; Yachnis, A.T.; Kupper, M.D.; Neal, D.; et al. The ZEB1 pathway links glioblastoma initiation, invasion and chemoresistance. *EMBO Moecular Med.* **2013**, 1196–1212. [CrossRef] [PubMed]
55. Elias, M.C.; Tozer, K.R.; Silber, J.R.; Mikheeva, S.; Deng, M.; Morrison, R.S.; Manning, T.C.; Silbergeld, D.L.; Glackin, C.A.; Reh, T.A.; et al. TWIST is Expressed in Human Gliomas and Promotes Invasion 1. *Neoplasia* **2005**, *7*, 824–837. [CrossRef] [PubMed]
56. Yang, H.W.; Menon, L.G.; Black, P.M.; Carroll, R.S.; Johnson, M.D. SNAI2/Slug promotes growth and invasion in human gliomas. *BMC Cancer* **2010**, *10*, 301. [CrossRef] [PubMed]
57 Mikheeva, S.A.; Mikheev, A.M.; Petit, A.; Beyer, R.; Oxford, R.G.; Khorasani, L.; Maxwell, J.; Glackin, C.A.; Wakimoto, H.; González-herrero, I.; et al. TWIST1 promotes invasion through mesenchymal change in human glioblastoma. *Mol. Cancer* **2010**, *9*, 194. [CrossRef] [PubMed]
58. Han, S.H.J.K.M. SNAI1 is Involved in the Proliferation and Migration of Glioblastoma Cells. *Cell Mol. Neurobiol.* **2011**, 489–496. [CrossRef] [PubMed]
59. Myung, J.K.; Choi, S.A.; Kim, S.; Wang, K.; Park, S. Snail plays an oncogenic role in glioblastoma by promoting epithelial mesenchymal transition. *Int. J. Clin. Exp. Pathol.* **2014**, *7*, 1977–1987. [PubMed]
60. Qi, S.; Song, Y.; Peng, Y.; Wang, H.; Long, H.; Yu, X.; Li, Z. ZEB2 Mediates Multiple Pathways Regulating Cell Proliferation, Migration, Invasion, and Apoptosis in Glioma. *PLoS ONE* **2012**, *7*, 1–12. [CrossRef] [PubMed]
61. Lewis-Tuffin, L.J.; Rodriguez, F.; Giannini, C.; Scheithauer, B.; Necela, B.M. Misregulated E-Cadherin Expression Associated with an Aggressive Brain Tumor Phenotype. *PLoS ONE* **2010**, *5*. [CrossRef] [PubMed]
62. Asano, K.; Duntsch, C.D.; Zhou, Q.; Weimar, J.D.; Bordelon, D.; Robertson, J.H.; Pourmotabbed, T. Correlation of N-cadherin expression in high grade gliomas with tissue invasion. *J. Neurooncol.* **2004**, *70*, 3–15. [CrossRef] [PubMed]
63. Camand, E.; Peglion, F.; Osmani, N.; Sanson, M.; Etienne-Manneville, S. N-cadherin expression level modulates integrin-mediated polarity and strongly impacts on the speed and directionality of glial cell migration. *J. Cell Sci.* **2012**, *125*, 844–857. [CrossRef] [PubMed]
64. Perego, C.; Vanoni, C.; Massari, S.; Raimondi, A.; Pola, S.; Cattaneo, M.G.; Francolini, M.; Vicentini, L.M.; Pietrini, G. Invasive behaviour of glioblastoma cell lines is associated with altered organisation of the cadherin- catenin adhesion system. *J. Cell Sci.* **2002**, *115*, 3331–3340. [PubMed]

65. Appolloni, I.; Barilari, M.; Caviglia, S.; Gambini, E.; Reisoli, E.; Malatesta, P. A cadherin switch underlies malignancy in high-grade gliomas. *Oncogene* **2014**, *34*, 1991. [CrossRef] [PubMed]
66. Halliday, J.; Helmy, K.; Pattwell, S.S.; Pitter, K.L.; Laplant, Q.; Ozawa, T. In vivo radiation response of proneural glioma characterized by protective p53 transcriptional program and proneural-mesenchymal shift. *Proc. Natl. Acad. Sci. USA* **2014**, *111*. [CrossRef] [PubMed]
67. Batchelor, T.T.; Reardon, D.A.; de Groot, J.F.; Wick, W.; Weller, M. Antiangiogenic Therapy for Glioblastoma: Current Status and Future Prospects. *Clin. Cancer Res.* **2014**, *20*, 5612–5619. [CrossRef] [PubMed]
68. Piao, Y.; Liang, J.; Holmes, L.; Henry, V.; Sulman, E.; Groot, J.F. De Acquired Resistance to Anti-VEGF Therapy in Glioblastoma Is Associated with a Mesenchymal Transition. *Clin. Cancer Res.* **2013**, *19*, 4392–4404. [CrossRef] [PubMed]
69. Kim, Y.; Yoo, K.; Cui, Y.; Uddin, N.; Lim, E.; Kim, M.; Nam, S.; Kim, I.; Suh, Y.; Lee, S. Radiation promotes malignant progression of glioma cells through HIF-1alpha stabilization. *Cancer Lett.* **2014**, *354*, 132–141. [CrossRef] [PubMed]
70. Meng, J.; Li, P.; Zhang, Q.; Yang, Z.; Fu, S. A radiosensitivity gene signature in predicting glioma prognostic via EMT pathway. *Oncotarget* **2014**, *5*, 4683. [CrossRef] [PubMed]
71. Timke, C.; Zieher, H.; Roth, A.; Hauser, K.; Lipson, K.E.; Weber, K.J.; Debus, J.; Abdollahi, A.; Huber, P.E. Combination of Vascular Endothelial Growth Factor Receptor/Platelet-Derived Growth Factor Receptor Inhibition Markedly Improves Radiation T umor Therapy. *Clin. Cancer Res.* **2008**, *14*, 2210–2220. [CrossRef] [PubMed]
72. Zhang, M.; Kleber, S.; Rohrich, M.; Timke, C.; Han, N.; Tuettenberg, J.; Martin-Villalba, A.; Debus, J.; Peschke, P.; Wirkner, U.; et al. Blockade of TGF-beta signaling by the TGFbetaR-I Kinase Inhibitor LY2109761 Enhances Radiation Response and Prolongs Survival in Glioblastoma. *Cancer Res.* **2011**, 7155–7168. [CrossRef] [PubMed]
73. Bhat, K.P.L.; Balasubramaniyan, V.; Vaillant, B.; Hummelink, K.; Hollingsworth, F.; Wani, K.; James, J.D.; Goodman, L.D.; Conroy, S.; Long, L.; et al. Mesenchymal Differentiation Mediated by NF-κB Promotes Radiation Resistance in Glioblastoma. *Cancer Cell* **2013**, *24*, 1–22. [CrossRef] [PubMed]
74. Klionsky, D.J. Autophagy revisited: A conversation with Christian de Duve. *Autophagy* **2008**, *4*, 740–743. [CrossRef] [PubMed]
75. Harnett, M.M.; Pineda, M.A.; Latré de Laté, P.; Eason, R.J.; Besteiro, S.; Harnett, W.; Langsley, G. From Christian de Duve to Yoshinori Ohsumi: More to autophagy than just dining at home. *Biomed. J.* **2017**, *40*, 9–22. [CrossRef] [PubMed]
76. Noboru, M. A brief history of autophagy from cell biology to physiology and disease. *Nat. Cell Biol.* **2018**, *20*, 521–527.
77. Tsukada, M.; Oshumi, Y. Isofation and characterization of autophagy-defective mutants of Saccaromyces cerevisiae. *FEBS* **1993**, *333*, 169–174. [CrossRef]
78. Boya, P.; Reggiori, F.; Codogno, P. Emerging regulation and functions of autophagy. *Nat. Cell Biol.* **2013**, *15*, 713. [CrossRef] [PubMed]
79. Feng, Y.; He, D.; Yao, Z.; Klionsky, D.J. The machinery of macroautophagy. *Cell Res.* **2014**, *24*, 24–41. [CrossRef] [PubMed]
80. Laplante, M.; Sabatini, D.M. mTOR signaling at a glance. *J. Cell Sci.* **2009**. [CrossRef] [PubMed]
81. Wong, P.M.; Puente, C.; Ganley, I.G.; Jiang, X. The ULK1 complex sensing nutrient signals for autophagy activation. *Autophagy* **2013**, *9*, 124–137. [CrossRef] [PubMed]
82. Maria Fimia, G.; Stoykova, A.; Romagnoli, A.; Giunta, L.; Di Bartolomeo, S.; Nardacci, R.; Corazzari, M.; Fuoco, C.; Ucar, A.; Schwartz, P.; et al. Ambra1 regulates autophagy and development of the nervous system. *Nature* **2007**, *447*. [CrossRef] [PubMed]
83. Di Bartolomeo, S.; Corazzari, M.; Nazio, F.; Oliverio, S.; Lisi, G.; Antonioli, M.; Pagliarini, V.; Matteoni, S.; Fuoco, C.; Giunta, L.; et al. The dynamic interaction of AMBRA1 with the dynein motor complex regulates mammalian autophagy. *J. Cell Biol.* **2010**, *191*, 155–168. [CrossRef] [PubMed]
84. Matsunaga, K.; Saitoh, T.; Tabata, K.; Omori, H.; Satoh, T.; Kurotori, N.; Maejima, I.; Shirahama-noda, K.; Ichimura, T.; Isobe, T.; et al. Two Beclin 1-binding proteins, Atg14L and Rubicon, reciprocally regulate autophagy at different stages. *Nat. Cell Biol.* **2009**, *11*. [CrossRef] [PubMed]

85. Zhong, Y.; Wang, Q.J.; Li, X.; Yan, Y.; Backer, J.M.; Brian, T.; Heintz, N.; Yue, Z. Distinct regulation of autophagic activity by Atg14L and Rubicon associated with Beclin 1- phosphatidylinositol 3-kinase complex. *Nat. Cell Biol.* **2009**, *11*, 468–476. [CrossRef] [PubMed]
86. Liang, C.; Feng, P.; Ku, B.; Dotan, I.; Canaani, D.; Oh, B.; Jung, J.U. Autophagic and tumour suppressor activity of a novel Beclin1-binding protein UVRAG. *Nat. Cell Biol.* **2006**, *8*. [CrossRef] [PubMed]
87. Mizushima, N. Autophagy: Process and function. *Genes Dev.* **2007**, *21*, 2861–2873. [CrossRef] [PubMed]
88. Di Bartolomeo, S.; Nazio, F.; Cecconi, F. The Role of Autophagy During Development in Higher Eukaryotes. *Traffic* **2010**, *11*, 1280–1289. [CrossRef] [PubMed]
89. Kroemer, G. Autophagy: A druggable process th a t is deregulated in aging and hum an disease. *J. Clin. Investig.* **2015**, *125*, 1–5. [CrossRef] [PubMed]
90. Galluzzi, L.; Pietrocola, F.; Pedro, J.M.B.; Ravi, K.; Maiuri, M.C.; Martin, S.J.; Penninger, J.; Piacentini, M. Autophagy in malignant transformation and cancer progression. *EMBO J.* **2015**, *34*, 856–880. [CrossRef] [PubMed]
91. Liang, X.H.; Jackson, S.; Seaman, M.; Brown, K.; Kempkes, B.; Hibshoosh, H.; Levine, B. Induction of autophagy and inhibition of tumorigenesis by beclin 1. *Nature* **1999**, *402*, 672. [CrossRef] [PubMed]
92. Qu, X.; Yu, J.; Bhagat, G.; Furuya, N.; Hibshoosh, H.; Troxel, A.; Rosen, J.; Eskelinen, E.; Mizushima, N.; Ohsumi, Y.; et al. Promotion of tumorigenesis by heterozygous disruption of the beclin 1 autophagy gene. *J. Clin. Investig.* **2003**, *112*, 1809–1820. [CrossRef] [PubMed]
93. Yue, Z.; Jin, S.; Yang, C.; Levine, A.J.; Heintz, N. Beclin 1, an autophagy gene essential for early embryonic development, is a haploinsufficient tumor suppressor. *Proc. Natl. Acad. Sci. USA* **2003**, *100*, 15077–15082. [CrossRef] [PubMed]
94. Cianfanelli, V.; Fuoco, C.; Lorente, M.; Salazar, M.; Quondamatteo, F.; Gherardini, P.F.; De Zio, D.; Nazio, F.; Antonioli, M.; D'Orazio, M.; et al. AMBRA1 links autophagy to cell proliferation and tumorigenesis by promoting c-Myc dephosphorylation and degradation. *Nat. Cell Biol.* **2015**, *17*. [CrossRef] [PubMed]
95. Takamura, A.; Komatsu, M.; Hara, T.; Sakamoto, A.; Kishi, C.; Waguri, S.; Eishi, Y.; Hino, O. Autophagy-deficient mice develop multiple liver tumors. *Genes Dev.* **2011**, *5*, 795–800. [CrossRef] [PubMed]
96. Marino, G.; Salvador-montoliu, N.; Fueyo, A.; Knecht, E.; Mizushima, N.; Lopez-Otin, C. Tissue-specific Autophagy Alterations and Increased Tumorigenesis in Mice Deficient in Atg4C/Autophagin-3. *J. Biol. Chem.* **2007**, *282*, 18573–18583. [CrossRef] [PubMed]
97. Strohecker, A.M.; Guo, J.Y.; Karsli-uzunbas, G.; Sandy, M.; Chen, G.J.; Mathew, R.; Mcmahon, M.; White, E. Autophagy Sustains Mitochondrial Glutamine Metabolism and Growth of BRAFV600E –Driven Lung Tumors. *Cancer Discov.* **2013**, *3*, 1–23. [CrossRef] [PubMed]
98. Rao, S.; Tortola, L.; Perlot, T.; Wirnsberger, G.; Novatchkova, M.; Nitsch, R.; Sykacek, P.; Frank, L.; Schramek, D.; Komnenovic, V.; et al. A dual role for autophagy in a murine model of lung cancer. *Nat. Commun.* **2014**. [CrossRef] [PubMed]
99. Takahashi, Y.; Hori, T.; Cooper, T.K.; Liao, J.; Desai, N.; Serfass, J.M.; Young, M.M.; Park, S.; Izu, Y.; Wang, H. Bif-1 haploinsufficiency promotes chromosomal instability and accelerates Myc -driven lymphomagenesis via suppression of mitophagy. *Lymphoid Neoplasia* **2016**, *121*, 1622–1633. [CrossRef] [PubMed]
100. Mathew, R.; Karp, C.; Beaudoin, B.; Vuong, N.; Chen, H.; Bray, K.; Reddy, A.; Bhanot, G.; Dipaola, R.S.; Karantza-wadsworth, V.; et al. Autophagy Suppresses Tumorigenesis Through Elimination of p62. *Cell* **2009**, *137*, 1062–1075. [CrossRef] [PubMed]
101. Karantza-wadsworth, V.; Patel, S.; Kravchuk, O.; Chen, G.; Mathew, R.; Jin, S.; White, E. Autophagy mitigates metabolic stress and genome damage in mammary tumorigenesis. *Genes Dev.* **2007**, 1621–1635. [CrossRef] [PubMed]
102. Mathew, R.; Kongara, S.; Beaudoin, B.; Karp, C.M.; Bray, K.; Degenhardt, K.; Chen, G.; Jin, S.; White, E. Autophagy suppresses tumor progression by limiting chromosomal instability. *Genes Dev.* **2007**, 1367–1381. [CrossRef] [PubMed]
103. Park, J.M.; Tougeron, D.; Huang, S.; Okamoto, K.; Sinicrope, F.A. Beclin 1 and UVRAG Confer Protection from Radiation- Induced DNA Damage and Maintain Centrosome Stability in Colorectal Cancer Cells. *PLoS ONE* **2014**, *9*, e100819.
104. Green, D.R.; Galluzzi, L.; Kroemer, G. Metabolic control of cell death. *Science* **2014**, *345*. [CrossRef] [PubMed]
105. Ma, Y.; Galluzzi, L.; Zitvogel, L.; Kroemer, G. Autophagy and Cellular Immune Responses. *Immunity* **2013**, *39*, 211–227. [CrossRef] [PubMed]

106. Salemi, S. Autophagy is required for self-renewal and differentiation of adult human stem cells. *Cell Res.* **2012**, *22*, 432–435. [CrossRef] [PubMed]
107. Garufi, A.; Pucci, D.; Orazi, V.D.; Cirone, M.; Bossi, G.; Avantaggiati, M.L.; Orazi, G.D. Degradation of mutant p53H175 protein by Zn (II) through autophagy. *Cell Death Dis.* **2014**, 1–9. [CrossRef] [PubMed]
108. Peng, Y.; Shi, Y.; Ding, Z.; Ke, A.; Gu, C.; Peng, Y.; Shi, Y.; Ding, Z.; Ke, A.; Gu, C.; et al. Autophagy inhibition suppresses pulmonary metastasis of HCC in mice via impairing anoikis resistance and colonization of HCC cells Autophagy inhibition suppresses pulmonary metastasis of HCC in mice via impairing anoikis resistance and colonization of HCC. *Autophagy* **2013**, *8627*. [CrossRef]
109. Boya, P.; Gonza, R.; Casares, N.; Perfettini, J.; Dessen, P.; Larochette, N.; Me, D.; Meley, D.; Souquere, S.; Yoshimori, T.; et al. Inhibition of Macroautophagy Triggers Apoptosis. *Mol. Cell. Biol.* **2005**, *25*, 1025–1040. [CrossRef] [PubMed]
110. Ko, A.; Kanehisa, A.; Martins, I.; Senovilla, L.; Chargari, C.; Dugue, D.; Marino, G.; Kepp, O.; Michaud, M.; et al. Autophagy inhibition radiosensitizes in vitro, yet reduces radioresponses in vivo due to deficient immunogenic signalling. *Cell Death Differ.* **2014**, 92–99. [CrossRef] [PubMed]
111. Levy, J.M.M.; Thompson, J.C.; Griesinger, A.M.; Amani, V.; Donson, A.M.; Birks, D.K.; Morgan, M.J.; Mirsky, D.M.; Michael, H. Autophagy Inhibition Improves Chemosensitivity in BRAFV600E Brain Tumors. *Cancer Discov.* **2014**, *4*, 773–780. [CrossRef] [PubMed]
112. Amaravadi, R.K.; Yu, D.; Lum, J.J.; Bui, T.; Christophorou, M.A.; Evan, G.I.; Thomas-tikhonenko, A.; Thompson, C.B. Autophagy inhibition enhances therapy-induced apoptosis in a Myc -induced model of lymphoma. *J. Clin. Investig.* **2007**, *117*, 326–336. [CrossRef] [PubMed]
113. Gugnoni, M.; Sancisi, V.; Manzotti, G.; Gandolfi, G.; Ciarrocchi, A. Autophagy and epithelial-mesenchymal transition: An intricate interplay in cancer. *Cell Death Dis.* **2016**, *7*, e2520. [CrossRef] [PubMed]
114. Pirtoli, L.; Cevenini, G.; Tini, P.; Vannini, M.; Oliveri, G.; Marsili, S.; Mourmouras, V.; Rubino, G.; Miracco, C.; Pirtoli, L.; et al. The prognostic role of Beclin 1 protein expression in high-grade gliomas. *Autophagy* **2009**, *8627*. [CrossRef]
115. Huang, X.; Bai, H.; Chen, L.; Li, B.; Lu, Y. Reduced expression of LC3B-II and Beclin 1 in glioblastoma multiforme indicates a down-regulated autophagic capacity that relates to the progression of astrocytic tumors. *J. Clin. Neurosci.* **2010**, *17*, 1515–1519. [CrossRef] [PubMed]
116. Galavotti, S.; Bartesaghi, S.; Faccenda, D.; Sanzone, S.; Mcevoy, A.; Dinsdale, D.; Condorelli, F.; Brandner, S. The autophagy-associated factors DRAM1 and p62 regulate cell migration and invasion in glioblastoma stem cells. *Oncogene* **2012**, *32*, 699–712. [CrossRef] [PubMed]
117. Singh, A.B. EGFR-Signaling and Autophagy: How they Fit in the Cancer Landscape. *J. Adenocarcinoma* **2016**, *1*, 1–9. [CrossRef]
118. Arico, S.; Petiot, A.; Bauvy, C.; Dubbelhuis, P.F.; Meijer, A.J.; Codogno, P.; Ogier-denis, E. The Tumor Suppressor PTEN positively Regulates Macroautophagy by Inhibiting the Phosphatidylinositol 3-Kinase/Protein Kinase B Pathway. *J. Biol. Chem. Biol. Chem.* **2001**, *276*, 35243–35247. [CrossRef] [PubMed]
119. Tasdemir, E.; Maiuri, M.C.; Galluzzi, L.; Vitale, I.; Djavaheri-mergny, M.; Amelio, M.D.; Criollo, A.; Morselli, E.; Zhu, C.; Harper, F.; et al. Regulation of autophagy by cytoplasmic p53. *Nat. Cell Biol.* **2008**, *10*, 676–687. [CrossRef] [PubMed]
120. Marcucci, F.; Ghezzi, P.; Rumio, C. The role of autophagy in the cross-talk between epithelial-mesenchymal transitioned tumor cells and cancer stem-like cells. *Mol. Cancer* **2017**, 1–8. [CrossRef] [PubMed]
121. Subramani, R.; Gonzalez, E.; Arumugam, A.; Ortega, A.; Bonkoungou, S.; Narayan, M.; Dwivedi, A. Nimbolide inhibits pancreatic cancer growth and metastasis through ROS-mediated apoptosis and inhibition of epithelial-to- mesenchymal transition. *Sci. Rep.* **2016**, 1–12. [CrossRef] [PubMed]
122. Ren, B.J.; Zhou, Z.-W.; Ju, Y.-L.; Ouyang, M.-Z.; Chen, X.-W.; Chen, S.-F. Alisertib Induces Cell Cycle Arrest, Apoptosis, Autophagy and Suppresses EMT in HT29 and Caco-2 cells. *Int. J. Mol. Sci.* **2015**, *17*, 41. [CrossRef] [PubMed]
123. Zi, D.; Zhou, Z.; Yang, Y.; Huang, L.; Zhou, Z.; He, S. Danusertib Induces Apoptosis, Cell Cycle Arrest, and Autophagy but Inhibits Epithelial to Mesenchymal Transition Involving PI3K/Akt/mTOR Signaling Pathway in Human Ovarian Cancer Cells. *Int. J. Mol. Sci.* **2015**, *16*, 27228–27251. [CrossRef] [PubMed]
124. Lv, Q.; Wang, W.; Xue, J.; Hua, F.; Mu, R.; Lin, H.; Yan, J.; Lv, X. DEDD Interacts with PI3KC3 to Activate Autophagy and Attenuate Epithelial—Mesenchymal Transition in Human Breast Cancer. *Cancer Res.* **2012**, *72*, 3238–3251. [CrossRef] [PubMed]

125. Qiang, L.; Zhao, B.; Ming, M.; Wang, N.; He, T.; Hwang, S.; Thorburn, A. Regulation of cell proliferation and migration by p62 through stabilization of Twist1. *Proc. Natl. Acad. Sci. USA* **2014**. [CrossRef] [PubMed]
126. Grassi, G.; Di Caprio, G.; Santangelo, L.; Fimia, G.M.; Cozzolino, A.M.; Komatsu, M.; Ippolito, G.; Tripodi, M.; Alonzi, T. Autophagy regulates hepatocyte identity and epithelial- transitions promoting Snail degradation. *Cell Death Dis.* **2015**, *6*, e1880. [CrossRef] [PubMed]
127. Li, J.; Yang, B.; Zhou, Q.; Wu, Y.; Shang, D.; Guo, Y.; Song, Z. Autophagy promotes hepatocellular carcinoma cell invasion through activation of epithelial—Mesenchymal transition. *Carcinogenesis* **2013**, *34*, 1343–1351. [CrossRef] [PubMed]
128. Chen, C.; Liang, Q.Y.U.; Chen, H.U.I.K.; Wu, P.I.N.F.E.I.; Feng, Z.Y.U. DRAM1 regulates the migration and invasion of hepatoblastoma cells via autophagy—EMT pathway. *Oncol. Lett.* **2018**, 2427–2433. [CrossRef] [PubMed]
129. Hu, S.; Wang, L.; Zhang, X.; Wu, Y.; Yang, J.; Li, J. Autophagy induces transforming growth factor-β—Dependent epithelial—Mesenchymal transition in hepatocarcinoma cells through cAMP response element binding signalling. *J. Cell Mol. Med.* **2018**, *22*, 5518–5532. [CrossRef] [PubMed]
130. Gulhati, P.; Bowen, K.A.; Liu, J. mTORC1 and mTORC2 Regulate EMT, Motility, and Metastasis of Colorectal Cancer via RhoA and Rac1 Signaling Pathways. *Cancer Res.* **2011**. [CrossRef] [PubMed]
131. Shen, H.; Yin, L.; Deng, G.; Guo, C.; Han, Y.; Li, Y.; Cai, C.; Fu, Y.; Liu, S.; Zeng, S. Knockdown of Beclin-1 impairs epithelial-mesenchymal transition of colon cancer cells. *J. Cell. Biochem.* **2018**, *119*, 7022–7030. [CrossRef] [PubMed]
132. Su, Z.; Li, G.; Ren, C.; Deng, T.; Zhang, S.; Tian, Y.; Liu, Y.; Qiu, Y. Autophagy inhibition impairs the epithelial-mesenchymal transition and enhances cisplatin sensitivity in nasopharyngeal carcinoma. *Oncol. Lett.* **2017**, *13*, 4147–4154. [CrossRef] [PubMed]
133. Alizadeh, J.; Glogowska, A.; Thliveris, J.; Kalantari, F.; Shojaei, S. Autophagy modulates transforming growth factor beta 1 induced epithelial to mesenchymal transition in non-small cell lung cancer cells. *BBA Mol. Cell Res.* **2018**, *1865*, 749–768. [CrossRef] [PubMed]
134. Luo, D.; Hu, S.; Tang, C.; Liu, G. Mesenchymal stem cells promote cell invasion and migration and autophagy-induced epithelial-mesenchymal transition in A549 lung adenocarcinoma cells. *Cell Biochem. Funct.* **2018**, *36*, 88–94. [CrossRef] [PubMed]
135. Liu, H.; Du, Y.; Zhang, Z.; Lv, L.; Xiong, W.; Zhang, L.; Li, N.; He, H.; Li, Q.; Liu, Y. Autophagy contributes to hypoxia-induced epithelial to mesenchymal transition of endometrial epithelial cells in endometriosis. *Biol. Reprod.* **2018**, *99*, 968–981. [CrossRef] [PubMed]
136. Zhao, S.; Li, D.; Bei, X.Y.; Zhu, Y.P.; Sun, W.L.; Shen, C.; Wood, K.; Han, B.M.; Jiang, J. Maternal exposure to di-n-butyl phthalate (DBP) promotes epithelial-mesenchymal transition via regulation of autophagy in uroepithelial cell. *Toxicology* **2018**, *406–407*, 114–122. [CrossRef] [PubMed]
137. Tong, H.; Yin, H.; Hossain, MA.; Wang, Y.; Wu, F.; Dong, X.; Gao, S.; Zhan, K.; He, W. Starvation-induced autophagy promotes the invasion and migration of human bladder cancer cells via TGF β1/Smad3-mediated epithelial-mesenchymal transition activation. *J. Cell. Biochem.* **2019**, *120*, 5118–5127. [CrossRef] [PubMed]
138. Li, G.; Li, C.; Xia, M.; Ritter, J.K.; Gehr, T.W.B.; Boini, K.; Li, P.; Commonwealth, V. Enhanced Epithelial-to-Mesenchymal Transition Associated with Lysosome Dysfunction in Podocytes: Role of p62/Sequestosome 1 as a Signaling Hub. *Cell Physiol. Biochem.* **2016**, *35*, 1773–1786. [CrossRef] [PubMed]
139. Qiang, L.; He, Y. Autophagy deficiency stabilizes TWIST1 to promote epithelial-mesenchymal transition. *Autophagy* **2014**, *8627*. [CrossRef] [PubMed]
140. Lv, Q.; Hua, F.; Hu, Z. DEDD, a novel tumor repressor, reverses epithelial-mesenchymal transition by activating selective autophagy. *Autophagy* **2012**, *8627*, 10–12. [CrossRef] [PubMed]
141. Qin, W.; Li, C.; Zheng, W.; Guo, Q.; Zhang, Y. Inhibition of autophagy promotes metastasis and glycolysis by inducing ROS in gastric cancer cells. *Oncotarget* **2015**, *6*, 39839. [CrossRef] [PubMed]
142. Macintosh, R.L.; Timpson, P.; Thorburn, J.; Anderson, K.I.; Thorburn, A.; Ryan, K.M. Inhibition of autophagy impairs tumor cell invasion in an organotypic model Do not distribute. *Landes Biosci.* **2012**, *11*, 2022–2029.
143. Zou, M.; Zhu, W.; Wang, L.; Shi, L.; Gao, R.; Ou, Y.; Chen, X. AEG-1/MTDH-activated autophagy enhances human malignant glioma susceptibility to TGF-β1-triggered epithelial-mesenchymal transition. *Oncotarget* **2016**, *7*, 13122. [CrossRef] [PubMed]

144. Lu, Y.; Xiao, L.; Liu, Y.; Wang, H.; Li, H.; Zhou, Q.; Pan, J.; Lei, B.; Huang, A.; Qi, S. MIR517C inhibits autophagy and the epithelial- to-mesenchymal (-like) transition phenotype in human glioblastoma through KPNA2-dependent disruption of TP53 nuclear translocation. *Autophagy* **2015**, *11*, 2213–2232. [CrossRef] [PubMed]
145. Liu, X.; Sun, K.; Wang, H.; Dai, Y. Inhibition of Autophagy by Chloroquine Enhances the Antitumor Efficacy of Sorafenib in Glioblastoma. *Cell. Mol. Neurobiol.* **2016**, *36*, 1197–1208. [CrossRef] [PubMed]
146. Catalano, M.; D'Alessandro, G.; Lepore, F.; Corazzari, M.; Caldarola, S.; Valacca, C.; Faienza, F.; Esposito, V.; Limatola, C.; Cecconi, F.; et al. Autophagy induction impairs migration and invasion by reversing EMT in glioblastoma cells. *Mol. Oncol.* **2015**, *9*. [CrossRef] [PubMed]
147. Colella, B.; Faienza, F.; Carinci, M.; Alessandro, G.D.; Catalano, M.; Santoro, A.; Cecconi, F.; Limatola, C.; Di Bartolomeo, S. Autophagy induction impairs Wnt/β-catenin signalling through β-catenin relocalisation in glioblastoma cells. *Cell. Signal.* **2019**, *53*, 357–364. [CrossRef] [PubMed]
148. Palumbo, S.; Pirtoli, L.; Tini, P.; Cevenini, G.; Calderaro, F.; Toscano, M.; Miracco, C.; Cominicini, S. Different involvement of autophagy in human malignant glioma cell lines undergoing irradiation and Temozolomide combined treatments. *J. Cell. Biochem.* **2012**, *113*, 2308–2318. [CrossRef] [PubMed]
149. Mecca, C.; Giambanco, I.; Bruscoli, S.; Bereshchenko, O.; Fioretti, B.; Riccardi, C.; Donato, R.; Arcuri, C.; Adamo, M.C.D. PP242 Counteracts Glioblastoma Cell Proliferation, Migration, Invasiveness and Stemness Properties by Inhibiting mTORC2/AKT. *Front. Cell. Neurosci.* **2018**, *12*, 1–18. [CrossRef] [PubMed]
150. MacDonald, B.T. Wnt/b-catenin signaling: Components, mechanisms, and diseases. *Dev. Biol.* **2010**, *17*, 9–26. [CrossRef] [PubMed]
151. Nusse, R.; Clevers, H. Wnt/β-Catenin Signaling, Disease, and Emerging Therapeutic Modalities. *Cell* **2017**, *169*, 985–999. [CrossRef] [PubMed]
152. Lee, Y.; Lee, J.; Ahn, S.H.; Lee, J.; Nam, D. WNT signaling in glioblastoma and therapeutic opportunities. *Lab. Investig.* **2016**, *96*, 137–150. [CrossRef] [PubMed]
153. Nager, M.; Bhardwaj, D.; Cant, C.; Medina, L.; Nogu, P.; Herreros, J. β -Catenin Signalling in Glioblastoma Multiforme and Glioma-Initiating Cells. *Chemother. Res. Pract.* **2012**, *2012*, 6–10. [CrossRef] [PubMed]
154. Sánchez-tilló, E.; De Barrios, O.; Siles, L.; Cuatrecasas, M.; Castells, A.; Postigo, A. β-catenin/TCF4 complex induces the epithelial-to-mesenchymal transition (EMT)-activator ZEB1 to regulate tumor invasiveness. *Proc. Natl. Acad. Sci. USA* **2011**, *108*, 19204–19209. [CrossRef] [PubMed]
155. Kim, K.; Lu, Z.; Hay, E.D. Direct evidence for a role of beta-catenin/LEF-1 signaling pathway in induction of EMT. *Cell Biol. Int.* **2002**, *26*, 463–476. [CrossRef] [PubMed]
156. Wang, X.; Wang, X. The critical role of EGF-β-catenin signaling in the epithelial-mesenchymal transition in human glioblastoma. *OncoTargets Ther.* **2017**, *10*, 2781–2789. [CrossRef] [PubMed]
157. Kanzawa, T.; Germano, I.M.; Komata, T.; Ito, H.; Kondo, Y.; Kondo, S. Role of autophagy in temozolomide-induced cytotoxicity for malignant glioma cells. *Cell Death Differ.* **2004**, *11*, 448–457. [CrossRef] [PubMed]
158. Sotelo, J.; Briceno, E.; Lopez-Gonzales, M.A. Annals of Internal Medicine Article Adding Chloroquine to Conventional Treatment for Glioblastoma Multiforme. *Ann. Intern. Med.* **2006**, *144*, 337–344. [CrossRef] [PubMed]
159. Rosenfeld, M.R.; Ye, X.; Supko, J.G.; Desideri, S.; Grossman, S.A.; Brem, S.; Mikkelson, T.; Wang, D.; Chang, Y.C.; Hu, J.; et al. A phase I/II trial of hydroxychloroquine in conjunction with radiation therapy and concurrent and adjuvant temozolomide in patients with newly diagnosed glioblastoma multiforme. *Autophagy* **2014**, *10*, 1359–1368. [CrossRef] [PubMed]
160. Mauthe, M.; Orhon, I.; Rocchi, C.; Zhou, X.; Luhr, M.; Hijlkema, K.; Coppes, R.P.; Engedal, N.; Mari, M.; Mauthe, M.; et al. Chloroquine inhibits autophagic flux by decreasing autophagosome-lysosome fusion. *Autophagy* **2018**, *14*, 1435–1455. [CrossRef] [PubMed]
161. Rebecca, V.; Amaravadi, R.K. Emerging strategies to effectively target autophagy in cancer. *Oncogene* **2016**, *35*, 1–11. [CrossRef] [PubMed]
162. Castilho, RM.; Squarize, CH.; Chodosh, LA.; Williams, B.O.; Gutkind, J. mTOR Mediates Wnt-Induced Epidermal Stem Cell Exhaustion and Aging. *Cell Stem Cell* **2009**, *5*, 279–289. [CrossRef] [PubMed]
163. Easley, C.A.; Ben-yehudah, A.; Redinger, C.J.; Oliver, S.L.; Varum, S.T.; Eisinger, V.M.; Carlisle, D.L.; Donovan, P.J.; Schatten, G.P. mTOR-Mediated Activation of p70 S6K Induces Differentiation of Pluripotent Human Embryonic Stem Cells. *Cell. Reprogram.* **2010**, *12*, 263–274. [CrossRef] [PubMed]

164. Josset, E.; Burckel, H.; Noël, G.; Bischoff, P. The mTOR Inhibitor RAD001 Potentiates Autophagic Cell Death Induced by Temozolomide in a Glioblastoma Cell Line. *Anticancer Res.* **2013**, *1852*, 1845–1851.
165. Alonso, M.M.; Jiang, H.; Yokoyama, T.; Xu, J.; Bekele, N.B.; Lang, F.F.; Kondo, S.; Gomez-manzano, C.; Fueyo, J. Delta-24-RGD in Combination With RAD001 Induces Enhanced Anti-glioma Effect via Autophagic Cell Death. *Mol. Ther.* **2008**, *16*, 487–493. [CrossRef] [PubMed]
166. Yokoyama, T.; Iwado, E.; Kondo, Y.; Aoki, H.; Hayashi, Y.; Georgescu, M.M.; Sawaya, R.; Hess, K.R.; Moi, O.; Moi, O.; et al. Autophagy-inducing agents augment the antitumor effect of telomerase-selective oncolytic adenovirus OBP-405 on glioblastoma cells. *Gene Ther.* **2008**, 1233–1239. [CrossRef] [PubMed]
167. Filippi-chiela, E.C.; Manssur, M.; Thom, M.P.; Lenz, G. Single-cell analysis challenges the connection between autophagy and senescence induced by DNA damage. *Autophagy* **2015**, *11*, 1099–1113. [CrossRef] [PubMed]
168. Gammoh, N.; Fraser, J.; Puente, C.; Syred, H.M.; Kang, H.; Ozawa, T.; Lam, D.; Acosta, J.C.; Finch, A.J.; Holland, E.; et al. Suppression of autophagy impedes glioblastoma development and induces senescence. *Autophagy* **2016**, *12*, 1431–1439. [CrossRef] [PubMed]
169. Harrington, L.S.; Findlay, G.M.; Gray, A.; Tolkacheva, T.; Wigfield, S.; Rebholz, H.; Barnett, J.; Leslie, N.R.; Cheng, S.; Shepherd, P.R.; et al. The TSC1-2 tumor suppressor controls insulin–PI3K signaling via regulation of IRS proteins. *J. Cell Biol.* **2004**, *166*, 213–223. [CrossRef] [PubMed]
170. Chiarini, F.; Evangelisti, C.; Mccubrey, J.A.; Martelli, A.M. Current treatment strategies for inhibiting mTOR in cancer. *Trends Pharmacol. Sci.* **2015**, *36*, 124–135. [CrossRef] [PubMed]
171. Jhanwar-uniyal, M.; Gillick, J.L.; Neil, J.; Tobias, M.; Thwing, Z.E.; Murali, R. Distinct signaling mechanisms of mTORC1 and mTORC2 in glioblastoma multiforme: A tale of two complexes. *Adv. Biol. Regul.* **2015**, *57*, 64–74. [CrossRef] [PubMed]
172. Roper, J.; Richardson, M.P.; Wang, W.V.; Richard, L.G.; Chen, W.; Erin, M.; Sinnamon, M.J.; Lee, L.; Chen, P.; Bronson, R.T.; et al. The Dual PI3K/mTOR Inhibitor NVP-BEZ235 Induces Tumor Regression in a Genetically Engineered Mouse Model of PIK3CA Wild-Type Colorectal Cancer. *PLoS ONE* **2011**, *6*, 1–10. [CrossRef] [PubMed]
173. Kanzawa, T.; Zhang, L.; Xiao, L.; Germano, I.M.; Kondo, Y.; Kondo, S. Arsenic trioxide induces autophagic cell death in malignant glioma cells by upregulation of mitochondrial cell death protein BNIP3. *Oncogene* **2005**, 980–991. [CrossRef] [PubMed]
174. Daido, S.; Kanzawa, T.; Yamamoto, A.; Takeuchi, H.; Kondo, Y.; Kondo, S. Pivotal Role of the Cell Death Factor BNIP3 in Pivotal Role of the Cell Death Factor BNIP3 in Ceramide-Induced Autophagic Cell Death in Malignant Glioma Cells. *Cancer Res.* **2004**, *1*, 4286–4293. [CrossRef] [PubMed]
175. Kim, E.H.; Sohn, S.; Kwon, H.J.; Kim, S.U.; Kim, M.; Lee, S.; Choi, K.S. Sodium Selenite Induces Superoxide-Mediated Sodium Selenite Induces Superoxide-Mediated Mitochondrial Damage and Subsequent Autophagic Cell Death in Malignant Glioma Cells. *Cancer Res.* **2007**, 6314–6325. [CrossRef] [PubMed]
176. Torres, S.; Lorente, M.; Rodriguez-Fomes, F.; Hernandez-Tiedra, S.; Salazar, M.; García-taboada, E.; Barcia, J.; Guzman, M.; Velasco, G. A Combined Preclinical Therapy of Cannabinoids and Temozolomide against Glioma. *Mol. Cancer Ther.* **2011**, *10*, 90–104. [CrossRef] [PubMed]
177. Hetschko, H.; Voss, V.; Senft, C.; Seifert, V.; Prehn, J.H.M.; Kögel, D. BH3 Mimetics Reactivate Autophagic Cell Death in Anoxia-Resistant Malignant Glioma Cells. *Neoplasia* **2008**, *10*, 873–885. [CrossRef] [PubMed]
178. Voss, V.; Senft, C.; Lang, V.; Ronellenfitsch, M.W.; Steinbach, J.P. The Pan-Bcl-2 Inhibitor (−)-Gossypol Triggers Autophagic Cell Death in Malignant Glioma. *Mol. Cancer Res.* **2010**, *8*, 1002–1017. [CrossRef] [PubMed]

cancers

MDPI

Review

Nitric Oxide Antagonism to Anti-Glioblastoma Photodynamic Therapy: Mitigation by Inhibitors of Nitric Oxide Generation

Jonathan M. Fahey and Albert W. Girotti *

Department of Biochemistry, Medical College of Wisconsin, Milwaukee, WI 53226, USA; jonfahey5@gmail.com
* Correspondence: agirotti@mcw.edu; Tel.: +1-414-955-8432; Fax: +1-424-955-6510

Received: 17 December 2018; Accepted: 9 February 2019; Published: 15 February 2019

Abstract: Many studies have shown that low flux nitric oxide (NO) produced by inducible NO synthase (iNOS/NOS2) in various tumors, including glioblastomas, can promote angiogenesis, cell proliferation, and migration/invasion. Minimally invasive, site-specific photodynamic therapy (PDT) is a highly promising anti-glioblastoma modality. Recent research in the authors' laboratory has revealed that iNOS-derived NO in glioblastoma cells elicits resistance to 5-aminolevulinic acid (ALA)-based PDT, and moreover endows PDT-surviving cells with greater proliferation and migration/invasion aggressiveness. In this contribution, we discuss iNOS/NO antagonism to glioblastoma PDT and how this can be overcome by judicious use of pharmacologic inhibitors of iNOS activity or transcription.

Keywords: nitric oxide (NO); inducible nitric oxide synthase (iNOS); photodynamic therapy (PDT); glioblastoma PDT; NO-mediated PDT resistance

1. Introduction

Malignant gliomas are the most prevalent primary brain tumors, and among these malignancies, glioblastoma multiforme (GBM), a grade-4 astrocytoma, is the most aggressive and resistant to a variety of therapeutic interventions [1–3]. Average patient survival even after the most advanced surgical treatments, or surgery combined with radiation or chemotherapy, remains dismal at only 12-18 months after initial diagnosis [3]. Cisplatin (CDDP), which intercalates into and cross-links DNA, has been widely used as a chemotherapeutic for GBM and other brain tumors [4,5]. Temozolimide (TMZ), a highly effective imidazotetrazine that was introduced more recently, alkylates DNA at guanine bases after hydrolysis, leading to single and double strand breaks [6,7]. Unfortunately, many tumors exhibit an inherent or acquired resistance to these chemotherapeutic agents, often necessitating doses that become cytotoxic to normal brain tissue [3]. Similar responses to radiotherapy may occur. Photodynamic therapy (PDT), which involves non-ionizing radiation, is less susceptible to pre-existing resistance and has emerged as one of the most attractive alternatives for treating brain malignancies [8–11]. An added advantage of PDT is that synergistic effects with chemotherapy (e.g., low-level cisplatin) are often possible, some of which are based on different subcellular sites of action. Whereas cisplatin and TMZ damage DNA in the nucleus, PDT typically causes cytoplasmic (mitochondrial, lysosomal, or ER) damage [12].

PDT was introduced about 45 years ago as a novel means of selectively eradicating a variety of solid tumors [13–15], many of which are refractory to conventional radiotherapy or chemotherapy. PDT is a minimally invasive modality, the classic version involving a pre-existing photosensitizing agent (PS), PS-exciting light in the visible-to-near infrared range, and molecular oxygen. All three of these factors must be engaged concurrently to produce cytotoxic reactive oxygen species (ROS). The most prominent of these ROS is singlet molecular oxygen ($^1O_2$), which can damage proteins, unsaturated lipids, and nucleic acids in target cancer cells [16]. A major advantage of PDT over chemotherapy or radiotherapy is that cytotoxic photodamage is typically limited to the tumor site itself and does not occur until PS, exciting light, and O_2 are all engaged. Thus, light alone or PS alone is usually ineffective on a tumor, nor does it have any significant effect on normal tissue. Precise light delivery to a tumor via fiber optic transmitters provides an additional element of site-specificity during PDT [13–15]. Photofrin®, an oligomeric hematoporphyrin derivative, was the first administrable photosensitizer to receive FDA approval for PDT about 20 years ago and is now used for a variety of solid tumors, including brain tumors [14,15,17].

In more recently developed 5-aminolevulinic acid (ALA)-based PDT, ALA itself or an ALA ester is taken up by cancer cells and metabolized to protoporphyin IX (PpIX), the active PS, via the heme biosynthetic pathway [18,19]. In order to provide heme for rapid growth and progression, this pathway is more active in tumor cells than normal counterparts, although iron typically becomes limiting, allowing relatively large levels of PpIX to accumulate [19]. In cancer cells, PpIX accumulates initially in mitochondria, making these organelles primary targets of ALA-PDT damage that can kill cells via intrinsic apoptosis [18,19]. In addition to sensitizing cytotoxic reactions, ALA-induced PpIX produces a striking red fluorescence under relatively low-intensity exciting light. Many oncologists have exploited this aspect for image-guided resection (IGR), i.e., to clearly define the extremities of a tumor prior to its surgical removal, thereby greatly improving procedural accuracy [19]. Thus, ALA-induced PpIX, which is largely localized in tumor cells, has the advantage over most other anti-tumor agents of serving as a surgical guide on the one hand and a cytotoxic PDT sensitizer on the other.

2. Nitric Oxide and Its Multifaceted Roles in Cancer

Nitric oxide (NO) is a short-lived bioactive free radical (<2 s in H_2O) that diffuses freely on its own and, like O_2, tends to partition into hydrophobic regions of cells, e.g., cell membranes [20]. NO is generated naturally by three different enzymes of the nitric oxide synthase (NOS) family: nNOS/NOS1 (neuronal), iNOS/NOS2 (inducible), and eNOS/NOS3 (endothelial). nNOS and eNOS usually function at relatively low constitutive levels, require Ca^{2+} for activation, and produce NO at nanomolar levels for short intervals. In contrast, iNOS can be induced by various stressors, does not require Ca^{2+} for activation, and can generate NO in the micromolar range for much longer periods [21,22]. All three NOS enzymes catalyze the conversion of L-arginine to citrulline and NO at the expense of NADPH and O_2 [22]. It well established that eNOS-derived NO at low steady-state levels (1–10 nM) stimulates cyclic-GMP formation in vascular smooth muscle cells, leading to blood vessel relaxation. On the other hand, in activated macrophages during an immune response, iNOS-derived NO at much higher levels (1 μM or greater) is cytotoxic and potentially oncogenic [23–25]. Such effects typically occur after NO reacts with superoxide radical (O_2^-) to give peroxynitrite ($ONOO^-$), a strong oxidant that can oxidatively damage DNA and membrane lipids. At intermediate levels (e.g., 50–300 nM), iNOS-derived NO can play a key role in cancer persistence and progression by activating oncogenic signaling pathways or inhibiting suppression pathways [26]. Thus, whether NO exhibits pro- or anti-tumor properties (Figure 1) depends largely on its steady-state levels, which are usually quite low (10–300 nM) in proliferating transformed cells. There is an increasing awareness that besides nNOS, most gliomas express iNOS and that iNOS-derived NO plays a major role in tumor cell survival, persistence, and progression [27–29]. For many malignancies, including gliomas, a direct correlation has been found between relatively high iNOS expression in tumor tissue and poor prognosis [29]. Proteins such as Survivin (which inhibits apoptosis [30]) and S100A4 (which stimulates invasion/metastasis [31]) can also serve as prognostic markers. It is important to note that in many cases, elevated expression of these markers might be a secondary effect of upregulated iNOS/NO, as demonstrated in a PDT challenge, for example (see below) [32].

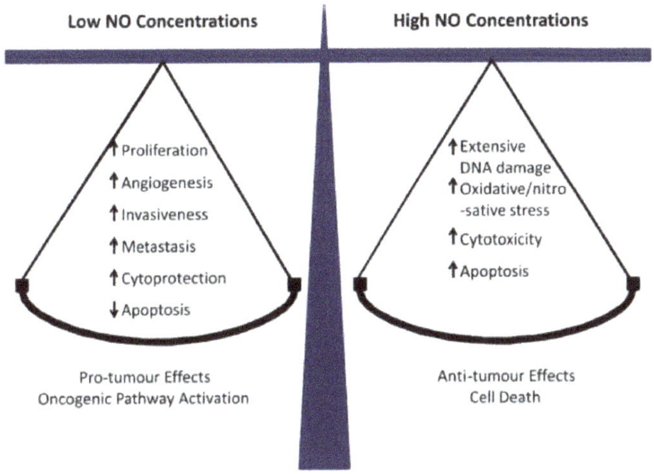

Figure 1. Concentration-dependent effects of NO in cancer: pro-tumor versus anti-tumor. Reproduced from Reference [24], with permission.

As already pointed out, sub-micromolar nitric oxide (NO) in tumors plays a key role in tumor cell survival, persistence, and progression [23–29]. Such NO can also signal for resistance to radiotherapy, chemotherapy, or PDT [33–37]. The NO-mediated PDT resistance was first observed by Henderson et al. [36] and Korbelik et al. [37], using various syngeneic mouse tumors (breast, cutaneous, RIF carcinomas) sensitized with Photofrin®. It was found that tumor regression could be greatly improved by administering an iNOS activity inhibitor (e.g., L-NAME) immediately after PDT [37]. The extent of improvement correlated with constitutive NO production, tumors with a high output responding much better than those with a low output. The resistance was attributed mainly to NO-mediated relaxation of tumor blood vessels acting in opposition to PDT-induced vasoconstriction, i.e., a vascular effect of NO as opposed to a pro-survival effect on tumor cells per se [37]. A similar mechanism was deduced in subsequent studies by Reeves et al. [38], using ALA-based PDT on mouse tumor models. However, until relatively recently, many other questions pertaining to NO's anti-PDT effects have remained unanswered, e.g., (i) the major cellular source of NO in any given tumor, e.g., cancer cells vs. stromal cells; (ii) the nitric oxide synthase (NOS) isoform that produces most of this NO, and (iii) the NO-mediated signaling events that lead to PDT resistance. Over the past eight years, these open questions have been addressed by the authors and colleagues using various cancer cell lines. Key findings from these studies will be discussed below, along with more recent findings dealing with NO's negative effects on glioblastoma PDT.

3. iNOS/NO-Mediated Antagonism to PDT

As indicated in Section 1, one of the advantages of PDT is that it can often overcome any resistance that a tumor may have to chemotherapy or radiotherapy. Like these other modalities, however, PDT may elicit a resistance response (an acquired resistance) in many cancer cell types. One type of resistance, which involves iNOS-generated NO, was discovered in the authors' laboratory. When subjected to an ALA-based photodynamic challenge in mitochondria, several human breast cancer lines (COH-BR1, MCF-7, MDA-MB-231) and prostate cancer lines (PC3, DU145) underwent intrinsic apoptosis that could be substantially enhanced by inhibitors of iNOS enzymatic activity (e.g., 1400W, GW274150) or by a selective NO scavenger (cPTIO) [39–41]. Seeing these enhancements implied that iNOS-derived NO was playing a key role in a hyper-resistance response. Moreover, this response was greatly attenuated when iNOS was depleted via shRNA-mediated knock down prior to ALA/light treatment [41]. More often than not, only iNOS (rather than other NOS isoforms) was upregulated by the challenge and this appeared to be more important in the hyper-resistance than any effect of the pre-existing enzyme [40–44]. As surviving cells continued to proliferate, a gradual decline in over-expressed iNOS was often observed, suggesting that elevated resistance was a transient phenomenon rather than selection for a relatively stable population of high iNOS expressing cells. Of added importance was our observation that breast and prostate cells surviving ALA/light treatment typically exhibited a more aggressive phenotype in terms of accelerated proliferation, migration, and invasion [43,44].

We recently extended the above in vitro findings to the in vivo level using immunodeficient female mice engrafted with breast MDA-MB-231 tumors [45]. After ALA administration, tumors were irradiated with red (633 nm) light, using an LED source. Treated animals exhibited a significant reduction in tumor growth compared with light-only controls over a two-day post-PDT period. However, 1400W in multiple post-PDT doses spaced one day apart reduced tumor growth much further, whereas it had no significant effect on control animals that were not ALA-treated before irradiation [45]. This suggested that pre-existing iNOS/NO had little, if any, effect on tumor resistance. Western blot analysis of tumor samples after PDT revealed a striking upregulation of iNOS and also a 1400W-inhibitable increase in the level of NO-derived nitrite [45]. This was the first reported evidence for iNOS upregulation by PDT in vivo using a human tumor model and for iNOS/NO-imposed resistance to tumor regression.

3.1. Post-PDT Upregulation of iNOS/NO in Glioma Cells: Increased Resistance to Photokilling

Like breast and prostate cancer cells [40–44], human glioblastoma U87MG cells (a.k.a. U87) that had been sensitized with ALA-induced PpIX underwent a progressive loss of MTT-assessed viability with increasing light fluence (Figure 2A). ALA alone or light alone without prior ALA treatment had no effect of viability. When sensitized cells were irradiated in the presence of 1400W or cPTIO, there was a striking increase in cytotoxicity with increasing light fluence, suggesting that endogenous iNOS/NO was signaling for resistance (Figure 2A). The extent of apoptosis after ALA/light treatment was also significantly greater when 1400W or L-NAME (a general NOS inhibitor) was present (Figure 2B), confirming iNOS/NO-elicited resistance [32]. As shown in Figure 2C(a), the level of iNOS protein in ALA/light (1 J/cm^2)-challenged U87 cells increased progressively during post-irradiation incubation, reaching ~3-times the control level after 5 h. Not-surprisingly, nNOS was also expressed in U87 cells; however, it remained at the same starting level after cells were photodynamically stressed (Figure 2C(b)). This suggests that unlike iNOS, nNOS made no significant contribution to cytoprotection in these cells. Another established glioblastoma line (U251) responded similarly to ALA/light, exhibiting 4–5-fold iNOS overexpression over 20 h of post-irradiation incubation (Figure 2D), and a dramatic iNOS/NO-mediated hyper-resistance [32]. Thus, stress-induced iNOS/NO appeared to play a crucial role in the photostress-enhanced resistance of both glioblastoma cell types, U87 and U251.

A fluorogenic probe for NO (DAF-2DA) was used to establish whether photostress-induced iNOS and associated hyper-resistance was in fact due to iNOS-derived NO. After DAF-2DA's cellular internalization and hydrolysis to DAF-2, the latter detects NO after its conversion to a nitrosating species such as N_2O_3 [46,47]. Within 4 h after an ALA/light challenge, U87 cells exhibited a strong NO-based fluorescence signal (>3-fold over an ALA-only background), which persisted for at least 20 h [32]. This signal was nearly abolished when 1400W or L-NAME was introduced immediately after irradiation, confirming that NO had been upregulated in photostressed U87 cells along with iNOS.

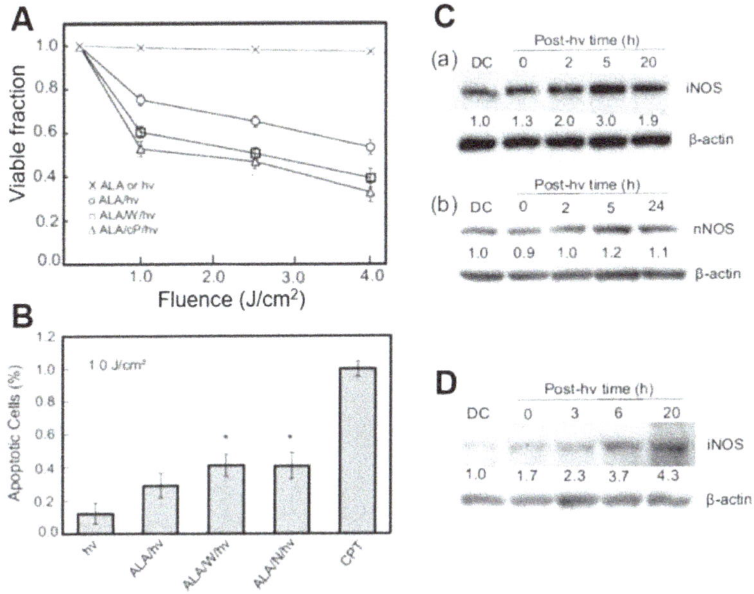

Figure 2. Upregulation of iNOS/NO in photodynamically-stressed glioblastoma cells confers resistance to photokilling. U87 or U251 cells at ~60% confluency were dark-incubated with 1 mM ALA for 30 min in serum-free medium, irradiated with broad-band visible light, then switched to serum-containing medium and dark-incubated for various periods. Where indicated, 25 μM 1400W (W) or cPTIO (cP) was introduced immediately after irradiation and kept at the same concentration throughout. (**A**) MTT-assessed U87 viability 20 h after ALA/hν exposure in absence vs. presence of W or cP; $n = 3$. (**B**) Annexin V-FITC-assessed U87 apoptosis 4 h after indicated treatments (CPT: camptothecin); $n = 3$, * $p < 0.05$ vs. ALA/hν. (**C**) Immunoblot of U87 iNOS (a) and nNOS (b) at indicated post-hν times. Numbers below NOS bands indicate intensity relative to actin and normalized to dark (ALA-alone) control (DC). (**D**) Immunoblot of iNOS in U251 cells at indicated post-hν times. Adapted from Reference [32].

3.2. Accelerated Proliferation of PDT-Surviving Glioma Cells: Role of iNOS/NO

Cancer cells often adapt to stressful conditions by acquiring a more aggressive proliferative and migratory phenotype [1–3]. This proved to be the case for photodynamically stressed glioblastoma cells and iNOS/NO played a key driving role. As shown in Figure 3A, U87 cells that survived an ALA/light challenge 24 h after it was delivered exhibited a striking 2-fold growth spurt over the next 24 h relative to non-irradiated controls. This spurt was strongly attenuated by 1400W and also by cPTIO, clearly indicating iNOS/NO involvement, particularly that upregulated by the photostress (Figure 2C). The growth of non-irradiated control cells was slowed somewhat by 1400W (Figure 3A), suggesting a constitutive stimulatory effect of pre-existing iNOS/NO in U87 cells. This confirms the findings of others using non-stressed glioblastoma cells [48,49]. However, observing a robust iNOS/NO-dependent growth stimulation in response to therapy-based oxidative stress, in this case PDT [32], had not been described previously.

3.3. Role of iNOS/NO in Accelerated Migration of PDT-Surviving Glioma Cells

One other manifestation of hyper-aggressiveness in photostressed glioblastoma cells is more rapid migration into a cell-depleted space. Two other manifestations of hyper-aggressiveness were also observed in photostressed glioblastoma cells: (i) accelerated migration into a cell-depleted space, and (ii) accelerated invasion through an extracellular matrix (ECM)-like membrane. A gap-closure or "wound-healing" assay is typically used to examine forward migration into a voided zone generated by a straight-line scratch on a culture dish [50]. In a recent study, we photographed ALA-treated U87 cells in a scratch zone before irradiation and at various times after irradiation up to 24 h, during which cells were kept in the incubator. 1400W or cPTIO was included in the medium of certain dishes to test for iNOS/NO involvement in any altered migration. As shown by the gap-closure data in Figure 3B, ALA/light-stressed cells migrated more rapidly than ALA-only control cells over a 24 h post-irradiation period. This response was substantially blunted by 1400W, signifying major iNOS/NO dependency. 1400W also slowed control cell migration, but to a much smaller extent than in photodynamically-stressed cells, demonstrating the greater importance of stress-upregulated iNOS over basal iNOS in stimulating migration.

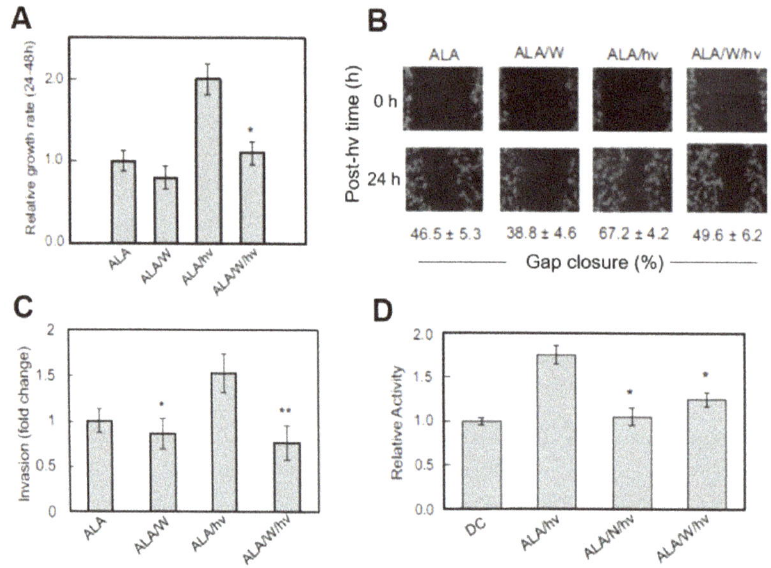

Figure 3. Enhanced aggressiveness of glioblastoma cells that survive a photodynamic challenge. U87 cells at ~40% confluence were sensitized with ALA-induced PpIX and irradiated (ALA/hv: light dose/fluence ~1 J/cm^2). Where indicated, 25 μM 1400W (W) or 1 mM L-NAME (N) was introduced immediately after irradiation and maintained as such thereafter. Dark controls (ALA or ALA/W) were run alongside. After 24 h of post-hv incubation, any detached/dead cells were washed off and aggressive properties of remaining live cells were determined. (**A**) MTT-assessed proliferation rate; means ± SEM, $n = 3$, * $p < 0.01$ vs. ALA/hv. (**B**) Gap-closure-assessed migration rate; means ± SEM, $n = 3$.(**C**) Trans well chamber-assessed invasion rate; means ± SEM, $n = 4$, * $p < 0.05$ vs. ALA; ** $p < 0.0001$ vs. ALA/hv. (**D**) Gel zymography-assessed MMP-9 activity; means ± SEM, $n = 3$, * $p < 0.01$ vs. ALA/hv. Adapted from Reference [32].

3.4. Role of iNOS/NO in Accelerated Invasion of PDT-Surviving Glioma Cells

A 96-place trans-well device was used to assess the invasiveness of U87 cells, i.e., ability to traverse a Matrigel-infused filter, moving from a serum-free upper well toward a serum-containing lower well [50]. Measurements commenced at 24 h after cell exposure to ALA alone (controls) or ALA plus irradiation. As shown in Figure 3C, 1400W inhibited the invasiveness of control cells by a small (barely significant) extent, suggesting iNOS/NO promotion of this basal activity. After irradiation, however, ALA-treated cells exhibited a striking 50% increase in invasion relative to control cells and 1400W abrogated this increase, demonstrating that iNOS/NO played a dominant role in the photostress-enhanced invasiveness.

Matrix metalloproteinases (MMPs) catalyze the degradation of ECM components such as collagen, laminin and fibronectin, and thus play a key role in cancer cell invasiveness and metastasis [51]. Type-9 MMP (MMP-9) is known to be actively involved in glioma cell migration/invasion [52] and this proved to be the case for U87 cells after an ALA/light challenge. Western blot analysis showed that there was no significant change in overall MMP-9 expression over a 24 h post-irradiation incubation period [32]. However, when in-gel zymography was used to monitor the activity status of externalized enzyme 24 h after irradiation, a dramatic 80% increase in activity was observed relative to a non-irradiated (ALA-only) control (Figure 3D). This activation was strongly depressed by L-NAME (>90%) and by 1400W (~70%), indicating that iNOS-derived NO played a prominent role in MMP-9 hyper-activation from its zymogen, pro-MMP-9 [32]. A fascinating sidelight to the MMP-9 response is that tissue inhibitor of metalloproteinase-1 (TIMP-1), which is known to be highly specific for MMP-9 [53], was progressively down-regulated in U87 cells after ALA/light treatment, as revealed by immunoblotting [32]. This response was strongly blunted by 1400W, pointing again to iNOS/NO dependency—in this case to prevent MMP-9 de-activation by TIMP-1. Of added interest is the fact that Survivin and S100A4, both known to play key roles in tumor cell growth, migration and invasion [30,31], were markedly upregulated in photostressed U87 cells, and in 1400W-inhibitable fashion [32]. Pro-metastatic S100A4 is particularly interesting because it was barely detectable in control cells, but reached a very high level 24 h after ALA/light, almost all of which was ablated by 1400W.

If occurring at the clinical PDT level, each of these negative responses to photostress-enhanced migration/invasion supported by MMP-9 activation and Survivin and S100A4 overexpression—would be problematic unless pharmacologically counteracted, e.g., by an inhibitor of iNOS expression or activity.

4. iNOS/NO-Induced Bystander Effects in PDT Models

Another dimension of post-PDT cancer cell aggressiveness was discovered recently: enhanced growth and migration of non-stressed bystander cells. In a tumor setting, bystander cells may lie in the general vicinity of cells targeted by a PDT or chemotherapeutic agent, but may not receive any (or enough) of it to elicit a therapeutic response, possibly due to insufficient vascular delivery. In the case of PDT, insufficient light delivery due to constraints of the light source and structural irregularities of the tumor can pose additional limitations. Most of the research relating to bystander effects has involved cancer-initiating vs. cancer-suppressing ionizing radiation (e.g., gamma rays, X-rays), which can produce effects ranging from DNA damage, mutations, and apoptosis to accelerated growth and migration of targeted cells [53]. Radiation-induced bystander effects can be transmitted via inter-cell gap junctions or via the medium, i.e., without physical contact between targeted and bystander cells [53]. Various signaling mediators have been proposed for ionizing radiation, including cytokines, H_2O_2, and NO, the latter receiving the greatest attention for bystander effects that occur independently of cell contact. NO produced specifically by radiation-targeted cells has been reported to elicit bystander effects ranging from DNA strand breaks and micronuclei formation to defective homologous recombination repair, leading to greater genetic instability, cell transformation, and accelerated proliferation [54]. Of special relevance here are early studies showing that X-ray-targeted glioblastoma cells overexpressed iNOS continuously over a 24 h post-radiation period, the resulting NO signaling for radioresistance in non-targeted bystander cells [55,56].

The possibility of bystander effects during non-ionizing PDT was first recognized about twenty years ago [57,58], but far less is known about this in mechanistic terms than its ionizing radiation counterpart. Hypothesizing that cells experiencing the greatest photodynamic (e.g., ALA/light) stress might send signals to non- or weakly-stressed bystanders, Bazak et al. [59,60] developed a novel approach for testing this. Two populations of sub-confluent cancer cells (initially prostate PC3 cells) were separated from one another on a large culture dish by two-to-four impermeable silicone-rimmed rings. The larger population (target cells, outside rings) was treated with ALA while the smaller population (bystander cells, inside rings) was not, after which the entire dish was irradiated. At some interval after irradiation, the rings were carefully removed, leaving a gap between the cell groups, and then both populations were analyzed for iNOS status, growth rate, and migration rate during dark incubation compared with non-irradiated (ALA-only) controls. Like surviving targeted cells, bystanders exhibited significant iNOS upregulation and accelerated growth and migration, each response being attributed to initial NO diffusing from the target compartment, based on the inhibitory effects of 1400W, cPTIO, or target cell iNOS knockdown [59]. Incubation of bystander cells with conditioned medium from ALA/light-targeted cells failed to induce the above responses, ruling out any involvement of relatively long-lived effectors such as cytokines, lipid peroxides, or NO-derived nitrite/nitrate. NO itself would not have survived in conditioned media because of its short lifetime (<2 s) [20]. Thus, it had to be continuously generated by targeted cells to elicit bystander effects. Screening for possible effector proteins associated with enhanced PC3 bystander aggressiveness revealed a strong transient activation of the Akt and ERK1/2 kinases and an induction of cyclooxygenase-2 (COX-2), each response being cPTIO-inhibitable [59]. Although NO-mediated bystander effects for glioblastoma cells have not yet been assessed in a PDT format, they would most likely occur, based on evidence obtained with ionizing radiation [55,56]. PDT is becoming an increasingly attractive treatment option for glioblastoma and other brain malignancies [8–11]. Therefore, the negative implications of NO-mediated bystander effects (e.g., promotion of tumor growth and metastatic expansion) deserve serious attention aimed at mitigating these effects and improving treatment outcomes. One can view NO-mediated bystander effects as a type of relay or "feed-forward" phenomenon in which NO produced by targeted cells induces iNOS/NO in bystanders, and that this is propagated through the population. Thus, cells that escape being targeted in any given tumor are not necessarily unaffected by NO diffusing from neighboring targeted cells, a possibility that has not been well recognized up to now.

5. Mechanism of Glioblastoma iNOS Induction by Photostress: Preventative Approaches

In examining the underlying mechanism of iNOS upregulation in glioblastoma U87 cells after ALA/light exposure, Fahey et al. [61] found that transcription factor NF-κB played a crucial role. Accordingly, the active p65/RelA subunit of NF-κB translocated from cytosol to nucleus after photostress, and Bay11, an inhibitor of IKK which phosphorylates and releases restraining subunit IκB on NF-κB, prevented p65 translocation. Importantly, Bay11 nearly abolished all iNOS expression (basal as well as photostress-induced), clearly linking this with p65-mediated iNOS gene transcription [61]. Similar findings were reported earlier for a breast cancer cell line [42]. Based on the non-glioblastoma work of others [62,63], we postulated that acetylation (ac) of specific lysine (K) residue(s) on p65 played a key driving role in iNOS transcription. Such acK modifications can be recognized by one or more proteins of the bromodomain and extra-terminal domain (BET) family, also known as epigenetic "readers" [64,65]. BET proteins are known to interact with histones and transcription factors at acK sites, thereby co-activating transcriptional processes [64–66]. In testing our hypothesis using pull-down assays, we found that BET-4 protein (Brd4), but not its BET-2 (Brd2) analogue, interacted strongly with NF-κB p65 in U87 cells that had been ALA/light-stressed [61]. Of added importance were our findings that the levels of Brd4 and acK310 on p65 each increased ~3-fold in photostressed cells, whereas the Brd2 and p65 levels were unaffected. Thus, stress-elevation of Brd4 and p65-ack310 would have promoted interaction of these proteins for augmented translational activity at the iNOS promoter.

Several BET protein inhibitors have recently emerged as highly potent and specific pharmacologic suppressors of cancer cell proliferation and invasive/metastatic expansion [67,68]. These inhibitors function by binding to acK recognition domains on BET proteins and preventing their interaction with histones or transcription factors. Several of these agents have advanced to clinical trials for various malignancies, including lymphomas, myelomas and triple negative breast cancers [67]. The thienodiazepine JQ1 is a prototypical BET inhibitor that has repressed cancer cell progression in many in vitro and in vivo models, including glioblastoma models [69]. It was recently reported that JQ1 blocks the TNF-α-stimulated expression of several pro-survival genes in lung cancer cells by binding and inactivating Brd4 [63]. However, iNOS was not included among the gene products described, prompting us to investigate this in our PDT studies on glioblastoma cells. At a very low concentration (0.3 μM), JQ1 by itself had little effect on U87 cell viability, but it enhanced ALA/light-induced apoptosis in synergistic fashion [61]. At the same time, JQ1 nearly abolished Brd4-p65 interaction after ALA/light treatment along with total iNOS expression - basal as well as stress-upregulated. However, it had no effect on the extent of Brd4 or p65-acK310 elevation, implying that these responses were not under JQ1-iinhibitable transcriptional control. A striking loss of photostress-enhanced U87 cell aggressiveness was also observed when JQ1 was present. For example, JQ1 slowed surviving cell proliferation rate by >80% compared with ~50% for 1400W, but more impressively, it did so at only ~1% the 1400W concentration in bulk cell suspension [61]. The BET inhibitor also suppressed the greater invasiveness of glioblastoma cells that survived an ALA/light challenge. Results for U87 and U251 cells were similar; the latter is represented in Figure 4. Active JQ1 reduced invasion rate of vehicle control (DMSO) by ~10% [61], whereas the inactive enantiomer [JQ1(-)] had no effect. PDT-surviving cells invaded ~40% faster than controls (Figure 4) and JQ1 (but not JQ1(-)) not only abolished this acceleration but reduced the rate to ~35% that of the control. Bay 11 at ~20-fold higher concentration than JQ1 also had a large inhibitory effect, but specificity could be an issue because Bay11 can inhibit other pro-survival effectors besides Iκκ [70]. Like JQ1, 1400W also eliminated stress-enhanced invasion but left residual activity closer to the control value (Figure 4). Consequently, the effects of an iNOS transcriptional inhibitor (JQ1) were clearly much more impressive than those of an iNOS activity inhibitor (1400W), the former operating at ~100-times lower concentration than the latter. Of added interest is the observation that photostress-upregulation of other NF-κB-regulated proteins besides iNOS (e.g., Survivin, Bcl-xL, MMP-9) was also blocked by JQ1, but whether this occurred directly or secondarily in some cases as a result of iNOS/NO suppression [61] is not yet clear, at least for glioblastoma cells. A scheme depicting Brd4-p65$_{ac}$-dependent transcriptional upregulation of iNOS/NO by ALA-PpIX/light-induced stress and its suppression by JQ1 is shown in Figure 5.

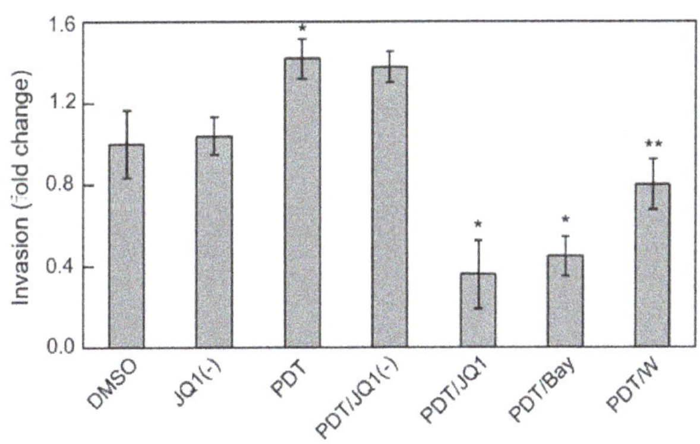

Figure 4. JQ1 abrogation of photostress-enhanced glioblastoma cell invasiveness. U251 cells at ~60% confluency were pre-incubated with 1 mM ALA. After PDT irradiation (1 J/cm^2), these cells, along with dark controls, were harvested and assessed for invasiveness in the absence vs. presence of 0.3 μM JQ1, 0.3 μM JQ1(−), 5 μM Bay11, or 25 μM 1400W. DMSO served as a vehicle control for JQ1 and JQ1(−). Cells crossing Matrigel-infused filters after an incubation period of 24 h were pelleted and quantified by CCK-8 assay. Plotted invasion values are means ± SEM (n = 3); * $p < 0.001$ vs. PDT; ** $p < 0.01$ vs. PDT. Reproduced from Reference [61], with permission.

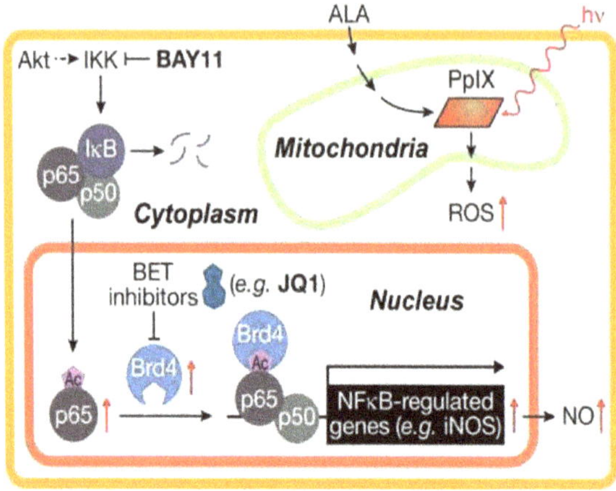

Figure 5. Scheme showing (i) photosensitized generation of ROS ($^1O_2$) by ALA-induced PpIX, (ii) activation of NF-κB, (iii) nuclear translocation and acetylation of p65, and (iv) JQ1-inhibitable transcription of iNOS as mediated by co-activator Brd4. Reproduced from Reference [61], with permission.

More recent studies have focused on events upstream of NF-κB activation and iNOS/NO upregulation in photodynamically-targeted glioblastoma cells. Some key findings were that (i) $^1O_2$ generated by photoactivation of ALA-induced PpIX in mitochondria was indirectly involved in activation of the cytosolic pro-survival PI3K/Akt kinase system in U87 cells; (ii) Tumor suppressor PTEN was rapidly inactivated by intramolecular disulfide bond formation during ALA/light treatment, thereby fostering PI3K activation; (iii) transacetylase p300 was phosphorylation-activated by Akt, leading to greater acK310 formation on NF-κB/p65; (iv) the latter was facilitated by greater p300-p65 interaction as demonstrated by pull-down assay; (v) the deacetylase Sirt1 was down-regulated in ALA/light-treated U87 cells, consistent with elevated p65-acK310 levels; and (vi) chemical inhibitors of upstream PI3K and p300 activity also inhibited iNOS upregulation under photodynamic stress [71].

6. Anti-Cancer Potential of NO at Relatively High Levels

Recent studies by Rapozzi et al. [72,73], using melanoma cells photosensitized with pheophorbide-a, have supported our findings regarding the cytoprotective effects of iNOS/NO on ALA/light-stressed glioma cells. Thus, low-level NO from low PDT light doses was found to be cytoprotective through anti-apoptotic activation of the NF-κB/Snail/RKIP pathway [72]. In contrast, higher NO levels from relatively high light doses proved to be cytotoxic due to pro-apoptotic activation of this pathway [72,73]. Whether a significant elevation of iNOS/NO via greater PDT pressure might produce a similar NO-enhanced proapoptotic response in glioblastoma cells has not yet been determined.

The anti-tumor potential of exogenous high-level NO, either alone or in combination with other treatments, is being explored by various investigators. NO donors such as glyceryl trinitrate, sodium nitroprusside, S-nitrosoglutathione (GSNO), diazeniumdiolates (NONOates), and Lopinavir-NO have been tested in this way with varying degrees of success, a key question being whether undesirable (off-target) effects occur [74–76]. The latter question has been addressed by the development of donors such as JS-K, which releases NO upon activation by glutathione S-transferases [77]. These enzymes are expressed at relatively high levels in many malignant cells, thus increasing the specificity of NO delivery. Cytotoxic NO donors have not yet been tested on gliomas. However, based on our own studies [32,39–45], some concerns arise, e.g., that any long-lasting release of NO in relatively low fluxes might actually enhance tumor aggressiveness rather than suppress it. Such concerns could be mitigated with the development of rapid NO release photosensitizing agents [78,79]. For example, the nitrosyl phthalocyanine ruthenium complex (Ru-NO) generates both $^1O_2$ and NO upon photoexcitation [78]. This complex was found to be more cytotoxic than a $^1O_2$-only control, suggesting that rapid photorelease of NO in high fluxes with little or no persistent low-flux release could provide major advantages for PDT [78,79]. Production of high dose NO by light-activated donors has also been shown to promote the chemotherapeutic effects of doxorubicin by inhibiting its export via ATP binding cassette (ABC) transporters [80]. A similar approach might benefit ALA-PDT for brain tumors, given the recent evidence that the ABCG2 transporter promotes efflux of ALA-induced PpIX from brain cancer cells [81].

7. Summary and Conclusions

Like two other gasotransmitters, carbon monoxide (CO) and hydrogen sulfide (H_2S), NO is known to have normo-physiologic as well as patho-physiologic roles, depending in large part on its generation rate and local concentration [82]. Many malignant tumors, including highly aggressive and lethal glioblastomas, exploit low flux NO to promote survival, stimulate proliferation and migration/invasion, and to resist eradication by PDT and other therapeutic interventions. PDT is now considered one of the most promising of the new anti-glioma therapies [8–11]. ALA-based PDT is attracting particular attention because tumor cells are hyperactive in heme biosynthesis and generate higher levels of ALA-derived PpIX than normal cells, thereby increasing tumor vulnerability to photodynamic action. Of added importance, PpIX fluorescence at low light intensities can be exploited for image-guided resection, i.e., to better define tumor boundaries during surgery [83]. In this review, we have discussed various aspects of iNOS/NO antagonism to ALA-PDT ranging from tumor cell resistance to photokilling to accelerated proliferation, migration and invasion of cells that withstand the challenge. Although these negative effects can be observed in directly targeted cells, they may also develop in non-targeted bystander cells as NO from the former diffuses to the latter and iNOS/NO is upregulated there as well [59]. Although we have emphasized glioblastoma PDT here, various human breast and prostate cancers respond similarly to this type of challenge, employing endogenous iNOS as the major source of signaling NO [84]. Based on substantial evidence, this NO typically derives from stress-upregulated iNOS rather than the pre-existing or basal enzyme. Other oxidative therapeutics may also give rise to iNOS/NO-dependent hyper-resistance and/or hyper-aggressive phenotypes, one example being cis-platin-treated prostate PC3 cells, which also exhibit bystander effects (Fahey and Girotti, unpublished observations).

How endogenous NO exerts anti-PDT effects is still not clear in chemical mechanistic terms. S-nitrosation of select cysteine residues on key effector proteins might initiate resistance signaling [23–26,85], but this modification can be transient and difficult to define in PDT systems such as we describe. Thus, much remains to be learned in terms of underlying chemical biology. Another key issue relates to how iNOS/NO's anti-therapeutic effects might be overcome by pharmacologic intervention. As discussed in the context of glioblastoma PDT, the BET inhibitor JQ1 suppressed iNOS transcription (Figure 5) and reduced acquired hyper-aggressiveness much more powerfully than an inhibitor of iNOS activity. Thus, JQ1, which has tested positively against glioblastoma on its own [61], might significantly improve therapeutic outcomes for this malignancy when combined with PDT or possibly some other modality that stimulates iNOS/NO. Transcriptional upregulation of pro-survival/expansion iNOS under therapeutic stress like PDT may occur more often than currently recognized, thus emphasizing the need for powerful inhibitors like JQ1 as therapeutic adjuvants.

Funding: The authors' research was supported by the following grants to AWG: USPHS Grant CA70823 from the National Cancer Institute, Grant No. 5520347 from the Advancing a Healthier Wisconsin Research and Education Foundation, and BSC Grant No. 3308239/FP12605 from the MCW Cancer Center.

Acknowledgments: Reshma Bhowmick and Jerzy Bazak are thanked for their valuable contributions to much of the work described. Also appreciated was the helpful advice and suggestions of Witold Korytowski, Neil Hogg, and Mladen Korbelik.

Conflicts of Interest: The authors have no conflicts of interest to declare.

References

1. Furnari, F.B.; Fenton, T.; Bachoo, R.M.; Mukasa, A.; Stommel, J.M.; Stegh, A.; Hahn, W.C.; Ligon, K.L.; Louis, D.N.; Brennan, C.; et al. Malignant astrocytic glioma: Genetics, biology, and paths to treatment. *Genes Dev.* **2007**, *21*, 2683–2710. [CrossRef] [PubMed]
2. Wen, P.Y.; Kesari, S. Malignant gliomas in adults. *N. Engl. J. Med.* **2008**, *359*, 492–507. [CrossRef] [PubMed]
3. Miranda, A.; Blanco-Prieto, M.; Sousa, J.; Pais, A.; Vitorino, C. Breaching barriers in glioblastoma. Part I: Molecular pathways and novel treatment approaches. *Int. J. Pharm.* **2017**, *531*, 373–388. [CrossRef] [PubMed]
4. Carrillo, J.A.; Munoz, C.A. Alternative chemotherapeutic agents: Nitrosoureas, cisplatin, irinotecan. *Neurosurg. Clin. N. Am.* **2012**, *23*, 297–306. [CrossRef] [PubMed]
5. Stewart, D.J.; Molep, J.M.; Eapen, L.; Montpetit, V.A.J.; Goel, R.; Wong, P.T.T.; Popovic, P.; Taylor, K.D.; Raaphorst, G.P. Cisplatin and radiation in the treatment of tumors of the central nervous system: Pharmacological considerations and results of early studies. *Int. J. Radiat. Oncol. Biol. Phys.* **1993**, *28*, 531–542. [CrossRef]
6. Friedman, H.S.; Kerby, T.; Calvert, H. Temozolomide and treatment of malignant glioma. *Clin. Cancer Res.* **2000**, *6*, 2585–2597. [PubMed]
7. Yang, L.J.; Zhou, C.F.; Lin, Z.X. Temozolomide and radiotherapy for newly diagnosed glioblastoma multiforme: A systematic review. *Cancer Investig.* **2014**, *32*, 31–36. [CrossRef]
8. Whelan, H.T. High-grade glioma/glioblastoma multiforme: Is there a role for photodynamic therapy? *J. Natl. Compr. Cancer Netw.* **2012**, *10*, S31–S34. [CrossRef]
9. Quirk, B.J.; Brandal, G.; Donlon, S.; Vera, J.C.; Mang, T.S.; Foy, A.B.; Lew, S.M.; Girotti, A.W.; Jogal, S.; LaViolette, P.S.; et al. Photodynamic therapy (PDT) for malignant brain tumors: Where do we stand? *Photodiagn. Photodyn. Ther.* **2012**, *12*, 530–544. [CrossRef]
10. Bechet, D.; Mordon, S.R.; Guillemin, F.; Barberi-Heyob, M.A. Photodynamic therapy of malignant brain tumours: A complementary approach to conventional therapies. *Cancer Treat. Rev.* **2014**, *40*, 229–241. [CrossRef]
11. Olzowy, B.; Hundt, C.S.; Stocker, S.; Bise, K.; Reulen, H.J.; Stummer, W. Photoirradiation therapy of experimental malignant glioma with 5-aminolevulinic acid. *J. Neurosurg.* **2002**, *97*, 970–976. [CrossRef] [PubMed]
12. Kessel, D.; Oleinick, N.L. Cell Death Pathways Associated with Photodynamic Therapy: An Update. Cell Death Pathways Associated with Photodynamic Therapy: An Update. *Photochem. Photobiol.* **2018**, *94*, 213–218. [CrossRef] [PubMed]
13. Dougherty, T.J.; Gomer, C.J.; Henderson, B.W.; Jori, G.; Kessel, D.; Korbelik, M.; Moan, J.; Peng, J. Photodynamic therapy. *J. Natl. Cancer Inst.* **1998**, *90*, 889–905. [CrossRef] [PubMed]
14. Agostinis, P.; Berg, K.; Cengel, K.A.; Foster, T.H.; Girotti, A.W.; Gollnick, S.O.; Hahn, S.M.; Hamblin, M.R.; Juzeniene, A.; Kessel, M.; et al. Photodynamic therapy of cancer: An update. *CA Cancer J. Clin.* **2011**, *61*, 250–281. [CrossRef] [PubMed]
15. Benov, L. Photodynamic therapy: Current status and future directions. *Med. Princ. Pract.* **2015**, *24*, 14–28. [CrossRef] [PubMed]
16. Foote, C.S. Photosensitized oxidation and singlet oxygen: Consequences in biological systems. In *Free Radicals in Biology*; Pryor, W.H., Ed.; Academic Press: New York, NY, USA, 1976; Volume II, pp. 85–133.
17. Tetard, M.C.; Bermandel, M.; Mordon, S.; Lejeune, J.P.; Reyns, N. Experimental use of photodynamic therapy in high grade gliomas: A review focused on 5-aminolevulinic acid. *Photodiagn. Photodyn. Ther.* **2014**, *11*, 319–330. [CrossRef] [PubMed]
18. Peng, Q.; Berg, K.; Moan, J.; Kongshaug, M.; Nesland, J.M. 5-Aminolevulinic acid-based photodynamic therapy: Principles and experimental research. *Photochem. Photobiol.* **1997**, *65*, 235–251. [CrossRef]
19. Yang, X.; Palasuberniam, P.; Kraus, D.; Chen, B. Aminolevulinic acid-based tumor detection and therapy: Molecular mechanisms and strategies for enhancement. *Int. J. Mol. Sci.* **2015**, *16*, 25865–25880. [CrossRef]
20. Thomas, D.D.; Liu, X.; Kantrow, S.P.; Lancaster, J.R., Jr. The biological lifetime of nitric oxide: Implications for the perivascular dynamics of NO and O_2. *Proc. Natl. Acad. Sci. USA* **2001**, *98*, 355–360. [CrossRef]

21. Geller, D.A.; Billiar, R.R. Molecular biology of nitric oxide synthases. *Cancer Metastasis Rev.* **1998**, *17*, 7–23. [CrossRef]
22. Alderton, W.K.; Cooper, C.E.; Knowles, R.G. Nitric oxide synthases: Structure, function and inhibition. *Biochem. J.* **2001**, *357*, 593–615. [CrossRef] [PubMed]
23. Ridnour, L.A.; Thomas, D.D.; Donzelli, S.; Espey, M.G.; Roberts, D.D.; Wink, D.A.; Isenberg, J.S. The biphasic nature of nitric oxide responses in tumor biology. *Antioxid. Redox Signal.* **2006**, *8*, 1329–1337. [CrossRef] [PubMed]
24. Burke, A.J.; Sullivan, F.J.; Giles, F.J.; Glynn, S.A. The yin and yang of nitric oxide in cancer progression. *Carcinogenesis* **2013**, *34*, 503–512. [CrossRef] [PubMed]
25. Vannini, F.; Kashfi, K.; Nath, N. The dual role of iNOS in cancer. *Redox Biol.* **2015**, *6*, 334–343. [CrossRef] [PubMed]
26. Switzer, C.H.; Glynn, S.A.; Ridnour, L.A.; Cheng, R.Y.; Vitek, M.P.; Ambs, S.; Wink, D.A. Nitric oxide and protein phosphatase 2A provide novel therapeutic opportunities in ER-negative breast cancer. *Trends Pharmacol. Sci.* **2011**, *32*, 644–651. [CrossRef] [PubMed]
27. Eyler, C.E.; Wu, Q.; Yan, K.; MacSwords, J.M.; Chandler-Militello, D.; Misuraca, K.L.; Lathia, J.D.; Forrester, M.T.; Lee, J.; Stamler, J.S.; et al. Glioma stem cell proliferation and tumor growth are promoted by nitric oxide synthase-2. *Cell* **2011**, *146*, 53–66. [CrossRef] [PubMed]
28. Jahani, A.; Bonni, A. iNOS: A potential therapeutic target for malignant glioma. *Curr. Mol. Med.* **2013**, *13*, 1241–1249. [CrossRef]
29. Tran, A.N.; Boyd, N.H.; Walker, K.; Hjelmeland, A.B. NOS expression and NO function in glioma and implications for patient therapies. *Antiox. Redox Signal.* **2017**, *26*, 986–999. [CrossRef]
30. Altieri, D.C. Survivin, cancer networks, and pathway-directed drug discovery. *Nat. Rev. Cancer* **2008**, *8*, 61–70. [CrossRef]
31. Boyle, K.; Maelandsmo, G.M. S100A4 and metastasis: A small actor playing many roles. *Am. J. Pathol.* **2010**, *176*, 528–535.
32. Fahey, J.M.; Emmer, J.V.; Korytowski, W.; Hogg, N.; Girotti, A.W. Antagonistic effects of endogenous nitric oxide in a glioblastoma photodynamic therapy model. *Photochem. Photobiol.* **2016**, *92*, 842–853. [CrossRef] [PubMed]
33. Crowell, J.A.; Steele, V.E.; Sigman, C.C.; Fay, J.R. Is inducible nitric oxide synthase a target for chemoprevention? *Mol. Cancer Ther.* **2003**, *2*, 815–823. [PubMed]
34. Matsunaga, T.; Yamaji, Y.; Tomokuni, T.; Morita, H.; Morikawa, Y.; Suzuki, A.; Yonezawa, A.; Endo, S.; Ikari, A.; Iguchi, K.; et al. Nitric oxide confers cisplatin resistance in human lung cancer cells through upregulation of aldo-keto reductase 1B10 and proteasome. *Free Radic. Res.* **2014**, *48*, 1371–1385. [CrossRef] [PubMed]
35. Kim, R.K.; Suh, Y.; Cui, Y.H.; Hwang, E.; Lim, E.J.; Yoo, K.C.; Lee, G.H.; Yi, J.M.; Kang, S.G.; Lee, S.J. Fractionated radiation-induced nitric oxide promotes expansion of glioma stem like cells. *Cancer Sci.* **2013**, *104*, 1172–1177. [CrossRef] [PubMed]
36. Henderson, B.W.; Sitnik-Busch, T.M.; Vaughan, L.A. Potentiation of photodynamic therapy antitumor activity in mice by nitric oxide synthase inhibition is fluence rate dependent. *Photochem. Photobiol.* **1999**, *70*, 64–71. [CrossRef]
37. Korbelik, M.; Parkins, C.S.; Shibuya, H.; Cecic, I.; Stratford, M.R.; Chaplin, D.J. Nitric oxide production by tumour tissue: Impact on the response to photodynamic therapy. *Br. J. Cancer* **2000**, *82*, 1835–1843. [CrossRef]
38. Reeves, K.J.; Reed, M.W.; Brown, N.J. The role of nitric oxide in the treatment of tumours with aminolaevulinic acid-induced photodynamic therapy. *J. Photochem. Photobiol. B* **2010**, *101*, 224–232. [CrossRef] [PubMed]
39. Bhowmick, R.; Girotti, A.W. Signaling events in apoptotic photokilling of 5-aminolevulinic acid-treated tumor cells: Inhibitory effects of nitric oxide. *Free Radic. Biol. Med.* **2009**, *47*, 731–740. [CrossRef] [PubMed]
40. Bhowmick, R.; Girotti, A.W. Cytoprotective induction of nitric oxide synthase in a cellular model of 5-aminolevulinic acid-based photodynamic therapy. *Free Radic. Biol. Med.* **2010**, *48*, 1296–1301. [CrossRef]
41. Bhowmick, R.; Girotti, A.W. Rapid upregulation of cytoprotective nitric oxide in breast tumor cells subjected to a photodynamic therapy-like oxidative challenge. *Photochem. Photobiol.* **2011**, *87*, 378–386. [CrossRef]
42. Bhowmick, R.; Girotti, A.W. Cytoprotective signaling associated with nitric oxide upregulation in tumor cells subjected to photodynamic therapy-like oxidative stress. *Free Radic. Biol. Med.* **2013**, *57*, 39–48. [CrossRef] [PubMed]

43. Bhowmick, R.; Girotti, A.W. Pro-survival and pro-growth effects of stress-induced nitric oxide in a prostate cancer photodynamic therapy model. *Cancer Lett.* **2014**, *343*, 115–122. [CrossRef] [PubMed]
44. Fahey, J.M.; Girotti, A.W. Accelerated migration and invasion of prostate cancer cells after a photodynamic therapy-like challenge: Role of nitric oxide. *Nitric Oxide* **2015**, *49*, 47–55. [CrossRef] [PubMed]
45. Fahey, J.M.; Girotti, A.W. Nitric oxide-mediated resistance to photodynamic therapy in a human breast tumor xenograft model: Improved outcome with NOS2 inhibitors. *Nitric Oxide* **2017**, *62*, 52–61. [CrossRef] [PubMed]
46. Wardman, P. Fluorescent and luminescent probes for measurement of oxidative and nitrosative species in cells and tissure: Progress, pitfalls, and prospects. *Free Radic. Biol. Med.* **2007**, *43*, 995–1022. [CrossRef] [PubMed]
47. Lancaster, J.R. The use of diaminofluorescein for nitric oxide detection: Conceptual and methodological distinction between NO and nitrosation. *Free Radic. Biol. Med.* **2010**, *49*, 1145. [CrossRef]
48. Belda-Iniesta, C.; de Castro Carpeño, J.; Casado Sáenz, E.; Cejas Guerrero, P.; Perona, R.; González Barón, M. Molecular biology of malignant gliomas. *Clin. Transl. Oncol.* **2006**, *8*, 635–641. [CrossRef]
49. Amberger-Murphy, V. Hypoxia helps glioma to fight therapy. *Curr. Cancer Drug Targets* **2009**, *9*, 381–390. [CrossRef]
50. Kramer, N.; Walzl, A.; Unger, C.; Rosner, M.; Krupitza, G.; Hengstschläger, M.; Dolznig, H. In vitro cell migration and invasion assays. *Mutat. Res.* **2013**, *752*, 10–24. [CrossRef]
51. Stamenkovic, I. Matrix metalloproteinases in tumor invasion and metastasis. *Semin. Cancer Biol.* **2000**, *10*, 415–433. [CrossRef]
52. Chintala, S.K.; Tonn, J.C.; Rao, J.S. Matrix metalloproteinases and their biological function in human gliomas. *Int. J. Dev. Neurosci.* **1999**, *17*, 495–502. [CrossRef]
53. Hei, T.K.; Zhou, H.; Chai, Y.; Ponnaiya, B.; Ivanov, V.N. Radiation induced non-targeted response: Mechanism and potential clinical implications. *Curr. Mol. Pharmacol.* **2011**, *4*, 96–105. [CrossRef] [PubMed]
54. Yakovlev, V.A. Role of nitric oxide in the radiation-induced bystander effect. *Redox Biol.* **2015**, *6*, 396–400. [CrossRef] [PubMed]
55. Matsumoto, H.; Hayashi, S.; Hatashita, M.; Ohnishi, K.; Shioura, H.; Ohtsubo, T.; Kitai, R.; Ohnishi, T.; Kano, E. Induction of radioresistance by a nitric oxide-mediated bystander effect. *Radiat. Res.* **2001**, *155*, 387–396. [CrossRef]
56. Matsumoto, H.; Takahashi, A.; Ohnishi, T. Nitric oxide radicals choreograph a radioadaptive response. *Cancer Res.* **2007**, *67*, 8574–8579. [CrossRef] [PubMed]
57. Dahle, J.; Bagdonas, S.; Kaalhus, O.; Olsen, G.; Steen, H.B.; Moan, J. The bystander effect in photodynamic inactivation of cells. *Biochim. Biophys. Acta* **2000**, *1475*, 273–280. [CrossRef]
58. Chakraborty, A.; Held, K.D.; Prise, K.M.; Liber, H.L.; Redmond, R.W. Bystander effects induced by diffusing mediators after photodynamic stress. *Radiat. Res.* **2009**, *172*, 74–81. [CrossRef]
59. Bazak, J.; Fahey, J.M.; Wawak, K.; Korytowski, W.; Girotti, A.W. Enhanced aggressiveness of bystander cells in an anti-tumor photodynamic therapy model: Role of nitric oxide produced by targeted cells. *Free Radic. Biol. Med.* **2017**, *102*, 111–121. [CrossRef]
60. Bazak, J.; Fahey, J.M.; Wawak, K.; Korytowski, W.; Girotti, A.W. Bystander effects of nitric oxide in anti-tumor photodynamic therapy. *Cancer Cell Microenviron.* **2017**, *4*, e1511. [CrossRef]
61. Fahey, J.M.; Stancill, J.S.; Smith, B.C.; Girotti, A.W. Nitric oxide antagonism to glioblastoma photodynamic therapy and mitigation thereof by BET bromodomain inhibitor JQ1. *J. Biol. Chem.* **2018**, *293*, 5345–5359. [CrossRef]
62. Huang, B.; Yang, X.D.; Zhou, M.M.; Ozato, K.; Chen, L.F. Brd4 coactivates transcriptional activation of NF-kappaB via specific binding to acetylated RelA. *Mol. Cell. Biol.* **2009**, *29*, 1375–1387. [CrossRef] [PubMed]
63. Zou, Z.; Huang, B.; Wu, X.; Zhang, H.; Qi, J.; Bradner, J.; Nair, S.; Chen, L.F. Brd4 maintains constitutively active NF-κB in cancer cells by binding to acetylated RelA. *Oncogene* **2014**, *33*, 2395–2404. [CrossRef] [PubMed]
64. Shu, S.; Polyak, K. BET bromodomain proteins as cancer therapeutic targets. *Cold Spring Harb. Symp. Quant. Biol.* **2016**, *81*, 123–129. [CrossRef] [PubMed]
65. Fujisawa, T.; Filippakopoulos, P. Functions of bromodomain-containing proteins and their roles in homeostasis and cancer. *Nat. Rev. Mol. Cell Biol.* **2017**, *18*, 246–262. [CrossRef] [PubMed]

66. Hajmirza, A.; Emadali, A.; Gauthier, A.; Casasnovas, O.; Gressin, R.; Callanan, M.B.; Hajmirza, A.; Emadali, A.; Gauthier, A.; Casasnovas, O.; et al. BET Family Protein BRD4: An Emerging Actor in NFκB Signaling in Inflammation and Cancer. *Biomedicines* **2018**, *6*, 16. [CrossRef] [PubMed]
67. Filippakopoulos, P.; Knapp, S. Targeting bromodomains: Epigenetic readers of lysine acetylation. *Nat. Rev.* **2014**, *13*, 337–356. [CrossRef] [PubMed]
68. Fu, L.L.; Tian, M.; Li, X.; Li, J.J.; Huang, J.; Ouyang, L.; Zhang, Y.; Liu, B. Inhibition of BET bromodomains as a therapeutic strategy for cancer drug discovery. *Oncotarget* **2015**, *6*, 5501–5516. [CrossRef] [PubMed]
69. Ishida, C.T.; Bianchetti, E.; Shu, C.; Halatsch, M.E.; Westhoff, M.A.; Karpel-Massler, G.; Siegelin, M.D. BH3-mimetics and BET-inhibitors elicit enhanced lethality in malignant glioma. *Oncotarget* **2017**, *8*, 29558–29573. [CrossRef] [PubMed]
70. Krishnan, N.; Bencze, G.; Cohen, P.; Tonks, N.K. The anti-inflammatory compound BAY-11-7082 is a potent inhibitor of protein tyrosine phosphatases. *FEBS J.* **2013**, *280*, 2830–2841. [CrossRef] [PubMed]
71. Fahey, J.M.; Girotti, A.W. Upstream signaling events leading to elevated production of pro-survival nitric oxide in photodynamically-challenged glioblastoma cells. *Free Radic. Biol. Med.* **2019**, in press.
72. Rapozzi, V.; Della Pietra, E.; Zorzet, M.; Zacchigna, B.; Bonavida, B.; Xodo, L.E. Nitric oxide-mediated activity in anti-cancer photodynamic therapy. *Nitric Oxide* **2013**, *30*, 26–35. [CrossRef] [PubMed]
73. Rapozzi, V.; Della Pietra, E.; Bonavida, B. Dual roles of nitric oxide in the regulation of tumor cell response and resistance to photodynamic theapy. *Redox Biol.* **2015**, *6*, 311–317. [CrossRef] [PubMed]
74. Bonavida, B.; Baritaki, S.; Huerta-Yepez, S.; Bega, M.I.; Jazirehi, R.; Berenson, J. Nitric oxide donors are a new class of anti-cancer therapeutics for the reversal of resistance and inhibition of metastasis. In *Nitric Oxide (NO) and Cancer*; Bonavida, B., Ed.; Springer: New York, NY, USA, 2010; pp. 459–477.
75. Huang, Z.; Fu, J.; Zhang, Y. Nitric oxide donor-based cancer therapy: Advances and prospects. *J. Med. Chem.* **2017**, *60*, 7617–7635. [CrossRef]
76. Basile, M.S.; Mazzon, E.; Krajnovic, T.; Draca, D.; Cavalli, E.; Al-Abed, Y.; Bramanti, P.; Nicoletti, F.; Mijatovic, S.; Maksimovic-Ivanic, D. Anticancer and Differentiation Properties of the Nitric Oxide Derivative of Lopinavir in Human Glioblastoma Cells. *Molecules* **2018**, *23*, 2463. [CrossRef] [PubMed]
77. Kiziltepe, T.; Hideshima, T.; Ishitsuka, K.; Ocio, E.M.; Raje, N.; Catley, L.; Li, C.Q.; Trudel, L.J.; Yasui, H.; Vallet, S.; et al. JS-K, a GST-activated nitric oxide generator, induces DNA double-strand breaks, activates DNA damage response pathways, and induces apoptosis in vitro and in vivo in human multiple myeloma cells. *Blood* **2007**, *110*, 709–718. [CrossRef] [PubMed]
78. Heinrich, T.A.; Tedesco, A.C.; Fukuto, J.M.; da Silva, R.S. Production of reactive oxygen and nitrogen species by light irradiation of a nitrosyl phthalocyanine ruthenium complex as a strategy for cancer treatment. *Dalton Trans.* **2014**, *43*, 4021–4025. [CrossRef] [PubMed]
79. Fraix, A.; Sortino, S. Combination of PDT photosensitizers with NO photodonors. *Photochem. Photobiol. Sci.* **2018**, *17*, 1709–1727. [CrossRef]
80. Chegaev, K.; Fraix, A.; Gazzano, E.; Abd-Ellatef, G.E.; Blangetti, M.; Rolando, B.; Conoci, S.; Riganti, C.; Fruttero, R.; Gasco, A.; et al. Light-regulated NO release as a novel strategy to overcome doxorubicin multidrug resistance. *ACS Med. Chem. Lett.* **2017**, *8*, 361–365. [CrossRef] [PubMed]
81. Ishikawa, T.; Kajimoto, Y.; Inoue, Y.; Ikegami, Y.; Kuroiwa, T. Critical role of ABCG2 in ALA-photodynamic diagnosis and therapy of human brain tumor. *Adv. Cancer Res.* **2015**, *125*, 97–216.
82. Fagone, P.; Masson, E.; Bramanti, P.; Bendzen, K.; Nicoletti, F. Gasotransmitters and the immune system: Mode of action and novel therapeutic targets. *Eur. J. Pharmacol.* **2018**, *834*, 92–102. [CrossRef]
83. Mahboob, S.O.; Eljamel, M. Intraoperative image-guided surgery in neuro-oncology with specific focus on high-grade gliomas. *Future Oncol.* **2017**, *13*, 2349–2361. [CrossRef] [PubMed]
84. Girotti, A.W. Upregulation of nitric oxide in tumor cells as a negative adaptation to photodynamic therapy. *Lasers Surg. Med.* **2018**, *50*, 590–598. [CrossRef] [PubMed]
85. Bignon, E.; Allega, M.F.; Lucchetta, M.; Tiberti, M.; Papaleo, E. Computational structural biology of S-nitrosylation of cancer targets. *Front. Oncol.* **2018**, *8*, 272. [CrossRef] [PubMed]

cancers

MDPI

Review

Protein Phosphatases—A Touchy Enemy in the Battle Against Glioblastomas: A Review

Arata Tomiyama [1,2,*], Tatsuya Kobayashi [1,2,3], Kentaro Mori [2] and Koichi Ichimura [1]

1 Division of Brain Tumor Translational Research, National Cancer Center Research Institute, 5-1-1 Tsukiji, Chuo-ku, Tokyo 104-0045, Japan; opera58840428@gmail.com (T.K.); kichimur@ncc.go.jp (K.I.)

2 Department of Neurosurgery, National Defense Medical College, 3-2 Namiki, Tokorozawa, Saitama 359-8513, Japan; kmori@ndmc.ac.jp

3 Department of Neurosurgery, Tokyo Women's Medical University, 8-1 Kawadacho, Shinjuku-ku, Tokyo 162-8666, Japan

* Correspondence: atomiyam@outlook.jp; Tel.: +81-3-3542-2511 (ext. 5983); Fax: +81-3-3542-8170

Received: 31 December 2018; Accepted: 16 February 2019; Published: 19 February 2019

Abstract: Glioblastoma (GBM) is the most common malignant tumor arising from brain parenchyma. Although many efforts have been made to develop therapies for GBM, the prognosis still remains poor, mainly because of the difficulty in total resection of the tumor mass from brain tissue and the resistance of the residual tumor against standard chemoradiotherapy. Therefore, novel adjuvant therapies are urgently needed. Recent genome-wide analyses of GBM cases have clarified molecular signaling mechanisms underlying GBM biology. However, results of clinical trials targeting phosphorylation-mediated signaling have been unsatisfactory to date. Protein phosphatases are enzymes that antagonize phosphorylation signaling by dephosphorylating phosphorylated signaling molecules. Recently, the critical roles of phosphatases in the regulation of oncogenic signaling in malignant tumor cells have been reported, and tumorigenic roles of deregulated phosphatases have been demonstrated in GBM. However, a detailed mechanism underlying phosphatase-mediated signaling transduction in the regulation of GBM has not been elucidated, and such information is necessary to apply phosphatases as a therapeutic target for GBM. This review highlights and summarizes the phosphatases that have crucial roles in the regulation of oncogenic signaling in GBM cells.

Keywords: protein phosphatase; glioblastoma; signaling; therapy

1. Introduction

Malignant gliomas are the most common primary intracranial malignant neoplasm arising from brain parenchyma. And, Glioblastoma (GBM) is the most malignant and aggressive form of gliomas (WHO grade IV) with the highly infiltrative phenotype, refractoriness, and genetical complexity [1,2]. Despite the efforts made over the past decades, therapeutic outcomes of GBM still remain poor [3,4], mainly because complete tumor mass resection is difficult because of its strong invasiveness into adjacent normal brain tissue and resistance by the residual tumor to standard chemoradiotherapy. Therefore, novel adjuvant therapies for GBM are urgently needed. The developments in high-throughput methods have allowed genome-wide analyses of malignant tumors, and landscapes of genetic mutations in GBMs have been reported [5–14]. In addition, molecular signaling networks involved in the regulation of GBM biology have been elucidated [5,6,15]. Concomitantly, clinical trials targeting signaling molecules identified in these studies have been started. However, most clinical trials have yielded unsatisfactory results [16,17].

Regulation of molecular signaling, including signaling related to tumor biology, generally involves post-translational modification of the signaling molecules. One of the most common modifications is phosphorylation. As a result of upstream signaling activation, kinases are activated and attach phosphate groups to certain amino-acid residues of their specific substrate molecules, resulting in downstream signaling activation. In GBM, kinases are often hyperactivated as a result of genetic alteration or enhanced upstream signaling activation, and, therefore, kinase hyperactivation has been focused on as a therapeutic target [5,6,18–20].

Protein phosphatases (PPPs), as negative regulators of phosphorylation-dependent signaling, remove phosphate groups of certain phosphorylated amino-acid residues of specific substrate molecules [21–28]. In tumor cells, most PPPs have been suggested to act as negative regulators of oncogenic signaling and, thus, to function as tumor suppressors, as oncogene-induced signal transduction is largely mediated by phosphorylation [22,23,29,30]. However, recent studies have demonstrated suppressed expression or activity of tumor-suppressive PPPs in certain tumors, including GBMs, and as such, PPPs might negatively regulate oncogene inhibition. However, our knowledge of the roles of PPPs in tumor signaling and oncogenic activity is still limited. In this review, we discuss the current knowledge on the roles of PPPs intertwined with kinase signaling in the regulation of GBM biology.

2. Protein Phosphatases

PPPs are classified as classical or atypical PPPs (Figure 1) [21,23,27]. Based on the amino-acid residue they dephosphorylate, classical and atypical PPPs are further grouped into protein serine/threonine phosphatases (PSPs), protein tyrosine phosphatases (PTPs), and dual-specificity phosphatases (DUSPs) (Figure 1) [21,27]. In classical PPPs, these groups are even more divided into subfamilies based on their chemical structures (Figure 1). Most of PPPs can suppress oncogenic signaling by dephosphorylating phosphorylated (activated) signaling molecules, such as mitogen-activated protein kinases (MAPKs), and are known to act as tumor suppressors in various tumors [21,31,32]. On the other hand, certain kinds of PPPs are known to be overexpressed in tumor cells and rather act as oncogenes [31,33–37]. Because both oncogenic signaling and tumor-suppressive signaling are simultaneously regulated by the same phosphatase, it is suggested that the oncogenic or tumor-suppressive function of PPPs in tumors probably depends on the balance between the two signaling effects of the PPPs.

To develop clinical GBM therapies targeting PPPs, it is important to understand the detailed molecular mechanisms underlying both tumor-suppressive and oncogenic PPPs in GBM cells. In what follows, we discuss PPPs that are suggested to have major roles in GBM biology in detail.

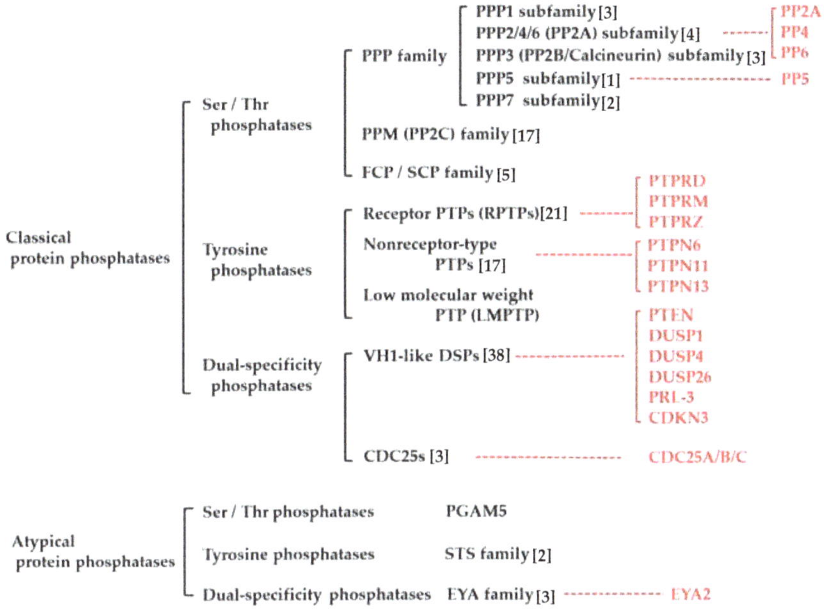

Figure 1. Protein phosphatases in human. Human phosphoprotein phosphatases (PPPs) are classified into two large classes as classical or atypical PPPs. And, these classes of PPPs are further grouped into families or subfamilies based on the amino-acid residue they dephosphorylate or their chemical structure. The PPPs focused in the text are written in red characters. Ser, Serine; Thr, Threonine; Tyr, Tyrosine; PPP, phosphoprotein phosphatase; PP2A, protein phosphatase 2A; PP2B protein phosphatase 2B; PP2C, protein phosphatase 2C; PPM, metal-dependent protein phosphatase; FCP, TFIIF-associating component of RNA polymerase II carboxy-terminal domain phosphatase; PTP, protein Tyr phosphatase; LMPTP, low molecular weight PTP; VH1-like DSP [38], Vaccinia virus gene H1-like dual specificity phosphatase; PGAM5, phosphoglycerate mutase family member 5; Sts, suppressor of T-cell receptor signaling; EYA, eyes absent.

2.1. Classical Protein Phosphatases

2.1.1. Protein Serine/Threonine Phosphatases

Protein Phosphatase 2A

As mentioned above, protein phosphatase 2A (PP2A) is one of the most major PSPs. PP2A is a heterotrimeric protein phosphatase complex which consists of the alpha (PPP2R1A) or beta (PPP2R1B) isoform of the structural A subunit, the alpha (PPP2CA) or beta (PPP2CB) isoform of the catalytic C subunit, and the regulatory B subunit. The A subunit and C subunit forms core heterodimer, and association of one of the multiple B subunits with the core dimer directs various substrate specificity (more than 60 combinations) of PP2A [39]. PP2A regulates various cellular signaling pathways, such as receptor tyrosine kinase (RTK) signaling, by dephosphorylating multiple substrates under physiological conditions, and ablation of PP2A expression or activity causes cardiovascular disorder, diabetes, and neurodegenerative disorder [26]. In cancer systems, involvement of genetic, epigenetic, or post-translational modification-mediated dysregulation of PP2A expression or activity in tumorigenesis are suggested, and dysregulated PP2A tumor cells cause an increase in cellular proliferation, formation of resistance against drug or irradiation, or impairment of tumor immunity [26,40–43]. However, the genetic alteration of PP2A subunits-encoding genes in GBMs are rare (about less than 1%) in The Cancer Genome Atlas (TCGA) datasets [5,43]. One of the mechanisms which is suggested to induce non-genetic dysregulation of PP2A in GBM is hyperactivation of RTKs, such as epidermal growth factor receptor (EGFR), by genetic alteration frequently observed in GBMs [5,6]. In a certain series of malignant tumors with RTK hyperactivation, downregulation of PP2A expression or activity has been reported, which would possibly relieve PP2A-mediated suppression of downstream signaling of RTK, resulting in further activation of RTK-mediated signaling [26,44–46]. In line herewith, downregulated expression of PP2A subunits—without genetic alteration—has been observed in glioma tissue [47,48]. And direct or indirect inhibition of PP2A resulted in enhanced oncogenic property of glioma cells [43,49–51], suggesting a role of PP2A as a tumor suppressor in GBMs. As the other non-genetic regulatory mechanisms of PP2A activity, the molecules which negatively regulate PP2A activity are also crucial. Among this group of proteins, cancerous inhibitor of PP2A (CIP2A), protein phosphatase methylesterase-1 (PME-1), and SE translocation (SET) oncoprotein, are well-characterized and known to downregulate PP2A activity by different biological processes [26]. CIP2A directly associates with and blocks the B56 regulatory subunits of PP2A complex [52], and importantly, high expression of CIP2A is correlated with overexpression of EGFR in the certain cancer systems [44–46]. PME-1 suppresses PP2Ac activity by the removal of metal ions from PP2Ac catalytic core and demethylation of the C-terminal lesion of PP2Ac, whereas SET directly associates and blocks the catalytic core of PP2Ac [53,54]. In GBMs, in vitro experiments revealed the possible role of PME-1 in the formation of GBM cell resistance against Ca^{2+}/calmodulin-dependent protein kinase inhibitor (H7), PI3K inhibitor (LY29644), and multi-RTKs inhibitor (sunitinib). These knowledges suggest not only expressional but also enzymatic inhibition of PP2A in GBM cells would be important for the maintenance of GBM malignancy, and the possible role of PP2A reactivation as the therapeutic strategy of GBM would also be considered (see below chapter 3.1. On the contrary, PP2A has also been suggested as a potent therapeutic target for GBMs. Treatment with PP2A inhibitor okadaic acid alone, without concomitant use of genotoxins, triggered mitotic cell death of GBM cells [55]. Treatment of GBM stem cells with a PP2A inhibitor LB100 resulted in induction of differentiation or cell death via dysregulation of nuclear receptor corepressor [56]. Treatment of GBM cells with the c-Jun N-terminal kinase (JNK) activator anisomycin induced cell death via suppression of PP2A subunit expression [57]. Pharmacological inhibition of PP2A activity by LB-1.2, LB-100, or Microcystin-LR under irradiation or genotoxin treatment resulted in enhanced cell death induction in GBM cells [58–60]. As mentioned above, these pro-survival effects of PP2A in GBM cell death might be due to the alteration of the balance between cell death-inducing and cell survival-inducing signaling under PP2A inhibition. In fact, PP2A regulates both cell death-inducing and cell survival-inducing signaling simultaneously in multiple sites of programmed cell death cascades (Figure 2) [26,38,61–63]. Collectively, these evidences suggest the potential of PP2A as a therapeutic target for GBMs not only through suppression but also through upregulation of its activity or expression.

Protein Phosphatase 4

Protein phosphatase 4 (PP4) is a member of the type 2A PSP family, which includes PP2A, PP4, and protein phosphatase 6 (PP6) [64,65]. These PSPs have approximately 60% sequence similarity, but they have different biological functions. In normal conditions, PP4 is reported to regulate the DNA-damage response as well as NF-kappaB and mTOR functions [66]. In several cancers, enhanced expression of PP4 subunits is observed in the cancer tissue and is suggested as a marker of poor prognosis in these cases [67–70]. Although alteration of *PPP4C* gene, which encodes the catalytic subunit of PP4 (PP4C), in the TCGA GBM dataset was not confirmed [5], high expression of PP4C was observed in a series of GBM cases, and the PP4C expression level was negatively correlated with prognosis [67]. In addition, knockdown of PP4C in cultured GBM cells resulted in a reduction in their oncogenic property [67]. Although the positive evidence of PP4 in GBM biology as noted above has been reported, the exact role of PP4 in GBM biology has not yet been elucidated, and more evidence of its potential as a novel therapeutic is needed.

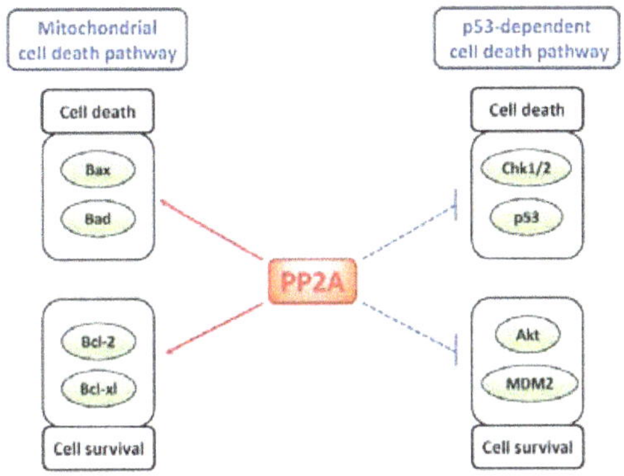

Figure 2. Multiple roles of protein phosphatase 2A (PP2A) in the regulation of cell death signaling. PP2A regulates both cell death-inducing and cell survival-inducing signaling simultaneously in not only mitochondria- but p53-dependent cell death cascade. In mitochondria-dependent cell death cascade, PP2A activates both pro-apopototic (Bax and Bad) and anti-apoptotic (Bcl-2 and Bcl-xl) Bcl-2 family proteins simultaneously by dephosphorylation. On the other hand, in p53-dependent cell death cascade, PP2A suppresses p53 inhibitor MDM2 and MDM2 activator Akt as well as p53 and p53 activator Chk1/2. Bcl-2, B cell lymphoma 2; Bcl-xl, B-cell lymphoma extra large; p53, tumor protein p53; Chk1/2, Serine/threonine-protein kinase Chk1/2; MDM2, Mouse double minute 2 homolog; Akt, protein kinase B.

Protein Phosphatase 5

Protein phosphatase 5 (PP5) is one of the classical PSPs encoded by *PPP5C* and highly expressed in central nervous system and neurons [71,72]. Physiologically, PP5 regulates cellular survival, differentiation, DNA damage repair, migration, and proliferation [73]. On the other hand, high expression of PP5 in the breast cancer or osteosarcoma tissue is observed [74,75]. However, no direct evidence about linking between aberrant expression of PP5 with tumorigenesis is reported yet. In GBMs, in vitro experiments revealed positive contribution of PP5 in regulation of GBM cell growth and migration [76]. Therefore, to investigate further about expression level of PP5 in GBM tissue and the detailed regulatory roles of PP5 in GBM oncogenicity would be suggested as meaningful.

Protein Phosphatase 6

Like PP4, Protein phosphatase 6 (PP6) is a member of the type 2A PSP family. It regulates various physiological processes, such as anti-inflammation, cell-cycle regulation, DNA-damage repair, and lymphocyte development. Furthermore, it is reportedly involved in the regulation of tumorigenesis. In melanomas, PP6 has been shown to act as a tumor suppressor [65]. On the other hand, anti-tumor activity of PP6 is reported in cervical cell carcinoma cell [77,78]. In GBMs, the PP6 catalytic subunit (PP6c) is overexpressed in around 0.5% of the TCGA GBM dataset [5], and siRNA knockdown of PP6c suppressed DNA-dependent protein kinase activity, resulting in an enhanced response of GBM cells to irradiation treatment in vitro and in vivo [64]. The representative PP2A inhibitor, okadaic acid, also inhibits PP6 activity [65], and okadaic acid has a tumor-suppressive effect in GBM cells [55,56], which is likely through suppression of not only PP2A but also PP6 activity. These evidences suggest PP6 rather contributes as an oncogene in GBMs, but the accumulation of further evidences is necessary for clarifying the exact roles of PP6 in GBM biology as well as PP4.

2.1.2. Protein Tyrosine Phosphatases

Tyrosine-Protein Phosphatase Non-Receptor Type 6

Tyrosine-protein phosphatase non-receptor type 6 (PTPN6; also known as SHP1) is a PTP primarily expressed in hematopoietic cells in normal tissue and regulates hematopoietic signaling, such as tyrosine-protein kinase Lyn-mediated pathway. In a series of glioma cases, high expression of SHP1 in glioma tissue was associated with poor prognosis [79]. Furthermore, in vitro study revealed expression of SHP1 in GBM cells resulted in an increase of chemoresistance, and it was also demonstrated that the expression level of SHP1 in glioma tissue was regulated by SHP1 promoter methylation status [79]. These results suggest SHP1 would have rather oncogenic roles in GBM cells, and epigenetic machinery is a key mechanism for regulation of SHP1 expression in GBMs.

Tyrosine-Protein Phosphatase Non-Receptor Type 11

Tyrosine-protein phosphatase non-receptor type 11 (PTPN11; also known as PTP-1D, PTP-2C, or SHP2) is a PTP activated mainly by RTKs and is one of the representative PTPs that have oncogenic roles in various cancers [24,33,80–85]. Dysregulation of SHP2 function because of germline mutations is involved in the pathogenesis of hereditary diseases, such as Noonan syndrome and Leopard syndrome, as well as in oncogenesis and malignancy of neoplasms [24,33,80–87]. The major signaling pathway regulated by SHP2 is the Ras-Raf-ERK (MAPK) cascade [33,80–84]. Both the hereditary diseases and oncogenic signaling triggered by dysregulation of SHP2 are known to be mediated by alterations of MAPK cascade activity [33,81,82,85–87]. Although SHP2 is reported to negatively regulate proto-oncogenes, such as STAT3, by dephosphorylation [88,89], SHP2 as a positive regulator of the MAPK cascade is reported to contribute as a proto-oncogene. There is evidence that SHP2 activates the MAPK cascade via adapter molecules, such as growth factor receptor-bound protein 2—although the detailed regulatory mechanism remains unknown—and through dephosphorylation of Ras [33,80,83]. In glioma cells, although MAPK cascade activation by oncogenic activation of other molecules is frequently reported, genetic alteration or enhanced activation of SHP2 is not common [5,6,13]. This is probably because SHP2 induces downregulation of other signaling factors, such as signal transducer and activator of transcription 3 (STAT3), which is also essential for glioma maintenance, and overactivation of SHP2 rather leads to repression of tumor growth, even when the MAPK pathway is activated. On the other hand, SHP2 has been reported to play an essential role in oncogenic signal transduction of EGFRviii, an activation mutation of EGFR frequently observed in GBMs [90]. In addition, reduced expression of SHP2 resulted in augmented radiosensitivity in glioma cells [91], suggesting that SHP2 has potential as a therapeutic target for GBM, even without oncogenic activation or overexpression. In line with this, the efficacy of SHP2 inhibitors in GBM treatment has been exhibited (see below chapter 3.2.).

Tyrosine-Protein Phosphatase Non-Receptor Type 13

Tyrosine-protein phosphatase non-receptor type 13 (PTPN13 or PTPL1) is a PTP that regulates various cellular functions, such as proliferation and differentiation, via the Ras-ERK cascade or the Rho-associated protein kinase (ROCK) pathway [24,92–95]. PTPN13 has shown both tumor-suppressive and oncogenic effects in several cancers [22,93,95–99]. As for GBMs, PTPTN13 is overexpressed in GBM cells compared with normal cells in GBM tissue [22,37]. An in vitro study revealed PTPN13 directly interacts with and dephosphorylates tyrosine-phosphorylated FAS, a cell membrane-localizing death receptor activated by interaction with FASL ligand, and blocks FASL-dependent cell death in GBM cells [37]. Because FAS and FASL-mediated cell death machinery are involved in irradiation- or genotoxin-induced cell death of GBM cells [37,100,101], these findings suggest that PTPN13 acts as an oncogene in GBM cells by inhibiting therapy-induced GBM cell death.

Receptor-Type Tyrosine-Protein Phosphatase Delta

Receptor-type tyrosine-protein phosphatase delta (PTPRD) is one of the receptor-type tyrosine kinases (RTPTPs). PTPRD is encoded by the *PTPRD* gene and is inactivated in about 2 to 3% of GBM cases in the TCGA dataset [5]. Physiologically, PTPRD regulates hippocampal memory by promoting neurite outgrowth or axon guidance in the central nervous system [102], and mutation of *PTPRD* triggers craniosynostosis, hearing loss, or intellectual disability [103,104]. In GBMs, suppression of PTPRD expression by missense or nonsense mutation or promoter hypermethylation of *PTPRD* is frequently observed [105,106]. Forced expression of PTPRD in cultured GBM cells resulted in growth arrest or cell death that might be because of inhibition of activation of STAT3, one of the substrates of PTPRD [105,107]. These findings suggest that PTPRD is a tumor suppressor in GBM, but when genetically or epigenetically inactivated, might contribute to tumorigenesis in GBM.

Receptor-Type Tyrosine-Protein Phosphatase Mu

Receptor-type tyrosine-protein phosphatase mu (PTPRM or RTPTPμ) is RPTPT expressed in endothelial, glial, and neuronal cells. PTPRM regulates various biological process, such as cell growth, differentiation, and mitosis. And importantly, PTPRM positively regulates cell–cell adhesion by association with another PTPRM expressed at adjacent cells via homophilic binding [108]. The role as the negative regulator of cancer biology is indicated [109,110]. In GBMs, decreased expression of PTPRM is reported in GBM tissue [111], and suppression of PTPRM expression in GBM cells resulted in enhancement of migration in vitro [111], suggesting decreased PTPRM expression contributes to tumorigenicity of GBM by augmentation of GBM cell migration. However, inhibition of PTPRM phosphatase activity by antagonizing peptide demonstrated inhibition of GBM cell migration [112]. Therefore, the therapeutic roles of PTPRM in GBM treatment should be further confirmed carefully.

Receptor-Type Tyrosine-Protein Phosphatase Zeta

Receptor-type tyrosine-protein phosphatase zeta (PTPRZ) is an RTPTP encoded by *PTPRZ1* [113,114]. In the central nervous system, PTPRZ regulates neurotransmission, endocytic transportation, and synapse motility, thereby controlling hippocampal memory under the physiological condition. Through alternative splicing, three variants of PTPRZ are generated: a long form, a short form, and phosphacan, which is the extracellular domain of PTPRZ and is secreted to the extracellular space [115,116]. Phosphacan is expressed mainly in matured glia in the adult human brain, whereas the other two variants are expressed mainly in glial precursors [115]. In GBM tissue, all variants are expressed, and high-level expression of PTPRZ is reported [117,118]. Importantly, single-cell RNA sequencing analysis of primary GBM cases revealed that PTPRZ positively regulates stemness of GBM cells [119]. In addition, PTPRZ knockdown resulted in a suppression of migration and tumor growth of GBM cells [117,120,121]. Although the exact role of the intrinsic phosphatase activity of PTPRZ in the regulation of GBM biology remains unclear, recent evidence revealed that re-expression of the extracellular domain in GBM cell lines after knockdown of PTPRZ resultthe ed in rescue of the migration, but not proliferation, in these cells [120]. Enhanced secretion of pleiotrophin, one of the extracellular ligands of PTPRZ abundantly expressed in GBM tissue, from tumor-associated macrophages in GBM tissue promotes tumor growth of GBM stem cells [122]. The biological and therapeutic roles of the extracellular domain, as well as the intracellular PTP domain of PTPRZ. should be further investigated to elucidate the oncogenic role of PTPRZ in GBM. Recently, the inhibitors against PTPRZ, SCB4380 and NAZ2329, have been developed, and efficacy of these inhibitors in tumorigenicity of GBM cells was demonstrated (see below chapter 3.3).

2.1.3. Dual-Specificity Phosphatases

Phosphatase and Tensin Homolog

Phosphatase and tensin homolog (PTEN) is one of the best known DUSPs in gliomas, because inactivation of PTEN by deletion or mutation was discovered in 41% of the TCGA GBM dataset [5]. In physiological conditions, PTEN catalyzes the dephosphorylation of phosphatidylinositol (3,4,5)-trisphosphate to phosphatidylinositol (4,5)-bisphosphate, which inhibits phosphatidylinositol-4,5-bisphosphate 3-kinase signaling [29,123,124]. This reaction consequently blocks RTK-mediated survival signaling, resulting in tumor suppression. In gliomas, PTEN has been highlighted for its tumorigenic role induced by genetic inactivation through chromosome 10 deletion. In addition to genetic inactivation of PTEN, recent evidence demonstrates that miR-26a, which is highly expressed in GBM tissues, targets PTEN and suppresses PTEN expression in GBMs with monoallelic PTEN deletion, resulting in further inhibition of PTEN [125]. *PTEN* promoter methylation-mediated PTEN silencing has also been detected in GBMs [126]. Because overactivation of RTKs, such as EGFR, by chromosomal mutations also frequently occurs in gliomas [5,6,15,127,128], it is suggested that PTEN inactivation results in enhanced RTK-mediated oncogenic signaling and. thus, to oncogenicity in glioma cells. This also implies that the phosphatidylinositol-4,5-bisphosphate 3-kinase pathway might be an effective therapeutic target for gliomas, although direct targeting of PTEN would be difficult. Interestingly, a recent finding demonstrated PTEN is exocytically secreted by exosomes from GBM cells and suppresses Akt activity of its recipient cells in vitro [129], and exosomal microRNA (miR)-21 and miR-26a enriched in GBM patients as the circulating microRNA suppress PTEN expression [130], suggesting the profound roles of exosomes in regulation of GBMs biology via modulation of PTEN expression. Therefore, understanding these epigenetical or exosomal regulations of PTEN are also suggested to be important for targeting PTEN-mediated signaling pathway in GBMs treatment.

Dual-Specificity Phosphatase 1/Mitogen-Activated Protein Kinase Phosphatase 1

Most DUSP family members regulate the phosphorylation (activation) status and subcellular localization of MAPKs [22,25,28,131]. Dephosphorylation of MAPKs by DUSPs usually results in the suppression of MAPK signaling induced by, e.g., RTKs [22,25,28,131]. Dual-specificity phosphatase 1 (DUSP1)/mitogen-activated protein kinase phosphatase 1 (MKP-1) is encoded by *DUSP1*, and DUSP1 expression is transcriptionally regulated by tumor suppressor p53 [132], which is frequently inactivated by genetic alteration in GBMs [5,6]. Thus, suppression of DUSP1 expression by p53 mutation would contribute to GBM tumorigenicity via upregulated MAPK signaling activation. Interestingly, one report demonstrated that DUSP1 expression is higher in GBM than in normal cells in glioma tissue [133]. Knockdown of DUSP1 in GBM cells resulted in enhanced induction of cell death by genotoxins, which was probably induced by JNK hyperactivation [134]. These findings suggest the potential of DUSP1 as a potent therapeutic target for GBMs, and that DUSP1 has both tumor-suppressive and oncogenic roles in GBMs, likely dependent on its expression level.

Dual-Specificity Phosphatase 4/Mitogen-Activated Protein Kinase Phosphatase 2

Dual-specificity phosphatase 4 (DUSP4)/mitogen-activated protein kinase phosphatase 2 (MKP-2) is encoded by *DUSP4* and is widely expressed in the nucleus in various tissues. Like other DUSPs, DUSP4 dephosphorylates MAPKs, such as ERK1, ERK2, and JNK, and regulates proliferation and differentiation [22,131]. In cancers, DUSP4 expression is either up- or downregulated [135–139], and in GBM tissues, DUSP4 is generally downregulated [140]. This downregulation of DUSP4 expression is often associated with and, thus, likely caused by *DUSP4* promoter methylation [140,141]. Importantly, in GBMs, *DUSP4* promoter methylation is reported to be associated with alteration of the *IDH1* gene encoding isocitrate dehydrogenase 1 (IDH1) [140], which catalyzes the oxidative decarboxylation of isocitrate to 2-oxoglutarate [142]. *IDH1* mutation is one of the most important genetic alteration in gliomas, which is not only necessary for molecular diagnosis of glioma cases, but is also essential for understanding glioma biology [143]. *IDH1* mutations in gliomas are gain-of-function mutations that lead to the generation of the oncometabolite 2-hydroxyglutarate instead of 2-oxoglutarate from isocitrate, which induces a wide range of epigenetic dysregulation collectively termed "glioma CpG island methylator phenotype" [144]. The presence of *IDH1* mutation is one of the most important determining factors for molecular diagnosis in the WHO 2016 classification of gliomas [145,146]. Therefore, downregulation of DUSP4 expression in gliomas is suggested to be caused by epigenetic regulation triggered by *IDH1* mutation and to contribute to glioma tumorigenesis through enhancement of MAPK signaling. Accordingly, the DUSP4 expression level might serve as a predictor of *IDH1* mutation, which currently is crucial for the molecular diagnosis of gliomas.

Dual-Specificity Phosphatase 26/Mitogen-Activated Protein Kinase Phosphatase 8

Dual-specificity phosphatase 26 (DUSP26) is encoded by *DUSP26* and dephosphorylates MAPKs, including ERK, JNK, and p38; however, its physiological function remains unclear. In tumor cells, DUSP26 is known to act as both oncogene and tumor suppressor. A recent study revealed that the oncogenic activity of DUSP26 depends on dephosphorylation and inactivation of tumor suppressor p53 [147]. In GBMs, DUSP26 expression is downregulated, and lower expression of DUSP26 predicts poor prognosis in GBM patients [148]. Another study reported downregulation of DUSP26 expression GBM tissues and DUSP26-mediated enhancement of cell–cell adhesion, which suppressed the invasive phenotype of GBM cells [149]. Collectively, these findings indicate that DUSP26 might act as a tumor suppressor in GBMs, and suppression of activity or expression of DUSP26 might contribute to tumorigenesis of GBMs.

Phosphatase of Regenerating Liver 3

Phosphatase of regenerating liver 3 (PRL-3) is a DUSP that positively regulates various cellular signaling factors, such as PI3K, Src, and Rho [35,150–152]. The C′-terminal region of PRL-3 is known to be prenylated, which might contribute to the intracellular membrane localization of PRL-3. The exact PRL-3-dependent regulatory molecular signaling and substrates of PRL-3 in normal cells are yet to be elucidated. It has been reported that PRL-3 contributes to growth induction, invasion, and metastasis of tumor cells [152–155]. In glioma cells, PRL-3 induces matrix metalloprotease expression via ERK, JNK, or other machinery, and this has been suggested to trigger invasion and metastasis [22,156–158]. More importantly, the PRL-3 expression level is elevated according to historical malignancy grade, and evidence demonstrates that PRL-3 expression is inversely correlated with the prognosis of GBM patients [22,156,157]. Based on these findings, it is suggested PRL-3 might be involved in the regulation of GBM malignancy; however, details on the molecular machinery behind PRL-3-mediated GBM oncogenesis required further investigation.

Cyclin-Dependent Kinase Inhibitor 3

Cyclin-dependent kinase inhibitor 3 (CDKN3, or KAP) is a DUSP that physiologically dephosphorylates cyclin-dependent kinase 2 (CDK2) and induces mitotic G1-S arrest [159,160]. In several cancers, KAP is reported to contribute as either a tumor suppressor or oncogene by de-phosphorylation of other targets. Although alteration of *CDKN3* gene, which encodes KAP, in the TCGA GBM dataset was not confirmed [5], KAP expression was increased, and a high KAP expression was associated with poor prognosis in a series of GBM cases [161]; however, KAP proteins expressed in GBMs were aberrantly spliced and acted dominant-negatively against wild-type KAP in the same cases [161]. An in vitro study revealed that wild-type KAP inhibits GBM cell migration in a CDC2-dependent manner [161]. In addition, a recent study revealed that KAP-ROCK2 pathway inhibition through overexpression of miR-26a which directly targets KAP and is often amplified in GBMs augments CDK2 activation in GBMs, resulting in disease progression [162]. Therefore, it is suggested downregulation of KAP activity by aberrant splicing or overexpression of microRNA facilitates GBM cell oncogenicity at least in part by promoting migration in a CDC2-dependent manner.

CDC25 Family

The cell division cycle 25 (CDC25) family of DUSPs is commonly involved in mitotic entry in the cell cycle through dephosphorylation and activation of cyclin-dependent kinases (CDKs). In many cancers, the upregulated expression of CDC25s and its role in tumor progression by accelerating proliferation has been confirmed. All three CDC25 variants (CDC25A, B, and C) are involved in the regulation of GBM biology. Although genetic alterations of the genes encoding CDC25s are not observed in the TCGA GBM dataset, in a certain series of human glioma samples, expression of the cell-cycle marker Ki67 is increased in parallel with increased CDC25A expression, and dephosphorylation of PKM2 by CDC25A induces glycolytic metabolism, resulting in the Warburg effect and tumorigenicity of GBM cells [163,164]. On the other hand, suppression of CDC25A expression during ionizing radiation in GBM cells resulted in enhanced invasion, and upregulation of CDC25A expression upon irradiation augmented ionizing radiation-induced GBM cell death [165,166]. These findings suggest that CDC25A is involved in positive regulation of GBM biology. However, CDC25A might also conversely contribute to radioresistance of GBMs. In the case of CDC25B, increased CDC25B expression in GBM cases is reported to associate with worsening of histological malignancy and poor prognosis [22,167,168]. Inhibition of the expression and activity of forkhead box protein M1 (FOXM1), an upstream regulator of CDC25B, by a natural compound plumbagin resulted in suppression of GBM cell growth [169,170]. Although one report indicated that protein expression of CDC25B in GBM tissue is not upregulated in cases with chemo- or ratio-resistance [171], these evidences suggest and an oncogenic rather than a tumor-suppressive role of CDC25B. In the case of CDC25C, genetic alteration of CDC25C and correlation of expression or activation of CDC25C with malignancy or prognosis is not common in GBMs; however, indirect pharmacological inhibition of CDC25C expression by demethocurcumin or ansamycins has been suggested to be effective in GBMs [172,173]. Collectively, the data suggest that all CDC variants act as proto-oncogenes. Although direct inhibition of CDCs is under development for clinical use, CDCs are strong candidates as a novel therapeutic target for GBMs.

2.2. *Atypical Protein Phosphatases*

Eyes Absent Transcriptional Coactivator and Phosphatase Homolog 2

The eyes absent (EYA) family (EYA1–4) is a group of atypical PPPs that function as transcriptional cofactors as well as phosphatases. As for the physiological functions of the EYA family, they are known to be involved in organ development by regulating RTK, transforming growth factor (TGF) and Hedgehog-mediated signaling [23,174–176]. In addition, EYAs contribute to DNA repair [23,177–179] and regulate cancer proliferation and metastasis in various cancers [23,180,181]. EYA2 has a notable function in the regulation of cancer biology and serves as a prognostic marker in cancer [181,182]. In gliomas, including GBMs, EYA2 is overexpressed in glioma cells in glioma tissues compared with normal cells and is associated with histological grading. In the TCGA GBM dataset, amplification of EYA2-encoding gene *EYA2* is observed in around 0.7% of the total cases [5]. An in vitro study revealed that EYA2 positively regulates GBM cell invasion by enhancing matrix metalloprotease 9 expression and proliferation [183]. Thus, although the knowledge on EYA2 in the context of GBM is still limited, EYA2 might serve as a therapeutic target for GBM, as has been suggested for other tumors.

3. Phosphatase Targeting in GBM

Recently, to establish the therapy targeting phosphatases based on the biological features of each phosphatase upon neoplasms, the small molecule inhibitors and activators against some phosphatases have been developed as the antitumor agents. The summary of these agents is shown in Table 1.

Table 1. Summary of pharmacological inhibitors and activators of protein phosphatases highlighted in the text. The number of relevant references (Ref), molecular targets or mechanisms, and other information, such as clinical trial of each agent, are also shown. The agents with anti-glioblastoma (GBM) effects (and the number of relevant references) and clinical trial in GBM cases are emphasized in red characters.

PPP	Agent (Reference)	Target / Mechanism	Notes
PP2A	(Inhibitor)		
	LB-100 [59,184]	PP2A (Binding partner: unknown) →	Clinical trial in recurrent GBM is ongoing (Phase 2, NCT03027388)
	LB-102 [56,58]		
	Microcystin-LR [60]	Catalytic subunit of PP2A	
	Okadaic acid [55]		
	Anisomycin [57]	Suppression of PP2A expression (JNK activator)	
	(Activator)		
	FTY720 [185,186]	SET (PP2A inhibitor protein)	
	OP440		
	DT-061 (SMAP) [41,187,188]	Binding to the PP2A Aα scaffold subunit (Driving conformational changes in PP2A)	
PP4			
PP5			
PP6			
PTPN6			
SHP2	(Inhibitor)		
	Cryptotanshinone [189,190,191,192,193]	Catalytic domain of SHP2	
	NSC-87877 [192,193,194]		
	PHPS1 [185,192,193]		
	RMC-4550 [195] SHP099 [196]	Allosteric inhibition of SHP2	
	TNO155 [197]	→	Clinical trial in advanced solid tumors is ongoing (Phase 1, NCT03114319)
	(Activator)		
	Geranylnaringenin (CG902)	Binding to N-SH2 domain of SHP2 (Suppression of autoinhibition)	
PTPN13			
PTPRD			
PTPRM			
PTTPRZ	(Inhibitor)		
	SCB4380 [198]	Catalytic domain of PTPRZ/G →	Cell permeability (-)
	NAZ2329 [199]	Allosteric inhibition of PTPRZ/G	
PTEN			
DUSP1	(Inhibitor)		
	Triptolide [200]	Unknown (Inhibition of DUSP1 expression)	
	BCI [201–203]	Allosteric inhibition of DUSP1/6	
DUSP4			
DUSP26			
PRL-3	(Inhibitor)		
	Thienopyridone [204]	Catalytic domain of PRL-3	
	Cmpd-43 [205]	Inhibition of trimerization of PRL-3	
	Monoclonal anti-PRL-3 antibody [206]		
CDKN3			
CDC25A			
CDC25B	(Inhibitor)		
	Plumbagin [169,170]	FOXM1 (Inhibition of CDC25B expression)	
	FDI-6 [207]		
CDC25C	(Inhibitor)		
	Demethocurcumin [172]	Unknown (Acceleration of CDC25C degradation)	
	Ansamycins [173]	Hsp90 (Acceleration of CDC25C degradation)	
EYA2	(Inhibitor)		
	MLS000544460 [21]	Allosteric inhibition of EYA2	

3.1. PP2A Inhibitors and Activators

As noted above, PP2A works as either oncogenic protein or tumor suppressor which is dependent on the type harboring tumor, and the inhibitor and activator have been the focused as novel anticancer drug against various malignancies previously. LB-100, a water-soluble small molecule inhibitor of

PP2A, and its lipid-soluble derivative LB-102 have been shown antitumor effects against various tumors [184]. And importantly, LB-102 potentiated the effect of genotoxins without increase of side effects in the mouse xenograft model of human GBM cell lines [58]. Based on these evidences, phase 2 study of LB-100 against relapsed GBMs has already been started (https://clinicaltrials.gov/ct2/show/NCT03027388). However, okadaic acid-induced dormancy of GBM stem-like cells was also reported [208]. This means PP2A inhibitor might also have the potency to induce resistance against treatment-induced GBM cell death. On the other hand, the usefulness of PP2A activators has also been validated in tumor treatment. Phenothiazine, a tricyclic neuroleptic, induces PP2A-mediated apoptosis in leukemia cells via targeting A subunit of PP2A [209]. OP440, a peptidyl inhibitor of PP2A targeting SET, also demonstrated antitumor activity against leukemia cells [194]. Anticancer activity of FTY720, a sphingosine-based classical PP2A reactivator which targets SET, against various cancer systems has been demonstrated [185]. In GBMs, in vitro study demonstrated apoptosis-inducing activity of FTY720 against human glioma cell lines [186]. In addition, ceramides are also known to inhibit SET, and anti-inflammatory drug indomethacin is reported to induce upregulation of PP2A activity and PP2A-mediated GBM cell death by an increase of intracellular ceramides [210]. Currently, a series of orally bioavailable small molecule activators of PP2A (SMAP), which have been obtained by reengineering of tricyclic neuroleptics and targets structural A subunit of PP2A, are highlighted as the potent PP2A activators [187]. Efficacy of DT-061, a lead molecule of SMAP, in treatment of rat sarcoma proto-oncogene (RAS)- or MAPK-driven tumors, such as lung cancers or prostate cancers, has been confirmed [41,188]; however, the evidences about the effectiveness of SMAP in the treatment of GBMs is not reported yet. Taken together, the potencies of both PP2A inhibitors and PP2A activators as the antitumor agents against GBMs have been shown especially in association with recent progress in drug development, and it is also suggested a further continuation of investigation and improvement about these PP2A-targeted drugs is still necessary for clinical application.

3.2. SHP2 Inhibitors and Activators

As mentioned above, not only tumor-suppressive roles but tumorigenic functions of SHP2 have been suggested in various cancer systems. Therefore, SHP2 inhibitors, as well as SHP2 activators, have been developed as the anticancer drugs. The classical SHP2 inhibitors, such as NSC-87877 or PHPS1 which targets the catalytic domain of SHP2 [211,212], demonstrated antitumor effect against GBM cells in the cell-based studies. Cryptotanshinone, another SHP2 inhibitor extracted from natural plants [189], has also been reported to show tumor-suppressive effects upon GBM cells via suppression of STAT3 activation [190,191]. However, these active site-targeted SHP2 inhibitors or cryptotanshinone are also revealed to have the potency to act on multiple molecular targets [192,193]. Recently, RMC-4550, SHP099, and TNO155, the allosteric inhibitors of SHP2, have been developed, and these allosteric SHP2 inhibitors block SHP2 activity by lower concentration and higher specificity [195–197]. Notably, phase I clinical trial of TNO155 as the therapeutic agent against solid tumors has already started (https://clinicaltrials.gov/ct2/show/NCT03114319); however, the efficacy of these allosteric inhibitors of SHP2 in GBM treatment has not yet been well-proven. On the other hand, the activator of SHP2, geranylnaringenin (CG902), is reported to inhibit proliferation of the prostate cell line via SHP2-dependent STAT3 suppression [213]. Although the efficacy of geranylnaringenin in GBM therapy is not validated yet, the possible role of geranylnaringenin in GBM treatment would be suggested because STAT3 is known to play a crucial role in the maintenance of GBM biology [214].

3.3. PTPRZ Inhibitor

As mentioned above, suppression of PTPRZ expression resulted in inhibition of GBM cell oncogenicity. SCB4380, a cell-impermeable small molecule PTPRZ inhibitor which also potentially blocks PTPRG activity, suppressed migration, proliferation, and tumorigenicity of the Rat GBM cell line by liposome-mediated intracellular loading [198]. NAZ2329, another cell-permeable allosteric inhibitor of both PTPRZ and PTPRG, demonstrated suppression of stem cell-like properties and tumorigenicity in GBM cells [199]. These results suggest the potency of PTPRZ inhibitors as the effective GBM-treating agents; however, especially from the viewpoint of target selectivity and drug delivery, further development of PTPRZ inhibitors would be necessary for actual clinical use.

3.4. DUSP1 Inhibitors

The various inhibitors targeting certain DUSPs (DUSP1, DUSP6, and DUSP26) have been developed for cancer therapy [25]. Among them, triptolide, a plant-derived DUSP1 inhibitor which suppresses DUSP1 expression, suppressed proliferation and invasion of GBM cells in vitro [200]. (E)-2-benzylidene-5-bromo-3-(cyclohexylamino)-2,3-dihydro-1H-inden-1-one (BCI), an allosteric inhibitor of DUSP1/6 was recently developed [201] and demonstrated anticancer effect against breast cancer and gastric cancer cells via modulation of ERK1/2 activity [202,203]. Although the efficacy of BCI in GBM treatment is also prospected, accumulation of further evidences would be necessary for therapeutic application.

3.5. PRL-3 Inhibitors

Inhibitors targeting PRL-3 have been developed as anti-cancer agents for years. Thienopyridone, a classical PRL-3 inhibitor targeting active site of PRL-3, blocks PRLs activity and has demonstrated anticancer effects [204]. In addition, Cmpd-43, a PRZ-3 inhibitor which interferes with PRL trimerization and possesses more specificity to PRZs, demonstrates anticancer activity against melanomas in vitro and in the xenograft model [205], and recently-developed monoclonal anti-PRL-3 antibody also exhibited anticancer activity in vitro and in vivo [206]; however, the effectiveness of these PRL-3 inhibitors in GBM treatment is not confirmed yet.

3.6. CDC25s Inhibitors

CDC25s play fundamental roles in regulation of GBM biology, and pharmacological inhibition of CDC25B or CDC25C expression resulted in growth arrest of GBM cells [170,172,173]. In addition, FDI-6, a recent developed small molecule inhibitor of FOXM1 which transcriptionally upregulates CDC25B expression, suppressed CDC25B expression in breast and ovarian cancer cells [207]. Although these CD25B or CDC25C inhibitors do not directly affect these phosphatases, the possibility of CDC25B or CDC25C as the pharmacological therapeutic target of GBM is suggested. Further development of direct specific inhibitors of these CDC25s would be necessary for GBM treatment.

3.7. EYA2 Inhibitor

As noted above, EYA2 is overexpressed and regulates growth and invasion of high-grade gliomas including GBMs [183]. MLS000544460, an allosteric Eya2 inhibitor, has been developed and inhibited migration of human epithelial cell line [215], and further investigation about effectiveness of MLS000544460 in GBM treatment is awaited.

Given current knowledge, it still remains difficult to determine which PPPs might be appropriate as a therapeutic target for GBM. Therefore, it is necessary to verify the effect and usefulness of PPP inactivation or activation in vitro before the efficacy of targeting PPPs which can be studies in in vivo models or clinical studies. As for inhibitors or activator of PPPs, substrate specificity and blood–brain barrier (BBB) permeability might pose issues, especially in the case of GBMs [216]. Now, as noted above, phase 2 clinical trial of PP2A inhibitor LB-100 against recurrent GBM cases has been already ongoing (https://clinicaltrials.gov/ct2/show/NCT03027388), and the drugs targeting PPPs with greater permeability of the BBB, such as SMAPs, have been developed recently [41,187,188]; clinical application of these agents in GBM therapy in the near future would be anticipated.

4. Conclusions

Recent studies on glioma biology have indicated the roles of PPPs in the regulation of GBM oncogenicity, and the modulation of PPP expression or activity for clinical GBM treatment is gradually getting closer. However, further in-depth studies are needed before the application of PPP modulation in clinical GBM treatment can be contemplated, as PPPs are more difficult as therapeutic targets than certain other molecules because a single PPP regulates multiple and complex signaling pathways. However, from another angle, PPP modulation might avoid relapse and the development of therapeutic insensitivity in GBM exactly because PPPs modulate diverse intracellular signaling pathways. In any case, targeting of PPPs in GBM therapy would be challenging but is worth further development. In future, highly specific and blood–brain barrier-permeable PPP inhibitors might have potential as novel therapeutic agents against refractory GBMs.

Author Contributions: Wrote the paper: A.T. Check the paper: A.T., T.K., K.M., K.I.

Funding: This research received no external funding.

Conflicts of Interest: The authors declare no conflicts of interest.

References

1. Thakkar, J.P.; Dolecek, T.A.; Horbinski, C.; Ostrom, Q.T.; Lightner, D.D.; Barnholtz-Sloan, J.S.; Villano, J.L. Epidemiologic and molecular prognostic review of glioblastoma. *Cancer Epidemiol. Biomark. Prev.* **2014**, *23*, 1985–1996. [CrossRef] [PubMed]
2. Diamandis, P.; Aldape, K. World Health Organization 2016 Classification of Central Nervous System Tumors. *Neurol. Clin.* **2018**, *36*, 439–447. [CrossRef] [PubMed]
3. Gilbert, M.R.; Wang, M.; Aldape, K.D.; Stupp, R.; Hegi, M.E.; Jaeckle, K.A.; Armstrong, T.S.; Wefel, J.S.; Won, M.; Blumenthal, D.T.; et al. Dose-dense temozolomide for newly diagnosed glioblastoma: A randomized phase III clinical trial. *J. Clin. Oncol.* **2013**, *31*, 4085–4091. [CrossRef]
4. Stupp, R.; Mason, W.P.; van den Bent, M.J.; Weller, M.; Fisher, B.; Taphoorn, M.J.; Belanger, K.; Brandes, A.A.; Marosi, C.; Bogdahn, U.; et al. Radiotherapy plus concomitant and adjuvant temozolomide for glioblastoma. *N. Engl. J. Med.* **2005**, *352*, 987–996. [CrossRef] [PubMed]
5. Brennan, C.W.; Verhaak, R.G.; McKenna, A.; Campos, B.; Noushmehr, H.; Salama, S.R.; Zheng, S.; Chakravarty, D.; Sanborn, J.Z.; Berman, S.H.; et al. The somatic genomic landscape of glioblastoma. *Cell* **2013**, *155*, 462–477. [CrossRef] [PubMed]
6. Cancer Genome Atlas Research, N. Comprehensive genomic characterization defines human glioblastoma genes and core pathways. *Nature* **2008**, *455*, 1061–1068.
7. Phillips, H.S.; Kharbanda, S.; Chen, R.; Forrest, W.F.; Soriano, R.H.; Wu, T.D.; Misra, A.; Nigro, J.M.; Colman, H.; Soroceanu, L.; et al. Molecular subclasses of high-grade glioma predict prognosis, delineate a pattern of disease progression, and resemble stages in neurogenesis. *Cancer Cell* **2006**, *9*, 157–173. [CrossRef] [PubMed]
8. Kim, H.; Zheng, S.; Amini, S.S.; Virk, S.M.; Mikkelsen, T.; Brat, D.J.; Grimsby, J.; Sougnez, C.; Muller, F.; Hu, J.; et al. Whole-genome and multisector exome sequencing of primary and post-treatment glioblastoma reveals patterns of tumor evolution. *Genome Res.* **2015**, *25*, 316–327. [CrossRef] [PubMed]
9. Johnson, B.E.; Mazor, T.; Hong, C.; Barnes, M.; Aihara, K.; McLean, C.Y.; Fouse, S.D.; Yamamoto, S.; Ueda, H.; Tatsuno, K.; et al. Mutational analysis reveals the origin and therapy-driven evolution of recurrent glioma. *Science* **2014**, *343*, 189–193. [CrossRef]
10. deCarvalho, A.C.; Kim, H.; Poisson, L.M.; Winn, M.E.; Mueller, C.; Cherba, D.; Koeman, J.; Seth, S.; Protopopov, A.; Felicella, M.; et al. Discordant inheritance of chromosomal and extrachromosomal DNA elements contributes to dynamic disease evolution in glioblastoma. *Nat. Genet.* **2018**, *50*, 708–717. [CrossRef]
11. Wang, Q.; Hu, B.; Hu, X.; Kim, H.; Squatrito, M.; Scarpace, L.; deCarvalho, A.C.; Lyu, S.; Li, P.; Li, Y.; et al. Tumor Evolution of Glioma-Intrinsic Gene Expression Subtypes Associates with Immunological Changes in the Microenvironment. *Cancer Cell* **2017**, *32*, 42–56. [CrossRef] [PubMed]

12. Kim, J.; Lee, I.H.; Cho, H.J.; Park, C.K.; Jung, Y.S.; Kim, Y.; Nam, S.H.; Kim, B.S.; Johnson, M.D.; Kong, D.S.; et al. Spatiotemporal Evolution of the Primary Glioblastoma Genome. *Cancer Cell* **2015**, *28*, 318–328. [CrossRef] [PubMed]
13. Wang, J.; Cazzato, E.; Ladewig, E.; Frattini, V.; Rosenbloom, D.I.; Zairis, S.; Abate, F.; Liu, Z.; Elliott, O.; Shin, Y.J.; et al. Clonal evolution of glioblastoma under therapy. *Nat. Genet.* **2016**, *48*, 768–776. [CrossRef] [PubMed]
14. Kwon, S.M.; Kang, S.H.; Park, C.K.; Jung, S.; Park, E.S.; Lee, J.S.; Kim, S.H.; Woo, H.G. Recurrent Glioblastomas Reveal Molecular Subtypes Associated with Mechanistic Implications of Drug-Resistance. *PLoS ONE* **2015**, *10*, e0140528. [CrossRef]
15. Verhaak, R.G.; Hoadley, K.A.; Purdom, E.; Wang, V.; Qi, Y.; Wilkerson, M.D.; Miller, C.R.; Ding, L.; Golub, T.; Mesirov, J.P.; et al. Integrated genomic analysis identifies clinically relevant subtypes of glioblastoma characterized by abnormalities in PDGFRA, IDH1, EGFR, and NF1. *Cancer Cell* **2010**, *17*, 98–110. [CrossRef]
16. De Witt Hamer, P.C. Small molecule kinase inhibitors in glioblastoma: A systematic review of clinical studies. *Neuro-Oncology* **2010**, *12*, 304–316. [CrossRef]
17. Ma, C.; Zhao, G.; Cruz, M.H.; Siden, A.; Yakisich, J.S. Translational gap in glioma research. *Anticancer Agents Med. Chem.* **2014**, *14*, 1110–1120. [CrossRef]
18. An, Z.; Aksoy, O.; Zheng, T.; Fan, Q.W.; Weiss, W.A. Epidermal growth factor receptor and EGFRvIII in glioblastoma: Signaling pathways and targeted therapies. *Oncogene* **2018**, *37*, 1561–1575. [CrossRef]
19. Zhao, H.F.; Wang, J.; Shao, W.; Wu, C.P.; Chen, Z.P.; To, S.T.; Li, W.P. Recent advances in the use of PI3K inhibitors for glioblastoma multiforme: Current preclinical and clinical development. *Mol. Cancer* **2017**, *16*, 100. [CrossRef]
20. Venkatesan, S.; Lamfers, M.L.; Dirven, C.M.; Leenstra, S. Genetic biomarkers of drug response for small-molecule therapeutics targeting the RTK/Ras/PI3K, p53 or Rb pathway in glioblastoma. *CNS Oncol.* **2016**, *5*, 77–90. [CrossRef]
21. Shi, Y. Serine/threonine phosphatases: Mechanism through structure. *Cell* **2009**, *139*, 468–484. [CrossRef]
22. Navis, A.C.; van den Eijnden, M.; Schepens, J.T.; Hooft van Huijsduijnen, R.; Wesseling, P.; Hendriks, W.J. Protein tyrosine phosphatases in glioma biology. *Acta Neuropathol.* **2010**, *119*, 157–175. [CrossRef] [PubMed]
23. Sadatomi, D.; Tanimura, S.; Ozaki, K.; Takeda, K. Atypical protein phosphatases: Emerging players in cellular signaling. *Int. J. Mol. Sci.* **2013**, *14*, 4596–4612. [CrossRef] [PubMed]
24. He, R.J.; Yu, Z.H.; Zhang, R.Y.; Zhang, Z.Y. Protein tyrosine phosphatases as potential therapeutic targets. *Acta Pharmacol. Sin.* **2014**, *35*, 1227–1246. [CrossRef]
25. Prabhakar, S.; Asuthkar, S.; Lee, W.; Chigurupati, S.; Zakharian, E.; Tsung, A.J.; Velpula, K.K. Targeting DUSPs in glioblastomas—Wielding a double-edged sword? *Cell Biol. Int.* **2014**, *38*, 145–153. [CrossRef] [PubMed]
26. Mazhar, S.; Taylor, S.E.; Sangodkar, J.; Narla, G. Targeting PP2A in cancer: Combination therapies. *Biochim. Biophys. Acta Mol. Cell Res.* **2019**, *1866*, 51–63. [CrossRef] [PubMed]
27. Alonso, A.; Sasin, J.; Bottini, N.; Friedberg, I.; Friedberg, I.; Osterman, A.; Godzik, A.; Hunter, T.; Dixon, J.; Mustelin, T. Protein tyrosine phosphatases in the human genome. *Cell* **2004**, *117*, 699–711. [CrossRef]
28. Pulido, R.; Hooft van Huijsduijnen, R. Protein tyrosine phosphatases: Dual-specificity phosphatases in health and disease. *FEBS J.* **2008**, *275*, 848–866. [CrossRef]
29. Chen, C.Y.; Chen, J.; He, L.; Stiles, B.L. PTEN: Tumor Suppressor and Metabolic Regulator. *Front. Endocrinol.* **2018**, *9*, 338. [CrossRef]
30. Thompson, J.J.; Williams, C.S. Protein Phosphatase 2A in the Regulation of Wnt Signaling, Stem Cells, and Cancer. *Genes* **2018**, *9*, 121. [CrossRef]
31. Bollu, L.R.; Mazumdar, A.; Savage, M.I.; Brown, P.H. Molecular Pathways: Targeting Protein Tyrosine Phosphatases in Cancer. *Clin. Cancer Res.* **2017**, *23*, 2136–2142. [CrossRef] [PubMed]
32. Grech, G.; Baldacchino, S.; Saliba, C.; Grixti, M.P.; Gauci, R.; Petroni, V.; Fenech, A.G.; Scerri, C. Deregulation of the protein phosphatase 2A, PP2A in cancer: Complexity and therapeutic options. *Tumour Biol.* **2016**, *37*, 11691–11700. [CrossRef] [PubMed]
33. Bunda, S.; Burrell, K.; Heir, P.; Zeng, L.; Alamsahebpour, A.; Kano, Y.; Raught, B.; Zhang, Z.Y.; Zadeh, G.; Ohh, M. Inhibition of SHP2-mediated dephosphorylation of Ras suppresses oncogenesis. *Nat. Commun.* **2015**, *6*, 8859. [CrossRef] [PubMed]

34. Julien, S.G.; Dube, N.; Read, M.; Penney, J.; Paquet, M.; Han, Y.; Kennedy, B.P.; Muller, W.J.; Tremblay, M.L. Protein tyrosine phosphatase 1B deficiency or inhibition delays ErbB2-induced mammary tumorigenesis and protects from lung metastasis. *Nat. Genet.* **2007**, *39*, 338–346. [CrossRef] [PubMed]
35. Fiordalisi, J.J.; Dewar, B.J.; Graves, L.M.; Madigan, J.P.; Cox, A.D. Src-mediated phosphorylation of the tyrosine phosphatase PRL-3 is required for PRL-3 promotion of Rho activation, motility and invasion. *PLoS ONE* **2013**, *8*, e64309. [CrossRef] [PubMed]
36. Cramer, J.M.; Zimmerman, M.W.; Thompson, T.; Homanics, G.E.; Lazo, J.S.; Lagasse, E. Deletion of Ptp4a3 reduces clonogenicity and tumor-initiation ability of colitis-associated cancer cells in mice. *Stem Cell Res.* **2014**, *13*, 164–171. [CrossRef] [PubMed]
37. Foehr, E.D.; Lorente, G.; Vincent, V.; Nikolich, K.; Urfer, R. FAS associated phosphatase (FAP-1) blocks apoptosis of astrocytomas through dephosphorylation of FAS. *J. Neurooncol.* **2005**, *74*, 241–248. [CrossRef] [PubMed]
38. Antony, R.; Lukiw, W.J.; Bazan, N.G. Neuroprotectin D1 induces dephosphorylation of Bcl-xL in a PP2A-dependent manner during oxidative stress and promotes retinal pigment epithelial cell survival. *J. Biol. Chem.* **2010**, *285*, 18301–18308. [CrossRef] [PubMed]
39. Eichhorn, P.J.; Creyghton, M.P.; Bernards, R. Protein phosphatase 2A regulatory subunits and cancer. *Biochim. Biophys. Acta* **2009**, *1795*, 1–15. [CrossRef] [PubMed]
40. Ruvolo, P.P. The broken "off" switch in cancer signaling: PP2A as a regulator of tumorigenesis, drug resistance, and immune surveillance. *BBA Clin.* **2016**, *6*, 87–99. [CrossRef]
41. Kauko, O.; O'Connor, C.M.; Kulesskiy, E.; Sangodkar, J.; Aakula, A.; Izadmehr, S.; Yetukuri, L.; Yadav, B.; Padzik, A.; Laajala, T.D.; et al. PP2A inhibition is a druggable MEK inhibitor resistance mechanism in KRAS-mutant lung cancer cells. *Sci. Transl. Med.* **2018**, *10*, eaaq1093. [CrossRef] [PubMed]
42. Allen-Petersen, B.L.; Risom, T.; Feng, Z.; Wang, Z.; Jenny, Z.P.; Thoma, M.C.; Pelz, K.R.; Morton, J.P.; Sansom, O.J.; Lopez, C.D.; et al. Activation of PP2A and Inhibition of mTOR Synergistically Reduce MYC Signaling and Decrease Tumor Growth in Pancreatic Ductal Adenocarcinoma. *Cancer Res.* **2019**, *79*, 209–219. [CrossRef] [PubMed]
43. Kaur, A.; Denisova, O.V.; Qiao, X.; Jumppanen, M.; Peuhu, E.; Ahmed, S.U.; Raheem, O.; Haapasalo, H.; Eriksson, J.; Chalmers, A.J.; et al. PP2A Inhibitor PME-1 Drives Kinase Inhibitor Resistance in Glioma Cells. *Cancer Res.* **2016**, *76*, 7001–7011. [CrossRef] [PubMed]
44. Roberts, K.G.; Smith, A.M.; McDougall, F.; Carpenter, H.; Horan, M.; Neviani, P.; Powell, J.A.; Thomas, D.; Guthridge, M.A.; Perrotti, D.; et al. Essential requirement for PP2A inhibition by the oncogenic receptor c-KIT suggests PP2A reactivation as a strategy to treat c-KIT+ cancers. *Cancer Res.* **2010**, *70*, 5438–5447. [CrossRef] [PubMed]
45. Smith, A.M.; Dun, M.D.; Lee, E.M.; Harrison, C.; Kahl, R.; Flanagan, H.; Panicker, N.; Mashkani, B.; Don, A.S.; Morris, J.; et al. Activation of protein phosphatase 2A in FLT3+ acute myeloid leukemia cells enhances the cytotoxicity of FLT3 tyrosine kinase inhibitors. *Oncotarget* **2016**, *7*, 47465–47478. [CrossRef] [PubMed]
46. Bockelman, C.; Lassus, H.; Hemmes, A.; Leminen, A.; Westermarck, J.; Haglund, C.; Butzow, R.; Ristimaki, A. Prognostic role of CIP2A expression in serous ovarian cancer. *Br. J. Cancer* **2011**, *105*, 989–995. [CrossRef]
47. Colella, S.; Ohgaki, H.; Ruediger, R.; Yang, F.; Nakamura, M.; Fujisawa, H.; Kleihues, P.; Walter, G. Reduced expression of the Aalpha subunit of protein phosphatase 2A in human gliomas in the absence of mutations in the Aalpha and Abeta subunit genes. *Int. J. Cancer* **2001**, *93*, 798–804. [CrossRef]
48. Fan, Y.L.; Chen, L.; Wang, J.; Yao, Q.; Wan, J.Q. Over expression of PPP2R2C inhibits human glioma cells growth through the suppression of mTOR pathway. *FEBS Lett.* **2013**, *587*, 3892–3897. [CrossRef]
49. Pitre, A.; Davis, N.; Paul, M.; Orr, A.W.; Skalli, O. Synemin promotes AKT-dependent glioblastoma cell proliferation by antagonizing PP2A. *Mol. Biol. Cell* **2012**, *23*, 1243–1253. [CrossRef]
50. Gursel, D.B.; Banu, M.A.; Berry, N.; Marongiu, R.; Burkhardt, J.K.; Kobylarz, K.; Kaplitt, M.G.; Rafii, S.; Boockvar, J.A. Tight regulation between cell survival and programmed cell death in GBM stem-like cells by EGFR/GSK3b/PP2A signaling. *J. Neurooncol.* **2015**, *121*, 19–29. [CrossRef]
51. Palanichamy, K.; Kanji, S.; Gordon, N.; Thirumoorthy, K.; Jacob, J.R.; Litzenberg, K.T.; Patel, D.; Chakravarti, A. NNMT Silencing Activates Tumor Suppressor PP2A, Inactivates Oncogenic STKs, and Inhibits Tumor Forming Ability. *Clin. Cancer Res.* **2017**, *23*, 2325–2334. [CrossRef] [PubMed]

52. Wang, J.; Okkeri, J.; Pavic, K.; Wang, Z.; Kauko, O.; Halonen, T.; Sarek, G.; Ojala, P.M.; Rao, Z.; Xu, W.; et al. Oncoprotein CIP2A is stabilized via interaction with tumor suppressor PP2A/B56. *EMBO Rep.* **2017**, *18*, 437–450. [CrossRef] [PubMed]
53. Kaur, A.; Westermarck, J. Regulation of protein phosphatase 2A (PP2A) tumor suppressor function by PME-1. *Biochem. Soc. Trans.* **2016**, *44*, 1683–1693. [CrossRef] [PubMed]
54. Arnaud, L.; Chen, S.; Liu, F.; Li, B.; Khatoon, S.; Grundke-Iqbal, I.; Iqbal, K. Mechanism of inhibition of PP2A activity and abnormal hyperphosphorylation of tau by I2(PP2A)/SET. *FEBS Lett.* **2011**, *585*, 2653–2659. [CrossRef]
55. Castigli, E.; Sciaccaluga, M.; Schiavoni, G.; Brozzi, F.; Fabiani, R.; Gorello, P.; Gianfranceschi, G.L. GL15 and U251 glioblastoma-derived human cell lines are peculiarly susceptible to induction of mitotic death by very low concentrations of okadaic acid. *Oncol. Rep.* **2006**, *15*, 463–470. [CrossRef]
56. Lu, J.; Zhuang, Z.; Song, D.K.; Mehta, G.U.; Ikejiri, B.; Mushlin, H.; Park, D.M.; Lonser, R.R. The effect of a PP2A inhibitor on the nuclear receptor corepressor pathway in glioma. *J. Neurosurg.* **2010**, *113*, 225–233. [CrossRef]
57. Li, J.Y.; Huang, J.Y.; Li, M.; Zhang, H.; Xing, B.; Chen, G.; Wei, D.; Gu, P.Y.; Hu, W.X. Anisomycin induces glioma cell death via down-regulation of PP2A catalytic subunit in vitro. *Acta Pharmacol. Sin.* **2012**, *33*, 935–940. [CrossRef]
58. Lu, J.; Kovach, J.S.; Johnson, F.; Chiang, J.; Hodes, R.; Lonser, R.; Zhuang, Z. Inhibition of serine/threonine phosphatase PP2A enhances cancer chemotherapy by blocking DNA damage induced defense mechanisms. *Proc. Natl. Acad. Sci. USA* **2009**, *106*, 11697–11702. [CrossRef]
59. Gordon, I.K.; Lu, J.; Graves, C.A.; Huntoon, K.; Frerich, J.M.; Hanson, R.H.; Wang, X.; Hong, C.S.; Ho, W.; Feldman, M.J.; et al. Protein Phosphatase 2A Inhibition with LB100 Enhances Radiation-Induced Mitotic Catastrophe and Tumor Growth Delay in Glioblastoma. *Mol. Cancer Ther.* **2015**, *14*, 1540–1547. [CrossRef]
60. Lankoff, A.; Bialczyk, J.; Dziga, D.; Carmichael, W.W.; Gradzka, I.; Lisowska, H.; Kuszewski, T.; Gozdz, S.; Piorun, I.; Wojcik, A. The repair of gamma-radiation-induced DNA damage is inhibited by microcystin-LR, the PP1 and PP2A phosphatase inhibitor. *Mutagenesis* **2006**, *21*, 83–90. [CrossRef]
61. Lin, S.S.; Bassik, M.C.; Suh, H.; Nishino, M.; Arroyo, J.D.; Hahn, W.C.; Korsmeyer, S.J.; Roberts, T.M. PP2A regulates BCL-2 phosphorylation and proteasome-mediated degradation at the endoplasmic reticulum. *J. Biol. Chem.* **2006**, *281*, 23003–23012. [CrossRef] [PubMed]
62. Scheidtmann, K.H.; Mumby, M.C.; Rundell, K.; Walter, G. Dephosphorylation of simian virus 40 large-T antigen and p53 protein by protein phosphatase 2A: Inhibition by small-t antigen. *Mol. Cell. Biol.* **1991**, *11*, 1996–2003. [CrossRef] [PubMed]
63. Li, H.H.; Cai, X.; Shouse, G.P.; Piluso, L.G.; Liu, X. A specific PP2A regulatory subunit, B56gamma, mediates DNA damage-induced dephosphorylation of p53 at Thr55. *EMBO J.* **2007**, *26*, 402–411. [CrossRef] [PubMed]
64. Shen, Y.; Wang, Y.; Sheng, K.; Fei, X.; Guo, Q.; Larner, J.; Kong, X.; Qiu, Y.; Mi, J. Serine/threonine protein phosphatase 6 modulates the radiation sensitivity of glioblastoma. *Cell Death Dis.* **2011**, *2*, e241. [CrossRef] [PubMed]
65. Ohama, T. The multiple functions of protein phosphatase 6. *Biochim. Biophys. Acta Mol. Cell Res.* **2019**, *1866*, 74–82. [CrossRef] [PubMed]
66. Cohen, P.T.; Philp, A.; Vazquez-Martin, C. Protein phosphatase 4—From obscurity to vital functions. *FEBS Lett.* **2005**, *579*, 3278–3286. [CrossRef]
67. Li, M.; Li, X.; Xu, S.; Xue, P.; Li, Q.; Lu, Q.; Jia, Q.; Zhang, L.; Li, X.; Li, X. Protein phosphatase 4 catalytic subunit is overexpressed in glioma and promotes glioma cell proliferation and invasion. *Tumour Biol.* **2016**, *37*, 11893–11901. [CrossRef] [PubMed]
68. Li, X.; Liang, L.; Huang, L.; Ma, X.; Li, D.; Cai, S. High expression of protein phosphatase 4 is associated with the aggressive malignant behavior of colorectal carcinoma. *Mol. Cancer* **2015**, *14*, 95. [CrossRef]
69. Weng, S.; Wang, H.; Chen, W.; Katz, M.H.; Chatterjee, D.; Lee, J.E.; Pisters, P.W.; Gomez, H.F.; Abbruzzese, J.L.; Fleming, J.B.; et al. Overexpression of protein phosphatase 4 correlates with poor prognosis in patients with stage II pancreatic ductal adenocarcinoma. *Cancer Epidemiol. Biomark. Prev.* **2012**, *21*, 1336–1343. [CrossRef]
70. Wang, B.; Zhao, A.; Sun, L.; Zhong, X.; Zhong, J.; Wang, H.; Cai, M.; Li, J.; Xu, Y.; Liao, J.; et al. Protein phosphatase PP4 is overexpressed in human breast and lung tumors. *Cell Res.* **2008**, *18*, 974–977. [CrossRef]

71. Becker, W.; Kentrup, H.; Klumpp, S.; Schultz, J.E.; Joost, H.G. Molecular cloning of a protein serine/threonine phosphatase containing a putative regulatory tetratricopeptide repeat domain. *J. Biol. Chem.* **1994**, *269*, 22586–22592. [PubMed]
72. Chinkers, M. Targeting of a distinctive protein-serine phosphatase to the protein kinase-like domain of the atrial natriuretic peptide receptor. *Proc. Natl. Acad. Sci. USA* **1994**, *91*, 11075–11079. [CrossRef] [PubMed]
73. Hinds, T.D., Jr.; Sanchez, E.R. Protein phosphatase 5. *Int. J. Biochem. Cell Biol.* **2008**, *40*, 2358–2362. [CrossRef] [PubMed]
74. Atiye, J.; Wolf, M.; Kaur, S.; Monni, O.; Bohling, T.; Kivioja, A.; Tas, E.; Serra, M.; Tarkkanen, M.; Knuutila, S. Gene amplifications in osteosarcoma-CGH microarray analysis. *Genes Chromosomes Cancer* **2005**, *42*, 158–163. [CrossRef]
75. Golden, T.; Swingle, M.; Honkanen, R.E. The role of serine/threonine protein phosphatase type 5 (PP5) in the regulation of stress-induced signaling networks and cancer. *Cancer Metastasis Rev.* **2008**, *27*, 169–178. [CrossRef]
76. Zhi, X.; Zhang, H.; He, C.; Wei, Y.; Bian, L.; Li, G. Serine/Threonine Protein Phosphatase-5 Accelerates Cell Growth and Migration in Human Glioma. *Cell. Mol. Neurobiol.* **2015**, *35*, 669–677. [CrossRef]
77. Kettenbach, A.N.; Schlosser, K.A.; Lyons, S.P.; Nasa, I.; Gui, J.; Adamo, M.E.; Gerber, S.A. Global assessment of its network dynamics reveals that the kinase Plk1 inhibits the phosphatase PP6 to promote Aurora A activity. *Sci. Signal.* **2018**, *11*, eaaq1441. [CrossRef]
78. Rusin, S.F.; Adamo, M.E.; Kettenbach, A.N. Identification of Candidate Casein Kinase 2 Substrates in Mitosis by Quantitative Phosphoproteomics. *Front. Cell Dev. Biol.* **2017**, *5*, 97. [CrossRef]
79. Sooman, L.; Ekman, S.; Tsakonas, G.; Jaiswal, A.; Navani, S.; Edqvist, P.H.; Ponten, F.; Bergstrom, S.; Johansson, M.; Wu, X.; et al. PTPN6 expression is epigenetically regulated and influences survival and response to chemotherapy in high-grade gliomas. *Tumour Biol.* **2014**, *35*, 4479–4488. [CrossRef]
80. Hanafusa, H.; Torii, S.; Yasunaga, T.; Matsumoto, K.; Nishida, E. Shp2, an SH2-containing protein-tyrosine phosphatase, positively regulates receptor tyrosine kinase signaling by dephosphorylating and inactivating the inhibitor Sprouty. *J. Biol. Chem.* **2004**, *279*, 22992–22995. [CrossRef]
81. Chan, G.; Kalaitzidis, D.; Neel, B.G. The tyrosine phosphatase Shp2 (PTPN11) in cancer. *Cancer Metastasis Rev.* **2008**, *27*, 179–192. [CrossRef] [PubMed]
82. Bondeson, M.L. Key insights into the protein tyrosine phosphatase PTPN11/SHP2 associated with noonan syndrome and cancer. *Hum. Mutat.* **2017**, *38*, 337. [CrossRef] [PubMed]
83. Jarvis, L.A.; Toering, S.J.; Simon, M.A.; Krasnow, M.A.; Smith-Bolton, R.K. Sprouty proteins are in vivo targets of Corkscrew/SHP-2 tyrosine phosphatases. *Development* **2006**, *133*, 1133–1142. [CrossRef] [PubMed]
84. Zhang, J.; Zhang, F.; Niu, R. Functions of Shp2 in cancer. *J. Cell. Mol. Med.* **2015**, *19*, 2075–2083. [CrossRef]
85. Bentires-Alj, M.; Paez, J.G.; David, F.S.; Keilhack, H.; Halmos, B.; Naoki, K.; Maris, J.M.; Richardson, A.; Bardelli, A.; Sugarbaker, D.J.; et al. Activating mutations of the noonan syndrome-associated SHP2/PTPN11 gene in human solid tumors and adult acute myelogenous leukemia. *Cancer Res.* **2004**, *64*, 8816–8820. [CrossRef] [PubMed]
86. Lauriol, J.; Jaffre, F.; Kontaridis, M.I. The role of the protein tyrosine phosphatase SHP2 in cardiac development and disease. *Semin. Cell Dev. Biol.* **2015**, *37*, 73–81. [CrossRef]
87. Zheng, H.; Yu, W.M.; Waclaw, R.R.; Kontaridis, M.I.; Neel, B.G.; Qu, C.K. Gain-of-function mutations in the gene encoding the tyrosine phosphatase SHP2 induce hydrocephalus in a catalytically dependent manner. *Sci. Signal.* **2018**, *11*, eaao1591. [CrossRef] [PubMed]
88. Zhang, W.; Chan, R.J.; Chen, H.; Yang, Z.; He, Y.; Zhang, X.; Luo, Y.; Yin, F.; Moh, A.; Miller, L.C.; et al. Negative regulation of Stat3 by activating PTPN11 mutants contributes to the pathogenesis of Noonan syndrome and juvenile myelomonocytic leukemia. *J. Biol. Chem.* **2009**, *284*, 22353–22363. [CrossRef] [PubMed]
89. Bard-Chapeau, E.A.; Li, S.; Ding, J.; Zhang, S.S.; Zhu, H.H.; Princen, F.; Fang, D.D.; Han, T.; Bailly-Maitre, B.; Poli, V.; et al. Ptpn11/Shp2 acts as a tumor suppressor in hepatocellular carcinogenesis. *Cancer Cell* **2011**, *19*, 629–639. [CrossRef]
90. Zhan, Y.; Counelis, G.J.; O'Rourke, D.M. The protein tyrosine phosphatase SHP-2 is required for EGFRvIII oncogenic transformation in human glioblastoma cells. *Exp. Cell Res.* **2009**, *315*, 2343–2357. [CrossRef] [PubMed]

91. Sun, Q.; Mu, L.; Qiao, W.; Li, H.; Tang, J.; Wang, C.; Hu, W.; Zhao, T.; Dong, B.; Song, Y.; et al. Inhibition of SHP-2 promotes radiosensitivity in glioma. *Mol. Med. Rep.* **2015**, *12*, 3563–3568. [CrossRef] [PubMed]
92. Saras, J.; Franzen, P.; Aspenstrom, P.; Hellman, U.; Gonez, L.J.; Heldin, C.H. A novel GTPase-activating protein for Rho interacts with a PDZ domain of the protein-tyrosine phosphatase PTPL1. *J. Biol. Chem.* **1997**, *272*, 24333–24338. [CrossRef] [PubMed]
93. Hoover, A.C.; Strand, G.L.; Nowicki, P.N.; Anderson, M.E.; Vermeer, P.D.; Klingelhutz, A.J.; Bossler, A.D.; Pottala, J.V.; Hendriks, W.J.; Lee, J.H. Impaired PTPN13 phosphatase activity in spontaneous or HPV-induced squamous cell carcinomas potentiates oncogene signaling through the MAP kinase pathway. *Oncogene* **2009**, *28*, 3960–3970. [CrossRef] [PubMed]
94. Spanos, W.C.; Hoover, A.; Harris, G.F.; Wu, S.; Strand, G.L.; Anderson, M.E.; Klingelhutz, A.J.; Hendriks, W.; Bossler, A.D.; Lee, J.H. The PDZ binding motif of human papillomavirus type 16 E6 induces PTPN13 loss, which allows anchorage-independent growth and synergizes with ras for invasive growth. *J. Virol.* **2008**, *82*, 2493–2500. [CrossRef] [PubMed]
95. Abaan, O.D.; Levenson, A.; Khan, O.; Furth, P.A.; Uren, A.; Toretsky, J.A. PTPL1 is a direct transcriptional target of EWS-FLI1 and modulates Ewing's Sarcoma tumorigenesis. *Oncogene* **2005**, *24*, 2715–2722. [CrossRef] [PubMed]
96. Revillion, F.; Puech, C.; Rabenoelina, F.; Chalbos, D.; Peyrat, J.P.; Freiss, G. Expression of the putative tumor suppressor gene PTPN13/PTPL1 is an independent prognostic marker for overall survival in breast cancer. *Int. J. Cancer* **2009**, *124*, 638–643. [CrossRef] [PubMed]
97. Wang, Z.; Shen, D.; Parsons, D.W.; Bardelli, A.; Sager, J.; Szabo, S.; Ptak, J.; Silliman, N.; Peters, B.A.; van der Heijden, M.S.; et al. Mutational analysis of the tyrosine phosphatome in colorectal cancers. *Science* **2004**, *304*, 1164–1166. [CrossRef] [PubMed]
98. Abaan, O.D.; Toretsky, J.A. PTPL1: A large phosphatase with a split personality. *Cancer Metastasis Rev.* **2008**, *27*, 205–214. [CrossRef] [PubMed]
99. Huang, W.; Bei, L.; Eklund, E.A. Inhibition of Fas associated phosphatase 1 (Fap1) facilitates apoptosis of colon cancer stem cells and enhances the effects of oxaliplatin. *Oncotarget* **2018**, *9*, 25891–25902. [CrossRef]
100. Shinoura, N.; Yamamoto, N.; Asai, A.; Kirino, T.; Hamada, H. Adenovirus-mediated transfer of Fas ligand gene augments radiation-induced apoptosis in U-373MG glioma cells. *Jpn. J. Cancer Res.* **2000**, *91*, 1044–1050. [CrossRef]
101. Maleniak, T.C.; Darling, J.L.; Lowenstein, P.R.; Castro, M.G. Adenovirus-mediated expression of HSV1-TK or Fas ligand induces cell death in primary human glioma-derived cell cultures that are resistant to the chemotherapeutic agent CCNU. *Cancer Gene Ther.* **2001**, *8*, 589–598. [CrossRef] [PubMed]
102. Gonzalez-Brito, M.R.; Bixby, J.L. Differential activities in adhesion and neurite growth of fibronectin type III repeats in the PTP-delta extracellular domain. *Int. J. Dev. Neurosci.* **2006**, *24*, 425–429. [CrossRef] [PubMed]
103. Choucair, N.; Mignon-Ravix, C.; Cacciagli, P.; Abou Ghoch, J.; Fawaz, A.; Megarbane, A.; Villard, L.; Chouery, E. Evidence that homozygous PTPRD gene microdeletion causes trigonocephaly, hearing loss, and intellectual disability. *Mol. Cytogenet.* **2015**, *8*, 39. [CrossRef] [PubMed]
104. Uetani, N.; Kato, K.; Ogura, H.; Mizuno, K.; Kawano, K.; Mikoshiba, K.; Yakura, H.; Asano, M.; Iwakura, Y. Impaired learning with enhanced hippocampal long-term potentiation in PTPdelta-deficient mice. *EMBO J.* **2000**, *19*, 2775–2785. [CrossRef] [PubMed]
105. Veeriah, S.; Brennan, C.; Meng, S.; Singh, B.; Fagin, J.A.; Solit, D.B.; Paty, P.B.; Rohle, D.; Vivanco, I.; Chmielecki, J.; et al. The tyrosine phosphatase PTPRD is a tumor suppressor that is frequently inactivated and mutated in glioblastoma and other human cancers. *Proc. Natl. Acad. Sci. USA* **2009**, *106*, 9435–9440. [CrossRef]
106. Solomon, D.A.; Kim, J.S.; Cronin, J.C.; Sibenaller, Z.; Ryken, T.; Rosenberg, S.A.; Ressom, H.; Jean, W.; Bigner, D.; Yan, H.; et al. Mutational inactivation of PTPRD in glioblastoma multiforme and malignant melanoma. *Cancer Res.* **2008**, *68*, 10300–10306. [CrossRef]
107. Ortiz, B.; Fabius, A.W.; Wu, W.H.; Pedraza, A.; Brennan, C.W.; Schultz, N.; Pitter, K.L.; Bromberg, J.F.; Huse, J.T.; Holland, E.C.; et al. Loss of the tyrosine phosphatase PTPRD leads to aberrant STAT3 activation and promotes gliomagenesis. *Proc. Natl. Acad. Sci. USA* **2014**, *111*, 8149–8154. [CrossRef]
108. Brady-Kalnay, S.M.; Tonks, N.K. Identification of the homophilic binding site of the receptor protein tyrosine phosphatase PTP mu. *J. Biol. Chem.* **1994**, *269*, 28472–28477.

109. Sudhir, P.R.; Lin, S.T.; Chia-Wen, C.; Yang, S.H.; Li, A.F.; Lai, R.H.; Wang, M.J.; Chen, Y.T.; Chen, C.F.; Jou, Y.S.; et al. Loss of PTPRM associates with the pathogenic development of colorectal adenoma-carcinoma sequence. *Sci. Rep.* **2015**, *5*, 9633. [CrossRef]
110. Sun, P.H.; Ye, L.; Mason, M.D.; Jiang, W.G. Protein tyrosine phosphatase micro (PTP micro or PTPRM), a negative regulator of proliferation and invasion of breast cancer cells, is associated with disease prognosis. *PLoS ONE* **2012**, *7*, e50183. [CrossRef]
111. Burgoyne, A.M.; Palomo, J.M.; Phillips-Mason, P.J.; Burden-Gulley, S.M.; Major, D.L.; Zaremba, A.; Robinson, S.; Sloan, A.E.; Vogelbaum, M.A.; Miller, R.H.; et al. PTPmu suppresses glioma cell migration and dispersal. *Neuro-Oncology* **2009**, *11*, 767–778. [CrossRef] [PubMed]
112. Burgoyne, A.M.; Phillips-Mason, P.J.; Burden-Gulley, S.M.; Robinson, S.; Sloan, A.E.; Miller, R.H.; Brady-Kalnay, S.M. Proteolytic cleavage of protein tyrosine phosphatase mu regulates glioblastoma cell migration. *Cancer Res.* **2009**, *69*, 6960–6968. [CrossRef] [PubMed]
113. Krueger, N.X.; Saito, H. A human transmembrane protein-tyrosine-phosphatase, PTP zeta, is expressed in brain and has an N-terminal receptor domain homologous to carbonic anhydrases. *Proc. Natl. Acad. Sci. USA* **1992**, *89*, 7417–7421. [CrossRef]
114. Ariyama, T.; Hasegawa, K.; Inazawa, J.; Mizuno, K.; Ogimoto, M.; Katagiri, T.; Yakura, H. Assignment of the human protein tyrosine phosphatase, receptor-type, zeta (PTPRZ) gene to chromosome band 7q31.3. *Cytogenet. Cell Genet.* **1995**, *70*, 52–54. [CrossRef]
115. Canoll, P.D.; Petanceska, S.; Schlessinger, J.; Musacchio, J.M. Three forms of RPTP-beta are differentially expressed during gliogenesis in the developing rat brain and during glial cell differentiation in culture. *J. Neurosci. Res.* **1996**, *44*, 199–215. [CrossRef]
116. Maurel, P.; Rauch, U.; Flad, M.; Margolis, R.K.; Margolis, R.U. Phosphacan, a chondroitin sulfate proteoglycan of brain that interacts with neurons and neural cell-adhesion molecules, is an extracellular variant of a receptor-type protein tyrosine phosphatase. *Proc. Natl. Acad. Sci. USA* **1994**, *91*, 2512–2516. [CrossRef] [PubMed]
117. Muller, S.; Kunkel, P.; Lamszus, K.; Ulbricht, U.; Lorente, G.A.; Nelson, A.M.; von Schack, D.; Chin, D.J.; Lohr, S.C.; Westphal, M.; et al. A role for receptor tyrosine phosphatase zeta in glioma cell migration. *Oncogene* **2003**, *22*, 6661–6668. [CrossRef] [PubMed]
118. Ulbricht, U.; Brockmann, M.A.; Aigner, A.; Eckerich, C.; Muller, S.; Fillbrandt, R.; Westphal, M.; Lamszus, K. Expression and function of the receptor protein tyrosine phosphatase zeta and its ligand pleiotrophin in human astrocytomas. *J. Neuropathol. Exp. Neurol.* **2003**, *62*, 1265–1275. [CrossRef]
119. Patel, A.P.; Tirosh, I.; Trombetta, J.J.; Shalek, A.K.; Gillespie, S.M.; Wakimoto, H.; Cahill, D.P.; Nahed, B.V.; Curry, W.T.; Martuza, R.L.; et al. Single-cell RNA-seq highlights intratumoral heterogeneity in primary glioblastoma. *Science* **2014**, *344*, 1396–1401. [CrossRef] [PubMed]
120. Bourgonje, A.M.; Navis, A.C.; Schepens, J.T.; Verrijp, K.; Hovestad, L.; Hilhorst, R.; Harroch, S.; Wesseling, P.; Leenders, W.P.; Hendriks, W.J. Intracellular and extracellular domains of protein tyrosine phosphatase PTPRZ-B differentially regulate glioma cell growth and motility. *Oncotarget* **2014**, *5*, 8690–8702. [CrossRef] [PubMed]
121. Ulbricht, U.; Eckerich, C.; Fillbrandt, R.; Westphal, M.; Lamszus, K. RNA interference targeting protein tyrosine phosphatase zeta/receptor-type protein tyrosine phosphatase beta suppresses glioblastoma growth in vitro and in vivo. *J. Neurochem.* **2006**, *98*, 1497–1506. [CrossRef] [PubMed]
122. Shi, Y.; Ping, Y.F.; Zhou, W.; He, Z.C.; Chen, C.; Bian, B.S.; Zhang, L.; Chen, L.; Lan, X.; Zhang, X.C.; et al. Tumour-associated macrophages secrete pleiotrophin to promote PTPRZ1 signalling in glioblastoma stem cells for tumour growth. *Nat. Commun.* **2017**, *8*, 15080. [CrossRef] [PubMed]
123. Tohma, Y.; Gratas, C.; Biernat, W.; Peraud, A.; Fukuda, M.; Yonekawa, Y.; Kleihues, P.; Ohgaki, H. PTEN (MMAC1) mutations are frequent in primary glioblastomas (de novo) but not in secondary glioblastomas. *J. Neuropathol. Exp. Neurol.* **1998**, *57*, 684–689. [CrossRef] [PubMed]
124. Li, J.; Yen, C.; Liaw, D.; Podsypanina, K.; Bose, S.; Wang, S.I.; Puc, J.; Miliaresis, C.; Rodgers, L.; McCombie, R.; et al. PTEN, a putative protein tyrosine phosphatase gene mutated in human brain, breast, and prostate cancer. *Science* **1997**, *275*, 1943–1947. [CrossRef] [PubMed]
125. Huse, J.T.; Brennan, C.; Hambardzumyan, D.; Wee, B.; Pena, J.; Rouhanifard, S.H.; Sohn-Lee, C.; le Sage, C.; Agami, R.; Tuschl, T.; et al. The PTEN-regulating microRNA miR-26a is amplified in high-grade glioma and facilitates gliomagenesis in vivo. *Genes Dev.* **2009**, *23*, 1327–1337. [CrossRef] [PubMed]

126. Mueller, S.; Phillips, J.; Onar-Thomas, A.; Romero, E.; Zheng, S.; Wiencke, J.K.; McBride, S.M.; Cowdrey, C.; Prados, M.D.; Weiss, W.A.; et al. PTEN promoter methylation and activation of the PI3K/Akt/mTOR pathway in pediatric gliomas and influence on clinical outcome. *Neuro-Oncology* **2012**, *14*, 1146–1152. [CrossRef] [PubMed]
127. Libermann, T.A.; Nusbaum, H.R.; Razon, N.; Kris, R.; Lax, I.; Soreq, H.; Whittle, N.; Waterfield, M.D.; Ullrich, A.; Schlessinger, J. Amplification, enhanced expression and possible rearrangement of EGF receptor gene in primary human brain tumours of glial origin. *Nature* **1985**, *313*, 144–147. [CrossRef] [PubMed]
128. Sugawa, N.; Ekstrand, A.J.; James, C.D.; Collins, V.P. Identical splicing of aberrant epidermal growth factor receptor transcripts from amplified rearranged genes in human glioblastomas. *Proc. Natl. Acad. Sci. USA* **1990**, *87*, 8602–8606. [CrossRef]
129. Putz, U.; Howitt, J.; Doan, A.; Goh, C.P.; Low, L.H.; Silke, J.; Tan, S.S. The tumor suppressor PTEN is exported in exosomes and has phosphatase activity in recipient cells. *Sci. Signal.* **2012**, *5*, ra70. [CrossRef]
130. Godlewski, J.; Krichevsky, A.M.; Johnson, M.D.; Chiocca, E.A.; Bronisz, A. Belonging to a network—MicroRNAs, extracellular vesicles, and the glioblastoma microenvironment. *Neuro-Oncology* **2015**, *17*, 652–662. [CrossRef]
131. Huang, C.Y.; Tan, T.H. DUSPs, to MAP kinases and beyond. *Cell Biosci.* **2012**, *2*, 24. [CrossRef]
132. Li, M.; Zhou, J.Y.; Ge, Y.; Matherly, L.H.; Wu, G.S. The phosphatase MKP1 is a transcriptional target of p53 involved in cell cycle regulation. *J. Biol. Chem.* **2003**, *278*, 41059–41068. [CrossRef] [PubMed]
133. Yu, H.; Park, J.; Lee, J.; Choi, K.; Choi, C. Constitutive Expression of MAP Kinase Phosphatase-1 Confers Multi-drug Resistance in Human Glioblastoma Cells. *Cancer Res. Treat.* **2012**, *44*, 195–201. [CrossRef] [PubMed]
134. Lomonaco, S.L.; Kahana, S.; Blass, M.; Brody, Y.; Okhrimenko, H.; Xiang, C.; Finniss, S.; Blumberg, P.M.; Lee, H.K.; Brodie, C. Phosphorylation of protein kinase Cdelta on distinct tyrosine residues induces sustained activation of Erk1/2 via down-regulation of MKP-1: Role in the apoptotic effect of etoposide. *J. Biol. Chem.* **2008**, *283*, 17731–17739. [CrossRef]
135. Yip-Schneider, M.T.; Lin, A.; Marshall, M.S. Pancreatic tumor cells with mutant K-ras suppress ERK activity by MEK-dependent induction of MAP kinase phosphatase-2. *Biochem. Biophys. Res. Commun.* **2001**, *280*, 992–997. [CrossRef]
136. Wang, H.Y.; Cheng, Z.; Malbon, C.C. Overexpression of mitogen-activated protein kinase phosphatases MKP1, MKP2 in human breast cancer. *Cancer Lett.* **2003**, *191*, 229–237. [CrossRef]
137. Gaedcke, J.; Grade, M.; Jung, K.; Camps, J.; Jo, P.; Emons, G.; Gehoff, A.; Sax, U.; Schirmer, M.; Becker, H.; et al. Mutated KRAS results in overexpression of DUSP4, a MAP-kinase phosphatase, and SMYD3, a histone methyltransferase, in rectal carcinomas. *Genes Chromosomes Cancer* **2010**, *49*, 1024–1034. [CrossRef] [PubMed]
138. Sieben, N.L.; Oosting, J.; Flanagan, A.M.; Prat, J.; Roemen, G.M.; Kolkman-Uljee, S.M.; van Eijk, R.; Cornelisse, C.J.; Fleuren, G.J.; van Engeland, M. Differential gene expression in ovarian tumors reveals Dusp 4 and Serpina 5 as key regulators for benign behavior of serous borderline tumors. *J. Clin. Oncol.* **2005**, *23*, 7257–7264. [CrossRef] [PubMed]
139. Chitale, D.; Gong, Y.; Taylor, B.S.; Broderick, S.; Brennan, C.; Somwar, R.; Golas, B.; Wang, L.; Motoi, N.; Szoke, J.; et al. An integrated genomic analysis of lung cancer reveals loss of DUSP4 in EGFR-mutant tumors. *Oncogene* **2009**, *28*, 2773–2783. [CrossRef] [PubMed]
140. Waha, A.; Felsberg, J.; Hartmann, W.; von dem Knesebeck, A.; Mikeska, T.; Joos, S.; Wolter, M.; Koch, A.; Yan, P.S.; Endl, E.; et al. Epigenetic downregulation of mitogen-activated protein kinase phosphatase MKP-2 relieves its growth suppressive activity in glioma cells. *Cancer Res.* **2010**, *70*, 1689–1699. [CrossRef] [PubMed]
141. Muller, T.; Gessi, M.; Waha, A.; Isselstein, L.J.; Luxen, D.; Freihoff, D.; Freihoff, J.; Becker, A.; Simon, M.; Hammes, J.; et al. Nuclear exclusion of TET1 is associated with loss of 5-hydroxymethylcytosine in IDH1 wild-type gliomas. *Am. J. Pathol.* **2012**, *181*, 675–683. [CrossRef] [PubMed]
142. Reitman, Z.J.; Yan, H. Isocitrate dehydrogenase 1 and 2 mutations in cancer: Alterations at a crossroads of cellular metabolism. *J. Natl. Cancer Inst.* **2010**, *102*, 932–941. [CrossRef] [PubMed]
143. Molenaar, R.J.; Radivoyevitch, T.; Maciejewski, J.P.; van Noorden, C.J.; Bleeker, F.E. The driver and passenger effects of isocitrate dehydrogenase 1 and 2 mutations in oncogenesis and survival prolongation. *Biochim. Biophys. Acta* **2014**, *1846*, 326–341. [CrossRef] [PubMed]
144. Malta, T.M.; de Souza, C.F.; Sabedot, T.S.; Silva, T.C.; Mosella, M.S.; Kalkanis, S.N.; Snyder, J.; Castro, A.V.B.; Noushmehr, H. Glioma CpG island methylator phenotype (G-CIMP): Biological and clinical implications. *Neuro-Oncology* **2018**, *20*, 608–620. [CrossRef] [PubMed]

145. Hoshide, R.; Jandial, R. 2016 World Health Organization Classification of Central Nervous System Tumors: An Era of Molecular Biology. *World Neurosurg.* **2016**, *94*, 561–562. [CrossRef] [PubMed]
146. Louis, D.N.; Perry, A.; Reifenberger, G.; von Deimling, A.; Figarella-Branger, D.; Cavenee, W.K.; Ohgaki, H.; Wiestler, O.D.; Kleihues, P.; Ellison, D.W. The 2016 World Health Organization Classification of Tumors of the Central Nervous System: A summary. *Acta Neuropathol.* **2016**, *131*, 803–820. [CrossRef] [PubMed]
147. Shang, X.; Vasudevan, S.A.; Yu, Y.; Ge, N.; Ludwig, A.D.; Wesson, C.L.; Wang, K.; Burlingame, S.M.; Zhao, Y.J.; Rao, P.H.; et al. Dual-specificity phosphatase 26 is a novel p53 phosphatase and inhibits p53 tumor suppressor functions in human neuroblastoma. *Oncogene* **2010**, *29*, 4938–4946. [CrossRef] [PubMed]
148. Bourgonje, A.M.; Verrijp, K.; Schepens, J.T.; Navis, A.C.; Piepers, J.A.; Palmen, C.B.; van den Eijnden, M.; Hooft van Huijsduijnen, R.; Wesseling, P.; Leenders, W.P.; et al. Comprehensive protein tyrosine phosphatase mRNA profiling identifies new regulators in the progression of glioma. *Acta Neuropathol. Commun.* **2016**, *4*, 96. [CrossRef] [PubMed]
149. Tanuma, N.; Nomura, M.; Ikeda, M.; Kasugai, I.; Tsubaki, Y.; Takagaki, K.; Kawamura, T.; Yamashita, Y.; Sato, I.; Sato, M.; et al. Protein phosphatase Dusp26 associates with KIF3 motor and promotes N-cadherin-mediated cell-cell adhesion. *Oncogene* **2009**, *28*, 752–761. [CrossRef]
150. Wang, H.; Quah, S.Y.; Dong, J.M.; Manser, E.; Tang, J.P.; Zeng, Q. PRL-3 down-regulates PTEN expression and signals through PI3K to promote epithelial-mesenchymal transition. *Cancer Res.* **2007**, *67*, 2922–2926. [CrossRef]
151. Abdollahi, P.; Vandsemb, E.N.; Hjort, M.A.; Misund, K.; Holien, T.; Sponaas, A.M.; Ro, T.B.; Slordahl, T.S.; Borset, M. Src Family Kinases Are Regulated in Multiple Myeloma Cells by Phosphatase of Regenerating Liver-3. *Mol. Cancer Res.* **2017**, *15*, 69–77. [CrossRef]
152. Zhang, Y.; Li, Z.; Fan, X.; Xiong, J.; Zhang, G.; Luo, X.; Li, K.; Jie, Z.; Cao, Y.; Huang, Z.; et al. PRL-3 promotes gastric cancer peritoneal metastasis via the PI3K/AKT signaling pathway in vitro and in vivo. *Oncol. Lett.* **2018**, *15*, 9069–9074. [CrossRef] [PubMed]
153. Li, B.H.; Wang, Y.; Wang, C.Y.; Zhao, M.J.; Deng, T.; Ren, X.Q. Up-Regulation of Phosphatase in Regenerating Liver-3 (PRL-3) Contributes to Malignant Progression of Hepatocellular Carcinoma by Activating Phosphatase and Tensin Homolog Deleted on Chromosome Ten (PTEN)/Phosphoinositide 3-Kinase (PI3K)/AKT Signaling Pathway. *Med. Sci. Monit.* **2018**, *24*, 8105–8114. [PubMed]
154. Xie, H.; Wang, H. PRL-3 promotes breast cancer progression by downregulating p14(ARF)-mediated p53 expression. *Oncol. Lett.* **2018**, *15*, 2795–2800. [PubMed]
155. Vandsemb, E.N.; Bertilsson, H.; Abdollahi, P.; Storkersen, O.; Vatsveen, T.K.; Rye, M.B.; Ro, T.B.; Borset, M.; Slordahl, T.S. Phosphatase of regenerating liver 3 (PRL-3) is overexpressed in human prostate cancer tissue and promotes growth and migration. *J. Transl. Med.* **2016**, *14*, 71. [CrossRef] [PubMed]
156. Mu, N.; Gu, J.; Liu, N.; Xue, X.; Shu, Z.; Zhang, K.; Huang, T.; Chu, C.; Zhang, W.; Gong, L.; et al. PRL-3 is a potential glioblastoma prognostic marker and promotes glioblastoma progression by enhancing MMP7 through the ERK and JNK pathways. *Theranostics* **2018**, *8*, 1527–1539. [CrossRef] [PubMed]
157. Soni, P.; Husain, N.; Chandra, A.; Ojha, B.K.; Bhatt, M.L.; Gupta, R.K. Do phosphatase of regenerating liver-3, matrix metalloproteinases-2, matrix metalloproteinases-9, and epidermal growth factor receptor-1 predict response to therapy and survival in glioblastoma multiforme? *Indian J. Pathol. Microbiol.* **2016**, *59*, 287–293. [PubMed]
158. Kong, L.; Li, Q.; Wang, L.; Liu, Z.; Sun, T. The value and correlation between PRL-3 expression and matrix metalloproteinase activity and expression in human gliomas. *Neuropathology* **2007**, *27*, 516–521. [CrossRef] [PubMed]
159. Gyuris, J.; Golemis, E.; Chertkov, H.; Brent, R. Cdi1, a human G1 and S phase protein phosphatase that associates with Cdk2. *Cell* **1993**, *75*, 791–803. [CrossRef]
160. Hannon, G.J.; Casso, D.; Beach, D. KAP: A dual specificity phosphatase that interacts with cyclin-dependent kinases. *Proc. Natl. Acad. Sci. USA* **1994**, *91*, 1731–1735. [CrossRef]
161. Yu, Y.; Jiang, X.; Schoch, B.S.; Carroll, R.S.; Black, P.M.; Johnson, M.D. Aberrant splicing of cyclin-dependent kinase-associated protein phosphatase KAP increases proliferation and migration in glioblastoma. *Cancer Res.* **2007**, *67*, 130–138. [CrossRef]
162. Li, H.; Jiang, X.; Yu, Y.; Huang, W.; Xing, H.; Agar, N.Y.; Yang, H.W.; Yang, B.; Carroll, R.S.; Johnson, M.D. KAP regulates ROCK2 and Cdk2 in an RNA-activated glioblastoma invasion pathway. *Oncogene* **2015**, *34*, 1432–1441. [CrossRef] [PubMed]

163. Yamashita, Y.; Kasugai, I.; Sato, M.; Tanuma, N.; Sato, I.; Nomura, M.; Yamashita, K.; Sonoda, Y.; Kumabe, T.; Tominaga, T.; et al. CDC25A mRNA levels significantly correlate with Ki-67 expression in human glioma samples. *J. Neurooncol.* **2010**, *100*, 43–49. [CrossRef] [PubMed]
164. Liang, J.; Cao, R.; Zhang, Y.; Xia, Y.; Zheng, Y.; Li, X.; Wang, L.; Yang, W.; Lu, Z. PKM2 dephosphorylation by Cdc25A promotes the Warburg effect and tumorigenesis. *Nat. Commun.* **2016**, *7*, 12431. [CrossRef] [PubMed]
165. Vanan, I.; Dong, Z.; Tosti, E.; Warshaw, G.; Symons, M.; Ruggieri, R. Role of a DNA damage checkpoint pathway in ionizing radiation-induced glioblastoma cell migration and invasion. *Cell. Mol. Neurobiol.* **2012**, *32*, 1199–1208. [CrossRef]
166. Li, Y.; Zhao, S.; Zhen, Y.; Li, Q.; Teng, L.; Asai, A.; Kawamoto, K. A miR-21 inhibitor enhances apoptosis and reduces G(2)-M accumulation induced by ionizing radiation in human glioblastoma U251 cells. *Brain Tumor Pathol.* **2011**, *28*, 209–214. [CrossRef] [PubMed]
167. Suren, D.; Isiksacan Ozen, O. CDC25B, Ki-67, and p53 expressions in reactive gliosis and astrocytomas. *J. BUON* **2013**, *18*, 1006–1011.
168. Nakabayashi, H.; Hara, M.; Shimizu, K. Prognostic significance of CDC25B expression in gliomas. *J. Clin. Pathol.* **2006**, *59*, 725–728. [CrossRef]
169. Niu, M.; Cai, W.; Liu, H.; Chong, Y.; Hu, W.; Gao, S.; Shi, Q.; Zhou, X.; Liu, X.; Yu, R. Plumbagin inhibits growth of gliomas in vivo via suppression of FOXM1 expression. *J. Pharmacol. Sci.* **2015**, *128*, 131–136. [CrossRef]
170. Liu, X.; Cai, W.; Niu, M.; Chong, Y.; Liu, H.; Hu, W.; Wang, D.; Gao, S.; Shi, Q.; Hu, J.; et al. Plumbagin induces growth inhibition of human glioma cells by downregulating the expression and activity of FOXM1. *J. Neurooncol.* **2015**, *121*, 469–477. [CrossRef]
171. Jung, Y.; Joo, K.M.; Seong, D.H.; Choi, Y.L.; Kong, D.S.; Kim, Y.; Kim, M.H.; Jin, J.; Suh, Y.L.; Seol, H.J.; et al. Identification of prognostic biomarkers for glioblastomas using protein expression profiling. *Int. J. Oncol.* **2012**, *40*, 1122–1132. [CrossRef]
172. Lal, N.; Nemaysh, V.; Luthra, P.M. Proteasome mediated degradation of CDC25C and Cyclin B1 in Demethoxycurcumin treated human glioma U87 MG cells to trigger G2/M cell cycle arrest. *Toxicol. Appl. Pharmacol.* **2018**, *356*, 76–89. [CrossRef]
173. Garcia-Morales, P.; Carrasco-Garcia, E.; Ruiz-Rico, P.; Martinez-Mira, R.; Menendez-Gutierrez, M.P.; Ferragut, J.A.; Saceda, M.; Martinez-Lacaci, I. Inhibition of Hsp90 function by ansamycins causes downregulation of cdc2 and cdc25c and G(2)/M arrest in glioblastoma cell lines. *Oncogene* **2007**, *26*, 7185–7193. [CrossRef] [PubMed]
174. Silver, S.J.; Davies, E.L.; Doyon, L.; Rebay, I. Functional dissection of eyes absent reveals new modes of regulation within the retinal determination gene network. *Mol. Cell. Biol.* **2003**, *23*, 5989–5999. [CrossRef]
175. Kumar, J.P.; Moses, K. EGF receptor and Notch signaling act upstream of Eyeless/Pax6 to control eye specification. *Cell* **2001**, *104*, 687–697. [CrossRef]
176. Kenyon, K.L.; Ranade, S.S.; Curtiss, J.; Mlodzik, M.; Pignoni, F. Coordinating proliferation and tissue specification to promote regional identity in the Drosophila head. *Dev. Cell* **2003**, *5*, 403–414. [CrossRef]
177. Matsuoka, S.; Ballif, B.A.; Smogorzewska, A.; McDonald, E.R., III; Hurov, K.E.; Luo, J.; Bakalarski, C.E.; Zhao, Z.; Solimini, N.; Lerenthal, Y.; et al. ATM and ATR substrate analysis reveals extensive protein networks responsive to DNA damage. *Science* **2007**, *316*, 1160–1166. [CrossRef] [PubMed]
178. Stokes, M.P.; Rush, J.; Macneill, J.; Ren, J.M.; Sprott, K.; Nardone, J.; Yang, V.; Beausoleil, S.A.; Gygi, S.P.; Livingstone, M.; et al. Profiling of UV-induced ATM/ATR signaling pathways. *Proc. Natl. Acad. Sci. USA* **2007**, *104*, 19855–19860. [CrossRef]
179. Cook, P.J.; Ju, B.G.; Telese, F.; Wang, X.; Glass, C.K.; Rosenfeld, M.G. Tyrosine dephosphorylation of H2AX modulates apoptosis and survival decisions. *Nature* **2009**, *458*, 591–596. [CrossRef] [PubMed]
180. Krishnan, N.; Jeong, D.G.; Jung, S.K.; Ryu, S.E.; Xiao, A.; Allis, C.D.; Kim, S.J.; Tonks, N.K. Dephosphorylation of the C-terminal tyrosyl residue of the DNA damage-related histone H2A.X is mediated by the protein phosphatase eyes absent. *J. Biol. Chem.* **2009**, *284*, 16066–16070. [CrossRef]
181. Farabaugh, S.M.; Micalizzi, D.S.; Jedlicka, P.; Zhao, R.; Ford, H.L. Eya2 is required to mediate the pro-metastatic functions of Six1 via the induction of TGF-beta signaling, epithelial-mesenchymal transition, and cancer stem cell properties. *Oncogene* **2012**, *31*, 552–562. [CrossRef]

182. Zhang, L.; Yang, N.; Huang, J.; Buckanovich, R.J.; Liang, S.; Barchetti, A.; Vezzani, C.; O'Brien-Jenkins, A.; Wang, J.; Ward, M.R.; et al. Transcriptional coactivator Drosophila eyes absent homologue 2 is up-regulated in epithelial ovarian cancer and promotes tumor growth. *Cancer Res.* **2005**, *65*, 925–932. [PubMed]
183. Wen, Z.; Liang, C.; Pan, Q.; Wang, Y. Eya2 overexpression promotes the invasion of human astrocytoma through the regulation of ERK/MMP9 signaling. *Int. J. Mol. Med.* **2017**, *40*, 1315–1322. [CrossRef]
184. Chung, V.; Mansfield, A.S.; Braiteh, F.; Richards, D.; Durivage, H.; Ungerleider, R.S.; Johnson, F.; Kovach, J.S. Safety, Tolerability, and Preliminary Activity of LB-100, an Inhibitor of Protein Phosphatase 2A, in Patients with Relapsed Solid Tumors: An Open-Label, Dose Escalation, First-in-Human, Phase I Trial. *Clin. Cancer Res.* **2017**, *23*, 3277–3284. [CrossRef] [PubMed]
185. White, C.; Alshaker, H.; Cooper, C.; Winkler, M.; Pchejetski, D. The emerging role of FTY720 (Fingolimod) in cancer treatment. *Oncotarget* **2016**, *7*, 23106–23127. [CrossRef] [PubMed]
186. Sonoda, Y.; Yamamoto, D.; Sakurai, S.; Hasegawa, M.; Aizu-Yokota, E.; Momoi, T.; Kasahara, T. FTY720, a novel immunosuppressive agent, induces apoptosis in human glioma cells. *Biochem. Biophys. Res. Commun.* **2001**, *281*, 282–288. [CrossRef] [PubMed]
187. Kastrinsky, D.B.; Sangodkar, J.; Zaware, N.; Izadmehr, S.; Dhawan, N.S.; Narla, G.; Ohlmeyer, M. Reengineered tricyclic anti-cancer agents. *Bioorg. Med. Chem.* **2015**, *23*, 6528–6534. [CrossRef] [PubMed]
188. Sangodkar, J.; Perl, A.; Tohme, R.; Kiselar, J.; Kastrinsky, D.B.; Zaware, N.; Izadmehr, S.; Mazhar, S.; Wiredja, D.D.; O'Connor, C.M.; et al. Activation of tumor suppressor protein PP2A inhibits KRAS-driven tumor growth. *J. Clin. Investig.* **2017**, *127*, 2081–2090. [CrossRef] [PubMed]
189. Liu, W.; Yu, B.; Xu, G.; Xu, W.R.; Loh, M.L.; Tang, L.D.; Qu, C.K. Identification of cryptotanshinone as an inhibitor of oncogenic protein tyrosine phosphatase SHP2 (PTPN11). *J. Med. Chem.* **2013**, *56*, 7212–7221. [CrossRef] [PubMed]
190. Lu, L.; Zhang, S.; Li, C.; Zhou, C.; Li, D.; Liu, P.; Huang, M.; Shen, X. Cryptotanshinone inhibits human glioma cell proliferation in vitro and in vivo through SHP-2-dependent inhibition of STAT3 activation. *Cell Death Dis.* **2017**, *8*, e2767. [CrossRef] [PubMed]
191. Lu, L.; Li, C.; Li, D.; Wang, Y.; Zhou, C.; Shao, W.; Peng, J.; You, Y.; Zhang, X.; Shen, X. Cryptotanshinone inhibits human glioma cell proliferation by suppressing STAT3 signaling. *Mol. Cell. Biochem.* **2013**, *381*, 273–282. [CrossRef] [PubMed]
192. Tsutsumi, R.; Ran, H.; Neel, B.G. Off-target inhibition by active site-targeting SHP2 inhibitors. *FEBS Open Bio* **2018**, *8*, 1405–1411. [CrossRef] [PubMed]
193. Chen, W.; Lu, Y.; Chen, G.; Huang, S. Molecular evidence of cryptotanshinone for treatment and prevention of human cancer. *Anticancer Agents Med. Chem.* **2013**, *13*, 979–987. [CrossRef]
194. Agarwal, A.; MacKenzie, R.J.; Pippa, R.; Eide, C.A.; Oddo, J.; Tyner, J.W.; Sears, R.; Vitek, M.P.; Odero, M.D.; Christensen, D.J.; et al. Antagonism of SET using OP449 enhances the efficacy of tyrosine kinase inhibitors and overcomes drug resistance in myeloid leukemia. *Clin. Cancer Res.*. **2014**, *20*, 2092–2103. [CrossRef] [PubMed]
195. Nichols, R.J.; Haderk, F.; Stahlhut, C.; Schulze, C.J.; Hemmati, G.; Wildes, D.; Tzitzilonis, C.; Mordec, K.; Marquez, A.; Romero, J.; et al. RAS nucleotide cycling underlies the SHP2 phosphatase dependence of mutant BRAF-, NF1- and RAS-driven cancers. *Nat. Cell Biol.* **2018**, *20*, 1064–1073. [CrossRef] [PubMed]
196. Chen, Y.N.; LaMarche, M.J.; Chan, H.M.; Fekkes, P.; Garcia-Fortanet, J.; Acker, M.G.; Antonakos, B.; Chen, C.H.; Chen, Z.; Cooke, V.G.; et al. Allosteric inhibition of SHP2 phosphatase inhibits cancers driven by receptor tyrosine kinases. *Nature* **2016**, *535*, 148–152. [CrossRef] [PubMed]
197. Dempke, W.C.M.; Uciechowski, P.; Fenchel, K.; Chevassut, T. Targeting SHP-1, 2 and SHIP Pathways: A Novel Strategy for Cancer Treatment? *Oncology* **2018**, *95*, 257–269. [CrossRef]
198. Fujikawa, A.; Nagahira, A.; Sugawara, H.; Ishii, K.; Imajo, S.; Matsumoto, M.; Kuboyama, K.; Suzuki, R.; Tanga, N.; Noda, M.; et al. Small-molecule inhibition of PTPRZ reduces tumor growth in a rat model of glioblastoma. *Sci. Rep.* **2016**, *6*, 20473. [CrossRef]
199. Fujikawa, A.; Sugawara, H.; Tanaka, T.; Matsumoto, M.; Kuboyama, K.; Suzuki, R.; Tanga, N.; Ogata, A.; Masumura, M.; Noda, M. Targeting PTPRZ inhibits stem cell-like properties and tumorigenicity in glioblastoma cells. *Sci. Rep.* **2017**, *7*, 5609. [CrossRef]
200. Zhang, H.; Zhu, W.; Su, X.; Wu, S.; Lin, Y.; Li, J.; Wang, Y.; Chen, J.; Zhou, Y.; Qiu, P.; et al. Triptolide inhibits proliferation and invasion of malignant glioma cells. *J. Neurooncol.* **2012**, *109*, 53–62. [CrossRef]

201. Molina, G.; Vogt, A.; Bakan, A.; Dai, W.; Queiroz de Oliveira, P.; Znosko, W.; Smithgall, T.E.; Bahar, I.; Lazo, J.S.; Day, B.W.; et al. Zebrafish chemical screening reveals an inhibitor of Dusp6 that expands cardiac cell lineages. *Nat. Chem. Biol.* **2009**, *5*, 680–687. [CrossRef] [PubMed]
202. Kaltenmeier, C.T.; Vollmer, L.L.; Vernetti, L.A.; Caprio, L.; Davis, K.; Korotchenko, V.N.; Day, B.W.; Tsang, M.; Hulkower, K.I.; Lotze, M.T.; et al. A Tumor Cell-Selective Inhibitor of Mitogen-Activated Protein Kinase Phosphatases Sensitizes Breast Cancer Cells to Lymphokine-Activated Killer Cell Activity. *J. Pharmacol. Exp. Ther.* **2017**, *361*, 39–50. [CrossRef] [PubMed]
203. Wu, Q.N.; Liao, Y.F.; Lu, Y.X.; Wang, Y.; Lu, J.H.; Zeng, Z.L.; Huang, Q.T.; Sheng, H.; Yun, J.P.; Xie, D.; et al. Pharmacological inhibition of DUSP6 suppresses gastric cancer growth and metastasis and overcomes cisplatin resistance. *Cancer Lett.* **2018**, *412*, 243–255. [CrossRef] [PubMed]
204. Daouti, S.; Li, W.H.; Qian, H.; Huang, K.S.; Holmgren, J.; Levin, W.; Reik, L.; McGady, D.L.; Gillespie, P.; Perrotta, A.; et al. A selective phosphatase of regenerating liver phosphatase inhibitor suppresses tumor cell anchorage-independent growth by a novel mechanism involving p130Cas cleavage. *Cancer Res.* **2008**, *68*, 1162–1169. [CrossRef] [PubMed]
205. Bai, Y.; Yu, Z.H.; Liu, S.; Zhang, L.; Zhang, R.Y.; Zeng, L.F.; Zhang, S.; Zhang, Z.Y. Novel Anticancer Agents Based on Targeting the Trimer Interface of the PRL Phosphatase. *Cancer Res.* **2016**, *76*, 4805–4815. [CrossRef] [PubMed]
206. Thura, M.; Al-Aidaroos, A.Q.; Yong, W.P.; Kono, K.; Gupta, A.; Lin, Y.B.; Mimura, K.; Thiery, J.P.; Goh, B.C.; Tan, P.; et al. PRL3-zumab, a first-in-class humanized antibody for cancer therapy. *JCI Insight* **2016**, *1*, e87607. [CrossRef] [PubMed]
207. Gormally, M.V.; Dexheimer, T.S.; Marsico, G.; Sanders, D.A.; Lowe, C.; Matak-Vinkovic, D.; Michael, S.; Jadhav, A.; Rai, G.; Maloney, D.J.; et al. Suppression of the FOXM1 transcriptional programme via novel small molecule inhibition. *Nat. Commun.* **2014**, *5*, 5165. [CrossRef] [PubMed]
208. Hofstetter, C.P.; Burkhardt, J.K.; Shin, B.J.; Gursel, D.B.; Mubita, L.; Gorrepati, R.; Brennan, C.; Holland, E.C.; Boockvar, J.A. Protein phosphatase 2A mediates dormancy of glioblastoma multiforme-derived tumor stem-like cells during hypoxia. *PLoS ONE* **2012**, *7*, e30059. [CrossRef] [PubMed]
209. Gutierrez, A.; Pan, L.; Groen, R.W.; Baleydier, F.; Kentsis, A.; Marineau, J.; Grebliunaite, R.; Kozakewich, E.; Reed, C.; Pflumio, F.; et al. Phenothiazines induce PP2A-mediated apoptosis in T cell acute lymphoblastic leukemia. *J. Clin. Investig.* **2014**, *124*, 644–655. [CrossRef] [PubMed]
210. Chang, C.Y.; Li, J.R.; Wu, C.C.; Wang, J.D.; Yang, C.P.; Chen, W.Y.; Wang, W.Y.; Chen, C.J. Indomethacin induced glioma apoptosis involving ceramide signals. *Exp. Cell Res.* **2018**, *365*, 66–77. [CrossRef] [PubMed]
211. Chen, L.; Sung, S.S.; Yip, M.L.; Lawrence, H.R.; Ren, Y.; Guida, W.C.; Sebti, S.M.; Lawrence, N.J.; Wu, J. Discovery of a novel shp2 protein tyrosine phosphatase inhibitor. *Mol. Pharmacol.* **2006**, *70*, 562–570. [CrossRef] [PubMed]
212. Hellmuth, K.; Grosskopf, S.; Lum, C.T.; Wurtele, M.; Roder, N.; von Kries, J.P.; Rosario, M.; Rademann, J.; Birchmeier, W. Specific inhibitors of the protein tyrosine phosphatase Shp2 identified by high-throughput docking. *Proc. Natl. Acad. Sci. USA* **2008**, *105*, 7275–7280. [CrossRef] [PubMed]
213. Jin, Y.; Yoon, Y.J.; Jeon, Y.J.; Choi, J.; Lee, Y.J.; Lee, J.; Choi, S.; Nash, O.; Han, D.C.; Kwon, B.M. Geranylnaringenin (CG902) inhibits constitutive and inducible STAT3 activation through the activation of SHP-2 tyrosine phosphatase. *Biochem. Pharmacol.* **2017**, *142*, 46–57. [CrossRef] [PubMed]
214. Tomiyama, A.; Ichimura, K. Signal transduction pathways and resistance to targeted therapies in glioma. *Semin. Cancer Biol.* **2019**.
215. Krueger, A.B.; Drasin, D.J.; Lea, W.A.; Patrick, A.N.; Patnaik, S.; Backos, D.S.; Matheson, C.J.; Hu, X.; Barnaeva, E.; Holliday, M.J.; et al. Allosteric inhibitors of the Eya2 phosphatase are selective and inhibit Eya2-mediated cell migration. *J. Biol. Chem.* **2014**, *289*, 16349–16361. [CrossRef] [PubMed]
216. Harder, B.G.; Blomquist, M.R.; Wang, J.; Kim, A.J.; Woodworth, G.F.; Winkles, J.A.; Loftus, J.C.; Tran, N.L. Developments in Blood-Brain Barrier Penetrance and Drug Repurposing for Improved Treatment of Glioblastoma. *Front. Oncol.* **2018**, *8*, 462. [CrossRef]

cancers

MDPI

Review

Lentiviral Vectors as Tools for the Study and Treatment of Glioblastoma

Claudia Del Vecchio, Arianna Calistri, Cristina Parolin and Carla Mucignat-Caretta *

Department of Molecular Medicine, University of Padova, 35121 Padova, Italy; claudia.delvecchio@unipd.it (C.D.V.); arianna.calistri@unipd.it (A.C.); cristina.parolin@unipd.it (C.P.)
* Correspondence: carla.mucignat@unipd.it; Tel.: +39-049-827-5304

Received: 4 February 2019; Accepted: 19 March 2019; Published: 24 March 2019

Abstract: Glioblastoma (GBM) has the worst prognosis among brain tumors, hence basic biology, preclinical, and clinical studies are necessary to design effective strategies to defeat this disease. Gene transfer vectors derived from the most-studied lentivirus—the Human Immunodeficiency Virus type 1—have wide application in dissecting GBM specific features to identify potential therapeutic targets. Last-generation lentiviruses (LV), highly improved in safety profile and gene transfer capacity, are also largely employed as delivery systems of therapeutic molecules to be employed in gene therapy (GT) approaches. LV were initially used in GT protocols aimed at the expression of suicide factors to induce GBM cell death. Subsequently, LV were adopted to either express small noncoding RNAs to affect different aspects of GBM biology or to overcome the resistance to both chemo- and radiotherapy that easily develop in this tumor after initial therapy. Newer frontiers include adoption of LV for engineering T cells to express chimeric antigen receptors recognizing specific GBM antigens, or for transducing specific cell types that, due to their biological properties, can function as carriers of therapeutic molecules to the cancer mass. Finally, LV allow the setting up of improved animal models crucial for the validation of GBM specific therapies.

Keywords: glioblastoma; lentiviral vectors; gene therapy; animal models

1. Introduction

Glioblastoma (GBM) is the most aggressive brain tumor, with poor prognosis and scarce progression-free survival time, usually leading to death within 20 months after initial diagnosis [1]. Currently, GBM classification by the World Health Organization couples clinical features to molecular phenotype, distinguishing isocitrate dehydrogenase (IDH) wildtype GBMs, mostly primary developed tumors accounting for nearly 90% of cases, from IDH mutant GBMs, usually arising as secondary tumors from lower grade gliomas and otherwise not specified GBMs [2]. However, additional molecular markers maybe found in specific subsets of patients. The blood–brain barrier (BBB) further complicates GBM treatment, limiting the access to the tumor mass when anticancer molecules are systemically administered. Moreover, targeting GBM growth is complicated by the presence, within the tumor mass, of at least two populations of cancer cells, self-renewing tumor stem cells as well as more differentiated ones that respond differently to the various chemo- and radiotherapies, and whose biology is still not completely unraveled. Hence, both basic and clinical research need new tools for wrecking GBM conundrum.

Due to the peculiar biology of the parental viruses, retroviral-based vectors (gamma-retroviral and lentiviral vectors) display some interesting features, including persistent gene expression due to integration of their genome into the host DNA. In the last three decades, lentiviral vectors (LV) have been widely used in the field of GBM research, because of their advantages over gammaretroviral vectors, which were initially developed [3]. In particular, LV are more stable and less prone to insertional mutagenesis with respect to gammaretroviral vectors. Twenty-five years have passed since the earlier studies demonstrated that the Human Immunodeficiency Virus type 1 (HIV-1), the best-known lentivirus, could be modified for efficient gene delivery to lymphocytes [4,5] and to nondividing cells [6]. Currently, third-generation HIV-based vectors, which are highly improved in their transduction efficiency and safety, are adopted in several clinical trials [7]. LV offer several advantages over different viral and nonviral gene delivery vehicles: (i) the ability of transducing both dividing and resting cells; (ii) high efficiency in delivering transgenes to primary and stem cells; (iii) the capability of integrating their genome into the host DNA, leading to a sustained transgene expression; in addition, as mentioned above, LV show a potentially safer integration site profile when compared to gammaretroviral vectors [8]; (iv) a tissue tropism that can be modified by providing the viral core particles with an heterologous glycoprotein envelope. This process, known as pseudotyping, can be exploited to extend the natural tropism of the parental virus, which otherwise would be restricted to CD4 positive T cells and/or to allow an efficient and possibly specific targeting of cell types of interest. Relevant to GBM, LV have been successfully pseudotyped with the envelope of the lymphocytic choriomeningitis virus, resulting in a preferential transduction of tumor cells over the normal surrounding tissue [9]; (v) a low immunogenicity due to the virtual lack of viral protein expression after transduction; (vi) the ability to efficiently deliver not only transgenes, but also non coding (nc)-RNAs, such as small interfering RNAs (siRNAs) and microRNAs (miRNAs); and (vii) effective standardized and relatively easy to adopt methods for vector manipulation and production. These features, all together, led to the adoption of LV in different basic and translational studies. Indeed, LV are currently employed for (i) gene silencing by RNA interference (RNAi) with the aim of both analyzing the biological role of a certain transcript or for its knock-out in therapeutic approaches; (ii) transgene overexpression in several cell types, including primary and stem cells, as well cell marking and tracing, in vitro and in vivo; (iii) strategies aimed at vaccine development; (iv) the generation of transgenic animal models; (v) the induction of pluripotent stem cells starting from adult cells (iPSCs); and (vi) genome editing purposes.

In the field of GBM research, LV were originally designed for gene and cell therapy, yet their use falls well beyond this initial application. Indeed, LV have been valuable tools for exploring the tumor biology and for dissecting functions of various cellular pathways and proteins, including their possible role as therapeutic targets. Furthermore, LV were adopted to create and develop GBM animal models that closely mimic the clinical proteiform manifestations of this tumor (Figure 1). In this review, we aim at summarizing how LV contributed to gain further insights in our understanding of GBM biology and to design new therapeutic strategies, by mainly focusing on studies performed in the animal model.

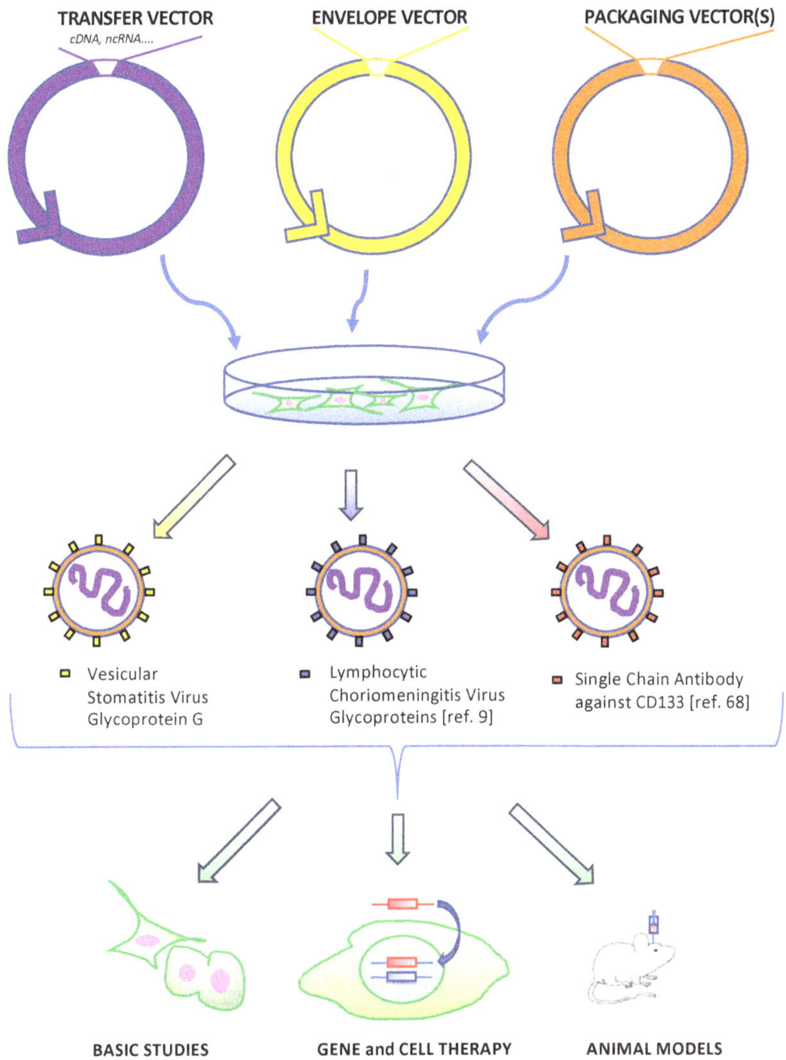

Figure 1. Overview of the main applications of lentiviruses LV in glioblastoma (GBM) research. Once transfected in appropriate cell lines, the packaging and envelope plasmids express the viral structural and enzymatic proteins (packaging vector(s)) along with the envelope glycoprotein (envelope vector) leading to the formation of viral particles that will incorporate the transgene encoding vector (transfer vector). The vesicular stomatitis virus glycoprotein G (VSV-G) is a widely employed envelope which confers to the recombinant particles the ability to infect a large range of target cells, including primary and stem cells. Additional envelopes can be adopted with the aim of restricting vector entry to cell types of interest, as explained in the text. In the context of GBM research, recombinant lentiviral particles were used in basic studies focused on tumor biology/new therapeutic target discovery, in gene and cell therapy approaches and to generate suitable animal models.

2. LV as Tools for Studying Tumor Biology and the Identification of Therapeutic Targets

Given the above described features, LV have been extensively used to explore basic biology of GBM. Indeed, LV have been mainly adopted to achieve effective gene silencing with the aim of exploring the role of different proteins and/or cellular pathways in crucial aspects of GBM development and progression, as well as resistance to chemotherapy and radiation.

RNAi is a natural process through which expression of a targeted gene can be knocked down with high specificity and selectivity. The silencing mechanisms either rely on the degradation of the target mRNA, as in the case of siRNAs and short hairpin RNAs (shRNAs), or on the suppression of mRNA translation, as induced by miRNAs. RNAi can be also accomplished by artificially delivering these small nc-RNA species to the target cells, through the adoption of viral as well as nonviral vectors. LV represent one of the most adopted transfer tools in the field of RNAi, being extremely efficient in transducing small nc-RNAs into different cell types, including primary and stem cells.

In the context of GBM research, LV versatility combined to RNAi potency have been widely exploited. For instance, LV have been valuable tools for the generation of shRNA libraries that allowed a genome-wide, high-throughput genetic screening for the identification of genes involved in tumor initiation, maintenance, and in cell growth inhibition (e.g., see the work by Thaker and colleagues [10]).

By adopting LV designed to transduce specific shRNAs (shRNA-LV), Li and coworkers showed the importance of the dopamine receptor D2 (DRD2) signaling pathway in GBM. Validating this result, DRD2 transcript and protein expression were found increased in clinical glioblastoma specimens when compared to matched non-neoplastic tissues [11]. On the other hand, Wanka and coworkers demonstrated, through LV-mediated shRNA suppression of p53 and of its downstream effector SCO2, that this pathway is involved in tumor resistance to hypoxia [12].

With the same strategy it was shown that the SNARE proteins, which are crucial for the functioning of the cellular machinery involved in the cell interaction with extracellular environment, and more specifically syntaxin1, are crucial for GBM proliferation [13].

LV-mediated expression of shRNAs has been also adopted to investigate the molecular mechanisms contributing to the invasive nature of GBM within the central nervous system (CNS), a feature that highly contributes to the poor prognosis of this tumor. By adopting this strategy, the role of the A Disintegrin And Metalloproteinase (ADAM) family of proteins in invasiveness of GBM cells has been unraveled. Specifically, ADAM-9 was identified as a mediator of tenascin-C-stimulated migration of a specific class of cancer cells, the so-called tumor initiating cells [14]. Tumor initiating cells (TIC), or glioblastoma stem cells (GSC), are defined as cells with self-renewal ability, tumor-initiating capacity, and ability to give rise to a more differentiated progeny. A fraction of these cells seems to be responsible not only for the onset and recurrence of GBM, but also for its characteristic resistance to currently employed treatment [15]. Auvergne and coworkers by LV silencing were able to show a role for the Protease Activated Receptor 1 (PAR1) in self-renewal of human TICs. Indeed, the knock-out of PAR1 expression led to a robust decrease in the tumorigenicity activity of these cells [15]. A similar role was found for the transcriptional modulator High Mobility Group AT-Hook 2 (HMGA2), which is involved in motility and self-renewal of normal and cancer cells. When this factor was targeted by shRNA-LV in a GBM neurosphere cell line (HSR-GBM1), a reduction in cell stemness, invasion capacity, and ability to induce tumors was observed [16]. Furthermore, shRNA-LV allowed the identification of the protooncogene protein tyrosine phosphatase SHP2 as a positive factor for GSC proliferation and tumorigenicity [17].

Different proteins have been connected to GBM cell proliferation by LV-mediated silencing, among these, the fatty acid regulator hydroxysteroid dehydrogenase-like 2 (HSDL2) [18], as well as the ribosomal protein s15a [19].

The tumor microenvironment is greatly involved in driving tumor growth through proinflammatory and tumorigenic molecules that enable communication between tumor and nontumor cells, as a result of NF-κB- and/or AP-1-induced genes. Upstream of these molecules is TRAF3

Interacting Protein 2, which was silenced via LV-mediated shRNA, resulting in a reduced growth of GBM [20].

The cytoskeletal dynamics and angiogenesis are two factors playing a significant role in tumor formation/progression. In the first case, there is a great interest towards the identification of proteins involved in the microtubule formation. Indeed, these cell factors would represent suitable therapeutic targets, given that most of the already approved and in use antitumoral drugs interfere with this process. Stathmin, a 17 kDa regulator of microtubule dynamics, was recently targeted via LV-shRNA in GBM cells, resulting in a reduction of tumor formation and growth in nude mice [21]. Importantly, a recent study employing a new murine model of platelet-derived growth factor receptor-α (PDGFRα)-driven GBM, demonstrated that PDGFRα activity synergizes with the microtubule stabilizer drug vinblastine via stathmin dephosphorylation. This work reveals a connection between vinblastine cytotoxic effect on GBM cells and stathmin, prompting additional investigations on this protein as therapeutic target [22]. As far as angiogenesis is concerned, the role of the endothelium and its interaction with GBM cells have been explored by LV-shRNA, disclosing the role of programmed cell death protein 10 (PDCD10) dysregulation in tumor angiogenesis and in GBM progression [23]. Furthermore, downmodulation by LV-mediated silencing of the vascular endothelial growth factor (VEGF)/VEGF receptor (VEGFR) signaling pathway allowed its association to decreased tumor size coupled to an increased tumor necrosis in orthotopic glioma xenograft models [24]. In a recent study, the methionine aminopeptidase MetAP2 was LV knocked down in tumor cells resulting in reduced angiogenesis and tumor growth, positively affecting survival of mice [25].

Finally, LV transduction of shRNAs were used in the identification of intracellular targets positively correlated with longer patient survival, like phosphodiesterase 5 [26] and thioredoxin-interacting protein [27].

Currently, LV are also adopted as platforms to deliver the novel genome editing machinery, such as CRISPR/Cas9, for the knock-out of specific genes with potential roles in GBM formation and development. In this context, LV have been mainly adopted for the development of therapeutic strategies, an aspect that will be further discussed in the next paragraph of this review.

Even though gene silencing represents one of the main applications of LV in the study of GBM biology, these vectors have been also exploited for their ability to lead to a stable and sustained overexpression of factors potentially involved in peculiar features of this tumor. Interestingly, in dissecting the role of HMGA2 in promoting GBM tumorigenicity, Kaur and colleagues not only silenced its transcript by shRNA-LV, as described above, by they also overexpressed HMGA2 in the HSR-GBM1 neurosphere cell line, further validating RNAi results [16]. Data from patients show that mutated isocitrate dehydrogenase 1 is associated with better prognosis. By overexpressing this protein in glioma stem cells by mean of an ad hoc designed LV, it was proven that GBM cell proliferation, migration and invasiveness were reduced. Identified mechanisms giving reason of this result were the induction of apoptosis and cell differentiation accompanied by a down-regulation of the Wnt/beta-catenin signaling pathway [28]. Furthermore, LV-mediated overexpression led to the identification of the SERCA $Ca2^{+}$-ATPase ATP2A2 as a beneficial factor for survival [29].

Finally, LV ability to efficiently transduce a wide range of cell types, including primary cells and cells in live organisms, has been exploited with the aim of dissecting the complex landscape of GBM protein expression patterns. In particular, Manricque and coworker adopted LV to optimize the methodology known as massively parallel reporter assay (MPRA) that facilitates the systematic analysis of transcriptional regulatory elements [30]. Thanks to the developed platform, those authors were able to demonstrate that local DNA sequence and regional chromatin affect regulatory activity, further advancing our vision of noncoding genome in GBM development [31].

3. LV as Tools for Gene and Cell Therapy of GBM

Due to its characteristics of aggressiveness, recurrence and resistance to traditional therapies (both chemo- and radiotherapy), GBM is considered among solid cancers one of the most suitable targets for innovative therapeutic approaches as gene and cell therapy. Indeed, GBM gene therapy history parallels gene therapy history itself, starting with approaches aimed at the delivery of either suicide or tumor-suppressor genes to the cancer mass, followed by more innovative strategies driven by the increased knowledge of the tumor biology. First of all, as already discussed in the previous paragraph, several proteins and cellular pathways were discovered to play crucial role in different aspects of GBM pathogenesis, and were targeted by gene silencing and, more recently, by genome editing approaches. In 2013, the journal Science selected as the breakthrough of the year cancer immunotherapy [32], a therapeutic strategy made possible by improved understanding of mechanisms accounting for the already well-known ability of cancer cells to evade the immune system control [33]. Viral vectors are employed also in immunotherapy approaches. Indeed, the idea of overcoming the immune suppression typical of the tumor microenvironment to increase the chance of therapeutic success was already exploited in earlier cancer gene therapy protocols, by combining immunomodulatory factors, as certain cytokines, to suicide proteins. In the case of GBM, for instance, Palù and coworkers generated a bicistronic retroviral vector expressing one of the most exploited suicide proteins, the thymidine kinase of herpes simplex virus type 1 (HSV1-tk), along with interleukin 2 (IL-2), which is widely adopted in cancer immunotherapy approaches. The developed vector was tested also in patients [34,35]. Currently, one of the most promising approaches of cancer immunotherapy is represented by viral vector mediated genetic modification of T cells to express chimeric antigen receptors (CAR) directed against specific cancer antigens [36].

Based on their features, LV have been widely used in gene and cell therapy approaches of GBM. In this context, LV were initially adopted mainly to deliver suicide genes, such as HSV1-tk. This viral protein works by activating through phosphorylation the prodrug ganciclovir, which, once activated, can block DNA replication leading to cell death. In the case of GBM, such a strategy resulted in significant tumor reduction (e.g., see [37,38]). Interestingly, the success of this approach was linked to a consistent bystander effect due to the presence of gap junctions among GBM cells [39].

Next, several preclinical and clinical studies adopted RNA interference as a strategy to treat GBM, especially by means of LV delivered shRNAs [40,41], targeting genes and pathways crucial for GBM biology, as previously described.

In addition to shRNAs, in the last years miRNAs have gained particular attention. Normally, these nc-RNAs are endogenously expressed and induce translational silencing/degradation by binding to target transcripts [42]. In this way, miRNAs play important regulatory roles both in physiological and in pathological conditions, cancer included. Indeed, almost all tumors are characterized by a peculiar pattern of miRNA expression that differs from the one found in the corresponding healthy tissue. This feature is the results of several complex mechanisms, such as an altered functioning in the cancer cell of the cellular machinery involved in the miRNA biogenesis [43]. Consequently, upregulation of oncogenes and/or downregulation of tumor suppressors occur that contribute to the development of tumors. GBM is not an exception, and different studies in the past years have clearly shown how miRNAs contribute to the phenotypic diversity of GBM subtypes and can be used as diagnostic and prognostic biomarkers, as well as therapeutic targets [44]. Specific miRNAs have been associated to the increase in GBM cell proliferation, resistance to apoptosis and cell death in general, invasiveness, induction of angiogenesis, and resistance to traditional treatments [45]. Not surprisingly, LV were used to investigate the involvement of specific miRNAs in different aspects of GBM biology. For instance, both overexpression and downregulation were obtained with an ad hoc designed LV, leading to the identification of miR-100 as a protective factor via interactions with the Fibroblast Growth Factor Receptor 3 (FGFR3) signaling pathway [46]. On the other hand, miR-297 was unraveled as a factor promoting survival of GBM cells, by targeting diacylglycerol kinase alpha [47].

Considering their crucial roles in tumor biology and their peculiar transcriptional signature in cancer cells, miRNAs have been considered a suitable target for developing novel therapeutic strategies to fight GBM. Antisense oligonucleotides and the so-called "miRNA sponges" [48] were used in strategies aimed at downregulating miRNAs playing a role in carcinogenesis [49]. In this context, LV have been adopted as a delivery system to achieve an affective and sustained expression of the anti-miRNA molecules also in vivo. For instance, Chen and coworkers adopted a LV to express a miRNA sponge able to block miR-23b function as an oncogene in GBM. Those authors showed that both in a glioma cell line and in an orthotopic mouse model miR-23b inhibition resulted in a significant reduction in tumor malignancy and angiogenesis, as well as in the invasiveness capacity of the tumor cells. As a consequence, cancer progression was affected [50]. On the other hand, the differential expression of specific miRNA in GBM cells with respect to healthy cells was exploited to achieve a targeted expression of suicide genes delivered by LV only within the cancer. Skalsky and coworkers generated an LV expressing HSV1-tk under the transcriptional control of miR-128, which is downregulated in GBM, obtaining a selective killing of cancer cells upon transduction and ganciclovir administration [51].

More recently, gene therapy approaches aimed at interfering with GBM biology have shifted on the use of the novel genome editing technologies. In particular, the Clustered Regularly Interspaced Short Palindromic Repeats (CRISPR) and CRISPR-associated (Cas) 9 system is currently one of the most exploited tools [52] and, as mentioned above, LV are among the most suitable viral vectors for its efficient delivery to several target cells in vitro and in vivo. Under this respect, Tome-Garcia and coworkers have achieved a strong impairment of cancer cell migration by LV-mediated transduction of an ad hoc-designed CRISPR-Cas9 system [53]. On the other hand, in a very recent study, Sun and coworkers demonstrated that GBM growth can be impaired increasing mice survival, by editing the sequence encoding the vascular laminin-411 that is overexpressed in higher grade GBM [54].

Finally, gene therapy approaches for the treatment of GBM based on LV vectors have also exploited their ability to lead to a sustained expression of heterologous genes in vitro and in vivo and their versatility to allocate expression cassettes. As an example, Sanchez-Hernandez and colleagues have recently designed and developed a LV to obtain equimolar expression of the Growth Arrest Specific 1 (GAS1) and of the tumor suppressor phosphatase and tensin homolog (PTEN). Indeed, the aim of the study was to investigate the potential additive effect of these proteins on cancer cell proliferation, considering that GAS1 is known to induce apoptosis in GBM cells, while PTEN blocks the phosphatidylinositol 3-Kinase (PI3K)/protein kinase B or the Akt pathway. To achieve a similar expression level of both proteins, the authors exploited the p2A peptide–base expression system, which has been demonstrated to work well in the context of LV, to allow the expression of both proteins under the transcriptional control of the same promoter. In this way, proliferation of GBM cells was significantly affected in vitro and upon LV inoculation into immunosuppressed mice [55].

An interesting application of LV as tool to overexpress proteins that might help in treating GBM was set up by Lamb and coworkers. The GBM standard therapy in clinical application is based on a combination of temozolomide with radiation therapy. Temozolomide (TMZ) is an alkylating agent that binds to DNA and interferes with replication, leading to breaks within the DNA and, as a consequence, cell death. However, TMZ has a limited activity mainly due to the overexpression in cancer cells of the DNA repair protein O6-methylguanine-DNA methyltransferase (MGMT). On the other hand, it is known that $\gamma\delta$ T cells, a subset of T cells, that are able to recognize stress-associated Natural Killer Group 2D (NKG2D) receptor ligands expressed by GBM and efficiently reduce tumor expansion. Noteworthy, TMZ induces the expression of these ligands on TMZ-resistant GBM cells making them more vulnerable to $\gamma\delta$ T cells. However, TMZ is also toxic to $\gamma\delta$ T cells. To overcome this issue, $\gamma\delta$ T cells were transduced with LV expressing MGMT, thus conferring them resistance to TMZ. Transduced $\gamma\delta$ T cells maintained their biological characteristics and displayed cytotoxicity toward TMZ resistant GBM cell lines, upon administration of TMZ, thus fostering their use to treat GBM [56].

3.1. LV-Based Gene Therapy Approaches Aimed at Overcoming GMB Resistance to Therapy

One of the peculiar features of GBM is the onset of resistance to both radiation and chemotherapy which leads to rapid recurrence after treatment.

By generating LV expressing an anti-MGMT shRNA, Viel and coworkers observed a specific inhibition of the MGMT expression in GBM cell lines as well as in the animal model. Furthermore, combining the developed LV with TMZ treatment those Authors were able to show a significant reduction in tumor formation [57].

In addition to MGMT, p53 is known to be involved in GBM resistance to TMZ [58]. Furthermore, protein phosphatase 1D magnesium-dependent PPM1D, a member of the protein phosphatase 2C (PP2C) family, seems to play a role as well by modulating p53 function [59]. Wang and colleagues showed that silencing of PPMD1 by LV-mediated expression of a specific shRNA improved the effect of TMZ by inducing apoptosis of GBM cells mainly through the PIK3R1/AKT pathway [59]. More recently, starting from the observation that SRY-Box 9 (SOX9) protein expression is linked to poor prognosis in GBM patients, its encoding transcript was targeted via shRNA-LV, demonstrating that this protein along with carbonic anhydrase 9 (CA9) are part of an oncogenic pathway. Furthermore, this study clearly showed that the inhibition of such a pathway leads to an increased sensitivity of GBM to TMZ [60].

On the other hand, not only resistance to TMZ is a major problem in GBM therapy, but also radioresistance contributes to scarce efficacy of therapy. The interaction of various proteins with radiation therapy has been exploited using LV. Shi and coworkers investigated the effect of K5, a kringle domain of plasminogen known to induce apoptosis of dermal microvessel endothelial cells, combined with the positive feedback effect of the Early growth response-1 (Egr1) promoter on sodium/iodide symporter (NIS) gene upon ^{131}I radiation. To this end, they generated a LV expressing K5 under the transcriptional control of the cytomegalovirus promoter and NIS under the control of the Egr1 promoter. The GBM cell line U87 was transduced with the developed LV and employed to induce tumor formation in nude mice. In this model, K5 and irradiation showed an addictive effect on glioma growth, cell proliferation and capillary density [61]. Oxidative stress may interact with cell survival: when Nox4 NADPH oxidase, which contributes to generate reactive oxygen species in GBM cells, was knocked-down via LV, tumor proliferation, invasion and radioresistance were diminished [62]. Stably expressing low-represented miRNA MiR-224 by means of LV was also shown to increase sensitivity of GBM cells to radiations [63]. Finally, the transcript encoding ATM protein, which is mutated in ataxia-telangiectasia development conferring radiosensitivity to the patients, was silenced by LV-mediated RNAi increasing tumor response to radiotherapy [64].

Interestingly, GSCs seem to play a crucial role in GBM resistance to both radio and chemotherapy due to cellular and microenviromental mechanisms [65–67]. Interestingly, the possibility to pseudotype LV offers the opportunity of specifically targeting a subset of GSCs that express on their surface the glycosylated pentaspan transmembrane protein CD133. Specifically, Bayin and coworkers were able to generate a LV displaying a single chain antibody against CD133 on its envelope. Those authors showed that such a LV selectively transduced CD133-expressing cells in GBM xenografts in NOD SCID mice, while sparing normal brain tissue [68]. This LV represents a valuable tool to dissect the role of CD133 positive GCSs in tumor biology in general and in its resistance to chemotherapy in particular. Furthermore, it can have an application in gene therapy approaches aimed at targeting this subset of cells in particular.

3.2. Immunotherapy of GBM via LV Expressing CAR Specific to GBM Antigens

As mentioned above, immunotherapy of cancer is emerging as one of the most promising strategies to fight cancer in general and GBM in particular. Among immunotherapy approaches, one of the most efficient is represented by the injection of genetically modification of autologous T cells to express chimeric antigen receptors (CARs) with defined specificities for tumor-associated antigens independent of MHC restriction. Typically, CARs are composed of domains of synthetic antibodies recognizing the selected antigen along with intracellular activation/stimulatory domains that transduce signals from the surface receptor inside the cell. Different generations of CARs do exist that differ in the number of intracellular stimulatory domains. First-generation CARs contain a single activation domain (i.e., the T cell receptor ζ chain CD3ζ). Second- and third-generation CARs consist of one or two additional costimulatory signaling domains, such as CD27, CD28, CD134 and CD137, leading to an increase in activated T cell overall survival, proliferation and persistence [69]. This innovative cell therapy approach has been shown to be particularly active in the case of hematological malignancies, while the results for solid tumors have not been as encouraging, likely due to their specific immunobiological features [70]. Interestingly, clinical studies have shown the safety and feasibility of this type of cell therapy in the treatment of GBM and signs of success at least in a subset of patients [71]. Different CARs directed against antigens expressed on the GBM cell surface have been generated and tested. In particular, major exploited targets include so far, the epidermal growth factor receptor variant III (EGFRvIII), the human epidermal growth factor receptor 2 (HER2), the IL13 receptor alpha 2 (IL13Rα2) and ephrin type A receptor 2 (EphA2). CAR T cell therapy against GBM targeting EGFRvIII and HER2 are already in clinical trials [72].

Even though retroviral vectors have been traditionally used in approaches of CAR therapy of GBM, LV were also used in this context. For instance, Kuramitsu and colleagues constructed a LV expressing a CAR directed against the EGFRvIII, to transduce CD3 positive (+) T cells, rendering them able to specifically lyse EGFRvIII positive GBM cells. Importantly CAR-expressing CD3(+) T cells reached intracranial xenografts of GBM in mice [73]. These results were further expanded by targeting both the mutated and the wild type (wt) form of EGFR. Specifically, natural killer (NK) lymphocytes were transduced with a LV expressing a second generation CAR targeting both wild type EGFR and EGFRvIII. In detail, the human NK cell line NK-92 was transduced and adopted to intracranially inject two orthotopic xenograft mice models, achieving efficient suppression of tumor growth and prolonged animal survival [74]. More recently Murakami and coworkers generated a novel type of NK cell line (KHYG-1) transduced with a LV to express a CAR targeting ECGRvIII. The authors were able to establish the transformed cell line and to show that it inhibited GBM cell growth via apoptosis depending on EGFRvIII expression [75].

On the other hand, LV have been also used to express CAR directed towards novel GBM specific antigens. In this context, Shiina and coworkers generated a LV expressing a third-generation CAR specifically recognizing podoplanin, a transmembrane mucin-like glycoprotein, expressed in the mesenchymal GBM subtype. The developed LV was used to obtain CAR-T cells that were active against podoplanin-positive GBM cells in vitro. Furthermore, upon injection of transduced T lymphocytes in immunodeficient mice, a significant growth inhibition of GBM xenograft was observed [76].

3.3. LV Transduced Cells as Carriers of Therapeutic Molecules for the Treatment of GBM

A different approach to cell therapy takes advantage of specific cell types that can be used as carriers for delivering therapeutic genes directly to the cancer mass. Different studies have clearly demonstrated that stem cells, in particular neural stem cells (NSCs) and mesenchymal stem cells (MSCs), have a tropism for brain tumors [77,78]. This feature has prompted the adoption of stem cells as vehicles for targeted therapeutic molecule delivery [79]. One of the main obstacles to the delivery of both viral vectors and stem cells to a tumor mass located within the CNS is represented by the BBB. Even though the direct injection into the brain works, the associated risks and the procedure complexity might render this method difficult to apply in clinical trials and eventually in a clinical setting. Much work has been carried on finding the best route of modified stem cells administration to overcome the blood–brain barrier [80,81]. In the case of GBM, intranasal delivery of cell carriers appears to be an efficient method [82]. Furthermore, LV, for their capacity of transducing stem cells, have been widely adopted as viral vectors to transform both MSCs and NSCs, also in the context of GBM cell therapy. In 2011, Balyasnikova and colleagues published a paper focused on the adoption LV-engineered NSCs to express membrane-bound tumor necrosis factor-α-related apoptosis-inducing ligand (NSCs-mTRAIL). Indeed, while it is well known that TRAIL can selectively induce apoptosis in GBM cells, some tumors were found resistant to the soluble form of this proapoptotic agent. The authors showed that NSCs-mTRAIL, improved the survival of mice bearing intracranial glial tumor xenografts [83].

A very interesting study, fully exploiting LV versatility, was carried on by Bagò and coworkers [84]. In their work these authors adopted LV both to set up a novel strategy to rapidly converting human skin fibroblasts into tumor-homing early-stage induced NSCs, named h-iNSC^{TE}. Next, they further adopted LV to engineer h-iNSC^{TE} to express reporter proteins and different therapeutic molecules. Finally, they showed the tumor-homing migration and therapeutic efficacy of transformed h-iNSC^{TE} in two different mouse models of GBM, one based on orthotopic established GBMs and the other on implantation of patient-derived cancer lines [84]. Specifically, stem cells transformed with LV expressing either TRAIL or HSV1-tk significantly affected the size of GBM xenografts and displayed a positive effect on animal survival. Furthermore, the authors were able to show that delivery of HSV1-tk expressing h-iNSC^{TE} into the postoperative surgical resection cavity delayed the regrowth of residual GMB in mice. Altogether these data demonstrate that h-iNSC^{TE} might represent a valid platform to generate tumor-homing cytotoxic cells to be applied for the treatment of GBM.

LV have been used to efficiently transduce also MSCs leading to a sustained expression of the transgene of interest. In the context of GBM, Bak and coworkers employed MSCs derived from human embryonic stem cells to deliver HSV1-tk to the tumor, showing that the transduced cells were able to migrate from the site of inoculum in one cerebral hemisphere towards the cancer located in the opposite side of the mouse brain, delivering the suicide gene [85].

LV-expressing cytosine deaminase (CD) or enhanced green fluorescent protein (eGFP) was constructed and transduced into rat MSCs (MSC-CD/eGFP). In addition to HSV1-tk, CD is another widely exploited suicide protein that activates the prodrug 5-fluocytosine into a cytotoxic agent, the 5-fluorouracyl (5-FU), which is phosphorylated inside the cells. The active metabolites can be incorporated into DNA and RNA resulting in the inhibition of their synthesis and cell death. As for HSV1-tk based strategy, also in this case, gap junctions mediate the bystander killing of tumor cells. The LV-mediated expression of the reporter protein allowed the tracing of the transduced MSCs in vivo, upon their intracranial injection in a GBM rat model. The cells were found to localize mainly at the interface between cancer and normal tissue. After administration of 5-FU, MSC-CD/eGFP determined a significant decrease in the tumor mass with increased cancer cell apoptosis and mice survival [86]. Balyasnikova and colleagues adopted LV engineered human bone marrow derived MSCs to express m-TRAIL. The authors were able to show that MSCs can migrate into the brain after nasal instillation and infiltrate intracranial glioma xenografts in a mouse model. Furthermore, MSCs expressing m-TRAIL increased the survival of mice treated with a combination of m-TRAIL expressing MSCs and irradiation [82]. An abundant and accessible source for MSCs is the adipose tissue. De Melo and colleagues succeeded in purifying MSCs from this source and transduced them with an LV expressing HSV1-tk. The Authors demonstrated that GBM cells died when co-cultured with these cells and upon GCV treatment. More importantly, they showed that when injected into the brains of nude mice bearing GBM, LV-transduced MSCs were chemoattracted to the tumor and killed cancer cells in situ, via a bystander effect, without significantly affecting normal brain tissue [87].

In an interesting approach, instead of NSCs or MSCs, patient-derived olfactory ensheathing cells (OECs) were modified by LV transduction to carry HSV1-tk to primary cultured human GBM, taking advantage of OECs natural tropism to the brain. Furthermore, these cells should reduce technical and ethical problems, being obtained from the olfactory mucosa, a source easily accessible from the patients themselves. The study showed that transduced cells affected survival of tumor cells after ganciclovir administration. This finding opened up the possibility to exploit OECs as vehicle to transfer anticancer molecules to brain tumors [88]. Indeed, more recently, Carvalho and coworkers studied OEC application in a cytotoxic therapeutic approach of GBM in vivo. Specifically, they transduced OECs with an LV-expressing CD fused in frame to the uracil phosphoribosyltransferase (CU) that should enhance the overall antitumor effect [89]. By injecting patient-derived GSCs in the striatum of nude mice brains, the authors were able to show that OECs migrated from the nasal cavity to the tumors, reached infiltrative GBM cells, and delivered the fusion protein, affecting tumor growth and mice survival. This work represents the first demonstration that OECs can efficiently target GBM in vivo and deliver therapeutic transgenes upon intranasal delivery [90].

Finally, a very interesting recent study demonstrated that it is possible to use genetically engineered macrophages to drive the expression of secreted proteins that can influence the GBM microenvironment, enhancing the immune system recognition of cancer cells. Importantly, to this end, a novel LV expression system was designed and tested that allowed the modification of macrophages [91]. This study clearly demonstrates the versatility of LV and their feasibility in approaches that require vectors easy to manipulate and efficient in the transduction of primary cells.

4. LV to Create Animal Models

One of the major challenges in studying GBM is the difficulty of reproducing in the laboratory all the features of this tumor, that has no clear molecular signature and whose effects may vary also according to the site of origin within the brain.

Different experimental GBM models are available in both invertebrates and vertebrates, from chemically induced to xeno- and allograft as well as viral mediated or genetically engineered animals [92], for the study of tumor biology and the development of new treatments. Despite easy manipulation and low-cost, few models have been created in Drosophila by manipulating glial progenitor cells [93]. More options are available for the zebrafish *Danio rerio*, which allows an easy tracing of single cells within the tumor mass induced via xenograft in animals [94], or in tumors originated by CRISPR/Cas9 genetic tailoring [95]. A model closer to human is the dog, which can develop spontaneous brain tumors with histological features similar to human counterpart, but ethical reasons and long timeframe for tumor development hamper its wide use as a model for GBM studies [96]. The most widely used animal models for the study of GBM are generated in rodents, which allow flexible and fast manipulations in a complex brain system. Initially, GBM-like tumors were induced in rodents by ethylnitrosurea injection or methylnitrosurea administration, but the complexity and low reproducibility of the procedure lead to its discontinuation [97]. Next, allograft and xenograft models were established, with the limits that the initial stages of tumor development are different from naturally occurring disease. Moreover, in the case of xenograft, immunodeficient mice are required so that tumors escape the immune destruction step, a phase taking place in naturally occurring cancers. Recently, however, an interesting approach was designed for mice orthotopic tumors, in which primary cells from patients or GBM cell lines were transduced with LV expressing green fluorescent protein (GFP)/firefly luciferase fusion proteins before implantation in the frontal lobe of mice. These proteins were detected in vivo via bioluminescence imaging [98], to allow the monitoring of GBM development also during chemotherapy administration.

LV were used in immunocompetent mice to overexpress oncogenes under Cre-loxP control, inducing tumors in some brain regions of the adult mouse: the expression in specific cell types or regions (subventricular zone or hippocampus) of the activated forms of different oncoproteins such as H-Ras and AKT successfully induced GBM growth [99,100], pointing at LV overexpressing these factors under the control of Cre-loxP as a powerful tool for GBM generation. Similarly, LV have been adopted to express Cre recombinase in mice carrying conditional alleles for different oncogenes and oncosuppressors successfully generating high-grade gliomas (e.g., see the work by de Vries and colleagues [101]). In a very elegant work, Niola and colleagues combined LV-mediated overexpression of oncogenic H-Ras, silencing of the oncosuppressor p53 and Cre-loxP technology to generate a mouse model of high-grade GBM and to study the role of Inhibitor of Differentiation (ID) proteins in TIC biology and tumor maintenance [102].

More recently, recombinant LV expressing a shRNA targeted to the Cyclin-dependent kinase Inhibitor 2A (Cdkn2a) and expressing the Platelet-Derived Growth Factor subunit B (PDGFB) were generated to induce proneural-type GBM growth in the dentate gyrus of immunocompetent mice, setting up an easy procedure for preclinical testing which does not require genetically engineered mice [103]. This model is powerful since it uses different version of the vector, for the deletion of floxed genes in transduced cells or for eGFP visualization, by direct injection into the dentate gyrus of immunocompetent adult mice, inducing specific gene expression signature reminiscent of proneural GBM. The rapid development of the tumor allows preclinical testing of drugs in immunocompetent animals.

Lynes and coworkers performed intracranial injection of LV expressing different factors involved in the receptor tyrosine kinase (RTK)/RAS/PI3K pathway to induce high-grade gliomas in rats, not only obtaining useful animal models, but also elucidating the role of the selected proteins in GBM development [104].

Lastly, a GBM model was obtained in the tree shrew *Tupaia belangeri*, which is phylogenetically closer to primates than rodents: a LV was used to transduce mutant H-Ras and a shRNA to silence p53, resulting in the fast development of tumors with histological features reminiscent of GBM, including aggressiveness and GBM histological features [105]. The genetic analysis suggested similarity to the mesenchymal subgroup of human GBM, with a better match with respect to comparable mouse models.

5. Conclusions

Currently, LV represent a valid tool to explore GBM biology, via the creation of improved animal models and the selective targeting of cellular specific pathways, and also in the complex tumor microenvironment. They can also support the design of innovative therapies, at the preclinical and clinical level (Table 1). Challenges and drawbacks still have to be faced in the development of safe and effective molecular approaches for GBM treatment. One of the main issues is represented by the difficulty for the treatments to gain access to the tumor mass due, for example, to the presence of the BBB. Furthermore, the need of highly specific therapeutic approaches combined with the biological characteristics of the CNS cell repertoire, complicate GBM therapy. Thus, the availability of versatile gene delivery systems, allowing, for instance, the targeting of selected cell types (pseudotyping) and highly efficient in transducing primary and stem cells, is extremely useful. In this context, LV, due to their peculiar features, are among the best vectors to be adopted in gene and cell therapy protocols based on the expression of small noncoding RNAs and on the engineering of primary cells to express CARs or to become carriers of therapeutic molecules to the tumor mass.

Table 1. Summary of the studies discussed in the text.

LV Main Applications in GBM Research	References
Basic Studies	
Dissecting the tumor biology	[10–14,18,19,23,26–31]
Identification of novel therapeutic targets	[15–17,20–22,24,25]
LV as Tools for Gene and Cell Therapy of GBM	
Gene Therapy	
Gene therapy approaches based on transgene expression	[37–39,55,56]
Gene therapy approaches based on silencing/gene editing	[46,47,50,51,53,54]
Gene Therapy approaches aimed at overcoming GMB resistance to therapy	[57–64,68]
Immunotherapy of GBM via LV expressing CARs	[72–76]
Cell Therapy	
LV transduced cells as carriers of therapeutic molecules	[82–88,90,91]
LV to Create Animal Models	
Murine models	[97–104]
Non-murine models	[93–96,105]

LV: Last-generation lentiviruses; GBM: Glioblastoma; CAR: chimeric antigen receptors.

Author Contributions: Conceptualization, A.C. and C.M.C.; Writing—Original Draft Preparation, C.D.V., A.C., C.P. and C.M.C.

Funding: This research received no external funding.

Conflicts of Interest: The authors declare no conflict of interest.

References

1. Stupp, R.; Taillibert, S.; Kanner, A.; Read, W.; Steinberg, D.; Lhermitte, B.; Toms, S.; Idbaih, A.; Ahluwalia, M.S.; Fink, K.; et al. Effect of Tumor-Treating Fields Plus Maintenance Temozolomide vs Maintenance Temozolomide Alone on Survival in Patients with Glioblastoma: A Randomized Clinical Trial. *JAMA* **2017**, *318*, 2306–2316. [CrossRef] [PubMed]
2. Louis, D.N.; Perry, A.; Reifenberger, G.; von Deimling, A.; Figarella-Branger, D.; Cavenee, W.K.; Ohgaki, H.; Wiestler, O.D.; Kleihues, P.; Ellison, D.W. The 2016 World Health Organization Classi- fication of tumors of the central nervous system: A summary. *Acta Neuropathol.* **2016**, *131*, 803–820. [CrossRef] [PubMed]
3. Mann, R.; Mulligan, R.C.; Baltimore, D. Construction of a retrovirus packaging mutant and its use to produce helper-free defective retrovirus. *Cell* **1983**, *33*, 153–159. [CrossRef]
4. Parolin, C.; Sodroski, J. A defective HIV-1 vector for gene transfer to human lymphocytes. *J. Mol. Med. (Berl.)* **1995**, *73*, 279–288. [CrossRef]
5. Parolin, C.; Dorfman, T.; Palú, G.; Göttlinger, H.; Sodroski, J. Analysis in human immunodeficiency virus type 1 vectors of cis-acting sequences that affect gene transfer into human lymphocytes. *J. Virol.* **1994**, *68*, 3888–3895.
6. Naldini, L.; Blömer, U.; Gallay, P.; Ory, D.; Mulligan, R.; Gage, F.H.; Verma, I.M.; Trono, D. In vivo gene delivery and stable transduction of nondividing cells by a lentiviral vector. *Science* **1996**, *272*, 263–267. [CrossRef]
7. Naldini, L. Genetic engineering of hematopoiesis: Current stage of clinical translation and future perspectives. *EMBO Mol. Med.* **2019**, e9958. [CrossRef] [PubMed]
8. Poletti, V.; Mavilio, F. Interactions between Retroviruses and the Host Cell Genome. *Mol. Ther. Methods Clin. Dev.* **2017**, *8*, 31–41. [CrossRef]
9. Miletic, H.; Fischer, Y.H.; Neumann, H.; Hans, V.; Stenzel, W.; Giroglou, T.; Hermann, M.; Deckert, M.; Von Laer, D. Selective transduction of malignant glioma by lentiviral vectors pseudotyped with lymphocytic choriomeningitis virus glycoproteins. *Hum. Gene Ther.* **2004**, *15*, 1091–1100. [CrossRef]
10. Thaker, N.G.; McDonald, P.R.; Zhang, F.; Kitchens, C.A.; Shun, T.Y.; Pollack, I.F.; Lazo, J.S. Designing, optimizing, and implementing high-throughput siRNA genomic screening with glioma cells for the discovery of survival genes and novel drug targets. *J. Neurosci. Methods* **2010**, *185*, 204–212. [CrossRef]
11. Li, J.; Zhu, S.; Kozono, D.; Ng, K.; Futalan, D.; Shen, Y.; Akers, J.C.; Steed, T.; Kushwaha, D.; Schlabach, M.; et al. Genome-wide shRNA screen revealed integrated mitogenic signaling between dopamine receptor D2 (DRD2) and epidermal growth factor receptor (EGFR) in glioblastoma. *Oncotarget* **2014**, *5*, 882–893. [CrossRef] [PubMed]
12. Wanka, C.; Brucker, D.P.; Bähr, O.; Ronellenfitsch, M.; Weller, M.; Steinbach, J.P.; Rieger, J. Synthesis of cytochrome C oxidase 2: A p53-dependent metabolic regulator that promotes respiratory function and protects glioma and colon cancer cells from hypoxia-induced cell death. *Oncogene* **2012**, *31*, 3764–3776. [CrossRef]
13. Ulloa, F.; Gonzàlez-Juncà, A.; Meffre, D.; Barrecheguren, P.J.; Martínez-Mármol, R.; Pazos, I.; Olivé, N.; Cotrufo, T.; Seoane, J.; Soriano, E. Blockade of the SNARE protein syntaxin 1 inhibits glioblastoma tumor growth. *PLoS ONE* **2015**, *10*, e0119707. [CrossRef] [PubMed]
14. Sarkar, S.; Zemp, F.J.; Senger, D.; Robbins, S.M.; Yong, V.W. ADAM-9 is a novel mediator of tenascin-C-stimulated invasiveness of brain tumor-initiating cells. *Neuro Oncol.* **2015**, *17*, 1095–1105. [CrossRef]
15. Auvergne, R.; Wu, C.; Connell, A.; Au, S.; Cornwell, A.; Osipovitch, M.; Benraiss, A.; Dangelmajer, S.; Guerrero-Cazares, H.; Quinones-Hinojosa, A.; et al. PAR1 inhibition suppresses the self-renewal and growth of A2B5-defined glioma progenitor cells and their derived gliomas in vivo. *Oncogene* **2016**, *35*, 3817–3828. [CrossRef] [PubMed]

16. Kaur, H.; Ali, S.Z.; Huey, L.; Hütt-Cabezas, M.; Taylor, I.; Mao, X.G.; Weingart, M.; Chu, Q.; Rodriguez, F.J.; Eberhart, C.G.; et al. The transcriptional modulator HMGA2 promotes stemness and tumorigenicity in glioblastoma. *Cancer Lett.* **2016**, *377*, 55–64. [CrossRef] [PubMed]
17. Roccograndi, L.; Binder, Z.A.; Zhang, L.; Aceto, N.; Zhang, Z.; Bentires-Alj, M.; Nakano, I.; Dahmane, N.; O'Rourke, D.M. SHP2 regulates proliferation and tumorigenicity of glioma stem cells. *J. Neurooncol.* **2017**, *135*, 487–496. [CrossRef]
18. Ruokun, C.; Yake, X.; Fengdong, Y.; Xinting, W.; Laijun, S.; Xianzhi, L. Lentivirus-mediated silencing of HSDL2 suppresses cell proliferation in human gliomas. *Tumour Biol.* **2016**, *37*, 15065–15077. [CrossRef]
19. Yao, Y.; Liu, Y.; Lv, X.; Dong, B.; Wang, F.; Li, J.; Zhang, Q.; Xu, R.; Xu, Y. Down-regulation of ribosomal protein S15A inhibits proliferation of human glioblastoma cells in vivo and in vitro via AKT pathway. *Tumour Biol.* **2016**, *37*, 4979–9490. [CrossRef]
20. Alt, E.U.; Barabadi, Z.; Pfnür, A.; Ochoa, J.E.; Daneshimehr, F.; Lang, L.M.; Lin, D.; Braun, S.E.; Chandrasekar, B.; Izadpanah, R. TRAF3IP2, a novel therapeutic target in glioblastoma multiforme. *Oncotarget* **2018**, *9*, 29772–29788. [CrossRef]
21. Liu, P.; Yu, J.; Tian, X.; Chang, J.; Zhang, Y.; Zhang, R.; Zhang, N.; Huang, R.; Li, L.; Qiao, X.; et al. The effect of downregulation of Stathmin gene on biological behaviors of U373 and U87-MG glioblastoma cells. *Biol. Res.* **2018**, *51*, 16. [CrossRef] [PubMed]
22. Jun, H.J.; Appleman, V.A.; Wu, H.J.; Rose, C.M.; Pineda, J.J.; Yeo, A.T.; Delcuze, B.; Lee, C.; Gyuris, A.; Zhu, H.; et al. A PDGFRα-driven mouse model of glioblastoma reveals a stathmin1-mediated mechanism of sensitivity to vinblastine. *Nat. Commun.* **2018**, *9*, 3116. [CrossRef] [PubMed]
23. Zhu, Y.; Zhao, K.; Prinz, A.; Keyvani, K.; Lambertz, N.; Kreitschmann-Andermahr, I.; Lei, T.; Sure, U. Loss of endothelial programmed cell death 10 activates glioblastoma cells and promotes tumor growth. *Neuro Oncol.* **2016**, *18*, 538–548. [CrossRef] [PubMed]
24. Szabo, E.; Schneider, H.; Seystahl, K.; Rushing, E.J.; Herting, F.; Weidner, K.M.; Weller, M. Autocrine VEGFR1 and VEGFR2 signaling promotes survival in human glioblastoma models in vitro and in vivo. *Neuro Oncol.* **2016**, *18*, 1242–1252. [CrossRef]
25. Lin, M.; Zhang, X.; Jia, B.; Guan, S. Suppression of glioblastoma growth and angiogenesis through molecular targeting of methionine aminopeptidase-2. *J. Neurooncol.* **2018**, *136*, 243–254. [CrossRef]
26. Cesarini, V.; Martini, M.; Vitiani, L.R.; Gravina, G.L.; Di Agostino, S.; Graziani, G.; D'Alessandris, Q.G.; Pallini, R.; Larocca, L.M.; Rossi, P.; et al. Type 5 phosphodiesterase regulates glioblastoma multiforme aggressiveness and clinical outcome. *Oncotarget* **2017**, *8*, 13223–13239. [CrossRef] [PubMed]
27. Zhang, P.; Gao, J.; Wang, X.; Wen, W.; Yang, H.; Tian, Y.; Liu, N.; Wang, Z.; Liu, H.; Zhang, Y.; et al. A novel indication of thioredoxin-interacting protein as a tumor suppressor gene in malignant glioma. *Oncol. Lett.* **2017**, *14*, 2053–2058. [CrossRef]
28. Yao, Q.; Cai, G.; Yu, Q.; Shen, J.; Gu, Z.; Chen, J.; Shi, W.; Shi, J. IDH1 mutation diminishes aggressive phenotype in glioma stem cells. *Int. J. Oncol.* **2018**, *52*, 270–278. [CrossRef]
29. Li, W.Q.; Zhong, N.Z.; He, J.; Li, Y.M.; Hou, L.J.; Liu, H.M.; Xia, C.Y.; Wang, L.Z.; Lu, Y.C. High ATP2A2 expression correlates with better prognosis of diffuse astrocytic tumor patients. *Oncol. Rep.* **2017**, *37*, 2865–2874. [CrossRef]
30. Melnikov, A.; Murugan, A.; Zhang, X.; Tesileanu, T.; Wang, L.; Rogov, P.; Feizi, S.; Gnirke, A.; Callan, C.G., Jr.; Kinney, J.B.; et al. Systematic dissection and optimization of inducible enhancers in human cells using a massively parallel reporter assay. *Nat. Biotechnol.* **2012**, *30*, 271–277. [CrossRef] [PubMed]
31. Maricque, B.B.; Dougherty, J.D.; Cohen, B.A. A genome-integrated massively parallel reporter assay reveals DNA sequence determinants of cis-regulatory activity in neural cells. *Nucleic Acids Res.* **2017**, *45*, e16. [CrossRef] [PubMed]
32. Couzin-Frankel, J. Breakthrough of the year 2013. Cancer immunotherapy. *Science* **2013**, *342*, 1432–1433. [CrossRef]
33. Yang, Y. Cancer immunotherapy: Harnessing the immune system to battle cancer. *J. Clin. Investig.* **2015**, *125*, 3335–3337. [CrossRef]
34. Palù, G.; Cavaggioni, A.; Calvi, P.; Franchin, E.; Pizzato, M.; Boschetto, R.; Parolin, C.; Chilosi, M.; Ferrini, S.; Zanusso, A.; et al. Gene therapy of glioblastoma multiforme via combined expression of suicide and cytokine genes: A pilot study in humans. *Gene Ther.* **1999**, *6*, 330–337. [CrossRef] [PubMed]

35. Colombo, F.; Barzon, L.; Franchin, E.; Pacenti, M.; Pinna, V.; Danieli, D.; Zanusso, M.; Palù, G. Combined HSV-TK/IL-2 gene therapy in patients with recurrent glioblastoma multiforme: Biological and clinical results. *Cancer Gene Ther.* **2005**, *12*, 835–848. [CrossRef] [PubMed]
36. Yu, S.; Li, A.; Liu, Q.; Li, T.; Yuan, X.; Han, X.; Wu, K. Chimeric antigen receptor T cells: A novel therapy for solid tumors. *J. Hematol. Oncol.* **2017**, *10*, 78. [CrossRef] [PubMed]
37. Miletic, H.; Fischer, Y.H.; Giroglou, T.; Rueger, M.A.; Winkeler, A.; Li, H.; Himmelreich, U.; Stenzel, W.; Jacobs, A.H.; von Laer, D. Normal brain cells contribute to the bystander effect in suicide gene therapy of malignant glioma. *Clin. Cancer Res.* **2007**, *13*, 6761–6768. [CrossRef] [PubMed]
38. Huszthy, P.C.; Giroglou, T.; Tsinkalovsky, O.; Euskirchen, P.; Skaftnesmo, K.O.; Bjerkvig, R.; von Laer, D.; Miletic, H. Remission of invasive, cancer stem-like glioblastoma xenografts using lentiviral vector-mediated suicide gene therapy. *PLoS ONE* **2009**, *4*, e6314. [CrossRef] [PubMed]
39. Cottin, S.; Gould, P.V.; Cantin, L.; Caruso, M. Gap junctions in human glioblastomas: Implications for suicide gene therapy. *Cancer Gene Ther.* **2011**, *18*, 674–681. [CrossRef] [PubMed]
40. Guo, D.; Wang, B.; Han, F.; Lei, T. RNA interference therapy for glioblastoma. *Expert Opin. Biol. Ther.* **2010**, *10*, 927–936. [CrossRef]
41. Lozada-Delgado, E.L.; Grafals-Ruiz, N.; Vivas-Mejía, P.E. RNA interference for glioblastoma therapy: Innovation ladder from the bench to clinical trials. *Life Sci.* **2017**, *188*, 26–36. [CrossRef]
42. Winter, J.; Jung, S.; Keller, S.; Gregory, R.I.; Diederichs, S. Many roads to maturity: MicroRNA biogenesis pathways and their regulation. *Nat. Cell Biol.* **2009**, *11*, 228–234. [CrossRef]
43. Visone, R.; Croce, C.M. MiRNAs and cancer. *Am. J. Pathol.* **2009**, *174*, 1131–1138. [CrossRef]
44. Ahir, B.K.; Ozer, H.; Engelhard, H.H.; Lakka, S.S. MicroRNAs in glioblastoma pathogenesis and therapy: A comprehensive review. *Crit. Rev. Oncol. Hematol.* **2017**, *120*, 22–33. [CrossRef] [PubMed]
45. Luo, J.W.; Wang, X.; Yang, Y.; Mao, Q. Role of microRNA (miRNA) in pathogenesis of glioblastoma. *Eur. Rev. Med. Pharmacol. Sci.* **2015**, *19*, 1630–1639. [PubMed]
46. Luan, Y.; Zhang, S.; Zuo, L.; Zhou, L. Overexpression of miR-100 inhibits cell proliferation, migration, and chemosensitivity in human glioblastoma through FGFR3. *Onco Targets Ther.* **2015**, *8*, 3391–3400. [CrossRef]
47. Kefas, B.; Floyd, D.H.; Comeau, L.; Frisbee, A.; Dominguez, C.; Dipierro, C.G.; Guessous, F.; Abounader, R.; Purow, B. A miR-297/hypoxia/DGK-α axis regulating glioblastoma survival. *Neuro Oncol.* **2013**, *15*, 1652–1663. [CrossRef] [PubMed]
48. Ebert, M.S.; Sharp, P.A. MicroRNA sponges: Progress and possibilities. *RNA* **2010**, *16*, 2043–2050. [CrossRef]
49. Shea, A.; Harish, V.; Afzal, Z.; Chijioke, J.; Kedir, H.; Dusmatova, S.; Roy, A.; Ramalinga, M.; Harris, B.; Blancato, J.; et al. MicroRNAs in glioblastoma multiforme pathogenesis and therapeutics. *Cancer Med.* **2016**, *5*, 1917–1946. [CrossRef]
50. Chen, L.; Zhang, K.; Shi, Z.; Zhang, A.; Jia, Z.; Wang, G.; Pu, P.; Kang, C.; Han, L. A lentivirus-mediated miR-23b sponge diminishes the malignant phenotype of glioma cells in vitro and in vivo. *Oncol. Rep.* **2014**, *31*, 1573–1580. [CrossRef]
51. Skalsky, R.L.; Cullen, B.R. Reduced expression of brain-enriched microRNAs in glioblastomas permits targeted regulation of a cell death gene. *PLoS ONE.* **2011**, *6*, e24248. [CrossRef] [PubMed]
52. Zhang, F.; Wen, Y.; Guo, X. CRISPR/Cas9 for genome editing: Progress, implications and challenges. *Hum. Mol. Genet.* **2014**, *23*, R40–R46. [CrossRef]
53. Tome-Garcia, J.; Erfani, P.; Nudelman, G.; Tsankov, A.M.; Katsyv, I.; Tejero, R.; Bin, Z.; Walsh, M.; Friedel, R.H.; Zaslavsky, E.; et al. Analysis of chromatin accessibility uncovers TEAD1 as a regulator of migration in human glioblastoma. *Nat. Commun.* **2018**, *9*, 4020. [CrossRef] [PubMed]
54. Sun, T.; Patil, R.; Galstyan, A.; Klymyshyn, D.; Ding, H.; Chesnokova, A.; Cavenee, W.K.; Furnari, F.B.; Ljubimov, V.A.; Shatalova, E.S.; et al. Blockade of a laminin-411—Notch axis with CRISPR/Cas9 or a nanobioconjugate inhibits glioblastoma growth through tumor-microenvironment crosstalk. *Cancer Res.* **2019**. [CrossRef]
55. Sánchez-Hernández, L.; Hernández-Soto, J.; Vergara, P.; González, R.O.; Segovia, J. Additive effects of the combined expression of soluble forms of GAS1 and PTEN inhibiting glioblastoma growth. *Gene Ther.* **2018**, *25*, 439–449. [CrossRef] [PubMed]
56. Lamb, L.S., Jr.; Bowersock, J.; Dasgupta, A.; Gillespie, G.Y.; Su, Y.; Johnson, A.; Spencer, H.T. Engineered drug resistant γδ T cells kill glioblastoma cell lines during a chemotherapy challenge: A strategy for combining chemo- and immunotherapy. *PLoS ONE* **2013**, *8*, e51805. [CrossRef] [PubMed]

57. Viel, T.; Monfared, P.; Schelhaas, S.; Fricke, I.B.; Kuhlmann, M.T.; Fraefel, C.; Jacobs, A.H. Optimizing glioblastoma temozolomide chemotherapy employing lentiviral-based anti-MGMT shRNA technology. *Mol. Ther.* **2013**, *21*, 570–579. [CrossRef]
58. Miao, W.; Liu, X.; Wang, H.; Fan, Y.; Lian, S.; Yang, X.; Wang, X.; Guo, G.; Li, Q.; Wang, S. p53 upregulated modulator of apoptosis sensitizes drug-resistant U251 glioblastoma stem cells to temozolomide through enhanced apoptosis. *Mol. Med. Rep.* **2015**, *11*, 4165–4173. [CrossRef]
59. Wang, P.; Ye, J.A.; Hou, C.X.; Zhou, D.; Zhan, S.Q. Combination of lentivirus-mediated silencing of PPM1D and temozolomide chemotherapy eradicates malignant glioma through cell apoptosis and cell cycle arrest. *Oncol. Rep.* **2016**, *36*, 2544–2552. [CrossRef] [PubMed]
60. Xu, X.; Wang, Z.; Liu, N.; Cheng, Y.; Jin, W.; Zhang, P.; Wang, X.; Yang, H.; Liu, H.; Zhang, Y.; et al. Association between SOX9 and CA9 in glioma, and its effects on chemosensitivity to TMZ. *Int. J. Oncol.* **2018**, *53*, 189–202. [CrossRef]
61. Shi, S.; Zhang, M.; Guo, R.; Zhang, M.; Hu, J.; Xi, Y.; Miao, Y.; Li, B. 131I therapy mediated by sodium/iodide symporter combined with kringle 5 has a synergistic therapeutic effect on glioma. *Oncol. Rep.* **2016**, *35*, 691–698. [CrossRef]
62. Li, Y.; Han, N.; Yin, T.; Huang, L.; Liu, S.; Liu, D.; Xie, C.; Zhang, M. Lentivirus-mediated Nox4 shRNA invasion and angiogenesis and enhances radiosensitivity in human glioblastoma. *Oxid. Med. Cell. Longev.* **2014**, *2014*, 581732. [CrossRef]
63. Upraity, S.; Kazi, S.; Padul, V.; Shirsat, N.V. MiR-224 expression increases radiation sensitivity of glioblastoma cells. *Biochem. Biophys. Res. Commun.* **2014**, *448*, 225–230. [CrossRef]
64. Chuah, T.L.; Walker, D.G.; Wei, M.; Scott, S.; Lavin, M.F. Approaches to sensitizing glioblastoma to radiotherapy: Use of lentiviral vectors. *Int. J. Oncol.* **2012**, *40*, 1963–1969. [CrossRef]
65. Chen, J.; Li, Y.; Yu, T.S.; McKay, R.M.; Burns, D.K.; Kernie, S.G.; Parada, L.F. A restricted cell population propagates glioblastoma growth after chemotherapy. *Nature* **2012**, *488*, 522–526. [CrossRef] [PubMed]
66. Hardee, M.E.; Marciscano, A.E.; Medina-Ramirez, C.M.; Zagzag, D.; Narayana, A.; Lonning, S.M.; Barcellos-Hoff, M.H. Resistance of glioblastoma-initiating cells to radiation mediated by the tumor microenvironment can be abolished by inhibiting transforming growth factor-beta. *Cancer Res.* **2012**, *72*, 4119–4129. [CrossRef]
67. Lathia, J.D.; Mack, S.C.; Mulkearns-Hubert, E.E.; Valentim, C.L.; Rich, J.N. Cancer stem cells in glioblastoma. *Genes Dev.* **2015**, *29*, 1203–1217. [CrossRef]
68. Bayin, N.S.; Modrek, A.S.; Dietrich, A.; Lebowitz, J.; Abel, T.; Song, H.R.; Schober, M. Selective lentiviral gene delivery to CD133-expressing human glioblastoma stem cells. *PLoS ONE* **2014**, *9*, e116114. [CrossRef] [PubMed]
69. Kalaitsidou, M.; Kueberuwa, G.; Schütt, A.; Gilham, D.E. CAR T-cell therapy: Toxicity and the relevance of preclinical models. *Immunotherapy* **2015**, *7*, 487–497. [CrossRef] [PubMed]
70. Kosti, P.; Maher, J.; Arnold, J.N. Perspectives on Chimeric Antigen Receptor T-Cell Immunotherapy for Solid Tumors. *Front. Immunol.* **2018**, *9*, 1104. [CrossRef] [PubMed]
71. Bagley, S.J.; Desai, A.S.; Linette, G.P.; June, C.H.; O'Rourke, D.M. CAR T-cell therapy for glioblastoma: Recent clinical advances and future challenges. *Neuro Oncol.* **2018**, *20*, 1429–1438. [CrossRef]
72. Petersen, C.T.; Krenciute, G. Next generation CAR T Cells for the immunotherapy of high-grade glioma. *Front. Oncol.* **2019**, *9*, 69. [CrossRef] [PubMed]
73. Kuramitsu, S.; Ohno, M.; Ohka, F.; Shiina, S.; Yamamichi, A.; Kato, A.; Tanahashi, K.; Motomura, K.; Kondo, G.; Kurimoto, M.; et al. Lenalidomide enhances the function of chimeric antigen receptor T cells against the epidermal growth factor receptor variant III by enhancing immune synapses. *Cancer Gene Ther.* **2015**, *22*, 487–495. [CrossRef] [PubMed]
74. Han, J.; Chu, J.; Keung Chan, W.; Zhang, J.; Wang, Y.; Cohen, J.B.; Victor, A.; Meisen, W.H.; Kim, S.H.; Grandi, P.; et al. CAR-Engineered NK Cells Targeting Wild-Type EGFR and EGFRvIII Enhance Killing of Glioblastoma and Patient-Derived Glioblastoma Stem Cells. *Sci. Rep.* **2015**, *5*, 11483. [CrossRef]
75. Murakami, T.; Nakazawa, T.; Natsume, A.; Nishimura, F.; Nakamura, M.; Matsuda, R.; Omoto, K.; Tanaka, Y.; Shida, Y.; Park, Y.S.; et al. Novel Human NK Cell Line Carrying CAR Targeting EGFRvIII Induces Antitumor Effects in Glioblastoma Cells. *Anticancer Res.* **2018**, *38*, 5049–5056. [CrossRef] [PubMed]

76. Shiina, S.; Ohno, M.; Ohka, F.; Kuramitsu, S.; Yamamichi, A.; Kato, A.; Motomura, K.; Tanahashi, K.; Yamamoto, T.; Watanabe, R.; et al. CAR T Cells Targeting Podoplanin Reduce Orthotopic Glioblastomas in Mouse Brains. *Cancer Immunol. Res.* **2016**, *4*, 259–268. [CrossRef] [PubMed]
77. Lee, D.H.; Ahn, Y.; Kim, S.U.; Wang, K.C.; Cho, B.K.; Phi, J.H.; Park, I.H.; Black, P.M.; Carroll, R.S.; Lee, J.; et al. Targeting rat brainstem glioma using human neural stem cells and human mesenchymal stem cells. *Clin. Cancer Res.* **2009**, *15*, 4925–4934. [CrossRef] [PubMed]
78. Nakamizo, A.; Marini, F.; Amano, T.; Khan, A.; Studeny, M.; Gumin, J.; Chen, J.; Hentschel, S.; Vecil, G.; Dembinski, J.; et al. Human bone marrow-derived mesenchymal stem cells in the treatment of gliomas. *Cancer Res.* **2005**, *65*, 3307–3318. [CrossRef]
79. Germano, I.M.; Binello, E. Stem cells and gliomas: Past, present, and future. *J. Neurooncol.* **2014**, *119*, 547–555. [CrossRef]
80. Danielyan, L.; Schäfer, R.; von Ameln-Mayerhofer, A.; Buadze, M.; Geisler, J.; Klopfer, T.; Burkhardt, U.; Proksch, B.; Verleysdonk, S.; Ayturan, M.; et al. Intranasal delivery of cells to the brain. *Eur. J. Cell Biol.* **2009**, *88*, 315–324. [CrossRef]
81. Li, L.; Jiang, Q.; Ding, G.; Zhang, L.; Zhang, Z.G.; Li, Q.; Panda, S.; Lu, M.; Ewing, J.R.; Chopp, M. Effects of administration route on migration and distribution of neural progenitor cells transplanted into rats with focal cerebral ischemia, an MRI study. *J. Cereb. Blood Flow Metab.* **2010**, *30*, 653–662. [CrossRef] [PubMed]
82. Balyasnikova, I.V.; Prasol, M.S.; Ferguson, S.D.; Han, Y.; Ahmed, A.U.; Gutova, M.; Tobias, A.L.; Mustafi, D.; Rincón, E.; Zhang, L.; et al. Intranasal delivery of mesenchymal stem cells significantly extends survival of irradiated mice with experimental brain tumors. *Mol. Ther.* **2014**, *22*, 140–148. [CrossRef]
83. Balyasnikova, I.V.; Ferguson, S.D.; Han, Y.; Liu, F.; Lesniak, M.S. Therapeutic effect of neural stem cells expressing TRAIL and bortezomib in mice with glioma xenografts. *Cancer Lett.* **2011**, *310*, 148–159. [CrossRef] [PubMed]
84. Bagó, J.R.; Okolie, O.; Dumitru, R.; Ewend, M.G.; Parker, J.S.; Werff, R.V.; Underhill, T.M.; Schmid, R.S.; Miller, C.R.; Hingtgen, S.D. Tumor-homing cytotoxic human induced neural stem cells for cancer therapy. *Sci. Transl. Med.* **2017**, *9*, 375. [CrossRef]
85. Bak, X.Y.; Lam, D.H.; Yang, J.; Ye, K.; Wei, E.L.; Lim, S.K.; Wang, S. Human embryonic stem cell-derived mesenchymal stem cells as cellular delivery vehicles for prodrug gene therapy of glioblastoma. *Hum. Gene Ther.* **2011**, *22*, 1365–1377. [CrossRef]
86. Fei, S.; Qi, X.; Kedong, S.; Guangchun, J.; Jian, L.; Wei, Q. The antitumor effect of mesenchymal stem cells transduced with a lentiviral vector expressing cytosine deaminase in a rat glioma model. *J. Cancer Res. Clin. Oncol.* **2012**, *138*, 347–357. [CrossRef]
87. de Melo, S.M.; Bittencourt, S.; Ferrazoli, E.G.; da Silva, C.S.; da Cunha, F.F.; da Silva, F.H.; Stilhano, R.S.; Denapoli, P.M.; Zanetti, B.F.; Martin, P.K.; et al. The Anti-Tumor Effects of Adipose Tissue Mesenchymal Stem Cell Transduced with HSV-Tk Gene on U-87-Driven Brain Tumor. *PLoS ONE* **2015**, *10*, e0128922. [CrossRef]
88. Hashemi, M.; Fallah, A.; Aghayan, H.R.; Arjmand, B.; Yazdani, N.; Verdi, J.; Ghodsi, S.M.; Miri, S.M.; Hadjighassem, M. A New Approach in Gene Therapy of Glioblastoma Multiforme: Human Olfactory Ensheathing Cells as a Novel Carrier for Suicide Gene Delivery. *Mol. Neurobiol.* **2016**, *53*, 5118–51128. [CrossRef]
89. Adachi, Y.; Tamiya, T.; Ichikawa, T.; Terada, K.; Ono, Y.; Matsumoto, K.; Furuta, T.; Hamada, H.; Ohmoto, T. Experimental gene therapy for brain tumors using adenovirus-mediated transfer of cytosine deaminase gene and uracil phosphoribosyltransferase gene with 5-fluorocytosine. *Hum. Gene Ther.* **2000**, *11*, 77–89. [CrossRef] [PubMed]
90. Carvalho, L.A.; Teng, J.; Fleming, R.L.; Tabet, E.I.; Zinter, M.; de Melo Reis, R.A.; Tannous, B.A. Olfactory Ensheathing Cells: A Trojan Horse for Glioma Gene Therapy. *J. Natl. Cancer Inst.* **2018**. [CrossRef]
91. Moyes, K.W.; Lieberman, N.A.; Kreuser, S.A.; Chinn, H.; Winter, C.; Deutsch, G.; Hoglund, V.; Watson, R.; Crane, C.A. Genetically Engineered Macrophages: A Potential Platform for Cancer Immunotherapy. *Hum. Gene Ther.* **2017**, *28*, 200–215. [CrossRef]
92. Miyai, M.; Tomita, H.; Soeda, A.; Yano, H.; Iwama, T.; Hara, A. Current trends in mouse models of glioblastoma. *J. Neurooncol.* **2017**, *135*, 423–432. [CrossRef] [PubMed]
93. Read, R.D. Drosophila melanogaster as a model system for human brain cancers. *Glia* **2011**, *59*, 1364–1376. [CrossRef] [PubMed]

94. Wehmas, L.C.; Tanguay, R.L.; Punnoose, A.; Greenwood, J.A. Developing a novel embryo-larval zebrafish xenograft assay to prioritize human glioblastoma therapeutics. *Zebrafish* **2016**, *13*, 317–329. [CrossRef] [PubMed]
95. Yen, J.; White, R.M.; Stemple, D.L. Zebrafish models of cancer: Progress and future challenges. *Curr. Opin. Genet. Dev.* **2014**, *24*, 38–45. [CrossRef]
96. Hicks, J.; Platt, S.; Kent, M.; Haley, A. Canine brain tumours: A model for the human disease? *Vet. Comp. Oncol.* **2015**, *15*, 252–272. [CrossRef] [PubMed]
97. Rushing, E.J.; Watson, M.L.; Schold, S.C.; Land, K.J.; Kokkinakis, D.M. Glial tumors in the MNU rat model: Induction of pure and mixed gliomas that do not require typical missense mutations of p53. *J. Neuropathol. Exp. Neurol.* **1998**, *57*, 1053–1060. [CrossRef]
98. Chang, E.; Pohling, C.; Natarajan, A.; Witney, T.H.; Kaur, J.; Xu, L.; Gowrishankar, G.; D'Souza, A.L.; Murty, S.; Schick, S.; et al. AshwaMAX and Withaferin A inhibits gliomas in cellular and murine orthotopic models. *J. Neurooncol.* **2016**, *126*, 253–264. [CrossRef] [PubMed]
99. Marumoto, T.; Tashiro, A.; Friedmann-Morvinski, D.; Scadeng, M.; Soda, Y.; Gage, F.H.; Verma, I.M. Development of a novel mouse glioma model using lentiviral vectors. *Nat. Med.* **2009**, *15*, 110–116. [CrossRef]
100. Friedmann-Morvinski, D.; Singer, O. Overexpression Models: Lentiviral Modeling of Brain Cancer. *Curr. Protoc. Mouse Biol.* **2013**, *3*, 121–139. [CrossRef]
101. de Vries, N.A.; Bruggeman, S.W.; Hulsman, D.; de Vries, H.I.; Zevenhoven, J.; Buckle, T.; Hamans, B.C.; Leenders, W.P.; Beijnen, J.H.; van Lohuizen, M.; et al. Rapid and robust transgenic high-grade glioma mouse models for therapy intervention studies. *Clin. Cancer Res.* **2010**, *16*, 3431–3441. [CrossRef] [PubMed]
102. Niola, F.; Zhao, X.; Singh, D.; Sullivan, R.; Castano, A.; Verrico, A.; Zoppoli, P.; Friedmann-Morvinski, D.; Sulman, E.; Barrett, L.; et al. Mesenchymal high-grade glioma is maintained by the ID-RAP1 axis. *J. Clin. Investig.* **2013**, *123*, 405–417. [CrossRef] [PubMed]
103. Rahme, G.J.; Luikart, B.W.; Cheng, C.; Israel, M.A. A recombinant lentiviral PDGF-driven mouse model of proneural glioblastoma. *Neuro Oncol.* **2018**, *20*, 332–342. [CrossRef] [PubMed]
104. Lynes, J.; Wibowo, M.; Koschmann, C.; Baker, G.J.; Saxena, V.; Muhammad, A.K.; Bondale, N.; Klein, J.; Assi, H.; Lieberman, A.P.; et al. Lentiviral-induced high-grade gliomas in rats: The effects of PDGFB, HRAS-G12V, AKT, and IDH1-R132H. *Neurotherapeutics* **2014**, *11*, 623–635. [CrossRef] [PubMed]
105. Tong, Y.; Hao, J.; Tu, Q.; Yu, H.; Yan, L.; Li, Y.; Lv, L.; Wang, F.; Iavarone, A.; Zhao, X. A tree shrew glioblastoma model recapitulates features of human glioblastoma. *Oncotarget* **2017**, *8*, 17897–17907. [CrossRef] [PubMed]

cancers

MDPI

Review

Cholesterol Metabolism: A Potential Therapeutic Target in Glioblastoma

Fahim Ahmad †, **Qian Sun** †, **Deven Patel and Jayne M. Stommel** *

National Institutes of Health, National Cancer Institute, Radiation Oncology Branch, Bethesda, MD 20892, USA; fahim.ahmad@nih.gov (F.A.); qiansun1128@gmail.com (Q.S.); devenpatel7a@gmail.com (D.P.)

* Correspondence: jayne.stommel@nih.gov

† These authors contributed equally to this work.

Received: 30 December 2018; Accepted: 24 January 2019; Published: 26 January 2019

Abstract: Glioblastoma is a highly lethal adult brain tumor with no effective treatments. In this review, we discuss the potential to target cholesterol metabolism as a new strategy for treating glioblastomas. Twenty percent of cholesterol in the body is in the brain, yet the brain is unique among organs in that it has no access to dietary cholesterol and must synthesize it de novo. This suggests that therapies targeting cholesterol synthesis in brain tumors might render their effects without compromising cell viability in other organs. We will describe cholesterol synthesis and homeostatic feedback pathways in normal brain and brain tumors, as well as various strategies for targeting these pathways for therapeutic intervention.

Keywords: glioblastoma; cholesterol; liver X receptor (LXR); brain; liver; metabolism; blood–brain barrier; low-density lipoprotein receptor (LDLR); sterol regulatory element binding protein (SREBP)

1. Pathological and Genetic Features of Glioblastoma

Glioblastoma (also called GBM) is the most common malignant primary brain tumor originating from glial cells [1]. Of the three types of gliomas (ependymomas, oligodendrogliomas, and astrocytomas), glioblastomas are WHO grade IV astrocytomas and have the worst prognosis. GBMs cause 225,000 deaths per year worldwide, and are diagnosed at an average age of 64 [2]. The median survival rate for GBM is 15 months from initial diagnosis even with the most current standard-of-care therapy, which consists of maximal surgical resection, radiation therapy, and adjuvant chemotherapy with temozolomide [3]. There are several obstacles to the development of efficient treatments against glioblastoma. Surgical resection of GBM is virtually impossible as these tumors are highly invasive and penetrate the normal brain. Full resection would require very fine tuned and precise imaging tools, which would enable the removal of invading tumor cells. GBM is highly resistant to cytotoxic drug regimens, including temozolomide, which only improves overall survival 2.5 months beyond radiation and surgery alone [4]. Therefore, new strategies are necessary to develop treatments that elicit durable responses in GBM patients.

Various rodent models have been developed for the accurate representation of preclinical GBM models, but those systems have multiple drawbacks such as immune system deficiency and incompatible stroma and microenvironment that might interfere with the testing new drugs [5]. Existing animal models fall under three categories. The first involves genetically engineered mouse models: for example, mice expressing v-src kinase driven by an astrocyte-specific Glial Fibrillary Acidic Protein (*GFAP*) promoter or *tp53*-null mice with astrocyte-specific loss of *NF1* both develop high-grade astrocytomas [6,7]. Mouse models of glioblastoma have also been generated using viruses expressing oncogenes injected into the mouse brain. For example, Pax3-Tv-a; Trp53 fl/fl mice injected with RCAS-PDGFB and RCAS-Cre virus, with or without RCAS-H3.3K27M, develop a tumor similar to diffuse pontine glioma [8]. A more recent technical development is the injection of patient-derived glioblastoma stem-like cells in immunocompromised mice. While many laboratories have adopted this technique for studying glioblastoma in vivo, two recent examples include injecting cells derived from isocitrate dehydrogenase 1 (*IDH1*)-mutant tumors into SCID (Severe Combined Immunodeficiency) mice to study vulnerability of this tumor genotype to 2-hydroxyglutarate depletion [9], and injecting cells from recurrent glioblastoma into NOD-SCID (Non-Obese Diabetic SCID) and NOG (NOD/Shi-scid/IL-2Rγnull) mice then treating them with a STAT3 (Signal Transducer and Activator of Transcription 3) inhibitor [10]. These animal models have led to many novel discoveries in glioma cell biology and metabolism, but thus far have not led to any new clinical advances.

Our most promising avenues for developing robust strategies to target GBM are likely to lie within recent efforts to genomically and proteomically catalog this disease. Indeed, a study published by The Cancer Genome Atlas (TCGA) Research Network in 2014 showed that the most frequently altered pathways in GBM are the RTK/PI3K/MAPK (90% of tumors), p53 (86%), and Rb pathways (79%) [11]. Because nearly all glioblastomas have at least one genomic alteration in the RTK/PI3K/MAPK axis, and because there are numerous small molecule inhibitors targeting this pathway that are already FDA-approved or at least in early-stage clinical trials, it appears inhibition of RTKs, PI3Ks, or MAPKs should be highly effective. Unfortunately, no RTK, PI3K, or MAPK pathway inhibitor thus far has improved patient survival beyond that of the current standard-of-care [4,12,13]. That said, the TCGA datasets have provided a trove of mutations, gene expression and proteomic profiles, and copy number variations that can be explored for novel therapeutic targets. A recent review published by Alphandery et al. has covered all the most recent glioblastoma treatments that are commercialized or under industrial development for glioblastoma treatment [2].

2. Dysregulated Metabolic Pathways in Glioblastoma

One strategy lacking from conventional GBM therapies is metabolic pathway targeting. Significant pre-clinical research has been done in this field to understand how tumor cells differ from their counterparts. For example, work from Calvert et al. using in silico and wet-bench analyses shows that non-mutated isocitrate dehydrogenase 1 (IDH1) is commonly overexpressed in the primary GBM and genetic and pharmacological inactivation leads to attenuation of GBM tumor growth [14]. Large-scale genomics studies have also revealed that IDH1 mutations occur in gliomas even before other mutations such as TP53 [15,16]. Glutamate and alpha-ketoglutarate are fundamental metabolites that are necessary for growth and proliferation of GBM cells. A study published by Franceschi et al., showed that glutamate dehydrogenase 2 (GLUD2, an enzyme responsible for glutamate oxidative deamination), inhibits GBM cell growth and could be a target to control tumor progression [17]. In contrast with previous work showing the importance of the tricarboxylic acid (TCA) cycle and amino acid metabolism in GBM, in this review we will discuss the role of cholesterol metabolism in influencing tumor cell growth and proliferation and as a potential therapeutic target.

3. Feedback Pathways in Cholesterol Biosynthesis

Cholesterol is critical for cell growth and function. It is important not only as a component of the plasma membrane and lipid rafts, it also serves as a precursor for steroid hormones, bile acids, and vitamin D [18–20]. In the last few years, researchers have explored the impact of cholesterol metabolism on immune response. For example, Yang et al. demonstrated that by modulating cholesterol metabolism in CD8 + T cells, higher antitumor activity could be achieved [21]. Another study published by Wang et al. showed that an analog of cholesterol can act as a negative regulator for TCR signaling [22]. Cells usually obtain cholesterol via different mechanisms. One is through direct synthesis via the transcriptional activity of sterol regulatory element binding proteins (SREBPs), which promote the transcription of enzymes involved in cholesterol and fatty acid biosynthesis pathways, including HMG-CoA reductase (HMGCR) [23]. Ground-breaking work by Brown and Goldstein, who later won the Nobel prize for the discovery, defined cholesterol as a key molecule regulating its own synthesis via the activation of a negative feedback mechanism (Figure 1) [24]. In the presence of cholesterol, SREBP is sequestered in the endoplasmic reticulum (ER) by the SREBP cleavage-activating protein (SCAP), whose function is inhibited by the ER-resident insulin-induced gene (*INSIG1, INSIG2*) [25]. When cholesterol levels drop below homeostatic levels, SCAP is separated from INSIGs via a conformational change [26] and carries SREBPs into the Golgi where their active sites are cleaved and activated as transcription factors [27]. SREBPs include three members: SREBP1a, SREBP1c, and SREBP2 [28]. SREBP2 is encoded by the *srebf2* gene and SREBP1a and SREBP1c are encoded by the *srebf1* gene. SREBP1c regulates the transcription of the genes that are associated with biosynthesis of fatty acids; SREBP2 mainly regulates genes involved with cholesterol biosynthesis. Activity of SREBP1a partially overlaps between SREBP1c and SREBP2.

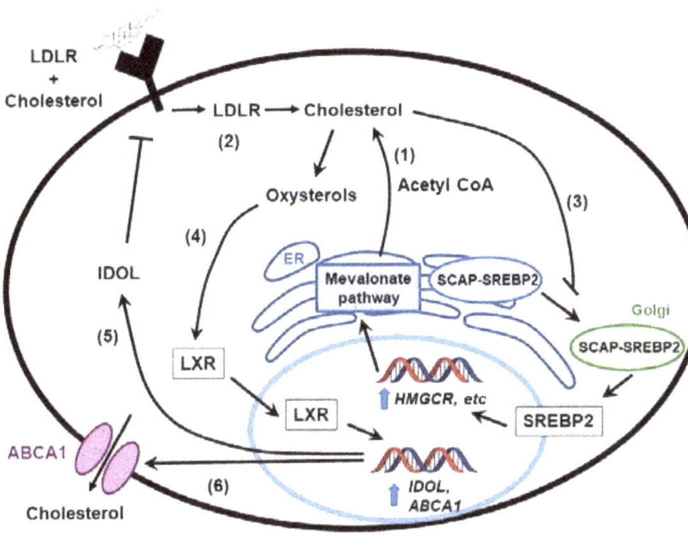

Figure 1. Cholesterol homeostasis in normal cells. Cells obtain cholesterol primarily through one of two mechanisms: (1) by synthesizing it de novo from acetyl CoA generated from glycolysis and (2) through exogenous uptake by low density lipoprotein receptors (LDLR). Cholesterol can negatively regulate its own levels through (3) the inhibition of proteolytic processing and nuclear import of sterol regulatory element binding proteins (SREBP2), leading to a decrease in activity in the mevalonate pathway or (4) through its conversion to oxysterols that activate liver X receptors (LXRs). LXRs lower cellular cholesterol levels by (5) inducing the transcription of the E3 ubiquitin ligase, *IDOL*, which ubiquitinates LDLR, and (6) by upregulating expression of the cholesterol efflux pump, *ABCA1*. SCAP: SREBP cleavage-activating protein; ER: endoplasmic reticulum.

Cholesterol can be enzymatically modified to form metabolites such as oxysterols. One species of oxysterol, 25-hydroxycholesterol (25-OHC), has been shown to be associated with suppression of proliferation [29]. The suppression in proliferation can be rescued by the addition of exogeneous cholesterol, indicating that the presence of 25-OHC inhibits SREBP activation and suppresses cholesterol biosynthesis [26]. 25-OHC also has anti-inflammatory effects: animal models have demonstrated that it inhibits transcription and inflammasome-mediated activation of interleukin-1β (IL1B) by inhibiting the activation of SREBPs [30]. Of interest for this review, recent studies have shown that 25-OHC also suppresses the immune response in the human glioblastoma cell line, U87-MG, and thus might increase tumor growth by modulating the immune system [31].

Oxysterols also regulate the activity of liver X receptors (LXRs), which are nuclear receptors that are activated by oxysterols. LXRα is expressed highly in liver, adrenal glands, intestine, adipose tissues, macrophages, lungs, and kidney, while LXRβ is ubiquitously expressed [32]. The LXRs maintain cholesterol homeostasis by maintaining the balance between biosynthesis, uptake via low density lipoprotein receptors (LDLRs), and efflux via the ATP-binding cassette transporters, ABCA1 and ABCG1 (Figure 1). LXRs inhibit LDLR protein expression through induction of an E3 ligase that ubiquitinates LDLRs called IDOL (inducible degrader of LDLR, which is encoded by the *MYLIP* gene [33]. The importance of LXRs in the central nervous system and in brain development was recently reviewed by Courtney et al. [34].

4. Cholesterol in the Normal Brain

The brain contains about 20% of the cholesterol of the whole body, rendering it the most cholesterol-rich organ [35]. Previous studies have shown the possibility of circulating cholesterol, in some manner, affecting the function of the central nervous system (CNS): for instance, low circulating cholesterol levels might be linked with violent behavior [36–38]. It is also postulated that brain development and intelligence is related to the levels of circulating cholesterol of a newborn infant [39,40]. However, a series of experiments conducted later provide no direct evidence for lipoprotein cholesterol crossing the blood–brain barrier (BBB) [41–44]. Thus, it is believed that the BBB prevents the entry of lipoproteins into the brain, and the accumulation of brain cholesterol is mainly achieved through de novo synthesis. In addition, several proteins related to cholesterol metabolism have been found in the brain, such as the apolipoproteins ApoE and ApoAI, LDLRs, scavenger receptor class B type I (SRB1, encoded by the *SCARB1* gene), and ABC transporters. Whether they play the same role in the brain as in other organs is still under investigation.

Cholesterol metabolism in the brain is well-regulated through the coordinating work of a series of proteins. The mechanisms of acquiring cholesterol include de novo synthesis and uptake of cholesterol from the external environment by LDLR, SRB1, and Niemann–Pick C1-like protein (NPC1L1) [45]. The synthesis of cholesterol in brain, as in other organs, starts from the conversion of acetyl-CoA to 3-hydroxy-3-methylglutaryl-CoA with HMG-CoA as the rate-limiting enzyme. SREBPs in the endoplasmic reticulum sense the levels of cholesterol and regulate the activity of HMG-CoA [46]. Meanwhile, the uptake of cholesterol can be achieved through taking up lipoproteins from the extracellular environment. One example is the binding of particles that contain ApoE to LDLR, which are then processed through the clathrin-coated pit pathway to endosomes and lysosomes [47]. Moreover, Niemann–Pick type C1 and C2 are also required to move cholesterol to the plasma membrane [48]. The excretion of cholesterol out of the cell may be driven by the chemical gradient between leaflet and lipoprotein receptors in the plasma membrane. Cholesterol can also be exported from the cells by ABC transporters. Hundreds of ABC transporters have been found in both prokaryotes and eukaryotes. Of the 48 ABC transporters in human genome, 13 ABC transporters (ABCA1, ABCA2, ABCA3, ABCA4, ABCA7, ABCA8, ABCB1, ABCB4, ABCD1, ABCD2, ABCG1, ABCG2, and ABCG4) have been studied in human brain [49]. As mentioned previously, LXRα and LXRβ can regulate the expression of ABCA1 and ABCG1 to control the efflux of cholesterol and phospholipids. It was found that LXR agonists enhance cholesterol efflux in astrocytes [50]. In addition, cholesterol in the brain and other organs can be hydroxylated by various enzymes to form hydroxylated sterol molecules and excreted from cells by diffusion [36]. Sterols in the brain, especially in the adult, are essentially non-esterified cholesterol. The presence of cholesteryl esters in the brain correlates with the occurrence of disease, such as multiple sclerosis [51].

5. Cholesterol Metabolism in Embryonic vs. Adult Brain

About 70–80% of the cholesterol in the brain is found in myelin sheaths, and the rest exists in the membranes of cell organelles. The total cholesterol levels in the brain are low during the perinatal period and at the time of birth, which is thought to limit the development of the CNS to allow the baby's head to pass through the pelvis and birth canal. The cholesterol concentration of the whole brain increases from about 6 mg/g at birth to 23 mg/g in young adults [47]. Animal experiments demonstrate that the rate of cholesterol synthesis in the brain correlates with the rate of cholesterol accumulation and the final concentration of cholesterol found in those regions [44,52–54]. Therefore, de novo synthesis, not exogenous lipoprotein cholesterol import, is the major pathway for the accumulation of cholesterol in the brain during early development. Furthermore, studies also show that cholesterol synthesis during early development is accompanied by the synthesis of myelin basic protein and cerebroside [54]. These findings indicate that the accumulation of cholesterol in the brain associates with myelination. The synthesis of brain cholesterol slows down when neuron myelination is complete in adulthood, and the half-life of cholesterol lengthens to between 6 months and 5 years [55]. In contrast with early development when cholesterol is synthesized de novo, in adulthood nerve cell bodies can take up cholesterol using LDLR-related protein. Cell culture studies have shown that media from glial cells which contains cholesterol and ApoE stimulates the extension of axons, and LDLR inhibitors prevent the extension [56]. Furthermore, the formation of synapses of retinal ganglion cells requires glial cells to produce cholesterol and ApoE [57,58]. Therefore, cholesterol is probably mainly synthesized in both neurons and glial cells in the adult brain.

6. Cholesterol Excretion from the Brain

To maintain cholesterol homeostasis in the brain, especially during adulthood, the excretion of cholesterol is important. 24(S)-hydroxycholesterol is the major hydroxylated sterol excreted from brain [59]. In addition, the enzyme that synthesizes this oxysterol, cholesterol 24-hydroxylase (encoded by the gene *CYP46A1*), is primarily expressed in brain compared to other organs. The hydroxylated cholesterol can cross the BBB and go to the liver to be converted to bile acids and excreted from the body.

7. Cholesterol Metabolism in the Liver vs. Brain

The liver plays an important role in cholesterol metabolism. In contrast with brain, hepatic cholesterol can be obtained from the diet. Dietary cholesterol is obtained by intestinal epithelial cells via endocytosis. Cholesterol can then be esterified and loaded into nascent chylomicrons together with triacylglycerol. The chylomicrons are released from intestinal cells into the circulation by the lymphatics [60]. Triacylglycerol in the chylomicrons is hydrolyzed by lipoprotein lipase in blood vessels, and the cholesterol left behind in the chylomicron remnants are taken up and utilized by the liver [60,61]. The cholesterol from liver and dietary origins can be packed into particles of very low-density lipoproteins (VLDLs), which leave the liver and transport cholesterol to other tissues [62]. Another difference between liver and brain cholesterol metabolism is that cholesterol can be recycled through enterohepatic circulation, which does not exist in the brain (Figure 2). Cholesterol can be oxidized in the liver to form bile acids which along with cholesterol is excreted from the liver into the bile [63,64]. The excretion of bile acid involves ABCB11 [65]. Only about 5% of bile acids are lost in the feces, and the rest are reabsorbed into enterocytes. Bile acid is important for the digestion and absorption of dietary fats.

In summary, cholesterol is involved in cell membrane formation and signaling, and is the precursor of many steroid molecules such as steroid hormones, vitamins, and bile salts. Thus, the metabolism of cholesterol is tightly regulated throughout the body. Although the liver is the primary organ regulating cholesterol homeostasis, the brain cannot uptake cholesterol from peripheral blood and diet due to the BBB (Figure 2). Brain cholesterol is primarily derived from de novo synthesis, and cholesterol levels start to accumulate after birth. Upon reaching adulthood, brain cholesterol levels are maintained at constant levels. Therefore, the excretion of cholesterol from the brain becomes more active in adulthood. Brain cholesterol can be hydroxylated and pass through the BBB to form bile acids in the liver. Moreover, for some types of nerve cells cholesterol must be acquired through the binding of low-density lipoproteins LDLs and LDLRs since cholesterol is mainly synthesized in glial cells and neurons in the adult brain. Disturbed homeostasis of brain cholesterol can lead to diseases such as dementia.

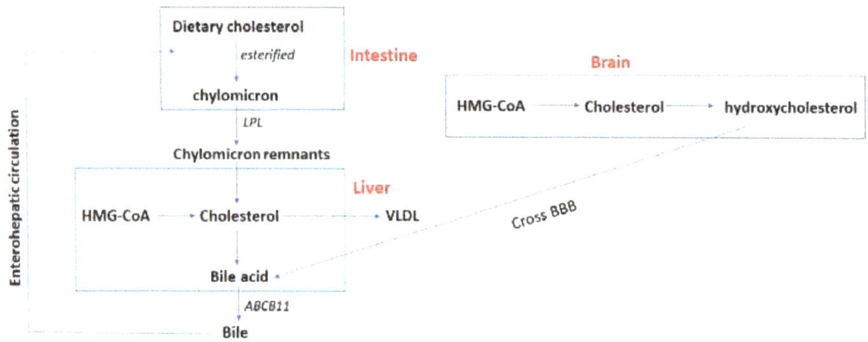

Figure 2. Cholesterol metabolism in liver vs. brain. The brain obtains cholesterol exclusively from de novo synthesis. On the contrary, hepatic cholesterol can be obtained by de novo synthesis and through dietary intake. Dietary cholesterol can be esterified and loaded into chylomicrons in the intestine. The chylomicrons are released into circulation and hydrolyzed by lipoprotein lipase (LPL) to form chylomicron remnants. Cholesterol left behind in the chylomicron remnants are taken up and utilized in the liver. The cholesterol synthesized in liver and from dietary origins can be packed into very low-density lipoprotein (VLDL) and exported from liver. Cholesterol can also be oxidized in the liver to form bile acids which excreted from liver into the bile via the ABCB11 transporter. Cholesterol in the brain can be hydrolyzed to form hydroxycholesterol which crosses the blood–brain barrier (BBB) and goes to the liver to be converted to bile acid. Cholesterol in the liver can be recycled through enterohepatic circulation, which does not exist in the brain. About 5% of the bile acids are lost in the feces, and the rest are reabsorbed into enterocytes.

8. Cholesterol Metabolism Pathways Are Altered in Brain Tumors

The brain has different ways to satisfy the requirements of cholesterol compared to peripheral organs. An epidemiological study investigated the relationship between dietary intake of cholesterol and the incidence of cancer, and found that high dietary cholesterol levels increase the risk of several cancers including stomach, colon, rectum, pancreas, lung, breast, kidney, bladder, and non-Hodgkin's lymphomas, but not brain tumors [66]. This result is not surprising: since cholesterol cannot pass the BBB, high plasma cholesterol levels are unlikely to affect cholesterol metabolism in the brain [66].

The brain obtains cholesterol primarily through de novo synthesis, which involves the mevalonate and Bloch and Kandutsch-Russell pathways [67–69]. Taking advantage of the Glioblastoma Bio Discovery Portal [70], our group found a correlation between upregulation of mevalonate and cholesterol pathway and poor survival of GBM patients [71]. Mechanistic studies demonstrated that densely-plated glioma cells increase the synthesis of cholesterol by enhancing oxygen consumption, glycolysis and the pentose phosphate pathway, and pharmacological inhibitors acting downstream of the mevalonate pathway induce glioma cell death [71]. Moreover, the study from our group also found that densely plated normal astrocytes but not tumor sphere glioma cells downregulate genes in the cholesterol biosynthetic pathway including farnesyl diphosphate synthase, farnesyl-diphosphate farnesyltransferase 1, and squalene epoxidase, (*FDPS*, *FDFT1*, and *SQLE*, Figure 3) [71]. A study conducted later by Kim et al. showed that inhibition of FDPS by pharmacological inhibitors and siRNA (small interfering RNA) prevents the formation of secondary spheres of glioma stem cells, and *FDPS* mRNA was associated with malignancy in glioblastoma patients [72]. In addition to mRNA and protein levels, Abate et al. demonstrate that the activity of FDPS is also upregulated in GBM tumor tissue [73].

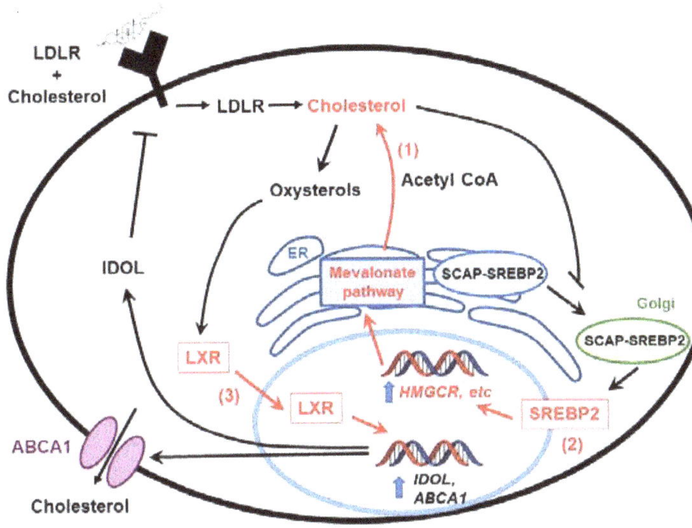

Figure 3. Cholesterol homeostasis in glioblastoma cells. Glioblastoma cells maintain cholesterol under conditions in which normal cells turn it off through multiple mechanisms of dysregulation (highlighted in red). They keep cholesterol biosynthesis on by constitutive activation of the mevalonate pathway (1), and by upregulating SREBPs under hypoxia (2). They are also highly dependent on appropriate levels of LXR activity—hyperactivating LXR with synthetic agonists overstimulates ABCA1 expression and cholesterol efflux, killing glioblastoma (GBM cells) (3). In sum, this provides them with cholesterol in an organ that is blocked from obtaining it from the circulation due to the blood-brain barrier.

Another study demonstrates that GBM is dependent proper cholesterol homeostasis for survival: instead of inhibiting the synthesis of cholesterol, Villa et al. tested the effect of LXRs on treating GBM [74]. LXR is a transcription factor that facilitates the efflux of cholesterol by increasing ABCA1 expression. Villa et al. showed that limiting cholesterol levels by treating cells with LXR agonists induced glioma cell death. In vivo experiments showed that LXR agonists inhibited GBM growth and prolonged the survival of mice [74]. As a key transcription factor in the regulation of sterol homeostasis, other groups have evaluated the effects of SREBPs on GBM development [75,76]. Lewis et al. showed that under hypoxia and serum-deprivation conditions, SREBP is upregulated to maintain the expression of fatty acid and cholesterol biosynthetic genes in GBM cells, and inhibiting SREBP activity under hypoxia led to GBM cell death [75]. These studies demonstrate that cholesterol metabolism pathways are upregulated in GBM patients and targeting cholesterol metabolism and/or homeostasis may be a promising strategy in treating GBM.

9. Targeting Cholesterol Metabolism as a Glioblastoma Therapy

Cancer cells have an increased demand for cholesterol and cholesterol precursors. Therefore, a reasonable assumption is that prevention of tumor-cell growth can be achieved by restricting either cholesterol availability or cholesterol synthesis [77]. Loss of cholesterol feedback inhibition mechanisms that regulate cholesterol synthesis is an important feature of malignant transformation. The cholesterol synthesis pathway has numerous proteins that are potential targets to disrupt cancer progression [78]. The therapeutic potential of targeting these cholesterol synthesis genes is under preclinical investigation [79,80]. The unique metabolic requirements of the brain might make glioblastoma particularly suitable for cholesterol pathway targeting [71].

Liver X-receptors (LXRs) act as transcriptional master switches that coordinate the regulation of genes such as *ABCA1* and *ABCG1*, which are involved in cellular cholesterol homeostasis [81,82]

LXR-623 is a synthetic ligand for LXRα and β that upregulates *ABCA1* and *ABCG1* expression in blood cells [83]. In a study published by the Mischel lab, LXR-623 killed GBM cell lines in an LXR β- and cholesterol-dependent fashion but not healthy brain cells. Upon further investigation of LXR-623, the group found that the drug penetrated the blood–brain barrier and retained its anticancer activity. In addition, mice harboring GBM tumors derived from human patients and treated with LXR-623 had substantially reduced tumor size and improved survival [74].

Statins are HMG-CoA reductase inhibitors that have anti-tumor effects and synergize with certain chemotherapeutic agents to decrease tumor development [80,84]. Several clinical trials have examined the potential chemo-preventive and therapeutic efficacy of different formulations of statins, including simvastatin, pitavastatin, and lovastatin [85–89]. For example, a short-term biomarker study showed lowered breast cancer recurrence in simvastatin-treated patients through the reduction of serum estrone sulfate levels [90]. Moreover, pitavastatin reduces GBM tumor burden in xenografts [86]. Bisphosphonates and tocotrienols are another class of drugs which act as downstream inhibitors of the cholesterol synthesis pathway. Early investigations have shown that they can slow down cancer cell and tumor growth similar to statins [91].

Protein geranylgeranylation, a branch of the cholesterol synthesis pathway, was also found to be essential for maintaining stemness of basal breast cancer cells and to promote human glioma cell growth. GGTI-288, an inhibitor of the geranylgeranyl transferase I (GGTI) reduced the cancer stem cell subpopulation in primary breast cancer xenografts [79,92]. Moreover, high dependency of malignant glioma cells on the isoprenoid pathway for post translational modification of intracellular signaling molecules make this pathway a potential candidate for drug targeting. Recent work published by Ciaglia et al. shows antitumor activity of N6-isopentenyladenosine (iPA) on glioma cells. Its mechanism of action is primarily driven through AMPK-dependent epidermal growth factor receptor (EGFR) degradation, which further adds value to its candidacy as a potent antitumor drug [93,94]. Thus, multiple preclinical studies demonstrate that targeting the cholesterol synthesis pathway could be useful for modulating cancer growth, either through directly inhibiting cholesterol synthesis, or though inhibiting the production of mevalonate pathway-derived moieties used in protein post-translational modifications of oncogenes.

10. Cholesterol and its Derivatives as Anti-Glioblastoma Agents

Steroids and their derivatives or triterpene precursors such as betulinic, oleanolic, and ursolic acids and stigmasterol have shown strong anti-cancerous properties [95–97]. Cholesterol derivatives, for example, sodium cholesteryl sulfate, cholesteryl chloride, cholesteryl bromide, and cholesteryl hemi-succinate, have been reported to possess inhibitory activities against DNA polymerase and DNA topoisomerase and to inhibit human cancer cell growth [98].

Another promising approach in use is a cholesterol-based anticancer agent containing carborane as the anticancer unit for boron neutron capture therapy (BNCT). The cholesteryl 1,12-dicarba-*closo*-dodecaborane 1-carboxylate (BCH) mimics the native cholesteryl ester in structure and was found to be effectively taken up by brain glioma cells in vitro. BNCT delivers boron-10 packed in liposomes. To increase the delivery with an enhanced specificity to tumor tissue, these boron-10 were packed inside cholesterol-anchored folate in EGFR-folate receptor targeted liposomes or consist of cholesterol-anchored anti EGFR antibodies. Once the boron-10 is delivered to its designated location, low energy radiation is passed in the form of a thermal neutron that triggers fission reactions, resulting in the production of high linear energy transfer (LET) α-particles which are highly lethal to the cells [99].

11. Cholesterol-Based Intracerebral Delivery of Chemotherapeutics in Brain Tumors

The absence of compelling treatment alternatives results in poor prognosis of glioblastoma. It is vital to deliver adequate amounts of therapeutic agents to the brain tumor site. However, delivery of therapeutic agents to the tumor site is technically very challenging due to the presence of the BBB [100].

Many attempts have been made to overcome this. It is well established that LDL receptors are present on the BBB capillary endothelial cells and, therefore, could be utilized for transporting cholesteryl-based or other compounds to the brain.

Conjugation of a cholesterol moiety to an active medicinal compound for either cancer diagnosis or treatment is an attractive approach for targeted drug delivery. Approaches to intracerebrally administer agents within the brain parenchyma through local delivery to tumor tissue are on the rise [101]. The advantage of this approach is it results in high drug concentrations at the tumor site with restricted exposure to non-neoplastic tissues and organs.

12. Conclusions

The unique metabolic demands and dysregulated metabolism of GBM makes it particularly suitable for cholesterol pathway targeting [71,97,102]. Genomic analyses performed by the TCGA provide correlative evidence suggesting an involvement of the cholesterol homeostasis pathways in cancer development, especially in glioblastoma [103]. A vast number of genetic and phenotypic alterations in cholesterol homeostasis pathways have been identified in cancer cells [104]. These include increases in gene copy numbers and upregulation of cholesterol synthesis gene expression, enhanced cholesterol import by LDL receptors, and decreased transport of cholesterol, all of which promote increased cellular cholesterol levels to aid cancer cell proliferation [104–107].

First and the foremost, the genetic alterations influencing the cholesterol pathways in cancer development need further investigation. Many cholesterol synthesis genes or mitochondrial cholesterol importers are upregulated, however their effects on cancer development remain unknown. Out of these, several genetic alterations were associated with known chromosomal amplification sites that harbor well-characterized oncogenes. For example, *HMGCS2* and *NOTCH2*, and *SQLE* and *MYC*, were co-localized to the same amplicons. Possibly, oncogenes and cholesterol synthesis genes act in tandem to promote disease progression. However, limited success has been achieved in restricting tumor growth by targeting these critical genes with their pharmacological inhibitors (Table 1). Due to a lack of evidence to support their efficacy in treating different forms of tumors, this needs further exploration. A second question needing attention is whether tumors could be classified into subclasses based on genetic abnormalities occurring in cholesterol homeostasis genes. This might facilitate development of precision medicine-based approaches for treating subgroups of cancer. For examples, the efficacy of statins, squalene synthesis inhibitors, farnesyl or geranylgeranyl transferase inhibitors might be more effective for certain patients with characteristic genetic signatures.

Table 1. Commercial drugs targeting cholesterol pathways.

Drug	Mechanism
Ciprofibrate	PPARα agonist
Clofibrates	PPARα agonist
Fenofibrate	PPARα agonist
Gemifibrozil	PPARα agonist
Anacetrapib	CETP inhibitor
Avasimibe	ACAT inhibitor
Berberine	Increases LDLR expression
Lapaquistat acetate	FDFT1 inhibitor
Ezetimibe	NPC1L1 inhibitor
Atorvastatin	HMGCR inhibitor
Fluvastatin	HMGCR inhibitor
Pitavastatin	HMGCR inhibitor
Rosuvastatin	HMGCR inhibitor
Simvastatin	HMGCR inhibitor
Pitavastatin	HMGCR inhibitor

PPARα = peroxisome proliferator activated receptor alpha; CETP = cholesteryl ester transfer protein; ACAT = sterol O-acyltransferase 1; LDLR = low density lipoprotein receptor; FDFT1 farnesyl-diphosphate farnesyltransferase 1; NPC1L1 = NPC1 like intracellular cholesterol transporter 1; HMGCR = 3-hydroxy-3-methylglutaryl-CoA reductase.

Conjugation of the cholesterol moiety to an active medicinal compound for cancer disease diagnosis and treatment is an attractive approach for targeted drug delivery. Several anticancer agents have also shown promise in cholesterol formulation for brain delivery. However, a greater understanding of the biodistribution and pharmacokinetics of these cholesteryl drug conjugates is essential for their practical use. For example, once active drug molecules are conjugated to cholesterol moieties, their chemical properties such as hydrophilicity/lipophilicity and molecular weight are significantly altered, which can change their biodistribution, pharmacokinetics, and efficacy. Hence, a thorough analysis of the interactions of these cholesteryl conjugates with cells, receptors, and membranes is necessary prior to their use in a clinical setting. Nevertheless, the use of cholesteryl drug conjugates for targeted delivery provides a novel approach for treating a variety of cancers, including glioblastoma.

In summary, it appears that deregulation of cholesterol homeostasis is an important contributing factor to cancer development and particularly to brain tumors. Studies are needed to link population-based epidemiological data, results from the TCGA database, and preclinical mechanistic evidence to more thoroughly dissect the involvement of cholesterol in cancer development, which would be helpful in devising new strategy for therapeutic intervention.

Author Contributions: Writing—original draft preparation, F.A., Q.S., D.P., and J.M.S.; writing—review and editing, F.A., Q.S., D.P., and J.M.S.; visualization, F.A., Q.S., and J.M.S.; supervision, J.M.S.; project administration, J.M.S.; funding acquisition, J.M.S.

Funding: This work was supported by the US National Cancer Institute Center for Cancer Research (NCI-CCR) Intramural Research Program ZIA BC 011441 (to J.M.S.).

Conflicts of Interest: The authors declare no conflict of interest.

References

1. Furnari, F.B.; Fenton, T.; Bachoo, R.M.; Mukasa, A.; Stommel, J.M.; Stegh, A.; Hahn, W.C.; Ligon, K.L.; Louis, D.N.; Brennan, C.; et al. Malignant astrocytic glioma: Genetics, biology, and paths to treatment. *Genes Dev.* **2007**, *21*, 2683–2710. [CrossRef] [PubMed]
2. Alphandery, E. Glioblastoma Treatments: An Account of Recent Industrial Developments. *Front. Pharmacol.* **2018**, *9*, 879. [CrossRef] [PubMed]
3. Thakkar, J.P.; Dolecek, T.A.; Horbinski, C.; Ostrom, Q.T.; Lightner, D.D.; Barnholtz-Sloan, J.S.; Villano, J.L. Epidemiologic and molecular prognostic review of glioblastoma. *Cancer Epidemiol. Biomark. Prev.* **2014**, *23*, 1985–1996. [CrossRef] [PubMed]
4. Prados, M.D.; Byron, S.A.; Tran, N.L.; Phillips, J.J.; Molinaro, A.M.; Ligon, K.L.; Wen, P.Y.; Kuhn, J.G.; Mellinghoff, I.K.; de Groot, J.F.; et al. Toward precision medicine in glioblastoma: The promise and the challenges. *Neuro Oncol.* **2015**, *17*, 1051–1063. [CrossRef] [PubMed]
5. Miyai, M.; Tomita, H.; Soeda, A.; Yano, H.; Iwama, T.; Hara, A. Current trends in mouse models of glioblastoma. *J. Neurooncol.* **2017**, *135*, 423–432. [CrossRef]
6. Weissenberger, J.; Steinbach, J.P.; Malin, G.; Spada, S.; Rulicke, T.; Aguzzi, A. Development and malignant progression of astrocytomas in GFAP-v-src transgenic mice. *Oncogene* **1997**, *14*, 2005–2013. [CrossRef]
7. Zhu, Y.; Guignard, F.; Zhao, D.; Liu, L.; Burns, D.K.; Mason, R.P.; Messing, A.; Parada, L.F. Early inactivation of p53 tumor suppressor gene cooperating with NF1 loss induces malignant astrocytoma. *Cancer Cell* **2005**, *8*, 119–130. [CrossRef]
8. Misuraca, K.L.; Hu, G.; Barton, K.L.; Chung, A.; Becher, O.J. A Novel Mouse Model of Diffuse Intrinsic Pontine Glioma Initiated in Pax3-Expressing Cells. *Neoplasia* **2016**, *18*, 60–70. [CrossRef]
9. Tateishi, K.; Wakimoto, H.; Iafrate, A.J.; Tanaka, S.; Loebel, F.; Lelic, N.; Wiederschain, D.; Bedel, O.; Deng, G.; Zhang, B.; et al. Extreme Vulnerability of IDH1 Mutant Cancers to NAD+ Depletion. *Cancer Cell* **2015**, *28*, 773–784. [CrossRef]
10. Ashizawa, T.; Miyata, H.; Iizuka, A.; Komiyama, M.; Oshita, C.; Kume, A.; Nogami, M.; Yagoto, M.; Ito, I.; Oishi, T.; et al. Effect of the STAT3 inhibitor STX-0119 on the proliferation of cancer stem-like cells derived from recurrent glioblastoma. *Int. J. Oncol.* **2013**, *43*, 219–227. [CrossRef]

11. Brennan, C.W.; Verhaak, R.G.; McKenna, A.; Campos, B.; Noushmehr, H.; Salama, S.R.; Zheng, S.; Chakravarty, D.; Sanborn, J.Z.; Berman, S.H.; et al. The somatic genomic landscape of glioblastoma. *Cell* **2013**, *155*, 462–477. [CrossRef]
12. De Witt Hamer, P.C. Small molecule kinase inhibitors in glioblastoma: A systematic review of clinical studies. *Neuro Oncol.* **2010**, *12*, 304–316. [CrossRef]
13. Thorne, A.H.; Zanca, C.; Furnari, F. Epidermal growth factor receptor targeting and challenges in glioblastoma. *Neuro Oncol.* **2016**, *18*, 914–918. [CrossRef]
14. Calvert, A.E.; Chalastanis, A.; Wu, Y.; Hurley, L.A.; Kouri, F.M.; Bi, Y.; Kachman, M.; May, J.L.; Bartom, E.; Hua, Y.; et al. Cancer-Associated IDH1 Promotes Growth and Resistance to Targeted Therapies in the Absence of Mutation. *Cell Rep.* **2017**, *19*, 1858–1873. [CrossRef]
15. Watanabe, T.; Nobusawa, S.; Kleihues, P.; Ohgaki, H. IDH1 mutations are early events in the development of astrocytomas and oligodendrogliomas. *Am. J. Pathol.* **2009**, *174*, 1149–1153. [CrossRef]
16. Lai, A.; Kharbanda, S.; Pope, W.B.; Tran, A.; Solis, O.E.; Peale, F.; Forrest, W.F.; Pujara, K.; Carrillo, J.A.; Pandita, A.; et al. Evidence for sequenced molecular evolution of IDH1 mutant glioblastoma from a distinct cell of origin. *J. Clin. Oncol.* **2011**, *29*, 4482–4490. [CrossRef]
17. Franceschi, S.; Corsinovi, D.; Lessi, F.; Tantillo, E.; Aretini, P.; Menicagli, M.; Scopelliti, C.; Civita, P.; Pasqualetti, F.; Naccarato, A.G.; et al. Mitochondrial enzyme GLUD2 plays a critical role in glioblastoma progression. *EBioMedicine* **2018**, *37*, 56–67. [CrossRef]
18. Janowski, B.A.; Willy, P.J.; Devi, T.R.; Falck, J.R.; Mangelsdorf, D.J. An oxysterol signalling pathway mediated by the nuclear receptor LXR alpha. *Nature* **1996**, *383*, 728–731. [CrossRef]
19. Umetani, M.; Domoto, H.; Gormley, A.K.; Yuhanna, I.S.; Cummins, C.L.; Javitt, N.B.; Korach, K.S.; Shaul, P.W.; Mangelsdorf, D.J. 27-Hydroxycholesterol is an endogenous SERM that inhibits the cardiovascular effects of estrogen. *Nat. Med.* **2007**, *13*, 1185–1192. [CrossRef]
20. Wei, W.; Schwaid, A.G.; Wang, X.; Wang, X.; Chen, S.; Chu, Q.; Saghatelian, A.; Wan, Y. Ligand Activation of ERRalpha by Cholesterol Mediates Statin and Bisphosphonate Effects. *Cell Metab.* **2016**, *23*, 479–491. [CrossRef]
21. Yang, W.; Bai, Y.; Xiong, Y.; Zhang, J.; Chen, S.; Zheng, X.; Meng, X.; Li, L.; Wang, J.; Xu, C.; et al. Potentiating the antitumour response of CD8(+) T cells by modulating cholesterol metabolism. *Nature* **2016**, *531*, 651–655. [CrossRef]
22. Wang, F.; Beck-Garcia, K.; Zorzin, C.; Schamel, W.W.; Davis, M.M. Inhibition of T cell receptor signaling by cholesterol sulfate, a naturally occurring derivative of membrane cholesterol. *Nat. Immunol.* **2016**, *17*, 844–850. [CrossRef] [PubMed]
23. Horton, J.D. Sterol regulatory element-binding proteins: Transcriptional activators of lipid synthesis. *Biochem. Soc. Trans.* **2002**, *30*, 1091–1095. [CrossRef]
24. Goldstein, J.L.; DeBose-Boyd, R.A.; Brown, M.S. Protein sensors for membrane sterols. *Cell* **2006**, *124*, 35–46. [CrossRef]
25. Gong, Y.; Lee, J.N.; Lee, P.C.; Goldstein, J.L.; Brown, M.S.; Ye, J. Sterol-regulated ubiquitination and degradation of Insig-1 creates a convergent mechanism for feedback control of cholesterol synthesis and uptake. *Cell Metab.* **2006**, *3*, 15–24. [CrossRef]
26. Adams, C.M.; Reitz, J.; De Brabander, J.K.; Feramisco, J.D.; Li, L.; Brown, M.S.; Goldstein, J.L. Cholesterol and 25-hydroxycholesterol inhibit activation of SREBPs by different mechanisms, both involving SCAP and Insigs. *J. Biol. Chem.* **2004**, *279*, 52772–52780. [CrossRef]
27. Sakai, J.; Duncan, E.A.; Rawson, R.B.; Hua, X.; Brown, M.S.; Goldstein, J.L. Sterol-regulated release of SREBP-2 from cell membranes requires two sequential cleavages, one within a transmembrane segment. *Cell* **1996**, *85*, 1037–1046. [CrossRef]
28. Horton, J.D.; Goldstein, J.L.; Brown, M.S. SREBPs: Activators of the complete program of cholesterol and fatty acid synthesis in the liver. *J. Clin. Investig.* **2002**, *109*, 1125–1131. [CrossRef]
29. Heiniger, H.J.; Marshall, J.D. Cholesterol synthesis in polyclonally activated cytotoxic lymphocytes and its requirement for differentiation and proliferation. *Proc. Natl. Acad. Sci. USA* **1982**, *79*, 3823–3827. [CrossRef]
30. Reboldi, A.; Dang, E.V.; McDonald, J.G.; Liang, G.; Russell, D.W.; Cyster, J.G. Inflammation. 25-Hydroxycholesterol suppresses interleukin-1-driven inflammation downstream of type I interferon. *Science* **2014**, *345*, 679–684. [CrossRef]

31. Tricarico, P.M.; Gratton, R.; Braga, L.; Celsi, F.; Crovella, S. 25-Hydroxycholesterol and inflammation in Lovastatin-deregulated mevalonate pathway. *Int. J. Biochem. Cell Biol.* **2017**, *92*, 26–33. [CrossRef] [PubMed]
32. Repa, J.J.; Mangelsdorf, D.J. The role of orphan nuclear receptors in the regulation of cholesterol homeostasis. *Annu. Rev. Cell Dev. Biol.* **2000**, *16*, 459–481. [CrossRef] [PubMed]
33. Zelcer, N.; Hong, C.; Boyadjian, R.; Tontonoz, P. LXR regulates cholesterol uptake through Idol-dependent ubiquitination of the LDL receptor. *Science* **2009**, *325*, 100–104. [CrossRef] [PubMed]
34. Courtney, R.; Landreth, G.E. LXR Regulation of Brain Cholesterol: From Development to Disease. *Trends Endocrinol. Metab.* **2016**, *27*, 404–414. [CrossRef] [PubMed]
35. Bjorkhem, I.; Meaney, S. Brain cholesterol: Long secret life behind a barrier. *Arterioscler. Thromb. Vasc. Biol.* **2004**, *24*, 806–815. [CrossRef] [PubMed]
36. Virkkunen, M.; Penttinen, H. Serum cholesterol in aggressive conduct disorder: A preliminary study. *Biol. Psychiatry* **1984**, *19*, 435–439. [PubMed]
37. Goldstein, M.R. Cholesterol and violence: Is there a connection? *Ann. Intern. Med.* **1998**, *129*, 668–669, author reply 669–670. [CrossRef]
38. Golomb, B.A.; Stattin, H.; Mednick, S. Low cholesterol and violent crime. *J. Psychiatry Res.* **2000**, *34*, 301–309. [CrossRef]
39. Lifshitz, F.; Moses, N. Growth failure. A complication of dietary treatment of hypercholesterolemia. *Am. J. Dis. Child.* **1989**, *143*, 537–542. [CrossRef]
40. Barness, L.A. Nutritional requirements of infants and children with respect to cholesterol and related compounds. *Am. J. Med. Genet.* **1994**, *50*, 353–354. [CrossRef]
41. Dietschy, J.M.; Kita, T.; Suckling, K.E.; Goldstein, J.L.; Brown, M.S. Cholesterol synthesis in vivo and in vitro in the WHHL rabbit, an animal with defective low density lipoprotein receptors. *J. Lipid Res.* **1983**, *24*, 469–480. [PubMed]
42. Osono, Y.; Woollett, L.A.; Herz, J.; Dietschy, J.M. Role of the low density lipoprotein receptor in the flux of cholesterol through the plasma and across the tissues of the mouse. *J. Clin. Investig.* **1995**, *95*, 1124–1132. [CrossRef] [PubMed]
43. Turley, S.D.; Burns, D.K.; Rosenfeld, C.R.; Dietschy, J.M. Brain does not utilize low density lipoprotein-cholesterol during fetal and neonatal development in the sheep. *J. Lipid Res.* **1996**, *37*, 1953–1961. [PubMed]
44. Quan, G.; Xie, C.; Dietschy, J.M.; Turley, S.D. Ontogenesis and regulation of cholesterol metabolism in the central nervous system of the mouse. *Brain Res. Dev. Brain Res.* **2003**, *146*, 87–98. [CrossRef] [PubMed]
45. Bloch, K. Sterol molecule: Structure, biosynthesis, and function. *Steroids* **1992**, *57*, 378–383. [CrossRef]
46. Geng, F.; Guo, D. Lipid droplets, potential biomarker and metabolic target in glioblastoma. *Intern. Med. Rev.* **2017**, *3*. [CrossRef]
47. Dietschy, J.M.; Turley, S.D. Thematic review series: Brain Lipids. Cholesterol metabolism in the central nervous system during early development and in the mature animal. *J. Lipid Res.* **2004**, *45*, 1375–1397. [CrossRef] [PubMed]
48. Betters, J.L.; Yu, L. NPC1L1 and cholesterol transport. *FEBS Lett.* **2010**, *584*, 2740–2747. [CrossRef]
49. Kim, W.S.; Weickert, C.S.; Garner, B. Role of ATP-binding cassette transporters in brain lipid transport and neurological disease. *J. Neurochem.* **2008**, *104*, 1145–1166. [CrossRef]
50. Whitney, K.D.; Watson, M.A.; Collins, J.L.; Benson, W.G.; Stone, T.M.; Numerick, M.J.; Tippin, T.K.; Wilson, J.G.; Winegar, D.A.; Kliewer, S.A. Regulation of cholesterol homeostasis by the liver X receptors in the central nervous system. *Mol. Endocrinol.* **2002**, *16*, 1378–1385. [CrossRef]
51. Davison, A.N.; Wajda, M. Cerebral lipids in multiple sclerosis. *J. Neurochem.* **1962**, *9*, 427–432. [CrossRef] [PubMed]
52. Jurevics, H.; Morell, P. Cholesterol for synthesis of myelin is made locally, not imported into brain. *J. Neurochem.* **1995**, *64*, 895–901. [CrossRef] [PubMed]
53. Cavender, C.P.; Turley, S.D.; Dietschy, J.M. Sterol metabolism in fetal, newborn, and suckled lambs and their response to cholesterol after weaning. *Am. J. Physiol.* **1995**, *269*, E331–E340. [CrossRef] [PubMed]
54. Muse, E.D.; Jurevics, H.; Toews, A.D.; Matsushima, G.K.; Morell, P. Parameters related to lipid metabolism as markers of myelination in mouse brain. *J. Neurochem.* **2001**, *76*, 77–86. [CrossRef] [PubMed]

55. Andersson, M.; Elmberger, P.G.; Edlund, C.; Kristensson, K.; Dallner, G. Rates of cholesterol, ubiquinone, dolichol and dolichyl-P biosynthesis in rat brain slices. *FEBS Lett.* **1990**, *269*, 15–18. [CrossRef]
56. Hayashi, H.; Campenot, R.B.; Vance, D.E.; Vance, J.E. Glial lipoproteins stimulate axon growth of central nervous system neurons in compartmented cultures. *J. Biol. Chem.* **2004**, *279*, 14009–14015. [CrossRef] [PubMed]
57. Mauch, D.H.; Nagler, K.; Schumacher, S.; Goritz, C.; Muller, E.C.; Otto, A.; Pfrieger, F.W. CNS synaptogenesis promoted by glia-derived cholesterol. *Science* **2001**, *294*, 1354–1357. [CrossRef]
58. Pfrieger, F.W. Outsourcing in the brain: Do neurons depend on cholesterol delivery by astrocytes? *Bioessays* **2003**, *25*, 72–78. [CrossRef]
59. Lutjohann, D.; Breuer, O.; Ahlborg, G.; Nennesmo, I.; Siden, A.; Diczfalusy, U.; Bjorkhem, I. Cholesterol homeostasis in human brain: Evidence for an age-dependent flux of 24S-hydroxycholesterol from the brain into the circulation. *Proc. Natl. Acad. Sci. USA* **1996**, *93*, 9799–9804. [CrossRef]
60. Dash, S.; Xiao, C.; Morgantini, C.; Lewis, G.F. New Insights into the Regulation of Chylomicron Production. *Annu. Rev. Nutr.* **2015**, *35*, 265–294. [CrossRef]
61. Dietschy, J.M.; Turley, S.D.; Spady, D.K. Role of liver in the maintenance of cholesterol and low density lipoprotein homeostasis in different animal species, including humans. *J. Lipid Res.* **1993**, *34*, 1637–1659.
62. Doonan, L.M.; Fisher, E.A.; Brodsky, J.L. Can modulators of apolipoproteinB biogenesis serve as an alternate target for cholesterol-lowering drugs? *Biochim. Biophys. Acta. Mol. Cell Biol. Lipids* **2018**, *1863*, 762–771. [CrossRef]
63. Russell, D.W. The enzymes, regulation, and genetics of bile acid synthesis. *Annu. Rev. Biochem.* **2003**, *72*, 137–174. [CrossRef] [PubMed]
64. Midzak, A.; Papadopoulos, V. Binding domain-driven intracellular trafficking of sterols for synthesis of steroid hormones, bile acids and oxysterols. *Traffic* **2014**, *15*, 895–914. [CrossRef] [PubMed]
65. Oude Elferink, R.P.; Groen, A.K. Mechanisms of biliary lipid secretion and their role in lipid homeostasis. *Semin. Liver Dis.* **2000**, *20*, 293–305. [CrossRef] [PubMed]
66. Hu, J.; La Vecchia, C.; de Groh, M.; Negri, E.; Morrison, H.; Mery, L.; Canadian Cancer Registries Epidemiology Research Group. Dietary cholesterol intake and cancer. *Ann. Oncol.* **2012**, *23*, 491–500. [CrossRef] [PubMed]
67. Kandutsch, A.A.; Russell, A.E. Preputial gland tumor sterols. 3. A metabolic pathway from lanosterol to cholesterol. *J. Biol. Chem.* **1960**, *235*, 2256–2261.
68. Bloch, K. The biological synthesis of cholesterol. *Science* **1965**, *150*, 19–28. [CrossRef]
69. Goldstein, J.L.; Brown, M.S. Regulation of the mevalonate pathway. *Nature* **1990**, *343*, 425–430. [CrossRef]
70. Celiku, O.; Johnson, S.; Zhao, S.; Camphausen, K.; Shankavaram, U. Visualizing molecular profiles of glioblastoma with GBM-BioDP. *PLoS ONE* **2014**, *9*, e101239. [CrossRef]
71. Kambach, D.M.; Halim, A.S.; Cauer, A.G.; Sun, Q.; Tristan, C.A.; Celiku, O.; Kesarwala, A.H.; Shankavaram, U.; Batchelor, E.; Stommel, J.M. Disabled cell density sensing leads to dysregulated cholesterol synthesis in glioblastoma. *Oncotarget* **2017**, *8*, 14860–14875. [CrossRef] [PubMed]
72. Kim, H.Y.; Kim, D.K.; Bae, S.H.; Gwak, H.; Jeon, J.H.; Kim, J.K.; Lee, B.I.; You, H.J.; Shin, D.H.; Kim, Y.H.; et al. Farnesyl diphosphate synthase is important for the maintenance of glioblastoma stemness. *Exp. Mol. Med.* **2018**, *50*, 137. [CrossRef] [PubMed]
73. Abate, M.; Laezza, C.; Pisanti, S.; Torelli, G.; Seneca, V.; Catapano, G.; Montella, F.; Ranieri, R.; Notarnicola, M.; Gazzerro, P.; et al. Deregulated expression and activity of Farnesyl Diphosphate Synthase (FDPS) in Glioblastoma. *Sci. Rep.* **2017**, *7*, 14123. [CrossRef] [PubMed]
74. Villa, G.R.; Hulce, J.J.; Zanca, C.; Bi, J.; Ikegami, S.; Cahill, G.L.; Gu, Y.; Lum, K.M.; Masui, K.; Yang, H.; et al. An LXR-Cholesterol Axis Creates a Metabolic Co-Dependency for Brain Cancers. *Cancer Cell* **2016**, *30*, 683–693. [CrossRef] [PubMed]
75. Lewis, C.A.; Brault, C.; Peck, B.; Bensaad, K.; Griffiths, B.; Mitter, R.; Chakravarty, P.; East, P.; Dankworth, B.; Alibhai, D.; et al. SREBP maintains lipid biosynthesis and viability of cancer cells under lipid- and oxygen-deprived conditions and defines a gene signature associated with poor survival in glioblastoma multiforme. *Oncogene* **2015**, *34*, 5128–5140. [CrossRef] [PubMed]
76. Geng, F.; Cheng, X.; Wu, X.; Yoo, J.Y.; Cheng, C.; Guo, J.Y.; Mo, X.; Ru, P.; Hurwitz, B.; Kim, S.H.; et al. Inhibition of SOAT1 Suppresses Glioblastoma Growth via Blocking SREBP-1-Mediated Lipogenesis. *Clin. Cancer Res.* **2016**, *22*, 5337–5348. [CrossRef]

77. Buchwald, H. Cholesterol inhibition, cancer, and chemotherapy. *Lancet* **1992**, *339*, 1154–1156. [CrossRef]
78. Brown, M.S.; Goldstein, J.L. A receptor-mediated pathway for cholesterol homeostasis. *Science* **1986**, *232*, 34–47. [CrossRef]
79. Ginestier, C.; Monville, F.; Wicinski, J.; Cabaud, O.; Cervera, N.; Josselin, E.; Finetti, P.; Guille, A.; Larderet, G.; Viens, P.; et al. Mevalonate metabolism regulates Basal breast cancer stem cells and is a potential therapeutic target. *Stem Cells* **2012**, *30*, 1327–1337. [CrossRef]
80. Cruz, P.M.; Mo, H.; McConathy, W.J.; Sabnis, N.; Lacko, A.G. The role of cholesterol metabolism and cholesterol transport in carcinogenesis: A review of scientific findings, relevant to future cancer therapeutics. *Front. Pharmacol.* **2013**, *4*, 119. [CrossRef]
81. Schultz, J.R.; Tu, H.; Luk, A.; Repa, J.J.; Medina, J.C.; Li, L.; Schwendner, S.; Wang, S.; Thoolen, M.; Mangelsdorf, D.J.; et al. Role of LXRs in control of lipogenesis. *Genes Dev.* **2000**, *14*, 2831–2838. [CrossRef]
82. Repa, J.J.; Turley, S.D.; Lobaccaro, J.A.; Medina, J.; Li, L.; Lustig, K.; Shan, B.; Heyman, R.A.; Dietschy, J.M.; Mangelsdorf, D.J. Regulation of absorption and ABC1-mediated efflux of cholesterol by RXR heterodimers. *Science* **2000**, *289*, 1524–1529. [CrossRef] [PubMed]
83. DiBlasio-Smith, E.A.; Arai, M.; Quinet, E.M.; Evans, M.J.; Kornaga, T.; Basso, M.D.; Chen, L.; Feingold, I.; Halpern, A.R.; Liu, Q.Y.; et al. Discovery and implementation of transcriptional biomarkers of synthetic LXR agonists in peripheral blood cells. *J. Transl. Med.* **2008**, *6*, 59. [CrossRef]
84. Warita, K.; Warita, T.; Beckwitt, C.H.; Schurdak, M.E.; Vazquez, A.; Wells, A.; Oltvai, Z.N. Statin-induced mevalonate pathway inhibition attenuates the growth of mesenchymal-like cancer cells that lack functional E-cadherin mediated cell cohesion. *Sci. Rep.* **2014**, *4*, 7593. [CrossRef] [PubMed]
85. Roeder, S.L.; University of Iowa. A Study of the Proper Dosage of Lovastatin and Docetaxel for Patients with Cancer. Available online: https://ClinicalTrials.gov/show/NCT00584012 (accessed on 2 January 2008).
86. Claes, A.; Wesseling, P.; Jeuken, J.; Maass, C.; Heerschap, A.; Leenders, W.P. Antiangiogenic compounds interfere with chemotherapy of brain tumors due to vessel normalization. *Mol. Cancer Ther.* **2008**, *7*, 71–78. [CrossRef] [PubMed]
87. Center, L.U.M.; Institute, N.C. Simvastatin and Panitumumab in Treating Patients with Advanced or Metastatic Colorectal Cancer. Available online: https://ClinicalTrials.gov/show/NCT01110785 (accessed on 27 April 2010).
88. Center, C.-S.M.; Institute, R.P.C. Cellular Effect of Cholesterol-Lowering Prior to Prostate Removal. Available online: https://ClinicalTrials.gov/show/NCT02534376 (accessed on 27 August 2015).
89. Centre, L.H.S. Metformin and Simvastatin Use in Bladder Cancer. Available online: https://ClinicalTrials.gov/show/NCT02360618 (accessed on 10 February 2015).
90. Higgins, M.J.; Prowell, T.M.; Blackford, A.L.; Byrne, C.; Khouri, N.F.; Slater, S.A.; Jeter, S.C.; Armstrong, D.K.; Davidson, N.E.; Emens, L.A.; et al. A short-term biomarker modulation study of simvastatin in women at increased risk of a new breast cancer. *Breast Cancer Res. Treat.* **2012**, *131*, 915–924. [CrossRef] [PubMed]
91. Mo, H.; Elson, C.E. Studies of the isoprenoid-mediated inhibition of mevalonate synthesis applied to cancer chemotherapy and chemoprevention. *Exp. Biol. Med.* **2004**, *229*, 567–585. [CrossRef]
92. Zhou, X.; Qian, J.; Hua, L.; Shi, Q.; Liu, Z.; Xu, Y.; Sang, B.; Mo, J.; Yu, R. Geranylgeranyltransferase I promotes human glioma cell growth through Rac1 membrane association and activation. *J. Mol. Neurosci.* **2013**, *49*, 130–139. [CrossRef]
93. Ciaglia, E.; Grimaldi, M.; Abate, M.; Scrima, M.; Rodriquez, M.; Laezza, C.; Ranieri, R.; Pisanti, S.; Ciuffreda, P.; Manera, C.; et al. The isoprenoid derivative N(6)-benzyladenosine CM223 exerts antitumor effects in glioma patient-derived primary cells through the mevalonate pathway. *Br. J. Pharmacol.* **2017**, *174*, 2287–2301. [CrossRef]
94. Ciaglia, E.; Abate, M.; Laezza, C.; Pisanti, S.; Vitale, M.; Seneca, V.; Torelli, G.; Franceschelli, S.; Catapano, G.; Gazzerro, P.; et al. Antiglioma effects of N6-isopentenyladenosine, an endogenous isoprenoid end product, through the downregulation of epidermal growth factor receptor. *Int. J. Cancer* **2017**, *140*, 959–972. [CrossRef]
95. Awad, A.B.; Fink, C.S. Phytosterols as anticancer dietary components: Evidence and mechanism of action. *J. Nutr.* **2000**, *130*, 2127–2130. [CrossRef] [PubMed]
96. Putz, M.V.; Lazea, M.; Sandjo, L.P. Quantitative Structure Inter-Activity Relationship (QSInAR). Cytotoxicity Study of Some Hemisynthetic and Isolated Natural Steroids and Precursors on Human Fibrosarcoma Cells HT1080. *Molecules* **2011**, *16*, 6603–6620. [CrossRef] [PubMed]

97. Ahmad, F.; Patrick, S.; Sheikh, T.; Sharma, V.; Pathak, P.; Malgulwar, P.B.; Kumar, A.; Joshi, S.D.; Sarkar, C.; Sen, E. Telomerase reverse transcriptase (TERT)—Enhancer of zeste homolog 2 (EZH2) network regulates lipid metabolism and DNA damage responses in glioblastoma. *J. Neurochem.* **2017**, *143*, 671–683. [CrossRef] [PubMed]
98. Ishimaru, C.; Yonezawa, Y.; Kuriyama, I.; Nishida, M.; Yoshida, H.; Mizushina, Y. Inhibitory effects of cholesterol derivatives on DNA polymerase and topoisomerase activities, and human cancer cell growth. *Lipids* **2008**, *43*, 373–382. [CrossRef]
99. Peacock, G.; Sidwell, R.; Pan, G.; Ole, S.; Lu, D.R. In vitro uptake of a new cholesteryl carborane ester compound by human glioma cell lines. *J. Pharm. Sci.* **2004**, *93*, 13–19. [CrossRef] [PubMed]
100. Zee-Cheng, R.K.; Cheng, C.C. Delivery of anticancer drugs. *Methods Find. Exp. Clin. Pharmacol.* **1989**, *11*, 439–529. [PubMed]
101. Lu, D.R.; Mehta, S.C.; Chen, W. Selective boron drug delivery to brain tumors for boron neutron capture therapy. *Adv. Drug Deliv. Rev.* **1997**, *26*, 231–247. [PubMed]
102. Ahmad, F.; Dixit, D.; Sharma, V.; Kumar, A.; Joshi, S.D.; Sarkar, C.; Sen, E. Nrf2-driven TERT regulates pentose phosphate pathway in glioblastoma. *Cell Death Dis.* **2016**, *7*, e2213. [CrossRef] [PubMed]
103. Cancer Genome Atlas Research, N.; Weinstein, J.N.; Collisson, E.A.; Mills, G.B.; Shaw, K.R.; Ozenberger, B.A.; Ellrott, K.; Shmulevich, I.; Sander, C.; Stuart, J.M. The Cancer Genome Atlas Pan-Cancer analysis project. *Nat. Genet.* **2013**, *45*, 1113–1120. [CrossRef]
104. Murai, T. Cholesterol lowering: Role in cancer prevention and treatment. *Biol. Chem.* **2015**, *396*, 1–11. [CrossRef]
105. Llaverias, G.; Danilo, C.; Mercier, I.; Daumer, K.; Capozza, F.; Williams, T.M.; Sotgia, F.; Lisanti, M.P.; Frank, P.G. Role of cholesterol in the development and progression of breast cancer. *Am. J. Pathol.* **2011**, *178*, 402–412. [CrossRef] [PubMed]
106. Smith, B.; Land, H. Anticancer activity of the cholesterol exporter ABCA1 gene. *Cell Rep.* **2012**, *2*, 580–590. [CrossRef] [PubMed]
107. Rama, A.R.; Alvarez, P.J.; Madeddu, R.; Aranega, A. ABC transporters as differentiation markers in glioblastoma cells. *Mol. Biol. Rep.* **2014**, *41*, 4847–4851. [CrossRef] [PubMed]

cancers

MDPI

Review

The Role of Platelets in Cancer Pathophysiology: Focus on Malignant Glioma

Sascha Marx [1], Yong Xiao [1,2], Marcel Baschin [3], Maximilian Splittstöhser [1,2], Robert Altmann [1,2], Eileen Moritz [2], Gabriele Jedlitschky [2], Sandra Bien-Möller [1,2], Henry W.S. Schroeder [1] and Bernhard H. Rauch [2,*]

[1] Department of Neurosurgery, University Medicine Greifswald, 17475 Greifswald, Germany; sascha.marx@uni-greifswald.de (S.M.); yungxiao@hotmail.com (Y.X.); msplittstoehser@yahoo.de (M.S.); robert.altmann@stud.uni-greifswald.de (R.A.); sandra.bien@uni-greifswald.de (S.B.-M.); henry.schroeder@uni-greifswald.de (H.W.S.S.)

[2] Department of Pharmacology, Center of Drug Absorption and Transport, University Medicine Greifswald, 17487 Greifswald, Germany; eileen.moritz@uni-greifswald.de (E.M.); gabriele.jedlitschky@uni-greifswald.de (G.J.)

[3] Department of Transfusion Medicine, University Medicine Greifswald, 17475 Greifswald, Germany; marcel.baschin@uni-greifswald.de

* Correspondence: Bernhard.Rauch@uni-greifswald.de; Tel.: +49-3834-86-5650

Received: 27 March 2019; Accepted: 18 April 2019; Published: 22 April 2019

Abstract: The link between thrombocytosis and malignancy has been well known for many years and its associations with worse outcomes have been reported mainly for solid tumors. Besides measuring platelet count, it has become popular to assess platelet function in the context of malignant diseases during the last decade. Malignant gliomas differ tremendously from malignancies outside the central nervous system because they virtually never form distant metastases. This review summarizes the current understanding of the platelet–immune cell communication and its potential role in glioma resistance and progression. Particularly, we focus on platelet-derived proinflammatory modulators, such as sphingosine-1-phosphate (S1P). The multifaceted interaction with immune cells puts the platelet into an interesting perspective regarding the recent advances in immunotherapeutic approaches in malignant glioma.

Keywords: glioblastoma multiforme; platelet; immune cell interaction; sphingosine-1-phosphate

1. Introduction

The glioblastoma multiforme (GBM) is the most common malignant brain tumor in adults. The mean overall survival is hardly beyond one year, despite standard treatment consisting of gross total resection of the contrast enhanced tumor segments followed by radiochemotherapy with temozolomide [1,2]. New treatment modalities such as tumor-treating fields (TTF) [3,4] may ultimately result in a survival benefit, and innovative immunotherapeutic approaches might bear the opportunity of curative interventions in the future [5–10]. One major problem for a successful systemic immunotherapy in GBM is the highly immunosuppressive tumor microenvironment which prevents the immune system from effectively attacking the tumor [11–13]. The predominant immune cell constituent in GBM tissue are tumor-associated macrophages (TAMs) [14,15], which can be derived from both microglia, the resident type of macrophages in the brain, as well as from bone marrow (BM)-derived peripheral monocytes/macrophages [16,17]. Comprehensive investigations from Pyonteck et al. have shown that about 40% of TAMs are of peripheral origin and invade the brain in a platelet-derived growth factor (PDGF)- and hypoxia-inducible factor-1α (HIF-1α)/stromal-cell-derived factor-1 (SDF-1)-dependent manner [18]. Strikingly, platelets are not only a major source of PDGF and of SDF-1α (CXCL12) [19,20], but also play a central role in both tumor angiogenesis [21,22] and in concerting the tumor microenvironment in solid tumors [23]. Platelets release a plethora of growth factors, inflammatory mediators, and chemokines into the microenvironment, and overwhelming evidence reveals platelets as key components in cancer biology and the challenges in targeting platelets for cancer treatment are currently discussed [21]. In comparison, the putative role of platelets in GBM pathophysiology and specifically in affecting immune cell functions such as conversion of macrophages into immunosuppressive TAMs in GBM is less clear to date.

The GBM is not only characterized by its highly infiltrative growth pattern and immunosuppressive properties but it is also a highly prothrombotic tumor entity. Typically, the balance between the procoagulant and anticoagulant system as well as the fibrinolytic system is severely impaired in favor of hypercoagulability in GBM patients [24]. High-grade gliomas, for instance, histologically exhibit areas of necrosis due to anoxia and microthrombosis despite diffuse microvascular proliferation and represent a significant risk factor for thromboembolism. As a consequence, this tumor is associated with a high risk for venous thromboembolism (VTE) and also cardiovascular events [25–27]. The heterogeneous morphology of GBM is characterized by its infiltrative growth, intense neoangiogenesis, and pseudopalisading necrosis [28]. In addition, striking endothelial cell proliferation and occlusive intravascular thrombosis can be observed in GBM tissue [29]. VTE is a common complication of cancer and it has been estimated that patients with cancer have a 4-fold to 20-fold increased risk of VTE, which is further accentuated by chemotherapy. The cumulative incidence of symptomatic VTE among patients with glioma has been estimated to be as high as 32.2% during their course of therapy [24,29,30]. Furthermore, inside the GBM a multitude of thrombosed vessels can be seen reflecting the state of hypercoagulability in these patients. However, the survival did not differ between GBM patients with and without VTE and was 53% after 12 months in both groups [31]. Interestingly, recent studies support the assumption that the plasma hypercoagulable profile seen in GBM patients is related to adverse outcomes [24]. Therefore, the significance and prognostic value of characterizing the coagulation profile in GBM patients as a novel approach to an individualized therapy of GBM is under discussion [24]. In addition, platelets release mediators which can directly or indirectly, i.e. by modulating immune responses, modify tumor cell activity, tumor growth and tumor angiogenesis (see Figure 1). The precise contribution of platelet-derived mediators in GBM in this context is not well studied to date.

2. Platelets and Solid Tumors

Platelets are well known for their classical function as a key player of the primary hemostasis, but platelets have a much broader range of other functions as well [32]. The relevance of platelets in oncological processes was first described in the 19th century by Leopold Riess and Theodor Billroth [33–35]. Malignant tumors have an impact on the platelet number and the functional state of platelets [32]. Platelets can be activated by tumor cells [36]. The daily use of aspirin can diminish the risk of dying due to a visceral malignancy [37–39]. Buergy et al. excellently reviewed the role of platelets in malignant diseases, which include platelet enhanced formation of metastasis, platelet-induced angiogenesis, and malignancy-induced thrombocytosis [33]. In recent years, a correlation of thrombocytosis and decreased overall survival could be shown for a magnitude of solid cancers such as lung, colon, breast, esophageal, gastric, renal, ovarian, and melanoma. However, since different tumor types induce thrombopoiesis, this may be an epiphenomenon rather than a causal relation. At least, the possible causality between thrombocytosis and malignancy is not fully understood to date [33]. Interestingly, decreased platelet reactivity in patients with cancer has been associated with a high risk of venous thromboembolism and poor prognosis, presumably as a consequence of continuous activation [40]. Thus, the status of platelets reactivity, which is modulated by solid tumors, appears to be of key relevance [40]. Platelets induce tumor angiogenesis by secreting proangiogenetic cytokines such as vascular endothelial growth factor (VEGF), PDGF, transforming growth factor (TGF), endothelial cell growth factor (ECGF), insulin-like growth factor (ILGF), basic fibroblast growth factor (bFGF, FGF-2), angiopoietin-1 as well as the lipid mediator sphingosine-1-phosphate [33]. Platelets cover circulating tumor cells in a P-selectin-dependent mechanism and thereby support their extravasation and prevent natural killer (NK) cell-mediated tumor cell lysis. Thus, platelets are a key player in the formation of distant metastasis in malignant diseases [41,42]. The inhibition of platelet binding to tumor cells by heparin can reduce the formation of metastasis [43]. Although it is well known that circulating GBM cells are present in patients suffering from GBM and tumor RNA can even be found in circulating platelets of these patients [44], GBM patients virtually never form metastases outside the central nervous system (CNS). This is a fundamental difference to malignancies outside the CNS, where distant metastases are a main determinant for disease progression and patient prognosis. As the CNS provides a unique microenvironment, i.e. determined by the blood–brain barrier and a distinct microglia, the contribution of platelets to GBM pathogenesis may vary from non-CNS tumors and is not well studied in this context. Thus, the functional interactions between platelets and the GBM may differ substantially from the role of platelets in tumors outside the CNS. Figure 1 summarizes the different mutual levels of interaction between tumor cells and platelets, and the role of platelets in concerting immune cell functions. The precise molecular levels of interaction of platelets with glioblastoma cells have to be clarified in future studies.

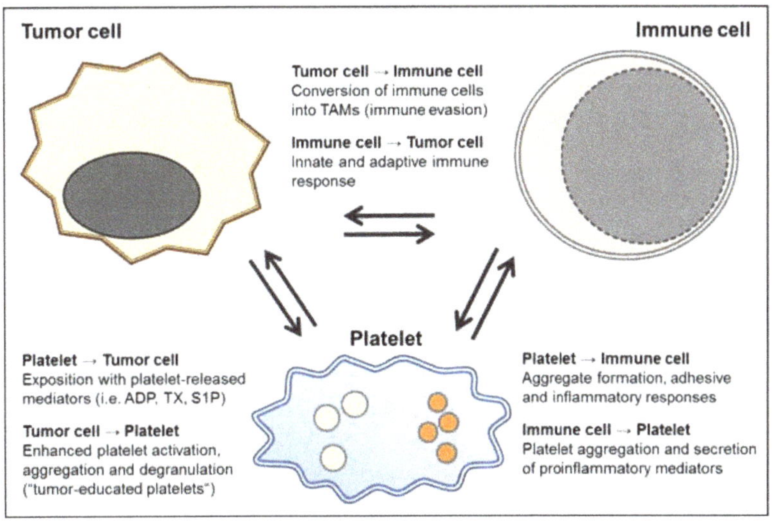

Figure 1. A scheme of the multiple and mutual interactions between tumor and immune cells and platelets. Platelets and tumor cells interact at various levels. Platelet-derived paracrine mediators such as adenosine diphosphate (ADP) as well as lipid signaling mediators like thromboxane (TX) and sphingosine-1-phosphate (S1P) are secreted upon platelet activation and can modulate tumor cell activity [39]. Tumor cells can in turn enhance platelet reactivity and educate platelets to release tumorigenic and angiogenic mediators and stimulate thrombopoiesis [45,46]. While the platelet itself can be seen as an immune cell, it can also interact in multiple ways with the different nucleated immune cells [47,48]. For example, platelet derived mediators such as S1P enhance monocyte activity levels via upregulation of the protease-activated thrombin receptors (PARs)-1 and -4 resulting in enhanced chemotactic capacity and amplifying cyclooxygenase-2 (COX-2) expression [49,50]. In turn, immune cells, such as monocytes, can also release tissue factor-rich microparticles to enhance fibrin formation, enhance clot stability, and ultimately thrombosis [51]. Activated immune cells such as microglia and also peripheral monocytes enter the tumor tissue [15]. In the case of the GBM, this results in the conversion of the infiltrating immune cells into immunosuppressive tumor-associated macrophage (TAMs), which are modulated to protect the tumor. Therapeutic approaches to target these immunosuppressive defense mechanisms are currently under discussion [14,15].

3. Platelets and Glioblastoma

More than 10 years ago, Brockmann et al. showed an association between thrombocytosis and a decreased overall survival in a group of 158 GBM patients [52]. They observed a significantly shorter median survival time in patients with preoperative thrombocytosis of 4 months (95% confidence interval (95% CI), 3–6 months) compared to 11 months survival time (95% CI, 8–13 months; $p = 0.0006$) in patients with normal platelet count [52]. The results suggested that a condition of preoperative thrombocytosis may represent a prognostic factor associated with shorter survival time in patients with glioblastoma. However, in a consecutive investigation involving 140 GBM patients, Lopes at al. could not confirm this observation [53]. Here, the authors found no correlation of neutrophil–lymphocyte ratio, platelets–lymphocyte ratio, or the absolute counts of neutrophils, lymphocytes, and platelets with overall survival in multivariate analyses [53]. Strikingly, the mean overall survival time in this study was 19.4 ± 14.3 months and the mean progression-free survival was 9.4 ± 8.7 months. Different adjuvant treatment regimens, which have changed over time, may have contributed to this difference in survival time. In agreement with this estimation, the rate of combined radiochemotherapy in the earlier study from Brockmann et al. [52] was below 40% in total, while in the more recent study from Lopes at al. [53] a large majority of patients of about 84% were treated according to the combined radiochemotherapy protocol established by Stupp et al. [1]. Besides this, different GBM subtypes exist, which were not further characterized in both of these studies, but should be distinguished. These diverse GBM variants have been described to exhibit substantial heterogeneity, including molecular, histopathological, and genomic features [54,55]. The understanding of the precise molecular pattern of these diverse tumor variants should be improved in the future. Both studies addressing the possible role of platelet count in GBM prognosis did not define the clinical features of molecular variants. Furthermore, the number of included patients—owing to the rare nature of the disease—was rather limited, making these controversial observations rather difficult to interpret.

Another aspect, which may be of relevance for the pathophysiological impact of platelet functions in GBM development is the extent of platelet reactivity. In a recent study from our group, we could show, for the first time in a homogenous cohort of primary diagnosed GBM patients, that the activation status of circulating platelets is increased [56]. However, in comparison to visceral malignancies, such as colon cancer, the available data on the clinical effects of platelet function inhibitors in GBM, e.g., with aspirin, appear sparse and inhomogeneous. A large epidemiological study with over 300,000 participants did not find any correlation between the intake of aspirin or other NSAID and the development of glioma and/or GBM [57]. However, the assessment of aspirin use was only according to self-questionnaire and there was no monitoring of treatment efficacy, i.e., by measuring thromboxane levels [57]. In addition, the definition of regular aspirin use as intake of more than two doses per week in this study [57] makes it highly unlikely, that aspirin treatment was efficacious in this study. Furthermore, the duration of aspirin treatment in the study was only one year. In comparison, the well documented effects of aspirin in colorectal cancer occur after a treatment time of at least 5 to 10 years [38,39]. To date, the precise underlying mode of action of the long-term effect of aspirin is neither in colorectal cancer nor in other tumor entities fully understood. Especially in colorectal cancer, the antitumor effects of aspirin occur at antiplatelet doses implicating a role of platelets in tumor pathogenesis. However, since many cancer cell types express both isoforms of cyclooxygenase, COX-1 and COX-2, the effects of aspirin may in part also be directly related to the inhibition of COX in cancer cells [37–39]. In comparison to the study from Daugherty et al. [57], where aspirin efficacy may be questionable (see above), another more recent study ascertained drug use in glioma patients more thoroughly through the evaluation of drug registries [58]. In their study, Gaist et al. observed that long-term continuous use of low-dose aspirin was associated with an OR (odds ratio) of 0.88 (95% CI: 0.54–1.42) and non-aspirin-NSAID (non-steroidal anti-inflammatory drugs) with an OR of 1.56 (95% CI: 0.71–3.46). According to the authors, their findings may be consistent with a small reduction in glioma risk associated with the long-term use of low-dose aspirin. However, they also point out that their findings should be interpreted cautiously due to limited statistical precision, notably in the duration and dose–response analyses [58]. Taken together, the clinical evidence of the role of platelets both concerning platelet numbers and the use of antiplatelet agents in GBM are limited. Therefore, future and ideally perspective studies are needed to clarify the possible direct role of platelets in GBM.

4. The Immunomodulatory Potential of Platelets in GBM

Besides their role in hemostasis and the above described mechanisms in tumor biology, platelets are well known to interact with the immune system [47,48]. Platelets express Toll-like receptors and, thereby, are sentinel cells for infection [59]. Furthermore, platelets interact in an activity-boosting manner with the complement system [60]. Platelet activation is accompanied with an increased formation of heterotypic platelet–leukocyte conjugates, which is well known as the surrogate of a proinflammatory function and described in inflammatory diseases as sepsis or atherosclerosis [61,62]. In agreement with these (patho)physiological functions, platelet inhibition is known to be anti-inflammatory and, for example, reduces the risk of pulmonary as well as other infections [59,63].

Since recently, immunotherapeutical approaches became of interest in GBM patients, the immunomodulatory functions of platelets might be of great interest as well. However, little is known about the role of platelets to modulate immune responses in GBM, yet. For the first time, we could recently show that the formation of heterotypic platelet–monocyte conjugates in the circulation of GBM patients is not increased, although the platelets had an increased activation status in these patients (including an increased P-selectin expression) [56]. This, at first sight a paradoxical finding, may be explained by a reduced expression of PSGL-1 on circulating monocytes in GBM patients [56], since the interaction between platelet P-selectin and monocyte PSGL-1 is the initial and key step in this conjugate formation [61]. In further experiments, we could show that the PSGL-1 expression is tremendously reduced on intratumoral macrophages in GBM as well (unpublished data). Thus, the PSGL-1 phenotype might be a yet unnoted biomarker for the GBM-induced immunosuppression.

Platelets contain and secrete a multitude of mediators that are known to participate in both hemostasis and inflammation [64]. These mediators predominantly originate from platelet α-granules [65]. Every platelet contains about 50–80 alpha granules. Inside these granules mediators and chemokines, such as P-selectin, CXCL1, platelet factor 4 (PF4) CXCL5, CXCL7, IL-8, and CXCL12, macrophage inflammatory protein-1 (CCL3) and RANTES (CCL5) are stored and can be released upon activation [65]. These mediators play a major role in the regulation of leukocyte migration into tissues and in other proinflammatory functions like phagocytosis, monocyte differentiation, and the generation of reactive oxygen species [66]. Activated platelets release IL-1, which plays a major role in the inflammatory cytokine cascade [67], and TGFβ, which is well known as key mediator of the GBM-induced immunosuppression [68–70]. Furthermore, soluble CD40 released from platelets was shown to inhibit regulatory T-cells in a glioma model [71]. As a note, in our own previous studies, we could show a tendency of an increased CD40 expression on platelets in GBM patients [56].

The role of platelets in the pathophysiology of GBM appears, however, to be two-edged. On the one hand, platelet activation may be beneficial to support immune responses, as activated platelets and their secretome can modulate immune responses. Recently, it was shown that an activated platelet-rich plasma clot can inhibit regulatory T-cell migration and prolong overall survival in a GBM model in mice [71]. On the other hand, platelet activation needs to be avoided, since GBM patients have an increased risk for systemic cardiovascular events and the intratumoral occlusion of numerous vessels leads to a hypoxia-induced tumor progression. A further key molecule, which is released from platelets and has been described as a modulator of inflammatory and oncological processes, is sphingosine-1-phosphate (S1P).

5. Sphingosine-1-Phosphate in Glioblastoma

The versatile lipid signaling mediator S1P has emerged as a regulator of a variety of cellular processes including proliferation, metastasis, inflammation, stem cell behavior, and the formation of microvascular networks, which provide nourishment to cancerous cells [72–74]. S1P has been recognized in numerous studies over the recent years as an important oncogenic factor involved in multiple cancer entities including breast, colorectal, kidney, lung, melanoma, and GBM [72,75]. Our group could show that the relative amount of S1P secreted from stimulated peripheral platelets of GBM patients is significantly elevated compared to healthy controls [56]. However, the absolute concentration of S1P in platelet-rich plasma of these patients was significantly lower than in the controls (Figure 2C) [56]. Furthermore, a significantly elevated S1P concentration was found in GBM tissue in comparison to the control brain specimens [76], and glioma cells, as well as GBM stem cells, are able to produce and release S1P [77,78]. Since S1P directs immune cell migration via concentration gradient-dependent mechanisms, a reduced peripheral, and in turn, elevated central S1P level in GBM tissue might foster monocyte migration from the peripheral blood into the brain. In addition, S1P has been reported as a key player in the transformation of intratumoral macrophages into ultimately immunosuppressive TAMs via S1P receptor-1 [79]. Thus, reduced peripheral blood S1P levels may drive monocyte invasion into the tumor tissue, and elevated S1P concentrations within the GBM consecutively could support intratumoral conversion of macrophage/microglia into TAMs (see also Figure 1). In addition, S1P is known as mediator between blood coagulation, platelets, and vascular inflammatory responses [49]. Not only the secretion of platelet-derived S1P can be stimulated by coagulation factors such as thrombin [80], but also enzymes which generate S1P in the vasculature can be upregulated by coagulation factors such as the activated factor-X (FXa) [81].

In general, the synthesis of sphingolipid and S1P in particular is tightly controlled by the metabolism of ceramide. Biosynthesis of cellular ceramide derives either de novo from serine, palmitoyl-CoA, and fatty acid, or from the breakdown of membrane-resident sphingomyelin [72,82,83]. S1P is ultimately generated by phosphorylation of sphingosine by the two isoenzymes sphingosine kinase-1 and -2 (SphK1 and SphK2) [49,82,83]. S1P degradation is achieved via dephosphorylation by two S1P-specific phosphatases (SGPP1 and SGPP2) or an irreversible hydrolysis by S1P lyase (SGPL) [84,85]. The formation of concentration gradients is essential in regulating the physiological effects of S1P in vivo [84]. The balance between S1P generation and degradation is also critical for the regulation of cell growth and plays a key role in pathological processes such as carcinogenesis [72,85]. Thus, the inhibition of S1P synthesis results in a sensitization of glioma stem cells against temozolomide as the standard chemotherapeutic in GBM treatment [78]. Interestingly, fingolimod (FTY720)—a sphingosine analogue approved for the treatment of multiple sclerosis—reduced the intracranial growth of brain tumors in a mouse model of GBM [86].

Besides elevated basal platelet activation and altered circulating peripheral S1P levels (see Figure 1), a severe dysregulation of the S1P signaling system was also found in tumor tissue samples of GBM patients in previous investigations [75]. Specifically, the S1P receptors $S1P_1$ and $S1P_2$ (Figure 2) and the S1P-generating enzyme, SPHK1, were significantly upregulated in GBM samples. Intriguingly, the expression levels of both $S1P_1$ and $S1P_2$ correlated significantly with patient survival time but in divergent ways. In GBM patients with high $S1P_1$ mRNA levels, a prolonged survival was observed, while patients with a high $S1P_2$ mRNA expression exhibited a shorter survival time [75]. In agreement with the literature, our results from this study implicated a complex interplay between S1P receptors, S1P signaling, and other tumor-promoting signaling cascades in GBM, i.e., opposing functions of $S1P_1$ and $S1P_2$ in the regulation of cell migration and proliferation. Going forward, a better understanding of the pathological basis of GBM tumors could lead to better diagnostic and treatment protocols, so that a tailored monitoring of platelet-derived proinflammatory modulators such as S1P could be easily incorporated into multiplex biomarker panels and guide clinicians in developing novel immunotherapeutic approaches to gliomas. Prospectively, a pharmacological modulation of S1P levels and/or its receptors may also represent a potential future therapeutic concept in GBM therapy.

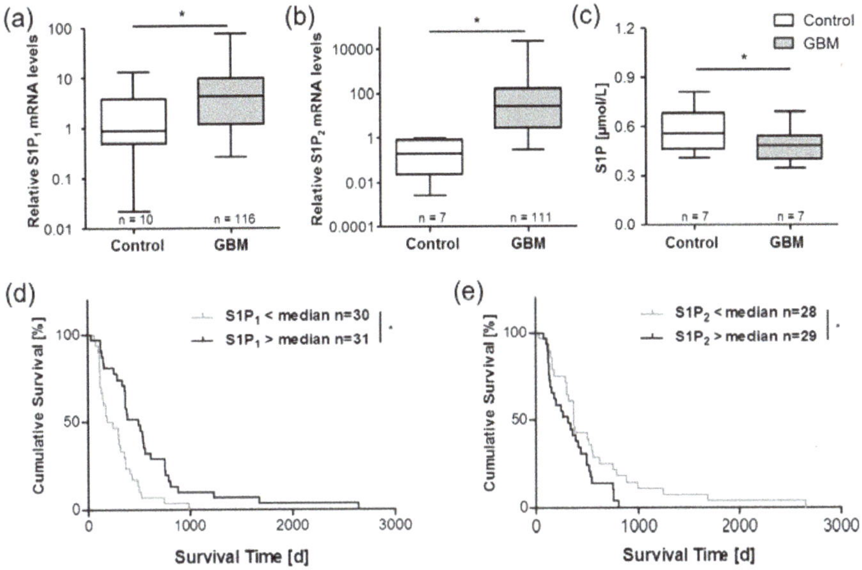

Figure 2. Expression of S1P receptors and circulating peripheral S1P levels are altered in GBM patients. In previous studies, we observed highly upregulated levels of mRNA for the S1P receptors $S1P_1$ (**a**) and$S1P_2$ (**b**) in GBM tissue [75]. The expression levels of both S1P receptors significantly correlated with the survival time of the respective GBM patients (**d**,**e**). In a later study, our data indicated reduced levels of S1P in platelet-rich plasma of GBM patients (**c**) [56]. These observations point towards a key role of S1P signaling system in the pathophysiology of GBM and are in agreement with studies from other groups [87,88]. Data are adapted from Bien-Möller et al. [75] and Marx et al. [56] and are shown as box plots representing the median as horizontal bars as well as the 5th and 95th percentile. * $p < 0.05$, Mann Whitney U test.

6. Summary and Conclusions

In summary, the contribution of platelets in tumor development, invasiveness, malignancy, and metastasis is widely acknowledged in the literature. However, the specific contribution of platelets to tumor pathophysiology in GBM is less clear. The preoperative platelet count has been suggested as a potential outcome predictor. However, the currently available data are inconclusive and future studies are needed to gain further insights into the role of platelet number as a possible biomarker. Studies on the reactivity level of platelets in GBM are also sparse to date. Our own data indicate elevated levels of P-selectin and impaired relative responsiveness to platelet stimulation ex vitro, possibly pointing towards an altered immunoregulatory function of platelets during GBM growth and progression. Since thrombosis is a hallmark of GBM, suitable antithrombotic and antiplatelet concepts may be a valuable addition to future individualized targeted therapeutic concepts.

Funding: This work was supported by grants from the Gerhard-Domagk scholarship program, the Jung-Stiftung für Wissenschaft und Forschung, and the Forschungsverbund Molekulare Medizin to Sascha Marx, and by the Forschungsverbund Neurowissenschaften to Sandra Bien-Möller.

Acknowledgments: We acknowledge support for the article processing charge from the DFG (German Research Foundation, 393148499) and the Open Access Publication Fund of the University of Greifswald. We are grateful to Sarah Polster for excellent technical assistance.

Conflicts of Interest: The authors declare no conflicts of interest.

References

1. Stupp, R.; Mason, W.P.; van den Bent, M.J.; Weller, M.; Fisher, B.; Taphoorn, M.J.; Belanger, K.; Brandes, A.A.; Marosi, C.; Bogdahn, U.; et al. Radiotherapy plus concomitant and adjuvant temozolomide for glioblastoma. *N. Engl. J. Med.* **2005**, *352*, 987–996. [CrossRef]
2. Wen, P.Y.; Kesari, S. Malignant gliomas in adults. *N. Engl. J. Med.* **2008**, *359*, 492–507. [CrossRef]
3. Stupp, R.; Taillibert, S.; Kanner, A.; Read, W.; Steinberg, D.; Lhermitte, B.; Toms, S.; Idbaih, A.; Ahluwalia, M.S.; Fink, K.; et al. Effect of Tumor-Treating Fields Plus Maintenance Temozolomide vs Maintenance Temozolomide Alone on Survival in Patients with Glioblastoma: A Randomized Clinical Trial. *JAMA* **2017**, *318*, 2306–2316. [CrossRef] [PubMed]
4. Stupp, R.; Taillibert, S.; Kanner, A.A.; Kesari, S.; Steinberg, D.M.; Toms, S.A.; Taylor, L.P.; Lieberman, F.; Silvani, A.; Fink, K.L.; et al. Maintenance Therapy with Tumor-Treating Fields Plus Temozolomide vs Temozolomide Alone for Glioblastoma: A Randomized Clinical Trial. *JAMA* **2015**, *314*, 2535–2543. [CrossRef] [PubMed]
5. Jackson, C.M.; Lim, M. Immunotherapy for Glioblastoma: Playing Chess, Not Checkers. *Clin. Cancer Res.* **2018**, *24*, 4059–4061. [CrossRef]
6. Keskin, D.B.; Anandappa, A.J.; Sun, J.; Tirosh, I.; Mathewson, N.D.; Li, S.; Oliveira, G.; Giobbie-Hurder, A.; Felt, K.; Gjini, E.; et al. Neoantigen vaccine generates intratumoral T cell responses in phase Ib glioblastoma trial. *Nature* **2019**, *565*, 234–239. [CrossRef]
7. Reardon, D.A.; Wucherpfennig, K.; Chiocca, E.A. Immunotherapy for glioblastoma: On the sidelines or in the game? *Discov. Med.* **2017**, *24*, 201–208. [PubMed]
8. Sayegh, E.T.; Oh, T.; Fakurnejad, S.; Bloch, O.; Parsa, A.T. Vaccine therapies for patients with glioblastoma. *J. Neurooncol.* **2014**, *119*, 531–546. [CrossRef]
9. Schijns, V.; Pretto, C.; Strik, A.M.; Gloudemans-Rijkers, R.; Deviller, L.; Pierre, D.; Chung, J.; Dandekar, M.; Carrillo, J.A.; Kong, X.T.; et al. Therapeutic Immunization against Glioblastoma. *Int. J. Mol. Sci.* **2018**, *19*, 2540. [CrossRef]
10. Suryadevara, C.M.; Verla, T.; Sanchez-Perez, L.; Reap, E.A.; Choi, B.D.; Fecci, P.E.; Sampson, J.H. Immunotherapy for malignant glioma. *Surg. Neurol. Int.* **2015**, *6*, S68–S77. [CrossRef]
11. Bloch, O.; Crane, C.A.; Kaur, R.; Safaee, M.; Rutkowski, M.J.; Parsa, A.T. Gliomas promote immunosuppression through induction of B7-H1 expression in tumor-associated macrophages. *Clin. Cancer Res.* **2013**, *19*, 3165–3175. [CrossRef]
12. Markovic, D.S.; Vinnakota, K.; Chirasani, S.; Synowitz, M.; Raguet, H.; Stock, K.; Sliwa, M.; Lehmann, S.; Kalin, R.; van Rooijen, N.; et al. Gliomas induce and exploit microglial MT1-MMP expression for tumor expansion. *Proc. Natl. Acad. Sci. USA* **2009**, *106*, 12530–12535. [CrossRef] [PubMed]
13. Ye, X.Z.; Xu, S.L.; Xin, Y.H.; Yu, S.C.; Ping, Y.F.; Chen, L.; Xiao, H.L.; Wang, B.; Yi, L.; Wang, Q.L.; et al. Tumor-associated microglia/macrophages enhance the invasion of glioma stem-like cells via TGF-beta1 signaling pathway. *J. Immunol.* **2012**, *189*, 444–453. [CrossRef]
14. Bowman, R.L.; Joyce, J.A. Therapeutic targeting of tumor-associated macrophages and microglia in glioblastoma. *Immunotherapy* **2014**, *6*, 663–666. [CrossRef]
15. Hussain, S.F.; Yang, D.; Suki, D.; Aldape, K.; Grimm, E.; Heimberger, A.B. The role of human glioma-infiltrating microglia/macrophages in mediating antitumor immune responses. *Neuro Oncol.* **2006**, *8*, 261–279. [CrossRef]
16. Parney, I.F.; Waldron, J.S.; Parsa, A.T. Flow cytometry and in vitro analysis of human glioma-associated macrophages. Laboratory investigation. *J. Neurosurg.* **2009**, *110*, 572–582. [CrossRef]
17. Qian, B.Z.; Pollard, J.W. Macrophage diversity enhances tumor progression and metastasis. *Cell* **2010**, *141*, 39–51. [CrossRef]
18. Pyonteck, S.M.; Akkari, L.; Schuhmacher, A.J.; Bowman, R.L.; Sevenich, L.; Quail, D.F.; Olson, O.C.; Quick, M.L.; Huse, J.T.; Teijeiro, V.; et al. CSF-1R inhibition alters macrophage polarization and blocks glioma progression. *Nat. Med.* **2013**, *19*, 1264–1272. [CrossRef]
19. Chatterjee, M.; Rath, D.; Gawaz, M. Role of chemokine receptors CXCR4 and CXCR7 for platelet function. *Biochem. Soc. Trans.* **2015**, *43*, 720–726. [CrossRef]
20. Stellos, K.; Gawaz, M. Platelets and stromal cell-derived factor-1 in progenitor cell recruitment. *Semin. Thromb. Hemost.* **2007**, *33*, 159–164. [CrossRef]

21. Haemmerle, M.; Stone, R.L.; Menter, D.G.; Afshar-Kharghan, V.; Sood, A.K. The Platelet Lifeline to Cancer: Challenges and Opportunities. *Cancer Cell* **2018**, *33*, 965–983. [CrossRef]
22. Wojtukiewicz, M.Z.; Sierko, E.; Hempel, D.; Tucker, S.C.; Honn, K.V. Platelets and cancer angiogenesis nexus. *Cancer Metastasis Rev.* **2017**, *36*, 249–262. [CrossRef]
23. Huong, P.T.; Nguyen, L.T.; Nguyen, X.B.; Lee, S.K.; Bach, D.H. The Role of Platelets in the Tumor-Microenvironment and the Drug Resistance of Cancer Cells. *Cancers* **2019**, *11*, 240. [CrossRef]
24. Navone, S.E.; Guarnaccia, L.; Locatelli, M.; Rampini, P.; Caroli, M.; la Verde, N.; Gaudino, C.; Bettinardi, N.; Riboni, L.; Marfia, G.; et al. Significance and Prognostic Value of The Coagulation Profile in Patients with Glioblastoma: Implications for Personalized Therapy. *World Neurosurg.* **2019**, *121*, e621–e629. [CrossRef]
25. Fisher, J.L.; Palmisano, S.; Schwartzbaum, J.A.; Svensson, T.; Lonn, S. Comorbid conditions associated with glioblastoma. *J. Neurooncol.* **2014**, *116*, 585–591. [CrossRef]
26. Smith, T.R.; Lall, R.R.; Graham, R.B.; McClendon, J.; Lall, R.R.; Nanney, A.D.; Adel, J.G.; Zakarija, A.; Chandler, J.P. Venous thromboembolism in high grade glioma among surgical patients: Results from a single center over a 10 year period. *J. Neurooncol.* **2014**, *120*, 347–352. [CrossRef]
27. Streiff, M.B.; Ye, X.; Kickler, T.S.; Desideri, S.; Jani, J.; Fisher, J.; Grossman, S.A. A prospective multicenter study of venous thromboembolism in patients with newly-diagnosed high-grade glioma: Hazard rate and risk factors. *J. Neurooncol.* **2015**, *124*, 299–305. [CrossRef]
28. Rong, Y.; Durden, D.L.; van Meir, E.G.; Brat, D.J. 'Pseudopalisading' necrosis in glioblastoma: A familiar morphologic feature that links vascular pathology, hypoxia, and angiogenesis. *J. Neuropathol. Exp. Neurol.* **2006**, *65*, 529–539. [CrossRef]
29. Edwin, N.C.; Khoury, M.N.; Sohal, D.; McCrae, K.R.; Ahluwalia, M.S.; Khorana, A.A. Recurrent venous thromboembolism in glioblastoma. *Thromb. Res.* **2016**, *137*, 184–188. [CrossRef]
30. Brat, D.J.; van Meir, E.G. Vaso-occlusive and prothrombotic mechanisms associated with tumor hypoxia, necrosis, and accelerated growth in glioblastoma. *Lab. Investig.* **2004**, *84*, 397–405. [CrossRef]
31. Simanek, R.; Vormittag, R.; Hassler, M.; Roessler, K.; Schwarz, M.; Zielinski, C.; Pabinger, I.; Marosi, C. Venous thromboembolism and survival in patients with high-grade glioma. *Neuro Oncol.* **2007**, *9*, 89–95. [CrossRef]
32. Meikle, C.K.; Kelly, C.A.; Garg, P.; Wuescher, L.M.; Ali, R.A.; Worth, R.G. Cancer and Thrombosis: The Platelet Perspective. *Front Cell Dev. Biol.* **2016**, *4*, 147. [CrossRef]
33. Buergy, D.; Wenz, F.; Groden, C.; Brockmann, M.A. Tumor-platelet interaction in solid tumors. *Int. J. Cancer* **2012**, *130*, 2747–2760. [CrossRef]
34. Riess, L. Zur pathologischen Anatomie des Blutes. *Arch. Anat. Physiol. Wissensch. Med.* **1872**, *39*, 237–249.
35. Billroth, T. *Lectures on Surgical Pathology and Therapeutics: A Handbook for Students and Practitionersed*; The New Sydenham Society: London, UK, 1878.
36. Heinmoller, E.; Weinel, R.J.; Heidtmann, H.H.; Salge, U.; Seitz, R.; Schmitz, I.; Muller, K.M.; Zirngibl, H. Studies on tumor-cell-induced platelet aggregation in human lung cancer cell lines. *J. Cancer Res. Clin. Oncol.* **1996**, *122*, 735–744. [CrossRef]
37. Jacobs, E.J.; Newton, C.C.; Gapstur, S.M.; Thun, M.J. Daily aspirin use and cancer mortality in a large US cohort. *J. Natl. Cancer Inst.* **2012**, *104*, 1208–1217. [CrossRef]
38. Rothwell, P.M.; Price, J.F.; Fowkes, F.G.; Zanchetti, A.; Roncaglioni, M.C.; Tognoni, G.; Lee, R.; Belch, J.F.; Wilson, M.; Mehta, Z.; et al. Short-term effects of daily aspirin on cancer incidence, mortality, and non-vascular death: Analysis of the time course of risks and benefits in 51 randomised controlled trials. *Lancet* **2012**, *379*, 1602–1612. [CrossRef]
39. Schrör, K.; Rauch, B.H. Aspirin and lipid mediators in the cardiovascular system. *Prostaglandins Other Lipid Mediat.* **2015**, *121*, 17–23. [CrossRef]
40. Riedl, J.; Kaider, A.; Marosi, C.; Prager, G.W.; Eichelberger, B.; Assinger, A.; Pabinger, I.; Panzer, S.; Ay, C. Decreased platelet reactivity in patients with cancer is associated with high risk of venous thromboembolism and poor prognosis. *Thromb. Haemost.* **2017**, *117*, 90–98. [CrossRef]
41. Placke, T.; Kopp, H.G.; Salih, H.R. Modulation of natural killer cell anti-tumor reactivity by platelets. *J. Innate Immun.* **2011**, *3*, 374–382. [CrossRef]
42. Zhang, W.; Dang, S.; Hong, T.; Tang, J.; Fan, J.; Bu, D.; Sun, Y.; Wang, Z.; Wisniewski, T. A humanized single-chain antibody against beta 3 integrin inhibits pulmonary metastasis by preferentially fragmenting activated platelets in the tumor microenvironment. *Blood* **2012**, *120*, 2889–2898. [CrossRef]

43. Borsig, L.; Wong, R.; Feramisco, J.; Nadeau, D.R.; Varki, N.M.; Varki, A. Heparin and cancer revisited: Mechanistic connections involving platelets, P-selectin, carcinoma mucins, and tumor metastasis. *Proc. Natl. Acad. Sci. USA* **2001**, *98*, 3352–3357. [CrossRef]
44. Nilsson, R.J.; Balaj, L.; Hulleman, E.; van Rijn, S.; Pegtel, D.M.; Walraven, M.; Widmark, A.; Gerritsen, W.R.; Verheul, H.M.; Vandertop, W.P.; et al. Blood platelets contain tumor-derived RNA biomarkers. *Blood* **2011**, *118*, 3680–3683. [CrossRef]
45. In't Veld, S.; Sjors, G.J.G.; Wurdinger, T. Tumor-educated platelets. *Blood* **2019**. [CrossRef]
46. Olsson, A.K.; Cedervall, J. The pro-inflammatory role of platelets in cancer. *Platelets* **2018**, *29*, 569–573. [CrossRef]
47. Koenen, R.R. The prowess of platelets in immunity and inflammation. *Thromb. Haemost.* **2016**, *116*, 605–612. [CrossRef]
48. Lam, F.W.; Vijayan, K.V.; Rumbaut, R.E. Platelets and Their Interactions with Other Immune Cells. *Compr. Physiol.* **2015**, *5*, 1265–1280.
49. Mahajan-Thakur, S.; Böhm, A.; Jedlitschky, G.; Schrör, K.; Rauch, B.H. Sphingosine-1-Phosphate and Its Receptors: A Mutual Link between Blood Coagulation and Inflammation. *Mediat. Inflamm.* **2015**, *2015*, 831059. [CrossRef]
50. Mahajan-Thakur, S.; Sostmann, B.D.; Fender, A.C.; Behrendt, D.; Felix, S.B.; Schrör, K.; Rauch, B.H. Sphingosine-1-phosphate induces thrombin receptor PAR-4 expression to enhance cell migration and COX-2 formation in human monocytes. *J. Leukoc. Biol.* **2014**, *96*, 611–618. [CrossRef]
51. Aleman, M.M.; Gardiner, C.; Harrison, P.; Wolberg, A.S. Differential contributions of monocyte- and platelet-derived microparticles towards thrombin generation and fibrin formation and stability. *J. Thromb. Haemost.* **2011**, *9*, 2251–2261. [CrossRef]
52. Brockmann, M.A.; Giese, A.; Mueller, K.; Kaba, F.J.; Lohr, F.; Weiss, C.; Gottschalk, S.; Nolte, I.; Leppert, J.; Tuettenberg, J.; et al. Preoperative thrombocytosis predicts poor survival in patients with glioblastoma. *Neuro Oncol.* **2007**, *9*, 335–342. [CrossRef] [PubMed]
53. Lopes, M.; Carvalho, B.; Vaz, R.; Linhares, P. Influence of neutrophil-lymphocyte ratio in prognosis of glioblastoma multiforme. *J. Neurooncol.* **2018**, *136*, 173–180. [CrossRef]
54. Karsy, M.; Gelbman, M.; Shah, P.; Balumbu, O.; Moy, F.; Arslan, E. Established and emerging variants of glioblastoma multiforme: Review of morphological and molecular features. *Folia Neuropathol.* **2012**, *50*, 301–321. [CrossRef]
55. Karsy, M.; Huang, T.; Kleinman, G.; Karpel-Massler, G. Molecular, histopathological, and genomic variants of glioblastoma. *Front Biosci.* **2014**, *19*, 1065–1087. [CrossRef]
56. Marx, S.; Splittstöhser, M.; Kinnen, F.; Moritz, E.; Joseph, C.; Paul, S.; Paland, H.; Seifert, C.; Marx, M.; Böhm, A.; et al. Platelet activation parameters and platelet-leucocyte-conjugate formation in glioblastoma multiforme patients. *Oncotarget* **2018**, *9*, 25860–25876. [CrossRef] [PubMed]
57. Daugherty, S.E.; Moore, S.C.; Pfeiffer, R.M.; Inskip, P.D.; Park, Y.; Hollenbeck, A.; Rajaraman, P. Nonsteroidal anti-inflammatory drugs and glioma in the NIH-AARP Diet and Health Study cohort. *Cancer Prev. Res.* **2011**, *4*, 2027–2034. [CrossRef]
58. Gaist, D.; Garcia-Rodriguez, L.A.; Sorensen, H.T.; Hallas, J.; Friis, S. Use of low-dose aspirin and non-aspirin nonsteroidal anti-inflammatory drugs and risk of glioma: A case-control study. *Br. J. Cancer* **2013**, *108*, 1189–1194. [CrossRef]
59. Thomas, M.R.; Storey, R.F. The role of platelets in inflammation. *Thromb. Haemost.* **2015**, *114*, 449–458.
60. Peerschke, E.I.; Yin, W.; Ghebrehiwet, B. Complement activation on platelets: Implications for vascular inflammation and thrombosis. *Mol. Immunol.* **2010**, *47*, 2170–2175. [CrossRef]
61. Lösche, W.; Heptinstall, S. Value of Platelet Activation Markers as Prothrombotic Risk Indicators. *Transfus Med. Hemother.* **2007**, *34*, 34–42. [CrossRef]
62. Zarbock, A.; Polanowska-Grabowska, R.K.; Ley, K. Platelet-neutrophil-interactions: Linking hemostasis and inflammation. *Blood Rev.* **2007**, *21*, 99–111. [CrossRef]
63. Storey, R.F.; James, S.K.; Siegbahn, A.; Varenhorst, C.; Held, C.; Ycas, J.; Husted, S.E.; Cannon, C.P.; Becker, R.C.; Steg, P.G.; et al. Lower mortality following pulmonary adverse events and sepsis with ticagrelor compared to clopidogrel in the PLATO study. *Platelets* **2014**, *25*, 517–525. [CrossRef]
64. Coppinger, J.A.; O'Connor, R.; Wynne, K.; Flanagan, M.; Sullivan, M.; Maguire, P.B.; Fitzgerald, D.J.; Cagney, G. Moderation of the platelet releasate response by aspirin. *Blood* **2007**, *109*, 4786–4792. [CrossRef]

65. Blair, P.; Flaumenhaft, R. Platelet alpha-granules: Basic biology and clinical correlates. *Blood Rev.* **2009**, *23*, 177–189. [CrossRef]
66. Scheuerer, B.; Ernst, M.; Durrbaum-Landmann, I.; Fleischer, J.; Grage-Griebenow, E.; Brandt, E.; Flad, H.D.; Petersen, F. The CXC-chemokine platelet factor 4 promotes monocyte survival and induces monocyte differentiation into macrophages. *Blood* **2000**, *95*, 1158–1166.
67. Loppnow, H.; Libby, P. Proliferating or interleukin 1-activated human vascular smooth muscle cells secrete copious interleukin 6. *J. Clin. Investig.* **1990**, *85*, 731–738. [CrossRef]
68. Kopp, H.G.; Placke, T.; Salih, H.R. Platelet-derived transforming growth factor-beta down-regulates NKG2D thereby inhibiting natural killer cell antitumor reactivity. *Cancer Res.* **2009**, *69*, 7775–7783. [CrossRef]
69. Lee, Y.L.; Lee, L.W.; Su, C.Y.; Hsiao, G.; Yang, Y.Y.; Leu, S.J.; Shieh, Y.H.; Burnouf, T. Virally inactivated human platelet concentrate lysate induces regulatory T cells and immunosuppressive effect in a murine asthma model. *Transfusion* **2013**, *53*, 1918–1928. [CrossRef]
70. Sengelov, L.; Kamby, C.; Schou, G.; von der Maase, H. Prognostic factors and significance of chemotherapy in patients with recurrent or metastatic transitional cell cancer of the urinary tract. *Cancer* **1994**, *74*, 123–133. [CrossRef]
71. Panek, W.K.; Pituch, K.C.; Miska, J.; Kim, J.W.; Rashidi, A.; Kanojia, D.; Lopez-Rosas, A.; Han, Y.; Yu, D.; Chang, C.L.; et al. Local Application of Autologous Platelet-Rich Fibrin Patch (PRF-P) Suppresses Regulatory T Cell Recruitment in a Murine Glioma Model. *Mol. Neurobiol.* **2018**. [CrossRef] [PubMed]
72. Mahajan-Thakur, S.; Bien-Moller, S.; Marx, S.; Schroeder, H.; Rauch, B.H. Sphingosine 1-phosphate (S1P) signaling in glioblastoma multiforme—A systematic review. *Int. J. Mol. Sci.* **2017**, *18*, 2448. [CrossRef]
73. Pyne, S.; Pyne, N.J. New perspectives on the role of sphingosine 1-phosphate in cancer. *Handb. Exp. Pharm.* **2013**, *216*, 55–71.
74. Pyne, N.J.; Ohotski, J.; Bittman, R.; Pyne, S. The role of sphingosine 1-phosphate in inflammation and cancer. *Adv. Biol. Regul.* **2014**, *54*, 121–129. [CrossRef] [PubMed]
75. Bien-Moller, S.; Lange, S.; Holm, T.; Bohm, A.; Paland, H.; Kupper, J.; Herzog, S.; Weitmann, K.; Havemann, C.; Vogelgesang, S.; et al. Expression of S1P metabolizing enzymes and receptors correlate with survival time and regulate cell migration in glioblastoma multiforme. *Oncotarget* **2016**, *7*, 13031–13046. [CrossRef] [PubMed]
76. Abuhusain, H.J.; Matin, A.; Qiao, Q.; Shen, H.; Kain, N.; Day, B.W.; Stringer, B.W.; Daniels, B.; Laaksonen, M.A.; Teo, C.; et al. A metabolic shift favoring sphingosine 1-phosphate at the expense of ceramide controls glioblastoma angiogenesis. *J. Biol. Chem.* **2013**, *288*, 37355–37364. [CrossRef]
77. Anelli, V.; Gault, C.R.; Cheng, A.B.; Obeid, L.M. Sphingosine kinase 1 is up-regulated during hypoxia in U87MG glioma cells. Role of hypoxia-inducible factors 1 and 2. *J. Biol. Chem.* **2008**, *283*, 3365–3375. [CrossRef] [PubMed]
78. Riccitelli, E.; Giussani, P.; di Vito, C.; Condomitti, G.; Tringali, C.; Caroli, M.; Galli, R.; Viani, P.; Riboni, L. Extracellular sphingosine-1-phosphate: A novel actor in human glioblastoma stem cell survival. *PLoS ONE* **2013**, *8*, e68229. [CrossRef]
79. Muller, J.; von Bernstorff, W.; Heidecke, C.D.; Schulze, T. Differential S1P Receptor Profiles on M1- and M2-Polarized Macrophages Affect Macrophage Cytokine Production and Migration. *Biomed. Res. Int.* **2017**, *2017*, 7584621. [CrossRef]
80. Ulrych, T.; Böhm, A.; Polzin, A.; Daum, G.; Nusing, R.M.; Geisslinger, G.; Hohlfeld, T.; Schrör, K.; Rauch, B.H. Release of sphingosine-1-phosphate from human platelets is dependent on thromboxane formation. *J. Thromb. Haemost.* **2011**, *9*, 790–798. [CrossRef]
81. Böhm, A.; Flösser, A.; Ermler, S.; Fender, A.C.; Luth, A.; Kleuser, B.; Schrör, K.; Rauch, B.H. Factor-Xa-induced mitogenesis and migration require sphingosine kinase activity and S1P formation in human vascular smooth muscle cells. *Cardiovasc. Res.* **2013**, *99*, 505–513. [CrossRef]
82. Obinata, H.; Hla, T. Sphingosine 1-phosphate in coagulation and inflammation. *Semin. Immunopathol.* **2012**, *34*, 73–91. [CrossRef]
83. Rauch, B.H. Sphingosine 1-phosphate as a link between blood coagulation and inflammation. *Cell Physiol. Biochem.* **2014**, *34*, 185–196. [CrossRef]
84. Olivera, A.; Allende, M.L.; Proia, R.L. Shaping the landscape: Metabolic regulation of S1P gradients. *Biochim. Biophys. Acta* **2013**, *1831*, 193–202. [CrossRef]
85. Heffernan-Stroud, L.A.; Obeid, L.M. Sphingosine kinase 1 in cancer. *Adv. Cancer Res.* **2013**, *117*, 201–235. [PubMed]

86. Estrada-Bernal, A.; Palanichamy, K.; Chaudhury, A.R.; van Brocklyn, J.R. Induction of brain tumor stem cell apoptosis by FTY720: A potential therapeutic agent for glioblastoma. *Neuro Oncol.* **2012**, *14*, 405–415. [CrossRef]
87. Cattaneo, M.G.; Vanetti, C.; Samarani, M.; Aureli, M.; Bassi, R.; Sonnino, S.; Giussani, P. Cross-talk between sphingosine-1-phosphate and EGFR signaling pathways enhances human glioblastoma cell invasiveness. *Febs Lett.* **2018**, *592*, 949–961. [CrossRef]
88. Quint, K.; Stiel, N.; Neureiter, D.; Schlicker, H.U.; Nimsky, C.; Ocker, M.; Strik, H.; Kolodziej, M.A. The role of sphingosine kinase isoforms and receptors S1P1, S1P2, S1P3, and S1P5 in primary, secondary, and recurrent glioblastomas. *Tumour Biol.* **2014**, *35*, 8979–8989. [CrossRef] [PubMed]

cancers

MDPI

Review

Virus-Based Immunotherapy of Glioblastoma

Miika Martikainen * and Magnus Essand

Department of Immunology, Genetics, and Pathology, Science for Life Laboratory, Uppsala University, 75185 Uppsala, Sweden; magnus.essand@igp.uu.se

* Correspondence: miika.martikainen@igp.uu.se

Received: 21 December 2018; Accepted: 2 February 2019; Published: 5 February 2019

Abstract: Glioblastoma (GBM) is the most common type of primary brain tumor in adults. Despite recent advances in cancer therapy, including the breakthrough of immunotherapy, the prognosis of GBM patients remains dismal. One of the new promising ways to therapeutically tackle the immunosuppressive GBM microenvironment is the use of engineered viruses that kill tumor cells via direct oncolysis and via stimulation of antitumor immune responses. In this review, we focus on recently published results of phase I/II clinical trials with different oncolytic viruses and the new interesting findings in preclinical models. From syngeneic preclinical GBM models, it seems evident that oncolytic virus-mediated destruction of GBM tissue coupled with strong adjuvant effect, provided by the robust stimulation of innate antiviral immune responses and adaptive anti-tumor T cell responses, can be harnessed as potent immunotherapy against GBM. Although clinical testing of oncolytic viruses against GBM is at an early stage, the promising results from these trials give hope for the effective treatment of GBM in the near future.

Keywords: oncolytic virotherapy; cancer immunotherapy; glioblastoma

---

1. Current Status of Glioblastoma (GBM) Immunotherapy

Glioblastoma (GBM) is the most common and most aggressive primary malignant brain tumor in humans [1]. GBM can arise and be diagnosed either "de novo" or by evolving from less malignant astrocytomas or oligodendrogliomas [2]. The current standard care for GBM consists of radiotherapy, chemotherapy (temozolomide), and surgery when possible [3]. However, none of these interventions are curative, and GBM recurrence is virtually 100% certain. Consequently, the prognosis for GBM patients is dismal, with a median survival of only 14–15 months after diagnosis [3,4]

The recent success of cancer immunotherapy using immune checkpoint inhibitors (CPIs) against many otherwise untreatable cancers has raised expectations that such approaches could also be successfully used against GBM. For example, combination of anti-PD-1, anti-TIM-3, and targeted radiation showed impressive results in the preclinical mouse GL261 glioma model [5]. The drawback of using carcinogen-induced mouse models such as GL261 [6] is their relatively high immunogenicity and mutational load as compared with the clinical GBMs [7]. The results obtained in such preclinical models might thus overestimate the efficacy of immunotherapy against GBM. A recent study by Genoud et al. indicates that CPIs are not effective against the novel and significantly less immunogenic (and highly tumorigenic) SB28 model [8], which may better mimic the immune landscape of GBM. In line with this, experiences in the clinics show that GBM presents a particularly difficult target for immunotherapy, and the currently ongoing clinical trials utilizing CPIs, chimeric antigen receptor (CAR) T cells, or different vaccination strategies have resulted in mostly disappointing results [9–15]. Through whole-exome and RNA sequencing of a patient's tumor and normal cells, neoantigens (derived from genetic mutations in tumor cells) can be predicted and evaluated as personalized vaccines. Two clinical trials have recently evaluated this approach for GBM and found that neoantigen-specific T cells can be generated and found inside GBM, showing that immunotherapy may be a fruitful way forward [16,17]. It is worth noting that PD-1 blockade has shown impressive results against hypermutated GBM [18,19], indicating that CPIs can be effective in a certain subset of patients. These cases can be associated with somatic mutations in the DNA mismatch-repair machinery [18,19]. Another glimmer of hope for effective GBM immunotherapy can be seen in the results of recent phase I/II clinical trials using cancer cell selectively replicating oncolytic viruses [20–22].

Oncolytic virotherapy is based on utilizing replicating viruses that can selectively kill the infected cancer cells. The virus-induced cell death can occur through a variety of different mechanisms, most (if not all) of which can be immunogenic (reviewed by [23]). During this process of immunogenic cell death (ICD), damage-associated molecular patterns (DAMPs) and tumor-associated antigens (TAAs), including patient-specific neoantigens caused by genomic mutations in cancer cells, are released from the disrupted cancer cells [24]. This in turn acts as a potent stimulus for the immune system and can, in the optimal case, lead into activation of effective antitumor immunity. In addition, the virus-induced antiviral innate immune response mediated by pathogen-associated molecular patterns (PAMPs) can act as a potent adjuvant to further boost antigen cross-presentation and consequent adaptive immune responses. Oncolytic viruses can thus be considered to function not only as directly cancer killing agents, but also as active anticancer vaccines [25,26]. Because of the recent appreciation of the immunostimulatory effects of oncolytic viruses, the focus of the field has clearly shifted from direct oncolysis to immunostimulatory properties of the viruses used. It is easy to hypothesize that virus-induced inflammation and ICD would be especially beneficial when treating heavily immunosuppressed (discussed later) tumors such as GBM. Importantly, use of oncolytic viruses together with radiation therapy or temozolomide has been shown to have synergistic activity [27–29], indicating that oncolytic virotherapy in combination with traditional forms of GBM therapy can be feasible.

Oncolytic viruses can also be used to transfer therapeutic payloads to the tumor. Viruses armed with immunoregulatory inserts such as interleukin 12 (IL-12) and OX40 ligand are currently being tested in clinical trials (Table 1). Other examples of clinically tested viruses with therapeutic payloads are gammaretrovirus "Toca 511" and vaccinia virus "TG6002", which carry the cytosine deaminase (CD) gene (Table 1). When active in infected tumor cells, CD can convert the subsequently given 5-fluorocytosine drug into chemotherapeutic fluorouracil [30].

The oncolytic viruses used against GBM in current trials are listed in Table 1. Of these, perhaps the most promising viruses are DNX-2401, PVS-RIPO, and Toca 511, all of which have shown complete durable responses in approximately 20% of GBM patients who received virus intratumorally [20–22]. Virus-related severe adverse events in these trials have been rare. In Toca 511 and DNX-2401 trials, no dose-limiting toxicities were observed [21,22]. In the PVSRIPO trial, one (possibly virus-related) death and one dose-limiting toxic effect were reported [20]. Despite the early status of these trials, it is clear that oncolytic virotherapy is among the most compelling new therapies for GBM. The encouraging results obtained with PVS-RIPO, Toca511, and DNX-2401 have granted them a fast track designation by the U.S. Food and Drug Administration (FDA) for expedited drug review process.

Table 1. Currently active or completed clinical trials utilizing replication-competent viruses against malignant gliomas. Search results from clinicaltrials.gov.

Currently Active or Recruiting					
Adenovirus	**Virus Construct**	**Phase**	**Therapy Regimen**	**Trial No.**	**Results**
DNX-2401 + pembrolizumab	Deletion in E1A. RGD-4C fiber modification	II	IT injection followed by Pembrolizumab (IV) every 3 weeks	NCT02798406	-
DNX-2440	DNX-2401 (above) armed with OX40L	I	IT injection	NCT03714334	-
CRad-S-pk7 (loaded into neural stem cells)	E1A under survivin promoter, pK7 fiber modification	I	injection into resection cavity	NCT03072134	-
HSV	**Virus Construct**	**Phase**	**Therapy Regimen**	**Trial No.**	**Results**
C134	γ34.5 deletion IRS1 under HCMV promoter	I	IT injection	NCT03657576	-
M032	γ34.5 deletion Armed with human IL-12	I	single IT infusion	NCT02062827 NSC733972	-
rQNestin 34.5	One copy of γ34.5 under nestin promoter With UL39 deletion	I	IT with/without preceding IV cyclophosphamide	NCT03152318	-
G207	γ34.5 deletion Inactivating insertion of lacZ in UL39 gene	I	IT infusion	NCT02457845	-
Other	**Virus Construct**	**Phase**	**Therapy Regimen**	**Trial No.**	**Results**
Vaccinia virus TG6002 + 5-FC	TK and RR deletion Expresses cytosine deaminase	I/II	3 weekly IV infusions, followed by oral 5-FC	NCT03294486	-
Measles Virus (MV-CEA)	expresses CEA	I	injection into resection cavity and/or IT	NCT00390299	-
Poliovirus (PVSRIPO)	attenuated (Sabin) poliovirus with IRES from HRV2	I/Ib	IT via convection- enhanced delivery	NCT01491893 NCT03043391	[20]
Reovirus (REOLYSIN) + sargramostim (rGM-CSF)	unmodified reovirus	I	repeated cycles of sargamostim followed by IV virus	NCT02444546	-
Toca 511 + Toca FC	described above	II/III	injection into resection cavity followed by oral Toca FC	NCT02414165	-
Completed					
Adenovirus	**Virus Construct**	**Phase**	**Therapy Regimen**	**Trial No.**	**Results**
DNX-2401 + Temozolomide	described above	I	injection in the brain parenchyma followed by temozolomide	NCT01956734	-
DNX-2401 + IFNγ	described above	I	IT injection followed by IFNγ	NCT02197169	-
DNX-2401	described above	I/II	intracerebral infusion	NCT01582516	[21]
Other	**Virus Construct**	**Phase**	**Therapy Regimen**	**Trial No.**	**Results**
HSV G207	described above	I/II	IT injection	NCT00028158	[31]
Parvovirus H-1PV	unmodified rat parvovirus	I/II	combinations of IV, IT, and resection cavity injections	NCT01301430	[32]
Reovirus (REOLYSIN)	unmodified human reovirus	I	IT infusion	NCT00528684	[33]
Toca 511 + Toca FC	described above	I	injection in resection cavity followed by oral Toca FC	NCT01470794	[22,34]

Abbreviations: IT: intratumoral, IV: intravenous, HSV: herpes simplex virus, HCMV: human cytomegalovirus, IL-12: interleukin 12, TK: thymidine kinase gene, RR: ribonucleotide reductase gene, 5-FC: 5-fluorocytosine, CEA: carcinoembryonic antigen, IRES: internal ribosomal entry site, HRV2: human rhinovirus 2, rGM-CSF: recombinant granulocyte-macrophage colony-stimulating factor, IFN-γ: interferon gamma.

2. Heating the Suppressive Tumor Microenvironment with Viruses

The GBM microenvironment has proven to be a particularly challenging target for immunotherapeutic approaches. This can be largely attributed to the naturally isolated and tightly controlled brain microenvironment, which GBM thrives upon and skews even more immunosuppressive. The overall T cell infiltration in GBM is relatively poor and has a low $CD8^+/CD4^+$ T cell ratio [35]. Although cytotoxic $CD8^+$ T cells can be found in patient GBM samples, they often display a PD-1^+, LAG-3^+, $TIGIT^+$, $CD39^+$, $KLRG1^-$, and $CD57^-$ profile, which is indicative of an exhausted phenotype of these cells [36]. GBM cells also attract microglia and macrophages that have been shown to promote immunosuppression and enhance tumor growth [37]. In fact, monocytes and macrophages can contribute to as much as 30% of the total amount of cells in gliomas [38], making them (in addition to the GBM cells themselves) a compelling target for therapy. In addition to the local immunosuppression, there is also evidence of GBM-mediated inhibition of peripheral immune responses. This is indicated by increased circulating regulatory T cells (Tregs) [39] and impaired cytokine response by peripheral blood lymphocytes [40]. It must also be noted that corticosteroids, which are used with standard GBM radiotherapy, chemotherapy (to alleviate swelling in the brain), also have notable immunosuppressive properties [41]. In preclinical studies, the corticosteroid dexamethasone has been shown to reduce serum neutralizing antibodies against oncolytic herpes simplex virus (HSV) G207 [42]. While having no effect on the direct oncolytic activity of the virus, the use of dexamethasone completely abolished virus-induced antitumor immunity against subcutaneous N18 neuroblastomas in A/J mice [42]. This indicates that temporal immunosuppression with corticosteroids could possibly be used early during oncolytic virotherapy to increase the efficacy of systemic virus delivery. The use of corticosteroids can, however, be clearly detrimental for the long-term antitumor immunotherapeutic effect.

Together with a low mutational burden and high heterogeneity, the immune evasive factors present in GBM (reviewed by [43]) are likely to render most immunotherapeutic approaches ineffective. There is, however, an increasing amount of evidence that oncolytic viruses can be used to disrupt a immunologically "cold" GBM microenvironment through induction of inflammation and ICD. One good example of ICD-driven oncolytic immunotherapy in an experimental mouse model is reported by Koks et al., using oncolytic Newcastle disease virus (NDV) in the syngeneic and fully immunocompetent GL261 model of orthotopic glioma [44]. Here, NDV was shown to induce ICD in GL261 cells, as detected by ectopic calreticulin surface expression, HMGB1 release, and increased cancer antigen expression. In line with the immunostimulatory potential seen in vitro, intratumoral administration of NDV was shown to lead to long-term survival, associated with elevated infiltration of cytotoxic T cells and reduced accumulation of myeloid-derived suppressor cells (MDSCs), in 50% of the treated mice. Similar evidence of an immunostimulatory therapeutic effect has been observed in the orthotopic GL261 model with adenovirus DNX-2401 (currently used in clinical trials). Here, intratumoral administration of DNX-2401 into GL261-OVA glioma-bearing mice enhanced the presentation of OVA epitopes to $CD8^+$ T cells and had the potency to induce anti-glioma immunity [45]. Of note in the case of NDV, the long-term antitumor effect was reported to be lost in immunodeficient Rag2 knockout and in CD8-depleted mice, stressing the important role of functional immune system in the observed therapeutic effect [44]. Adding to the list of viruses that have shown survival benefit in mouse immunocompetent glioma models are the modified Semliki forest virus (SFV), VSV Δ51, and mouse-adapted Zika virus [46–49]. Interestingly, Zika virus has been reported to efficiently target and replicate in glioma stem cells [46], while viruses such as γ34.5-deleted HSV show restricted replication in this cell type [50]. Glioma stem cells, characterized as a highly self-renewing capable population of cells in the GBM tissue, have been proposed to be one of the main contributors to the high therapeutic resistance of GBM [51]. These cells, therefore, present a highly relevant therapeutic target, and the use of oncolytic viruses that are capable of efficiently infecting and killing these cells would be favorable.

The ability of oncolytic viruses to induce therapeutically relevant immune responses against GBM (at least in subsets of patients) has been demonstrated in clinical trials [20,21,52]. In fact, the immunostimulatory effect of oncolytic viruses used in the clinics can be speculated to be more relevant than the direct oncolytic power of the virus. Therapy with adenovirus DNX-2401 was shown to increase $CD8^+$, T-bet$^+$ T cell infiltration together with a reduction of TIM-3 expression [21]. Parvovirus H-1 therapy was noted to activate GBM-associated microglia (Cathepsin B expression) and, similar to DNX-2401 therapy, increase cytotoxic T cell infiltration [52]. In addition, PVS-RIPO therapy has been reported to induce notable GBM tissue inflammation, as evidenced by the "soap bubble" appearance in magnetic resonance imaging [20]. One classic sign of immunotherapy-induced inflammation is also an apparent initial growth of the tumor. This so-called pseudoprogression is caused by increased immune cell infiltration to the tumor, not by the growth of tumor cells [53]. Pseudoprogression has been observed, for example, after PVS-RIPO infusions as increased cerebral edema. Although this type of tumor swelling is therapeutically favorable, it proposes challenges for the interpretation of radiologic images and patient care.

Dendritic cells (DCs) are professional antigen presenting cells that are essential in the induction of adaptive immunity. Owing to their capability to activate T cells, DCs function in the interface between the innate and adaptive immune system. DC activation has been shown to have an especially potent adjuvant effect in the promotion and development of antitumor immunity against GBM. For example, DC stimulation with Toll-like receptor 3 (TLR3) agonists enhances the anti-tumor immune response to anti-PD-1 therapy in the orthotopic GL261 glioma model [54]. A potent adjuvant effect has also been observed in the subcutaneous B16 melanoma model using heat-inactivated vaccinia virus, which can activate the cytosolic DNA-sensing cGAS-STING-pathway in DCs [55]. It is likely that the improved therapeutic activity of inactivated virus in this case can be explained by the profound immunosuppressive effect of fully replicative vaccinia. This study also points out that Batf3-dependent $CD103^+/CD8\alpha^+$ DCs, also known as type-1 conventional DCs, which excel at cross presentation of TAAs to $CD8^+$ T cells [56], are crucial for the observed therapeutic effect.

Interestingly, PVS-RIPO was also recently reported to be able to directly infect dendritic cells and macrophages [57]. Although PVS-RIPO propagation in these cells was shown to be non-lethal and only marginally productive, it resulted in a robust IFN-I response (shown by STAT1 Y701 phosphorylation and expression of interferon-stimulated genes IFIT1, ISG15, and PD-L1), accompanied by the expression of costimulatory molecules and cytokines [57]. It is, however, not clear how big a role this effect has in the therapeutic effects seen in patients treated with PVS-RIPO.

Taken together, the increasing amount of evidence suggests that oncolytic viruses can be used as potent immunotherapy against GBM. Given the immunologically cold GBM microenvironment, it is likely that potent "de novo" induction of antitumor T cell immunity is needed for effective therapy. In the context, oncolytic virotherapy can be seen to have an especially favorable effect, both disrupting the system by destruction of cancer cells and re-stimulating both the innate and adaptive immunity against the infected cancer cells.

3. The Prospect of Combining Oncolytic Viruses with Checkpoint Inhibitors in GBM Therapy

During recent years, the use of immune checkpoint inhibitors (CPIs) has shown great success in inducing long-term complete remissions in patients with otherwise refractory cancers [58]. Since 2011, when the first immune CPI was approved by the FDA for treatment of melanoma, these novel drugs have been rapidly approved for many different types of cancers including melanoma, head and neck squamous cell carcinoma, non-small cell lung cancer, renal cell carcinoma, Hodgkin's lymphoma, and urothelial carcinoma [58]. The use of CPIs can also result in therapeutic response against primary or metastatic brain tumors [18,19,59]. It is unclear to what extent the observed responses are the result of CPIs' ability to get across the blood–brain barrier, as peripheral immune activation might be effective enough to induce antitumor immunity. There are currently several ongoing clinical trials of CPIs in glioblastoma, including phase III trials with Ipilimumab (blocking CTLA-4) and Nivolumab (blocking PD-1) (NCT02017717, NCT02617589).

CPIs work by blocking the negative regulators of T cell function, thereby sustaining T cell activity. Consequently, tumors with low T cell infiltration (such as GBM) are unlikely to get significant benefit from CPI therapy. In these cases, it would be important to induce T cell infiltration into the tumor prior to CPI therapy. As oncolytic viruses have a potent ability to induce $CD8^+$ T cell responses against tumor cells, it can be expected that precursory oncolytic virotherapy would enhance the effectiveness of CPIs. The combination of checkpoint blockade with oncolytic virotherapy is also an attractive option because virus-induced inflammatory response in the tumor can lead to upregulation of PD1 on T cells and PD-L1 on tumor cells [60]. Taken together, there is a clear rationale for combining

CPIs with precursory therapy with oncolytic viruses for synergistic and effective immunotherapy of immunologically cold tumors, such as GBM.

Many different viruses have been tested together with CPIs in different preclinical tumor models with encouraging results [61]. For example, intratumoral injection of Newcastle disease virus (NDV) into B16 melanomas has been shown to induce CD4$^+$ and CD8$^+$ T cell infiltration in both infected and distant (non-infected) tumors, rendering them vulnerable to anti-CTLA-4 blockade [62]. As another good example of synergy between oncolytic viruses and CPIs, recombinant adenovirus (hTertAd) infection/oncolysis combined with anti PD-1 therapy was shown to broaden the spectrum of tumor neoantigen-specific T cell response in the subcutaneous CMT64 lung adenocarcinoma model [63]. A notable added benefit of combining oncolytic virotherapy and checkpoint blockade has also been observed in immunocompetent mouse glioma models. For example, measles virus together with anti-PD-1 [64], VSV (expressing TAAs HIF-2α, Sox-10, and c-Myc) together with anti-PD-1 [65], adenovirus Delta-24-RGDOX (expressing the immune costimulatory OX40 ligand) combined with anti-PD-L1 [66], and reovirus combined with anti-PD-1 antibodies [60] have shown therapeutic benefits against the GL261 glioma model. In addition, HSV (G47Δ-mIL12) together with anti-PD-1 and anti-CTLA-4 has shown therapeutic benefits in the mouse 005 glioma stem cell model [67]. These two immunotherapies have also been combined by engineering oncolytic herpes simplex virus, which expresses single-chain antibody fragments (scFv) against PD-1 [68]. According to the results by Passaro et al., GBM cells infected with this virus express and secrete scFvPD-1, which can bind to mouse PD-1 without affecting the oncolysis. It remains to be seen whether this design of oncolytic viral constructs can also be functional under clinical settings.

Because of the success of checkpoint inhibitors against other types of tumors and the evident synergy with oncolytic viruses, it is not a big surprise that many virus/antibody combinations are currently investigated in clinical trials [61,69]. Of note, a phase 2 clinical trial with DNX-2401 + pembrolizumab (anti-PD-1 antibody) is currently recruiting patients with recurrent GBM or gliosarcoma (NCT02798406, Table 1).

4. Feasibility of Systemic Virus Delivery into the GBM

Because of the blood–brain barrier, the brain is a notoriously difficult target for efficient drug delivery. GBM are characterized by highly abnormal tumor vessels, but although GBM vasculature can be considered leaky, regions with an intact blood–brain barrier are also present (reviewed by [70]) and can limit effective systemic delivery of drugs into the tumor. In most of the clinical trials, the virus is administered intratumorally or into the resection cavity. The clear rationale for this strategy is to ensure efficient direct tumor cell infection. Systemic delivery of oncolytic virus, if done safely, could, however, present a less technically challenging option. Systemic delivery would also be more broadly applicable in disseminated cases or if the tumor cannot be reached directly. A peripheral administration route could be used to stimulate therapeutically beneficial systemic immune responses.

Some viruses (such as SFV) have natural tropism to the central nervous system (CNS), which can be harnessed to achieve potent delivery of virus into the brain. At least, SFV [47,48,71], vaccinia virus [72,73], chimeric vesicular stomatitis virus (VSV) [74], parvovirus H-1 [75], and picornavirus SVV-001 [76] have shown the ability to infect intracranial tumors in animal models when administered systemically. Notably, parvovirus H-1 and reovirus have been shown to be able to reach GBM tumors following intravenous injection in clinical trials [32,52,60], indicating that effective systemic delivery of the virus can also be achieved in human patients.

Classically, induction of virus neutralizing antibodies (NAbs) is considered to limit repeated virus injections or the use of virus strains that the patients' immune system has previously encountered. Interestingly, studies by Berkeley et al. challenge this paradigm by showing that monocytes can uptake and internalize reovirus–NAb complexes (mediated by FcγRIII) and release functional replication competent reovirus to tumor [77]. As GBM is naturally heavily infiltrated by peripheral monocytoid cells, this "trojan horse" strategy might present an interesting (if possibly limited to certain viruses/strains) new way of efficient systemic virus delivery.

5. Innate Antiviral Response as a Challenge and Opportunity for Virus-Based Therapy of GBM

The first line of defense against viral infections is innate antiviral response, which is largely orchestrated by type-I interferons (IFN-I). IFN-Is are secreted by infected cells in response to PAMPs (e.g., viral nucleic acids) detected by host cell pattern recognition receptors (PRRs). Generally speaking, the IFN-I response plays a major role in inducing an antiviral state (via signaling through the IFNAR–JAK–STAT pathway) in the surrounding cells, thereby limiting viral spread. In addition, IFN-Is promote antigen presentation, natural killer (NK) cell function, and the development of antigen-specific adaptive immune responses [78]. In the GBM context, however, the effect of IFN-Is seems to be more contradictory or unclear. On the one hand, studies in mouse models show that IFNAR1-deficiency accelerates gliomagenesis by impairing immunosurveillance (seen as increased infiltration of $CD11b^+/Ly6G^+$ and $CD4^+/FoxP3^+$ cells, but decreased Tc1 effector cells and $CD11c^+$ DCs) [79]. Also supporting the antitumor effect of IFN-I signaling, high expression of negative IFN-I regulatory factor ATM (ataxia-telangiectasia mutated kinase) in GBM has been shown to correlate with poor patient survival [80]. On the other hand, the constitutively activated IFN-I signaling pathway [81] and constitutive activation of STAT proteins have been observed in GBM [82]. Autocrine IFN-I signaling has also been suggested to contribute to the immune evasion of glioma cells [81], probably by negatively regulating the antigen-presenting capacity of glial cells [83]. Interestingly, chronically elevated levels of IFN-I in the brain have been shown to be related to ageing [83]. As GBM mostly affects people at an older age (median age of diagnosis is 64) [4], it would then also be possible that these tumors are evolved to tolerate or even to employ active IFN-I signaling to their benefit.

Many of the currently used oncolytic viruses are either natural or the result of genetic engineering sensitive to the antiviral effect of IFN-Is. The rationale for use of such viruses is that cancer cells in many cases have dysfunctional IFN-I signaling [84]. Notably, human GBM samples have been reported to both respond to and produce IFN-I [85–87]. This, together with possibly abnormal effects of IFN-Is on GBM immunosurveillance (described above), would indicate that, although considered an "Achilles heel" of cancer cells, a defective IFN-I system cannot be taken for granted in GBM. Thereby, viruses that have at least some resistance to innate antiviral signaling would be favorable. Indeed, viruses such as myxomavirus, certain versions of SFV, and clinically tested poliovirus PVS-RIPO show resistance/insensitivity to IFN-I mediated antiviral signaling [48,88,89]. Importantly, SFV and PVS-RIPO do, however, activate IFN-I response in the infected cells. The use of IFN-I-tolerant viruses raises safety issues by possibly enhancing the risk of an uncontrolled spread of the virus in healthy cells. In the case of PVS-RIPO, unwanted replication in neurons is inhibited by replacing the poliovirus internal ribosomal entry site (IRES) with the IRES from human rhinovirus type 2, which is not functional in normal neuronal cells [90]. Neuronal replication of SFV (another (+)ssRNA virus, studied for its oncolytic properties by us and others) can be controlled by microRNA-mediated detargeting [47,48,91].

The virus infection-induced release of IFN-I and other DAMPs attracts innate immune cells to eat and destroy the infected cells. While this is an important step in priming adaptive antitumor (and antiviral) immunity, the premature killing of infected cells can also lead to suboptimal virus spread in the tumor. For example, HSV therapy has been shown to increase infiltration of macrophages and microglia in the mouse GBM model [92]. Although shown to be polarized toward favorable proinflammatory M1 phenotype (with high levels of CD86, MHCII, and Ly6C), increased tumor necrosis factor alpha (TNFα) produced by these cells induced apoptosis in infected tumor cells and ultimately inhibited viral spread. A negative effect of the innate immune system attacking the HSV-infected GBM cell in terms of replication has been shown in another study, in which temporal NK-cell depletion by administration of transforming growth factor beta (TGF-β) improved therapeutic effect of the virus [93]. NK cells and infiltrating monocytes/macrophages have also been implicated to play an important role in the early clearance of oncolytic Myxoma virus and HSV in syngeneic glioma models [94,95]. Notably, tumor-associated microglia (and astrocytes) have been shown to attenuate the replication of the oncolytic vaccinia virus LIVP 1.1.1 in murine GL261 gliomas by acting as vaccinia virus traps [96].

On the basis of the findings in preclinical GBM models, temporal inhibition of innate responses in the early stage of viral therapy could lead to increased viral spread in the GBM microenvironment. This would in turn lead to more potent oncolysis and subsequent immunotherapeutic effect. Alternatively, viruses that can tolerate and escape the innate antiviral effect could be used (if safety can be assured). It is likely that preclinical GBM models do not perfectly recapitulate the innate immune microenvironment of human GBM, making direct translation of results difficult. In addition, human GBMs show high heterogeneity both within the same tumors and between different patients [97]. Because of these limitations, it is reasonable to hypothesize that viruses that potently activate the innate immune system (inducing robust IFN-I response) and resist the following antiviral response (without risk of unwanted spread in healthy tissues) would have the best chance of replicating in GBM and, consequently, the best chance of inducing curative therapeutic response.

6. Conclusions

The use of viruses as oncolytic agents is not a particularly new idea. While the focus used to be on the direct oncolytic effect, the current working model of oncolytic virotherapy rather relies on utilizing viruses as immunotherapeutic agents. As oncolytic viruses can replicate in the target cells, they provide an intriguing possibility for one shot self-amplifying cancer therapy. On the other hand, it is also likely that repeated dosing could be beneficial (if not crucial) for the therapeutic effect through immunostimulatory effect as therapeutic vaccines. Intratumoral administration clearly has the benefit of delivering the virus directly to the tumor, thereby mostly evading virus neutralization by systemic antiviral antibodies, while the systemically delivered virus has the potential to break through the blood–brain barrier and reach tumor cells and sites not reachable by intratumoral injections. The effect on antitumor immune response from stimulating systemic antiviral immune response must be further studied before conclusions can be made. It is possible that combinations of intratumoral and intravenous doses of oncolytic viruses would produce synergistic therapeutic benefits by inducing both local and systemic immune responses. Direct side-by-side comparison of immune-related characteristics of the different viruses being developed and used in preclinical and clinical settings would also be valuable in order to evaluate the best candidates for GBM therapy. Nevertheless, recent phase I/II clinical trials demonstrate that the use of oncolytic viruses can significantly improve the survival of some GBM patients [20–22]. Most of the therapeutic benefits so far in the trials can be associated with the potent stimulation of antitumor immune responses, thereby warming the naturally cold GBM immune microenvironment (Figure 1A). Later phase clinical trials will demonstrate whether therapy with oncolytic viruses such as PVS-RIPO, DNX-2401, and Toca511 (currently on fast track status by the FDA) can lead to truly curative GBM therapy.

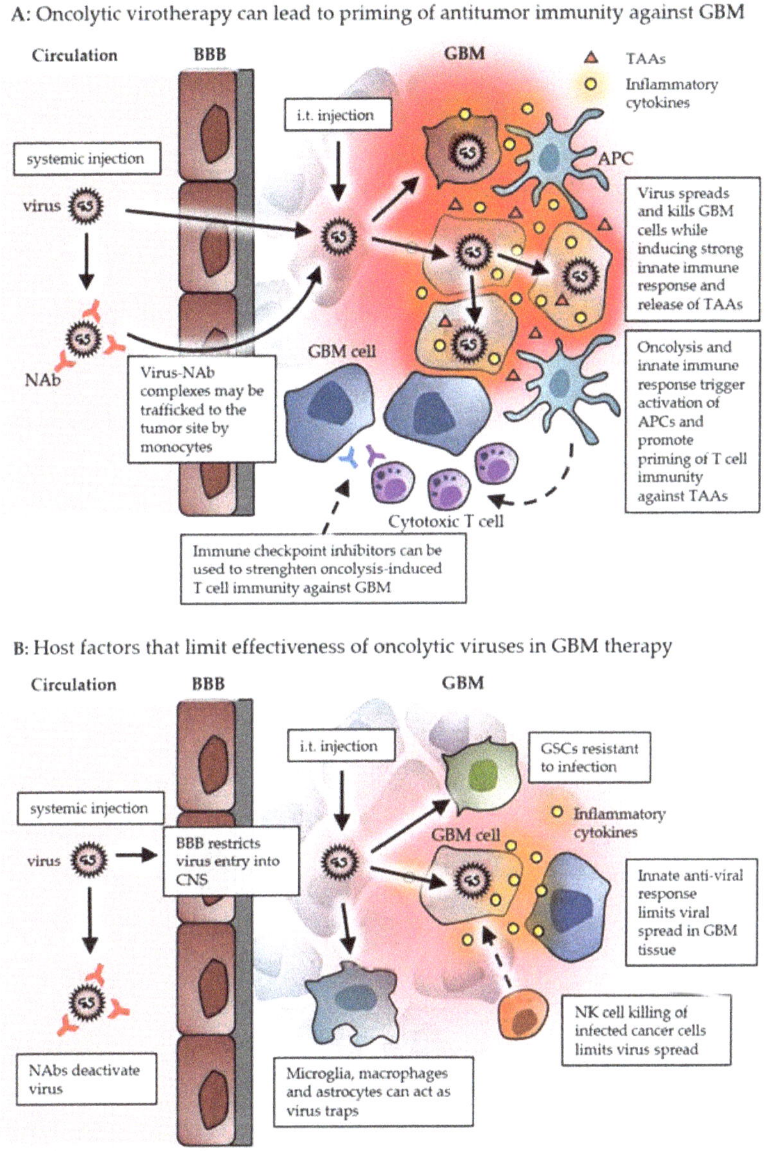

Figure 1. Schematic picture summarizing the events and factors related to successful oncolytic immunotherapy (**A**), and to limited effectiveness of oncolytic viruses (**B**) in GBM. Abbreviations: BBB: blood–brain barrier, GBM: glioblastoma, TAA: tumor-associated antigen, NAb: neutralizing antibody, APC: antigen-presenting cell, CNS: central nervous system, GSC: glioma stem cell, NK: natural killer cell.

The addition of therapeutic payloads to oncolytic viruses to boost antitumor potency is an attractive option. The increased therapeutic effect of such an armed virus construct is naturally dependent on the efficient replication of the virus in the target cells. Robust replication-induced cell death can, however, lower the amount of transgene produced by the infected cell. However, adding genes or replacing viral genes with therapeutic ones can hamper viral replication potency. Therefore, any oncolytic virus construct that depends on delivery and expression of therapeutic genes must be carefully optimized. It must also be considered that excessive expression of immunoregulatory transgenes might have unwanted toxic side-effects [98].

The GBM microenvironment presents a challenge for oncolytic virus delivery into the tumor, as well as effective viral replication and spread within the tumor (Figure 1B). Although the functionality and exact role of the innate immune system in GBM is still elusive, it is reasonable to speculate that at least some replicative power, despite antiviral signaling, would be beneficial for viral spread in the tumor. In addition, the heavy infiltration of innate immune cells (in both human and mouse models) and the presence glioma stem cells can render the GBM tissue resistant to effective viral spread, pointing out the need for robustly replicating viruses in order to disrupt the suppressive GBM microenvironment. In optimal cases, oncolytic viruses can serve as versatile innate adjuvants for boosting adaptive anti-tumor immunity. Notable overlap between virotherapy and other forms of immunotherapy, together with the possibility of turning an immunosuppressive tumor microenvironment into an immune-vulnerable one, opens interesting possibilities for combination therapies. It is likely that synergistic combination therapy with, for example, immune checkpoint inhibitors will ultimately be needed for a curative GBM therapy.

More work is needed to fully unlock the potential of viruses as effective anticancer agents. However, virus-based immunotherapy clearly shows promise in leading to curative treatment of currently incurable tumors such as GBM in the near future.

Funding: This research was funded by The Swedish Cancer Society (CAN 2016/318), The Swedish Children Cancer Foundation (PR2015-0049) and by the Swedish Research Council (2015-03688).

Acknowledgments: In this section you can acknowledge any support given which is not covered by the author contribution or funding sections. This may include administrative and technical support, or donations in kind (e.g., materials used for experiments).

Conflicts of Interest: The authors declare no conflict of interest.

References

1. Louis, D.N.; Perry, A.; Reifenberger, G.; von Deimling, A.; Figarella-Branger, D.; Cavenee, W.K.; Ohgaki, H.; Wiestler, O.D.; Kleihues, P.; Ellison, D.W. The 2016 World Health Organization Classification of Tumors of the Central Nervous System: A summary. *Acta Neuropathol.* **2016**, *131*, 803–820. [CrossRef]
2. Ohgaki, H.; Kleihues, P. The definition of primary and secondary glioblastoma. *Clin. Cancer Res.* **2013**, *19*, 764–772. [CrossRef] [PubMed]
3. Stupp, R.; Mason, W.P.; van den Bent, M.J.; Weller, M.; Fisher, B.; Taphoorn, M.J.B.; Belanger, K.; Brandes, A.A.; Marosi, C.; Bogdahn, U.; et al. Radiotherapy plus concomitant and adjuvant temozolomide for glioblastoma. *N. Engl. J. Med.* **2005**, *352*, 987–996. [CrossRef]
4. Ostrom, Q.T.; Gittleman, H.; Liao, P.; Vecchione-Koval, T.; Wolinsky, Y.; Kruchko, C.; Barnholtz-Sloan, J.S. CBTRUS Statistical Report: Primary brain and other central nervous system tumors diagnosed in the United States in 2010–2014. *Neuro-Oncology* **2017**, *19*, v1–v88. [CrossRef]
5. Kim, J.E.; Patel, M.A.; Mangraviti, A.; Kim, E.S.; Theodros, D.; Velarde, E.; Liu, A.; Sankey, E.W.; Tam, A.; Xu, H.; et al. Combination Therapy with Anti-PD-1, Anti-TIM-3, and Focal Radiation Results in Regression of Murine Gliomas. *Clin. Cancer Res.* **2017**, *23*, 124–136. [CrossRef] [PubMed]
6. Oh, T.; Fakurnejad, S.; Sayegh, E.T.; Clark, A.J.; Ivan, M.E.; Sun, M.Z.; Safaee, M.; Bloch, O.; James, C.D.; Parsa, A.T. Immunocompetent murine models for the study of glioblastoma immunotherapy. *J. Transl. Med.* **2014**, *12*, 107. [CrossRef] [PubMed]
7. Johanns, T.M.; Ward, J.P.; Miller, C.A.; Wilson, C.; Kobayashi, D.K.; Bender, D.; Fu, Y.; Alexandrov, A.; Mardis, E.R.; Artyomov, M.N.; et al. Endogenous Neoantigen-Specific CD8 T Cells Identified in Two Glioblastoma Models Using a Cancer Immunogenomics Approach. *Cancer Immunol. Res.* **2016**, *4*, 1007–1015. [CrossRef] [PubMed]
8. Genoud, V.; Marinari, E.; Nikolaev, S.I.; Castle, J.C.; Bukur, V.; Dietrich, P.-Y.; Okada, H.; Walker, P.R. Responsiveness to anti-PD-1 and anti-CTLA-4 immune checkpoint blockade in SB28 and GL261 mouse glioma models. *Oncoimmunology* **2018**, *7*, e1501137. [CrossRef] [PubMed]

9. Reardon, D.A.; Omuro, A.; Brandes, A.A.; Rieger, J.; Wick, A.; Sepulveda, J.; Phuphanich, S.; de Souza, P.; Ahluwalia, M.S.; Lim, M.; et al. OS10.3 Randomized Phase 3 Study Evaluating the Efficacy and Safety of Nivolumab vs Bevacizumab in Patients with Recurrent Glioblastoma: CheckMate 143. *Neuro-Oncology* **2017**, *19*, iii21. [CrossRef]
10. Wen, P.Y.; Reardon, D.A.; Phuphanich, S.; Aiken, R.; Landolfi, J.C.; Curry, W.T.; Zhu, J.-J.; Glantz, M.J.; Peereboom, D.M.; Markert, J.; et al. A randomized, double-blind, placebo-controlled phase 2 trial of dendritic cell (DC) vaccination with ICT-107 in newly diagnosed glioblastoma (GBM) patients. *J. Clin. Oncol.* **2014**, *32*, 2005. [CrossRef]
11. Rampling, R.; Peoples, S.; Mulholland, P.J.; James, A.; Al-Salihi, O.; Twelves, C.J.; McBain, C.; Jefferies, S.; Jackson, A.; Stewart, W.; et al. A Cancer Research UK First Time in Human Phase I Trial of IMA950 (Novel Multipeptide Therapeutic Vaccine) in Patients with Newly Diagnosed Glioblastoma. *Clin. Cancer Res.* **2016**, *22*, 4776–4785. [CrossRef] [PubMed]
12. Weller, M.; Butowski, N.; Tran, D.D.; Recht, L.D.; Lim, M.; Hirte, H.; Ashby, L.; Mechtler, L.; Goldlust, S.A.; Iwamoto, F.; et al. Rindopepimut with temozolomide for patients with newly diagnosed, EGFRvIII-expressing glioblastoma (ACT IV): A randomised, double-blind, international phase 3 trial. *Lancet Oncol.* **2017**, *18*, 1373–1385. [CrossRef]
13. Brown, C.E.; Alizadeh, D.; Starr, R.; Weng, L.; Wagner, J.R.; Naranjo, A.; Ostberg, J.R.; Blanchard, M.S.; Kilpatrick, J.; Simpson, J.; et al. Regression of Glioblastoma after Chimeric Antigen Receptor T-Cell Therapy. *N. Engl. J. Med.* **2016**, *375*, 2561–2569. [CrossRef] [PubMed]
14. O'Rourke, D.M.; Nasrallah, M.P.; Desai, A.; Melenhorst, J.J.; Mansfield, K.; Morrissette, J.J.D.; Martinez-Lage, M.; Brem, S.; Maloney, E.; Shen, A.; et al. A single dose of peripherally infused EGFRvIII-directed CAR T cells mediates antigen loss and induces adaptive resistance in patients with recurrent glioblastoma. *Sci. Transl. Med.* **2017**, *9*, eaaa0984. [CrossRef]
15. Lim, M.; Xia, Y.; Bettegowda, C.; Weller, M. Current state of immunotherapy for glioblastoma. *Nat. Rev. Clin. Oncol.* **2018**, *15*, 422–442. [CrossRef]
16. Hilf, N.; Kuttruff-Coqui, S.; Frenzel, K.; Bukur, V.; Stevanović, S.; Gouttefangeas, C.; Platten, M.; Tabatabai, G.; Dutoit, V.; van der Burg, S.H.; et al. Actively personalized vaccination trial for newly diagnosed glioblastoma. *Nature* **2019**, *565*, 240–245. [CrossRef]
17. Keskin, D.B.; Anandappa, A.J.; Sun, J.; Tirosh, I.; Mathewson, N.D.; Li, S.; Oliveira, G.; Giobbie-Hurder, A.; Felt, K.; Gjini, E.; et al. Neoantigen vaccine generates intratumoral T cell responses in phase Ib glioblastoma trial. *Nature* **2019**, *565*, 234–239. [CrossRef]
18. Johanns, T.M.; Miller, C.A.; Dorward, I.G.; Tsien, C.; Chang, E.; Perry, A.; Uppaluri, R.; Ferguson, C.; Schmidt, R.E.; Dahiya, S.; et al. Immunogenomics of Hypermutated Glioblastoma: A Patient with Germline POLE Deficiency Treated with Checkpoint Blockade Immunotherapy. *Cancer Discov.* **2016**, *6*, 1230–1236. [CrossRef]
19. Bouffet, E.; Larouche, V.; Campbell, B.B.; Merico, D.; de Borja, R.; Aronson, M.; Durno, C.; Krueger, J.; Cabric, V.; Ramaswamy, V.; et al. Immune Checkpoint Inhibition for Hypermutant Glioblastoma Multiforme Resulting from Germline Biallelic Mismatch Repair Deficiency. *J. Clin. Oncol.* **2016**, *34*, 2206–2211. [CrossRef]
20. Desjardins, A.; Gromeier, M.; Herndon, J.E.; Beaubier, N.; Bolognesi, D.P.; Friedman, A.H.; Friedman, H.S.; McSherry, F.; Muscat, A.M.; Nair, S.; et al. Recurrent Glioblastoma Treated with Recombinant Poliovirus. *N. Engl. J. Med.* **2018**, *379*, 150–161. [CrossRef]
21. Lang, F.F.; Conrad, C.; Gomez-Manzano, C.; Yung, W.K.A.; Sawaya, R.; Weinberg, J.S.; Prabhu, S.S.; Rao, G.; Fuller, G.N.; Aldape, K.D.; et al. Phase I Study of DNX-2401 (Delta-24-RGD) Oncolytic Adenovirus: Replication and Immunotherapeutic Effects in Recurrent Malignant Glioma. *J. Clin. Oncol.* **2018**, *36*, 1419–1427. [CrossRef] [PubMed]
22. Cloughesy, T.F.; Landolfi, J.; Vogelbaum, M.A.; Ostertag, D.; Elder, J.B.; Bloomfield, S.; Carter, B.; Chen, C.C.; Kalkanis, S.N.; Kesari, S.; et al. Durable complete responses in some recurrent high-grade glioma patients treated with Toca 511 + Toca FC. *Neuro-Oncology* **2018**, *20*, 1383–1392. [CrossRef] [PubMed]
23. Aurelian, L. Oncolytic viruses as immunotherapy: Progress and remaining challenges. *Onco Targets Ther.* **2016**, *9*, 2627–2637. [CrossRef] [PubMed]
24. Russell, S.J.; Barber, G.N. Oncolytic Viruses as Antigen-Agnostic Cancer Vaccines. *Cancer Cell* **2018**, *33*, 599–605. [CrossRef] [PubMed]

25. Bartlett, D.L.; Liu, Z.; Sathaiah, M.; Ravindranathan, R.; Guo, Z.; He, Y.; Guo, Z.S. Oncolytic viruses as therapeutic cancer vaccines. *Mol. Cancer* **2013**, *12*, 103. [CrossRef] [PubMed]
26. Chiocca, E.A.; Rabkin, S.D. Oncolytic viruses and their application to cancer immunotherapy. *Cancer Immunol. Res.* **2014**, *2*, 295–300. [CrossRef]
27. Geletneky, K.; Hartkopf, A.D.; Krempien, R.; Rommelaere, J.; Schlehofer, J.R. Improved killing of human high-grade glioma cells by combining ionizing radiation with oncolytic parvovirus H-1 infection. *J. Biomed. Biotechnol.* **2010**, *2010*, 350748. [CrossRef]
28. Liu, C.; Sarkaria, J.N.; Petell, C.A.; Paraskevakou, G.; Zollman, P.J.; Schroeder, M.; Carlson, B.; Decker, P.A.; Wu, W.; James, C.D.; et al. Combination of measles virus virotherapy and radiation therapy has synergistic activity in the treatment of glioblastoma multiforme. *Clin. Cancer Res.* **2007**, *13*, 7155–7165. [CrossRef]
29. Bai, Y.; Chen, Y.; Hong, X.; Liu, X.; Su, X.; Li, S.; Dong, X.; Zhao, G.; Li, Y. Newcastle disease virus enhances the growth-inhibiting and proapoptotic effects of temozolomide on glioblastoma cells in vitro and in vivo. *Sci. Rep.* **2018**, *8*, 11470. [CrossRef]
30. Nakamura, H.; Mullen, J.T.; Chandrasekhar, S.; Pawlik, T.M.; Yoon, S.S.; Tanabe, K.K. Multimodality therapy with a replication-conditional herpes simplex virus 1 mutant that expresses yeast cytosine deaminase for intratumoral conversion of 5-fluorocytosine to 5-fluorouracil. *Cancer Res.* **2001**, *61*, 5447–5452.
31. Markert, J.M.; Medlock, M.D.; Rabkin, S.D.; Gillespie, G.Y.; Todo, T.; Hunter, W.D.; Palmer, C.A.; Feigenbaum, F.; Tornatore, C.; Tufaro, F.; et al. Conditionally replicating herpes simplex virus mutant, G207 for the treatment of malignant glioma: Results of a phase I trial. *Gene Ther.* **2000**, *7*, 867–874. [CrossRef] [PubMed]
32. Geletneky, K.; Hajda, J.; Angelova, A.L.; Leuchs, B.; Capper, D.; Bartsch, A.J.; Neumann, J.-O.; Schöning, T.; Hüsing, J.; Beelte, B.; et al. Oncolytic H-1 Parvovirus Shows Safety and Signs of Immunogenic Activity in a First Phase I/IIa Glioblastoma Trial. *Mol. Ther.* **2017**, *25*, 2620–2634. [CrossRef] [PubMed]
33. Forsyth, P.; Roldán, G.; George, D.; Wallace, C.; Palmer, C.A.; Morris, D.; Cairncross, G.; Matthews, M.V.; Markert, J.; Gillespie, Y.; et al. A phase I trial of intratumoral administration of reovirus in patients with histologically confirmed recurrent malignant gliomas. *Mol. Ther.* **2008**, *16*, 627–632. [CrossRef] [PubMed]
34. Cloughesy, T.F.; Landolfi, J.; Hogan, D.J.; Bloomfield, S.; Carter, B.; Chen, C.C.; Elder, J.B.; Kalkanis, S.N.; Kesari, S.; Lai, A.; et al. Phase 1 trial of vocimagene amiretrorepvec and 5-fluorocytosine for recurrent high-grade glioma. *Sci. Transl. Med.* **2016**, *8*, 341ra75. [CrossRef] [PubMed]
35. Han, S.; Ma, E.; Wang, X.; Yu, C.; Dong, T.; Zhan, W.; Wei, X.; Liang, G.; Feng, S. Rescuing defective tumor-infiltrating T-cell proliferation in glioblastoma patients. *Oncol. Lett.* **2016**, *12*, 2924–2929. [CrossRef] [PubMed]
36. Woroniecka, K.; Chongsathidkiet, P.; Rhodin, K.; Kemeny, H.; Dechant, C.; Farber, S.H.; Elsamadicy, A.A.; Cui, X.; Koyama, S.; Jackson, C.; et al. T-Cell Exhaustion Signatures Vary with Tumor Type and Are Severe in Glioblastoma. *Clin. Cancer Res.* **2018**, *24*, 4175–4186. [CrossRef] [PubMed]
37. Hambardzumyan, D.; Gutmann, D.H.; Kettenmann, H. The role of microglia and macrophages in glioma maintenance and progression. *Nat. Neurosci.* **2016**, *19*, 20–27. [CrossRef]
38. Graeber, M.B.; Scheithauer, B.W.; Kreutzberg, G.W. Microglia in brain tumors. *Glia* **2002**, *40*, 252–259. [CrossRef]
39. Crane, C.A.; Ahn, B.J.; Han, S.J.; Parsa, A.T. Soluble factors secreted by glioblastoma cell lines facilitate recruitment, survival, and expansion of regulatory T cells: Implications for immunotherapy. *Neuro-Oncology* **2012**, *14*, 584–595. [CrossRef]
40. Mohme, M.; Schliffke, S.; Maire, C.L.; Rünger, A.; Glau, L.; Mende, K.C.; Matschke, J.; Gehbauer, C.; Akyüz, N.; Zapf, S.; et al. Immunophenotyping of Newly Diagnosed and Recurrent Glioblastoma Defines Distinct Immune Exhaustion Profiles in Peripheral and Tumor-infiltrating Lymphocytes. *Clin. Cancer Res.* **2018**, *24*, 4187–4200. [CrossRef]
41. Pitter, K.L.; Tamagno, I.; Alikhanyan, K.; Hosni-Ahmed, A.; Pattwell, S.S.; Donnola, S.; Dai, C.; Ozawa, T.; Chang, M.; Chan, T.A.; et al. Corticosteroids compromise survival in glioblastoma. *Brain* **2016**, *139*, 1458–1471. [CrossRef] [PubMed]
42. Todo, T.; Rabkin, S.D.; Chahlavi, A.; Martuza, R.L. Corticosteroid administration does not affect viral oncolytic activity, but inhibits antitumor immunity in replication-competent herpes simplex virus tumor therapy. *Hum. Gene Ther.* **1999**, *10*, 2869–2878. [CrossRef] [PubMed]

43. Razavi, S.-M.; Lee, K.E.; Jin, B.E.; Aujla, P.S.; Gholamin, S.; Li, G. Immune Evasion Strategies of Glioblastoma. *Front. Surg.* **2016**, *3*, 11. [CrossRef] [PubMed]
44. Koks, C.A.; Garg, A.D.; Ehrhardt, M.; Riva, M.; Vandenberk, L.; Boon, L.; De Vleeschouwer, S.; Agostinis, P.; Graf, N.; Van Gool, S.W. Newcastle disease virotherapy induces long-term survival and tumor-specific immune memory in orthotopic glioma through the induction of immunogenic cell death. *Int. J. Cancer* **2015**, *136*, E313–E325. [CrossRef] [PubMed]
45. Jiang, H.; Clise-Dwyer, K.; Ruisaard, K.E.; Fan, X.; Tian, W.; Gumin, J.; Lamfers, M.L.; Kleijn, A.; Lang, F.F.; Yung, W.-K.A.; et al. Delta-24-RGD oncolytic adenovirus elicits anti-glioma immunity in an immunocompetent mouse model. *PLoS ONE* **2014**, *9*, e97407. [CrossRef] [PubMed]
46. Zhu, Z.; Gorman, M.J.; McKenzie, L.D.; Chai, J.N.; Hubert, C.G.; Prager, B.C.; Fernandez, E.; Richner, J.M.; Zhang, R.; Shan, C.; et al. Zika virus has oncolytic activity against glioblastoma stem cells. *J. Exp. Med.* **2017**, *214*, 2843–2857. [CrossRef] [PubMed]
47. Ramachandran, M.; Yu, D.; Dyczynski, M.; Baskaran, S.; Zhang, L.; Lulla, A.; Lulla, V.; Saul, S.; Nelander, S.; Dimberg, A.; et al. Safe and Effective Treatment of Experimental Neuroblastoma and Glioblastoma Using Systemically Delivered Triple MicroRNA-Detargeted Oncolytic Semliki Forest Virus. *Clin. Cancer Res.* **2017**, *23*, 1519–1530. [CrossRef] [PubMed]
48. Martikainen, M.; Niittykoski, M.; von und zu Fraunberg, M.; Immonen, A.; Koponen, S.; van Geenen, M.; Vähä-Koskela, M.; Ylösmäki, E.; Jääskeläinen, J.E.; Saksela, K.; et al. MicroRNA-Attenuated Clone of Virulent Semliki Forest Virus Overcomes Antiviral Type I Interferon in Resistant Mouse CT-2A Glioma. *J. Virol.* **2015**, *89*, 10637–10647. [CrossRef]
49. Balathasan, L.; Tang, V.A.; Yadollahi, B.; Brun, J.; Labelle, M.; Lefebvre, C.; Swift, S.L.; Stojdl, D.F. Activating Peripheral Innate Immunity Enables Safe and Effective Oncolytic Virotherapy in the Brain. *Mol. Ther. Oncolytics* **2017**, *7*, 45–56. [CrossRef]
50. Peters, C.; Paget, M.; Tshilenge, K.-T.; Saha, D.; Antoszczyk, S.; Baars, A.; Frost, T.; Martuza, R.L.; Wakimoto, H.; Rabkin, S.D. Restriction of γ34.5-Deleted Oncolytic Herpes Simplex Virus Replication in Glioblastoma Stem-Like Cells. *J. Virol.* **2018**, *92*. [CrossRef]
51. Jackson, M.; Hassiotou, F.; Nowak, A. Glioblastoma stem-like cells: At the root of tumor recurrence and a therapeutic target. *Carcinogenesis* **2015**, *36*, 177–185. [CrossRef] [PubMed]
52. Angelova, A.L.; Barf, M.; Geletneky, K.; Unterberg, A.; Rommelaere, J. Immunotherapeutic Potential of Oncolytic H-1 Parvovirus: Hints of Glioblastoma Microenvironment Conversion towards Immunogenicity. *Viruses* **2017**, *9*, 382. [CrossRef] [PubMed]
53. Chiou, V.L.; Burotto, M. Pseudoprogression and Immune-Related Response in Solid Tumors. *J. Clin. Oncol.* **2015**, *33*, 3541–3543. [CrossRef] [PubMed]
54. Garzon-Muvdi, T.; Theodros, D.; Luksik, A.S.; Maxwell, R.; Kim, E.; Jackson, C.M.; Belcaid, Z.; Ganguly, S.; Tyler, B.; Brem, H.; et al. Dendritic cell activation enhances anti-PD-1 mediated immunotherapy against glioblastoma. *Oncotarget* **2018**, *9*, 20681–20697. [CrossRef] [PubMed]
55. Dai, P.; Wang, W.; Yang, N.; Serna-Tamayo, C.; Ricca, J.M.; Zamarin, D.; Shuman, S.; Merghoub, T.; Wolchok, J.D.; Deng, L. Intratumoral delivery of inactivated modified vaccinia virus Ankara (iMVA) induces systemic antitumor immunity via STING and Batf3-dependent dendritic cells. *Sci. Immunol.* **2017**, *2*, eaal1713. [CrossRef]
56. Sánchez-Paulete, A.R.; Teijeira, A.; Cueto, F.J.; Garasa, S.; Pérez-Gracia, J.L.; Sánchez-Arráez, A.; Sancho, D.; Melero, I. Antigen cross-presentation and T-cell cross-priming in cancer immunology and immunotherapy. *Ann. Oncol.* **2017**, *28*, xii44–xii55. [CrossRef] [PubMed]
57. Brown, M.C.; Holl, E.K.; Boczkowski, D.; Dobrikova, E.; Mosaheb, M.; Chandramohan, V.; Bigner, D.D.; Gromeier, M.; Nair, S.K. Cancer immunotherapy with recombinant poliovirus induces IFN-dominant activation of dendritic cells and tumor antigen-specific CTLs. *Sci. Transl. Med.* **2017**, *9*, eaan4220. [CrossRef]
58. Wei, S.C.; Duffy, C.R.; Allison, J.P. Fundamental Mechanisms of Immune Checkpoint Blockade Therapy. *Cancer Discov.* **2018**, *8*, 1069–1086. [CrossRef]
59. Goldberg, S.B.; Gettinger, S.N.; Mahajan, A.; Chiang, A.C.; Herbst, R.S.; Sznol, M.; Tsiouris, A.J.; Cohen, J.; Vortmeyer, A.; Jilaveanu, L.; et al. Pembrolizumab for patients with melanoma or non-small-cell lung cancer and untreated brain metastases: Early analysis of a non-randomised, open-label, phase 2 trial. *Lancet Oncol.* **2016**, *17*, 976–983. [CrossRef]

60. Samson, A.; Scott, K.J.; Taggart, D.; West, E.J.; Wilson, E.; Nuovo, G.J.; Thomson, S.; Corns, R.; Mathew, R.K.; Fuller, M.J.; et al. Intravenous delivery of oncolytic reovirus to brain tumor patients immunologically primes for subsequent checkpoint blockade. *Sci. Transl. Med.* **2018**, *10*, eaam7577. [CrossRef]
61. Chen, C.-Y.; Hutzen, B.; Wedekind, M.F.; Cripe, T.P. Oncolytic virus and PD-1/PD-L1 blockade combination therapy. *Oncolytic Virother.* **2018**, *7*, 65–77. [CrossRef] [PubMed]
62. Zamarin, D.; Holmgaard, R.B.; Subudhi, S.K.; Park, J.S.; Mansour, M.; Palese, P.; Merghoub, T.; Wolchok, J.D.; Allison, J.P. Localized oncolytic virotherapy overcomes systemic tumor resistance to immune checkpoint blockade immunotherapy. *Sci. Transl. Med.* **2014**, *6*, 226ra32. [CrossRef] [PubMed]
63. Woller, N.; Gürlevik, E.; Fleischmann-Mundt, B.; Schumacher, A.; Knocke, S.; Kloos, A.M.; Saborowski, M.; Geffers, R.; Manns, M.P.; Wirth, T.C.; et al. Viral Infection of Tumors Overcomes Resistance to PD-1-immunotherapy by Broadening Neoantigenome-directed T-cell Responses. *Mol. Ther.* **2015**, *23*, 1630–1640. [CrossRef] [PubMed]
64. Hardcastle, J.; Mills, L.; Malo, C.S.; Jin, F.; Kurokawa, C.; Geekiyanage, H.; Schroeder, M.; Sarkaria, J.; Johnson, A.J.; Galanis, E. Immunovirotherapy with measles virus strains in combination with anti-PD-1 antibody blockade enhances antitumor activity in glioblastoma treatment. *Neuro. Oncol.* **2017**, *19*, 493–502. [CrossRef] [PubMed]
65. Cockle, J.V.; Rajani, K.; Zaidi, S.; Kottke, T.; Thompson, J.; Diaz, R.M.; Shim, K.; Peterson, T.; Parney, I.F.; Short, S.; et al. Combination viroimmunotherapy with checkpoint inhibition to treat glioma, based on location-specific tumor profiling. *Neuro-Oncology* **2016**, *18*, 518–527. [CrossRef] [PubMed]
66. Jiang, H.; Rivera-Molina, Y.; Gomez-Manzano, C.; Clise-Dwyer, K.; Bover, L.; Vence, L.M.; Yuan, Y.; Lang, F.F.; Toniatti, C.; Hossain, M.B.; et al. Oncolytic Adenovirus and Tumor-Targeting Immune Modulatory Therapy Improve Autologous Cancer Vaccination. *Cancer Res.* **2017**, *77*, 3894–3907. [CrossRef] [PubMed]
67. Saha, D.; Martuza, R.L.; Rabkin, S.D. Macrophage Polarization Contributes to Glioblastoma Eradication by Combination Immunovirotherapy and Immune Checkpoint Blockade. *Cancer Cell* **2017**, *32*, 253–267.e5. [CrossRef]
68. Passaro, C.; Alayo, Q.; De Laura, I.; McNulty, J.; Grauwet, K.; Ito, H.; Bhaskaran, V.; Mineo, M.; Lawler, S.E.; Shah, K.; et al. Arming an Oncolytic Herpes Simplex Virus Type 1 with a Single-chain Fragment Variable Antibody against PD-1 for Experimental Glioblastoma Therapy. *Clin. Cancer Res.* **2019**, *25*, 290–299. [CrossRef]
69. LaRocca, C.J.; Warner, S.G. Oncolytic viruses and checkpoint inhibitors: Combination therapy in clinical trials. *Clin. Transl. Med.* **2018**, *7*, 35. [CrossRef]
70. Sarkaria, J.N.; Hu, L.S.; Parney, I.F.; Pafundi, D.H.; Brinkmann, D.H.; Laack, N.N.; Giannini, C.; Burns, T.C.; Kizilbash, S.H.; Laramy, J.K.; et al. Is the blood-brain barrier really disrupted in all glioblastomas? A critical assessment of existing clinical data. *Neuro-Oncology* **2018**, *20*, 184–191. [CrossRef]
71. Heikkilä, J.E.; Vähä-Koskela, M.J.V.; Ruotsalainen, J.J.; Martikainen, M.W.; Stanford, M.M.; McCart, J.A.; Bell, J.C.; Hinkkanen, A.E. Intravenously administered alphavirus vector VA7 eradicates orthotopic human glioma xenografts in nude mice. *PLoS ONE* **2010**, *5*, e8603. [CrossRef] [PubMed]
72. Lun, X.Q.; Jang, J.-H.; Tang, N.; Deng, H.; Head, R.; Bell, J.C.; Stojdl, D.F.; Nutt, C.L.; Senger, D.L.; Forsyth, P.A.; et al. Efficacy of systemically administered oncolytic vaccinia virotherapy for malignant gliomas is enhanced by combination therapy with rapamycin or cyclophosphamide. *Clin. Cancer Res.* **2009**, *15*, 2777–2788. [CrossRef] [PubMed]
73. Advani, S.J.; Buckel, L.; Chen, N.G.; Scanderbeg, D.J.; Geissinger, U.; Zhang, Q.; Yu, Y.A.; Aguilar, R.J.; Mundt, A.J.; Szalay, A.A. Preferential replication of systemically delivered oncolytic vaccinia virus in focally irradiated glioma xenografts. *Clin. Cancer Res.* **2012**, *18*, 2579–2590. [CrossRef] [PubMed]
74. Muik, A.; Stubbert, L.J.; Jahedi, R.Z.; Geiβ, Y.; Kimpel, J.; Dold, C.; Tober, R.; Volk, A.; Klein, S.; Dietrich, U.; et al. Re-engineering vesicular stomatitis virus to abrogate neurotoxicity, circumvent humoral immunity, and enhance oncolytic potency. *Cancer Res.* **2014**, *74*, 3567–3578. [CrossRef] [PubMed]
75. Geletneky, K.; Kiprianova, I.; Ayache, A.; Koch, R.; Herrero, Y.; Calle, M.; Deleu, L.; Sommer, C.; Thomas, N.; Rommelaere, J.; et al. Regression of advanced rat and human gliomas by local or systemic treatment with oncolytic parvovirus H-1 in rat models. *Neuro-Oncology* **2010**, *12*, 804–814. [CrossRef] [PubMed]

76. Liu, Z.; Zhao, X.; Mao, H.; Baxter, P.A.; Huang, Y.; Yu, L.; Wadhwa, L.; Su, J.M.; Adesina, A.; Perlaky, L.; et al. Intravenous injection of oncolytic picornavirus SVV-001 prolongs animal survival in a panel of primary tumor-based orthotopic xenograft mouse models of pediatric glioma. *Neuro-Oncology* **2013**, *15*, 1173–1185. [CrossRef]
77. Berkeley, R.A.; Steele, L.P.; Mulder, A.A.; van den Wollenberg, D.J.M.; Kottke, T.J.; Thompson, J.; Coffey, M.; Hoeben, R.C.; Vile, R.G.; Melcher, A.; et al. Antibody-Neutralized Reovirus Is Effective in Oncolytic Virotherapy. *Cancer Immunol. Res.* **2018**, *6*, 1161–1173. [CrossRef]
78. Ivashkiv, L.B.; Donlin, L.T. Regulation of type I interferon responses. *Nat. Rev. Immunol.* **2014**, *14*, 36–49. [CrossRef]
79. Fujita, M.; Scheurer, M.E.; Decker, S.A.; McDonald, H.A.; Kohanbash, G.; Kastenhuber, E.R.; Kato, H.; Bondy, M.L.; Ohlfest, J.R.; Okada, H. Role of type 1 IFNs in antiglioma immunosurveillance–using mouse studies to guide examination of novel prognostic markers in humans. *Clin. Cancer Res.* **2010**, *16*, 3409–3419. [CrossRef]
80. Yang, C.H.; Wang, Y.; Sims, M.; Cai, C.; He, P.; Häcker, H.; Yue, J.; Cheng, J.; Boop, F.A.; Pfeffer, L.M. MicroRNA203a suppresses glioma tumorigenesis through an ATM-dependent interferon response pathway. *Oncotarget* **2017**, *8*, 112980–112991. [CrossRef]
81. Silginer, M.; Nagy, S.; Happold, C.; Schneider, H.; Weller, M.; Roth, P. Autocrine activation of the IFN signaling pathway may promote immune escape in glioblastoma. *Neuro-Oncology* **2017**, *19*, 1338–1349. [CrossRef] [PubMed]
82. Swiatek-Machado, K.; Kaminska, B. STAT signaling in glioma cells. *Adv. Exp. Med. Biol.* **2013**, *986*, 189–208. [PubMed]
83. Baruch, K.; Deczkowska, A.; David, E.; Castellano, J.M.; Miller, O.; Kertser, A.; Berkutzki, T.; Barnett-Itzhaki, Z.; Bezalel, D.; Wyss-Coray, T.; et al. Aging. Aging-induced type I interferon response at the choroid plexus negatively affects brain function. *Science* **2014**, *346*, 89–93. [CrossRef] [PubMed]
84. Critchley-Thorne, R.J.; Simons, D.L.; Yan, N.; Miyahira, A.K.; Dirbas, F.M.; Johnson, D.L.; Swetter, S.M.; Carlson, R.W.; Fisher, G.A.; Koong, A.; et al. Impaired interferon signaling is a common immune defect in human cancer. *Proc. Natl. Acad. Sci. USA* **2009**, *106*, 9010–9015. [CrossRef] [PubMed]
85. Alain, T.; Lun, X.; Martineau, Y.; Sean, P.; Pulendran, B.; Petroulakis, E.; Zemp, F.J.; Lemay, C.G.; Roy, D.; Bell, J.C.; et al. Vesicular stomatitis virus oncolysis is potentiated by impairing mTORC1-dependent type I IFN production. *Proc. Natl. Acad. Sci. USA* **2010**, *107*, 1576–1581. [CrossRef] [PubMed]
86. Cosset, É.; Petty, T.J.; Dutoit, V.; Cordey, S.; Padioleau, I.; Otten-Hernandez, P.; Farinelli, L.; Kaiser, L.; Bruyère-Cerdan, P.; Tirefort, D.; et al. Comprehensive metagenomic analysis of glioblastoma reveals absence of known virus despite antiviral-like type I interferon gene response. *Int. J. Cancer* **2014**, *135*, 1381–1389. [CrossRef] [PubMed]
87. Duarte, C.W.; Willey, C.D.; Zhi, D.; Cui, X.; Harris, J.J.; Vaughan, L.K.; Mehta, T.; McCubrey, R.O.; Khodarev, N.N.; Weichselbaum, R.R.; et al. Expression signature of IFN/STAT1 signaling genes predicts poor survival outcome in glioblastoma multiforme in a subtype-specific manner. *PLoS ONE* **2012**, *7*, e29653. [CrossRef]
88. Zemp, F.J.; McKenzie, B.A.; Lun, X.; Maxwell, L.; Reilly, K.M.; McFadden, G.; Yong, V.W.; Forsyth, P.A. Resistance to oncolytic myxoma virus therapy in nf1(−/−)/trp53(−/−) syngeneic mouse glioma models is independent of anti-viral type-I interferon. *PLoS ONE* **2013**, *8*, e65801. [CrossRef]
89. Walton, R.W.; Brown, M.C.; Sacco, M.T.; Gromeier, M. Engineered Oncolytic Poliovirus PVSRIPO Subverts MDA5-Dependent Innate Immune Responses in Cancer Cells. *J. Virol.* **2018**, *92*, e00879-18. [CrossRef]
90. Gromeier, M.; Bossert, B.; Arita, M.; Nomoto, A.; Wimmer, E. Dual stem loops within the poliovirus internal ribosomal entry site control neurovirulence. *J. Virol.* **1999**, *73*, 958–964.
91. Ylösmäki, E.; Martikainen, M.; Hinkkanen, A.; Saksela, K. Attenuation of Semliki Forest virus neurovirulence by microRNA-mediated detargeting. *J. Virol.* **2013**, *87*, 335–344. [CrossRef] [PubMed]
92. Meisen, W.H.; Wohleb, E.S.; Jaime-Ramirez, A.C.; Bolyard, C.; Yoo, J.Y.; Russell, L.; Hardcastle, J.; Dubin, S.; Muili, K.; Yu, J.; et al. The Impact of Macrophage- and Microglia-Secreted TNFα on Oncolytic HSV-1 Therapy in the Glioblastoma Tumor Microenvironment. *Clin. Cancer Res.* **2015**, *21*, 3274–3285. [CrossRef] [PubMed]
93. Han, J.; Chen, X.; Chu, J.; Xu, B.; Meisen, W.H.; Chen, L.; Zhang, L.; Zhang, J.; He, X.; Wang, Q.-E.; et al. TGFβ Treatment Enhances Glioblastoma Virotherapy by Inhibiting the Innate Immune Response. *Cancer Res.* **2015**, *75*, 5273–5282. [CrossRef] [PubMed]

94. Fulci, G.; Breymann, L.; Gianni, D.; Kurozomi, K.; Rhee, S.S.; Yu, J.; Kaur, B.; Louis, D.N.; Weissleder, R.; Caligiuri, M.A.; et al. Cyclophosphamide enhances glioma virotherapy by inhibiting innate immune responses. *Proc. Natl. Acad. Sci. USA* **2006**, *103*, 12873–12878. [CrossRef] [PubMed]
95. Zemp, F.J.; McKenzie, B.A.; Lun, X.; Reilly, K.M.; McFadden, G.; Yong, V.W.; Forsyth, P.A. Cellular factors promoting resistance to effective treatment of glioma with oncolytic myxoma virus. *Cancer Res.* **2014**, *74*, 7260–7273. [CrossRef] [PubMed]
96. Kober, C.; Rohn, S.; Weibel, S.; Geissinger, U.; Chen, N.G.; Szalay, A.A. Microglia and astrocytes attenuate the replication of the oncolytic vaccinia virus LIVP 1.1.1 in murine GL261 gliomas by acting as vaccinia virus traps. *J. Transl. Med.* **2015**, *13*, 216. [CrossRef]
97. Wick, W.; Kessler, T. New glioblastoma heterogeneity atlas—A shared resource. *Nat. Rev. Neurol.* **2018**, *14*, 453–454. [CrossRef]
98. Cohen, J. IL-12 deaths: Explanation and a puzzle. *Science* **1995**, *270*, 908. [CrossRef]

cancers

MDPI

Review

Recent Advances in Oncolytic Virotherapy and Immunotherapy for Glioblastoma: A Glimmer of Hope in the Search for an Effective Therapy?

Aleksei A. Stepanenko [1,*] **and Vladimir P. Chekhonin** [1,2]

[1] Department of Fundamental and Applied Neurobiology, V. P. Serbsky National Medical Research Center for Psychiatry and Narcology, the Ministry of Health of the Russian Federation, Kropotkinsky lane 23, 119034 Moscow, Russia; chekhoninnew@yandex.ru

[2] Department of Medical Nanobiotechnologies, Medico-Biological Faculty, N. I. Pirogov Russian National Research Medical University, the Ministry of Health of the Russian Federation, Ostrovitianov str. 1, 117997 Moscow, Russia

* Correspondence: a.a.stepanenko@gmail.com

Received: 7 November 2018; Accepted: 29 November 2018; Published: 5 December 2018

Abstract: To date, no targeted drugs, antibodies or combinations of chemotherapeutics have been demonstrated to be more efficient than temozolomide, or to increase efficacy of standard therapy (surgery, radiotherapy, temozolomide, steroid dexamethasone). According to recent phase III trials, standard therapy may ensure a median overall survival of up to 18–20 months for adult patients with newly diagnosed glioblastoma. These data explain a failure of positive non-controlled phase II trials to predict positive phase III trials and should result in revision of the landmark Stupp trial as a historical control for median overall survival in non-controlled trials. A high rate of failures in clinical trials and a lack of effective chemotherapy on the horizon fostered the development of conceptually distinct therapeutic approaches: dendritic cell/peptide immunotherapy, chimeric antigen receptor (CAR) T-cell therapy and oncolytic virotherapy. Recent early phase trials with the recombinant adenovirus DNX-2401 (Ad5-delta24-RGD), polio-rhinovirus chimera (PVSRIPO), parvovirus H-1 (ParvOryx), Toca 511 retroviral vector with 5-fluorocytosine, heat shock protein-peptide complex-96 (HSPPC-96) and dendritic cell vaccines, including DCVax-L vaccine, demonstrated that subsets of patients with glioblastoma/glioma may benefit from oncolytic virotherapy/immunotherapy (>3 years of survival after treatment). However, large controlled trials are required to prove efficacy of next-generation immunotherapeutics and oncolytic vectors.

Keywords: immunotherapy; oncolytic virotherapy; temozolomide; targeted drugs; glioma; dendritic cell vaccine; radiotherapy; TTFields; PD-L1; bevacizumab

1. Introduction

Despite aggressive multimodal therapy (surgery, radiation, a genotoxic drug temozolomide, steroid dexamethasone, TTFields, lomustine, bevacizumab, re-irradiation, etc.), survival of adult patients with newly diagnosed and recurrent glioblastoma (grade IV malignant glioma) is usually less than 18–20 months [1–3] and 8–12 months [4–7], respectively. Over the last decade, high expectations were placed on targeted drug therapy, which was hoped to provide tumor growth control and further improvement in survival rates. However, to date, no molecularly targeted drug/antibody or combinations of small molecule inhibitors have been demonstrated to be more efficient than temozolomide (TMZ) or to increase efficacy of standard therapy in patients with primary/recurrent glioblastoma. A high rate of failures in clinical trials (Figure 1) and a lack of effective targeted therapy on the horizon have fostered the development of conceptually distinct therapeutic approaches such as cellular/peptide immunotherapy and oncolytic virotherapy. Next-generation immunotherapeutics and replication-competent genetically engineered oncolytic viruses demonstrated high efficacy in preclinical models (e.g., [8–12]). Recent early phase trials with the recombinant adenovirus DNX-2401 (Ad5-delta24-RGD) [13], polio-rhinovirus chimera (PVSRIPO) [14], parvovirus H-1 (ParvOryx) [15], Toca 511 retroviral vector with 5-fluorocytosine [16], heat shock protein-peptide complex-96 (HSPPC-96) vaccine [17,18], cytomegalovirus pp65 RNA-pulsed dendritic cells [8,19], and a large phase III trial of an autologous tumor lysate-pulsed dendritic cell vaccine (DCVax-L) [20], have demonstrated that subsets of recurrent glioblastoma/glioma patients may significantly benefit from oncolytic virotherapy or dendritic cell-/peptide-based vaccines and survive >3 years after treatment. These encouraging clinical data raise a glimmer of hope in fighting glioblastoma after many years of intensive search for a cure.

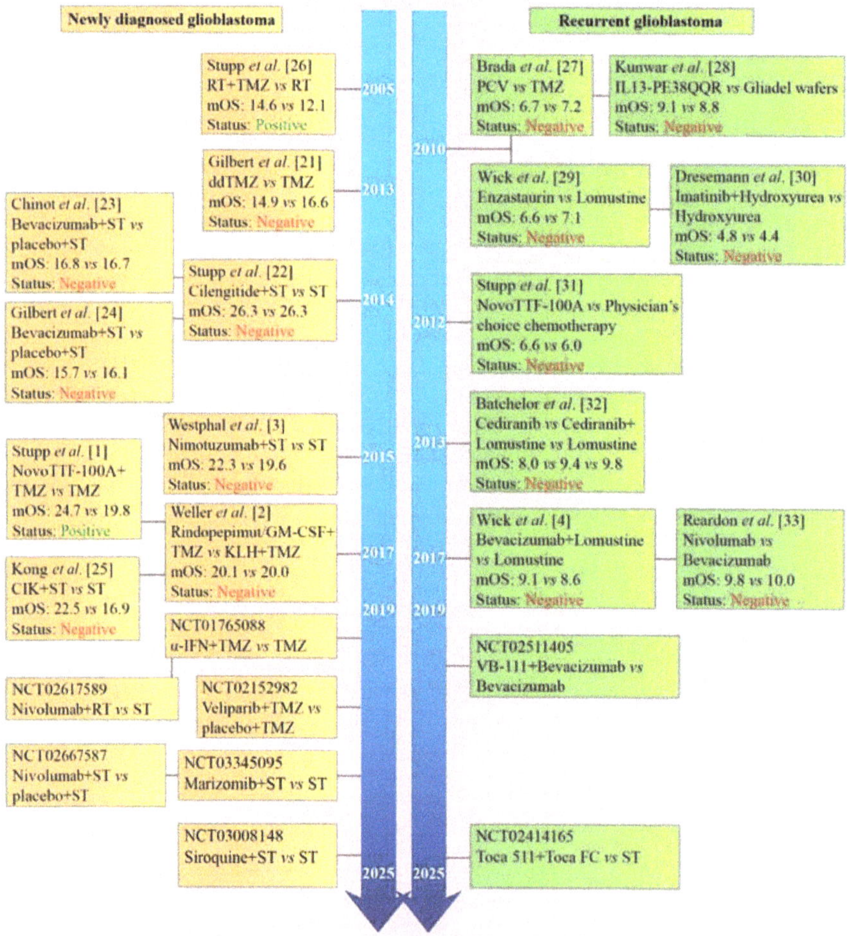

Figure 1. Timeline of phase III clinical trials in patients with newly diagnosed or recurrent glioblastoma. Almost all trials have been negative and failed to predict positive outcomes of the preceding phase II trials. α-IFN: interferon-alpha; Bevacizumab: anti-vascular endothelial growth factor (VEGF) antibody; Cediranib: an inhibitor of VEGF receptor; CIK: autologous cytokine-induced killer cells; Enzastaurin: an inhibitor of protein kinase Cβ (PKCβ, as well as PKCα, PKCγ, and PKCε at higher concentrations);

IL13-PE38QQR, also known as Cintredekin besudotox: a recombinant chimeric cytotoxin composed of human interleukin 13 (IL-13) fused to a truncated, mutated form of Pseudomonas aeruginosa exotoxin A (PE38QQR); Marizomib: an irreversible proteasome inhibitor; mOS: median overall survival; Nimotuzumab: an anti-epidermal growth factor receptor (EGFR) antibody; Nivolumab: an anti-programmed cell death protein 1 (PD-1) antibody; NovoTTF-100A System™, or Optune™ generates Tumor Treating Fields (TTFields); PCV: procarbazine, lomustine, and vincristine; Rapamycin: an inhibitor of the mechanistic target of rapamycin (mTOR) protein kinase; RT: radiotherapy; Siroquine: sirolimus (rapamycin) plus hydroxychloroquine sulfate; ST: standard therapy; TMZ/ddTMZ: temozolomide/dose dense temozolomide; Toca 511: a gammaretroviral replicating vector encoding cytosine deaminase that converts the antifungal drug 5-fluorocytosine (FC) into the antineoplastic drug 5-fluorouracil (FU); VB-111: a non-replicating adenovirus 5 carrying a proapoptotic Fas-chimera transgene under the control of an endothelial cell-specific promoter; Veliparib: a poly(ADP-ribose) polymerase (PARP) 1 and 2 inhibitor. Cited literature: [1–4,21–33].

2. Standard Therapy for Adult Glioblastoma: 15 Years of Experience and TTFields

Since 2003, only two chemotherapeutic agents were approved as first-line drugs for the treatment of newly diagnosed glioblastoma: BCNU/carmustine (Gliadel® wafers intracranially implanted local chemotherapy composed of a biodegradable copolymer prolifeprospan 20 impregnated with the alkylating agent carmustine) in 2003 and the alkylating agent temozolomide/TMZ (Temodal®, oral systemic chemotherapy) in 2005 [34,35]. However, the widespread use of Gliadel wafers was limited for different reasons including contradictory survival benefit and a high complication rate [35–39]. The current standard first-line treatment for adult patients (<65 years of age) with newly diagnosed glioblastoma is maximum safe tumor resection, followed by concurrent radiotherapy (RT, 60 Gy delivered in 30 fractions, five times a week for 6 weeks, fractions of 2 Gy each) and daily oral TMZ (75 mg/m^2 per day, given 7 days per week), and then, after 4 treatment-free weeks, adjuvant TMZ up to 6 cycles (150–200 mg/m^2 per day, for five consecutive days, 28-day cycle) [40]. In a landmark phase III trial, this treatment regimen (the Stupp protocol) resulted in a median overall survival of 14.6 months in the RT-TMZ group (n = 287) versus 12.1 months in the RT only group (n = 286), 12.6 months versus 11.8 months in the *MGMT*-unmethylated subgroup, and 23.4 versus 15.3 months in the *MGMT*-methylated subgroup [40]. Since then, it has been established that, in addition to patient age and Karnofsky Performance Status, an extent of tumor resection is an independent prognostic factor for survival [41–43]; the use of anesthetic isoflurane, desflurane or a propofol infusion during glioblastoma surgery was not associated with overall survival [44]; irradiation doses above 60 Gy did not result in any survival prolongation regardless of exploited RT technique [45], and more than six cycles of TMZ also did not increase overall survival, including the *MGMT*-methylated subgroup [46–48], although no conclusive evidence exists [49], highlighting the need of a prospective randomized controlled trial to reconcile contradictory conclusions. A randomized, multicenter phase IIB trial of TMZ 12 cycles versus 6 cycles in patients with glioblastoma is ongoing (NCT02209948).

In addition to standard RT-TMZ, many patients routinely receive the standard corticosteroid dexamethasone to control peritumoral vasogenic cerebral edema and ameliorate neurological symptoms [50,51]. In clinical practice, steroids are administered more often to patients with aggressive tumor growth. Dexamethasone use (steroid dependency) was associated with shorter survival in primary and recurrent glioblastoma patients [52–62]. Corticosteroids are known to cause many adverse systemic effects, including immunosuppression and lymphopenia, which was also independently associated with shorter survival [63–69]. Thus, the negative association between dexamethasone use (steroid dependency) and survival may be due to more aggressive tumor growth in patients (such patients receive dexamethasone more likely), or due to dexamethasone-induced immunosuppression, or both, or due to another unknown reason. Since there are no randomized trials prospectively evaluating an association between dexamethasone use (its total daily dose, duration of application) and survival, no definitive answer can exist.

Patients with glioblastoma also take anticonvulsants/anti-epileptic drugs (40–60% of patients in different studies) to reduce tumor-associated seizures [70–72]. Levetiracetam is now the most frequently prescribed drug for brain tumor-related epilepsy, followed by valproic acid [70,71]. Although multiple retrospective clinical studies stated improved outcome of patients with newly diagnosed glioblastoma after the addition of valproic acid [73–81] or levetiracetam [82] to standard therapy, while others did not reveal a significant effect on overall survival [70,75,83–87], the analysis of the prospective phase III clinical trials with a pooled cohort of 1869 patients demonstrated that levetiracetam or valproic acid did not influence median overall survival on multivariate analysis [88]. No significant impact on overall survival of patients with glioblastoma was also documented for other anti-epileptic drug [70].

Venous thromboembolism, including deep vein thrombosis and pulmonary embolism, is a common complication in patients with newly diagnosed glioblastoma, particularly in the first six months after diagnosis/during the course of treatment, associated with significant morbidity and shorter survival [89,90]. About 20–25% of patients require post-operative long-term curative treatment to manage symptomatic venous thromboembolism [91–93]. Low-molecular-weight heparin is preferred to other anticoagulants due to its excellent therapeutic index [89,90,92]. The application of low-molecular-weight heparin in glioblastoma patients did not affect overall survival in clinical trials [94].

Finally, prophylaxis against *Pneumocystis jiroveci* pneumonia is recommended for newly diagnosed glioblastoma patients receiving RT-TMZ, especially in combination with the chronic use of corticosteroids [95]. Interestingly, a retrospective analysis of 127 glioblastoma patients treated with standard therapy who did not receive prophylaxis against *Pneumocystis jiroveci* pneumonia revealed that only one patient suffered from pneumonia [96]. It was proposed to reconsider the administration of prophylactic drugs against pneumonia in every glioblastoma patient treated with RT-TMZ in favor of avoiding potentially unnecessary toxic prophylaxis [95].

Altogether, an extent of tumor resection, RT and TMZ are the main well-established modulators of patients' survival. In 2015, following a phase III trial (EF-14) [1], a new electric-physical cancer treatment modality (low-intensity, intermediate-frequency, alternating electric fields (TTFields) generated by the NovoTTF-100A device/Optune®) was approved by the FDA for the treatment of newly diagnosed glioblastoma patients [97,98]. In this trial, patients (median age 56 years) were randomized 2:1 to the TTFields plus adjuvant TMZ group (n = 466) or the TMZ only group (n = 229). Median overall survival was 20.9 months in the TTFields-TMZ group versus 16.0 months in the TMZ only group from randomization (plus median time from diagnosis to randomization 3.8 months). In exploratory analyses, the percentage of patients alive at 2 years (from randomization) was 43%, 26% at 3 years, and 13% at 5 years in the TTFields-TMZ group versus 31%, 16%, and 5%, respectively, in the TMZ only group. No significant differences in the incidence, distribution, and severity of systemic adverse effects were observed between groups [1]. The meta-analysis data of primary and recurrent glioblastoma patients (n = 1769) also indicated that the addition of TTFields to standard therapy was associated with a better median overall survival after 1 and 2 years [99]. However, due to relatively small numbers of 3-year survivors, a 3-year survival rate could not be estimated reliably for TTFields-treated patients [99]. Integrating EF-14 trial data with glioblastoma epidemiology data, Guzauskas et al. estimated the conditional survival rates at 3, 5, 10, and 15 years for the EF-14 trial patients alive at year 2 [100]. The authors concluded that patients alive at year 2 after starting TTFields with adjuvant TMZ had 59.6%, 29.4%, 20.7%, and 17.4% probability of surviving to year 3, 5, 10, and 15 versus 53.1%, 14.7%, 10.3%, and 8.7% probability of surviving for patients alive at year 2 after starting maintenance TMZ only [100]. These estimations of conditional survival should be confirmed by further monitoring survival of the EF-14 trial patients and in additional large randomized controlled TTFields trials with long follow-up.

3. Revision of the Landmark Stupp Trial as a Historical Control for Median Overall Survival in Non-Controlled Clinical Trials

Over the last decade, the results of Stupp's EORTC/NCIC study [40] have been considered a historical control in non-controlled phase II trials assessing the efficacy of investigational drugs and for making phase III go/no-go decision. However, the follow-up phase III trials in adult patients with newly diagnosed glioblastoma demonstrated a trending increase in median overall survival in the control cohorts/arms receiving standard therapy (from ≈14 to ≈20.0) (Table 1). In addition, a Korean single-institution retrospective report on outcomes of 252 patients with newly diagnosed glioblastoma who received standard therapy between 2005 and 2013 indicated that median overall survival was 20.8 months [101]. It has been suggested that a general trend in increase of median overall survival (at least >2–4 months) is a result of more aggressive glioma management, improvement in surgery, RT, and toxicity management rather than a selection bias [3,45]. However, a trend of increasing median overall survival has not translated into an increase in the 3- or 5-year survival rates so far (Table 2). Importantly, an increase in median overall survival in the control cohorts/arms is one of the reasons why all positive non-controlled (compared with a historical control) phase II trials of investigational treatments in glioblastoma failed to predict positive phase III trials [2,102–105]. To the point, about 85% of glioblastoma trials registered in ClinicalTrials.gov and conducted from 2005 to 2016 were non-controlled [106]. Altogether, the landmark Stupp's trial as a historical control for median overall survival in non-controlled trials should be revisited. The available data highlight the importance of designing controlled randomized phase II clinical trials due to the failure to an adequate estimate of therapeutic efficacy based on a historical control even after matching for patients eligibility [104,105].

Table 1. A trending increase in median overall survival in the control cohorts/arms receiving standard therapy in phase III clinical trials.

Trial	Number of Patients in the Control Arm	Median Overall Survival in the Landmark Stupp Study and in the Control Standard Therapy Arms/Cohorts of the Follow-Up Phase III Trials	Year of Publication	References
TMZ + RT versus RT	287	14.6 months; 12.6 months (*MGMT*-unmethylated subgroup), 23.4 months (*MGMT*-methylated subgroup)	2009	[40]
Dose-dense TMZ versus standard TMZ	411	16.6 months; 14.6 months (*MGMT*-unmethylated subgroup); 21.4 months (*MGMT*-methylated subgroup)	2013	[21]
Cilengitide + ST versus ST	273	26.3 months (*MGMT* methylated cohort)	2014	[22]
Bevacizumab + ST versus ST + placebo	463	16.7 months	2014	[23]
Bevacizumab + ST versus ST + placebo	317	16.1 months; in the pooled analysis of both arms (n = 621): 14.3 months (*MGMT*-unmethylated cohort); 23.2 months (*MGMT*-methylated cohort)	2014	[24]
Nimotuzumab + ST versus ST	71	19.6 months; 15.5 months (*MGMT*-unmethylated subgroup); 33.8 months (*MGMT*-methylated subgroup)	2015	[3]
Rindopepimut/GM-CSF + TMZ versus KLH + TMZ	374	20.0 months	2017	[2]
TTFields + TMZ versus TMZ	229	19.8 months (16.0 months from randomization plus median time from diagnosis to randomization 3.8 months)	2017	[1]

* Bevacizumab: anti-VEGF antibody; cilengitide: αVβ3/αVβ5 integrin inhibitor; KLH: the keyhole limpet hemocyanin, a large copper-containing immunogenic carrier glycoprotein; nimotuzumab: anti-EGFR antibody; rindopepimut: anti-EGFRvIII vaccine; ST: standard therapy; TMZ: temozolomide; TTFields: tumor-treating alternating electric fields.

Table 2. The 2-, 3-, 5-, and 10-year overall survival (OS) rates for glioblastoma patients in clinical trials, registries and systematic reviews.

Patient Groups/Registries	2-Year OS Rate	3-Year OS Rate	5-Year OS Rate	10-Year OS Rate	References
RT-TMZ group versus RT only group	27.2% vs. 10.9%	16.0% vs. 4.4%	9.8% vs. 1.9%		[40]
RT-TMZ group versus RT only group in MGMT-unmethylated subgroup	14.8% vs. 1.8%	11.1% vs. 0%	8.3% vs. 0%		
RT-TMZ group versus RT only group in MGMT-methylated subgroup	48.9% vs. 23.9%	27.6% vs. 7.8%	13.8% vs. 5.2%		
RT-TMZ MGMT-methylated group	56%				[22]
RT-TMZ group (exploratory analysis)	31%	16%	5%		[1]
RT-TMZ plus placebo group	30.1%				[23]
The National Cancer Institute's SEER Program (1985–2005, n = 5991)	9.5%	5.4%	3.6%	2.9%	[107]
The National Cancer Institute's SEER Program (2005–2007)	24%				[108]
The Central Brain Tumor Registry of the United States (CBTRUS) (1995–2011, n = 30611)	14.8%	8.7%	5%	2.6%	[109]
Systematic reviews		2-5%		<1%	[110]

4. No Molecularly Targeted Drug(s) for Glioblastoma on the Horizon

About 190 phase II and 25 phase III glioblastoma clinical trials were launched between 2005 and 2015. In total, 100 different agents (43 clinically approved and 57 with investigational status: 67 small molecules, 32 biologicals, and one unclassified substance) were tested in those trials [111]. The systematic reviews and meta-analyses studies evaluating efficacy of the addition of molecularly targeted drugs to RT or standard therapy for newly diagnosed or recurrent glioblastoma showed no improvement in overall survival but increased risks of severe adverse events [112–115]. An overview of 100 ongoing Phase I/II glioma chemotherapy trials is given in [116]. Comparing ongoing trials with 29 phase I/II trials published in 2011, it was found that there is an increase in the number of trials using two drugs (from 24.1% to 44.9%) and an increase in the number of drugs able to pass the blood–brain barrier (7.14% versus 64.29%) [116].

It is worth mentioning a currently ongoing individualized screening trial of innovative glioblastoma therapy (INSIGhT) designed as a randomized, multi-arm phase II trial for patients with newly diagnosed glioblastoma and unmethylated *MGMT* promoter (NCT02977780) [117]. In this trial, patients are assigned to experimental arms according to their specific genetic aberrations. INSIGhT compares experimental arms to standard therapy. Three experimental arms consist of neratinib (EGFR, HER2, and HER4 inhibitor), abemaciclib (CDK4/6 inhibitor) or CC-115 (TORC1/2 and DNA-PK inhibitor) added to radio- or radiochemotherapy [117]. In addition, a randomized controlled phase III trial in patients with recurrent glioblastoma comparing standard chemotherapy versus chemotherapy chosen by cancer stem cell chemosensitivity testing by the ChemoID drug response assay (NCT03632135) has been launched. However, the inspiring concept of personalized medicine based on a patient-specific combination of targeted drugs has been challenged recently [118–120]. In the preliminary proof-of-concept trials, efficacy of targeted therapy matched to genomic alterations has not been proved in advanced carcinomas [121,122]. Thus, targeted drugs matched to specific mutant kinases might also not result in any benefit in personalized precision medicine trials in glioblastoma patients.

Glioblastoma is highly vascularized, critically dependent on angiogenesis brain neoplasm that provides a rationale for targeting a formation of blood vessels. However, in two phase III trials (RTOG 0825 and AVAglio), the addition of bevacizumab, a humanized monoclonal antibody targeting vascular endothelial growth factor (VEGF), to standard therapy for newly diagnosed glioblastoma failed to demonstrate improvement in overall survival [23,24]. Further, in a randomized phase II trial of hypofractionated RT (40 Gy in 15 fractions) with bevacizumab (n = 50) or without bevacizumab (n = 25) in elderly patients (≥65 years) with newly diagnosed glioblastoma, overall survival in two arms was similar (12.1 versus 12.2 months, ARTE trial, NCT01443676) [123]. In general, a meta-analysis of fourteen randomized clinical trials demonstrated that seven tested drugs with antiangiogenic potential did not improve overall survival in glioblastoma patients, either as first or second-line treatment, and either as single agent or in combination with conventional chemotherapy [124].

Beyond its costly purpose for brain edema reduction, the role of bevacizumab as a therapeutic anti-tumor agent remains uncertain even for recurrent glioblastoma [125,126]. Bevacizumab was approved by the FDA for the treatment of recurrent glioblastoma relying upon the results of two non-controlled phase II trials without the completion of a controlled randomized phase III trial [127]. In contrast, the European Medicines Agency has not approved bevacizumab, since a phase III trial comparing lomustine plus bevacizumab versus lomustine in progressive glioblastoma revealed no significant difference between groups (9.1 versus 8.6 months, respectively) [4]. The systematic review and meta-analysis of randomized controlled studies combining bevacizumab with chemotherapy versus single-agent therapy in recurrent glioblastoma indicated no overall survival benefit from combination [5,128]. In a randomized controlled phase II trial (n = 155, TAVAREC trial), the addition of bevacizumab to TMZ in patients with first recurrence of grade II/III glioma without 1p/19q co-deletion did not also improve overall survival [129]. In different randomized trials, patients with recurrent glioblastoma receiving bevacizumab alone had a median overall survival of about 7–10 months from recurrence, and this efficacy was comparable to lomustine monotherapy [5–7]. Moreover, in the retrospective studies, hypofractionated stereotactic re-irradiation therapy alone also demonstrated comparable survival benefit (a median survival time of about 9–11 months from recurrence) [130–133]. These observations are corroborated by data from a randomized phase II trial of APG101 (a CD95 ligand-binding fusion protein) plus reirradiation versus reirradiation only in progressive glioblastoma (a median overall survival of 11.5 months in each group) [134]. A recent retrospective study suggests that a combination of bevacizumab and re-irradiation (fractionated stereotactic RT) for progressive or recurrent high-grade gliomas may moderately increase median overall survival (>13 months from recurrence) [135]. A randomized phase II trial of concurrent bevacizumab and re-irradiation versus bevacizumab only for recurrent glioblastoma patients is ongoing (NCT01730950). In addition, several trials including NovoTTF-100A with bevacizumab (NCT01894061) and NovoTTF-100A with bevacizumab and hypofractionated stereotactic irradiation (NCT01925573) for patients with recurrent glioblastoma have been launched. The results of a phase II study of pembrolizumab (anti-programmed cell death protein 1 antibody, PD-1) with and without bevacizumab for recurrent glioblastoma (NCT02337491) are awaited. Finally, a phase II non-randomized trial of pembrolizumab and reirradiation in bevacizumab-naïve and bevacizumab-resistant recurrent glioblastoma has been announced (NCT03661723). It should be noted that after tumor progression on bevacizumab, there is no effective therapeutic option, and an estimated median overall survival on bevacizumab progression and in post-bevacizumab salvage studies is 3.36 months and 4.46 months, respectively [7].

There are many ongoing clinical trials with immune checkpoint inhibitors in patients with primary and recurrent glioma/glioblastoma [136,137]. In a large randomized clinical trial for recurrent glioblastoma (CheckMate 143, NCT02017717), the anti-PD-1 antibody, nivolumab, did not demonstrate better efficacy in comparison to bevacizumab [6,136]. Two phase III trials comparing nivolumab versus TMZ, each in combination with RT, in patients with newly diagnosed *MGMT*-unmethylated glioblastoma (CheckMate-498, NCT02617589) and TMZ plus RT combined with nivolumab or placebo in patients with newly diagnosed *MGMT*-methylated glioblastoma (CheckMate-548, NCT02667587) are ongoing. There is emerging evidence that cancer patients with high tumor mutational load (a total number of nonsynonymous point mutations) or the total number of mutations per coding area and associated neoantigen burden show a much better response to immune checkpoint inhibitors. To evaluate the correlation between tumor mutational load and objective response rate, Yarchoan et al. plotted the objective response rate for PD-1 or anti-PD-L1 (PD ligand 1) therapy against median tumor mutational load across 27 tumor types/subtypes. The authors observed a strong correlation between them, with glioma/glioblastoma predicted to be among tumor types with the lowest chances to respond [138]. Although the tumor mutational load is correlated with response to immune checkpoint inhibitors, it is neither necessary nor sufficient to drive it [139–141]. In case studies, glioblastoma patients with DNA repair deficiency, which results in increased tumor mutational load, demonstrated significant clinical and immunological responses to immune checkpoint inhibition [142,143]. However, the analysis of the tumor mutational load, mismatch repair (MMR) and immune checkpoint expression in glioblastoma (n = 198) revealed that only 3.5% of glioblastoma samples (seven of 198) had high tumor mutational load (DNA MMR mutations) associated with the loss of MLH1, MSH2, MSH6, and/or PMS2 expression [144]. Neither glioblastomas with high and moderate tumor mutational load nor IDH1 mutant gliomas exhibited increased PD-1+ T cell infiltrate or PD-L1 expression by tumor cells in comparison to samples with low tumor mutational load [144]. This analysis suggests that only some glioblastoma patients (mainly with DNA repair deficiency) might benefit from immune checkpoint inhibitors.

Altogether, an extremely high rate of failures in clinical trials and a lack of effective molecularly targeted drug(s) on the horizon have encouraged the development of conceptually distinct therapeutic approaches such as cellular immunotherapy and oncolytic virotherapy.

5. Dendritic Cell/Peptide Vaccines and CAR T-cells for Glioblastoma Treatment: A Need for Large Controlled Trials to Prove Efficacy

Systematic reviews and meta-analysis studies of phase I-II clinical trials demonstrated that the addition of dendritic cell vaccines to standard therapy improved the median overall survival and 2- and 3-year survival rates of patients with newly diagnosed or recurrent high-grade gliomas [145–149]. Recent early phase clinical trials largely supported these conclusions (Table 3). However, until now, encouraging results derived from small controlled or non-controlled and/or non-randomized early phase clinical trials of vaccines have not been confirmed in multicenter, controlled, randomized phase II/III trials (Table 3). Only minor subgroups of patients benefit from dendritic cell/peptide vaccines. Recently, Liau et al. reported the interim results of a large randomized controlled phase III trial of DCVax-L vaccine for newly diagnosed glioblastoma (NCT00045968) [20]. Patients were randomized 2:1 to standard therapy plus DCVax-L (n = 232) or standard therapy plus placebo (n = 99). However, 86.4% of patients received dendritic cell vaccine at some point during the trial because of the cross-over study design. DCVax-L was administered by intra-dermal injection in the arm, six times in year one and twice per year thereafter. Among the patients (n = 223) who have lived $\geq$30 months past their surgery, 67 (30.0%) have a median overall survival of 46.5 months. Among the patients (n = 182) who have lived $\geq$36 months past their surgery, 24.2% (n = 44) have a median overall survival of 88.2 months. For patients with methylated *MGMT* (n = 131), median overall survival was 34.7 months from surgery, with a 3-year survival rate of 46.4% [20]. Altogether, these preliminary results seem very encouraging. However, the presentation of these interim immature data derived from a highly selected patient population as well as the trial design have been criticized [150]. The currently ongoing or recently announced phase II/III trials of dendritic cell/peptide vaccines are presented in Table 4.

Table 3. Completed phase I–III clinical trials of vaccines for glioblastoma with primary outcomes.

Investigational Treatment versus Comparator Treatment	N of Patients	Newly Diagnosed/ Recurrent	Results for Primary Outcome	ClinicalTrials.gov Identifier	References
Phase III trials					
Rindopepimut * plus GM-CSF and TMZ versus KLH plus TMZ	745	Newly diagnosed	mOS: 20.1 versus 20.0 months (HR 1.01, 95% CI 0.79–1.30; $p = 0.93$)	NCT01480479	[2]
Autologous cytokine-induced killer cells plus ST versus ST	180	Newly diagnosed	mOS: 22.5 versus 16.9 months ($p = 0.5237$) mPFS: 8.1 versus 5.4 months (HR 0.693, 90% CI 0.512–0.937, $p = 0.0218$)	NA	[25]
Autologous DC * vaccine versus autologous PBMC	331	Newly diagnosed	Pending	NCT00045968	[20]
Phase II trials					
Rindopepimut plus GM-CSF and standard or dose-intensified TMZ versus a historical control	22	Newly diagnosed	mOS: 23.6 versus 15.0 months (HR = 0.23; 95% CI 0.07–0.79; $p = 0.019$); mPFS: 15.2 versus 6.3 months (HR = 0.35, 95% CI 0.14–0.87; $p = 0.024$)	NCT00643097	[151]
Rindopepimut plus GM-CSF and adjuvant TMZ	65	Newly diagnosed	mOS: 21.8 months; mPFS: 9.2 months	NCT00458601	[152]
Rindopepimut and GM-CSF plus bevacizumab versus KLH plus bevacizumab	73	Recurrent	mOS: 11.6 versus 9.3 months (HR = 0.57, 95% CI 0.33–0.98, $p = 0.039$)	NCT01498328	[153]
ICT-107 * versus unpulsed autologous DC vaccine	124	Newly diagnosed	mOS: 18.3 versus 16.7 months ($p > 0.05$); PFS: 11.2 versus 9.0 months ($p = 0.010$)	NCT01280552	[154]
Autologous DC vaccine plus ST versus ST	76	Newly diagnosed	mOS: 564 versus 568 days ($p = 0.99$); PFS: 204 versus 210 days ($p = 0.83$)	EudraCT number 2009-015979-27	[155]
Autologous DC vaccine plus ST versus ST	34	Newly diagnosed	mOS: 31.9 versus 15.0 months ($p < 0.002$); mPFS: 8.5 versus 8.0 months ($p = 0.075$)	NA	[156]
Autologous DC vaccine plus ST	27	Newly diagnosed	mOS: 23.4 months; mPFS: 12.7 months	NCT01006044	[157]
Autologous DC vaccine plus ST versus ST plus placebo	43	Newly diagnosed or recurrent	mOS: 13.7 versus 10.7 months ($p = 0.05$); mPFS: 7.7 versus 6.9 months ($p = 0.75$)	NA	[158]
HSPPC-96 * plus TMZ	46	Newly diagnosed	mOS: 23.8 months; mPFS: 18.0 months	NCT00905060	[17]
ERC1671 */GM-CSF/cyclophosphamide plus bevacizumab versus placebo plus bevacizumab	9	Recurrent	Interim mOS: 12.0 versus 7.5 months	NCT01903330	[159]
LAK cells *	33	Newly diagnosed	mOS: 20.5 months	NCT00331526	[160]
Phase I and I/II trials					
IMA950 * vaccine with poly ICLC plus ST	16	Newly diagnosed	mOS: 21.2 months	NCT01920191	[161]
Autologous DC vaccine plus ST	23	Newly diagnosed	mOS: 31.4 months	NA	[162]
Autologous DC vaccine plus ST versus ST	25	Newly diagnosed	mOS: 17.0 versus 10.5 months ($p < 0.05$)	NA	[163]
Autologous DC vaccine plus ST versus a historical control	11	Newly diagnosed	mOS: 759 days versus 585 days	NCT00846456	[164]
Autologous DC vaccine plus ST	77	Newly diagnosed	mOS: 18.3 months in ITT analysis; mPFS: 10.4 months in the ITT group versus 20.4 months in the PP group	NA	[165]

Table 3. *Cont.*

Investigational Treatment versus Comparator Treatment	N of Patients	Newly Diagnosed/ Recurrent	Results for Primary Outcome	ClinicalTrials.gov Identifier	References
		Phase I and I/II trials			
Autologous DC vaccine versus RT plus nitrosourea	45	Recurrent	mOS: 480 versus 400 days (p = 0.010)	NA	[166]
Autologous DC vaccine	56	Recurrent	mOS: 9.6 months; mPFS: 3 months	NA	[167]
Autologous DC vaccine pulsed with pp65 RNA plus tetanus/diphtheria (Td) toxoid or unpulsed autologous DCs	12	Newly diagnosed	mOS: 18.5 months; mPFS: 10.8 months	NA	[8]
Autologous DC vaccine pulsed with pp65 RNA plus GM-CSF and dose-intensified TMZ	11	Newly diagnosed	mOS: 41.1 months; mPFS: 25.3 months	NA	[19]
HSPPC-96 vaccine plus ST	20	Newly diagnosed	mOS: 31.4 months	NA	[18]

* DC: dendritic cells; ERC1671 vaccine is composed of primary irradiated/inactivated whole tumor cells and lysates from the patient to be treated and from three other allogeneic and autologous glioblastoma patients; GM-CSF: granulocyte-macrophage colony-stimulating factor; Rindopepimut, or CDX-110-KLH peptide vaccine, presents a 14-amino acid peptide corresponding to the fusion junction of EGFRvIII, which is linked to the keyhole limpet hemocyanin (KLH), a large copper-containing immunogenic carrier glycoprotein; HSPPC-96 peptide vaccine is comprised of autologous antigenic peptides chaperoned by heat shock glycoprotein-96; ICLC: an immunostimulating adjuvant consisting of double-stranded RNAs of polyinosinic-polycytidylic acid stabilized with poly L-lysine in carboxymethylcellulose; ICT-107 vaccine presents an autologous dendritic cell vaccine pulsed with peptides from six glioma-associated antigens (HER2, TRP-2, gp100, MAGE-1, IL13Rα2, and AIM-2); IMA950 peptide vaccine consists of 11 glioma-associated peptides; ITT: intention-to-treat; LAK cells: lymphokine-activated killer cells; mOS: median overall survival; mPFS: median progression-free survival; PP: per protocol; ST: standard therapy; TMZ: temozolomide.

Table 4. Ongoing phase II/III clinical trials of dendritic cell/peptide vaccines in newly diagnosed or recurrent glioblastoma patients.

ClinicalTrials.gov Identifier	Trial Title	Estimated Sample Size	Satus
NCT03395587	Phase II multicenter open label, randomized trial of vaccination with lysate-loaded, mature dendritic cells integrated into standard therapy in newly diagnosed glioblastoma (GlioVax)	136	Recruiting
NCT02465268	A phase II randomized, blinded, and placebo-controlled trial of CMV RNA-pulsed dendritic cells with tetanus-diphtheria toxoid vaccine in patients with newly diagnosed glioblastoma (ATTAC-II)	150	Recruiting
NCT02366728	A randomized phase II study of evaluation of overcoming limited migration and enhancing cytomegalovirus (CMV)-specific dendritic cell vaccines with adjuvant tetanus pre-conditioning in patients with newly diagnosed glioblastoma	100	Active, not recruiting
NCT03548571	Open label randomized phase II/III trial of dendritic cell immunotherapy against cancer stem cells in glioblastoma patients receiving standard therapy (DEN-STEM)	60	Recruiting
NCT03018288	A randomized, double blind phase II trial of surgery, RT plus TMZ and pembrolizumab with and without heat shock protein-peptide complex-96 (HSPPC-96) in newly diagnosed glioblastoma	108	Recruiting
NCT01814813	A phase II randomized trial comparing the efficacy of heat shock protein–peptide complex-96 (HSPPC-96) vaccine given with bevacizumab versus bevacizumab alone in the treatment of surgically resectable recurrent glioblastoma	90	Active, not recruiting
NCT02455557	A phase II Study of the safety and efficacy of SVN53-67/M57-KLH (SurVaxM) peptide vaccine in survinin-positive newly diagnosed glioblastoma	64	Active, not recruiting
NCT01204684	A phase II clinical trial evaluating autologous dendritic cells pulsed with tumor lysate antigen +/− toll-like receptor agonists for the treatment of malignant glioma	60	Active, not recruiting
NCT02799238	An open label, randomized, phase II study to investigate the efficacy and safety of autologous lymphoid effector cells specific against tumor (ALECSAT) treatment as an add-on therapy to RT-TMZ in patients with newly diagnosed glioblastoma	87	Recruiting
NCT03650257	A large-scale research for immunotherapy of glioblastoma with autologous heat shock protein gp96	150	Not yet recruiting

Genetically engineered chimeric antigen receptor-expressing T-cells (CAR T-cells) present recently advanced immunotherapy technology, which showed significant antiglioma activity in preclinical models. Initial experience with CAR T-cells targeting EGFRvIII, epidermal growth factor receptor 2 (HER2), and interleukin 13 receptor α2 (IL13Rα2) has demonstrated their safety and antitumor activity in some glioblastoma patients [168–170]. However, it is too early to judge about the efficacy of CAR T-cell immune-therapeutics for which heterogeneous antigen expression and the immunosuppressive tumor microenvironment are considered the major barriers [171,172]. Glioblastoma antigens that are targeted by CAR T-cell therapy in ongoing clinical trials include EGFRvIII (NCT01454596, NCT02209376, NCT02844062, and NCT03283631), HER2 (NCT02442297, NCT01109095, and NCT03389230), IL-13Rα2 (NCT02208362), ephrin type-A receptor 2 (EphA2) (NCT02575261), and programmed death-ligand 1 (PD-L1) (NCT02937844).

6. Oncolytic Virotherapy for Glioma/Glioblastoma Treatment at Recurrence: Feasibility and Safety in Phase I Trials with Promising Efficacy in Subsets of Patients

Different oncolytic viruses have been tested in progressive/recurrent glioblastoma/glioma and proved feasibility and safety, but not efficacy, in terms of median overall survival in randomized trials until now [173]. Here, we discuss the oncolytic viruses that have been recently advanced to phase I/II trials in recurrent glioma patients and demonstrated remarkable efficacy in subsets of patients.

DNX-2401 (Ad5-Delta-24-RGD; tasadenoturev) is an infectivity-enhanced, replication-competent, tumor-selective oncolytic adenovirus 5 (Ad5)-based vector [174]. In a phase I dose-escalation trial of DNX-2401 in 37 patients with recurrent malignant glioma [13], 25 patients received a single intratumoral injection (eight dose levels: 1×10^7–3×10^{10} vp in 1 mL) of DNX-2401 through the biopsy needle into recurrent tumor, while 12 patients underwent intratumoral injection (1×10^7–3×10^8 vp) through a permanently implanted catheter to mark the injection site. Two weeks later, the tumor and catheter were resected. In a first cohort, 20% of patients survived >3 years from treatment and three patients showed >3 years of progression-free survival from treatment. The analysis of post-treatment surgical specimens from a second cohort revealed that DNX-2401 replicated and spread within the tumor and induced CD8+ and T-bet+ cells infiltration. However, some patients did not demonstrate evidence of infection and therapeutic response. No dose-limiting toxicities were observed and no maximum tolerated dose was identified in this trial [13]. In a randomized phase Ib trial of DNX-2401 versus DNX-2401 plus interferon gamma (IFN-γ) for recurrent glioblastoma (first or second recurrence, n = 27; TARGET-I trial, NCT02197169), the 1- and 1.5-year overall survival rates for all patients enrolled (regardless of treatment assignment) was 33% and 22%, respectively, and the addition of IFN γ did not improve survival upon a preliminary intent-to-treat analysis [175]. The combination of DNX-2401 delivered directly into the tumor with anti-PD-1 antibodies (pembrolizumab) administered intravenously is under evaluation in a phase II trial for recurrent glioblastoma/gliosarcoma (CAPTIVE/KEYNOTE-192, NCT02798406). Further, a phase I trial (NCT03330197) of intratumoral Ad-RTS-hIL-12, an inducible adenoviral vector engineered to express human interleukin 12 (hIL-12) in the presence of the activator ligand veledimex in pediatric patients with recurrent or progressive grade III/IV glioma (n = 25) demonstrated good tolerability of controlled local IL-12 expression [176]. A similar phase I trial in adults with glioblastoma/glioma is ongoing (NCT02026271).

The polio–rhinovirus chimera (PVSRIPO) targeting the poliovirus receptor CD155 represents a replication-competent attenuated poliovirus type 1 (Sabin) with its internal ribosome entry site substituted for that of human rhinovirus type 2. This internal ribosome entry site replacement ablates neurovirulence of PVSRIPO preventing from propagation in neurons. In a phase I trial (NCT01491893) with a dose-escalation phase (ranging between 10^7–10^{10} 50% tissue-culture infectious doses ($TCID_{50}$) and then a dose-expansion phase (5.0×10^7 $TCID_{50}$), a total of 61 recurrent supratentorial grade IV malignant glioma patients were treated with intratumoral infusion of PVSRIPO by convection-enhanced delivery via a catheter [14]. Sixty nine percent of the patients had grade 1 or 2 PVSRIPO infusion-related adverse events. In the dose-expansion phase, 19% of the patients had grade 3 or higher PVSRIPO-related adverse events. No neuropathogenicity or virus shedding was observed. Median overall survival among all 61 patients was 12.5 months. Overall survival reached a plateau at 24 months, with overall survival rate being 21% at 24 and 36 months. A few patients remained alive >57–70 months after the PVSRIPO infusion. It should be noted that some patients received additional treatments after the administration of PVSRIPO: 37 patients were treated with TMZ, lomustine, or other agents and 34 patients received bevacizumab to mitigate peritumoral inflammation [14]. A randomized phase II trial of PVSRIPO only or in combination with single-cycle lomustine in patients with recurrent grade IV malignant glioma (NCT02986178) is ongoing.

Oncolytic H-1 parvovirus (ParvOryx) whose natural host is the rat was tested in a phase I/IIa dose-escalating trial with different routes of administration in patients with recurrent glioblastoma [15]. Eighteen patients were enrolled. ParvOryx was administered via intratumoral or intravenous injection, then tumors were resected 9 days after treatment, and parvovirus was re-administered around the resection cavity. Median overall survival was 464 days (≈15.5 months) after first ParvOryx treatment. Eight patients survived >12 months and three patients >24 months after first administration of ParvOryx. Clinical response did not depend on the dose or route of ParvOryx administration. A maximum tolerated dose was not identified. Tumors from six ParvOryx-treated patients displayed strong CD8+ and CD4+ T lymphocytes infiltration [15].

Toca 511 *is* a retroviral replication-competent vector based on murine leukemia virus encoding the yeast cytosine deaminase that converts the antifungal drug 5-fluorocytosine into the antineoplastic drug 5-fluorouracil. In a phase I trial (NCT01470794) of Toca 511 injected into resection cavity of patients with recurrent high-grade gliomas (n = 56), followed by cycles of oral 5-fluorocytosine, 23 patients matched the recommended phase III Toca 511 dose [16]. In this patient subgroup (the phase III-eligible subgroup), which included both IDH1-mutant and wild type tumors, median overall survival was 14.4 months, the 1- and 2-year survival rates were 65.2% (15/23) and 34.8% (8/23). The estimated probability of a 3-year survival rate was 26.1% (6/23). Five patients demonstrated complete response and have been alive 33.9–52.2 months after Toca 511 administration [16]. A randomized phase II/III trial of Toca 511 combined with 5-fluorocytosine versus standard of care in patients undergoing planned resection for recurrent glioblastoma or anaplastic astrocytoma is ongoing (NCT02414165). Several other ongoing clinical trials in adult patients with recurrent glioblastoma/glioma include: a phase I/II trial of oncolytic vaccinia virus TG6002 combined with flucytosine (ONCOVIRAC, NCT03294486); a phase I trial of a measles virus derivative producing carcinoembryonic antigen (NCT00390299); a phase I trial of M032, a genetically engineered herpes simplex virus (HSV-1) expressing IL-12 (NCT02062827); and a phase I trial of a genetically engineered HSV-1, rQNestin34.5v.2, with cyclophosphamide (NCT03152318).

In general, early-phase clinical trials discussed above demonstrated that oncolytic virotherapy might markedly improve survival in subsets of patients. However, a pooled analysis of recent virotherapy trials for recurrent glioblastoma revealed that the 2- and 3-year survival rates were comparable to non-virotherapy clinical trials (2-year survival: 15% versus 12%; 3-year survival rate: 9% versus 6%) [177]. Thus, a benefit of oncolytic virotherapy has yet to be proven in the large randomized controlled phase II/III trials.

7. Is a Benefit Derived from Immunotherapy/Oncolytic Virotherapy Correlated with a Degree of Immunosuppression?

Only subsets of patients benefit from vaccination/oncolytic virotherapy. In clinical trials, immunotherapies/oncolytic viruses are assessed concurrently with or after standard therapy. Importantly, arguing against its putative positive immunomodulatory role, standard therapy induces systemic immunosuppression and long-lasting severe lymphopenia [178–187], and may interfere with the immunotherapy/oncolytic virotherapy efficacy, which is critically dependent on the activity of the host's own immune cells [10,159,188–197]. In support of this, recent studies have reported that high blood CD3+/CD4+ T cells counts [159] and tumor-infiltrating lymphocyte density [188] were correlated with better overall survival in glioblastoma patients receiving dendritic cell vaccination, while adjuvant TMZ hampered a CD8+ T cell count increase and the generation of CD8+ T cell-associated antitumor memory promoted by dendritic cell vaccination [198]. Standard therapy differentially affects the immune system of each patient [198–200], and patients with less severe standard therapy-induced immune suppression might derive more benefit from immunotherapy/oncolytic virotherapy than severely immunosuppressed patients. Detailed blood/tumor immunophenotyping should be incorporated in immunotherapy/oncolytic virotherapy trials to correlate immune cell subsets (counts, phenotype), immune responses to therapy and survival of patients to find immune-related predictive/prognostic markers.

8. Conclusions

Since 2003, only two chemotherapeutic agents have been approved for the treatment of newly diagnosed glioblastoma: BCNU/carmustine and temozolomide. In 2015, tumor-treating alternating electric fields (TTFields) generated by the NovoTTF-100A device was approved by the FDA as a new glioblastoma treatment modality concurrently with standard therapy. However, more randomized controlled studies with long follow-up are required to assess the real clinical efficiency of the addition of TTFields to standard therapy in terms of overall survival and the 2-, 3- and 5-year survival rates. To date, no cytotoxic chemotherapeutic agent, antibody, molecularly targeted drug or combinations of small molecule inhibitors have been demonstrated to be more efficient than TMZ or to increase survival when combined with standard therapy. The addition of vaccines or oncolytic virotherapy to standard therapy has markedly improved the survival of subsets of patients in early-phase clinical trials. Large controlled trials are required to prove the efficacy of next-generation immunotherapeutics and oncolytic vectors. Since immunotherapy efficacy critically depends on the activity of the host's own immune cells, blood cell counts and immunophenotyping may potentially help find immune-related predictive/prognostic markers in immunotherapy/oncolytic virotherapy trials.

Funding: The work was financially supported by the Russian Foundation for Basic Research (RFBR), Grant No. 18-29-01009.

Conflicts of Interest: The authors declare no conflict of interest.

References

1. Stupp, R.; Taillibert, S.; Kanner, A.; Read, W.; Steinberg, D.; Lhermitte, B.; Toms, S.; Idbaih, A.; Ahluwalia, M.S.; Fink, K.; et al. Effect of Tumor-Treating Fields Plus Maintenance Temozolomide vs Maintenance Temozolomide Alone on Survival in Patients with Glioblastoma: A Randomized Clinical Trial. *JAMA* **2017**, *318*, 2306–2316. [CrossRef] [PubMed]
2. Weller, M.; Butowski, N.; Tran, D.D.; Recht, L.D.; Lim, M.; Hirte, H.; Ashby, L.; Mechtler, L.; Goldlust, S.A.; Iwamoto, F.; et al. Rindopepimut with temozolomide for patients with newly diagnosed, EGFRvIII-expressing glioblastoma (ACT IV): A randomised, double-blind, international phase 3 trial. *Lancet Oncol.* **2017**, *18*, 1373–1385. [CrossRef]
3. Westphal, M.; Heese, O.; Steinbach, J.P.; Schnell, O.; Schackert, G.; Mehdorn, M.; Schulz, D.; Simon, M.; Schlegel, U.; Senft, C.; et al. A randomised, open label phase III trial with nimotuzumab, an anti-epidermal growth factor receptor monoclonal antibody in the treatment of newly diagnosed adult glioblastoma. *Eur. J. Cancer* **2015**, *51*, 522–532. [CrossRef] [PubMed]
4. Wick, W.; Gorlia, T.; Bendszus, M.; Taphoorn, M.; Sahm, F.; Harting, I.; Brandes, A.A.; Taal, W.; Domont, J.; Idbaih, A.; et al. Lomustine and Bevacizumab in Progressive Glioblastoma. *N. Engl. J. Med.* **2017**, *377*, 1954–1963. [CrossRef] [PubMed]
5. Seystahl, K.; Wick, W.; Weller, M. Therapeutic options in recurrent glioblastoma—An update. *Crit. Rev. Oncol. Hematol.* **2016**, *99*, 389–408. [CrossRef] [PubMed]

6. Reardon, D.A.; Omuro, A.; Brandes, A.A.; Rieger, J.; Wick, A.; Sepulveda, J.; Phuphanich, S.; de Souza, P.; Ahluwalia, M.S.; Lim, M.; et al. OS10.3 Randomized Phase 3 Study Evaluating the Efficacy and Safety of Nivolumab vs. Bevacizumab in Patients with Recurrent Glioblastoma: CheckMate 143. *Neuro Oncol.* **2017**, *19*, iii21. [CrossRef]
7. Tipping, M.; Eickhoff, J.; Ian Robins, H. Clinical outcomes in recurrent glioblastoma with bevacizumab therapy: An analysis of the literature. *J. Clin. Neurosci.* **2017**, *44*, 101–106. [CrossRef]
8. Mitchell, D.A.; Batich, K.A.; Gunn, M.D.; Huang, M.-N.; Sanchez-Perez, L.; Nair, S.K.; Congdon, K.L.; Reap, E.A.; Archer, G.E.; Desjardins, A.; et al. Tetanus toxoid and CCL3 improve dendritic cell vaccines in mice and glioblastoma patients. *Nature* **2015**, *519*, 366–369. [CrossRef]
9. Angelova, A.L.; Geletneky, K.; Nüesch, J.P.F.; Rommelaere, J. Tumor Selectivity of Oncolytic Parvoviruses: From in vitro and Animal Models to Cancer Patients. *Front. Bioeng. Biotechnol.* **2015**, *3*, 55. [CrossRef]
10. Garg, A.D.; Vandenberk, L.; Koks, C.; Verschuere, T.; Boon, L.; Van Gool, S.W.; Agostinis, P. Dendritic cell vaccines based on immunogenic cell death elicit danger signals and T cell-driven rejection of high-grade glioma. *Sci. Transl. Med.* **2016**, *8*, 328ra27. [CrossRef]
11. Stepanenko, A.A.; Chekhonin, V.P. A compendium of adenovirus genetic modifications for enhanced replication, oncolysis, and tumor immunosurveillance in cancer therapy. *Gene* **2018**, *679*, 11–18. [CrossRef] [PubMed]
12. Stepanenko, A.A.; Chekhonin, V.P. Tropism and transduction of oncolytic adenovirus 5 vectors in cancer therapy: Focus on fiber chimerism and mosaicism, hexon and pIX. *Virus Res.* **2018**, *257*, 40–51. [CrossRef] [PubMed]
13. Lang, F.F.; Conrad, C.; Gomez-Manzano, C.; Yung, W.K.A.; Sawaya, R.; Weinberg, J.S.; Prabhu, S.S.; Rao, G.; Fuller, G.N.; Aldape, K.D.; et al. Phase I Study of DNX-2401 (Delta-24-RGD) Oncolytic Adenovirus: Replication and Immunotherapeutic Effects in Recurrent Malignant Glioma. *J. Clin. Oncol.* **2018**. [CrossRef] [PubMed]
14. Desjardins, A.; Gromeier, M.; Herndon, J.E.; Beaubier, N.; Bolognesi, D.P.; Friedman, A.H.; Friedman, H.S.; McSherry, F.; Muscat, A.M.; Nair, S.; et al. Recurrent Glioblastoma Treated with Recombinant Poliovirus. *N. Engl. J. Med.* **2018**, *379*, 150–161. [CrossRef] [PubMed]
15. Geletneky, K.; Hajda, J.; Angelova, A.L.; Leuchs, B.; Capper, D.; Bartsch, A.J.; Neumann, J.-O.; Schöning, T.; Hüsing, J.; Beelte, B.; et al. Oncolytic H-1 Parvovirus Shows Safety and Signs of Immunogenic Activity in a First Phase I/IIa Glioblastoma Trial. *Mol. Ther.* **2017**, *25*, 2620–2634. [CrossRef] [PubMed]
16. Cloughesy, T.F.; Landolfi, J.; Vogelbaum, M.A.; Ostertag, D.; Elder, J.B.; Bloomfield, S.; Carter, B.; Chen, C.C.; Kalkanis, S.N.; Kesari, S.; et al. Durable complete responses in some recurrent high-grade glioma patients treated with Toca 511 + Toca FC. *Neuro Oncol.* **2018**, *20*, 1383–1392. [CrossRef]
17. Bloch, O.; Lim, M.; Sughrue, M.E.; Komotar, R.J.; Abrahams, J.M.; O'Rourke, D.M.; D'Ambrosio, A.; Bruce, J.N.; Parsa, A.T. Autologous Heat Shock Protein Peptide Vaccination for Newly Diagnosed Glioblastoma: Impact of Peripheral PD-L1 Expression on Response to Therapy. *Clin. Cancer Res.* **2017**, *23*, 3575–3584. [CrossRef]
18. Ji, N.; Zhang, Y.; Liu, Y.; Xie, J.; Wang, Y.; Hao, S.; Gao, Z. Heat shock protein peptide complex-96 vaccination for newly diagnosed glioblastoma: A phase I, single-arm trial. *JCI Insight* **2018**, *3*. [CrossRef]
19. Batich, K.A.; Reap, E.A.; Archer, G.E.; Sanchez-Perez, L.; Nair, S.K.; Schmittling, R.J.; Norberg, P.; Xie, W.; Herndon, J.E.; Healy, P.; et al. Long-term Survival in Glioblastoma with Cytomegalovirus pp65-Targeted Vaccination. *Clin. Cancer Res.* **2017**, *23*, 1898–1909. [CrossRef]
20. Liau, L.M.; Ashkan, K.; Tran, D.D.; Campian, J.L.; Trusheim, J.E.; Cobbs, C.S.; Heth, J.A.; Salacz, M.; Taylor, S.; D'Andre, S.D.; et al. First results on survival from a large Phase 3 clinical trial of an autologous dendritic cell vaccine in newly diagnosed glioblastoma. *J. Transl. Med.* **2018**, *16*, 142. [CrossRef]
21. Gilbert, M.R.; Wang, M.; Aldape, K.D.; Stupp, R.; Hegi, M.E.; Jaeckle, K.A.; Armstrong, T.S.; Wefel, J.S.; Won, M.; Blumenthal, D.T.; et al. Dose-Dense Temozolomide for Newly Diagnosed Glioblastoma: A Randomized Phase III Clinical Trial. *J. Clin. Oncol.* **2013**, *31*, 4085–4091. [CrossRef] [PubMed]
22. Stupp, R.; Hegi, M.E.; Gorlia, T.; Erridge, S.C.; Perry, J.; Hong, Y.-K.; Aldape, K.D.; Lhermitte, B.; Pietsch, T.; Grujicic, D.; et al. Cilengitide combined with standard treatment for patients with newly diagnosed glioblastoma with methylated MGMT promoter (CENTRIC EORTC 26071-22072 study): A multicentre, randomised, open-label, phase 3 trial. *Lancet Oncol.* **2014**, *15*, 1100–1108. [CrossRef]

23. Chinot, O.L.; Wick, W.; Mason, W.; Henriksson, R.; Saran, F.; Nishikawa, R.; Carpentier, A.F.; Hoang-Xuan, K.; Kavan, P.; Cernea, D.; et al. Bevacizumab plus Radiotherapy–Temozolomide for Newly Diagnosed Glioblastoma. *N. Engl. J. Med.* **2014**, *370*, 709–722. [CrossRef] [PubMed]
24. Gilbert, M.R.; Dignam, J.J.; Armstrong, T.S.; Wefel, J.S.; Blumenthal, D.T.; Vogelbaum, M.A.; Colman, H.; Chakravarti, A.; Pugh, S.; Won, M.; et al. A Randomized Trial of Bevacizumab for Newly Diagnosed Glioblastoma. *N. Engl. J. Med.* **2014**, *370*, 699–708. [CrossRef] [PubMed]
25. Kong, D.-S.; Nam, D.-H.; Kang, S.-H.; Lee, J.W.; Chang, J.-H.; Kim, J.-H.; Lim, Y.-J.; Koh, Y.-C.; Chung, Y.-G.; Kim, J.-M.; et al. Phase III randomized trial of autologous cytokine-induced killer cell immunotherapy for newly diagnosed glioblastoma in Korea. *Oncotarget* **2017**, *8*, 7003–7013. [CrossRef] [PubMed]
26. Stupp, R.; Mason, W.P.; van den Bent, M.J.; Weller, M.; Fisher, B.; Taphoorn, M.J.B.; Belanger, K.; Brandes, A.A.; Marosi, C.; Bogdahn, U.; et al. Radiotherapy plus Concomitant and Adjuvant Temozolomide for Glioblastoma. *N. Engl. J. Med.* **2005**, *352*, 987–996. [CrossRef] [PubMed]
27. Brada, M.; Stenning, S.; Gabe, R.; Thompson, L.C.; Levy, D.; Rampling, R.; Erridge, S.; Saran, F.; Gattamaneni, R.; Hopkins, K.; et al. Temozolomide Versus Procarbazine, Lomustine, and Vincristine in Recurrent High-Grade Glioma. *J. Clin. Oncol.* **2010**, *28*, 4601–4608. [CrossRef]
28. Kunwar, S.; Chang, S.; Westphal, M.; Vogelbaum, M.; Sampson, J.; Barnett, G.; Shaffrey, M.; Ram, Z.; Piepmeier, J.; Prados, M.; et al. Phase III randomized trial of CED of IL13-PE38QQR vs Gliadel wafers for recurrent glioblastoma. *Neuro. Oncol.* **2010**, *12*, 871–881. [CrossRef]
29. Wick, W.; Puduvalli, V.K.; Chamberlain, M.C.; van den Bent, M.J.; Carpentier, A.F.; Cher, L.M.; Mason, W.; Weller, M.; Hong, S.; Musib, L.; et al. Phase III study of enzastaurin compared with lomustine in the treatment of recurrent intracranial glioblastoma. *J. Clin. Oncol.* **2010**, *28*, 1168–1174. [CrossRef]
30. Dresemann, G.; Weller, M.; Rosenthal, M.A.; Wedding, U.; Wagner, W.; Engel, E.; Heinrich, B.; Mayer-Steinacker, R.; Karup-Hansen, A.; Fluge, Ø.; et al. Imatinib in combination with hydroxyurea versus hydroxyurea alone as oral therapy in patients with progressive pretreated glioblastoma resistant to standard dose temozolomide. *J. Neurooncol.* **2010**, *96*, 393–402. [CrossRef]
31. Stupp, R.; Wong, E.T.; Kanner, A.A.; Steinberg, D.; Engelhard, H.; Heidecke, V.; Kirson, E.D.; Taillibert, S.; Liebermann, F.; Dbalý, V.; et al. NovoTTF-100A versus physician's choice chemotherapy in recurrent glioblastoma: A randomised phase III trial of a novel treatment modality. *Eur. J. Cancer* **2012**, *48*, 2192–2202. [CrossRef] [PubMed]
32. Batchelor, T.T.; Mulholland, P.; Neyns, B.; Nabors, L.B.; Campone, M.; Wick, A.; Mason, W.; Mikkelsen, T.; Phuphanich, S.; Ashby, L.S.; et al. Phase III Randomized Trial Comparing the Efficacy of Cediranib As Monotherapy, and in Combination With Lomustine, Versus Lomustine Alone in Patients With Recurrent Glioblastoma. *J. Clin. Oncol.* **2013**, *31*, 3212–3218. [CrossRef] [PubMed]
33. Reardon, D.A.; Omuro, A.; Brandes, A.A.; Rieger, J.; Wick, A.; Sepulveda, J.; Phuphanich, S.; de Souza, P.; Ahluwalia, M.S.; Lim, M.; et al. OS10.3 Randomized Phase 3 Study Evaluating the Efficacy and Safety of Nivolumab vs Bevacizumab in Patients With Recurrent Glioblastoma: CheckMate 143. *Neuro. Oncol.* **2017**, *19*, iii21. [CrossRef]
34. Lillehei, K.O.; Kalkanis, S.N.; Liau, L.M.; Mydland, D.E.; Olson, J.; Paleologos, N.A.; Ryken, T.; Johnson, T.; Scullin, E. Rationale and design of the 500-patient, 3-year, and prospective Vigilant ObservatIon of GlIadeL WAfer ImplaNT registry. *CNS Oncol.* **2018**, *7*, CNS08. [CrossRef] [PubMed]
35. Ashby, L.S.; Smith, K.A.; Stea, B. Gliadel wafer implantation combined with standard radiotherapy and concurrent followed by adjuvant temozolomide for treatment of newly diagnosed high-grade glioma: A systematic literature review. *World J. Surg. Oncol.* **2016**, *14*, 225. [CrossRef] [PubMed]
36. Bregy, A.; Shah, A.H.; Diaz, M.V.; Pierce, H.E.; Ames, P.L.; Diaz, D.; Komotar, R.J. The role of Gliadel wafers in the treatment of high-grade gliomas. *Expert Rev. Anticancer Ther.* **2013**, *13*, 1453–1461. [CrossRef] [PubMed]
37. Chowdhary, S.A.; Ryken, T.; Newton, H.B. Survival outcomes and safety of carmustine wafers in the treatment of high-grade gliomas: A meta-analysis. *J. Neurooncol.* **2015**, *122*, 367–382. [CrossRef]
38. Nagpal, S. The Role of BCNU Polymer Wafers (Gliadel) in the Treatment of Malignant Glioma. *Neurosurg. Clin.* **2012**, *23*, 289–295. [CrossRef]
39. Xing, W.; Shao, C.; Qi, Z.; Yang, C.; Wang, Z. The role of Gliadel wafers in the treatment of newly diagnosed GBM: A meta-analysis. *Drug Des. Dev. Ther.* **2015**, *9*, 3341–3348. [CrossRef]

40. Stupp, R.; Hegi, M.E.; Mason, W.P.; van den Bent, M.J.; Taphoorn, M.J.B.; Janzer, R.C.; Ludwin, S.K.; Allgeier, A.; Fisher, B.; Belanger, K.; et al. Effects of radiotherapy with concomitant and adjuvant temozolomide versus radiotherapy alone on survival in glioblastoma in a randomised phase III study: 5-year analysis of the EORTC-NCIC trial. *Lancet Oncol.* **2009**, *10*, 459–466. [CrossRef]
41. Brown, T.J.; Brennan, M.C.; Li, M.; Church, E.W.; Brandmeir, N.J.; Rakszawski, K.L.; Patel, A.S.; Rizk, E.B.; Suki, D.; Sawaya, R.; et al. Association of the Extent of Resection with Survival in Glioblastoma: A Systematic Review and Meta-analysis. *JAMA Oncol.* **2016**, *2*, 1460–1469. [CrossRef] [PubMed]
42. Almenawer, S.A.; Badhiwala, J.H.; Alhazzani, W.; Greenspoon, J.; Farrokhyar, F.; Yarascavitch, B.; Algird, A.; Kachur, E.; Cenic, A.; Sharieff, W.; et al. Biopsy versus partial versus gross total resection in older patients with high-grade glioma: A systematic review and meta-analysis. *Neuro Oncol.* **2015**, *17*, 868–881. [CrossRef] [PubMed]
43. Xia, L.; Fang, C.; Chen, G.; Sun, C. Relationship between the extent of resection and the survival of patients with low-grade gliomas: A systematic review and meta-analysis. *BMC Cancer* **2018**, *18*, 48. [CrossRef] [PubMed]
44. Cata, J.P.; Hagan, K.B.; Bhavsar, S.D.O.; Arunkumar, R.; Grasu, R.; Dang, A.; Carlson, R.; Arnold, B.; Potylchansky, Y.; Lipski, I.; et al. The use of isoflurane and desflurane as inhalational agents for glioblastoma surgery. A survival analysis. *J. Clin. Neurosci.* **2017**, *35*, 82–87. [CrossRef] [PubMed]
45. Kazda, T.; Dziacky, A.; Burkon, P.; Pospisil, P.; Slavik, M.; Rehak, Z.; Jancalek, R.; Slampa, P.; Slaby, O.; Lakomy, R. Radiotherapy of glioblastoma 15 years after the landmark Stupp's trial: More controversies than standards? *Radiol. Oncol.* **2018**, *52*, 121–128. [CrossRef]
46. Skardelly, M.; Dangel, E.; Gohde, J.; Noell, S.; Behling, F.; Lepski, G.; Borchers, C.; Koch, M.; Schittenhelm, J.; Bisdas, S.; et al. Prolonged Temozolomide Maintenance Therapy in Newly Diagnosed Glioblastoma. *Oncologist* **2017**, *22*, 570–575. [CrossRef]
47. Blumenthal, D.T.; Gorlia, T.; Gilbert, M.R.; Kim, M.M.; Burt Nabors, L.; Mason, W.P.; Hegi, M.E.; Zhang, P.; Golfinopoulos, V.; Perry, J.R.; et al. Is more better? The impact of extended adjuvant temozolomide in newly diagnosed glioblastoma: A secondary analysis of EORTC and NRG Oncology/RTOG. *Neuro Oncol.* **2017**, *19*, 1119–1126. [CrossRef]
48. Gramatzki, D.; Kickingereder, P.; Hentschel, B.; Felsberg, J.; Herrlinger, U.; Schackert, G.; Tonn, J.-C.; Westphal, M.; Sabel, M.; Schlegel, U.; et al. Limited role for extended maintenance temozolomide for newly diagnosed glioblastoma. *Neurology* **2017**, *88*, 1422–1430. [CrossRef]
49. Xu, W.; Li, T.; Gao, L.; Zheng, J.; Shao, A.; Zhang, J. Efficacy and safety of long-term therapy for high-grade glioma with temozolomide: A meta-analysis. *Oncotarget* **2017**, *8*, 51758–51765. [CrossRef]
50. Deutsch, M.B.; Panageas, K.S.; Lassman, A.B.; DeAngelis, L.M. Steroid management in newly diagnosed glioblastoma. *J. Neurooncol.* **2013**, *113*, 111–116. [CrossRef]
51. Ly, K.I.; Wen, P.Y. Clinical Relevance of Steroid Use in Neuro-Oncology. *Curr. Neurol. Neurosci. Rep.* **2017**, *17*, 5. [CrossRef] [PubMed]
52. Hohwieler Schloss, M.; Freidberg, S.R.; Heatley, G.J.; Lo, T.C. Glucocorticoid dependency as a prognostic factor in radiotherapy for cerebral gliomas. *Acta Oncol.* **1989**, *28*, 51–55. [CrossRef] [PubMed]
53. Watne, K.; Hannisdal, E.; Nome, O.; Hager, B.; Hirschberg, H. Prognostic factors in malignant gliomas with special reference to intra-arterial chemotherapy. *Acta Oncol.* **1993**, *32*, 307–310. [CrossRef]
54. Tieu, M.T.; Lovblom, L.E.; McNamara, M.G.; Mason, W.; Laperriere, N.; Millar, B.-A.; Ménard, C.; Kiehl, T.-R.; Perkins, B.A.; Chung, C. Impact of glycemia on survival of glioblastoma patients treated with radiation and temozolomide. *J. Neurooncol.* **2015**, *124*, 119–126. [CrossRef] [PubMed]
55. Michaelsen, S.R.; Christensen, I.J.; Grunnet, K.; Stockhausen, M.-T.; Broholm, H.; Kosteljanetz, M.; Poulsen, H.S. Clinical variables serve as prognostic factors in a model for survival from glioblastoma multiforme: An observational study of a cohort of consecutive non-selected patients from a single institution. *BMC Cancer* **2013**, *13*, 402. [CrossRef] [PubMed]
56. Shields, L.B.E.; Shelton, B.J.; Shearer, A.J.; Chen, L.; Sun, D.A.; Parsons, S.; Bourne, T.D.; LaRocca, R.; Spalding, A.C. Dexamethasone administration during definitive radiation and temozolomide renders a poor prognosis in a retrospective analysis of newly diagnosed glioblastoma patients. *Radiat. Oncol.* **2015**, *10*, 222. [CrossRef] [PubMed]

57. Kostopoulou, O.N.; Mohammad, A.-A.; Bartek, J.; Winter, J.; Jung, M.; Stragliotto, G.; Söderberg-Nauclér, C.; Landázuri, N. Glucocorticoids promote a glioma stem cell-like phenotype and resistance to chemotherapy in human glioblastoma primary cells: Biological and prognostic significance. *Int. J. Cancer* **2017**. [CrossRef] [PubMed]
58. Díez Valle, R.; Becerra Castro, V.; Marigil Sánchez, M.; Gállego Pérez-Larraya, J.; Núñez-Córdoba, J.M.; Tejada Solis, S. Results of a Policy of Fast Tapering of Steroids After Resection Surgery in Glioblastoma. *World Neurosurg.* **2018**, *109*, e845–e852. [CrossRef]
59. Pitter, K.L.; Tamagno, I.; Alikhanyan, K.; Hosni-Ahmed, A.; Pattwell, S.S.; Donnola, S.; Dai, C.; Ozawa, T.; Chang, M.; Chan, T.A.; et al. Corticosteroids compromise survival in glioblastoma. *Brain* **2016**, *139*, 1458–1471. [CrossRef]
60. Van Linde, M.E.; Brahm, C.G.; de Witt Hamer, P.C.; Reijneveld, J.C.; Bruynzeel, A.M.E.; Vandertop, W.P.; van de Ven, P.M.; Wagemakers, M.; van der Weide, H.L.; Enting, R.H.; et al. Treatment outcome of patients with recurrent glioblastoma multiforme: A retrospective multicenter analysis. *J. Neurooncol.* **2017**. [CrossRef]
61. Duerinck, J.; Du Four, S.; Bouttens, F.; Andre, C.; Verschaeve, V.; Van Fraeyenhove, F.; Chaskis, C.; D'Haene, N.; Le Mercier, M.; Rogiers, A.; et al. Randomized phase II trial comparing axitinib with the combination of axitinib and lomustine in patients with recurrent glioblastoma. *J. Neurooncol.* **2017**. [CrossRef] [PubMed]
62. Wong, E.T.; Lok, E.; Gautam, S.; Swanson, K.D. Dexamethasone exerts profound immunologic interference on treatment efficacy for recurrent glioblastoma. *Br. J. Cancer* **2015**, *113*, 232–241. [CrossRef] [PubMed]
63. Coutinho, A.E.; Chapman, K.E. The anti-inflammatory and immunosuppressive effects of glucocorticoids, recent developments and mechanistic insights. *Mol. Cell. Endocrinol.* **2011**, *335*, 2–13. [CrossRef] [PubMed]
64. Zen, M.; Canova, M.; Campana, C.; Bettio, S.; Nalotto, L.; Rampudda, M.; Ramonda, R.; Iaccarino, L.; Doria, A. The kaleidoscope of glucorticoid effects on immune system. *Autoimmun. Rev.* **2011**, *10*, 305–310. [CrossRef] [PubMed]
65. Cain, D.W.; Cidlowski, J.A. Immune regulation by glucocorticoids. *Nat. Rev. Immunol.* **2017**, *17*, 233–247. [CrossRef] [PubMed]
66. Sengupta, S.; Thaci, B.; Crawford, A.C.; Sampath, P. Interleukin-13 Receptor Alpha 2-Targeted Glioblastoma Immunotherapy. *Biomed. Res. Int.* **2014**, *2014*, 1–8. [CrossRef] [PubMed]
67. Gustafson, M.P.; Lin, Y.; New, K.C.; Bulur, P.A.; O'Neill, B.P.; Gastineau, D.A.; Dietz, A.B. Systemic immune suppression in glioblastoma: The interplay between CD14+HLA-DRlo/neg monocytes, tumor factors, and dexamethasone. *Neuro Oncol.* **2010**, *12*, 631–644. [CrossRef] [PubMed]
68. Chitadze, G.; Flüh, C.; Quabius, E.S.; Freitag-Wolf, S.; Peters, C.; Lettau, M.; Bhat, J.; Wesch, D.; Oberg, H.-H.; Luecke, S.; et al. In-depth immunophenotyping of patients with glioblastoma multiforme: Impact of steroid treatment. *Oncoimmunology* **2017**, *6*, e1358839. [CrossRef] [PubMed]
69. Cook, A.M.; McDonnell, A.M.; Lake, R.A.; Nowak, A.K. Dexamethasone co-medication in cancer patients undergoing chemotherapy causes substantial immunomodulatory effects with implications for chemo-immunotherapy strategies. *Oncoimmunology* **2016**, *5*, e1066062. [CrossRef] [PubMed]
70. Knudsen-Baas, K.M.; Engeland, A.; Gilhus, N.E.; Storstein, A.M.; Owe, J.F. Does the choice of antiepileptic drug affect survival in glioblastoma patients? *J. Neurooncol.* **2016**, *129*, 461–469. [CrossRef]
71. Vecht, C.J.; Kerkhof, M.; Duran-Pena, A. Seizure Prognosis in Brain Tumors: New Insights and Evidence-Based Management. *Oncologist* **2014**, *19*, 751–759. [CrossRef] [PubMed]
72. Bruna, J.; Miró, J.; Velasco, R. Epilepsy in glioblastoma patients: Basic mechanisms and current problems in treatment. *Expert Rev. Clin. Pharmacol.* **2013**, *6*, 333–344. [CrossRef] [PubMed]
73. Guthrie, G.D.; Eljamel, S. Impact of particular antiepileptic drugs on the survival of patients with glioblastoma multiforme. *J. Neurosurg.* **2013**, *118*, 859–865. [CrossRef] [PubMed]
74. Watanabe, S.; Kuwabara, Y.; Suehiro, S.; Yamashita, D.; Tanaka, M.; Tanaka, A.; Ohue, S.; Araki, H. Valproic acid reduces hair loss and improves survival in patients receiving temozolomide-based radiation therapy for high-grade glioma. *Eur. J. Clin. Pharmacol.* **2017**, *73*, 357–363. [CrossRef] [PubMed]
75. Tsai, H.-C.; Wei, K.-C.; Tsai, C.-N.; Huang, Y.-C.; Chen, P.-Y.; Chen, S.-M.; Lu, Y.-J.; Lee, S.-T. Effect of valproic acid on the outcome of glioblastoma multiforme. *Br. J. Neurosurg.* **2012**, *26*, 347–354. [CrossRef] [PubMed]
76. Kerkhof, M.; Dielemans, J.C.M.; van Breemen, M.S.; Zwinkels, H.; Walchenbach, R.; Taphoorn, M.J.; Vecht, C.J. Effect of valproic acid on seizure control and on survival in patients with glioblastoma multiforme. *Neuro Oncol.* **2013**, *15*, 961–967. [CrossRef]

77. Barker, C.A.; Bishop, A.J.; Chang, M.; Beal, K.; Chan, T.A. Valproic Acid Use During Radiation Therapy for Glioblastoma Associated With Improved Survival. *Int. J. Radiat. Oncol.* **2013**, *86*, 504–509. [CrossRef]
78. Rudà, R.; Pellerino, A.; Soffietti, R. Does valproic acid affect tumor growth and improve survival in glioblastomas? *CNS Oncol.* **2016**, *5*, 51–53. [CrossRef]
79. Weller, M.; Gorlia, T.; Cairncross, J.G.; van den Bent, M.J.; Mason, W.; Belanger, K.; Brandes, A.A.; Bogdahn, U.; Macdonald, D.R.; Forsyth, P.; et al. Prolonged survival with valproic acid use in the EORTC/NCIC temozolomide trial for glioblastoma. *Neurology* **2011**, *77*, 1156–1164. [CrossRef]
80. Redjal, N.; Reinshagen, C.; Le, A.; Walcott, B.P.; McDonnell, E.; Dietrich, J.; Nahed, B.V. Valproic acid, compared to other antiepileptic drugs, is associated with improved overall and progression-free survival in glioblastoma but worse outcome in grade II/III gliomas treated with temozolomide. *J. Neurooncol.* **2016**, *127*, 505–514. [CrossRef]
81. Yuan, Y.; Xiang, W.; Qing, M.; Yanhui, L.; Jiewen, L.; Yunhe, M. Survival analysis for valproic acid use in adult glioblastoma multiforme: A meta-analysis of individual patient data and a systematic review. *Seizure* **2014**, *23*, 830–835. [CrossRef] [PubMed]
82. Kim, Y.-H.; Kim, T.; Joo, J.-D.; Han, J.H.; Kim, Y.J.; Kim, I.A.; Yun, C.-H.; Kim, C.-Y. Survival benefit of levetiracetam in patients treated with concomitant chemoradiotherapy and adjuvant chemotherapy with temozolomide for glioblastoma multiforme. *Cancer* **2015**, *121*, 2926–2932. [CrossRef] [PubMed]
83. Valiyaveettil, D.; Malik, M.; Joseph, D.; Ahmed, S.; Kothwal, S.; Vijayasaradhi, M. Effect of valproic acid on survival in glioblastoma: A prospective single-arm study. *South Asian J. Cancer* **2018**, *7*, 159. [CrossRef] [PubMed]
84. Felix, F.H.C.; de Araujo, O.L.; da Trindade, K.M.; Trompieri, N.M.; Fontenele, J.B. Survival of children with malignant brain tumors receiving valproate: A retrospective study. *Childs. Nerv. Syst.* **2013**, *29*, 195–197. [CrossRef] [PubMed]
85. Breemen, M.S.M.; Rijsman, R.M.; Taphoorn, M.J.B.; Walchenbach, R.; Zwinkels, H.; Vecht, C.J. Efficacy of anti-epileptic drugs in patients with gliomas and seizures. *J. Neurol.* **2009**, *256*, 1519–1526. [CrossRef] [PubMed]
86. Berendsen, S.; Varkila, M.; Kroonen, J.; Seute, T.; Snijders, T.J.; Kauw, F.; Spliet, W.G.M.; Willems, M.; Poulet, C.; Broekman, M.L.; et al. Prognostic relevance of epilepsy at presentation in glioblastoma patients. *Neuro Oncol.* **2016**, *18*, 700–706. [CrossRef] [PubMed]
87. Toledo, M.; Sarria-Estrada, S.; Quintana, M.; Maldonado, X.; Martinez-Ricarte, F.; Rodon, J.; Auger, C.; Salas-Puig, J.; Santamarina, E.; Martinez-Saez, E. Prognostic implications of epilepsy in glioblastomas. *Clin. Neurol. Neurosurg.* **2015**, *139*, 166–171. [CrossRef]
88. Happold, C.; Gorlia, T.; Chinot, O.; Gilbert, M.R.; Nabors, L.B.; Wick, W.; Pugh, S.L.; Hegi, M.; Cloughesy, T.; Roth, P.; et al. Does Valproic Acid or Levetiracetam Improve Survival in Glioblastoma? A Pooled Analysis of Prospective Clinical Trials in Newly Diagnosed Glioblastoma. *J. Clin. Oncol.* **2016**, *34*, 731–739. [CrossRef]
89. Cote, D.J.; Smith, T.R. Venous thromboembolism in brain tumor patients. *J. Clin. Neurosci.* **2016**, *25*, 13–18. [CrossRef]
90. Morgan, E.R.; Mason, W.P.; Maurice, C. A critical balance: Managing coagulation in patients with glioma. *Expert Rev. Neurother.* **2016**, *16*, 803–814. [CrossRef]
91. Streiff, M.B.; Ye, X.; Kickler, T.S.; Desideri, S.; Jani, J.; Fisher, J.; Grossman, S.A. A prospective multicenter study of venous thromboembolism in patients with newly-diagnosed high-grade glioma: Hazard rate and risk factors. *J. Neurooncol.* **2015**, *124*, 299–305. [CrossRef] [PubMed]
92. Taillibert, S.; Taillandier, L.; Le Rhun, E. Venous thrombosis in patients with high-grade glioma. *Curr. Opin. Oncol.* **2015**, *27*, 516–521. [CrossRef] [PubMed]
93. Edwin, N.C.; Khoury, M.N.; Sohal, D.; McCrae, K.R.; Ahluwalia, M.S.; Khorana, A.A. Recurrent venous thromboembolism in glioblastoma. *Thromb. Res.* **2016**, *137*, 184–188. [CrossRef] [PubMed]
94. Schnoor, R.; Maas, S.L.N.; Broekman, M.L.D. Heparin in malignant glioma: Review of preclinical studies and clinical results. *J. Neurooncol.* **2015**, *124*, 151–156. [CrossRef] [PubMed]
95. De Vos, F.Y.; Gijtenbeek, J.M.; Bleeker-Rovers, C.P.; van Herpen, C.M. Pneumocystis jirovecii pneumonia prophylaxis during temozolomide treatment for high-grade gliomas. *Crit. Rev. Oncol. Hematol.* **2013**, *85*, 373–382. [CrossRef] [PubMed]

96. Neuwelt, A.J.; Nguyen, T.M.; Fu, R.; Bubalo, J.; Tyson, R.M.; Lacy, C.; Gahramanov, S.; Nasseri, M.; Barnes, P.D.; Neuwelt, E.A. Incidence of Pneumocystis jirovecii pneumonia after temozolomide for CNS malignancies without prophylaxis. *CNS Oncol.* **2014**, *3*, 267–273. [CrossRef] [PubMed]
97. Mehta, M.; Wen, P.; Nishikawa, R.; Reardon, D.; Peters, K. Critical review of the addition of tumor treating fields (TTFields) to the existing standard of care for newly diagnosed glioblastoma patients. *Crit. Rev. Oncol. Hematol.* **2017**, *111*, 60–65. [CrossRef]
98. Zhu, P.; Zhu, J.-J. Tumor treating fields: A novel and effective therapy for glioblastoma: Mechanism, efficacy, safety and future perspectives. *Chin. Clin. Oncol.* **2017**, *6*, 41. [CrossRef]
99. Magouliotis, D.E.; Asprodini, E.K.; Svokos, K.A.; Tasiopoulou, V.S.; Svokos, A.A.; Toms, S.A. Tumor-treating fields as a fourth treating modality for glioblastoma: A meta-analysis. *Acta Neurochir. (Wien)* **2018**, *160*, 1167–1174. [CrossRef]
100. Guzauskas, G.F.; Salzberg, M.; Wang, B.C. Estimated lifetime survival benefit of tumor treating fields and temozolomide for newly diagnosed glioblastoma patients. *CNS Oncol.* **2018**. [CrossRef]
101. Roh, T.H.; Park, H.H.; Kang, S.-G.; Moon, J.H.; Kim, E.H.; Hong, C.-K.; Ahn, S.S.; Choi, H.J.; Cho, J.; Kim, S.H.; et al. Long-term outcomes of concomitant chemoradiotherapy with temozolomide for newly diagnosed glioblastoma patients. *Medicine (Baltimore).* **2017**, *96*, e7422. [CrossRef] [PubMed]
102. Mandel, J.J.; Yust-Katz, S.; Patel, A.J.; Cachia, D.; Liu, D.; Park, M.; Yuan, Y.; Kent, T.A.; de Groot, J.F. Inability of positive phase II clinical trials of investigational treatments to subsequently predict positive phase III clinical trials in glioblastoma. *Neuro Oncol.* **2018**, *20*, 113–122. [CrossRef] [PubMed]
103. Mandel, J.J.; Youssef, M.; Ludmir, E.; Yust-Katz, S.; Patel, A.J.; De Groot, J.F. Highlighting the need for reliable clinical trials in glioblastoma. *Expert Rev. Anticancer Ther.* **2018**, *18*, 1031–1040. [CrossRef] [PubMed]
104. Weller, M.; Butowski, N.; Tran, D.D.; Recht, L.D.; Lim, M.; Hirte, H.; Ashby, L.; Mechtler, L.; Goldlust, S.A.; Iwamoto, F.; et al. Go, no-go decision making for phase 3 clinical trials: ACT IV revisited—Authors' reply. *Lancet Oncol.* **2017**, *18*, e708. [CrossRef]
105. Nguyen, H.T.N.; Grogan, P.; Robins, H.I. Go, no-go decision making for phase 3 clinical trials: ACT IV revisited. *Lancet Oncol.* **2017**, *18*, e708. [CrossRef]
106. Vanderbeek, A.M.; Rahman, R.; Fell, G.; Ventz, S.; Chen, T.; Redd, R.; Parmigiani, G.; Cloughesy, T.F.; Wen, P.Y.; Trippa, L.; et al. The clinical trials landscape for glioblastoma: Is it adequate to develop new treatments? *Neuro Oncol.* **2018**, *20*, 1034–1043. [CrossRef]
107. Porter, K.R.; McCarthy, B.J.; Berbaum, M.L.; Davis, F.G. Conditional survival of all primary brain tumor patients by age, behavior, and histology. *Neuroepidemiology* **2011**, *36*, 230–239. [CrossRef]
108. Darefsky, A.S.; King, J.T.; Dubrow, R. Adult glioblastoma multiforme survival in the temozolomide era: A population-based analysis of Surveillance, Epidemiology, and End Results registries. *Cancer* **2012**, *118*, 2163–2172. [CrossRef]
109. Ostrom, Q.T.; Gittleman, H.; Liao, P.; Rouse, C.; Chen, Y.; Dowling, J.; Wolinsky, Y.; Kruchko, C.; Barnholtz-Sloan, J. CBTRUS Statistical Report: Primary Brain and Central Nervous System Tumors Diagnosed in the United States in 2007–2011. *Neuro Oncol.* **2014**, *16*, iv1–iv63. [CrossRef]
110. Tykocki, T.; Eltayeb, M. Ten-year survival in glioblastoma. A systematic review. *J. Clin. Neurosci.* **2018**, *54*, 7–13. [CrossRef]
111. Cihoric, N.; Tsikkinis, A.; Minniti, G.; Lagerwaard, F.J.; Herrlinger, U.; Mathier, E.; Soldatovic, I.; Jeremic, B.; Ghadjar, P.; Elicin, O.; et al. Current status and perspectives of interventional clinical trials for glioblastoma—Analysis of ClinicalTrials.gov. *Radiat. Oncol.* **2017**, *12*. [CrossRef] [PubMed]
112. Dos Santos, M.A.; Pignon, J.-P.; Blanchard, P.; Lefeuvre, D.; Levy, A.; Touat, M.; Louvel, G.; Dhermain, F.; Soria, J.-C.; Deutsch, E.; et al. Systematic review and meta-analysis of phase I/II targeted therapy combined with radiotherapy in patients with glioblastoma multiforme: Quality of report, toxicity, and survival. *J. Neurooncol.* **2015**, *123*, 307–314. [CrossRef] [PubMed]
113. Su, J.; Cai, M.; Li, W.; Hou, B.; He, H.; Ling, C.; Huang, T.; Liu, H.; Guo, Y. Molecularly Targeted Drugs Plus Radiotherapy and Temozolomide Treatment for Newly Diagnosed Glioblastoma: A Meta-Analysis and Systematic Review. *Oncol. Res. Featur. Preclin. Clin. Cancer Ther.* **2016**, *24*, 117–128. [CrossRef] [PubMed]
114. Sim, H.-W.; Morgan, E.R.; Mason, W.P. Contemporary management of high-grade gliomas. *CNS Oncol.* **2018**, *7*, 51–65. [CrossRef] [PubMed]
115. Yin, A.; Cheng, J.; Zhang, X.; Liu, B. The treatment of glioblastomas: A systematic update on clinical Phase III trials. *Crit. Rev. Oncol. Hematol.* **2013**, *87*, 265–282. [CrossRef] [PubMed]

116. Guishard, A.F.; Yakisich, J.S.; Azad, N.; Iyer, A.K.V. Translational gap in ongoing clinical trials for glioma. *J. Clin. Neurosci.* **2018**, *47*, 28–42. [CrossRef] [PubMed]
117. Alexander, B.M.; Trippa, L.; Gaffey, S.C.; Arrillaga, I.; Lee, E.Q.; Tanguturi, S.K.; Ahluwalia, M.S.; Colman, H.; Galanis, E.; De Groot, J.F.; et al. Individualized screening trial of innovative glioblastoma therapy (INSIGhT). *J. Clin. Oncol.* **2017**, *35*, TPS2079. [CrossRef]
118. Prasad, V. Perspective: The precision-oncology illusion. *Nature* **2016**, *537*, S63. [CrossRef]
119. Joyner, M.J.; Paneth, N.; Ioannidis, J.P.A. What Happens When Underperforming Big Ideas in Research Become Entrenched? *JAMA* **2016**, *316*, 1355–1356. [CrossRef]
120. Tannock, I.F.; Hickman, J.A. Limits to Personalized Cancer Medicine. *N. Engl. J. Med.* **2016**, *375*, 1289–1294. [CrossRef]
121. Massard, C.; Michiels, S.; Ferté, C.; Le Deley, M.-C.; Lacroix, L.; Hollebecque, A.; Verlingue, L.; Ileana, E.; Rosellini, S.; Ammari, S.; et al. High-Throughput Genomics and Clinical Outcome in Hard-to-Treat Advanced Cancers: Results of the MOSCATO 01 Trial. *Cancer Discov.* **2017**, *7*, 586–595. [CrossRef] [PubMed]
122. Le Tourneau, C.; Delord, J.-P.; Gonçalves, A.; Gavoille, C.; Dubot, C.; Isambert, N.; Campone, M.; Trédan, O.; Massiani, M.-A.; Mauborgne, C.; et al. Molecularly targeted therapy based on tumour molecular profiling versus conventional therapy for advanced cancer (SHIVA): A multicentre, open-label, proof-of-concept, randomised, controlled phase 2 trial. *Lancet Oncol.* **2015**, *16*, 1324–1334. [CrossRef]
123. Wirsching, H.-G.; Tabatabai, G.; Roelcke, U.; Hottinger, A.F.; Jörger, F.; Schmid, A.; Plasswilm, L.; Schrimpf, D.; Mancao, C.; Capper, D.; et al. Bevacizumab plus hypofractionated radiotherapy versus radiotherapy alone in elderly patients with glioblastoma: The randomized, open-label, phase II ARTE trial. *Ann. Oncol.* **2018**, *29*, 1423–1430. [CrossRef] [PubMed]
124. Lombardi, G.; Pambuku, A.; Bellu, L.; Farina, M.; Della Puppa, A.; Denaro, L.; Zagonel, V. Effectiveness of antiangiogenic drugs in glioblastoma patients: A systematic review and meta-analysis of randomized clinical trials. *Crit. Rev. Oncol. Hematol.* **2017**, *111*, 94–102. [CrossRef] [PubMed]
125. Song, J.; Xue, Y.-Q.; Zhao, M.-M.; Xu, P. Effectiveness of lomustine and bevacizumab in progressive glioblastoma: A meta-analysis. *Onco. Targets Ther.* **2018**, *11*, 3435–3439. [CrossRef] [PubMed]
126. Taal, W.; Oosterkamp, H.M.; Walenkamp, A.M.E.; Dubbink, H.J.; Beerepoot, L.V.; Hanse, M.C.J.; Buter, J.; Honkoop, A.H.; Boerman, D.; de Vos, F.Y.F.; et al. Single-agent bevacizumab or lomustine versus a combination of bevacizumab plus lomustine in patients with recurrent glioblastoma (BELOB trial): A randomised controlled phase 2 trial. *Lancet Oncol.* **2014**, *15*, 943–953. [CrossRef]
127. Kim, M.M.; Umemura, Y.; Leung, D. Bevacizumab and Glioblastoma. *Cancer J.* **2018**, *24*, 180–186. [CrossRef]
128. Chen, Z.; Xu, N.; Zhao, C.; Xue, T.; Wu, X.; Wang, Z. Bevacizumab combined with chemotherapy vs single-agent therapy in recurrent glioblastoma: Evidence from randomized controlled trials. *Cancer Manag. Res.* **2018**, *10*, 2193–2205. [CrossRef] [PubMed]
129. Van den Bent, M.J.; Klein, M.; Smits, M.; Reijneveld, J.C.; French, P.J.; Clement, P.; de Vos, F.Y.F.; Wick, A.; Mulholland, P.J.; Taphoorn, M.J.B.; et al. Bevacizumab and temozolomide in patients with first recurrence of WHO grade II and III glioma, without 1p/19q co-deletion (TAVAREC): A randomised controlled phase 2 EORTC trial. *Lancet Oncol.* **2018**, *19*, 1170–1179. [CrossRef]
130. Yazici, G.; Cengiz, M.; Ozyigit, G.; Eren, G.; Yildiz, F.; Akyol, F.; Gurkaynak, M.; Zorlu, F. Hypofractionated stereotactic reirradiation for recurrent glioblastoma. *J. Neurooncol.* **2014**, *120*, 117–123. [CrossRef]
131. Fogh, S.E.; Andrews, D.W.; Glass, J.; Curran, W.; Glass, C.; Champ, C.; Evans, J.J.; Hyslop, T.; Pequignot, E.; Downes, B.; et al. Hypofractionated Stereotactic Radiation Therapy: An Effective Therapy for Recurrent High-Grade Gliomas. *J. Clin. Oncol.* **2010**, *28*, 3048–3053. [CrossRef] [PubMed]
132. Ciammella, P.; Podgornii, A.; Galeandro, M.; D'Abbiero, N.; Pisanello, A.; Botti, A.; Cagni, E.; Iori, M.; Iotti, C. Hypofractionated stereotactic radiation therapy for recurrent glioblastoma: Single institutional experience. *Radiat. Oncol.* **2013**, *8*, 222. [CrossRef] [PubMed]
133. Fokas, E.; Wacker, U.; Gross, M.W.; Henzel, M.; Encheva, E.; Engenhart-Cabillic, R. Hypofractionated Stereotactic Reirradiation of Recurrent Glioblastomas. *Strahlenther. Onkol.* **2009**, *185*, 235–240. [CrossRef] [PubMed]
134. Wick, W.; Fricke, H.; Junge, K.; Kobyakov, G.; Martens, T.; Heese, O.; Wiestler, B.; Schliesser, M.G.; von Deimling, A.; Pichler, J.; et al. A Phase II, Randomized, Study of Weekly APG101+Reirradiation versus Reirradiation in Progressive Glioblastoma. *Clin. Cancer Res.* **2014**, *20*, 6304–6313. [CrossRef] [PubMed]

135. Palmer, J.D.; Bhamidipati, D.; Song, A.; Eldredge-Hindy, H.B.; Siglin, J.; Dan, T.D.; Champ, C.E.; Zhang, I.; Bar-Ad, V.; Kim, L.; et al. Bevacizumab and re-irradiation for recurrent high grade gliomas: Does sequence matter? *J. Neurooncol.* **2018**. [CrossRef] [PubMed]
136. Filley, A.C.; Henriquez, M.; Dey, M. Recurrent glioma clinical trial, CheckMate-143: The game is not over yet. *Oncotarget* **2017**, *8*, 91779–91794. [CrossRef]
137. Wilcox, J.A.; Ramakrishna, R.; Magge, R. Immunotherapy in Glioblastoma. *World Neurosurg.* **2018**, *116*, 518–528. [CrossRef]
138. Yarchoan, M.; Hopkins, A.; Jaffee, E.M. Tumor Mutational Burden and Response Rate to PD-1 Inhibition. *N. Engl. J. Med.* **2017**, *377*, 2500–2501. [CrossRef]
139. Naboush, A.; Roman, C.A.J.; Shapira, I. Immune checkpoint inhibitors in malignancies with mismatch repair deficiency: A review of the state of the current knowledge. *J. Investig. Med.* **2017**, *65*, 754–758. [CrossRef]
140. Mouw, K.W.; Goldberg, M.S.; Konstantinopoulos, P.A.; D'Andrea, A.D. DNA Damage and Repair Biomarkers of Immunotherapy Response. *Cancer Discov.* **2017**, *7*, 675–693. [CrossRef]
141. Viale, G.; Trapani, D.; Curigliano, G. Mismatch Repair Deficiency as a Predictive Biomarker for Immunotherapy Efficacy. *Biomed. Res. Int.* **2017**, *2017*, 1–7. [CrossRef] [PubMed]
142. Johanns, T.M.; Miller, C.A.; Dorward, I.G.; Tsien, C.; Chang, E.; Perry, A.; Uppaluri, R.; Ferguson, C.; Schmidt, R.E.; Dahiya, S.; et al. Immunogenomics of Hypermutated Glioblastoma: A Patient with Germline POLE Deficiency Treated with Checkpoint Blockade Immunotherapy. *Cancer Discov.* **2016**, *6*, 1230–1236. [CrossRef] [PubMed]
143. Bouffet, E.; Larouche, V.; Campbell, B.B.; Merico, D.; de Borja, R.; Aronson, M.; Durno, C.; Krueger, J.; Cabric, V.; Ramaswamy, V.; et al. Immune Checkpoint Inhibition for Hypermutant Glioblastoma Multiforme Resulting From Germline Biallelic Mismatch Repair Deficiency. *J. Clin. Oncol.* **2016**, *34*, 2206–2211. [CrossRef] [PubMed]
144. Hodges, T.R.; Ott, M.; Xiu, J.; Gatalica, Z.; Swensen, J.; Zhou, S.; Huse, J.T.; de Groot, J.; Li, S.; Overwijk, W.W.; et al. Mutational burden, immune checkpoint expression, and mismatch repair in glioma: Implications for immune checkpoint immunotherapy. *Neuro Oncol.* **2017**, *19*, 1047–1057. [CrossRef] [PubMed]
145. Wang, X.; Zhao, H.-Y.; Zhang, F.-C.; Sun, Y.; Xiong, Z.-Y.; Jiang, X.-B. Dendritic Cell-Based Vaccine for the Treatment of Malignant Glioma: A Systematic Review. *Cancer Investig.* **2014**, *32*, 451–457. [CrossRef] [PubMed]
146. Cao, J.-X.; Zhang, X.-Y.; Liu, J.-L.; Li, D.; Li, J.-L.; Liu, Y.-S.; Wang, M.; Xu, B.-L.; Wang, H.-B.; Wang, Z.-X. Clinical efficacy of tumor antigen-pulsed DC treatment for high-grade glioma patients: Evidence from a meta-analysis. *PLoS ONE* **2014**, *9*, e107173. [CrossRef] [PubMed]
147. Artene, S.-A.; Turcu-Stiolica, A.; Ciurea, M.E.; Folcuti, C.; Tataranu, L.G.; Alexandru, O.; Purcaru, O.S.; Tache, D.E.; Boldeanu, M.V.; Silosi, C.; et al. Comparative effect of immunotherapy and standard therapy in patients with high grade glioma: A meta-analysis of published clinical trials. *Sci. Rep.* **2018**, *8*, 11800. [CrossRef] [PubMed]
148. Hanaei, S.; Afshari, K.; Hirbod-Mobarakeh, A.; Mohajer, B.; Amir Dastmalchi, D.; Rezaei, N. Therapeutic efficacy of specific immunotherapy for glioma: A systematic review and meta-analysis. *Rev. Neurosci.* **2018**, *29*, 443–461. [CrossRef]
149. Sokratous, G.; Polyzoidis, S.; Ashkan, K. Immune infiltration of tumor microenvironment following immunotherapy for glioblastoma multiforme. *Hum. Vaccin. Immunother.* **2017**, *13*, 2575–2582. [CrossRef]
150. Wick, W.; van den Bent, M.J. First results on the DCVax phase III trial: Raising more questions than providing answers. *Neuro Oncol.* **2018**, *20*, 1283–1284. [CrossRef]
151. Sampson, J.H.; Aldape, K.D.; Archer, G.E.; Coan, A.; Desjardins, A.; Friedman, A.H.; Friedman, H.S.; Gilbert, M.R.; Herndon, J.E.; McLendon, R.E.; et al. Greater chemotherapy-induced lymphopenia enhances tumor-specific immune responses that eliminate EGFRvIII-expressing tumor cells in patients with glioblastoma. *Neuro Oncol.* **2011**, *13*, 324–333. [CrossRef]
152. Schuster, J.; Lai, R.K.; Recht, L.D.; Reardon, D.A.; Paleologos, N.A.; Groves, M.D.; Mrugala, M.M.; Jensen, R.; Baehring, J.M.; Sloan, A.; et al. A phase II, multicenter trial of rindopepimut (CDX-110) in newly diagnosed glioblastoma: The ACT III study. *Neuro Oncol.* **2015**, *17*, 854–861. [CrossRef] [PubMed]
153. Reardon, D.A.; Desjardins, A.; Schuster, J.; Tran, D.D.; Fink, K.L.; Nabors, L.B.; Li, G.; Bota, D.A.; Lukas, R.V.; Ashby, L.S.; et al. IMCT-08ReACT: long-term survival from a randomized phase ii study of rindopepimut (CDX-110) plus bevacizumab in relapsed glioblastoma. *Neuro Oncol.* **2015**, *17*. [CrossRef]

154. Wen, P.; Reardon, D.; Phuphanich, S.; Aiken, R.; Landolfi, J.; Curry, W.; Zhu, J.-J.; Glantz, M.; Peereboom, D.; Markert, J.; et al. A randomized, double-blind, placebo-controlled phase 2 trial of dendritic cell (DC) vaccination with ICT-107 in newly diagnosed glioblastoma (GBM) patients. *J. Clin. Oncol.* **2014**, *32*, 2005.
155. Buchroithner, J.; Erhart, F.; Pichler, J.; Widhalm, G.; Preusser, M.; Stockhammer, G.; Nowosielski, M.; Iglseder, S.; Freyschlag, C.; Oberndorfer, S.; et al. Audencel Immunotherapy Based on Dendritic Cells Has No Effect on Overall and Progression-Free Survival in Newly Diagnosed Glioblastoma: A Phase II Randomized Trial. *Cancers* **2018**, *10*, 372. [CrossRef] [PubMed]
156. Cho, D.-Y.; Yang, W.-K.; Lee, H.-C.; Hsu, D.-M.; Lin, H.-L.; Lin, S.-Z.; Chen, C.-C.; Harn, H.-J.; Liu, C.-L.; Lee, W.-Y.; et al. Adjuvant immunotherapy with whole-cell lysate dendritic cells vaccine for glioblastoma multiforme: A phase II clinical trial. *World Neurosurg.* **2012**, *77*, 736–744. [CrossRef] [PubMed]
157. Inogés, S.; Tejada, S.; de Cerio, A.L.-D.; Gállego Pérez-Larraya, J.; Espinós, J.; Idoate, M.A.; Domínguez, P.D.; de Eulate, R.G.; Aristu, J.; Bendandi, M.; et al. A phase II trial of autologous dendritic cell vaccination and radiochemotherapy following fluorescence-guided surgery in newly diagnosed glioblastoma patients. *J. Transl. Med.* **2017**, *15*, 104. [CrossRef]
158. Yao, Y.; Luo, F.; Tang, C.; Chen, D.; Qin, Z.; Hua, W.; Xu, M.; Zhong, P.; Yu, S.; Chen, D.; et al. Molecular subgroups and B7-H4 expression levels predict responses to dendritic cell vaccines in glioblastoma: An exploratory randomized phase II clinical trial. *Cancer Immunol. Immunother.* **2018**. [CrossRef]
159. Bota, D.A.; Chung, J.; Dandekar, M.; Carrillo, J.A.; Kong, X.-T.; Fu, B.D.; Hsu, F.P.; Schönthal, A.H.; Hofman, F.M.; Chen, T.C.; et al. Phase II study of ERC1671 plus bevacizumab versus bevacizumab plus placebo in recurrent glioblastoma: Interim results and correlations with $CD4^+$ T-lymphocyte counts. *CNS Oncol.* **2018**, cns-2018-0009. [CrossRef]
160. Dillman, R.O.; Duma, C.M.; Ellis, R.A.; Cornforth, A.N.; Schiltz, P.M.; Sharp, S.L.; DePriest, M.C. Intralesional lymphokine-activated killer cells as adjuvant therapy for primary glioblastoma. *J. Immunother.* **2009**, *32*, 914–919. [CrossRef]
161. Dutoit, V.; Migliorini, D.; Patrikidou, A.; Mayer-Mokler, A.; Hilf, N.; Walker, P.R.; Dietrich, P.-Y. IMA950 multipeptide vaccine adjuvanted with poly-ICLC in combination with standard therapy in newly diagnosed HLA-A2 glioblastoma patients. *Ann. Oncol.* **2017**, *28*, B148. [CrossRef]
162. Prins, R.M.; Soto, H.; Konkankit, V.; Odesa, S.K.; Eskin, A.; Yong, W.H.; Nelson, S.F.; Liau, L.M. Gene Expression Profile Correlates with T-Cell Infiltration and Relative Survival in Glioblastoma Patients Vaccinated with Dendritic Cell Immunotherapy. *Clin. Cancer Res.* **2011**, *17*, 1603–1615. [CrossRef] [PubMed]
163. Jie, X.; Hua, L.; Jiang, W.; Feng, F.; Feng, G.; Hua, Z. Clinical application of a dendritic cell vaccine raised against heat-shocked glioblastoma. *Cell Biochem. Biophys.* **2012**, *62*, 91–99. [CrossRef] [PubMed]
164. Vik-Mo, E.O.; Nyakas, M.; Mikkelsen, B.V.; Moe, M.C.; Due-Tønnesen, P.; Suso, E.M.I.; Sæbøe-Larssen, S.; Sandberg, C.; Brinchmann, J.E.; Helseth, E.; et al. Therapeutic vaccination against autologous cancer stem cells with mRNA-transfected dendritic cells in patients with glioblastoma. *Cancer Immunol. Immunother.* **2013**, *62*, 1499–1509. [CrossRef]
165. Ardon, H.; Van Gool, S.W.; Verschuere, T.; Maes, W.; Fieuws, S.; Sciot, R.; Wilms, G.; Demaerel, P.; Goffin, J.; Van Calenbergh, F.; et al. Integration of autologous dendritic cell-based immunotherapy in the standard of care treatment for patients with newly diagnosed glioblastoma: Results of the HGG-2006 phase I/II trial. *Cancer Immunol. Immunother.* **2012**, *61*, 2033–2044. [CrossRef] [PubMed]
166. Yamanaka, R.; Homma, J.; Yajima, N.; Tsuchiya, N.; Sano, M.; Kobayashi, T.; Yoshida, S.; Abe, T.; Narita, M.; Takahashi, M.; et al. Clinical evaluation of dendritic cell vaccination for patients with recurrent glioma: Results of a clinical phase I/II trial. *Clin. Cancer Res.* **2005**, *11*, 4160–4167. [CrossRef] [PubMed]
167. De Vleeschouwer, S.; Fieuws, S.; Rutkowski, S.; Van Calenbergh, F.; Van Loon, J.; Goffin, J.; Sciot, R.; Wilms, G.; Demaerel, P.; Warmuth-Metz, M.; et al. Postoperative Adjuvant Dendritic Cell-Based Immunotherapy in Patients with Relapsed Glioblastoma Multiforme. *Clin. Cancer Res.* **2008**, *14*, 3098–3104. [CrossRef]
168. Brown, C.E.; Alizadeh, D.; Starr, R.; Weng, L.; Wagner, J.R.; Naranjo, A.; Ostberg, J.R.; Blanchard, M.S.; Kilpatrick, J.; Simpson, J.; et al. Regression of Glioblastoma after Chimeric Antigen Receptor T-Cell Therapy. *N. Engl. J. Med.* **2016**, *375*, 2561–2569. [CrossRef]
169. O'Rourke, D.M.; Nasrallah, M.P.; Desai, A.; Melenhorst, J.J.; Mansfield, K.; Morrissette, J.J.D.; Martinez-Lage, M.; Brem, S.; Maloney, E.; Shen, A.; et al. A single dose of peripherally infused EGFRvIII-directed CAR T cells mediates antigen loss and induces adaptive resistance in patients with recurrent glioblastoma. *Sci. Transl. Med.* **2017**, *9*, eaaa0984. [CrossRef]

170. Ahmed, N.; Brawley, V.; Hegde, M.; Bielamowicz, K.; Kalra, M.; Landi, D.; Robertson, C.; Gray, T.L.; Diouf, O.; Wakefield, A.; et al. HER2-Specific Chimeric Antigen Receptor–Modified Virus-Specific T Cells for Progressive Glioblastoma. *JAMA Oncol.* **2017**, *3*, 1094. [CrossRef]
171. Migliorini, D.; Dietrich, P.-Y.; Stupp, R.; Linette, G.P.; Posey, A.D.; June, C.H. CAR T-Cell Therapies in Glioblastoma: A First Look. *Clin. Cancer Res.* **2018**, *24*, 535–540. [CrossRef] [PubMed]
172. Prinzing, B.L.; Gottschalk, S.M.; Krenciute, G. CAR T-cell therapy for glioblastoma: Ready for the next round of clinical testing? *Expert Rev. Anticancer Ther.* **2018**, *18*, 451–461. [CrossRef] [PubMed]
173. Eissa, I.; Bustos-Villalobos, I.; Ichinose, T.; Matsumura, S.; Naoe, Y.; Miyajima, N.; Morimoto, D.; Mukoyama, N.; Zhiwen, W.; Tanaka, M.; et al. The Current Status and Future Prospects of Oncolytic Viruses in Clinical Trials against Melanoma, Glioma, Pancreatic, and Breast Cancers. *Cancers* **2018**, *10*, 356. [CrossRef] [PubMed]
174. Fueyo, J.; Alemany, R.; Gomez-Manzano, C.; Fuller, G.N.; Khan, A.; Conrad, C.A.; Liu, T.-J.; Jiang, H.; Lemoine, M.G.; Suzuki, K.; et al. Preclinical characterization of the antiglioma activity of a tropism-enhanced adenovirus targeted to the retinoblastoma pathway. *J. Natl. Cancer Inst.* **2003**, *95*, 652–660. [CrossRef] [PubMed]
175. Lang, F.F.; Tran, N.D.; Puduvalli, V.K.; Elder, J.B.; Fink, K.L.; Conrad, C.A.; Yung, W.K.A.; Penas-Prado, M.; Gomez-Manzano, C.; Peterkin, J.; et al. Phase 1b open-label randomized study of the oncolytic adenovirus DNX-2401 administered with or without interferon gamma for recurrent glioblastoma. *J. Clin. Oncol.* **2017**, *35*, 2002. [CrossRef]
176. Chiocca, E.A.; Yu, J.; Phuphanich, S.; Lukas, R.V.; Kumthekar, P.; Yang, Y.; Zhou, Q.; Buck, J.Y.; Deary, A.; Cai, H.; Barrett, J.A.; et al. Expanded phase I study of intratumoral Ad-RTS-hIL-12 plus oral veledimex: Tolerability and survival in recurrent glioblastoma. *J. Clin. Oncol.* **2017**, *35*, 2044. [CrossRef]
177. Chiocca, E.A.; Nassiri, F.; Wang, J.; Peruzzi, P.; Zadeh, G. Viral and other therapies for recurrent GBM: Is a 24-month durable response unusual? *Neuro Oncol.* **2018**. [CrossRef]
178. Rudra, S.; Hui, C.; Rao, Y.J.; Samson, P.; Lin, A.J.; Chang, X.; Tsien, C.; Fergus, S.; Mullen, D.; Yang, D.; et al. Effect of Radiation Treatment Volume Reduction on Lymphopenia in Patients Receiving Chemoradiotherapy for Glioblastoma. *Int. J. Radiat. Oncol. Biol. Phys.* **2018**, *101*, 217–225. [CrossRef]
179. Ellsworth, S.G. Field size effects on the risk and severity of treatment-induced lymphopenia in patients undergoing radiation therapy for solid tumors. *Adv. Radiat. Oncol.* **2018**, *3*, 512–519. [CrossRef]
180. Yovino, S.; Kleinberg, L.; Grossman, S.A.; Narayanan, M.; Ford, E. The Etiology of Treatment-related Lymphopenia in Patients with Malignant Gliomas: Modeling Radiation Dose to Circulating Lymphocytes Explains Clinical Observations and Suggests Methods of Modifying the Impact of Radiation on Immune Cells. *Cancer Investig.* **2013**, *31*. [CrossRef]
181. Gupta, T.; Mohanty, S.; Moiyadi, A.; Jalali, R. Factors predicting temozolomide induced clinically significant acute hematologic toxicity in patients with high-grade gliomas: A clinical audit. *Clin. Neurol. Neurosurg.* **2013**, *115*, 1814–1819. [CrossRef] [PubMed]
182. Ishikawa, E.; Yamamoto, T.; Sakamoto, N.; Nakai, K.; Akutsu, H.; Tsuboi, K.; Takano, S.; Matsumura, A. Low peripheral lymphocyte count before focal radiotherapy plus concomitant temozolomide predicts severe lymphopenia during malignant glioma treatment. *Neurol. Med. Chir. (Tokyo)* **2010**, *50*, 638–644. [CrossRef] [PubMed]
183. Gerber, D.E.; Grossman, S.A.; Zeltzman, M.; Parisi, M.A.; Kleinberg, L. The impact of thrombocytopenia from temozolomide and radiation in newly diagnosed adults with high-grade gliomas1. *Neuro Oncol.* **2007**, *9*, 47–52. [CrossRef] [PubMed]
184. Mendez, J.S.; Govindan, A.; Leong, J.; Gao, F.; Huang, J.; Campian, J.L. Association between treatment-related lymphopenia and overall survival in elderly patients with newly diagnosed glioblastoma. *J. Neurooncol.* **2016**, *127*, 329–335. [CrossRef] [PubMed]
185. Rahman, R.; Catalano, P.J.; Arvold, N.D.; Aizer, A.A.; Weiss, S.E.; Pinnell, N.; Horvath, M.C.; Christianson, L.; Reardon, D.A.; Lee, E.Q.; et al. Chemoradiation-Related Lymphopenia Is Common Among Glioblastoma Patients and Is Associated with Worse Progression-Free and Overall Survival. *Int. J. Radiat. Oncol.* **2016**, *96*, E123. [CrossRef]
186. Mariucci, S.; Rovati, B.; Manzoni, M.; Della Porta, M.G.; Comolli, G.; Delfanti, S.; Danova, M. Lymphocyte subpopulation and dendritic cell phenotyping during antineoplastic therapy in human solid tumors. *Clin. Exp. Med.* **2011**, *11*, 199–210. [CrossRef]

187. Sengupta, S.; Marrinan, J.; Frishman, C.; Sampath, P. Impact of temozolomide on immune response during malignant glioma chemotherapy. *Clin. Dev. Immunol.* **2012**, *2012*, 831090. [CrossRef]
188. Hsu, M.; Sedighim, S.; Wang, T.; Antonios, J.P.; Everson, R.G.; Tucker, A.M.; Du, L.; Emerson, R.; Yusko, E.; Sanders, C.; et al. TCR Sequencing Can Identify and Track Glioma-Infiltrating T Cells after DC Vaccination. *Cancer Immunol. Res.* **2016**, *4*, 412–418. [CrossRef]
189. Tysome, J.R.; Li, X.; Wang, S.; Wang, P.; Gao, D.; Du, P.; Chen, D.; Gangeswaran, R.; Chard, L.S.; Yuan, M.; et al. A novel therapeutic regimen to eradicate established solid tumors with an effective induction of tumor-specific immunity. *Clin. Cancer Res.* **2012**, *18*, 6679–6689. [CrossRef]
190. Wang, P.; Li, X.; Wang, J.; Gao, D.; Li, Y.; Li, H.; Chu, Y.; Zhang, Z.; Liu, H.; Jiang, G.; et al. Re-designing Interleukin-12 to enhance its safety and potential as an anti-tumor immunotherapeutic agent. *Nat. Commun.* **2017**, *8*, 1395. [CrossRef]
191. Kleijn, A.; van den Bossche, W.; Haefner, E.S.; Belcaid, Z.; Burghoorn-Maas, C.; Kloezeman, J.J.; Pas, S.D.; Leenstra, S.; Debets, R.; de Vrij, J.; et al. The Sequence of Delta24-RGD and TMZ Administration in Malignant Glioma Affects the Role of CD8(+)T Cell Anti-tumor Activity. *Mol. Ther. Oncol.* **2017**, *5*, 11–19. [CrossRef] [PubMed]
192. Koks, C.A.; Garg, A.D.; Ehrhardt, M.; Riva, M.; Vandenberk, L.; Boon, L.; De Vleeschouwer, S.; Agostinis, P.; Graf, N.; Van Gool, S.W. Newcastle disease virotherapy induces long-term survival and tumor-specific immune memory in orthotopic glioma through the induction of immunogenic cell death. *Int. J. Cancer* **2015**, *136*, E313–E325. [CrossRef] [PubMed]
193. Saha, D.; Martuza, R.L.; Rabkin, S.D. Macrophage Polarization Contributes to Glioblastoma Eradication by Combination Immunovirotherapy and Immune Checkpoint Blockade. *Cancer Cell* **2017**, *32*, 253–267. [CrossRef] [PubMed]
194. Fritzell, S.; Sandén, E.; Eberstål, S.; Visse, E.; Darabi, A.; Siesjö, P. Intratumoral temozolomide synergizes with immunotherapy in a T cell-dependent fashion. *Cancer Immunol. Immunother.* **2013**, *62*, 1463–1474. [CrossRef] [PubMed]
195. Curtin, J.F.; Liu, N.; Candolfi, M.; Xiong, W.; Assi, H.; Yagiz, K.; Edwards, M.R.; Michelsen, K.S.; Kroeger, K.M.; Liu, C.; et al. HMGB1 mediates endogenous TLR2 activation and brain tumor regression. *PLoS Med.* **2009**, *6*, e10. [CrossRef] [PubMed]
196. Curtin, J.F.; Candolfi, M.; Fakhouri, T.M.; Liu, C.; Alden, A.; Edwards, M.; Lowenstein, P.R.; Castro, M.G. Treg depletion inhibits efficacy of cancer immunotherapy: Implications for clinical trials. *PLoS ONE* **2008**, *3*, e1983. [CrossRef] [PubMed]
197. Li, X.; Wang, P.; Li, H.; Du, X.; Liu, M.; Huang, Q.; Wang, Y.; Wang, S. The Efficacy of Oncolytic Adenovirus Is Mediated by T-cell Responses against Virus and Tumor in Syrian Hamster Model. *Clin. Cancer Res.* **2017**, *23*, 239–249. [CrossRef]
198. Pellegatta, S.; Eoli, M.; Cuccarini, V.; Anghileri, E.; Pollo, B.; Pessina, S.; Frigerio, S.; Servida, M.; Cuppini, L.; Antozzi, C.; et al. Survival gain in glioblastoma patients treated with dendritic cell immunotherapy is associated with increased NK but not CD8+ T cell activation in the presence of adjuvant temozolomide. *Oncoimmunology* **2018**, *7*, e1412901. [CrossRef]
199. Fadul, C.E.; Fisher, J.L.; Gui, J.; Hampton, T.H.; Cote, A.L.; Ernstoff, M.S. Immune modulation effects of concomitant temozolomide and radiation therapy on peripheral blood mononuclear cells in patients with glioblastoma multiforme. *Neuro Oncol.* **2011**, *13*, 393–400. [CrossRef]
200. Ellsworth, S.; Balmanoukian, A.; Kos, F.; Nirschl, C.J.; Nirschl, T.R.; Grossman, S.A.; Luznik, L.; Drake, C.G. Sustained $CD4^+$ T cell-driven lymphopenia without a compensatory IL-7/IL-15 response among high-grade glioma patients treated with radiation and temozolomide. *Oncoimmunology* **2014**, *3*, e27357. [CrossRef]

cancers

MDPI

Review

Glioblastoma in Elderly Patients: Current Management and Future Perspectives

Giuseppe Minniti [1,*], Giuseppe Lombardi [2] and Sergio Paolini [3]

1 Radiation Oncology Unit, UPMC Hillman Cancer Center, San Pietro Hospital FBF, 00189 Rome, Italy
2 Department of Oncology, Veneto Institute of Oncology IOV-IRCCS, 35128 Padua, Italy; giuseppe.lombardi@iov.veneto.it
3 IRCCS Neuromed, 86077 Pozzilli (IS), Italy; nch@neuromed.it
* Correspondence: minnitig@upmc.edu; Tel.: +39-06-88846800

Received: 31 January 2019; Accepted: 5 March 2019; Published: 8 March 2019

Abstract: The incidence of glioblastoma (GBM) in the elderly population is slowly increasing in Western countries. Current management includes surgery, radiation therapy (RT) and chemotherapy; however, survival is significantly worse than that observed in younger patients and the optimal treatment in terms of efficacy and safety remains a matter of debate. Surgical resection is often employed as initial treatment for elderly patients with GBM, although the survival benefit is modest. Better survival has been reported in elderly patients treated with RT compared with those receiving supportive care alone, with similar survival outcome for patients undergoing standard RT (60 Gy over 6 weeks) and hypofractionated RT (25–40 Gy in 5–15 daily fractions). Temozolomide, an alkylating agent, may represent an effective and safe therapy in patients with promoter methylation of O^6-methylguanine-DNA-methyltransferase (*MGMT*) gene which is predictor of responsiveness to alkylating agents. An abbreviated course of RT, 40 Gy in 15 daily fractions in combination with adjuvant and concomitant temozolomide has emerged as an effective treatment for patients aged 65 years old or over with GBM. Results of the National Cancer Institute of Canada Clinical Trials Group (NCIC CTG CE6) and European Organization for Research and Treatment of Cancer (EORTC 26062/22061) randomized study of short-course RT with or without concurrent and adjuvant temozolomide have demonstrated a significant improvement in progression-free survival and overall survival for patients receiving RT and temozolomide over RT alone, without impairing either quality of life or functional status. Although combined chemoradiation has become the recommended treatment in fit elderly patients with GBM, several questions remain unanswered, including the survival impact of chemoradiation in patients with impaired neurological status, advanced age (>75–80 years old), or for those with severe comorbidities. In addition, the efficacy and safety of alternative therapeutic approaches according to the methylation status of the O^6-methylguanine-DNA methyl-transferase (*MGMT*) gene promoter need to be explored in future trials.

Keywords: glioblastoma; elderly; surgery; radiotherapy; chemotherapy; temozolomide

1. Introduction

Gliomas account for almost 80% of all primary malignant brain tumors. Glioblastoma (GBM) is the most frequent histology and accounts for more than 50% of gliomas in all age groups, with an incidence rate among elderly patients of 70 years and older of 17.5 per 100,000 person-years, and a relative risk of 3–4 times compared with young adults [1–3]. Considering that the population of 65 years or older is expected to increase in the next two decades in USA, Canada, Australia, and Europe, this age group will account for the majority of GBM cases in these nations, representing an important aspect of public health.

Based on the results of the EORTC-NCIC CTG phase III study showing a significant improvement in the median survival for patients aged 18–70 years who received chemoradiation over radiation therapy (RT) alone from 12.1 to 14.6 months, and an improvement in 2-year survival from 10% to 26%, respectively, the standard of care for adult patients with GBM is represented by post-operative standard RT (60 Gy in 30 fractions) with concurrent and adjuvant temozolomide [4,5]. However, the majority of elderly patients with GBM are less likely to receive standard chemoradiation because aggressive combined approaches are associated with lower survival benefit and increased toxicity.

Elderly patients with GBM have historically been treated with standard or hypofractionated RT, with a reported median survival in the range of 5 to 9 months [6–17]. For elderly patients with O_6-methylguanine-DNA methyl-transferase (*MGMT*) gene promoter methylation, the alkylating agent temozolomide has emerged as an effective treatment option associated with a survival benefit [18–20]. Published results from the EORTC (26981-22981)/NCIC CTG (CE.3) randomized study of short-course RT with or without concurrent and adjuvant temozolomide have made a significant contribution to the management of GBM [21] showing that the addition of temozolomide to short-course RT in patients 65 years of age or older with newly diagnosed GBM resulted in significantly longer survival compared to short-course RT alone. Although this regimen is likely to become the new standard of care in the elderly population with GBM, questions if all older patients may receive combined chemoradiation regardless of advanced age (>75–80 years old), impaired neurological status, presence of comorbidities, or molecular profiling of the tumor remain open.

The purpose of this review is to summarize the published literature on the efficacy of RT and chemotherapy given alone or in combination in elderly patients with GBM, and to address important issues such as the importance of molecular profiling in predicting response to treatments, the impact of treatments on quality of life and neurocognitive outcomes, and future research priorities for this population.

2. Methods

There is no generally agreed criterion for definition of "older people". A cut-off of over 60 or 65 years is often used (http://www.who.int/healthinfo/survey/ageingdefnolder/en/). Most developed world countries have accepted the chronological age of 65 years as a definition of "elderly" or older person; however, in the developed countries, the most relevant geriatric syndromes (e.g., insomnia, urinary incontinence, severe hearing/vision problem, functional decline, fall and depressive disorder) are most common over 70–75 years. In order to include all published studies reporting on older patients with GBM, we have used an age cutoff of 60 years old for defining older patients in the current research.

A literature search was conducted in MEDLINE PubMed evaluating older people with GBM. The search focused on randomized, prospective and retrospective studies published in English. The literature search was performed using a combination of medical subject headings (MeSH) "glioblastoma" and free text terms ("radiation therapy" or "hypofractionated radiotherapy" or "chemotherapy" or "chemoradiation" or "elderly"). We included relevant studies published from 1990 to 2017. Studies published in languages other than English or not involving human subjects were not reviewed. A total of 178 potentially relevant studies were identified, including 19 prospective/randomized studies and 159 retrospective studies. The results of the literature research were used and included if appropriate.

3. Overview of Treatments

3.1. Surgery

Surgical resection is the first step in treating patients with GBM. The goals of the surgical procedure include histologic diagnosis, relief of the tumor mass effect, safe tumor cytoreduction, and possibly prolong patient survival. Although several reports have found that extensive surgical resection is associated with longer survival [22–32], aggressive surgery is still a controversial issue in neurosurgery. Simpson et al. [22] reviewed the impact of the extent of surgery on the survival outcome in 645 patients with GBM who were enrolled in three consecutive randomized Radiation Therapy Oncology Group (RTOG) trials. Patients undergoing gross total resection had a significant longer median survival than those receiving a biopsy only (11.3 vs. 6.6 months; $p < 0.001$); notably, a significant difference in median survival times was also found for partial resection compared with biopsy only treatment (10.4 vs. 6.6 months; $p < 0.01$). In a series of 788 patients who underwent surgery for a malignant glioma between 1997 and 2001 in North America, Laws et al. [23] showed that total/subtotal surgical resection was an independent favorable prognostic factor compared with biopsy only ($p < 0.0001$), with no difference between older and younger patients. In another series of 1229 patients aged less than 80 years with histologically verified GBM undergoing surgery at the University of Texas MD Anderson Cancer Center, Li et al. [32] reported a median survival time of 15.2 months after complete resection and 9.8 months after incomplete resection ($p < 0.001$). This survival advantage was achieved without an increased risk of overall or neurological postoperative deficits even after correcting for prognostic factors including age, Karnofsky Performance Scale (KPS) score, and preoperative contrast-enhancing tumor volume. The favorable impact of complete resection on survival has been confirmed in published series reporting the use of modern surgical techniques, such as intraoperative magnetic resonance imaging, intraoperative ultrasonography, and fluorescence-guided surgery [28,29,33]. In a series of 243 patients enrolled in a phase III trial assessing the efficacy of fluorescence-guided surgery with 5-aminolevulinic acid for resection of malignant gliomas, Stummer et al. [29] showed that patients who underwent complete resection survived longer than those who had incomplete resection (16.7 vs. 11.8 months, $p < 0.0001$), even in elderly patients older than 60 years of age with GBM.

A few studies have evaluated the efficacy of surgery in elderly patients with GBM [34–40]. In a systematic review and meta-analysis of biopsy versus partial versus gross total resection in patients older than 60 years of age with high-grade glioma, Almenawer et al. [37] have compared overall survival, KPS, progression-free survival, mortality, and morbidity amongst 12607 patients who were included in 34 studies. The overall survival was 5.71 months (95% CI 5.04–6.36) in patients undergoing biopsy, 8.68 months (95% CI 7.87–9.48) in those having subtotal resection, and 14.04 months (95% CI 12.8–15.2) in those subjected to gross total resection. For the entire population, patients undergoing overall resection (of any extent) had a significant benefit compared with those having biopsy, with mean difference in overall survival of 3.88 months (95% CI 2.14–5.62, $p < 0.001$). Mean difference in postoperative KPS was 10.4 (95% CI 6.58–14.22, $p < 0.001$) and mean difference in progression-free survival was 2.44 months (95% CI 1.45–3.43, $p < 0.001$). Overall, the analysis showed longer survival time, delayed tumor progression rate, and improved functional recovery with decreasing trends of mortality and morbidity rates in the order of biopsy, sub-total resection, and complete resection, suggesting a progressive improvement in clinical outcomes with greater degrees of resection. In a small randomized study of 30 patients older than 65 years with malignant glioma who received stereotactic biopsy or surgical resection, Vuorinen et al. [35] reported median survival times of 171 and 85 days after surgical resection or biopsy, respectively ($p = 0.035$). Similar results have been reported in other retrospective series showing significantly improved survival in patients with GBM receiving subtotal/complete resection compared with those undergoing stereotactic biopsy [36–40].

Biopsy, which has limited mortality and serious morbidity in the range of 1–5%, is usually performed to assess histological and molecular characteristics of the tumor when surgical resection carries high risks [41–44]. Tumor tissue is in fact fundamental to assess the molecular profile of GBMs and consequently to tailor the appropriate treatment. In this contest biopsy can be avoided only when clinical and radiological data provide an accurate diagnosis of GBM and results will not affect treatment choices.

Overall, findings from these studies support the general principle of considering maximal degrees of tumor removal when the operative option is indicated, regardless of age. The main limitation of these studies is a lack of detail about surgical complications, functional recovery time, and neurocognitive outcome following variable levels of resections. Preventing new permanent neurological deficits and maintaining good quality of life are essential factors for guiding surgical management of these patients. In absence of randomized studies, in current clinical practice the optimal surgical approach in elderly patients should be individually based on the carefully evaluation of known established safety measures, as those included in the neurologic assessment in neuro-oncology (NANO) scale [45] and in geriatric assessment models [46].

3.2. Radiotherapy

Postoperative RT, either standard RT or abbreviated courses of RT, has been historically employed for elderly patients with GBM [6–17]. A summary of randomized controlled trials addressing the efficacy and safety of RT in the elderly population with GBM is shown in Table 1 [14,15,17,19,20,40].

The superiority of RT (50 Gy in 28 daily fractions) over best supportive care has been demonstrated in a French multi-institutional randomized trial of 85 elderly patients with GBM aged 70 years or older with a KPS score of 70 or higher [47]. Median overall survival and progression-free survival times were 29.1 and 14.9 weeks for patients receiving RT plus supportive care and 16.9 and 5.4 weeks for those receiving supportive care alone, respectively. Notably, RT did not cause further deterioration in the KPS, health-related quality of life and cognitive functions compared with supportive care. The efficacy of RT versus supportive care alone has been demonstrated in other few studies [6,8,10–12,14,16,48]. Using the Surveillance, Epidemiology, and End Results (SEER) registry (1988–2004) as data source of 10987 patients with GBM aged 70 years or older, Scott and colleagues [41] showed that the overall survival time was significantly improved by RT after adjusting for surgery, tumor size, gender, ethnicity, and age at diagnosis. In another retrospective series of 202 patients with GBM treated between 1990 and 2000 at Leiden University, Marijnen et al. [16] reported a significant longer survival of 10.6 months in patients treated with RT compared with 1.9 months in non-irradiated patients ($p < 0.0001$), with no significant survival difference between elderly and young patients.

Few randomized studies have addressed the efficacy and safety of either radical RT or abbreviated courses of RT in elderly patients with GBM [15,17,19] (Table 1). In a randomized trial of 100 patients with GBM aged 60 years or older who received postoperative standard RT or short-course RT (40 Gy in 15 fractions over 3 weeks), Roa et al. [15] showed no survival differences between the two groups. Median survival times and 6-month survival rates were 5.1 months and 44.7% for patients treated with standard RT and 5.6 months and 41.7% for those receiving short-course RT, respectively.

The Nordic randomized, phase III trial enrolled 291 patients older than 60 years of age with newly diagnosed GBM who were assigned to receive three different treatments: temozolomide (200 mg/m^2 on days 1–5 of every 28 days for up to six cycles), hypofractionated RT (34 Gy given in 3.4 Gy fractions over two weeks), or standard RT (60 Gy given in 2 Gy fractions over 6 weeks) [19]. The efficacy of radiation treatments was similar between the two radiation groups. The median survival time was 7.5 months for patients treated with hypofractionated RT and 6.0 months for those receiving standard RT; however, in patients older than 70 years hypofractionated RT resulted in significantly longer survival than standard RT (7.0 months vs. 5.2 months, $p = 0.02$).

Roa et al. [17] conducted a randomized trial of 98 frail and/or elderly patients aged 65 years and older with GBM who were randomized to receive two different hypofractionated radiation schedules. Median overall survival and progression-free survival times were 7.9 and 4.2 months in patients who received 25 Gy in five daily fractions and 6.4 and 4.2 months in those subjected to 40 Gy in 15 daily fractions over three weeks, respectively ($p = 0.9$). Neurological outcome and quality of life were similar between the two groups at 4 weeks and 8 weeks after treatment.

The assessment of neurocognitive status and quality of life is of particular relevance in elderly patients with GBM (Table 1). In Roa et al. [15] trial comparing standard RT versus short-course RT, KPS scores varied markedly over time but were not significantly different between groups; notably, 20% of elderly patients receiving standard RT interrupted the treatment because of acute toxicity. In the

Nordic study, global health status and several functioning scales, including physical, role, emotional, social, functioning and cognitive (assessed by the European Organisation for Research and Treatment of Cancer (EORTC)) quality of life questionnaire Core 30 (QLQ-C30) did not change significantly between patients receiving standard or hypofractionated RT [19]. However, data need be interpreted with caution due to the low number of patients who completed questionnaires. A similar improvement of quality of life and performance status have been reported using hypofractionated schedules of 25–40 Gy in 5–15 daily fractions [6,8,9,14,17].

In summary, RT is associated with an improved survival in elderly patients with GBM with no significant detrimental effects on neurocognitive function and quality of life. Although standard RT may represent a feasible treatment option for elderly patients of 60–70 years old with a good performance status, results of randomized controlled studies comparing standard and hypofractionated radiation schedules clearly indicate that short-course RT should be recommended in elderly patients because it offers similar survival benefit and shortens the time of treatment. In addition, advanced radiation techniques, including intensity-modulated radiotherapy (IMRT), volumetric-modulated arc therapy (VMAT), and image-guided RT (IGRT), which allow the delivery of more precise radiation doses to the target while minimizing exposure of the surrounding normal brain tissues, should be routinely used in clinical practice to treat these patients with the objective to reduce the risk of neurocognitive deficits.

3.3. Chemotherapy

Historically, the role of chemotherapy as initial treatment for elderly patients with GBM has been poorly investigated in the past, mainly because of the concern about the limited efficacy [49,50] and the severity of side effects of nitrosurea-based regimens [51,52]. Recently, the use of temozolomide as an alternative to RT in older patients with malignant gliomas has been addressed in prospective and randomized studies (Table 1) [18–20,53,54].

The German Neuro-oncology Working Group (NOA) phase 3 trial (NOA-08) has compared the efficacy and safety of RT to temozolomide in elderly patients with anaplastic astrocytoma or GBM; 373 patients aged 65 years and older with histologically confirmed tumors, and a KPS score $\geq$ 60, were randomly assigned to receive dose-dense temozolomide (one week on, one week off cycles) or standard RT [20]. Median survival times and one-year overall survival rates were 8.6 months and 34.4% for patients receiving temozolomide, and 9.6 months and 37.4% for those treated with standard RT, respectively, indicating that chemotherapy was non-inferior to standard RT. Similarly, median event-free survival times (progression or death as event) were not different: 3.3 months for patients treated with temozolomide and 4.7 months for those treated with standard RT. The major novelty of the study was the strong predictive role of *MGMT* promoter methylation status on event-free survival outcome. Event-free survival rates were significantly better in patients with *MGMT* methylated promoter than in those with *MGMT* unmethylated promoter, although this favorable impact was seen only in patients receiving temozolomide. For patients with a methylated *MGMT* promoter, event-free survival rates were better with temozolomide than RT, while the opposite was true for patients with an unmethylated *MGMT* promoter. Analysis of health-related quality of life scales showed no significant differences between the two groups; however, grade 2–4 adverse events were more frequent in patients receiving temozolomide.

Table 1. Selected prospective studies on radiotherapy or chemotherapy in older patients with glioblastoma.

Authors	Type of Study	pts	Age Yrs	RT Dose Gy/fr	CHT	Median PFS Months	Median OS Months	Toxicity	Neurological Outcome and Quality of Life (QoL)
McAleese JJ et al., 2003 [14]	Prospective	30 29	65–70 ≥70	30/6 30/6	no no	NR NR	6 m 37% 6 m 41%	Neurological deterioration. Occurred in 3% of patients.	68% of patients improved or remained stable, as assessed by Barthel score.
Chinot O et al., 2004 [53]	Prospective	32	≥70	no	TMZ*	5 1-yr 15%	6.4 1-yr 25%	Any grade 3–4 hematological toxicity 15%.	NR
Roa W et al., 2004 [15]	Randomized	51 49	≥60 ≥60	60/30 40/15	no no	NR NR	5.1 5.6	26% of patients receiving standard RT and 10% receiving short course RT discontinued RT for clinical deterioration.	No significant differences in KPS scores between groups; insufficient number of completed questionnaires for QoL evaluation.
Keime-Guiber F et al., 2007 [47]	Randomized	39 39	≥70 ≥70	50/28 no	no no	3.6 1.5	7 4	No grade 3–4 toxicity reported.	QoL (QLQ-BN20) and neurological function by mini-mental state examination (MMSE) showed no differences between groups.
Gallego Perez-Larraya et al., 2011 [18]	Prospective	70	≥70	no	TMZ*	4 1-yr 6.5%	6 1-yr 11.4%	Any grade 3–4 hematological toxicities 25%.	33% of patients improved their KPS by 10 or more points, and 18 (26%) became capable of self-care (KPS ≥ 70). MMSE and QLQ C30-BN20 improved.
Malmstrom et al., 2012 [19]	Randomized	100 98 93	>60 >60 >60	60/30 34/10 no	no no TMZ*	NA NA NA	6 (1-yr 17%) 7.5 (1-yr 23%) 8.3 (1-yr 27%)	72% completed standard RT and 95% hypofractionated RT; Grade 3–4 hematological toxicity in 19% of patients receiving TMZ.	Global health status between groups; better cognitive and phisical functioning in TMZ group at 3 months (QLQC30-BN20).
Wick et al., 2012 [20]	Randomized	178* 195*	>65 >65	60/30 no	no TMZ⁺	4.7 (1-yr 9.3%) 3.3 (1-yr 12%)	9.6 (1-yr 37.4%) 8.6 (1-yr 34.4%)	Grade 2–4 toxicities were more frequent in TMZ than RT group in all categories except for cutaneous adverse events.	QoL scales were siilar between groups (QLQC30-BN20), except for communication deficits, greater in RT group.
Roa et al., 2015 [17]	Randomized	48* 50*	≥65 ≥65	40/15 25/5	no no	4.2 4.2	7.9 6.4	No grade 3–4 acute toxicity.	Similar mean global QoL scores at 8 weeks.
Reyes-Botero, 2018 [54]	Prospective	66	≥70	no	TMZ* + Bev	4 months	5.8 months	Grade ≥ 3 hematological toxicity 20%, high blood pressure 24%, venous thromboembolism 4.5%, cerebral hemorrhage 3%.	Twenty-two (33%) patients became transiently capable of self-care (i.e., KPS > 70). Cognition and quality of life significantly improved over time during treament.

RT, radiotherapy; CHT, chemotherapy; OS, overall survival; PFS, progression-free survival; NR, not reported. TMZ, Temozolomide; *TMZ (200 mg/m^2 on days 1–5) every 4 weeks; ⁺TMZ (200 mg/m^2 1 week on/1 week off); Bev, bevacizumab.

In the Nordic trial, elderly patients receiving temozolomide had better survival outcome than those having RT. The median survival time was 8.3 months after temozolomide and 6.0 months after standard RT (HR 0.7, 95% CI 0.52–0.93, $p = 0.01$); however, with no significant differences between patients receiving standard RT and hypofractionated RT (6.0 months vs. 7.5 months, HR 0.82, 95% CI 0.63–1.06, $p = 0.12$) [19]. As for the NOA-8, a striking finding of the study was the predictive value of *MGMT* promoter methylation status. For patients receiving temozolomide, the median survival was 9.7 months in patients with *MGMT* promoter methylation and 6.8 months in those without methylation ($p = 0.02$). In contrast, *MGMT* methylation status did not affect the survival in patients having RT (methylated tumors, 8.2 months; unmethylated tumors, 6.8 months; $p = 0.81$). EORTC QLQ-C30 data evaluating global health status and cognitive functioning were generally better in patients receiving temozolomide than those having standard RT.

A French phase II trial including 70 patients aged 70 years and older with newly diagnosed GBM and a postoperative KPS score < 70, showed that temozolomide alone, given at doses of 150 to 200 mg/m^2/die for 5 days every 4 weeks until disease progression, resulted in improved functional status and quality of life, with a substantial proportion of patients who became capable of self-care, especially those with an *MGMT* methylated tumor [18]. The median and 6-month survival rates were 31 weeks and 69.2% in patients with *MGMT* promoter methylation and 18.7 weeks and 28% in those without methylation ($p = 0.03$), respectively. The same ANOCEF French group has recently published the results of another phase II trial exploring the combination of temozolomide and bevacizumab in 66 patients aged 70 years and older with a KPS < 70 and histologically confirmed GBM [54]. With a median overall survival (OS) of 23.9 weeks, cognition and quality of life significantly improved over time during treatment, and 33% of patients became transiently capable of self-care. Grade 3 or 4 hematologic toxicity and high blood pressure occurred in 20% of patients. Other toxicities, including venous thromboembolism, cerebral hemorrhage, and intestinal perforation were less common (<5%).

In summary, temozolomide is an effective and tolerated treatment for elderly patients with GBM associated with a significant improvement of functional status and quality of life. Response to treatment is significantly associated with *MGMT* promoter methylation status. For patients with *MGMT* methylated tumors, temozolomide results in longer compared with standard RT; in contrast, there is no evidence of a survival benefit in patients with *MGMT* unmethylated tumor. In clinical practice, this means that postoperative temozolomide should be considered only in patients with *MGMT* promoter methylated tumors, whereas its use is not recommended in those with *MGMT* unmethylated tumors.

3.4. Combined Chemoradiation

The use of standard or hypofractionated RT in combination with concomitant and/or adjuvant TMZ has been evaluated in several studies [55–69]. Results from published prospective series are shown in Table 2 [21,56,57,60,64].

Table 2. Selected studies on combined radiochemotherapy in older patients with glioblastoma.

Authors	Type of Study	Pts	Age yrs	RT Dose Gy/fr	CHT	Median PFS Months	Median OS Months	Toxicity	Neurological Outcomeand Quality of Life (QoL)
Minniti G et al., 2008 [56]	Prospective	32	≥70	60/30	TMZ	6.7 (1-yr 16%)	10.8 (1-yr 7%)	Neurological deterioration in 40%; grade 3–4 hematological toxicity 24%.	NR
Brandes et al., 2009 [57]	Prospective	58	≥65	60/30	TMZ	9.5 (1-yr 35%)	13.7 (2-yr 31.4%)	Grade 2 neurological deterioration, 31%; grade 3, 25%; grade 3–4 hematological toxicity, 9%.	NR
Minniti et al., 2009 [60]	Prospective	43	≥70	30/6	TMZ	6.3 (1-yr 12%)	9.3 (1-yr 35%)	Neurological deterioration in 16%; Grade 3–4 hematological toxicity 27%.	No significant decline in functioning scales and global health status (QLQC30-BN20) in patients free of disease progression.
Minniti et al., 2012 [64]	Prospective	70	≥70	40/15	TMZ	6 (1-yr 20%)	12.4 (1-yr 58%)	Grade 2/3 neurological toxicity, 10%; Grade 3–4 hematological toxicity, 29%.	Global health, social and cognitive functioning, and motor dysfunction improved over time (QLQC30-BN20); MMSE score improved or remained stable in 89% of patients free of disease progression.
Perry et al., 2016 [21]	Randomized	178*	>65	40/15	no	4.7 (1-yr 9.3%)	9.6 (1-yr 37.4%)	Grade 3–4 hematological toxicity in 25% and 9% of patients receiving RT plus TMZ or RT alone, respectively.	Changes from baseline scores during treatment and follow-up were similar by groups (QLQC30-BN20), with the exception of nausea and vomiting being worse in the RT + TMZ group.
		195*	>65	40/15	TMZ	3.3 (1-yr 12%)	8.6 (1-yr 34.4%)		

RT, radiotherapy; CHT, chemotherapy; OS, overall survival; PFS, progression-free survival; NR, not reporded TMZ, temozolomide given concomitantly (75 mg/m^2/day) and adjuvantly (200 mg/m^2 on days 1–5 every four weeks).

In a small prospective series of 32 patients aged 70 years and older with a good KPS receiving standard RT with adjuvant and concomitant temozolomide at Sant'Andrea Hospital (University of Rome Sapienza, Rome, Italy) the median survival time and one-year survival rates were 10.6 months and 37%, respectively [56]; grade 3 or 4 hematologic toxic effects occurred in 24% of patients. In another prospective study of 58 patients aged 65 years and older with GBM treated with standard RT and concomitant and adjuvant temozolomide, Brandes et al. [57] observed a median survival of 13.7 months. *MGMT* methylation status was an independent prognostic factor for survival; the two-year survival rates were 83% for patients with methylated tumors and 56% for those with unmethylated tumors. Grade 3 or 4 hematological toxicity occurred in 10% of patients and grade 3 neurocognitive deterioration in 25% of patients. A similar incidence of neurological toxicity up to one third of elderly patients with GBM treated standard chemoradiation has been reported in few series [58,59].

The efficacy and safety of hypofractionated RT with or without temozolomide has been recently evaluated by the intergroup EORTC 26062-22061/NCIC CTG (CE.3) randomized trial comparing an abbreviated course of RT (40 Gy in 15 fractions) plus concomitant and adjuvant temozolomide versus abbreviated RT alone in 562 patients older than 65 years old with newly diagnosed GBM [21]. The median survival time (9.3 vs. 7.6 months, $p < 0.0001$) and progression-free survival time (5.3 vs. 3.9 months, $p < 0.0001$) were significantly better in patients receiving combined chemoradiation over RT alone. *MGMT* promoter methylation status was a strong predictor for survival. Amongst 165 patients with *MGMT* methylated promoter, the overall survival was 13.5 months in patients receiving RT and temozolomide and 7.7 months in those receiving RT alone ($p = 0.0001$); in patients with *MGMT* unmethylated promoter, the respective overall survival times were 10.0 months and 7.9 months ($p = 0.055$). Quality of life analysis assessed by the EORTC QLQ-C30 questionnaire and the EORTC brain module (QLQ-BN20) showed that changes in global health status, functioning and symptom scales were similar in the two groups, although nausea and constipation were worse in patients receiving temozolomide.

A similar survival benefit has been reported in a phase 2 trial of 70 patients aged 70 years and older with newly diagnosed GBM treated with the same regimen at the University of Rome [64]. For the whole population, the median overall survival time and 1-year survival rates were 12.4 months and 58%, respectively. According to the *MGMT* promoter methylation status, the 1-year and 2-year survival rates were 81% and 20% in patients with *MGMT* methylated tumors, and 32% and 0% in those with *MGMT* unmethylated tumors, respectively ($p = 0.0001$). The treatment was well tolerated and resulted in a significant improvement or stability in global health, social functioning, and cognitive functioning scores between baseline and 6-month follow-up.

In summary, combined chemoradiation represents an effective therapeutic strategy for elderly patients with GBM. Standard RT and concomitant and adjuvant temozolomide remain a feasible treatment in patients aged less than 70 years with good KPS, although treatment-related neurotoxicity leading to serious disability and worsening of the quality of life represents a major concern. Based on the results of the EORTC 26062-22061/NCIC CTG (CE.3) randomized trial, elderly patients aged 70 year and older with *MGMT* promoter methylation who are considered eligible for combined modality treatment should be offered a short-course RT, 40 Gy in 15 fractions, in combination with concomitant and adjuvant temozolomide. Elderly patients not considered candidates for combined chemoradiation should be treated with short-course RT or temozolomide based on *MGMT* promoter methylation status [70].

4. Future Perspectives

Combined chemoradiation remains a matter of concern in frail patients presenting with functional deficits, multiple comorbidities and geriatric syndromes, like gait imbalance, malnutrition, delirium, and incontinence that make them more vulnerable to treatment-related toxicities [71–73]. Identification of frail, vulnerable, or fit patients is essential for making more appropriate treatment decisions for elderly patients. A comprehensive geriatric assessment which includes the evaluation of functional status, cognitive function, nutritional status, comorbidities, polipharmacy, and socioeconomic status needs to be incorporated in future clinical trials with the aim of improving treatment outcome and reducing the risk of adverse events [74,75]. Several instruments may be used to assess the different domains of geriatric assessment, including performance status scales, daily activities, cognitive function, presence of comorbidities, psycological status, health and nutritional status, and socioeconomic status [46,74,76]. For daily clinical practice, several screening tests have been developed, including the abbreviated comprehensive geriatric assessment (aCGA) [77], the Groningen frailty indicator (GFI), as the G8 [78], and the vulnerable elders survey-13 (VES-13) [79] to help neurooncologists to guide cancer treatment decision-making and improve quality of life and functional independence of elderly patients with GBM. Based on geriatric assessment, patients with a higher frailty index score would receive less aggressive treatments; e.g., temozolomide or RT alone according to *MGMT* methylation promoter status.

Recently, experts from United States, Canada, and Europe have developed the Neurologic Assessment in Neuro-Oncology (NANO) scale which is an objective and simple tool for an accurate assessment of neurological function [45]. The NANO scale evaluates nine major domains of neurologic function that are most relevant to patients with supratentorial, infratentorial, and brainstem tumors, including gait, strength, upper extremity ataxia, sensation, visual fields, facial strength, language, level of consciousness, and behavior. It will provide an accurate neurological evaluation of elderly patients with GBM in both clinical trials and daily practice.

A novel treatment modality for patients with GBM is represented by the tumor-treating fields (TTFields) device (Optune®, Novocure Ltd., Novocure Inc, Israel) which is a portable, battery-operated device that generates TTFields. Results from a prospective phase 3 trial, EF-14, comparing TTFields plus temozolomide versus temozolomide alone after standard chemoradiation in patients with GBM has shown significant longer survival and clinical improvement in those having TTFields plus temozolomide [80]. Based on these results, Optune® has received FDA approval for adult patients with newly diagnosed supratentorial glioblastoma, in addition to standard postoperative chemotherapy, or as monotherapy for the treatment of recurrent GBM.

For patient of all ages, surgery, radiation and systemic therapies have been employed for recurrent tumors, although standards of care are not well defined. Currently, there are no prospective studies evaluating the management of older patients with recurrent GBM. Outside of the context of clinical trials, systemic therapy with either lomustine or bevacizumab may represent a feasible treatment option for fit patients with recurrent tumors. Randomized studies evaluating the efficacy of lomustine given alone or in combination with other agents have observed a median survival in the range of 8–10 months [81–83]. In a phase III trial of 437 patients assigned to receive lomustine plus bevacizumab or lomustine alone, Wick et al. [83] showed a similar survival between the two groups (9.1 months versus 8.6 months; $p = 0.65$). Grade 3 to 5 adverse events occurred in 63.6% of the patients in the combination group and 38.1% of the patients in the monotherapy group; this means that the use of lomustine should be carefully considered in older and frail patients at increased risk of toxicity. The antiangiogenic agent bevacizumab has been approved at recurrence in various countries, but not in the European Union. Although bevacizumab did not have superior efficacy compared with lomustine [82,83], it produces evident symptom relief and steroid-sparing effects. Future clinical trials need to evaluate the efficacy and safety of different treatment approaches in elderly patients with recurrent GBM and good functional status.

Although data from EORTC 26062-22061/NCIC CTG (CE.3) randomized trial support the combination of hypofractionated RT with temozolomide in patients older than 65 years of age, the survival benefit in patients with unmethylated tumors remains questionable. In addition, no trials have compared the survival and neurocognitive outcomes of combined chemoradiation versus temozolomide alone in patients with *MGMT* promoter methylation. Thus, future trials are needed to explore these important issues. Finally, molecular profiling contributes to identify prognostic subgroups of elderly patients with GBM who may benefit from new therapies. Although recent trials have failed to demonstrate the efficacy of new agents in patients with GBM, including bevacizumab and cilengitide, the discovery of targeted agents and immunotherapy strategies which can translate to a survival benefit in patients with GBM remain an area of continued research.

5. Conclusions

For elderly patients with newly diagnosed GBM, current management includes surgery, RT and chemotherapy; however, survival is significantly worse than that observed in younger patients. Standard RT with concomitant and adjuvant temozolomide, which represent the standard of care for newly diagnosed adult patients with GBM in good general and neurological condition, may be considered in selected fit patients aged between 65 and 70 years. Elderly patients aged 70 years and older who are considered eligible for combined modality treatment should receive a short-course RT with concomitant and adjuvant temozolomide up to 12 cycles. In the absence of comparative data between temozolomide alone and chemoradiation, elderly patients with *MGMT* promoter methylation may be considered for temozolomide alone, especially those presenting with functional impairment and geriatric syndromes. Temozolomide alone is not associated with survival advantages in elderly patients with unmethylated tumors; patients who are not considered eligible for combined chemoradiation should receive hypofractionated RT. Supportive and palliative care may represent an appropriate approach for frail patients with large or multifocal tumors and low KPS at an increased of treatment-related toxicity.

Author Contributions: G.M., G.L., and S.P. participated in article preparation, data analysis and wrote the manuscript. All authors have approved the final article.

Funding: This research received no external funding.

Conflicts of Interest: All authors declare that they have no competing interests.

References

1. Jukich, P.J.; McCarthy, B.J.; Surawicz, T.S.; Freels, S.; Davis, F.G. Trends in incidence of primary brain tumors in the United States, 1985–1994. *Neuro Oncol.* **2001**, *3*, 141–151. [CrossRef]
2. Parkin, D.M. Global cancer statistics in the year 2000. *Lancet Oncol.* **2001**, *2*, 533–543. [CrossRef]
3. Ostrom, Q.T.; Gittleman, H.; Truitt, G.; Boscia, A.; Kruchko, C.; Barnholtz-Sloan, J.S. CBTRUS Statistical Report: Primary Brain and Other Central Nervous System Tumors Diagnosed in the United States in 2011–2015. *Neuro Oncol.* **2018**, *20*, iv1–iv86. [CrossRef]
4. Stupp, R.; Mason, W.P.; van den Bent, M.J.; Weller, M.; Fisher, B.; Taphoorn, M.J.; Belanger, K.; Brandes, A.A.; Marosi, C.; Bogdahn, U.; et al. Radiotherapy plus concomitant and adjuvant Temozolomide for glioblastoma. *N. Engl. J. Med.* **2005**, *352*, 987–996. [CrossRef]
5. Stupp, R.; Hegi, M.E.; Mason, W.P.; van den Bent, M.J.; Taphoorn, M.J.; Janzer, R.C.; Ludwin, S.K.; Allgeier, A.; Fisher, B.; Belanger, K.; et al. Effects of radiotherapy with concomitant and adjuvant temozolomide versus radiotherapy alone on survival in glioblastoma in a randomised phase III study: 5-year analysis of the EORTC-NCIC trial. *Lancet Oncol.* **2009**, *10*, 459–466. [CrossRef]
6. Bauman, G.S.; Gaspar, L.E.; Fisher, B.J.; Halperin, E.C.; Macdonald, D.R.; Cairncross, J.G. A prospective study of short course RT in poor prognosis glioblastoma multiforme. *Int. J. Radiat. Oncol. Biol. Phys.* **1994**, *29*, 835–839. [CrossRef]
7. Thomas, R.; James, N.; Guerrero, D.; Ashley, S.; Gregor, A.; Brada, M. Hypofractionated radiotherapy as palliative treatment in poor prognosis patients with high grade glioma. *Radiother. Oncol.* **1994**, *33*, 113–116. [CrossRef]
8. Ford, J.M.; Stenning, S.P.; Boote, D.J.; Counsell, R.; Falk, S.J.; Flavin, A.; Laurence, V.M.; Bleehen, N.M. A short fractionation radiotherapy treatment for poor prognosis patients with high grade glioma. *Clin. Oncol.* **1997**, *9*, 20–24. [CrossRef]
9. Hoegler, D.B.; Davey, P. A prospective study of short course radiotherapy in elderly patients with malignant glioma. *J. Neurooncol.* **1997**, *33*, 201–214. [CrossRef] [PubMed]

10. Mohan, D.S.; Suh, J.H.; Phan, J.L.; Kupelian, P.A.; Cohen, B.H.; Barnett, G.H. Outcome in elderly patients undergoing definitive surgery and radiation therapy for supratentorial glioblastoma multiforme at a tertiary care institution. *Int. J. Radiat. Oncol. Biol. Phys.* **1998**, *42*, 981–987. [CrossRef]
11. Villà, S.; Viñolas, N.; Verger, E.; Yaya, R.; Martínez, A.; Gil, M.; Moreno, V.; Caral, L.; Graus, F. Efficacy of radiotherapy for malignant gliomas in elderly patients. *Int. J. Radiat. Oncol. Biol. Phys.* **1998**, *42*, 977–980. [CrossRef]
12. Pierga, J.Y.; Hoang-Xuan, K.; Feuvret, L.; Simon, J.M.; Cornu, P.; Baillet, F.; Mazeron, J.J.; Delattre, J.Y. Treatment of malignant gliomas in the elderly. *J. Neurooncol.* **1999**, *43*, 187–193. [CrossRef] [PubMed]
13. Whittle, I.R.; Basu, N.; Grant, R.; Walker, M.; Gregor, A. Management of patients aged >60 years with malignant glioma: Good clinical status and radiotherapy determine outcome. *Br. J. Neurosurg.* **2002**, *16*, 343–347. [CrossRef] [PubMed]
14. McAleese, J.J.; Stenning, S.P.; Ashley, S.; Traish, D.; Hines, F.; Sardell, S.; Guerrero, D.; Brada, M. Hypofractionated radiotherapy for poor prognosis malignant glioma: Matched pair survival analysis with MRC controls. *Radiother. Oncol.* **2003**, *67*, 177–182. [CrossRef]
15. Roa, W.; Brasher, P.M.; Bauman, G.; Anthes, M.; Bruera, E.; Chan, A.; Fisher, B.; Fulton, D.; Gulavita, S.; Hao, C.; et al. Abbreviated Course of Radiation Therapy in Older Patients with Glioblastoma Multiforme: A Prospective Randomized Clinical Trial. *J. Clin. Oncol.* **2004**, *22*, 1593–1598.
16. Marijnen, C.A.; van den Berg, S.M.; van Duinen, S.G.; Voormolen, J.H.; Noordijk, E.M. Radiotherapy is effective in patients with glioblastoma multiforme with a limited prognosis and in patients above 70 years of age: A retrospective single institution analysis. *Radiother. Oncol.* **2005**, *75*, 210–216. [CrossRef]
17. Roa, W.; Kepka, L.; Kumar, N.; Sinaika, V.; Matiello, J.; Lomidze, D.; Hentati, D.; Guedes de Castro, D.; Dyttus-Cebulok, K.; Drodge, S.; et al. International Atomic Energy Agency Randomized Phase III Study of Radiation Therapy in Elderly and/or Frail Patients with Newly Diagnosed Glioblastoma Multiforme. *J. Clin. Oncol.* **2015**, *33*, 4145–4150. [CrossRef]
18. Gállego Pérez-Larraya, J.; Ducray, F.; Chinot, O.; Catry-Thomas, I.; Taillandier, L.; Guillamo, J.S.; Campello, C.; Monjour, A.; Cartalat-Carel, S.; Barrie, M.; et al. Temozolomide in elderly patients with newly diagnosed glioblastoma and poor performance status: An ANOCEF phase II trial. *J. Clin. Oncol.* **2011**, *29*, 3050–3055. [CrossRef]
19. Malmström, A.; Grønberg, B.H.; Marosi, C.; Stupp, R.; Frappaz, D.; Schultz, H.; Abacioglu, U.; Tavelin, B.; Lhermitte, B.; Hegi, M.E.; et al. Temozolomide versus standard 6-week radiotherapy versus hypofractionated radiotherapy in patients older than 60 years with glioblastoma: The Nordic randomised, phase 3 trial. *Lancet Oncol.* **2012**, *13*, 916–926. [CrossRef]
20. Wick, W.; Platten, M.; Meisner, C.; Felsberg, J.; Tabatabai, G.; Simon, M.; Nikkhah, G.; Papsdorf, K.; Steinbach, J.P.; Sabel, M.; et al. Temozolomide chemotherapy alone versus radiotherapy alone for malignant astrocytoma in the elderly: The NOA-08 randomised, phase 3 trial. *Lancet Oncol.* **2012**, *13*, 707–715. [CrossRef]
21. Perry, J.R.; Laperriere, N.; O'Callaghan, C.J.; Brandes, A.A.; Menten, J.; Phillips, C.; Fay, M.; Nishikawa, R.; Cairncross, J.G.; Roa, W.; et al. Trial Investigators. Short-Course Radiation plus Temozolomide in Elderly Patients with Glioblastoma. *N. Engl. J. Med.* **2017**, *376*, 1027–1037. [CrossRef] [PubMed]
22. Simpson, J.R.; Horton, J.; Scott, C.; Curran, W.J.; Rubin, P.; Fischbach, J.; Isaacson, S.; Rotman, M.; Asbell, S.O.; Nelson, J.S.; et al. Influence of location and extent of surgical resection on survival of patients with glioblastoma multiforme: Results of three consecutive Radiation Therapy Oncology Group (RTOG) clinical trials. *Int. J. Radiat. Oncol. Biol. Phys.* **1993**, *26*, 239–244. [CrossRef]
23. Lacroix, M.; Abi-Said, D.; Fourney, D.R.; Gokaslan, Z.L.; Shi, W.; DeMonte, F.; Lang, F.F.; McCutcheon, I.E.; Hassenbusch, S.J.; Holland, E.; et al. A multivariate analysis of 416 patients with glioblastoma multiforme: Prognosis, extent of resection, and survival. *J. Neurosurg.* **2001**, *95*, 190–198. [CrossRef]
24. Laws, E.R.; Shaffrey, M.E.; Morris, A.; Anderson, F.A., Jr. Surgical management of intracranial gliomas—Does radical resection improve outcome? *Acta Neurochir. Suppl.* **2003**, *85*, 47–53. [PubMed]
25. Brown, P.D.; Maurer, M.J.; Rummans, T.A.; Clark, M.M.; Frost, M.H.; Ballman, K.V.; Arusell, R.M.; Buckner, J.C. A prospective study of quality of life in adults with newly diagnosed high-grade gliomas: The impact of the extent of resection on quality of life and survival. *Neurosurgery* **2005**, *57*, 495–504. [CrossRef]
26. Mitchell, P.; Ellison, D.W.; Mendelow, A.D. Surgery for malignant gliomas: Mechanistic reasoning and slippery statistics. *Lancet Neurol.* **2005**, *4*, 413–422. [CrossRef]

27. Litofsky, N.S.; Bauer, A.M.; Kasper, R.S.; Sullivan, C.M.; Dabbous, O.H. Image-guided resection of high-grade glioma: Patient selection factors and outcome. *Neurosurg. Focus* **2006**, *20*, E16. [CrossRef] [PubMed]
28. Stummer, W.; Pichlmeier, U.; Meinel, T.; Wiestler, O.D.; Zanella, F.; Reulen, H.; ALA-Glioma Study Group. Fluorescence-guided surgery with 5-aminolevulinic acid for resection of malignant glioma: A randomised controlled multicentre phase III trial. *Lancet Oncol.* **2006**, *7*, 392–401. [CrossRef]
29. Stummer, W.; Reulen, H.J.; Meinel, T.; Pichlmeier, U.; Schumacher, W.; Tonn, J.C.; Rohde, V.; Oppel, F.; Turowski, B.; Woiciechowsky, C.; et al. Extent of resection and survival in glioblastoma multiforme: Identification of and adjustment for bias. *Neurosurgery* **2008**, *62*, 564–576. [PubMed]
30. Mineo, J.F.; Bordron, A.; Baroncini, M.; Ramirez, C.; Maurage, C.A.; Blond, S.; Dam-Hieu, P. Prognosis factors of survival time in patients with glioblastoma multiforme: A multivariate analysis of 340 patients. *Acta Neurochir.* **2007**, *149*, 245–252. [CrossRef] [PubMed]
31. Brown, T.J.; Brennan, M.C.; Li, M.; Church, E.W.; Brandmeir, N.J.; Rakszawski, K.L.; Patel, A.S.; Rizk, E.B.; Suki, D.; Sawaya, R.; et al. Association of the Extent of Resection with Survival in Glioblastoma: A Systematic Review and Meta-analysis. *JAMA Oncol.* **2016**, *2*, 1460–1469. [CrossRef]
32. Li, Y.M.; Suki, D.; Hess, K.; Sawaya, R. The influence of maximum safe resection of glioblastoma on survival in 1229 patients: Can we do better than gross-total resection? *J. Neurosurg.* **2016**, *124*, 977–988. [CrossRef] [PubMed]
33. Jenkinson, M.D.; Barone, D.G.; Bryant, A.; Vale, L.; Bulbeck, H.; Lawrie, T.A.; Hart, M.G.; Watts, C. Intraoperative imaging technology to maximise extent of resection for glioma. *Cochrane Database Syst. Rev.* **2018**, *1*, CD012788. [CrossRef] [PubMed]
34. Kelly, P.J.; Hunt, C. The limited value of cytoreductive surgery in elderly patients with malignant gliomas. *Neurosurgery* **1994**, *34*, 62–66. [PubMed]
35. Vuorinen, V.; Hinkka, S.; Färkkilä, M.; Jääskeläinen, J. Debulking or biopsy of malignant glioma in elderly people—A randomised study. *Acta Neurochir.* **2003**, *145*, 5–10. [CrossRef] [PubMed]
36. Ewelt, C.; Goeppert, M.; Rapp, M.; Steiger, H.J.; Stummer, W.; Sabel, M. Glioblastoma multiforme of the elderly: The prognostic effect of resection on survival. *J. Neurooncol.* **2011**, *103*, 611–618. [CrossRef] [PubMed]
37. Almenawer, S.A.; Badhiwala, J.H.; Alhazzani, W.; Greenspoon, J.; Farrokhyar, F.; Yarascavitch, B.; Algird, A.; Kachur, E.; Cenic, A.; Sharieff, W.; et al. Biopsy versus partial versus gross total resection in older patients with high-grade glioma: A systematic review and meta-analysis. *Neuro Oncol.* **2015**, *17*, 868–881. [CrossRef]
38. Babu, R.; Komisarow, J.M.; Agarwal, V.J.; Rahimpour, S.; Iyer, A.; Britt, D.; Karikari, I.O.; Grossi, P.M.; Thomas, S.; Friedman, A.H.; et al. Glioblastoma in the elderly: The effect of aggressive and modern therapies on survival. *J. Neurosurg.* **2016**, *124*, 998–1007. [CrossRef]
39. Hager, J.; Herrmann, E.; Kammerer, S.; Dinc, N.; Won, S.Y.; Senft, C.; Seifert, V.; Marquardt, G.; Quick-Weller, J. Impact of resection on overall survival of recurrent Glioblastoma in elderly patients. *Clin. Neurol. Neurosurg.* **2018**, *174*, 21–25. [CrossRef]
40. Karsy, M.; Yoon, N.; Boettcher, L.; Jensen, R.; Shah, L.; MacDonald, J.; Menacho, S.T. Surgical treatment of glioblastoma in the elderly: The impact of complications. *J. Neurooncol.* **2018**, *138*, 123–132. [CrossRef]
41. Chaichana, K.L.; Garzon-Muvdi, T.; Parker, S.; Weingart, J.D.; Olivi, A.; Bennett, R.; Brem, H.; Quiñones-Hinojosa, A. Supratentorial glioblastoma multiforme: The role of surgical resection versus biopsy among older patients. *Ann. Surg. Oncol.* **2011**, *18*, 239–245. [CrossRef]
42. Metcalfe, S.E.; Grant, R. Biopsy versus resection for malignant glioma. *Cochrane Database Syst. Rev.* **2001**, *3*, CD002034.
43. McGirt, M.J.; Woodworth, G.F.; Coon, A.L.; Frazier, J.M.; Amundson, E.; Garonzik, I.; Olivi, A.; Weingart, J.D. Independent predictors of morbidity after image-guided stereotactic brain biopsy: A risk assessment of 270 cases. *J. Neurosurg.* **2005**, *102*, 897–901. [CrossRef]
44. Kongkham, P.N.; Knifed, E.; Tamber, M.S.; Bernstein, M. Complications in 622 cases of frame-based stereotactic biopsy, a decreasing procedure. *Can. J. Neurol. Sci.* **2008**, *35*, 79–84. [CrossRef]
45. Nayak, L.; DeAngelis, L.M.; Brandes, A.A.; Peereboom, D.M.; Galanis, E.; Lin, N.U.; Soffietti, R.; Macdonald, D.R.; Chamberlain, M.; Perry, J.; et al. The Neurologic Assessment in Neuro-Oncology (NANO) scale: A tool to assess neurologic function for integration into the Response Assessment in Neuro-Oncology (RANO) criteria. *Neuro Oncol.* **2017**, *19*, 625–635. [CrossRef] [PubMed]
46. Minniti, G.; Filippi, A.R.; Osti, M.F.; Ricardi, U. Radiation therapy for older patients with brain tumors. *Radiat. Oncol.* **2017**, *12*, 101. [CrossRef]

47. Keime-Guibert, F.; Chinot, O.; Taillandier, L.; Cartalat-Carel, S.; Frenay, M.; Kantor, G.; Guillamo, J.S.; Jadaud, E.; Colin, P.; Bondiau, P.Y.; et al. Association of French-Speaking Neuro-Oncologists. Radiotherapy for glioblastoma in the elderly. *N. Engl. J. Med.* **2007**, *356*, 1527–1535. [CrossRef] [PubMed]
48. Scott, J.; Tsai, Y.Y.; Chinnaiyan, P.; Yu, H.H. Effectiveness of radiotherapy for elderly patients with glioblastoma. *Int. J. Radiat. Oncol. Biol. Phys.* **2011**, *81*, 206–210. [CrossRef]
49. Rosenblum, M.L.; Gerosa, M.; Dougherty, D.V.; Reese, C.; Barger, G.R.; Davis, R.L.; Levin, V.A.; Wilson, C.B. Age-related chemosensitivity of stem cells from human malignant brain tumours. *Lancet* **1982**, *1*, 885–887. [CrossRef]
50. Shapiro, W.R.; Shapiro, J.R. Biology and treatment of malignant glioma. *Oncology* **1998**, *12*, 233–240.
51. Yellen, S.B.; Cella, D.F.; Leslie, W.T. Age and clinical decision making in oncology patients. *J. Natl. Cancer Inst.* **1994**, *86*, 1766–1770. [CrossRef]
52. Hutchins, L.F.; Unger, J.M.; Crowley, J.J.; Coltman, C.A., Jr.; Albain, K.S. Underrepresentation of patients 65 years of age or older in cancer-treatment trials. *N. Engl. J. Med.* **1999**, *341*, 2061–2067. [CrossRef] [PubMed]
53. Chinot, O.L.; Barrie, M.; Frauger, E.; Dufour, H.; Figarella-Branger, D.; Palmari, J.; Braguer, D.; Hoang-Xuan, K.; Moktari, K.; Peragut, J.C.; et al. Phase II study of temozolomide without radiotherapy in newly diagnosed glioblastoma multiforme in an elderly populations. *Cancer* **2004**, *100*, 2208–2214. [CrossRef]
54. Reyes-Botero, G.; Cartalat-Carel, S.; Chinot, O.L.; Barrie, M.; Taillandier, L.; Beauchesne, P.; Catry-Thomas, I.; Barrière, J.; Guillamo, J.S.; Fabbro, M.; et al. Temozolomide Plus Bevacizumab in Elderly Patients with Newly Diagnosed Glioblastoma and Poor Performance Status: An ANOCEF Phase II Trial (ATAG). *Oncologist* **2018**, *23*, 524-e44. [CrossRef] [PubMed]
55. Combs, S.E.; Wagner, J.; Bischof, M.; Welzel, T.; Wagner, F.; Debus, J.; Schulz-Ertner, D. Postoperative treatment of primary glioblastoma multiforme with radiation and concomitant temozolomide in elderly patients. *Int. J. Radiat. Oncol. Biol. Phys.* **2008**, *70*, 987–992. [CrossRef] [PubMed]
56. Minniti, G.; De Sanctis, V.; Muni, R.; Filippone, F.; Bozzao, A.; Valeriani, M.; Osti, M.F.; De Paula, U.; Lanzetta, G.; Tombolini, V.; et al. Radiotherapy plus concomitant and adjuvant temozolomide for glioblastoma in elderly patients. *J. Neurooncol.* **2008**, *88*, 97–103. [CrossRef]
57. Brandes, A.A.; Franceschi, E.; Tosoni, A.; Benevento, F.; Scopece, L.; Mazzocchi, V.; Bacci, A.; Agati, R.; Calbucci, F.; Ermani, M. Temozolomide concomitant and adjuvant to radiotherapy in elderly patients with glioblastoma: Correlation with MGMT promoter methylation status. *Cancer* **2009**, *115*, 3512–3518. [CrossRef]
58. Gerstner, E.R.; Yip, S.; Wang, D.L.; Louis, D.N.; Iafrate, A.J.; Batchelor, T.T. Mgmt methylation is a prognostic biomarker in elderly patients with newly diagnosed glioblastoma. *Neurology* **2009**, *73*, 1509–1510. [CrossRef]
59. Gerstein, J.; Franz, K.; Steinbach, J.P.; Seifert, V.; Fraunholz, I.; Weiss, C.; Rödel, C. Postoperative radiotherapy and concomitant temozolomide for elderly patients withglioblastoma. *Radiother. Oncol.* **2010**, *97*, 382–386. [CrossRef]
60. Minniti, G.; De Sanctis, V.; Muni, R.; Rasio, D.; Lanzetta, G.; Bozzao, A.; Osti, M.F.; Salvati, M.; Valeriani, M.; Cantore, G.P.; et al. Hypofractionated radiotherapy followed by adjuvant chemotherapy with temozolomide in elderly patients with glioblastoma. *J. Neurooncol.* **2009**, *91*, 95–100. [CrossRef]
61. Minniti, G.; Salvati, M.; Arcella, A.; Buttarelli, F.; D'Elia, A.; Lanzetta, G.; Esposito, V.; Scarpino, S.; Maurizi Enrici, R.; Giangaspero, F. Correlation between O6-methylguanine-DNA methyltransferase and survival in elderly patients with glioblastoma treated with radiotherapy plus concomitant and adjuvant temozolomide. *J. Neurooncol.* **2011**, *102*, 311–316. [CrossRef] [PubMed]
62. Barker, C.A.; Chang, M.; Chou, J.F.; Zhang, Z.; Beal, K.; Gutin, P.H.; Iwamoto, F.M. Radiotherapy and concomitant temozolomide may improve survival of elderly patients with glioblastoma. *J. Neurooncol.* **2012**, *109*, 391–397. [CrossRef]
63. Niyazi, M.; Schwarz, S.B.; Suchorska, B.; Belka, C. Radiotherapy with and without temozolomide in elderly patients with glioblastoma. *Strahlenther. Onkol.* **2012**, *188*, 154–159. [CrossRef] [PubMed]
64. Minniti, G.; Lanzetta, G.; Scaringi, C.; Caporello, P.; Salvati, M.; Arcella, A.; De Sanctis, V.; Giangaspero, F.; Enrici, R.M. Phase II study of short-course radiotherapy plus concomitant and adjuvant temozolomide in elderly patients with glioblastoma. *Int. J. Radiat. Oncol. Biol. Phys.* **2012**, *83*, 93–99. [CrossRef] [PubMed]
65. Behm, T.; Horowski, A.; Schneider, S.; Bock, H.C.; Mielke, D.; Rohde, V.; Stockhammer, F. Concomitant and adjuvant temozolomide of newly diagnosed glioblastoma in elderly patients. *Clin. Neurol. Neurosurg.* **2013**, *115*, 2142–2146. [CrossRef] [PubMed]

66. Minniti, G.; Scaringi, C.; Baldoni, A.; Lanzetta, G.; De Sanctis, V.; Esposito, V.; Enrici, R.M. Health-related quality of life in elderly patients with newly diagnosed glioblastoma treated with short-course radiation therapy plus concomitant and adjuvant temozolomide. *Int. J. Radiat. Oncol. Biol. Phys.* **2013**, *86*, 285–291. [CrossRef] [PubMed]
67. Nguyen, L.T.; Touch, S.; Nehme-Schuster, H.; Antoni, D.; Eav, S.; Clavier, J.B.; Bauer, N.; Vigneron, C.; Schott, R.; Kehrli, P.; et al. Outcomes in newly diagnosed elderly glioblastoma patients after concomitant temozolomide administration and hypofractionated radiotherapy. *Cancers* **2013**, *5*, 1177–1198. [CrossRef]
68. Uzuka, T.; Asano, K.; Sasajima, T.; Sakurada, K.; Kumabe, T.; Beppu, T.; Ichikawa, M.; Kitanaka, C.; Aoki, H.; Saito, K.; et al. Treatment outcomes in glioblastoma patients aged 76 years or older: A multicenter retrospective cohort study. *J. Neurooncol.* **2014**, *116*, 299–306. [CrossRef]
69. Lombardi, G.; Pace, A.; Pasqualetti, F.; Rizzato, S.; Faedi, M.; Anghileri, E.; Nicolotto, E.; Bazzoli, E.; Bellu, L.; Villani, V.; et al. Predictors of survival and effect of short (40 Gy) or standard-course (60 Gy) irradiation plus concomitant temozolomide in elderly patients with glioblastoma: A multicenter retrospective study of AINO (Italian Association of Neuro-Oncology). *J. Neurooncol.* **2015**, *125*, 359–367. [CrossRef]
70. Weller, M.; van den Bent, M.; Tonn, J.C.; Stupp, R.; Preusser, M.; Cohen-Jonathan-Moyal, E.; Henriksson, R.; Le Rhun, E.; Balana, C.; Chinot, O.; et al. European Association for Neuro-Oncology (EANO) Task Force on Gliomas. European Association for Neuro-Oncology (EANO) guideline on the diagnosis and treatment of adult astrocytic and oligodendroglial gliomas. *Lancet Oncol.* **2017**, *18*, e315–e329. [CrossRef]
71. Hoffe, S.; Balducci, L. Cancer and age: General considerations. *Clin. Geriatr. Med.* **2012**, *28*, 1–18. [CrossRef] [PubMed]
72. Hamaker, M.E.; Jonker, J.M.; de Rooij, S.E.; Vos, A.G.; Smorenburg, C.H.; van Munster, B.C. Frailty screening methods for predicting outcome of a comprehensive geriatric assessment in elderly patients with cancer: A systematic review. *Lancet Oncol.* **2012**, *13*, e437–e444. [CrossRef]
73. Ferrat, E.; Paillaud, E.; Caillet, P.; Laurent, M.; Tournigand, C.; Lagrange, J.L.; Droz, J.P.; Balducci, L.; Audureau, E.; Canouï-Poitrine, F.; et al. Performance of Four Frailty Classifications in Older Patients with Cancer: Prospective Elderly Cancer Patients Cohort Study. *J. Clin. Oncol.* **2017**, *35*, 766–777. [CrossRef] [PubMed]
74. Extermann, M.; Hurria, A. Comprehensive geriatric assessment for older patients with cancer. *J. Clin. Oncol.* **2007**, *25*, 1824–1831. [CrossRef] [PubMed]
75. Wildiers, H.; Heeren, P.; Puts, M.; Topinkova, E.; Janssen-Heijnen, M.L.; Extermann, M.; Falandry, C.; Artz, A.; Brain, E.; Colloca, G.; et al. International Society of Geriatric Oncology Consensus on Geriatric Assessment in Older Patients with Cancer. *J. Clin. Oncol.* **2014**, *32*, 2595–2603. [CrossRef] [PubMed]
76. Extermann, M.; Aapro, M.; Bernabei, R.; Cohen, H.J.; Droz, J.P.; Lichtman, S.; Mor, V.; Monfardini, S.; Repetto, L.; Sørbye, L.; et al. Task Force on CGA of the International Society of Geriatric Oncology. Use of comprehensive geriatric assessment in older cancer patients: Recommendations from the task force on CGA of the International Society of Geriatric Oncology (SIOG). *Crit. Rev. Oncol. Hematol.* **2005**, *55*, 241–252. [CrossRef] [PubMed]
77. Overcash, J.A.; Beckstead, J.; Moody, L.; Extermann, M.; Cobb, S. The abbreviated comprehensive geriatric assessment (aCGA) for use in the older cancer patient as a prescreen: Scoring and interpretation. *Crit. Rev. Oncol. Hematol.* **2006**, *59*, 205–210. [CrossRef] [PubMed]
78. Bielderman, A.; Van Der Schans, C.P.; van Lieshout, M.R.; de Greef, M.H.; Boersma, F.; Krijnen, W.P.; Steverink, N. Multidimensional structure of the Groningen Frailty Indicator in community-dwelling older People. *BMC Geriatr.* **2013**, *13*, 86. [CrossRef]
79. Saliba, D.; Elliott, M.; Rubenstein, L.Z.; Solomon, D.H.; Young, R.T.; Kamberg, C.J.; Roth, C.; MacLean, C.H.; Shekelle, P.G.; Sloss, E.M.; et al. The vulnerable elders survey: A tool for identifying vulnerable older People in the community. *J. Am. Geriatr. Soc.* **2001**, *49*, 1691–1699. [CrossRef]
80. Mehta, M.; Wen, P.; Nishikawa, R.; Reardon, D.; Peters, K. Critical review of the addition of tumor treating fields (TTFields) to the existing standard of care for newly diagnosed glioblastoma patients. *Crit. Rev. Oncol. Hematol.* **2017**, *111*, 60–65. [CrossRef]
81. Batchelor, T.T.; Mulholland, P.; Neyns, B.; Nabors, L.B.; Campone, M.; Wick, A.; Mason, W.; Mikkelsen, T.; Phuphanich, S.; Ashby, L.S.; et al. Phase III randomized trial comparing the efficacy of cediranib as monotherapy, and in combination with lomustine, versus lomustine alone in patients with recurrent glioblastoma. *J. Clin. Oncol.* **2013**, *31*, 3212–3218. [CrossRef] [PubMed]

82. Taal, W.; Oosterkamp, H.M.; Walenkamp, A.M.; Dubbink, H.J.; Beerepoot, L.V.; Hanse, M.C.; Buter, J.; Honkoop, A.H.; Boerman, D.; de Vos, F.Y.; et al. Single-agent bevacizumab or lomustine versus combination of bevacizumab plus lomustine in patients with recurrent glioblastoma (BELOB trial): A randomised controlled phase 2 trial. *Lancet Oncol.* **2014**, *15*, 943–953. [CrossRef]
83. Wick, W.; Gorlia, T.; Bendszus, M.; Taphoorn, M.; Sahm, F.; Harting, I.; Brandes, A.A.; Taal, W.; Domont, J.; Idbaih, A.; et al. Lomustine and Bevacizumab in Progressive Glioblastoma. *N. Engl. J. Med.* **2017**, *377*, 1954–1963. [CrossRef] [PubMed]

cancers

MDPI

Review

Allergic Signs in Glioma Pathology: Current Knowledge and Future Perspectives

Massimo Costanza * and Gaetano Finocchiaro

Unit of Molecular Neuro-Oncology, Fondazione IRCCS Istituto Neurologico Carlo Besta, 20133 Milan, Italy; gaetano.finocchiaro@istituto-besta.it

* Correspondence: massimo.costanza@istituto-besta.it; Tel.: +39-02-2394-4654

Received: 21 February 2019; Accepted: 19 March 2019; Published: 22 March 2019

Abstract: Historically restrained to immune defense against parasite infections, allergic inflammation has been recently rediscovered to protect from a wide array of environmental triggers, such as xenobiotics and carcinogens, which can induce DNA damage and ultimately lead to cancer development. Moreover, cells and mediators typical of allergic responses can importantly modulate the tissue inflammatory milieu, which represents a crucial gatekeeper towards the acquisition of malignancy by cancer cells through immune escape. Numerous studies have described an inverse association between allergies and glioma development. Mast cells, key players of allergic reactions, have been recently found at increased numbers in glioblastoma multiforme (GBM), the most common and lethal primary brain tumor, and they have been implicated in GBM pathogenesis. In this review, we summarize epidemiological studies and discuss the main evidence highlighting a potential interplay between allergic responses, and glioma formation and progression. Last, we draw future lines of research for better clarification whether and through which mechanisms allergic inflammation might impact on gliomagenesis. The comprehension of the immune mechanisms favoring or counteracting tumor growth might open the path to novel immunotherapy approaches.

Keywords: allergy; glioma; GBM; IgE; IL-4; mast cells

---

1. Introduction

Gliomas are the most common primary brain tumors of adults and are classified into different types and grades, based on histologic and molecular patterns [1]. Glioblastoma multiforme (GBM) accounts for approximately 70% of malignant gliomas and it is characterized by severe morbidity and high mortality [2]. Recent data suggest that GBMs arise from driver mutations of neural stem cells in the subventricular zone of the adult human brain [3]. Divergent genetic evolution may then be one factor underlying GBM progression [4]. The best-established risk factor for malignant gliomas is the exposure to ionizing radiation, however, a large body of literature has suggested that subjects with atopy have a diminished probability of developing gliomas [2]. The reason for this inverse association is still unknown. Type 2 immune reactions have long been ascribed to host defense against parasite infections, with allergic inflammation considered as an off-targeted side-effect of these protective responses [5]. Interestingly, recent lines of evidence have clearly shown that type 2 immunity exerts crucial functions in a complex array of responses, including venom detoxification as well as protection from hematophagous fluids, toxic xenobiotics and carcinogens [5]. These responses are driven by T helper (Th)2 cells, which produce mainly interleukin (IL)-4 and IL-13, that in turn promote immunoglobulin (Ig)E production by B cells, and IL-5, important mediator for eosinophil activation [6]. Allergic inflammation includes also IgG1 and components of innate immunity, such as mast cells, basophils, eosinophils, alternatively-activated macrophages and innate lymphoid cells [5].

The relationship between allergic inflammation and cancer has been the subject of several investigations, leading to opposite views regarding the impact of these responses on tumor growth.

According to one current of thought, allergic reactions might foster tumor development by impairing Th1 and cytotoxic T cell functions, and by supporting myeloid cells with a suppressive phenotype [7]. Indeed, in a model of breast cancer, IL-4-producing Th2 cells have been shown to sustain metastasis by promoting effector functions of tumor-associated macrophages [8]. Conversely, according to an alternative viewpoint, allergic responses might be important for tumor-surveillance [5]. In a model of skin cancer induced by exposure to the carcinogen dimethylbenz[a]anthracene (DMBA), it was demonstrated that IL-4-deficient mice develop tumors of increased size and greater incidence compared to wild-type mice [9]. Moreover, sensitization to DMBA was associated with the establishment of a strong autoreactive IgE response that was protective against carcinogenesis [9]. A similarly puzzling scenario emerges from studies analyzing the contribution of mast cells to cancer, as they have been suggested as either protective or detrimental [10]. Most probably, depending on the specific neoplastic condition and the experimental model utilized, allergic inflammation can promote or counteract tumor development [10]. Very recently, the long controversy regarding the activity of eosinophils in tumor development has gained new insight from the study of Holland and colleagues [11,12]. In this work, it has been shown that in several solid tumors such as hepatocellular carcinoma, breast cancer and prostatic carcinoma, eosinophils can dampen tumor growth by inducing tumor cell-cytotoxicity. In particular, tumor-derived IL-33 was shown to promote chemokine CCL11-mediated eosinophil infiltration and degranulation [11,12].

In recent years, numerous studies have investigated the contribution of allergic components in the pathogenesis of gliomas and GBM. In this review, we summarize main epidemiological, histopathological and genetic studies that have suggested the involvement of allergic inflammation in glioma pathology and we discuss experimental approaches to further characterize in future the relationship between this specific arm of the immune system and brain tumorigenesis.

2. Epidemiology of Gliomas and GBM in Allergic Subjects

Most of the research on glioma and allergy has relied on epidemiological investigations, which have shown an inverse association between allergies and the risk of developing gliomas since the early 1990s [13]. In an international population-based case-control study performed in six different countries and including 1178 cases, it was reported that there is a reduced risk of 30–40% of developing gliomas among subjects with asthma, eczema and other allergic conditions [14]. In two subsequent case-control studies performed by Wiemels and colleagues on subjects enrolled in the San Francisco Bay Area, a significant inverse association for both self- and proxy-reported histories of allergic conditions with a diagnosis of adult glioma was shown [15,16]. A particularly decreased risk was found among subjects with allergies to pollen, dairy products and nuts [15]. Additionally, a strong association with diminished risk of developing gliomas has been detected in cases with late-onset (>12 years) respiratory allergies [16]. Another case-control study enrolling 489 cases of gliomas among three different hospitals in the United States (US), confirmed an inverse association between the risk of glioma and history of any allergy (e.g., eczema, allergy to insects, chemicals, etc.) [17]. Of note, in this work, people affected by both allergies and autoimmune diseases were at particularly low risk of developing glioma (odds ratio (OR) = 0.24, 95% CI, 0.14–0.42) [17]. In addition, a retrospective study has shown in two independent cohorts selected from Swedish twin registries that high-grade gliomas (grades III and IV) are inversely associated with allergic conditions that comprised of eczema, asthma, hay fever, and allergic rhinitis (hazard ratio, HR = 0.45 and 0.46, 95% CI) [18]. In a large, population-based case-control study in the UK, history of asthma, eczema, hay fever or other types of allergy was inversely correlated to the risk of developing glioma (OR = 0.63, 95% CI, 0.53–0.76) [19]. Association with allergies was not increased or decreased in grade I/II compared with grade III/IV tumors in this work [19]. Moreover, there was no evidence for a gradient of risk with the age of onset, time since onset or with the number of conditions presented [19]. A meta-analysis of both case-control and cohort studies published between 1979 and 2007 and including 3450 cases, has shown that the risk of glioma was decreased by 40% among subjects with a history of allergy, 30% diminished among those with a history of eczema, and 30% reduced among those with a history of asthma [20]. Similar findings have been obtained in a large population-based case-control study published in 2007 and comprising 1527 cases of gliomas recruited in Denmark, Norway, Finland, Sweden and Southeast England [21]. This epidemiological trend has been confirmed also in another case-control study involving 366 cases of glioma recruited among the German population [22]. In an alternative approach, Il'yasova and colleagues have compared 388 glioma cases with three different groups of control—siblings, friends and clinic-based controls—to better evaluate the contribution of genotypic (for siblings) or environmental (for friends) effects on the relationship with allergic inflammation [23]. Allergies were always found significantly inversely associated with gliomas: ORs were 0.53 (95% CI, 0.15–1.84), 0.54 (95% CI, 0.28–1.07) and 0.34 (95% CI, 0.23–0.50) with siblings, friends and clinic-based controls, respectively [23]. A case-control study recruited 855 high-grade glioma patients from four different geographic regions of the US and belonging to five inherited glioma risk variants: 5p15.3 (*TERT*), 8q24.21 (*CCDC26/MLZE*), 9p21.3 (*CDKN2B*), 11q23.3 (*PHLDB1/DDX6*) and 20q13.3 (*RTEL1*) [24]. A significantly stronger inverse association of allergy history with glioma was detected among subjects who do not carry the glioma risk allele in the 9p21.3 region, while it was suggestively higher among those carrying the glioma risk allele in the 20q13.3 region [24]. In 2014, Cahoon and co-workers analyzed a cohort of 4.5 million male US veterans comprising 4383 cases of patients with a discharge diagnosis of malignant neoplasm of the brain, which included mainly gliomas, expected to cover 95% of cases, and some rare childhood tumors such as medulloblastoma. In this study, allergy/atopy of long latency ($\geq$10 years) was associated with a reduced risk of developing brain cancer (rate ratio = 0.60, 95% CI, 0.43–0.83) [25]. Last, a recent multicenter case-control study carried out in four areas in France in 2004–2010 has explored the relationship between allergy (e.g., asthma, eczema, rhinitis/hay fever and other allergic conditions) and the risk of glioma in 273 glioma cases and 982 matched controls [26]. In addition to confirming previous findings, this work highlighted a dose-effect relationship between the number of allergic conditions and the inverse association with glioma risk and a stronger relationship in women [26]. To avoid the possibility that allergic inflammation might be somehow influenced by (or be the effect of) brain tumor itself, Schwartzbaum and colleagues have investigated germline polymorphisms as potential biomarkers of GBM susceptibility, selecting five single nucleotide polymorphisms (SNPs) strongly associated with asthma and allergic conditions [27]. Authors found that in subjects carrying three of these polymorphisms, located on the *IL-4 receptor α* (*IL4RA*) and *IL13* genes, the odds ratios for GBM were in the opposite direction with those for asthma [27]. Interestingly, pre-diagnostic serum levels of IL-4 and soluble IL-4RA have been later found inversely associated with gliomas and GBM in a nested case-control study including 487 glioma cases and 487 matched controls [28]. Of note, this association was present >20 years before glioma diagnosis [28]. Epidemiological data are summarized in Table 1.

Table 1. Epidemiological studies of allergy and risk of glioma.

Study Type	Number of Glioma Cases	Number of Control Subjects	Association with Allergy (95% CI)	Year [Ref.]
Case-control	115	418	RR = 0.71 (0.5–1.0)	1992 [13]
Case-control	1178	2493	RR = 0.59 (0.49–0.71)	1999 [14]
Case-control	405	402	OR = 0.5 (0.3–0.7)	2002 [15]
Case-control	489	799	OR = 0.67 (0.52–0.86)	2002 [17]
Cohort study	Cohort I (14535 subjects/33 glioma cases)		HR = 0.38 (0.05–3.13) [1] HR = 0.46 (0.18–1.21) [2]	2003 [18]
	Cohort II (29573 subjects/42 glioma cases)		HR = 2.60 (0.86–7.81) [1] HR = 0.45 (0.11–1.92) [2]	
Case-control	965	1716	OR = 0.63 (0.53–0.76)	2006 [19]
Meta-Analysis	Participants (53223 subjects/3450 glioma cases)		RR = 0.61 (0.55–0.67)	2007 [20]
Case-control	1527	3309	OR = 0.70 (0.61–0.80)	2007 [21]
Case-control	366	1494	OR = 0.92 (0.70–1.22)	2009 [22]
Case-control	388	80 (siblings)	OR = 0.53 (0.15–1.84)	2009 [23]
		191 (friends)	OR = 0.54 (0.28–1.07)	
		177 (clinic-based controls)	OR = 0.34 (0.23–0.50)	
Case-control	855	1160	OR = 0.62 (0.51–0.76)	2011 [24]
Cohort study	4.5 million subjects/4383 malignant neoplasm brain [3]		Rate ratio = 0.60 (0.43–0.83)	2014 [25]
Case-control	273	982	OR = 0.52 (0.36–0.75)	2018 [26]

Confidence interval (CI); relative risk (RR); odds ratio (OR); hazard ratio (HR). [1] Association of low-grade (I and II) glioma cases. [2] Association of high-grade (III and IV) glioma cases. [3] Brain tumors included rare childhood tumors, but mainly gliomas, expected to cover 95% of cases. Ref., reference.

3. Allergic Mediators in Glioma and GBM

Much effort has been made to correlate the above described epidemiological observations with the prototypical biomarker of allergic inflammation, that is IgE. In 2004, Wiemels et al. found significantly lower IgE levels in glioma patients compared to controls (OR = 0.37, 95% CI, 0.22–0.64), with a more striking inverse association for IgE specific to food allergens (OR = 0.12, 95% CI, 0.04–0.41) [16]. However, a follow-up of this work by the same group has suggested that this inverse relationship is detectable only among cases receiving temozolomide [29]. Conversely, two later works have actually confirmed a relationship between serum IgE levels and gliomas. Indeed, a nested case-control study combining data from four prospective cohort studies have found a statistically significant inverse association between "borderline-elevated" total IgE levels (25–100 kU/L) and glioma (with 169 cases) (OR = 0.63, 95% CI, 0.42–0.93), even though no association was detected between high IgE (>100 kU/L) and glioma (OR = 0.98, 95% CI, 0.61–1.56) [30]. A prospective case-control study with a nested design including a higher number of cases (n = 275) has also demonstrated that the risk of glioma is inversely correlated to IgE response to inhalant allergens (OR = 0.73, 95% CI, 0.51–1.06) [31]. This relationship is particularly pronounced in women (OR = 0.53, 95% CI, 0.30–0.95) and the lowest OR was found in samples with the highest serum IgE levels [31]. Last, a nested case-control study with serum samples from 594 glioma and 374 GBM cases has shown that high levels of total IgE are associated with a significantly reduced risk of glioma, while allergen-specific IgE levels are correlated with a decreased risk of GBM specifically in women, but not in men [32]. Of note, this inverse association is present at least 20 years before tumor diagnosis [32].

A few studies have addressed the interplay between allergic inflammation and gliomagenesis with strategies different from the epidemiological approach. First experimental evidence can be found in initial attempts of immunotherapy. In a mouse model of glioma elicited in nude mice by the injection of U87 human glioma cells, the co-transplantation of an IL-4-secreting cell line promoted a significantly increased survival and a massive infiltration by eosinophils [33]. In line with these data, a phase I/II clinical trial has shown that the intracavitary injection of IL-2 and lymphokine-activated killer (LAK) cells in GBM patients leads to improved survival, associated with eosinophilic infiltration of the central nervous system (CNS) [34]. Several gene therapy approaches by our group and others have shown that IL-4 delivery in immunocompetent hosts with malignant gliomas results in a significant impairment of tumor growth and prolonged survival [35–37]. In particular, we found that local production of IL-4 at the tumor site is associated with increased infiltration of $CD8^+$ T and $CD4^+$ T cells, B lymphocytes and macrophages [37]. Later, we have shown that transplantation of neural stem/progenitor cells retrovirally engineered to produce IL-4 in C57BL/6 mice with established GBM leads to the survival of most tumor-bearing mice [38]. These data, obtained in both experimental models and human disease, suggest that the induction of an IL-4 and/or eosinophil-mediated immune response might be beneficial to counteract tumor development.

In addition, mast cells (MCs) have been implicated in the pathogenesis of glioma. MCs constitutively express the high affinity receptor for IgE (FcεRI) and represent the key effector players in allergic reactions, when the re-exposure to a previously encountered antigen induces IgE-mediated anaphylactic degranulation and the consequent release of massive amounts of pre-formed mediators, such as histamine, cytokines (e.g., tumor necrosis factor (TNF), IL-4), chemokines, leukotrienes and growth factors (e.g., vascular endothelial growth factor, VEGF) [39]. Furthermore, monomeric IgE in the absence of antigen, can be even more effective than IgE plus antigen to induce MC-mediated release of inflammatory cytokines such as IL-4, IL-6 and TNF [40]. Together with microglia, MCs represent a key immune subset resident within the CNS, where they are mainly localized in the thalamus, hypothalamus and leptomeninges [41]. Significantly increased numbers of MCs have been detected in GBM tissue specimens in comparison to low-grade glioma (grade II), displaying often a perivascular localization and staining for CXCL12 and CXCR4 [42]. Similar findings have been obtained in an immunocompetent mouse model of glioma, elicited with the retroviral system RCAS/TV-a of gene delivery. In detail, the RCAS retrovirus encoding for two oncogenes (*KRas* and *Akt*) has been administered postnatally in *Ntv-a* and *Gtv-a* transgenic mice, which express the TV-a receptor for RCAS retrovirus under the control of Nestin or GFAP promoters, respectively, to obtain retroviral infection (and consequent oncogene expression) specifically in neural/glial progenitor cells or astrocytes. These *Ntv-a* and *Gtv-a* mouse lines carried at the same time deficiencies in tumor suppressor genes *Ink4a* or *Arf*, which are often deleted in human gliomas and developed gliomas of various grades and types [42]. The mouse strain with high-grade glioma ($Arf^{-/-}$) displayed a significantly higher infiltration of MCs than $Ink4a^{-/-}$ mice developing low-grade gliomas and exhibited also a strong expression of stem cell factor (SCF), the main growth factor for MCs, around tumor blood vessels [42]. Interestingly, primary human gliomas have been shown to express SCF in a grade-dependent manner [43]. The same research group has later extended these findings to a larger cohort of patients (n = 188), confirming an accumulation of MCs in high-grade (i.e., grade III and IV) vs. low-grade (grade II) gliomas [44]. By taking advantage of a transwell migration assay, human glioma cell lines were shown to potently attract MCs by a mechanism partially dependent on the chemoattractant macrophage migration inhibitory factor (MIF), plasminogen activator inhibitor 1 (PAI-1) and the phosphorylation of STAT5 [44,45]. Of note, a significantly positive correlation was observed between the number of MCs and the cytoplasmic intensity of MIF staining in glioma tissue samples [44].

Notably, a recent analysis from lower grade gliomas showed that allergy is a robust, prognostic factor independent from other major prognostic molecular markers [46].

4. Conclusions and Future Perspectives

Type 2 immunity is recently emerging as a specific arm of the immune system physiologically deputed to defense against toxic and carcinogenic agents [5]. Data so far have suggested that at least some components of allergic inflammation might counteract the development of gliomas and GBM (Figure 1).

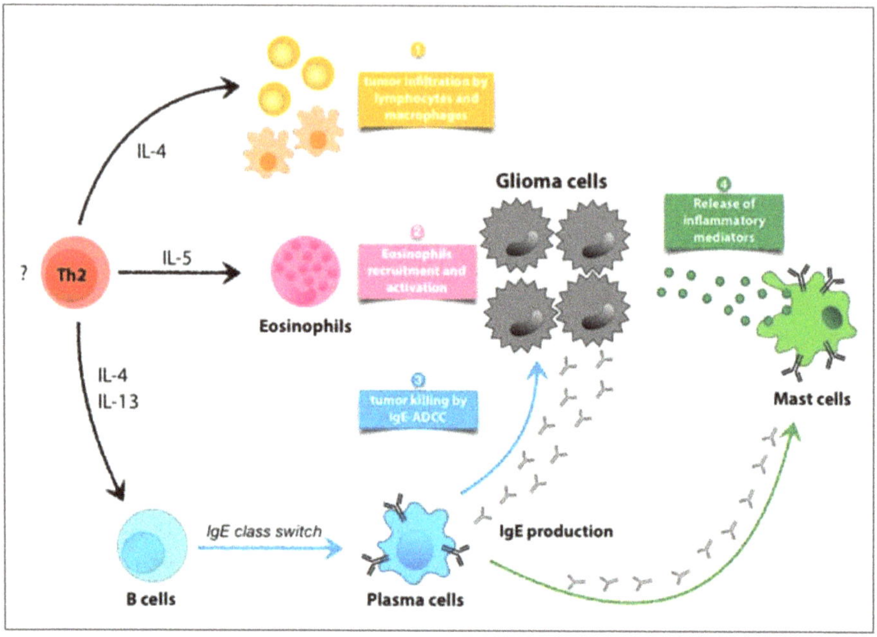

Figure 1. Schematic representation of the components of allergic inflammation with potentially anti-glioma activity. The establishment of a Th2-driven immune response against the tumor might support resistance to glioma growth through several modalities. The overexpression of the prototypical Th2-cytokine interleukin (IL)-4 within tumor microenvironment has been shown to counteract gliomagenesis [33,35–38,47]. This anti-tumor effect is associated with enhanced infiltration of immune cells such as CD8$^+$ T lymphocytes, CD4$^+$ T cells, B cells and monocytes at the tumor site (1) [37]. Th2 cells also secrete IL-5, a key cytokine for the recruitment and activation of eosinophils. Eosinophilic infiltrates within glioma tissue have been detected in early attempts of immunotherapy displaying a positive outcome [33,34] (2). IL-4 and IL-13 derived from Th2 cells can promote IgE isotype class switch recombination in B cells and their maturation to plasma cells. Theoretically, antigen-specific IgE might promote direct tumor killing through ADCC (3), similar to what was observed in a mouse model of ovarian cancer [48]. Alternatively, the engagement of FcεRI on mast cells (MCs) by monomeric IgE is known to induce the release of selective inflammatory cytokines [40], such as IL-4 and tumor necrosis factor (TNF), both endowed with anti-glioma activity [38,49] (4).

Epidemiological investigations have provided a huge amount of data showing an inverse association between allergy and gliomas, even though they do not provide any information on the causative relationship between these two conditions. In principle, it is possible that the tumor itself promotes immune-suppression of type 1 immunity and indirectly favors an imbalance of systemic immunity toward a Th2/allergic profile. However, the timing of most epidemiological observations, anticipating the clinical emergence of gliomas of many years, is in some contradiction with this hypothesis, at least in cases of primary GBM that are known to be formed in months rather than years. Moreover, some genetic variants in *IL13* and *IL4RA* genes confer opposite odds ratios for asthma and GBM development [27]. Given that germline polymorphisms were utilized as biomarkers of susceptibility to asthma in this study, these results cannot be ascribed to an effect of GBM on the immune system [27]. According to the cancer immunoediting hypothesis, the interaction between the immune system and tumor cells undergoes three different stages termed respectively "elimination", "equilibrium" and "escape" [50]. The elimination phase is characterized by the interplay of innate and adaptive immune mechanisms, which drives the destruction of tumor cells from the host. However, if this process is not complete, rare cancer cell variants might enter an equilibrium stage, in which tumor growth is still prevented by adaptive immunity and the tumor is subject to constant editing of its immunogenicity. During this stage, immune pressure on genetically-unstable tumor cells might result in the selection of tumor variants endowed with the ability to escape immune surveillance and form clinically visible tumors [50]. In the context of brain tumors, it's possible to speculate that the genetic background favoring the development of allergic inflammation might support the elimination phase and reduce the probability of developing glioma over the course of time.

Immunotherapeutic approaches have shown that IL-4 can counteract glioma growth and eosinophilic infiltration of tumor tissue correlates with a positive outcome of immunotherapy. Nonetheless, the mechanisms through which these typical components of allergic inflammation might exert an anti-tumoral effect are not clearly understood.

Higher levels of IgE are associated with a reduced risk of glioma. IgE is endowed with tumor-killing properties by mediating antibody-dependent cell-mediated cytotoxicity. Its effect has been shown in a mouse model of ovarian cancer [48], but no study has investigated whether IgE is reactive against tumor antigens in glioma patients. Furthermore, IgE alone or in combination with an antigen can stimulate MC functions.

Studies describing the accumulation of MCs in GBM have offered important evidence of the potential involvement of these crucial effector cells of allergy in gliomagenesis, however, whether MCs promote or counteract tumor growth is not known and should be subject to further investigation. According to the modality of activation, MCs have been shown to promote or break immune tolerance. This concept is illustrated in models of allotransplantation and $CD8^+$ T cell-mediated graft rejection. Indeed, IL-9 secreted by regulatory T (Treg) cells has been shown to recruit and activate MCs to promote regional immune suppression and acceptance of allograft skin transplants [51]. Conversely, in the same model when MCs undergo anaphylactic degranulation, they break Treg-mediated suppression and induce allograft rejection [52]. To understand the contribution of MCs to gliomagenesis, it might be worthwhile to evaluate the development of gliomas in mouse models of MC-deficiency [53]. Interestingly, recent work has shown that $Hdc^{-/-}$ mice, that genetically lack histamine, exhibit increased glioma growth and reduced survival, associated with the accumulation of immunosuppressive immature myeloid cells [54]. Of note, MC granules represent the main immunological source of histamine and $Hdc^{-/-}$ mice display several deficits in MC numbers, morphology and granular contents [55]. In addition, there are data supporting the contribution of MCs to tumor formation in the context of neurofibromatosis type 1 [56].

Author Contributions: Both authors contributed in the conceptualization, writing and final revision of the manuscript.

Conflicts of Interest: The authors declare no conflict of interest.

References

1. Louis, D.N.; Perry, A.; Reifenberger, G.; von Deimling, A.; Figarella-Branger, D.; Cavenee, W.K.; Ohgaki, H.; Wiestler, O.D.; Kleihues, P.; Ellison, D.W. The 2016 World Health Organization Classification of Tumors of the Central Nervous System: A summary. *Acta Neuropathol.* **2016**, *131*, 803–820. [CrossRef]
2. Wen, P.Y.; Kesari, S. Malignant gliomas in adults. *N. Engl. J. Med.* **2008**, *359*, 492–507. [CrossRef]
3. Lee, J.H.; Lee, J.E.; Kahng, J.Y.; Kim, S.H.; Park, J.S.; Yoon, S.J.; Um, J.-Y.; Kim, W.K.; Lee, J.-K.; Park, J.; et al. Human glioblastoma arises from subventricular zone cells with low-level driver mutations. *Nature* **2018**, *560*, 243–247. [CrossRef] [PubMed]
4. Orzan, F.; De Bacco, F.; Crisafulli, G.; Pellegatta, S.; Mussolin, B.; Siravegna, G.; D'Ambrosio, A.; Comoglio, P.M.; Finocchiaro, G.; Boccaccio, C. Genetic Evolution of Glioblastoma Stem-Like Cells From Primary to Recurrent Tumor. *Stem Cells* **2017**, *35*, 2218–2228. [CrossRef] [PubMed]
5. Palm, N.W.; Rosenstein, R.K.; Medzhitov, R. Allergic host defences. *Nature* **2012**, *484*, 465–472. [CrossRef]
6. Galli, S.J.; Tsai, M.; Piliponsky, A.M. The development of allergic inflammation. *Nature* **2008**, *454*, 445–454. [CrossRef]

7. Disis, M.L. Immune regulation of cancer. *J. Clin. Oncol.* **2010**, *28*, 4531–4538. [CrossRef] [PubMed]
8. DeNardo, D.G.; Barreto, J.B.; Andreu, P.; Vasquez, L.; Tawfik, D.; Kolhatkar, N.; Coussens, L.M. CD4(+) T cells regulate pulmonary metastasis of mammary carcinomas by enhancing protumor properties of macrophages. *Cancer Cell* **2009**, *16*, 91–102. [CrossRef]
9. Crawford, G.; Hayes, M.D.; Seoane, R.C.; Ward, S.; Dalessandri, T.; Lai, C.; Healy, E.; Kipling, D.; Proby, C.; Moyes, C.; et al. Epithelial damage and tissue $\gamma\delta$ T cells promote a unique tumor-protective IgE response. *Nat. Immunol.* **2018**, *19*, 859–870. [CrossRef]
10. Rigoni, A.; Colombo, M.P.; Pucillo, C. Mast cells, basophils and eosinophils: From allergy to cancer. *Semin. Immunol.* **2018**, *35*, 29–34. [CrossRef]
11. Hollande, C.; Boussier, J.; Ziai, J.; Nozawa, T.; Bondet, V.; Phung, W.; Lu, B.; Duffy, D.; Paradis, V.; Mallet, V.; et al. Inhibition of the dipeptidyl peptidase DPP4 (CD26) reveals IL-33-dependent eosinophil-mediated control of tumor growth. *Nat. Immunol.* **2019**, *20*, 257–264. [CrossRef]
12. Munitz, A.; Hogan, S.P. Alarming eosinophils to combat tumors. *Nat. Immunol.* **2019**, *20*, 250–252. [CrossRef]
13. Schlehofer, B.; Blettner, M.; Becker, N.; Martinsohn, C.; Wahrendorf, J. Medical risk factors and the development of brain tumors. *Cancer* **1992**, *69*, 2541–2547. [CrossRef]
14. Schlehofer, B.; Blettner, M.; Preston-Martin, S.; Niehoff, D.; Wahrendorf, J.; Arslan, A.; Ahlbom, A.; Choi, W.N.; Giles, G.G.; Howe, G.R.; et al. Role of medical history in brain tumour development. Results from the international adult brain tumour study. *Int. J. Cancer* **1999**, *82*, 155–160. [CrossRef]
15. Wiemels, J.L.; Wiencke, J.K.; Sison, J.D.; Miike, R.; McMillan, A.; Wrensch, M. History of allergies among adults with glioma and controls. *Int. J. Cancer* **2002**, *98*, 609–615. [CrossRef]
16. Wiemels, J.L.; Wiencke, J.K.; Patoka, J.; Moghadassi, M.; Chew, T.; McMillan, A.; Miike, R.; Barger, G.; Wrensch, M. Reduced immunoglobulin E and allergy among adults with glioma compared with controls. *Cancer Res.* **2004**, *64*, 8468–8473. [CrossRef]
17. Brenner, A.V.; Linet, M.S.; Fine, H.A.; Shapiro, W.R.; Selker, R.G.; Black, P.M.; Inskip, P.D. History of allergies and autoimmune diseases and risk of brain tumors in adults. *Int. J. Cancer* **2002**, *99*, 252–259. [CrossRef]
18. Schwartzbaum, J.; Jonsson, F.; Ahlbom, A.; Preston-Martin, S.; Lönn, S.; Söderberg, K.C.; Feychting, M. Cohort studies of association between self-reported allergic conditions, immune-related diagnoses and glioma and meningioma risk. *Int. J. Cancer* **2003**, *106*, 423–428. [CrossRef]
19. Schoemaker, M.J.; Swerdlow, A.J.; Hepworth, S.J.; McKinney, P.A.; van Tongeren, M.; Muir, K.R. History of allergies and risk of glioma in adults. *Int. J. Cancer* **2006**, *119*, 2165–2172. [CrossRef]
20. Linos, E.; Raine, T.; Alonso, A.; Michaud, D. Atopy and risk of brain tumors: A meta-analysis. *J. Natl. Cancer Inst.* **2007**, *99*, 1544–1550. [CrossRef]
21. Wigertz, A.; Lönn, S.; Schwartzbaum, J.; Hall, P.; Auvinen, A.; Christensen, H.C.; Johansen, C.; Klaeboe, L.; Salminen, T.; Schoemaker, M.J.; et al. Allergic conditions and brain tumor risk. *Am. J. Epidemiol.* **2007**, *166*, 941–950. [CrossRef] [PubMed]
22. Berg-Beckhoff, G.; Schüz, J.; Blettner, M.; Münster, E.; Schlaefer, K.; Wahrendorf, J.; Schlehofer, B. History of allergic disease and epilepsy and risk of glioma and meningioma (INTERPHONE study group, Germany). *Eur. J. Epidemiol.* **2009**, *24*, 433–440. [CrossRef] [PubMed]
23. Il'yasova, D.; McCarthy, B.; Marcello, J.; Schildkraut, J.M.; Moorman, P.G.; Krishnamachari, B.; Ali-Osman, F.; Bigner, D.D.; Davis, F. Association between glioma and history of allergies, asthma and eczema: A case-control study with three groups of controls. *Cancer Epidemiol. Biomark. Prev.* **2009**, *18*, 1232–1238. [CrossRef] [PubMed]
24. Lachance, D.H.; Yang, P.; Johnson, D.R.; Decker, P.A.; Kollmeyer, T.M.; McCoy, L.S.; Rice, T.; Xiao, Y.; Ali-Osman, F.; Wang, F.; et al. Associations of high-grade glioma with glioma risk alleles and histories of allergy and smoking. *Am. J. Epidemiol.* **2011**, *174*, 574–581. [CrossRef]
25. Cahoon, E.K.; Inskip, P.D.; Gridley, G.; Brenner, A.V. Immune-related conditions and subsequent risk of brain cancer in a cohort of 4.5 million male US veterans. *Br. J. Cancer* **2014**, *110*, 1825–1833. [CrossRef] [PubMed]
26. Pouchieu, C.; Raherison, C.; Piel, C.; Migault, L.; Carles, C.; Fabbro-Perray, P.; Loiseau, H.; Guillamo, J.-S.; Lebailly, P.; Baldi, I. Allergic conditions and risk of glioma and meningioma in the CERENAT case-control study. *J. Neurooncol.* **2018**, *138*, 271–281. [CrossRef]
27. Schwartzbaum, J.; Ahlbom, A.; Malmer, B.; Lönn, S.; Brookes, A.J.; Doss, H.; Debinski, W.; Henriksson, R.; Feychting, M. Polymorphisms associated with asthma are inversely related to glioblastoma multiforme. *Cancer Res.* **2005**, *65*, 6459–6465. [CrossRef] [PubMed]

28. Schwartzbaum, J.; Seweryn, M.; Holloman, C.; Harris, R.; Handelman, S.K.; Rempala, G.A.; Huang, R.-P.; Burkholder, B.; Brandemihl, A.; Kallberg, H.; et al. Association between Prediagnostic Allergy-Related Serum Cytokines and Glioma. *PLoS ONE* **2015**, *10*, e0137503. [CrossRef] [PubMed]
29. Wiemels, J.L.; Wilson, D.; Patil, C.; Patoka, J.; McCoy, L.; Rice, T.; Schwartzbaum, J.; Heimberger, A.; Sampson, J.H.; Chang, S.; et al. IgE, allergy and risk of glioma: Update from the San Francisco Bay Area Adult Glioma Study in the temozolomide era. *Int. J. Cancer* **2009**, *125*, 680–687. [CrossRef] [PubMed]
30. Calboli, F.C.F.; Cox, D.G.; Buring, J.E.; Gaziano, J.M.; Ma, J.; Stampfer, M.; Willett, W.C.; Tworoger, S.S.; Hunter, D.J.; Camargo, C.A.; et al. Prediagnostic plasma IgE levels and risk of adult glioma in four prospective cohort studies. *J. Natl. Cancer Inst.* **2011**, *103*, 1588–1595. [CrossRef] [PubMed]
31. Schlehofer, B.; Siegmund, B.; Linseisen, J.; Schüz, J.; Rohrmann, S.; Becker, S.; Michaud, D.; Melin, B.; Bas Bueno-de-Mesquita, H.; Peeters, P.H.M.; et al. Primary brain tumours and specific serum immunoglobulin E: A case-control study nested in the European Prospective Investigation into Cancer and Nutrition cohort. *Allergy* **2011**, *66*, 1434–1441. [CrossRef]
32. Schwartzbaum, J.; Ding, B.; Johannesen, T.B.; Osnes, L.T.N.; Karavodin, L.; Ahlbom, A.; Feychting, M.; Grimsrud, T.K. Association between prediagnostic IgE levels and risk of glioma. *J. Natl. Cancer Inst.* **2012**, *104*, 1251–1259. [CrossRef]
33. Yu, J.S.; Wei, M.X.; Chiocca, E.A.; Martuza, R.L.; Tepper, R.I. Treatment of glioma by engineered interleukin 4-secreting cells. *Cancer Res.* **1993**, *53*, 3125–3128.
34. Hayes, R.L.; Koslow, M.; Hiesiger, E.M.; Hymes, K.B.; Hochster, H.S.; Moore, E.J.; Pierz, D.M.; Chen, D.K.; Budzilovich, G.N.; Ransohoff, J. Improved long term survival after intracavitary interleukin-2 and lymphokine-activated killer cells for adults with recurrent malignant glioma. *Cancer* **1995**, *76*, 840–852. [CrossRef]
35. Benedetti, S.; Dimeco, F.; Pollo, B.; Cirenei, N.; Colombo, B.M.; Bruzzone, M.G.; Cattaneo, E.; Vescovi, A.; Didonato, S.; Colombo, M.P.; et al. Limited efficacy of the HSV-TK/GCV system for gene therapy of malignant gliomas and perspectives for the combined transduction of the interleukin-4 gene. *Hum. Gene Ther.* **1997**, *8*, 1345–1353. [CrossRef]
36. Andreansky, S.; He, B.; van Cott, J.; McGhee, J.; Markert, J.M.; Gillespie, G.Y.; Roizman, B.; Whitley, R.J. Treatment of intracranial gliomas in immunocompetent mice using herpes simplex viruses that express murine interleukins. *Gene Ther.* **1998**, *5*, 121–130. [CrossRef] [PubMed]
37. Benedetti, S.; Bruzzone, M.G.; Pollo, B.; DiMeco, F.; Magrassi, L.; Pirola, B.; Cirenei, N.; Colombo, M.P.; Finocchiaro, G. Eradication of rat malignant gliomas by retroviral-mediated, in vivo delivery of the interleukin 4 gene. *Cancer Res.* **1999**, *59*, 645–652. [PubMed]
38. Benedetti, S.; Pirola, B.; Pollo, B.; Magrassi, L.; Bruzzone, M.G.; Rigamonti, D.; Galli, R.; Selleri, S.; Di Meco, F.; De Fraja, C.; et al. Gene therapy of experimental brain tumors using neural progenitor cells. *Nat. Med.* **2000**, *6*, 447–450. [CrossRef] [PubMed]
39. Mukai, K.; Tsai, M.; Saito, H.; Galli, S.J. Mast cells as sources of cytokines, chemokines and growth factors. *Immunol. Rev.* **2018**, *282*, 121–150. [CrossRef]
40. Kalesnikoff, J.; Huber, M.; Lam, V.; Damen, J.E.; Zhang, J.; Siraganian, R.P.; Krystal, G. Monomeric IgE stimulates signaling pathways in mast cells that lead to cytokine production and cell survival. *Immunity* **2001**, *14*, 801–811. [CrossRef]
41. Sayed, B.A.; Christy, A.; Quirion, M.R.; Brown, M.A. The master switch: The role of mast cells in autoimmunity and tolerance. *Annu. Rev. Immunol.* **2008**, *26*, 705–739. [CrossRef]
42. Põlajeva, J.; Sjösten, A.M.; Lager, N.; Kastemar, M.; Waern, I.; Alafuzoff, I.; Smits, A.; Westermark, B.; Pejler, G.; Uhrbom, L.; et al. Mast cell accumulation in glioblastoma with a potential role for stem cell factor and chemokine CXCL12. *PLoS ONE* **2011**, *6*, e25222. [CrossRef] [PubMed]
43. Sun, L.; Hui, A.-M.; Su, Q.; Vortmeyer, A.; Kotliarov, Y.; Pastorino, S.; Passaniti, A.; Menon, J.; Walling, J.; Bailey, R.; et al. Neuronal and glioma-derived stem cell factor induces angiogenesis within the brain. *Cancer Cell* **2006**, *9*, 287–300. [CrossRef]
44. Põlajeva, J.; Bergström, T.; Edqvist, P.-H.; Lundequist, A.; Sjösten, A.; Nilsson, G.; Smits, A.; Bergqvist, M.; Pontén, F.; Westermark, B.; et al. Glioma-derived macrophage migration inhibitory factor (MIF) promotes mast cell recruitment in a STAT5-dependent manner. *Mol. Oncol.* **2014**, *8*, 50–58. [CrossRef]

45. Roy, A.; Coum, A.; Marinescu, V.D.; Põlajeva, J.; Smits, A.; Nelander, S.; Uhrbom, L.; Westermark, B.; Forsberg-Nilsson, K.; Pontén, F.; et al. Glioma-derived plasminogen activator inhibitor-1 (PAI-1) regulates the recruitment of LRP1 positive mast cells. *Oncotarget* **2015**, *6*, 23647–23661. [CrossRef] [PubMed]
46. Lehrer, S.; Rheinstein, P.H.; Rosenzweig, K.E. Allergy may confer better survival on patients with gliomas. *Clin. Neurol. Neurosurg.* **2018**, *177*, 63–67. [CrossRef]
47. Tepper, R.I.; Pattengale, P.K.; Leder, P. Murine interleukin-4 displays potent anti-tumor activity in vivo. *Cell* **1989**, *57*, 503–512. [CrossRef]
48. Karagiannis, S.N.; Wang, Q.; East, N.; Burke, F.; Riffard, S.; Bracher, M.G.; Thompson, R.G.; Durham, S.R.; Schwartz, L.B.; Balkwill, F.R.; et al. Activity of human monocytes in IgE antibody-dependent surveillance and killing of ovarian tumor cells. *Eur. J. Immunol.* **2003**, *33*, 1030–1040. [CrossRef]
49. Ehtesham, M.; Samoto, K.; Kabos, P.; Acosta, F.L.; Gutierrez, M.A.R.; Black, K.L.; Yu, J.S. Treatment of intracranial glioma with in situ interferon-gamma and tumor necrosis factor-alpha gene transfer. *Cancer Gene Ther.* **2002**, *9*, 925–934. [CrossRef]
50. Schreiber, R.D.; Old, L.J.; Smyth, M.J. Cancer immunoediting: Integrating immunity's roles in cancer suppression and promotion. *Science* **2011**, *331*, 1565–1570. [CrossRef]
51. Lu, L.-F.; Lind, E.F.; Gondek, D.C.; Bennett, K.A.; Gleeson, M.W.; Pino-Lagos, K.; Scott, Z.A.; Coyle, A.J.; Reed, J.L.; Van Snick, J.; et al. Mast cells are essential intermediaries in regulatory T-cell tolerance. *Nature* **2006**, *442*, 997–1002. [CrossRef]
52. de Vries, V.C.; Wasiuk, A.; Bennett, K.A.; Benson, M.J.; Elgueta, R.; Waldschmidt, T.J.; Noelle, R.J. Mast cell degranulation breaks peripheral tolerance. *Am. J. Transplant.* **2009**, *9*, 2270–2280. [CrossRef] [PubMed]
53. Cildir, G.; Pant, H.; Lopez, A.F.; Tergaonkar, V. The transcriptional program, functional heterogeneity and clinical targeting of mast cells. *J. Exp. Med.* **2017**, *214*, 2491–2506. [CrossRef] [PubMed]
54. Ahn, B.; Kohanbash, G.; Ohkuri, T.; Kosaka, A.; Chen, X.; Ikeura, M.; Wang, T.C.; Okada, H. Histamine deficiency promotes accumulation of immunosuppressive immature myeloid cells and growth of murine gliomas. *Oncoimmunology* **2015**, *4*, e1047581. [CrossRef]
55. Ohtsu, H.; Tanaka, S.; Terui, T.; Hori, Y.; Makabe-Kobayashi, Y.; Pejler, G.; Tchougounova, E.; Hellman, L.; Gertsenstein, M.; Hirasawa, N.; et al. Mice lacking histidine decarboxylase exhibit abnormal mast cells. *FEBS Lett.* **2001**, *502*, 53–56. [CrossRef]
56. Karmakar, S.; Reilly, K.M. The role of the immune system in neurofibromatosis type 1-associated nervous system tumors. *CNS Oncol.* **2017**, *6*, 45–60. [CrossRef]

cancers

MDPI

Review

Combined Amino Acid Positron Emission Tomography and Advanced Magnetic Resonance Imaging in Glioma Patients

Philipp Lohmann [1,*], Jan-Michael Werner [2], N. Jon Shah [1,3,4], Gereon R. Fink [1,2], Karl-Josef Langen [1,5] and Norbert Galldiks [1,2,6]

1 Institute of Neuroscience and Medicine (INM-3, -4, -5, -11), Forschungszentrum Juelich, 52425 Juelich, Germany; n.j.shah@fz-juelich.de (N.J.S.); gereon.fink@uk-koeln.de (G.R.F.); k.j.langen@fz-juelich.de (K.-J.L.); n.galldiks@fz-juelich.de (N.G.)
2 Department of Neurology, Faculty of Medicine and University Hospital Cologne, University of Cologne, 50937 Cologne, Germany; jan-michael.werner@uk-koeln.de
3 JARA-BRAIN-Translational Medicine, 52074 Aachen, Germany
4 Department of Neurology, RWTH Aachen University, 52074 Aachen, Germany
5 Department of Nuclear Medicine, RWTH Aachen University, 52074 Aachen, Germany
6 Center of Integrated Oncology (CIO), Universities of Cologne and Bonn, 50937 Cologne, Germany
* Correspondence: p.lohmann@fz-juelich.de; Tel.: +49-2461-61-96357

Received: 4 January 2019; Accepted: 25 January 2019; Published: 29 January 2019

Abstract: Imaging techniques such as positron emission tomography (PET) and magnetic resonance imaging (MRI) provide valuable information about brain tumor patients. Particularly amino acid PET, advanced MRI techniques, and combinations thereof are of great interest for the non-invasive assessment of biological characteristics in patients with primary or secondary brain cancer. A methodological innovation that potentially advances research in patients with brain tumors is the increasing availability of hybrid PET/MRI systems, which enables the simultaneous acquisition of both imaging modalities. Furthermore, the advent of ultra-high field MRI scanners operating at magnetic field strengths of 7 T or more will allow further development of metabolic MR imaging at higher resolution. This review focuses on the combination of amino acid PET with MR spectroscopic imaging, perfusion- and diffusion-weighted imaging, as well as chemical exchange saturation transfer in patients with high-grade gliomas, especially glioblastomas.

Keywords: [^{11}C]-methyl-L-methionine (MET); *O*-(2-[^{18}F]-fluoroethyl)-L-tyrosine (FET); 3,4-dihydroxy-6-[^{18}F]-fluoro-L-phenylalanine (FDOPA); magnetic resonance spectroscopy; perfusion-weighted imaging; diffusion-weighted imaging; chemical exchange saturation transfer; brain tumors; high-grade glioma; hybrid PET/MRI scanner

1. Introduction

At present, contrast-enhanced magnetic resonance imaging (MRI) is the method of choice for brain tumor diagnostics since MRI provides excellent soft tissue contrast, comparatively high resolution, and is widely available. On the downside, its specificity for neoplastic tissue is low, hampering the evaluation of tumor extent in both enhancing and non-enhancing gliomas as well as the differentiation of tumor progression from non-specific treatment-related changes [1–3].

[^{18}F]-2-Fluoro-2-deoxy-D-glucose (FDG) is the most extensively used positron emission tomography (PET) tracer to date and has hence gained exceptional importance in general oncology. In stark contrast, its applicability in brain tumor diagnostics is hindered by high levels of physiological glucose uptake in the cerebral cortex resulting in diminished contrast between tumor and background. On the contrary, the cerebral uptake of radiolabeled amino acids is low while it is typically increased in brain tumors, resulting in an enhanced tumor-to-background contrast [4,5]. Importantly, the uptake of amino acid tracers is independent of the blood-brain barrier integrity, allowing the evaluation of amino acid uptake in non-enhancing gliomas [1,6]. Additionally, amino acid PET has demonstrated its usefulness for the differentiation of tumor progression from treatment-related changes [7–12] and for various other indications, e.g., treatment monitoring, prognostication or tumor grading [4–6,13–17]. Among amino acid tracers labelled with carbon-11 used for PET imaging in neuro-oncology, [^{11}C]-methyl-L-methionine (MET) is currently best evaluated [18,19]. However, the short half-life of carbon-11 (20 min) limits its use to centers with an on-site cyclotron. To overcome such logistical limitations, amino acid tracers have been labeled with fluorine-18, which has a longer half-life of 110 min [4,6]. For that reason, O-(2-[^{18}F]-fluoroethyl)-L-tyrosine (FET) has replaced MET in many neuro-oncological centers. Subsequently, FET has become the most widely used radiotracer for brain tumor diagnostics, especially in Western Europe. Another fluorine-18 labeled amino acid tracer is 3,4-dihydroxy-6-[^{18}F]-fluoro-L-phenylalanine (FDOPA), originally developed to evaluate the dopamine synthesis in the basal ganglia, which also shows increased uptake in brain tumors [20,21]. Although brain tumors distant to the striatum are depicted with high contrast, the high striatal uptake of FDOPA limits its usefulness for tumor delineation especially near the basal ganglia [22,23]. The Response Assessment in Neuro-Oncology (RANO) working group considers the additional value of amino acid PET for brain tumor diagnostics as highly relevant and recommends its use at every stage of brain tumor management [24].

The combination of MRI and PET, especially if hybrid PET/MR scanners are used, offers great potential for brain tumor diagnostics. This technology is particularly attractive when ultra-high field MRI scanners operating at magnetic field strengths of 7 T or more are widely available, allowing comparative anatomical and metabolic imaging at high resolution. Currently, amino acid PET tracers are the best evaluated and most frequently used PET tracers for brain tumor diagnostics. Therefore, this review focuses on the combination of amino acid PET with MR spectroscopy (MRS), perfusion- and diffusion-weighted imaging (PWI, DWI), and chemical exchange saturation transfer (CEST) in patients with glioblastoma (GBM).

2. Search Strategy

A PubMed search of the published literature with the combination of the search terms "glioblastoma", "brain tumors", "high-grade glioma", "positron emission tomography", "magnetic resonance imaging", "magnetic resonance spectroscopy", "perfusion-weighted imaging", "diffusion-weighted imaging", "chemical exchange saturation transfer", "PET", "amino acid PET", "MRI", "advanced MRI", "MRS", "PWI", "DWI", "CEST", and "hybrid PET/MR" before and inclusive of December 2018 was performed. Additional literature was retrieved from the reference lists of all identified articles. Furthermore, articles identified through searches of the authors' files were included. Only papers in English were considered.

3. Amino Acid PET and MRS

MR spectroscopy (MRS) is a non-invasive method to detect selected water-soluble metabolites in vivo. In magnetic resonance, an externally applied magnetic field experienced by the nucleus is modulated by the distribution of surrounding electrons. Given that every molecule presents with a slightly different distribution of electrons, every nucleus experiences a slightly different magnetic field.

Consequently, every molecule has its characteristic magnetic field 'signature' that, combined with interactions amongst the nuclei, results in slightly different resonance frequencies leading to differential signals. These signal differences are used in MRS to identify the metabolites of interest. Every nucleus possessing a non-zero spin can be theoretically used for MRS, e.g., protons, carbon-13 and phosphorous-31. However, due to the low abundance of the latter and limited resolution of metabolite profiles in vivo, mainly proton spectra are used in clinical practice.

Metabolites that are frequently assessed by MRS are lactate, lipids, alanine, N-acetylaspartate (NAA), glutamine, glutamate, 2-hydroxyglutarate (2-HG), citrate, creatine, choline, and myo-inositol. In gliomas, with increasing grade of malignancy, NAA and creatine are usually decreased, whereas choline, lipids and lactate are increased [25–27]. Furthermore, the accumulation of 2-HG caused by gene mutations encoding for the enzyme isocitrate dehydrogenase (IDH) can be detected by 2-HG spectroscopy at 3 T [28]. Although several studies demonstrated the feasibility of 2-HG spectroscopy in a clinical setting [29–32], 2-HG spectroscopy remains challenging due to a small and complex signal and it is therefore not yet established in clinical routine [33].

Two different MRS methods exist using either data from a single voxel or multiple voxels in a single slice or multiple slices of the investigated organ. Single-voxel spectroscopy is the most widely used method due to simple data acquisition and relatively short scanning time. Moreover, the comparatively high signal-to-noise ratio achieved by single-voxel spectroscopy results in high-quality spectra enabling a quantitative analysis and straightforward interpretation. However, the manually preselected region of interest (ROI) based on information from T2-weighted or contrast-enhanced MRI may represent only a fraction of the tumor, potentially resulting in an incomplete evaluation of tumor biology.

Multi-voxel spectroscopy techniques may overcome these limitations by covering a more extended two- or three-dimensional ROI at higher spatial resolution. Especially in heterogenous lesions, this allows detection of subtle intratumoral signal alterations [34]. Another advantage of smaller voxels used in multi-voxel spectroscopy is the reduction of partial volume effects as structures such as cerebrospinal fluid or fat may diminish the quality of spectra and can be excluded from ROI definition. However, because of the difficulty of achieving good enough shimming over a large enough region, multi-voxel spectroscopy can have a lower signal-to-noise ratio, a reduced spectral quality for individual voxels, requires longer scanning times, and is technically more demanding than single-voxel spectroscopy [35].

In a few studies, MRS combined with amino acid PET in glioma patients was used. D´Souza and colleagues [36] evaluated in 29 high-grade glioma patients the diagnostic performance of MET PET and single-voxel MRS at 3 T for the correct diagnosis of tumor recurrence after neurooncological treatment including radiotherapy and chemotherapy. Metrics derived from both MET PET (i.e., tumor/brain ratios) and single-voxel MRS (i.e., choline/creatine ratios) suggested a high diagnostic accuracy of 85–90% for the detection of high-grade glioma recurrence. Floeth and co-workers [37] explored the use of combined FET PET and single-voxel MRS at 1.5 T in 50 patients with newly diagnosed lesions suspicious for gliomas on contrast-enhanced MRI. Based on the neuropathological assessment, gliomas could be identified with an accuracy of 68% using conventional MRI alone. Accuracy could be increased to 97% when conventional MR imaging was used in combination with FET PET and single-voxel MRS. In that study, FET lesion/brain ratios and NAA/choline ratios were identified as significant independent predictors for the identification of glioma tissue.

Stadlbauer et al. [38] used two-dimensional multi-voxel spectroscopy at 1.5 T in combination with FET PET to spatially correlate concentrations of choline, creatine, and NAA with FET uptake in 15 glioma patients. The authors found significant correlations between increased FET uptake and parameters derived from MRS. More precisely, increased FET uptake was significantly correlated with the extent of neuronal loss (NAA) and partially with membrane proliferation (choline), indicating that both PET and MRS provide complementary information. Yet, the number of patients included in that study was relatively small and two-dimensional MRS allowed only limited coverage of the tumor, especially in comparison with amino acid PET, which provided three-dimensional information concerning tracer distribution.

To overcome these limitations, Mauler and colleagues [39] investigated the spatial correlation between FET PET and three-dimensional MRS covering the whole brain. Forty-one glioma patients were investigated using a 3 T hybrid PET/MR scanner. Importantly, the authors reported that the FET uptake and increased choline/NAA ratios were not consistently spatially congruent (Figure 1).

Unfortunately, however, no spatial correlation with neuropathological information obtained via stereotactic biopsy was performed to further explore the differences between amino acid uptake and the metabolites detected by MRS in gliomas.

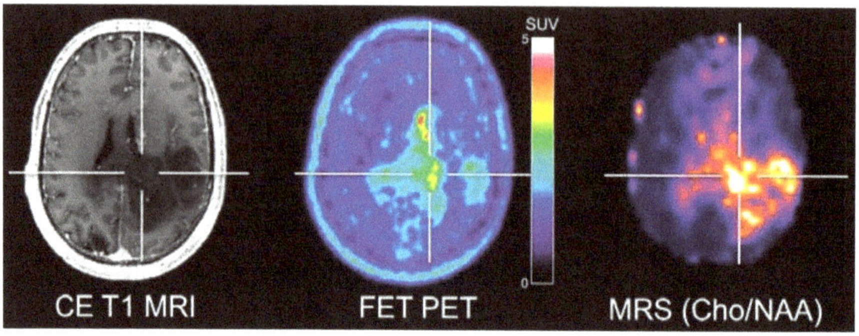

Figure 1. Simultaneously acquired contrast-enhanced (CE) T1-weighted MRI, FET PET and whole-brain MRS shows elevated choline/N-acetylaspartate (Cho/NAA) ratios in a patient with neuropathologically confirmed glioblastoma. Areas of contrast enhancement, FET uptake and increased Cho/NAA are spatially incongruent. Image courtesy of Joerg Mauler, Institute of Neuroscience and Medicine, Forschungszentrum Juelich, Germany, and Andrew Maudsley, Department of Radiology, University of Miami Medical School, Miami, FL, USA.

4. Amino Acid PET and PWI

PWI is a non-invasive MRI technique to measure blood flow quantitatively. In Neuro-Oncology, the parameters relative cerebral blood volume (rCBV) and cerebral blood flow (rCBF) are frequently assessed. Most commonly, a gadolinium-based contrast agent is used to assess tissue perfusion. After i.v. injection, the passage of the contrast agent leads to i) a local magnetic field distortion (susceptibility effect) in the vicinity of the vessels causing a signal drop in T2- or T2*-weighted MRI, also called dynamic susceptibility contrast (DSC), or ii) a shortening of T1-relaxation time causing a signal increase in T1-weighted MRI, also called dynamic contrast-enhanced (DCE) MRI. While DSC data requires only the first pass of intravascular contrast agent for assessing tissue perfusion, DCE additionally evaluates information about the continuous accumulation of the contrast agent in the extracellular space. Consequently, the acquisition times for DCE are longer than for DSC.

Another PWI method called arterial spin labelling (ASL) does not require a contrast agent. Here, endogenous water molecules in blood vessels are magnetically labeled by applying a specific radiofrequency pulse. Passage of these labeled molecules through the tissue of interest leads to a reduction of signal intensity in proportion to the perfusion. However, by dispensing with the administration of a contrast agent, the signal-to-noise ratio of ASL is inherently low, so that repetitive signal averaging is mandatory resulting in prolonged acquisition times. ASL significantly benefits from higher magnetic field strengths. Thus, ultra-high field MRI scanners operating at magnetic field strengths of up to 7 T may boost its clinical utility in the future.

Several studies combined PWI and amino acid PET in patients with high-grade gliomas for the delineation of the glioma extent. Filss and colleagues [40] compared DSC PWI and FET PET acquired concurrently using a 3 T hybrid PET/MR scanner in 56 glioma patients (24 patients with GBM). One main finding was that FET PET tumor volumes were significantly larger than tumor volumes delineated by rCBV maps. Furthermore, the spatial congruence of the two methods was poor, and the localization of tumor hotspots identified by PWI and FET PET yielded inconsistent results indicating that the delineation of glioma extent is not appropriately reflected by PWI and hence cannot replace amino acid PET (Figure 2). These findings were confirmed subsequently [41,42]. Besides, similar results were reported by Cicone and co-workers for the comparison of DSC PWI at 1.5 T and FDOPA PET [22].

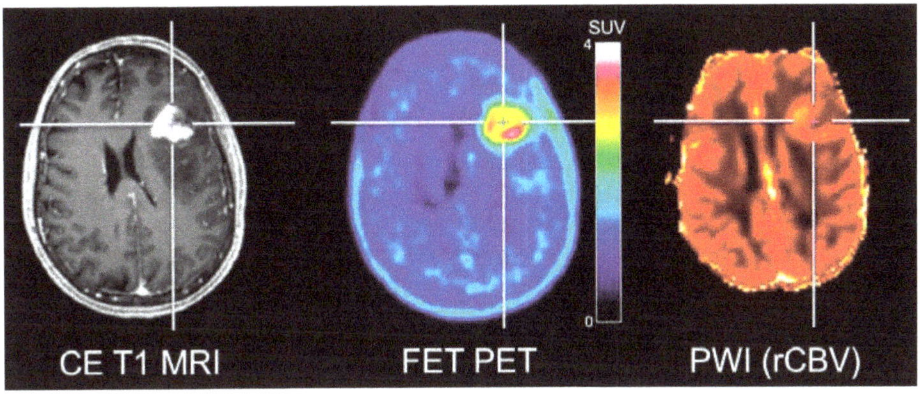

Figure 2. Simultaneously acquired contrast-enhanced (CE) T1-weighted MRI, FET PET and relative cerebral blood volume (rCBV) obtained from PWI in a patient with neuropathologically confirmed glioblastoma. FET PET identifies metabolically active tumor areas without an increased rCBV. Image courtesy of Christian Filss, Institute of Neuroscience and Medicine, Forschungszentrum Juelich, Juelich, Germany.

Verger et al. [43] investigated the usefulness of FET PET and DSC PWI at 3 T for the grading of gliomas in 72 patients with newly diagnosed glioma. The diagnostic accuracy for glioma grading was comparable for both FET PET and rCBV with an area under the receiver-operating characteristic curve of about 0.80. However, neuropathological diagnoses were not based on the revised WHO classification from 2016 [44], which limits the applicability of the results. The authors also found in 78% of patients a considerable spatial disparity between the local hotspots delineated using the two methods, consistent with previous studies [22,40,41].

Dandois and co-workers [45] compared the diagnostic accuracy of DSC PWI at 1.5 T and amino acid PET using MET for the differentiation of recurrent glioma from radiation necrosis in 28 high-grade glioma patients. The diagnostic performance was comparable between rCBV maps and MET PET for this clinically relevant question. These findings were reinforced by other studies [36,46,47]. However, in the majority of the cases, the diagnosis of either recurrent glioma or radiation necrosis was based on clinical observation and radiological follow-up rather than neuropathology. In stark contrast, Verger and colleagues [48] reported that FET PET seems to be superior to PWI at 3 T to diagnose recurrent glioma. Importantly, and in contrast to the latter studies mentioned above, the definite diagnosis was based on neuropathology in about 80% of the cases. These findings were confirmed by another study [49]. Subsequently, Roodakker et al. [50] compared regional MET PET and PWI at 3 T with local neuropathology in patients with en bloc-resected oligodendrogliomas. They showed that MET uptake correlated with tumor cell density throughout the entire tumor volume in all patients. Interestingly, tumor perfusion (rCBV) did not correlate with MET uptake or with any other histological marker.

Morana and colleagues [51] compared FDOPA PET and ASL at 1.5 T for the grading and prediction of tumor progression in 26 pediatric gliomas. Authors reported a better diagnostic performance for FDOPA PET with an area under the receiver-operating characteristic curve of 0.95 compared to 0.88 for rCBF derived from ASL.

The reported differences across the studies highlight the need for prospective studies with larger cohorts and neuropathological validation to further evaluate the differences between perfusion MRI and amino acid PET. Furthermore, a more recent meta-analysis regarding PWI in high-grade gliomas by Patel and co-workers [52] revealed a considerable heterogeneity of applied MR sequences, acquisition and post-processing parameters utilized in PWI studies of brain tumors thereby hampering comparability and reproducibility of results.

5. Amino Acid PET and DWI

Diffusion-weighted imaging (DWI) is an MRI technique based on the measurement of Brownian motion of water molecules to generate image contrast. DWI contrast uses two opposing gradient pulses; the first one induces a phase shift in water molecules, leading to a signal reduction. Subsequently, a second opposed gradient pulse is applied, which rephases the water molecules in the region of interest, leading to a recovery of the water signal. If water molecules moved out of the region of interest within the time between the two pulses, the number of water molecules contributing to the water signal is reduced. Therefore, the signal intensity is decreased. By variation of the amplitude, pulse duration, and time between the pulses, the degree of diffusion-weighting, represented by the b-value, can be altered. The apparent diffusion coefficient (ADC) is calculated from different b-values (typical range of b-values, 0–1000 s/mm^2), and represents a quantitative measure of diffusivity [53]. In brain tumors, hypointense signals in ADC maps (indicating lower ADC values) are suspicious for neoplastic tissue due to high tumor cellularity leading to a restriction of diffusivity [6].

Several studies addressed the spatial congruency of DWI and amino acid PET in glioma patients with mixed results. Karavaeva and colleagues [54] demonstrated a significant spatial relationship of FDOPA uptake and ADC signal alterations at 1.5 T in patients with recurrent high-grade glioma. Furthermore, a significant spatial correlation of both ADC values and FDOPA uptake with the cellular proliferation rate was reported. However, it remains unclear whether the cellular proliferation rate was associated with tumor progression or treatment-related tissue reactions such as inflammatory processes. On the other hand, several other studies reported contradictory results concerning the spatial congruency of amino acid uptake and ADC abnormalities [55–57]. Recently, Popp and co-workers [57] showed that tumor volumes in patients with recurrent glioblastoma delineated using ADC maps acquired at 1.5 T were frequently localized outside of areas with increased FET uptake. Another study by Kinoshita [58] and colleagues reported that MET PET demonstrated a more robust and reliable estimation of cell density in gliomas compared to ADC maps acquired at 3 T. These controversial study results concerning the spatial congruency are most likely related to the fact that amino acid PET and DWI encode different biological properties [57,59] (Figure 3).

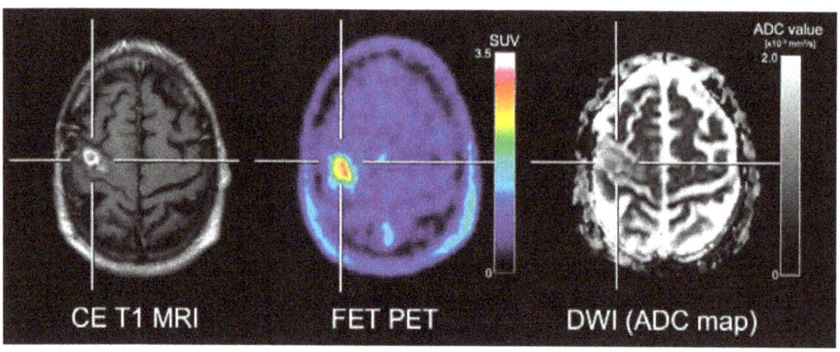

Figure 3. Contrast-enhanced (CE) T1-weighted MRI, FET PET and the apparent diffusion coefficient (ADC) map obtained from DWI in a patient with neuropathologically confirmed glioblastoma. In contrast to the ADC map, FET PET shows the metabolically active tumor tissue with high contrast. Furthermore, the ADC map shows slightly increased diffusivity in various subregions, erroneously indicating treatment-related changes.

Pyka and colleagues [49] compared FET PET and DWI at 3 T for the differentiation of tumor progression from treatment-related changes in 47 high-grade glioma patients. The authors reported a superior performance for FET PET compared to ADC (AUC, 0.86 vs. 0.69). The combination of FET PET and ADC with other MRI metrics such as rCBV resulted in an AUC improvement to 0.89.

It should be pointed out that DWI has a low spatial resolution and a poor signal-to-noise ratio. Although stronger gradients and higher magnetic field strengths can increase signal-to-noise ratios, susceptibility artifacts would become concurrently more pronounced. Another limitation of DWI is the compromised reproducibility and comparability of ADC values, even when the same MR scanner is used [60]. Moreover, diffusivity can also be altered by ischemia, infection, inflammation, gliosis, necrosis, or vascular proliferation, suggesting non-specificity of DWI changes [6,61,62].

6. Amino Acid PET and CEST

Chemical exchange saturation transfer (CEST) is an MRI technique for the indirect detection of different metabolites using the MR free water signal [63]. CEST contrasts are based on the spontaneous exchange of free water protons and protons bound to metabolites (solute-bound protons). After saturation of solute-bound protons by the application of a radio-frequency pulse with a specific frequency depending on the metabolite of interest, these protons exchange with free water protons and cause a signal reduction of free water. Thus, the CEST signal depends upon the exchange rate between free water and solute-bound protons and the concentration of the metabolite of interest [64,65].

CEST has already been applied for the measurement of glycogen (GlycoCEST) [66], glucose (GlucoCEST) [67], glutamate (GluCEST) [68], amide proton transfer (APT) [69], and intracellular pH [69–71]. The signal used to generate CEST contrasts is complex and small. Therefore, CEST imaging may benefit from higher magnetic field strengths (> 3 T) due to an increased signal-to-noise ratio and spectral dispersion. However, complex data processing steps are required to evaluate metabolite concentrations from CEST contrasts [72–75].

Although several studies evaluated CEST in patients with GBM [75–77], only few studies directly compared amino acid PET and CEST in more than 5 patients. Da Silva et al. [78] compared FET PET and APT CEST MRI in 8 high-grade gliomas patients using a 3 T hybrid PET/MR scanner. The authors showed that FET PET and APT CEST are spatially incongruent and reflect different biological information. However, although CEST and amino acid PET were acquired simultaneously, the number of patients was low, and no correlation with local neuropathology was obtained.

7. Conclusions and Outlook

The combination of amino acid PET and advanced MRI methods offers a variety of new vistas for the assessment of brain tumors with the potential to overcome the limitations of conventional MRI. Amino acid PET and advanced MRI methods already provided promising results for the clinical management of patients with GBM. Both the complementary and additive value of these methods as well as their incongruent findings, suggesting differential biological information, warrant further investigation including neuropathological validation. The increasing advent of hybrid PET/MR scanners offers the great potential of comparative studies using both amino acid PET and advanced MRI in a single session. Furthermore, the clinical benefits of simultaneous PET/MR imaging need to be balanced against the relatively high cost of such an approach. Of note, the increasing availability of ultra-high field MRI scanners with magnetic field strengths of 7 T will help to develop novel techniques thereby promoting advanced MRI in neuro-oncology as almost all contrasts benefit from the higher spatial resolution related to the increased signal-to-noise ratio.

Amino acid PET is a robust and attractive approach for clinicians for many reasons including easy scan reading. Furthermore, most studies using amino acid PET in neuro-oncology provide comparable results across different scanners, which is also a consequence of national and international efforts concerning the standardization of amino acid PET acquisition and evaluation in brain tumor imaging [24,79]. Recently, joint practice guidelines were developed by the European Association of Nuclear Medicine (EANM), the Society of Nuclear Medicine and Molecular Imaging (SNMMI), the European Association of Neuro-Oncology (EANO), and the working group for Response Assessment in Neuro-Oncology with PET (PET-RANO) [80].

In contrast, the yet missing standardization of advanced MRI methods is—at least in part—likely to account for conflicting data. In a significant number of advanced MRI studies, self-developed or self-optimized MR sequence protocols and acquisition parameters, as well as post-processing tools, were used, thereby impeding comparability and reproducibility of the results. Besides, vendors of MRI and PET scanners should strive for further standardization of imaging protocols and data processing since better comparability and easier reproducibility of the data is likely to boost their clinical value. Finally, further implementation of various advanced MRI methods as well as amino acid PET in clinical routine requires the validation of neuroimaging findings through neuropathology.

To conclude, advanced MRI in combination with amino acid PET has the potential to become an essential diagnostic tool for improving the clinical management of patients with high-grade gliomas, as well as in light of emerging high-throughput analysis methods of large-scale data sets such as radiomics and machine learning [16,81,82].

Funding: This research received no external funding.

Conflicts of Interest: The authors declare no conflict of interest.

References

1. Lohmann, P.; Stavrinou, P.; Lipke, K.; Bauer, E.K.; Ceccon, G.; Werner, J.M.; Neumaier, B.; Fink, G.R.; Shah, N.J.; Langen, K.J.; et al. FET PET reveals considerable spatial differences in tumour burden compared to conventional MRI in newly diagnosed glioblastoma. *Eur. J. Nucl. Med. Mol. Imaging* **2018**. [CrossRef] [PubMed]
2. Dhermain, F.G.; Hau, P.; Lanfermann, H.; Jacobs, A.H.; van den Bent, M.J. Advanced MRI and PET imaging for assessment of treatment response in patients with gliomas. *Lancet Neurol.* **2010**, *9*, 906–920. [CrossRef]
3. Ahluwalia, M.S.; Wen, P.Y. Antiangiogenic therapy for patients with glioblastoma: Current challenges in imaging and future directions. *Expert Rev. Anticancer Ther.* **2011**, *11*, 653–656. [CrossRef] [PubMed]
4. Herholz, K. Brain Tumors: An Update on Clinical PET Research in Gliomas. *Semin. Nucl. Med.* **2017**, *47*, 5–17. [CrossRef]
5. Galldiks, N.; Langen, K.J.; Pope, W.B. From the clinician's point of view—What is the status quo of positron emission tomography in patients with brain tumors? *Neuro Oncol.* **2015**, *17*, 1434–1444. [CrossRef] [PubMed]
6. Langen, K.J.; Galldiks, N.; Hattingen, E.; Shah, N.J. Advances in neuro-oncology imaging. *Nat. Rev. Neurol.* **2017**, *13*, 279–289. [CrossRef] [PubMed]
7. Galldiks, N.; Dunkl, V.; Stoffels, G.; Hutterer, M.; Rapp, M.; Sabel, M.; Reifenberger, G.; Kebir, S.; Dorn, F.; Blau, T.; et al. Diagnosis of pseudoprogression in patients with glioblastoma using *O*-(2-[18F]fluoroethyl)-L-tyrosine PET. *Eur. J. Nucl. Med. Mol. Imaging* **2015**, *42*, 685–695. [CrossRef] [PubMed]
8. Belliveau, J.G.; Bauman, G.; Macdonald, D.R. Detecting tumor progression in glioma: Current standards and new techniques. *Expert Rev. Anticancer Ther.* **2016**, *16*, 1177–1188. [CrossRef]
9. Galldiks, N.; Stoffels, G.; Filss, C.; Rapp, M.; Blau, T.; Tscherpel, C.; Ceccon, G.; Dunkl, V.; Weinzierl, M.; Stoffel, M.; et al. The use of dynamic *O*-(2-18F-fluoroethyl)-L-tyrosine PET in the diagnosis of patients with progressive and recurrent glioma. *Neuro Oncol.* **2015**, *17*, 1293–1300. [CrossRef]
10. Ceccon, G.; Lohmann, P.; Stoffels, G.; Judov, N.; Filss, C.P.; Rapp, M.; Bauer, E.; Hamisch, C.; Ruge, M.I.; Kocher, M.; et al. Dynamic *O*-(2-18F-fluoroethyl)-L-tyrosine positron emission tomography differentiates brain metastasis recurrence from radiation injury after radiotherapy. *Neuro Oncol.* **2017**, *19*, 281–288. [CrossRef]
11. Galldiks, N.; Kocher, M.; Langen, K.-J. Pseudoprogression after glioma therapy: An update. *Expert Rev. Neurother.* **2017**, *17*, 1109–1115. [CrossRef] [PubMed]
12. Muoio, B.; Giovanella, L.; Treglia, G. Recent developments of 18F-FET PET in neuro-oncology. *Curr. Med. Chem.* **2018**, *25*, 3061–3073. [CrossRef]

13. Suchorska, B.; Albert, N.L.; Tonn, J.C. Role of amino-tracer PET for decision-making in neuro-oncology. *Curr. Opin. Neurol.* **2018**, *31*, 720–726. [CrossRef] [PubMed]
14. Langen, K.J.; Galldiks, N. Update on amino acid PET of brain tumours. *Curr. Opin. Neurol.* **2018**, *31*, 354–361. [CrossRef] [PubMed]
15. Galldiks, N.; Law, I.; Pope, W.B.; Arbizu, J.; Langen, K.J. The use of amino acid PET and conventional MRI for monitoring of brain tumor therapy. *Neuroimage Clin.* **2017**, *13*, 386–394. [CrossRef] [PubMed]
16. Lohmann, P.; Kocher, M.; Steger, J.; Galldiks, N. Radiomics derived from amino-acid PET and conventional MRI in patients with high-grade gliomas. *Q J. Nucl. Med. Mol. Imaging* **2018**, *62*, 272–280. [CrossRef] [PubMed]
17. Galldiks, N.; Albert, N.L.; Sommerauer, M.; Grosu, A.L.; Ganswindt, U.; Law, I.; Preusser, M.; Le Rhun, E.; Vogelbaum, M.A.; Zadeh, G.; et al. PET imaging in patients with meningioma-report of the RANO/PET Group. *Neuro Oncol.* **2017**, *19*, 1576–1587. [CrossRef]
18. Singhal, T.; Narayanan, T.K.; Jain, V.; Mukherjee, J.; Mantil, J. 11C-L-methionine positron emission tomography in the clinical management of cerebral gliomas. *Mol. Imaging Biol.* **2008**, *10*, 1–18. [CrossRef]
19. Galldiks, N.; Ullrich, R.; Schroeter, M.; Fink, G.R.; Jacobs, A.H.; Kracht, L.W. Volumetry of [(11)C]-methionine PET uptake and MRI contrast enhancement in patients with recurrent glioblastoma multiforme. *Eur. J. Nucl. Med. Mol. Imaging* **2010**, *37*, 84–92. [CrossRef]
20. Becherer, A.; Karanikas, G.; Szabo, M.; Zettinig, G.; Asenbaum, S.; Marosi, C.; Henk, C.; Wunderbaldinger, P.; Czech, T.; Wadsak, W.; et al. Brain tumour imaging with PET: A comparison between [18F]fluorodopa and [11C]methionine. *Eur. J. Nucl. Med. Mol. Imaging* **2003**, *30*, 1561–1567. [CrossRef]
21. Herrmann, K.; Czernin, J.; Cloughesy, T.; Lai, A.; Pomykala, K.L.; Benz, M.R.; Buck, A.K.; Phelps, M.E.; Chen, W. Comparison of visual and semiquantitative analysis of 18F-FDOPA-PET/CT for recurrence detection in glioblastoma patients. *Neuro Oncol.* **2014**, *16*, 603–609. [CrossRef]
22. Cicone, F.; Filss, C.P.; Minniti, G.; Rossi-Espagnet, C.; Papa, A.; Scaringi, C.; Galldiks, N.; Bozzao, A.; Shah, N.J.; Scopinaro, F.; et al. Volumetric assessment of recurrent or progressive gliomas: Comparison between F-DOPA PET and perfusion-weighted MRI. *Eur. J. Nucl. Med. Mol. Imaging* **2015**, *42*, 905–915. [CrossRef]
23. Galldiks, N.; Langen, K.J. Applications of PET imaging of neurological tumors with radiolabeled amino acids. *Q J. Nucl. Med. Mol. Imaging* **2015**, *59*, 70–82. [PubMed]
24. Albert, N.L.; Weller, M.; Suchorska, B.; Galldiks, N.; Soffietti, R.; Kim, M.M.; la Fougere, C.; Pope, W.; Law, I.; Arbizu, J.; et al. Response Assessment in Neuro-Oncology working group and European Association for Neuro-Oncology recommendations for the clinical use of PET imaging in gliomas. *Neuro Oncol.* **2016**, *18*, 1199–1208. [CrossRef] [PubMed]
25. Negendank, W.G.; Sauter, R.; Brown, T.R.; Evelhoch, J.L.; Falini, A.; Gotsis, E.D.; Heerschap, A.; Kamada, K.; Lee, B.C.; Mengeot, M.M.; et al. Proton magnetic resonance spectroscopy in patients with glial tumors: A multicenter study. *J. Neurosurg.* **1996**, *84*, 449–458. [CrossRef] [PubMed]
26. Glunde, K.; Bhujwalla, Z.M.; Ronen, S.M. Choline metabolism in malignant transformation. *Nat. Rev. Cancer* **2011**, *11*, 835–848. [CrossRef]
27. Herminghaus, S.; Pilatus, U.; Moller-Hartmann, W.; Raab, P.; Lanfermann, H.; Schlote, W.; Zanella, F.E. Increased choline levels coincide with enhanced proliferative activity of human neuroepithelial brain tumors. *NMR Biomed.* **2002**, *15*, 385–392. [CrossRef]
28. Choi, C.; Ganji, S.K.; DeBerardinis, R.J.; Hatanpaa, K.J.; Rakheja, D.; Kovacs, Z.; Yang, X.L.; Mashimo, T.; Raisanen, J.M.; Marin-Valencia, I.; et al. 2-hydroxyglutarate detection by magnetic resonance spectroscopy in IDH-mutated patients with gliomas. *Nat. Med.* **2012**, *18*, 624–629. [CrossRef] [PubMed]
29. Natsumeda, M.; Igarashi, H.; Nomura, T.; Ogura, R.; Tsukamoto, Y.; Kobayashi, T.; Aoki, H.; Okamoto, K.; Kakita, A.; Takahashi, H.; et al. Accumulation of 2-hydroxyglutarate in gliomas correlates with survival: A study by 3.0-tesla magnetic resonance spectroscopy. *Acta Neuropathol. Commun.* **2014**, *2*, 158. [CrossRef]
30. Andronesi, O.C.; Loebel, F.; Bogner, W.; Marjanska, M.; Vander Heiden, M.G.; Iafrate, A.J.; Dietrich, J.; Batchelor, T.T.; Gerstner, E.R.; Kaelin, W.G.; et al. Treatment Response Assessment in IDH-Mutant Glioma Patients by Noninvasive 3D Functional Spectroscopic Mapping of 2-Hydroxyglutarate. *Clin. Cancer Res.* **2016**, *22*, 1632–1641. [CrossRef]

31. De la Fuente, M.I.; Young, R.J.; Rubel, J.; Rosenblum, M.; Tisnado, J.; Briggs, S.; Arevalo-Perez, J.; Cross, J.R.; Campos, C.; Straley, K.; et al. Integration of 2-hydroxyglutarate-proton magnetic resonance spectroscopy into clinical practice for disease monitoring in isocitrate dehydrogenase-mutant glioma. *Neuro Oncol.* **2016**, *18*, 283–290. [CrossRef] [PubMed]
32. Choi, C.; Raisanen, J.M.; Ganji, S.K.; Zhang, S.; McNeil, S.S.; An, Z.; Madan, A.; Hatanpaa, K.J.; Vemireddy, V.; Sheppard, C.A.; et al. Prospective Longitudinal Analysis of 2-Hydroxyglutarate Magnetic Resonance Spectroscopy Identifies Broad Clinical Utility for the Management of Patients With IDH-Mutant Glioma. *J. Clin. Oncol.* **2016**, *34*, 4030–4039. [CrossRef]
33. Andronesi, O.C.; Rapalino, O.; Gerstner, E.; Chi, A.; Batchelor, T.T.; Cahill, D.P.; Sorensen, A.G.; Rosen, B.R. Detection of oncogenic IDH1 mutations using magnetic resonance spectroscopy of 2-hydroxyglutarate. *J. Clin. Investig.* **2013**, *123*, 3659–3663. [CrossRef] [PubMed]
34. Nelson, S.J. Multivoxel magnetic resonance spectroscopy of brain tumors. *Mol. Cancer Ther.* **2003**, *2*, 497–507. [PubMed]
35. Rabinov, J.D.; Lee, P.L.; Barker, F.G.; Louis, D.N.; Harsh, G.R.; Cosgrove, G.R.; Chiocca, E.A.; Thornton, A.F.; Loeffler, J.S.; Henson, J.W.; et al. In vivo 3-T MR spectroscopy in the distinction of recurrent glioma versus radiation effects: Initial experience. *Radiology* **2002**, *225*, 871–879. [CrossRef] [PubMed]
36. D'Souza, M.M.; Sharma, R.; Jaimini, A.; Panwar, P.; Saw, S.; Kaur, P.; Mondal, A.; Mishra, A.; Tripathi, R.P. 11C-MET PET/CT and advanced MRI in the evaluation of tumor recurrence in high-grade gliomas. *Clin. Nucl. Med.* **2014**, *39*, 791–798. [CrossRef] [PubMed]
37. Floeth, F.W.; Pauleit, D.; Wittsack, H.J.; Langen, K.J.; Reifenberger, G.; Hamacher, K.; Messing-Junger, M.; Zilles, K.; Weber, F.; Stummer, W.; et al. Multimodal metabolic imaging of cerebral gliomas: Positron emission tomography with [18F]fluoroethyl-L-tyrosine and magnetic resonance spectroscopy. *J. Neurosurg.* **2005**, *102*, 318–327. [CrossRef] [PubMed]
38. Stadlbauer, A.; Prante, O.; Nimsky, C.; Salomonowitz, E.; Buchfelder, M.; Kuwert, T.; Linke, R.; Ganslandt, O. Metabolic imaging of cerebral gliomas: Spatial correlation of changes in *O*-(2-18F-fluoroethyl)-L-tyrosine PET and proton magnetic resonance spectroscopic imaging. *J. Nucl. Med.* **2008**, *49*, 721–729. [CrossRef] [PubMed]
39. Mauler, J.; Maudsley, A.A.; Langen, K.J.; Nikoubashman, O.; Stoffels, G.; Sheriff, S.; Lohmann, P.; Filss, C.; Galldiks, N.; Kops, E.R.; et al. Spatial Relationship of Glioma Volume Derived from (18)F-FET PET and Volumetric MR Spectroscopy Imaging: A Hybrid PET/MRI Study. *J. Nucl. Med.* **2018**, *59*, 603–609. [CrossRef]
40. Filss, C.P.; Galldiks, N.; Stoffels, G.; Sabel, M.; Wittsack, H.J.; Turowski, B.; Antoch, G.; Zhang, K.; Fink, G.R.; Coenen, H.H.; et al. Comparison of 18F-FET PET and perfusion-weighted MR imaging: A PET/MR imaging hybrid study in patients with brain tumors. *J. Nucl. Med.* **2014**, *55*, 540–545. [CrossRef]
41. Henriksen, O.M.; Larsen, V.A.; Muhic, A.; Hansen, A.E.; Larsson, H.B.; Poulsen, H.S.; Law, I. Simultaneous evaluation of brain tumour metabolism, structure and blood volume using [(18)F]-fluoroethyltyrosine (FET) PET/MRI: Feasibility, agreement and initial experience. *Eur. J. Nucl. Med. Mol. Imaging* **2016**, *43*, 103–112. [CrossRef] [PubMed]
42. Gottler, J.; Lukas, M.; Kluge, A.; Kaczmarz, S.; Gempt, J.; Ringel, F.; Mustafa, M.; Meyer, B.; Zimmer, C.; Schwaiger, M.; et al. Intra-lesional spatial correlation of static and dynamic FET-PET parameters with MRI-based cerebral blood volume in patients with untreated glioma. *Eur. J. Nucl. Med. Mol. Imaging* **2017**, *44*, 392–397. [CrossRef] [PubMed]
43. Verger, A.; Filss, C.P.; Lohmann, P.; Stoffels, G.; Sabel, M.; Wittsack, H.J.; Kops, E.R.; Galldiks, N.; Fink, G.R.; Shah, N.J.; et al. Comparison of (18)F-FET PET and perfusion-weighted MRI for glioma grading: A hybrid PET/MR study. *Eur. J. Nucl. Med. Mol. Imaging* **2017**, *44*, 2257–2265. [CrossRef] [PubMed]
44. Louis, D.N.; Perry, A.; Reifenberger, G.; von Deimling, A.; Figarella-Branger, D.; Cavenee, W.K.; Ohgaki, H.; Wiestler, O.D.; Kleihues, P.; Ellison, D.W. The 2016 World Health Organization Classification of Tumors of the Central Nervous System: A summary. *Acta Neuropathol.* **2016**, *131*, 803–820. [CrossRef] [PubMed]
45. Dandois, V.; Rommel, D.; Renard, L.; Jamart, J.; Cosnard, G. Substitution of 11C-methionine PET by perfusion MRI during the follow-up of treated high-grade gliomas: Preliminary results in clinical practice. *J. Neuroradiol.* **2010**, *37*, 89–97. [CrossRef] [PubMed]
46. Kim, Y.H.; Oh, S.W.; Lim, Y.J.; Park, C.K.; Lee, S.H.; Kang, K.W.; Jung, H.W.; Chang, K.H. Differentiating radiation necrosis from tumor recurrence in high-grade gliomas: Assessing the efficacy of 18F-FDG PET, 11C-methionine PET and perfusion MRI. *Clin. Neurol. Neurosurg.* **2010**, *112*, 758–765. [CrossRef]

47. Jena, A.; Taneja, S.; Gambhir, A.; Mishra, A.K.; D'Souza, M.M.; Verma, S.M.; Hazari, P.P.; Negi, P.; Jhadav, G.K.; et al. Glioma Recurrence Versus Radiation Necrosis: Single-Session Multiparametric Approach Using Simultaneous *O*-(2-18F-Fluoroethyl)-L-Tyrosine PET/MRI. *Clin. Nucl. Med.* **2016**, *41*, e228–e236. [CrossRef] [PubMed]
48. Verger, A.; Filss, C.P.; Lohmann, P.; Stoffels, G.; Sabel, M.; Wittsack, H.J.; Kops, E.R.; Galldiks, N.; Fink, G.R.; Shah, N.J.; et al. Comparison of *O*-(2-(18)F-Fluoroethyl)-L-Tyrosine Positron Emission Tomography and Perfusion-Weighted Magnetic Resonance Imaging in the Diagnosis of Patients with Progressive and Recurrent Glioma: A Hybrid Positron Emission Tomography/Magnetic Resonance Study. *World Neurosurg.* **2018**, *113*, e727–e737. [CrossRef]
49. Pyka, T.; Hiob, D.; Preibisch, C.; Gempt, J.; Wiestler, B.; Schlegel, J.; Straube, C.; Zimmer, C. Diagnosis of glioma recurrence using multiparametric dynamic 18F-fluoroethyl-tyrosine PET-MRI. *Eur. J. Radiol.* **2018**, *103*, 32–37. [CrossRef]
50. Roodakker, K.R.; Alhuseinalkhudhur, A.; Al-Jaff, M.; Georganaki, M.; Zetterling, M.; Berntsson, S.G.; Danfors, T.; Strand, R.; Edqvist, P.H.; Dimberg, A.; et al. Region-by-region analysis of PET, MRI, and histology in en bloc-resected oligodendrogliomas reveals intra-tumoral heterogeneity. *Eur. J. Nucl. Med. Mol. Imaging* **2018**. [CrossRef]
51. Morana, G.; Piccardo, A.; Tortora, D.; Puntoni, M.; Severino, M.; Nozza, P.; Ravegnani, M.; Consales, A.; Mascelli, S.; Raso, A.; et al. Grading and outcome prediction of pediatric diffuse astrocytic tumors with diffusion and arterial spin labeling perfusion MRI in comparison with 18F-DOPA PET. *Eur. J. Nucl. Med. Mol. Imaging* **2017**, *44*, 2084–2093. [CrossRef] [PubMed]
52. Patel, P.; Baradaran, H.; Delgado, D.; Askin, G.; Christos, P.; John Tsiouris, A.; Gupta, A. MR perfusion-weighted imaging in the evaluation of high-grade gliomas after treatment: A systematic review and meta-analysis. *Neuro Oncol.* **2017**, *19*, 118–127. [CrossRef]
53. Koh, D.M.; Padhani, A.R. Diffusion-weighted MRI: A new functional clinical technique for tumour imaging. *Br. J. Radiol.* **2006**, *79*, 633–635. [CrossRef] [PubMed]
54. Karavaeva, E.; Harris, R.J.; Leu, K.; Shabihkhani, M.; Yong, W.H.; Pope, W.B.; Lai, A.; Nghiemphu, P.L.; Liau, L.M.; Chen, W.; et al. Relationship Between [18F]FDOPA PET Uptake, Apparent Diffusion Coefficient (ADC), and Proliferation Rate in Recurrent Malignant Gliomas. *Mol. Imaging Biol.* **2015**, *17*, 434–442. [CrossRef] [PubMed]
55. Rose, S.; Fay, M.; Thomas, P.; Bourgeat, P.; Dowson, N.; Salvado, O.; Gal, Y.; Coulthard, A.; Crozier, S. Correlation of MRI-derived apparent diffusion coefficients in newly diagnosed gliomas with [18F]-fluoro-L-dopa PET: What are we really measuring with minimum ADC? *AJNR Am. J. Neuroradiol.* **2013**, *34*, 758–764. [CrossRef] [PubMed]
56. Choi, H.; Paeng, J.C.; Cheon, G.J.; Park, C.K.; Choi, S.H.; Min, H.S.; Kang, K.W.; Chung, J.K.; Kim, E.E.; Lee, D.S. Correlation of 11C-methionine PET and diffusion-weighted MRI: Is there a complementary diagnostic role for gliomas? *Nucl. Med. Commun.* **2014**, *35*, 720–726. [CrossRef] [PubMed]
57. Popp, I.; Bott, S.; Mix, M.; Oehlke, O.; Schimek-Jasch, T.; Nieder, C.; Nestle, U.; Bock, M.; Yuh, W.T.C.; Meyer, P.T.; et al. Diffusion-weighted MRI and ADC versus FET-PET and GdT1w-MRI for gross tumor volume (GTV) delineation in re-irradiation of recurrent glioblastoma. *Radiother. Oncol.* **2018**. [CrossRef]
58. Kinoshita, M.; Arita, H.; Okita, Y.; Kagawa, N.; Kishima, H.; Hashimoto, N.; Tanaka, H.; Watanabe, Y.; Shimosegawa, E.; Hatazawa, J.; et al. Comparison of diffusion tensor imaging and (11)C-methionine positron emission tomography for reliable prediction of tumor cell density in gliomas. *J. Neurosurg.* **2016**, *125*, 1136–1142. [CrossRef]
59. Tietze, A.; Boldsen, J.K.; Mouridsen, K.; Ribe, L.; Dyve, S.; Cortnum, S.; Ostergaard, L.; Borghammer, P. Spatial distribution of malignant tissue in gliomas: Correlations of 11C-L-methionine positron emission tomography and perfusion- and diffusion-weighted magnetic resonance imaging. *Acta Radiol.* **2015**, *56*, 1135–1144. [CrossRef]
60. Sasaki, M.; Yamada, K.; Watanabe, Y.; Matsui, M.; Ida, M.; Fujiwara, S.; Shibata, E.; Acute Stroke Imaging Standardization Group-Japan, I. Variability in absolute apparent diffusion coefficient values across different platforms may be substantial: A multivendor, multi-institutional comparison study. *Radiology* **2008**, *249*, 624–630. [CrossRef]

61. Huang, R.Y.; Neagu, M.R.; Reardon, D.A.; Wen, P.Y. Pitfalls in the neuroimaging of glioblastoma in the era of antiangiogenic and immuno/targeted therapy—Detecting illusive disease, defining response. *Front. Neurol.* **2015**, *6*, 33. [CrossRef] [PubMed]
62. Sundgren, P.C.; Fan, X.; Weybright, P.; Welsh, R.C.; Carlos, R.C.; Petrou, M.; McKeever, P.E.; Chenevert, T.L. Differentiation of recurrent brain tumor versus radiation injury using diffusion tensor imaging in patients with new contrast-enhancing lesions. *Magn. Reson. Imaging* **2006**, *24*, 1131–1142. [CrossRef]
63. Wolff, S.D.; Balaban, R.S. NMR imaging of labile proton exchange. *J. Magn. Reson.* **1990**, *86*, 164–169. [CrossRef]
64. Ward, K.M.; Aletras, A.H.; Balaban, R.S. A new class of contrast agents for MRI based on proton chemical exchange dependent saturation transfer (CEST). *J. Magn. Reson.* **2000**, *143*, 79–87. [CrossRef] [PubMed]
65. Ward, K.M.; Balaban, R.S. Determination of pH using water protons and chemical exchange dependent saturation transfer (CEST). *Magn. Reson. Med.* **2000**, *44*, 799–802. [CrossRef]
66. Van Zijl, P.C.; Jones, C.K.; Ren, J.; Malloy, C.R.; Sherry, A.D. MRI detection of glycogen in vivo by using chemical exchange saturation transfer imaging (glycoCEST). *Proc. Natl. Acad. Sci. USA* **2007**, *104*, 4359–4364. [CrossRef]
67. Walker-Samuel, S.; Ramasawmy, R.; Torrealdea, F.; Rega, M.; Rajkumar, V.; Johnson, S.P.; Richardson, S.; Goncalves, M.; Parkes, H.G.; Arstad, E.; et al. In vivo imaging of glucose uptake and metabolism in tumors. *Nat. Med.* **2013**, *19*, 1067–1072. [CrossRef] [PubMed]
68. Cai, K.; Haris, M.; Singh, A.; Kogan, F.; Greenberg, J.H.; Hariharan, H.; Detre, J.A.; Reddy, R. Magnetic resonance imaging of glutamate. *Nat. Med.* **2012**, *18*, 302–306. [CrossRef]
69. Zhou, J. *Amide Proton Transfer Imaging of the Human Brain*; Modo, M., Bulte, J.W.M., Eds.; Humana Press: Totowa, NJ, USA, 2011; Volume 711, pp. 227–237.
70. Zhou, J.; Payen, J.F.; Wilson, D.A.; Traystman, R.J.; van Zijl, P.C. Using the amide proton signals of intracellular proteins and peptides to detect pH effects in MRI. *Nat. Med.* **2003**, *9*, 1085–1090. [CrossRef]
71. Jones, C.K.; Schlosser, M.J.; van Zijl, P.C.; Pomper, M.G.; Golay, X.; Zhou, J. Amide proton transfer imaging of human brain tumors at 3T. *Magn. Reson. Med.* **2006**, *56*, 585–592. [CrossRef]
72. Van Zijl, P.C.; Yadav, N.N. Chemical exchange saturation transfer (CEST): What is in a name and what isn't? *Magn. Reson. Med.* **2011**, *65*, 927–948. [CrossRef] [PubMed]
73. Wu, B.; Warnock, G.; Zaiss, M.; Lin, C.; Chen, M.; Zhou, Z.; Mu, L.; Nanz, D.; Tuura, R.; Delso, G. An overview of CEST MRI for non-MR physicists. *EJNMMI Phys.* **2016**, *3*, 19. [CrossRef] [PubMed]
74. Zaiss, M.; Bachert, P. Chemical exchange saturation transfer (CEST) and MR Z-spectroscopy in vivo: A review of theoretical approaches and methods. *Phys. Med. Biol.* **2013**, *58*, R221–R269. [CrossRef] [PubMed]
75. Dreher, C.; Oberhollenzer, J.; Meissner, J.E.; Windschuh, J.; Schuenke, P.; Regnery, S.; Sahm, F.; Bickelhaupt, S.; Bendszus, M.; Wick, W.; et al. Chemical exchange saturation transfer (CEST) signal intensity at 7T MRI of WHO IV degrees gliomas is dependent on the anatomic location. *J. Magn. Reson. Imaging* **2018**. [CrossRef] [PubMed]
76. Regnery, S.; Adeberg, S.; Dreher, C.; Oberhollenzer, J.; Meissner, J.E.; Goerke, S.; Windschuh, J.; Deike-Hofmann, K.; Bickelhaupt, S.; Zaiss, M.; et al. Chemical exchange saturation transfer MRI serves as predictor of early progression in glioblastoma patients. *Oncotarget* **2018**, *9*, 28772–28783. [CrossRef]
77. Sagiyama, K.; Mashimo, T.; Togao, O.; Vemireddy, V.; Hatanpaa, K.J.; Maher, E.A.; Mickey, B.E.; Pan, E.; Sherry, A.D.; Bachoo, R.M.; et al. In vivo chemical exchange saturation transfer imaging allows early detection of a therapeutic response in glioblastoma. *Proc. Natl. Acad. Sci. USA* **2014**, *111*, 4542–4547. [CrossRef]
78. Da Silva, N.A.; Lohmann, P.; Fairney, J.; Magill, A.W.; Oros Peusquens, A.M.; Choi, C.H.; Stirnberg, R.; Stoffels, G.; Galldiks, N.; Golay, X.; et al. Hybrid MR-PET of brain tumours using amino acid PET and chemical exchange saturation transfer MRI. *Eur. J. Nucl. Med. Mol. Imaging* **2018**, *45*, 1031–1040. [CrossRef] [PubMed]
79. Vander Borght, T.; Asenbaum, S.; Bartenstein, P.; Halldin, C.; Kapucu, O.; Van Laere, K.; Varrone, A.; Tatsch, K.; European Association of Nuclear Medicine (EANM). EANM procedure guidelines for brain tumour imaging using labelled amino acid analogues. *Eur. J. Nucl. Med. Mol. Imaging* **2006**, *33*, 1374–1380. [CrossRef]

80. Law, I.; Albert, N.L.; Arbizu, J.; Boellaard, R.; Drzezga, A.; Galldiks, N.; la Fougere, C.; Langen, K.J.; Lopci, E.; Lowe, V.; et al. Joint EANM/EANO/RANO practice guidelines/SNMMI procedure standards for imaging of gliomas using PET with radiolabelled amino acids and [(18)F]FDG: Version 1.0. *Eur. J. Nucl. Med. Mol. Imaging* **2018**. [CrossRef]
81. Lambin, P.; Rios-Velazquez, E.; Leijenaar, R.; Carvalho, S.; Van Stiphout, R.G.P.M.; Granton, P.; Zegers, C.M.L.; Gillies, R.; Boellard, R.; Dekker, A.; et al. Radiomics: Extracting more information from medical images using advanced feature analysis. *Eur. J. Cancer* **2012**, *48*, 441–446. [CrossRef]
82. Lohmann, P.; Kocher, M.; Ceccon, G.; Bauer, E.K.; Stoffels, G.; Viswanathan, S.; Ruge, M.I.; Neumaier, B.; Shah, N.J.; Fink, G.R.; et al. Combined FET PET/MRI radiomics differentiates radiation injury from recurrent brain metastasis. *Neuroimage Clin.* **2018**, *20*, 537–542. [CrossRef] [PubMed]

cancers

MDPI

Review

Treatment of Glioblastoma (GBM) with the Addition of Tumor-Treating Fields (TTF): A Review

Denise Fabian [1], Maria del Pilar Guillermo Prieto Eibl [2], Iyad Alnahhas [2], Nikhil Sebastian [1], Pierre Giglio [2], Vinay Puduvalli [2], Javier Gonzalez [2] and Joshua D. Palmer [2,*]

1 Department of Radiation Oncology, The Ohio State University, Columbus, OH 43210, USA; denise.fabian@osumc.edu (D.F.); Nikhil.Sebastian@osumc.edu (N.S.)

2 Department of Neuro-Oncology, The Ohio State University, Columbus, OH 43210, USA; Pilar.Prieto@osumc.edu (M.d.P.G.P.E.); Iyad.Alnahhas@osumc.edu (I.A.); Pierre.Giglio@osumc.edu (P.G.); Vinay.Puduvalli@osumc.edu (V.P.); Javier.GonzalezAlarcon@osumc.edu (J.G.)

* Correspondence: Joshua.Palmer@osumc.edu; Tel.: +1-614-293-5066

Received: 14 January 2019; Accepted: 31 January 2019; Published: 2 February 2019

Abstract: Glioblastoma (GBM) is the most common primary brain tumor. Despite aggressive treatment, GBM almost always recurs. The current standard-of-care for treatment of newly diagnosed GBM has remained relatively unchanged since 2005: maximal safe resection followed by concomitant chemoradiation (CRT) with temozolomide (TMZ), and subsequent adjuvant TMZ. In 2011, the first-generation tumor treating fields (TTF) device, known at the time as the NovoTTF-100A System (renamed Optune), was approved by the Food and Drug Administration (FDA) for treatment of recurrent GBM. The TTF device was subsequently approved as an adjuvant therapy for newly-diagnosed GBM in 2015. The following is a review of the TTF device, including evidence supporting its use and limitations.

Keywords: glioblastoma; GBM; tumor treating fields; TTF

1. Background

Glioblastoma (GBM) is the most aggressive and common primary brain tumor. Treatment remains challenging, as GBM inevitably recurs despite surgical debulking, radiation and chemotherapy. The current standard-of-care is comprised of maximal safe resection—gross total resection if feasible—followed by chemoradiation (CRT). Radiation therapy consists of 60 Gray (Gy) in 30 fractions over a period of 6 weeks with concomitant daily temozolomide (TMZ), followed by adjuvant TMZ (days 1–5 every 28 days). Median survival ranges from twelve to fifteen months following diagnosis and treatment [1]. The current five-year survival rate is about five percent in the United States [2].

Since 2005, several clinical trials have been conducted attempting to improve outcomes for GBM patients. For example, the Radiation Therapy Oncology Group (RTOG, Philadelphia, PA, USA) 0525 was a phase III trial that compared conventional adjuvant TMZ with dose-dense (dd) TMZ. Despite confirming the prognostic significance of *MGMT* promotor methylation, survival did not improve with dd TMZ [3]. The addition of Bevacizumab in RTOG 0825 demonstrated improvement in progression free survival (PFS), however, it did not yield changes on overall survival (OS) [4,5]. The addition of Everolimus, an oral mammialian target of rapamycin (mTOR) inhibitor, to chemoradiation, increased treatment-related toxicities, and did not have any impact on progression-free survival (PFS) and it even shortened the OS [4]. Additional trials are being conducted to look into the use of checkpoint inhibitors, such as Ipilimumab and Nivolumab (NRG-BN002) and radiation dose-escalation with photon Intensity modulated radiotherapy (IMRT) or Proton Beam Therapy (NRG-BN001) NRG Oncology is a National Clinical Trials Network group created through the efforts of the National Surgical Adjuvant Breast and Bowel Project (NSABP), the RTOG, and the Gynecologic Oncology Group (GOG). Recently, a Phase

II Trial of Neoadjuvant TMZ followed by accelerated hypofractionated radiation therapy (60 Gy in 20 fractions) demonstrated a median OS of twenty-two months with a PFS of 13.2 months, comparing favorably to OS previously reported in other clinical trials [6].

In 2011, the United States Food and Drug Administration (FDA) approved a tumor treating fields (TTF) device for treatment of recurrent or refractory GBM. More recently, the FDA approved the TTF device as adjuvant treatment for newly-diagnosed patients after completing standard-of-care surgery and chemoradiation. The National Comprehensive Cancer Network (NCCN) added the TTF device as an option for treatment of newly-diagnosed GBM. Despite FDA approval, skepticism remains regarding this therapy. In this review we discuss the current evidence supporting treatment with the TTF device and its limitations.

2. Materials and Methods

We conducted a comprehensive literary investigation utilizing PubMed and Google search engines. Approximately 50 journal articles, newspaper articles, and abstracts were reviewed. Ultimately, 43 sources were selected for relevance and impact. Relevance of topics was selected based on talking points at the 2018 American Society for Radiation Oncology (ASTRO) conference and common questions proposed by the patients and practicing clinicians at our institution.

2.1. Tumor Treating Fields Device Proposed Mechanism

The TTF device includes four transducer arrays, each consisting of nine insulated electrodes which are applied to the patient's scalp to deliver low-intensity, intermediate-frequency (100–300 kHz) alternating electric fields [7,8]. In 2004, a preclinical model demonstrated the inhibitory effect of the TTF device on proliferating cells whereas nonproliferating cells remained unaffected. Treatment with the TTF device is thought to interfere with normal polymerization and depolymerization of microtubules of the mitotic spindle by positioning tubulin dimers further away from the growing end of the microtubules [8]. This leads to mitotic disruption, which leads to mitotic catastrophe and ultimately to mitotic cell death. The investigators demonstrated this by setting up melanoma cell cultures in vitro with TTFs generated by pairs of insulated wires. In cells exposed to TTF, significant inhibition of growth was seen after 24 h exposure. This effect was also seen beyond the exposure time [8]. To explore the effects of TTF on molecular processes, the investigators used time-lapse microphotography. In cells treated with TTF, mitosis began normally, but was prolonged [8]. Additionally, a quarter of cells in the TTF cultures were destroyed during the formation of the mitotic cleavage furrow [8]. Finally, nuclear rotation was seen in the TTF cultures [8]. The investigators explain that microtubules in dividing cells have electric dipole moments, which may be altered by the forces exerted by TTF [8]. They showed this by comparing the movement of cellular microtubules by fixing the cells after 24 h of TTF vs. no treatment; when the fixed cells were viewed under fluorescence microscopy, more than 50% of the TTF treated cells had abnormal mitosis compared to less than 5% of the control cells [8]. The investigators defined two mechanisms of action: (1) disruption of the polar tubulin molecule orientation, pushing tubulin dimers further than 14 nm away from the growing end of the microtubule and thereby interfering with proper microtubule assembly; and (2) cell destruction by pulling of all intracellular and polar particles towards the mitotic cleavage furrow, resulting in a pile-up that interferes with cytokinesis [8].

Further laboratory studies demonstrated similar cancer growth inhibition in multiple in vitro cell lines and animal tumor models. A single-arm pilot clinical trial was performed in ten patients shortly thereafter [9]. Novocure Ltd. (St. Helier, NJ, USA) manufactured a clinical TTF device, named Optune (formally Novo TTF-100A, St. Helierl, NJ, USA). It utilizes electrodes configured as a cap that is placed on the patient's shaved scalp and powered by a battery package [10]. At present, the proposed mechanism also includes interference with Septin fibers of proliferating cells as well as endoplasmic reticulum stress leading to cellular stress and autophagy [11,12]. This demonstrates that the effect of TTF on interactions of intercellular may be more complex than originally thought; similarly, the effects on many other intracellular proteins are unknown. Further studies are warranted in both cancer and normal cell lines.

2.2. Treatment of Glioblastoma with Tumor Treating Fields

The first phase III clinical trial testing the TTF device (EF-11) was published in 2012 and included 237 patients with recurrent GBM [13]. This study compared the first generation of the TFF device alone—worn 18–24 h a day—to chemotherapy of the treating physician's choice. The median survival was 6.6 vs. 6.0 months ($p = 0.27$) for patients subject to treatment utilizing the TTF device or chemotherapy, respectively. Although the trial did not demonstrate its primary endpoint of improved overall survival, efficacy was similar to commonly used chemotherapy regimens. Given the localized nature of the treatment, a lower toxicity profile and better quality of life—including improved cognition and emotional well-being—were reported in the TTF arm. These patients also experienced less chemotherapy-related side effects, such as hematologic toxicity, gastrointestinal adverse events, and infections. Further analysis found that compliance with therapy was linked to outcomes; more specifically device compliance above 75% correlated with improved OS [14]. It is important to note that the trials were not powered to look into these subgroup analyses. The limitations of the study included absence of a placebo control arm and a heterogeneous patient population who had received various chemotherapy treatments prior to the trial. The results from this trial lead to the 2011 FDA approval of the first generation of the TFF device for treatment of patients with recurrent GBM or GBM that has not responded to traditional therapy [15]. Approval of this device was contingent on Novocure conducting additional studies and clinical trials to demonstrate device efficacy [16].

Following the trial, the Patient Registry Data set (PRiDe)—a registry of all patients with recurrent GBM who received therapy with the TTF device—was used to analyze the clinical outcomes of TTF device use in 91 cancer centers across the United States [17]. Compared to patients in the EF-11 trial, the 457 recurrent GBM patients in the PRiDe data set were more likely to use the TTF device for their first recurrence (33% vs. 9%). Median OS was improved in clinical practice when compared to the EF-11 trial (9.6 months vs. 6.6 months, hazard ratio (HR) 0.66, $p = 0.003$). As seen previously, device compliance rate of 75% or greater was associated with a significantly improved overall survival when compared to device compliance rate of less than 75% (7.7 vs. 4.5 months, $p = 0.042$). There were no significant adverse events, and the most common side effect was skin reaction [17].

In 2009, a multi-center phase III clinical trial lead by Dr. Roger Stupp was initiated for newly diagnosed GBM, studying the addition of TTF device treatment to maintenance TMZ (EF-14 trial) [7]. This study randomized 695 GBM patients in a 2:1 format to receive TTF device treatment plus maintenance TMZ or TMZ alone, following standard-of-care surgery and concurrent chemoradiotherapy. Patients with a Karnofsky performance score (KPS) of less than seventy, evidence of progressive disease following chemoradiotherapy, infratentorial tumor location or severe comorbidities were excluded from the study. An interim analysis presented in 2015 demonstrated improved PFS (7.1 months in the TTF device plus TMZ group and 4.0 in the TMZ alone group ($p = 0.001$) [7]. The final study was published in 2017, which reported an improved median overall survival of 20.9 months in the TTF device plus TMZ group vs. 16.0 months in the TMZ only group ($p < 0.001$). Of note, the reported survival times were measured from time of randomization, which was done after completion of chemoradiation, and which was about 3.8 months from original diagnosis. About half of patients who received treatment with the TFF device experienced mild to moderate skin toxicity [7]. Accordingly, the FDA approved the use of the TTF device for use in newly diagnosed GBM on 5 October 2015 [15]. Both Phase III clinical trials using TTF for GBM are summarized in Table 1.

This publication lead to the National Comprehensive Cancer Network (NCCN) adaptation of TTF treatment to the Clinical Practice Guidelines in Oncology for Central Nervous System. Additional analysis of the EF-14 trial has shown that the overall survival (OS) is improved at five years, regardless of other prognostic factors, such as age, performance status, extent of resection, and neurologic status.

An abstract presented at ASTRO 2018 showed that that use of the TTF device and TMZ improved overall survival out to five years in all three recursive partitioning analysis (RPA) classes [18].

As expected, further clinical investigations using TFF are well underway (Table 2). In November 2018, a phase II trial opened comparing a combination treatment with the TTF and nivolumab with or without ipilimumab in patients with bevacizumab-naive recurrent GBM [19]. The trial is expected to close in 2021 and has an accrual goal of sixty (60) patients. Table 2 is a summary of ongoing trials using TTF in GBM. Research on utilization of TTF in cancer care is also not limited to GBM. A phase 2 pilot study on the safety and efficacy of the TTF device concomitant with pemetrexed and cisplatin or carboplatin in malignant pleural mesothelioma (STELLER) recently closed in April 2018. In this study, 150 kHz electrode arrays were placed on the thorax of patients with previously treated malignant pleural mesothelioma in addition to chemotherapy. The preliminary efficacy will be compared to historical cohorts [20]. TFF devices are being explored in disease sites such brain metastasis secondary to non-small cell lung carcinoma, ovarian carcinoma, pancreatic carcinoma, meningioma and even as an alternative to prophylactic cranial irradiation in small cell lung cancer [21,22].

Table 1. Completed clinical trials using tumor treating fields (TTF) for glioblastoma (GBM), published [1].

Trial Name	Patient Cohort	Number of Patients	Study Design	Intervention	Endpoints	Toxicity	Ref.
EF-11 (NCT00379470)	GBM, progressed on prior therapy	237	Prospective, randomized (1:1) Phase III	Standard: best available CHT alone; Experimental: TTF alone (20–24 h/d)	Median OS 6.6 vs. 6.0 months (primary endpoint); 1-y OS 20% vs. 20%; 6-month PFS 21.4% vs. 15.1% (NS)	Severe adverse events 6% vs. 16% ($p = 0.022$). TTF-related adverse events were mild (14%) to moderate (2%) skin rash.	[12]
EF-14 (NCT00916409)	Newly diagnosed GBM after completion of concurrent TMZ and RT.	695	Prospective, randomized (2:1) Phase III	Standard: Maintenance TMZ (150–200 mg/m^2/d for 5 days every 28 days for 6–12 cycles); Experimental: Maintenance TMZ with TTF (>18 h/d)	Median PFS 20.9 months vs. 16.0 months ($p < 0.001$)	Grade 1–2 skin toxicity 52% vs. 0%.	[6]

[1] Abbreviations: GBM: glioblastoma multiforme, CHT: chemotherapy, TMZ: temozolomide, TTF: tumor treating fields, OS: overall survival, PFS: progression-free survival.

Table 2. Ongoing trials using TTF for GBM as of December 2018 [1].

ClinicalTrials.gov Identifier	Patient Cohort	Study Design	Intervention	Primary Outcome Measures	Secondary Outcome Measures	Estimated Enrollment	Duration	Institution
NCT03430791	Bevacizumab-naïve, recurrent GBM	Prospective, randomized Phase II	TTF + nivolumab +/− ipilimumab	ORR	-	60	Nov 2018–Aug 2021	Miami Cancer Institute
NCT03405792	Newly diagnosed GBM after resection and CRT.	Phase II, single arm	TMZ + TTF + pembrolizumab	PFS	AEs, OS, Augmentation of TTF immune reaction	29	Feb 2018–Feb 2023	University of Florida
NCT03477110	Newly diagnosed GBM	Phase I, single arm, single institution	CRT + TTF (up front at initiation of CRT)	TTF-discontinuation rate due to toxicity	PFS, PS, EFS	35	May 2018–Mar 2020	Thomas Jefferson University
NCT02663271	Bevacizumab-refractory recurrent GBM	Phase II multicenter, single-arm	TTF + bevacizumab	PFS	AEs, KPS/MMSE change from baseline, imaging response	18	Aug 2016–Mar 2019	University of Florida, Washington University
NCT02743078	Bevacizumab-refractory recurrent GBM	Phase II, multicenter, single-arm	TTF + bevacizumab	OS	PFS, ORR, AEs	85	Apr 2017–Aug 2022	Multiple
NCT01894061	Bevacizumab-naïve, recurrent GBM	Phase II, single-arm	TTF + bevacizumab	PFS	ORR, AEs, neuro-cognition, QOL	40	June 2013–May 2019	Case Western, Cleveland Clinic, University of Cincinnati
NCT03223103	Newly diagnosed GBM, after CRT	Phase I single-arm, single-institution	TTF + mutation-derived tumor vaccine	DLT	Toxicity, PFS, OS, ORR	20	Mar 2018–May 2020	Mt. Sinai
NCT02903069	Newly diagnosed GBM	Phase I, multicenter	TTF + proteasome inhibitor	MTD/DLT	AEs, OS, PFS	72	Apr 2016–May 2020	Multiple

[1] Abbreviations: GBM: glioblastoma multiforme, TTF: tumor treating fields, ORR: objective response rate, OS: overall survival, PFS: progression-free survival, TTP: time to disease progression, QOL: quality of life, AEs: adverse events, CRT: chemoradiation, EFS: event-free survival, MTD: maximum tolerated dose, DLT: dose-limiting toxicities.

2.3. Existing Skepticism

The TFF device has been described as "polarizing" amongst neuro-oncology experts. To date, the TTF device is not covered by Medicare or many other insurance companies, on the grounds that the therapy is still experimental. More recently, Novocure has applied for a reconsideration request, which has been accepted for local coverage determination [15].

Despite FDA-approval in the United States (US), existing skepticism regarding the use of the TTF device persists. Primary criticism includes the unblinded nature of the TTF clinical trials. A "sham" device—to better discern a potential placebo-effect of wearing the device—wasn't used. The lack of blinding becomes more of an issue when the primary outcome is PFS. Other concerns arise from the lack of understanding of the TTF device's manner of operation, specifically across a variety of tissues and in combination with other treatments [23]. Furthermore, the fact that randomization in the EF-14 trial occurred over two months after diagnosis suggests a selection bias of patients who did not have progression after the initial treatment and would therefore likely have better survival regardless. Hence, it is difficult to consider the device as "standard-of-care" for future patients who may either: (1) be receiving a shorter course of radiotherapy; or (2) experience an interruption chemotherapy during initial treatment [23].

Several of these points were addressed during a round table discussion of the EF-14 trial at ASCO 2015. This meeting included 5 neuro-oncology experts who did not participate in the trial [21]. In response to the time of randomization question, the EF-14 trial compared favorably in efficacy to other GBM trials when adjustments were made for this difference. Specifically, the improvement in the 2-year survival rate, which was 43% with TTF vs. 29% without TTF, along with median overall survival, (19.6 months with TTF device treatment vs. 16.6 months without TTF device treatment, HR = 0.744, p = 0.0038) were considered clinically meaningful [21]. Omission of a placebo-control device in the trial was also a topic of consideration. Several arguments were made in support of this decision, including that: (1) the panel thought it was unlikely that an objective endpoint like OS would be a result of placebo effect; (2) the magnitude of benefit was beyond what would be expected for a placebo effect (HR of 0.75 for OS); and (3) previous trials that lacked a placebo, such as RTOG 0525, did not show improved survival [21]. Ultimately, the group decided that that the TTF device should be considered a treatment option for patients with GBM who were willing to undergo therapy and did not have contraindications. Furthermore, future studies were recommended to identify which subset of patients benefits most from TTF therapy [21].

Despite criticism, most medical device evaluations traditionally lack randomized control groups [24]. Although this may be attributed to the lesser sophistication of clinical trial design by device manufacturers, it's also likely secondary to properties of the device. In this case, patients sense heat from the TTF device, which would be impractical for a sham device. Furthermore, ethical issues arise. For example, it would be difficult and unethical to observe a placebo heart valve replacement [24]. Similarly, it is argued that having a patient shave their head and wear a sham cap for greater than eighteen hours a day would also be unethical. Nevertheless, long-term observational studies will be necessary to observe device efficacy.

2.4. Safety

Use of a TFF device was not associated with systemic toxicity in either the EF-11 or EF-14 trials. In the EF-11 trial, typical systemic side effects were not seen; whereas in the EF-14 trial, systemic side effect rates did not show a significant difference from the TMZ-alone cohort (48% vs. 44%) [7,13]. Some events in EF-14 had a slightly higher incidence in the TTF group, which was attributed to longer use of TMZ in this group due to delayed progression [7]. A common toxicity was a moderate skin reaction on the scalp below the transducers; dermatitis was observed in 16% of patients wearing the TFF device on EF-11 and 52% of patients in EF-14 [7,13]. Severe skin toxicity was seen in 2% of patients on EF-14 [7]. TFF therapy did not increase the incidence of seizures in either study. The mild to moderate skin reaction patients may experience is reversible and does not require treatment discontinuation [25].

Recently, a phase I clinical trial of 10 patients exploring TTF device treatment with concomitant radiation therapy reported preliminary results of similar rate of TTF-related skin toxicity (40%) to that of the EF-14 trial [26].

2.5. Compliance and Quality of Life

The efficacy of the TFF device has been correlated with patient compliance, and a compliance rate of over 75% has been linked to higher OS [14]. A recently published subgroup analysis showed that a threshold value of 50% compliance, defined as percent usage per month, improved PFS (HR = 0.70) and OS (HR 0.67) [27]. This was seen independent of gender, extent of resection, *MGMT* status, age or performance status [27]. Interestingly, patients with a compliance rate of >90% showed a prolonged median survival of 24.9 months and a five-year survival rate of 29.3% [27].

Consequently, patients need to wear the TFF device continuously with minimal interruption. This inevitably requires lifestyle modifications. Enrollment in the trial was self-selecting and the cohort was not representative of the entire GBM population. Some patients may find the device confining, as it potentially interferes with daily activities. Despite improved survival in the EF-14 trial, patients in clinic may be reluctant to use the TFF device for social or cultural reasons [22]. Others may be averse to shaving their heads. Studies on chemotherapy-induced alopecia have shown that the psychosocial impact of hair loss can be devastating [28]. Quality of life is certainly a priority for this patient population.

A planned interim analysis of the EF-14 trial analyzing health-related quality of life (HRQoL) was performed on 315 patients. HRQoL was slightly improved for patients in the TTF group at six months; physical and social functioning showed no difference at nine and twelve month time points. Patients in the TTF group complained of itchy skin. There was no preliminary evidence that HRQoL, functional status, or cognitive functioning was altered by use of TTF [29]. A final analysis, published in 2018, found that HRQoL did not differ significantly between the two arms of the trial with the exception of itchy skin in the TTF arm [29]. The analysis found that deterioration-free survival was longer in patients receiving TTF therapy for global health, physical and emotional functioning [29]. However, it is important to note that adherence to the HRQoL survey questionnaire decreased significantly over time; survey completion-rate was 91.9% at baseline, 65.8% at three months, and 41.7% at twelve months [29]. Remaining HRQoL scales not previously reported were subjected to an exploratory analysis, which was presented at ASTRO 2018. This study reported a larger proportion of patients having stable/improved bladder control and diarrhea when compared to baseline. The deterioration-free survival for diarrhea, future uncertainty, and headaches was delayed in patients receiving TTF and TMZ treatment when compared to treatment consisting of TMZ alone (HR 0.68, $p < 0.001$). No negative impact on HRQoL was seen in this study [30].

Efforts are underway to improve convenience of utilizing the TTF device. The first-generation Optune TTF device weighed 6 pounds and consisted of a field generator, transducers, lithium batteries and a carry bag. In 2016, a second-generation TFF device weighing only 2.7 pounds was approved [31]. The goal of the second-generation TFF device was to improve patient convenience and compliance. Additionally, the newer device contains an objective log that records treatments compliance [31]. There is hope that the growing popularity of wearable devices, such activity trackers and other smart devices, may encourage patients to be more open about wearing the TTF device [22].

2.6. Cost of the Device

As mentioned previously, the specific payments for this device have not yet been established by Medicare and other insurance providers. Medicare's Ambulatory Payment Classification covers the technical component of the device; Medicare's Physician Fee reimburses the professional component of therapy [29]. The total monthly therapy cost is about $21,000, which translates to about $86,000 for the average patient using the device for 4.1 months [30]. Novocure's executive chairman William F.

Doyle reported that the company had been providing the device for free for to patients without health insurance in a 2015 New York Times Article [32].

There have been few studies evaluating the cost-effectiveness of TTF. A French group measured the cost-effectiveness of the device and measured it to be "far beyond conventional thresholds due to the prohibitive announced cost of the device." [33]. This report suggested that there is 0% chance of achieving the cost-effectiveness threshold, which was chosen arbitrarily, of €100,000 ($114,213 in US dollars) per year life gained [33]. Total cost for TTF therapy and conventional therapy strategies were €243,141 and €57,665, respectively. After applying a 4% annual discount according to French national guidelines, the analysis resulted in a total incremental-cost effectiveness ratio of €596 411 per year life gained [33]. A US group presented an abstract at the America Association for Cancer Research (AACR) meeting in 2018 projected that TFF plus TMZ compared to TMZ alone resulted in an undiscounted mean survival of 1.8 life years based on 5 survival results of EF-14 trial [34]. The incremental cost effectiveness ratio based on this survival was $150,638 per life year gained and $198,032 per quality life year gained. This abstract reported a high probability of cost-effectiveness at a threshold of $200,000 [34]. Regardless, the device is certainly expensive and the cost-effectiveness threshold will likely vary considerably between, and even within, countries [35].

How does TTF therapy compare to other novel cancer therapies available in the US? The CheckMate 067 trial evaluating dual checkpoint inhibitors nivolumab and ipilimumab in patients with advanced melanoma showed an increase in PFS from 6.9 months with nivolumab alone to 11.5 months with combination therapy [36]. In 2015, the estimated cost per patient per year for nivolumab was $103,220 and ipilimumab was $158,252. The combination therapy was estimated to run $295,566 [37]. The cost-effectiveness of nivolumab in patients with advanced renal cell carcinoma treated in the US found this treatment to have a 91.7% probability of being cost-effective at a $150,000 threshold per quality adjusted life year. Other treatments like chimeric antigen receptor (CAR) T-cell therapy are costly with a price tag of $475,000 for Tisagenlecleucel and $373,000 for Axicabtagene ciloleucel [38].

Currently, the average direct cost of therapy for GBM patients in the US is about $8500 per month, and there is regional variability [39]. Mayo clinic reported a median total direct cost of $91,000 from 1987 to 1992, before the routine use of TMZ: radiotherapy and imaging cost contributed the most to this number [39,40]. It has since been reported that the addition of TMZ, in both the concomitant and adjuvant settings, has led to an eightfold increase to the direct cost of GBM therapy [39,41]. Nevertheless, administration of TTF therapy significantly increases the overall cost of GBM treatment.

3. Discussion

Since the publication of the Stupp trial in 2005, there have been minimal advancements in GBM therapy. Trials that studied targeted molecular inhibitors, immunotherapy agents, vascular growth factor inhibitors, and radiotherapy dose escalation have fallen short. The EF-11 and EF-14 trials have demonstrated that TTF therapy is a promising alternative to chemotherapy in patients with recurrent GBM and adjuvant therapy for patients with newly diagnosed GBM. Most interestingly, the EF-14 trial reported an improved median overall survival of 20.9 months in the TTF plus TMZ group vs. 16.0 months in the TMZ only group. This led to the NCCN endorsement of routine adaptation of TTF as an upfront therapy for patients with newly diagnosed GBM. Since clinical implementation, there have been several reviews on TTF therapy for GBM [21,42]. Given the amount of newer studies done within the last year on this topic, especially in the realm of quality-of-life and cost-effectiveness, we conducted an updated review.

The implementation of the device has polarized many neuro-oncology experts. Some are enthusiastic about this development while others remain skeptical, perhaps understandably. There is a general lack of understanding of the TTF device's mechanism of action beyond what is proposed at the cellular level. Of note, most of the pre-clinical data used to drive the trials was pioneered by Novocure scientists, like Dr. Eilon Kirson. Additionally, the absence of a placebo-control "sham" device in the trials makes some experts less eager to recommend the therapy to their patients. However, it has been discussed that the inclusion of such a device in the trial would have created an ethical dilemma. Moreover, for better or worse, a more lenient precedence for medical devices exists: these devices become approved without randomized, placebo-controlled studies. Again, it is noted that both the EF-11 and EF-14 trials were sponsored by Novocure. Despite these considerations, the EF-14 trial certainly produced compelling results and clinicians echo the phrase "it is difficult to argue with overall survival."

One of the attractive properties of TTF therapy is the minimally-invasive nature and lack of systemic side-effects. This is especially important in the realm of recurrent disease, where patients undergo a variety of treatments from chemotherapy to additional surgery and/or re-irradiation. There are lifestyle drawbacks, as the device needs to be worn as continuously as possible, especially given the recent analyses showing correlation between device compliance and OS. Device compliance of >90% notably correlated with a 5-year OS reaching 30%. It is reassuring that most of the quality of life studies have reported similar results between patients who received TTF and those who did not. Neither health-related quality of life, functional status, nor cognitive functioning have been shown to be altered by TTF use.

The cost of therapy continues to be high, which may be prohibitive for certain patients. Willingness-to-pay for the device varies significantly between countries and even between regions. However, it is important to consider that other novel cancer therapies—such as dual checkpoint blockade with nivolumab and ipilimumab and CAR T-cell therapy—also carry a very high price tag. Hopefully prices will become more manageable as these therapies mature and become more widespread. It's important to make prices manageable for patients while also maintaining incentive for product innovation.

An additional obstacle is that the device is not routinely available in many centers. An electronic survey sent to an international group of radiation oncologists, neurosurgeons, and neuro-oncologists between January 2015 and July 2015 found that only 41% of surveyed practitioners had TTF available to offer their patients [43].

4. Conclusions

Ultimately, long-term, real-world results are necessary to make clinicians more comfortable with the device. Furthermore, positive clinical trial results in disease sites other than GBM may make the device more acceptable and familiar. It remains important to define which subset of patients benefit the most from TTF therapy, through identifying markers that predict improved outcome with TTF therapy. Reluctant patients may be more willing to try the device if it is more convenient and less intrusive to daily life. Further industry efforts to improve the device will help with patient compliance, such as the lower weight in second generation devices. Finally, it is important to define the benefits and potential toxicities of TFF therapy in combination with radiation therapy and chemoradiation therapy via larger, randomized clinical trials. Should we expect to hear much more about TTF therapy in oncology? We'll bet our best hat.

Author Contributions: Conceptualization, D.F. and J.D.P.; methodology, D.F.; investigation, D.F.; data curation, D.F. and N.S.; writing—original draft preparation, D.F., M.d.P.G.P.E., I.A., P.G., and J.D.P.; writing—review and editing, D.F., M.d.P.G.P.E., I.A., N.S., J.G., V.P., P.G., and J.D.P.; visualization, D.F., M.d.P.G.P.E., I.A.; supervision, J.D.P.; project administration, J.D.P.

Funding: This research received no external funding.

Acknowledgments: The Ohio State University Department of Radiation Oncology and The Ohio State University Department of Neuro-Oncology.

Conflicts of Interest: The authors declare no conflict of interest.

References

1. Stupp, R.; Hegi, M.E.; Mason, W.P.; van den Bent, M.J.; Taphoorn, M.J.B.; Janzer, R.C.; Ludwin, S.K.; Allgeier, A.; Fisher, B.; Belanger, K.; et al. Effects of radiotherapy with concomitant and adjuvant temozolomide versus radiotherapy alone on survival in glioblastoma in a randomised phase III study: 5-year analysis of the EORTC-NCIC trial. *Lancet Oncol.* **2009**, *10*, 459–466. [CrossRef]
2. Thakkar, J.P.; Dolecek, T.A.; Horbinski, C.; Ostrom, Q.T.; Lightner, D.D.; Barnholtz-Sloan, J.S.; Villano, J.L. Epidemiologic and Molecular Prognostic Review of Glioblastoma. *Cancer Epidemiol. Biomark. Prev.* **2014**, *23*, 1985–1996. [CrossRef] [PubMed]
3. Gilbert, M.R.; Wang, M.; Aldape, K.D.; Stupp, R.; Hegi, M.E.; Jaeckle, K.A.; Armstrong, T.S.; Wefel, J.S.; Won, M.; Blumenthal, D.T.; et al. Dose-dense temozolomide for newly diagnosed glioblastoma: A randomized phase III clinical trial. *J. Clin. Oncol.* **2013**, *31*, 4085–4091. [CrossRef] [PubMed]
4. Gilbert, M.R.; Dignam, J.J.; Armstrong, T.S.; Wefel, J.S.; Blumenthal, D.T.; Vogelbaum, M.A.; Colman, H.; Chakravarti, A.; Pugh, S.; Won, M.; et al. A Randomized Trial of Bevacizumab for Newly Diagnosed Glioblastoma. *N. Engl. J. Med.* **2014**, *370*, 699–708. [CrossRef] [PubMed]
5. Chinnaiyan, P.; Won, M.; Wen, P.Y.; Rojiani, A.M.; Werner-Wasik, M.; Shih, H.A.; Ashby, L.S.; Michael Yu, H.H.; Stieber, V.W.; Malone, S.C.; et al. A randomized phase II study of everolimus in combination with chemoradiation in newly diagnosed glioblastoma: Results of NRG Oncology RTOG 0913. *Neuro-Oncology* **2018**, *20*, 666–673. [CrossRef] [PubMed]
6. Shenouda, G.; Souhami, L.; Petrecca, K.; Owen, S.; Panet-Raymond, V.; Guiot, M.-C.; Corredor, A.G.; Abdulkarim, B. A Phase 2 Trial of Neoadjuvant Temozolomide Followed by Hypofractionated Accelerated Radiation Therapy With Concurrent and Adjuvant Temozolomide for Patients With Glioblastoma. *Int. J. Radiat. Oncol. Biol. Phys.* **2017**, *97*, 487–494. [CrossRef] [PubMed]
7. Stupp, R.; Taillibert, S.; Kanner, A.; Read, W.; Steinberg, D.; Lhermitte, B.; Toms, S.; Idbaih, A.; Ahluwalia, M.S.; Fink, K.; et al. Effect of Tumor-Treating Fields Plus Maintenance Temozolomide vs Maintenance Temozolomide Alone on Survival in Patients With Glioblastoma: A Randomized Clinical Trial. *JAMA* **2017**, *318*, 2306–2316. [CrossRef]
8. Kirson, E.D.; Gurvich, Z.; Schneiderman, R.; Dekel, E.; Itzhaki, A.; Wasserman, Y.; Schatzberger, R.; Palti, Y. Disruption of cancer cell replication by alternating electric fields. *Cancer Res.* **2004**, *64*, 3288–3295. [CrossRef]
9. Kirson, E.D.; Dbalý, V.; Tovarys, F.; Vymazal, J.; Soustiel, J.F.; Itzhaki, A.; Mordechovich, D.; Steinberg-Shapira, S.; Gurvich, Z.; Schneiderman, R.; et al. Alternating electric fields arrest cell proliferation in animal tumor models and human brain tumors. *Proc. Natl. Acad. Sci. USA* **2007**, *104*, 10152–10157. [CrossRef]
10. Optune: Official Patient Site. Available online: https://www.optune.com/ (accessed on 30 January 2019).
11. Shteingauz, A.; Porat, Y.; Voloshin, T.; Schneiderman, R.S.; Munster, M.; Zeevi, E.; Kaynan, N.; Gotlib, K.; Giladi, M.; Kirson, E.D.; et al. AMPK-dependent autophagy upregulation serves as a survival mechanism in response to Tumor Treating Fields (TTFields). *Cell Death Dis.* **2018**, *9*, 1074. [CrossRef]
12. Gera, N.; Yang, A.; Holtzman, T.S.; Lee, S.X.; Wong, E.T.; Swanson, K.D. Tumor Treating Fields Perturb the Localization of Septins and Cause Aberrant Mitotic Exit. *PLoS ONE* **2015**, *10*, e0125269. [CrossRef] [PubMed]
13. Stupp, R.; Wong, E.T.; Kanner, A.A.; Steinberg, D.; Engelhard, H.; Heidecke, V.; Kirson, E.D.; Taillibert, S.; Liebermann, F.; Dbalý, V.; et al. NovoTTF-100A versus physician's choice chemotherapy in recurrent glioblastoma: A randomised phase III trial of a novel treatment modality. *Eur. J. Cancer* **2012**, *48*, 2192–2202. [CrossRef] [PubMed]
14. Kanner, A.A.; Wong, E.T.; Villano, J.L.; Ram, Z. No-065. Tumor Treating Fields (ttfields) In Recurrent Gbm. An Updated Subgroup Analysis Of The Phase Iii Data. *Neuro-Oncology* **2013**, *15*, iii114.
15. FDA Approves Expanded Indication for Medical Device to Treat Glioblastoma Multiforme | ESMO. Available online: https://www.esmo.org/Oncology-News/FDA-Approves-Expanded-Indication-for-Medical-Device-to-Treat-Glioblastoma-Multiforme (accessed on 30 January 2019).
16. Foreman, C. RE: P100034 Novo TTF-100A System. Available online: https://www.accessdata.fda.gov/cdrh_docs/pdf10/p100034a.pdf (accessed on 30 January 2019).

17. Mrugala, M.M.; Engelhard, H.H.; Dinh Tran, D.; Kew, Y.; Cavaliere, R.; Villano, J.L.; Annenelie Bota, D.; Rudnick, J.; Love Sumrall, A.; Zhu, J.-J.; et al. Clinical practice experience with NovoTTF-100A™ system for glioblastoma: The Patient Registry Dataset (PRiDe). *Semin. Oncol.* **2014**, *41* (Suppl. 6), S4–S13. [CrossRef] [PubMed]
18. Choe, K. Analysis of RTOG-RPA Scores in the Phase 3 EF-14 Trial of Tumor Treating Fields with Temozolomide (TTFields/TMZ) Versus TMZ Alone in Newly Diagnosed Glioblastoma. *Int. J. Radiat. Oncol. Biol. Phys.* **2018**, *102*, S48. [CrossRef]
19. Trial of Combination TTF(Optune), Nivolumab Plus/Minus Ipilimumab for Bevacizumab-naïve, Recurrent Glioblastoma—Full Text View—ClinicalTrials.gov. Available online: https://clinicaltrials.gov/ct2/show/NCT03430791 (accessed on 30 January 2019).
20. Safety and Efficacy of TTFields (150 kHz) Concomitant With Pemetrexed and Cisplatin or Carboplatin in Malignant Pleural Mesothelioma (STELLAR)—Full Text View—ClinicalTrials.gov. Available online: https://clinicaltrials.gov/ct2/show/NCT02397928 (accessed on 30 January 2019).
21. Mehta, M.; Wen, P.; Nishikawa, R.; Reardon, D.; Peters, K. Critical review of the addition of tumor treating fields (TTFields) to the existing standard of care for newly diagnosed glioblastoma patients. *Crit. Rev. Oncol. Hematol.* **2017**, *111*, 60–65. [CrossRef] [PubMed]
22. Halasz, L.M.; Mitin, T. Tumor-Treating Fields: Answering the Concern About Quality of Life. *JAMA Oncol.* **2018**, *4*, 504–505. [CrossRef]
23. Wick, W. TTFields: Where does all the skepticism come from? *Neuro-Oncology* **2015**, *18*, 303–305. [CrossRef]
24. Institute of Medicine (US) Committee on Technological Innovation in Medicine; Gelijns, A.C. *Comparing the Development of Drugs, Devices, and Clinical Procedures*; National Academies Press (US): Washington, DC, USA, 1990.
25. Optune for Healthcare Professionals: Safety. Available online: https://www.optune.com/hcp/clinical-data/tolerability (accessed on 30 January 2019).
26. Grossman, R.; Bukstein, F.; Blumenthal, D.T.; Ben Harush, C.; Limon, D.; Ram, Z. Safety of tumor treating fields and concomitant radiotherapy for newly diagnosed glioblastoma. *J. Clin. Oncol.* **2018**, *36*, e14078. [CrossRef]
27. Toms, S.A.; Kim, C.Y.; Nicholas, G.; Ram, Z. Increased compliance with tumor treating fields therapy is prognostic for improved survival in the treatment of glioblastoma: A subgroup analysis of the EF-14 phase III trial. *J. Neurooncol.* **2018**. [CrossRef]
28. Hesketh, P.J.; Batchelor, D.; Golant, M.; Lyman, G.H.; Rhodes, N.; Yardley, D. Chemotherapy-induced alopecia: Psychosocial impact and therapeutic approaches. *Support Care Cancer* **2004**, *12*, 543–549. [CrossRef] [PubMed]
29. Taphoorn, M.J.B.; Dirven, L.; Kanner, A.A.; Lavy-Shahaf, G.; Weinberg, U.; Taillibert, S.; Toms, S.A.; Honnorat, J.; Chen, T.C.; Sroubek, J.; et al. Influence of Treatment With Tumor-Treating Fields on Health-Related Quality of Life of Patients With Newly Diagnosed Glioblastoma: A Secondary Analysis of a Randomized Clinical Trial. *JAMA Oncol.* **2018**, *4*, 495–504. [CrossRef] [PubMed]
30. Toms, S.A. Effects of Tumor Treating Fields (TTFields) on Health-Related Quality of Life (HRQOL) in Newly Diagnosed Glioblastoma: An Exploratory Analysis of the EF-14 Randomized Phase III Trial. *Int. J. Radiat. Oncol. Biol. Phys.* **2018**, *102*, S170–S171. [CrossRef]
31. Second-Generation Optune® Tumor Treating Fields System Enhances Compliance in Patients with Glioblastoma vs. the First-Generation Device. Available online: https://www.practiceupdate.com/content/second-generation-optunereg-tumor-treating-fields-system-enhances-compliance-in-patients-with-glioblastoma-vs-the-first-generation-device/52701 (accessed on 27 December 2018).
32. Electrical Scalp Device Can Slow Progression of Deadly Brain Tumors—The New York Times. Available online: https://www.nytimes.com/2014/11/16/health/electrical-scalp-device-can-slow-progression-of-deadly-brain-tumors.html (accessed on 27 December 2018).
33. Bernard-Arnoux, F.; Lamure, M.; Ducray, F.; Aulagner, G.; Honnorat, J.; Armoiry, X. The cost-effectiveness of tumor-treating fields therapy in patients with newly diagnosed glioblastoma. *Neuro-Oncology* **2016**, *18*, 1129–1136. [CrossRef] [PubMed]
34. Guzauskas, G.F.; Pollom, E.L.; Stieber, V.W.; Wang, B.C.; Garrison, L. Abstract LB-257: Tumor treating fields treatment for patients with newly diagnosed glioblastoma: A cost-effectiveness analysis. *Cancer Res.* **2018**, *78*, LB-257. [CrossRef]

35. Musgrove, P.; Fox-Rushby, J. Cost-Effectiveness Analysis for Priority Setting. In *Disease Control Priorities in Developing Countries*, 2nd ed.; Jamison, D.T., Breman, J.G., Measham, A.R., Alleyne, G., Claeson, M., Evans, D.B., Jha, P., Mills, A., Musgrove, P., Eds.; World Bank: Washington, DC, USA, 2006.
36. Wolchok, J.D.; Chiarion-Sileni, V.; Gonzalez, R.; Rutkowski, P.; Grob, J.-J.; Cowey, C.L.; Lao, C.D.; Wagstaff, J.; Schadendorf, D.; Ferrucci, P.F.; et al. Overall Survival with Combined Nivolumab and Ipilimumab in Advanced Melanoma. *N. Engl. J. Med.* **2017**, *377*, 1345–1356. [CrossRef] [PubMed]
37. Andrews, A. Treating with Checkpoint Inhibitors—Figure $1 Million per Patient. *Am. Health Drug Benefits* **2015**, *8*, 9.
38. Weighing the Cost and Value of CAR T-Cell Therapy—The ASCO Post. Available online: http://www.ascopost.com/issues/may-25-2018/weighing-the-cost-and-value-of-car-t-cell-therapy/ (accessed on 30 January 2019).
39. Cagney, D.N.; Alexander, B.M. The cost and value of glioblastoma therapy. *Expert Rev. Anticancer Ther.* **2017**, *17*, 657–659. [CrossRef]
40. Silverstein, M.D.; Cascino, T.L.; Harmsen, W.S. High-grade astrocytomas: Resource use, clinical outcomes, and cost of care. *Mayo Clin. Proc.* **1996**, *71*, 936–944. [CrossRef]
41. Wasserfallen, J.-B.; Ostermann, S.; Leyvraz, S.; Stupp, R. Cost of temozolomide therapy and global care for recurrent malignant gliomas followed until death. *Neuro-Oncology* **2005**, *7*, 189–195. [CrossRef]
42. Rick, J.; Chandra, A.; Aghi, M.K. Tumor treating fields: A new approach to glioblastoma therapy. *J. Neurooncol.* **2018**, *137*, 447–453. [CrossRef] [PubMed]
43. Palmer, J.D.; Bhamidipati, D.; Mehta, M.; Williams, N.L.; Dicker, A.P.; Werner-Wasik, M.; Shi, W. Treatment recommendations for elderly patients with newly diagnosed glioblastoma lack worldwide consensus. *J. Neurooncol.* **2018**, *140*, 421–426. [CrossRef] [PubMed]

cancers MDPI

Perspective

Finding the Right Way to Target EGFR in Glioblastomas; Lessons from Lung Adenocarcinomas

Ya Gao , Wies R. Vallentgoed and Pim J. French *

Department of Neurology, Erasmus MC Cancer Institute; 3015 CD Rotterdam, The Netherlands; y.gao@erasmusmc.nl (Y.G.); w.vallentgoed@erasmusmc.nl (W.R.V.)

* Correspondence: p.french@erasmusmc.nl; Tel.: +31-10-704-4333

Received: 9 November 2018; Accepted: 30 November 2018; Published: 4 December 2018

Abstract: The *EGFR* gene is one of the most frequently mutated and/or amplified gene both in lung adenocarcinomas (LUAD) and in glioblastomas (GBMs). Although both tumor types depend on the mutation for growth, clinical benefit of EGFR tyrosine kinase inhibitors (TKIs) has only been observed in LUAD patients and, thus-far, not in GBM patients. Also in LUAD patients however, responses are restricted to specific *EGFR* mutations only and these 'TKI-sensitive' mutations hardly occur in GBMs. This argues for mutation-specific (as opposed to tumor-type specific) responses to EGFR-TKIs. We here discuss potential reasons for the differences in mutation spectrum and highlight recent evidence for specific functions of different *EGFR* mutations. These mutation-specific effects likely underlie the differential treatment response between LUAD and GBMs and provide new insights into how to target EGFR in GBM patients.

Keywords: EGFR; glioblastoma; glioma; pulmonary adenocarcinoma; lung cancer; erlotinib; gefitinib; lapatinib

1. Introduction

More than fifty years ago, Stanley Cohen reported on the isolation of a heat-stable protein from mouse salivary glands that was able to induce the eruption of incisors and separation of the eyelids [1]. Later studies showed that the eyelid separation was actually a consequence of enhanced keratinization and growth of the epidermis and, based on this function, the protein was coined epidermal growth factor (EGF) [2,3]. Around a decade later, the same group found evidence for a receptor for the growth factor on human fibroblasts and another decade later, the receptor was cloned and sequenced [4,5].

The *EGF* receptor (*EGFR*) gene encodes for a protein of 1210 amino acids which is built up of an extracellular ligand binding domain, a single transmembrane domain, a juxtamembrane region, an intracellular tyrosine kinase domain and a C-terminal regulatory domain. The extracellular region can be divided into four subdomains, domains I–IV, that contain two leucine rich regions (I and III, also known as L1 and L2) and two cysteine rich domains (II and IV, also known as CR1 and CR2). Proteins with a similar domain structure include ERBB2, ERBB3 and ERBB4 (which are also known as HER2-4), and are collectively known as the ERBB protein family, which are all members of the receptor tyrosine kinase superfamily.

In the absence of ligand, EGFR exists as a monomer (or as an inactive dimer) on the plasma membrane. When ligands are available, they bind to the extracellular leucine rich domains I and III, and this association results in the exposure of the cysteine rich domain II, which allows for receptor dimerization. The dimerization then induces phosphorylation of tyrosine residues on the C-terminal intracellular domain of the dimerization partner (trans-phosphorylation) [6]. Phosphorylation occurs on multiple sites: the intracellular domain contains twenty tyrosine residues, twelve of which can be phosphorylated following ligand binding [7]. This phosphorylation results in the recruitment of specific adaptor proteins and subsequent activation of signal transduction cascades. These cascades include the RAS-RAF-MEK-ERK, PI3kinase-AKT-mTOR, SRC, JNK, PLC-γ-PKC and STAT pathways and ultimately result in cellular proliferation, migration and survival [8,9].

Although the EGF receptor (EGFR) was identified based on its affinity for EGF, later studies showed it can be also activated by six other, related- ligands: heparin-binding EGF-like growth factor (HB-EGF), amphiregulin, epiregulin, epigen, betacellulin and transforming growth factor-alpha (TGF-α) [10]. EGF, TGF-α, HB-EGF and betacellulin are the ligands that bind with high affinity to EGFR; amphiregulin, epiregulin and epigen bind with 10–100 fold less affinity. Each ligand can induce quantitative differences in responses, but can also elicit ligand-specific responses [10,11].

2. *EGFR* Mutations in LUAD and Gliomas

Cancer is caused by the accumulation of acquired genetic changes in oncogenes and tumor suppressor genes. The epidermal growth factor receptor (*EGFR*) gene is a key oncogene that is mutated in many tumors including lung adenocarcinomas (LUAD) and glioblastomas (GBM). In LUAD, around 90% of all *EGFR* mutations comprise of either short in-frame deletions in exon 19 (in particular around residues 747–750) or the *L858R* missense mutation in exon 21 [12,13]. Other, less common mutations in LUAD include *G719X* missense mutations (~3%) and in-frame insertions in exon 20 (~3%) [14]. All these mutations result in increased and continuous EGFR phosphorylation and activation [15,16].

In GBMs, the initial driving event is thought to be high copy number amplification of the *EGFR* gene, present in tumor cells as double minutes (extrachromosomal circular DNA fragments) with levels ranging from >5 to more than 100 copies per cell [17]. These double minutes likely increase the copy number (and RNA expression) of the oncogene more effectively compared to chromosomal amplification. Double minutes are unevenly distributed across the two daughter cells during cell division which enhances tumor heterogeneity and plasticity [17]. Amplification of the *EGFR* gene is followed by the acquisition of a plethora of mutations that include intragenic deletions, point mutations and gene-fusions [18]. Multiple *EGFR* mutations may be present within the same tumor which also contributes to tumor heterogeneity [19]. The most common *EGFR* mutation in GBM is the in-frame deletion of exon 2–7, coined *EGFRvIII* and occurs in ~50% of all *EGFR*-amplified GBM cases. EGFRvIII has impaired ligand binding and is constitutively active, though its activity is only ~10% of endogenous EGFR signaling [20]. Missense mutations that are commonly found in GBM are often located on the extracellular domain and include A289X, G598X and R108K mutations. These mutations also result in a constitutively active protein and increase the tumorigenic potential of cells [19].

Both in lung cancer and in GBMs, *EGFR* mutations are driver mutations and the tumors remain dependent on this oncogene for growth [21–24]. Despite the similarities in activity, the most prominent mutations in LUAD, exon-19 deletions and L858R, have, to date, never been identified in GBMs and the most common mutation in GBMs, EGFRvIII, has never been identified in LUAD. This indicates that each tumor type has an almost (see below), unique mutation spectrum [14,20].

3. Clinical Activity of EGFR-TKIs in LUAD, But Not GBM Patients

It is well known that EGFR-TKIs provide remarkable survival benefit to patients with *EGFR*-mutated LUAD. These benefits were initially discovered by research on clinical trials that examined the efficacy of EGFR-TKIs in LUAD patients, in which antitumor activity and an increase in survival was observed in patients who failed on prior chemotherapy [25,26]: translational research showed that the clinical responses were correlated to mutations in *EGFR* [27,28]. Landmark studies such as the IPASS study showed that gefitinib improved the 12 months progression free survival in advanced, previously untreated pulmonary adenocarcinoma patients, but only in those patients where activating mutations in the *EGFR* gene were identified. Similar improvements were observed in *EGFR*-mutated, metastatic non-small cell lung cancer patients [29,30]. A phase III study that included only *EGFR*-mutated lung cancer patients confirmed these observations [29].

Because of the clinical responses observed in *EGFR*-mutated LUAD patients, and because of the high frequency of *EGFR* mutations in GBMs, it was logical to test the clinical efficacy of EGFR-TKIs in GBM patients. Although several such trials have been conducted (Table 1), thus-far none of these demonstrated a clear clinical benefit of the inhibitors, despite inhibitors showing target inhibition on the various EGFR mutations in preclinical models [19]. For example, two studies conducted in primary gliomas showed no additional clinical benefit of adding gefitinib after radiotherapy [31,32]. Similar disappointing data were obtained in two studies on recurrent gliomas where single agent erlotinib did not improve the 6 months progression-free survival [33,34].

Table 1. clinical trials of EGFR-TKIs in gliomas.

Drug	Phase	Clinical Trial ID	Comparator	Histology	*n*	Ref.
Erl + TMZ/RT	I/II	NCT00039494	single arm	GBM	97	[35]
Erl + Bev	II	NCT00671970	single arm	GBM, AG	57	[36]
Erl + Soraf	II	NCT00445588	single arm	rGBM	56	[37]
Erl + Bev + TMZ	II	NCT00525525	single arm	GBM, GSC	74	[38]
Erl + TMZ	II	NCT00187486	single arm	GBM, GLS	65	[39]
Erl	I/II	NCT00301418	single arm	GBM, AA	11	[40]
Erl + Sirol	II	NCT00672243	single arm	rGBM	32	[41]
Erl + Sirol	I/II	NCT00112736	single arm	rGlioma	69	[42]
Erl	II	NCT00250887	TMZ/BCNU	rGBM	110	[34]
Gef	II	NCT00250887	single arm	rGBM	22	[43]
Gef + Cedir	II	NCT01310855	Cedir	rGBM	38	[44]
Gef	II	NCT00016991	Single arm	rGBM	57	[45]
Lap	I/II	NCT00099060	single arm	rGBM	17	[46]
Afa	I/II	NCT00727506	TMZ/afa, TMZ	rGBM	119	[47]
Dac	II	NCT01520870	single arm	rGBM	49	[48]
Sun	II	NCT00923117	Bev naïve/resistant	GBM	72	[49]

rGBM: recurrent or progressive GBM; AG: anaplastic glioma, rGlioma: recurrent glioma; GLS: gliosarcoma; AA: anaplastic astrocytoma. Erl: erlotinib; TMZ: temozolomide; RT: radiotherapy; soraf: sorafenib; sirol: temsirolimus; cedir: cediranib; bev: bevacizumab; lap: lapatinib; afa: afatinib; Dac: dacomitinib; Sun: sunitinib.

There are several possibilities why those trials failed to improve on their primary endpoint. First of all, the trials may not have included the right patient population. While this seems trivial, trials testing EGFR-TKIs selected patients based on histological criteria and did not select for *EGFR*-mutated tumors [31,33,43,45,50]. However, the frequency of *EGFR* amplification and mutations are sufficiently high in GBM patients (estimated to be ~50% of all GBMs) to show some signal of efficacy. Moreover, translational research on the material from these trials also failed to show clinical improvement in the samples that had *EGFR*-mutations [31,34,51,52]. A second possibility for trial failure is that the concentration of drug does not reach sufficiently high concentrations in the tumor, for example by lack of penetration through the blood brain barrier. Indeed, intratumoral drug concentrations for erlotinib were much lower of that in plasma and erlotinib did not affect intratumoral EGFR signaling [33]. In contrast, a phase II trial in which the intratumoral levels of gefitinib was measured in 22 patients showed concentrations sufficiently high to inhibit the phosphorylation of EGFR [43].

A third option for therapy refractoriness of GBMs to EGFR-TKIs is that these tumors no longer depend on the oncogene for growth. However, biological experiments conducted in mice and using primary patient cell lines showed that these tumors do remain dependent on EGFR and therefore do not explain why GBMs do not respond to EGFR-TKIs [21,22]. It is possible that GBMs have an innate resistance to EGFR-TKIs, such as the upregulation (or coactivation) of PDGFRA and cMET [53]. Such an innate resistance has been identified in colon carcinomas where inhibition of one tyrosine kinase (BRAF) is bypassed by the activation of another (EGFR) [54]. Alternatively, perturbation of downstream pathways such as PTEN deletion may also confer resistance to EGFR TKIs [40]. It remains however to be determined if, and if so which-, changes (genetic or epi-genetic) underlie the therapy refractoriness of GBMs.

The dynamics of *EGFR* mutations can also play a role in treatment resistance of GBMs. Since *EGFR* is present as double minutes, cell division can result in the asymmetric distribution of *EGFR* copynumber and variants, and thus in rapid selection of potentially resistant clones [17]. An example of dynamics is the expression of *EGFRvIII* which can be restricted to certain regions of the tumor only and can change over time [55–57]. Our unpublished observations show that temporal dynamics is not restricted to *EGFRvIII* but also true for other *EGFR* mutations. If only one of those mutations is resistant to the TKI, the dynamics of double minutes likely results in a rapidly acquired resistance to EGFR-TKIs.

4. Different Mutations Activate Different Pathways and May Explain Refractoriness to EGFR-TKIs

On the functional level, several lines of evidence suggest that different mutations in *EGFR* activate unique signal transduction pathways. For example, EGFRvIII and EGFRwt have differential activation of the JNK, STAT and MAPK signaling pathways and induce the expression of a unique set of genes [58–62]. We have also demonstrated that EGFRwt, EGFRL858R and EGFRvIII each bind to a unique set of proteins and activate different molecular pathways [63]. If mutation-specific pathways active in GBMs are independent of EGFR phosphorylation, they explain why GBMs do not respond to EGFR-TKIs. For example, EGFR contains a functional nuclear localization signal and has been found present at high levels in the nucleus where it associates with the promoter of cell proliferation genes [64,65]. However, the nuclear accumulation is mutation dependent: mutations found in GBMs (e.g., EGFRvIII) accumulate in the nucleus whereas LUAD-specific mutations such as EGFR L858R do not [66–68]. Other studies have shown that the nuclear accumulation is independent of EGFR inhibitors [65,69]. It is therefore possible that activation of mutation-specific and TKI-independent pathways contribute to the lack of response to EGFR-TKIs. Which pathways are differentially activated and whether (co-) targeting these pathways will ultimately benefit GBM patients remains to be determined.

However, one argument against activation of mutation-specific pathway activation is the observation that some mutations have been found in both tumor types. Yes, the type of mutations vastly differs between LUAD and GBMs: exon 19 deletions or the *EGFR L858R* point mutations have never been found in GBMs and *EGFRvIII* is not present in LUAD [18,20]. Sporadically however, exon 20 mutations, though rare in LUAD, have also been found in GBM patients [18]. Indeed, in a recent sequencing effort of ~200 GBM samples performed by our group, we identified two patients with mutations near identical to those found in LUAD (p.H773_V774insAH and p.H773dup) (Figure 1). Conversely, sporadic R108K, A289A/T and G598V mutations that are common in GBMs have been identified in LUAD (Figure 1) [70–72]. This overlap in mutational spectrum, though rare, may suggest that LUAD and GBMs are unlikely to activate mutation-specific pathways. However, since multiple mutations accumulate in the *EGFR* gene in GBMs, other activating mutations may confer potential mutation-specific effects.

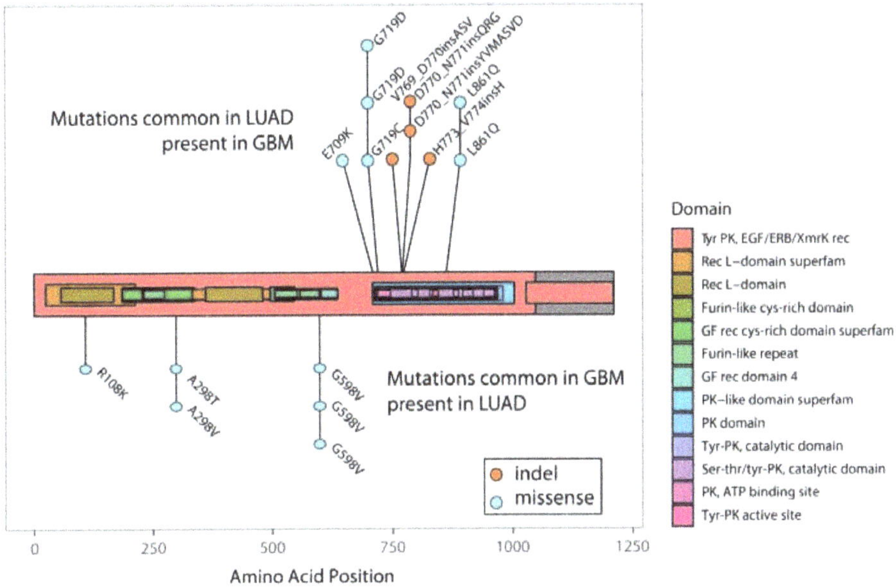

Figure 1. Mutations common to LUAD are sporadically identified in GBMs and vice versa. Each dot reflects a single sample identified with the specific mutation. Domains of EGFR are highlighted in color. PK: protein kinase; GF: growth factor; rec: receptor. The transmembrane is localized around amino acids 622–644.

5. Lessons from LUAD: Not all LUAD Respond to EGFR-TKIs and Not All TKIs are Effective in LUAD

Mutations that show response to EGFR-TKIs in LUAD have sporadically been identified in other tumor types. In various tumors harboring 'classical' exon 19 deletions or the L858R point mutation, EGFR-TKIs have shown clinical responses [73–75]. Such responses have also been documented for the less common, but responsive, G719X mutations, also in different tumor types [76,77]. The fact that sensitive mutations in LUAD appear also sensitive to TKIs in other tumor types indicates that response to EGFR-TKIs is not specific to the type of tumor, but is specific for the mutation present. This can have important consequences for GBM patients, as sporadic G719X mutations have been identified (Figure 1) [72,78].

This mutation-specific response is also conversely observed in LUAD patients: while the most common mutations in LUAD are sensitive to EGFR-TKIs, LUAD patients harboring exon 20 mutations are largely insensitive [79,80]. This lack of responsiveness was also observed with these mutations in various preclinical model systems [81,82]. Although rare, they comprise of ~3% of all *EGFR* mutations in LUAD and are considered to be activating mutations. Interestingly, and despite the lack of a clinical response, addition of TKIs does result in EGFR dephosphorylation in preclinical model systems [83,84]. The observation that not all *EGFR*-mutated LUAD respond to EGFR-TKIs indicate that response is limited to a defined set of mutations only.

Apart from the mutation-specificity of the response, there also is evidence for drug-specificity: although several EGFR-TKIs (erlotinib, gefitinib, afatinib, dacomitinib and osimertinib) have provided clinical benefit to *EGFR*-mutated LUAD patients, a phase II study on lapatinib did not show any signs of clinical activity [85]. This lack of clinical activity is surprising as lapatinib, similar to the other listed TKIs, is a highly potent inhibitor of EGFR phosphorylation. Erlotinib and gefitinib were the first to be developed and are both reversible inhibitors. Afatinib and dacomitinib are second generation irreversible inhibitors and osimertinib is a third generation inhibitor that also inhibits EGFR containing the T790M resistance mutation. Although all inhibitors inhibit EGFR phosphorylation, each drug has its own unique binding pocket, kinetics and properties (see e.g., [86]). Of note in this is that drugs like erlotinib associate with the active conformation while lapatinib traps the protein in an inactive conformation [87–89]. Perhaps this drug-specific conformational association underlies the marked difference in treatment response.

Nevertheless, the lack of clinical activity to lapatinib indicates that response to EGFR-TKIs is specific to the drug that was used. If so, response to inhibitors is a highly specific event, in which only specific mutations (exon 19 deletions, L858R and G719X missense mutations), and in the context of specific drugs only provide clinical benefit. The absence of either will result in clinical inactivity. Importantly, this would mean a shift in research paradigms from understanding why certain tumors do not respond to research into why only specific mutations do respond. Understanding why tumors respond will then lead to better design of novel targeted treatments.

6. Conclusions

Only specific mutations in EGFR respond to treatment, and this response seems independent of the type of tumor. Conversely, activating mutations with lack of clinical response to EGFR-TKIs, such as those found in GBMs, also do not respond to treatment in LUAD patients. Treatment response therefore is not tumor-type dependent, but mutation dependent. Moreover, only specific drugs seem to provide clinical benefit, despite showing similar target inhibition. Response to inhibitors therefore appears to be a highly specific event: only specific mutations in the context of specific drugs provide clinical benefit. The absence of either will result in clinical inactivity. The specificity of response warrants further investigation into its mechanisms as understanding why tumors respond will lead to better design of novel targeted treatments.

Author Contributions: Conceptualization: Y.G. and P.J.F.; Writing-Original Draft Preparation: P.J.F.; Writing-Review and Editing: Y.G., W.R.V. and P.J.F.; Funding Acquisition: P.J.F.

Funding: This work was funded by the Dutch Cancer Foundation, KWF kankerbestrijding, grant numbers 11125 and 11026. P.J.F. received grant support from AbbVie.

Conflicts of Interest: The authors declare no conflict of interest.

References

1. Cohen, S. Isolation of a mouse submaxillary gland protein accelerating incisor eruption and eyelid opening in the new-born animal. *J. Biol. Chem.* **1962**, *237*, 1555–1562. [PubMed]
2. Cohen, S.; Elliott, G.A. The stimulation of epidermal keratinization by a protein isolated from the submaxillary gland of the mouse. *J. Investig. Dermatol.* **1963**, *40*, 1–5. [CrossRef] [PubMed]
3. Cohen, S. The stimulation of epidermal proliferation by a specific protein (EGF). *Dev. Biol.* **1965**, *12*, 394–407. [CrossRef]
4. Carpenter, G.; Lembach, K.J.; Morrison, M.M.; Cohen, S. Characterization of the binding of 125-I-labeled epidermal growth factor to human fibroblasts. *J. Biol. Chem.* **1975**, *250*, 4297–4304. [PubMed]
5. Ullrich, A.; Coussens, L.; Hayflick, J.S.; Dull, T.J.; Gray, A.; Tam, A.W.; Lee, J.; Yarden, Y.; Libermann, T.A.; Schlessinger, J.; et al. Human epidermal growth factor receptor cDNA sequence and aberrant expression of the amplified gene in A431 epidermoid carcinoma cells. *Nature* **1984**, *309*, 418–425. [CrossRef] [PubMed]
6. Ferguson, K.M. Structure-based view of epidermal growth factor receptor regulation. *Annu. Rev. Biophys.* **2008**, *37*, 353–373. [CrossRef] [PubMed]
7. Maruyama, I.N. Mechanisms of activation of receptor tyrosine kinases: Monomers or dimers. *Cells* **2014**, *3*, 304–330. [CrossRef] [PubMed]
8. Weinberg, R.A. *The Biology of Cancer*, 2nd ed.; Ww Norton & Co.: New York, NY, USA, 2013.
9. Wee, P.; Wang, Z. Epidermal Growth Factor Receptor Cell Proliferation Signaling Pathways. *Cancers (Basel)* **2017**, *9*. [CrossRef]
10. Singh, B.; Carpenter, G.; Coffey, R.J. EGF receptor ligands: Recent advances. *F1000Research* **2016**, *5*. [CrossRef]

11. Wilson, K.J.; Gilmore, J.L.; Foley, J.; Lemmon, M.A.; Riese, D.J., 2nd. Functional selectivity of EGF family peptide growth factors: Implications for cancer. *Pharmacol. Ther.* **2009**, *122*, 1–8. [CrossRef]
12. Campbell, J.D.; Alexandrov, A.; Kim, J.; Wala, J.; Berger, A.H.; Pedamallu, C.S.; Shukla, S.A.; Guo, G.; Brooks, A.N.; Murray, B.A.; et al. Distinct patterns of somatic genome alterations in lung adenocarcinomas and squamous cell carcinomas. *Nat. Genet.* **2016**, *48*, 607–616. [CrossRef] [PubMed]
13. Jordan, E.J.; Kim, H.R.; Arcila, M.E.; Barron, D.; Chakravarty, D.; Gao, J.; Chang, M.T.; Ni, A.; Kundra, R.; Jonsson, P.; et al. Prospective Comprehensive Molecular Characterization of Lung Adenocarcinomas for Efficient Patient Matching to Approved and Emerging Therapies. *Cancer Discov.* **2017**, *7*, 596–609. [CrossRef] [PubMed]
14. Sondka, Z.; Bamford, S.; Cole, C.G.; Ward, S.A.; Dunham, I.; Forbes, S.A. The COSMIC Cancer Gene Census: Describing genetic dysfunction across all human cancers. *Nat. Rev. Cancer* **2018**. [CrossRef] [PubMed]
15. Gazdar, A.F. Activating and resistance mutations of EGFR in non-small-cell lung cancer: Role in clinical response to EGFR tyrosine kinase inhibitors. *Oncogene* **2009**, *28* (Suppl. 1), S24–S31. [CrossRef] [PubMed]
16. Chen, Y.R.; Fu, Y.N.; Lin, C.H.; Yang, S.T.; Hu, S.F.; Chen, Y.T.; Tsai, S.F.; Huang, S.F. Distinctive activation patterns in constitutively active and gefitinib-sensitive EGFR mutants. *Oncogene* **2006**, *25*, 1205–1215. [CrossRef] [PubMed]
17. Turner, K.M.; Deshpande, V.; Beyter, D.; Koga, T.; Rusert, J.; Lee, C.; Li, B.; Arden, K.; Ren, B.; Nathanson, D.A.; et al. Extrachromosomal oncogene amplification drives tumour evolution and genetic heterogeneity. *Nature* **2017**, *543*, 122–125. [CrossRef]
18. Brennan, C.W.; Verhaak, R.G.; McKenna, A.; Campos, B.; Noushmehr, H.; Salama, S.R.; Zheng, S.; Chakravarty, D.; Sanborn, J.Z.; Berman, S.H.; et al. The somatic genomic landscape of glioblastoma. *Cell* **2013**, *155*, 462–477. [CrossRef]
19. Lee, J.C.; Vivanco, I.; Beroukhim, R.; Huang, J.H.; Feng, W.L.; DeBiasi, R.M.; Yoshimoto, K.; King, J.C.; Nghiemphu, P.; Yuza, Y.; et al. Epidermal growth factor receptor activation in glioblastoma through novel missense mutations in the extracellular domain. *PLoS Med.* **2006**, *3*, e485. [CrossRef]
20. Gan, H.K.; Cvrljevic, A.N.; Johns, T.G. The epidermal growth factor receptor variant III (EGFRvIII): Where wild things are altered. *FEBS J.* **2013**, *280*, 5350–5370. [CrossRef]
21. Klingler, S.; Guo, B.; Yao, J.; Yan, H.; Zhang, L.; Vaseva, A.V.; Chen, S.; Canoll, P.; Horner, J.W.; Wang, Y.A.; et al. Development of Resistance to EGFR-Targeted Therapy in Malignant Glioma Can Occur through EGFR-Dependent and -Independent Mechanisms. *Cancer Res.* **2015**, *75*, 2109–2119. [CrossRef]
22. Vivanco, I.; Robins, H.I.; Rohle, D.; Campos, C.; Grommes, C.; Nghiemphu, P.L.; Kubek, S.; Oldrini, B.; Chheda, M.G.; Yannuzzi, N.; et al. Differential sensitivity of glioma- versus lung cancer-specific EGFR mutations to EGFR kinase inhibitors. *Cancer Discov.* **2012**, *2*, 458–471. [CrossRef] [PubMed]
23. Politi, K.; Zakowski, M.F.; Fan, P.D.; Schonfeld, E.A.; Pao, W.; Varmus, H.E. Lung adenocarcinomas induced in mice by mutant EGF receptors found in human lung cancers respond to a tyrosine kinase inhibitor or to down-regulation of the receptors. *Genes Dev.* **2006**, *20*, 1496–1510. [CrossRef] [PubMed]
24. Ji, H.; Li, D.; Chen, L.; Shimamura, T.; Kobayashi, S.; McNamara, K.; Mahmood, U.; Mitchell, A.; Sun, Y.; Al-Hashem, R.; et al. The impact of human EGFR kinase domain mutations on lung tumorigenesis and in vivo sensitivity to EGFR-targeted therapies. *Cancer Cell* **2006**, *9*, 485–495. [CrossRef] [PubMed]
25. Shepherd, F.A.; Rodrigues Pereira, J.; Ciuleanu, T.; Tan, E.H.; Hirsh, V.; Thongprasert, S.; Campos, D.; Maoleekoonpiroj, S.; Smylie, M.; Martins, R.; et al. Erlotinib in previously treated non-small-cell lung cancer. *N. Engl. J. Med.* **2005**, *353*, 123–132. [CrossRef] [PubMed]
26. Fukuoka, M.; Yano, S.; Giaccone, G.; Tamura, T.; Nakagawa, K.; Douillard, J.Y.; Nishiwaki, Y.; Vansteenkiste, J.; Kudoh, S.; Rischin, D.; et al. Multi-institutional randomized phase II trial of gefitinib for previously treated patients with advanced non-small-cell lung cancer (The IDEAL 1 Trial) [corrected]. *J. Clin. Oncol.* **2003**, *21*, 2237–2246. [CrossRef] [PubMed]
27. Paez, J.G.; Janne, P.A.; Lee, J.C.; Tracy, S.; Greulich, H.; Gabriel, S.; Herman, P.; Kaye, F.J.; Lindeman, N.; Boggon, T.J.; et al. EGFR mutations in lung cancer: Correlation with clinical response to gefitinib therapy. *Science* **2004**, *304*, 1497–1500. [CrossRef]
28. Lynch, T.J.; Bell, D.W.; Sordella, R.; Gurubhagavatula, S.; Okimoto, R.A.; Brannigan, B.W.; Harris, P.L.; Haserlat, S.M.; Supko, J.G.; Haluska, F.G.; et al. Activating mutations in the epidermal growth factor receptor underlying responsiveness of non-small-cell lung cancer to gefitinib. *N. Engl. J. Med.* **2004**, *350*, 2129–2139. [CrossRef]

29. Maemondo, M.; Inoue, A.; Kobayashi, K.; Sugawara, S.; Oizumi, S.; Isobe, H.; Gemma, A.; Harada, M.; Yoshizawa, H.; Kinoshita, I.; et al. Gefitinib or chemotherapy for non-small-cell lung cancer with mutated EGFR. *N. Engl. J. Med.* **2010**, *362*, 2380–2388. [CrossRef]
30. Mok, T.S.; Wu, Y.L.; Thongprasert, S.; Yang, C.H.; Chu, D.T.; Saijo, N.; Sunpaweravong, P.; Han, B.; Margono, B.; Ichinose, Y.; et al. Gefitinib or carboplatin-paclitaxel in pulmonary adenocarcinoma. *N. Engl. J. Med.* **2009**, *361*, 947–957. [CrossRef]
31. Uhm, J.H.; Ballman, K.V.; Wu, W.; Giannini, C.; Krauss, J.C.; Buckner, J.C.; James, C.D.; Scheithauer, B.W.; Behrens, R.J.; Flynn, P.J.; et al. Phase II evaluation of gefitinib in patients with newly diagnosed Grade 4 astrocytoma: Mayo/North Central Cancer Treatment Group Study N0074. *Int. J. Radiat. Oncol. Biol. Phys.* **2011**, *80*, 347–353. [CrossRef]
32. Chakravarti, A.; Wang, M.; Robins, H.I.; Lautenschlaeger, T.; Curran, W.J.; Brachman, D.G.; Schultz, C.J.; Choucair, A.; Dolled-Filhart, M.; Christiansen, J.; et al. RTOG 0211: A phase 1/2 study of radiation therapy with concurrent gefitinib for newly diagnosed glioblastoma patients. *Int. J. Radiat. Oncol. Biol. Phys.* **2013**, *85*, 1206–1211. [CrossRef]
33. Raizer, J.J.; Abrey, L.E.; Lassman, A.B.; Chang, S.M.; Lamborn, K.R.; Kuhn, J.G.; Yung, W.K.; Gilbert, M.R.; Aldape, K.A.; Wen, P.Y.; et al. A phase II trial of erlotinib in patients with recurrent malignant gliomas and nonprogressive glioblastoma multiforme postradiation therapy. *Neuro. Oncol.* **2010**, *12*, 95–103. [CrossRef]
34. van den Bent, M.J.; Brandes, A.A.; Rampling, R.; Kouwenhoven, M.C.; Kros, J.M.; Carpentier, A.F.; Clement, P.M.; Frenay, M.; Campone, M.; Baurain, J.F.; et al. Randomized phase II trial of erlotinib versus temozolomide or carmustine in recurrent glioblastoma: EORTC brain tumor group study 26034. *J. Clin. Oncol.* **2009**, *27*, 1268–1274. [CrossRef] [PubMed]
35. Brown, P.D.; Krishnan, S.; Sarkaria, J.N.; Wu, W.; Jaeckle, K.A.; Uhm, J.H.; Geoffroy, F.J.; Arusell, R.; Kitange, G.; Jenkins, R.B.; et al. Phase I/II trial of erlotinib and temozolomide with radiation therapy in the treatment of newly diagnosed glioblastoma multiforme: North Central Cancer Treatment Group Study N0177. *J. Clin. Oncol.* **2008**, *26*, 5603–5609. [CrossRef] [PubMed]
36. Sathornsumetee, S.; Desjardins, A.; Vredenburgh, J.J.; McLendon, R.E.; Marcello, J.; Herndon, J.E.; Mathe, A.; Hamilton, M.; Rich, J.N.; Norfleet, J.A.; et al. Phase II trial of bevacizumab and erlotinib in patients with recurrent malignant glioma. *Neuro. Oncol.* **2010**, *12*, 1300–1310. [CrossRef] [PubMed]
37. Peereboom, D.M.; Ahluwalia, M.S.; Ye, X.; Supko, J.G.; Hilderbrand, S.L.; Phuphanich, S.; Nabors, L.B.; Rosenfeld, M.R.; Mikkelsen, T.; Grossman, S.A.; et al. NABTT 0502: A phase II and pharmacokinetic study of erlotinib and sorafenib for patients with progressive or recurrent glioblastoma multiforme. *Neuro. Oncol.* **2013**, *15*, 490–496. [CrossRef]
38. Clarke, J.L.; Molinaro, A.M.; Phillips, J.J.; Butowski, N.A.; Chang, S.M.; Perry, A.; Costello, J.F.; DeSilva, A.A.; Rabbitt, J.E.; Prados, M.D. A single-institution phase II trial of radiation, temozolomide, erlotinib, and bevacizumab for initial treatment of glioblastoma. *Neuro. Oncol.* **2014**, *16*, 984–990. [CrossRef]
39. Prados, M.D.; Chang, S.M.; Butowski, N.; DeBoer, R.; Parvataneni, R.; Carliner, H.; Kabuubi, P.; Ayers-Ringler, J.; Rabbitt, J.; Page, M.; et al. Phase II study of erlotinib plus temozolomide during and after radiation therapy in patients with newly diagnosed glioblastoma multiforme or gliosarcoma. *J. Clin. Oncol.* **2009**, *27*, 579–584. [CrossRef]
40. Mellinghoff, I.K.; Wang, M.Y.; Vivanco, I.; Haas-Kogan, D.A.; Zhu, S.; Dia, E.Q.; Lu, K.V.; Yoshimoto, K.; Huang, J.H.; Chute, D.J.; et al. Molecular determinants of the response of glioblastomas to EGFR kinase inhibitors. *N. Engl. J. Med.* **2005**, *353*, 2012–2024. [CrossRef]
41. Reardon, D.A.; Desjardins, A.; Vredenburgh, J.J.; Gururangan, S.; Friedman, A.H.; Herndon, J.E., 2nd; Marcello, J.; Norfleet, J.A.; McLendon, R.E.; Sampson, J.H.; et al. Phase 2 trial of erlotinib plus sirolimus in adults with recurrent glioblastoma. *J. Neurooncol.* **2010**, *96*, 219–230. [CrossRef]
42. Wen, P.Y.; Chang, S.M.; Lamborn, K.R.; Kuhn, J.G.; Norden, A.D.; Cloughesy, T.F.; Robins, H.I.; Lieberman, F.S.; Gilbert, M.R.; Mehta, M.P.; et al. Phase I/II study of erlotinib and temsirolimus for patients with recurrent malignant gliomas: North American Brain Tumor Consortium trial 04-02. *Neuro. Oncol.* **2014**, *16*, 567–578. [CrossRef] [PubMed]
43. Hegi, M.E.; Diserens, A.C.; Bady, P.; Kamoshima, Y.; Kouwenhoven, M.C.; Delorenzi, M.; Lambiv, W.L.; Hamou, M.F.; Matter, M.S.; Koch, A.; et al. Pathway analysis of glioblastoma tissue after preoperative treatment with the EGFR tyrosine kinase inhibitor gefitinib–a phase II trial. *Mol. Cancer Ther.* **2011**, *10*, 1102–1112. [CrossRef] [PubMed]

44. Brown, N.; McBain, C.; Nash, S.; Hopkins, K.; Sanghera, P.; Saran, F.; Phillips, M.; Dungey, F.; Clifton-Hadley, L.; Wanek, K.; et al. Multi-Center Randomized Phase II Study Comparing Cediranib plus Gefitinib with Cediranib plus Placebo in Subjects with Recurrent/Progressive Glioblastoma. *PLoS ONE* **2016**, *11*, e0156369. [CrossRef] [PubMed]
45. Rich, J.N.; Reardon, D.A.; Peery, T.; Dowell, J.M.; Quinn, J.A.; Penne, K.L.; Wikstrand, C.J.; Van Duyn, L.B.; Dancey, J.E.; McLendon, R.E.; et al. Phase II trial of gefitinib in recurrent glioblastoma. *J. Clin. Oncol.* **2004**, *22*, 133–142. [CrossRef] [PubMed]
46. Thiessen, B.; Stewart, C.; Tsao, M.; Kamel-Reid, S.; Schaiquevich, P.; Mason, W.; Easaw, J.; Belanger, K.; Forsyth, P.; McIntosh, L.; et al. A phase I/II trial of GW572016 (lapatinib) in recurrent glioblastoma multiforme: Clinical outcomes, pharmacokinetics and molecular correlation. *Cancer Chemother. Pharmacol.* **2010**, *65*, 353–361. [CrossRef] [PubMed]
47. Reardon, D.A.; Nabors, L.B.; Mason, W.P.; Perry, J.R.; Shapiro, W.; Kavan, P.; Mathieu, D.; Phuphanich, S.; Cseh, A.; Fu, Y.; et al. Phase I/randomized phase II study of afatinib, an irreversible ErbB family blocker, with or without protracted temozolomide in adults with recurrent glioblastoma. *Neuro. Oncol.* **2015**, *17*, 430–439. [CrossRef]
48. Sepulveda-Sanchez, J.M.; Vaz, M.A.; Balana, C.; Gil-Gil, M.; Reynes, G.; Gallego, O.; Martinez-Garcia, M.; Vicente, E.; Quindos, M.; Luque, R.; et al. Phase II trial of dacomitinib, a pan-human EGFR tyrosine kinase inhibitor, in recurrent glioblastoma patients with EGFR amplification. *Neuro. Oncol.* **2017**, *19*, 1522–1531. [CrossRef] [PubMed]
49. Kreisl, T.N.; Smith, P.; Sul, J.; Salgado, C.; Iwamoto, F.M.; Shih, J.H.; Fine, H.A. Continuous daily sunitinib for recurrent glioblastoma. *J. Neurooncol.* **2013**, *111*, 41–48. [CrossRef]
50. Reardon, D.A.; Quinn, J.A.; Vredenburgh, J.J.; Gururangan, S.; Friedman, A.H.; Desjardins, A.; Sathornsumetee, S.; Herndon, J.E., 2nd; Dowell, J.M.; McLendon, R.E.; et al. Phase 1 trial of gefitinib plus sirolimus in adults with recurrent malignant glioma. *Clin. Cancer Res.* **2006**, *12*, 860–868. [CrossRef]
51. Franceschi, E.; Cavallo, G.; Lonardi, S.; Magrini, E.; Tosoni, A.; Grosso, D.; Scopece, L.; Blatt, V.; Urbini, B.; Pession, A.; et al. Gefitinib in patients with progressive high-grade gliomas: A multicentre phase II study by Gruppo Italiano Cooperativo di Neuro-Oncologia (GICNO). *Br. J. Cancer* **2007**, *96*, 1047–1051. [CrossRef]
52. Yung, W.K.; Vredenburgh, J.J.; Cloughesy, T.F.; Nghiemphu, P.; Klencke, B.; Gilbert, M.R.; Reardon, D.A.; Prados, M.D. Safety and efficacy of erlotinib in first-relapse glioblastoma: A phase II open-label study. *Neuro. Oncol.* **2010**, *12*, 1061–1070. [CrossRef]
53. Stommel, J.M.; Kimmelman, A.C.; Ying, H.; Nabioullin, R.; Ponugoti, A.H.; Wiedemeyer, R.; Stegh, A.H.; Bradner, J.E.; Ligon, K.L.; Brennan, C.; et al. Coactivation of receptor tyrosine kinases affects the response of tumor cells to targeted therapies. *Science* **2007**, *318*, 287–290. [CrossRef] [PubMed]
54. Prahallad, A.; Sun, C.; Huang, S.; Di Nicolantonio, F.; Salazar, R.; Zecchin, D.; Beijersbergen, R.L.; Bardelli, A.; Bernards, R. Unresponsiveness of colon cancer to BRAF(V600E) inhibition through feedback activation of EGFR. *Nature* **2012**, *483*, 100–103. [CrossRef]
55. Del Vecchio, C.A.; Giacomini, C.P.; Vogel, H.; Jensen, K.C.; Florio, T.; Merlo, A.; Pollack, J.R.; Wong, A.J. EGFRvIII gene rearrangement is an early event in glioblastoma tumorigenesis and expression defines a hierarchy modulated by epigenetic mechanisms. *Oncogene* **2013**, *32*, 2670–2681. [CrossRef] [PubMed]
56. van den Bent, M.J.; Gao, Y.; Kerkhof, M.; Kros, J.M.; Gorlia, T.; van Zwieten, K.; Prince, J.; van Duinen, S.; Sillevis Smitt, P.A.; Taphoorn, M.; et al. Changes in the EGFR amplification and EGFRvIII expression between paired primary and recurrent glioblastomas. *Neuro. Oncol.* **2015**, *17*, 935–941. [CrossRef] [PubMed]
57. Nathanson, D.A.; Gini, B.; Mottahedeh, J.; Visnyei, K.; Koga, T.; Gomez, G.; Eskin, A.; Hwang, K.; Wang, J.; Masui, K.; et al. Targeted therapy resistance mediated by dynamic regulation of extrachromosomal mutant EGFR DNA. *Science* **2014**, *343*, 72–76. [CrossRef] [PubMed]
58. Antonyak, M.A.; Moscatello, D.K.; Wong, A.J. Constitutive activation of c-Jun N-terminal kinase by a mutant epidermal growth factor receptor. *J. Biol. Chem.* **1998**, *273*, 2817–2822. [CrossRef] [PubMed]
59. Chu, C.T.; Everiss, K.D.; Wikstrand, C.J.; Batra, S.K.; Kung, H.J.; Bigner, D.D. Receptor dimerization is not a factor in the signalling activity of a transforming variant epidermal growth factor receptor (EGFRvIII). *Biochem. J.* **1997**, *324 Pt 3*, 855–861. [CrossRef] [PubMed]
60. Chumbalkar, V.; Latha, K.; Hwang, Y.; Maywald, R.; Hawley, L.; Sawaya, R.; Diao, L.; Baggerly, K.; Cavenee, W.K.; Furnari, F.B.; et al. Analysis of phosphotyrosine signaling in glioblastoma identifies STAT5 as a novel downstream target of DeltaEGFR. *J. Proteome Res.* **2011**, *10*, 1343–1352. [CrossRef]

61. Latha, K.; Li, M.; Chumbalkar, V.; Gururaj, A.; Hwang, Y.; Dakeng, S.; Sawaya, R.; Aldape, K.; Cavenee, W.K.; Bogler, O.; et al. Nuclear EGFRvIII-STAT5b complex contributes to glioblastoma cell survival by direct activation of the Bcl-XL promoter. *Int. J. Cancer* **2012**. [CrossRef]
62. Pedersen, M.W.; Pedersen, N.; Damstrup, L.; Villingshoj, M.; Sonder, S.U.; Rieneck, K.; Bovin, L.F.; Spang-Thomsen, M.; Poulsen, H.S. Analysis of the epidermal growth factor receptor specific transcriptome: Effect of receptor expression level and an activating mutation. *J. Cell. Biochem.* **2005**, *96*, 412–427. [CrossRef] [PubMed]
63. Erdem-Eraslan, L.; Gao, Y.; Kloosterhof, N.K.; Atlasi, Y.; Demmers, J.; Sacchetti, A.; Kros, J.M.; Sillevis Smitt, P.; Aerts, J.; French, P.J. Mutation specific functions of EGFR result in a mutation-specific downstream pathway activation. *Eur. J. Cancer* **2015**, *51*, 893–903. [CrossRef]
64. Lin, S.Y.; Makino, K.; Xia, W.; Matin, A.; Wen, Y.; Kwong, K.Y.; Bourguignon, L.; Hung, M.C. Nuclear localization of EGF receptor and its potential new role as a transcription factor. *Nat. Cell. Biol.* **2001**, *3*, 802–808. [CrossRef] [PubMed]
65. Mikula, M.; Skrzypczak, M.; Goryca, K.; Paczkowska, K.; Ledwon, J.K.; Statkiewicz, M.; Kulecka, M.; Grzelak, M.; Dabrowska, M.; Kuklinska, U.; et al. Genome-wide co-localization of active EGFR and downstream ERK pathway kinases mirrors mitogen-inducible RNA polymerase 2 genomic occupancy. *Nucleic. Acids Res.* **2016**, *44*, 10150–10164. [CrossRef] [PubMed]
66. Liccardi, G.; Hartley, J.A.; Hochhauser, D. EGFR nuclear translocation modulates DNA repair following cisplatin and ionizing radiation treatment. *Cancer Res.* **2011**, *71*, 1103–1114. [CrossRef] [PubMed]
67. Lo, H.W.; Hsu, S.C.; Ali-Seyed, M.; Gunduz, M.; Xia, W.; Wei, Y.; Bartholomeusz, G.; Shih, J.Y.; Hung, M.C. Nuclear interaction of EGFR and STAT3 in the activation of the iNOS/NO pathway. *Cancer Cell.* **2005**, *7*, 575–589. [CrossRef] [PubMed]
68. Marti, U.; Ruchti, C.; Kampf, J.; Thomas, G.A.; Williams, E.D.; Peter, H.J.; Gerber, H.; Burgi, U. Nuclear localization of epidermal growth factor and epidermal growth factor receptors in human thyroid tissues. *Thyroid* **2001**, *11*, 137–145. [CrossRef] [PubMed]
69. Gururaj, A.E.; Gibson, L.; Panchabhai, S.; Bai, M.; Manyam, G.; Lu, Y.; Latha, K.; Rojas, M.L.; Hwang, Y.; Liang, S.; et al. Access to the nucleus and functional association with c-Myc is required for the full oncogenic potential of DeltaEGFR/EGFRvIII. *J. Biol. Chem.* **2013**, *288*, 3428–3438. [CrossRef] [PubMed]
70. Bollig-Fischer, A.; Chen, W.; Gadgeel, S.M.; Wenzlaff, A.S.; Cote, M.L.; Schwartz, A.G.; Bepler, G. Racial diversity of actionable mutations in non-small cell lung cancer. *J. Thorac. Oncol.* **2015**, *10*, 250–255. [CrossRef]
71. Seo, J.S.; Ju, Y.S.; Lee, W.C.; Shin, J.Y.; Lee, J.K.; Bleazard, T.; Lee, J.; Jung, Y.J.; Kim, J.O.; Shin, J.Y.; et al. The transcriptional landscape and mutational profile of lung adenocarcinoma. *Genome Res.* **2012**, *22*, 2109–2119. [CrossRef]
72. Zehir, A.; Benayed, R.; Shah, R.H.; Syed, A.; Middha, S.; Kim, H.R.; Srinivasan, P.; Gao, J.; Chakravarty, D.; Devlin, S.M.; et al. Mutational landscape of metastatic cancer revealed from prospective clinical sequencing of 10,000 patients. *Nat. Med.* **2017**, *23*, 703–713. [CrossRef] [PubMed]
73. Iyevleva, A.G.; Novik, A.V.; Moiseyenko, V.M.; Imyanitov, E.N. EGFR mutation in kidney carcinoma confers sensitivity to gefitinib treatment: A case report. *Urol. Oncol.* **2009**, *27*, 548–550. [CrossRef] [PubMed]
74. Masago, K.; Asato, R.; Fujita, S.; Hirano, S.; Tamura, Y.; Kanda, T.; Mio, T.; Katakami, N.; Mishima, M.; Ito, J. Epidermal growth factor receptor gene mutations in papillary thyroid carcinoma. *Int. J. Cancer* **2009**, *124*, 2744–2749. [CrossRef] [PubMed]
75. Ali, S.M.; Alpaugh, R.K.; Buell, J.K.; Stephens, P.J.; Yu, J.Q.; Wu, H.; Hiemstra, C.N.; Miller, V.A.; Lipson, D.; Palmer, G.A.; et al. Antitumor response of an ERBB2 amplified inflammatory breast carcinoma with EGFR mutation to the EGFR-TKI erlotinib. *Clin. Breast Cancer* **2014**, *14*, e14–e16. [CrossRef] [PubMed]
76. Voss, J.S.; Holtegaard, L.M.; Kerr, S.E.; Fritcher, E.G.; Roberts, L.R.; Gores, G.J.; Zhang, J.; Highsmith, W.E.; Halling, K.C.; Kipp, B.R. Molecular profiling of cholangiocarcinoma shows potential for targeted therapy treatment decisions. *Hum. Pathol.* **2013**, *44*, 1216–1222. [CrossRef]
77. Agatsuma, N.; Yasuda, Y.; Ozasa, H. Malignant Pleural Mesothelioma Harboring Both G719C and S768I Mutations of EGFR Successfully Treated with Afatinib. *J. Thorac. Oncol.* **2017**, *12*, e141–e143. [CrossRef]
78. Chi, A.S.; Batchelor, T.T.; Dias-Santagata, D.; Borger, D.; Stiles, C.D.; Wang, D.L.; Curry, W.T.; Wen, P.Y.; Ligon, K.L.; Ellisen, L.; et al. Prospective, high-throughput molecular profiling of human gliomas. *J. Neurooncol.* **2012**, *110*, 89–98. [CrossRef] [PubMed]

79. Byeon, S.; Kim, Y.; Lim, S.W.; Cho, J.H.; Park, S.; Lee, J.; Sun, J.M.; Choi, Y.L.; Lee, S.H.; Ahn, J.S.; et al. Clinical Outcomes of EGFR Exon 20 Insertion Mutations in Advanced Non-small Cell Lung Cancer in Korea. *Cancer Res. Treat.* **2018**. [CrossRef]
80. Yasuda, H.; Kobayashi, S.; Costa, D.B. EGFR exon 20 insertion mutations in non-small-cell lung cancer: Preclinical data and clinical implications. *Lancet Oncol.* **2012**, *13*, e23–e31. [CrossRef]
81. Hirano, T.; Yasuda, H.; Tani, T.; Hamamoto, J.; Oashi, A.; Ishioka, K.; Arai, D.; Nukaga, S.; Miyawaki, M.; Kawada, I.; et al. In vitro modeling to determine mutation specificity of EGFR tyrosine kinase inhibitors against clinically relevant EGFR mutants in non-small-cell lung cancer. *Oncotarget* **2015**, *6*, 38789–38803. [CrossRef]
82. Yang, M.; Xu, X.; Cai, J.; Ning, J.; Wery, J.P.; Li, Q.X. NSCLC harboring EGFR exon-20 insertions after the regulatory C-helix of kinase domain responds poorly to known EGFR inhibitors. *Int. J. Cancer* **2016**, *139*, 171–176. [CrossRef]
83. Ruan, Z.; Kannan, N. Altered conformational landscape and dimerization dependency underpins the activation of EGFR by alphaC-beta4 loop insertion mutations. *Proc. Natl. Acad. Sci. USA* **2018**, *115*, E8162–E8171. [CrossRef] [PubMed]
84. Hasako, S.; Terasaka, M.; Abe, N.; Uno, T.; Ohsawa, H.; Hashimoto, A.; Fujita, R.; Tanaka, K.; Okayama, T.; Wadhwa, R.; et al. TAS6417, A Novel EGFR Inhibitor Targeting Exon 20 Insertion Mutations. *Mol. Cancer Ther.* **2018**, *17*, 1648–1658. [CrossRef] [PubMed]
85. Ross, H.J.; Blumenschein, G.R., Jr.; Aisner, J.; Damjanov, N.; Dowlati, A.; Garst, J.; Rigas, J.R.; Smylie, M.; Hassani, H.; Allen, K.E.; et al. Randomized phase II multicenter trial of two schedules of lapatinib as first- or second-line monotherapy in patients with advanced or metastatic non-small cell lung cancer. *Clin. Cancer Res.* **2010**, *16*, 1938–1949. [CrossRef] [PubMed]
86. Jia, Y.; Yun, C.H.; Park, E.; Ercan, D.; Manuia, M.; Juarez, J.; Xu, C.; Rhee, K.; Chen, T.; Zhang, H.; et al. Overcoming EGFR(T790M) and EGFR(C797S) resistance with mutant-selective allosteric inhibitors. *Nature* **2016**, *534*, 129–132. [CrossRef]
87. Stamos, J.; Sliwkowski, M.X.; Eigenbrot, C. Structure of the epidermal growth factor receptor kinase domain alone and in complex with a 4-anilinoquinazoline inhibitor. *J. Biol. Chem.* **2002**, *277*, 46265–46272. [CrossRef]
88. Zhang, X.; Gureasko, J.; Shen, K.; Cole, P.A.; Kuriyan, J. An allosteric mechanism for activation of the kinase domain of epidermal growth factor receptor. *Cell.* **2006**, *125*, 1137–1149. [CrossRef]
89. Wood, E.R.; Truesdale, A.T.; McDonald, O.B.; Yuan, D.; Hassell, A.; Dickerson, S.H.; Ellis, B.; Pennisi, C.; Horne, E.; Lackey, K.; et al. A unique structure for epidermal growth factor receptor bound to GW572016 (Lapatinib): Relationships among protein conformation, inhibitor off-rate, and receptor activity in tumor cells. *Cancer Res.* **2004**, *64*, 6652–6659. [CrossRef] [PubMed]

cancers

MDPI

Perspective

The 'Ins and Outs' of Early Preclinical Models for Brain Tumor Research: Are They Valuable and Have We Been Doing It Wrong?

Ola Rominiyi [1,2,*], Yahia Al-Tamimi [1,3] and Spencer J. Collis [2]

1 Department of Neurosurgery, Sheffield Teaching Hospitals NHS Foundation Trust, Sheffield S10 2JF, UK; y.al-tamimi@sheffield.ac.uk

2 Department of Oncology & Metabolism, The University of Sheffield Medical School, Sheffield S10 2RX, UK; s.collis@sheffield.ac.uk

3 NIHR Sheffield Biomedical Research Centre, Sheffield Teaching Hospitals NHS Foundation Trust, Sheffield S10 2JF, UK

* Correspondence: o.rominiyi@sheffield.ac.uk

Received: 26 February 2019; Accepted: 23 March 2019; Published: 25 March 2019

Abstract: In this perspective, we congratulate the international efforts to highlight critical challenges in brain tumor research through a recent Consensus Statement. We also illustrate the importance of developing more accurate and clinically relevant early translational in vitro brain tumor models—a perspective given limited emphasis in the Consensus Statement, despite in vitro models being widely used to prioritize candidate therapeutic strategies prior to in vivo studies and subsequent clinical trials. We argue that successful translation of effective novel treatments into the clinic will require investment into the development of more predictive early pre-clinical models. It is in the interest of researchers, clinicians, and ultimately, patients that the most promising therapeutic candidates are identified and translated toward use in the clinic. Highlighting the value of early pre-clinical brain tumor models and debating how such models can be improved is of the utmost importance to the neuro-oncology research community and cancer research more broadly.

Keywords: pre-clinical; research models; 3D models; brain tumors; cancer stem cells; glioblastoma; DNA damage response (DDR)

1. Introduction

The recent Consensus Statement 'Challenges to curing primary brain tumors [1]', produced by a Cancer Research UK convened panel of international experts, represents an important 'call-to-arms' to the neuro-oncology research community and should be congratulated for highlighting areas worthy of investment to improve patient outcomes. Agreement exists within the cancer research community that more accurate genetically engineered mouse models (GEMMs) and orthotopic patient-derived xenografts (PDXs), along with greater sharing of these resources, will help translate new treatments into the clinic. However, it is important to consider that, in some contexts, these in vivo models may only be as successful as the therapeutic strategies they are used to evaluate. Consequently, greater emphasis should also be directed toward enhancing and validating early pre-clinical in vitro culture systems used to model brain tumors.

2. The Value of Early Pre-Clinical Models in Brain Tumor Research

Using the example of targeting the DNA damage response (DDR) [2], effective future treatment strategies may need to incorporate novel therapeutic combinations (such as combined PARP and ATR inhibition with Olaparib and AZD6738, respectively) to overcome the extensive intra-tumoral heterogeneity brain tumors can exhibit, and functional redundancy governed by the interconnectedness of many biological processes [3]. At least 19 novel drugs that target the DDR in cancer are currently approved or in clinical trials [2]. Conservative estimates provide at least 147 potential two-drug or 939 three-drug DDR inhibitor combinations (excluding strategies that combine drugs of the same class). In the quest to improve patient outcomes, all of these combinations may warrant evaluation, however, testing every combination using rodent models would be impractical and potentially unethical. Therefore, robust validation of biologically relevant in vitro models that are able to recapitulate key disease features and predict clinical response may represent one of the most urgent and important challenges facing our community.

3. A Need to Validate and Contrast In Vitro Models

Whilst organoids represent a promising 3D culture system highlighted in the Consensus Statement, other clinically relevant 3D co-culture systems do exist [4,5], and the relative value of each to prospectively determine which therapies extend survival for brain tumor patients in Phase III clinical trials remains unclear. Furthermore, the Consensus Statement alludes to the blood–brain barrier (BBB) representing a key obstacle to the efficient delivery of chemotherapy and novel therapeutics to brain tumors [1]. In recent years, numerous advances in modeling the BBB have been made including microfluidic 'BBB on a chip' systems [6] and multi-cellular BBB spheroids, based on non-adherent co-culture of human astrocytes, pericytes, and brain endothelial cells [7]. Refining the ability of in vitro models such as these to reproduce in vivo selectivity of both the BBB and blood-tumor barrier remains critically important to accurately predict the ability of novel therapeutics to reach brain tumors. Increasingly, the 'gold-standard' assessment of each should perhaps be based on the ability to predict tumoral or peri-tumoral drug concentrations and intended target effects in Phase 0 clinical trials, rather than animal studies. To prioritize novel therapies for the future, many researchers would benefit from the clear validation of which in vitro BBB models best predict drug delivery in human subjects. Greater investment in these areas is crucial to avoid the loss of talent within the research community and ensure candidate therapies with the greatest likelihood of success are selected for in vivo studies and subsequent clinical trials.

4. Toward Early Translational Models of Post-Surgical Disease

Additionally, worthy of consideration are the neurosurgical features of deriving human brain tumor cells used in the majority of pre-clinical models, whether in vitro (2D, 3D, and organoid culture systems) or in vivo (PDXs). Generally, these cells come from the resected tumor taken out at surgery, even though the infiltrative residual cells left inside patients after surgery are responsible for disease progression. For example, considering glioblastoma, which is characterized by extensive spatial and temporal heterogeneity [8,9], it should not be assumed that the more invasive cells left behind after surgery respond to treatment in an identical fashion to those cells that are removed. Consequently, greater emphasis should be placed on the incorporation of typical residual cells into patient-derived pre-clinical models. Our group and others are currently working on such models of clinically relevant post-surgical residual disease, which we envisage will be valuable to the research community and help prioritize the most promising novel therapies in the context of current multi-modal treatment.

5. Conclusions

Highlighting the value of early pre-clinical brain tumor models and debating how such models can be improved is essential to enable the identification and translation of novel treatment strategies that are most likely to be successful in clinical trials for patients with brain tumors. In cancer research, developing more clinically relevant in vitro models to better prioritize novel therapies worthy of further investigation is critical. The translational 'pipeline' may only ever be as successful as those treatments which enter it.

Author Contributions: Conceptualization, O.R. and S.J.C.; Writing—Original Draft Preparation, O.R.; Writing—Review and Editing, O.R., Y.A.-T., and S.J.C.

Funding: O.R. is funded by a Royal College of Surgeons England and Neurocare Clinical Research Fellowship (Clinical Research Fellowship Neurocare: 181911). Y.A.-T. is supported by the NIHR Sheffield Biomedical Research Centre/NIHR Sheffield Clinical Research Facility. The views expressed are those of the authors and not necessarily those of the RCS, Neurocare, NHS, NIHR, or the Department of Health and Social Care.

Conflicts of Interest: The authors declare no conflict of interest.

References

1. Aldape, K.; Brindle, K.M.; Chesler, L.; Chopra, R.; Gajjar, A.; Gilbert, M.R.; Gottardo, N.; Gutmann, D.H.; Hargrave, D.; Holland, E.C.; et al. Challenges to curing primary brain tumours. *Nat. Rev. Clin. Oncol.* **2019**. [CrossRef] [PubMed]
2. Pilie, P.G.; Tang, C.; Mills, G.B.; Yap, T.A. State-of-the-art strategies for targeting the DNA damage response in cancer. *Nat. Rev. Clin. Oncol.* **2019**, *16*, 81–104. [CrossRef]
3. Pearl, L.H.; Schierz, A.C.; Ward, S.E.; Al-Lazikani, B.; Pearl, F.M. Therapeutic opportunities within the DNA damage response. *Nat. Rev. Cancer* **2015**, *15*, 166–180. [CrossRef]
4. Caragher, S.; Chalmers, A.J.; Gomez-Roman, N. Glioblastoma's Next Top Model: Novel Culture Systems for Brain Cancer Radiotherapy Research. *Cancers* **2019**, *11*, 44. [CrossRef] [PubMed]
5. Rominiyi, O.; Gomez-Roman, N.; Lad, N.; Al-Tamimi, Y.; Jellinek, D.; Chalmers, A.; Carroll, T.; Chen, B.; Collis, S. P04.74 Preclinical evaluation of combinations targeting the DNA damage response in 2D and 3D models of glioblastoma stem cells . *Neuro-Oncology* **2018**, *20*, iii297. [CrossRef]
6. Griep, L.M.; Wolbers, F.; de Wagenaar, B.; ter Braak, P.M.; Weksler, B.B.; Romero, I.A.; Couraud, P.O.; Vermes, I.; van der Meer, A.D.; van den Berg, A. BBB on chip: Microfluidic platform to mechanically and biochemically modulate blood-brain barrier function. *Biomed. Microdevices* **2013**, *15*, 145–150. [CrossRef] [PubMed]
7. Cho, C.F.; Wolfe, J.M.; Fadzen, C.M.; Calligaris, D.; Hornburg, K.; Chiocca, E.A.; Agar, N.Y.R.; Pentelute, B.L.; Lawler, S.E. Blood-brain-barrier spheroids as an in vitro screening platform for brain-penetrating agents. *Nat. Commun.* **2017**, *8*, 15623. [CrossRef] [PubMed]
8. Smith, S.J.; Diksin, M.; Chhaya, S.; Sairam, S.; Estevez-Cebrero, M.A.; Rahman, R. The Invasive Region of Glioblastoma Defined by 5ALA Guided Surgery Has an Altered Cancer Stem Cell Marker Profile Compared to Central Tumour. *Int. J. Mol. Sci.* **2017**, *18*, 2452. [CrossRef]
9. Spiteri, I.; Caravagna, G.; Cresswell, G.D.; Vatsiou, A.; Nichol, D.; Acar, A.; Ermini, L.; Chkhaidze, K.; Werner, B.; Mair, R.; et al. Evolutionary dynamics of residual disease in human glioblastoma. *Ann. Oncol.* **2018**. [CrossRef] [PubMed]

MDPI
St. Alban-Anlage 66
4052 Basel
Switzerland
Tel. +41 61 683 77 34
Fax +41 61 302 89 18
www.mdpi.com

Cancers Editorial Office
E-mail: cancers@mdpi.com
www.mdpi.com/journal/cancers

www.ingramcontent.com/pod-product-compliance
Lightning Source LLC
Chambersburg PA
CBHW041014270926
51850CB00024B/10
* 9 7 8 3 0 3 9 2 8 2 6 0 9 *